Applied Calculus & Regression Analysis

University of Georgia Custom Edition

Soo T. Tan | Terry E. Dielman

CENGAGE Learning

Applied Calculus & Regression Analysis: University of Georgia Custom Edition

Senior Manager, Student Engagement:
Linda deStefano

Manager, Student Engagement:
Julie Dierig

Marketing Manager:
Rachael Kloos

Manager, Premedia:
Kim Fry

Manager, Intellectual Property Project Manager:
Brian Methe

Senior Manager, Production:
Donna M. Brown

Manager, Production:
Terri Daley

Sources:

Applied Calculus for the Managerial, Life, and Social Sciences, 9th Edition
Soo T. Tan
© 2014 Cengage Learning. All rights reserved.

Applied Regression Analysis: A Second Course in Business and Economic Statistics, 4th Edition
Terry E. Dielman
© 2005 Cengage Learning. All rights reserved.

ALL RIGHTS RESERVED. No part of this work covered by the copyright herein may be reproduced, transmitted, stored or used in any form or by any means graphic, electronic, or mechanical, including but not limited to photocopying, recording, scanning, digitizing, taping, Web distribution, information networks, or information storage and retrieval systems, except as permitted under Section 107 or 108 of the 1976 United States Copyright Act, without the prior written permission of the publisher.

> For product information and technology assistance, contact us at
> **Cengage Learning Customer & Sales Support, 1-800-354-9706**
> For permission to use material from this text or product,
> submit all requests online at **cengage.com/permissions**
> Further permissions questions can be emailed to
> **permissionrequest@cengage.com**

This book contains select works from existing Cengage Learning resources and was produced by Cengage Learning Custom Solutions for collegiate use. As such, those adopting and/or contributing to this work are responsible for editorial content accuracy, continuity and completeness.

Compilation © 2014 Cengage Learning

ISBN: 978-1-305-28144-8

WCN: 01-100-101

Cengage Learning
20 Channel Center Street
Boston, MA 02210
USA

Cengage Learning is a leading provider of customized learning solutions with office locations around the globe, including Singapore, the United Kingdom, Australia, Mexico, Brazil, and Japan. Locate your local office at:
international.cengage.com/region.

Cengage Learning products are represented in Canada by Nelson Education, Ltd.
For your lifelong learning solutions, visit **www.cengage.com/custom.**
Visit our corporate website at **www.cengage.com.**

Brief Contents

From:
Applied Calculus for the Managerial, Life, and Social Sciences
9th Edition
Soo T. Tan

Chapter 1	Preliminaries	1
Chapter 2	Functions, Limits, and the Derivative	49
Chapter 3	Differentiation	157
Chapter 4	Applications of the Derivative	245
Chapter 5	Exponential and Logarithmic Functions	329
Chapter 8	Calculus of Several Variables	533

From:
Applied Regression Analysis: A Second Course in Business and Economic Statistics, 4th Edition
Terry E. Dielman

Chapter 1	An Introduction to Regression Analysis	1
Chapter 2	Review of Basic Statistical Concepts	5
Chapter 3	Simple Regression Analysis	63
Chapter 4	Multiple Regression Analysis	133
Chapter 5	Fitting Curves to Data	179
Chapter 6	Assessing the Assumptions of the Regression Model	205
Chapter 7	Using Indicator and Interaction Variables	273
Chapter 8	Variable Selection	311
Chapter 10	Qualitative Dependent Variables: An Introduction to Discriminant Analysis and Logistic Regression	373

1 PRELIMINARIES

How much money is needed to purchase at least 100,000 shares of the Starr Communications Company? Corbyco, a giant conglomerate, wishes to purchase a minimum of 100,000 shares of the company. In Example 11, page 21, you will see how Corbyco's management determines how much money they will need for the acquisition.

THE FIRST TWO sections of this chapter contain a brief review of algebra. We then introduce the Cartesian coordinate system, which allows us to represent points in the plane in terms of ordered pairs of real numbers. This in turn enables us to compute the distance between two points algebraically. This chapter also covers straight lines. The slope of a straight line plays an important role in the study of calculus.

Use this test to diagnose any weaknesses that you might have in the algebra that you will need for the calculus material that follows. The review section and examples that will help you brush up on the skills necessary to work the problem are indicated after each exercise. The answers follow the test.

Diagnostic Test

1. **a.** Evaluate the expression:

   (i) $\left(\dfrac{16}{9}\right)^{3/2}$ **(ii)** $\sqrt[3]{\dfrac{27}{125}}$

   b. Rewrite the expression using positive exponents only: $(x^{-2}y^{-1})^3$

   (Exponents and radicals, Examples 1 and 2, pages 6–7)

2. Rationalize the numerator: $\sqrt[3]{\dfrac{x^2}{yz^3}}$

   (Rationalization, Example 5, page 7)

3. Simplify the following expressions:
   a. $(3x^4 + 10x^3 + 6x^2 + 10x + 3) + (2x^4 + 10x^3 + 6x^2 + 4x)$
   b. $(3x - 4)(3x^2 - 2x + 3)$

   (Operations with algebraic expressions, Examples 6 and 7, pages 8–9)

4. Factor completely:
   a. $6a^4b^4c - 3a^3b^2c - 9a^2b^2$ **b.** $6x^2 - xy - y^2$

   (Factoring, Examples 8–10, pages 9–11)

5. Use the quadratic formula to solve the following equation: $9x^2 - 12x = 4$

   (The quadratic formula, Example 11, pages 12–13)

6. Simplify the following expressions:

   a. $\dfrac{2x^2 + 3x - 2}{2x^2 + 5x - 3}$ **b.** $\dfrac{(t^2 + 4)(2t - 4) - (t^2 - 4t + 4)(2t)}{(t^2 + 4)^2}$

   (Rational expressions, Example 1, page 16)

7. Perform the indicated operations and simplify:

   a. $\dfrac{2x - 6}{x + 3} \cdot \dfrac{x^2 + 6x + 9}{x^2 - 9}$ **b.** $\dfrac{3x}{x^2 + 2} + \dfrac{3x^2}{x^3 + 1}$

   (Rational expressions, Examples 2 and 3, pages 16–18)

8. Perform the indicated operations and simplify:

   a. $\dfrac{1 + \dfrac{1}{x + 2}}{x - \dfrac{9}{x}}$ **b.** $\dfrac{x(3x^2 + 1)}{x - 1} \cdot \dfrac{3x^3 - 5x^2 + x}{x(x - 1)(3x^2 + 1)^{1/2}}$

   (Rational expressions, Examples 4 and 5, pages 18–19)

9. Rationalize the denominator: $\dfrac{3}{1 + 2\sqrt{x}}$

   (Rationalizing algebraic fractions, Example 6, page 19)

10. Solve the inequalities:
    a. $x^2 + x - 12 \leq 0$

    (Inequalities, Example 9, page 20)

    b. $|3x - 4| \leq 2$

    (Absolute value, Example 14, page 22)

ANSWERS:

1. a. (i) $\dfrac{64}{27}$ (ii) $\dfrac{3}{5}$ b. $\dfrac{1}{x^6 y^3}$ 2. $\dfrac{x}{z\sqrt[3]{xy}}$

3. a. $5x^4 + 20x^3 + 12x^2 + 14x + 3$ b. $9x^3 - 18x^2 + 17x - 12$

4. a. $3a^2 b^2 (2a^2 b^2 c - ac - 3)$ b. $(2x - y)(3x + y)$

5. $\dfrac{2}{3}(1 - \sqrt{2}); \dfrac{2}{3}(1 + \sqrt{2})$ 6. a. $\dfrac{x+2}{x+3}$ b. $\dfrac{4(t^2 - 4)}{(t^2 + 4)^2}$

7. a. 2 b. $\dfrac{3x(2x^3 + 2x + 1)}{(x^2 + 2)(x^3 + 1)}$

8. a. $\dfrac{x}{(x+2)(x-3)}$ b. $\dfrac{x\sqrt{1 + 3x^2}(3x^2 - 5x + 1)}{(x-1)^2}$

9. $\dfrac{3(1 - 2\sqrt{x})}{1 - 4x}$ 10. a. $[-4, 3]$ b. $\left[\dfrac{2}{3}, 2\right]$

1.1 Precalculus Review I

Sections 1.1 and 1.2 review some basic concepts and techniques of algebra that are essential in the study of calculus. The material in this review will help you work through the examples and exercises in this book. You can read through this material now and do the exercises in areas where you feel a little "rusty," or you can review the material on an as-needed basis as you study the text. The self-diagnostic test that precedes this section will help you pinpoint the areas where you might have any weaknesses.

The Real Number Line

The real number system is made up of the set of real numbers together with the usual operations of addition, subtraction, multiplication, and division.

We can represent real numbers geometrically by points on a **real number,** or **coordinate, line.** This line can be constructed as follows. Arbitrarily select a point on a straight line to represent the number 0. This point is called the **origin.** If the line is horizontal, then a point at a convenient distance to the right of the origin is chosen to represent the number 1. This determines the scale for the number line. Each positive real number lies at an appropriate distance to the right of the origin, and each negative real number lies at an appropriate distance to the left of the origin (Figure 1).

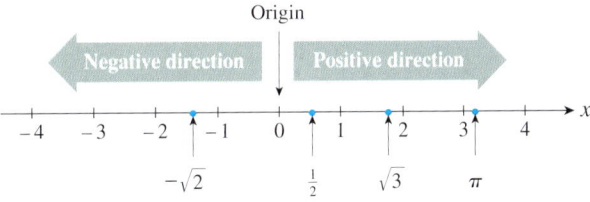

FIGURE 1
The real number line

A *one-to-one correspondence* is set up between the set of all real numbers and the set of points on the number line; that is, exactly one point on the line is associated with each real number. Conversely, exactly one real number is associated with each point on the line. The real number that is associated with a point on the real number line is called the **coordinate** of that point.

Intervals

Throughout this book, we will often restrict our attention to subsets of the set of real numbers. For example, if x denotes the number of cars rolling off a plant assembly line each day, then x must be nonnegative—that is, $x \geq 0$. Further, suppose management decides that the daily production must not exceed 200 cars. Then, x must satisfy the inequality $0 \leq x \leq 200$.

More generally, we will be interested in the following subsets of real numbers: open intervals, closed intervals, and half-open intervals. The set of all real numbers that lie *strictly* between two fixed numbers a and b is called an **open interval** (a, b). It consists of all real numbers x that satisfy the inequalities $a < x < b$, and it is called "open" because neither of its endpoints is included in the interval. A **closed interval** contains *both* of its endpoints. Thus, the set of all real numbers x that satisfy the inequalities $a \leq x \leq b$ is the closed interval $[a, b]$. Notice that square brackets are used to indicate that the endpoints are included in this interval. **Half-open intervals** contain only *one* of their endpoints. Thus, the interval $[a, b)$ is the set of all real numbers x that satisfy $a \leq x < b$, whereas the interval $(a, b]$ is described by the inequalities $a < x \leq b$. Examples of these **finite intervals** are illustrated in Table 1.

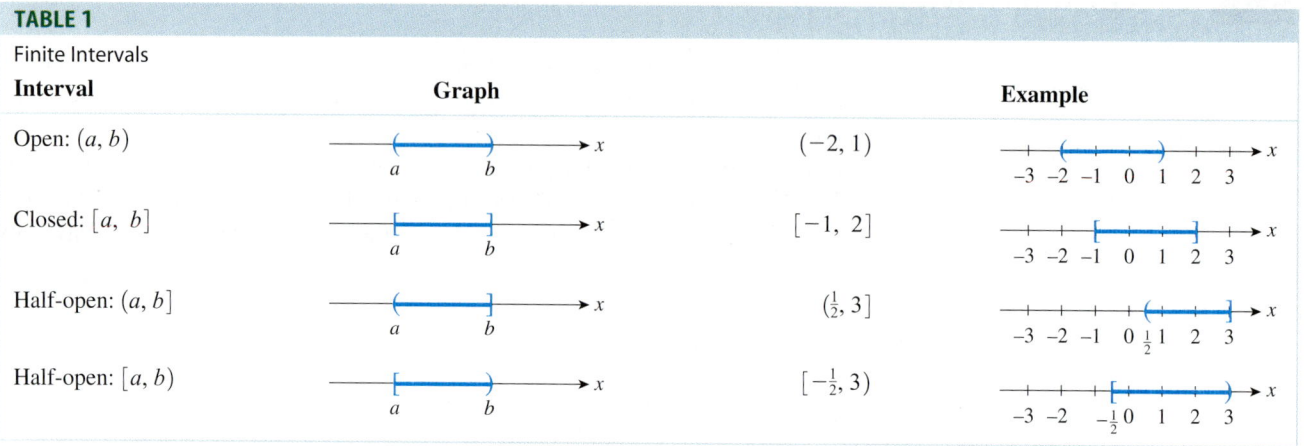

TABLE 1 Finite Intervals

In addition to finite intervals, we will encounter **infinite intervals**. Examples of infinite intervals are the half-lines (a, ∞), $[a, \infty)$, $(-\infty, a)$, and $(-\infty, a]$ defined by the set of all real numbers that satisfy $x > a$, $x \geq a$, $x < a$, and $x \leq a$, respectively. The symbol ∞, called *infinity*, is not a real number. It is used here only for notational purposes. The notation $(-\infty, \infty)$ is used for the set of all real numbers x, since by definition, the inequalities $-\infty < x < \infty$ hold for any real number x. Infinite intervals are illustrated in Table 2.

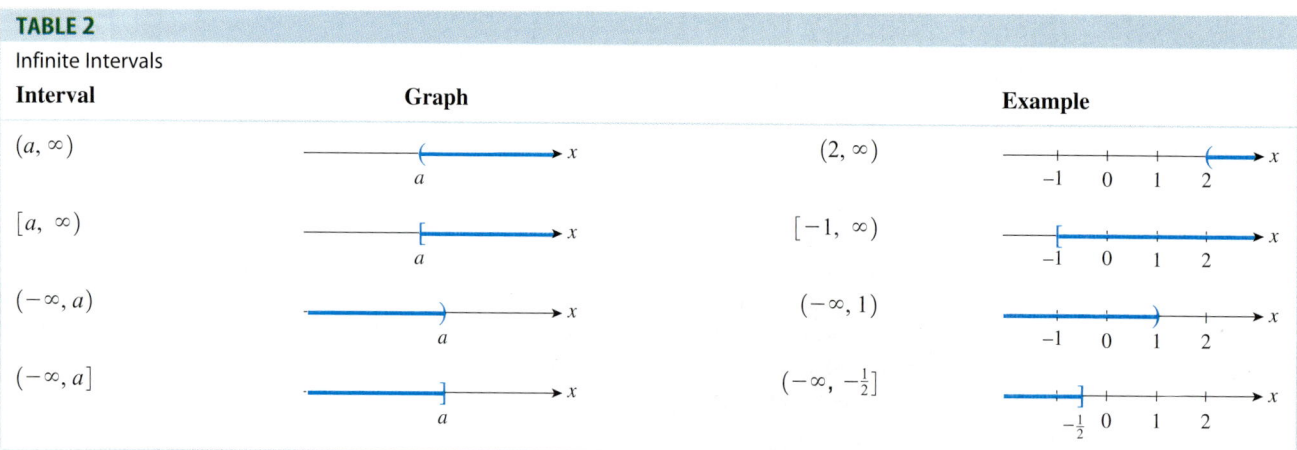

TABLE 2 Infinite Intervals

Exponents and Radicals

Recall that if b is any real number and n is a positive integer, then the expression b^n (read "b to the power n") is defined as the number

$$b^n = \underbrace{b \cdot b \cdot b \cdot \cdots \cdot b}_{n \text{ factors}}$$

The number b is called the **base,** and the superscript n is called the **power** of the exponential expression b^n. For example,

$$2^5 = 2 \cdot 2 \cdot 2 \cdot 2 \cdot 2 = 32 \quad \text{and} \quad \left(\frac{2}{3}\right)^3 = \left(\frac{2}{3}\right)\left(\frac{2}{3}\right)\left(\frac{2}{3}\right) = \frac{8}{27}$$

If $b \neq 0$, we define

$$b^0 = 1$$

For example, $2^0 = 1$ and $(-\pi)^0 = 1$, but the expression 0^0 is undefined.

Next, recall that if n is a positive integer, then the expression $b^{1/n}$ is defined to be the number that, when raised to the nth power, is equal to b. Thus,

$$(b^{1/n})^n = b$$

Such a number, if it exists, is called the **nth root of b,** also written $\sqrt[n]{b}$.

 If n is even, the nth root of a negative number is not defined. For example, the square root of -2 ($n = 2$) is not defined because there is no real number b such that $b^2 = -2$. Also, given a number b, more than one number might satisfy our definition of the nth root. For example, both 3 and -3 squared equal 9, and each is a square root of 9. So to avoid ambiguity, we define $b^{1/n}$ to be the positive nth root of b whenever it exists. Thus, $\sqrt{9} = 9^{1/2} = 3$. That's why your calculator will give the answer 3 when you use it to evaluate $\sqrt{9}$.

Next, recall that if p/q (where p and q are positive integers and $q \neq 0$) is a rational number in lowest terms, then the expression $b^{p/q}$ is defined as the number $(b^{1/q})^p$ or, equivalently, $\sqrt[q]{b^p}$, whenever it exists. For example,

$$2^{3/2} = (2^{1/2})^3 \approx (1.4142)^3 \approx 2.8283$$

Expressions involving negative rational exponents are taken care of by the definition

$$b^{-p/q} = \frac{1}{b^{p/q}}$$

Thus,

$$4^{-5/2} = \frac{1}{4^{5/2}} = \frac{1}{(4^{1/2})^5} = \frac{1}{2^5} = \frac{1}{32}$$

The rules defining the exponential expression a^n, where $a > 0$, for all rational values of n are given in Table 3.

The first three definitions in Table 3 are also valid for negative values of a. The fourth definition holds for all values of a if n is odd but only for nonnegative values of a if n is even. Thus,

$$(-8)^{1/3} = \sqrt[3]{-8} = -2 \quad \text{n is odd.}$$
$$(-8)^{1/2} \text{ has no real value} \quad \text{n is even.}$$

Finally, it can be shown that a^n has meaning for *all* real numbers n. For example, using a calculator with a $\boxed{y^x}$ key, we see that $2^{\sqrt{2}} \approx 2.665144$.

TABLE 3
Rules for Defining a^n

Definition of a^n ($a > 0$)	Example	Definition of a^n ($a > 0$)	Example
Integer exponent: If n is a positive integer, then $$a^n = a \cdot a \cdot a \cdots a$$ (n factors of a)	$2^5 = 2 \cdot 2 \cdot 2 \cdot 2 \cdot 2$ (5 factors) $= 32$	**Fractional exponent:** **a.** If n is a positive integer, then $$a^{1/n} \quad \text{or} \quad \sqrt[n]{a}$$ denotes the nth root of a.	$16^{1/2} = \sqrt{16}$ $= 4$
Zero exponent: If n is equal to zero, then $$a^0 = 1$$ (0^0 is not defined.)	$7^0 = 1$	**b.** If m and n are positive integers, then $$a^{m/n} = \sqrt[n]{a^m} = (\sqrt[n]{a})^m$$	$8^{2/3} = (\sqrt[3]{8})^2$ $= 4$
Negative exponent: If n is a positive integer, then $$a^{-n} = \frac{1}{a^n} \quad (a \neq 0)$$	$6^{-2} = \frac{1}{6^2}$ $= \frac{1}{36}$	**c.** If m and n are positive integers, then $$a^{-m/n} = \frac{1}{a^{m/n}} \quad (a \neq 0)$$	$9^{-3/2} = \frac{1}{9^{3/2}}$ $= \frac{1}{27}$

The five laws of exponents are listed in Table 4.

TABLE 4
Laws of Exponents

Law	Example
1. $a^m \cdot a^n = a^{m+n}$	$x^2 \cdot x^3 = x^{2+3} = x^5$
2. $\dfrac{a^m}{a^n} = a^{m-n} \quad (a \neq 0)$	$\dfrac{x^7}{x^4} = x^{7-4} = x^3$
3. $(a^m)^n = a^{m \cdot n}$	$(x^4)^3 = x^{4 \cdot 3} = x^{12}$
4. $(ab)^n = a^n \cdot b^n$	$(2x)^4 = 2^4 \cdot x^4 = 16x^4$
5. $\left(\dfrac{a}{b}\right)^n = \dfrac{a^n}{b^n} \quad (b \neq 0)$	$\left(\dfrac{x}{2}\right)^3 = \dfrac{x^3}{2^3} = \dfrac{x^3}{8}$

These laws are valid for any real numbers a, b, m, and n whenever the quantities are defined.

⚠️ Remember, $(x^2)^3 \neq x^5$. The correct equation is $(x^2)^3 = x^{2 \cdot 3} = x^6$.

The next several examples illustrate the use of the laws of exponents.

EXAMPLE 1 Simplify the expressions:

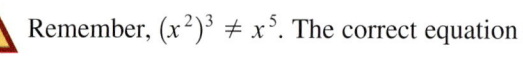

a. $(3x^2)(4x^3)$ **b.** $\dfrac{16^{5/4}}{16^{1/2}}$ **c.** $(6^{2/3})^3$ **d.** $(x^3 y^{-2})^{-2}$ **e.** $\left(\dfrac{y^{3/2}}{x^{1/4}}\right)^{-2}$

Solution

a. $(3x^2)(4x^3) = 12x^{2+3} = 12x^5$ Law 1

b. $\dfrac{16^{5/4}}{16^{1/2}} = 16^{5/4 - 1/2} = 16^{3/4} = (\sqrt[4]{16})^3 = 2^3 = 8$ Law 2

c. $(6^{2/3})^3 = 6^{(2/3)(3)} = 6^2 = 36$ Law 3

d. $(x^3 y^{-2})^{-2} = (x^3)^{-2}(y^{-2})^{-2} = x^{(3)(-2)} y^{(-2)(-2)} = x^{-6} y^4 = \dfrac{y^4}{x^6}$ Law 4

e. $\left(\dfrac{y^{3/2}}{x^{1/4}}\right)^{-2} = \dfrac{y^{(3/2)(-2)}}{x^{(1/4)(-2)}} = \dfrac{y^{-3}}{x^{-1/2}} = \dfrac{x^{1/2}}{y^3}$ Law 5

We can also use the laws of exponents to simplify expressions involving radicals, as illustrated in the next example.

EXAMPLE 2 Simplify the expressions. (Assume that x, y, m, and n are positive.)

a. $\sqrt[4]{16 x^4 y^8}$ **b.** $\sqrt{12 m^3 n} \cdot \sqrt{3 m^5 n}$ **c.** $\dfrac{\sqrt[3]{-27 x^6}}{\sqrt[3]{8 y^3}}$

Solution

a. $\sqrt[4]{16 x^4 y^8} = (16 x^4 y^8)^{1/4} = 16^{1/4} \cdot x^{4/4} y^{8/4} = 2xy^2$

b. $\sqrt{12 m^3 n} \cdot \sqrt{3 m^5 n} = \sqrt{36 m^8 n^2} = (36 m^8 n^2)^{1/2} = 36^{1/2} \cdot m^4 n = 6 m^4 n$

c. $\dfrac{\sqrt[3]{-27 x^6}}{\sqrt[3]{8 y^3}} = \dfrac{(-27 x^6)^{1/3}}{(8 y^3)^{1/3}} = \dfrac{-27^{1/3} x^2}{8^{1/3} y} = -\dfrac{3 x^2}{2 y}$

If a radical appears in the numerator or denominator of an algebraic expression, we often try to simplify the expression by eliminating the radical from the numerator or denominator. This process, called **rationalization**, is illustrated in the next two examples.

EXAMPLE 3 Rationalize the denominator of the expression $\dfrac{3x}{2\sqrt{x}}$.

Solution

$$\dfrac{3x}{2\sqrt{x}} = \dfrac{3x}{2\sqrt{x}} \cdot \dfrac{\sqrt{x}}{\sqrt{x}} = \dfrac{3x\sqrt{x}}{2\sqrt{x^2}} = \dfrac{3x\sqrt{x}}{2x} = \dfrac{3}{2}\sqrt{x}$$

EXAMPLE 4 Express $\dfrac{1}{2} x^{-1/2}$ as a radical and rationalize the denominator of the expression that you obtain.

Solution

$$\dfrac{1}{2} x^{-1/2} = \dfrac{1}{2\sqrt{x}} \cdot \dfrac{\sqrt{x}}{\sqrt{x}} = \dfrac{\sqrt{x}}{2x}$$

EXAMPLE 5 Rationalize the numerator of the expression $\dfrac{3\sqrt{x}}{2x}$.

Solution

$$\dfrac{3\sqrt{x}}{2x} = \dfrac{3\sqrt{x}}{2x} \cdot \dfrac{\sqrt{x}}{\sqrt{x}} = \dfrac{3\sqrt{x^2}}{2x\sqrt{x}} = \dfrac{3x}{2x\sqrt{x}} = \dfrac{3}{2\sqrt{x}}$$

Operations with Algebraic Expressions

In calculus, we often work with algebraic expressions such as

$$2x^{4/3} - x^{1/3} + 1 \qquad 2x^2 - x - \frac{2}{\sqrt{x}} \qquad \frac{3xy + 2}{x + 1} \qquad 2x^3 + 2x + 1$$

An algebraic expression of the form $ax^m y^n$, where the coefficient a is a real number and m and n are nonnegative integers, is called a **monomial,** meaning it consists of one term. For example, $7x^2$ is a monomial. A **polynomial** is a monomial or the sum of two or more monomials. For example,

$$x^2 + 4x + 4 \qquad x^3 + 5 \qquad x^4 + 3x^2 + 3 \qquad x^2 y + xy + y$$

are all polynomials. The degree of a polynomial is the highest power $(m + n)$ of the variables that appears in the polynomial.

Constant terms and terms containing the same variable factor are called **like,** or **similar, terms.** Like terms may be combined by adding or subtracting their numerical coefficients. For example,

$$3x + 7x = 10x \quad \text{and} \quad \frac{1}{2}xy + 3xy = \frac{7}{2}xy$$

The distributive property of the real number system,

$$ab + ac = a(b + c)$$

is used to justify this procedure.

To add or subtract two or more algebraic expressions, first remove the parentheses and then combine like terms. The resulting expression is written in order of nonincreasing degree from left to right.

EXAMPLE 6

a. $(2x^4 + 3x^3 + 4x + 6) - (3x^4 + 9x^3 + 3x^2)$

$\qquad = 2x^4 + 3x^3 + 4x + 6 - 3x^4 - 9x^3 - 3x^2$ Remove parentheses.

$\qquad = 2x^4 - 3x^4 + 3x^3 - 9x^3 - 3x^2 + 4x + 6$

$\qquad = -x^4 - 6x^3 - 3x^2 + 4x + 6$ Combine like terms.

b. $2t^3 - \{t^2 - [t - (2t - 1)] + 4\}$

$\qquad = 2t^3 - \{t^2 - [t - 2t + 1] + 4\}$

$\qquad = 2t^3 - \{t^2 - [-t + 1] + 4\}$ Remove parentheses and combine like terms within brackets.

$\qquad = 2t^3 - \{t^2 + t - 1 + 4\}$ Remove brackets.

$\qquad = 2t^3 - \{t^2 + t + 3\}$ Combine like terms within braces.

$\qquad = 2t^3 - t^2 - t - 3$ Remove braces.

Observe that when the algebraic expression in Example 6b was simplified, the innermost grouping symbols were removed first; that is, the parentheses ( ) were removed first, the brackets [ ] second, and the braces { } third.

When algebraic expressions are multiplied, each term of one algebraic expression is multiplied by each term of the other. The resulting algebraic expression is then simplified.

EXAMPLE 7 Perform the indicated operations:

a. $(x^2 + 1)(3x^2 + 10x + 3)$ **b.** $x\left(300 - \frac{1}{4}x - \frac{1}{8}y\right) + y\left(240 - \frac{1}{8}x - \frac{3}{8}y\right)$

c. $(e^t + e^{-t})e^t - e^t(e^t - e^{-t})$

Solution

a. $(x^2 + 1)(3x^2 + 10x + 3) = x^2(3x^2 + 10x + 3) + 1(3x^2 + 10x + 3)$
$= 3x^4 + 10x^3 + 3x^2 + 3x^2 + 10x + 3$
$= 3x^4 + 10x^3 + 6x^2 + 10x + 3$

b. $x\left(300 - \frac{1}{4}x - \frac{1}{8}y\right) + y\left(240 - \frac{1}{8}x - \frac{3}{8}y\right)$
$= 300x - \frac{1}{4}x^2 - \frac{1}{8}xy + 240y - \frac{1}{8}xy - \frac{3}{8}y^2$
$= -\frac{1}{4}x^2 - \frac{3}{8}y^2 - \frac{1}{4}xy + 300x + 240y$

c. $(e^t + e^{-t})e^t - e^t(e^t - e^{-t}) = e^{2t} + e^0 - e^{2t} + e^0$
$= e^{2t} - e^{2t} + e^0 + e^0$
$= 1 + 1$ Recall that $e^0 = 1$.
$= 2$

Certain product formulas that are frequently used in algebraic computations are given in Table 5.

TABLE 5
Some Useful Product Formulas

Formula	Example
$(a + b)^2 = a^2 + 2ab + b^2$	$(2x + 3y)^2 = (2x)^2 + 2(2x)(3y) + (3y)^2$
	$= 4x^2 + 12xy + 9y^2$
$(a - b)^2 = a^2 - 2ab + b^2$	$(4x - 2y)^2 = (4x)^2 - 2(4x)(2y) + (2y)^2$
	$= 16x^2 - 16xy + 4y^2$
$(a + b)(a - b) = a^2 - b^2$	$(2x + y)(2x - y) = (2x)^2 - (y)^2$
	$= 4x^2 - y^2$

Factoring

Factoring is the process of expressing an algebraic expression as a product of other algebraic expressions. For example, by applying the distributive property, we may write

$$3x^2 - x = x(3x - 1)$$

To factor an algebraic expression, first check to see whether any of its terms have common factors. If they do, then factor out the greatest common factor. For example, the common factor of the algebraic expression $2a^2x + 4ax + 6a$ is $2a$ because

$$2a^2x + 4ax + 6a = 2a \cdot ax + 2a \cdot 2x + 2a \cdot 3 = 2a(ax + 2x + 3)$$

EXAMPLE 8 Factor out the greatest common factor in each expression:

a. $-3t^2 + 3t$ **b.** $2x^{3/2} - 3x^{1/2}$ **c.** $2ye^{xy^2} + 2xy^3e^{xy^2}$

d. $4x(x + 1)^{1/2} - 2x^2\left(\frac{1}{2}\right)(x + 1)^{-1/2}$

Solution

a. $-3t^2 + 3t = -3t(t - 1)$
b. $2x^{3/2} - 3x^{1/2} = x^{1/2}(2x - 3)$
c. $2ye^{xy^2} + 2xy^3e^{xy^2} = 2ye^{xy^2}(1 + xy^2)$

d. $4x(x+1)^{1/2} - 2x^2\left(\dfrac{1}{2}\right)(x+1)^{-1/2} = 4x(x+1)^{1/2} - x^2(x+1)^{-1/2}$

$\qquad = x(x+1)^{-1/2}[4(x+1)^{1/2}(x+1)^{1/2} - x]$
$\qquad = x(x+1)^{-1/2}[4(x+1) - x]$
$\qquad = x(x+1)^{-1/2}(4x + 4 - x) = x(x+1)^{-1/2}(3x+4)$

Here we select $(x+1)^{-1/2}$ as the greatest common factor because it is the highest power of $(x+1)$ in each algebraic term. In particular, observe that

$$(x+1)^{-1/2}(x+1)^{1/2}(x+1)^{1/2} = (x+1)^{-1/2+1/2+1/2} = (x+1)^{1/2}$$ ∎

Sometimes an algebraic expression may be factored by regrouping and rearranging its terms so that a common term can be factored out. This technique is illustrated in Example 9.

EXAMPLE 9 Factor:

a. $2ax + 2ay + bx + by$ **b.** $3x\sqrt{y} - 4 - 2\sqrt{y} + 6x$

Solution

a. First, factor the common term $2a$ from the first two terms and the common term b from the last two terms. Thus,

$$2ax + 2ay + bx + by = 2a(x+y) + b(x+y)$$

Since $(x+y)$ is common to both terms of the polynomial, we may factor it out. Hence

$$2a(x+y) + b(x+y) = (2a+b)(x+y)$$

b. $3x\sqrt{y} - 4 - 2\sqrt{y} + 6x = 3x\sqrt{y} - 2\sqrt{y} + 6x - 4$ *Rearrange terms.*
$\qquad\qquad\qquad\qquad\quad = \sqrt{y}(3x-2) + 2(3x-2)$ *Factor out common factors.*
$\qquad\qquad\qquad\qquad\quad = (3x-2)(\sqrt{y}+2)$ ∎

As we have seen, the first step in factoring a polynomial is to find the common factors. The next step is to express the polynomial as the product of a constant and/or one or more prime polynomials.

Certain product formulas that are useful in factoring binomials and trinomials are listed in Table 6.

TABLE 6
Product Formulas Used in Factoring

Formula	Example
Difference of two squares:	
$x^2 - y^2 = (x+y)(x-y)$	$x^2 - 36 = (x+6)(x-6)$
	$8x^2 - 2y^2 = 2(4x^2 - y^2)$
	$\qquad\qquad\quad = 2(2x+y)(2x-y)$
	$9 - a^6 = (3+a^3)(3-a^3)$
Perfect-square trinomial:	
$x^2 + 2xy + y^2 = (x+y)^2$	$x^2 + 8x + 16 = (x+4)^2$
$x^2 - 2xy + y^2 = (x-y)^2$	$4x^2 - 4xy + y^2 = (2x-y)^2$
Sum of two cubes:	
$x^3 + y^3 = (x+y)(x^2 - xy + y^2)$	$z^3 + 27 = z^3 + (3)^3$
	$\qquad\quad = (z+3)(z^2 - 3z + 9)$
Difference of two cubes:	
$x^3 - y^3 = (x-y)(x^2 + xy + y^2)$	$8x^3 - y^6 = (2x)^3 - (y^2)^3$
	$\qquad\quad\; = (2x - y^2)(4x^2 + 2xy^2 + y^4)$

The factors of the second-degree polynomial with integral coefficients

$$px^2 + qx + r$$

are $(ax + b)(cx + d)$, where $ac = p$, $ad + bc = q$, and $bd = r$. Since only a limited number of choices are possible, we can use a trial-and-error method to factor polynomials having this form.

For example, to factor $x^2 - 2x - 3$, we first observe that the only possible first-degree terms are

$$(x \quad)(x \quad) \quad \text{Since the coefficient of } x^2 \text{ is 1}$$

Next, we observe that the product of the constant terms is (-3). This gives us the following possible factors:

$$(x - 1)(x + 3)$$
$$(x + 1)(x - 3)$$

Looking once again at the polynomial $x^2 - 2x - 3$, we see that the coefficient of x is -2. Checking to see which set of factors yields -2 for the coefficient of x, we find that

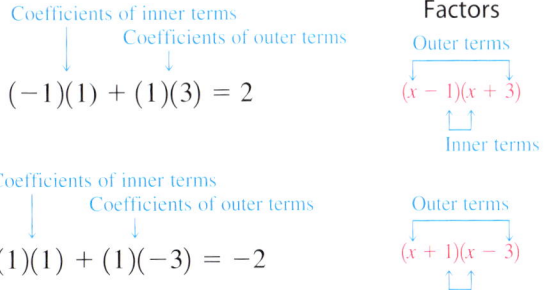

and we conclude that the correct factorization is

$$x^2 - 2x - 3 = (x + 1)(x - 3)$$

With practice, you will soon find that you can perform many of these steps mentally and the need to write out each step will be eliminated.

EXAMPLE 10 Factor:

a. $3x^2 + 4x - 4$ **b.** $3x^2 - 6x - 24$ **c.** $-3t^2 + 192t + 195$

Solution

a. Using trial and error, we find that the correct factorization is

$$3x^2 + 4x - 4 = (3x - 2)(x + 2)$$

b. Since each term has the common factor 3, we have

$$3x^2 - 6x - 24 = 3(x^2 - 2x - 8)$$

Using the trial-and-error method of factorization, we find that

$$x^2 - 2x - 8 = (x - 4)(x + 2)$$

Thus, we have

$$3x^2 - 6x - 24 = 3(x - 4)(x + 2)$$

c. Since each term has the common factor -3, we have

$$-3t^2 + 192t + 195 = -3(t^2 - 64t - 65)$$

Using the trial-and-error method of factorization, we find that

$$(t^2 - 64t - 65) = (t - 65)(t + 1)$$

Therefore,

$$-3t^2 + 192t + 195 = -3(t - 65)(t + 1)$$

Roots of Polynomial Equations

A polynomial equation of degree n in the variable x is an equation of the form

$$a_n x^n + a_{n-1} x^{n-1} + \cdots + a_0 = 0$$

where n is a nonnegative integer and $a_0, a_1, \ldots, a_n$ are real numbers with $a_n \neq 0$. For example, the equation

$$-2x^5 + 8x^3 - 6x^2 + 3x + 1 = 0$$

is a polynomial equation of degree 5 in x.

The **roots of a polynomial equation** are precisely the values of x that satisfy the given equation.* One way to find the roots of a polynomial equation is to factor the polynomial and then solve the resulting equation. For example, the polynomial equation

$$x^3 - 3x^2 + 2x = 0$$

may be rewritten in the form

$$x(x^2 - 3x + 2) = 0 \quad \text{or} \quad x(x-1)(x-2) = 0$$

Since the product of two real numbers can be equal to zero if and only if one (or both) of the factors is equal to zero, we have

$$x = 0 \qquad x - 1 = 0 \quad \text{or} \quad x - 2 = 0$$

from which we see that the desired roots are $x = 0, 1,$ and 2.

The Quadratic Formula

In general, the problem of finding the roots of a polynomial equation is a difficult one. But the roots of a quadratic equation (a polynomial equation of degree 2) are easily found either by factoring or by using the following quadratic formula.

> **Quadratic Formula**
>
> The solutions of the equation $ax^2 + bx + c = 0$ $(a \neq 0)$ are given by
>
> $$x = \frac{-b \pm \sqrt{b^2 - 4ac}}{2a}$$

Note If you use the quadratic formula to solve a quadratic equation, first make sure that the equation is in the *standard form* $ax^2 + bx + c = 0$. ∎

EXAMPLE 11 Solve each of the following quadratic equations:

a. $2x^2 + 5x - 12 = 0$ **b.** $x^2 = -3x + 8$

Solution

a. The equation is in standard form, with $a = 2$, $b = 5$, and $c = -12$. Using the quadratic formula, we find

$$x = \frac{-b \pm \sqrt{b^2 - 4ac}}{2a} = \frac{-5 \pm \sqrt{5^2 - 4(2)(-12)}}{2(2)}$$

$$= \frac{-5 \pm \sqrt{121}}{4} = \frac{-5 \pm 11}{4}$$

$$= -4 \quad \text{or} \quad \frac{3}{2}$$

---

*In this book, we are interested only in the *real* roots of an equation.

1.1 PRECALCULUS REVIEW I

This equation can also be solved by factoring. Thus,
$$2x^2 + 5x - 12 = (2x - 3)(x + 4) = 0$$
from which we see that the desired roots are $x = \frac{3}{2}$ or $x = -4$, as obtained earlier.

b. We first rewrite the given equation in the standard form $x^2 + 3x - 8 = 0$, from which we see that $a = 1$, $b = 3$, and $c = -8$. Using the quadratic formula, we find
$$x = \frac{-b \pm \sqrt{b^2 - 4ac}}{2a} = \frac{-3 \pm \sqrt{3^2 - 4(1)(-8)}}{2(1)}$$
$$= \frac{-3 \pm \sqrt{41}}{2}$$
That is, the solutions are
$$\frac{-3 + \sqrt{41}}{2} \approx 1.7 \quad \text{and} \quad \frac{-3 - \sqrt{41}}{2} \approx -4.7$$
In this case, the quadratic formula proves quite handy!

1.1 Exercises

In Exercises 1–6, show the interval on a number line.

1. $(3, 6)$ **2.** $(-2, 5]$ **3.** $[-1, 4)$

4. $\left[-\frac{6}{5}, -\frac{1}{2}\right]$ **5.** $(0, \infty)$ **6.** $(-\infty, 5]$

In Exercises 7–22, evaluate the expression.

7. $27^{2/3}$ **8.** $8^{-4/3}$

9. $\left(\frac{1}{\sqrt{5}}\right)^0$ **10.** $(7^{1/2})^6$

11. $\left[\left(\frac{1}{8}\right)^{1/3}\right]^{-2}$ **12.** $\left[\left(-\frac{1}{3}\right)^2\right]^{-3}$

13. $\left(\frac{8^{-5} \cdot 8^2}{8^{-2}}\right)^{-1}$ **14.** $\left(\frac{9}{16}\right)^{-1/2}$

15. $(125^{2/3})^{-1/2}$ **16.** $\sqrt[3]{2^6}$

17. $\frac{\sqrt{72}}{\sqrt{18}}$ **18.** $\sqrt[3]{\frac{-8}{27}}$

19. $\frac{16^{5/8} 16^{1/2}}{16^{7/8}}$ **20.** $\left(\frac{9^{-3.5} \cdot 9^{2.5}}{9^{-2}}\right)^{-0.5}$

21. $16^{1/4} \cdot 8^{-1/3}$ **22.** $\frac{6^{2.5} \cdot 6^{-1.9}}{6^{-1.4}}$

In Exercises 23–32, determine whether the statement is true or false. Give a reason for your choice.

23. $x^4 + 2x^4 = 3x^4$ **24.** $3^2 \cdot 2^2 = 6^2$

25. $x^3 \cdot 2x^2 = 2x^6$ **26.** $3^3 + 3^2 = 3^5$

27. $\frac{2^{4x}}{1^{3x}} = 2^{4x-3x}$ **28.** $(2^2 \cdot 3^2)^2 = 6^4$

29. $\frac{1}{4^{-3}} = \frac{1}{64}$ **30.** $\frac{4^{3/2}}{2^4} = \frac{1}{2}$

31. $(1.2^{1/2})^{-1/2} = 1$ **32.** $5^{2/3} \cdot (25)^{2/3} = 25$

In Exercises 33–38, rewrite the expression using positive exponents only.

33. $(xy)^{-2}$ **34.** $3s^{1/3} \cdot 2s^{-7/3}$

35. $\frac{3x^{-1/3}}{x^{1/2}}$ **36.** $\sqrt{4x^{-1}} \cdot \sqrt{9x^{-3}}$

37. $12^0(s + t)^{-3}$ **38.** $(x - y)(x^{-1} + y^{-1})$

In Exercises 39–54, simplify the expression. (Assume that x, y, r, s, and t are positive.)

39. $\frac{x^{7/3}}{x^{-2}}$ **40.** $(49x^{-2})^{-1/2}$

41. $(x^2 y^{-3})(x^{-5} y^3)$ **42.** $\frac{5x^{5/2} y^{3/2}}{2x^{3/2} y^{7/4}}$

43. $\frac{x^{3/4}}{x^{-1/4}}$ **44.** $\left(\frac{x^3 y^2}{z^2}\right)^2$

45. $\left(\frac{x^3}{-27y^{-6}}\right)^{-2/3}$ **46.** $\left(16 \frac{e^x}{e^{x-2}}\right)^{-1/2}$

47. $\left(\frac{x^{-3}}{y^{-2}}\right)^2 \left(\frac{y}{x}\right)^4$ **48.** $\frac{8^{2/3}(r^n)^4}{2^4 r^{5-2n}}$

49. $\sqrt[3]{x^{-2}} \cdot \sqrt{4x^5}$ **50.** $\sqrt{121x^6 y^{-4}}$

51. $-\sqrt[4]{16x^4 y^8}$ **52.** $\sqrt[3]{x^{3a+b}}$

53. $\sqrt[6]{64x^8 y^3}$ **54.** $\sqrt[3]{27r^6} \cdot \sqrt{9s^2 t^4}$

In Exercises 55–58, use the fact that $2^{1/2} \approx 1.414$ and $3^{1/2} \approx 1.732$ to evaluate the expression without using a calculator.

55. $2^{3/2}$ **56.** $8^{1/2}$ **57.** $9^{3/4}$ **58.** $6^{1/2}$

In Exercises 59–62, use the fact that $10^{1/2} \approx 3.162$ and $10^{1/3} \approx 2.154$ to evaluate the expression without using a calculator.

59. $10^{3/2}$ **60.** $1000^{3/2}$

61. $10^{2.5}$ **62.** $(0.0001)^{-1/3}$

In Exercises 63–68, rationalize the denominator.

63. $\dfrac{3}{2\sqrt{x}}$ **64.** $\dfrac{5}{2\sqrt{xy}}$ **65.** $\dfrac{2y}{\sqrt{3y}}$

66. $\dfrac{5x^2}{\sqrt{5x}}$ **67.** $\dfrac{1}{\sqrt[3]{x}}$ **68.** $\sqrt{\dfrac{2x}{y}}$

In Exercises 69–74, rationalize the numerator.

69. $\dfrac{2\sqrt{x}}{3}$ **70.** $\dfrac{\sqrt[3]{8x}}{24}$ **71.** $\sqrt{\dfrac{2y}{3x}}$

72. $\sqrt[3]{\dfrac{2x}{3y}}$ **73.** $\dfrac{\sqrt[3]{x^2 z}}{y}$ **74.** $\dfrac{\sqrt[3]{x^2 y}}{2x}$

In Exercises 75–98, perform the indicated operations and/or simplify each expression.

75. $(7x^2 - 3x + 5) + (2x^2 + 5x - 4)$

76. $(3x^2 + 5xy + 2y) + (4 - 3xy - 2x^2)$

77. $(5y^2 - 2y + 1) - (y^2 - 3y - 7)$

78. $3(2a - b) - 4(b - 2a)$

79. $x - \{2x - [-x - (1 - x)]\}$

80. $3x^2 - \{x^2 + 1 - x[x - (2x - 1)]\} + 2$

81. $\left(\dfrac{1}{3} - 1 + e\right) - \left(-\dfrac{1}{3} - 1 + e^{-1}\right)$

82. $-\dfrac{3}{4}y - \dfrac{1}{4}x + 100 + \dfrac{1}{2}x + \dfrac{1}{4}y - 120$

83. $3\sqrt{8} + 8 - 2\sqrt{y} + \dfrac{1}{2}\sqrt{x} - \dfrac{3}{4}\sqrt{y}$

84. $\dfrac{8}{9}x^2 + \dfrac{2}{3}x + \dfrac{16}{3}x^2 - \dfrac{16}{3}x - 2x + 2$

85. $(3x + 8)(x - 2)$ **86.** $(5x + 2)(3x - 4)$

87. $(a + 5)^2$ **88.** $(3a - 4b)^2$

89. $(x + 2y)^2$ **90.** $(6y - 3x)^2$

91. $(2x + y)(2x - y)$ **92.** $(3x + 2)(2 - 3x)$

93. $(2x^2 - 1)(3x^2) + (x^2 + 3)(4x)$

94. $(x^2 - 1)(2x) - x^2(2x)$

95. $6x\left(\dfrac{1}{2}\right)(2x^2 + 3)^{-1/2}(4x) + 6(2x^2 + 3)^{1/2}$

96. $(x^{1/2} + 1)\left(\dfrac{1}{2}x^{-1/2}\right) - (x^{1/2} - 1)\left(\dfrac{1}{2}x^{-1/2}\right)$

97. $100(-10te^{-0.1t} - 100e^{-0.1t})$

98. $2(t + \sqrt{t})^2 - 2t^2$

In Exercises 99–106, factor out the greatest common factor from each expression.

99. $4x^5 - 12x^4 - 6x^3$

100. $4x^2y^2z - 2x^5y^2 + 6x^3y^2z^2$

101. $7a^4 - 42a^2b^2 + 49a^3b$

102. $3x^{2/3} - 2x^{1/3}$

103. $3e^{-x} - 9xe^{-x}$

104. $2ye^{xy^2} + 2xy^3 e^{xy^2}$

105. $2x^{-5/2}y^2 - \dfrac{3}{2}x^{-3/2}y^4$

106. $\dfrac{1}{2}\left(\dfrac{2}{3}u^{5/2} - 2u^{3/2}\right)$

In Exercises 107–120, factor each expression completely.

107. $6ac + 3bc - 4ad - 2bd$

108. $3x^3 - x^2 + 3x - 1$

109. $4a^2 - 9b^2$ **110.** $12x^2 - 3y^2$

111. $10 - 14x - 12x^2$ **112.** $6x^2 - 7x - 20$

113. $3x^2 - 6x - 24$ **114.** $15x^2 - 4x - 4$

115. $12x^2 - 2x - 30$ **116.** $(x + y)^2 - 1$

117. $9x^2 - 16y^2$ **118.** $8a^2 - 2ab - 6b^2$

119. $8x^6 + 125$ **120.** $27x^3 - 64$

In Exercises 121–128, perform the indicated operations and simplify each expression.

121. $(x^2 + y^2)x - xy(2y)$

122. $2kr(R - r) - kr^2$

123. $2(x - 1)(2x + 2)^3[4(x - 1) + (2x + 2)]$

124. $5x^2(3x^2 + 1)^4(6x) + (3x^2 + 1)^5(2x)$

125. $4(x - 1)^2(2x + 2)^3(2) + (2x + 2)^4(2)(x - 1)$

126. $(x^2 + 1)(4x^3 - 3x^2 + 2x) - (x^4 - x^3 + x^2)(2x)$

127. $(x^2 + 2)^2[5(x^2 + 2)^2 - 3](2x)$

128. $(x^2 - 4)(x^2 + 4)(2x + 8) - (x^2 + 8x - 4)(4x^3)$

In Exercises 129–134, find the real roots of each equation by factoring.

129. $x^2 + x - 12 = 0$ **130.** $3x^2 - x - 4 = 0$

131. $4t^2 + 2t - 2 = 0$ **132.** $-6x^2 + x + 12 = 0$

133. $\frac{1}{4}x^2 - x + 1 = 0$ **134.** $\frac{1}{2}a^2 + a - 12 = 0$

In Exercises 135–140, solve the equation by using the quadratic formula.

135. $4x^2 + 5x - 6 = 0$ **136.** $3x^2 - 4x + 1 = 0$

137. $8x^2 - 8x - 3 = 0$ **138.** $x^2 - 6x + 6 = 0$

139. $2x^2 + 4x - 3 = 0$ **140.** $2x^2 + 7x - 15 = 0$

141. DISTRIBUTION OF INCOMES The distribution of income in a certain city can be described by the mathematical model $y = (5.6 \cdot 10^{11})(x)^{-1.5}$, where y is the number of families with an income of x or more dollars.

a. How many families in this city have an income of $30,000 or more?
b. How many families have an income of $60,000 or more?
c. How many families have an income of $150,000 or more?

In Exercises 142–144, determine whether the statement is true or false. If it is true, explain why it is true. If it is false, give an example to show why it is false.

142. If $b^2 - 4ac > 0$, then $ax^2 + bx + c = 0$ ($a \neq 0$) has two real roots.

143. If $b^2 - 4ac < 0$, then $ax^2 + bx + c = 0$ ($a \neq 0$) has no real roots.

144. $\sqrt{(a+b)(b-a)} = \sqrt{b^2 - a^2}$ for all real numbers a and b.

1.2 Precalculus Review II

Rational Expressions

Quotients of polynomials are called **rational expressions**. Examples of rational expressions are

$$\frac{6x - 1}{2x + 3} \qquad \frac{3x^2y^3 - 2xy}{4x} \qquad \frac{2}{5ab}$$

Since rational expressions are quotients in which the variables represent real numbers, the properties of real numbers apply to rational expressions as well, and operations with rational fractions are performed in the same manner as operations with arithmetic fractions. For example, using the properties of the real number system, we may write

$$\frac{ac}{bc} = \frac{a}{b} \cdot \frac{c}{c} = \frac{a}{b} \cdot 1 = \frac{a}{b}$$

where a, b, and c are any real numbers and b and c are not zero.

Similarly, using the same properties of real numbers, we may write

$$\frac{(x+2)(x-3)}{(x-2)(x-3)} = \frac{x+2}{x-2} \qquad (x \neq 2, 3)$$

after "canceling" the common factors.

 An example of incorrect cancellation is

$$\frac{\cancel{3} + 4x}{\cancel{3}} \neq 1 + 4x$$

because 3 is not a factor of the numerator. Instead, we need to write

$$\frac{3 + 4x}{3} = \frac{3}{3} + \frac{4x}{3} = 1 + \frac{4x}{3}$$

An algebraic fraction is simplified, or in lowest terms, when the numerator and denominator have no common factors other than 1 and -1 and the fraction contains no negative exponents.

EXAMPLE 1 Simplify the following expressions:

a. $\dfrac{x^2 + 2x - 3}{x^2 + 4x + 3}$ b. $\dfrac{(x^2 + 1)^2(-2) + (2x)(2)(x^2 + 1)(2x)}{(x^2 + 1)^4}$

Solution

a. $\dfrac{x^2 + 2x - 3}{x^2 + 4x + 3} = \dfrac{(x + 3)(x - 1)}{(x + 3)(x + 1)} = \dfrac{x - 1}{x + 1}$

b. $\dfrac{(x^2 + 1)^2(-2) + (2x)(2)(x^2 + 1)(2x)}{(x^2 + 1)^4}$

$= \dfrac{(x^2 + 1)[(x^2 + 1)(-2) + (2x)(2)(2x)]}{(x^2 + 1)^4}$ Factor out $(x^2 + 1)$.

$= \dfrac{(x^2 + 1)(-2x^2 - 2 + 8x^2)}{(x^2 + 1)^4}$ Carry out indicated multiplication.

$= \dfrac{(x^2 + 1)(6x^2 - 2)}{(x^2 + 1)^4}$ Combine like terms.

$= \dfrac{(6x^2 - 2)}{(x^2 + 1)^3}$ Cancel the common factors.

$= \dfrac{2(3x^2 - 1)}{(x^2 + 1)^3}$ Factor out 2 from the numerator.

The operations of multiplication and division are performed with algebraic fractions in the same manner as with arithmetic fractions (Table 7).

TABLE 7
Rules of Multiplication and Division: Algebraic Fractions

Operation	Example
If $P, Q, R,$ and S are polynomials, then	
Multiplication:	
$\dfrac{P}{Q} \cdot \dfrac{R}{S} = \dfrac{PR}{QS}$ $(Q, S \neq 0)$	$\dfrac{2x}{y} \cdot \dfrac{(x + 1)}{(y - 1)} = \dfrac{2x(x + 1)}{y(y - 1)} = \dfrac{2x^2 + 2x}{y^2 - y}$
Division:	
$\dfrac{P}{Q} \div \dfrac{R}{S} = \dfrac{P}{Q} \cdot \dfrac{S}{R} = \dfrac{PS}{QR}$ $(Q, R, S \neq 0)$	$\dfrac{x^2 + 3}{y} \div \dfrac{y^2 + 1}{x} = \dfrac{x^2 + 3}{y} \cdot \dfrac{x}{y^2 + 1} = \dfrac{x^3 + 3x}{y^3 + y}$

When rational expressions are multiplied and divided, the resulting expressions should be simplified if possible.

EXAMPLE 2 Perform the indicated operations and simplify:

$$\dfrac{2x - 8}{x + 2} \cdot \dfrac{x^2 + 4x + 4}{x^2 - 16}$$

Solution

$$\frac{2x-8}{x+2} \cdot \frac{x^2+4x+4}{x^2-16} = \frac{2(x-4)}{x+2} \cdot \frac{(x+2)^2}{(x+4)(x-4)}$$

$$= \frac{2(x-4)(x+2)(x+2)}{(x+2)(x+4)(x-4)}$$

$$= \frac{2(x+2)}{x+4} \quad \text{Cancel the common factors } (x+2)(x-4).$$

For rational expressions, the operations of addition and subtraction are performed by finding a common denominator of the fractions and then adding or subtracting the numerators. Table 8 shows the rules for fractions with equal denominators.

TABLE 8
Rules of Addition and Subtraction: Fractions with Equal Denominators

Operation	Example
If P, Q, and R are polynomials, then	
Addition: $\dfrac{P}{R} + \dfrac{Q}{R} = \dfrac{P+Q}{R} \quad (R \neq 0)$	$\dfrac{2x}{x+2} + \dfrac{6x}{x+2} = \dfrac{2x+6x}{x+2} = \dfrac{8x}{x+2}$
Subtraction: $\dfrac{P}{R} - \dfrac{Q}{R} = \dfrac{P-Q}{R} \quad (R \neq 0)$	$\dfrac{3y}{y-x} - \dfrac{y}{y-x} = \dfrac{3y-y}{y-x} = \dfrac{2y}{y-x}$

$$\frac{x}{2+y} \neq \frac{x}{2} + \frac{x}{y}$$

To add or subtract fractions that have different denominators, first find a common denominator, preferably the least common denominator (LCD). Then carry out the indicated operations following the procedure described in Table 8.

To find the LCD of two or more rational expressions:

1. *Find the prime factors of each denominator.*
2. *Form the product of the different prime factors that occur in the denominators. Each prime factor in this product should be raised to the highest power of that factor appearing in the denominators.*

EXAMPLE 3 Simplify:

a. $\dfrac{2x}{x^2+1} + \dfrac{6(3x^2)}{x^3+2}$ b. $\dfrac{1}{x+h} - \dfrac{1}{x}$

Solution

a. $\dfrac{2x}{x^2+1} + \dfrac{6(3x^2)}{x^3+2} = \dfrac{2x(x^3+2) + 6(3x^2)(x^2+1)}{(x^2+1)(x^3+2)}$ LCD $= (x^2+1)(x^3+2)$

$$= \frac{2x^4 + 4x + 18x^4 + 18x^2}{(x^2+1)(x^3+2)} \quad \text{Carry out the indicated multiplication.}$$

$$= \frac{20x^4 + 18x^2 + 4x}{(x^2+1)(x^3+2)} \quad \text{Combine like terms.}$$

$$= \frac{2x(10x^3 + 9x + 2)}{(x^2+1)(x^3+2)} \quad \text{Factor.}$$

b. $\dfrac{1}{x+h} - \dfrac{1}{x} = \dfrac{x-(x+h)}{x(x+h)}$ LCD $= x(x+h)$

$$= \dfrac{x-x-h}{x(x+h)}$$ Remove parentheses.

$$= \dfrac{-h}{x(x+h)}$$ Combine like terms.

Other Algebraic Fractions

The techniques used to simplify rational expressions may also be used to simplify algebraic fractions in which the numerator and denominator are not polynomials, as illustrated in Example 4.

EXAMPLE 4 Simplify:

a. $\dfrac{1 + \dfrac{1}{x+1}}{x - \dfrac{4}{x}}$ **b.** $\dfrac{x^{-1} + y^{-1}}{x^{-2} - y^{-2}}$

Solution

a. $\dfrac{1 + \dfrac{1}{x+1}}{x - \dfrac{4}{x}} = \dfrac{\dfrac{x+1+1}{x+1}}{\dfrac{x^2-4}{x}}$ LCD for numerator is $x+1$ and LCD for denominator is x.

$$= \dfrac{x+2}{x+1} \cdot \dfrac{x}{x^2-4} = \dfrac{x+2}{x+1} \cdot \dfrac{x}{(x+2)(x-2)}$$

$$= \dfrac{x}{(x+1)(x-2)}$$

b. $\dfrac{x^{-1} + y^{-1}}{x^{-2} - y^{-2}} = \dfrac{\dfrac{1}{x} + \dfrac{1}{y}}{\dfrac{1}{x^2} - \dfrac{1}{y^2}} = \dfrac{\dfrac{y+x}{xy}}{\dfrac{y^2-x^2}{x^2y^2}}$ $x^{-n} = \dfrac{1}{x^n}$

$$= \dfrac{y+x}{xy} \cdot \dfrac{x^2y^2}{y^2-x^2} = \dfrac{y+x}{xy} \cdot \dfrac{x^2y^2}{(y+x)(y-x)}$$

$$= \dfrac{xy}{y-x}$$

EXAMPLE 5 Perform the given operations and simplify:

a. $\dfrac{x^2(2x^2+1)^{1/2}}{x-1} \cdot \dfrac{4x^3-6x^2+x-2}{x(x-1)(2x^2+1)}$ **b.** $\dfrac{12x^2}{\sqrt{2x^2+3}} + 6\sqrt{2x^2+3}$

Solution

a. $\dfrac{x^2(2x^2+1)^{1/2}}{x-1} \cdot \dfrac{4x^3-6x^2+x-2}{x(x-1)(2x^2+1)} = \dfrac{x(4x^3-6x^2+x-2)}{(x-1)^2(2x^2+1)^{1-1/2}}$

$$= \dfrac{x(4x^3-6x^2+x-2)}{(x-1)^2(2x^2+1)^{1/2}}$$

b. $\dfrac{12x^2}{\sqrt{2x^2+3}} + 6\sqrt{2x^2+3} = \dfrac{12x^2}{(2x^2+3)^{1/2}} + 6(2x^2+3)^{1/2}$ Write radicals in exponential form.

$$= \dfrac{12x^2 + 6(2x^2+3)^{1/2}(2x^2+3)^{1/2}}{(2x^2+3)^{1/2}}$$ LCD is $(2x^2+3)^{1/2}$.

$$= \dfrac{12x^2 + 6(2x^2+3)}{(2x^2+3)^{1/2}}$$

$$= \dfrac{24x^2 + 18}{(2x^2+3)^{1/2}} = \dfrac{6(4x^2+3)}{\sqrt{2x^2+3}}$$

Rationalizing Algebraic Fractions

When the denominator of an algebraic fraction contains sums or differences involving radicals, we may **rationalize the denominator**—that is, transform the fraction into an equivalent one with a denominator that does not contain radicals. In doing so, we make use of the fact that

$$(\sqrt{a} + \sqrt{b})(\sqrt{a} - \sqrt{b}) = (\sqrt{a})^2 - (\sqrt{b})^2$$
$$= a - b$$

This procedure is illustrated in Example 6.

EXAMPLE 6 Rationalize the denominator: $\dfrac{1}{1+\sqrt{x}}$.

Solution Upon multiplying the numerator and the denominator by $(1-\sqrt{x})$, we obtain

$$\dfrac{1}{1+\sqrt{x}} = \dfrac{1}{1+\sqrt{x}} \cdot \dfrac{1-\sqrt{x}}{1-\sqrt{x}} \quad \dfrac{1-\sqrt{x}}{1-\sqrt{x}} = 1$$

$$= \dfrac{1-\sqrt{x}}{1-(\sqrt{x})^2}$$

$$= \dfrac{1-\sqrt{x}}{1-x}$$

In other situations, it may be necessary to rationalize the numerator of an algebraic expression. In calculus, for example, one encounters the following problem.

EXAMPLE 7 Rationalize the numerator: $\dfrac{\sqrt{1+h}-1}{h}$.

Solution

$$\dfrac{\sqrt{1+h}-1}{h} = \dfrac{\sqrt{1+h}-1}{h} \cdot \dfrac{\sqrt{1+h}+1}{\sqrt{1+h}+1}$$

$$= \dfrac{(\sqrt{1+h})^2 - (1)^2}{h(\sqrt{1+h}+1)}$$

$$= \dfrac{1+h-1}{h(\sqrt{1+h}+1)} \quad (\sqrt{1+h})^2 = \sqrt{1+h} \cdot \sqrt{1+h}$$
$$= 1+h$$

$$= \dfrac{h}{h(\sqrt{1+h}+1)}$$

$$= \dfrac{1}{\sqrt{1+h}+1}$$

Inequalities

The following properties may be used to solve one or more inequalities involving a variable.

> **Properties of Inequalities**
> If a, b, and c are any real numbers, then
>
> | | | Example |
> |---|---|---|
> | **Property 1** | If $a < b$ and $b < c$, then $a < c$. | $2 < 3$ and $3 < 8$, so $2 < 8$. |
> | **Property 2** | If $a < b$, then $a + c < b + c$. | $-5 < -3$, so $-5 + 2 < -3 + 2$; that is, $-3 < -1$. |
> | **Property 3** | If $a < b$ and $c > 0$, then $ac < bc$. | $-5 < -3$, and since $2 > 0$, we have $(-5)(2) < (-3)(2)$; that is, $-10 < -6$. |
> | **Property 4** | If $a < b$ and $c < 0$, then $ac > bc$. | $-2 < 4$, and since $-3 < 0$, we have $(-2)(-3) > (4)(-3)$; that is, $6 > -12$. |

Similar properties hold if each inequality sign, $<$, between a and b and between b and c is replaced by $\geq$, $>$, or $\leq$. Note that Property 4 says that an inequality sign is reversed if the inequality is multiplied by a negative number.

A real number is a *solution of an inequality* involving a variable if a true statement is obtained when the variable is replaced by that number. The set of all real numbers satisfying the inequality is called the *solution set*. We often use interval notation to describe the solution set.

EXAMPLE 8 Find the set of real numbers that satisfy $-1 \leq 2x - 5 < 7$.

Solution Add 5 to each member of the given double inequality, obtaining
$$4 \leq 2x < 12$$
Next, multiply each member of the resulting double inequality by $\frac{1}{2}$, yielding
$$2 \leq x < 6$$
Thus, the solution is the set of all values of x lying in the interval $[2, 6)$.

EXAMPLE 9 Solve the inequality $x^2 + 2x - 8 < 0$.

Solution Observe that $x^2 + 2x - 8 = (x + 4)(x - 2)$, so the given inequality is equivalent to the inequality $(x + 4)(x - 2) < 0$. Since the product of two real numbers is negative if and only if the two numbers have opposite signs, we solve the inequality $(x + 4)(x - 2) < 0$ by studying the signs of the two factors $x + 4$ and $x - 2$. Now, $x + 4 > 0$ when $x > -4$, and $x + 4 < 0$ when $x < -4$. Similarly, $x - 2 > 0$ when $x > 2$, and $x - 2 < 0$ when $x < 2$. These results are summarized graphically in Figure 2.

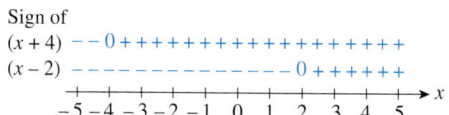

FIGURE 2
Sign diagram for $(x + 4)(x - 2)$

From Figure 2, we see that the two factors $x + 4$ and $x - 2$ have opposite signs when and only when x lies strictly between -4 and 2. Therefore, the required solution is the interval $(-4, 2)$.

EXAMPLE 10 Solve the inequality $\dfrac{x + 1}{x - 1} \geq 0$.

Solution The quotient $(x + 1)/(x - 1)$ is strictly positive if and only if both the numerator and the denominator have the same sign. The signs of $x + 1$ and $x - 1$ are shown in Figure 3.

FIGURE 3
Sign diagram for $\dfrac{x + 1}{x - 1}$

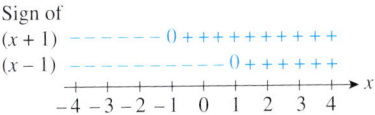

From Figure 3, we see that $x + 1$ and $x - 1$ have the same sign if $x < -1$ or $x > 1$. The quotient $(x + 1)/(x - 1)$ is equal to zero if $x = -1$. Therefore, the required solution is the set of all x in the intervals $(-\infty, -1]$ and $(1, \infty)$.

APPLIED EXAMPLE 11 Stock Purchase The management of Corbyco, a giant conglomerate, has estimated that x thousand dollars is needed to purchase

$$100{,}000(-1 + \sqrt{1 + 0.001x})$$

shares of common stock of Starr Communications. Determine how much money Corbyco needs to purchase at least 100,000 shares of Starr's stock.

Solution The amount of money Corbyco needs to purchase at least 100,000 shares is found by solving the inequality

$$100{,}000(-1 + \sqrt{1 + 0.001x}) \geq 100{,}000$$

Proceeding, we find

$$-1 + \sqrt{1 + 0.001x} \geq 1$$
$$\sqrt{1 + 0.001x} \geq 2$$
$$1 + 0.001x \geq 4 \quad \text{Square both sides.}$$
$$0.001x \geq 3$$
$$x \geq 3000$$

so Corbyco needs at least $3,000,000. (Recall that x is measured in thousands of dollars.)

Absolute Value

Absolute Value
The **absolute value** of a number a is denoted by $|a|$ and is defined by

$$|a| = \begin{cases} a & \text{if } a \geq 0 \\ -a & \text{if } a < 0 \end{cases}$$

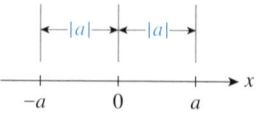

FIGURE 4
The absolute value of a number

Since $-a$ is a positive number when a is negative, it follows that the absolute value of a number is always nonnegative. For example, $|5| = 5$ and $|-5| = -(-5) = 5$. Geometrically, $|a|$ is the distance between the origin and the point on the number line that represents the number a (Figure 4).

> **Absolute Value Properties**
> If a and b are any real numbers, then
>
> | | | Example |
> |---|---|---|
> | **Property 5** | $\|-a\| = \|a\|$ | $\|-3\| = -(-3) = 3 = \|3\|$ |
> | **Property 6** | $\|ab\| = \|a\|\|b\|$ | $\|(2)(-3)\| = \|-6\| = 6 = (2)(3)$ $= \|2\|\|-3\|$ |
> | **Property 7** | $\left\|\dfrac{a}{b}\right\| = \dfrac{\|a\|}{\|b\|} \quad (b \neq 0)$ | $\left\|\dfrac{(-3)}{(-4)}\right\| = \left\|\dfrac{3}{4}\right\| = \dfrac{3}{4} = \dfrac{\|-3\|}{\|-4\|}$ |
> | **Property 8** | $\|a + b\| \leq \|a\| + \|b\|$ | $\|8 + (-5)\| = \|3\| = 3$ $\leq \|8\| + \|-5\| = 13$ |

Property 8 is called the **triangle inequality**.

EXAMPLE 12 Evaluate each of the following expressions:

a. $|\pi - 5| + 3$ **b.** $|\sqrt{3} - 2| + |2 - \sqrt{3}|$

Solution

a. Since $\pi - 5 < 0$, we see that $|\pi - 5| = -(\pi - 5)$. Therefore,
$$|\pi - 5| + 3 = -(\pi - 5) + 3 = 8 - \pi$$

b. Since $\sqrt{3} - 2 < 0$, we see that $|\sqrt{3} - 2| = -(\sqrt{3} - 2)$. Next, observe that $2 - \sqrt{3} > 0$, so $|2 - \sqrt{3}| = 2 - \sqrt{3}$. Therefore,
$$|\sqrt{3} - 2| + |2 - \sqrt{3}| = -(\sqrt{3} - 2) + (2 - \sqrt{3})$$
$$= 4 - 2\sqrt{3} = 2(2 - \sqrt{3})$$

EXAMPLE 13 Solve the inequalities $|x| \leq 5$ and $|x| \geq 5$.

Solution First, we consider the inequality $|x| \leq 5$. If $x \geq 0$, then $|x| = x$, so $|x| \leq 5$ implies $x \leq 5$ in this case. On the other hand, if $x < 0$, then $|x| = -x$, so $|x| \leq 5$ implies $-x \leq 5$ or $x \geq -5$. Thus, $|x| \leq 5$ means $-5 \leq x \leq 5$ (Figure 5a). To obtain an alternative solution, observe that $|x|$ is the distance from the point x to zero, so the inequality $|x| \leq 5$ implies immediately that $-5 \leq x \leq 5$.

FIGURE 5

(a) (b)

Next, the inequality $|x| \geq 5$ states that the distance from x to zero is greater than or equal to 5. This observation yields the result $x \geq 5$ or $x \leq -5$ (Figure 5b).

EXAMPLE 14 Solve the inequality $|2x - 3| \leq 1$.

Solution The inequality $|2x - 3| \leq 1$ is equivalent to the inequalities $-1 \leq 2x - 3 \leq 1$ (see Example 13). Thus, $2 \leq 2x \leq 4$ and $1 \leq x \leq 2$. The solution is therefore given by the set of all x in the interval $[1, 2]$ (Figure 6).

FIGURE 6
$|2x - 3| \leq 1$

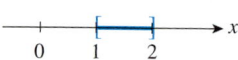

1.2 Exercises

In Exercises 1–6, simplify the expression.

1. $\dfrac{x^2 - x - 6}{x^2 - 4}$

2. $\dfrac{2a^2 - 3ab - 9b^2}{2ab^2 + 3b^3}$

3. $\dfrac{12t^2 + 12t + 3}{4t^2 - 1}$

4. $\dfrac{x^4 + 2x^3 - 3x^2}{-2x^2 - x + 3}$

5. $\dfrac{(4x - 1)(3) - (3x + 1)(4)}{(4x - 1)^2}$

6. $\dfrac{(1 + x^2)^2(2) - 2x(2)(1 + x^2)(2x)}{(1 + x^2)^4}$

In Exercises 7–28, perform the indicated operations and simplify each expression.

7. $\dfrac{2a^2 - 2b^2}{b - a} \cdot \dfrac{6a + 6b}{a^2 + 2ab + b^2}$

8. $\dfrac{x^2 - 6x + 9}{x^2 - x - 6} \cdot \dfrac{3x + 6}{2x^2 - 7x + 3}$

9. $\dfrac{3x^2 + 2x - 1}{2x + 6} \div \dfrac{x^2 - 1}{x^2 + 2x - 3}$

10. $\dfrac{3x^2 - 4xy - 4y^2}{x^2 y} \div \dfrac{(2y - x)^2}{x^3 y}$

11. $\dfrac{58}{3(3t + 2)} + \dfrac{1}{3}$

12. $\dfrac{a + 1}{3a} + \dfrac{b - 2}{5b}$

13. $\dfrac{2x}{2x - 1} - \dfrac{3x}{2x + 5}$

14. $\dfrac{-xe^x}{x + 1} + e^x$

15. $\dfrac{4}{x^2 - 9} - \dfrac{5}{x^2 - 6x + 9}$

16. $\dfrac{x}{1 - x} + \dfrac{2x + 3}{x^2 - 1}$

17. $\dfrac{1 + \dfrac{1}{x}}{1 - \dfrac{1}{x}}$

18. $\dfrac{\dfrac{1}{x} + \dfrac{1}{y}}{1 - \dfrac{1}{xy}}$

19. $\dfrac{4x^2}{2\sqrt{2x^2 + 7}} + \sqrt{2x^2 + 7}$

20. $6(2x + 1)^3 \sqrt{x^2 + x} + \dfrac{(2x + 1)^5}{2\sqrt{x^2 + x}}$

21. $5\left[\dfrac{(t^2 + 1)(1) - t(2t)}{(t^2 + 1)^2}\right]$

22. $\dfrac{2x(x + 1)^{-1/2} - (x + 1)^{1/2}}{x^2}$

23. $\dfrac{(x^2 + 1)^2(-2) + (2x)2(x^2 + 1)(2x)}{(x^2 + 1)^4}$

24. $\dfrac{(x^2 + 1)^{1/2} - 2x^2(x^2 + 1)^{-1/2}}{1 - x^2}$

25. $3\left(\dfrac{2x + 1}{3x + 2}\right)^2 \left[\dfrac{(3x + 2)(2) - (2x + 1)(3)}{(3x + 2)^2}\right]$

26. $\dfrac{(2x + 1)^{1/2} - (x + 2)(2x + 1)^{-1/2}}{2x + 1}$

27. $100\left[\dfrac{(t^2 + 20t + 100)(2t + 10) - (t^2 + 10t + 100)(2t + 20)}{(t^2 + 20t + 100)^2}\right]$

28. $\dfrac{2(2x - 3)^{1/3} - (x - 1)(2x - 3)^{-2/3}}{(2x - 3)^{2/3}}$

In Exercises 29–34, rationalize the denominator of each expression.

29. $\dfrac{1}{\sqrt{3} - 1}$

30. $\dfrac{1}{\sqrt{x} + 5}$

31. $\dfrac{1}{\sqrt{x} - \sqrt{y}}$

32. $\dfrac{a}{1 - \sqrt{3a}}$

33. $\dfrac{\sqrt{a} + \sqrt{b}}{\sqrt{a} - \sqrt{b}}$

34. $\dfrac{2\sqrt{a} + \sqrt{b}}{2\sqrt{a} - \sqrt{b}}$

In Exercises 35–40, rationalize the numerator of each expression.

35. $\dfrac{\sqrt{x}}{3}$

36. $\dfrac{\sqrt[3]{y}}{x}$

37. $\dfrac{1 - \sqrt{3}}{3}$

38. $\dfrac{\sqrt{x} - 1}{x}$

39. $\dfrac{1 + \sqrt{x + 2}}{\sqrt{x + 2}}$

40. $\dfrac{\sqrt{x + 3} - \sqrt{x}}{3}$

In Exercises 41–44, determine whether the statement is true or false.

41. $-3 < -20$

42. $-\pi \leq -\pi$

43. $\dfrac{2}{3} > \dfrac{5}{6}$

44. $-\dfrac{5}{6} < \dfrac{11}{12}$

In Exercises 45–62, find the values of x that satisfy the inequality (inequalities).

45. $2x + 4 < 8$

46. $-6 > 4 + 5x$

47. $-\dfrac{1}{4}x \geq 20$

48. $-12 \leq -\dfrac{1}{3}x$

49. $-6 < x - 2 < 4$

50. $0 \leq x + 1 \leq 4$

51. $x + 1 > 4$ or $x + 2 < -1$

52. $x + 1 > 2$ or $x - 1 < -2$

53. $x + 3 > 1$ and $x - 2 < 1$

54. $x - 4 \leq 1$ and $x + 3 > 2$

55. $(x + 3)(x - 5) \leq 0$

56. $(2x - 4)(x + 2) \geq 0$

57. $(2x - 3)(x - 1) \geq 0$

58. $(3x - 4)(2x + 2) \leq 0$

59. $\dfrac{x + 3}{x - 2} \geq 0$

60. $\dfrac{2x - 3}{x + 1} \geq 4$

61. $\dfrac{x - 2}{x - 1} \leq 2$

62. $\dfrac{2x - 1}{x + 2} \leq 4$

In Exercises 63–72, evaluate the expression.

63. $|-6 - 2|$

64. $4 + |-4|$

65. $\dfrac{|-12 + 4|}{|12 - 18|}$

66. $\left|\dfrac{0.2 - 1.4}{1.6 - 2.4}\right|$

67. $\sqrt{3}\,|-2| + 3\,|-\sqrt{3}\,|$

68. $|-1| + \sqrt{2}\,|-2|$

69. $|\pi - 1| + 2$

70. $|\pi - 6| - e$

71. $|\sqrt{2} - 1| + |3 - \sqrt{2}\,|$

72. $|2\sqrt{3} - 3| - |\sqrt{3} - 4|$

In Exercises 73–78, suppose a and b are real numbers other than zero and that $a > b$. State whether the inequality is true or false for all real numbers a and b.

73. $b - a > 0$

74. $\dfrac{a}{b} > 1$

75. $a^2 > b^2$

76. $\dfrac{1}{a} > \dfrac{1}{b}$

77. $a^3 > b^3$

78. $-a < -b$

In Exercises 79–84, determine whether the statement is true or false for all real numbers a and b.

79. $|-a| = a$

80. $|b^2| = b^2$

81. $|a - 4| = |4 - a|$

82. $|a + 1| = |a| + 1$

83. $|a + b| = |a| + |b|$

84. $|a - b| = |a| - |b|$

85. **Driving Range of a Car** An advertisement for a certain car states that the EPA fuel economy is 20 mpg city and 27 mpg highway and that the car's fuel-tank capacity is 18.1 gal. Assuming ideal driving conditions, determine the driving range for the car from the foregoing data.

86. Find the minimum cost C (in dollars), given that
$$5(C - 25) \geq 1.75 + 2.5C$$

87. Find the maximum profit P (in dollars) given that
$$6(P - 2500) \leq 4(P + 2400)$$

88. **Celsius and Fahrenheit Temperatures** The relationship between Celsius (°C) and Fahrenheit (°F) temperatures is given by the formula
$$C = \dfrac{5}{9}(F - 32)$$
   a. If the average temperature range for Montreal during the month of January is $-15° < C < -5°$, find the range in degrees Fahrenheit in Montreal for the same period.

   b. If the average temperature range for New York City during the month of June is $63° < F < 80°$, find the range in degrees Celsius in New York City for the same period.

89. **Meeting Sales Targets** A salesman's monthly commission is 15% on all sales over $12,000. If his goal is to make a commission of at least $6000/month, what minimum monthly sales figures must he attain?

90. **Markup on a Car** The markup on a used car was at least 30% of its current wholesale price. If the car was sold for $11,200, what was the maximum wholesale price?

91. **Quality Control** PAR Manufacturing manufactures steel rods. Suppose the rods ordered by a customer are manufactured to a specification of 0.5 in. and are acceptable only if they are within the *tolerance limits* of 0.49 in. and 0.51 in. Letting x denote the diameter of a rod, write an inequality using absolute value signs to express a criterion involving x that must be satisfied in order for a rod to be acceptable.

92. **Quality Control** The diameter x (in inches) of a batch of ball bearings manufactured by PAR Manufacturing satisfies the inequality
$$|x - 0.1| \leq 0.01$$
What is the smallest diameter a ball bearing in the batch can have? The largest diameter?

93. **Meeting Profit Goals** A manufacturer of a certain commodity has estimated that her profit in thousands of dollars is given by the expression
$$-6x^2 + 30x - 10$$
where x (in thousands) is the number of units produced. What production range will enable the manufacturer to realize a profit of at least $14,000 on the commodity?

94. **Concentration of a Drug in the Bloodstream** The concentration (in milligrams/cubic centimeter) of a certain drug in a patient's bloodstream t hr after injection is given by
$$\dfrac{0.2t}{t^2 + 1}$$
Find the interval of time when the concentration of the drug is greater than or equal to 0.08 mg/cc.

95. **Cost of Removing Toxic Pollutants** A city's main well was recently found to be contaminated with trichloroethylene (a cancer-causing chemical) as a result of an abandoned chemical dump that leached chemicals into the water. A proposal submitted to the city council indicated that the cost, in millions of dollars, of removing x% of the toxic pollutants is
$$\dfrac{0.5x}{100 - x}$$
If the city could raise between $25 million and $30 million for the purpose of removing the toxic pollutants, what is the range of pollutants that could be expected to be removed?

96. **Average Speed of a Vehicle** The average speed of a vehicle in miles per hour on a stretch of Route 134 between

6 A.M. and 10 A.M. on a typical weekday is approximated by the expression

$$20t - 40\sqrt{t} + 50 \qquad (0 \leq t \leq 4)$$

where t is measured in hours, with $t = 0$ corresponding to 6 A.M. Over what interval of time is the average speed of a vehicle less than or equal to 35 mph?

97. **AIR POLLUTION** Nitrogen dioxide is a brown gas that impairs breathing. The amount of nitrogen dioxide present in the atmosphere on a certain May day in the city of Long Beach measured in PSI (pollutant standard index) at time t, where t is measured in hours and $t = 0$ corresponds to 7 A.M., is approximated by

$$\frac{136}{1 + 0.25(t - 4.5)^2} + 28 \qquad (0 \leq t \leq 11)$$

Find the time of the day when the amount of nitrogen dioxide is greater than or equal to 128 PSI.
Source: Los Angeles Times.

In Exercises 98–102, determine whether the statement is true or false. If it is true, explain why it is true. If it is false, give an example to show why it is false.

98. $\dfrac{a}{b + c} = \dfrac{a}{b} + \dfrac{a}{c}$

99. If $a < b$, then $a - c > b - c$.

100. $|a - b| = |b - a|$

101. $|a - b| \leq |b| + |a|$

102. $\sqrt{a^2 - b^2} = |a| - |b|$

1.3 The Cartesian Coordinate System

The Cartesian Coordinate System

In Section 1.1, we saw how a one-to-one correspondence between the set of real numbers and the points on a straight line leads to a coordinate system on a line (a one-dimensional space).

A similar representation for points in a plane (a two-dimensional space) is realized through the **Cartesian coordinate system**, which is constructed as follows: Take two perpendicular lines, one of which is normally chosen to be horizontal. These lines intersect at a point O, called the **origin** (Figure 7). The horizontal line is called the **x-axis,** and the vertical line is called the **y-axis.** A number scale is set up along the x-axis, with the positive numbers lying to the right of the origin and the negative numbers lying to the left of it. Similarly, a number scale is set up along the y-axis, with the positive numbers lying above the origin and the negative numbers lying below it.

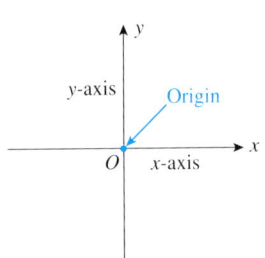

FIGURE 7
The Cartesian coordinate system

The number scales on the two axes need not be the same. Indeed, in many applications, different quantities are represented by x and y. For example, x may represent the number of cell phones sold, and y may represent the total revenue resulting from the sales. In such cases, it is often desirable to choose different number scales to represent the different quantities. Note, however, that the zeros of both number scales coincide at the origin of the two-dimensional coordinate system.

We can now uniquely represent a point in the plane in this coordinate system by an **ordered pair** of numbers—that is, a pair (x, y), where x is the first number and y is the second. To see this, let P be any point in the plane (Figure 8). Draw perpendiculars from P to the x-axis and to the y-axis. Then the number x is precisely the number that corresponds to the point on the x-axis at which the perpendicular through P crosses the x-axis. Similarly, y is the number that corresponds to the point on the y-axis at which the perpendicular through P crosses the y-axis.

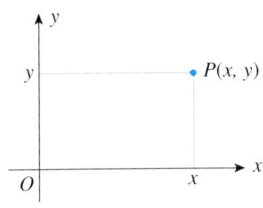

FIGURE 8
An ordered pair (x, y)

Conversely, given an ordered pair (x, y) with x as the first number and y the second, a point P in the plane is uniquely determined as follows: Locate the point on the x-axis represented by the number x and draw a line through that point parallel to the y-axis. Next, locate the point on the y-axis represented by the number y and draw a line through that point parallel to the x-axis. The point of intersection of these two lines is the point P (see Figure 8).

In the ordered pair (x, y), x is called the **abscissa,** or **x-coordinate;** y is called the **ordinate,** or **y-coordinate;** and x and y together are referred to as the **coordinates** of the point P.

Letting $P(a, b)$ denote the point with x-coordinate a and y-coordinate b, we plot the points $A(2, 3)$, $B(-2, 3)$, $C(-2, -3)$, $D(2, -3)$, $E(3, 2)$, $F(4, 0)$, and $G(0, -5)$ in Figure 9. The fact that, in general, $P(x, y) \neq P(y, x)$ is clearly illustrated by points A and E.

The axes divide the xy-plane into four quadrants. Quadrant I consists of the points $P(x, y)$ that satisfy $x > 0$ and $y > 0$; Quadrant II, the points $P(x, y)$, where $x < 0$ and $y > 0$; Quadrant III, the points $P(x, y)$, where $x < 0$ and $y < 0$; and Quadrant IV, the points $P(x, y)$, where $x > 0$ and $y < 0$ (Figure 10).

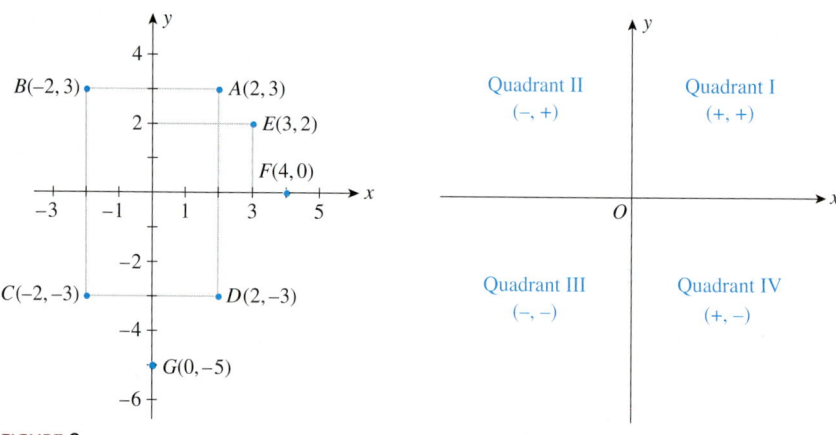

FIGURE 9
Several points in the Cartesian plane

FIGURE 10
The four quadrants in the Cartesian plane

The Distance Formula

One immediate benefit that arises from using the Cartesian coordinate system is that the distance between any two points in the plane may be expressed solely in terms of their coordinates. Suppose, for example, that (x_1, y_1) and (x_2, y_2) are any two points in the plane (Figure 11). Then the distance between these two points can be computed by using the following formula.

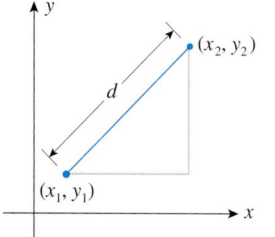

FIGURE 11
The distance d between the points (x_1, y_1) and (x_2, y_2)

Distance Formula

The distance d between two points $P_1(x_1, y_1)$ and $P_2(x_2, y_2)$ in the plane is given by

$$d = \sqrt{(x_2 - x_1)^2 + (y_2 - y_1)^2} \tag{1}$$

For a proof of this result, see Exercise 48, page 32.

In what follows, we give several applications of the distance formula.

EXAMPLE 1 Find the distance between the points $(-4, 3)$ and $(2, 6)$.

Solution Let $P_1(-4, 3)$ and $P_2(2, 6)$ be points in the plane. Then we have

$$x_1 = -4 \quad y_1 = 3 \quad x_2 = 2 \quad y_2 = 6$$

Using Formula (1), we have

$$\begin{aligned} d &= \sqrt{[2 - (-4)]^2 + (6 - 3)^2} \\ &= \sqrt{6^2 + 3^2} \\ &= \sqrt{45} = 3\sqrt{5} \approx 6.7 \end{aligned}$$

Explore & Discuss

Refer to Example 1. Suppose we label the point $(2, 6)$ as P_1 and the point $(-4, 3)$ as P_2.
(1) Show that the distance d between the two points is the same as that obtained earlier.
(2) Prove that, in general, the distance d in Formula (1) is independent of the way we label the two points.

EXAMPLE 2 Let $P(x, y)$ denote a point lying on the circle with radius r and center $C(h, k)$ (Figure 12). Find a relationship between x and y.

Solution By the definition of a circle, the distance between $C(h, k)$ and $P(x, y)$ is r. Using Formula (1), we have

$$\sqrt{(x - h)^2 + (y - k)^2} = r$$

which, upon squaring both sides, gives the equation

$$(x - h)^2 + (y - k)^2 = r^2$$

that must be satisfied by the variables x and y.

FIGURE 12
A circle with radius r and center $C(h, k)$

A summary of the result obtained in Example 2 follows.

> **Equation of a Circle**
> An equation of the circle with center $C(h, k)$ and radius r is given by
> $$(x - h)^2 + (y - k)^2 = r^2 \qquad (2)$$

EXAMPLE 3 Find an equation of the circle with

a. Radius 2 and center $(-1, 3)$.
b. Radius 3 and center located at the origin.

Solution

a. We use Formula (2) with $r = 2$, $h = -1$, and $k = 3$, obtaining

$$[x - (-1)]^2 + (y - 3)^2 = 2^2 \quad \text{or} \quad (x + 1)^2 + (y - 3)^2 = 4$$

(Figure 13a).

b. Using Formula (2) with $r = 3$ and $h = k = 0$, we obtain

$$x^2 + y^2 = 3^2 \quad \text{or} \quad x^2 + y^2 = 9$$

(Figure 13b).

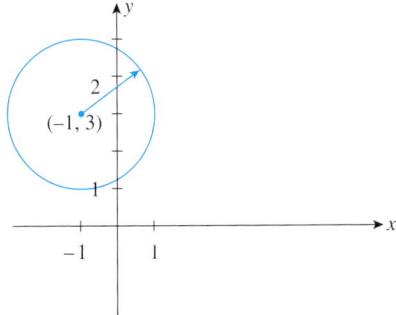
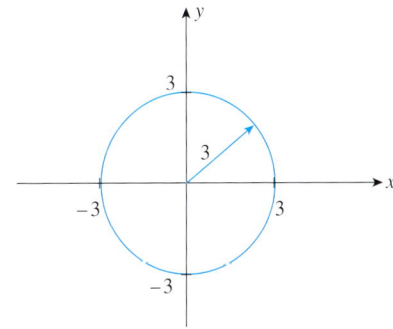

FIGURE 13

(a) The circle with radius 2 and center $(-1, 3)$

(b) The circle with radius 3 and center $(0, 0)$

Explore & Discuss

1. Use the distance formula to help you describe the set of points in the xy-plane satisfying each of the following inequalities.
   a. $(x - h)^2 + (y - k)^2 \leq r^2$
   b. $(x - h)^2 + (y - k)^2 < r^2$
   c. $(x - h)^2 + (y - k)^2 \geq r^2$
   d. $(x - h)^2 + (y - k)^2 > r^2$

2. Consider the equation $x^2 + y^2 = 4$.
   a. Show that $y = \pm\sqrt{4 - x^2}$.
   b. Describe the set of points (x, y) in the xy-plane satisfying the following equations:
      (i) $y = \sqrt{4 - x^2}$
      (ii) $y = -\sqrt{4 - x^2}$

APPLIED EXAMPLE 4 Cost of Laying Cable In Figure 14, S represents the position of a power relay station located on a straight coastal highway, and M shows the location of a marine biology experimental station on an island. A cable is to be laid connecting the relay station with the experimental station. If the cost of running the cable on land is $3.00 per running foot and the cost of running the cable under water is $5.00 per running foot, find the total cost for laying the cable.

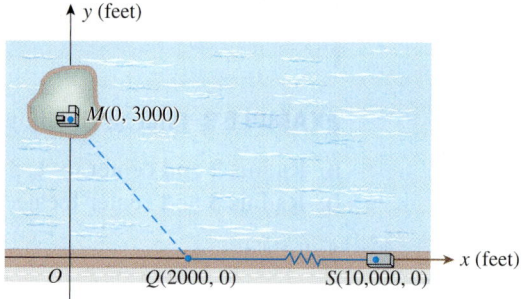

FIGURE 14
Cable connecting relay station S to experimental station M

Solution The length of cable required on land is given by the distance from S to Q. This distance is $(10{,}000 - 2000)$, or 8000 feet. Next, we see that the length of cable required underwater is given by the distance from M to Q. This distance is

$$\sqrt{(0 - 2000)^2 + (3000 - 0)^2} = \sqrt{2000^2 + 3000^2}$$
$$= \sqrt{13{,}000{,}000}$$
$$\approx 3606$$

or approximately 3606 feet. Therefore, the total cost for laying the cable is approximately

$$3(8000) + 5(3606) = 42{,}030$$

dollars.

Explore & Discuss

In the Cartesian coordinate system, the two axes are perpendicular to each other. Consider a coordinate system in which the x- and y-axes are not collinear and are not perpendicular to each other (see the accompanying figure).

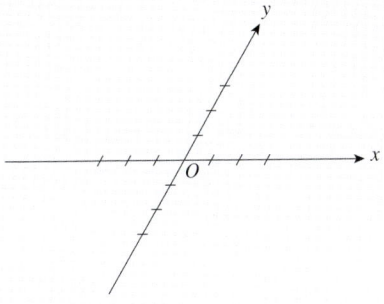

1. Describe how a point is represented in this coordinate system by an ordered pair (x, y) of real numbers. Conversely, show how an ordered pair (x, y) of real numbers uniquely determines a point in the plane.

2. Suppose you want to find a formula for the distance between two points $P_1(x_1, y_1)$ and $P_2(x_2, y_2)$ in the plane. What is the advantage that the Cartesian coordinate system has over the coordinate system under consideration? Comment on your answer.

1.3 Self-Check Exercises

1. a. Plot the points $A(4, -2)$, $B(2, 3)$, and $C(-3, 1)$.
    b. Find the distance between the points A and B; between B and C; between A and C.
    c. Use the Pythagorean Theorem to show that the triangle with vertices A, B, and C is a right triangle.

2. The figure opposite shows the location of cities A, B, and C. Suppose a pilot wishes to fly from City A to City C but must make a mandatory stopover in City B. If the single-engine light plane has a range of 650 miles, can she make the trip without refueling in City B?

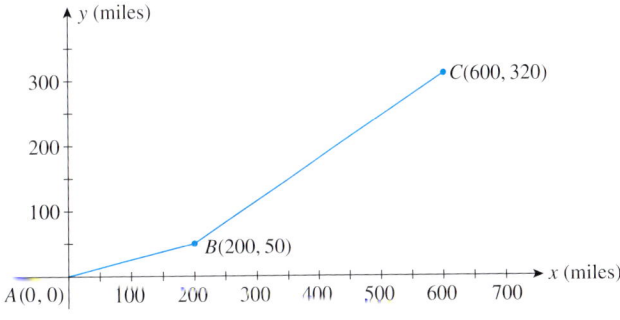

Solutions to Self-Check Exercises 1.3 can be found on page 32.

1.3 Concept Questions

1. What can you say about the signs of a and b if the point $P(a, b)$ lies in (a) the second quadrant? (b) The third quadrant? (c) The fourth quadrant?

2. a. What is the distance between $P_1(x_1, y_1)$ and $P_2(x_2, y_2)$?
    b. When you use the distance formula, does it matter which point is labeled P_1 and which point is labeled P_2? Explain.

1.3 Exercises

In Exercises 1–6, refer to the following figure and determine the coordinates of each point and the quadrant in which it is located.

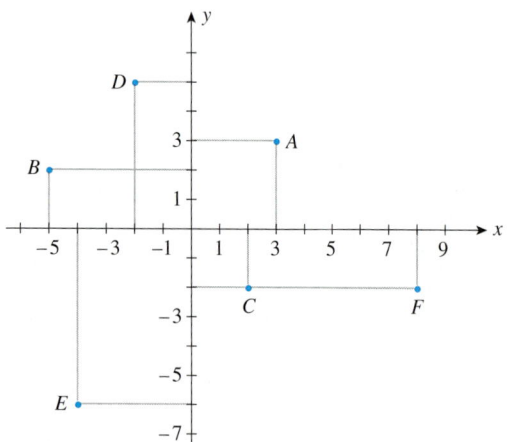

1. A
2. B
3. C
4. D
5. E
6. F

In Exercises 7–12, refer to the following figure.

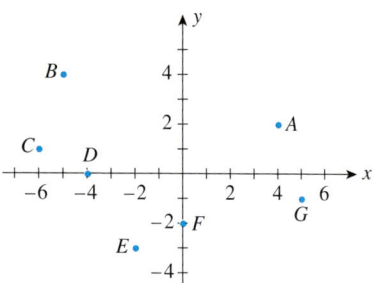

7. Which point has coordinates (4, 2)?
8. What are the coordinates of point B?
9. Which points have negative y-coordinates?
10. Which point has a negative x-coordinate and a negative y-coordinate?
11. Which point has an x-coordinate that is equal to zero?
12. Which point has a y-coordinate that is equal to zero?

In Exercises 13–20, sketch a set of coordinate axes and plot each point.

13. $(-2, 5)$
14. $(1, 3)$
15. $(3, -1)$
16. $(3, -4)$
17. $\left(8, -\dfrac{7}{2}\right)$
18. $\left(-\dfrac{5}{2}, \dfrac{3}{2}\right)$
19. $(4.5, -4.5)$
20. $(1.2, -3.4)$

In Exercises 21–24, find the distance between the given points.

21. $(1, 3)$ and $(4, 7)$
22. $(1, 0)$ and $(4, 4)$
23. $(-2, 3)$ and $(4, 9)$
24. $(-2, 1)$ and $(10, 6)$

25. Find the coordinates of the points that are 10 units away from the origin and have a y-coordinate equal to -6.
26. Find the coordinates of the points that are 5 units away from the origin and have an x-coordinate equal to 3.
27. Show that the points $(3, 4)$, $(-3, 7)$, $(-6, 1)$, and $(0, -2)$ form the vertices of a square.
28. Show that the triangle with vertices $(-5, 2)$, $(-2, 5)$, and $(5, -2)$ is a right triangle.

In Exercises 29–34, find an equation of the circle that satisfies the given conditions.

29. Radius 4 and center $(2, -3)$
30. Radius 3 and center $(-2, -4)$
31. Radius 6 and center at the origin
32. Center at the origin and passes through $(2, 3)$
33. Center $(2, -3)$ and passes through $(5, 2)$
34. Center $(-a, a)$ and radius $2a$

35. **Distance Traveled** A grand tour of four cities begins at City A and makes successive stops at cities B, C, and D before returning to City A. If the cities are located as shown in the following figure, find the total distance covered on the tour.

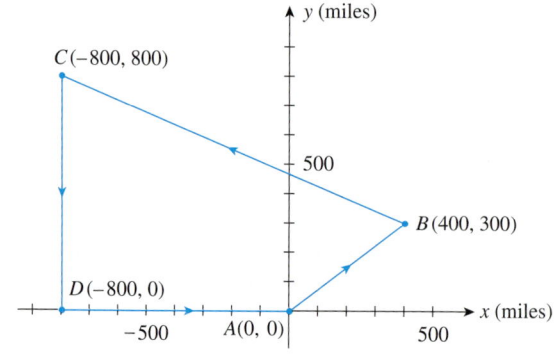

36. Delivery Charges A furniture store offers free setup and delivery services to all points within a 25-mi radius of its warehouse distribution center. If you live 20 mi east and 14 mi south of the warehouse, will you incur a delivery charge? Justify your answer.

37. Optimizing Travel Time Towns A, B, C, and D are located as shown in the following figure. Two highways link town A to town D. Route 1 runs from Town A to Town D via Town B, and Route 2 runs from Town A to Town D via Town C. If a salesman wishes to drive from Town A to Town D and traffic conditions are such that he could expect to average the same speed on either route, which highway should he take to arrive in the shortest time?

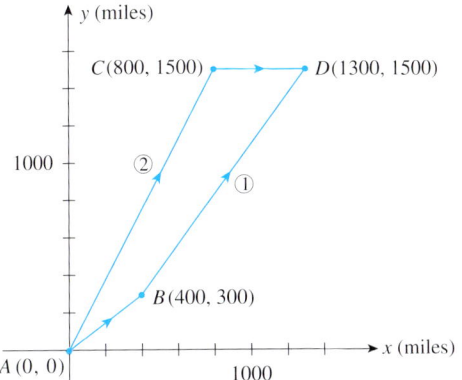

38. Minimizing Shipping Costs Refer to the figure for Exercise 37. Suppose a fleet of 100 automobiles are to be shipped from an assembly plant in Town A to Town D. They may be shipped either by freight train along Route 1 at a cost of 66¢/mile per automobile or by truck along Route 2 at a cost of 62¢/mile per automobile. Which means of transportation minimizes the shipping cost? What is the net savings?

39. Consumer Decisions Ivan wishes to determine which HDTV antenna he should purchase for his home. The TV store has supplied him with the following information:

Range in Miles		Model	Price
VHF	UHF		
30	20	A	$50
45	35	B	$60
60	40	C	$70
75	55	D	$80

Ivan wishes to receive Channel 17 (VHF), which is located 25 mi east and 35 mi north of his home, and Channel 38 (UHF), which is located 20 mi south and 32 mi west of his home. Which model will allow him to receive both channels at the least cost? (Assume that the terrain between Ivan's home and both broadcasting stations is flat.)

40. Cost of Laying Cable In the following diagram, S represents the position of a power relay station located on a straight coastal highway, and M shows the location of a marine biology experimental station on an island. A cable is to be laid connecting the relay station with the experimental station. If the cost of running the cable on land is $3.00/running foot and the cost of running cable under water is $5.00/running foot, find an expression in terms of x that gives the total cost for laying the cable. What is the total cost when $x = 2500$? When $x = 3000$?

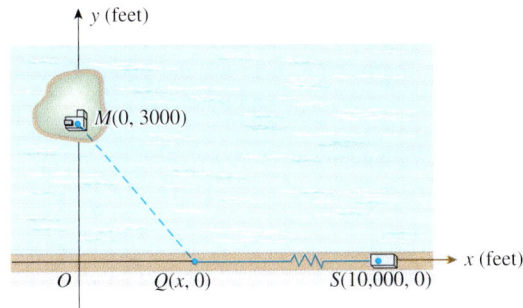

41. Two ships leave port at the same time. Ship A sails north at a speed of 20 mph while Ship B sails east at a speed of 30 mph.
 a. Find an expression in terms of the time t (in hours) giving the distance between the two ships.
 b. Using the expression obtained in part (a), find the distance between the two ships 2 hr after leaving port.

42. Ship A leaves port sailing north at a speed of 25 mph. A half hour later, Ship B leaves the same port sailing east at a speed of 20 mph. Let t (in hours) denote the time Ship B has been at sea.
 a. Find an expression in terms of t giving the distance between the two ships.
 b. Use the expression obtained in part (a) to find the distance between the two ships 2 hr after ship A has left port.

43. Watching a Rocket Launch At a distance of 4000 ft from the launch site, a spectator is observing a rocket being launched. Suppose the rocket lifts off vertically and reaches an altitude of x ft (see the accompanying figure).

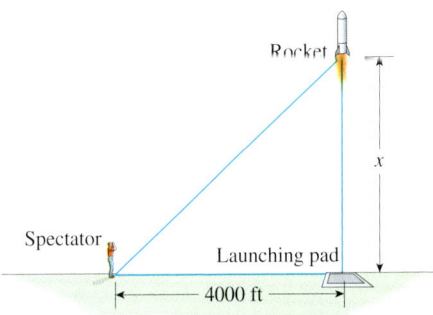

 a. Find an expression giving the distance between the spectator and the rocket.
 b. What is the distance between the spectator and the rocket when the rocket reaches an altitude of 20,000 ft?

In Exercises 44–47, determine whether the statement is true or false. If it is true, explain why it is true. If it is false, give an example to show why it is false.

44. The point $(-a, b)$ is symmetric to the point (a, b) with respect to the y-axis.

45. The point $(-a, -b)$ is symmetric to the point (a, b) with respect to the origin.

46. If the distance between the points $P_1(a, b)$ and $P_2(c, d)$ is D, then the distance between the points $P_1(a, b)$ and $P_3(kc, kd)$, $(k \neq 0)$, is given by $|k|D$.

47. The circle with equation $kx^2 + ky^2 = a^2$ lies inside the circle with equation $x^2 + y^2 = a^2$, provided $k > 1$.

48. Let (x_1, y_1) and (x_2, y_2) be two points lying in the xy-plane. Show that the distance between the two points is given by

$$d = \sqrt{(x_2 - x_1)^2 + (y_2 - y_1)^2}$$

Hint: Refer to the accompanying figure and use the Pythagorean Theorem.

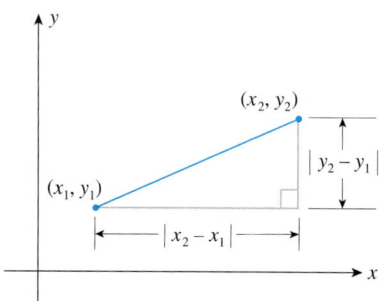

49. **a.** Show that the midpoint of the line segment joining the points $P_1(x_1, y_1)$ and $P_2(x_2, y_2)$ is

$$\left(\frac{x_1 + x_2}{2}, \frac{y_1 + y_2}{2}\right)$$

b. Use the result of part (a) to find the midpoint of the line segment joining the points $(-3, 2)$ and $(4, -5)$.

50. Show that an equation of a circle can be written in the form

$$x^2 + y^2 + Cx + Dy + E = 0$$

where C, D, and E are constants. This is called the general form of an equation of a circle.

1.3 Solutions to Self-Check Exercises

1. **a.** The points are plotted in the following figure:

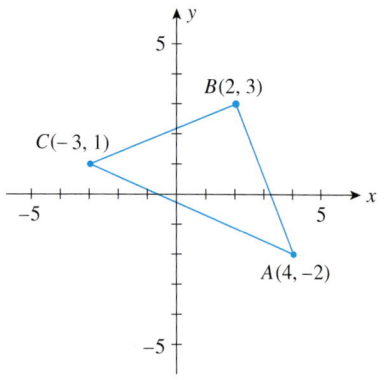

b. The distance between A and B is

$$d(A, B) = \sqrt{(2-4)^2 + [3-(-2)]^2}$$
$$= \sqrt{(-2)^2 + 5^2} = \sqrt{4 + 25} = \sqrt{29}$$

The distance between B and C is

$$d(B, C) = \sqrt{(-3-2)^2 + (1-3)^2}$$
$$= \sqrt{(-5)^2 + (-2)^2} = \sqrt{25 + 4} = \sqrt{29}$$

The distance between A and C is

$$d(A, C) = \sqrt{(-3-4)^2 + [1-(-2)]^2}$$
$$= \sqrt{(-7)^2 + 3^2} = \sqrt{49 + 9} = \sqrt{58}$$

c. We will show that

$$[d(A, C)]^2 = [d(A, B)]^2 + [d(B, C)]^2$$

From part (b), we see that $[d(A, B)]^2 = 29$, $[d(B, C)]^2 = 29$, and $[d(A, C)]^2 = 58$, and the desired result follows.

2. The distance between City A and City B is

$$d(A, B) = \sqrt{200^2 + 50^2} \approx 206$$

or 206 mi. The distance between City B and City C is

$$d(B, C) = \sqrt{(600 - 200)^2 + (320 - 50)^2}$$
$$= \sqrt{400^2 + 270^2} \approx 483$$

or 483 mi. Therefore, the total distance the pilot would have to cover is about 689 mi, so she must refuel in City B.

1.4 Straight Lines

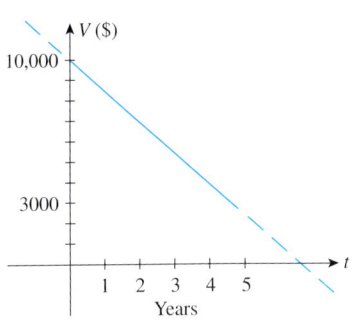

FIGURE 15
Linear depreciation of an asset

In computing income tax, business firms are allowed by law to depreciate certain assets, such as buildings, machines, furniture, and automobiles, over a period of time. Linear depreciation, or the straight-line method, is often used for this purpose. The graph of the straight line shown in Figure 15 describes the book value V of a network server that has an initial value of $10,000 and that is being depreciated linearly over 5 years with a scrap value of $3,000. Note that only the solid portion of the straight line is of interest here.

The book value of the server at the end of year t, where t lies between 0 and 5, can be read directly from the graph. But there is one shortcoming in this approach: The result depends on how accurately you draw and read the graph. A better and more accurate method is based on finding an *algebraic* representation of the depreciation line.

Slope of a Line

To see how a straight line in the *xy*-plane may be described algebraically, we need to first recall certain properties of straight lines. Let L denote the unique straight line that passes through the two distinct points (x_1, y_1) and (x_2, y_2). If $x_1 \neq x_2$, we define the slope of L as follows.

> **Slope of a Nonvertical Line**
> If (x_1, y_1) and (x_2, y_2) are any two distinct points on a nonvertical line L, then the slope m of L is given by
> $$m = \frac{\Delta y}{\Delta x} = \frac{y_2 - y_1}{x_2 - x_1} \qquad (3)$$

See Figure 16.

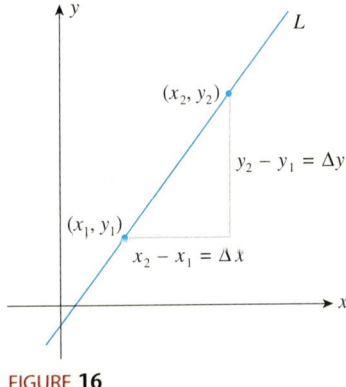

FIGURE 16

If $x_1 = x_2$, then L is a vertical line (Figure 17). Its slope is undefined since the denominator in Equation (3) will be zero and division by zero is proscribed.

Observe that the slope of a straight line is a constant whenever it is defined. The number $\Delta y = y_2 - y_1$ (Δy is read "delta y") is a measure of the vertical change in y, and $\Delta x = x_2 - x_1$ is a measure of the horizontal change in x, as shown in Figure 16. From this figure, we can see that the slope m of a straight line L is a measure of the *rate of change of y with respect to x*.

Figure 18a shows a straight line L_1 with slope 2. Observe that L_1 has the property that a 1-unit increase in x results in a 2-unit increase in y. To see this, let $\Delta x = 1$ in

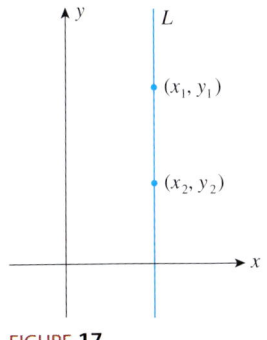

FIGURE 17
m is undefined.

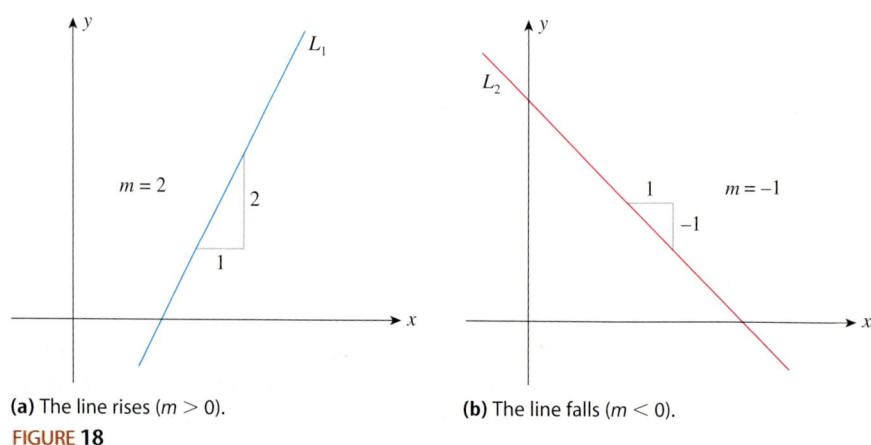

(a) The line rises ($m > 0$).   (b) The line falls ($m < 0$).

FIGURE 18

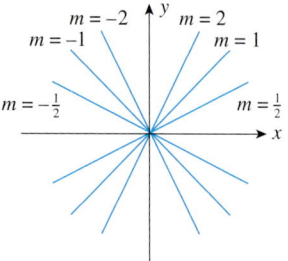

FIGURE 19
A family of straight lines

Formula (3); then $m = \Delta y$. Since $m = 2$, we conclude that $\Delta y = 2$. Similarly, Figure 18b shows a line L_2 with slope -1. Observe that a straight line with positive slope slants upward from left to right (y increases as x increases), whereas a line with negative slope slants downward from left to right (y decreases as x increases). Figure 19 shows a family of straight lines passing through the origin with indicated slopes.

EXAMPLE 1 Sketch the straight line that passes through the point $(-2, 5)$ and has slope $-\frac{4}{3}$.

Solution First, plot the point $(-2, 5)$ (Figure 20).

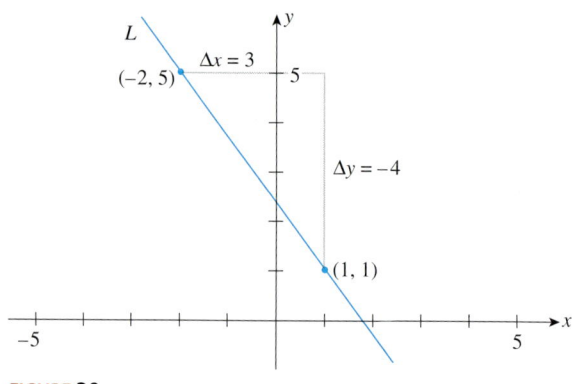

FIGURE 20
L has slope $-\frac{4}{3}$ and passes through $(-2, 5)$.

Next, recall that a slope of $-\frac{4}{3}$ indicates that an increase of 1 unit in the x-direction produces a decrease of $\frac{4}{3}$ units in the y-direction, or equivalently, a 3-unit increase in the x-direction produces a $3(\frac{4}{3})$, or 4, units decrease in the y-direction. Using this information, we plot the point $(1, 1)$ and draw the line through the two points. ∎

> ### Explore & Discuss
>
> Show that the slope of a nonvertical line is independent of the two distinct points $P_1(x_1, y_1)$ and $P_2(x_2, y_2)$ used to compute it.
>
> **Hint:** Suppose we pick two other distinct points, $P_3(x_3, y_3)$ and $P_4(x_4, y_4)$ lying on L. Draw a picture and use similar triangles to demonstrate that using P_3 and P_4 gives the same value as that obtained using P_1 and P_2.

1.4 STRAIGHT LINES 35

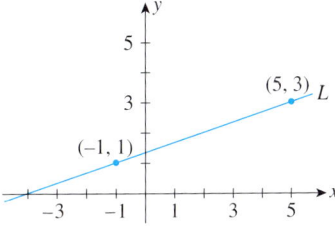

FIGURE 21
L passes through (5, 3) and (−1, 1).

EXAMPLE 2 Find the slope m of the line that passes through the points $(-1, 1)$ and $(5, 3)$.

Solution Choose (x_1, y_1) to be the point $(-1, 1)$ and (x_2, y_2) to be the point $(5, 3)$. Then, with $x_1 = -1$, $y_1 = 1$, $x_2 = 5$, and $y_2 = 3$, we find

$$m = \frac{y_2 - y_1}{x_2 - x_1} = \frac{3 - 1}{5 - (-1)} = \frac{1}{3} \quad \text{Use Formula (3).}$$

(Figure 21). Verify that the result obtained would have been the same had we chosen the point $(-1, 1)$ to be (x_2, y_2) and the point $(5, 3)$ to be (x_1, y_1).

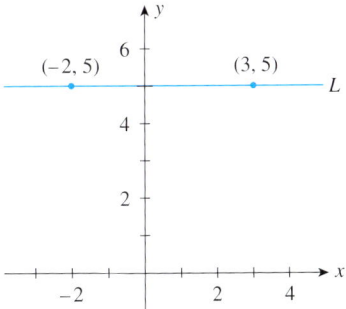

FIGURE 22
The slope of the horizontal line L is 0.

EXAMPLE 3 Find the slope of the line that passes through the points $(-2, 5)$ and $(3, 5)$.

Solution The slope of the required line is given by

$$m = \frac{5 - 5}{3 - (-2)} = \frac{0}{5} = 0 \quad \text{Use Formula (3).}$$

(Figure 22).

Note In general, the slope of a horizontal line is zero.

We can use the slope of a straight line to determine whether a line is parallel to another line.

> **Parallel Lines**
>
> Two distinct lines are **parallel** if and only if their slopes are equal or their slopes are undefined.

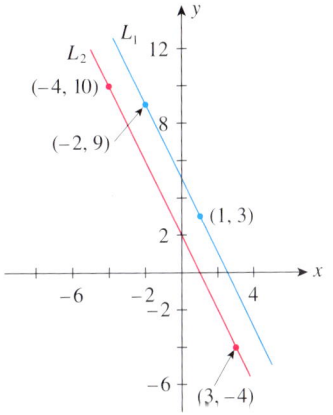

FIGURE 23
L_1 and L_2 have the same slope and hence are parallel.

EXAMPLE 4 Let L_1 be a line that passes through the points $(-2, 9)$ and $(1, 3)$, and let L_2 be the line that passes through the points $(-4, 10)$ and $(3, -4)$. Determine whether L_1 and L_2 are parallel.

Solution The slope m_1 of L_1 is given by

$$m_1 = \frac{3 - 9}{1 - (-2)} = -2$$

The slope m_2 of L_2 is given by

$$m_2 = \frac{-4 - 10}{3 - (-4)} = -2$$

Since $m_1 = m_2$, the lines L_1 and L_2 are in fact parallel (Figure 23).

Equations of Lines

We will now show that every straight line lying in the xy-plane may be represented by an equation involving the variables x and y. One immediate benefit of this is that problems involving straight lines may be solved algebraically.

Let L be a straight line parallel to the y-axis (perpendicular to the x-axis) (Figure 24). Then L crosses the x-axis at some point $(a, 0)$ with the x-coordinate given by $x = a$, where a is some real number. Any other point on L has the form (a, y),

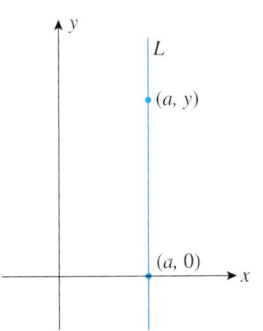

FIGURE 24
The vertical line $x = a$

where y is an appropriate number. Therefore, the vertical line L is described by the sole condition

$$x = a$$

and this is, accordingly, an equation of L. For example, the equation $x = -2$ represents a vertical line 2 units to the left of the y-axis, and the equation $x = 3$ represents a vertical line 3 units to the right of the y-axis (Figure 25).

Next, suppose L is a nonvertical line, so that it has a well-defined slope m. Suppose (x_1, y_1) is a fixed point lying on L and (x, y) is a variable point on L distinct from (x_1, y_1) (Figure 26).

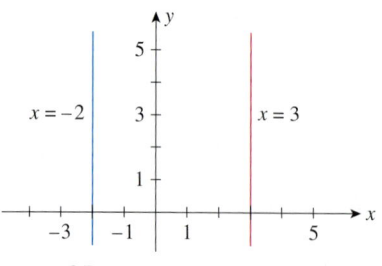

FIGURE 25
The vertical lines $x = -2$ and $x = 3$

FIGURE 26
L passes through (x_1, y_1) and has slope m.

Using Formula (3) with the point $(x_2, y_2) = (x, y)$, we find that the slope of L is given by

$$m = \frac{y - y_1}{x - x_1}$$

Upon multiplying both sides of the equation by $x - x_1$, we obtain Formula (4).

> **Point-Slope Form of an Equation of a Line**
>
> An equation of the line that has slope m and passes through the point (x_1, y_1) is given by
>
> $$y - y_1 = m(x - x_1) \qquad (4)$$

Equation (4) is called the **point-slope form of an equation of a line,** since it utilizes a given point (x_1, y_1) on a line and the slope m of the line.

EXAMPLE 5 Find an equation of the line that passes through the point $(1, 3)$ and has slope 2.

Solution Using the point-slope form of the equation of a line with the point $(1, 3)$ and $m = 2$, we obtain

$$y - 3 = 2(x - 1) \qquad y - y_1 = m(x - x_1)$$

which, when simplified, becomes

$$2x - y + 1 = 0$$

(Figure 27).

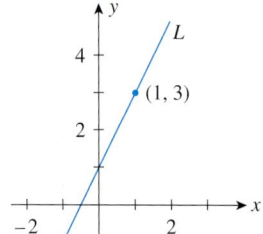

FIGURE 27
L passes through $(1, 3)$ and has slope 2.

EXAMPLE 6 Find an equation of the line that passes through the points $(-3, 2)$ and $(4, -1)$.

Solution The slope of the line is given by

$$m = \frac{-1 - 2}{4 - (-3)} = -\frac{3}{7}$$

Using the point-slope form of an equation of a line with the point $(4, -1)$ and the slope $m = -\frac{3}{7}$, we have

$$y + 1 = -\frac{3}{7}(x - 4) \quad \text{\color{red}{$y - y_1 = m(x - x_1)$}}$$

$$7y + 7 = -3x + 12$$

$$3x + 7y - 5 = 0$$

(Figure 28).

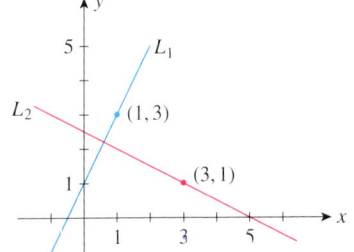

FIGURE 28
L passes through $(-3, 2)$ and $(4, -1)$.

We can use the slope of a straight line to determine whether a line is perpendicular to another line.

Perpendicular Lines

If L_1 and L_2 are two distinct nonvertical lines that have slopes m_1 and m_2, respectively, then L_1 is **perpendicular** to L_2 (written $L_1 \perp L_2$) if and only if

$$m_1 = -\frac{1}{m_2}$$

If the line L_1 is vertical (so that its slope is undefined), then L_1 is perpendicular to another line, L_2, if and only if L_2 is horizontal (so that its slope is zero). For a proof of these results, see Exercise 84, page 45.

EXAMPLE 7 Find an equation of the line that passes through the point $(3, 1)$ and is perpendicular to the line of Example 5.

Solution Since the slope of the line in Example 5 is 2, the slope of the required line is given by $m = -\frac{1}{2}$, the negative reciprocal of 2. Using the point-slope form of the equation of a line, we obtain

$$y - 1 = -\frac{1}{2}(x - 3) \quad \text{\color{red}{$y - y_1 = m(x - x_1)$}}$$

$$2y - 2 = -x + 3$$

$$x + 2y - 5 = 0$$

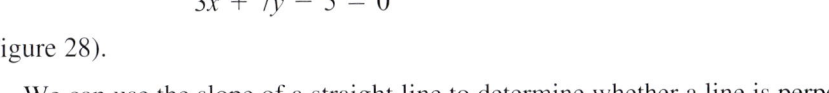

FIGURE 29
L_2 is perpendicular to L_1 and passes through $(3, 1)$.

(Figure 29).

A straight line L that is neither horizontal nor vertical cuts the x-axis and the y-axis at, say, points $(a, 0)$ and $(0, b)$, respectively (Figure 30). The numbers a and b are called the **x-intercept** and **y-intercept,** respectively, of L.

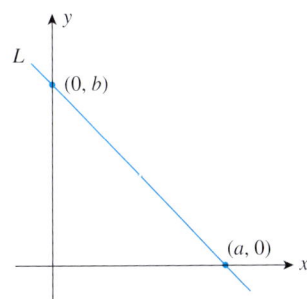

FIGURE 30
The line L has x-intercept a and y-intercept b.

> **Exploring with TECHNOLOGY**
>
> 1. Use a graphing utility to plot the straight lines L_1 and L_2 with equations $2x + y - 5 = 0$ and $41x + 20y - 11 = 0$ on the same set of axes, using the standard viewing window.
>   a. Can you tell whether the lines L_1 and L_2 are parallel to each other?
>   b. Verify your observations by computing the slopes of L_1 and L_2 algebraically.
> 2. Use a graphing utility to plot the straight lines L_1 and L_2 with equations $x + 2y - 5 = 0$ and $5x - y + 5 = 0$ on the same set of axes, using the standard viewing window.
>   a. Can you tell whether the lines L_1 and L_2 are perpendicular to each other?
>   b. Verify your observation by computing the slopes of L_1 and L_2 algebraically.

Now let L be a line with slope m and y-intercept b. Using Formula (4), the point-slope form of the equation of a line, with the point $(0, b)$ and slope m, we have

$$y - b = m(x - 0)$$
$$y = mx + b$$

> **Slope-Intercept Form of an Equation of a Line**
>
> An equation of the line that has slope m and intersects the y-axis at the point $(0, b)$ is given by
>
> $$y = mx + b \tag{5}$$

EXAMPLE 8 Find an equation of the line that has slope 3 and y-intercept -4.

Solution Using Equation (5) with $m = 3$ and $b = -4$, we obtain the required equation

$$y = 3x - 4$$

EXAMPLE 9 Determine the slope and y-intercept of the line whose equation is $3x - 4y = 8$.

Solution Rewrite the given equation in the slope-intercept form. Thus,

$$3x - 4y = 8$$
$$-4y = -3x + 8$$
$$y = \frac{3}{4}x - 2$$

Comparing this result with Equation (5), we find $m = \frac{3}{4}$ and $b = -2$, and we conclude that the slope and y-intercept of the given line are $\frac{3}{4}$ and -2, respectively.

> *Explore & Discuss*
>
> Consider the slope-intercept form of an equation of a straight line $y = mx + b$. Describe the family of straight lines obtained by keeping
>
> 1. The value of m fixed and allowing the value of b to vary.
> 2. The value of b fixed and allowing the value of m to vary.

Exploring with TECHNOLOGY

1. Use a graphing utility to plot the straight lines with equations $y = -2x + 3$, $y = -x + 3$, $y = x + 3$, and $y = 2.5x + 3$ on the same set of axes, using the standard viewing window. What effect does changing the coefficient m of x in the equation $y = mx + b$ have on its graph?
2. Use a graphing utility to plot the straight lines with equations $y = 2x - 2$, $y = 2x - 1$, $y = 2x$, $y = 2x + 1$, and $y = 2x + 4$ on the same set of axes, using the standard viewing window. What effect does changing the constant b in the equation $y = mx + b$ have on its graph?
3. Describe in words the effect of changing both m and b in the equation $y = mx + b$.

APPLIED EXAMPLE 10 Sales of a Sporting Goods Store The sales manager of a local sporting goods store plotted sales versus time for the last 5 years and found the points to lie approximately along a straight line (Figure 31). By using the points corresponding to the first and fifth years, find an equation of the trend line. What sales figure can be predicted for the sixth year?

Solution Using Formula (3) with the points (1, 20) and (5, 60), we find that the slope of the required line is given by

$$m = \frac{60 - 20}{5 - 1} = 10$$

Next, using the point-slope form of the equation of a line with the point (1, 20) and $m = 10$, we obtain

$$y - 20 = 10(x - 1) \quad y - y_1 = m(x - x_1)$$
$$y = 10x + 10$$

as the required equation.

The sales figure for the sixth year is obtained by letting $x = 6$ in the last equation, giving

$$y = 10(6) + 10 = 70$$

or $700,000.

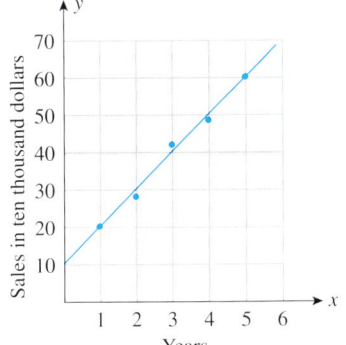

FIGURE 31
Sales of a sporting goods store

APPLIED EXAMPLE 11 Appreciation in Value of an Art Object Suppose an art object purchased for $50,000 is expected to appreciate in value at a constant rate of $5000 per year for the next 5 years. Use Formula (5) to write an equation predicting the value of the art object in the next several years. What will be its value 3 years from the date of purchase?

Solution Let x denote the time (in years) that has elapsed since the date the object was purchased, and let y denote the object's value (in dollars). Then $y = 50{,}000$ when $x = 0$. Furthermore, the slope of the required equation is given by $m = 5000$, since each unit increase in x (1 year) implies an increase of 5000 units (dollars) in y. Using (5) with $m = 5000$ and $b = 50{,}000$, we obtain

$$y = 5000x + 50{,}000 \quad y = mx + b$$

Three years from the date of purchase, the value of the object will be given by

$$y = 5000(3) + 50{,}000$$

or $65,000.

Explore & Discuss

Refer to Example 11. Can the equation predicting the value of the art object be used to predict long-term growth?

General Form of an Equation of a Line

We have considered several forms of an equation of a straight line in the plane. These different forms of the equation are equivalent to each other. In fact, each is a special case of the following equation.

> **General Form of a Linear Equation**
> The equation
> $$Ax + By + C = 0 \qquad (6)$$
> where A, B, and C are constants and A and B are not both zero, is called the general form of a linear equation in the variables x and y.

We will now state (without proof) an important result concerning the algebraic representation of straight lines in the plane.

> **THEOREM 1**
> An equation of a straight line is a linear equation; conversely, every linear equation represents a straight line.

This result justifies the use of the adjective *linear* describing Equation (6).

EXAMPLE 12 Sketch the straight line represented by the equation
$$3x - 4y - 12 = 0$$

Solution Since every straight line is uniquely determined by two distinct points, we need find only two such points through which the line passes in order to sketch it. For convenience, let's compute the x- and y-intercepts. Setting $y = 0$, we find $x = 4$; thus, the x-intercept is 4. Setting $x = 0$ gives $y = -3$, and the y-intercept is -3. A sketch of the line appears in Figure 32. ∎

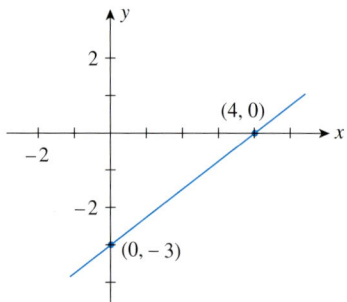

FIGURE 32
To sketch $3x - 4y - 12 = 0$, first find the x-intercept, 4, and the y-intercept, -3.

Following is a summary of the common forms of the equations of straight lines discussed in this section.

> **Equations of Straight Lines**
> Vertical line: $\quad x = a$
> Horizontal line: $\quad y = b$
> Point-slope form: $\quad y - y_1 = m(x - x_1)$
> Slope-intercept form: $\quad y = mx + b$
> General form: $\quad Ax + By + C = 0$

1.4 Self-Check Exercises

1. Determine the number a such that the line passing through the points $(a, 2)$ and $(3, 6)$ is parallel to a line with slope 4.

2. Find an equation of the line that passes through the point $(3, -1)$ and is perpendicular to a line with slope $-\frac{1}{2}$.

3. Does the point $(3, -3)$ lie on the line with equation $2x - 3y - 12 = 0$? Sketch the graph of the line.

4. **SATELLITE TV SUBSCRIBERS** The following table gives the number of satellite TV subscribers in the United States (in millions) from 2004 through 2008 ($t = 0$ corresponds to the beginning of 2004).

Year, t	0	1	2	3	4
Number, y	22.5	24.8	27.1	29.1	30.7

a. Plot the number of satellite TV subscribers in the United States (y) versus the year (t).
b. Draw the line L through the points $(0, 22.5)$ and $(4, 30.7)$.
c. Find an equation of the line L.
d. Assuming that this trend continued, estimate the number of satellite TV subscribers in the United States in 2010.

Sources: National Cable & Telecommunications Association, Federal Communications Commission.

Solutions to Self-Check Exercises 1.4 can be found on page 45.

1.4 Concept Questions

1. What is the slope of a nonvertical line? What can you say about the slope of a vertical line?

2. Give (a) the point-slope form, (b) the slope-intercept form, and (c) the general form of an equation of a line.

3. Let L_1 have slope m_1 and L_2 have slope m_2. State the conditions on m_1 and m_2 if (a) L_1 is parallel to L_2 and (b) L_1 is perpendicular to L_2.

1.4 Exercises

In Exercises 1–6, match the statement with one of the graphs (a)–(f).

1. The slope of the line is zero.
2. The slope of the line is undefined.
3. The slope of the line is positive, and its y-intercept is positive.
4. The slope of the line is positive, and its y-intercept is negative.
5. The slope of the line is negative, and its x-intercept is negative.
6. The slope of the line is negative, and its x-intercept is positive.

(a)
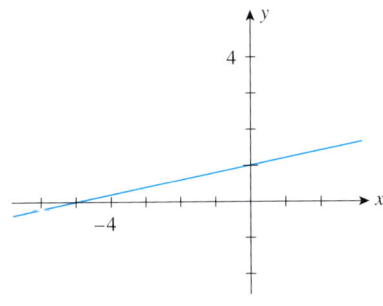

(b)
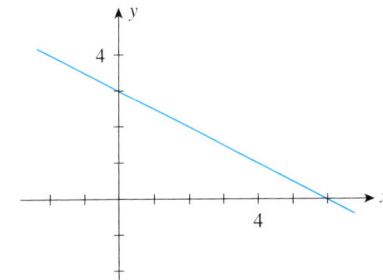

(c)
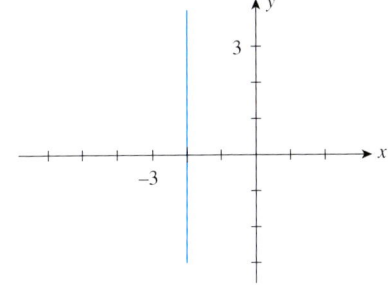

(d)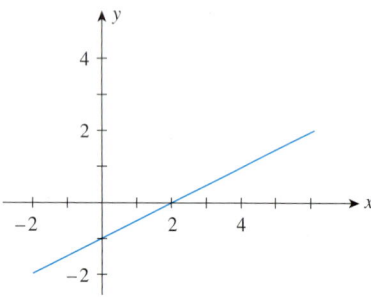

(e)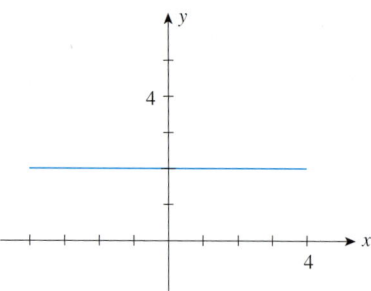

(f)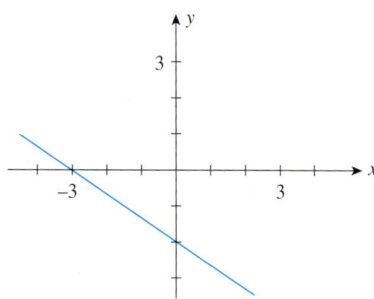

In Exercises 7–10, find the slope of the line shown in each figure.

7.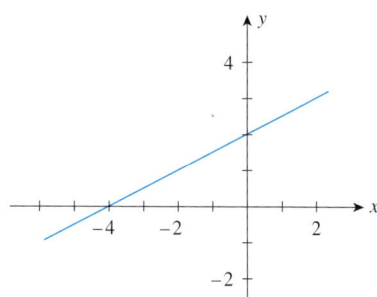

8.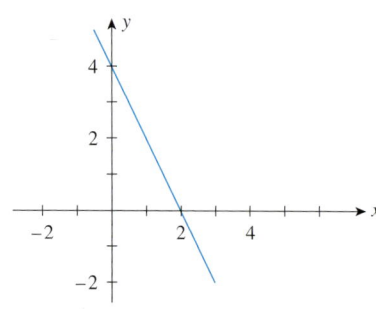

9.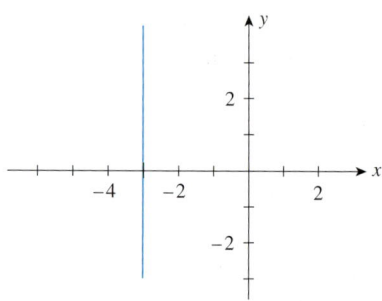

10.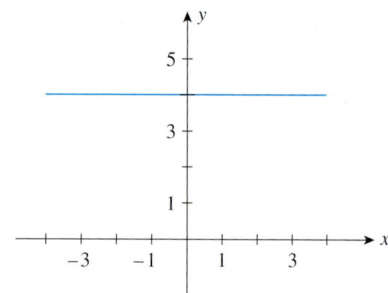

In Exercises 11–16, find the slope of the line that passes through each pair of points.

11. $(4, 3)$ and $(5, 8)$ 12. $(4, 5)$ and $(3, 8)$

13. $(-2, 3)$ and $(4, 8)$ 14. $(-2, -2)$ and $(4, -4)$

15. (a, b) and (c, d)

16. $(-a + 1, b - 1)$ and $(a + 1, -b)$

17. Given the equation $y = 4x - 3$, answer the following questions:
    a. If x increases by 1 unit, what is the corresponding change in y?
    b. If x decreases by 2 units, what is the corresponding change in y?

18. Given the equation $2x + 3y = 4$, answer the following questions:
    a. Is the slope of the line described by this equation positive or negative?
    b. As x increases in value, does y increase or decrease?
    c. If x decreases by 2 units, what is the corresponding change in y?

In Exercises 19 and 20, determine whether the lines AB and CD are parallel.

19. $A(1, -2)$, $B(-3, -10)$ and $C(1, 5)$, $D(-1, 1)$

20. $A(2, 3)$, $B(2, -2)$ and $C(-2, 4)$, $D(-2, 5)$

In Exercises 21 and 22, determine whether the lines AB and CD are perpendicular.

21. $A(-2, 5)$, $B(4, 2)$ and $C(-1, -2)$, $D(3, 6)$

22. $A(2, 0)$, $B(1, -2)$ and $C(4, 2)$, $D(-8, 4)$

23. If the line passing through the points (1, a) and (4, -2) is parallel to the line passing through the points (2, 8) and (-7, $a + 4$), what is the value of a?

24. If the line passing through the points (a, 1) and (5, 8) is parallel to the line passing through the points (4, 9) and ($a + 2$, 1), what is the value of a?

25. Find an equation of the horizontal line that passes through (-4, -3).

26. Find an equation of the vertical line that passes through (0, 5).

In Exercises 27–30, find an equation of the line that passes through the point and has the indicated slope m.

27. (3, -4); $m = 2$
28. (2, 4); $m = -1$
29. (-3, 2); $m = 0$
30. (1, 2); $m = -\dfrac{1}{2}$

In Exercises 31–34, find an equation of the line that passes through the points.

31. (2, 4) and (3, 7)
32. (2, 1) and (2, 5)
33. (1, 2) and (-3, -2)
34. (-1, -2) and (3, -4)

In Exercises 35–38, find an equation of the line that has slope m and y-intercept b.

35. $m = 3$; $b = 4$
36. $m = -2$; $b = -1$
37. $m = 0$; $b = 5$
38. $m = -\dfrac{1}{2}$; $b = \dfrac{3}{4}$

In Exercises 39–44, write the equation in the slope-intercept form and then find the slope and y-intercept of the corresponding line.

39. $x - 2y = 0$
40. $y - 2 = 0$
41. $2x - 3y - 9 = 0$
42. $3x - 4y + 8 = 0$
43. $2x + 4y = 14$
44. $5x + 8y - 24 = 0$

45. Find an equation of the line that passes through the point (-2, 2) and is parallel to the line $2x - 4y - 8 = 0$.

46. Find an equation of the line that passes through the point (2, 4) and is perpendicular to the line $3x + 4y - 22 = 0$.

In Exercises 47–52, find an equation of the line that satisfies the given condition.

47. The line parallel to the x-axis and 4 units below it

48. The line passing through the origin and parallel to the line joining the points (2, 4) and (4, 7)

49. The line passing through the point (a, b) with slope equal to zero

50. The line passing through (-3, 8) and parallel to the x-axis

51. The line passing through (-5, -4) and parallel to the line joining (-3, 2) and (6, 8)

52. The line passing through (a, b) with undefined slope

53. Given that the point $P(-3, 5)$ lies on the line $kx + 3y + 9 = 0$, find k.

54. Given that the point $P(2, -3)$ lies on the line $-2x + ky + 10 = 0$, find k.

In Exercises 55–60, sketch the straight line defined by the given linear equation by finding the x- and y-intercepts.
Hint: See Example 12, page 40.

55. $3x - 2y + 6 = 0$
56. $2x - 5y + 10 = 0$
57. $x + 2y - 4 = 0$
58. $2x + 3y - 15 = 0$
59. $y + 5 = 0$
60. $-2x - 8y + 24 = 0$

61. Show that an equation of a line through the points (a, 0) and (0, b) with $a \neq 0$ and $b \neq 0$ can be written in the form

$$\dfrac{x}{a} + \dfrac{y}{b} = 1$$

(Recall that the numbers a and b are the x- and y-intercepts, respectively, of the line. This form of an equation of a line is called the *intercept form*.)

In Exercises 62–65, use the results of Exercise 61 to find an equation of a line with the given x- and y-intercepts.

62. x-intercept 3; y-intercept 4
63. x-intercept -3; y-intercept -5
64. x-intercept $-\dfrac{1}{2}$; y-intercept $\dfrac{3}{4}$
65. x-intercept 4; y-intercept $-\dfrac{1}{2}$

In Exercises 66 and 67, determine whether the given points lie on a straight line.

66. $A(-1, 7)$, $B(2, -2)$, and $C(5, -9)$
67. $A(-2, 1)$, $B(1, 7)$, and $C(4, 13)$

68. **Temperature Conversion** The relationship between the temperature in degrees Fahrenheit (°F) and the temperature in degrees Celsius (°C) is

$$F = \dfrac{9}{5}C + 32$$

a. Sketch the line with the given equation.
b. What is the slope of the line? What does it represent?
c. What is the F-intercept of the line? What does it represent?

69. NUCLEAR PLANT UTILIZATION The United States is not building many nuclear plants, but the ones it has are running full tilt. The output (as a percent of total capacity) of nuclear plants is described by the equation

$$y = 1.9467t + 70.082$$

where t is measured in years, with $t = 0$ corresponding to the beginning of 1990.
a. Sketch the line with the given equation.
b. What are the slope and the y-intercept of the line found in part (a)?
c. Give an interpretation of the slope and the y-intercept of the line found in part (a).
d. If the utilization of nuclear power continued to grow at the same rate and the total capacity of nuclear plants in the United States remained constant, by what year were the plants generating at maximum capacity?
Source: Nuclear Energy Institute.

70. SOCIAL SECURITY CONTRIBUTIONS For wages less than the maximum taxable wage base, Social Security contributions (including those for Medicare) by employees are 7.65% of the employee's wages.
a. Find an equation that expresses the relationship between the wages earned (x) and the Social Security taxes paid (y) by an employee who earns less than the maximum taxable wage base.
b. For each additional dollar that an employee earns, by how much is his or her Social Security contribution increased? (Assume that the employee's wages remain less than the maximum taxable wage base.)
c. What Social Security contributions will an employee who earns $65,000 (which is less than the maximum taxable wage base) be required to make?
Source: Social Security Administration.

71. COLLEGE ADMISSIONS Using data compiled by the Admissions Office at Faber University, college admissions officers estimate that 55% of the students who are offered admission to the freshman class at the university will actually enroll.
a. Find an equation that expresses the relationship between the number of students who actually enroll (y) and the number of students who are offered admission to the university (x).
b. If the desired freshman class size for the upcoming academic year is 1100 students, how many students should be admitted?

72. WEIGHT OF WHALES The equation $W = 3.51L - 192$, expressing the relationship between the length L (in feet) and the expected weight W (in British tons) of adult blue whales, was adopted in the late 1960s by the International Whaling Commission.
a. What is the expected weight of an 80-ft blue whale?
b. Sketch the straight line that represents the equation.

73. THE NARROWING GENDER GAP Since the founding of the Equal Employment Opportunity Commission and the passage of equal-pay laws, the gulf between men's and women's earnings has continued to close gradually. At the beginning of 1990 ($t = 0$), women's wages were 68% of men's wages, and by the beginning of 2000 ($t = 10$), women's wages were projected to be 80% of men's wages. If this gap between women's and men's wages continued to narrow *linearly*, what percentage of men's wages are women's wages expected to be at the beginning of 2008?
Source: Journal of Economic Perspectives.

74. SALES GROWTH Metro Department Store's annual sales (in millions of dollars) during the past 5 years were

Annual Sales, y	5.8	6.2	7.2	8.4	9.0
Year, x	1	2	3	4	5

a. Plot the annual sales (y) versus the year (x).
b. Draw a straight line L through the points corresponding to the first and fifth years.
c. Derive an equation of the line L.
d. Using the equation found in part (c), estimate Metro's annual sales 4 years from now ($x = 9$).

75. SALES OF GPS EQUIPMENT The annual sales (in billions of dollars) of global positioning system (GPS) equipment from the year 2000 through 2006 follow. (Sales for 2004–2006 were projections.) Here $x = 0$ corresponds to 2000.

Annual Sales, y	7.9	9.6	11.5	13.3	15.2	17	18.8
Year, x	0	1	2	3	4	5	6

a. Plot the annual sales (y) versus the year (x).
b. Draw a straight line L through the points corresponding to 2000 and 2006.
c. Derive an equation of the line L.
d. Use the equation found in part (c) to estimate the annual sales of GPS equipment for 2005. Compare this figure with the projected sales for that year.
Source: ABI Research.

76. DIGITAL TV SERVICES The percentage of homes with digital TV services, which stood at 5% at the beginning of 1999 ($t = 0$) was projected to grow linearly so that at the beginning of 2003 ($t = 4$) the percentage of such homes was projected to be 25%.
a. Derive an equation of the line passing through the points $A(0, 5)$ and $B(4, 25)$.
b. Plot the line with the equation found in part (a).
c. Using the equation found in part (a), find the percentage of homes with digital TV services at the beginning of 2001.
Source: Paul Kagan Associates.

In Exercises 77–81, determine whether the statement is true or false. If it is true, explain why it is true. If it is false, give an example to show why it is false.

77. Suppose the slope of a line L is $-\frac{1}{2}$ and P is a given point on L. If Q is the point on L lying 4 units to the left of P, then Q is situated 2 units above P.

78. The line with equation $Ax + By + C = 0$, $(B \neq 0)$, and the line with equation $ax + by + c = 0$, $(b \neq 0)$, are parallel if $Ab - aB = 0$.

79. If the slope of the line L_1 is positive, then the slope of a line L_2 perpendicular to L_1 may be positive or negative.

80. The lines with equations $ax + by + c_1 = 0$ and $bx - ay + c_2 = 0$, where $a \neq 0$ and $b \neq 0$, are perpendicular to each other.

81. If L is the line with equation $Ax + By + C = 0$, where $A \neq 0$, then L crosses the x-axis at the point $(-C/A, 0)$.

82. Is there a difference between the statements "The slope of a straight line is zero" and "The slope of a straight line does not exist (is not defined)"? Explain your answer.

83. Show that two distinct lines with equations $a_1 x + b_1 y + c_1 = 0$ and $a_2 x + b_2 y + c_2 = 0$, respectively, are parallel if and only if $a_1 b_2 - b_1 a_2 = 0$.
Hint: Write each equation in the slope-intercept form and compare.

84. Prove that if a line L_1 with slope m_1 is perpendicular to a line L_2 with slope m_2, then $m_1 m_2 = -1$.
Hint: Refer to the following figure. Show that $m_1 = b$ and $m_2 = c$. Next, apply the Pythagorean Theorem to triangles OAC, OCB, and OBA to show that $1 = -bc$.

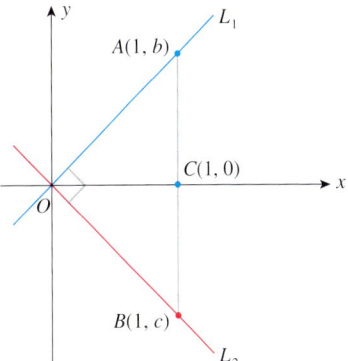

1.4 Solutions to Self-Check Exercises

1. The slope of the line that passes through the points $(a, 2)$ and $(3, 6)$ is
$$m = \frac{6 - 2}{3 - a}$$
$$= \frac{4}{3 - a}$$
Since this line is parallel to a line with slope 4, m must be equal to 4; that is,
$$\frac{4}{3 - a} = 4$$
or, upon multiplying both sides of the equation by $3 - a$,
$$4 = 4(3 - a)$$
$$4 = 12 - 4a$$
$$4a = 8$$
$$a = 2$$

2. Since the required line L is perpendicular to a line with slope $-\frac{1}{2}$, the slope of L is
$$m = \frac{-1}{-\frac{1}{2}} = 2$$
Next, using the point-slope form of the equation of a line, we have
$$y - (-1) = 2(x - 3)$$
$$y + 1 = 2x - 6$$
$$y = 2x - 7$$

3. Substituting $x = 3$ and $y = -3$ into the left-hand side of the given equation, we find
$$2(3) - 3(-3) - 12 = 3$$
which is not equal to zero (the right-hand side). Therefore, $(3, -3)$ does not lie on the line with equation $2x - 3y - 12 = 0$. (See the accompanying figure.)
Setting $x = 0$, we find $y = -4$, the y-intercept. Next, setting $y = 0$ gives $x = 6$, the x-intercept. We now draw the line passing through the points $(0, -4)$ and $(6, 0)$ as shown.

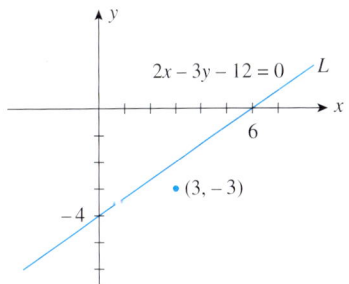

4. a. and b. See the accompanying figure.

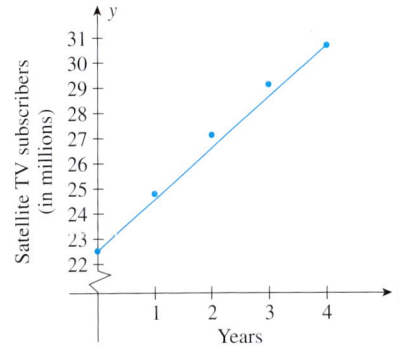

c. The slope of L is

$$m = \frac{30.7 - 22.5}{4 - 0} = 2.05$$

Using the point-slope form of the equation of a line with the point $(0, 22.5)$, we find

$$y - 22.5 = 2.05(t - 0)$$
$$y = 2.05t + 22.5$$

d. The estimated number of satellite TV subscribers in the United States in 2010 is

$$y = 2.05(6) + 22.5 = 34.8$$

or 34.8 million.

CHAPTER 1 Summary of Principal Formulas and Terms

FORMULAS

1.	Quadratic formula	$x = \dfrac{-b \pm \sqrt{b^2 - 4ac}}{2a}$
2.	Distance between two points	$d = \sqrt{(x_2 - x_1)^2 + (y_2 - y_1)^2}$
3.	Equation of a circle with center $C(h, k)$ and radius r	$(x - h)^2 + (y - k)^2 = r^2$
4.	Slope of a line	$m = \dfrac{y_2 - y_1}{x_2 - x_1}$
5.	Equation of a vertical line	$x = a$
6.	Equation of a horizontal line	$y = b$
7.	Point-slope form of the equation of a line	$y - y_1 = m(x - x_1)$
8.	Slope-intercept form of the equation of a line	$y = mx + b$
9.	General equation of a line	$Ax + By + C = 0$

TERMS

real number (coordinate) line (3)
open interval (4)
closed interval (4)
half-open interval (4)
finite interval (4)

infinite interval (4)
polynomial (8)
roots of a polynomial equation (12)
absolute value (21)
triangle inequality (22)

Cartesian coordinate system (25)
ordered pair (25
parallel lines (35)
perpendicular lines (37)

CHAPTER 1 Concept Review Questions

Fill in the blanks.

1. A point in the plane can be represented uniquely by a/an _____ pair of numbers. The first number of the pair is called the _____, and the second number of the pair is called the _____.

2. a. The point $P(a, 0)$ lies on the _____-axis, and the point $P(0, b)$ lies on the _____-axis.
   b. If the point $P(a, b)$ lies in the fourth quadrant, then the point $P(-a, b)$ lies in the _____ quadrant.

3. The distance between two points $P_1(a, b)$ and $P_2(c, d)$ is _____.

4. An equation of the circle with center $C(a, b)$ and radius r is given by _____.

5. a. If $P_1(x_1, y_1)$ and $P_2(x_2, y_2)$ are any two distinct points on a nonvertical line L, then the slope of L is $m = $ _____.
   b. The slope of a vertical line is _____.
   c. The slope of a horizontal line is _____.
   d. The slope of a line that slants upward from left to right is _____.

6. If L_1 and L_2 are nonvertical lines with slopes m_1 and m_2, respectively, then L_1 is parallel to L_2 if and only if _____ and L_1 is perpendicular to L_2 if and only if _____.

7. a. An equation of the line passing through the point $P(x_1, y_1)$ and having slope m is _____. This form of an equation of a line is called the _____ _____.
   b. An equation of the line that has slope m and y-intercept b is _____. It is called the _____ form of an equation of a line.

8. a. The general form of an equation of a line is _____.
   b. If a line has equation $ax + by + c = 0$ $(b \neq 0)$, then its slope is _____.

CHAPTER 1 Review Exercises

In Exercises 1–4, find the values of x that satisfy the inequality (inequalities).

1. $-x + 3 \leq 2x + 9$
2. $-2 \leq 3x + 1 \leq 7$
3. $x - 3 > 2$ or $x + 3 < -1$
4. $2x^2 > 50$

In Exercises 5–8, evaluate the expression.

5. $|-5 + 7| + |-2|$
6. $\left|\dfrac{5 - 12}{-4 - 3}\right|$
7. $|2\pi - 6| - \pi$
8. $|\sqrt{3} - 4| + |4 - 2\sqrt{3}|$

In Exercises 9–14, evaluate the expression.

9. $\left(\dfrac{9}{4}\right)^{3/2}$
10. $\dfrac{5^6}{25^4}$
11. $(3 \cdot 4)^{-2}$
12. $(-8)^{5/3}$
13. $\dfrac{(3 \cdot 2^{-3})(4 \cdot 3^5)}{2 \cdot 9^3}$
14. $\dfrac{3\sqrt[3]{54}}{\sqrt[3]{18}}$

In Exercises 15–20, simplify the expression.

15. $\dfrac{4(x^2 + y)^3}{x^2 + y}$
16. $\dfrac{a^6 b^{-5}}{(a^3 b^{-2})^{-3}}$
17. $\dfrac{\sqrt[4]{16x^5 yz}}{\sqrt[4]{81xyz^5}}$; $x > 0, y > 0, z > 0$
18. $(2x^3)(-3x^{-2})\left(\dfrac{1}{6}x^{-1/2}\right)$
19. $\left(\dfrac{3xy^2}{4x^3 y}\right)^{-2}\left(\dfrac{3xy^3}{2x^2}\right)^3$
20. $\sqrt[3]{81x^5 y^{10}} \sqrt[3]{9xy^2}$

In Exercises 21–24, factor each expression completely.

21. $-2\pi^2 r^3 + 100\pi r^2$
22. $2v^3 w + 2vw^3 + 2u^2 vw$
23. $16 - x^2 y^4$
24. $12t^3 - 6t^2 - 18t$

In Exercises 25–28, solve the equation by factoring.

25. $8x^2 + 2x - 3 = 0$
26. $-6x^2 - 10x + 4 = 0$
27. $-x^3 - 2x^2 + 3x = 0$
28. $2x^4 + x^2 = 1$

In Exercises 29–32, find the value(s) of x that satisfy the expression.

29. $2x^2 + 3x - 2 \leq 0$
30. $\dfrac{1}{x + 2} > 2$
31. $|2x - 3| < 5$
32. $\left|\dfrac{x + 1}{x - 1}\right| = 5$

In Exercises 33 and 34, use the quadratic formula to solve the quadratic equation.

33. $x^2 - 2x - 5 = 0$
34. $2x^2 + 8x + 7 = 0$

In Exercises 35–38, perform the indicated operations and simplify the expression.

35. $\dfrac{(t + 6)(60) - (60t + 180)}{(t + 6)^2}$
36. $\dfrac{6x}{2(3x^2 + 2)} + \dfrac{1}{4(x + 2)}$
37. $\dfrac{2}{3}\left(\dfrac{4x}{2x^2 - 1}\right) + 3\left(\dfrac{3}{3x - 1}\right)$
38. $\dfrac{-2x}{\sqrt{x + 1}} + 4\sqrt{x + 1}$

39. Rationalize the numerator: $\dfrac{\sqrt{x} - 1}{x - 1}$.

40. Rationalize the denominator: $\dfrac{\sqrt{x} - 1}{2\sqrt{x}}$.

In Exercises 41 and 42, find the distance between the two points.

41. $(-2, -3)$ and $(1, -7)$
42. $\left(\dfrac{1}{2}, \sqrt{3}\right)$ and $\left(-\dfrac{1}{2}, 2\sqrt{3}\right)$

In Exercises 43–48, find an equation of the line L that passes through the point $(-2, 4)$ and satisfies the condition.

43. L is a vertical line.
44. L is a horizontal line.
45. L passes through the point $(3, \tfrac{7}{2})$.
46. The x-intercept of L is 3.
47. L is parallel to the line $5x - 2y = 6$.
48. L is perpendicular to the line $4x + 3y = 6$.

49. Find an equation of the straight line that passes through the point $(2, 3)$ and is parallel to the line with equation $3x + 4y - 8 = 0$.

50. Find an equation of the straight line that passes through the point $(-1, 3)$ and is parallel to the line passing through the points $(-3, 4)$ and $(2, 1)$.

51. Find an equation of the line that passes through the point $(-3, -2)$ and is parallel to the line passing through the points $(-2, -4)$ and $(1, 5)$.

52. Find an equation of the line that passes through the point $(-2, -4)$ and is perpendicular to the line with equation $2x - 3y - 24 = 0$.

53. Sketch the graph of the equation $3x - 4y = 24$.

54. Sketch the graph of the line that passes through the point $(3, 2)$ and has slope $-\frac{2}{3}$.

55. Find the minimum cost C (in dollars) given that
$$2(1.5C + 80) \le 2(2.5C - 20)$$

56. Find the maximum revenue R (in dollars) given that
$$12(2R - 320) \le 4(3R + 240)$$

57. **A Falling Stone** A stone is thrown straight up from the roof of an 80-ft building, and the height (in feet) of the stone any time t later (in seconds), measured from the ground, is given by
$$-16t^2 + 64t + 80$$
Find the interval of time when the stone is at or greater than a height of 128 ft from the ground.

58. **Sales of Navigation Systems** The estimated number of navigation systems (in millions) sold in North America, Europe, and Japan from 2002 through 2006 follow. Here $t = 0$ corresponds to 2002.

Systems Installed, y	3.9	4.7	5.8	6.8	7.8
Year, t	0	1	2	3	4

a. Plot the annual sales (y) versus the year (t).
b. Draw a straight line L through the points corresponding to 2002 and 2006.
c. Derive an equation of the line L.
d. Use the equation found in part (c) to find the number of navigation systems installed for 2005. Compare this figure with the estimated sales for that year.

Source: ABI Research.

The problem-solving skills that you learn in each chapter are building blocks for the rest of the course. Therefore, it is a good idea to make sure that you have mastered these skills before moving on to the next chapter. The Before Moving On exercises that follow are designed for that purpose.

CHAPTER 1 Before Moving On . . .

1. Evaluate:

   a. $|\pi - 2\sqrt{3}| - |\sqrt{3} - \sqrt{2}|$ b. $\left[\left(-\dfrac{1}{3}\right)^{-3}\right]^{1/3}$

2. Simplify:

   a. $\sqrt[3]{64x^6} \cdot \sqrt{9y^2x^6}$; $x > 0$ b. $\left(\dfrac{a^{-3}}{b^{-4}}\right)^2 \left(\dfrac{b}{a}\right)^{-3}$

3. Rationalize the denominator:

   a. $\dfrac{2x}{3\sqrt{y}}$ b. $\dfrac{x}{\sqrt{x} - 4}$

4. Perform each operation and simplify:

   a. $\dfrac{(x^2 + 1)(\frac{1}{2}x^{-1/2}) - x^{1/2}(2x)}{(x^2 + 1)^2}$

   b. $-\dfrac{3x}{\sqrt{x + 2}} + 3\sqrt{x + 2}$

5. Rationalize the numerator: $\dfrac{\sqrt{x} + \sqrt{y}}{\sqrt{x} - \sqrt{y}}$.

6. Factor completely:

   a. $12x^3 - 10x^2 - 12x$ b. $2bx - 2by + 3cx - 3cy$

7. Solve each equation:

   a. $12x^2 - 9x - 3 = 0$ b. $3x^2 - 5x + 1 = 0$

8. Find the distance between $(-2, 4)$ and $(6, 8)$.

9. Find an equation of the line that passes through $(-1, -2)$ and $(4, 5)$.

10. Find an equation of the line that has slope $-\frac{1}{3}$ and y-intercept $\frac{4}{3}$.

2 FUNCTIONS, LIMITS, AND THE DERIVATIVE

Unless changes are made, when is the current Social Security system expected to go broke? In Example 3, page 79, we use a mathematical model constructed from data from the Social Security Administration to predict the year in which the assets of the current system will be depleted.

IN THIS CHAPTER, we define a *function,* a special relationship between two variables. The concept of a function enables us to describe many relationships that exist in applications. We also begin the study of differential calculus. Historically, differential calculus was developed in response to the problem of finding the tangent line to an arbitrary curve. But it quickly became apparent that solving this problem provided mathematicians with a method for solving many practical problems involving the rate of change of one quantity with respect to another. The basic tool used in differential calculus is the *derivative* of a function. The concept of the derivative is based, in turn, on a more fundamental notion—that of the *limit* of a function.

2.1 Functions and Their Graphs

Functions

A manufacturer would like to know how his company's profit is related to its production level; a biologist would like to know how the size of the population of a certain culture of bacteria will change over time; a psychologist would like to know the relationship between the learning time of an individual and the length of a vocabulary list; and a chemist would like to know how the initial speed of a chemical reaction is related to the amount of substrate used. In each instance, we are concerned with the same question: How does one quantity depend upon another? The relationship between two quantities is conveniently described in mathematics by using the concept of a function.

> **Function**
> A **function** is a rule that assigns to each element in a set A one and only one element in a set B.

FIGURE 1
A function machine

FIGURE 2
The function f viewed as a mapping

The set A is called the **domain** of the function. It is customary to denote a function by a letter of the alphabet, such as the letter f. If x is an element in the domain of a function f, then the element in B that f associates with x is written $f(x)$ (read "f of x") and is called the value of f at x. The set comprising all the values assumed by $y = f(x)$ as x takes on all possible values in its domain is called the **range** of the function f.

We can think of a function f as a machine. The domain is the set of inputs (raw material) for the machine, the rule describes how the input is to be processed, and the values of the function are the outputs of the machine (Figure 1).

We can also think of a function f as a mapping in which an element x in the domain of f is mapped onto a unique element $f(x)$ in B (Figure 2).

Notes

1. The output $f(x)$ associated with an input x is unique. To appreciate the importance of this uniqueness property, consider a rule that associates with each item x in a department store its selling price y. Then, each x must correspond to *one and only one y*. Notice, however, that different x's may be associated with the same y. In the context of the present example, this says that different items may have the same price.
2. Although the sets A and B that appear in the definition of a function may be quite arbitrary, in this book they will denote sets of real numbers.

An example of a function may be taken from the familiar relationship between the area of a circle and its radius. Letting x and y denote the radius and area of a circle, respectively, we have, from elementary geometry,

$$y = \pi x^2 \tag{1}$$

Equation (1) defines y as a function of x since for each admissible value of x (that is, for each nonnegative number representing the radius of a certain circle), there corresponds precisely one number $y = \pi x^2$ that gives the area of the circle. The rule defining this "area function" may be written as

$$f(x) = \pi x^2 \tag{2}$$

To compute the area of a circle of radius 5 inches, we simply replace x in Equation (2) with the number 5. Thus, the area of the circle is

$$f(5) = \pi 5^2 = 25\pi$$

or 25π square inches.

In general, to evaluate a function at a specific value of x, we replace x with that value, as illustrated in Examples 1 and 2.

EXAMPLE 1 Let the function f be defined by the rule $f(x) = 2x^2 - x + 1$. Find:

a. $f(1)$ **b.** $f(-2)$ **c.** $f(a)$ **d.** $f(a + h)$

Solution

a. $f(1) = 2(1)^2 - (1) + 1 = 2 - 1 + 1 = 2$
b. $f(-2) = 2(-2)^2 - (-2) + 1 = 8 + 2 + 1 = 11$
c. $f(a) = 2(a)^2 - (a) + 1 = 2a^2 - a + 1$
d. $f(a + h) = 2(a + h)^2 - (a + h) + 1 = 2a^2 + 4ah + 2h^2 - a - h + 1$

APPLIED EXAMPLE 2 Profit Functions ThermoMaster manufactures an indoor–outdoor thermometer at its Mexican subsidiary. Management estimates that the profit (in dollars) realizable by ThermoMaster in the manufacture and sale of x thermometers per week is

$$P(x) = -0.001x^2 + 8x - 5000$$

Find ThermoMaster's weekly profit if its level of production is (a) 1000 thermometers per week and (b) 2000 thermometers per week.

Solution

a. The weekly profit when the level of production is 1000 units per week is found by evaluating the profit function P at $x = 1000$. Thus,

$$P(1000) = -0.001(1000)^2 + 8(1000) - 5000 = 2000$$

or $2000.
b. When the level of production is 2000 units per week, the weekly profit is given by

$$P(2000) = -0.001(2000)^2 + 8(2000) - 5000 = 7000$$

or $7000.

Determining the Domain of a Function

Suppose we are given the function $y = f(x)$.* Then, the variable x is called the **independent variable**. The variable y, whose value depends on x, is called the **dependent variable**.

To determine the domain of a function, we need to find what restrictions, if any, are to be placed on the independent variable x. In many practical applications, the domain of a function is dictated by the nature of the problem, as illustrated in Example 3.

---

*It is customary to refer to a function f as $f(x)$ or by the equation $y = f(x)$ defining it.

APPLIED EXAMPLE 3 Packaging An open box is to be made from a rectangular piece of cardboard 16 inches long and 10 inches wide by cutting away identical squares (x inches by x inches) from each corner and folding up the resulting flaps (Figure 3). Find an expression that gives the volume V of the box as a function of x. What is the domain of the function?

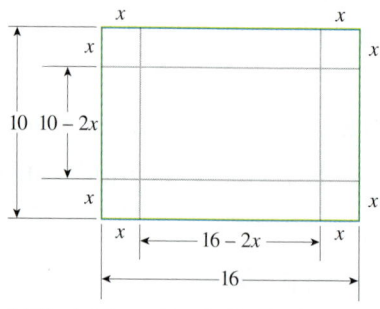
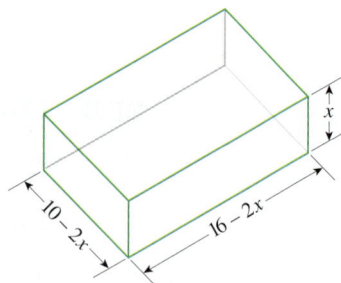

(a) The box is constructed by cutting x'' by x'' squares from each corner.

(b) The dimensions of the resulting box are $(10 - 2x)''$ by $(16 - 2x)''$ by x''.

FIGURE 3

Solution The dimensions of the box are $(10 - 2x)$ inches by $(16 - 2x)$ inches by x inches, so its volume (in cubic inches) is given by

$$V = f(x) = (16 - 2x)(10 - 2x)x \quad \text{Length} \cdot \text{width} \cdot \text{height}$$
$$= (160 - 52x + 4x^2)x$$
$$= 4x^3 - 52x^2 + 160x$$

Since the length of each side of the box must be greater than or equal to zero, we see that

$$16 - 2x \geq 0 \qquad 10 - 2x \geq 0 \qquad x \geq 0$$

simultaneously; that is,

$$x \leq 8 \qquad x \leq 5 \qquad x \geq 0$$

All three inequalities are satisfied simultaneously provided that $0 \leq x \leq 5$. Thus, the domain of the function f is the interval $[0, 5]$.

In general, if a function is defined by a rule relating x to $f(x)$ without specific mention of its domain, it is understood that the domain will consist of all values of x for which $f(x)$ is a real number. In this connection, you should keep in mind that (1) division by zero is not permitted and (2) the even root of a negative number is not a real number.

EXAMPLE 4 Find the domain of each function.

a. $f(x) = \sqrt{x - 1}$ **b.** $f(x) = \dfrac{1}{x^2 - 4}$ **c.** $f(x) = x^2 + 3$

Solution

a. Since the square root of a negative number is not a real number, it is necessary that $x - 1 \geq 0$. The inequality is satisfied by the set of real numbers $x \geq 1$. Thus, the domain of f is the interval $[1, \infty)$.
b. The only restriction on x is that $x^2 - 4$ be different from zero, since division by zero is not allowed. But $(x^2 - 4) = (x + 2)(x - 2) = 0$ if $x = -2$ or $x = 2$. Thus, the domain of f in this case consists of the intervals $(-\infty, -2)$, $(-2, 2)$, and $(2, \infty)$.
c. Here, any real number satisfies the equation, so the domain of f is the set of all real numbers.

Graphs of Functions

If f is a function with domain A, then corresponding to each real number x in A, there is precisely one real number $f(x)$. We can also express this fact by using **ordered pairs** of real numbers. Write each number x in A as the first member of an ordered pair and each number $f(x)$ corresponding to x as the second member of the ordered pair. This gives exactly one ordered pair $(x, f(x))$ for each x in A.

Observe that the condition that there be one and only one number $f(x)$ corresponding to each number x in A translates into the requirement that *no two ordered pairs have the same first number*.

Since ordered pairs of real numbers correspond to points in the plane, we have found a way to exhibit a function graphically.

> **Graph of a Function of One Variable**
>
> The **graph of a function** f is the set of all points (x, y) in the xy-plane such that x is in the domain of f and $y = f(x)$.

Figure 4 shows the graph of a function f. Observe that the y-coordinate of the point (x, y) on the graph of f gives the height of that point (the distance above the x-axis), if $f(x)$ is positive. If $f(x)$ is negative, then $-f(x)$ gives the depth of the point (x, y) (the distance below the x-axis). Also, observe that the domain of f is a set of real numbers lying on the x-axis, whereas the range of f lies on the y-axis.

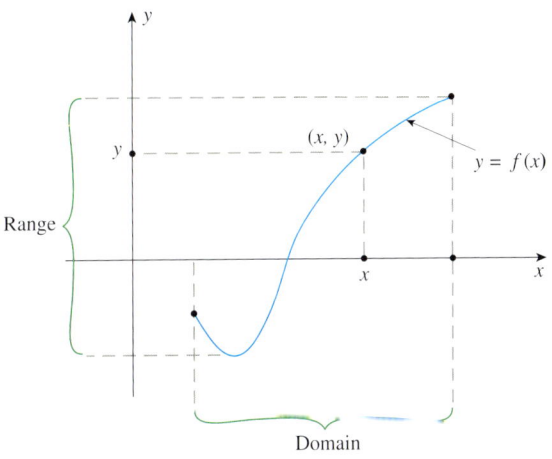

FIGURE 4
The graph of f

EXAMPLE 5 The graph of a function f is shown in Figure 5.

a. What is the value of $f(3)$? The value of $f(5)$?
b. What is the height or depth of the point $(3, f(3))$ from the x-axis? The point $(5, f(5))$ from the x-axis?
c. What is the domain of f? The range of f?

Solution

a. From the graph of f, we see that $y = -2$ when $x = 3$, and we conclude that $f(3) = -2$. Similarly, we see that $f(5) = 3$.

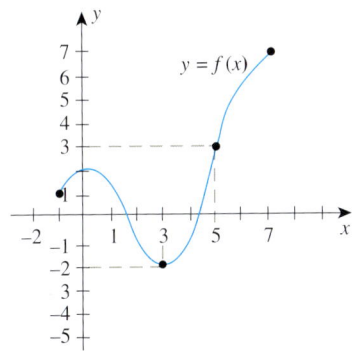

FIGURE 5
The graph of f

b. Since the point $(3, -2)$ lies below the x-axis, we see that the depth of the point $(3, f(3))$ is $-f(3) = -(-2) = 2$ units below the x-axis. The point $(5, f(5))$ lies above the x-axis and is located at a height of $f(5)$, or 3 units above the x-axis.

c. Observe that x may take on all values between $x = -1$ and $x = 7$, inclusive, and so the domain of f is $[-1, 7]$. Next, observe that as x takes on all values in the domain of f, $f(x)$ takes on all values between -2 and 7, inclusive. (You can easily see this by running your index finger along the x-axis from $x = -1$ to $x = 7$ and observing the corresponding values assumed by the y-coordinate of each point of the graph of f.) Therefore, the range of f is $[-2, 7]$. ■

Much information about the graph of a function can be gained by plotting a few points on its graph. Later on, we will develop more systematic and sophisticated techniques for graphing functions.

EXAMPLE 6 Sketch the graph of the function defined by the equation $y = x^2 + 1$. What is the range of f?

Solution The domain of the function is the set of all real numbers. By assigning several values to the variable x and computing the corresponding values for y, we obtain the following solutions to the equation $y = x^2 + 1$:

x	-3	-2	-1	0	1	2	3
y	10	5	2	1	2	5	10

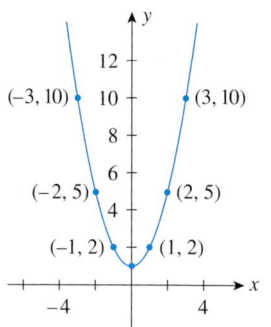

FIGURE 6
The graph of $y = x^2 + 1$ is a parabola.

By plotting these points and then connecting them with a smooth curve, we obtain the graph of $y = f(x)$, which is a parabola (Figure 6). To determine the range of f, we observe that $x^2 \geq 0$ if x is any real number, and so $x^2 + 1 \geq 1$ for all real numbers x. We conclude that the range of f is $[1, \infty)$. The graph of f confirms this result visually. ■

> **Exploring with TECHNOLOGY**
>
> Let $f(x) = x^2$.
>
> 1. Plot the graphs of $F(x) = x^2 + c$ on the same set of axes for $c = -2, -1, -\frac{1}{2}, 0, \frac{1}{2}, 1, 2$.
> 2. Plot the graphs of $G(x) = (x + c)^2$ on the same set of axes for $c = -2, -1, -\frac{1}{2}, 0, \frac{1}{2}, 1, 2$.
> 3. Plot the graphs of $H(x) = cx^2$ on the same set of axes for $c = -2, -1, -\frac{1}{2}, -\frac{1}{4}, 0, \frac{1}{4}, \frac{1}{2}, 1, 2$.
> 4. Study the family of graphs in parts 1–3, and describe the relationship between the graph of a function f and the graphs of the functions defined by (a) $y = f(x) + c$, (b) $y = f(x + c)$, and (c) $y = cf(x)$, where c is a constant.

Sometimes a function is defined by giving different formulas for different parts of its domain. Such a function is said to be a **piecewise-defined function.**

2.1 FUNCTIONS AND THEIR GRAPHS

EXAMPLE 7 Sketch the graph of the function f defined by

$$f(x) = \begin{cases} -x & \text{if } x < 0 \\ \sqrt{x} & \text{if } x \geq 0 \end{cases}$$

Solution The function f is defined in a piecewise fashion on the set of all real numbers. In the subdomain $(-\infty, 0)$, the rule for f is given by $f(x) = -x$. The equation $y = -x$ is a linear equation in the slope-intercept form (with slope -1 and intercept 0). Therefore, the graph of f corresponding to the subdomain $(-\infty, 0)$ is the half-line shown in Figure 7. Next, in the subdomain $[0, \infty)$, the rule for f is given by $f(x) = \sqrt{x}$. The values of $f(x)$ corresponding to $x = 0, 1, 2, 3,$ and 4 are shown in the following table:

x	0	1	2	3	4
$f(x)$	0	1	$\sqrt{2}$	$\sqrt{3}$	2

Using these values, we sketch the graph of the function f as shown in Figure 7.

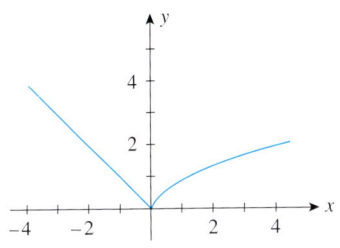

FIGURE 7
The graph of $y = f(x)$ is obtained by graphing $y = -x$ over $(-\infty, 0)$ and $y = \sqrt{x}$ over $[0, \infty)$.

APPLIED EXAMPLE 8 Bank Deposits Madison Finance Company plans to open two branch offices 2 years from now in two separate locations: an industrial complex and a newly developed commercial center in the city. As a result of these expansion plans, Madison's total deposits during the next 5 years are expected to grow in accordance with the rule

$$f(x) = \begin{cases} \sqrt{2x} + 20 & \text{if } 0 \leq x \leq 2 \\ \dfrac{1}{2}x^2 + 20 & \text{if } 2 < x \leq 5 \end{cases}$$

where $y = f(x)$ gives the total amount of money (in millions of dollars) on deposit with Madison in year x ($x = 0$ corresponds to the present). Sketch the graph of the function f.

Solution The function f is defined in a piecewise fashion on the interval $[0, 5]$. In the subdomain $[0, 2]$, the rule for f is given by $f(x) = \sqrt{2x} + 20$. The values of $f(x)$ corresponding to $x = 0, 1,$ and 2 may be tabulated as follows:

x	0	1	2
$f(x)$	20	21.4	22

Next, in the subdomain $(2, 5]$, the rule for f is given by $f(x) = \frac{1}{2}x^2 + 20$. The values of $f(x)$ corresponding to $x = 3, 4,$ and 5 are shown in the following table:

x	3	4	5
$f(x)$	24.5	28	32.5

Using the values of $f(x)$ in this table, we sketch the graph of the function f as shown in Figure 8.

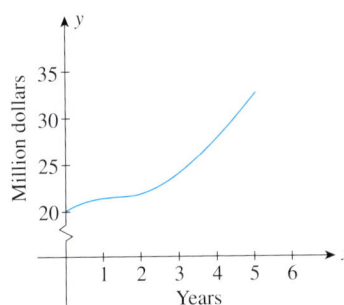

FIGURE 8
We obtain the graph of the function $y = f(x)$ by graphing $y = \sqrt{2x} + 20$ over $[0, 2]$ and $y = \frac{1}{2}x^2 + 20$ over $(2, 5]$.

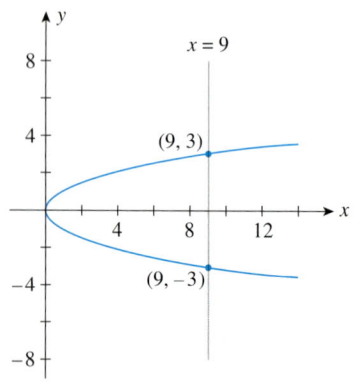

FIGURE 9
Since a vertical line passes through the curve at more than one point, we deduce that the curve is *not* the graph of a function.

The Vertical Line Test

Although it is true that every function f of a variable x has a graph in the xy-plane, it is not true that every curve in the xy-plane is the graph of a function. For example, consider the curve depicted in Figure 9. This is the graph of the equation $y^2 = x$. In general, the **graph of an equation** is the set of all ordered pairs (x, y) that satisfy the given equation. Observe that the points $(9, -3)$ and $(9, 3)$ both lie on the curve. This implies that the number $x = 9$ is associated with *two* numbers: $y = -3$ and $y = 3$. But this clearly violates the uniqueness property of a function. Thus, we conclude that the curve under consideration cannot be the graph of a function.

This example suggests the following **Vertical Line Test** for determining whether a curve is the graph of a function.

> **Vertical Line Test**
>
> A curve in the xy-plane is the graph of a function $y = f(x)$ if and only if each vertical line intersects it in at most one point.

EXAMPLE 9 Determine which of the curves shown in Figure 10 are the graphs of functions of x.

Solution The curves depicted in Figure 10a, c, and d are graphs of functions because each curve satisfies the requirement that each vertical line intersects the curve in at most one point. Note that the vertical line shown in Figure 10c does *not* intersect the graph because the point on the x-axis through which this line passes does not lie in the domain of the function. The curve depicted in Figure 10b is *not* the graph of a function of x because the vertical line shown there intersects the graph at three points.

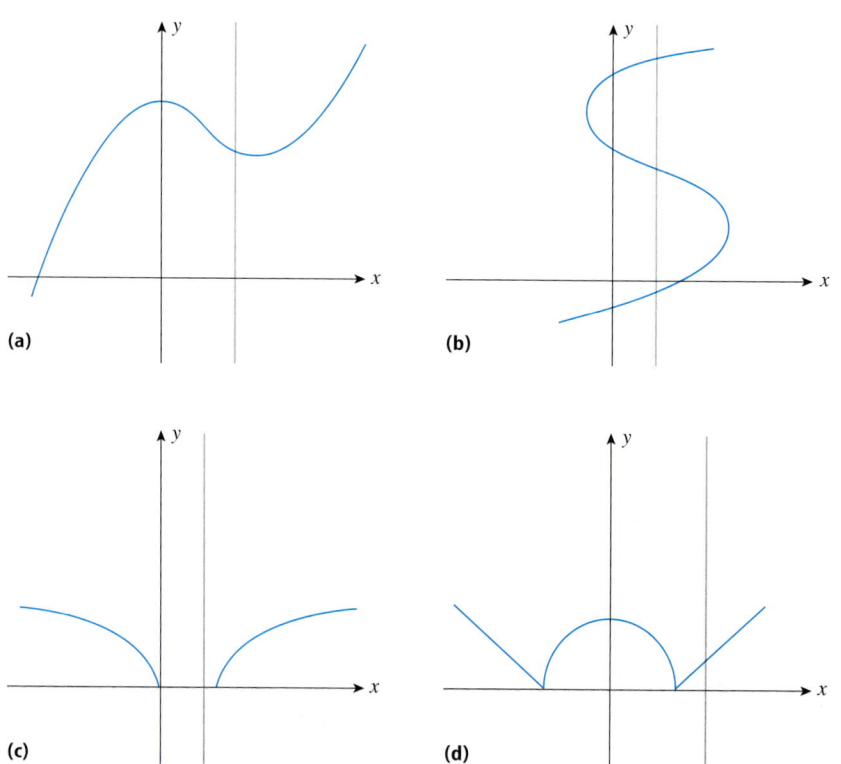

FIGURE 10
The Vertical Line Test can be used to determine which of these curves are graphs of functions.

2.1 Self-Check Exercises

1. Let f be the function defined by
$$f(x) = \frac{\sqrt{x+1}}{x}$$
   a. Find the domain of f.   b. Compute $f(3)$.
   c. Compute $f(a + h)$.

2. Let
$$f(x) = \begin{cases} -x + 1 & \text{if } -1 \leq x < 1 \\ \sqrt{x-1} & \text{if } 1 \leq x \leq 5 \end{cases}$$

   a. Find $f(0)$ and $f(2)$.
   b. Sketch the graph of f.

3. Let $f(x) = \sqrt{2x+1} + 2$. Determine whether the point $(4, 6)$ lies on the graph of f.

Solutions to Self-Check Exercises 2.1 can be found on page 63.

2.1 Concept Questions

1. a. What is a function?
   b. What is the domain of a function? The range of a function?
   c. What is an independent variable? A dependent variable?

2. a. What is the graph of a function? Use a drawing to illustrate the graph, the domain, and the range of a function.
   b. If you are given a curve in the xy-plane, how can you tell whether the graph is that of a function f defined by $y = f(x)$?

3. Are the following graphs of functions? Explain.

a.
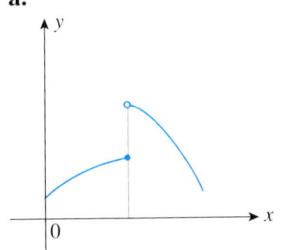

b.
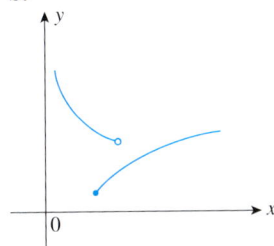

c.
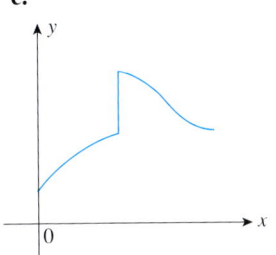

d.
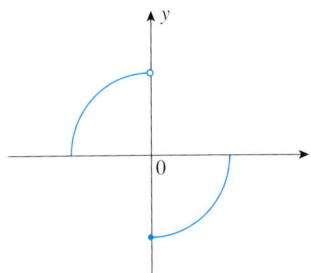

4. What are the domain and range of the function f with the following graph?

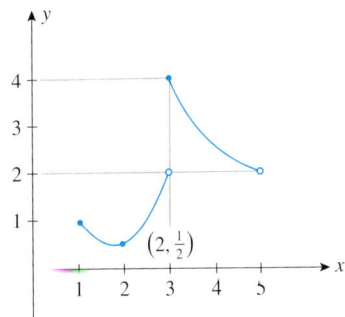

2.1 Exercises

1. Let f be the function defined by $f(x) = 5x + 6$. Find $f(3)$, $f(-3)$, $f(a)$, $f(-a)$, and $f(a + 3)$.

2. Let f be the function defined by $f(x) = 4x - 3$. Find $f(4)$, $f(\frac{1}{4})$, $f(0)$, $f(a)$, and $f(a + 1)$.

3. Let g be the function defined by $g(x) = 3x^2 - 6x - 3$. Find $g(0)$, $g(-1)$, $g(a)$, $g(-a)$, and $g(x + 1)$.

4. Let h be the function defined by $h(x) = x^3 - x^2 + x + 1$. Find $h(-5)$, $h(0)$, $h(a)$, and $h(-a)$.

5. Let f be the function defined by $f(x) = 2x + 5$. Find $f(a + h)$, $f(-a)$, $f(a^2)$, $f(a - 2h)$, and $f(2a - h)$.

6. Let g be the function defined by $g(x) = -x^2 + 2x$. Find $g(a + h)$, $g(-a)$, $g(\sqrt{a})$, $a + g(a)$, and $\dfrac{1}{g(a)}$.

7. Let s be the function defined by $s(t) = \dfrac{2t}{t^2 - 1}$. Find $s(4)$, $s(0)$, $s(a)$, $s(2 + a)$, and $s(t + 1)$.

8. Let g be the function defined by $g(u) = (3u - 2)^{3/2}$. Find $g(1), g(6), g(\frac{11}{3})$, and $g(u + 1)$.

9. Let f be the function defined by $f(t) = \dfrac{2t^2}{\sqrt{t-1}}$. Find $f(2)$, $f(a), f(x + 1)$, and $f(x - 1)$.

10. Let f be the function defined by $f(x) = 2 + 2\sqrt{5 - x}$. Find $f(-4), f(1), f(\frac{11}{4})$, and $f(x + 5)$.

11. Let f be the function defined by
$$f(x) = \begin{cases} x^2 + 1 & \text{if } x \leq 0 \\ \sqrt{x} & \text{if } x > 0 \end{cases}$$
Find $f(-2), f(0)$, and $f(1)$.

12. Let g be the function defined by
$$g(x) = \begin{cases} -\dfrac{1}{2}x + 1 & \text{if } x < 2 \\ \sqrt{x - 2} & \text{if } x \geq 2 \end{cases}$$
Find $g(-2), g(0), g(2)$, and $g(4)$.

13. Let f be the function defined by
$$f(x) = \begin{cases} -\dfrac{1}{2}x^2 + 3 & \text{if } x < 1 \\ 2x^2 + 1 & \text{if } x \geq 1 \end{cases}$$
Find $f(-1), f(0), f(1)$, and $f(2)$.

14. Let f be the function defined by
$$f(x) = \begin{cases} 2 + \sqrt{1 - x} & \text{if } x \leq 1 \\ \dfrac{1}{1 - x} & \text{if } x > 1 \end{cases}$$
Find $f(0), f(1)$, and $f(2)$.

15. Refer to the graph of the function f in the following figure.

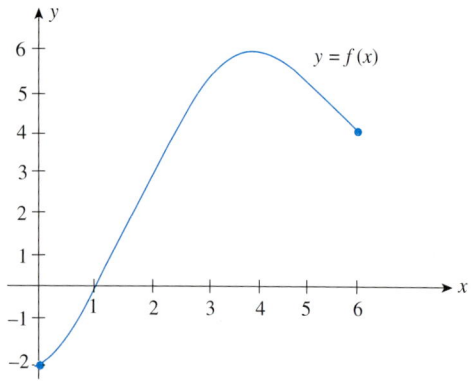

a. Find the value of $f(0)$.
b. Find the value of x for which (i) $f(x) = 3$ and (ii) $f(x) = 0$.
c. Find the domain of f.
d. Find the range of f.

16. Refer to the graph of the function f in the following figure.

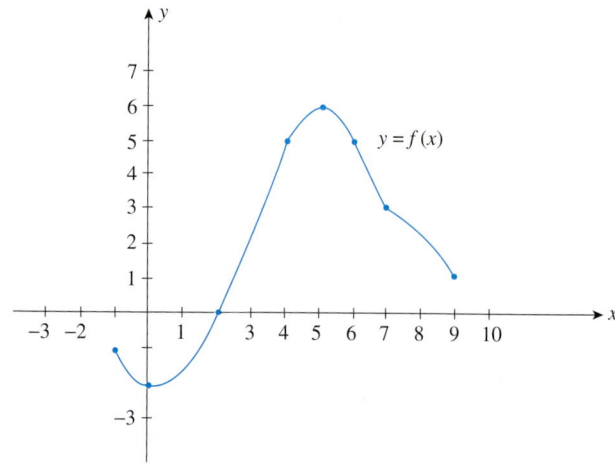

a. Find the value of $f(7)$.
b. Find the values of x corresponding to the point(s) on the graph of f located at a height of 5 units from the x-axis.
c. Find the point on the x-axis at which the graph of f crosses it. What is the value of $f(x)$ at this point?
d. Find the domain and range of f.

In Exercises 17–20, determine whether the point lies on the graph of the function.

17. $(2, \sqrt{3}); g(x) = \sqrt{x^2 - 1}$

18. $(3, 3); f(x) = \dfrac{x + 1}{\sqrt{x^2 + 7}} + 2$

19. $(-2, -3); f(t) = \dfrac{|t - 1|}{t + 1}$

20. $\left(-3, -\dfrac{1}{13}\right); h(t) = \dfrac{|t + 1|}{t^3 + 1}$

In Exercises 21 and 22, find the value of c such that the point $P(a, b)$ lies on the graph of the function f.

21. $f(x) = 2x^2 - 4x + c; P(1, 5)$

22. $f(x) = x\sqrt{9 - x^2} + c; P(2, 4)$

In Exercises 23–36, find the domain of the function.

23. $f(x) = x^2 + 3$

24. $f(x) = 7 - x^2$

25. $f(x) = \dfrac{3x + 1}{x^2}$

26. $g(x) = \dfrac{2x + 1}{x - 1}$

27. $f(x) = \sqrt{x^2 + 1}$

28. $f(x) = \sqrt{x - 5}$

29. $f(x) = \sqrt{5 - x}$

30. $g(x) = \sqrt{2x^2 + 3}$

31. $f(x) = \dfrac{x}{x^2 - 1}$

32. $f(x) = \dfrac{1}{x^2 + x - 2}$

33. $f(x) = (x + 3)^{3/2}$

34. $g(x) = 2(x - 1)^{5/2}$

35. $f(x) = \dfrac{\sqrt{1-x}}{x^2 - 4}$ 36. $f(x) = \dfrac{\sqrt{x-1}}{(x+2)(x-3)}$

37. Let f be the function defined by the rule $f(x) = x^2 - x - 6$.
    a. Find the domain of f.
    b. Compute $f(x)$ for $x = -3, -2, -1, 0, \tfrac{1}{2}, 1, 2, 3$.
    c. Use the results obtained in parts (a) and (b) to sketch the graph of f.

38. Let f be the function defined by the rule $f(x) = 2x^2 + x - 3$.
    a. Find the domain of f.
    b. Compute $f(x)$ for $x = -3, -2, -1, -\tfrac{1}{2}, 0, 1, 2, 3$.
    c. Use the results obtained in parts (a) and (b) to sketch the graph of f.

In Exercises 39–50, sketch the graph of the function with the given rule. Find the domain and range of the function.

39. $f(x) = 2x^2 + 1$ 40. $f(x) = 9 - x^2$

41. $f(x) = 2 + \sqrt{x}$ 42. $g(x) = 4 - \sqrt{x}$

43. $f(x) = \sqrt{1-x}$ 44. $f(x) = \sqrt{x-1}$

45. $f(x) = |x| - 1$ 46. $f(x) = |x| + 1$

47. $f(x) = \begin{cases} x & \text{if } x < 0 \\ 2x + 1 & \text{if } x \geq 0 \end{cases}$

48. $f(x) = \begin{cases} 4 - x & \text{if } x < 2 \\ 2x - 2 & \text{if } x \geq 2 \end{cases}$

49. $f(x) = \begin{cases} -x + 1 & \text{if } x \leq 1 \\ x^2 - 1 & \text{if } x > 1 \end{cases}$

50. $f(x) = \begin{cases} -x - 1 & \text{if } x < -1 \\ 0 & \text{if } -1 \leq x \leq 1 \\ x + 1 & \text{if } x > 1 \end{cases}$

In Exercises 51–58, use the Vertical Line Test to determine whether the graph represents y as a function of x.

51.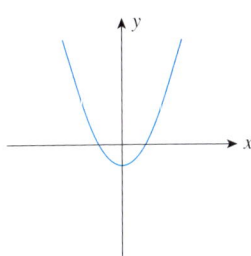

52.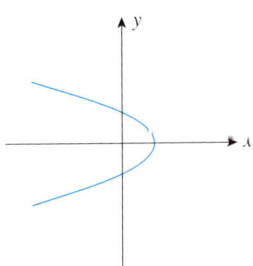

53.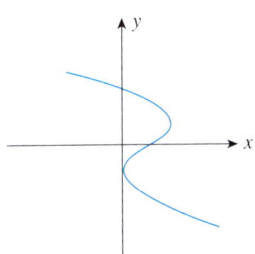

54.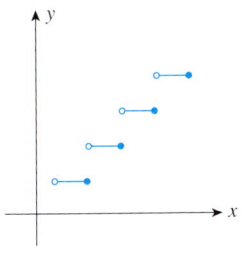

55.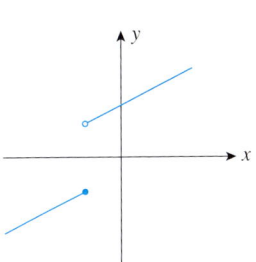

56.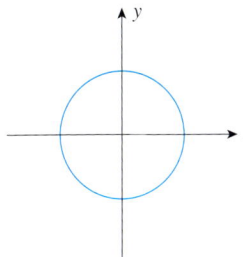

57.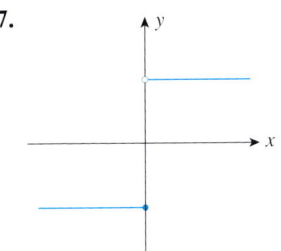

58.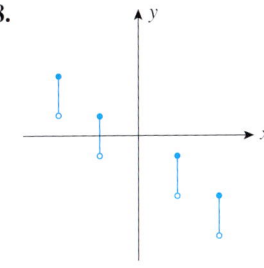

59. The circumference of a circle is given by

$$C(r) = 2\pi r$$

where r is the radius of the circle. What is the circumference of a circle with a 5-in. radius?

60. The volume of a sphere of radius r is given by

$$V(r) = \tfrac{4}{3}\pi r^3$$

Compute $V(2.1)$ and $V(2)$. What does the quantity $V(2.1) - V(2)$ measure?

61. **GROWTH OF A CANCEROUS TUMOR** The volume of a spherical cancerous tumor is given by the function

$$V(r) = \frac{4}{3}\pi r^3$$

where r is the radius of the tumor in centimeters. By what factor is the volume of the tumor increased if its radius is doubled?

62. **LIFE EXPECTANCY AFTER AGE 65** The average life expectancy after age 65 is soaring, putting pressure on the Social Security Administration's resources. According to the Social Security Trustees, the average life expectancy after age 65 is given by

$$L(t) = 0.056t + 18.1 \qquad (0 \leq t \leq 7)$$

where t is measured in years, with $t = 0$ corresponding to 2003.
    a. How fast is the average life expectancy after age 65 changing at any time during the period under consideration?
    b. What was the average life expectancy after age 65 in 2013?

Source: Social Security Trustees.

63. Sales of Prerecorded Music The following graphs show the sales y of prerecorded music (in billions of dollars) by format as a function of time t (in years), with $t = 0$ corresponding to 1985.

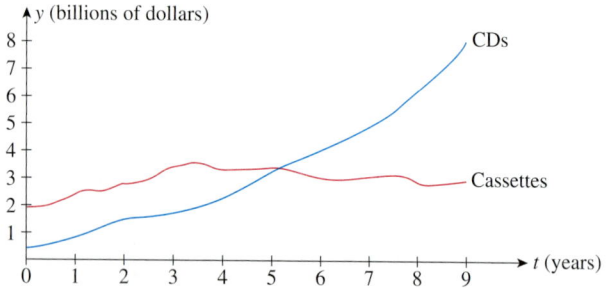

a. In what years were the sales of prerecorded cassettes greater than those of prerecorded CDs?
b. In what years were the sales of prerecorded CDs greater than those of prerecorded cassettes?
c. In what year were the sales of prerecorded cassettes the same as those of prerecorded CDs? Estimate the level of sales in each format at that time.
Source: Recording Industry Association of America.

64. The Gender Gap The following graph shows the ratio of women's earnings to men's from 1960 through 2000.

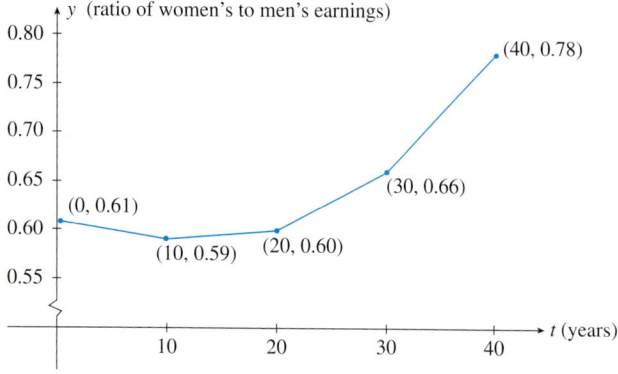

a. Write the rule for the function f giving the ratio of women's earnings to men's in year t, with $t = 0$ corresponding to 1960.
Hint: The function f is defined piecewise and is linear over each of four subintervals.
b. In what decade(s) was the gender gap expanding? Shrinking?
c. Refer to part (b). How fast was the gender gap expanding or shrinking in each of these decades?
Source: U.S. Bureau of Labor Statistics.

65. Closing the Gender Gap in Education The following graph shows the ratio of the number of bachelor's degrees earned by women to that of men from 1960 through 1990.

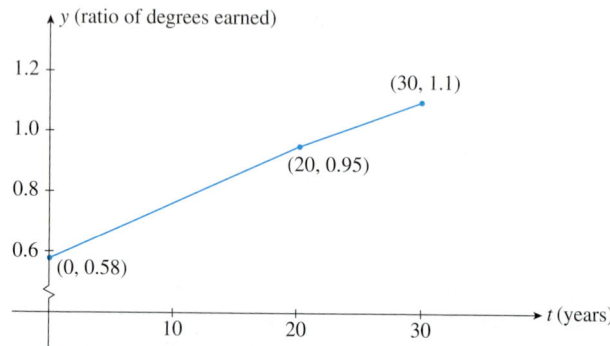

a. Write the rule for the function f giving the ratio of the number of bachelor's degrees earned by women to that of men in year t, with $t = 0$ corresponding to 1960.
Hint: The function f is defined piecewise and is linear over each of two subintervals.
b. How fast was the ratio changing in the period from 1960 to 1980? From 1980 to 1990?
c. In what year (approximately) was the number of bachelor's degrees earned by women equal for the first time to the number earned by men?
Source: Department of Education

66. Consumption Function The consumption function in a certain economy is given by the equation

$$C(y) = 0.75y + 6$$

where $C(y)$ is the personal consumption expenditure, y is the disposable personal income, and both $C(y)$ and y are measured in billions of dollars. Find $C(0)$, $C(50)$, and $C(100)$.

67. Sales Taxes In a certain state, the sales tax T on the amount of taxable goods is 6% of the value of the goods purchased x, where both T and x are measured in dollars.
a. Express T as a function of x.
b. Find $T(200)$ and $T(5.65)$.

68. Surface Area of a Single-Celled Organism The surface area S of a single-celled organism may be found by multiplying 4π times the square of the radius r of the cell. Express S as a function of r.

69. Friend's Rule Friend's Rule, a method for calculating pediatric drug dosages, is based on a child's age. If a denotes the adult dosage (in milligrams) and if t is the age of the child (in years), then the child's dosage is given by

$$D(t) = \frac{2}{25}ta$$

If the adult dose of a substance is 500 mg, how much should a 4-year-old child receive?

70. COLAs Social Security recipients receive an automatic cost-of-living adjustment (COLA) once each year. Their monthly benefit is increased by the amount that consumer prices increased during the preceding year. Suppose that consumer prices increased by 5.3% during the preceding year.

a. Express the adjusted monthly benefit of a Social Security recipient as a function of his or her current monthly benefit.
b. If Harrington's monthly Social Security benefit is now $1520, what will be his adjusted monthly benefit?

71. GLOBAL DEFENSE SPENDING Global defense spending stood at $1.44 trillion in 2009 and is projected to grow at the rate of $0.058 trillion per year through 2018.
a. Find a function $f(t)$ giving the projected global defense spending in year t, where $t = 0$ corresponds to 2009.
Hint: The graph of f lies on a straight line.
b. What is the projected global defense spending in 2018?
Source: Homeland Security Research.

72. COST OF RENTING A TRUCK Ace Truck leases its 10-ft box truck at $30/day and $0.45/mi, whereas Acme Truck leases a similar truck at $25/day and $0.50/mi.
a. Find the daily cost of leasing from each company as a function of the number of miles driven.
b. Sketch the graphs of the two functions on the same set of axes.
c. Which company should a customer rent a truck from for 1 day if she plans to drive at most 70 mi and wishes to minimize her cost?

73. LINEAR DEPRECIATION A new machine was purchased by National Textile for $120,000. For income tax purposes, the machine is depreciated linearly over 10 years; that is, the book value of the machine decreases at a constant rate, so that at the end of 10 years the book value is zero.
a. Express the book value of the machine V as a function of the age, in years, of the machine n.
b. Sketch the graph of the function in part (a).
c. Find the book value of the machine at the end of the sixth year.
d. Find the rate at which the machine is being depreciated each year.

74. LINEAR DEPRECIATION Refer to Exercise 73. An office building worth $1 million when completed in 1997 was depreciated linearly over 50 years. What was the book value of the building in 2012? What will be the book value in 2016? In 2021? (Assume that the book value of the building will be zero at the end of the 50th year.)

75. BOYLE'S LAW As a consequence of Boyle's Law, the pressure P of a fixed sample of gas held at a constant temperature is related to the volume V of the gas by the rule

$$P = f(V) = \frac{k}{V}$$

where k is a constant. What is the domain of the function f? Sketch the graph of the function f.

76. POISEUILLE'S LAW According to a law discovered by the nineteenth century physician Poiseuille, the velocity (in centimeters/second) of blood r cm from the central axis of an artery is given by

$$v(r) = k(R^2 - r^2)$$

where k is a constant and R is the radius of the artery. Suppose that for a certain artery, $k = 1000$ and $R = 0.2$, so that $v(r) = 1000(0.04 - r^2)$.
a. What is the domain of the function v?
b. Compute $v(0)$, $v(0.1)$, and $v(0.2)$ and interpret your results.
c. Sketch the graph of the function v on the interval $[0, 0.2]$.
d. What can you say about the velocity of blood as we move away from the central axis toward the artery wall?

77. CANCER SURVIVORS The number of living Americans who have had a cancer diagnosis has increased drastically since 1971. In part, this is due to more testing for cancer and better treatment for some cancers. In part, it is because the population is older, and cancer is largely a disease of the elderly. The number of cancer survivors (in millions) between 1975 ($t = 0$) and 2005 ($t = 30$) is approximately

$$N(t) = 0.0031t^2 + 0.16t + 3.6 \quad (0 \le t \le 30)$$

a. How many living Americans had a cancer diagnosis in 1975? In 2005?
b. Assuming that the trend continued, how many cancer survivors were there in 2010?
Source: National Cancer Institute.

78. PREVALENCE OF ALZHEIMER'S PATIENTS Based on a study conducted in 1997, the percentage of the U.S. population by age afflicted with Alzheimer's disease is given by the function

$$P(x) = 0.0726x^2 + 0.7902x + 4.9623 \quad (0 \le x \le 25)$$

where x is measured in years, with $x = 0$ corresponding to age 65. What percentage of the U.S. population at age 65 is expected to have Alzheimer's disease? At age 90?
Source: Alzheimer's Association.

79. WORKER EFFICIENCY An efficiency study conducted for Elektra Electronics showed that the number of Space Commander walkie-talkies assembled by the average worker t hr after starting work at 8 A.M. is given by

$$N(t) = -t^3 + 6t^2 + 15t \quad (0 \le t \le 4)$$

How many walkie-talkies can an average worker be expected to assemble between 8 and 9 A.M.? Between 9 and 10 A.M.?

80. POLITICS Political scientists have discovered the following empirical rule, known as the "cube rule," which gives the relationship between the proportion of seats in the House of Representatives won by Democratic candidates $s(x)$ and the proportion of popular votes x received by the Democratic presidential candidate:

$$s(x) = \frac{x^3}{x^3 + (1 - x)^3} \quad (0 \le x \le 1)$$

Compute $s(0.6)$, and interpret your result.

81. U.S. Health-Care Information Technology Spending As health-care costs increase, payers are turning to technology and outsourced services to keep a lid on expenses. The amount of health-care information technology (IT) spending by payer is projected to be

$$S(t) = -0.03t^3 + 0.2t^2 + 0.23t + 5.6 \quad (0 \le t \le 4)$$

where $S(t)$ is measured in billions of dollars and t is measured in years, with $t = 0$ corresponding to 2004. What was the amount spent by payers on health-care IT in 2004? Assuming that the projection held true, what amount was spent by payers in 2008?
Source: U.S. Department of Commerce.

82. Hotel Rates The average daily rate of U.S. hotels from 2006 through 2009 is approximated by the function

$$f(t) = \begin{cases} 0.88t^2 + 3.21t + 96.75 & \text{if } 0 \le t < 2 \\ -5.58t + 117.85 & \text{if } 2 \le t \le 3 \end{cases}$$

where $f(t)$ is measured in dollars and $t = 0$ corresponds to 2006.
a. What was the average daily rate of U.S. hotels in 2006? In 2007? In 2008?
b. Sketch the graph of f.
Source: Smith Travel Research.

83. Investments in Hedge Funds Investments in hedge funds have increased along with their popularity. The assets of hedge funds (in trillions of dollars) from 2002 through 2007 are modeled by the function

$$f(t) = \begin{cases} 0.6 & \text{if } 0 \le t < 1 \\ 0.6t^{0.43} & \text{if } 1 \le t \le 5 \end{cases}$$

where t is measured in years, with $t = 0$ corresponding to the beginning of 2002.
a. What were the assets in hedge funds at the beginning of 2002? At the beginning of 2003?
b. What were the assets in hedge funds at the beginning of 2005? At the beginning of 2007?
Source: Hennessee Group.

84. Postal Regulations In 2012, the postage for parcels sent by first-class mail was raised to $1.95 for any parcel weighing less than 4 oz or fraction thereof and 17¢ for each additional ounce or fraction thereof. Any parcel not exceeding 13 oz may be sent by first-class mail. Letting x denote the weight of a parcel in ounces and $f(x)$ the postage in dollars, complete the following description of the "postage function" f:

$$f(x) = \begin{cases} \$1.95 & \text{if } 0 < x < 4 \\ \$2.12 & \text{if } 4 \le x < 5 \\ \vdots & \\ ? & \text{if } x = 13 \end{cases}$$

a. What is the domain of f?
b. Sketch the graph of f.

85. Harbor Cleanup The amount of solids discharged from the MWRA (Massachusetts Water Resources Authority) sewage treatment plant on Deer Island (in Boston Harbor) is given by the function

$$f(t) = \begin{cases} 130 & \text{if } 0 \le t \le 1 \\ -30t + 160 & \text{if } 1 < t \le 2 \\ 100 & \text{if } 2 < t \le 4 \\ -5t^2 + 25t + 80 & \text{if } 4 < t \le 6 \\ 1.25t^2 - 26.25t + 162.5 & \text{if } 6 < t \le 10 \end{cases}$$

where $f(t)$ is measured in tons per day and t is measured in years, with $t = 0$ corresponding to 1989.
a. What amount of solids were discharged per day in 1989? In 1992? In 1996?
b. Sketch the graph of f.
Source: Metropolitan District Commission.

86. Rising Median Age Increased longevity and the aging of the baby boom generation—those born between 1946 and 1965—are the primary reasons for a rising median age. The median age (in years) of the U.S. population from 1900 through 2000 is approximated by the function

$$f(t) = \begin{cases} 1.3t + 22.9 & \text{if } 0 \le t \le 3 \\ -0.7t^2 + 7.2t + 11.5 & \text{if } 3 < t \le 7 \\ 2.6t + 9.4 & \text{if } 7 < t \le 10 \end{cases}$$

where t is measured in decades, with $t = 0$ corresponding to the beginning of 1900.
a. What was the median age of the U.S. population at the beginning of 1900? At the beginning of 1950? At the beginning of 1990?
b. Sketch the graph of f.
Source: U.S. Census Bureau.

87. Distance Between Ships A passenger ship leaves port sailing east at 14 mph. Two hours later, a cargo ship leaves the same port heading north at 10 mph.
a. Find a function giving the distance between the two ships t hr after the passenger ship leaves port.
b. How far apart are the two ships 3 hr after the cargo ship leaves port?

In Exercises 88–94, determine whether the statement is true or false. If it is true, explain why it is true. If it is false, give an example to show why it is false.

88. If $a = b$, then $f(a) = f(b)$.

89. If $f(a) = f(b)$, then $a = b$.

90. If f is a function, then $f(a + b) = f(a) + f(b)$.

91. A vertical line must intersect the graph of $y = f(x)$ at exactly one point.

92. The domain of $f(x) = \sqrt{x + 2} + \sqrt{2 - x}$ is $[-2, 2]$.

93. If f is a function defined on $(-\infty, \infty)$ and k is a real number, then $f(kx) = kf(x)$.

94. If f is a linear function, then $f(cx + y) = cf(x) + f(y)$, where c is a real number.

2.1 Solutions to Self-Check Exercises

1. **a.** The expression under the radical sign must be nonnegative, so $x + 1 \geq 0$ or $x \geq -1$. Also, $x \neq 0$ because division by zero is not permitted. Therefore, the domain of f is $[-1, 0) \cup (0, \infty)$.

   b. $f(3) = \dfrac{\sqrt{3+1}}{3} = \dfrac{\sqrt{4}}{3} = \dfrac{2}{3}$

   c. $f(a+h) = \dfrac{\sqrt{(a+h)+1}}{a+h} = \dfrac{\sqrt{a+h+1}}{a+h}$

2. **a.** The function f is defined in a piecewise fashion. For $x = 0$, the rule is $f(x) = -x + 1$, and so $f(0) = 1$. For $x = 2$, the rule is $f(x) = \sqrt{x - 1}$, and so $f(2) = \sqrt{2 - 1} = 1$.

   b. In the subdomain $[-1, 1)$, the graph of f is the line segment $y = -x + 1$, which is a linear equation with slope -1 and y-intercept 1. In the subdomain $[1, 5]$, the graph of f is given by the rule $f(x) = \sqrt{x - 1}$. From the table below,

x	1	2	3	4	5
$f(x)$	0	1	$\sqrt{2}$	$\sqrt{3}$	2

   we obtain the following graph of f.

   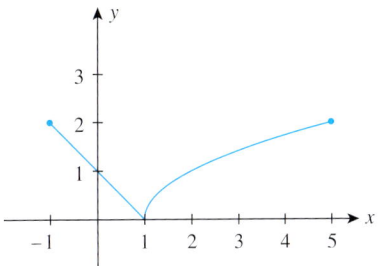

3. A point (x, y) lies on the graph of the function f if and only if the coordinates satisfy the equation $y = f(x)$. Now,

   $$f(4) = \sqrt{2(4) + 1} + 2 = \sqrt{9} + 2 = 5 \neq 6$$

   and we conclude that the given point does *not* lie on the graph of f.

USING TECHNOLOGY

Graphing a Function

Most of the graphs of functions in this book can be plotted with the help of a graphing utility. Furthermore, a graphing utility can be used to analyze the nature of a function. However, the amount and accuracy of the information obtained by using a graphing utility depend on the experience and sophistication of the user. As you progress through this book, you will see that the more knowledge of calculus you gain, the more effective the graphing utility will prove to be as a tool in problem solving.

Finding a Suitable Viewing Window

The first step in plotting the graph of a function with a graphing utility is to select a suitable viewing window. We usually do this by experimenting. For example, you might first plot the graph using the *standard viewing window* $[-10, 10]$ by $[-10, 10]$. If necessary, you then might adjust the viewing window by enlarging it or reducing it to obtain a sufficiently complete view of the graph or at least the portion of the graph that is of interest.

(continued)

EXAMPLE 1 Plot the graph of $f(x) = 2x^2 - 4x - 5$ in the standard viewing window.

Solution The graph of f, shown in Figure T1a, is a parabola. From our previous work (Example 6, Section 2.1), we know that the figure does give a good view of the graph.

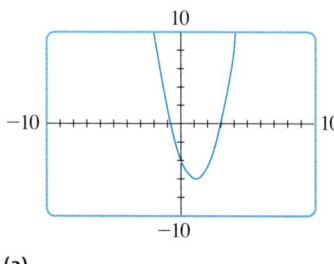

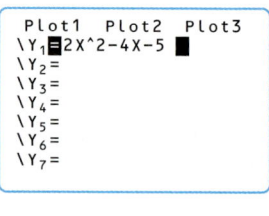

(a) **(b)** **(c)**

FIGURE T1
(a) The graph of $f(x) = 2x^2 - 4x - 5$ on $[-10, 10] \times [-10, 10]$; (b) the TI-83/84 window screen for (a); (c) the TI-83/84 equation screen

EXAMPLE 2 Let $f(x) = x^3(x - 3)^4$.

a. Plot the graph of f in the standard viewing window.
b. Plot the graph of f in the window $[-1, 5] \times [-40, 40]$.

Solution

a. The graph of f in the standard viewing window is shown in Figure T2a. Since the graph does not appear to be complete, we need to adjust the viewing window.

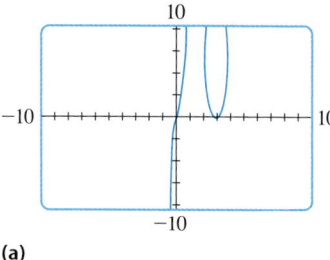

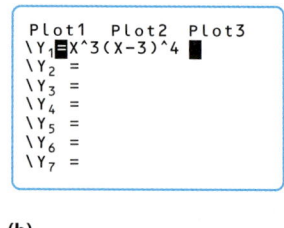

(a) **(b)**

FIGURE T2
(a) An incomplete sketch of $f(x) = x^3(x - 3)^4$ on $[-10, 10] \times [-10, 10]$; (b) the TI-83/84 equation screen

b. The graph of f in the window $[-1, 5] \times [-40, 40]$, shown in Figure T3a, is an improvement over the previous graph. (Later we will be able to show that the figure does in fact give a rather complete view of the graph of f.)

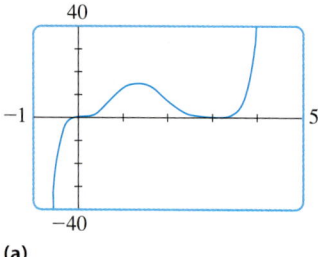

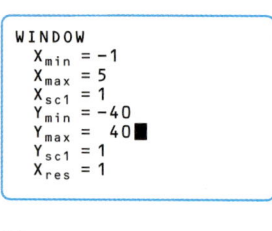

(a) **(b)**

FIGURE T3
(a) A complete sketch of $f(x) = x^3(x - 3)^4$ is shown using the window $[-1, 5] \times [-40, 40]$; (b) the TI-83/84 window screen

Evaluating a Function

A graphing utility can be used to find the value of a function with minimal effort, as the next example shows.

EXAMPLE 3 Let $f(x) = x^3 - 4x^2 + 4x + 2$.

a. Plot the graph of f in the standard viewing window.
b. Find $f(3)$ and verify your result by direct computation.
c. Find $f(4.215)$.

Solution

a. The graph of f is shown in Figure T4a.

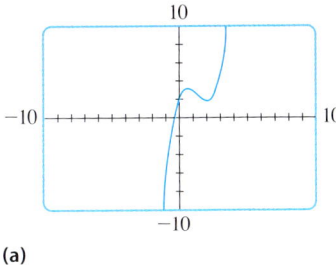
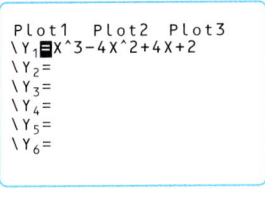

FIGURE T4
(a) The graph of $f(x) = x^3 - 4x^2 + 4x + 2$ in the standard viewing window; (b) the TI-83/84 equation screen

(a) (b)

b. Using the evaluation function of the graphing utility and the value 3 for x, we find $y = 5$. This result is verified by computing

$$f(3) = 3^3 - 4(3^2) + 4(3) + 2 = 27 - 36 + 12 + 2 = 5$$

c. Using the evaluation function of the graphing utility and the value 4.215 for x, we find $y = 22.679738$. Thus, $f(4.215) = 22.679738$. The efficacy of the graphing utility is clearly demonstrated here!

APPLIED EXAMPLE 4 Number of Alzheimer's Patients The number of Alzheimer's patients in the United States is approximated by

$$f(t) = 0.142t^3 - 0.557t^2 + 1.340t + 3.8 \qquad (0 \leq t \leq 4)$$

where $f(t)$ is measured in millions and t is measured in decades, with $t = 0$ corresponding to the beginning of 1990.

a. Use a graphing utility to plot the graph of f in the viewing window $[0, 4] \times [0, 12]$.
b. What is the projected number of Alzheimer's patients in the United States at the beginning of 2020 ($t = 3$)?

Source: Alzheimer's Association.

Solution

a. The graph of f in the viewing window $[0, 4] \times [0, 12]$ is shown in Figure T5a.

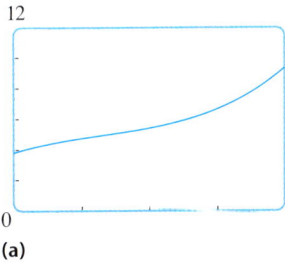
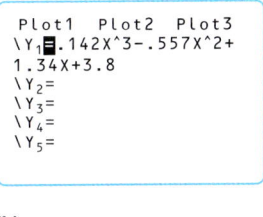

FIGURE T5
(a) The graph of f in the viewing window $[0, 4] \times [0, 12]$; (b) the TI-83/84 equation screen

(a) (b)

(continued)

b. Using the evaluation function of the graphing utility and the value 3 for x, we see that the anticipated number of Alzheimer's patients at the beginning of 2020 is given by $f(3) \approx 6.64$, or approximately 6.6 million. ∎

TECHNOLOGY EXERCISES

In Exercises 1–4, plot the graph of the function f in (a) the standard viewing window and (b) the indicated window.

1. $f(x) = x^4 - 2x^2 + 8$; $[-2, 2] \times [6, 10]$

2. $f(x) = x^3 - 20x^2 + 8x - 10$; $[-20, 20] \times [-1200, 100]$

3. $f(x) = x\sqrt{4 - x^2}$; $[-3, 3] \times [-2, 2]$

4. $f(x) = \dfrac{4}{x^2 - 8}$; $[-5, 5] \times [-5, 5]$

In Exercises 5–8, plot the graph of the function f in an appropriate viewing window. (*Note:* The answer is *not* unique.)

5. $f(x) = 2x^4 - 3x^3 + 5x^2 - 20x + 40$

6. $f(x) = -2x^4 + 5x^2 - 4$

7. $f(x) = \dfrac{x^3}{x^3 + 1}$

8. $f(x) = \dfrac{2x^4 - 3x}{x^2 - 1}$

In Exercises 9–12, use the evaluation function of your graphing utility to find the value of f at the indicated value of x. Express your answer accurate to four decimal places.

9. $f(x) = 3x^3 - 2x^2 + x - 4$; $x = 2.145$

10. $f(x) = 5x^4 - 2x^2 + 8x - 3$; $x = 1.28$

11. $f(x) = \dfrac{2x^3 - 3x + 1}{3x - 2}$; $x = 2.41$

12. $f(x) = \sqrt{2x^2 + 1} + \sqrt{3x^2 - 1}$; $x = 0.62$

13. **Lobbyists' Spending** Lobbyists try to persuade legislators to propose, pass, or defeat legislation or to change existing laws. The amount (in billions of dollars) spent by lobbyists from 2003 through 2009, where $t = 0$ corresponds to 2003, is given by

$$f(t) = -0.0056t^3 + 0.112t^2 + 0.51t + 8 \quad (0 \leq t \leq 6)$$

a. Plot the graph of f in the viewing window $[0, 6] \times [0, 15]$.
b. What amount was spent by lobbyists in the year 2005? In 2009?

Source: OpenSecrets.org.

14. **Surveillance Cameras** Research reports indicate that surveillance cameras at major intersections dramatically reduce the number of drivers who barrel through red lights. The cameras automatically photograph vehicles that drive into intersections after the light turns red. Vehicle owners are then mailed citations instructing them to pay a fine or sign an affidavit that they weren't driving at the time. The function

$$N(t) = 6.08t^3 - 26.79t^2 + 53.06t + 69.5 \quad (0 \leq t \leq 4)$$

gives the number, $N(t)$, of U.S. communities using surveillance cameras at intersections in year t, with $t = 0$ corresponding to 2003.
a. Plot the graph of N in the viewing window $[0, 4] \times [0, 250]$.
b. How many communities used surveillance cameras at intersections in 2004? In 2006?

Source: Insurance Institute for Highway Safety.

15. **Keeping With the Traffic Flow** By driving at a speed to match the prevailing traffic speed, you decrease the chances of an accident. According to data obtained in a university study, the number of accidents per 100 million vehicle miles, y, is related to the deviation from the mean speed, x, in miles per hour by

$$y = 1.05x^3 - 21.95x^2 + 155.9x - 327.3 \quad (6 \leq x \leq 11)$$

a. Plot the graph of y in the viewing window $[6, 11] \times [20, 150]$.
b. What is the number of accidents per 100 million vehicle miles if the deviation from the mean speed is 6 mph, 8 mph, and 11 mph?

Source: University of Virginia School of Engineering and Applied Science.

16. **Safe Drivers** The fatality rate in the United States (per 100 million miles traveled) by age of driver (in years) is given by the function

$$f(x) = 0.00000304x^4 - 0.0005764x^3 + 0.04105x^2 - 1.30366x + 16.579 \quad (18 \leq x \leq 82)$$

a. Plot the graph of f in the viewing window $[18, 82] \times [0, 8]$.
b. What is the fatality rate for 18-year-old drivers? For 50-year-old drivers? For 80-year-old drivers?

Source: National Highway Traffic Safety Administration.

2.2 The Algebra of Functions

The Sum, Difference, Product, and Quotient of Functions

Let $S(t)$ and $R(t)$ denote the federal government's spending and revenue, respectively, at any time t, measured in billions of dollars. The graphs of these functions for the period between 2004 and 2009 are shown in Figure 11.

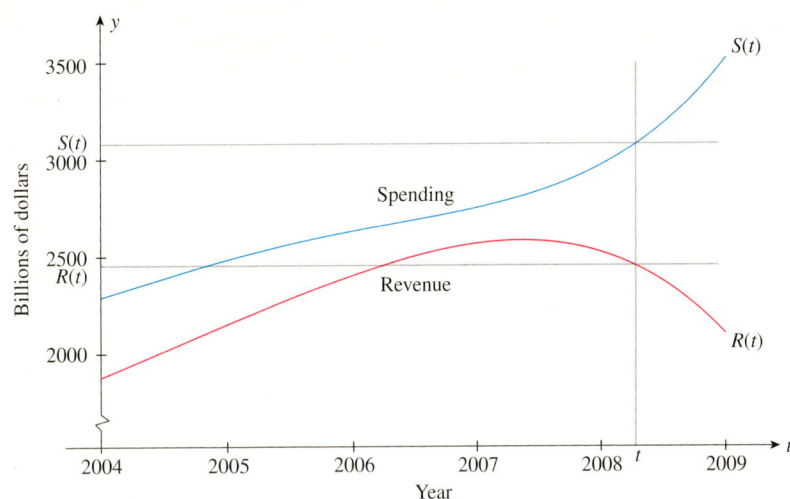

FIGURE 11
$R(t) - S(t)$ gives the federal budget deficit (surplus) at any time t.

Source: Office of Management and Budget.

The difference $R(t) - S(t)$ gives the deficit (surplus) in billions of dollars at any time t if $R(t) - S(t)$ is negative (positive). This observation suggests that we can define a function D whose value at any time t is given by $R(t) - S(t)$. The function D, the *difference* of the two functions R and S, is written $D = R - S$ and may be called the "deficit (surplus) function," since it gives the budget deficit or surplus at any time t. It has the same domain as the functions S and R. The graph of the function D is shown in Figure 12.

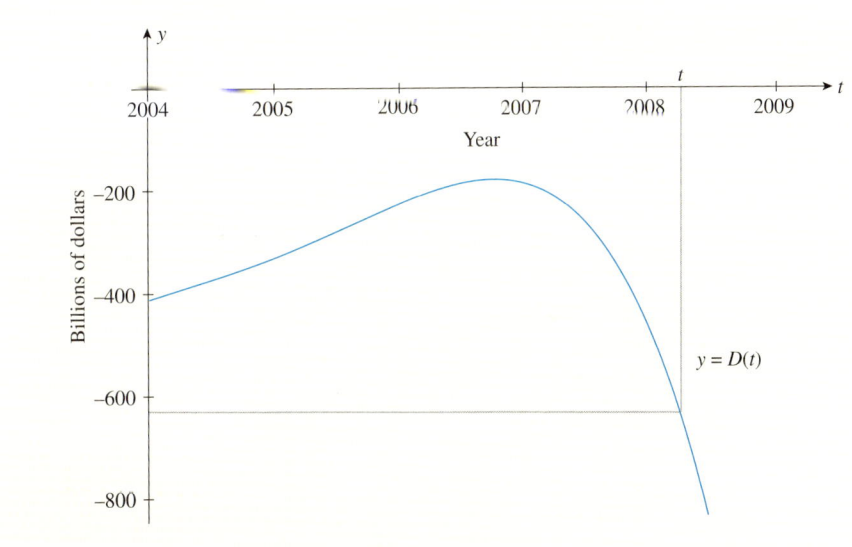

FIGURE 12
The graph of $D(t)$

Source: Office of Management and Budget.

Most functions are built up from other, generally simpler, functions. For example, we may view the function $f(x) = 2x + 4$ as the sum of the two functions $g(x) = 2x$ and $h(x) = 4$. The function $g(x) = 2x$ may in turn be viewed as the product of the functions $p(x) = 2$ and $q(x) = x$.

In general, given the functions f and g, we define the sum $f + g$, the difference $f - g$, the product fg, and the quotient f/g of f and g as follows.

The Sum, Difference, Product, and Quotient of Functions

Let f and g be functions with domains A and B, respectively. Then the **sum** $f + g$, **difference** $f - g$, and **product** fg of f and g are functions with domain $A \cap B$* and rule given by

$$(f + g)(x) = f(x) + g(x) \quad \text{Sum}$$
$$(f - g)(x) = f(x) - g(x) \quad \text{Difference}$$
$$(fg)(x) = f(x)g(x) \quad \text{Product}$$

The **quotient** f/g of f and g has domain $A \cap B$ excluding all numbers x such that $g(x) = 0$ and rule given by

$$\left(\frac{f}{g}\right)(x) = \frac{f(x)}{g(x)} \quad \text{Quotient}$$

---

*$A \cap B$ is read "A intersected with B" and denotes the set of all points common to both A and B.

EXAMPLE 1 Let $f(x) = \sqrt{x + 1}$ and $g(x) = 2x + 1$. Find the sum s, the difference d, the product p, and the quotient q of the functions f and g, and give their domains.

Solution Since the domain of f is $A = [-1, \infty)$ and the domain of g is $B = (-\infty, \infty)$, we see that the domain of s, d, and p is $A \cap B = [-1, \infty)$. The rules follow.

$$s(x) = (f + g)(x) = f(x) + g(x) = \sqrt{x + 1} + 2x + 1$$
$$d(x) = (f - g)(x) = f(x) - g(x) = \sqrt{x + 1} - (2x + 1) = \sqrt{x + 1} - 2x - 1$$
$$p(x) = (fg)(x) = f(x)g(x) = \sqrt{x + 1}(2x + 1) = (2x + 1)\sqrt{x + 1}$$

The quotient function q has rule

$$q(x) = \left(\frac{f}{g}\right)(x) = \frac{f(x)}{g(x)} = \frac{\sqrt{x + 1}}{2x + 1}$$

Its domain is $[-1, \infty)$ together with the restriction $x \neq -\frac{1}{2}$. We denote this by $[-1, -\frac{1}{2}) \cup (-\frac{1}{2}, \infty)$. ∎

The mathematical formulation of a problem arising from a practical situation often leads to an expression that involves the combination of functions. Consider, for example, the costs incurred in operating a business. Costs that remain more or less constant regardless of the firm's level of activity are called **fixed costs**. Examples of fixed costs are rental fees and executive salaries. On the other hand, costs that vary with production or sales are called **variable costs**. Examples of variable costs are wages and costs of raw materials. The **total cost** of operating a business is thus given by the *sum* of the variable costs and the fixed costs, as illustrated in the next example.

APPLIED EXAMPLE 2 Cost Functions Suppose Puritron, a manufacturer of water filters, has a monthly fixed cost of $10,000 and a variable cost of

$$-0.0001x^2 + 10x \quad (0 \leq x \leq 40,000)$$

dollars, where x denotes the number of filters manufactured per month. Find a function C that gives the total monthly cost incurred by Puritron in the manufacture of x filters.

Solution Puritron's monthly fixed cost is always $10,000, regardless of the level of production, and it is described by the constant function $F(x) = 10,000$. Next, the variable cost is described by the function $V(x) = -0.0001x^2 + 10x$. Since the total cost incurred by Puritron at any level of production is the sum of the variable cost and the fixed cost, we see that the required total cost function is given by

$$C(x) = V(x) + F(x)$$
$$= -0.0001x^2 + 10x + 10,000 \qquad (0 \le x \le 40,000)$$

Next, the **total profit** realized by a firm in operating a business is the *difference* between the total revenue realized and the total cost incurred; that is,

$$P(x) = R(x) - C(x)$$

APPLIED EXAMPLE 3 Profit Functions Refer to Example 2. Suppose the total revenue realized by Puritron from the sale of x water filters is given by the total revenue function

$$R(x) = -0.0005x^2 + 20x \qquad (0 \le x \le 40,000)$$

a. Find the total profit function—that is, the function that describes the total profit Puritron realizes in manufacturing and selling x water filters per month.
b. What is the profit when the level of production is 10,000 filters per month?

Solution

a. The total profit realized by Puritron in manufacturing and selling x water filters per month is the difference between the total revenue realized and the total cost incurred. Thus, the required total profit function is given by

$$P(x) = R(x) - C(x)$$
$$= (-0.0005x^2 + 20x) - (-0.0001x^2 + 10x + 10,000)$$
$$= -0.0004x^2 + 10x - 10,000$$

b. The profit realized by Puritron when the level of production is 10,000 filters per month is

$$P(10,000) = -0.0004(10,000)^2 + 10(10,000) - 10,000 = 50,000$$

or $50,000 per month.

Composition of Functions

Another way to build up a function from other functions is through a process known as the *composition of functions*. Consider, for example, the function h, whose rule is given by $h(x) = \sqrt{x^2 - 1}$. Let f and g be functions defined by the rules $f(x) = x^2 - 1$ and $g(x) = \sqrt{x}$. Evaluating the function g at the point $f(x)$ [remember that for each real number x in the domain of f, $f(x)$ is simply a real number], we find that

$$g(f(x)) = \sqrt{f(x)} = \sqrt{x^2 - 1}$$

which is just the rule defining the function h!

In general, the composition of a function g with a function f is defined as follows.

> **The Composition of Two Functions**
>
> Let f and g be functions. Then the composition of g and f is the function $g \circ f$ defined by
>
> $$(g \circ f)(x) = g(f(x))$$
>
> The domain of $g \circ f$ is the set of all x in the domain of f such that $f(x)$ lies in the domain of g.

The function $g \circ f$ (read "g circle f") is also called a **composite function**. The interpretation of the function $h = g \circ f$ as a machine is illustrated in Figure 13, and its interpretation as a mapping is shown in Figure 14.

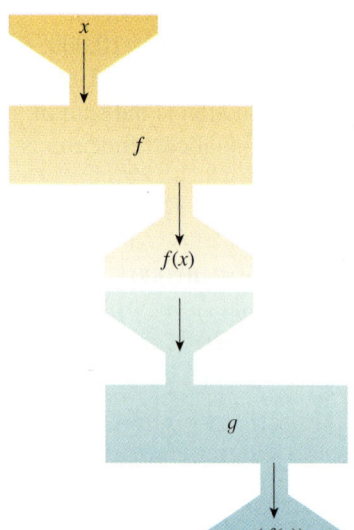

FIGURE 13
The composite function $h = g \circ f$ viewed as a machine

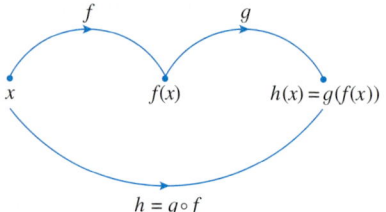

FIGURE 14
The function $h = g \circ f$ viewed as a mapping

EXAMPLE 4 Let $f(x) = x^2 - 1$ and $g(x) = \sqrt{x} + 1$. Find:

a. The rule for the composite function $g \circ f$.
b. The rule for the composite function $f \circ g$.

Solution

a. To find the rule for the composite function $g \circ f$, evaluate the function g at $f(x)$. We obtain

$$(g \circ f)(x) = g(f(x)) = \sqrt{f(x)} + 1 = \sqrt{x^2 - 1} + 1$$

b. To find the rule for the composite function $f \circ g$, evaluate the function f at $g(x)$. Thus,

$$(f \circ g)(x) = f(g(x)) = (g(x))^2 - 1 = (\sqrt{x} + 1)^2 - 1$$
$$= x + 2\sqrt{x} + 1 - 1 = x + 2\sqrt{x}$$

 Example 4 shows us that in general $g \circ f$ is different from $f \circ g$, so care must be taken in finding the rule for a composite function.

> *Explore & Discuss*
>
> Let $f(x) = \sqrt{x} + 1$ for $x \geq 0$, and let $g(x) = (x - 1)^2$ for $x \geq 1$.
>
> 1. Show that $(g \circ f)(x)$ and $(f \circ g)(x) = x$. (*Note:* The function g is said to be the *inverse* of f and vice versa.)
>
> 2. Plot the graphs of f and g together with the straight line $y = x$. Describe the relationship between the graphs of f and g.

APPLIED EXAMPLE 5 Automobile Pollution An environmental impact study conducted for the city of Oxnard indicates that under existing environmental protection laws, the level of carbon monoxide (CO) present in the air due to pollution from automobile exhaust will be $0.01x^{2/3}$ parts per million when the number of motor vehicles is x thousand. A separate study conducted by a state government agency estimates that t years from now, the number of motor vehicles in Oxnard will be $0.2t^2 + 4t + 64$ thousand.

a. Find an expression for the concentration of CO in the air due to automobile exhaust t years from now.
b. What will be the level of concentration 5 years from now?

Solution

a. The level of CO present in the air due to pollution from automobile exhaust is described by the function $g(x) = 0.01x^{2/3}$, where x is the number (in thousands) of motor vehicles. But the number of motor vehicles x (in thousands) t years from now may be estimated by the rule $f(t) = 0.2t^2 + 4t + 64$. Therefore, the concentration of CO due to automobile exhaust t years from now is given by

$$C(t) = (g \circ f)(t) = g(f(t)) = 0.01(0.2t^2 + 4t + 64)^{2/3}$$

parts per million.

b. The level of concentration 5 years from now will be

$$C(5) = 0.01[0.2(5)^2 + 4(5) + 64]^{2/3}$$
$$= (0.01)89^{2/3} \approx 0.20$$

or approximately 0.20 parts per million.

2.2 Self-Check Exercises

1. Let f and g be functions defined by the rules

$$f(x) = \sqrt{x} + 1 \quad \text{and} \quad g(x) = \frac{x}{1+x}$$

respectively. Find the rules for
a. The sum s, the difference d, the product p, and the quotient q of f and g.
b. The composite functions $f \circ g$ and $g \circ f$.

2. Health-care spending per person by the private sector includes payments by individuals, corporations, and their insurance companies and is approximated by the function

$$f(t) = 2.48t^2 + 18.47t + 509 \quad (0 \leq t \leq 6)$$

where $f(t)$ is measured in dollars and t is measured in years, with $t = 0$ corresponding to the beginning of 1994. The corresponding government spending—including expenditures for Medicaid, Medicare, and other federal, state, and local government public health care—is

$$g(t) = -1.12t^2 + 29.09t + 429 \quad (0 \leq t \leq 6)$$

where t has the same meaning as before.
a. Find a function that gives the difference between private and government health-care spending per person at any time t.
b. What was the difference between private and government expenditures per person at the beginning of 1995? At the beginning of 2000?

Source: Health Care Financing Administration.

Solutions to Self-Check Exercises 2.2 can be found on page 74.

2.2 Concept Questions

1. The figure opposite shows the graphs of a total cost function and a total revenue function. Let P, defined by $P(x) = R(x) - C(x)$, denote the total profit function.
a. Find an expression for $P(x_1)$. Explain its significance.
b. Find an expression for $P(x_2)$. Explain its significance.

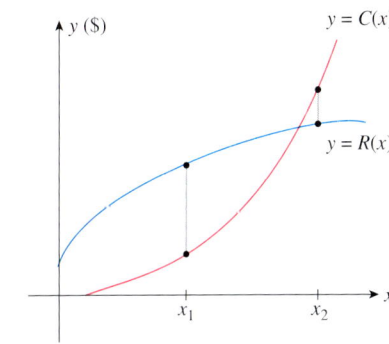

2. a. Explain what is meant by the sum, difference, product, and quotient of the functions f and g with domains A and B, respectively.
  b. If $f(2) = 3$ and $g(2) = -2$, what is $(f+g)(2)$? $(f-g)(2)$? $(fg)(2)$? $(f/g)(2)$?

3. Let f and g be functions, and suppose that (x, y) is a point on the graph of h. What is the value of y for $h = f + g$? $h = f - g$? $h = fg$? $h = f/g$?

4. a. What is the composition of the functions f and g? The functions g and f?
  b. If $f(2) = 3$ and $g(3) = 8$, what is $(g \circ f)(2)$? Can you conclude from the given information what $(f \circ g)(3)$ is? Explain.

5. Let f be a function with domain A, and let g be a function whose domain contains the range of f. If a is any number in A, must $(g \circ f)(a)$ be defined? Explain with an example.

2.2 Exercises

In Exercises 1–8, let $f(x) = x^3 + 5$, $g(x) = x^2 - 2$, and $h(x) = 2x + 4$. Find the rule for each function.

1. $f + g$ **2.** $f - g$ **3.** fg **4.** gf

5. $\dfrac{f}{g}$ **6.** $\dfrac{f-g}{h}$ **7.** $\dfrac{fg}{h}$ **8.** fgh

In Exercises 9–18, let $f(x) = x - 1$, $g(x) = \sqrt{x+1}$, and $h(x) = 2x^3 - 1$. Find the rule for each function.

9. $f + g$ **10.** $g - f$ **11.** fg **12.** gf

13. $\dfrac{g}{h}$ **14.** $\dfrac{h}{g}$ **15.** $\dfrac{fg}{h}$ **16.** $\dfrac{fh}{g}$

17. $\dfrac{f-h}{g}$ **18.** $\dfrac{gh}{g-f}$

In Exercises 19–24, find the functions $f + g$, $f - g$, fg, and f/g.

19. $f(x) = x^2 + 5$; $g(x) = \sqrt{x} - 2$

20. $f(x) = \sqrt{x-1}$; $g(x) = x^3 + 1$

21. $f(x) = \sqrt{x+3}$; $g(x) = \dfrac{1}{x-1}$

22. $f(x) = \dfrac{1}{x^2+1}$; $g(x) = \dfrac{1}{x^2-1}$

23. $f(x) = \dfrac{x+1}{x-1}$; $g(x) = \dfrac{x+2}{x-2}$

24. $f(x) = x^2 + 1$; $g(x) = \sqrt{x+1}$

In Exercises 25–30, find the rules for the composite functions $f \circ g$ and $g \circ f$.

25. $f(x) = x^2 + x + 1$; $g(x) = x^2$

26. $f(x) = 3x^2 + 2x + 1$; $g(x) = x + 3$

27. $f(x) = \sqrt{x} + 1$; $g(x) = x^2 - 1$

28. $f(x) = 2\sqrt{x} + 3$; $g(x) = x^2 + 1$

29. $f(x) = \dfrac{x}{x^2+1}$; $g(x) = \dfrac{1}{x}$

30. $f(x) = \sqrt{x+1}$; $g(x) = \dfrac{1}{x-1}$

In Exercises 31–34, evaluate $h(2)$, where $h = g \circ f$.

31. $f(x) = x^2 + x + 1$; $g(x) = x^2$

32. $f(x) = \sqrt[3]{x^2 - 1}$; $g(x) = 3x^3 + 1$

33. $f(x) = \dfrac{1}{2x+1}$; $g(x) = \sqrt{x}$

34. $f(x) = \dfrac{1}{x-1}$; $g(x) = x^2 + 1$

In Exercises 35–42, find functions f and g such that $h = g \circ f$. (Note: The answer is not unique.)

35. $h(x) = (2x^3 + x^2 + 1)^5$ **36.** $h(x) = (3x^2 - 4)^{-3}$

37. $h(x) = \sqrt{x^2 - 1}$ **38.** $h(x) = (2x - 3)^{3/2}$

39. $h(x) = \dfrac{1}{x^2 - 1}$ **40.** $h(x) = \dfrac{1}{\sqrt{x^2 - 4}}$

41. $h(x) = \dfrac{1}{(3x^2 + 2)^{3/2}}$ **42.** $h(x) = \dfrac{1}{\sqrt{2x+1}} + \sqrt{2x+1}$

In Exercises 43–46, find $f(a + h) - f(a)$ for each function. Simplify your answer.

43. $f(x) = 3x + 4$ **44.** $f(x) = -\dfrac{1}{2}x + 3$

45. $f(x) = 4 - x^2$ **46.** $f(x) = x^2 - 2x + 1$

In Exercises 47–52, find and simplify

$$\dfrac{f(a+h) - f(a)}{h} \quad (h \neq 0)$$

for each function.

47. $f(x) = x^2 + 1$ **48.** $f(x) = 2x^2 - x + 1$

49. $f(x) = x^3 - x$ **50.** $f(x) = 2x^3 - x^2 + 1$

51. $f(x) = \dfrac{1}{x}$ **52.** $f(x) = \sqrt{x}$

53. Restaurant Revenue Nicole owns and operates two restaurants. The revenue of the first restaurant at time t is $f(t)$ dollars, and the revenue of the second restaurant at time t is $g(t)$ dollars. What does the function $F(t) = f(t) + g(t)$ represent?

54. Birthrate of Endangered Species The birthrate of an endangered species of whales in year t is $f(t)$ whales/year. This species of whales is dying at the rate of $g(t)$ whales/year in year t. What does the function $F(t) = f(t) - g(t)$ represent?

55. Value of an Investment The number of AAPL shares that Nancy owns is given by $f(t)$. The price per share of the stock of AAPL at time t is $g(t)$ dollars. What does the function $f(t)g(t)$ represent?

56. Production Costs The total cost incurred by time t in the production of a certain commodity is $f(t)$ dollars. The number of products produced by time t is $g(t)$ units. What does the function $f(t)/g(t)$ represent?

57. Carbon Monoxide Pollution The number of cars running in the business district of a town at time t is given by $f(t)$. Carbon monoxide pollution coming from these cars is given by $g(x)$ parts per million, where x is the number of cars being operated in the district. What does the function $g \circ f$ represent?

58. Effect of Advertising on Revenue The revenue of Leisure Travel is given by $f(x)$ dollars, where x is the dollar amount spent by the company on advertising. The amount spent by Leisure at time t on advertising is given by $g(t)$ dollars. What does the function $f \circ g$ represent?

59. Manufacturing Costs TMI, a manufacturer of blank DVDs, has a monthly fixed cost of $12,100 and a variable cost of $.60/disc. Find a function C that gives the total cost incurred by TMI in the manufacture of x discs/month.

60. Spam Messages The total number of email messages per day (in billions) between 2003 and 2007 is approximated by

$$f(t) = 1.54t^2 + 7.1t + 31.4 \quad (0 \le t \le 4)$$

where t is measured in years, with $t = 0$ corresponding to 2003. Over the same period, the total number of spam messages per day (in billions) is approximated by

$$g(t) = 1.21t^2 + 6t + 14.5 \quad (0 \le t \le 4)$$

a. Find the rule for the function $D = f - g$. Compute $D(4)$ and explain what it measures.
b. Find the rule for the function $P = f/g$. Compute $P(4)$ and explain what it means.

Source: Technology Review.

61. Public Transportation Budget Deficit According to the Massachusetts Bay Transportation Authority (MBTA), the projected cumulative MBTA budget deficit with a $160 million rescue package (in billions of dollars) is given by

$$D_1(t) = 0.0275t^2 + 0.081t + 0.07 \quad (0 \le t \le 3)$$

and the budget deficit without the rescue package is given by

$$D_2(t) = 0.035t^2 + 0.21t + 0.24 \quad (0 \le t \le 3)$$

Find the function $D = D_2 - D_1$, and interpret your result.

Source: MBTA Review.

62. Motorcycle Deaths Suppose the fatality rate (deaths per 100 million miles traveled) of motorcyclists is given by $g(x)$, where x is the percentage of motorcyclists who wear helmets. Next, suppose the percent of motorcyclists who wear helmets at time t (t measured in years) is $f(t)$, with $t = 0$ corresponding to 2000.
a. If $f(0) = 0.64$ and $g(0.64) = 26$, find $(g \circ f)(0)$ and interpret your result.
b. If $f(6) = 0.51$ and $g(0.51) = 42$, find $(g \circ f)(6)$ and interpret your result.
c. Comment on the results of parts (a) and (b).

Source: National Highway Traffic Safety Administration.

63. Fighting Crime Suppose the reported serious crimes (crimes that include homicide, rape, robbery, aggravated assault, burglary, and car theft) that end in arrests or in the identification of suspects is $g(x)$ percent, where x denotes the total number of detectives. Next, suppose the total number of detectives in year t is $f(t)$, with $t = 0$ corresponding to 2001.
a. If $f(1) = 406$ and $g(406) = 23$, find $(g \circ f)(1)$ and interpret your result.
b. If $f(6) = 326$ and $g(326) = 18$, find $(g \circ f)(6)$ and interpret your result.
c. Comment on the results of parts (a) and (b).

Source: Boston Police Department.

64. Cost of Producing Smartphones Apollo manufactures smartphones at a variable cost of

$$V(x) = 0.000003x^3 - 0.03x^2 + 200x$$

dollars, where x denotes the number of units manufactured per month. The monthly fixed cost attributable to the division that produces them is $100,000. Find a function C that gives the total cost incurred by the manufacture of x smartphones. What is the total cost incurred in producing 2000 units/month?

65. Profit From Sale of Smartphones Refer to Exercise 64. Suppose the total revenue realized by Apollo from the sale of x smartphones is given by the total revenue function

$$R(x) = -0.1x^2 + 500x \quad (0 \le x \le 5000)$$

where $R(x)$ is measured in dollars.
a. Find the total profit function.
b. What is the profit when 1500 units are produced and sold each month?

66. Profit From Sale of Pagers A division of Chapman Corporation manufactures a pager. The weekly fixed cost for the division is $20,000, and the variable cost for producing x pagers/week is

$$V(x) = 0.000001x^3 - 0.01x^2 + 50x$$

dollars. The company realizes a revenue of

$$R(x) = -0.02x^2 + 150x \quad (0 \le x \le 7500)$$

dollars from the sale of x pagers/week.
a. Find the total cost function.
b. Find the total profit function.
c. What is the profit for the company if 2000 units are produced and sold each week?

67. Overcrowding of Prisons The 1980s saw a trend toward old-fashioned punitive deterrence as opposed to the more liberal penal policies and community-based corrections popular in the 1960s and early 1970s. As a result, prisons became more crowded, and the gap between the number of people in prison and the prison capacity widened. The number of prisoners (in thousands) in federal and state prisons is approximated by the function

$$N(t) = 3.5t^2 + 26.7t + 436.2 \quad (0 \le t \le 10)$$

where t is measured in years, with $t = 0$ corresponding to 1983. The number of inmates for which prisons were designed is given by

$$C(t) = 24.3t + 365 \quad (0 \le t \le 10)$$

where $C(t)$ is measured in thousands and t has the same meaning as before.
a. Find an expression that shows the gap between the number of prisoners and the number of inmates for which the prisons were designed at any time t.
b. Find the gap at the beginning of 1983 and at the beginning of 1986.

Source: U.S. Department of Justice.

68. Effect of Mortgage Rates on Housing Starts A study prepared for the National Association of Realtors estimated that the number of housing starts per year over the next 5 years will be

$$N(r) = \frac{7}{1 + 0.02r^2}$$

million units, where r (percent) is the mortgage rate. Suppose the mortgage rate t months from now will be

$$r(t) = \frac{5t + 75}{t + 10} \quad (0 \le t \le 24)$$

percent/year.
a. Find an expression for the number of housing starts per year as a function of t, t months from now.
b. Using the result from part (a), determine the number of housing starts at present, 12 months from now, and 18 months from now.

69. Hotel Occupancy Rate The occupancy rate of the all-suite Wonderland Hotel, located near an amusement park, is given by the function

$$r(t) = \frac{10}{81}t^3 - \frac{10}{3}t^2 + \frac{200}{9}t + 55 \quad (0 \le t \le 11)$$

percent, where t is measured in months and $t = 0$ corresponds to the beginning of January. Management has estimated that the monthly revenue (in thousands of dollars) is approximated by the function

$$R(r) = -\frac{3}{5000}r^3 + \frac{9}{50}r^2 \quad (0 \le r \le 100)$$

where r (percent) is the occupancy rate.
a. What is the hotel's occupancy rate at the beginning of January? At the beginning of June?
b. What is the hotel's monthly revenue at the beginning of January? At the beginning of June?
Hint: Compute $R(r(0))$ and $R(r(5))$.

70. Housing Starts and Construction Jobs The president of a major housing construction firm reports that the number of construction jobs (in millions) created is given by

$$N(x) = 1.42x$$

where x denotes the number of housing starts. Suppose the number of housing starts in the next t months is expected to be

$$x(t) = \frac{7(t + 10)^2}{(t + 10)^2 + 2(t + 15)^2}$$

million units. Find an expression for the number of jobs created per year in the next t months. How many jobs/year will have been created 6 months and 12 months from now?

71. a. Let f, g, and h be functions. How would you define the "sum" of f, g, and h?
b. Give a real-life example involving the sum of three functions. (*Note:* The answer is not unique.)

72. a. Let f, g, and h be functions. How would you define the "composition" of h, g, and f, in that order?
b. Give a real-life example involving the composition of these functions. (*Note:* The answer is not unique.)

In Exercises 73–76, determine whether the statement is true or false. If it is true, explain why it is true. If it is false, give an example to show why it is false.

73. If f and g are functions with domain D, then $f + g = g + f$.

74. If $g \circ f$ is defined at $x = a$, then $f \circ g$ must also be defined at $x = a$.

75. If f and g are functions, then $f \circ g = g \circ f$.

76. If f is a function, then $(f \circ f)(x) = [f(x)]^2$.

2.2 Solutions to Self-Check Exercises

1. a. $s(x) = f(x) + g(x) = \sqrt{x} + 1 + \dfrac{x}{1 + x}$

$d(x) = f(x) - g(x) = \sqrt{x} + 1 - \dfrac{x}{1 + x}$

$p(x) = f(x)g(x) = (\sqrt{x} + 1) \cdot \dfrac{x}{1 + x} = \dfrac{x(\sqrt{x} + 1)}{1 + x}$

$q(x) = \dfrac{f(x)}{g(x)} = \dfrac{\sqrt{x} + 1}{\dfrac{x}{1 + x}} = \dfrac{(\sqrt{x} + 1)(1 + x)}{x}$

b. $(f \circ g)(x) = f(g(x)) = \sqrt{\dfrac{x}{1 + x}} + 1$

$(g \circ f)(x) = g(f(x)) = \dfrac{\sqrt{x} + 1}{1 + (\sqrt{x} + 1)} = \dfrac{\sqrt{x} + 1}{\sqrt{x} + 2}$

2. a. The difference between private and government health-care spending per person at any time t is given by the function d with the rule

$d(t) = f(t) - g(t) = (2.48t^2 + 18.47t + 509)$
$\qquad - (-1.12t^2 + 29.09t + 429)$
$\qquad = 3.6t^2 - 10.62t + 80$

b. The difference between private and government expenditures per person at the beginning of 1995 is given by

$$d(1) = 3.6(1)^2 - 10.62(1) + 80$$

or $72.98/person.

The difference between private and government expenditures per person at the beginning of 2000 is given by

$$d(6) = 3.6(6)^2 - 10.62(6) + 80$$

or $145.88/person.

2.3 Functions and Mathematical Models

Mathematical Models

One of the fundamental goals in this book is to show how mathematics and, in particular, calculus can be used to solve real-world problems such as those arising from the world of business and the social, life, and physical sciences. You have already seen some of these problems earlier. Here are a few more examples of real-world phenomena that we will analyze in this and ensuing chapters.

- Global warming (page 78)
- The solvency of the U.S. Social Security trust fund (page 79)
- The total cost of the health-care bill (page 88)
- The Case-Shiller Home Price Index (page 279)
- Pharmaceutical theft (page 366)
- Mexico's hedging tactic (page 463)

Regardless of the field from which the real-world problem is drawn, the problem is analyzed by using a process called **mathematical modeling.** The four steps in this process are illustrated in Figure 15.

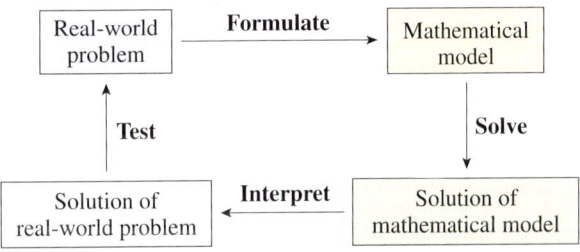

FIGURE 15

1. **Formulate** Given a real-world problem, our first task is to formulate the problem, using the language of mathematics. The many techniques used in constructing mathematical models range from theoretical consideration of the problem on the one extreme to an interpretation of data associated with the problem on the other. For example, the mathematical model giving the accumulated amount at any time when a certain sum of money is deposited in the bank can be derived theoretically (see Chapter 5). On the other hand, many of the mathematical models in this book are constructed by studying the data associated with the problem (see Using Technology, pages 93–96). In calculus, we are primarily concerned with how one (dependent) variable depends on one or more (independent) variables. Consequently, most of our mathematical models will involve functions of one or more variables or equations defining these functions (implicitly).
2. **Solve** Once a mathematical model has been constructed, we can use the appropriate mathematical techniques, which we will develop throughout the book, to solve the problem.
3. **Interpret** Bearing in mind that the solution obtained in Step 2 is just the solution of the mathematical model, we need to interpret these results in the context of the original real-world problem.

4. Test Some mathematical models of real-world applications describe the situations with complete accuracy. For example, the model describing a deposit in a bank account gives the exact accumulated amount in the account at any time. But other mathematical models give, at best, an approximate description of the real-world problem. In this case, we need to test the accuracy of the model by observing how well it describes the original real-world problem and how well it predicts past and/or future behavior. If the results are unsatisfactory, then we may have to reconsider the assumptions made in the construction of the model or, in the worst case, return to Step 1.

Many real-world phenomena, including those mentioned at the beginning of this section, are modeled by an appropriate function.

In what follows, we will recall some familiar functions and give examples of real-world phenomena that are modeled by using these functions.

Polynomial Functions

A **polynomial function** of degree n is a function of the form

$$f(x) = a_n x^n + a_{n-1} x^{n-1} + \cdots + a_2 x^2 + a_1 x + a_0 \qquad (a_n \neq 0)$$

where n is a nonnegative integer and the numbers $a_0, a_1, \ldots, a_n$ are constants, called the **coefficients** of the polynomial function. For example, the functions

$$f(x) = 2x^5 - 3x^4 + \frac{1}{2}x^3 + \sqrt{2}x^2 - 6$$

$$g(x) = 0.001x^3 - 0.2x^2 + 10x + 200$$

are polynomial functions of degrees 5 and 3, respectively. Observe that a polynomial function is defined for every value of x and so its domain is $(-\infty, \infty)$.

A polynomial function of degree 1 ($n = 1$) has the form

$$y = f(x) = a_1 x + a_0 \qquad (a_1 \neq 0)$$

and is an equation of a straight line in the slope-intercept form with slope $m = a_1$ and y-intercept $b = a_0$ (see Section 2.1). For this reason, a polynomial function of degree 1 is called a **linear function**.

Linear functions are used extensively in mathematical modeling for two important reasons. First, some models are *linear* by nature. For example, the formula for converting temperature from Celsius (°C) to Fahrenheit (°F) is $F = \frac{9}{5}C + 32$, and F is a linear function of C. Second, some natural phenomena exhibit linear characteristics over a small range of values and can therefore be modeled by a linear function restricted to a small interval.

The following example uses a linear function to model the bank revenue from overdraft fees from 2004 through 2009. In Section 8.4, we will show how this model is constructed using the *least-squares technique*. In Using Technology on pages 92–96, you will be asked to use a graphing calculator to construct other mathematical models from raw data.

APPLIED EXAMPLE 1 Bounced-Check Charges Overdraft fees have become an important piece of a bank's total fee income. The following table gives the bank revenue from overdraft fees (in billions of dollars) from 2004 through 2009.

Year, t	0	1	2	3	4	5
Revenue, y	27.5	29	31	34	36	38

where t is measured in years, with $t = 0$ corresponding to 2004. A mathematical model giving the approximate projected bank revenue from overdraft fees over the period under consideration is given by

$$f(t) = 2.19t + 27.12 \quad (0 \leq t \leq 5)$$

a. Plot the six data points and sketch the graph of the function f on the same set of axes.

b. Assuming that the projection held and the trend continued, what was the projected bank revenue from overdraft fees in 2010 ($t = 6$)?

c. What was the rate of increase of the bank revenue from overdraft fees over the period from 2004 through 2009?

Source: New York Times.

Solution

a. The graph of f is shown in Figure 16.

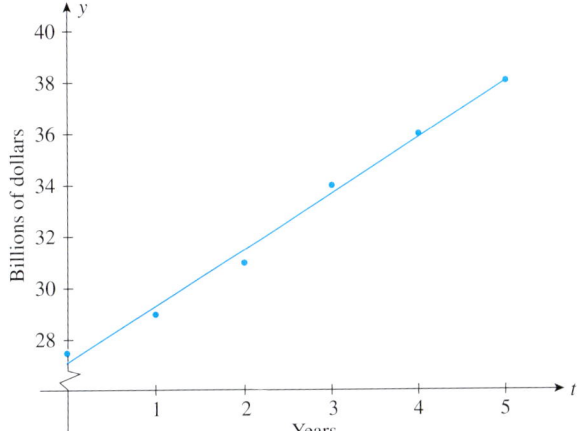

FIGURE 16
Bank revenue from overdraft fees from 2004 to 2009

b. The projected bank revenue from overdraft fees in 2010 was

$$f(6) = 2.19(6) + 27.12 = 40.26$$

or $40.26 billion.

c. The rate of increase of the bank revenue from overdraft fees over the period from 2004 through 2009 is $2.19 billion per year.

A polynomial function of degree 2 has the form

$$y = f(x) = a_2 x^2 + a_1 x + a_0 \quad (a_2 \neq 0)$$

or, more simply, $y = ax^2 + bx + c$, and is called a **quadratic function.** The graph of a quadratic function is a parabola (see Figure 17).

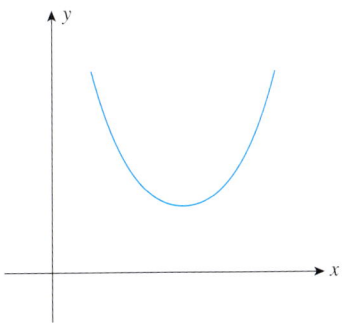
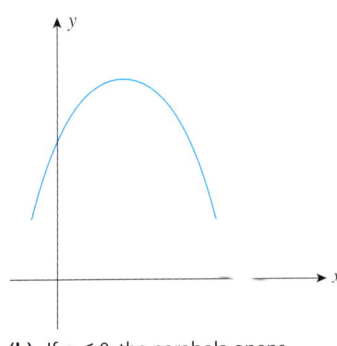

FIGURE 17
The graph of a quadratic function is a parabola.

(a) If $a > 0$, the parabola opens upward.

(b) If $a < 0$, the parabola opens downward.

The parabola opens upward if $a > 0$ and downward if $a < 0$. To see this, we rewrite the equation for y obtaining

$$f(x) = ax^2 + bx + c = x^2\left(a + \frac{b}{x} + \frac{c}{x^2}\right) \quad (x \neq 0)$$

Observe that if x is large in absolute value, then the expression inside the parentheses is close to a, so $f(x)$ behaves like ax^2 for large values of x. Thus, $y = f(x)$ is large and positive (the parabola opens upward) if $a > 0$ and is large in magnitude and negative if $a < 0$ (the parabola opens downward).

Quadratic functions serve as mathematical models for many phenomena, as Example 2 shows.

APPLIED EXAMPLE 2 Global Warming The increase in carbon dioxide (CO_2) in the atmosphere is a major cause of global warming. The Keeling curve, named after Charles David Keeling, a professor at Scripps Institution of Oceanography, gives the average amount of CO_2, measured in parts per million volume (ppmv), in the atmosphere from 1958 through 2010. Even though data were available for every year in this time interval, we'll construct the curve based only on the following randomly selected data points.

Year	1958	1970	1974	1978	1985	1991	1998	2003	2007	2010
Amount	315	325	330	335	345	355	365	375	380	390

The **scatter plot** associated with these data is shown in Figure 18a. A mathematical model giving the approximate amount of CO_2 in the atmosphere during this period is given by

$$A(t) = 0.012313t^2 + 0.7545t + 313.9 \quad (1 \leq t \leq 53)$$

where t is measured in years, with $t = 1$ corresponding to 1958. The graph of A is shown in Figure 18b.

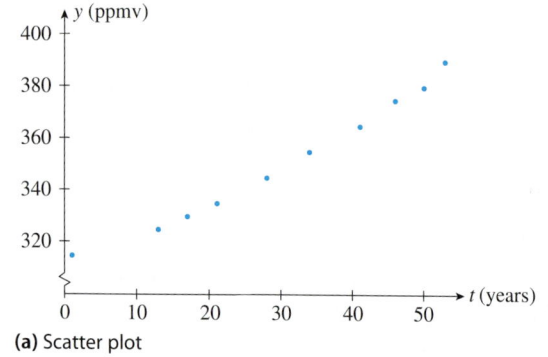
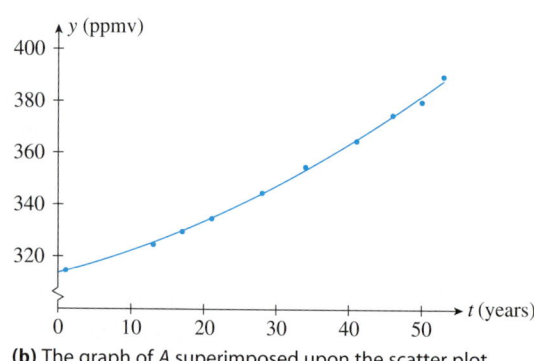

FIGURE 18
(a) Scatter plot
(b) The graph of A superimposed upon the scatter plot

a. Use the model to estimate the average amount of atmospheric CO_2 in 1980 ($t = 23$).
b. Assume that the trend continued, and use the model to estimate the average amount of atmospheric CO_2 in 2013.
Source: Scripps Institution of Oceanography.

Solution

a. The average amount of atmospheric carbon dioxide in 1980 is given by

$$A(23) = 0.012313(23)^2 + 0.7545(23) + 313.9 \approx 337.77$$

or approximately 338 ppmv.

b. Assuming that the trend continued, the average amount of atmospheric CO_2 in 2013 will be

$$A(56) = 0.012313(56)^2 + 0.7545(56) + 313.9 \approx 394.77$$

or approximately 395 ppmv.

The next example uses a polynomial of degree 4 to help us construct a model that describes the projected assets of the Social Security trust fund.

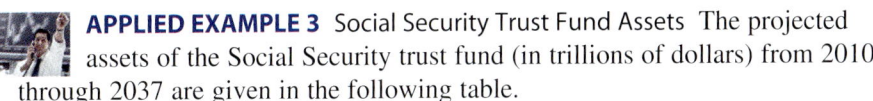

APPLIED EXAMPLE 3 Social Security Trust Fund Assets The projected assets of the Social Security trust fund (in trillions of dollars) from 2010 through 2037 are given in the following table.

Year	2010	2015	2020	2025	2030	2035	2037
Assets	2.69	3.56	4.22	4.24	3.24	0.87	0

The scatter plot associated with these data are shown in Figure 19a, where $t = 0$ corresponds to 2010. A mathematical model giving the approximate value of the assets in the trust fund $A(t)$, in trillions of dollars in year t is

$$A(t) = 0.0000263t^4 - 0.0017501t^3 + 0.0206t^2 + 0.0999t + 2.7 \qquad (0 \le t \le 27)$$

The graph of $A(t)$ is shown in Figure 19b. (You will be asked to construct this model in Exercise 22, Using Technology Exercises 2.3.)

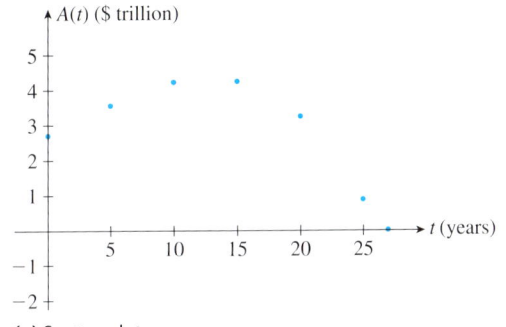

FIGURE 19   (a) Scatter plot   (b) The graph of A together with the scatter plot

a. The first baby boomers will turn 65 in 2011. What will be the assets of the Social Security system trust fund at that time? The last of the baby boomers will turn 65 in 2029. What will the assets of the trust fund be at that time?

b. Unless payroll taxes are increased significantly and/or benefits are scaled back dramatically, it is a matter of time before the assets of the current system are depleted. Use the graph of the function $A(t)$ to estimate the year in which the current Social Security system is projected to go broke.

Source: Social Security Administration.

Solution

a. The assets of the Social Security trust fund in 2011 ($t = 1$) will be

$$A(1) = 0.0000263(1)^4 - 0.0017501(1)^3 + 0.0206(1)^2 + 0.0999(1) + 2.7 \approx 2.82$$

or approximately \$2.82 trillion. The assets of the trust fund in 2029 ($t = 19$) will be

$$A(19) = 0.0000263(19)^4 - 0.0017501(19)^3 + 0.0206(19)^2 \\ + 0.0999(19) + 2.7 \approx 3.46$$

or approximately \$3.46 trillion.

b. From Figure 19b, we see that the graph of A crosses the t-axis at approximately $t = 27$. So unless the current system is changed, it is projected to go broke in 2037. (At this time, the first of the baby boomers will be 91, and the last of the baby boomers will be 73.)

Rational and Power Functions

Another important class of functions is rational functions. A **rational function** is simply the quotient of two polynomials. Examples of rational functions are

$$F(x) = \frac{3x^3 + x^2 - x + 1}{x - 2}$$

$$G(x) = \frac{x^2 + 1}{x^2 - 1}$$

In general, a rational function has the form

$$R(x) = \frac{f(x)}{g(x)}$$

where $f(x)$ and $g(x)$ are polynomial functions. Since division by zero is not allowed, we conclude that the domain of a rational function is the set of all real numbers except the zeros of g—that is, the roots of the equation $g(x) = 0$. Thus, the domain of the function F is the set of all numbers except $x = 2$, whereas the domain of the function G is the set of all numbers except those that satisfy $x^2 - 1 = 0$, or $x = \pm 1$.

Functions of the form

$$f(x) = x^r$$

where r is any real number, are called **power functions.** We encountered examples of power functions earlier in our work. For example, the functions

$$f(x) = \sqrt{x} = x^{1/2} \quad \text{and} \quad g(x) = \frac{1}{x^2} = x^{-2}$$

are power functions.

Many of the functions that we encounter later will involve combinations of the functions introduced here. For example, the following functions may be viewed as combinations of such functions:

$$f(x) = \sqrt{\frac{1 - x^2}{1 + x^2}}$$

$$g(x) = \sqrt{x^2 - 3x + 4}$$

$$h(x) = (1 + 2x)^{1/2} + \frac{1}{(x^2 + 2)^{3/2}}$$

As with polynomials of degree 3 or greater, analyzing the properties of these functions is facilitated by using the tools of calculus, to be developed later.

In the next example, we use a power function to construct a model that describes the driving costs of a car.

 APPLIED EXAMPLE 4 Driving Costs A study of driving costs based on a 2008 medium-sized sedan found the following average costs (car payments, gas, insurance, upkeep, and depreciation), measured in cents per mile.

Miles/year	5000	10,000	15,000	20,000
Cost/mile, y (¢)	147.52	71.9	55.2	46.9

A mathematical model (using least-squares techniques) giving the average cost in cents per mile is

$$C(x) = \frac{1735.2}{x^{1.72}} + 38.6$$

where x (in thousands) denotes the number of miles the car is driven in each year. The scatter plot associated with these data and the graph of C are shown in Figure 20. Using this model, estimate the average cost of driving a 2008 medium-sized sedan 8000 miles per year and 18,000 miles per year.

Source: American Automobile Association.

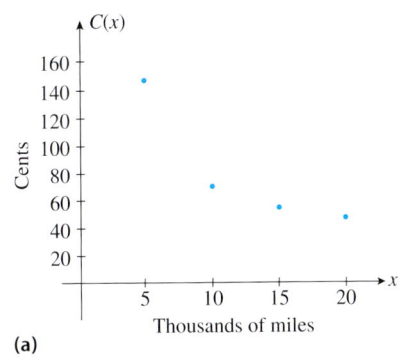
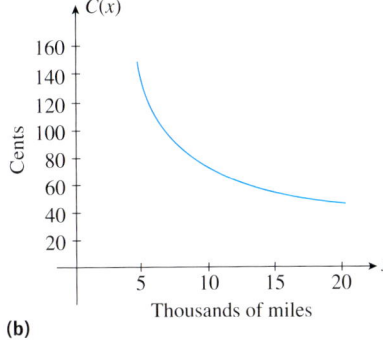

FIGURE 20
(a) The scatter plot and (b) the graph of the model for driving costs

Solution The average cost for driving a car 8000 miles per year is

$$C(8) = \frac{1735.2}{8^{1.72}} + 38.6 \approx 87.1$$

or approximately 87.1¢/mile. The average cost for driving it 18,000 miles per year is

$$C(18) = \frac{1735.2}{18^{1.72}} + 38.6 \approx 50.6$$

or approximately 50.6¢/mile.

Some Economic Models

In the remainder of this section, we look at some economic models.

In a free-market economy, consumer demand for a particular commodity depends on the commodity's unit price. A **demand equation** expresses the relationship between the unit price and the quantity demanded. The graph of the demand equation is called a **demand curve**. In general, the quantity demanded of a commodity decreases as the commodity's unit price increases, and vice versa. Accordingly, a **demand function** defined by $p = f(x)$, where p measures the unit price and x measures the number of units of the commodity in question, is generally characterized as a decreasing function of x; that is, $p = f(x)$ decreases as x increases. Since both x and p assume only nonnegative values, the demand curve is that part of the graph of $f(x)$ that lies in the first quadrant (Figure 21).

In a competitive market, a relationship also exists between the unit price of a commodity and the commodity's availability in the market. In general, an increase in the commodity's unit price induces the producer to increase the supply of the commodity. Conversely, a decrease in the unit price generally leads to a drop in the supply. The equation that expresses the relation between the unit price and the quantity supplied is called a **supply equation**, and its graph is called a **supply curve**. A **supply function** defined by $p = f(x)$ is generally characterized as an increasing function of x; that is, $p = f(x)$ increases as x increases. Since both x and p assume only nonnegative values, the supply curve is that part of the graph of $f(x)$ that lies in the first quadrant (Figure 22).

FIGURE 21
A demand curve

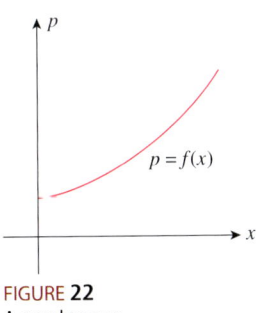

FIGURE 22
A supply curve

PORTFOLIO Todd Kodet

TITLE   Senior Vice-President of Supply
INSTITUTION   Earthbound Farm

Earthbound Farm is America's largest grower of organic produce, offering more than 100 varieties of organic salads, vegetables, fruits, and herbs on 34,000 crop acres. As Senior Vice-President of Supply, I am responsible for getting our products into and out of Earthbound Farm. A major part of my work is scheduling plantings for upcoming seasons, matching projected supply to projected demand for any given day and season. I use applied mathematics in every step of my planning to create models for predicting supply and demand.

After the sales department provides me with information about projected demand, I take their estimates, along with historical data for expected yields, to determine how much of each organic product we need to plant. There are several factors that I have to think about when I make these determinations. For example, I not only have to consider gross yield per acre of farmland, but also have to calculate average trimming waste per acre, to arrive at net pounds needed per customer.

Some of the other variables I consider are the amount of organic land available, the location of the farms, seasonal information (because days to maturity for each of our crops varies greatly depending on the weather), and historical information relating to weeds, pests, and diseases.

I emphasize the importance of understanding the mathematics that drives our business plans when I work with my team to analyze the reports they have generated. They need to recognize when the information they have gathered does not make sense so that they can spot errors that could skew our projections. With a sound understanding of mathematics, we are able to create more accurate predictions to help us meet our company's goals.

Alli Pura, Earthbound Farm; (inset) © istockphoto.com/Dan Moore

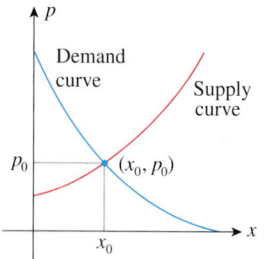

FIGURE 23
Market equilibrium corresponds to (x_0, p_0), the point at which the supply and demand curves intersect.

Under pure competition, the price of a commodity will eventually settle at a level dictated by the following condition: The supply of the commodity will be equal to the demand for it. If the price is too high, the consumer will not buy; if the price is too low, the supplier will not produce. **Market equilibrium** prevails when the quantity produced is equal to the quantity demanded. The quantity produced at market equilibrium is called the **equilibrium quantity**, and the corresponding price is called the **equilibrium price**.

Market equilibrium corresponds to the point at which the demand curve and the supply curve intersect. In Figure 23, x_0 represents the equilibrium quantity, and p_0 represents the equilibrium price. The point (x_0, p_0) lies on the supply curve and therefore satisfies the supply equation. At the same time, it also lies on the demand curve and therefore satisfies the demand equation. Thus, to find the point (x_0, p_0), and hence the equilibrium quantity and price, we solve the demand and supply equations simultaneously for x and p. For meaningful solutions, x and p must both be positive.

 APPLIED EXAMPLE 5 Supply-Demand The demand function for a certain brand of Bluetooth wireless headsets is given by

$$p = d(x) = -0.025x^2 - 0.5x + 60$$

and the corresponding supply function is given by

$$p = s(x) = 0.02x^2 + 0.6x + 20$$

where p is expressed in dollars and x is measured in units of a thousand. Find the equilibrium quantity and price.

Solution   We solve the following system of equations:

$$p = -0.025x^2 - 0.5x + 60$$
$$p = 0.02x^2 + 0.6x + 20$$

Substituting the first equation into the second yields

$$-0.025x^2 - 0.5x + 60 = 0.02x^2 + 0.6x + 20$$

which is equivalent to

$$0.045x^2 + 1.1x - 40 = 0$$
$$45x^2 + 1100x - 40{,}000 = 0 \quad \text{Multiply by 1000.}$$
$$9x^2 + 220x - 8000 = 0 \quad \text{Divide by 5.}$$
$$(9x + 400)(x - 20) = 0$$

Thus, $x = -\frac{400}{9}$ or $x = 20$. Since x must be nonnegative, the root $x = -\frac{400}{9}$ is rejected. Therefore, the equilibrium quantity is 20,000 headsets. The equilibrium price is given by

$$p = 0.02(20)^2 + 0.6(20) + 20 = 40$$

or $40 per headset (Figure 24).

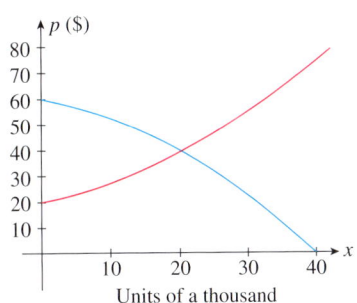

FIGURE 24
The supply curve and the demand curve intersect at the point (20, 40).

Exploring with TECHNOLOGY

1. **a.** Use a graphing utility to plot the straight lines L_1 and L_2 with equations $y = 2x - 1$ and $y = 2.1x + 3$, respectively, on the same set of axes, using the standard viewing window. Do the lines appear to intersect?
   b. Plot the straight lines L_1 and L_2, using the viewing window $[-100, 100] \times [-100, 100]$. Do the lines appear to intersect? Can you find the point of intersection using TRACE and ZOOM? Using the "intersection" function of your graphing utility?
   c. Find the point of intersection of L_1 and L_2 algebraically.
   d. Comment on the effectiveness of the methods of solutions in parts (b) and (c).

2. **a.** Use a graphing utility to plot the straight lines L_1 and L_2 with equations $y = 3x - 2$ and $y = -2x + 3$, respectively, on the same set of axes, using the standard viewing window. Then use TRACE and ZOOM to find the point of intersection of L_1 and L_2. Repeat using the "intersection" function of your graphing utility.
   b. Find the point of intersection of L_1 and L_2 algebraically.
   c. Comment on the effectiveness of the methods.

Constructing Mathematical Models

We close this section by showing how some mathematical models can be constructed by using elementary geometric and algebraic arguments.

The following guidelines can be used to construct mathematical models.

Guidelines for Constructing Mathematical Models

1. Assign a letter to each variable mentioned in the problem. If appropriate, draw and label a figure.
2. Find an expression for the quantity sought.
3. Use the conditions given in the problem to write the quantity sought as a function f of one variable. Note any restrictions to be placed on the domain of f from physical considerations of the problem.

FIGURE 25
The rectangular grazing land has width x and length y.

APPLIED EXAMPLE 6 Enclosing an Area The owner of Rancho Los Feliz has 3000 yards of fencing with which to enclose a rectangular piece of grazing land along the straight portion of a river. Fencing is not required along the river. Letting x denote the width of the rectangle, find a function f in the variable x giving the area of the grazing land if she uses all of the fencing (Figure 25).

Solution

1. This information was given.
2. The area of the rectangular grazing land is $A = xy$. Next, observe that the amount of fencing is $2x + y$ and this must be equal to 3000, since all the fencing is used; that is,

$$2x + y = 3000$$

3. From the equation, we see that $y = 3000 - 2x$. Substituting this value of y into the expression for A gives

$$A = xy = x(3000 - 2x) = 3000x - 2x^2$$

Finally, observe that both x and y must be nonnegative, since they represent the width and length of a rectangle, respectively. Thus, $x \geq 0$ and $y \geq 0$. But the latter is equivalent to $3000 - 2x \geq 0$, or $x \leq 1500$. So the required function is $f(x) = 3000x - 2x^2$ with domain $0 \leq x \leq 1500$.

Note Observe that if we view the function $f(x) = 3000x - 2x^2$ strictly as a mathematical entity, then its domain is the set of all real numbers. But physical considerations dictate that its domain should be restricted to the interval $[0, 1500]$.

APPLIED EXAMPLE 7 Charter-Flight Revenue If exactly 200 people sign up for a charter flight, Leisure World Travel Agency charges $300 per person. However, if more than 200 people sign up for the flight (assume that this is the case), then each fare is reduced by $1 for each additional person. Letting x denote the number of passengers above 200, find a function giving the revenue realized by the company.

Solution

1. This information was given.
2. If there are x passengers above 200, then the number of passengers signing up for the flight is $200 + x$. Furthermore, the fare will be $(300 - x)$ dollars per passenger.
3. The revenue will be

$$R = (200 + x)(300 - x) \qquad \text{Number of passengers} \times$$
$$= -x^2 + 100x + 60{,}000 \qquad \text{the fare per passenger}$$

Clearly, x must be nonnegative, and $300 - x \geq 0$, or $x \leq 300$. So the required function is $f(x) = -x^2 + 100x + 60{,}000$ with domain $[0, 300]$.

2.3 Self-Check Exercises

1. Thomas Young has suggested the following rule for calculating the dosage of medicine for children from 1 to 12 years of age. If a denotes the adult dosage (in milligrams) and t is the age of the child (in years), then the child's dosage is given by

$$D(t) = \frac{at}{t + 12}$$

If the adult dose of a substance is 500 mg, how much should a 4-year-old child receive?

2. The demand function for Mrs. Baker's cookies is given by

$$d(x) = -\frac{2}{15}x + 4$$

where $d(x)$ is the wholesale price in dollars per pound and x is the quantity demanded each week, measured in thousands of pounds. The supply function for the cookies is given by

$$s(x) = \frac{1}{75}x^2 + \frac{1}{10}x + \frac{3}{2}$$

where $s(x)$ is the wholesale price in dollars per pound and x is the quantity, in thousands of pounds, that will be made available in the market each week by the supplier.

a. Sketch the graphs of the functions d and s.
b. Find the equilibrium quantity and price.

Solutions to Self-Check Exercises 2.3 can be found on page 91.

2.3 Concept Questions

1. Describe mathematical modeling in your own words.
2. Define (a) a polynomial function and (b) a rational function. Give an example of each.
3. a. What is a demand function? A supply function?
   b. What is market equilibrium? Describe how you would go about finding the equilibrium quantity and equilibrium price given the demand and supply equations associated with a commodity.

2.3 Exercises

In Exercises 1–8, determine whether the equation defines y as a linear function of x. If so, write it in the form $y = mx + b$.

1. $2x + 3y = 6$
2. $-2x + 4y = 7$
3. $x = 2y - 4$
4. $2x = 3y + 8$
5. $2x - 4y + 9 = 0$
6. $3x - 6y + 7 = 0$
7. $2x^2 - 8y + 4 = 0$
8. $3\sqrt{x} + 4y = 0$

In Exercises 9–14, determine whether the given function is a polynomial function, a rational function, or some other function. State the degree of each polynomial function.

9. $f(x) = 3x^6 - 2x^2 + 1$
10. $f(x) = \dfrac{x^2 - 9}{x - 3}$
11. $G(x) = 2(x^2 - 3)^3$
12. $H(x) = 2x^{-3} + 5x^{-2} + 6$
13. $f(t) = 2t^2 + 3\sqrt{t}$
14. $f(r) = \dfrac{6r}{r^3 - 8}$

15. Find the constants m and b in the linear function $f(x) = mx + b$ such that $f(0) = 2$ and $f(3) = -1$.

16. Find the constants m and b in the linear function $f(x) = mx + b$ such that $f(2) = 4$ and the straight line represented by f has slope -1.

17. A manufacturer has a monthly fixed cost of $40,000 and a production cost of $8 for each unit produced. The product sells for $12/unit.
    a. What is the cost function?
    b. What is the revenue function?
    c. What is the profit function?
    d. Compute the profit (loss) corresponding to production levels of 8000 and 12,000 units.

18. A manufacturer has a monthly fixed cost of $100,000 and a production cost of $14 for each unit produced. The product sells for $20/unit.
    a. What is the cost function?
    b. What is the revenue function?
    c. What is the profit function?
    d. Compute the profit (loss) corresponding to production levels of 12,000 and 20,000 units.

19. **DISPOSABLE INCOME** Economists define the *disposable annual income* for an individual by the equation $D = (1 - r)T$, where T is the individual's total income and r is the net rate at which he or she is taxed. What is the disposable income for an individual whose income is $60,000 and whose net tax rate is 28%?

20. **DRUG DOSAGES** A method sometimes used by pediatricians to calculate the dosage of medicine for children is based on the child's surface area. If a denotes the adult dosage (in milligrams) and S is the surface area of the child (in square meters), then the child's dosage is given by

$$D(S) = \frac{Sa}{1.7}$$

If the adult dose of a substance is 500 mg, how much should a child whose surface area is 0.4 m² receive?

21. **COWLING'S RULE** Cowling's Rule is a method for calculating pediatric drug dosages. If a denotes the adult dosage (in milligrams) and t is the age of the child (in years), then the child's dosage is given by

$$D(t) = \left(\frac{t + 1}{24}\right)a$$

If the adult dose of a substance is 500 mg, how much should a 4-year-old child receive?

22. **Worker Efficiency** An efficiency study showed that the average worker at Delphi Electronics assembled cordless telephones at the rate of

$$f(t) = -\frac{3}{2}t^2 + 6t + 10 \quad (0 \leq t \leq 4)$$

phones/hour, t hr after starting work during the morning shift. At what rate does the average worker assemble telephones 2 hr after starting work?

23. **Effect of Advertising on Sales** The quarterly profit of Cunningham Realty depends on the amount of money x spent on advertising per quarter according to the rule

$$P(x) = -\frac{1}{8}x^2 + 7x + 30 \quad (0 \leq x \leq 50)$$

where $P(x)$ and x are measured in thousands of dollars. What is Cunningham's profit when its quarterly advertising budget is $28,000?

24. **Instant Messaging Accounts** The number of enterprise instant messaging (IM) accounts is projected to grow according to the function

$$N(t) = 2.96t^2 + 11.37t + 59.7 \quad (0 \leq t \leq 5)$$

where $N(t)$ is measured in millions and t in years, with $t = 0$ corresponding to 2006.
a. How many enterprise IM accounts were there in 2006?
b. What was the expected number of enterprise IM accounts in 2010?

Source: The Radical Group.

25. **Solar Power** More and more businesses and homeowners are installing solar panels on their roofs to draw energy from the sun's rays. According to the U.S. Department of Energy, the solar cell kilowatt-hour use in the United States (in millions) is projected to be

$$S(t) = 0.73t^2 + 15.8t + 2.7 \quad (0 \leq t \leq 8)$$

in year t, with $t = 0$ corresponding to 2000. What was the projected solar cell kilowatt-hours used in the United States for 2006? For 2008?

Source: U.S. Department of Energy.

26. **Average Single-Family Property Tax** Based on data from 298 of 351 cities and towns in Massachusetts, the average single-family tax bill from 1997 through 2007 is approximated by the function

$$T(t) = 7.26t^2 + 91.7t + 2360 \quad (0 \leq t \leq 10)$$

where $T(t)$ is measured in dollars and t in years, with $t = 0$ corresponding to 1997.
a. What was the average property tax on a single-family home in Massachusetts in 1997?
b. If the trend continued, what was the average property tax in 2010?

Source: Massachusetts Department of Revenue.

27. **Revenue of Polo Ralph Lauren** Citing strong sales and benefits from a new arm that will design lifestyle brands for department and specialty stores, the company projects revenue (in billions of dollars) to be

$$R(t) = -0.06t^2 + 0.69t + 3.25 \quad (0 \leq t \leq 3)$$

in year t, where $t = 0$ corresponds to 2005.
a. What was the revenue of the company in 2005?
b. Find $R(1)$, $R(2)$, and $R(3)$ and interpret your results.
c. Sketch the graph of R.

Source: Company reports.

28. **Aging Drivers** The number of fatalities due to car crashes, based on the number of miles driven, begins to climb after the driver is past age 65. Aside from declining ability as one ages, the older driver is more fragile. The number of fatalities per 100 million vehicle miles driven is approximately

$$N(x) = 0.0336x^3 - 0.118x^2 + 0.215x + 0.7 \quad (0 \leq x \leq 7)$$

where x denotes the age group of drivers, with $x = 0$ corresponding to those aged 50–54, $x = 1$ corresponding to those aged 55–59, $x = 2$ corresponding to those aged 60–64, ..., and $x = 7$ corresponding to those aged 85–89. What is the fatality rate per 100 million vehicle miles driven for an average driver in the 50–54 age group? In the 85–89 age group?

Source: U.S. Department of Transportation.

29. **Total Global Mobile Data Traffic** In a 2009 report, equipment maker Cisco forecast the total global mobile data traffic to be

$$f(t) = 0.021t^3 + 0.015t^2 + 0.12t + 0.06 \quad (0 \leq t \leq 5)$$

million terabytes per month in year t, where $t = 0$ corresponds to 2009.
a. What was the total global mobile data traffic in 2009?
b. According to Cisco, what will the total global mobile data traffic be in 2014?

Source: Cisco.

30. **Gift Cards** Gift cards have increased in popularity in recent years. Consumers appreciate gift cards because they get to select the present they like. The U.S. sales of gift cards (in billions of dollars) is approximated by

$$S(t) = -0.6204t^3 + 4.671t^2 + 3.354t + 47.4 \quad (0 \leq t \leq 5)$$

in year t, where $t = 0$ corresponds to 2003.
a. What were the sales of gift cards for 2003?
b. What were the sales of gift cards for 2008?

Source: The Tower Group.

31. **BlackBerry Subscribers** According to a study conducted in 2004, the number of subscribers of BlackBerry, the handheld email devices manufactured by Research in Motion Ltd., is approximated by

$$N(t) = -0.0675t^4 + 0.5083t^3 - 0.893t^2 + 0.66t + 0.32$$
$$(0 \leq t \leq 4)$$

where $N(t)$ is measured in millions and t in years, with $t = 0$ corresponding to the beginning of 2002.
a. How many BlackBerry subscribers were there at the beginning of 2002?
b. How many BlackBerry subscribers were there at the beginning of 2006?

Source: ThinkEquity Partners.

32. INFANT MORTALITY RATES IN MASSACHUSETTS The deaths of children less than 1 year old per 1000 live births is modeled by the function

$$R(t) = 162.8t^{-3.025} \quad (1 \le t \le 3)$$

where t is measured in 50-year intervals, with $t = 1$ corresponding to 1900.
a. Find $R(1)$, $R(2)$, and $R(3)$ and use your result to sketch the graph of the function R over the domain $[1, 3]$.
b. What was the infant mortality rate in 1900? in 1950? in 2000?

Source: Massachusetts Department of Public Health.

33. ONLINE VIDEO VIEWERS As broadband Internet grows more popular, video services such as YouTube will continue to expand. The number of online video viewers (in millions) is projected to grow according to the rule

$$N(t) = 52t^{0.531} \quad (1 \le t \le 10)$$

where $t = 1$ corresponds to 2003.
a. Sketch the graph of N.
b. How many online video viewers were there in 2012?

Source: eMarketer.com.

34. CHIP SALES The worldwide sales of flash memory chip (in billions of dollars) is approximated by

$$S(t) = 4.3(t + 2)^{0.94} \quad (0 \le t \le 8)$$

where t is measured in years, with $t = 0$ corresponding to 2002. Flash chips are used in cell phones, digital cameras, and other products.
a. What were the worldwide flash memory chip sales in 2002?
b. What were the estimated sales for 2010?

Source: Web-Feet Research, Inc.

35. OUTSOURCING OF JOBS According to a study conducted in 2003, the total number of U.S. jobs (in millions) that are projected to leave the country by year t, where $t = 0$ corresponds to 2000, is

$$N(t) = 0.0018425(t + 5)^{2.5} \quad (0 \le t \le 15)$$

What was the projected number of outsourced jobs for 2005 ($t = 5$)? For 2013 ($t = 13$)?

Source: Forrester Research.

36. IMMIGRATION TO THE UNITED STATES Immigration to the United States from Europe, as a percentage of total immigration, is approximately

$$P(t) = 0.767t^3 - 0.636t^2 - 19.17t + 52.7 \quad (0 \le t \le 4)$$

where t is measured in decades, with $t = 0$ corresponding to the decade of the 1950s.
a. Complete the following table:

t	0	1	2	3	4
$P(t)$					

b. Use the result of part (a) to sketch the graph of P.
c. Use the result of part (b) to estimate the decade when immigration, as a percentage of total immigration, was the greatest and the decade when it was the smallest.

Source: Jeffrey Williamson, Harvard University.

37. SELLING PRICE OF DVD RECORDERS The rise of digital music and the improvement to the DVD format are part of the reasons why the average selling price of stand-alone DVD recorders will drop in the coming years. The function

$$A(t) = \frac{699}{(t + 1)^{0.94}} \quad (0 \le t \le 5)$$

gives the average selling price (in dollars) of stand-alone DVD recorders in year t, where $t = 0$ corresponds to the beginning of 2002. What was the average selling price of stand-alone DVD recorders at the beginning of 2002? at the beginning of 2007?

Source: Consumer Electronics Association.

38. REACTION OF A FROG TO A DRUG Experiments conducted by A. J. Clark suggest that the response $R(x)$ of a frog's heart muscle to the injection of x units of acetylcholine (as a percent of the maximum possible effect of the drug) may be approximated by the rational function

$$R(x) = \frac{100x}{b + x} \quad (x \ge 0)$$

where b is a positive constant that depends on the particular frog.
a. If a concentration of 40 units of acetylcholine produces a response of 50% for a certain frog, find the "response function" for this frog.
b. Using the model found in part (a), find the response of the frog's heart muscle when 60 units of acetylcholine are administered.

39. DIGITAL VERSUS FILM CAMERAS The sales of digital cameras (in millions of units) in year t are given by the function

$$f(t) = 3.05t + 6.85 \quad (0 \le t \le 3)$$

where $t = 0$ corresponds to 2001. Over that same period, the sales of film cameras (in millions of units) are given by

$$g(t) = -1.85t + 16.58 \quad (0 \le t \le 3)$$

a. Show that more film cameras than digital cameras were sold in 2001.
b. When did the sales of digital cameras first exceed those of film cameras?

Source: Popular Science.

40. WALKING VERSUS RUNNING The oxygen consumption (in milliliter per pound per minute) for a person walking at x mph is approximated by the function

$$f(x) = \frac{5}{3}x^2 + \frac{5}{3}x + 10 \quad (0 \leq x \leq 9)$$

whereas the oxygen consumption for a runner at x mph is approximated by the function

$$g(x) = 11x + 10 \quad (4 \leq x \leq 9)$$

a. Sketch the graphs of f and g.
b. At what speed is the oxygen consumption the same for a walker as it is for a runner? What is the level of oxygen consumption at that speed?
c. What happens to the oxygen consumption of the walker and the runner at speeds beyond that found in part (b)?

Source: William McArdley, Frank Katch, and Victor Katch, *Exercise Physiology.*

41. PRICE OF AUTOMOBILE PARTS For years, automobile manufacturers had a monopoly on the replacement-parts market, particularly for sheet metal parts such as fenders, doors, and hoods, the parts most often damaged in a crash. Beginning in the late 1970s, however, competition appeared on the scene. In a report conducted by an insurance company to study the effects of the competition, the price of an OEM (original equipment manufacturer) fender for a particular 1983 model car was found to be

$$f(t) = \frac{110}{\frac{1}{2}t + 1} \quad (0 \leq t \leq 2)$$

where $f(t)$ is measured in dollars and t is in years. Over the same period of time, the price of a non-OEM fender for the car was found to be

$$g(t) = 26\left(\frac{1}{4}t^2 - 1\right)^2 + 52 \quad (0 \leq t \leq 2)$$

where $g(t)$ is also measured in dollars. Find a function $h(t)$ that gives the difference in price between an OEM fender and a non-OEM fender. Compute $h(0)$, $h(1)$, and $h(2)$. What does the result of your computation seem to say about the price gap between OEM and non-OEM fenders over the 2 years?

42. CRICKET CHIRPING AND TEMPERATURE Entomologists have discovered that a linear relationship exists between the number of chirps by crickets of a certain species and the air temperature. When the temperature is 70°F, the crickets chirp at the rate of 120 times/minute, and when the temperature is 80°F, they chirp at the rate of 160 times/minute.
a. Find an equation giving the relationship between the air temperature T and the number of chirps/minute, N, of the crickets.
b. Find N as a function of T, and use this formula to determine the rate at which the crickets chirp when the temperature is 102°F.

43. LINEAR DEPRECIATION In computing income tax, businesses are allowed by law to depreciate certain assets such as buildings, machines, furniture, and automobiles over a period of time. Linear depreciation, or the straight-line method, is often used for this purpose. Suppose an asset has an initial value of $\$C$ and is to be depreciated linearly over n years with a scrap value of $\$S$. Show that the book value of the asset at any time t ($0 \leq t \leq n$) is given by the linear function

$$V(t) = C - \frac{C-S}{n}t$$

Hint: Find an equation of the straight line that passes through the points $(0, C)$ and (n, S). Then rewrite the equation in the slope-intercept form.

44. LINEAR DEPRECIATION Using the linear depreciation model of Exercise 43, find the book value of a printing machine at the end of the second year if its initial value is $100,000 and it is depreciated linearly over 5 years with a scrap value of $30,000.

45. COST OF THE HEALTH-CARE BILL The Congressional Budget Office estimates that the health-care bill passed by the Senate in November 2009, combined with a package of revisions known as the reconciliation bill, will result in a cost by year t (in billions of dollars) of

$$f(t) = \begin{cases} 5 & \text{if } 0 \leq t < 2 \\ -0.5278t^3 + 3.012t^2 + 49.23t - 103.29 & \text{if } 2 \leq t \leq 8 \end{cases}$$

where t is measured in years, with $t = 0$ corresponding to 2010. What will be the cost of the health-care bill in 2011? By 2015?

Source: U.S. Congressional Budget Office.

46. OBESE CHILDREN IN THE UNITED STATES The percentage of obese children aged 12–19 in the United States is approximately

$$P(t) = \begin{cases} 0.04t + 4.6 & \text{if } 0 \leq t < 10 \\ -0.01005t^2 + 0.945t - 3.4 & \text{if } 10 \leq t \leq 30 \end{cases}$$

where t is measured in years, with $t = 0$ corresponding to the beginning of 1970. What was the percentage of obese children aged 12–19 at the beginning of 1970? At the beginning of 1985? At the beginning of 2000?

Source: Centers for Disease Control and Prevention.

47. PRICE OF IVORY According to the World Wildlife Fund, a group in the forefront of the fight against illegal ivory trade, the price of ivory (in dollars per kilo) compiled from a variety of legal and black market sources is approximated by the function

$$f(t) = \begin{cases} 8.37t + 7.44 & \text{if } 0 \leq t \leq 8 \\ 2.84t + 51.68 & \text{if } 8 < t \leq 30 \end{cases}$$

where t is measured in years, with $t = 0$ corresponding to the beginning of 1970.
a. Sketch the graph of the function f.

b. What was the price of ivory at the beginning of 1970? At the beginning of 1990?
Source: World Wildlife Fund.

48. Working-Age Population The ratio of working-age population to the elderly in the United States (including projections after 2000) is given by

$$f(t) = \begin{cases} 4.1 & \text{if } 0 \le t < 5 \\ -0.03t + 4.25 & \text{if } 5 \le t < 15 \\ -0.075t + 4.925 & \text{if } 15 \le t \le 35 \end{cases}$$

with $t = 0$ corresponding to the beginning of 1995.
a. Sketch the graph of f.
b. What was the ratio at the beginning of 2005? What will the ratio be at the beginning of 2020?
c. Over what years is the ratio constant?
d. Over what years is the decline of the ratio greatest?
Source: U.S. Census Bureau.

49. Senior Citizens' Health Care According to a study, the out-of-pocket cost to senior citizens for health care, $f(t)$ (as a percentage of income), in year t where $t = 0$ corresponds to 1977 is given by

$$f(t) = \begin{cases} \dfrac{2}{7}t + 12 & \text{if } 0 \le t \le 7 \\ t + 7 & \text{if } 7 < t \le 10 \\ \dfrac{1}{3}t + \dfrac{41}{3} & \text{if } 10 < t \le 25 \end{cases}$$

a. Sketch the graph of f.
b. What was the out-of-pocket cost, as a percentage of income, to senior citizens for health care in 1982? In 2002?
Source: Senate Select Committee on Aging, AARP.

50. Sales of DVD Players Versus VCRs The sales of DVD players in year t (in millions of units) is given by the function

$$f(t) = 5.6(1 + t) \quad (0 \le t \le 3)$$

where $t = 0$ corresponds to 2001. Over the same period, the sales of VCRs (in millions of units) is given by

$$g(t) = \begin{cases} -9.6t + 22.5 & \text{if } 0 \le t \le 1 \\ -0.5t + 13.4 & \text{if } 1 < t \le 2 \\ -7.8t + 28 & \text{if } 2 < t \le 3 \end{cases}$$

a. Show that more VCRs than DVD players were sold in 2001.
b. When did the sales of DVD players first exceed those of VCRs?
Source: Popular Science.

For the demand equations in Exercises 51–54, where x represents the quantity demanded in units of a thousand and p is the unit price in dollars, (a) sketch the demand curve and (b) determine the quantity demanded when the unit price is set at $\$p$.

51. $p = -x^2 + 16;\ p = 7$ **52.** $p = -x^2 + 36;\ p = 11$
53. $p = \sqrt{18 - x^2};\ p = 3$ **54.** $p = \sqrt{9 - x^2};\ p = 2$

For the supply equations in Exercises 55–58, where x is the quantity supplied in units of a thousand and p is the unit price in dollars, (a) sketch the supply curve and (b) determine the price at which the supplier will make 2000 units of the commodity available in the market.

55. $p = x^2 + 16x + 40$ **56.** $p = 2x^2 + 18$
57. $p = x^3 + 2x + 3$ **58.** $p = x^3 + x + 10$

59. Demand for Smoke Alarms The demand function for the Sentinel smoke alarm is given by

$$p = \frac{30}{0.02x^2 + 1} \quad (0 \le x \le 10)$$

where x (measured in units of a thousand) is the quantity demanded per week and p is the unit price in dollars.
a. Sketch the graph of the demand function.
b. What is the unit price that corresponds to a quantity demanded of 10,000 units?

60. Demand for Commodities Assume that the demand function for a certain commodity has the form

$$p = \sqrt{-ax^2 + b} \quad (a \ge 0, b \ge 0)$$

where x is the quantity demanded, measured in units of a thousand and p is the unit price in dollars. Suppose the quantity demanded is 6000 ($x = 6$) when the unit price is $8.00 and 8000 ($x = 8$) when the unit price is $6.00. Determine the demand equation. What is the quantity demanded when the unit price is set at $7.50?

61. Supply of Desk Lamps The supply function for the Luminar desk lamp is given by

$$p = 0.1x^2 + 0.5x + 15$$

where x is the quantity supplied (in thousands) and p is the unit price in dollars.
a. Sketch the graph of the supply function.
b. What unit price will induce the supplier to make 5000 lamps available in the marketplace?

62. Supply of Satellite Radios Suppliers of satellite radios will market 10,000 units when the unit price is $20 and 62,500 units when the unit price is $35. Determine the supply function if it is known to have the form

$$p = a\sqrt{x} + b \quad (a > 0, b > 0)$$

where x is the quantity supplied and p is the unit price in dollars. Sketch the graph of the supply function. What unit price will induce the supplier to make 40,000 satellite radios available in the marketplace?

63. Suppose the demand and supply equations for a certain commodity are given by $p = ax + b$ and $p = cx + d$, respectively, where $a < 0, c > 0$, and $b > d > 0$ (see the figure on the following page).
a. Find the equilibrium quantity and equilibrium price in terms of $a, b, c,$ and d.
b. Use part (a) to determine what happens to the market equilibrium if c is increased while $a, b,$ and d remain fixed. Interpret your answer in economic terms.

c. Use part (a) to determine what happens to the market equilibrium if b is decreased while a, c, and d remain fixed. Interpret your answer in economic terms.

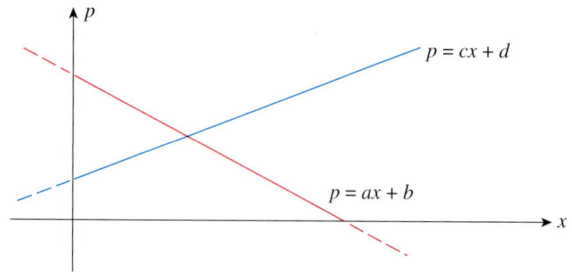

For each pair of supply and demand equations in Exercises 64–67, where x represents the quantity demanded in units of a thousand and p the unit price in dollars, find the equilibrium quantity and the equilibrium price.

64. $p = -x^2 - 2x + 100$ and $p = 8x + 25$

65. $p = -2x^2 + 80$ and $p = 15x + 30$

66. $p = 60 - 2x^2$ and $p = x^2 + 9x + 30$

67. $11p + 3x - 66 = 0$ and $2p^2 + p - x = 10$

68. MARKET EQUILIBRIUM The weekly demand and supply functions for Sportsman 5×7 tents are given by

$$p = -0.1x^2 - x + 40$$
$$p = 0.1x^2 + 2x + 20$$

respectively, where p is measured in dollars and x is measured in units of a hundred. Find the equilibrium quantity and price.

69. MARKET EQUILIBRIUM The management of Titan Tire Company has determined that the weekly demand and supply functions for their Super Titan tires are given by

$$p = 144 - x^2$$
$$p = 48 + \frac{1}{2}x^2$$

respectively, where p is measured in dollars and x is measured in units of a thousand. Find the equilibrium quantity and price.

70. ENCLOSING AN AREA Patricia wishes to have a rectangular garden in her backyard. She has 80 ft of fencing with which to enclose her garden. Letting x denote the width of the garden, find a function f in the variable x giving the area of the garden. What is its domain?

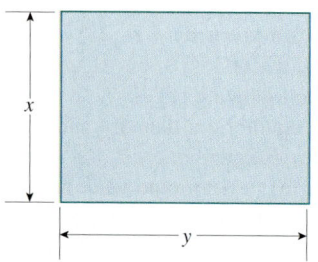

71. ENCLOSING AN AREA Juanita wishes to have a rectangular garden in her backyard. But Juanita wants her garden to have an area of 250 ft². Letting x denote the width of the garden, find a function f in the variable x giving the length of the fencing required to construct the garden. What is the domain of the function?

Hint: Refer to the figure for Exercise 70. The amount of fencing required is equal to the perimeter of the rectangle, which is twice the width plus twice the length of the rectangle.

72. PACKAGING By cutting away identical squares from each corner of a rectangular piece of cardboard and folding up the resulting flaps, an open box can be made. If the cardboard is 15 in. long and 8 in. wide and the square cutaways have dimensions of x in. by x in., find a function giving the volume of the resulting box.

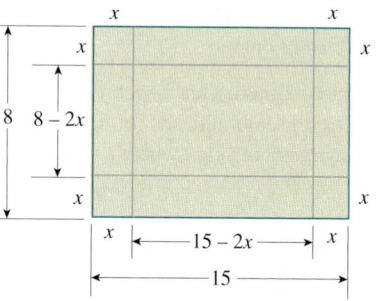

73. CONSTRUCTION COSTS A rectangular box is to have a square base and a volume of 20 ft³. The material for the base costs 30¢/ft², the material for the sides costs 10¢/ft², and the material for the top costs 20¢/ft². Letting x denote the length of one side of the base, find a function in the variable x giving the cost of constructing the box.

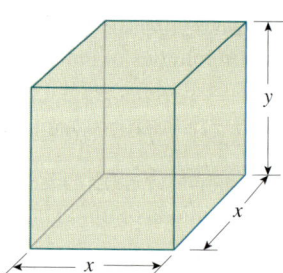

74. AREA OF A NORMAN WINDOW A Norman window has the shape of a rectangle surmounted by a semicircle (see the accompanying figure). Suppose a Norman window is to have a perimeter of 28 ft. Find a function in the variable x giving the area of the window.

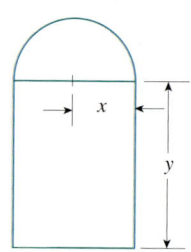

75. **Yield of an Apple Orchard** An apple orchard has an average yield of 36 bushels of apples/tree if tree density is 22 trees/acre. For each unit increase in tree density, the yield decreases by 2 bushels/tree. Letting x denote the number of trees beyond 22/acre, find a function in x that gives the yield of apples.

76. **Book Design** A book designer has decided that the pages of a book should have 1-in. margins at the top and bottom and $\frac{1}{2}$-in. margins on the sides. She further stipulated that each page should have a total area of 50 in.^{2}. Find a function in the variable x, giving the area of the printed part of the page. What is the domain of the function?

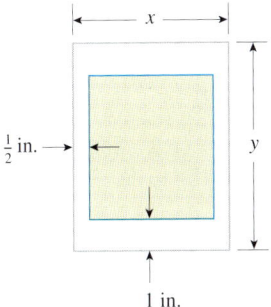

77. **Profit of a Vineyard** Phillip, the proprietor of a vineyard, estimates that if 10,000 bottles of wine were produced this season, then the profit would be $5/bottle. But if more than 10,000 bottles were produced, then the profit per bottle for the entire lot would drop by $0.0002 for each additional bottle sold. Assume that at least 10,000 bottles of wine are produced and sold, and let x denote the number of bottles produced and sold in excess of 10,000.

a. Find a function P giving the profit in terms of x.
b. What is the profit Phillip can expect from the sale of 16,000 bottles of wine from his vineyard?

78. **Charter Revenue** The owner of a luxury motor yacht that sails among the 4000 Greek islands charges $600/person/day if exactly 20 people sign up for the cruise. However, if more than 20 people sign up for the cruise (up to the maximum capacity of 90), the fare for all the passengers is reduced by $4 per person for each additional passenger. Assume that at least 20 people sign up for the cruise, and let x denote the number of passengers above 20.
a. Find a function R giving the revenue per day realized from the charter.
b. What is the revenue per day if 60 people sign up for the cruise?
c. What is the revenue per day if 80 people sign up for the cruise?

In Exercises 79–82, determine whether the statement is true or false. If it is true, explain why it is true. If it is false, give an example to show why it is false.

79. A polynomial function is a sum of constant multiples of power functions.

80. A polynomial function is a rational function, but the converse is false.

81. If $r > 0$, then the power function $f(x) = x^r$ is defined for all values of x.

82. The function $f(x) = 2^x$ is a power function.

2.3 Solutions to Self-Check Exercises

1. Since the adult dose of the substance is 500 mg, $a = 500$; thus, the rule in this case is

$$D(t) = \frac{500t}{t + 12}$$

A 4-year-old should receive

$$D(4) = \frac{500(4)}{4 + 12}$$

or 125 mg of the substance.

2. a. The graphs of the functions d and s are shown in the following figure:

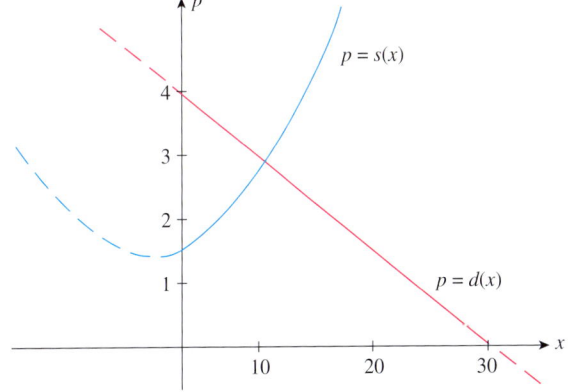

b. Solve the following system of equations:

$$p = -\frac{2}{15}x + 4$$

$$p = \frac{1}{75}x^2 + \frac{1}{10}x + \frac{3}{2}$$

Substituting the first equation into the second yields

$$\frac{1}{75}x^2 + \frac{1}{10}x + \frac{3}{2} = -\frac{2}{15}x + 4$$

$$\frac{1}{75}x^2 + \left(\frac{1}{10} + \frac{2}{15}\right)x - \frac{5}{2} = 0$$

$$\frac{1}{75}x^2 + \frac{7}{30}x - \frac{5}{2} = 0$$

Multiplying both sides of the last equation by 150, we have

$$2x^2 + 35x - 375 = 0$$

$$(2x - 15)(x + 25) = 0$$

Thus, $x = -25$ or $x = 15/2 = 7.5$. Since x must be non-negative, we take $x = 7.5$, and the equilibrium quantity is 7500 lb. The equilibrium price is given by

$$p = -\frac{2}{15}\left(\frac{15}{2}\right) + 4$$

or $3/lb.

USING TECHNOLOGY

Finding the Points of Intersection of Two Graphs and Modeling

A graphing utility can be used to find the point(s) of intersection of the graphs of two functions.

EXAMPLE 1 Find the points of intersection of the graphs of

$$f(x) = 0.3x^2 - 1.4x - 3 \quad \text{and} \quad g(x) = -0.4x^2 + 0.8x + 6.4$$

Solution The graphs of both f and g in the standard viewing window are shown in Figure T1a. Using the function for finding the points of intersection of two graphs on a graphing utility, we find the point(s) of intersection, accurate to four decimal places, to be $(-2.4158, 2.1329)$ (Figure T1b) and $(5.5587, -1.5125)$ (Figure T1c). To access this function on the TI-83/84, select **5: intersect** on the Calc menu.

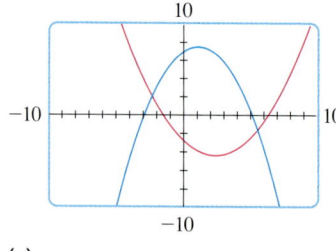
(a)

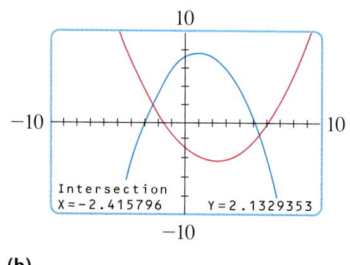
(b)

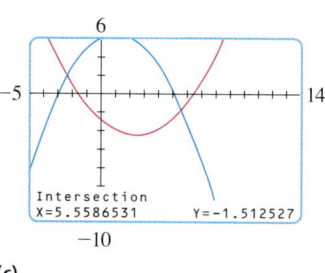
(c)

FIGURE T1
(a) The graphs of f and g in the standard viewing window; (b) and (c) the TI-83/84 intersection screens

EXAMPLE 2 Consider the demand and supply functions

$$p = d(x) = -0.01x^2 - 0.2x + 8 \quad \text{and} \quad p = s(x) = 0.01x^2 + 0.1x + 3$$

a. Plot the graphs of d and s in the viewing window $[0, 15] \times [0, 10]$.
b. Verify that the equilibrium point is $(10, 5)$.

Solution

a. The graphs of d and s are shown in Figure T2a.

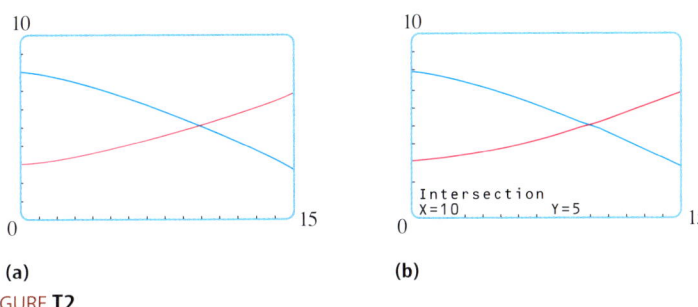

(a)   (b)

FIGURE T2
(a) The graphs of d and s in the window $[0, 15] \times [0, 10]$; (b) the TI-83/84 intersection screen

b. Using the function for finding the point of intersection of two graphs, we see that $x = 10$ and $y = 5$ (Figure T2b), so the equilibrium point is $(10, 5)$.

Constructing Mathematical Models from Raw Data

A graphing utility can sometimes be used to construct mathematical models from sets of data. For example, if the points corresponding to the given data are scattered about a straight line, then use **LinReg(ax+b)** (linear regression) from the statistical calculations menu of the graphing utility to obtain a function (model) that approximates the data at hand. If the points seem to be scattered along a parabola (the graph of a quadratic function), then use **QuadReg** (second-degree polynomial regression), and so on. (These are functions on the TI-83/84 calculator.)

APPLIED EXAMPLE 3 Indian Gaming Industry The following table gives the estimated gross revenues (in billions of dollars) from the Indian gaming industries from 2000 ($t = 0$) to 2008 ($t = 8$).

Year	0	1	2	3	4	5	6	7	8
Revenue	11.0	12.8	14.7	16.8	19.5	22.7	25.1	26.4	26.8

a. Use a graphing utility to find a polynomial function f of degree 4 that models the data.
b. Plot the graph of the function f, using the viewing window $[0, 8] \times [0, 30]$.
c. Use the function evaluation capability of the graphing utility to compute $f(0)$, $f(1), \ldots, f(8)$, and compare these values with the original data.
d. If the trend continued, what was the gross revenue for 2009 ($t = 9$)?

Source: National Indian Gaming Association.

Solution

a. First, enter the data using the statistical menu. Then choose **QuartReg** (fourth-degree polynomial regression) from the statistical calculations menu of a graphing utility. We find

$$f(t) = -0.00737t^4 + 0.0655t^3 - 0.008t^2 + 1.61t + 11$$

b. The graph of f is shown in Figure T3.

(continued)

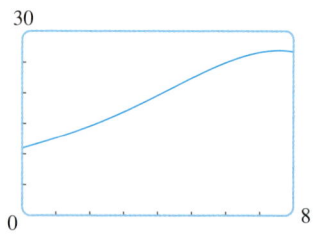

FIGURE T3
The graph of f in the viewing window [0, 8] × [0, 30]

c. The required values, which compare favorably with the given data, follow:

t	0	1	2	3	4	5	6	7	8
f(t)	11.0	12.7	14.6	16.9	19.6	22.4	25.0	26.6	26.7

d. The gross revenue for 2009 ($t = 9$) is given by

$$f(9) = -0.00737(9)^4 + 0.0655(9)^3 - 0.008(9)^2 + 1.61(9) + 11 \approx 24.24$$

or approximately $24.2 billion.

TECHNOLOGY EXERCISES

In Exercises 1–6, find the points of intersection of the graphs of the functions. Express your answer accurate to four decimal places.

1. $f(x) = 1.2x + 3.8$; $g(x) = -0.4x^2 + 1.2x + 7.5$

2. $f(x) = 0.2x^2 - 1.3x - 3$; $g(x) = -1.3x + 2.8$

3. $f(x) = 0.3x^2 - 1.7x - 3.2$; $g(x) = -0.4x^2 + 0.9x + 6.7$

4. $f(x) = -0.3x^2 + 0.6x + 3.2$; $g(x) = 0.2x^2 - 1.2x - 4.8$

5. $f(x) = 0.3x^3 - 1.8x^2 + 2.1x - 2$; $g(x) = 2.1x - 4.2$

6. $f(x) = -0.2x^3 + 1.2x^2 - 1.2x + 2$; $g(x) = -0.2x^2 + 0.8x + 2.1$

7. **MARKET EQUILIBRIUM** The monthly demand and supply functions for a certain brand of wall clock are given by

$$p = -0.2x^2 - 1.2x + 50$$
$$p = 0.1x^2 + 3.2x + 25$$

respectively, where p is measured in dollars and x is measured in units of a hundred.
   a. Plot the graphs of both functions in an appropriate viewing window.
   b. Find the equilibrium quantity and price.

8. **MARKET EQUILIBRIUM** The quantity demanded x (in units of a hundred) of Mikado miniature cameras per week is related to the unit price p (in dollars) by

$$p = -0.2x^2 + 80$$

The quantity x (in units of a hundred) that the supplier is willing to make available in the market is related to the unit price p (in dollars) by

$$p = 0.1x^2 + x + 40$$

   a. Plot the graphs of both functions in an appropriate viewing window.
   b. Find the equilibrium quantity and price.

In Exercises 9–22, use the statistical calculations menu to construct a mathematical model associated with the given data.

9. **CONSUMPTION OF BOTTLED WATER** The annual per-capita consumption of bottled water (in gallons) and the scatter plot for these data follow:

Year	2001	2002	2003	2004	2005	2006
Consumption	18.8	20.9	22.4	24	26.1	28.3

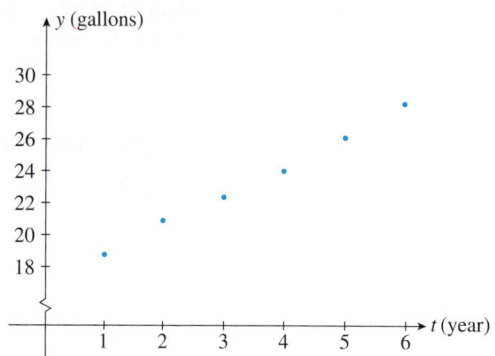

   a. Use **LinReg(ax+b)** to find a first-degree (linear) polynomial regression model for the data. Let $t = 1$ correspond to 2001.
   b. Plot the graph of the function f found in part (a), using the viewing window [1, 6] × [0, 30].
   c. Compute the values for $t = 1, 2, 3, 4, 5$, and 6. How do your figures compare with the given data?
   d. If the trend continued, what was the annual per-capita consumption of bottled water in 2008?

   Source: Beverage Marketing Corporation.

10. **WEB CONFERENCING** Web conferencing is a big business, and it's growing rapidly. The amount (in billions of dollars) spent on Web conferencing from the beginning of 2003 through 2010, and the scatter diagram for these data follow:

Year	2003	2004	2005	2006	2007	2008	2009	2010
Amount	0.50	0.63	0.78	0.92	1.16	1.38	1.60	1.90

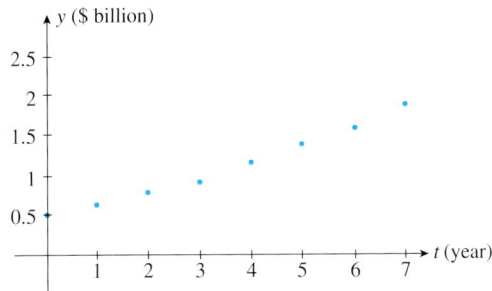

a. Let $t = 0$ correspond to the beginning of 2003 and use **QuadReg** to find a second-degree polynomial regression model based on the given data.
b. Plot the graph of the function f found in part (a) using the window $[0, 7] \times [0, 2]$.
c. Compute $f(0), f(3), f(6),$ and $f(7)$. Compare these values with the given data.

Source: Gartner Dataquest.

11. **STUDENT POPULATION** The projected total number of students in elementary schools, secondary schools, and colleges (in millions) from 1995 through 2015 is given in the following table:

Year	1995	2000	2005	2010	2015
Number	64.8	68.7	72.6	74.8	78

a. Use **QuadReg** to find a second-degree polynomial regression model for the data. Let t be measured in 5-year intervals, with $t = 0$ corresponding to the beginning of 1995.
b. Plot the graph of the function f found in part (a), using the viewing window $[0, 4] \times [0, 85]$.
c. Using the model found in part (a), what will be the projected total number of students (all categories) enrolled in 2015?

Source: U.S. National Center for Education Statistics.

12. **DIGITAL TV SHIPMENTS** The estimated number of digital TV shipments between the year 2000 and 2006 (in millions of units) and the scatter plot for these data follow:

Year	2000	2001	2002	2003	2004	2005	2006
Units Shipped	0.63	1.43	2.37	4.1	6	8.1	10

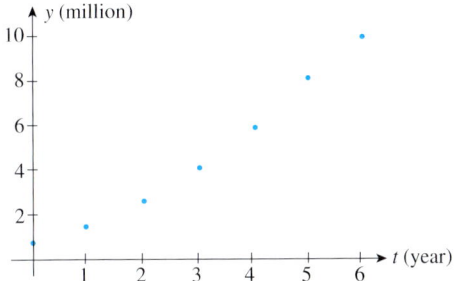

a. Use **CubicReg** to find a third-degree polynomial regression model for the data. Let $t = 0$ correspond to the beginning of 2000.
b. Plot the graph of the function f found in part (a), using the viewing window $[0, 6] \times [0, 11]$.
c. Compute the values of $f(t)$ for $t = 0, 1, 2, 3, 4, 5,$ and 6.

Source: Consumer Electronics Manufacturers Association.

13. **HEALTH-CARE SPENDING** Health-care spending by business (in billions of dollars) from the year 2000 through 2006 is summarized below:

Year	2000	2001	2002	2003	2004	2005	2006
Number	185	235	278	333	389	450	531

a. Plot the scatter diagram for the above data. Let $t = 0$ correspond to the beginning of 2000.
b. Use **QuadReg** to find a second-degree polynomial regression model for the data.
c. If the trend continued, what was the spending in 2010?

Source: Centers for Medicine and Medicaid Services.

14. **TiVo OWNERS** The projected number of households (in millions) with digital video recorders that allow viewers to record shows onto a server and skip commercials are given in the following table:

Year	2006	2007	2008	2009	2010
Households	31.2	49.0	71.6	97.0	130.2

a. Let $t = 0$ correspond to the beginning of 2006, and use **QuadReg** to find a second-degree polynomial regression model based on the given data.
b. Obtain the scatter plot and the graph of the function f found in part (a), using the viewing window $[0, 4] \times [0, 140]$.

Source: Strategy Analytics.

15. **TELECOMMUNICATIONS INDUSTRY REVENUE** Telecommunications industry revenue is expected to grow in the coming years, fueled by the demand for broadband and high-speed data services. The worldwide revenue for the industry (in trillions of dollars) and the scatter diagram for these data follow:

Year	2000	2002	2004	2006	2008	2010
Revenue	1.7	2.0	2.5	3.0	3.6	4.2

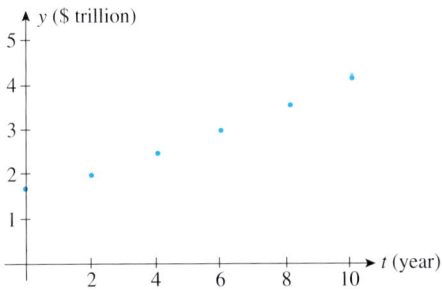

a. Let $t = 0$ correspond to the beginning of 2000 and use **CubicReg** to find a third-degree polynomial regression model based on the given data.
b. Plot the graph of the function f found in part (a), using the viewing window $[0, 10] \times [0, 5]$.
c. Find the worldwide revenue for the industry in 2001 and 2005 and find the projected revenue for 2010.

Source: Telecommunication Industry Association.

(continued)

16. Population Growth in Clark County Clark County in Nevada—dominated by greater Las Vegas—is one of the fastest-growing metropolitan areas in the United States. The population of the county from 1970 through 2000 is given in the following table:

Year	1970	1980	1990	2000
Population	273,288	463,087	741,459	1,375,765

a. Use **CubicReg** to find a third-degree polynomial regression model for the data. Let t be measured in decades, with $t = 0$ corresponding to the beginning of 1970.
b. Plot the graph of the function f found in part (a), using the viewing window $[0, 3] \times [0, 1{,}500{,}000]$.
c. Compare the values of f at $t = 0, 1, 2$, and 3, with the given data.

Source: U.S. Census Bureau.

17. Lobbyists' Spending Lobbyists try to persuade legislators to propose, pass, or defeat legislation or to change existing laws. The amount (in billions of dollars) spent by lobbyists from 2003 through 2009 is shown in the following table:

Year	2003	2004	2005	2006	2007	2008	2009
Amount	8.0	8.5	9.7	10.2	11.3	12.9	13.8

a. Use **CubicReg** to find a third-degree polynomial regression model for the data, letting $t = 0$ correspond to 2003.
b. Plot the scatter diagram and the graph of the function f found in part (a), using the viewing window $[0, 6] \times [0, 15]$.
c. Compare the values of f at $t = 0, 3$, and 6 with the given data.

Source: Center for Public Integrity.

18. Mobile Enterprise IM Accounts The projected number of mobile enterprise instant messaging (IM) accounts (in millions) from 2006 through 2010 is given in the following table ($t = 0$ corresponds to the beginning of 2006):

Year	0	1	2	3	4
Accounts	2.3	3.6	5.8	8.7	14.9

a. Use **CubicReg** to find a third-degree polynomial regression model based on the given data.
b. Plot the graph of the function f found in part (a), using the viewing window $[0, 5] \times [0, 16]$.
c. Compute $f(0), f(1), f(2), f(3)$, and $f(4)$.

Source: The Radical Group.

19. Measles Deaths Measles is still a leading cause of vaccine-preventable death among children, but because of improvements in immunizations, measles deaths have dropped globally. The following table gives the number of measles deaths (in thousands) in sub-Saharan Africa from 1999 through 2005:

Year	1999	2001	2003	2005
Amount	506	338	250	126

a. Use **CubicReg** to find a third-degree polynomial regression model for the data, letting $t = 0$ correspond to the beginning of 1999.
b. Plot the scatter diagram and the graph of the function f found in part (a).
c. Compute the values of f for $t = 0, 2$, and 6.

Source: Centers for Disease Control and Prevention and World Health Organization.

20. Office Vacancy Rate The total vacancy rate (as a percent) of offices in Manhattan from 2000 through 2006 is shown in the following table:

Year	2000	2001	2002	2003	2004	2005	2006
Vacancy Rate	3.8	8.9	12	12.5	11	8.4	6.7

a. Use **CubicReg** to find a third-degree polynomial regression model for the data, letting $t = 0$ correspond to the beginning of 2000.
b. Plot the scatter diagram and the graph of the function f found in part (a).
c. Compute the values for $t = 1, 2, 3, 4, 5$, and 6.

Source: Cushman and Wakefield.

21. Nicotine Content of Cigarettes Even as measures to discourage smoking have been growing more stringent in recent years, the nicotine content of cigarettes has been rising, making it more difficult for smokers to quit. The following table gives the average amount of nicotine in cigarette smoke from 1999 through 2004:

Year	1999	2000	2001	2002	2003	2004
Yield per Cigarette (mg)	1.71	1.81	1.85	1.84	1.83	1.89

a. Use **QuartReg** to find a fourth-degree polynomial regression model for the data. Let $t = 0$ correspond to the beginning of 1999.
b. Plot the graph of the function f found in part (a), using the viewing window $[0, 5] \times [0, 2]$.
c. Compute the values of $f(t)$ for $t = 0, 1, 2, 3, 4$, and 5.
d. If the trend continued, what was the average amount of nicotine in cigarettes in 2005?

Source: Massachusetts Tobacco Control Program.

22. Social Security Trust Fund Assets The projected assets of the Social Security trust fund (in trillions of dollars) from 2010 through 2037 are given in the following table:

Year	2010	2015	2020	2025	2030	2035	2037
Assets	2.69	3.56	4.22	4.24	3.24	0.87	0

Use **QuartReg** to find a fourth-degree polynomial regression model for the data. Let $t = 0$ correspond to 2010.

Source: Social Security Administration.

2.4 Limits

Introduction to Calculus

Historically, the development of calculus by Isaac Newton (1642–1727) and Gottfried Wilhelm Leibniz (1646–1716) resulted from the investigation of the following problems:

1. Finding the tangent line to a curve at a given point on the curve (Figure 26a)
2. Finding the area of a planar region bounded by an arbitrary curve (Figure 26b)

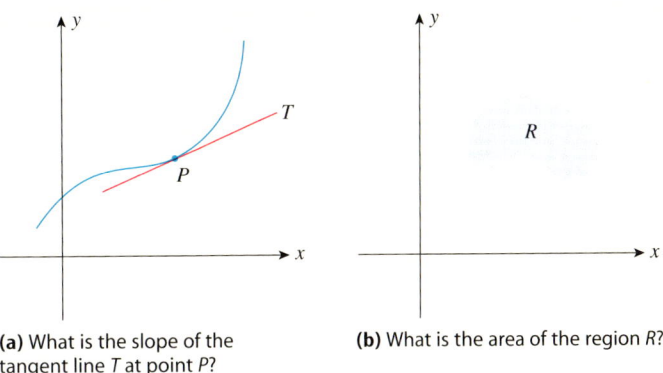

(a) What is the slope of the tangent line T at point P?

(b) What is the area of the region R?

FIGURE 26

The tangent-line problem might appear to be unrelated to any practical applications of mathematics, but as you will see later, the problem of finding the *rate of change* of one quantity with respect to another is mathematically equivalent to the geometric problem of finding the slope of the *tangent line* to a curve at a given point on the curve. It is precisely the discovery of the relationship between these two problems that spurred the development of calculus in the seventeenth century and made it such an indispensable tool for solving practical problems. The following are a few examples of such problems:

- Finding the velocity of an object
- Finding the rate of change of a bacteria population with respect to time
- Finding the rate of change of a company's profit with respect to time
- Finding the rate of change of a travel agency's revenue with respect to the agency's expenditure for advertising

The study of the tangent-line problem led to the creation of *differential calculus*, which relies on the concept of the *derivative* of a function. The study of the area problem led to the creation of *integral calculus*, which relies on the concept of the *antiderivative*, or *integral*, of a function. (The derivative of a function and the integral of a function are intimately related, as you will see in Section 6.4.) Both the derivative of a function and the integral of a function are defined in terms of a more fundamental concept—the limit—our next topic.

A Real-Life Example

From data obtained in a test run conducted on a prototype of a maglev (magnetic levitation train), which moves along a straight monorail track, engineers have determined that the position of the maglev (in feet) from the origin at time t (in seconds) is given by

$$s = f(t) = 4t^2 \qquad (0 \le t \le 30) \tag{3}$$

where f is called the **position function** of the maglev. The position of the maglev at time $t = 0, 1, 2, 3, \ldots, 10$, measured from its initial position, is

$$f(0) = 0 \quad f(1) = 4 \quad f(2) = 16 \quad f(3) = 36, \ldots \quad f(10) = 400$$

feet (Figure 27).

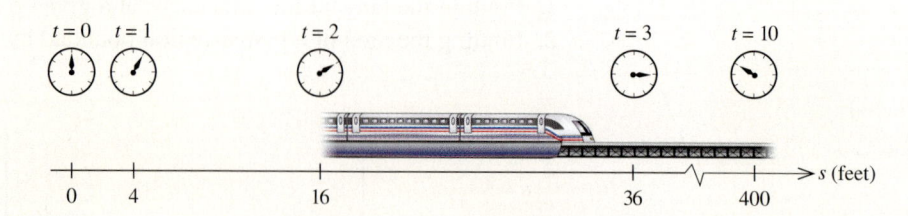

FIGURE 27
A maglev moving along an elevated monorail track

Suppose we want to find the velocity of the maglev at $t = 2$. This is just the velocity of the maglev as shown on its speedometer at that precise instant of time. Offhand, calculating this quantity using only Equation (3) appears to be an impossible task; but consider what quantities we *can* compute using this relationship. Obviously, we can compute the position of the maglev at any time t as we did earlier for some selected values of t. Using these values, we can then compute the *average velocity* of the maglev over an interval of time. For example, the average velocity of the train over the time interval [2, 4] is given by

$$\frac{\text{Distance covered}}{\text{Time elapsed}} = \frac{f(4) - f(2)}{4 - 2}$$

$$= \frac{4(4^2) - 4(2^2)}{2}$$

$$= \frac{64 - 16}{2} = 24$$

or 24 feet/second.

Although this is not quite the velocity of the maglev at $t = 2$, it does provide us with an approximation of its velocity at that time.

Can we do better? Intuitively, the smaller the time interval we pick (with $t = 2$ as the left endpoint), the better the average velocity over that time interval will approximate the actual velocity of the maglev at $t = 2$.*

Now, let's describe this process in general terms. Let $t > 2$. Then, the average velocity of the maglev over the time interval $[2, t]$ is given by

$$\frac{f(t) - f(2)}{t - 2} = \frac{4t^2 - 4(2^2)}{t - 2} = \frac{4(t^2 - 4)}{t - 2} \quad \quad (4)$$

By choosing the values of t closer and closer to 2, we obtain a sequence of numbers that give the average velocities of the maglev over smaller and smaller time intervals. As we observed earlier, this sequence of numbers should approach the *instantaneous velocity* of the train at $t = 2$.

Let's try some sample calculations. Using Equation (4) and taking the sequence $t = 2.5, 2.1, 2.01, 2.001,$ and 2.0001, which approaches 2, we find the following:

---

*Actually, any interval containing $t = 2$ will do.

The average velocity over [2, 2.5] is $\dfrac{4(2.5^2 - 4)}{2.5 - 2} = 18$, or 18 feet/second.

The average velocity over [2, 2.1] is $\dfrac{4(2.1^2 - 4)}{2.1 - 2} = 16.4$, or 16.4 feet/second.

and so forth. These results are summarized in Table 1.

TABLE 1

			t approaches 2 from the right.		
t	2.5	2.1	2.01	2.001	2.0001
Average Velocity over $[2, t]$	18	16.4	16.04	16.004	16.0004

Average velocity approaches 16 from the right.

From Table 1, we see that the average velocity of the maglev seems to approach the number 16 as it is computed over smaller and smaller time intervals. These computations suggest that the instantaneous velocity of the train at $t = 2$ is 16 feet/second.

Note Notice that we cannot obtain the instantaneous velocity for the maglev at $t = 2$ by substituting $t = 2$ into Equation (4) because this value of t is not in the domain of the average velocity function. ■

Intuitive Definition of a Limit

Consider the function g defined by

$$g(t) = \frac{4(t^2 - 4)}{t - 2}$$

which gives the average velocity of the maglev [see Equation (4)]. Suppose we are required to determine the value that $g(t)$ approaches as t approaches the (fixed) number 2. If we take the sequence of values of t approaching 2 from the right-hand side, as we did earlier, we see that $g(t)$ approaches the number 16. Similarly, if we take a sequence of values of t approaching 2 from the left, such as $t = 1.5, 1.9, 1.99, 1.999,$ and 1.9999, we obtain the results shown in Table 2.

TABLE 2

			t approaches 2 from the left.		
t	1.5	1.9	1.99	1.999	1.9999
$g(t)$	14	15.6	15.96	15.996	15.9996

Average velocity approaches 16 from the left.

Observe that $g(t)$ approaches the number 16 as t approaches 2—this time from the left-hand side. In other words, as t approaches 2 from *either* side of 2, $g(t)$ approaches 16. In this situation, we say that the limit of $g(t)$ as t approaches 2 is 16, written

$$\lim_{t \to 2} g(t) = \lim_{t \to 2} \frac{4(t^2 - 4)}{t - 2} = 16$$

The graph of the function g, shown in Figure 28, confirms this observation.

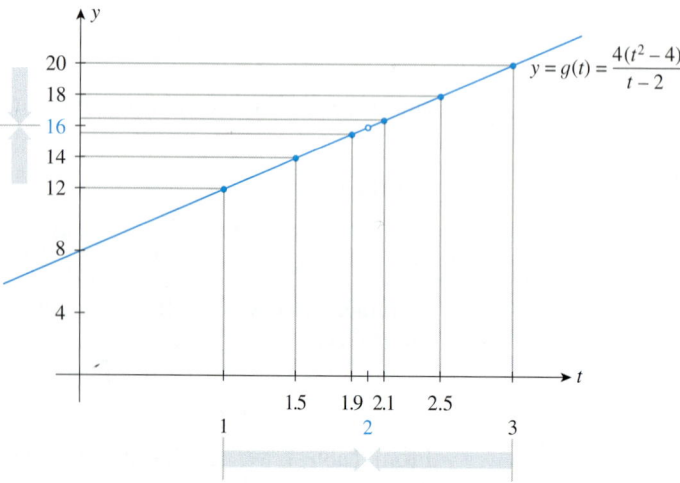

FIGURE 28
As t approaches t = 2 from either direction, g(t) approaches y = 16.

Observe that the point $t = 2$ is not in the domain of the function g [for this reason, the point (2, 16) is missing from the graph of g]. This, however, is inconsequential because the value, if any, of $g(t)$ at $t = 2$ plays no role in computing the limit.

This example leads to the following informal definition.

> **Limit of a Function**
> The function f has the **limit** L as x approaches a, written
> $$\lim_{x \to a} f(x) = L$$
> if the value of $f(x)$ can be made as close to the number L as we please by taking x sufficiently close to (but not equal to) a.

Exploring with TECHNOLOGY

1. Use a graphing utility to plot the graph of
$$g(x) = \frac{4(x^2 - 4)}{x - 2}$$
in the viewing window $[0, 3] \times [0, 20]$.
2. Use ZOOM and TRACE to describe what happens to the values of $g(x)$ as x approaches 2, first from the right and then from the left.
3. What happens to the y-value when you try to evaluate $g(x)$ at $x = 2$? Explain.
4. Reconcile your results with those of the preceding example.

Evaluating the Limit of a Function

Let's now consider some examples involving the computation of limits.

2.4 LIMITS

EXAMPLE 1 Let $f(x) = x^3$ and evaluate $\lim_{x \to 2} f(x)$.

Solution The graph of f is shown in Figure 29. You can see that $f(x)$ can be made as close to the number 8 as we please by taking x sufficiently close to 2. Therefore,

$$\lim_{x \to 2} x^3 = 8$$

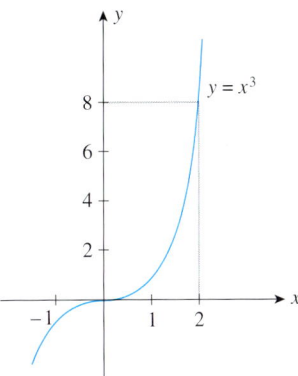

FIGURE 29
$f(x)$ is close to 8 whenever x is close to 2.

EXAMPLE 2 Let

$$g(x) = \begin{cases} x + 2 & \text{if } x \neq 1 \\ 1 & \text{if } x = 1 \end{cases}$$

Evaluate $\lim_{x \to 1} g(x)$.

Solution The domain of g is the set of all real numbers. From the graph of g shown in Figure 30, we see that $g(x)$ can be made as close to 3 as we please by taking x sufficiently close to 1. Therefore,

$$\lim_{x \to 1} g(x) = 3$$

Observe that $g(1) = 1$, which is not equal to the limit of the function g as x approaches 1. [Once again, the value of $g(x)$ at $x = 1$ has no bearing on the existence or value of the limit of g as x approaches 1.]

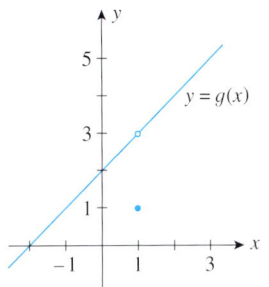

FIGURE 30
$\lim_{x \to 1} g(x) = 3$

EXAMPLE 3 Evaluate the limit of the following functions as x approaches the indicated point.

a. $f(x) = \begin{cases} -1 & \text{if } x < 0 \\ 1 & \text{if } x \geq 0 \end{cases}$; $x = 0$ **b.** $g(x) = \dfrac{1}{x^2}$; $x = 0$

Solution The graphs of the functions f and g are shown in Figure 31.

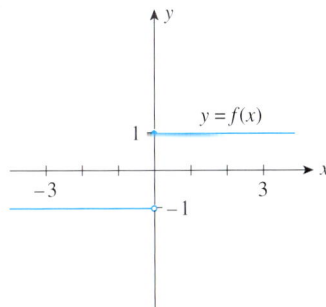

FIGURE 31

(a) $\lim_{x \to 0} f(x)$ does not exist.

(b) $\lim_{x \to 0} g(x)$ does not exist.

a. Referring to Figure 31a, we see that no matter how close x is to zero, $f(x)$ takes on the values 1 or -1, depending on whether x is positive or negative. Thus, there is no *single* real number L that $f(x)$ approaches as x approaches zero. We conclude that the limit of $f(x)$ does *not* exist as x approaches zero.

b. Referring to Figure 31b, we see that as x approaches zero (from either side), $g(x)$ increases without bound and thus does not approach any specific real number. We conclude, accordingly, that the limit of $g(x)$ does *not* exist as x approaches zero.

Explore & Discuss

Consider the graph of the function h shown in the following figure.

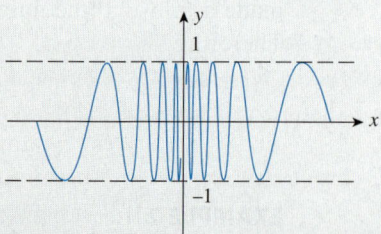

It has the property that as x approaches zero from either the right or the left, the curve oscillates more and more frequently between the lines $y = -1$ and $y = 1$.

1. Explain why $\lim_{x \to 0} h(x)$ does not exist.
2. Compare this function with those in Example 3. More specifically, discuss the different ways the functions fail to have a limit at $x = 0$.

Until now, we have relied on knowing the actual values of a function or the graph of a function near $x = a$ to help us evaluate the limit of the function $f(x)$ as x approaches a. The following properties of limits, which we list without proof, enable us to evaluate limits of functions algebraically.

THEOREM 1
Properties of Limits

Suppose
$$\lim_{x \to a} f(x) = L \quad \text{and} \quad \lim_{x \to a} g(x) = M$$

Then

1. $\lim_{x \to a} [f(x)]^r = [\lim_{x \to a} f(x)]^r = L^r$ r, a positive constant
2. $\lim_{x \to a} cf(x) = c \lim_{x \to a} f(x) = cL$ c, a real number
3. $\lim_{x \to a} [f(x) \pm g(x)] = \lim_{x \to a} f(x) \pm \lim_{x \to a} g(x) = L \pm M$
4. $\lim_{x \to a} [f(x)g(x)] = [\lim_{x \to a} f(x)][\lim_{x \to a} g(x)] = LM$
5. $\lim_{x \to a} \frac{f(x)}{g(x)} = \frac{\lim_{x \to a} f(x)}{\lim_{x \to a} g(x)} = \frac{L}{M}$ Provided that $M \neq 0$

EXAMPLE 4 Use Theorem 1 to evaluate the following limits.

a. $\lim_{x \to 2} x^3$ b. $\lim_{x \to 4} 5x^{3/2}$ c. $\lim_{x \to 1}(5x^4 - 2)$

d. $\lim_{x \to 3} 2x^3 \sqrt{x^2 + 7}$ e. $\lim_{x \to 2} \frac{2x^2 + 1}{x + 1}$

Solution

a. $\lim_{x \to 2} x^3 = [\lim_{x \to 2} x]^3$ Property 1

       $= 2^3 = 8$ $\lim_{x \to 2} x = 2$

b. $\lim\limits_{x \to 4} 5x^{3/2} = 5[\lim\limits_{x \to 4} x^{3/2}]$ Property 2

$= 5(4)^{3/2} = 40$ Property 1

c. $\lim\limits_{x \to 1}(5x^4 - 2) = \lim\limits_{x \to 1} 5x^4 - \lim\limits_{x \to 1} 2$ Property 3

To evaluate $\lim\limits_{x \to 1} 2$, observe that the constant function $g(x) = 2$ has value 2 for all values of x. Therefore, $g(x)$ must approach the limit 2 as x approaches 1 (or any other point for that matter!). Therefore,

$$\lim\limits_{x \to 1}(5x^4 - 2) = 5(1)^4 - 2 = 3$$

d. $\lim\limits_{x \to 3} 2x^3 \sqrt{x^2 + 7} = 2 \lim\limits_{x \to 3} x^3 \sqrt{x^2 + 7}$ Property 2

$= 2 \lim\limits_{x \to 3} x^3 \lim\limits_{x \to 3} \sqrt{x^2 + 7}$ Property 4

$= 2(3)^3 \sqrt{3^2 + 7}$ Properties 1 and 3

$= 2(27)\sqrt{16} = 216$

e. $\lim\limits_{x \to 2} \dfrac{2x^2 + 1}{x + 1} = \dfrac{\lim\limits_{x \to 2}(2x^2 + 1)}{\lim\limits_{x \to 2}(x + 1)}$ Property 5

$= \dfrac{2(2)^2 + 1}{2 + 1} = \dfrac{9}{3} = 3$

Indeterminate Forms

Let's emphasize once again that Property 5 of limits is valid only when the limit of the function that appears in the denominator is not equal to zero at the number in question.

If the numerator has a limit different from zero and the denominator has a limit equal to zero, then the limit of the quotient does not exist at the number in question. This is the case with the function $g(x) = 1/x^2$ in Example 3b. Here, as x approaches zero, the numerator approaches 1 but the denominator approaches zero, so the quotient becomes arbitrarily large. Thus, as was observed earlier, the limit does not exist.

Next, consider

$$\lim\limits_{x \to 2} \dfrac{4(x^2 - 4)}{x - 2}$$

which we evaluated earlier by looking at the values of the function for x near $x = 2$. If we attempt to evaluate this expression by applying Property 5 of limits, we see that both the numerator and denominator of the function

$$\dfrac{4(x^2 - 4)}{x - 2}$$

approach zero as x approaches 2; that is, we obtain an expression of the form 0/0. In this event, we say that the limit of the quotient $f(x)/g(x)$ as x approaches 2 has the **indeterminate form 0/0**.

We need to evaluate limits of this type when we discuss the derivative of a function, a fundamental concept in the study of calculus. As the name suggests, the meaningless expression 0/0 does not provide us with a solution to our problem. One strategy that can be used to solve this type of problem follows.

> **Strategy for Evaluating Indeterminate Forms**
> 1. Replace the given function with an appropriate one that takes on the same values as the original function everywhere except at $x = a$.
> 2. Evaluate the limit of this function as x approaches a.

Examples 5 and 6 illustrate this strategy.

EXAMPLE 5 Evaluate:
$$\lim_{x \to 2} \frac{4(x^2 - 4)}{x - 2}$$

Solution Since both the numerator and the denominator of this expression approach zero as x approaches 2, we have the indeterminate form $0/0$. We rewrite
$$\frac{4(x^2 - 4)}{x - 2} = \frac{4(x - 2)(x + 2)}{(x - 2)}$$
which, upon cancellation of the common factors, is equivalent to $4(x + 2)$, provided that $x \neq 2$. Next, we replace $4(x^2 - 4)/(x - 2)$ with $4(x + 2)$ and find that
$$\lim_{x \to 2} \frac{4(x^2 - 4)}{x - 2} = \lim_{x \to 2} 4(x + 2) = 16$$

The graphs of the functions
$$f(x) = \frac{4(x^2 - 4)}{x - 2} \quad \text{and} \quad g(x) = 4(x + 2)$$
are shown in Figure 32. Observe that the graphs are identical except when $x = 2$. The function g is defined for all values of x and, in particular, its value at $x = 2$ is $g(2) = 4(2 + 2) = 16$. Thus, the point $(2, 16)$ is on the graph of g. However, the function f is not defined at $x = 2$. Since $f(x) = g(x)$ for all values of x except $x = 2$, it follows that the graph of f must look exactly like the graph of g, with the exception that the point $(2, 16)$ is missing from the graph of f. This illustrates graphically why we can evaluate the limit of f by evaluating the limit of the "equivalent" function g.

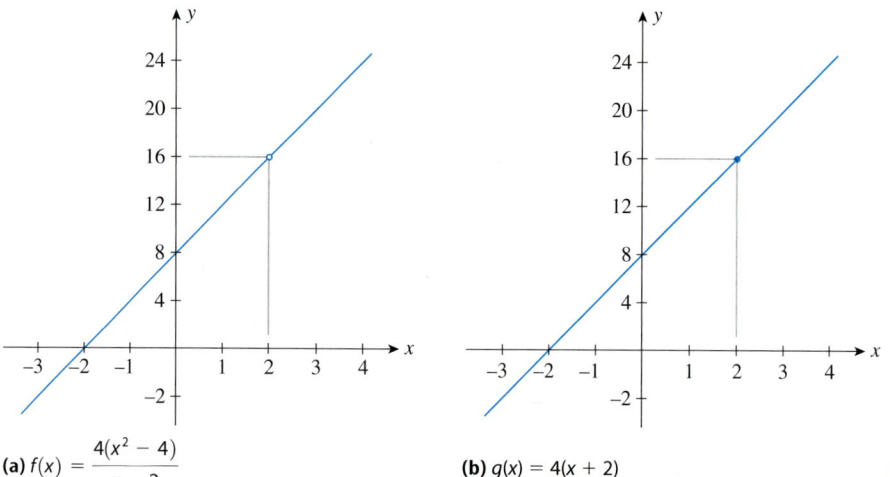

FIGURE 32
The graphs of $f(x)$ and $g(x)$ are identical except at the point $(2, 16)$.

(a) $f(x) = \dfrac{4(x^2 - 4)}{x - 2}$ (b) $g(x) = 4(x + 2)$

Note Notice that the limit in Example 5 is the same limit that we evaluated earlier when we discussed the instantaneous velocity of a maglev at a specified time.

Exploring with TECHNOLOGY

1. Use a graphing utility to plot the graph of
$$f(x) = \frac{4(x^2 - 4)}{x - 2}$$
in the viewing window $[0, 3] \times [0, 20]$. Then use ZOOM and TRACE to find
$$\lim_{x \to 2} \frac{4(x^2 - 4)}{x - 2}$$

2. Use a graphing utility to plot the graph of $g(x) = 4(x + 2)$ in the viewing window $[0, 3] \times [0, 20]$. Then use ZOOM and TRACE to find $\lim_{x \to 2} 4(x + 2)$.

   What happens to the y-value when you try to evaluate $f(x)$ at $x = 2$? Explain.

3. Can you distinguish between the graphs of f and g?

4. Reconcile your results with those of Example 5.

EXAMPLE 6 Evaluate:
$$\lim_{h \to 0} \frac{\sqrt{1 + h} - 1}{h}$$

(x²) The algebra icon is used to indicate that the algebraic computation or problem-solving skill used in the example is reviewed on the referenced page. For instance, in Example 6, if you refer to page 19 you will find a review of the process of rationalizing an algebraic fraction. This is followed by a worked example in which the numerator of the expression $\frac{\sqrt{1 + h} - 1}{h}$ is rationalized.

Solution Letting h approach zero, we obtain the indeterminate form $0/0$. Next, we rationalize the numerator of the quotient by multiplying both the numerator and the denominator by the expression $(\sqrt{1 + h} + 1)$, obtaining

$$\frac{\sqrt{1 + h} - 1}{h} = \frac{(\sqrt{1 + h} - 1)(\sqrt{1 + h} + 1)}{h(\sqrt{1 + h} + 1)} \quad \text{(x²) See page 19.}$$
$$= \frac{1 + h - 1}{h(\sqrt{1 + h} + 1)} \quad (\sqrt{a} - \sqrt{b})(\sqrt{a} + \sqrt{b}) = a - b$$
$$= \frac{h}{h(\sqrt{1 + h} + 1)}$$
$$= \frac{1}{\sqrt{1 + h} + 1}$$

Therefore,
$$\lim_{h \to 0} \frac{\sqrt{1 + h} - 1}{h} = \lim_{h \to 0} \frac{1}{\sqrt{1 + h} + 1} = \frac{1}{\sqrt{1} + 1} = \frac{1}{2}$$

Exploring with TECHNOLOGY

1. Use a graphing utility to plot the graph of
$$g(x) = \frac{\sqrt{1 + x} - 1}{x}$$
in the viewing window $[-1, 2] \times [0, 1]$. Then use ZOOM and TRACE to find
$$\lim_{x \to 0} \frac{\sqrt{1 + x} - 1}{x}$$
by observing the values of $g(x)$ as x approaches zero from the left and from the right.

2. Use a graphing utility to plot the graph of
$$f(x) = \frac{1}{\sqrt{1 + x} + 1}$$
in the viewing window $[-1, 2] \times [0, 1]$. Then use ZOOM and TRACE to find
$$\lim_{x \to 0} \frac{1}{\sqrt{1 + x} + 1}$$
What happens to the y-value when x takes on the value zero? Explain.

3. Can you distinguish between the graphs of f and g?

4. Reconcile your results with those of Example 6.

Limits at Infinity

Up to now, we have studied the limit of a function as x approaches a (finite) number a. There are occasions, however, when we want to know whether $f(x)$ approaches a unique number as x increases without bound. Consider, for example, the function P, giving the number of fruit flies (*Drosophila*) in a container under controlled laboratory conditions, as a function of a time t. The graph of P is shown in Figure 33. You can see from the graph of P that as t increases without bound (gets larger and larger), $P(t)$ approaches the number 400. This number, called the *carrying capacity* of the environment, is determined by the amount of living space and food available, as well as other environmental factors.

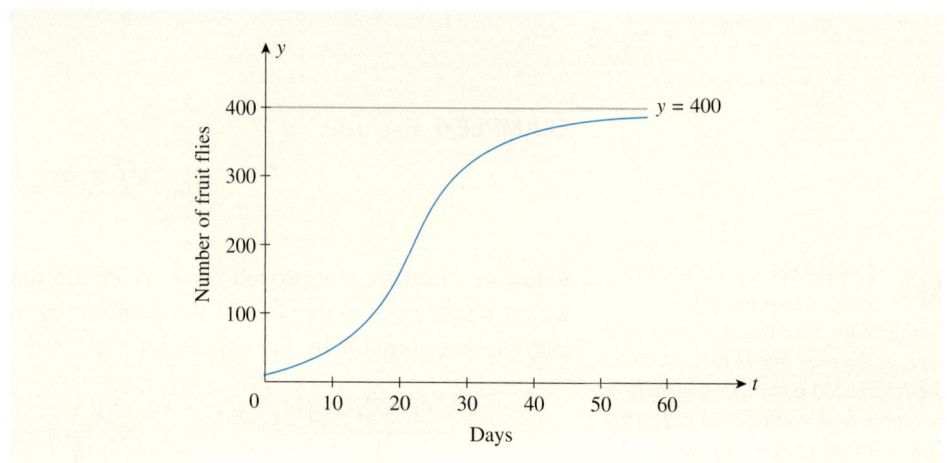

FIGURE 33
The graph of $P(t)$ gives the population of fruit flies in a laboratory experiment.

As another example, suppose we are given the function

$$f(x) = \frac{2x^2}{1 + x^2}$$

and we want to determine what happens to $f(x)$ as x gets larger and larger. Picking the sequence of numbers 1, 2, 5, 10, 100, and 1000 and computing the corresponding values of $f(x)$, we obtain the following table of values:

x	1	2	5	10	100	1000
$f(x)$	1	1.6	1.92	1.98	1.9998	1.999998

From the table, we see that as x gets larger and larger, $f(x)$ gets closer and closer to 2. The graph of the function f shown in Figure 34 confirms this observation. We call the line $y = 2$ a **horizontal asymptote.*** In this situation, we say that the limit of the function $f(x)$ as x increases without bound is 2, written

$$\lim_{x \to \infty} \frac{2x^2}{1 + x^2} = 2$$

FIGURE 34
The graph of
$$y = \frac{2x^2}{1 + x^2}$$
has a horizontal asymptote at $y = 2$.

In the general case, the following definition for a **limit of a function at infinity** is applicable.

---

*We will discuss asymptotes in greater detail in Section 4.3.

Limit of a Function at Infinity

The function f has the limit L as x increases without bound (or, as x approaches infinity), written

$$\lim_{x \to \infty} f(x) = L$$

if $f(x)$ can be made arbitrarily close to L by taking x large enough.

Similarly, the function f has the limit M as x decreases without bound (or as x approaches negative infinity), written

$$\lim_{x \to -\infty} f(x) = M$$

if $f(x)$ can be made arbitrarily close to M by taking x to be negative and sufficiently large in absolute value.

EXAMPLE 7 Let f and g be the functions

$$f(x) = \begin{cases} -1 & \text{if } x < 0 \\ 1 & \text{if } x \geq 0 \end{cases} \quad \text{and} \quad g(x) = \frac{1}{x^2}$$

Evaluate:

a. $\lim_{x \to \infty} f(x)$ and $\lim_{x \to -\infty} f(x)$ **b.** $\lim_{x \to \infty} g(x)$ and $\lim_{x \to -\infty} g(x)$

Solution The graphs of $f(x)$ and $g(x)$ are shown in Figure 35. Referring to the graphs of the respective functions, we see that

a. $\lim_{x \to \infty} f(x) = 1$ and $\lim_{x \to -\infty} f(x) = -1$ **b.** $\lim_{x \to \infty} \frac{1}{x^2} = 0$ and $\lim_{x \to -\infty} \frac{1}{x^2} = 0$

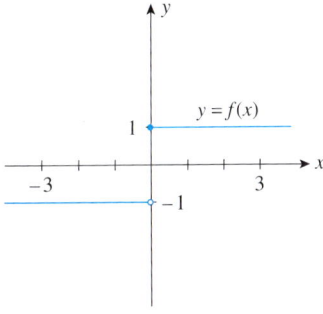
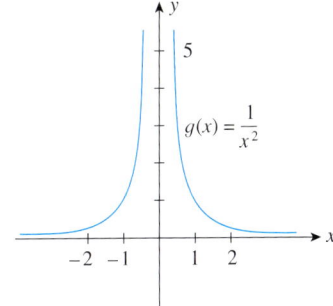

FIGURE 35

(a) $\lim_{x \to \infty} f(x) = 1$ and $\lim_{x \to -\infty} f(x) = -1$ (b) $\lim_{x \to \infty} g(x) = 0$ and $\lim_{x \to -\infty} g(x) = 0$

All the properties of limits listed in Theorem 1 are valid when a is replaced by ∞ or $-\infty$. In addition, we have the following property for the limit at infinity.

THEOREM 2
For all $n > 0$,

$$\lim_{x \to \infty} \frac{1}{x^n} = 0 \quad \text{and} \quad \lim_{x \to -\infty} \frac{1}{x^n} = 0$$

provided that $\frac{1}{x^n}$ is defined.

> **Exploring with TECHNOLOGY**
>
> 1. Use a graphing utility to plot the graphs of
>
> $$y_1 = \frac{1}{x^{0.5}} \quad y_2 = \frac{1}{x} \quad y_3 = \frac{1}{x^{1.5}}$$
>
> in the viewing window $[0, 200] \times [0, 0.5]$. What can you say about $\lim_{x \to \infty} \frac{1}{x^n}$ if $n = 0.5$, $n = 1$, and $n = 1.5$? Are these results predicted by Theorem 2?
>
> 2. Use a graphing utility to plot the graphs of
>
> $$y_1 = \frac{1}{x} \quad \text{and} \quad y_2 = \frac{1}{x^{5/3}}$$
>
> in the viewing window $[-50, 0] \times [-0.5, 0]$. What can you say about $\lim_{x \to -\infty} \frac{1}{x^n}$ if $n = 1$ and $n = \frac{5}{3}$? Are these results predicted by Theorem 2?
>
> *Hint:* To graph y_2, write it in the form y2 = 1/(x^(1/3))^5.

We often use the following technique to evaluate the limit at infinity of a rational function: *Divide the numerator and denominator of the expression by x^n, where n is the highest power present in the denominator of the expression.*

EXAMPLE 8 Evaluate:

$$\lim_{x \to \infty} \frac{x^2 - x + 3}{2x^3 + 1}$$

Solution Since the limits of both the numerator and the denominator do not exist as x approaches infinity, the property pertaining to the limit of a quotient (Property 5) is not applicable. Let's divide the numerator and denominator of the rational expression by x^3, obtaining

$$\lim_{x \to \infty} \frac{x^2 - x + 3}{2x^3 + 1} = \lim_{x \to \infty} \frac{\frac{1}{x} - \frac{1}{x^2} + \frac{3}{x^3}}{2 + \frac{1}{x^3}}$$

$$= \frac{0 - 0 + 0}{2 + 0} = \frac{0}{2} \quad \text{Use Theorem 2.}$$

$$= 0$$

EXAMPLE 9 Let

$$f(x) = \frac{3x^2 + 8x - 4}{2x^2 + 4x - 5}$$

Compute $\lim_{x \to \infty} f(x)$ if it exists.

Solution Again, we see that Property 5 is not applicable. Dividing the numerator and the denominator by x^2, we obtain

$$\lim_{x \to \infty} \frac{3x^2 + 8x - 4}{2x^2 + 4x - 5} = \lim_{x \to \infty} \frac{3 + \dfrac{8}{x} - \dfrac{4}{x^2}}{2 + \dfrac{4}{x} - \dfrac{5}{x^2}}$$

$$= \frac{\lim\limits_{x \to \infty} 3 + 8 \lim\limits_{x \to \infty} \dfrac{1}{x} - 4 \lim\limits_{x \to \infty} \dfrac{1}{x^2}}{\lim\limits_{x \to \infty} 2 + 4 \lim\limits_{x \to \infty} \dfrac{1}{x} - 5 \lim\limits_{x \to \infty} \dfrac{1}{x^2}}$$

$$= \frac{3 + 0 - 0}{2 + 0 - 0} \qquad \text{Use Theorem 2.}$$

$$= \frac{3}{2}$$

EXAMPLE 10 Let $f(x) = \dfrac{2x^3 - 3x^2 + 1}{x^2 + 2x + 4}$ and evaluate:

a. $\lim\limits_{x \to \infty} f(x)$ **b.** $\lim\limits_{x \to -\infty} f(x)$

Solution

a. Dividing the numerator and the denominator of the rational expression by x^2, we obtain

$$\lim_{x \to \infty} \frac{2x^3 - 3x^2 + 1}{x^2 + 2x + 4} = \lim_{x \to \infty} \frac{2x - 3 + \dfrac{1}{x^2}}{1 + \dfrac{2}{x} + \dfrac{4}{x^2}}$$

Since the numerator becomes arbitrarily large, whereas the denominator approaches 1 as x approaches infinity, we see that the quotient $f(x)$ gets larger and larger as x approaches infinity. In other words, the limit does not exist. In this case we indicate this by writing

$$\lim_{x \to \infty} \frac{2x^3 - 3x^2 + 1}{x^2 + 2x + 4} = \infty$$

b. Once again, dividing both the numerator and the denominator by x^2, we obtain

$$\lim_{x \to -\infty} \frac{2x^3 - 3x^2 + 1}{x^2 + 2x + 4} = \lim_{x \to -\infty} \frac{2x - 3 + \dfrac{1}{x^2}}{1 + \dfrac{2}{x} + \dfrac{4}{x^2}}$$

In this case, the numerator becomes arbitrarily large in magnitude but negative in sign, whereas the denominator approaches 1 as x approaches negative infinity. Therefore, the quotient $f(x)$ decreases without bound, and the limit does not exist. In this case we indicate this by writing

$$\lim_{x \to -\infty} \frac{2x^3 - 3x^2 + 1}{x^2 + 2x + 4} = -\infty$$

Example 11 gives an application of the concept of the limit of a function at infinity.

APPLIED EXAMPLE 11 Average Cost Functions Custom Office makes a line of executive desks. It is estimated that the total cost of making x Senior Executive Model desks is $C(x) = 100x + 200{,}000$ dollars per year, so the average cost of making x desks is given by

$$\overline{C}(x) = \frac{C(x)}{x}$$

$$= \frac{100x + 200{,}000}{x} = 100 + \frac{200{,}000}{x}$$

dollars per desk. Evaluate $\lim_{x \to \infty} \overline{C}(x)$, and interpret your results.

Solution

$$\lim_{x \to \infty} \overline{C}(x) = \lim_{x \to \infty} \left(100 + \frac{200{,}000}{x} \right)$$

$$= \lim_{x \to \infty} 100 + \lim_{x \to \infty} \frac{200{,}000}{x} = 100$$

A sketch of the graph of the function $\overline{C}(x)$ appears in Figure 36. The result we obtained is fully expected if we consider its economic implications. Note that as the level of production increases, the fixed cost per desk produced, represented by the term $(200{,}000/x)$, drops steadily. The average cost should approach a constant unit cost of production—$100 in this case.

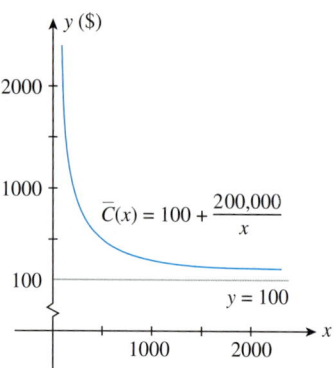

FIGURE 36
As the level of production increases, the average cost approaches $100 per desk.

Explore & Discuss

Consider the graph of the function f depicted in the following figure:

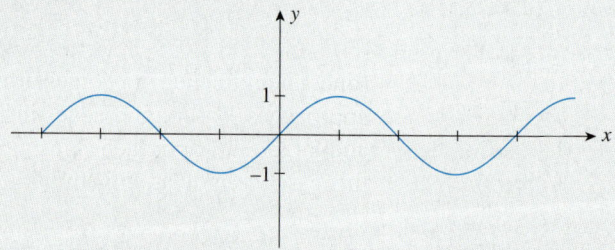

It has the property that the curve oscillates between $y = -1$ and $y = 1$ indefinitely in either direction.

1. Explain why $\lim_{x \to -\infty} f(x)$ and $\lim_{x \to \infty} f(x)$ do not exist.
2. Compare this function with those of Example 10. More specifically, discuss the different ways each function fails to have a limit at infinity or minus infinity.

2.4 Self-Check Exercises

1. Find the indicated limit if it exists.

   a. $\lim_{x \to 3} \dfrac{\sqrt{x^2 + 7} + \sqrt{3x - 5}}{x + 2}$

   b. $\lim_{x \to -1} \dfrac{x^2 - x - 2}{2x^2 - x - 3}$

2. The average cost per disc (in dollars) incurred by Herald Records in pressing x CDs is given by the average cost function

$$\overline{C}(x) = 1.8 + \frac{3000}{x}$$

Evaluate $\lim_{x \to \infty} \overline{C}(x)$, and interpret your result.

Solutions to Self-Check Exercises 2.4 can be found on page 114.

2.4 Concept Questions

1. Explain what is meant by the statement $\lim_{x \to 2} f(x) = 3$.

2. a. If $\lim_{x \to 3} f(x) = 5$, what can you say about $f(3)$? Explain.
   b. If $f(2) = 6$, what can you say about $\lim_{x \to 2} f(x)$? Explain.

3. Evaluate the following and state the property of limits that you use at each step.
   a. $\lim_{x \to 4} \sqrt{x}(2x^2 + 1)$
   b. $\lim_{x \to 1} \left(\dfrac{2x^2 + x + 5}{x^4 + 1} \right)^{3/2}$

4. What is an indeterminate form? Illustrate with an example.

5. Explain in your own words the meaning of $\lim_{x \to \infty} f(x) = L$ and $\lim_{x \to -\infty} f(x) = M$.

2.4 Exercises

In Exercises 1–8, use the graph of the given function f to determine $\lim_{x \to a} f(x)$ at the indicated value of a, if it exists.

1.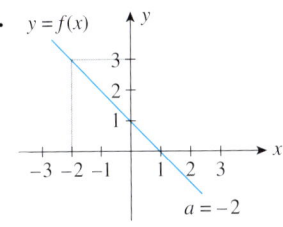

2.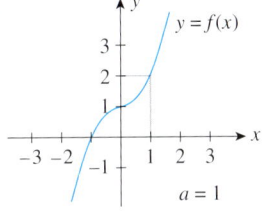

3.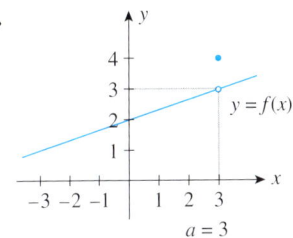

4.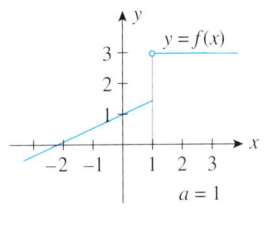

5.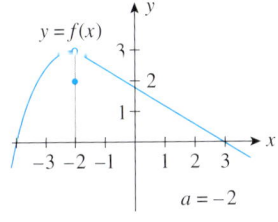

6.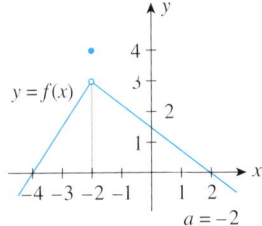

7.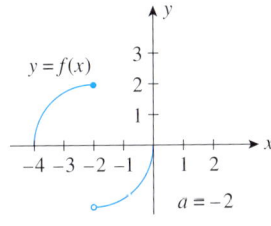

8.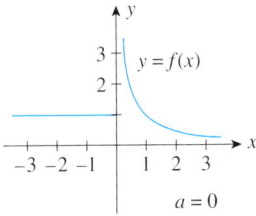

In Exercises 9–16, complete the table by computing $f(x)$ at the given values of x. Use these results to estimate the indicated limit (if it exists).

9. $f(x) = x^2 + 1$; $\lim_{x \to 2} f(x)$

x	1.9	1.99	1.999	2.001	2.01	2.1
$f(x)$						

10. $f(x) = 2x^2 - 1$; $\lim_{x \to 1} f(x)$

x	0.9	0.99	0.999	1.001	1.01	1.1
$f(x)$						

11. $f(x) = \dfrac{|x|}{x}$; $\lim_{x \to 0} f(x)$

x	-0.1	-0.01	-0.001	0.001	0.01	0.1
$f(x)$						

12. $f(x) = \dfrac{|x - 1|}{x - 1}$; $\lim_{x \to 1} f(x)$

x	0.9	0.99	0.999	1.001	1.01	1.1
$f(x)$						

13. $f(x) = \dfrac{1}{(x - 1)^2}$; $\lim_{x \to 1} f(x)$

x	0.9	0.99	0.999	1.001	1.01	1.1
$f(x)$						

14. $f(x) = \dfrac{1}{x - 2}$; $\lim_{x \to 2} f(x)$

x	1.9	1.99	1.999	2.001	2.01	2.1
$f(x)$						

15. $f(x) = \dfrac{x^2 + x - 2}{x - 1}$; $\lim_{x \to 1} f(x)$

x	0.9	0.99	0.999	1.001	1.01	1.1
$f(x)$						

16. $f(x) = \dfrac{x-1}{x-1}$; $\lim_{x \to 1} f(x)$

x	0.9	0.99	0.999	1.001	1.01	1.1
$f(x)$						

In Exercises 17–22, sketch the graph of the function f and evaluate $\lim_{x \to a} f(x)$, if it exists, for the given value of a.

17. $f(x) = \begin{cases} x-1 & \text{if } x \leq 0 \\ -1 & \text{if } x > 0 \end{cases}$ $(a = 0)$

18. $f(x) = \begin{cases} x-1 & \text{if } x \leq 3 \\ -2x+8 & \text{if } x > 3 \end{cases}$ $(a = 3)$

19. $f(x) = \begin{cases} x & \text{if } x < 1 \\ 0 & \text{if } x = 1 \\ -x+2 & \text{if } x > 1 \end{cases}$ $(a = 1)$

20. $f(x) = \begin{cases} -2x+4 & \text{if } x < 1 \\ 4 & \text{if } x = 1 \\ x^2+1 & \text{if } x > 1 \end{cases}$ $(a = 1)$

21. $f(x) = \begin{cases} |x| & \text{if } x \neq 0 \\ 1 & \text{if } x = 0 \end{cases}$ $(a = 0)$

22. $f(x) = \begin{cases} |x-1| & \text{if } x \neq 1 \\ 0 & \text{if } x = 1 \end{cases}$ $(a = 1)$

In Exercises 23–40, find the indicated limit.

23. $\lim_{x \to 2} 3$

24. $\lim_{x \to -2} -3$

25. $\lim_{x \to 3} x$

26. $\lim_{x \to -2} -3x$

27. $\lim_{x \to 1} (1 - 3x^2)$

28. $\lim_{t \to 3} (4t^2 - 2t + 1)$

29. $\lim_{x \to 1} (2x^3 - 3x^2 + x + 2)$

30. $\lim_{x \to 0} (4x^5 - 20x^2 + 2x + 1)$

31. $\lim_{s \to 0} (2s^2 - 1)(2s + 4)$

32. $\lim_{x \to 2} (x^2 + 1)(x^2 - 4)$

33. $\lim_{x \to 2} \dfrac{2x+1}{x+4}$

34. $\lim_{x \to 1} \dfrac{x^3+1}{2x^3+2}$

35. $\lim_{x \to 2} \sqrt{x+2}$

36. $\lim_{x \to -2} \sqrt[3]{5x+2}$

37. $\lim_{x \to -3} \sqrt{2x^4 + x^2}$

38. $\lim_{x \to 2} \sqrt{\dfrac{2x^3+4}{x^2+1}}$

39. $\lim_{x \to -1} \dfrac{\sqrt{x^2+8}}{2x+5}$

40. $\lim_{x \to 3} \dfrac{x\sqrt{x^2+7}}{2x - \sqrt{2x+3}}$

In Exercises 41–48, find the indicated limit given that $\lim_{x \to a} f(x) = 3$ and $\lim_{x \to a} g(x) = 4$.

41. $\lim_{x \to a} [f(x) - g(x)]$

42. $\lim_{x \to a} 2f(x)$

43. $\lim_{x \to a} [4f(x) - 3g(x)]$

44. $\lim_{x \to a} [f(x)g(x)]$

45. $\lim_{x \to a} \sqrt{g(x)}$

46. $\lim_{x \to a} \sqrt[3]{5f(x) + 3g(x)}$

47. $\lim_{x \to a} \dfrac{2f(x) - g(x)}{f(x)g(x)}$

48. $\lim_{x \to a} \dfrac{g(x) - f(x)}{f(x) + \sqrt{g(x)}}$

In Exercises 49–62, find the indicated limit, if it exists.

49. $\lim_{x \to 1} \dfrac{x^2-1}{x-1}$

50. $\lim_{x \to -2} \dfrac{x^2-4}{x+2}$

51. $\lim_{x \to 0} \dfrac{x^2-x}{2x}$

52. $\lim_{x \to 0} \dfrac{2x^2-3x}{x}$

53. $\lim_{x \to -5} \dfrac{x^2-25}{x+5}$

54. $\lim_{b \to -3} \dfrac{b+1}{b+3}$

55. $\lim_{x \to 1} \dfrac{x}{x-1}$

56. $\lim_{x \to 2} \dfrac{x+2}{x-2}$

57. $\lim_{x \to -2} \dfrac{x^2-x-6}{x^2+x-2}$

58. $\lim_{z \to 2} \dfrac{z^3-8}{z-2}$

59. $\lim_{x \to 1} \dfrac{\sqrt{x}-1}{x-1}$

60. $\lim_{x \to 4} \dfrac{x-4}{\sqrt{x}-2}$

Hint: Multiply by $\dfrac{\sqrt{x}+1}{\sqrt{x}+1}$.

Hint: See Exercise 59.

61. $\lim_{x \to 1} \dfrac{2x-2}{x^3+x^2-2x}$

62. $\lim_{x \to -2} \dfrac{4-x^2}{2x^2+x^3}$

In Exercises 63–68, use the graph of the function f to determine $\lim_{x \to \infty} f(x)$ and $\lim_{x \to -\infty} f(x)$, if they exist.

63.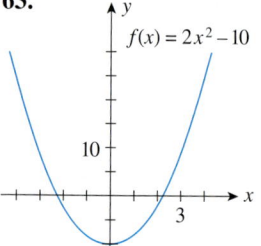

64.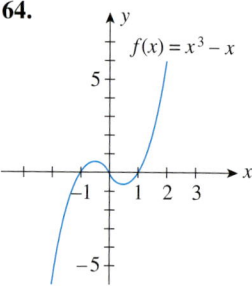

65.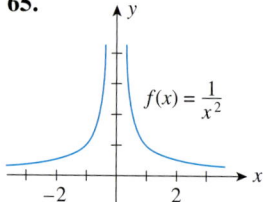

66.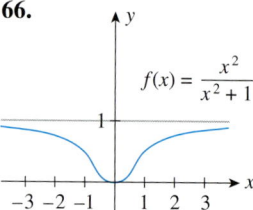

67.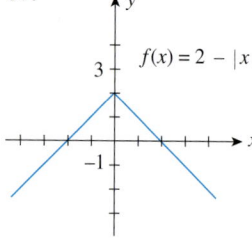

68.

$$f(x) = \begin{cases} \sqrt{-x} & \text{if } x \leq 0 \\ \dfrac{x}{x+1} & \text{if } x > 0 \end{cases}$$

In Exercises 69–72, complete the table by computing f(x) at the given values of x. Use the results to guess at the indicated limits, if they exist.

69. $f(x) = \dfrac{1}{x^2 + 1}$; $\lim\limits_{x \to \infty} f(x)$ and $\lim\limits_{x \to -\infty} f(x)$

x	1	10	100	1000
f(x)				

x	−1	−10	−100	−1000
f(x)				

70. $f(x) = \dfrac{2x}{x+1}$; $\lim\limits_{x \to \infty} f(x)$ and $\lim\limits_{x \to -\infty} f(x)$

x	1	10	100	1000
f(x)				

x	−5	−10	−100	−1000
f(x)				

71. $f(x) = 3x^3 - x^2 + 10$; $\lim\limits_{x \to \infty} f(x)$ and $\lim\limits_{x \to -\infty} f(x)$

x	1	5	10	100	1000
f(x)					

x	−1	−5	−10	−100	−1000
f(x)					

72. $f(x) = \dfrac{|x|}{x}$; $\lim\limits_{x \to \infty} f(x)$ and $\lim\limits_{x \to -\infty} f(x)$

x	1	10	100	−1	−10	−100
f(x)						

In Exercises 73–80, find the indicated limits, if they exist.

73. $\lim\limits_{x \to \infty} \dfrac{3x + 2}{x - 5}$

74. $\lim\limits_{x \to -\infty} \dfrac{4x^2 - 1}{x + 2}$

75. $\lim\limits_{x \to -\infty} \dfrac{3x^3 + x^2 + 1}{x^3 + 1}$

76. $\lim\limits_{x \to \infty} \dfrac{2x^2 + 3x + 1}{x^4 - x^2}$

77. $\lim\limits_{x \to -\infty} \dfrac{x^4 + 1}{x^3 - 1}$

78. $\lim\limits_{x \to \infty} \dfrac{4x^4 - 3x^2 + 1}{2x^4 + x^3 + x^2 + x + 1}$

79. $\lim\limits_{x \to \infty} \dfrac{x^5 - x^3 + x - 1}{x^6 + 2x^2 + 1}$

80. $\lim\limits_{x \to \infty} \dfrac{2x^2 - 1}{x^3 + x^2 + 1}$

81. Toxic Waste A city's main well was recently found to be contaminated with trichloroethylene, a cancer-causing chemical, as a result of an abandoned chemical dump leaching chemicals into the water. A proposal submitted to city council members indicates that the cost, measured in millions of dollars, of removing x% of the toxic pollutant is given by

$$C(x) = \dfrac{0.5x}{100 - x} \quad (0 < x < 100)$$

a. Find the cost of removing 50%, 60%, 70%, 80%, 90%, and 95% of the pollutant.
b. Evaluate

$$\lim\limits_{x \to 100} \dfrac{0.5x}{100 - x}$$

and interpret your result.

82. A Doomsday Situation The population of a certain breed of rabbits introduced into an isolated island is given by

$$P(t) = \dfrac{72}{9 - t} \quad (0 \leq t < 9)$$

where t is measured in months.
a. Find the number of rabbits present in the island initially (at t = 0).
b. Show that the population of rabbits is increasing without bound.
c. Sketch the graph of the function P.
(Comment: This phenomenon is referred to as a doomsday situation.)

83. Average Cost The average cost/disc in dollars incurred by Herald Records in pressing x DVDs is given by the average cost function

$$\overline{C}(x) = 2.2 + \dfrac{2500}{x}$$

Evaluate $\lim\limits_{x \to \infty} \overline{C}(x)$ and interpret your result.

84. Concentration of a Drug in the Bloodstream The concentration of a certain drug in a patient's bloodstream t hr after injection is given by

$$C(t) = \dfrac{0.2t}{t^2 + 1}$$

mg/cm³. Evaluate $\lim\limits_{t \to \infty} C(t)$ and interpret your result.

85. Box-Office Receipts The total worldwide box-office receipts for a long-running blockbuster movie are approximated by the function

$$T(x) = \dfrac{120x^2}{x^2 + 4}$$

where T(x) is measured in millions of dollars and x is the number of months since the movie's release.
a. What are the total box-office receipts after the first month? The second month? The third month?
b. What will the movie gross in the long run (when x is very large)?

86. POPULATION GROWTH A major corporation is building a 4325-acre complex of homes, offices, stores, schools, and churches in the rural community of Glen Cove. As a result of this development, the planners have estimated that Glen Cove's population (in thousands) t years from now will be given by

$$P(t) = \frac{25t^2 + 125t + 200}{t^2 + 5t + 40}$$

a. What is the current population of Glen Cove?
b. What will be the population in the long run?

87. DRIVING COSTS A study of driving costs of 1992 model subcompact (four-cylinder) cars found that the average cost (car payments, gas, insurance, upkeep, and depreciation), measured in cents/mile, is approximated by the function

$$C(x) = \frac{2010}{x^{2.2}} + 17.8$$

where x denotes the number of miles (in thousands) the car is driven in a year.
a. What is the average cost of driving a subcompact car 5000 mi/year? 10,000 mi/year? 15,000 mi/year? 20,000 mi/year? 25,000 mi/year?
b. Use part (a) to sketch the graph of the function C.
c. What happens to the average cost as the number of miles driven increases without bound?

Source: American Automobile Association.

88. PHOTOSYNTHESIS The rate of production R in photosynthesis is related to the light intensity I by the function

$$R(I) = \frac{aI}{b + I^2}$$

where a and b are positive constants.
a. Taking $a = b = 1$, compute $R(I)$ for $I = 0, 1, 2, 3, 4,$ and 5.
b. Evaluate $\lim_{I \to \infty} R(I)$.
c. Use the results of parts (a) and (b) to sketch the graph of R. Interpret your results.

In Exercises 89–94, determine whether the statement is true or false. If it is true, explain why it is true. If it is false, give an example to show why it is false.

89. If $\lim_{x \to a} f(x)$ exists, then f is defined at $x = a$.

90. If $\lim_{x \to 0} f(x) = 4$ and $\lim_{x \to 0} g(x) = 0$, then $\lim_{x \to 0} f(x)g(x) = 0$.

91. If $\lim_{x \to 2} f(x) = 3$ and $\lim_{x \to 2} g(x) = 0$, then $\lim_{x \to 2} [f(x)]/[g(x)]$ does not exist.

92. If $\lim_{x \to 3} f(x) = 0$ and $\lim_{x \to 3} g(x) = 0$, then $\lim_{x \to 3} [f(x)]/[g(x)]$ does not exist.

93. $\lim_{x \to 2} \left(\dfrac{x}{x+1} + \dfrac{3}{x-1} \right) = \lim_{x \to 2} \dfrac{x}{x+1} + \lim_{x \to 2} \dfrac{3}{x-1}$

94. $\lim_{x \to 1} \left(\dfrac{2x}{x-1} - \dfrac{2}{x-1} \right) = \lim_{x \to 1} \dfrac{2x}{x-1} - \lim_{x \to 1} \dfrac{2}{x-1}$

95. SPEED OF A CHEMICAL REACTION Certain proteins, known as enzymes, serve as catalysts for chemical reactions in living things. In 1913 Leonor Michaelis and L. M. Menten discovered the following formula giving the initial speed V (in moles per liter per second) at which the reaction begins in terms of the amount of substrate x (the substance being acted upon, measured in moles per liters) present:

$$V = \frac{ax}{x + b}$$

where a and b are positive constants. Evaluate

$$\lim_{x \to \infty} \frac{ax}{x + b}$$

and interpret your result.

96. Show by means of an example that $\lim_{x \to a} [f(x) + g(x)]$ may exist even though neither $\lim_{x \to a} f(x)$ nor $\lim_{x \to a} g(x)$ exists. Does this example contradict Theorem 1?

97. Show by means of an example that $\lim_{x \to a} [f(x)g(x)]$ may exist even though neither $\lim_{x \to a} f(x)$ nor $\lim_{x \to a} g(x)$ exists. Does this example contradict Theorem 1?

98. Show by means of an example that $\lim_{x \to a} f(x)/g(x)$ may exist even though neither $\lim_{x \to a} f(x)$ nor $\lim_{x \to a} g(x)$ exists. Does this example contradict Theorem 1?

2.4 Solutions to Self-Check Exercises

1. a. $\displaystyle\lim_{x \to 3} \frac{\sqrt{x^2 + 7} + \sqrt{3x - 5}}{x + 2} = \frac{\sqrt{9 + 7} + \sqrt{3(3) - 5}}{3 + 2}$

$= \dfrac{\sqrt{16} + \sqrt{4}}{5}$

$= \dfrac{6}{5}$

b. Letting x approach -1 leads to the indeterminate form $0/0$. Thus, we proceed as follows:

$\displaystyle\lim_{x \to -1} \frac{x^2 - x - 2}{2x^2 - x - 3} = \lim_{x \to -1} \frac{(x+1)(x-2)}{(x+1)(2x-3)}$

$= \displaystyle\lim_{x \to -1} \frac{x - 2}{2x - 3}$ Cancel the common factors.

$= \dfrac{-1 - 2}{2(-1) - 3}$

$= \dfrac{3}{5}$

2. $\lim_{x\to\infty} \overline{C}(x) = \lim_{x\to\infty}\left(1.8 + \frac{3000}{x}\right)$

$= \lim_{x\to\infty} 1.8 + \lim_{x\to\infty} \frac{3000}{x}$

$= 1.8$

Our computation reveals that, as the production of CDs increases "without bound," the average cost drops and approaches a unit cost of $1.80/disc.

USING TECHNOLOGY

Finding the Limit of a Function

A graphing utility can be used to help us find the limit of a function, if it exists, as illustrated in the following examples.

EXAMPLE 1 Let $f(x) = \dfrac{x^3 - 1}{x - 1}$.

a. Plot the graph of f in the viewing window $[-2, 2] \times [0, 4]$.

b. Use ZOOM to find $\lim_{x\to 1} \dfrac{x^3 - 1}{x - 1}$.

c. Verify your result by evaluating the limit algebraically.

Solution

a. The graph of f in the viewing window $[-2, 2] \times [0, 4]$ is shown in Figure T1a.

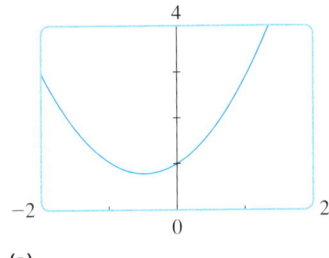
(a)

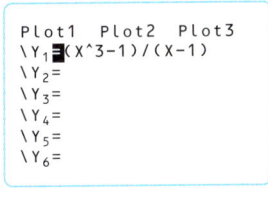
(b)

FIGURE T1
(a) The graph of
$f(x) = \dfrac{x^3 - 1}{x - 1}$
in the viewing window $[-2, 2] \times [0, 4]$;
(b) the TI-83/84 equation screen

b. Using ZOOM IN repeatedly, we see that the y-value approaches 3 as the x-value approaches 1. We conclude, accordingly, that

$$\lim_{x\to 1} \frac{x^3 - 1}{x - 1} = 3$$

c. We compute

$$\lim_{x\to 1} \frac{x^3 - 1}{x - 1} = \lim_{x\to 1} \frac{(x - 1)(x^2 + x + 1)}{x - 1}$$

$$= \lim_{x\to 1} (x^2 + x + 1) = 3$$

Note If you attempt to find the limit in Example 1 by using the evaluation function of your graphing utility to find the value of $f(x)$ when $x = 1$, you will see that the graphing utility does not display the y-value. This happens because $x = 1$ is not in the domain of f.

(continued)

EXAMPLE 2 Use ZOOM to find $\lim_{x \to 0}(1 + x)^{1/x}$.

Solution We first plot the graph of $f(x) = (1 + x)^{1/x}$ in a suitable viewing window. Figure T2a shows a plot of f in the window $[-1, 1] \times [0, 4]$. Using ZOOM-IN repeatedly, we see that $\lim_{x \to 0}(1 + x)^{1/x} \approx 2.71828$.

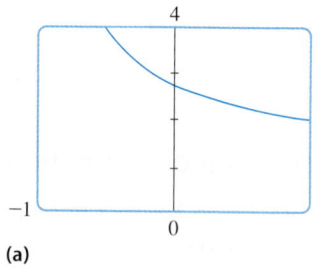
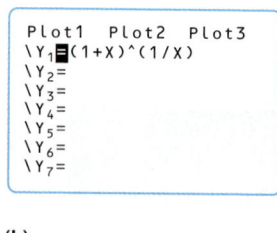

FIGURE T2
(a) The graph of $f(x) = (1 + x)^{1/x}$ in the viewing window $[-1, 1] \times [0, 4]$; (b) the TI-83/84 equation screen

(a) (b)

The limit of $f(x) = (1 + x)^{1/x}$ as x approaches zero, denoted by the letter e, plays a very important role in the study of mathematics and its applications (see Section 5.6). Thus,

$$\lim_{x \to 0}(1 + x)^{1/x} = e$$

where, as we have just seen, $e \approx 2.71828$.

APPLIED EXAMPLE 3 Oxygen Content of a Pond When organic waste is dumped into a pond, the oxidation process that takes place reduces the pond's oxygen content. However, given time, nature will restore the oxygen content to its natural level. Suppose the oxygen content t days after the organic waste has been dumped into the pond is given by

$$f(t) = 100\left(\frac{t^2 + 10t + 100}{t^2 + 20t + 100}\right)$$

percent of its normal level.

a. Plot the graph of f in the viewing window $[0, 200] \times [70, 100]$.
b. What can you say about $f(t)$ when t is very large?
c. Verify your observation in part (b) by evaluating $\lim_{t \to \infty} f(t)$.

Solution

a. The graph of f is shown in Figure T3a.
b. From the graph of f, it appears that $f(t)$ approaches 100 steadily as t gets larger and larger. This observation tells us that eventually the oxygen content of the pond will be restored to its natural level.

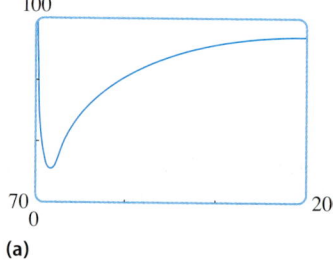
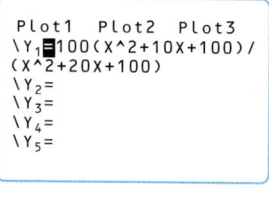

FIGURE T3
(a) The graph of f in the viewing window $[0, 200] \times [70, 100]$; (b) the TI-83/84 equation screen

(a) (b)

c. To verify the observation made in part (b), we compute

$$\lim_{t \to \infty} f(t) = \lim_{t \to \infty} 100 \left(\frac{t^2 + 10t + 100}{t^2 + 20t + 100} \right)$$

$$= 100 \lim_{t \to \infty} \left(\frac{1 + \dfrac{10}{t} + \dfrac{100}{t^2}}{1 + \dfrac{20}{t} + \dfrac{100}{t^2}} \right) = 100$$

TECHNOLOGY EERCISES

In Exercises 1–8, find the indicated limit by first plotting the graph of the function in a suitable viewing window and then using the ZOOM-IN feature of the calculator.

1. $\lim\limits_{x \to 1} \dfrac{2x^3 - 2x^2 + 3x - 3}{x - 1}$

2. $\lim\limits_{x \to -2} \dfrac{2x^3 + 3x^2 - x + 2}{x + 2}$

3. $\lim\limits_{x \to -1} \dfrac{x^3 + 1}{x + 1}$

4. $\lim\limits_{x \to -1} \dfrac{x^4 - 1}{x - 1}$

5. $\lim\limits_{x \to 1} \dfrac{x^3 - x^2 - x + 1}{x^3 - 3x + 2}$

6. $\lim\limits_{x \to 0} \dfrac{\sqrt{x + 1} - 1}{x}$

7. $\lim\limits_{x \to 0} (1 + 2x)^{1/x}$

8. $\lim\limits_{x \to 0} \dfrac{2^x - 1}{x}$

9. Show that $\lim\limits_{x \to 3} \dfrac{2}{x - 3}$ does not exist.

10. Show that $\lim\limits_{x \to 2} \dfrac{x^3 - 2x + 1}{x - 2}$ does not exist.

11. **CITY PLANNING** A major developer is building a 5000-acre complex of homes, offices, stores, schools, and churches in the rural community of Marlboro. As a result of this development, the planners have estimated that Marlboro's population (in thousands) t years from now will be given by

$$P(t) = \frac{25t^2 + 125t + 200}{t^2 + 5t + 40}$$

a. Plot the graph of P in the viewing window $[0, 50] \times [0, 30]$.
b. What will be the population of Marlboro in the long run?
Hint: Find $\lim\limits_{t \to \infty} P(t)$.

12. **AMOUNT OF RAINFALL** The total amount of rain (in inches) after t hr during a rainfall is given by

$$T(t) = \frac{0.8t}{t + 4.1}$$

a. Plot the graph of T in the viewing window $[0, 30] \times [0, 0.8]$.
b. What is the total amount of rain during this rainfall?
Hint: Find $\lim\limits_{t \to \infty} T(t)$.

2.5 One-Sided Limits and Continuity

One-Sided Limits

Consider the function f defined by

$$f(x) = \begin{cases} x - 1 & \text{if } x < 0 \\ x + 1 & \text{if } x \geq 0 \end{cases}$$

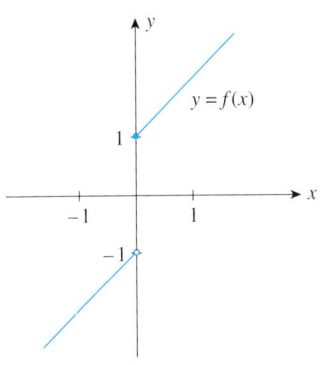

FIGURE 37
The function f does not have a limit as x approaches zero.

From the graph of f shown in Figure 37, we see that the function f does not have a limit as x approaches zero because, no matter how close x is to zero, $f(x)$ takes on values that are close to 1 if x is positive and values that are close to -1 if x is negative. Therefore, $f(x)$ cannot be close to a single number L—no matter how close x is to zero. Now, if we restrict x to be greater than zero (to the right of zero), then we see that $f(x)$ can be made as close to 1 as we please by taking x sufficiently close to zero. In this situation we say that the right-hand limit of f as x approaches zero (from the right) is 1, written

$$\lim_{x \to 0^+} f(x) = 1$$

Similarly, we see that $f(x)$ can be made as close to -1 as we please by taking x sufficiently close to, but to the left of, zero. In this situation we say that the left-hand limit of f as x approaches zero (from the left) is -1, written

$$\lim_{x \to 0^-} f(x) = -1$$

These limits are called **one-sided limits.** More generally, we have the following informal definitions.

One-Sided Limits

The function f has the **right-hand limit** L as x approaches a from the right, written

$$\lim_{x \to a^+} f(x) = L$$

if the values of $f(x)$ can be made as close to L as we please by taking x sufficiently close to (but not equal to) a and to the right of a.

Similarly, the function f has the **left-hand limit** M as x approaches a from the left, written

$$\lim_{x \to a^-} f(x) = M$$

if the values of $f(x)$ can be made as close to M as we please by taking x sufficiently close to (but not equal to) a and to the left of a.

The connection between one-sided limits and the two-sided limit defined earlier is given by the following theorem.

THEOREM 3

Let f be a function that is defined for all values of x close to $x = a$ with the possible exception of a itself. Then

$$\lim_{x \to a} f(x) = L \quad \text{if and only if} \quad \lim_{x \to a^+} f(x) = \lim_{x \to a^-} f(x) = L$$

Thus, the two-sided limit exists if and only if the one-sided limits exist and are equal.

EXAMPLE 1 Let

$$f(x) = \begin{cases} -x & \text{if } x \leq 0 \\ \sqrt{x} & \text{if } x > 0 \end{cases} \quad \text{and} \quad g(x) = \begin{cases} -1 & \text{if } x < 0 \\ 1 & \text{if } x \geq 0 \end{cases}$$

a. Show that $\lim_{x \to 0} f(x)$ exists by studying the one-sided limits of f as x approaches $x = 0$.
b. Show that $\lim_{x \to 0} g(x)$ does not exist.

Solution

a. For $x \leq 0$,

$$\lim_{x \to 0^-} f(x) = \lim_{x \to 0^-} (-x) = 0$$

and for $x > 0$, we find

$$\lim_{x \to 0^+} f(x) = \lim_{x \to 0^+} \sqrt{x} = 0$$

Thus,
$$\lim_{x \to 0} f(x) = 0$$
(Figure 38a).

b. We have
$$\lim_{x \to 0^-} g(x) = -1 \quad \text{and} \quad \lim_{x \to 0^+} g(x) = 1$$
and since these one-sided limits are not equal, we conclude that $\lim_{x \to 0} g(x)$ does not exist (Figure 38b).

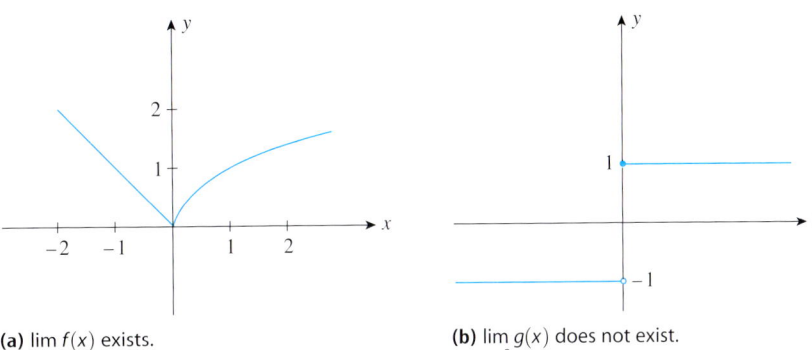

FIGURE 38

(a) $\lim_{x \to 0} f(x)$ exists.

(b) $\lim_{x \to 0} g(x)$ does not exist.

Continuous Functions

Continuous functions will play an important role throughout most of our study of calculus. Loosely speaking, a function is continuous at a point if the graph of the function at that point is devoid of holes, gaps, jumps, or breaks. Consider, for example, the graph of the function f depicted in Figure 39.

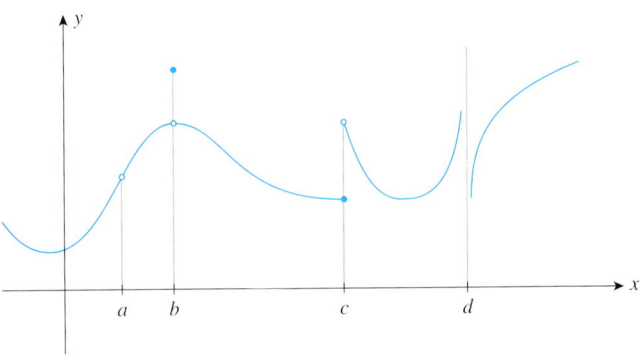

FIGURE 39
The graph of this function is not continuous at $x = a$, $x = b$, $x = c$, and $x = d$.

Let's take a closer look at the behavior of f at or near $x = a$, $x = b$, $x = c$, and $x = d$. First, note that f is not defined at $x = a$; that is, $x = a$ is not in the domain of f, thereby resulting in a "hole" in the graph of f. Next, observe that the value of f at b, $f(b)$, is not equal to the limit of $f(x)$ as x approaches b, resulting in a "jump" in the graph of f at $x = b$. The function f does not have a limit at $x = c$ since the left-hand and right-hand limits of $f(x)$ are not equal, also resulting in a jump in the graph of f at $x = c$. Finally, the limit of f does not exist at $x = d$, resulting in a break in the graph of f. The function f is *discontinuous* at each of these numbers. It is *continuous* everywhere else.

Continuity of a Function at a Number

A function f is **continuous at a number** $x = a$ if the following conditions are satisfied.

1. $f(a)$ is defined.   **2.** $\lim_{x \to a} f(x)$ exists.   **3.** $\lim_{x \to a} f(x) = f(a)$

Thus, a function f is continuous at $x = a$ if the limit of f at $x = a$ exists and has the value $f(a)$. Geometrically, f is continuous at $x = a$ if the proximity of x to a implies the proximity of $f(x)$ to $f(a)$.

If f is not continuous at $x = a$, then f is said to be **discontinuous** at $x = a$. Also, f is **continuous on an interval** if f is continuous at every number in the interval.

Figure 40 depicts the graph of a continuous function on the interval (a, b). Notice that the graph of the function over the stated interval can be sketched without lifting one's pencil from the paper.

FIGURE 40
The graph of f is continuous on the interval (a, b).

EXAMPLE 2 Find the values of x for which each function is continuous.

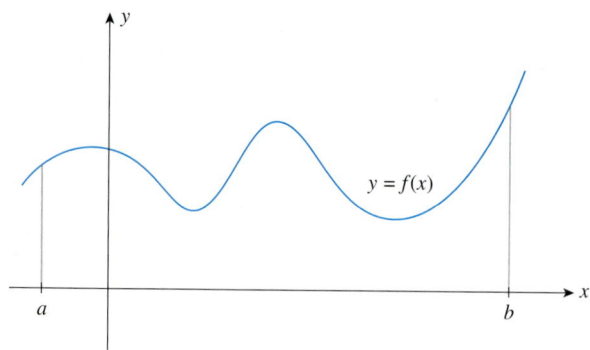

a. $f(x) = x + 2$ **b.** $g(x) = \dfrac{x^2 - 4}{x - 2}$ **c.** $h(x) = \begin{cases} x + 2 & \text{if } x \neq 2 \\ 1 & \text{if } x = 2 \end{cases}$

d. $F(x) = \begin{cases} -1 & \text{if } x < 0 \\ 1 & \text{if } x \geq 0 \end{cases}$ **e.** $G(x) = \begin{cases} \dfrac{1}{x} & \text{if } x > 0 \\ -1 & \text{if } x \leq 0 \end{cases}$

The graph of each function is shown in Figure 41.

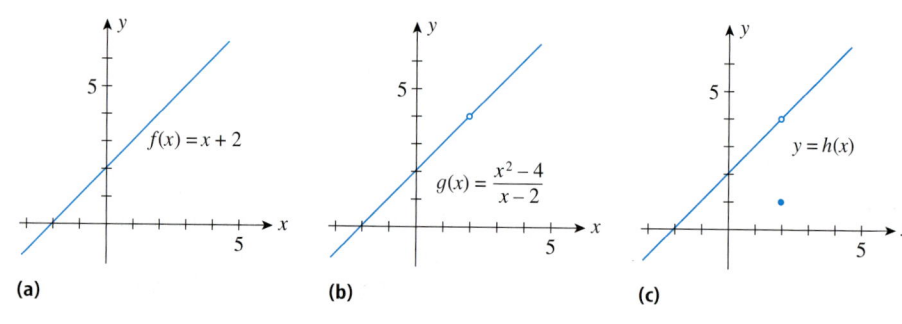

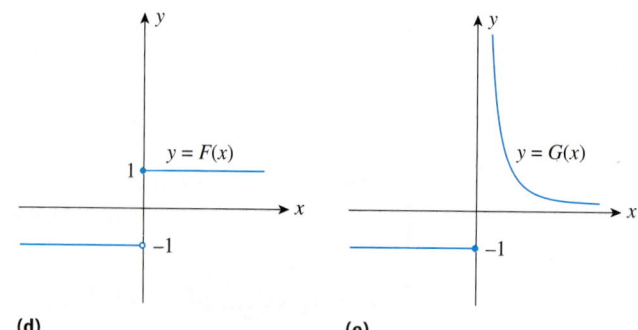

FIGURE 41

Solution

a. The function f is continuous everywhere because the three conditions for continuity are satisfied for all values of x.
b. The function g is discontinuous at $x = 2$ because g is not defined at that number. It is continuous everywhere else.
c. The function h is discontinuous at $x = 2$ because the third condition for continuity is violated; the limit of $h(x)$ as x approaches 2 exists and has the value 4, but this limit is not equal to $h(2) = 1$. It is continuous for all other values of x.
d. The function F is continuous everywhere except at $x = 0$, where the limit of $F(x)$ fails to exist as x approaches zero (see Example 3a, Section 2.4).
e. Since the limit of $G(x)$ does not exist as x approaches zero, we conclude that G fails to be continuous at $x = 0$. The function G is continuous everywhere else. ■

Properties of Continuous Functions

The following properties of continuous functions follow directly from the definition of continuity and the corresponding properties of limits. They are stated without proof.

> **Properties of Continuous Functions**
> 1. The constant function $f(x) = c$ is continuous everywhere.
> 2. The identity function $f(x) = x$ is continuous everywhere.
>
> *If f and g are continuous at $x = a$, then*
>
> 3. $[f(x)]^n$, where n is a real number, is continuous at $x = a$ whenever it is defined at that number.
> 4. $f \pm g$ is continuous at $x = a$.
> 5. fg is continuous at $x = a$.
> 6. f/g is continuous at $x = a$ provided that $g(a) \neq 0$.

Using these properties of continuous functions, we can prove the following results. (A proof is sketched in Exercise 100, page 130.)

> **Continuity of Polynomial and Rational Functions**
> 1. A polynomial function $y = P(x)$ is continuous at every value of x.
> 2. A rational function $R(x) = p(x)/q(x)$ is continuous at every value of x where $q(x) \neq 0$.

EXAMPLE 3 Find the values of x for which each function is continuous.

a. $f(x) = 3x^3 + 2x^2 - x + 10$ b. $g(x) = \dfrac{8x^{10} - 4x + 1}{x^2 + 1}$

c. $h(x) = \dfrac{4x^3 - 3x^2 + 1}{x^2 - 3x + 2}$

Solution

a. The function f is a polynomial function of degree 3, so $f(x)$ is continuous for all values of x.
b. The function g is a rational function. Observe that the denominator of g—namely, $x^2 + 1$—is never equal to zero. Therefore, we conclude that g is continuous for all values of x.

c. The function h is a rational function. In this case, however, the denominator of h is equal to zero at $x = 1$ and $x = 2$, which can be seen by factoring it. Thus,

$$x^2 - 3x + 2 = (x - 2)(x - 1)$$

We therefore conclude that h is continuous everywhere except at $x = 1$ and $x = 2$, where it is discontinuous.

Up to this point, most of the applications we have discussed involved functions that are continuous everywhere. In Example 4, we consider an application from the field of educational psychology that involves a discontinuous function.

APPLIED EXAMPLE 4 Learning Curves Figure 42 depicts the learning curve associated with a certain individual. Beginning with no knowledge of the subject being taught, the individual makes steady progress toward understanding it over the time interval $0 \leq t < t_1$. In this instance, the individual's progress slows as we approach time t_1 because he fails to grasp a particularly difficult concept. All of a sudden, a breakthrough occurs at time t_1, propelling his knowledge of the subject to a higher level. The curve is discontinuous at t_1.

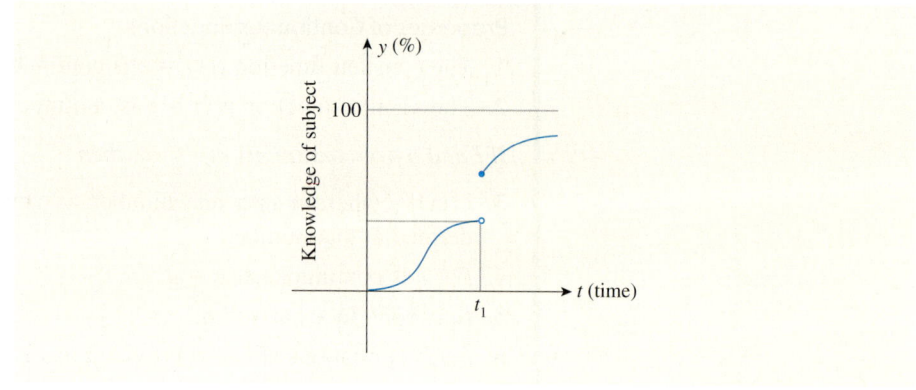

FIGURE 42
A learning curve that is discontinuous at $t = t_1$

Intermediate Value Theorem

Let's look again at our model of the motion of the maglev on a straight stretch of track. We know that the train cannot vanish at any instant of time and it cannot skip portions of the track and reappear someplace else. To put it another way, the train cannot occupy the positions s_1 and s_2 without at least, at some time, occupying an intermediate position (Figure 43).

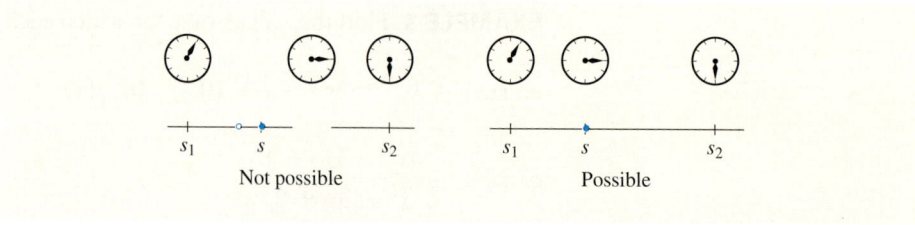

FIGURE 43
The position of the maglev

To state this fact mathematically, recall that the position of the maglev as a function of time is described by

$$f(t) = 4t^2 \quad (0 \leq t \leq 10)$$

Suppose the position of the maglev is s_1 at some time t_1 and its position is s_2 at some time t_2 (Figure 44). Then, if s_3 is any number between s_1 and s_2 giving an intermediate position of the maglev, there must be at least one t_3 between t_1 and t_2 giving the time at which the train is at s_3—that is, $f(t_3) = s_3$.

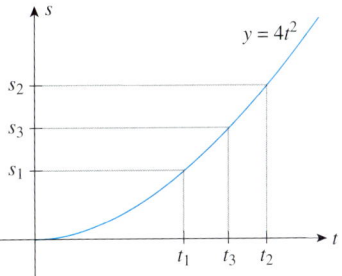

FIGURE 44
If $s_1 \leq s_3 \leq s_2$, then there must be at least one t_3 ($t_1 \leq t_3 \leq t_2$) such that $f(t_3) = s_3$.

This discussion carries the gist of the Intermediate Value Theorem. The proof of this theorem can be found in most advanced calculus texts.

> **THEOREM 4**
> **The Intermediate Value Theorem**
> If f is a continuous function on a closed interval $[a, b]$ and M is any number between $f(a)$ and $f(b)$, then there is at least one number c in $[a, b]$ such that $f(c) = M$ (Figure 45).
>
>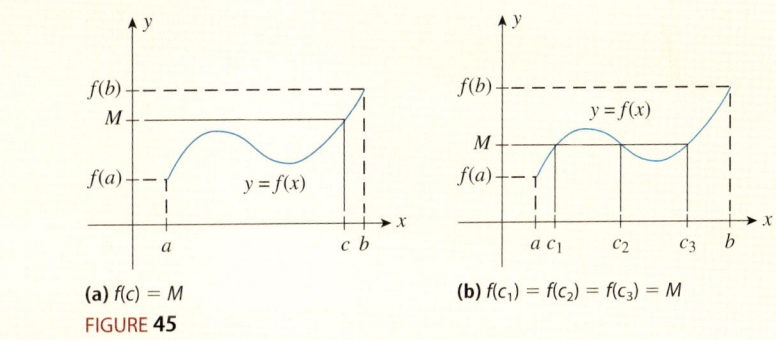
>
> **(a)** $f(c) = M$　　　　　　　　　**(b)** $f(c_1) = f(c_2) = f(c_3) = M$
> **FIGURE 45**

To illustrate the Intermediate Value Theorem, let's look at the example involving the motion of the maglev again (see Figure 27, page 98). Notice that the initial position of the train is $f(0) = 0$ and the position at the end of its test run is $f(10) = 400$. Furthermore, the function f is continuous on $[0, 10]$. So the Intermediate Value Theorem guarantees that if we arbitrarily pick a number between 0 and 400—say, 100—giving the position of the maglev, there must be a $\bar{t}$ (read "t bar") between 0 and 10 at which time the train is at the position $s = 100$. To find the value of $\bar{t}$, we solve the equation $f(\bar{t}) = s$, or

$$4\bar{t}^2 = 100$$

giving $\bar{t} = 5$ (t must lie between 0 and 10).

 It is important to remember when we use Theorem 4 that the function f must be continuous. The conclusion of the Intermediate Value Theorem may not hold if f is not continuous (see Exercise 101, page 130).

The next theorem is an immediate consequence of the Intermediate Value Theorem. It not only tells us when a **zero of a function** f [root of the equation $f(x) = 0$] exists but also provides the basis for a method of approximating it.

> **THEOREM 5**
> **Existence of Zeros of a Continuous Function**
> If f is a continuous function on a closed interval $[a, b]$, and if $f(a)$ and $f(b)$ have opposite signs, then there is at least one solution of the equation $f(x) = 0$ in the interval (a, b) (Figure 46).
>
>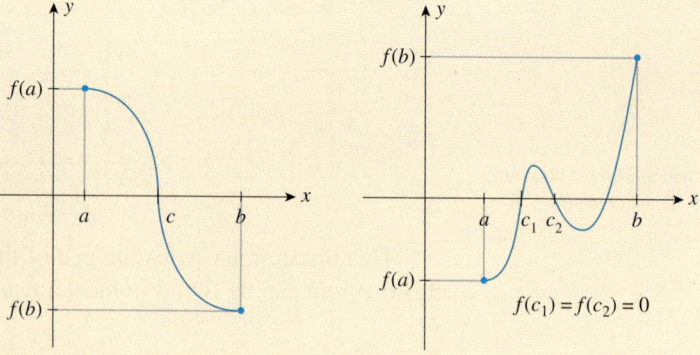
>
> **FIGURE 46**
> If $f(a)$ and $f(b)$ have opposite signs, there must be at least one number c ($a < c < b$) such that $f(c) = 0$.

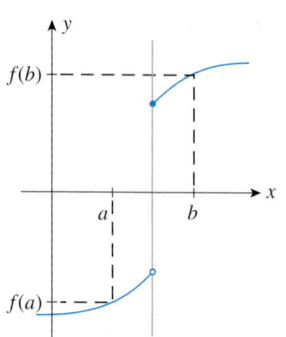

FIGURE 47
$f(a) < 0$ and $f(b) > 0$, but the graph of f does not cross the x-axis between a and b because f is discontinuous.

Geometrically, this property states that if the graph of a continuous function goes from above the x-axis to below the x-axis or vice versa, it must *cross* the x-axis. This is not necessarily true if the function is discontinuous (Figure 47).

EXAMPLE 5 Let $f(x) = x^3 + x + 1$.

a. Show that f is continuous for all values of x.
b. Compute $f(-1)$ and $f(1)$, and use the results to deduce that there must be at least one number $x = c$, where c lies in the interval $(-1, 1)$ and $f(c) = 0$.

Solution

a. The function f is a polynomial function of degree 3 and is therefore continuous everywhere.
b. $f(-1) = (-1)^3 + (-1) + 1 = -1$ and $f(1) = 1^3 + 1 + 1 = 3$

Since $f(-1)$ and $f(1)$ have opposite signs, Theorem 5 tells us that there must be at least one number $x = c$ with $-1 < c < 1$ such that $f(c) = 0$.

The next example shows how the Intermediate Value Theorem can be used to help us find a zero of a function.

EXAMPLE 6 Let $f(x) = x^3 + x - 1$. Since f is a polynomial function, it is continuous everywhere. Observe that $f(0) = -1$ and $f(1) = 1$, so Theorem 5 guarantees the existence of at least one root of the equation $f(x) = 0$ in $(0, 1)$.*

We can locate the root more precisely by using Theorem 5 once again as follows: Evaluate $f(x)$ at the midpoint of $[0, 1]$, obtaining

$$f(0.5) = -0.375$$

---

*It can be shown that f has precisely one zero in $(0, 1)$ (see Exercise 105, Section 4.1).

Because $f(0.5) < 0$ and $f(1) > 0$, Theorem 5 now tells us that the root must lie in $(0.5, 1)$.

Repeat the process: Evaluate $f(x)$ at the midpoint of $[0.5, 1]$, which is

$$\frac{0.5 + 1}{2} = 0.75$$

Thus,

$$f(0.75) \approx 0.1719$$

Because $f(0.5) < 0$ and $f(0.75) > 0$, Theorem 5 tells us that the root is in $(0.5, 0.75)$. This process can be continued. Table 3 summarizes the results of our computations through nine steps.

From Table 3, we see that the root is approximately 0.68, accurate to two decimal places. By continuing the process through a sufficient number of steps, we can obtain as accurate an approximation to the root as we please.

TABLE 3

Step	Root of $f(x) = 0$ Lies in
1	(0, 1)
2	(0.5, 1)
3	(0.5, 0.75)
4	(0.625, 0.75)
5	(0.625, 0.6875)
6	(0.65625, 0.6875)
7	(0.671875, 0.6875)
8	(0.6796875, 0.6875)
9	(0.6796875, 0.6835937)

Note The process of finding the root of $f(x) = 0$ used in Example 6 is called the **method of bisection.** It is crude but effective.

2.5 Self-Check Exercises

1. Evaluate $\lim_{x \to -1^-} f(x)$ and $\lim_{x \to -1^+} f(x)$, where

$$f(x) = \begin{cases} 1 & \text{if } x < -1 \\ 1 + \sqrt{x + 1} & \text{if } x \geq -1 \end{cases}$$

Does $\lim_{x \to -1} f(x)$ exist?

2. Determine the values of x for which the function is discontinuous. At each number where the function is discontinuous, indicate which condition(s) for continuity are violated. Sketch the graph of the function.

a. $f(x) = \begin{cases} -x^2 + 1 & \text{if } x \leq 1 \\ x - 1 & \text{if } x > 1 \end{cases}$

b. $g(x) = \begin{cases} -x + 1 & \text{if } x < -1 \\ 2 & \text{if } -1 < x \leq 1 \\ -x + 3 & \text{if } x > 1 \end{cases}$

Solutions to Self-Check Exercises 2.5 can be found on page 130.

2.5 Concept Questions

1. Explain what is meant by the statement $\lim_{x \to 3^-} f(x) = 2$ and $\lim_{x \to 3^+} f(x) = 4$.

2. Suppose $\lim_{x \to 1^-} f(x) = 3$ and $\lim_{x \to 1^+} f(x) = 4$.
   a. What can you say about $\lim_{x \to 1} f(x)$? Explain.
   b. What can you say about $f(1)$? Explain.

3. Explain what it means for a function f to be continuous (a) at a number a and (b) on an interval I.

4. Determine whether each function f is continuous or discontinuous. Explain your answer.

   a. $f(t)$ gives the altitude of an airplane at time t.
   b. $f(t)$ measures the total amount of rainfall at time t at the Municipal Airport.
   c. $f(s)$ measures the fare as a function of the distance s for taking a cab from Kennedy Airport to downtown Manhattan.
   d. $f(t)$ gives the interest rate charged by a financial institution at time t.

5. Explain the Intermediate Value Theorem in your own words.

2.5 Exercises

In Exercises 1–8, use the graph of the function f to find $\lim_{x \to a^-} f(x)$, $\lim_{x \to a^+} f(x)$, and $\lim_{x \to a} f(x)$ at the indicated value of a, if the limit exists.

1.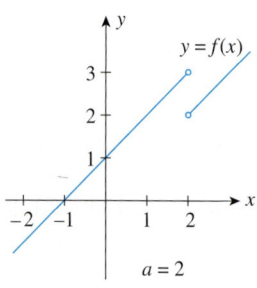
$a = 2$

2.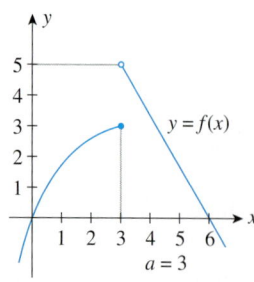
$a = 3$

3.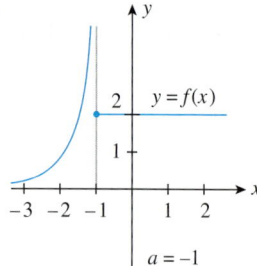
$a = -1$

4.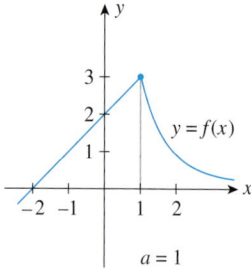
$a = 1$

5.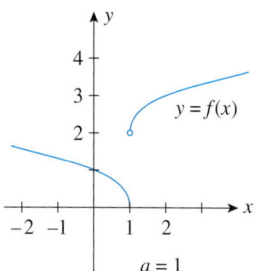
$a = 1$

6.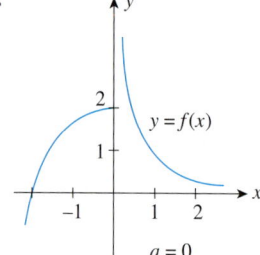
$a = 0$

7.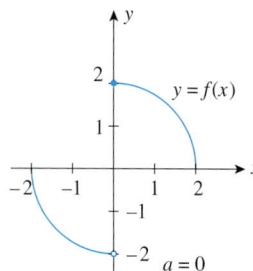
$a = 0$

8.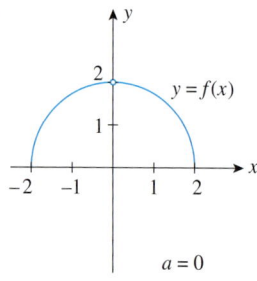
$a = 0$

In Exercises 9–14, refer to the graph of the function f and determine whether each statement is true or false.

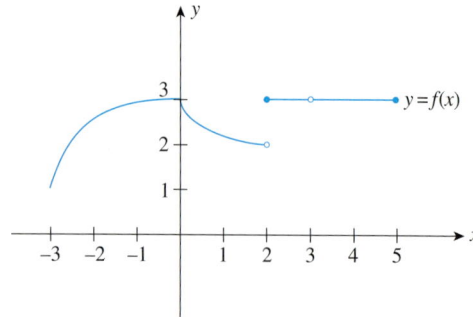

9. $\lim_{x \to -3^+} f(x) = 1$

10. $\lim_{x \to 0} f(x) = f(0)$

11. $\lim_{x \to 2^-} f(x) = 2$

12. $\lim_{x \to 2^+} f(x) = 3$

13. $\lim_{x \to 3} f(x)$ does not exist.

14. $\lim_{x \to 5^-} f(x) = 3$

In Exercises 15–20, refer to the graph of the function f and determine whether each statement is true or false.

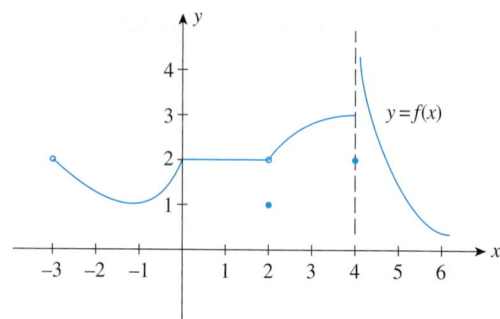

15. $\lim_{x \to -3^+} f(x) = 2$

16. $\lim_{x \to 0} f(x) = 2$

17. $\lim_{x \to 2^-} f(x) = 1$

18. $\lim_{x \to 4^-} f(x) = 3$

19. $\lim_{x \to 4^+} f(x)$ does not exist.

20. $\lim_{x \to 4} f(x) = 2$

In Exercises 21–38, find the indicated one-sided limit, if it exists.

21. $\lim_{x \to 1^+} (2x + 5)$

22. $\lim_{x \to 1^-} (3x - 4)$

23. $\lim_{x \to 2^-} \dfrac{x - 4}{x + 2}$

24. $\lim_{x \to 1^+} \dfrac{x + 2}{x + 1}$

25. $\lim_{x \to 0^+} \dfrac{1}{x}$

26. $\lim_{x \to 0^-} \dfrac{1}{x}$

27. $\lim_{x \to 0^+} \dfrac{x - 1}{x^2 + 1}$

28. $\lim_{x \to 2^+} \dfrac{x + 1}{x^2 - 2x + 3}$

29. $\lim_{x \to 0^+} \sqrt{x}$

30. $\lim_{x \to 2^+} 2\sqrt{x - 2}$

31. $\lim_{x \to -2^+} (2x + \sqrt{2 + x})$

32. $\lim_{x \to -5^+} x(1 + \sqrt{5 + x})$

33. $\lim_{x \to 1^-} \dfrac{1 + x}{1 - x}$

34. $\lim_{x \to 1^+} \dfrac{1 + x}{1 - x}$

35. $\lim_{x \to 2^-} \dfrac{x^2 - 4}{x - 2}$

36. $\lim_{x \to -3^+} \dfrac{\sqrt{x + 3}}{x^2 + 1}$

37. $\lim_{x \to 0^+} f(x)$ and $\lim_{x \to 0^-} f(x)$, where

$$f(x) = \begin{cases} 2x & \text{if } x < 0 \\ x^2 & \text{if } x \geq 0 \end{cases}$$

38. $\lim_{x \to 0^+} f(x)$ and $\lim_{x \to 0^-} f(x)$, where

$$f(x) = \begin{cases} -x + 1 & \text{if } x \leq 0 \\ 2x + 3 & \text{if } x > 0 \end{cases}$$

2.5 ONE-SIDED LIMITS AND CONTINUITY

In Exercises 39–44, determine the values of x, if any, at which each function is discontinuous. At each number where f is discontinuous, state the condition(s) for continuity that are violated.

39.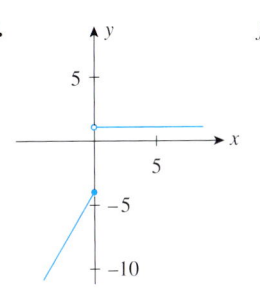
$$f(x) = \begin{cases} 2x - 4 & \text{if } x \leq 0 \\ 1 & \text{if } x > 0 \end{cases}$$

40.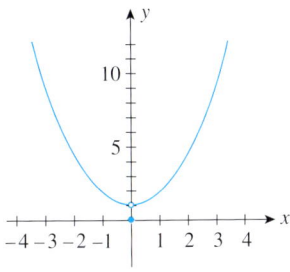
$$f(x) = \begin{cases} x^2 + 1 & \text{if } x \neq 0 \\ 0 & \text{if } x = 0 \end{cases}$$

41.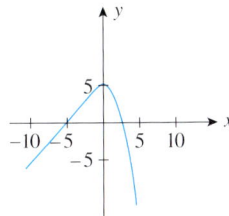
$$f(x) = \begin{cases} x + 5 & \text{if } x \leq 0 \\ -x^2 + 5 & \text{if } x > 0 \end{cases}$$

42.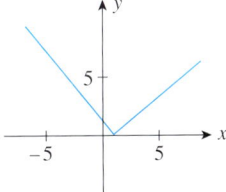
$$f(x) = |x - 1|$$

43.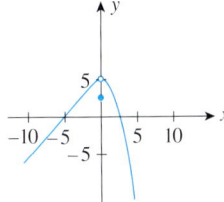
$$f(x) = \begin{cases} x + 5 & \text{if } x < 0 \\ 2 & \text{if } x = 0 \\ -x^2 + 5 & \text{if } x > 0 \end{cases}$$

44.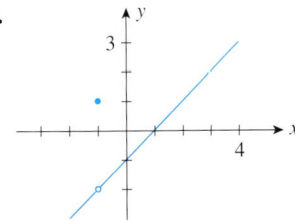
$$f(x) = \begin{cases} \dfrac{x^2 - 1}{x + 1} & \text{if } x \neq -1 \\ 1 & \text{if } x = -1 \end{cases}$$

In Exercises 45–56, find the values of x for which each function is continuous.

45. $f(x) = 2x^2 + x - 1$

46. $f(x) = x^3 - 2x^2 + x - 1$

47. $f(x) = \dfrac{2}{x^2 + 1}$

48. $f(x) = \dfrac{x}{2x^2 + 1}$

49. $f(x) = \dfrac{2}{2x - 1}$

50. $f(x) = \dfrac{x + 1}{x - 1}$

51. $f(x) = \dfrac{2x + 1}{x^2 + x - 2}$

52. $f(x) = \dfrac{x - 1}{x^2 + 2x - 3}$

53. $f(x) = \begin{cases} x & \text{if } x \leq 1 \\ 2x - 1 & \text{if } x > 1 \end{cases}$

54. $f(x) = \begin{cases} -2x + 1 & \text{if } x < 0 \\ x^2 + 1 & \text{if } x \geq 0 \end{cases}$

55. $f(x) = |x + 1|$

56. $f(x) = \dfrac{|x - 1|}{x - 1}$

In Exercises 57–60, determine all values of x at which the function is discontinuous.

57. $f(x) = \dfrac{2x}{x^2 - 1}$

58. $f(x) = \dfrac{1}{(x - 1)(x - 2)}$

59. $f(x) = \dfrac{x^2 - 2x}{x^2 - 3x + 2}$

60. $f(x) = \dfrac{x^2 - 3x + 2}{x^2 - 2x}$

61. **THE POSTAGE FUNCTION** The graph of the "postage function" for 2012,
$$f(x) = \begin{cases} 195 & \text{if } 0 < x < 4 \\ 212 & \text{if } 4 \leq x < 5 \\ \vdots & \\ 348 & \text{if } 12 \leq x < 13 \\ 365 & \text{if } x = 13 \end{cases}$$
where x denotes the weight of a package in ounces and $f(x)$ the postage in cents, is shown in the accompanying figure. Determine the values of x for which f is discontinuous.

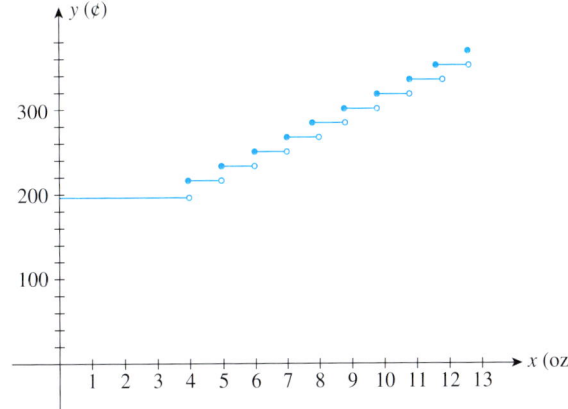

62. **Inventory Control** As part of an optimal inventory policy, the manager of an office supply company orders 500 reams of photocopy paper every 20 days. The accompanying graph shows the *actual* inventory level of paper in an office supply store during the first 60 business days of 2013. Determine the values of t for which the "inventory function" is discontinuous, and give an interpretation of the graph.

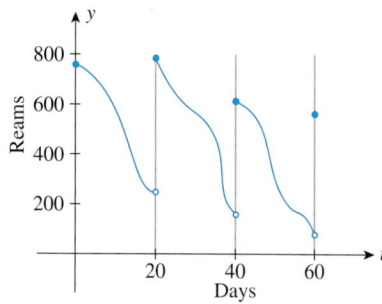

63. **Learning Curves** The following graph describes the progress Michael made in solving a problem correctly during a mathematics quiz. Here, y denotes the percentage of work completed, and x is measured in minutes. Give an interpretation of the graph.

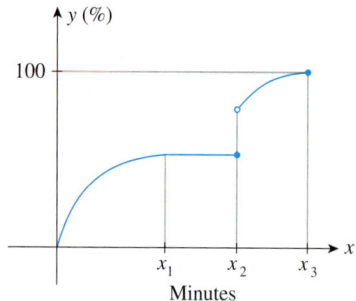

64. **Ailing Financial Institutions** Franklin Savings and Loan acquired two ailing financial institutions in 2012. One of them was acquired at time $t = T_1$, and the other was acquired at time $t = T_2$ ($t = 0$ corresponds to the beginning of 2012). The following graph shows the total amount of money on deposit with Franklin. Explain the significance of the discontinuities of the function at T_1 and T_2.

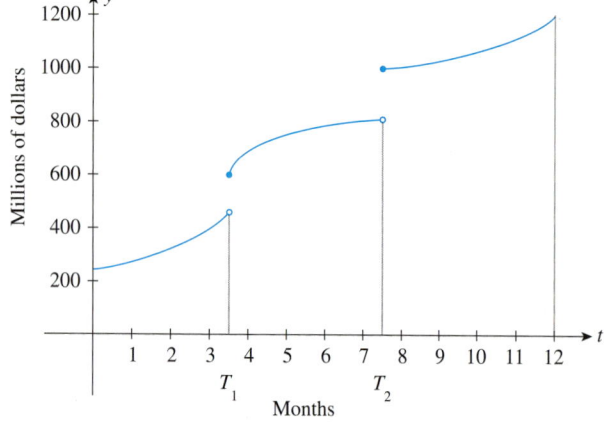

65. **Energy Consumption** The following graph shows the amount of home heating oil remaining in a 200-gal tank over a 120-day period ($t = 0$ corresponds to October 1). Explain why the function is discontinuous at $t = 40, 70, 95$, and 110.

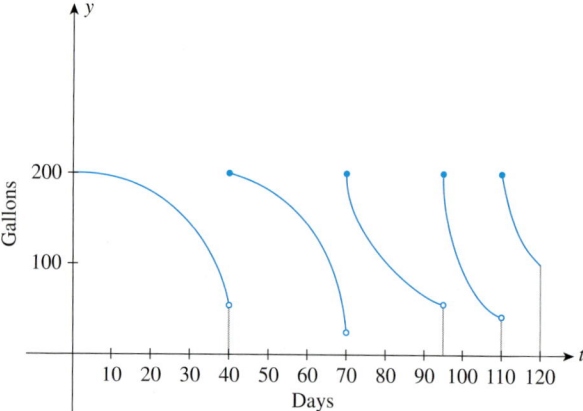

66. **Prime Interest Rate** The function P, whose graph follows, gives the prime rate (the interest rate banks charge their best corporate customers) for a certain country as a function of time for the first 32 weeks in 2012. Determine the values of t for which P is discontinuous, and interpret your results.

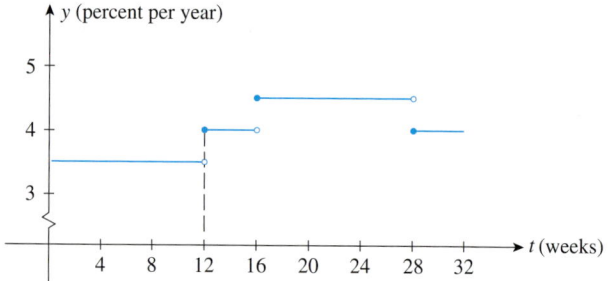

67. **Administration of an Intravenous Solution** A dextrose solution is being administered to a patient intravenously. The 1-liter (L) bottle holding the solution is removed and replaced by another as soon as the contents drop to approximately 5% of the initial (1-L) amount. The rate of discharge is constant, and it takes 6 hr to discharge 95% of the contents of a full bottle. Draw a graph showing the amount of dextrose solution in a bottle in the IV system over a 24-hr period, assuming that we started with a full bottle.

68. **Commissions** The base monthly salary of a salesman working on commission is $22,000. For each $50,000 of sales beyond $100,000, he is paid a $1000 commission. Sketch a graph showing his earnings as a function of the level of his sales x. Determine the values of x for which the function f is discontinuous.

69. **Parking Fees** The fee charged per car in a downtown parking lot is $2.00 for the first half hour and $1.00 for each additional half hour or part thereof, subject to a maximum of $10.00. Derive a function f relating the parking fee to the length of time a car is left in the lot. Sketch the graph of f and determine the values of x for which the function f is discontinuous.

70. COMMODITY PRICES The function that gives the cost of a certain commodity is defined by

$$C(x) = \begin{cases} 5x & \text{if } 0 < x < 10 \\ 4x & \text{if } 10 \le x < 30 \\ 3.5x & \text{if } 30 \le x < 60 \\ 3.25x & \text{if } x \ge 60 \end{cases}$$

where x is the number of pounds of a certain commodity sold and $C(x)$ is measured in dollars. Sketch the graph of the function C and determine the values of x for which the function C is discontinuous.

71. WEISS'S LAW According to Weiss's law of excitation of tissue, the strength S of an electric current is related to the time t the current takes to excite tissue by the formula

$$S(t) = \frac{a}{t} + b \qquad (t > 0)$$

where a and b are positive constants.

a. Evaluate $\lim_{t \to 0^+} S(t)$ and interpret your result.

b. Evaluate $\lim_{t \to \infty} S(t)$ and interpret your result.

(*Note:* The limit in part (b) is called the threshold strength of the current. Why?)

72. ENERGY EXPENDED BY A FISH Suppose a fish swimming a distance of L ft at a speed of v ft/sec relative to the water and against a current flowing at the rate of u ft/sec ($u < v$) expends a total energy given by

$$E(v) = \frac{aLv^3}{v - u}$$

where E is measured in foot-pounds (ft-lb) and a is a constant.

a. Evaluate $\lim_{v \to u^+} E(v)$, and interpret your result.

b. Evaluate $\lim_{v \to \infty} E(v)$, and interpret your result.

73. Let

$$f(x) = \begin{cases} x + 2 & \text{if } x \le 1 \\ kx^2 & \text{if } x > 1 \end{cases}$$

Find the value of k that will make f continuous on $(-\infty, \infty)$.

74. Let

$$f(x) = \begin{cases} \dfrac{x^2 - 4}{x + 2} & \text{if } x \ne -2 \\ k & \text{if } x = -2 \end{cases}$$

For what value of k will f be continuous on $(-\infty, \infty)$?

75. a. Suppose f is continuous at a and g is discontinuous at a. Is the sum $f + g$ necessarily discontinuous at a? Explain.
b. Suppose f and g are both discontinuous at a. Is the sum $f + g$ necessarily discontinuous at a? Explain.

76. a. Suppose f is continuous at a and g is discontinuous at a. Is the product fg necessarily discontinuous at a? Explain.
b. Suppose f and g are both discontinuous at a. Is the product fg necessarily discontinuous at a? Explain.

In Exercises 77–80, (a) show that the function f is continuous for all values of x in the interval $[a, b]$ and (b) prove that f must have at least one zero in the interval (a, b) by showing that $f(a)$ and $f(b)$ have opposite signs.

77. $f(x) = x^2 - 6x + 8$; $a = 1$, $b = 3$

78. $f(x) = 2x^3 - 3x^2 - 36x + 14$; $a = 0$, $b = 1$

79. $f(x) = x^3 - 2x^2 + 3x + 2$; $a = -1$, $b = 1$

80. $f(x) = 2x^{5/3} - 5x^{4/3}$; $a = 14$, $b = 16$

In Exercises 81 and 82, use the Intermediate Value Theorem to show that there exists a number c in the given interval such that $f(c) = M$. Then find its value.

81. $f(x) = x^2 - 4x + 6$ on $[0, 3]$; $M = 4$

82. $f(x) = x^2 - x + 1$ on $[-1, 4]$; $M = 7$

83. Use the method of bisection (see Example 6) to find the root of the equation $x^5 + 2x - 7 = 0$ accurate to two decimal places.

84. Use the method of bisection (see Example 6) to find the root of the equation $x^3 - x + 1 = 0$ accurate to two decimal places.

85. FALLING OBJECT Joan is looking straight out a window of an apartment building at a height of 32 ft from the ground. A boy on the ground throws a tennis ball straight up by the side of the building where the window is located. Suppose the height of the ball (measured in feet) from the ground at time t is $h(t) = 4 + 64t - 16t^2$.

a. Show that $h(0) = 4$ and $h(2) = 68$.
b. Use the Intermediate Value Theorem to conclude that the ball must cross Joan's line of sight at least once.
c. At what time(s) does the ball cross Joan's line of sight? Interpret your results.

86. OXYGEN CONTENT OF A POND The oxygen content t days after organic waste has been dumped into a pond is given by

$$f(t) = 100 \left(\frac{t^2 + 10t + 100}{t^2 + 20t + 100} \right)$$

percent of its normal level.

a. Show that $f(0) = 100$ and $f(10) = 75$.
b. Use the Intermediate Value Theorem to conclude that the oxygen content of the pond must have been at a level of 80% at some time.
c. At what time(s) is the oxygen content at the 80% level?
Hint: Use the quadratic formula.

In Exercises 87–96, determine whether the statement is true or false. If it is true, explain why it is true. If it is false, give an example to show why it is false.

87. If $f(2) = 4$, then $\lim_{x \to 2} f(x) = 4$.

88. If $\lim_{x \to 0} f(x) - 3$, then $f(0) = 3$.

89. If $\lim_{x \to 2^+} f(x) = 3$ and $f(2) = 3$ then $\lim_{x \to 2^-} f(x) = 3$.

90. If $\lim_{x \to 3^-} f(x)$ and $\lim_{x \to 3^+} f(x)$ both exist, then $\lim_{x \to 3} f(x)$ exists.

91. If $f(5)$ is not defined, then $\lim_{x \to 5^-} f(x)$ does not exist.

92. Suppose the function f is defined on the interval $[a, b]$. If $f(a)$ and $f(b)$ have the same sign, then f has no zero in $[a, b]$.

93. If $\lim_{x \to a^-} f(x) = L$ and $\lim_{x \to a^+} f(x) = L$, then $f(a) = L$.

94. If $\lim_{x \to a} f(x) = L$, then $\lim_{x \to a^+} f(x) - \lim_{x \to a^-} f(x) \neq 0$.

95. If f is continuous for all $x \neq 0$ and $f(0) = 0$, then $\lim_{x \to 0} f(x) = 0$.

96. If $\lim_{x \to a} f(x) = L$ and $g(a) = M$, then $\lim_{x \to a} f(x)g(x) = LM$.

97. Is the following statement true or false? Suppose f is continuous on $[a, b]$ and $f(a) < f(b)$. If M is a number that lies outside the interval $[f(a), f(b)]$, then there does not exist a number $a < c < b$ such that $f(c) = M$. Does this contradict the intermediate value theorem?

98. Let $f(x) = \dfrac{x^2}{x^2 + 1}$.

  a. Show that f is continuous for all values of x.
  b. Show that $f(x)$ is nonnegative for all values of x.
  c. Show that f has a zero at $x = 0$. Does this contradict Theorem 5?

99. Let $f(x) = x - \sqrt{1 - x^2}$.

  a. Show that f is continuous for all values of x in the interval $[-1, 1]$.
  b. Show that f has at least one zero in $[-1, 1]$.
  c. Find the zeros of f in $[-1, 1]$ by solving the equation $f(x) = 0$.

100. a. Prove that a polynomial function $y = P(x)$ is continuous at every number x. Follow these steps:
  (i) Use Properties 2 and 3 of continuous functions to establish that the function $g(x) = x^n$, where n is a positive integer, is continuous everywhere.
  (ii) Use Properties 1 and 5 to show that $f(x) = cx^n$, where c is a constant and n is a positive integer, is continuous everywhere.
  (iii) Use Property 4 to complete the proof of the result.
  b. Prove that a rational function $R(x) = p(x)/q(x)$ is continuous at every point x where $q(x) \neq 0$.
  Hint: Use the result of part (a) and Property 6.

101. Show that the conclusion of the intermediate value theorem does not necessarily hold if f is discontinuous on $[a, b]$.

2.5 Solutions to Self-Check Exercises

1. For $x < -1$, $f(x) = 1$, and so
$$\lim_{x \to -1^-} f(x) = \lim_{x \to -1^-} 1 = 1$$

For $x \geq -1$, $f(x) = 1 + \sqrt{x + 1}$, and so
$$\lim_{x \to -1^+} f(x) = \lim_{x \to -1^+} (1 + \sqrt{x + 1}) = 1$$

Since the left-hand and right-hand limits of f exist as x approaches -1 and both are equal to 1, we conclude that
$$\lim_{x \to -1} f(x) = 1$$

2. a. The graph of f follows:

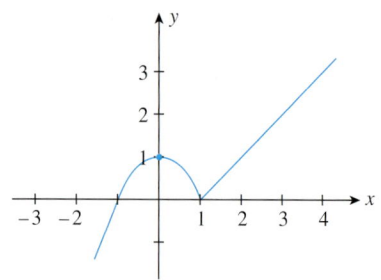

We see that f is continuous everywhere.

b. The graph of g follows:

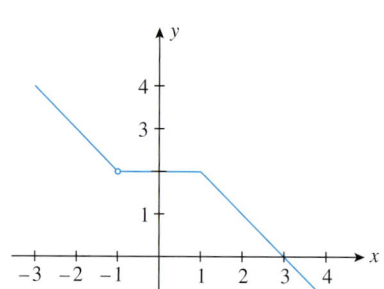

Since g is not defined at $x = -1$, it is discontinuous there. It is continuous everywhere else.

USING TECHNOLOGY

Finding the Points of Discontinuity of a Function

You can very often recognize the points of discontinuity of a function f by examining its graph. For example, Figure T1a shows the graph of $f(x) = x/(x^2 - 1)$ obtained using a graphing utility. It is evident that f is discontinuous at $x = -1$ and $x = 1$. This observation is also borne out by the fact that both points are not in the domain of f.

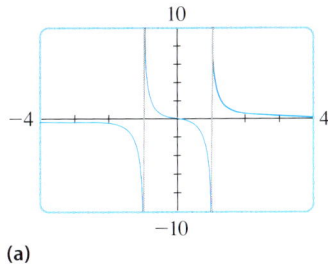
(a)

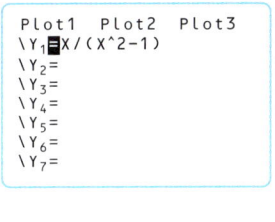
(b)

FIGURE T1
(a) The graph of
$$f(x) = \frac{x}{x^2 - 1}$$
in the viewing window $[-4, 4] \times [-10, 10]$; (b) the TI-83/84 equation screen

Consider the function
$$g(x) = \frac{2x^3 + x^2 - 7x - 6}{x^2 - x - 2}$$

Using a graphing utility, we obtain the graph of g shown in Figure T2a.

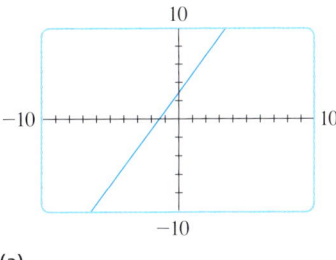
(a)

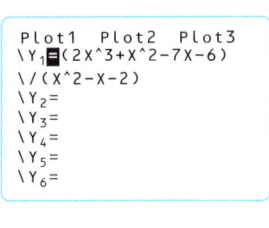
(b)

FIGURE T2
(a) The graph of
$$g(x) = \frac{2x^3 + x^2 - 7x - 6}{x^2 - x - 2}$$
in the standard viewing window; (b) the TI-83/84 equation screen

An examination of this graph does not reveal any points of discontinuity. However, if we factor both the numerator and the denominator of the rational expression, we see that

$$g(x) = \frac{(x + 1)(x - 2)(2x + 3)}{(x + 1)(x - 2)}$$
$$= 2x + 3$$

provided that $x \neq -1$ and $x \neq 2$, so its graph in fact looks like that shown in Figure T3.

This example shows the limitation of the graphing utility and reminds us of the importance of studying functions analytically!

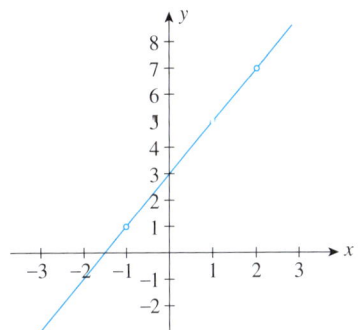

FIGURE T3
The graph of g has holes at $(-1, 1)$ and $(2, 7)$.

Graphing Functions Defined Piecewise

The following example illustrates how to plot the graphs of functions defined in a piecewise manner on a graphing utility.

EXAMPLE 1 Plot the graph of
$$f(x) = \begin{cases} x + 1 & \text{if } x \leq 1 \\ \dfrac{2}{x} & \text{if } x > 1 \end{cases}$$

(continued)

Solution We enter the function

$$y1 = (x + 1)(x \le 1) + (2/x)(x > 1)$$

Figure T4a shows the graph of the function in the viewing window $[-5, 5] \times [-2, 4]$.

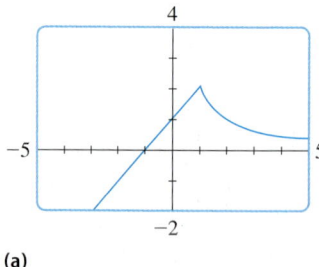
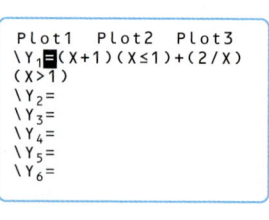

FIGURE T4
(a) The graph of f in the viewing window $[-5, 5] \times [-2, 4]$; (b) the TI-83/84 equation screen

(a)   (b)

APPLIED EXAMPLE 2 TV Viewing Patterns The percent of U.S. households, $P(t)$, watching television during weekdays between the hours of 4 P.M. and 4 A.M. is given by

$$P(t) = \begin{cases} 0.01354t^4 - 0.49375t^3 + 2.58333t^2 + 3.8t + 31.60704 & \text{if } 0 \le t \le 8 \\ 1.35t^2 - 33.05t + 208 & \text{if } 8 < t \le 12 \end{cases}$$

where t is measured in hours, with $t = 0$ corresponding to 4 P.M. Plot the graph of P in the viewing window $[0, 12] \times [0, 80]$.
Source: A. C. Nielsen Co.

Solution We enter the function

$$y1 = (.01354x^4 - .49375x^3 + 2.58333x^2 + 3.8x + 31.60704)(x \ge 0)(x \le 8)$$
$$+ (1.35x^2 - 33.05x + 208)(x > 8)(x \le 12)$$

Figure T5a shows the graph of P.

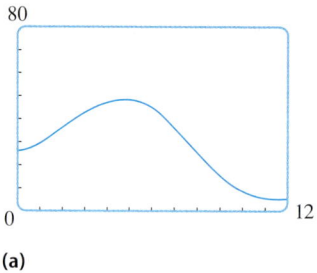

FIGURE T5
(a) The graph of P in the viewing window $[0, 12] \times [0, 80]$; (b) the TI-83/84 equation screen

(a)   (b)

TECHNOLOGY EXERCISES

In Exercises 1–8, plot the graph of f and find the points of discontinuity of f. Then use analytical means to verify your observation and find all numbers where f is discontinuous.

1. $f(x) = \dfrac{2}{x^2 - x}$

2. $f(x) = \dfrac{3}{\sqrt{x}(x + 1)}$

3. $f(x) = \dfrac{6x^3 + x^2 - 2x}{2x^2 - x}$

4. $f(x) = \dfrac{2x^3 - x^2 - 13x - 6}{2x^2 - 5x - 3}$

5. $f(x) = \dfrac{2x^4 - 3x^3 - 2x^2}{2x^2 - 3x - 2}$

6. $f(x) = \dfrac{6x^4 - x^3 + 5x^2 - 1}{6x^2 - x - 1}$

7. $f(x) = \dfrac{x^3 + x^2 - 2x}{x^4 + 2x^3 - x - 2}$
Hint: $x^4 + 2x^3 - x - 2 = (x^3 - 1)(x + 2)$

8. $f(x) = \dfrac{x^3 - x}{x^{4/3} - x + x^{1/3} - 1}$
Hint: $x^{4/3} - x + x^{1/3} - 1 = (x^{1/3} - 1)(x + 1)$
Can you explain why part of the graph is missing?

In Exercises 9 and 10, plot the graph of f in the indicated viewing window.

9. $f(x) = \begin{cases} 2 & \text{if } x \leq 0 \\ \sqrt{4-x^2} & \text{if } x > 0 \end{cases}; [-2, 2] \times [-4, 4]$

10. $f(x) = \begin{cases} -x^2 + x + 2 & \text{if } x \leq 1 \\ 2x^3 - x^2 - 4 & \text{if } x > 1 \end{cases}; [-4, 4] \times [-5, 5]$

11. FLIGHT PATH OF A PLANE The function

$$f(x) = \begin{cases} 0 & \text{if } 0 \leq x < 1 \\ -0.00411523x^3 + 0.0679012x^2 \\ \quad -0.123457x + 0.0596708 & \text{if } 1 \leq x < 10 \\ 1.5 & \text{if } 10 \leq x \leq 100 \end{cases}$$

where both x and $f(x)$ are measured in units of 1000 ft, describes the flight path of a plane taking off from the origin and climbing to an altitude of 15,000 ft. Plot the graph of f to visualize the trajectory of the plane.

12. HOME SHOPPING INDUSTRY According to industry sources, revenue from the home shopping industry for the years since its inception may be approximated by the function

$$R(t) = \begin{cases} -0.03t^3 + 0.25t^2 - 0.12t & \text{if } 0 \leq t \leq 3 \\ 0.57t - 0.63 & \text{if } 3 < t \leq 11 \end{cases}$$

where $R(t)$ measures the revenue in billions of dollars and t is measured in years, with $t = 0$ corresponding to the beginning of 1984. Plot the graph of R.

Source: Paul Kagan Associates.

2.6 The Derivative

An Intuitive Example

We mentioned in Section 2.4 that the problem of finding the *rate of change* of one quantity with respect to another is mathematically equivalent to the problem of finding the *slope of the tangent line* to a curve at a given point on the curve. Before going on to establish this relationship, let's show its plausibility by looking at it from an intuitive point of view.

Consider the motion of the maglev discussed in Section 2.4. Recall that the position of the maglev at any time t is given by

$$s = f(t) = 4t^2 \quad (0 \leq t \leq 30)$$

where s is measured in feet and t in seconds. The graph of the function f is sketched in Figure 48.

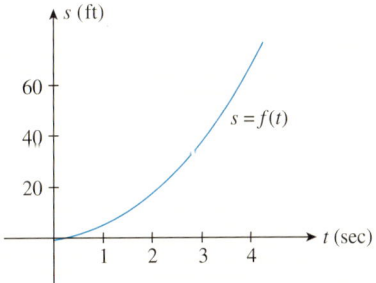

FIGURE 48
Graph showing the position s of a maglev at time t

Observe that the graph of f rises slowly at first but more rapidly as t increases, reflecting the fact that the speed of the maglev is increasing with time. This observation suggests a relationship between the speed of the maglev at any time t and the *steepness* of the curve at the point corresponding to this value of t. Thus, it would appear that we can solve the problem of finding the speed of the maglev at any time if we can find a way to measure the steepness of the curve at any point on the curve.

To discover a yardstick that will measure the steepness of a curve, consider the graph of a function f such as the one shown in Figure 49a. Think of the curve as representing a stretch of roller coaster track (Figure 49b). When the car is at the point

P on the curve, a passenger sitting erect in the car and looking straight ahead will have a line of sight that is parallel to the line T, the tangent to the curve at P.

As Figure 49a suggests, the steepness of the curve—that is, the rate at which y is increasing or decreasing with respect to x—is given by the slope of the tangent line to the graph of f at the point $P(x, f(x))$. But for now we will show how this relationship can be used to estimate the rate of change of a function from its graph.

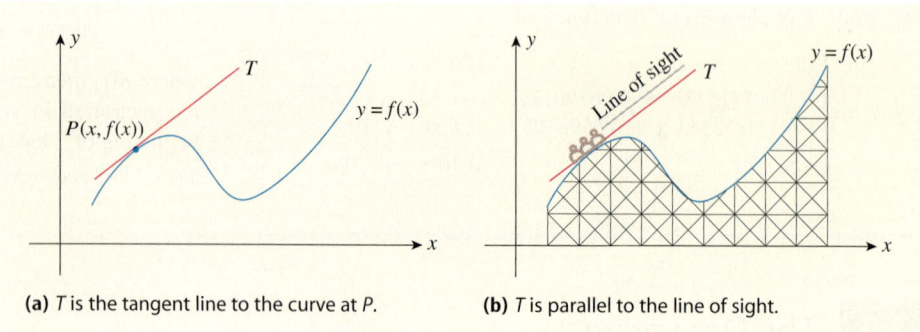

(a) T is the tangent line to the curve at P.

(b) T is parallel to the line of sight.

FIGURE 49

APPLIED EXAMPLE 1 Social Security Beneficiaries The graph of the function $y = N(t)$, shown in Figure 50, gives the number of Social Security beneficiaries from the beginning of 1990 ($t = 0$) through the year 2045 ($t = 55$).

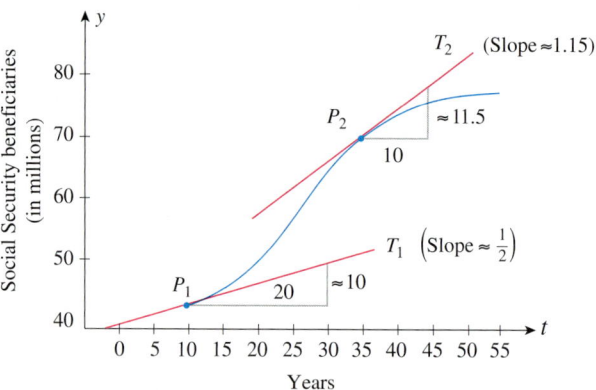

FIGURE 50
The number of Social Security beneficiaries from 1990 through 2045. We can use the slope of the tangent line at the indicated points to estimate the rate at which the number of Social Security beneficiaries will be changing.

Use the graph of $y = N(t)$ to estimate the rate at which the number of Social Security beneficiaries was growing at the beginning of the year 2000 ($t = 10$). How fast will the number be growing at the beginning of 2025 ($t = 35$)? [Assume that the rate of change of the function N at any value of t is given by the slope of the tangent line at the point $P(t, N(t))$.]
Source: Social Security Administration.

Solution From the figure, we see that the slope of the tangent line T_1 to the graph of $y = N(t)$ at $P_1(10, 44.7)$ is approximately 0.5. This tells us that the quantity y is increasing at the rate of $\frac{1}{2}$ unit per unit increase in t, when $t = 10$. In other words, at the beginning of the year 2000, the number of Social Security beneficiaries was increasing at the rate of approximately 0.5 million, or 500,000, per year.

The slope of the tangent line T_2 at $P_2(35, 71.9)$ is approximately 1.15. This tells us that at the beginning of 2025 the number of Social Security beneficiaries will be growing at the rate of approximately 1.15 million, or 1,150,000, per year.

Slope of a Tangent Line

In Example 1, we answered the questions raised by drawing the graph of the function N and estimating the position of the tangent lines. Ideally, however, we would like to solve a problem analytically whenever possible. To do this, we need a precise definition of the slope of a tangent line to a curve.

To define the tangent line to a curve C at a point P on the curve, fix P and let Q be any point on C distinct from P (Figure 51). The straight line passing through P and Q is called a **secant line**.

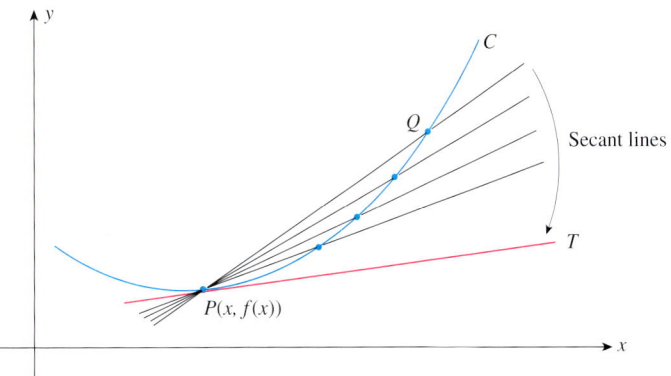

FIGURE 51
As Q approaches P along the curve C, the secant lines approach the tangent line T.

Now, as the point Q is allowed to move toward P along the curve, the secant line through P and Q rotates about the fixed point P and approaches a fixed line through P. This fixed line, which is the limiting position of the secant lines through P and Q as Q approaches P, is the **tangent line to the graph of f** at the point P.

We can describe the process more precisely as follows. Suppose the curve C is the graph of a function f defined by $y = f(x)$. Then the point P is described by $P(x, f(x))$ and the point Q by $Q(x + h, f(x + h))$, where h is some appropriate nonzero number (Figure 52a). Observe that we can make Q approach P along the curve C by letting h approach zero (Figure 52b).

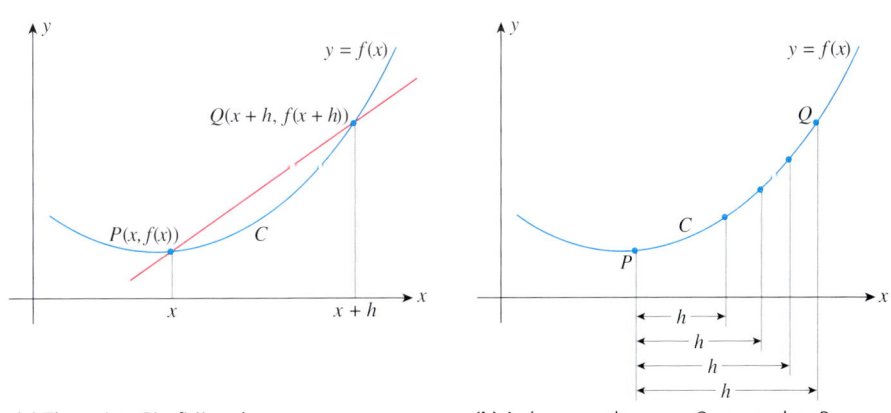

FIGURE 52

(a) The points $P(x, f(x))$ and $Q(x + h, f(x + h))$

(b) As h approaches zero, Q approaches P.

Next, using the formula for the slope of a line, we can write the slope of the secant line passing through $P(x, f(x))$ and $Q(x + h, f(x + h))$ as

$$\frac{f(x + h) - f(x)}{(x + h) - x} = \frac{f(x + h) - f(x)}{h} \tag{5}$$

As we observed earlier, Q approaches P, and therefore the secant line through P and Q approaches the tangent line T as h approaches zero. Consequently, we might expect that the slope of the secant line would approach the slope of the tangent line T as h approaches zero. This leads to the following definition.

> **Slope of a Tangent Line**
>
> The slope of the tangent line to the graph of f at the point $P(x, f(x))$ is given by
>
> $$\lim_{h \to 0} \frac{f(x+h) - f(x)}{h} \tag{6}$$
>
> if it exists.

Rates of Change

We now show that the problem of finding the slope of the tangent line to the graph of a function f at the point $P(x, f(x))$ is mathematically equivalent to the problem of finding the rate of change of f at x. To see this, suppose we are given a function f that describes the relationship between the two quantities x and y—that is, $y = f(x)$. The number $f(x + h) - f(x)$ measures the change in y that corresponds to a change h in x (Figure 53).

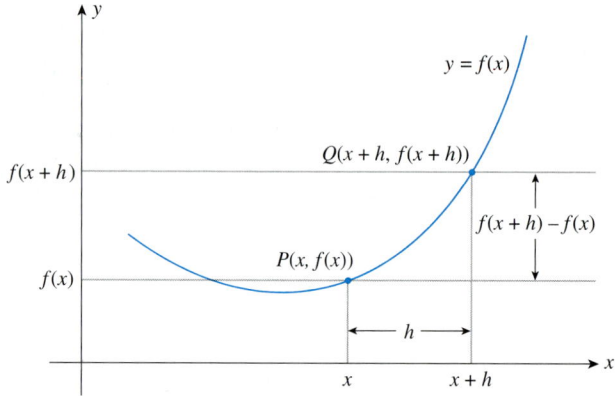

FIGURE 53
$f(x + h) - f(x)$ is the change in y that corresponds to a change h in x.

Then, the **difference quotient**

$$\frac{f(x+h) - f(x)}{h} \tag{7}$$

measures the **average rate of change of y with respect to x** over the interval $[x, x + h]$. For example, if y measures the position of a car at time x, then quotient (7) gives the average velocity of the car over the time interval $[x, x + h]$.

Observe that the difference quotient (7) is the same as (5). We conclude that the difference quotient (7) also measures the slope of the secant line that passes through the two points $P(x, f(x))$ and $Q(x + h, f(x + h))$ lying on the graph of $y = f(x)$. Next, by taking the limit of the difference quotient (7) as h goes to zero—that is, by evaluating

$$\lim_{h \to 0} \frac{f(x+h) - f(x)}{h} \tag{8}$$

we obtain the **rate of change of f at x**. For example, if y measures the position of a car at time x, then the limit (8) gives the velocity of the car at time x. For emphasis, the rate of change of a function f at x is often called the **instantaneous rate of change**

of f at x. This distinguishes it from the average rate of change of f, which is computed over an *interval* $[x, x + h]$ rather than at a *number* x.

Observe that the limit (8) is the same as (6). Therefore, the limit of the difference quotient also measures the slope of the tangent line to the graph of $y = f(x)$ at the point $(x, f(x))$. The following summarizes this discussion.

> **Average and Instantaneous Rates of Change**
>
> The **average rate of change** of f over the interval $[x, x + h]$ or **slope of the secant line** to the graph of f through the points $(x, f(x))$ and $(x + h, f(x + h))$ is
>
> $$\frac{f(x + h) - f(x)}{h} \tag{9}$$
>
> The **instantaneous rate of change** of f at x or **slope of the tangent line** to the graph of f at $(x, f(x))$ is
>
> $$\lim_{h \to 0} \frac{f(x + h) - f(x)}{h} \tag{10}$$

Explore & Discuss

Explain the difference between the average rate of change of a function and the instantaneous rate of change of a function.

The Derivative

The limit (6) or (10), which measures both the slope of the tangent line to the graph of $y = f(x)$ at the point $P(x, f(x))$ and the (instantaneous) rate of change of f at x, is given a special name: the **derivative of f at x.**

> **Derivative of a Function**
>
> The derivative of a function f with respect to x is the function f' (read "f prime"),
>
> $$f'(x) = \lim_{h \to 0} \frac{f(x + h) - f(x)}{h} \tag{11}$$
>
> The domain of f' is the set of all x for which the limit exists.

Thus, the derivative of a function f is a function f' that gives the slope of the tangent line to the graph of f at *any* point $(x, f(x))$ and also the rate of change of f at x (Figure 54).

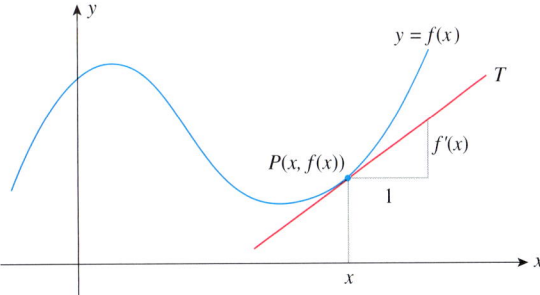

FIGURE 54
The slope of the tangent line at $P(x, f(x))$ is $f'(x)$; f changes at the rate of $f'(x)$ units per unit change in x at x.

Other notations for the derivative of f include:

$D_x f(x)$ Read "d sub x of f of x"

$\dfrac{dy}{dx}$ Read "$d\, y\, d\, x$"

y' Read "y prime"

The last two are used when the rule for f is written in the form $y = f(x)$.

The calculation of the derivative of f is facilitated by using the following four-step process.

> **Four-Step Process for Finding $f'(x)$**
> 1. Compute $f(x + h)$.
> 2. Form the difference $f(x + h) - f(x)$.
> 3. Form the quotient $\dfrac{f(x + h) - f(x)}{h}$.
> 4. Compute the limit $f'(x) = \lim\limits_{h \to 0} \dfrac{f(x + h) - f(x)}{h}$.

EXAMPLE 2 Find the slope of the tangent line to the graph of $f(x) = 3x + 5$ at any point $(x, f(x))$.

Solution The slope of the tangent line at any point on the graph of f is given by the derivative of f at x. To find the derivative, we use the four-step process:

Step 1 $\quad f(x + h) = 3(x + h) + 5 = 3x + 3h + 5$

Step 2 $\quad f(x + h) - f(x) = (3x + 3h + 5) - (3x + 5) = 3h$

Step 3 $\quad \dfrac{f(x + h) - f(x)}{h} = \dfrac{3h}{h} = 3$

Step 4 $\quad f'(x) = \lim\limits_{h \to 0} \dfrac{f(x + h) - f(x)}{h} = \lim\limits_{h \to 0} 3 = 3$

We expect this result since the tangent line to any point on a straight line must coincide with the line itself and therefore must have the same slope as the line. In this case, the graph of f is a straight line with slope 3.

EXAMPLE 3 Let $f(x) = x^2$.

a. Find $f'(x)$.
b. Compute $f'(2)$ and interpret your result.

Solution

a. To find $f'(x)$, we use the four-step process:

Step 1 $\quad f(x + h) = (x + h)^2 = x^2 + 2xh + h^2$

Step 2 $\quad f(x + h) - f(x) = x^2 + 2xh + h^2 - x^2 = 2xh + h^2 = h(2x + h)$

Step 3 $\quad \dfrac{f(x + h) - f(x)}{h} = \dfrac{h(2x + h)}{h} = 2x + h$

Step 4 $\quad f'(x) = \lim\limits_{h \to 0} \dfrac{f(x + h) - f(x)}{h} = \lim\limits_{h \to 0} (2x + h) = 2x$

b. $f'(2) = 2(2) = 4$. This result tells us that the slope of the tangent line to the graph of f at the point $(2, 4)$ is 4. It also tells us that the function f is changing at the rate of 4 units per unit change in x at $x = 2$. The graph of f and the tangent line at $(2, 4)$ are shown in Figure 55.

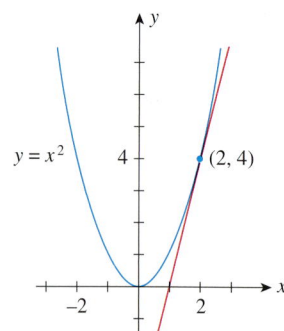

FIGURE 55
The tangent line to the graph of $f(x) = x^2$ at $(2, 4)$

Exploring with TECHNOLOGY

1. Consider the function $f(x) = x^2$ of Example 3. Suppose we want to compute $f'(2)$, using Equation (11). Thus,

$$f'(2) = \lim_{h \to 0} \frac{f(2+h) - f(2)}{h} = \lim_{h \to 0} \frac{(2+h)^2 - 2^2}{h}$$

Use a graphing utility to plot the graph of

$$g(x) = \frac{(2+x)^2 - 4}{x}$$

in the viewing window $[-3, 3] \times [-2, 6]$.

2. Use ZOOM and TRACE to find $\lim_{x \to 0} g(x)$.
3. Explain why the limit found in part 2 is $f'(2)$.

VIDEO **EXAMPLE 4** Let $f(x) = x^2 - 4x$.

a. Find $f'(x)$.
b. Find the point on the graph of f where the tangent line to the curve is horizontal.
c. Sketch the graph of f and the tangent line to the curve at the point found in part (b).
d. What is the rate of change of f at this point?

Solution

a. To find $f'(x)$, we use the four-step process:

Step 1 $f(x + h) = (x + h)^2 - 4(x + h) = x^2 + 2xh + h^2 - 4x - 4h$

Step 2 $f(x + h) - f(x) = x^2 + 2xh + h^2 - 4x - 4h - (x^2 - 4x)$
$= 2xh + h^2 - 4h = h(2x + h - 4)$

Step 3 $\dfrac{f(x + h) - f(x)}{h} = \dfrac{h(2x + h - 4)}{h} = 2x + h - 4$

Step 4 $f'(x) = \lim_{h \to 0} \dfrac{f(x + h) - f(x)}{h} = \lim_{h \to 0} (2x + h - 4) = 2x - 4$

b. At a point on the graph of f where the tangent line to the curve is horizontal and hence has slope zero, the derivative f' of f is zero. Accordingly, to find such point(s), we set $f'(x) = 0$, which gives $2x - 4 = 0$, or $x = 2$. The corresponding value of y is given by $y = f(2) = -4$, and the required point is $(2, -4)$.

c. The graph of f and the tangent line are shown in Figure 56.
d. The rate of change of f at $x = 2$ is zero.

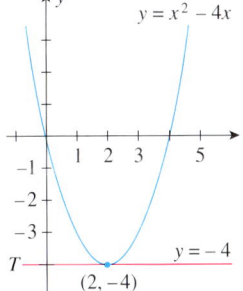

FIGURE 56
The tangent line to the graph of $y = x^2 - 4x$ at $(2, -4)$ is $y = -4$.

Explore & Discuss

Can the tangent line to the graph of a function intersect the graph at more than one point? Explain your answer using illustrations.

EXAMPLE 5 Let $f(x) = \dfrac{1}{x}$.

a. Find $f'(x)$.
b. Find the slope of the tangent line T to the graph of f at the point where $x = 1$.
c. Find an equation of the tangent line T in part (b).

Solution

a. To find $f'(x)$, we use the four-step process:

Step 1 $f(x + h) = \dfrac{1}{x + h}$

Step 2 $f(x + h) - f(x) = \dfrac{1}{x + h} - \dfrac{1}{x} = \dfrac{x - (x + h)}{x(x + h)} = -\dfrac{h}{x(x + h)}$

Step 3 $\dfrac{f(x + h) - f(x)}{h} = -\dfrac{h}{x(x + h)} \cdot \dfrac{1}{h} = -\dfrac{1}{x(x + h)}$ (x^2) See page 18.

Step 4 $f'(x) = \lim\limits_{h \to 0} \dfrac{f(x + h) - f(x)}{h} = \lim\limits_{h \to 0} -\dfrac{1}{x(x + h)} = -\dfrac{1}{x^2}$

b. The slope of the tangent line T to the graph of f where $x = 1$ is given by $f'(1) = -1$.

c. When $x = 1$, $y = f(1) = 1$ and T is tangent to the graph of f at the point $(1, 1)$. From part (b), we know that the slope of T is -1. Thus, an equation of T is

$$y - 1 = -1(x - 1)$$
$$y = -x + 2$$

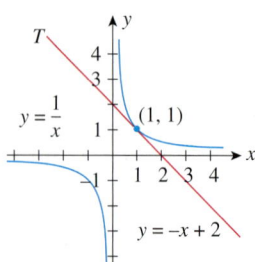

FIGURE 57
The tangent line to the graph of $f(x) = 1/x$ at $(1, 1)$.

(Figure 57).

Exploring with TECHNOLOGY

1. Use the results of Example 5 to draw the graph of $f(x) = 1/x$ and its tangent line at the point $(1, 1)$ by plotting the graphs of $y_1 = 1/x$ and $y_2 = -x + 2$ in the viewing window $[-4, 4] \times [-4, 4]$.

2. Some graphing utilities draw the tangent line to the graph of a function at a given point automatically—you need only specify the function and give the x-coordinate of the point of tangency. If your graphing utility has this feature, verify the result of part 1 without finding an equation of the tangent line.

Explore & Discuss

Consider the following alternative approach to the definition of the derivative of a function: Let h be a positive number and suppose $P(x - h, f(x - h))$ and $Q(x + h, f(x + h))$ are two points on the graph of f.

1. Give a geometric and a physical interpretation of the quotient

$$\dfrac{f(x + h) - f(x - h)}{2h}$$

Make a sketch to illustrate your answer.

2. Give a geometric and a physical interpretation of the limit

$$\lim\limits_{h \to 0} \dfrac{f(x + h) - f(x - h)}{2h}$$

Make a sketch to illustrate your answer.

3. Explain why it makes sense to define

$$f'(x) = \lim\limits_{h \to 0} \dfrac{f(x + h) - f(x - h)}{2h}$$

4. Using the definition given in part 3, formulate a four-step process for finding $f'(x)$ similar to that given on page 138, and use it to find the derivative of $f(x) = x^2$. Compare your answer with that obtained in Example 3 on page 138.

APPLIED EXAMPLE 6 Velocity of a Car Suppose the distance (in feet) covered by a car moving along a straight road t seconds after starting from rest is given by the function $f(t) = 2t^2$ $(0 \leq t \leq 30)$.

a. Calculate the average velocity of the car over the time intervals [22, 23], [22, 22.1], and [22, 22.01].
b. Calculate the (instantaneous) velocity of the car when $t = 22$.
c. Compare the results obtained in part (a) with that obtained in part (b).

Solution

a. We first compute the average velocity (average rate of change of f) over the interval $[t, t + h]$ using Formula (9). We find

$$\frac{f(t + h) - f(t)}{h} = \frac{2(t + h)^2 - 2t^2}{h}$$

$$= \frac{2t^2 + 4th + 2h^2 - 2t^2}{h}$$

$$= 4t + 2h$$

Next, using $t = 22$ and $h = 1$, we find that the average velocity of the car over the time interval [22, 23] is

$$4(22) + 2(1) = 90$$

or 90 feet per second. Similarly, using $t = 22$, $h = 0.1$, and $h = 0.01$, we find that its average velocities over the time intervals [22, 22.1] and [22, 22.01] are 88.2 and 88.02 feet per second, respectively.

b. Using the limit (10), we see that the instantaneous velocity of the car at any time t is given by

$$\lim_{h \to 0} \frac{f(t + h) - f(t)}{h} = \lim_{h \to 0} (4t + 2h) \quad \text{Use the results from part (a).}$$

$$= 4t$$

In particular, the velocity of the car 22 seconds from rest ($t = 22$) is given by

$$v = 4(22)$$

or 88 feet per second.

c. The computations in part (a) show that, as the time intervals over which the average velocity of the car are computed become smaller and smaller, the average velocities over these intervals do approach 88 feet per second, the instantaneous velocity of the car at $t = 22$.

APPLIED EXAMPLE 7 Demand for Tires The management of Titan Tire Company has determined that the weekly demand function of their Super Titan tires is given by

$$p = f(x) = 144 - x^2$$

where p, the price per tire, is measured in dollars and x is measured in units of a thousand (Figure 58).

a. Find the average rate of change in the unit price of a tire if the quantity demanded is between 5000 and 6000 tires, between 5000 and 5100 tires, and between 5000 and 5010 tires.
b. What is the instantaneous rate of change of the unit price when the quantity demanded is 5000 units?

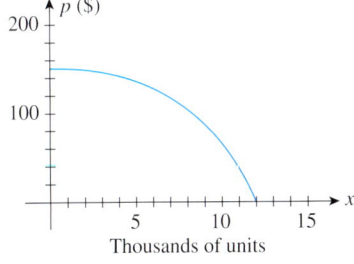

FIGURE 58
The graph of the demand function
$p = 144 - x^2$

Solution

a. The average rate of change of the unit price of a tire if the quantity demanded (in thousands) is between x and $x + h$ is

$$\frac{f(x + h) - f(x)}{h} = \frac{[144 - (x + h)^2] - (144 - x^2)}{h}$$

$$= \frac{144 - x^2 - 2xh - h^2 - 144 + x^2}{h}$$

$$= -2x - h$$

To find the average rate of change of the unit price of a tire when the quantity demanded is between 5000 and 6000 tires (that is, over the interval [5, 6]), we take $x = 5$ and $h = 1$, obtaining

$$-2(5) - 1 = -11$$

or $-\$11$ per 1000 tires. (Remember, x is measured in units of a thousand.) Similarly, taking $h = 0.1$ and $h = 0.01$ with $x = 5$, we find that the average rates of change of the unit price when the quantities demanded are between 5000 and 5100 and between 5000 and 5010 are $-\$10.10$ and $-\$10.01$ per 1000 tires, respectively.

b. The instantaneous rate of change of the unit price of a tire when the quantity demanded is x units is given by

$$\lim_{h \to 0} \frac{f(x + h) - f(x)}{h} = \lim_{h \to 0} (-2x - h) \quad \text{Use the results from part (a).}$$

$$= -2x$$

In particular, the instantaneous rate of change of the unit price per tire when the quantity demanded is 5000 is given by $-2(5)$, or $-\$10$ per 1000 tires. ∎

The derivative of a function provides us with a tool for measuring the rate of change of one quantity with respect to another. Table 4 lists several other applications involving this limit.

TABLE 4
Applications Involving Rate of Change

x stands for	y stands for	$\dfrac{f(a + h) - f(a)}{h}$ measures	$\lim\limits_{h \to 0} \dfrac{f(a + h) - f(a)}{h}$ measures
Time	**Concentration of a drug** in the bloodstream at time x	Average rate of change in the concentration of the drug over the time interval $[a, a + h]$	Instantaneous rate of change in the concentration of the drug in the bloodstream at time $x = a$
Number of items sold	**Revenue** at a sales level of x units	Average rate of change in the revenue when the sales level is between $x = a$ and $x = a + h$	Instantaneous rate of change in the revenue when the sales level is a units
Time	**Volume of sales** at time x	Average rate of change in the volume of sales over the time interval $[a, a + h]$	Instantaneous rate of change in the volume of sales at time $x = a$
Time	**Population** of *Drosophila* (fruit flies) at time x	Average rate of growth of the fruit fly population over the time interval $[a, a + h]$	Instantaneous rate of change of the fruit fly population at time $x = a$
Temperature in a chemical reaction	**Amount of product formed in the chemical reaction** when the temperature is x degrees	Average rate of formation of chemical product over the temperature range $[a, a + h]$	Instantaneous rate of formation of chemical product when the temperature is a degrees

Differentiability and Continuity

In practical applications one encounters continuous functions that fail to be **differentiable**—that is, do not have a derivative—at certain values in the domain of the function f. It can be shown that a continuous function f fails to be differentiable at $x = a$ when the graph of f makes an abrupt change of direction at $(a, f(a))$. We call such a point a "corner." A function also fails to be differentiable at a point where the tangent line is vertical, since the slope of a vertical line is undefined. These cases are illustrated in Figure 59.

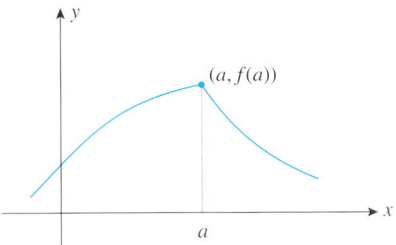
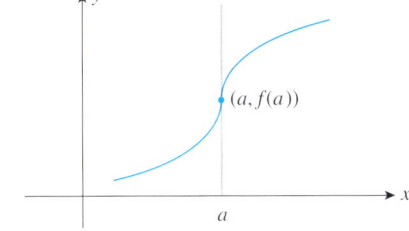

(a) The graph makes an abrupt change of direction at $x = a$.

(b) The slope at $x = a$ is undefined.

FIGURE 59

The next example illustrates a function that is not differentiable at a point.

APPLIED EXAMPLE 8 Wages Mary works at the B&O department store, where, on a weekday, she is paid $8 an hour for the first 8 hours and $12 an hour for overtime. The function

$$f(x) = \begin{cases} 8x & \text{if } 0 \leq x \leq 8 \\ 12x - 32 & \text{if } 8 < x \end{cases}$$

gives Mary's earnings on a weekday in which she worked x hours. Sketch the graph of the function f, and explain why it is not differentiable at $x = 8$.

Solution The graph of f is shown in Figure 60. Observe that the graph of f has a corner at $x = 8$ and consequently is not differentiable at $x = 8$.

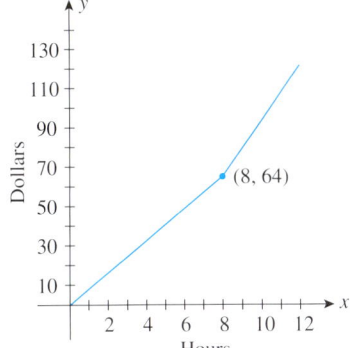

FIGURE 60
The function f is not differentiable at (8, 64).

We close this section by mentioning the connection between the continuity and the differentiability of a function at a given value $x = a$ in the domain of f. By reexamining the function of Example 8, it becomes clear that f is continuous everywhere and, in particular, when $x = 8$. This shows that in general the continuity of a function at $x = a$ does not necessarily imply the differentiability of the function at that number. The converse, however, is true: If a function f is differentiable at $x = a$, then it is continuous there.

> **Differentiability and Continuity**
> If a function is differentiable at $x = a$, then it is continuous at $x = a$.

For a proof of this result, see Exercise 62, page 149.

Explore & Discuss

Suppose a function f is differentiable at $x = a$. Can there be two tangent lines to the graphs of f at the point $(a, f(a))$? Explain your answer.

Exploring with TECHNOLOGY

1. Use a graphing utility to plot the graph of $f(x) = x^{1/3}$ in the viewing window $[-2, 2] \times [-2, 2]$.
2. Use a graphing utility to draw the tangent line to the graph of f at the point $(0, 0)$. Can you explain why the process breaks down?

EXAMPLE 9 Figure 61 depicts a portion of the graph of a function. Explain why the function fails to be differentiable at each of the numbers $x = a, b, c, d, e, f,$ and g.

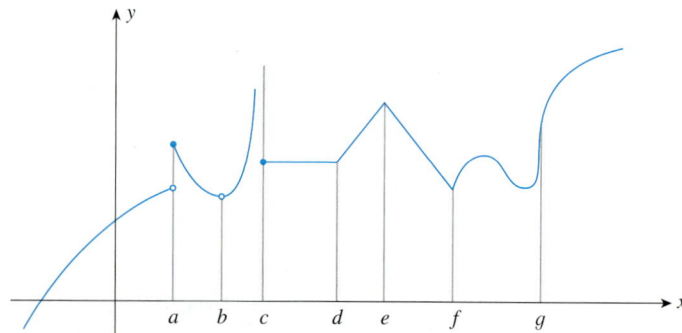

FIGURE 61
The graph of this function is not differentiable at the numbers a–g.

Solution The function fails to be differentiable at $x = a, b,$ and c because it is discontinuous at each of these numbers. The derivative of the function does not exist at $x = d, e,$ and f because it has a kink at each point on the graph corresponding to these numbers. Finally, the function is not differentiable at $x = g$ because the tangent line is vertical at the corresponding point on the graph.

2.6 Self-Check Exercises

1. Let $f(x) = -x^2 - 2x + 3$.
   a. Find the derivative f' of f, using the definition of the derivative.
   b. Find the slope of the tangent line to the graph of f at the point $(0, 3)$.
   c. Find the rate of change of f when $x = 0$.
   d. Find an equation of the tangent line to the graph of f at the point $(0, 3)$.
   e. Sketch the graph of f and the tangent line to the curve at the point $(0, 3)$.

2. The losses (in millions of dollars) due to bad loans extended chiefly in agriculture, real estate, shipping, and energy by the Franklin Bank are estimated to be

$$A = f(t) = -t^2 + 10t + 30 \quad (0 \leq t \leq 10)$$

where t is the time in years ($t = 0$ corresponds to the beginning of 2005). How fast were the losses mounting at the beginning of 2008? At the beginning of 2010? At the beginning of 2012?

Solutions to Self-Check Exercises 2.6 can be found on page 149.

2.6 Concept Questions

For Questions 1 and 2, refer to the following figure.

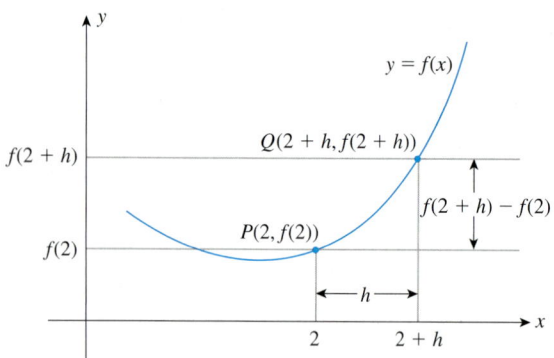

1. Let $P(2, f(2))$ and $Q(2 + h, f(2 + h))$ be points on the graph of a function f.
   a. Find an expression for the slope of the secant line passing through P and Q.
   b. Find an expression for the slope of the tangent line passing through P.

2. Refer to Question 1.
   a. Find an expression for the average rate of change of f over the interval $[2, 2 + h]$.
   b. Find an expression for the instantaneous rate of change of f at 2.
   c. Compare your answers for parts (a) and (b) with those of Question 1.

3. a. Give a geometric and a physical interpretation of the expression
   $$\frac{f(x + h) - f(x)}{h}$$
   b. Give a geometric and a physical interpretation of the expression
   $$\lim_{h \to 0} \frac{f(x + h) - f(x)}{h}$$

4. Under what conditions does a function fail to have a derivative at a number? Illustrate your answer with sketches.

2.6 Exercises

1. **AVERAGE WEIGHT OF AN INFANT** The following graph shows the weight measurements of the average infant from the time of birth ($t = 0$) through age 2 ($t = 24$). By computing the slopes of the respective tangent lines, estimate the rate of change of the average infant's weight when $t = 3$ and when $t = 18$. What is the average rate of change in the average infant's weight over the first year of life?

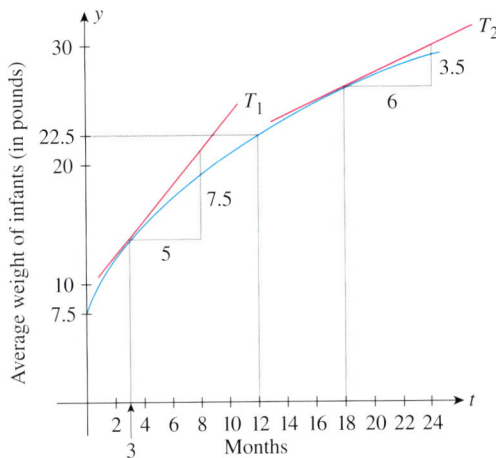

2. **FORESTRY** The following graph shows the volume of wood produced in a single-species forest. Here, $f(t)$ is measured in cubic meters per hectare, and t is measured in years. By computing the slopes of the respective tangent lines, estimate the rate at which the wood grown is changing at the beginning of year 10 and at the beginning of year 30.
   Source: The Random House Encyclopedia.

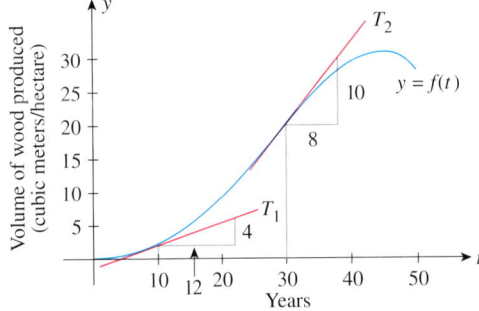

3. **TV-VIEWING PATTERNS** The following graph shows the percentage of U.S. households watching television during a 24-hr period on a weekday ($t = 0$ corresponds to 6 A.M.). By computing the slopes of the respective tangent lines, estimate the rate of change of the percent of households watching television at 4 P.M. and 11 P.M.
   Source: A. C. Nielsen Company.

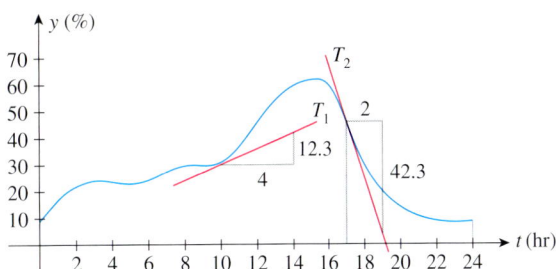

4. **CROP YIELD** Productivity and yield of cultivated crops are often reduced by insect pests. The following graph shows the relationship between the yield of a certain crop, $f(x)$, as a function of the density of aphids x. (Aphids are small insects that suck plant juices.) Here, $f(x)$ is measured in kilograms per 4000 square meters, and x is measured in hundreds of aphids per bean stem. By computing the slopes of the respective tangent lines, estimate the rate of change of the crop yield with respect to the density of aphids when that density is 200 aphids/bean stem and when it is 800 aphids/bean stem.
   Source: The Random House Encyclopedia.

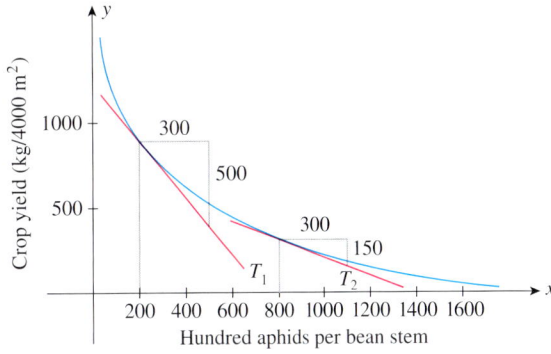

5. The positions of Car A and Car B, starting out side by side and traveling along a straight road, are given by $s = f(t)$ and $s = g(t)$, respectively, where s is measured in feet and t is measured in seconds (see the accompanying figure).

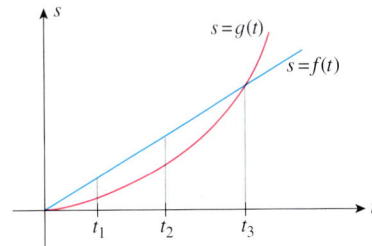

   a. Which car is traveling faster at t_1?
   b. What can you say about the speed of the cars at t_2?
      Hint: Compare tangent lines.
   c. Which car is traveling faster at t_3?
   d. What can you say about the positions of the cars at t_3?

6. The velocities of Car A and Car B, which start out side by side and travel along a straight road, are given by $v = f(t)$ and $v = g(t)$, respectively, where v is measured in feet per second and t is measured in seconds (see the accompanying figure).

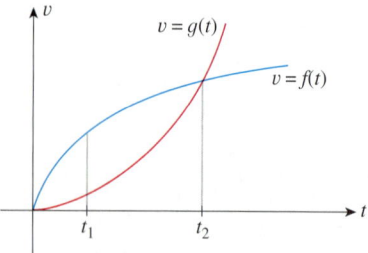

a. What can you say about the velocity and acceleration of the two cars at t_1? (Acceleration is the rate of change of velocity.)
b. What can you say about the velocity and acceleration of the two cars at t_2?

7. **Effect of a Bactericide on Bacteria** In the following figure, $f(t)$ gives the population P_1 of a certain bacteria culture at time t after a portion of Bactericide A was introduced into the population at $t = 0$. The graph of g gives the population P_2 of a similar bacteria culture at time t after a portion of Bactericide B was introduced into the population at $t = 0$.
a. Which population is decreasing faster at t_1?
b. Which population is decreasing faster at t_2?

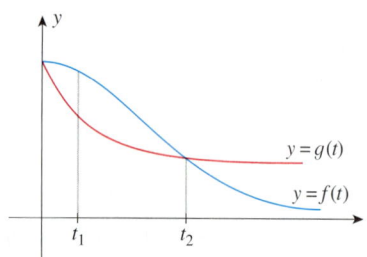

c. Which bactericide is more effective in reducing the population of bacteria in the short run? In the long run?

8. **Market Share** The following figure shows the devastating effect the opening of a new discount department store had on an established department store in a small town. The revenue of the discount store at time t (in months) is given by $f(t)$ million dollars, whereas the revenue of the established department store at time t is given by $g(t)$ million dollars. Answer the following questions by giving the value of t at which the specified event took place.

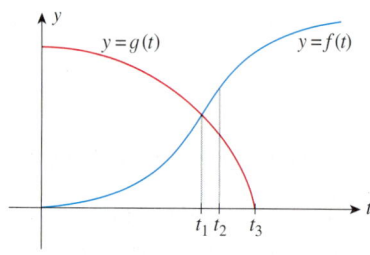

a. The revenue of the established department store is decreasing at the slowest rate.
b. The revenue of the established department store is decreasing at the fastest rate.
c. The revenue of the discount store first overtakes that of the established store.
d. The revenue of the discount store is increasing at the fastest rate.

In Exercises 9–16, use the four-step process to find the slope of the tangent line to the graph of the function at any point.

9. $f(x) = 13$
10. $f(x) = -6$
11. $f(x) = 2x + 7$
12. $f(x) = 8 - 4x$
13. $f(x) = 3x^2$
14. $f(x) = -\frac{1}{2}x^2$
15. $f(x) = -x^2 + 3x$
16. $f(x) = 2x^2 + 5x$

In Exercises 17–22, find the slope of the tangent line to the graph of the function at the given point, and determine an equation of the tangent line.

17. $f(x) = 2x + 7$ at $(2, 11)$
18. $f(x) = -3x + 4$ at $(-1, 7)$
19. $f(x) = 3x^2$ at $(1, 3)$
20. $f(x) = 3x - x^2$ at $(-2, -10)$
21. $f(x) = -\frac{1}{x}$ at $\left(3, -\frac{1}{3}\right)$
22. $f(x) = \frac{3}{2x}$ at $\left(1, \frac{3}{2}\right)$

23. Let $f(x) = 2x^2 + 1$.
a. Find the derivative f' of f.
b. Find an equation of the tangent line to the curve at the point $(1, 3)$.
c. Sketch the graph of f and its tangent line at $(1, 3)$.

24. Let $f(x) = x^2 + 6x$.
a. Find the derivative f' of f.
b. Find the point on the graph of f where the tangent line to the curve is horizontal.
  Hint: Find the value of x for which $f'(x) = 0$.
c. Sketch the graph of f and the tangent line to the curve at the point found in part (b).

25. Let $f(x) = x^2 - 2x + 1$.
a. Find the derivative f' of f.
b. Find the point on the graph of f where the tangent line to the curve is horizontal.
c. Sketch the graph of f and the tangent line to the curve at the point found in part (b).
d. What is the rate of change of f at this point?

26. Let $f(x) = \dfrac{1}{x-1}$.
   a. Find the derivative f' of f.
   b. Find an equation of the tangent line to the curve at the point $(-1, -\tfrac{1}{2})$.
   c. Sketch the graph of f and the tangent line to the curve at $(-1, -\tfrac{1}{2})$.

27. Let $y = f(x) = x^2 + x$.
   a. Find the average rate of change of y with respect to x in the interval from $x = 2$ to $x = 3$, from $x = 2$ to $x = 2.5$, and from $x = 2$ to $x = 2.1$.
   b. Find the (instantaneous) rate of change of y at $x = 2$.
   c. Compare the results obtained in part (a) with the result of part (b).

28. Let $y = f(x) = x^2 - 4x$.
   a. Find the average rate of change of y with respect to x in the interval from $x = 3$ to $x = 4$, from $x = 3$ to $x = 3.5$, and from $x = 3$ to $x = 3.1$.
   b. Find the (instantaneous) rate of change of y at $x = 3$.
   c. Compare the results obtained in part (a) with the result of part (b).

29. **VELOCITY OF A CAR** Suppose the distance s (in feet) covered by a car moving along a straight road after t sec is given by the function $s = f(t) = 2t^2 + 48t$.
   a. Calculate the average velocity of the car over the time intervals [20, 21], [20, 20.1], and [20, 20.01].
   b. Calculate the (instantaneous) velocity of the car when $t = 20$.
   c. Compare the results of part (a) with the result of part (b).

30. **VELOCITY OF A BALL THROWN INTO THE AIR** A ball is thrown straight up with an initial velocity of 128 ft/sec, so that its height (in feet) after t sec is given by $s(t) = 128t - 16t^2$.
   a. What is the average velocity of the ball over the time intervals [2, 3], [2, 2.5], and [2, 2.1]?
   b. What is the instantaneous velocity at time $t = 2$?
   c. What is the instantaneous velocity at time $t = 5$? Is the ball rising or falling at this time?
   d. When will the ball hit the ground?

31. **VELOCITY OF A FALLING OBJECT** During the construction of a high-rise building, a worker accidentally dropped his portable electric screwdriver from a height of 400 ft. After t sec, the screwdriver had fallen a distance of $s = 16t^2$ ft.
   a. How long did it take the screwdriver to reach the ground?
   b. What was the average velocity of the screwdriver between the time it was dropped and the time it hit the ground?
   c. What was the velocity of the screwdriver at the time it hit the ground?

32. **VELOCITY OF A HOT-AIR BALLOON** A hot-air balloon rises vertically from the ground so that its height after t sec is $h = \tfrac{1}{2}t^2 + \tfrac{1}{2}t$ ft $(0 \le t \le 60)$.
   a. What is the height of the balloon at the end of 40 sec?
   b. What is the average velocity of the balloon between $t = 0$ and $t = 40$?
   c. What is the velocity of the balloon at the end of 40 sec?

33. At a temperature of 20°C, the volume V (in liters) of 1.33 g of O_2 is related to its pressure p (in atmospheres) by the formula $V = 1/p$.
   a. What is the average rate of change of V with respect to p as p increases from $p = 2$ to $p = 3$?
   b. What is the rate of change of V with respect to p when $p = 2$?

34. **COST OF PRODUCING SURFBOARDS** The total cost $C(x)$ (in dollars) incurred by Aloha Company in manufacturing x surfboards a day is given by
$$C(x) = -10x^2 + 300x + 130 \qquad (0 \le x \le 15)$$
   a. Find $C'(x)$.
   b. What is the rate of change of the total cost when the level of production is ten surfboards a day?

35. **EFFECT OF ADVERTISING ON PROFIT** The quarterly profit (in thousands of dollars) of Cunningham Realty is given by
$$P(x) = -\tfrac{1}{3}x^2 + 7x + 30 \qquad (0 \le x \le 50)$$
where x (in thousands of dollars) is the amount of money Cunningham spends on advertising per quarter.
   a. Find $P'(x)$.
   b. What is the rate of change of Cunningham's quarterly profit if the amount it spends on advertising is $10,000/quarter $(x = 10)$ and $30,000/quarter $(x = 30)$?

36. **DEMAND FOR TENTS** The demand function for Sportsman 5 × 7 tents is given by
$$p = f(x) = -0.1x^2 - x + 40$$
where p is measured in dollars and x is measured in units of a thousand.
   a. Find the average rate of change in the unit price of a tent if the quantity demanded is between 5000 and 5050 tents; between 5000 and 5010 tents.
   b. What is the rate of change of the unit price if the quantity demanded is 5000?

37. **A COUNTRY'S GDP** The gross domestic product (GDP) of a certain country is projected to be
$$N(t) = t^2 + 2t + 50 \qquad (0 \le t \le 5)$$
billion dollars t years from now. What will be the rate of change of the country's GDP 2 years and 4 years from now?

38. **GROWTH OF BACTERIA** Under a set of controlled laboratory conditions, the size of the population of a certain bacteria culture at time t (in minutes) is described by the function
$$P = f(t) = 3t^2 + 2t + 1$$
Find the rate of population growth at $t = 10$ min.

39. **AIR TEMPERATURE** The air temperature at a height of h ft from the surface of the earth is $T = f(h)$ degrees Fahrenheit.
   a. Give a physical interpretation of $f'(h)$. Give units.
   b. Generally speaking, what do you expect the sign of $f'(h)$ to be?
   c. If you know that $f'(1000) = -0.05$, estimate the change in the air temperature if the altitude changes from 1000 ft to 1001 ft.

40. Revenue of a Travel Agency Suppose that the total revenue realized by the Odyssey Travel Agency is $R = f(x)$ thousand dollars if x thousand dollars are spent on advertising.
a. What does
$$\frac{f(b) - f(a)}{b - a} \quad (0 < a < b)$$
measure? What are the units?
b. What does $f'(x)$ measure? Give units.
c. Given that $f'(20) = 3$, what is the approximate change in the revenue if Odyssey increases its advertising budget from \$20,000 to \$21,000?

In Exercises 41–46, let x and $f(x)$ represent the given quantities. Fix $x = a$ and let h be a small positive number. Give an interpretation of the quantities
$$\frac{f(a+h) - f(a)}{h} \quad \text{and} \quad \lim_{h \to 0} \frac{f(a+h) - f(a)}{h}$$

41. x denotes time and $f(x)$ denotes the population of seals at time x.

42. x denotes time and $f(x)$ denotes the prime interest rate at time x.

43. x denotes time and $f(x)$ denotes a country's industrial production.

44. x denotes the level of production of a certain commodity, and $f(x)$ denotes the total cost incurred in producing x units of the commodity.

45. x denotes altitude and $f(x)$ denotes atmospheric pressure.

46. x denotes the speed of a car (in mph), and $f(x)$ denotes the fuel consumption of the car measured in miles per gallon (mpg).

In each of Exercises 47–52, the graph of a function is shown. For each function, state whether or not (a) $f(x)$ has a limit at $x = a$, as x approaches a (b) $f(x)$ is continuous at $x = a$, and (c) $f(x)$ is differentiable at $x = a$. Justify your answers.

47.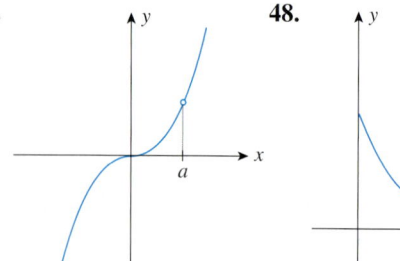

48.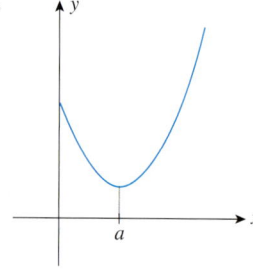

49.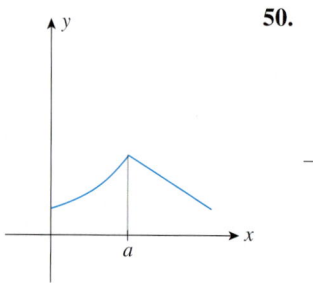

50.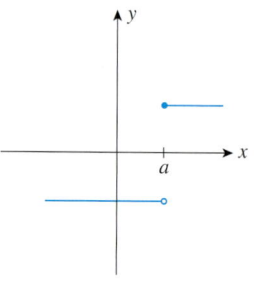

51. **52.**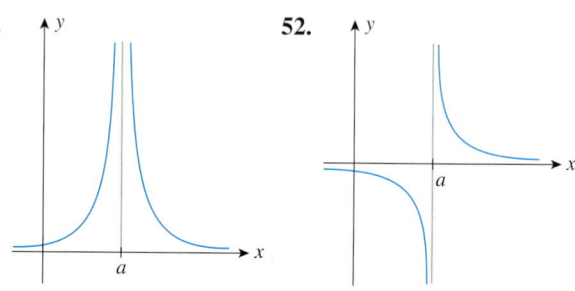

53. Velocity of a Motorcycle The distance s (in feet) covered by a motorcycle traveling in a straight line and starting from rest in t sec is given by the function
$$s(t) = -0.1t^3 + 2t^2 + 24t \quad (0 \leq t \leq 3)$$
Calculate the motorcycle's average velocity over the time interval $[2, 2 + h]$ for $h = 1, 0.1, 0.01, 0.001, 0.0001$, and 0.00001 and use your results to guess at the motorcycle's instantaneous velocity at $t = 2$.

54. Rate of Change of Production Costs The daily total cost $C(x)$ incurred by Trappee and Sons for producing x cases of TexaPep hot sauce is given by
$$C(x) = 0.000002x^3 + 5x + 400$$
Calculate
$$\frac{C(100 + h) - C(100)}{h}$$
for $h = 1, 0.1, 0.01, 0.001$, and 0.0001, and use your results to estimate the rate of change of the total cost function when the level of production is 100 cases/day.

In Exercises 55 and 56, determine whether the statement is true or false. If it is true, explain why it is true. If it is false, give an example to show why it is false.

55. If f is continuous at $x = a$, then f is differentiable at $x = a$.

56. If f is continuous at $x = a$ and g is differentiable at $x = a$, then $\lim_{x \to a} f(x)g(x) = f(a)g(a)$.

57. Sketch the graph of the function $f(x) = |x + 1|$, and show that the function does not have a derivative at $x = -1$.

58. Sketch the graph of the function $f(x) = 1/(x - 1)$, and show that the function does not have a derivative at $x = 1$.

59. Let
$$f(x) = \begin{cases} x^2 & \text{if } x \leq 1 \\ ax + b & \text{if } x > 1 \end{cases}$$
Find the values of a and b such that f is continuous and has a derivative at $x = 1$. Sketch the graph of f.

60. Sketch the graph of the function $f(x) = x^{2/3}$. Is the function continuous at $x = 0$? Does $f'(0)$ exist? Why or why not?

61. Prove that the derivative of the function $f(x) = |x|$ for $x \neq 0$ is given by
$$f'(x) = \begin{cases} 1 & \text{if } x > 0 \\ -1 & \text{if } x < 0 \end{cases}$$
Hint: Recall the definition of the absolute value of a number.

62. Show that if a function f is differentiable at $x = a$, then f must be continuous at $x = a$.

Hint: Write

$$f(x) - f(a) = \left[\frac{f(x) - f(a)}{x - a}\right](x - a)$$

Use the product rule for limits and the definition of the derivative to show that

$$\lim_{x \to a} [f(x) - f(a)] = 0$$

2.6 Solutions to Self-Check Exercises

1. a.
$$f'(x) = \lim_{h \to 0} \frac{f(x + h) - f(x)}{h}$$
$$= \lim_{h \to 0} \frac{[-(x + h)^2 - 2(x + h) + 3] - (-x^2 - 2x + 3)}{h}$$
$$= \lim_{h \to 0} \frac{-x^2 - 2xh - h^2 - 2x - 2h + 3 + x^2 + 2x - 3}{h}$$
$$= \lim_{h \to 0} \frac{h(-2x - h - 2)}{h}$$
$$= \lim_{h \to 0} (-2x - h - 2) = -2x - 2$$

b. From the result of part (a), we see that the slope of the tangent line to the graph of f at any point $(x, f(x))$ is given by

$$f'(x) = -2x - 2$$

In particular, the slope of the tangent line to the graph of f at $(0, 3)$ is

$$f'(0) = -2$$

c. The rate of change of f when $x = 0$ is given by $f'(0) = -2$, or -2 units/unit change in x.

d. Using the result from part (b), we see that an equation of the required tangent line is

$$y - 3 = -2(x - 0)$$
$$y = -2x + 3$$

e.

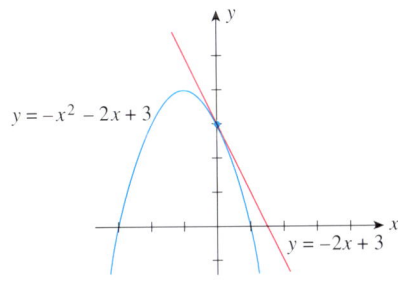

2. The rate of change of the losses at any time t is given by $f'(t)$

$$= \lim_{h \to 0} \frac{f(t + h) - f(t)}{h}$$
$$= \lim_{h \to 0} \frac{[-(t + h)^2 + 10(t + h) + 30] - (-t^2 + 10t + 30)}{h}$$
$$= \lim_{h \to 0} \frac{-t^2 - 2th - h^2 + 10t + 10h + 30 + t^2 - 10t - 30}{h}$$
$$= \lim_{h \to 0} \frac{h(-2t - h + 10)}{h}$$
$$= \lim_{h \to 0} (-2t - h + 10)$$
$$= -2t + 10$$

Therefore, the rate of change of the losses suffered by the bank at the beginning of 2008 ($t = 3$) was

$$f'(3) = -2(3) + 10 = 4$$

In other words, the losses were increasing at the rate of $4 million/year. At the beginning of 2010 ($t = 5$),

$$f'(5) = -2(5) + 10 = 0$$

and we see that the growth in losses due to bad loans was zero at this point. At the beginning of 2012 ($t = 7$),

$$f'(7) = -2(7) + 10 = -4$$

and we conclude that the losses were decreasing at the rate of $4 million/year.

USING TECHNOLOGY

Graphing a Function and Its Tangent Line

We can use a graphing utility to plot the graph of a function f and the tangent line at any point on the graph.

EXAMPLE 1 Let $f(x) = x^2 - 4x$.

a. Find an equation of the tangent line to the graph of f at the point $(3, -3)$.
b. Plot both the graph of f and the tangent line found in part (a) on the same set of axes.

(continued)

Solution

a. The slope of the tangent line at any point on the graph of f is given by $f'(x)$. But from Example 4 (page 139), we find $f'(x) = 2x - 4$. Using this result, we see that the slope of the required tangent line is

$$f'(3) = 2(3) - 4 = 2$$

Finally, using the point-slope form of the equation of a line, we find that an equation of the tangent line is

$$y - (-3) = 2(x - 3)$$
$$y + 3 = 2x - 6$$
$$y = 2x - 9$$

b. Figure T1a shows the graph of f in the standard viewing window and the tangent line of interest.

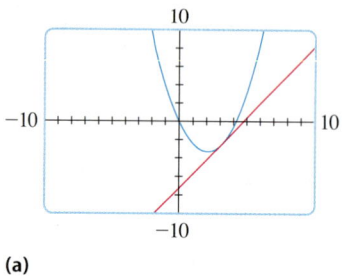
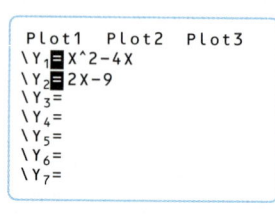

(a)   (b)

FIGURE T1
(a) The graph of $f(x) = x^2 - 4x$ and the tangent line $y = 2x - 9$ in the standard viewing window; (b) the TI-83/84 equation screen

Note Some graphing utilities will draw both the graph of a function f and the tangent line to the graph of f at a specified point when the function and the specified value of x are entered. If you use a TI-83 or TI-84 to plot the tangent line, proceed as follows:

1. Plot the graph of f in the usual manner.
2. Select the function **5:Tangent** (from the Draw menu). (Press 2ND followed by DRAW.)
3. Enter the x-coordinate of the point of tangency. In the present example, we press the key 3 to obtain $x = 3$.
4. Press ENTER to obtain the desired result.

Finding the Derivative of a Function at a Given Point

The numerical derivative operation of a graphing utility can be used to give an approximate value of the derivative of a function for a given value of x.

EXAMPLE 2 Let $f(x) = \sqrt{x}$.

a. Use the numerical derivative operation of a graphing utility to find the derivative of f at $(4, 2)$.
b. Find an equation of the tangent line to the graph of f at $(4, 2)$.
c. Plot the graph of f and the tangent line on the same set of axes.

Solution

a. Using the numerical derivative operation of a graphing utility, we find that

$$f'(4) = \frac{1}{4}$$

(Figure T2).

```
nDeriv(X^.5,X,4)
          .250000002
```

FIGURE T2
The TI-83/84 numerical derivative screen

b. An equation of the required tangent line is

$$y - 2 = \frac{1}{4}(x - 4)$$

$$y = \frac{1}{4}x + 1$$

c. Figure T3a shows the graph of f and the tangent line in the viewing window $[0, 15] \times [0, 5]$.

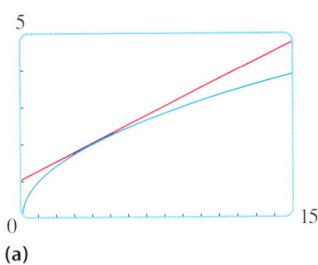
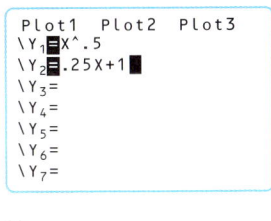

FIGURE T3
(a) The graph of $f(x) = \sqrt{x}$ and the tangent line $y = \frac{1}{4}x + 1$ in the viewing window $[0, 15] \times [0, 5]$; (b) the TI-83/84 equation screen

(a) (b)

TECHNOLOGY EXERCISES

In Exercises 1–4, (a) find an equation of the tangent line to the graph of f at the indicated point and (b) plot the graph of f and the tangent line on the same set of axes. Use a suitable viewing window.

1. $f(x) = 2x^2 + x - 3; (2, 7)$
2. $f(x) = x + \dfrac{1}{x}; (1, 2)$
3. $f(x) = x^{1/3}; (8, 2)$
4. $f(x) = \dfrac{1}{\sqrt{x}}; \left(4, \dfrac{1}{2}\right)$

In Exercises 5–8, (a) use the numerical derivative operation to find the derivative of f for the given value of x (to two decimal places of accuracy), (b) find an equation of the tangent line to the graph of f at the indicated point, and (c) plot the graph of f and the tangent line on the same set of axes. Use a suitable viewing window.

5. $f(x) = x^3 + x + 1; x = 1; (1, 3)$
6. $f(x) = \dfrac{1}{x + 1}; x = 1; \left(1, \dfrac{1}{2}\right)$
7. $f(x) = x\sqrt{x^2 + 1}; x = 2; (2, 2\sqrt{5})$
8. $f(x) = \dfrac{x}{\sqrt{x^2 + 1}}; x = 1; \left(1, \dfrac{\sqrt{2}}{2}\right)$

9. **Driving Costs** The average cost of owning and operating a car in the United States from 1991 through 2001 is approximated by the function

$$C(t) = 0.06t^2 + 0.74t + 37.3 \quad (0 \leq t \leq 11)$$

where $C(t)$ is measured in cents/mile and t is measured in years, with $t = 0$ corresponding to the beginning of 1991.

a. Plot the graph of C in the viewing window $[0, 10] \times [35, 52]$.
b. What was the average cost of driving a car at the beginning of 1995?
c. How fast was the average cost of driving a car changing at the beginning of 1995?

Source: Automobile Association of America.

10. **Modeling With Data** Annual retail sales in the United States from 2000 through the year 2008 (in billions of dollars) are given in the following table:

Year	2000	2001	2002	2003	2004
Sales	2.988	3.068	3.134	3.267	3.480

Year	2005	2006	2007	2008
Sales	3.698	3.882	4.005	3.959

a. Let $t = 0$ correspond to 2000, and use **QuadReg** to find a second-degree polynomial regression model based on the given data.
b. Plot the graph of the function found in part (a) in the viewing window $[0, 8] \times [0, 5]$.
c. What were the annual retail sales in the United States in 2006 ($t = 6$)?
d. Approximately, how fast were the retail sales changing in 2006 ($t = 6$)?

Source: U.S. Census Bureau.

CHAPTER 2 Summary of Principal Formulas and Terms

FORMULAS

1. Average rate of change of f over $[x, x+h]$ or
   Slope of the secant line to the graph of f through $(x, f(x))$ and $(x+h, f(x+h))$ or
   Difference quotient

   $$\frac{f(x+h) - f(x)}{h}$$

2. Instantaneous rate of change of f at $(x, f(x))$ or
   Slope of tangent line to the graph of f at $(x, f(x))$ at x or
   Derivative of f

   $$\lim_{h \to 0} \frac{f(x+h) - f(x)}{h}$$

TERMS

function (50)
domain (50)
range (50)
independent variable (51)
dependent variable (51)
ordered pairs (53)
graph of a function (53)
graph of an equation (56)
Vertical Line Test (56)
composite function (70)
polynomial function (76)

linear function (76)
quadratic function (77)
rational function (80)
power function (80)
demand function (81)
supply function (81)
market equilibrium (82)
equilibrium quantity (82)
equilibrium price (82)
limit of a function (100)

indeterminate form (103)
limit of a function at infinity (106)
right-hand limit of a function (118)
left-hand limit of a function (118)
continuity of a function at a number (119)
zero of a function (123)
secant line (135)
tangent line to the graph of f (135)
differentiable function (143)

CHAPTER 2 Concept Review Questions

Fill in the blanks.

1. If f is a function from the set A to the set B, then A is called the _____ of f, and the set of all values of $f(x)$ as x takes on all possible values in A is called the _____ of f. The range of f is contained in the set _____.

2. The graph of a function is the set of all points (x, y) in the xy-plane such that x is in the _____ of f and $y =$ _____. The Vertical Line Test states that a curve in the xy-plane is the graph of a function $y = f(x)$ if and only if each _____ line intersects it in at most one _____.

3. If f and g are functions with domains A and B, respectively, then (a) $(f \pm g)(x) =$ _____, (b) $(fg)(x) =$ _____, and (c) $\left(\frac{f}{g}\right)(x) =$ _____. The domain of $f + g$ is _____. The domain of $\frac{f}{g}$ is _____ with the additional condition that $g(x)$ is never _____.

4. The composition of g and f is the function with rule $(g \circ f)(x) =$ _____. Its domain is the set of all x in the domain of _____ such that _____ lies in the domain of _____.

5. a. A polynomial function of degree n is a function of the form _____.
   b. A polynomial function of degree 1 is called a/an _____ function; one of degree 2 is called a/an _____ function; one of degree 3 is called a/an _____ function.
   c. A rational function is a/an _____ of two _____.
   d. A power function has the form $f(x) =$ _____.

6. The statement $\lim_{x \to a} f(x) = L$ means that the values of _____ can be made as close to _____ as we please by taking x sufficiently close to _____.

7. If $\lim_{x \to a} f(x) = L$ and $\lim_{x \to a} g(x) = M$, then
   a. $\lim_{x \to a} [f(x)]^r =$ _____, where r is a positive constant.
   b. $\lim_{x \to a} [f(x) \pm g(x)] =$ _____.
   c. $\lim_{x \to a} [f(x)g(x)] =$ _____.
   d. $\lim_{x \to a} \frac{f(x)}{g(x)} =$ _____ provided that _____.

8. a. The statement $\lim_{x \to \infty} f(x) = L$ means that $f(x)$ can be made arbitrarily close to _____ by taking _____ large enough.
   b. The statement $\lim_{x \to -\infty} f(x) = M$ means that $f(x)$ can be made arbitrarily close to _____ by taking x to be _____ and sufficiently large in _____ value.

9. a. The statement $\lim_{x \to a^+} f(x) = L$ is similar to the statement $\lim_{x \to a} f(x) = L$, but here x is required to lie to the _____ of a.
   b. The statement $\lim_{x \to a^-} f(x) = L$ is similar to the statement $\lim_{x \to a} f(x) = L$, but here x is required to lie to the _____ of a.
   c. $\lim_{x \to a} f(x) = L$ if and only if both $\lim_{x \to a^-} f(x) =$ _____ and $\lim_{x \to a^+} f(x) =$ _____.

10. a. If $f(a)$ is defined, $\lim_{x \to a} f(x)$ exists and $\lim_{x \to a} f(x) = f(a)$, then f is _____ at a.
    b. If f is not continuous at a, then it is _____ at a.
    c. f is continuous on an interval I if f is continuous at _____ number in the interval.

11. a. If f and g are continuous at a, then $f \pm g$ and fg are continuous at _____. Also, $\frac{f}{g}$ is continuous at _____, provided _____ $\ne 0$.
    b. A polynomial function is continuous _____.
    c. A rational function $R = \frac{P}{Q}$ is continuous everywhere except at values of x for which _____ $= 0$.

12. a. Suppose f is continuous on $[a, b]$ and $f(a) < M < f(b)$. Then the intermediate value theorem guarantees the existence of at least one number c in _____ such that _____.
    b. If f is continuous on $[a, b]$ and $f(a)f(b) < 0$, then there must be at least one solution of the equation _____ in the interval _____.

13. a. The tangent line at $P(a, f(a))$ to the graph of f is the line passing through P and having slope _____.
    b. If the slope of the tangent line at $P(a, f(a))$ is m, then an equation of the tangent line at P is _____.

14. a. The slope of the secant line passing through $P(a, f(a))$ and $Q(a + h, f(a + h))$ and the average rate of change of f over the interval $[a, a + h]$ are both given by _____.
    b. The slope of the tangent line at $P(a, f(a))$ and the instantaneous rate of change of f at a are both given by _____.

CHAPTER 2 Review Exercises

1. Find the domain of the function.
   a. $f(x) = \sqrt{9 - x}$
   b. $f(x) = \dfrac{x + 3}{2x^2 - x - 3}$

2. Find the domain of the function.
   a. $f(x) = \dfrac{\sqrt{2 - x}}{x + 3}$
   b. $f(x) = \dfrac{x^2 + 3x + 4}{\sqrt{x^2 + 1}}$

3. Let $f(x) = 3x^2 + 5x - 2$. Find:
   a. $f(-2)$
   b. $f(a + 2)$
   c. $f(2a)$
   d. $f(a + h)$

4. Let $f(x) = 2x^2 - x + 1$. Find:
   a. $f(x - 1) + f(x + 1)$
   b. $f(x + 2h)$

5. Let $y^2 = 2x + 1$.
   a. Sketch the graph of this equation.
   b. Is y a function of x? Why?
   c. Is x a function of y? Why?

6. Sketch the graph of the function defined by
$$f(x) = \begin{cases} x + 1 & \text{if } x < 1 \\ -x^2 + 4x - 1 & \text{if } x \ge 1 \end{cases}$$

7. Let $f(x) = 1/x$ and $g(x) = 2x + 3$. Find:
   a. $f(x)g(x)$
   b. $f(x)/g(x)$
   c. $f(g(x))$
   d. $g(f(x))$

8. Find the rules for the composite functions $f \circ g$ and $g \circ f$.
   a. $f(x) = 2x - 1; g(x) = x^2 + 4$
   b. $f(x) = 1 - x; g(x) = \dfrac{1}{3x + 4}$
   c. $f(x) = x - 3; g(x) = \dfrac{1}{\sqrt{x + 1}}$

9. Find functions f and g such that $h = g \circ f$. (Note: The answer is not unique.)
   a. $h(x) = \dfrac{1}{(2x^2 + x + 1)^3}$
   b. $h(x) = \sqrt{x^2 + x + 4}$

10. Find the value of c such that the point $(4, 2)$ lies on the graph of $f(x) = cx^2 + 3x - 4$.

In Exercises 11–24, find the indicated limits, if they exist.

11. $\lim_{x \to 0} (5x - 3)$
12. $\lim_{x \to 1} (x^2 + 1)$
13. $\lim_{x \to -1} (3x^2 + 4)(2x - 1)$
14. $\lim_{x \to 3} \dfrac{x - 3}{x + 4}$
15. $\lim_{x \to 2} \dfrac{x + 3}{x^2 - 9}$
16. $\lim_{x \to -2} \dfrac{x^2 - 2x - 3}{x^2 + 5x + 6}$
17. $\lim_{x \to 3} \sqrt{2x^3 - 5}$
18. $\lim_{x \to 3} \dfrac{4x - 3}{\sqrt{x + 1}}$
19. $\lim_{x \to 1^+} \dfrac{x - 1}{x(x - 1)}$

20. $\lim_{x \to 1^-} \dfrac{\sqrt{x}-1}{x-1}$

21. $\lim_{x \to \infty} \dfrac{x^2}{x^2-1}$

22. $\lim_{x \to -\infty} \dfrac{x+1}{x}$

23. $\lim_{x \to \infty} \dfrac{3x^2+2x+4}{2x^2-3x+1}$

24. $\lim_{x \to -\infty} \dfrac{x^2}{x+1}$

25. Sketch the graph of the function
$$f(x) = \begin{cases} 2x - 3 & \text{if } x \leq 2 \\ -x + 3 & \text{if } x > 2 \end{cases}$$
and evaluate $\lim_{x \to a^+} f(x)$, $\lim_{x \to a^-} f(x)$, and $\lim_{x \to a} f(x)$ at the point $a = 2$, if the limits exist.

26. Sketch the graph of the function
$$f(x) = \begin{cases} 4 - x & \text{if } x \leq 2 \\ x + 2 & \text{if } x > 2 \end{cases}$$
and evaluate $\lim_{x \to a^+} f(x)$, $\lim_{x \to a^-} f(x)$, and $\lim_{x \to a} f(x)$ at the point $a = 2$, if the limits exist.

In Exercises 27–30, determine all values of *x* for which each function is discontinuous.

27. $g(x) = \begin{cases} x + 3 & \text{if } x \neq 2 \\ 0 & \text{if } x = 2 \end{cases}$

28. $f(x) = \dfrac{3x+4}{4x^2-2x-2}$

29. $f(x) = \begin{cases} \dfrac{1}{(x+1)^2} & \text{if } x \neq -1 \\ 2 & \text{if } x = -1 \end{cases}$

30. $f(x) = \dfrac{|2x|}{x}$

31. Let $y = x^2 + 2$.
    a. Find the average rate of change of *y* with respect to *x* over the intervals [1, 2], [1, 1.5], and [1, 1.1].
    b. Find the (instantaneous) rate of change of *y* at $x = 1$.

32. Use the definition of the derivative to find the slope of the tangent line to the graph of the function $f(x) = 4x + 5$ at any point $P(x, f(x))$ on the graph.

33. Use the definition of the derivative to find the slope of the tangent line to the graph of the function $f(x) = -1/x$ at any point $P(x, f(x))$ on the graph.

34. Use the definition of the derivative to find the slope of the tangent line to the graph of the function $f(x) = \frac{3}{2}x + 5$ at the point $(-2, 2)$ and determine an equation of the tangent line.

35. Use the definition of the derivative to find the slope of the tangent line to the graph of the function $f(x) = -x^2$ at the point $(2, -4)$ and determine an equation of the tangent line.

36. The graph of the function *f* is shown in the accompanying figure.
    a. Is *f* continuous at $x = a$? Why or why not?
    b. Is *f* differentiable at $x = a$? Justify your answers.

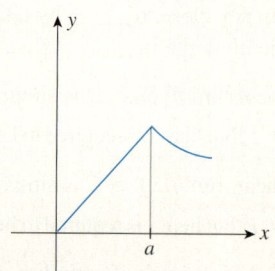

37. **Sales of Clock Radios** Sales of a certain stereo clock radio are approximated by the relationship $S(x) = 6000x + 30{,}000$ $(0 \leq x \leq 5)$, where $S(x)$ denotes the number of clock radios sold in year *x* ($x = 0$ corresponds to the year 2008). Find the number of clock radios expected to be sold in 2012.

38. **Sales of a Company** A company's total sales (in millions of dollars) are approximately linear as a function of *t* in years ($t = 0$ corresponds to the year 2007). Sales in 2007 were $2.4 million, whereas sales in 2012 amounted to $7.4 million.
    a. Find an equation that gives the company's sales as a function of time.
    b. What were the sales in 2010?

39. **Profit Functions** A company has a fixed cost of $30,000 and a production cost of $6 for each unit it manufactures. A unit sells for $10.
    a. What is the cost function?
    b. What is the revenue function?
    c. What is the profit function?
    d. Compute the profit (loss) corresponding to production levels of 6000, 8000, and 12,000 units, respectively.

40. Find the point of intersection of the two straight lines having the equations $y = \frac{3}{4}x + 6$ and $3x - 2y + 3 = 0$.

41. The cost and revenue functions for a certain firm are given by $C(x) = 12x + 20{,}000$ and $R(x) = 20x$, respectively. Find the company's profit function.

42. **Market Equilibrium** Given the demand equation $3x + p - 40 = 0$ and the supply equation $2x - p + 10 = 0$, where *p* is the unit price in dollars and *x* represents the quantity in units of a thousand, determine the equilibrium quantity and the equilibrium price.

43. **Clark's Rule** Clark's rule is a method for calculating pediatric drug dosages based on a child's weight. If *a* denotes the adult dosage (in milligrams) and *w* is the weight of the child (in pounds), then the child's dosage is given by
$$D(w) = \dfrac{aw}{150}$$
If the adult dose of a substance is 500 mg, how much should a child who weighs 35 lb receive?

44. **REVENUE FUNCTIONS** The revenue (in dollars) realized by Apollo from the sale of its ink-jet printers is given by
$$R(x) = -0.1x^2 + 500x$$
where x denotes the number of units manufactured each month. What is Apollo's revenue when 1000 units are produced?

45. **REVENUE FUNCTIONS** The monthly revenue R (in hundreds of dollars) realized from the sale of Royal electric shavers is related to the unit price p (in dollars) by the equation
$$R(p) = -\frac{1}{2}p^2 + 30p$$
Find the revenue when an electric shaver is priced at $30.

46. **HEALTH CLUB MEMBERSHIP** The membership of the newly opened Venus Health Club is approximated by the function
$$N(x) = 200(4 + x)^{1/2} \quad (1 \leq x \leq 24)$$
where $N(x)$ denotes the number of members x months after the club's grand opening. Find $N(0)$ and $N(12)$, and interpret your results.

47. **POPULATION GROWTH** A study prepared for a Sunbelt town's Chamber of Commerce projected that the population of the town in the next 3 years will grow according to the rule
$$P(x) = 50{,}000 + 30x^{3/2} + 20x$$
where $P(x)$ denotes the population x months from now. By how much will the population increase during the next 9 months? During the next 16 months?

48. **THURSTONE LEARNING CURVE** Psychologist L. L. Thurstone discovered the following model for the relationship between the learning time T and the length of a list n:
$$T = f(n) = An\sqrt{n - b}$$
where A and b are constants that depend on the person and the task. Suppose that, for a certain person and a certain task, $A = 4$ and $b = 4$. Compute $f(4), f(5), \ldots, f(12)$, and use this information to sketch the graph of the function f. Interpret your results.

49. **FORECASTING SALES** The annual sales of Crimson Drug Store are expected to be given by
$$S_1(t) = 2.3 + 0.4t$$
million dollars t years from now, whereas the annual sales of Cambridge Drug Store are expected to be given by
$$S_2(t) = 1.2 + 0.6t$$
million dollars t years from now. When will the annual sales of Cambridge first surpass the annual sales of Crimson?

50. **MARKET EQUILIBRIUM** The monthly demand and supply functions for the Luminar desk lamp are given by
$$p = d(x) = -1.1x^2 + 1.5x + 40$$
$$p = s(x) = 0.1x^2 + 0.5x + 15$$
respectively, where p is measured in dollars and x in units of a thousand. Find the equilibrium quantity and price.

51. **TESTOSTERONE USE** Fueled by the promotion of testosterone as an antiaging elixir, use of the hormone by middle-age and older men grew dramatically. The total number of prescriptions for testosterone from 1999 through 2002 is given by
$$N(t) = -35.8t^3 + 202t^2 + 87.8t + 648 \quad (0 \leq t \leq 3)$$
where $N(t)$ is measured in thousands and t is measured in years, with $t = 0$ corresponding to the beginning of 1999. Find the total number of prescriptions for testosterone in 1999, 2000, 2001, and 2002.
Source: IMS Health.

52. **U.S. NUTRITIONAL SUPPLEMENTS MARKET** The size of the U.S. nutritional supplements market from 1999 through 2003 is approximated by the function
$$A(t) = 16.4(t + 1)^{0.1} \quad (0 \leq t \leq 4)$$
where $A(t)$ is measured in billions of dollars and t is measured in years, with $t = 0$ corresponding to the beginning of 1999.
   a. Compute $A(0), A(1), A(2), A(3)$, and $A(4)$. Interpret your results.
   b. Use the results of part (a) to sketch the graph of A.
Source: Nutrition Business Journal.

53. **GLOBAL SUPPLY OF PLUTONIUM** The global stockpile of plutonium for military applications between 1990 ($t = 0$) and 2003 ($t = 13$) stood at a constant 267 tons. On the other hand, the global stockpile of plutonium for civilian use was
$$2t^2 + 46t + 733$$
tons in year t over the same period.
   a. Find the function f giving the global stockpile of plutonium for military use from 1990 through 2003 and the function g giving the global stockpile of plutonium for civilian use over the same period.
   b. Find the function h giving the total global stockpile of plutonium between 1990 and 2003.
   c. What was the total global stockpile of plutonium in 2003?
Source: Institute for Science and International Security.

54. **HOTEL OCCUPANCY RATE** A forecast released by PricewaterhouseCoopers in June of 2004 estimated the occupancy rate of U.S. hotels between 2001 ($t = 0$) and 2005 ($t = 4$) to be
$$P(t) = \begin{cases} -0.9t + 59.8 & \text{if } 0 \leq t < 1 \\ 0.3t + 58.6 & \text{if } 1 \leq t < 2 \\ 56.79 t^{0.06} & \text{if } 2 \leq t \leq 4 \end{cases}$$
percent.
   a. Compute $P(0), P(1), P(2), P(3)$, and $P(4)$.
   b. Sketch the graph of P.
   c. What was the estimated occupancy rate of hotels for 2004?
Source: PricewaterhouseCoopers LLP Hospitality & Leisure Research.

55. **OIL SPILLS** The oil spilling from the ruptured hull of a grounded tanker spreads in all directions in calm waters.

Suppose the area polluted is a circle of radius r and the radius is increasing at the rate of 2 ft/sec.
  a. Find a function f giving the area polluted in terms of r.
  b. Find a function g giving the radius of the polluted area in terms of t.
  c. Find a function h giving the area polluted in terms of t.
  d. What is the size of the polluted area 30 sec after the hull was ruptured?

56. **PACKAGING** By cutting away identical squares from each corner of a 20-in. × 20-in. piece of cardboard and folding up the resulting flaps, an open box can be made. Denoting the length of a side of a cutaway by x, find a function of x giving the volume of the resulting box.

57. **CONSTRUCTION COSTS** The length of a rectangular box is to be twice its width, and its volume is to be 30 ft³. The material for the base costs 30¢/ft², the material for the sides costs 15¢/ft², and the material for the top costs 20¢/ft². Letting x denote the width of the box, find a function in the variable x giving the cost of constructing the box.

58. **FILM CONVERSION PRICES** PhotoMart transfers movie films to DVDs. The fees charged for this service are shown in the following table. Find a function C relating the cost $C(x)$ to the number of feet x of film transferred. Sketch the graph of the function C and discuss its continuity.

Length of Film in Feet, x	Price ($) for Conversion
$1 \le x \le 100$	5.00
$100 < x \le 200$	9.00
$200 < x \le 300$	12.50
$300 < x \le 400$	15.00
$x > 400$	$7 + 0.02x$

59. **AVERAGE PRICE OF A COMMODITY** The average cost (in dollars) of producing x units of a certain commodity is given by

$$\overline{C}(x) = 20 + \frac{400}{x}$$

Evaluate $\lim_{x \to \infty} \overline{C}(x)$, and interpret your results.

60. **MANUFACTURING COSTS** Suppose that the total cost of manufacturing x units of a certain product is $C(x)$ dollars.
  a. What does $C'(x)$ measure? Give units.
  b. What can you say about the sign of C'?
  c. Given that $C'(1000) = 20$, estimate the additional cost to be incurred by the company in producing the 1001st unit of the product.

CHAPTER 2 Before Moving On . . .

1. Let
$$f(x) = \begin{cases} -2x + 1 & \text{if } -1 \le x < 0 \\ x^2 + 2 & \text{if } 0 \le x \le 2 \end{cases}$$
Find (a) $f(-1)$, (b) $f(0)$, and (c) $f(\frac{3}{2})$.

2. Let $f(x) = \dfrac{1}{x+1}$ and $g(x) = x^2 + 1$. Find the rules for (a) $f + g$, (b) fg, (c) $f \circ g$, and (d) $g \circ f$.

3. Postal regulations specify that a parcel sent by priority mail may have a combined length and girth of no more than 108 in. Suppose a rectangular package that has a square cross section of x in. × x in. is to have a combined length and girth of exactly 108 in. Find a function in terms of x giving the volume of the package.
Hint: The length plus the girth is $4x + h$ (see the accompanying figure).

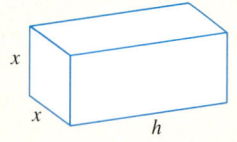

4. Find $\lim_{x \to -1} \dfrac{x^2 + 4x + 3}{x^2 + 3x + 2}$.

5. Let
$$f(x) = \begin{cases} x^2 - 1 & \text{if } -2 \le x < 1 \\ x^3 & \text{if } 1 \le x \le 2 \end{cases}$$
Find (a) $\lim_{x \to 1^-} f(x)$ and (b) $\lim_{x \to 1^+} f(x)$. Is f continuous at $x = 1$? Explain.

6. Find the slope of the tangent line to the graph of $f(x) = x^2 - 3x + 1$ at the point $(1, -1)$. What is an equation of the tangent line?

3 DIFFERENTIATION

What happens to the sales of a DVD recording of a certain hit movie over a 10-year period after it is first released into the market? In Example 6, page 174, you will see how to find the rate of change of sales for the DVD over the first 10 years after its release.

THIS CHAPTER GIVES several rules that will greatly simplify the task of finding the derivative of a function, thus enabling us to study how fast one quantity is changing with respect to another in many real-world situations. For example, we will be able to find how fast the population of an endangered species of whales grows after certain conservation measures have been implemented, how fast an economy's consumer price index (CPI) is changing at any time, and how fast the time taken to learn the items on a list changes with respect to the length of a list. We also see how these rules of differentiation facilitate the study of the rate of change of economic quantities—that is, the study of marginal analysis. Finally, we introduce the notion of the differential of a function. Differentials are used to approximate the change in one quantity due to a small change in a related quantity.

3.1 Basic Rules of Differentiation

Four Basic Rules

The method used in Chapter 2 for computing the derivative of a function is based on a faithful interpretation of the definition of the derivative as the limit of a quotient. To find the rule for the derivative f' of a function f, we first computed the difference quotient

$$\frac{f(x+h) - f(x)}{h}$$

and then evaluated its limit as h approached zero. As you have probably observed, this method is tedious even for relatively simple functions.

The main purpose of this chapter is to derive certain rules that will simplify the process of finding the derivative of a function. We will use the notation

$$\frac{d}{dx}[f(x)]$$ Read "d, dx of f of x"

to mean "the derivative of f with respect to x at x."

Rule 1: Derivative of a Constant

$$\frac{d}{dx}(c) = 0 \quad (c, \text{ a constant})$$

The derivative of a constant function is equal to zero.

We can see this from a geometric viewpoint by recalling that the graph of a constant function is a straight line parallel to the x-axis (Figure 1). Since the tangent line to a straight line at any point on the line coincides with the straight line itself, its slope [as given by the derivative of $f(x) = c$] must be zero. We can also use the definition of the derivative to prove this result by computing

$$f'(x) = \lim_{h \to 0} \frac{f(x+h) - f(x)}{h}$$

$$= \lim_{h \to 0} \frac{c - c}{h}$$

$$= \lim_{h \to 0} 0 = 0$$

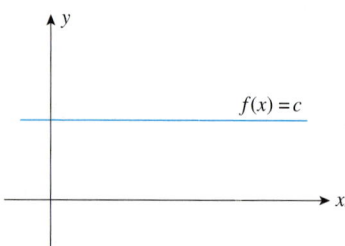

FIGURE 1
The slope of the tangent line to the graph of $f(x) = c$, where c is a constant, is zero.

EXAMPLE 1

a. If $f(x) = 28$, then

$$f'(x) = \frac{d}{dx}(28) = 0$$

b. If $f(x) = \pi^2$, then

$$f'(x) = \frac{d}{dx}(\pi^2) = 0$$

> **Rule 2: The Power Rule**
>
> If n is any real number, then $\dfrac{d}{dx}(x^n) = nx^{n-1}$.

Let's verify the Power Rule for the special case $n = 2$. If $f(x) = x^2$, then

$$\begin{aligned}
f'(x) &= \frac{d}{dx}(x^2) = \lim_{h \to 0} \frac{f(x+h) - f(x)}{h} \\
&= \lim_{h \to 0} \frac{(x+h)^2 - x^2}{h} \\
&= \lim_{h \to 0} \frac{x^2 + 2xh + h^2 - x^2}{h} \\
&= \lim_{h \to 0} \frac{2xh + h^2}{h} = \lim_{h \to 0} \frac{h(2x + h)}{h} \\
&= \lim_{h \to 0} (2x + h) = 2x
\end{aligned}$$

as we set out to show.

The Power Rule for the general case is not easy to prove and the proof will be omitted. However, you will be asked to prove the rule for the special case $n = 3$ in Exercise 79, page 169.

EXAMPLE 2

a. If $f(x) = x$, then

$$f'(x) = \frac{d}{dx}(x) = 1 \cdot x^{1-1} = x^0 = 1$$

b. If $f(x) = x^8$, then

$$f'(x) = \frac{d}{dx}(x^8) = 8x^7$$

c. If $f(x) = x^{5/2}$, then

$$f'(x) = \frac{d}{dx}(x^{5/2}) = \frac{5}{2}x^{3/2}$$

To differentiate a function whose rule involves a radical, we first rewrite the rule using fractional powers. The resulting expression can then be differentiated by using the Power Rule.

EXAMPLE 3 Find the derivative of the following functions:

a. $f(x) = \sqrt{x}$ **b.** $g(x) = \dfrac{1}{\sqrt[3]{x}}$

Solution

a. Rewriting $\sqrt{x}$ in the form $x^{1/2}$, we obtain   *(x^2) See page 6.*

$$\begin{aligned}
f'(x) &= \frac{d}{dx}(x^{1/2}) \\
&= \frac{1}{2}x^{-1/2} = \frac{1}{2x^{1/2}} = \frac{1}{2\sqrt{x}}
\end{aligned}$$

b. Rewriting $\dfrac{1}{\sqrt[3]{x}}$ in the form $x^{-1/3}$, we obtain

$$g'(x) = \frac{d}{dx}(x^{-1/3})$$

$$= -\frac{1}{3}x^{-4/3} = -\frac{1}{3x^{4/3}}$$

In stating the remaining rules of differentiation, we assume that the functions f and g are differentiable.

Rule 3: Derivative of a Constant Multiple of a Function

$$\frac{d}{dx}[cf(x)] = c\frac{d}{dx}[f(x)] \qquad (c, \text{ a constant})$$

The derivative of a constant times a differentiable function is equal to the constant times the derivative of the function.

This result follows from the following computations:

If $g(x) = cf(x)$, then

$$g'(x) = \lim_{h \to 0} \frac{g(x+h) - g(x)}{h} = \lim_{h \to 0} \frac{cf(x+h) - cf(x)}{h}$$

$$= c \lim_{h \to 0} \frac{f(x+h) - f(x)}{h}$$

$$= cf'(x)$$

EXAMPLE 4

a. If $f(x) = 5x^3$, then

$$f'(x) = \frac{d}{dx}(5x^3) = 5\frac{d}{dx}(x^3)$$

$$= 5(3x^2) = 15x^2$$

b. If $f(x) = \dfrac{3}{\sqrt{x}}$, then

$$f'(x) = \frac{d}{dx}(3x^{-1/2})$$

$$= 3\left(-\frac{1}{2}x^{-3/2}\right) = -\frac{3}{2x^{3/2}}$$

Rule 4: The Sum Rule

$$\frac{d}{dx}[f(x) \pm g(x)] = \frac{d}{dx}[f(x)] \pm \frac{d}{dx}[g(x)]$$

The derivative of the sum (difference) of two differentiable functions is equal to the sum (difference) of their derivatives.

This result may be extended to the sum and difference of any finite number of differentiable functions. Let's verify the rule for a sum of two functions.

If $s(x) = f(x) + g(x)$, then

$$\begin{aligned}
s'(x) &= \lim_{h \to 0} \frac{s(x+h) - s(x)}{h} \\
&= \lim_{h \to 0} \frac{[f(x+h) + g(x+h)] - [f(x) + g(x)]}{h} \\
&= \lim_{h \to 0} \frac{[f(x+h) - f(x)] + [g(x+h) - g(x)]}{h} \\
&= \lim_{h \to 0} \frac{f(x+h) - f(x)}{h} + \lim_{h \to 0} \frac{g(x+h) - g(x)}{h} \\
&= f'(x) + g'(x)
\end{aligned}$$

VIDEO

EXAMPLE 5 Find the derivatives of the following functions:

a. $f(x) = 4x^5 + 3x^4 - 8x^2 + x + 3$ **b.** $g(t) = \dfrac{t^2}{5} + \dfrac{5}{t^3}$

Solution

a. $f'(x) = \dfrac{d}{dx}(4x^5 + 3x^4 - 8x^2 + x + 3)$

$= \dfrac{d}{dx}(4x^5) + \dfrac{d}{dx}(3x^4) - \dfrac{d}{dx}(8x^2) + \dfrac{d}{dx}(x) + \dfrac{d}{dx}(3)$

$= 20x^4 + 12x^3 - 16x + 1$

b. Here, the independent variable is t instead of x, so we differentiate with respect to t. Thus,

$$g'(t) = \frac{d}{dt}\left(\frac{1}{5}t^2 + 5t^{-3}\right) \qquad \text{Rewrite } \frac{1}{t^3} \text{ as } t^{-3}.$$

$$= \frac{2}{5}t - 15t^{-4} = \frac{2}{5}t - \frac{15}{t^4} \qquad \text{Rewrite } t^{-4} \text{ as } \frac{1}{t^4}.$$

$$= \frac{2t^5 - 75}{5t^4}$$

EXAMPLE 6 Find the slope and an equation of the tangent line to the graph of $f(x) = 2x + 1/\sqrt{x}$ at the point $(1, 3)$.

Solution The slope of the tangent line at any point on the graph of f is given by

$$f'(x) = \frac{d}{dx}\left(2x + \frac{1}{\sqrt{x}}\right)$$

$$= \frac{d}{dx}(2x + x^{-1/2}) \qquad \text{Rewrite } \frac{1}{\sqrt{x}} \text{ as } \frac{1}{x^{1/2}} = x^{-1/2}.$$

$$= 2 - \frac{1}{2}x^{-3/2} \qquad \text{Use the Sum Rule.}$$

$$= 2 - \frac{1}{2x^{3/2}} \qquad \text{Rewrite } \frac{1}{2}x^{-3/2} \text{ as } \frac{1}{2x^{3/2}}.$$

In particular, the slope of the tangent line to the graph of f at $(1, 3)$ (where $x = 1$) is

$$f'(1) = 2 - \frac{1}{2(1^{3/2})} = 2 - \frac{1}{2} = \frac{3}{2}$$

Using the point-slope form of the equation of a line with slope $\frac{3}{2}$ and the point $(1, 3)$, we see that an equation of the tangent line is

$$y - 3 = \frac{3}{2}(x - 1) \qquad y - y_1 = m(x - x_1) \qquad (x^2) \text{ See page 36.}$$

or, upon simplification,

$$y = \frac{3}{2}x + \frac{3}{2}$$

(see Figure 2).

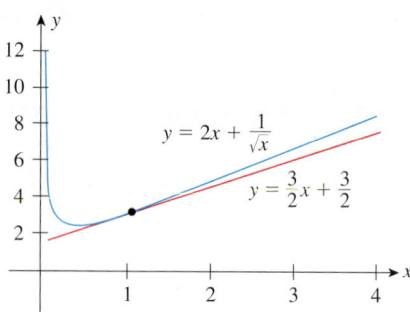

FIGURE 2
The tangent line to the graph of $f(x) = 2x + 1/\sqrt{x}$ at $(1, 3)$.

APPLIED EXAMPLE 7 Conservation of a Species A group of marine biologists at the Neptune Institute of Oceanography recommended that a series of conservation measures be carried out over the next decade to save a certain species of whale from extinction. After implementation of the conservation measures, the population of this species is expected to be

$$N(t) = 3t^3 + 2t^2 - 10t + 600 \qquad (0 \le t \le 10)$$

where $N(t)$ denotes the population at the end of year t. Find the rate of growth of the whale population when $t = 2$ and $t = 6$. How large will the whale population be 8 years after implementing the conservation measures?

Solution The rate of growth of the whale population at any time t is given by

$$N'(t) = 9t^2 + 4t - 10$$

In particular, when $t = 2$ and $t = 6$, we have

$$N'(2) = 9(2)^2 + 4(2) - 10$$
$$= 34$$
$$N'(6) = 9(6)^2 + 4(6) - 10$$
$$= 338$$

Thus, the whale population's rate of growth will be 34 whales per year after 2 years and 338 per year after 6 years.

The whale population at the end of the eighth year will be

$$N(8) = 3(8)^3 + 2(8)^2 - 10(8) + 600$$
$$= 2184$$

The graph of the function N appears in Figure 3. Note the rapid growth of the population in the later years, as the conservation measures begin to pay off, compared with the growth in the early years.

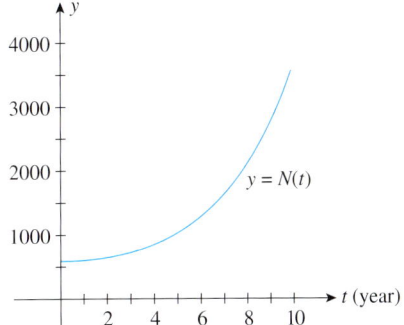

FIGURE 3
The whale population at the end of year t is given by $N(t)$.

APPLIED EXAMPLE 8 Altitude of a Rocket An experimental rocket lifts off vertically. Its altitude (in feet) t seconds into flight is given by

$$s = f(t) = -t^3 + 96t^2 + 5 \qquad (t \geq 0)$$

a. Find an expression v for the rocket's velocity at any time t.
b. Compute the rocket's velocity when $t = 0, 30, 50, 64,$ and 70. Interpret your results.
c. Using the results from the solution to part (b) and the observation that at the highest point in its trajectory the rocket's velocity is zero, find the maximum altitude attained by the rocket.

Solution

a. The rocket's velocity at any time t is given by

$$v = f'(t) = -3t^2 + 192t$$

b. The rocket's velocity when $t = 0, 30, 50, 64,$ and 70 is given by

$$f'(0) = -3(0)^2 + 192(0) = 0$$
$$f'(30) = -3(30)^2 + 192(30) = 3060$$
$$f'(50) = -3(50)^2 + 192(50) = 2100$$
$$f'(64) = -3(64)^2 + 192(64) = 0$$
$$f'(70) = -3(70)^2 + 192(70) = -1260$$

or $0, 3060, 2100, 0,$ and -1260 feet per second (ft/sec).

Thus, the rocket has an initial velocity of 0 ft/sec at $t = 0$ and accelerates to a velocity of 3060 ft/sec at $t = 30$. Fifty seconds into the flight, the rocket's velocity is 2100 ft/sec, which is less than the velocity at $t = 30$. This means that the rocket begins to decelerate after an initial period of acceleration. (Later on, we will learn how to determine the rocket's maximum velocity.)

The deceleration continues: The velocity is 0 ft/sec at $t = 64$ and -1260 ft/sec when $t = 70$. This result tells us that 70 seconds into flight, the rocket is heading back to the earth with a speed of 1260 ft/sec.

c. The results of part (b) show that the rocket's velocity is zero when $t = 64$. At this instant, the rocket's maximum altitude is

$$s = f(64) = -(64)^3 + 96(64)^2 + 5$$
$$= 131{,}077$$

or 131,077 feet. A sketch of the graph of f appears in Figure 4.

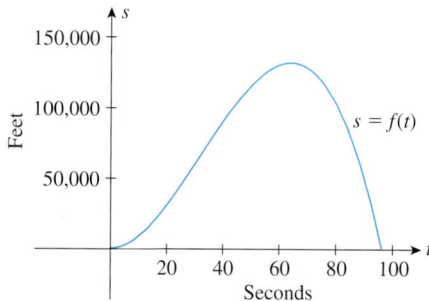

FIGURE 4
The rocket's altitude t seconds into flight is given by $f(t)$.

You may have observed that the domain of the function f in Example 8 is restricted, for practical reasons, to the interval $[0, \infty)$. Since the definition of the derivative of a function f at a number a requires that f be defined in an open interval containing a, the derivative of f is not, strictly speaking, defined at 0. But notice that the function f can, in fact, be defined for all values of t, and hence it makes sense to calculate $f'(0)$. You will encounter situations such as this throughout the book, especially in exercises pertaining to real-world applications. The nature of the functions appearing in these applications obviates the necessity to consider "one-sided" derivatives.

Exploring with TECHNOLOGY

Refer to Example 8.

1. Use a graphing utility to plot the graph of the velocity function
$$v = f'(t) = -3t^2 + 192t$$
using the viewing window $[0, 120] \times [-5000, 5000]$. Then, using ZOOM and TRACE or the root-finding capability of your graphing utility, verify that $f'(64) = 0$.

2. Plot the graph of the position function of the rocket
$$s = f(t) = -t^3 + 96t^2 + 5$$
using the viewing window $[0, 120] \times [0, 150{,}000]$. Then, using ZOOM and TRACE repeatedly, verify that the maximum altitude of the rocket is 131,077 feet.

3. Use ZOOM and TRACE or the root-finding capability of your graphing utility to find when the rocket returns to the earth.

3.1 Self-Check Exercises

1. Find the derivative of each function using the rules of differentiation.
   a. $f(x) = 1.5x^2 + 2x^{1.5}$
   b. $g(x) = 2\sqrt{x} + \dfrac{3}{\sqrt{x}}$

2. Let $f(x) = 2x^3 - 3x^2 + 2x - 1$.
   a. Compute $f'(x)$.
   b. What is the slope of the tangent line to the graph of f when $x = 2$?
   c. What is the rate of change of the function f at $x = 2$?

3. A certain country's gross domestic product (GDP) (in billions of dollars) is described by the function
$$G(t) = -2t^3 + 45t^2 + 20t + 6000 \quad (0 \le t \le 11)$$
where $t = 0$ corresponds to the beginning of 2002.
   a. At what rate was the GDP changing at the beginning of 2007? At the beginning of 2009? At the beginning of 2012?
   b. What was the average rate of growth of the GDP from the beginning of 2007 to the beginning of 2012?

Solutions to Self-Check Exercises 3.1 can be found on page 169.

3.1 Concept Questions

1. State the following rules of differentiation in your own words.
   a. The rule for differentiating a constant function
   b. The power rule
   c. The constant multiple rule
   d. The sum rule

2. If $f'(2) = 3$ and $g'(2) = -2$, find
   a. $h'(2)$ if $h(x) = 2f(x)$
   b. $F'(2)$ if $F(x) = 3f(x) - 4g(x)$

3. Suppose f and g are differentiable functions and a and b are nonzero numbers. Find $F'(x)$ if
   a. $F(x) = af(x) + bg(x)$
   b. $F(x) = \dfrac{f(x)}{a}$

3.1 Exercises

In Exercises 1–34, find the derivative of the function f by using the rules of differentiation.

1. $f(x) = -3$
2. $f(x) = 365$
3. $f(x) = x^5$
4. $f(x) = x^7$
5. $f(x) = x^{3.1}$
6. $f(x) = x^{0.8}$
7. $f(x) = 3x^2$
8. $f(x) = -2x^3$
9. $f(r) = \pi r^2$
10. $f(r) = \dfrac{4}{3}\pi r^3$
11. $f(x) = 9x^{1/3}$
12. $f(x) = \dfrac{5}{4}x^{4/5}$
13. $f(x) = 3\sqrt{x}$
14. $f(u) = \dfrac{2}{\sqrt{u}}$
15. $f(x) = 7x^{-12}$
16. $f(x) = 0.3x^{-1.2}$
17. $f(x) = 5x^2 - 3x + 7$
18. $f(x) = x^3 - 3x^2 + 1$
19. $f(x) = -x^3 + 2x^2 - 6$
20. $f(x) = x^4 - 2x^2 + 5$
21. $f(x) = 0.03x^2 - 0.4x + 10$
22. $f(x) = 0.002x^3 - 0.05x^2 + 0.1x - 20$
23. $f(x) = \dfrac{2x^3 - 4x^2 + 3}{x}$
24. $f(x) = \dfrac{x^3 + 2x^2 + x - 1}{x}$
25. $f(x) = 4x^4 - 3x^{5/2} + 2$
26. $f(x) = 5x^{4/3} - \dfrac{2}{3}x^{3/2} + x^2 - 3x + 1$
27. $f(x) = 5x^{-1} + 4x^{-2}$
28. $f(x) = -\dfrac{1}{3}(x^{-3} - x^6)$
29. $f(t) = -\dfrac{4}{t^4} - \dfrac{3}{t^3} + \dfrac{2}{t}$
30. $f(x) = \dfrac{5}{x^3} - \dfrac{2}{x^2} - \dfrac{1}{x} + 200$
31. $f(x) = 3x - 5\sqrt{x}$
32. $f(t) = 2t^2 + \sqrt{t^3}$
33. $f(x) = \dfrac{2}{x^2} - \dfrac{3}{x^{1/3}}$
34. $f(x) = \dfrac{3}{x^3} + \dfrac{4}{\sqrt{x}} + 1$

35. Let $f(x) = 2x^3 - 4x$. Find:
    a. $f'(-2)$
    b. $f'(0)$
    c. $f'(2)$

36. Let $f(x) = 4x^{5/4} + 2x^{3/2} + x$. Find:
    a. $f'(4)$
    b. $f'(16)$

In Exercises 37–40, find each limit by evaluating the derivative of a suitable function at an appropriate point.
Hint: Look at the definition of the derivative.

37. $\displaystyle\lim_{h \to 0} \dfrac{(1+h)^3 - 1}{h}$

38. $\displaystyle\lim_{x \to 1} \dfrac{x^5 - 1}{x - 1}$
    Hint: Let $h = x - 1$.

39. $\displaystyle\lim_{h \to 0} \dfrac{3(2+h)^2 - (2+h) - 10}{h}$

40. $\displaystyle\lim_{t \to 0} \dfrac{1 - (1+t)^2}{t(1+t)^2}$

In Exercises 41–44, find the slope and an equation of the tangent line to the graph of the function f at the specified point.

41. $f(x) = 2x^2 - 3x + 4;\ (2, 6)$
42. $f(x) = -\dfrac{5}{3}x^2 + 2x + 2;\ \left(-1, -\dfrac{5}{3}\right)$
43. $f(x) = x^4 - 3x^3 + 2x^2 - x + 1;\ (2, -1)$
44. $f(x) = \sqrt{x} + \dfrac{1}{\sqrt{x}};\ \left(4, \dfrac{5}{2}\right)$

45. Let $f(x) = x^3$.
    a. Find the point on the graph of f where the tangent line is horizontal.
    b. Sketch the graph of f and draw the horizontal tangent line.

46. Let $f(x) = x^3 - 4x^2$. Find the points on the graph of f where the tangent line is horizontal.

47. Let $f(x) = x^3 + 1$.
    a. Find the points on the graph of f where the slope of the tangent line is equal to 12.
    b. Find the equation(s) of the tangent line(s) of part (a).
    c. Sketch the graph of f showing the tangent line(s).

48. Let $f(x) = \frac{2}{3}x^3 + x^2 - 12x + 6$. Find the values of x for which:
    a. $f'(x) = -12$ b. $f'(x) = 0$
    c. $f'(x) = 12$

49. Let $f(x) = \frac{1}{4}x^4 - \frac{1}{3}x^3 - x^2$. Find the points on the graph of f where the slope of the tangent line is equal to:
    a. $-2x$ b. 0   c. $10x$

50. A straight line perpendicular to and passing through the point of tangency of the tangent line is called the *normal* to the curve at that point. Find an equation of the tangent line and the normal to the curve $y = x^3 - 3x + 1$ at the point $(2, 3)$.

51. **GROWTH OF A CANCEROUS TUMOR** The volume of a spherical cancerous tumor is given by the function
$$V(r) = \frac{4}{3}\pi r^3$$
where r is the radius of the tumor in centimeters. Find the rate of change in the volume of the tumor with respect to its radius when
    a. $r = \frac{2}{3}$ cm   b. $r = \frac{5}{4}$ cm

52. **VELOCITY OF BLOOD IN AN ARTERY** The velocity (in centimeters/second) of blood r cm from the central axis of an artery is given by
$$v(r) = k(R^2 - r^2)$$
where k is a constant and R is the radius of the artery (see the accompanying figure). Suppose $k = 1000$ and $R = 0.2$ cm. Find $v(0.1)$ and $v'(0.1)$, and interpret your results.

Blood vessel

53. **SALES OF DIGITAL CAMERAS** The worldwide shipments of digital point-and-shoot cameras are given by
$$N(t) = 16.3t^{0.8766} \quad (1 \le t \le 8)$$
where $N(t)$ is measured in millions and t is measured in years, with $t = 1$ corresponding to 2001.

a. How many digital cameras were sold in 2001 ($t = 1$)?
b. How fast were sales increasing in 2001?
c. What were the sales in 2005?
d. How fast did the sales grow in 2005?

Source: International Data Corp.

54. **ONLINE BUYERS** As use of the Internet grows, so does the number of consumers who shop online. The number of online buyers, as a percentage of net users, is expected to be
$$P(t) = 53t^{0.12} \quad (1 \le t \le 7)$$
where t is measured in years, with $t = 1$ corresponding to the beginning of 2002.

a. How many online buyers, as a percentage of net users, were there at the beginning of 2007?
b. How fast was the number of online buyers, as a percentage of net users, changing at the beginning of 2007?

Source: Strategy Analytics.

55. **MARRIED COUPLES WITH CHILDREN** The percentage of families that were married couples with children between 1970 and 2000 is approximately
$$P(t) = \frac{49.6}{t^{0.27}} \quad (1 \le t \le 4)$$
where t is measured in decades, with $t = 1$ corresponding to 1970.

a. What percentage of families were married couples with children in 1970? In 1980? In 1990? In 2000?
b. How fast was the percentage of families that were married couples with children changing in 1980? In 1990?

Source: U.S. Census Bureau.

56. **EFFECT OF STOPPING ON AVERAGE SPEED** According to data from a study, the average speed of your trip A (in miles per hour) is related to the number of stops per mile you make on the trip x by the equation.
$$A = \frac{26.5}{x^{0.45}}$$
Compute dA/dx for $x = 0.25$ and $x = 2$. How is the rate of change with respect to x of the average speed of your trip affected by the number of stops per mile?

Source: General Motors.

57. **ONLINE VIDEO VIEWERS** As broadband Internet grows more popular, video services such as YouTube will continue to expand. The number of online video viewers (in millions) is projected to grow according to the rule
$$N(t) = 52t^{0.531} \quad (1 \le t \le 10)$$
where $t = 1$ corresponds to 2003.

a. What was the projected number of online video viewers in 2010?
b. How fast was the projected number of online video viewers changing in 2010?

Source: eMarketer.com.

58. Demand Functions The demand function for the Luminar desk lamp is given by
$$p = f(x) = -0.1x^2 - 0.4x + 35$$
where x is the quantity demanded in thousands and p is the unit price in dollars.
a. Find $f'(x)$.
b. What is the rate of change of the unit price when the quantity demanded is 10,000 units ($x = 10$)? What is the unit price at that level of demand?

59. Stopping Distance of a Racing Car During a test by the editors of an auto magazine, the distance s (in feet) traveled by the MacPherson X-2 racing car t seconds after the brakes were applied conformed to the rule
$$s = f(t) = 120t - 15t^2 \quad (t \geq 0)$$
a. Find an expression for the car's velocity v at any time t.
b. What was the car's velocity when the brakes were first applied?
c. What was the car's stopping distance for that particular test?
Hint: The stopping time is found by setting $v = 0$.

60. Instant Messaging Accounts Mobile instant messaging (IM) is a small portion of total IM usage, but it is growing sharply. The function
$$P(t) = 0.257t^2 + 0.57t + 3.9 \quad (0 \leq t \leq 4)$$
gives the mobile IM accounts as a percentage of total enterprise IM accounts from 2006 ($t = 0$) through 2010 ($t = 4$).
a. What percentage of total enterprise IM accounts were the mobile accounts in 2008?
b. How fast was this percentage changing in 2008?
Source: The Radical Group.

61. Child Obesity The percentage of obese children, ages 12–19 years, in the United States has grown dramatically in recent years. The percentage of obese children from 1980 through the year 2000 is approximated by the function
$$P(t) = -0.0105t^2 + 0.735t + 5 \quad (0 \leq t \leq 20)$$
where t is measured in years, with $t = 0$ corresponding to the beginning of 1980.
a. What percentage of children were obese at the beginning of 1980? At the beginning of 1990? At the beginning of the year 2000?
b. How fast was the percentage of obese children changing at the beginning of 1985? At the beginning of 1990?
Source: Centers for Disease Control and Prevention.

62. Spending on Medicare Based on the current eligibility requirement, a study conducted in 2004 showed that federal spending on entitlement programs, particularly Medicare, would grow enormously in the future. The study predicted that spending on Medicare, as a percentage of the gross domestic product (GDP), will be
$$P(t) = 0.27t^2 + 1.4t + 2.2 \quad (0 \leq t \leq 5)$$
percent in year t, where t is measured in decades, with $t = 0$ corresponding to 2000.
a. How fast was the spending on Medicare, as a percentage of the GDP, growing in 2010? How fast will it be growing in 2020?
b. What was the predicted spending on Medicare in 2010? What will it be in 2020?
Source: Congressional Budget Office.

63. Fisheries The total groundfish population on Georges Bank in New England between 1989 and 1999 is approximated by the function
$$f(t) = 5.303t^2 - 53.977t + 253.8 \quad (0 \leq t \leq 10)$$
where $f(t)$ is measured in thousands of metric tons and t in years, with $t = 0$ corresponding to the beginning of 1989.
a. What was the rate of change of the groundfish population at the beginning of 1994? At the beginning of 1996?
b. Fishing restrictions were imposed on Dec. 7, 1994. Were the conservation measures effective?
Source: New England Fishery Management Council.

64. Worker Efficiency An efficiency study conducted for Elektra Electronics showed that the number of Space Commander walkie-talkies assembled by the average worker during the morning shift t hr after starting work at 8 A.M. is given by
$$N(t) = -t^3 + 6t^2 + 15t \quad (0 \leq t \leq 4)$$
a. Find the rate at which the average worker will be assembling walkie-talkies t hr after starting work.
b. At what rate will the average worker be assembling walkie-talkies at 10 A.M.? At 11 A.M.?
c. How many walkie-talkies will the average worker assemble between 10 A.M. and 11 A.M.?

65. Consumer Price Index An economy's consumer price index (CPI) is described by the function
$$I(t) = -0.2t^3 + 3t^2 + 100 \quad (0 \leq t \leq 10)$$
where $t = 0$ corresponds to 2002.
a. At what rate was the CPI changing in 2007? In 2009? In 2012?
b. What was the average rate of increase in the CPI over the period from 2007 to 2012?

66. Effect of Advertising on Sales The relationship between the amount of money x that Cannon Precision Instruments spends on advertising and the company's total sales $S(x)$ is given by the function
$$S(x) = -0.002x^3 + 0.6x^2 + x + 500 \quad (0 \leq x \leq 200)$$
where x is measured in thousands of dollars. Find the rate of change of the sales with respect to the amount of money spent on advertising. Are Cannon's total sales increasing at a faster rate when the amount of money spent on advertising is (a) $100,000 or (b) $150,000?

67. Supply Functions The supply function for a certain make of satellite radio is given by

$$p = f(x) = 0.0001x^{5/4} + 10$$

where x is the quantity supplied and p is the unit price in dollars.
a. Find $f'(x)$.
b. What is the rate of change of the unit price if the quantity supplied is 10,000 satellite radios?

68. Population Growth A study prepared for a Sunbelt town's chamber of commerce projected that the town's population in the next 3 years will grow according to the rule

$$P(t) = 50{,}000 + 30t^{3/2} + 20t$$

where $P(t)$ denotes the population t months from now. How fast will the population be increasing 9 months and 16 months from now?

69. Average Speed of a Vehicle on a Highway The average speed of a vehicle on a stretch of Route 134 between 6 A.M. and 10 A.M. on a typical weekday is approximated by the function

$$f(t) = 20t - 40\sqrt{t} + 50 \qquad (0 \le t \le 4)$$

where $f(t)$ is measured in miles per hour and t is measured in hours, with $t = 0$ corresponding to 6 A.M.
a. Compute $f'(t)$.
b. What is the average speed of a vehicle on that stretch of Route 134 at 6 A.M.? At 7 A.M.? At 8 A.M.?
c. How fast is the average speed of a vehicle on that stretch of Route 134 changing at 6:30 A.M.? At 7 A.M.? At 8 A.M.?

70. Curbing Population Growth Five years ago, the government of a Pacific Island country launched an extensive propaganda campaign aimed toward curbing the country's population growth. According to the Census Department, the population (measured in thousands of people) for the following 4 years was

$$P(t) = -\frac{1}{3}t^3 + 64t + 3000$$

where t is measured in years and $t = 0$ corresponds to the start of the campaign. Find the rate of change of the population at the end of years 1, 2, 3, and 4. Was the plan working?

71. Conservation of Species A certain species of turtle faces extinction because dealers collect truckloads of turtle eggs to be sold as aphrodisiacs. After severe conservation measures are implemented, it is hoped that the turtle population will grow according to the rule

$$N(t) = 2t^3 + 3t^2 - 4t + 1000 \qquad (0 \le t \le 10)$$

where $N(t)$ denotes the population at the end of year t. Find the rate of growth of the turtle population when $t = 2$ and $t = 8$. What will be the population 10 years after the conservation measures are implemented?

72. Flight of a Model Rocket The altitude (in feet) of a model rocket t sec into a trial flight is given by

$$s = f(t) = -2t^3 + 12t^2 + 5 \qquad (t \ge 0)$$

a. Find an expression v for the rocket's velocity at any time t.
b. Compute the rocket's vertical velocity when $t = 0, 2, 4,$ and 6. Interpret your results.
c. Using the results from the solution to part (b), find the maximum altitude attained by the rocket.
Hint: At its highest point, the velocity of the rocket is zero.

73. Obesity in America The body mass index (BMI) measures body weight in relation to height. A BMI of 25 to 29.9 is considered overweight, a BMI of 30 or more is considered obese, and a BMI of 40 or more is morbidly obese. The percentage of the U.S. population that is obese is approximated by the function

$$P(t) = 0.0004t^3 + 0.0036t^2 + 0.8t + 12 \qquad (0 \le t \le 13)$$

where t is measured in years, with $t = 0$ corresponding to the beginning of 1991.
a. What percentage of the U.S. population was deemed obese at the beginning of 1991? At the beginning of 2004?
b. How fast was the percentage of the U.S. population that is deemed obese changing at the beginning of 1991? At the beginning of 2004?

(*Note:* A formula for calculating the BMI of a person is given in Exercise 33, page 542.)
Source: Centers for Disease Control and Prevention.

74. Health-Care Spending Despite efforts at cost containment, the cost of the Medicare program is increasing. Two major reasons for this increase are an aging population and extensive use by physicians of new technologies. Based on data from the Health Care Financing Administration and the U.S. Census Bureau, health-care spending through the year 2000 may be approximated by the function

$$S(t) = 0.02836t^3 - 0.05167t^2 + 9.60881t + 41.9$$
$$(0 \le t \le 35)$$

where $S(t)$ is the spending in billions of dollars and t is measured in years, with $t = 0$ corresponding to the beginning of 1965.
a. Find an expression for the rate of change of health-care spending at any time t.
b. How fast was health-care spending changing at the beginning of 1980? At the beginning of 2000?
c. What was the amount of health-care spending at the beginning of 1980? At the beginning of 2000?
Source: Health Care Financing Administration and U.S. Census Bureau.

75. Aging Population The population age 65 and over (in millions) of developed countries from 2005 through 2034 is projected to be

$$f(t) = 3.567t + 175.2 \qquad (5 \le t \le 35)$$

where t is measured in years and $t = 5$ corresponds to 2005. On the other hand, the population age 65 and over of underdeveloped/emerging countries over the same period is projected to be

$$g(t) = 0.46t^2 + 0.16t + 287.8 \qquad (5 \le t \le 35)$$

a. What does the function $D = g + f$ represent?
b. Find D' and $D'(10)$, and interpret your results.
Source: U.S. Census Bureau, United Nations.

76. SHORTAGE OF NURSES The projected number of nurses (in millions) from 2000 through 2015 is given by

$$N(t) = \begin{cases} 1.9 & \text{if } 0 \le t < 5 \\ -0.0004t^2 + 0.038t + 1.72 & \text{if } 5 \le t \le 15 \end{cases}$$

where $t = 0$ corresponds to 2000. The projected number of nursing jobs (in millions) over the same period is

$$J(t) = \begin{cases} -0.0002t^2 + 0.032t + 2 & \text{if } 0 \le t < 10 \\ -0.0016t^2 + 0.12t + 1.26 & \text{if } 10 \le t \le 15 \end{cases}$$

a. Find the rule for the function $G = J - N$ giving the gap between the supply and the demand of nurses from 2000 through 2015.
b. How fast was the gap between the supply and the demand of nurses changing in 2008? In 2012?

Source: U.S. Department of Health and Human Services.

In Exercises 77 and 78, determine whether the statement is true or false. If it is true, explain why it is true. If it is false, give an example to show why it is false.

77. If f and g are differentiable, then

$$\frac{d}{dx}[2f(x) - 5g(x)] = 2f'(x) - 5g'(x)$$

78. If $f(x) = \pi^x$, then $f'(x) = x\pi^{x-1}$.

79. Prove the Power Rule (Rule 2) for the special case $n = 3$.

Hint: Compute $\lim\limits_{h \to 0} \left[\dfrac{(x + h)^3 - x^3}{h} \right]$.

3.1 Solutions to Self-Check Exercises

1. a. $f'(x) = \dfrac{d}{dx}(1.5x^2) + \dfrac{d}{dx}(2x^{1.5})$

$= (1.5)(2x) + (2)(1.5x^{0.5})$

$= 3x + 3x^{0.5}$

b. $g'(x) = \dfrac{d}{dx}(2x^{1/2}) + \dfrac{d}{dx}(3x^{-1/2})$

$= (2)\left(\dfrac{1}{2}x^{-1/2}\right) + (3)\left(-\dfrac{1}{2}x^{-3/2}\right)$

$= x^{-1/2} - \dfrac{3}{2}x^{-3/2} = \dfrac{1}{\sqrt{x}} - \dfrac{3}{2\sqrt{x^3}}$

2. a. $f'(x) = \dfrac{d}{dx}(2x^3) - \dfrac{d}{dx}(3x^2) + \dfrac{d}{dx}(2x) - \dfrac{d}{dx}(1)$

$= (2)(3x^2) - (3)(2x) + 2$

$= 6x^2 - 6x + 2$

b. The slope of the tangent line to the graph of f when $x = 2$ is given by

$$f'(2) = 6(2)^2 - 6(2) + 2 = 14$$

c. The rate of change of f at $x = 2$ is given by $f'(2)$. Using the results of part (b), we see that the required rate of change is 14 units/unit change in x.

3. a. The rate at which the GDP was changing at any time t $(0 < t < 11)$ is given by

$$G'(t) = -6t^2 + 90t + 20$$

In particular, the rates of change of the GDP at the beginning of the years 2007 ($t = 5$), 2009 ($t = 7$), and 2012 ($t = 10$) are given by

$$G'(5) = 320 \qquad G'(7) = 356 \qquad G'(10) = 320$$

respectively—that is, by $320 billion/year, $356 billion/year, and $320 billion/year, respectively.

b. The average rate of growth of the GDP over the period from the beginning of 2007 ($t = 5$) to the beginning of 2012 ($t = 10$) is given by

$$\dfrac{G(10) - G(5)}{10 - 5} = \dfrac{[-2(10)^3 + 45(10)^2 + 20(10) + 6000]}{5}$$

$$- \dfrac{[-2(5)^3 + 45(5)^2 + 20(5) + 6000]}{5}$$

$$= \dfrac{8700 - 6975}{5}$$

or $345 billion/year.

USING TECHNOLOGY

Finding the Rate of Change of a Function

We can use the numerical derivative operation of a graphing utility to obtain the value of the derivative at a given value of x. Since the derivative of a function f measures the rate of change of the function with respect to x, the numerical derivative operation can be used to answer questions pertaining to the rate of change of one quantity y with respect to another quantity x, where $y = f(x)$, for a specific value of x.

(continued)

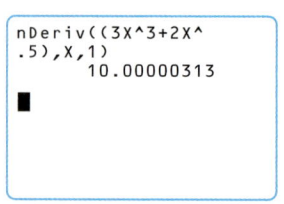

FIGURE T1
The TI-83/84 numerical derivative screen for computing $f'(1)$

EXAMPLE 1 Let $y = 3t^3 + 2\sqrt{t}$.

a. Use the numerical derivative operation of a graphing utility to find how fast y is changing with respect to t when $t = 1$.
b. Verify the result of part (a), using the rules of differentiation of this section.

Solution

a. Write $f(t) = 3t^3 + 2\sqrt{t}$. Using the numerical derivative operation of a graphing utility, we find that the rate of change of y with respect to t when $t = 1$ is given by $f'(1) = 10$ (Figure T1).
b. Here, $f(t) = 3t^3 + 2t^{1/2}$ and

$$f'(t) = 9t^2 + 2\left(\frac{1}{2}t^{-1/2}\right) = 9t^2 + \frac{1}{\sqrt{t}}$$

Using this result, we see that when $t = 1$, y is changing at the rate of

$$f'(1) = 9(1^2) + \frac{1}{\sqrt{1}} = 10$$

units per unit change in t, as obtained earlier.

APPLIED EXAMPLE 2 Fuel Economy of Cars According to data obtained from the U.S. Department of Energy and the Shell Development Company, a typical car's fuel economy depends on the speed it is driven and is approximated by the function

$$f(x) = 0.00000310315x^4 - 0.000455174x^3 + 0.00287869x^2 + 1.25986x \quad (0 \le x \le 75)$$

where x is measured in miles per hour and $f(x)$ is measured in miles per gallon (mpg).

a. Use a graphing utility to graph the function f on the interval $[0, 75]$.
b. Find the rate of change of f when $x = 20$ and when $x = 50$.
c. Interpret your results.

Source: U.S. Department of Energy and the Shell Development Company.

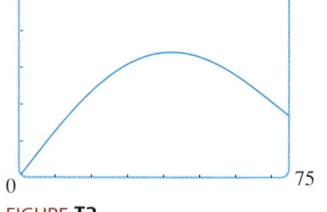

FIGURE T2
The graph of the function f on the interval $[0, 75]$

Solution

a. The graph is shown in Figure T2.
b. Using the numerical derivative operation of a graphing utility, we see that $f'(20) = 0.9280996$. The rate of change of f when $x = 50$ is given by $f'(50) = -0.3145009995$. (See Figure T3a and T3b.)

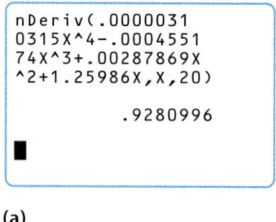

(a)   (b)

FIGURE T3
The TI-83/84 numerical derivative screen for computing (a) $f'(20)$ and (b) $f'(50)$

c. The results of part (b) tell us that when a typical car is being driven at 20 mph, its fuel economy increases at the rate of approximately 0.9 mpg per 1 mph increase in its speed. At a speed of 50 mph, its fuel economy decreases at the rate of approximately 0.3 mpg per 1 mph increase in its speed.

TECHNOLOGY EXERCISES

In Exercises 1–6, use the numerical derivative operation to find the rate of change of f at the given value of x. Give your answer accurate to four decimal places.

1. $f(x) = 4x^5 - 3x^3 + 2x^2 + 1$; $x = 0.5$
2. $f(x) = -x^5 + 4x^2 + 3$; $x = 0.4$
3. $f(x) = x - 2\sqrt{x}$; $x = 3$
4. $f(x) = \dfrac{\sqrt{x} - 1}{x}$; $x = 2$
5. $f(x) = x^{1/2} - x^{1/3}$; $x = 1.2$
6. $f(x) = 2x^{5/4} + x$; $x = 2$

7. **CARBON MONOXIDE IN THE ATMOSPHERE** The projected average global atmospheric concentration of carbon monoxide is approximated by the function

$$f(t) = 0.881443t^4 - 1.45533t^3 + 0.695876t^2 + 2.87801t + 293 \quad (0 \le t \le 4)$$

where t is measured in 40-year intervals, with $t = 0$ corresponding to the beginning of 1860, and $f(t)$ is measured in parts per million by volume.
   a. Plot the graph of f in the viewing window $[0, 4] \times [280, 400]$.
   b. Use a graphing utility to estimate how fast the projected average global atmospheric concentration of carbon monoxide was changing at the beginning of 1900 ($t = 1$) and at the beginning of 2010 ($t = 4$).

   Source: Meadows et al., *Beyond the Limits.*

8. **SPREAD OF HIV** The estimated number of children newly infected with HIV through mother-to-child contact worldwide is given by

$$f(t) = -0.2083t^3 + 3.0357t^2 + 44.0476t + 200.2857 \quad (0 \le t \le 12)$$

where $f(t)$ is measured in thousands and t is measured in years, with $t = 0$ corresponding to the beginning of 1990.
   a. Plot the graph of f in the viewing window $[0, 12] \times [0, 800]$.
   b. How fast was the estimated number of children newly infected with HIV through mother-to-child contact worldwide increasing at the beginning of the year 2000?

   Source: United Nations.

9. **MODELING WITH DATA** A hedge fund is a lightly regulated pool of professionally managed money. The assets (in billions of dollars) of hedge funds from the beginning of 1999 ($t = 0$) through the beginning of 2004 are given in the following table:

Year	1999	2000	2001	2002	2003	2004
Assets ($ billions)	472	517	594	650	817	950

   a. Use **CubicReg** to find a third-degree polynomial function for the data, letting $t = 0$ correspond to the beginning of 1999.
   b. Plot the graph of the function found in part (a).
   c. Use the numerical derivative capability of your graphing utility to find the rate at which the assets of hedge funds were increasing at the beginning of 2000 and the beginning of 2003.

   Sources: Hennessee Group; Institutional Investor.

10. **MODELING WITH DATA** The number of people (in millions) enrolled in HMOs from 1994 through 2002 is given in the following table:

Year	1994	1995	1996	1997	1998	1999	2000	2001	2002
People	45.4	50.6	58.7	67.0	76.4	81.3	80.9	80.0	74.2

   a. Use **QuartReg** to find a fourth-degree polynomial regression model for this data. Let $t = 0$ correspond to 1994.
   b. Use the model to estimate the number of people enrolled in HMOs in 2000. How does this number compare with the actual number?
   c. How fast was the number of people receiving their care in an HMO changing in 2001?

   Source: Group Health Association of America.

3.2 The Product and Quotient Rules

In this section, we study two more rules of differentiation: the **Product Rule** and the **Quotient Rule**.

The Product Rule

The derivative of the product of two differentiable functions is given by the following rule:

> **Rule 5: The Product Rule**
>
> $$\frac{d}{dx}[f(x)g(x)] = f(x)g'(x) + g(x)f'(x)$$

The derivative of the product of two functions is the first function times the derivative of the second plus the second function times the derivative of the first.

The Product Rule may be extended to the case involving the product of any finite number of functions (see Exercise 68, p. 180). We prove the Product Rule at the end of this section.

 The derivative of the product of two functions is *not* given by the product of the derivatives of the functions; that is, in general

$$\frac{d}{dx}[f(x)g(x)] \neq f'(x)g'(x)$$

For example, if $f(x) = x$ and $g(x) = 2x^2$. Then

$$\frac{d}{dx}[f(x)g(x)] = \frac{d}{dx}[x(2x^2)] = \frac{d}{dx}(2x^3) = 6x^2$$

On the other hand, $f'(x)g'(x) = (1)(4x) = 4x$. So

$$\frac{d}{dx}[f(x)g(x)] \neq f'(x)g'(x)$$

VIDEO   **EXAMPLE 1** Find the derivative of the function

$$f(x) = (2x^2 - 1)(x^3 + 3)$$

Solution   By the Product Rule,

$$f'(x) = (2x^2 - 1)\frac{d}{dx}(x^3 + 3) + (x^3 + 3)\frac{d}{dx}(2x^2 - 1)$$

$$= (2x^2 - 1)(3x^2) + (x^3 + 3)(4x) \qquad \text{(x²) See page 8.}$$

$$= 6x^4 - 3x^2 + 4x^4 + 12x$$

$$= 10x^4 - 3x^2 + 12x \qquad \text{Combine like terms.}$$

$$= x(10x^3 - 3x + 12) \qquad \text{Factor out } x.$$

EXAMPLE 2 Differentiate (that is, find the derivative of) the function

$$f(x) = x^3(\sqrt{x} + 1)$$

Solution   First, we express the function in exponential form, obtaining

$$f(x) = x^3(x^{1/2} + 1)$$

By the Product Rule,

$$f'(x) = x^3 \frac{d}{dx}(x^{1/2} + 1) + (x^{1/2} + 1)\frac{d}{dx}x^3$$

$$= x^3\left(\frac{1}{2}x^{-1/2}\right) + (x^{1/2} + 1)(3x^2)$$

$$= \frac{1}{2}x^{5/2} + 3x^{5/2} + 3x^2$$

$$= \frac{7}{2}x^{5/2} + 3x^2$$

Note   We can also solve the problem by first expanding the product before differentiating f. Examples for which this is not possible will be considered in Section 3.3, where the true value of the Product Rule will be appreciated.

The Quotient Rule

The derivative of the quotient of two differentiable functions is given by the following rule:

> **Rule 6: The Quotient Rule**
> $$\frac{d}{dx}\left[\frac{f(x)}{g(x)}\right] = \frac{g(x)f'(x) - f(x)g'(x)}{[g(x)]^2} \qquad (g(x) \neq 0)$$

As an aid to remembering this expression, observe that it has the following form:

$$\frac{d}{dx}\left[\frac{f(x)}{g(x)}\right] = \frac{(\text{Denominator})\begin{pmatrix}\text{Derivative of}\\ \text{numerator}\end{pmatrix} - (\text{Numerator})\begin{pmatrix}\text{Derivative of}\\ \text{denominator}\end{pmatrix}}{(\text{Square of denominator})}$$

For a proof of the Quotient Rule, see Exercise 69, page 180.

 The derivative of a quotient is *not* equal to the quotient of the derivatives; that is,

$$\frac{d}{dx}\left[\frac{f(x)}{g(x)}\right] \neq \frac{f'(x)}{g'(x)}$$

For example, if $f(x) = x^3$ and $g(x) = x^2$, then

$$\frac{d}{dx}\left[\frac{f(x)}{g(x)}\right] = \frac{d}{dx}\left(\frac{x^3}{x^2}\right) = \frac{d}{dx}(x) = 1$$

which is *not* equal to

$$\frac{f'(x)}{g'(x)} = \frac{\dfrac{d}{dx}(x^3)}{\dfrac{d}{dx}(x^2)} = \frac{3x^2}{2x} = \frac{3}{2}x$$

EXAMPLE 3 Find $f'(x)$ if $f(x) = \dfrac{x}{2x - 4}$.

Solution Using the Quotient Rule, we obtain

$$f'(x) = \frac{(2x - 4)\dfrac{d}{dx}(x) - x\dfrac{d}{dx}(2x - 4)}{(2x - 4)^2}$$

$$= \frac{(2x - 4)(1) - x(2)}{(2x - 4)^2}$$

$$= \frac{2x - 4 - 2x}{(2x - 4)^2} = -\frac{4}{(2x - 4)^2}$$

EXAMPLE 4 Find $f'(x)$ if $f(x) = \dfrac{x^2 + 1}{x^2 - 1}$.

Solution By the Quotient Rule,

$$f'(x) = \frac{(x^2 - 1)\dfrac{d}{dx}(x^2 + 1) - (x^2 + 1)\dfrac{d}{dx}(x^2 - 1)}{(x^2 - 1)^2}$$

$$= \frac{(x^2 - 1)(2x) - (x^2 + 1)(2x)}{(x^2 - 1)^2}$$

$$= \frac{2x^3 - 2x - 2x^3 - 2x}{(x^2 - 1)^2}$$

$$= -\frac{4x}{(x^2 - 1)^2}$$

EXAMPLE 5 Find $h'(x)$ if $h(x) = \dfrac{\sqrt{x}}{x^2 + 1}$.

Solution Rewrite $h(x)$ in the form $h(x) = \dfrac{x^{1/2}}{x^2 + 1}$. By the Quotient Rule, we find

$$h'(x) = \frac{(x^2 + 1)\dfrac{d}{dx}(x^{1/2}) - x^{1/2}\dfrac{d}{dx}(x^2 + 1)}{(x^2 + 1)^2}$$

$$= \frac{(x^2 + 1)(\tfrac{1}{2}x^{-1/2}) - x^{1/2}(2x)}{(x^2 + 1)^2}$$

$$= \frac{\tfrac{1}{2}x^{-1/2}(x^2 + 1 - 4x^2)}{(x^2 + 1)^2} \quad \text{Factor out } \tfrac{1}{2}x^{-1/2} \text{ from the numerator.} \quad \text{(x²) See page 9.}$$

$$= \frac{1 - 3x^2}{2\sqrt{x}(x^2 + 1)^2}$$

APPLIED EXAMPLE 6 Rate of Change of DVD Sales The annual sales (in millions of dollars per year) of a DVD recording of a hit movie t years from the date of release is given by

$$S(t) = \frac{5t}{t^2 + 1}$$

a. Find the rate at which the annual sales are changing at time t.
b. How fast are the annual sales changing at the time the DVDs are released ($t = 0$)? Two years from the date of release?

Solution

a. The rate at which the annual sales are changing at time t is given by $S'(t)$. Using the Quotient Rule, we obtain

$$S'(t) = \frac{d}{dt}\left[\frac{5t}{t^2 + 1}\right] = 5\frac{d}{dt}\left[\frac{t}{t^2 + 1}\right]$$

$$= 5\left[\frac{(t^2 + 1)(1) - t(2t)}{(t^2 + 1)^2}\right] \quad \text{(x²) See page 16.}$$

$$= 5\left[\frac{t^2 + 1 - 2t^2}{(t^2 + 1)^2}\right] = \frac{5(1 - t^2)}{(t^2 + 1)^2}$$

b. The rate at which the annual sales are changing at the time the DVDs are released is given by

$$S'(0) = \frac{5(1-0)}{(0+1)^2} = 5$$

That is, they are increasing at the rate of $5 million per year per year.

Two years from the date of release, the annual sales are changing at the rate of

$$S'(2) = \frac{5(1-4)}{(4+1)^2} = -\frac{3}{5} = -0.6$$

That is, they are decreasing at the rate of $600,000 per year per year.
The graph of the function S is shown in Figure 5.

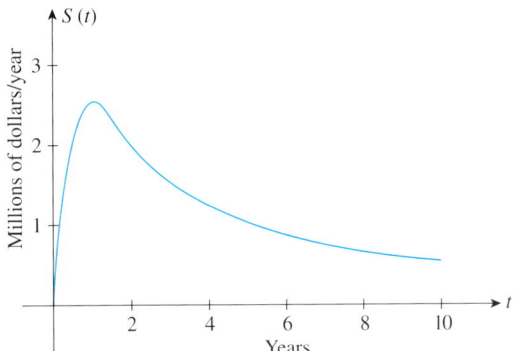

FIGURE 5
After a spectacular rise, the annual sales begin to taper off.

Exploring with TECHNOLOGY

Refer to Example 6.

1. Use a graphing utility to plot the graph of the function S, using the viewing window $[0, 10] \times [0, 3]$.
2. Use TRACE and ZOOM to determine the coordinates of the highest point on the graph of S in the interval $[0, 10]$. Interpret your results.

Explore & Discuss

Suppose the revenue of a company is given by $R(x) = xp(x)$, where x is the number of units of the product sold at a unit price of $p(x)$ dollars.

1. Compute $R'(x)$ and explain, in words, the relationship between $R'(x)$ and $p(x)$ and/or its derivative.
2. What can you say about $R'(x)$ if $p(x)$ is constant? Is this expected?

APPLIED EXAMPLE 7 Oxygen-Restoration Rate in a Pond When organic waste is dumped into a pond, the oxidation process that takes place reduces the pond's oxygen content. However, given time, nature will restore the oxygen content to its natural level. Suppose the oxygen content t days after organic waste has been dumped into the pond is given by

$$f(t) = 100 \left[\frac{t^2 + 10t + 100}{t^2 + 20t + 100} \right] \qquad (0 < t < \infty)$$

percent of its normal level.

a. Derive a general expression that gives the rate of change of the pond's oxygen level at any time t.

b. How fast is the pond's oxygen content changing 1 day, 10 days, and 20 days after the organic waste has been dumped?

Solution

a. The rate of change of the pond's oxygen level at any time t is given by the derivative of the function f. Thus, the required expression is

$$f'(t) = 100 \frac{d}{dt}\left[\frac{t^2 + 10t + 100}{t^2 + 20t + 100}\right]$$

$$= 100 \left[\frac{(t^2 + 20t + 100)\frac{d}{dt}(t^2 + 10t + 100) - (t^2 + 10t + 100)\frac{d}{dt}(t^2 + 20t + 100)}{(t^2 + 20t + 100)^2}\right]$$

$$= 100 \left[\frac{(t^2 + 20t + 100)(2t + 10) - (t^2 + 10t + 100)(2t + 20)}{(t^2 + 20t + 100)^2}\right] \quad \text{(x^2) See page 16.}$$

$$= 100 \left[\frac{2t^3 + 10t^2 + 40t^2 + 200t + 200t + 1000 - 2t^3 - 20t^2 - 20t^2 - 200t - 200t - 2000}{(t^2 + 20t + 100)^2}\right]$$

$$= 100 \left[\frac{10t^2 - 1000}{(t^2 + 20t + 100)^2}\right] \quad \text{Combine like terms in the numerator.}$$

b. The rate at which the pond's oxygen content is changing 1 day after the organic waste has been dumped is given by

$$f'(1) = 100\left[\frac{10 - 1000}{(1 + 20 + 100)^2}\right] \approx -6.76$$

That is, it is dropping at the rate of 6.8% per day. After 10 days, the rate is

$$f'(10) = 100\left[\frac{10(10)^2 - 1000}{(10^2 + 20(10) + 100)^2}\right] = 0$$

That is, it is neither increasing nor decreasing. After 20 days, the rate is

$$f'(20) = 100\left[\frac{10(20)^2 - 1000}{(20^2 + 20(20) + 100)^2}\right] \approx 0.37$$

That is, the oxygen content is increasing at the rate of 0.37% per day, and the restoration process has indeed begun. ∎

Verification of the Product Rule

We will now verify the Product Rule. If $p(x) = f(x)g(x)$, then

$$p'(x) = \lim_{h \to 0} \frac{p(x + h) - p(x)}{h} = \lim_{h \to 0} \frac{f(x + h)g(x + h) - f(x)g(x)}{h}$$

By adding $-f(x + h)g(x) + f(x + h)g(x)$ (which is zero!) to the numerator and factoring, we have

$$p'(x) = \lim_{h \to 0} \frac{f(x+h)[g(x+h) - g(x)] + g(x)[f(x+h) - f(x)]}{h}$$

$$= \lim_{h \to 0} \left\{ f(x+h) \left[\frac{g(x+h) - g(x)}{h} \right] + g(x) \left[\frac{f(x+h) - f(x)}{h} \right] \right\}$$

$$= \lim_{h \to 0} f(x+h) \left[\frac{g(x+h) - g(x)}{h} \right] + \lim_{h \to 0} g(x) \left[\frac{f(x+h) - f(x)}{h} \right] \quad \text{By Property 3 of limits}$$

$$= \lim_{h \to 0} f(x+h) \cdot \lim_{h \to 0} \frac{g(x+h) - g(x)}{h}$$

$$+ \lim_{h \to 0} g(x) \cdot \lim_{h \to 0} \frac{f(x+h) - f(x)}{h} \quad \text{By Property 4 of limits}$$

$$= f(x)g'(x) + g(x)f'(x)$$

Observe that in the last link in the chain of equalities, we have used the fact that $\lim_{h \to 0} f(x+h) = f(x)$ because f is continuous at x.

3.2 Self-Check Exercises

1. Find the derivative of $f(x) = \dfrac{2x + 1}{x^2 - 1}$.

2. What is the slope of the tangent line to the graph of
$$f(x) = (x^2 + 1)(2x^3 - 3x^2 + 1)$$
at the point $(2, 25)$? How fast is the function f changing when $x = 2$?

3. The total sales of Security Products in its first 2 years of operation are given by

$$S = f(t) = \frac{0.3t^3}{1 + 0.4t^2} \quad (0 \le t \le 2)$$

where S is measured in millions of dollars and $t = 0$ corresponds to the date Security Products began operations. How fast were the sales increasing at the beginning of the company's second year of operation?

Solutions to Self-Check Exercises 3.2 can be found on page 180.

3.2 Concept Questions

1. State the rule of differentiation in your own words.
   a. Product Rule   b. Quotient Rule

2. If $f(1) = 3$, $g(1) = 2$, $f'(1) = -1$, and $g'(1) = 4$, find
   a. $h'(1)$ if $h(x) = f(x)g(x)$ b. $F'(1)$ if $F(x) = \dfrac{f(x)}{g(x)}$

3.2 Exercises

In Exercises 1–30, find the derivative of each function.

1. $f(x) = 2x(x^2 + 1)$

2. $f(x) = 3x^2(x - 1)$

3. $f(t) = (t - 1)(2t + 1)$

4. $f(x) = (2x + 3)(3x - 4)$

5. $f(x) = (3x + 1)(x^2 - 2)$

6. $f(x) = (x + 1)(2x^2 - 3x + 1)$

7. $f(x) = (x^3 - 1)(x + 1)$

8. $f(x) = (x^3 - 12x)(3x^2 + 2x)$

9. $f(w) = (w^3 - w^2 + w - 1)(w^2 + 2)$

10. $f(x) = \dfrac{1}{5}x^5 + (x^2 + 1)(x^2 - x - 1) + 28$

11. $f(x) = (5x^2 + 1)(2\sqrt{x} - 1)$

12. $f(t) = (1 + \sqrt{t})(2t^2 - 3)$

13. $f(x) = (x^2 - 5x + 2)\left(x - \dfrac{2}{x}\right)$

14. $f(x) = (x^3 + 2x + 1)\left(2 + \dfrac{1}{x^2}\right)$

15. $f(x) = \dfrac{1}{x - 2}$

16. $g(x) = \dfrac{3}{2x + 4}$

17. $f(x) = \dfrac{2x - 1}{2x + 1}$

18. $f(t) = \dfrac{1 - 2t}{1 + 3t}$

19. $f(x) = \dfrac{1}{x^2 + 1}$

20. $f(u) = \dfrac{u}{u^2 + 1}$

21. $f(s) = \dfrac{s^2 - 4}{s + 1}$

22. $f(x) = \dfrac{x^3 - 2}{x^2 + 1}$

23. $f(x) = \dfrac{\sqrt{x} + 1}{x^2 + 1}$

24. $f(x) = \dfrac{x^2 + 1}{\sqrt{x}}$

25. $f(x) = \dfrac{x^2 + 2}{x^2 + x + 1}$

26. $f(x) = \dfrac{x + 1}{2x^2 + 2x + 3}$

27. $f(x) = \dfrac{(x + 1)(x^2 + 1)}{x - 2}$

28. $f(x) = (3x^2 - 1)\left(x^2 - \dfrac{1}{x}\right)$

29. $f(x) = \dfrac{x}{x^2 - 4} - \dfrac{x - 1}{x^2 + 4}$

30. $f(x) = \dfrac{x + \sqrt{3x}}{3x - 1}$

In Exercises 31–34, suppose f and g are functions that are differentiable at $x = 1$ and that $f(1) = 2$, $f'(1) = -1$, $g(1) = -2$, and $g'(1) = 3$. Find the value of $h'(1)$.

31. $h(x) = f(x)g(x)$

32. $h(x) = (x^2 + 1)g(x)$

33. $h(x) = \dfrac{xf(x)}{x + g(x)}$

34. $h(x) = \dfrac{f(x)g(x)}{f(x) - g(x)}$

In Exercises 35–38, find the derivative of each function and evaluate $f'(x)$ at the given value of x.

35. $f(x) = (2x - 1)(x^2 + 3); x = 1$

36. $f(x) = \dfrac{2x + 1}{2x - 1}; x = 2$

37. $f(x) = \dfrac{x}{x^4 - 2x^2 - 1}; x = -1$

38. $f(x) = (\sqrt{x} + 2x)(x^{3/2} - x); x = 4$

In Exercises 39–42, find the slope and an equation of the tangent line to the graph of the function f at the specified point.

39. $f(x) = (x^3 + 1)(x^2 - 2); (2, 18)$

40. $f(x) = \dfrac{x^2}{x + 1}; \left(2, \dfrac{4}{3}\right)$

41. $f(x) = \dfrac{x + 1}{x^2 + 1}; (1, 1)$

42. $f(x) = \dfrac{1 + 2x^{1/2}}{1 + x^{3/2}}; \left(4, \dfrac{5}{9}\right)$

43. Suppose $g(x) = x^2 f(x)$ and it is known that $f(2) = 3$ and $f'(2) = -1$. Evaluate $g'(2)$.

44. Suppose $g(x) = (x^2 + 1)f(x)$ and it is known that $f(2) = 3$ and $f'(2) = -1$. Evaluate $g'(2)$.

45. Find an equation of the tangent line to the graph of the function $f(x) = (x^3 + 1)(3x^2 - 4x + 2)$ at the point $(1, 2)$.

46. Find an equation of the tangent line to the graph of the function $f(x) = \dfrac{3x}{x^2 - 2}$ at the point $(2, 3)$.

47. Let $f(x) = (x^2 + 1)(2 - x)$. Find the point(s) on the graph of f where the tangent line is horizontal.

48. Let $f(x) = \dfrac{x}{x^2 + 1}$. Find the point(s) on the graph of f where the tangent line is horizontal.

49. Find the point(s) on the graph of the function $f(x) = (x^2 + 6)(x - 5)$ where the slope of the tangent line is equal to -2.

50. Find the point(s) on the graph of the function $f(x) = \dfrac{x + 1}{x - 1}$ where the slope of the tangent line is equal to $-\dfrac{1}{2}$.

51. A straight line perpendicular to and passing through the point of tangency of the tangent line is called the *normal* to the curve at that point. Find the equation of the tangent line and the normal to the curve

$$y = \dfrac{1}{1 + x^2}$$

at the point $(1, \tfrac{1}{2})$.

52. **CONCENTRATION OF A DRUG IN THE BLOODSTREAM** The concentration of a certain drug in a patient's bloodstream t hr after injection is given by

$$C(t) = \dfrac{0.2t}{t^2 + 1}$$

a. Find the rate at which the concentration of the drug is changing with respect to time.
b. How fast is the concentration changing $\tfrac{1}{2}$ hr, 1 hr, and 2 hr after the injection?

53. **COST OF REMOVING TOXIC WASTE** A city's main water reservoir was recently found to be contaminated with trichloroethylene, a cancer-causing chemical, as a result of an abandoned chemical dump leaching chemicals into the water. A proposal submitted to the city's council members indicates that the cost, measured in millions of dollars, of removing x% of the toxic pollutant is given by

$$C(x) = \dfrac{0.5x}{100 - x}$$

Find $C'(80)$, $C'(90)$, $C'(95)$, and $C'(99)$. What does your result tell you about the cost of removing *all* of the pollutant?

54. Drug Dosages Thomas Young has suggested the following rule for calculating the dosage of medicine for children 1 to 12 years old. If a denotes the adult dosage (in milligrams) and if t is the child's age (in years), then the child's dosage is given by

$$D(t) = \frac{at}{t + 12}$$

Suppose the adult dosage of a substance is 500 mg. Find an expression that gives the rate of change of a child's dosage with respect to the child's age. What is the rate of change of a child's dosage with respect to his or her age for a 6-year-old child? A 10-year-old child?

55. Effect of Bactericide The number of bacteria $N(t)$ in a certain culture t min after an experimental bactericide is introduced is given by

$$N(t) = \frac{10{,}000}{1 + t^2} + 2000$$

Find the rate of change of the number of bacteria in the culture 1 min and 2 min after the bactericide is introduced. What is the population of the bacteria in the culture 1 min and 2 min after the bactericide is introduced?

56. Demand Functions The demand function for the Sicard sports watch is given by

$$d(x) = \frac{50}{0.01x^2 + 1} \quad (0 \le x \le 20)$$

where x (measured in units of a thousand) is the quantity demanded per week and $d(x)$ is the unit price in dollars.
a. Find $d'(x)$.
b. Find $d'(5)$, $d'(10)$, and $d'(15)$, and interpret your results.

57. Revenue Functions Refer to Exercise 56.
a. Find an expression for the revenue function R for the Sicard sports watch.
Hint: $R(x) = xd(x)$
b. Find $R'(x)$.
c. Find $R'(8)$, $R'(10)$, and $R'(12)$, and interpret your results.

58. Profit Functions Refer to Exercise 56. The total profit function P (in thousands of dollars) for the Sicard sports watch is given by

$$P(x) = \frac{50x}{0.01x^2 + 1} - 0.025x^3 + 0.35x^2 - 10x - 30$$
$$(0 \le x \le 20)$$

where x is measured in units of a thousand.
a. Find $P(0)$, and interpret your result.
b. Find $P'(5)$ and $P'(10)$, and interpret your results.

59. Learning Curves From experience, Emory Secretarial School knows that the average student taking Advanced Typing will progress according to the rule

$$N(t) = \frac{60t + 180}{t + 6} \quad (t \ge 0)$$

where $N(t)$ measures the number of words per minute the student can type after t weeks in the course.
a. Find an expression for $N'(t)$.
b. Compute $N'(t)$ for $t = 1, 3, 4,$ and 7, and interpret your results.
c. Sketch the graph of the function N. Does it confirm the results obtained in part (b)?
d. What will be the average student's typing speed at the end of the 12-week course?

60. Box-Office Receipts The total worldwide box-office receipts for a long-running movie are approximated by the function

$$T(x) = \frac{120x^2}{x^2 + 4}$$

where $T(x)$ is measured in millions of dollars and x is the number of years since the movie's release. How fast are the total receipts changing 1 year, 3 years, and 5 years after its release?

61. Formaldehyde Levels A study on formaldehyde levels in 900 homes indicates that emissions of various chemicals can decrease over time. The formaldehyde level (parts per million) in an average home in the study is given by

$$f(t) = \frac{0.055t + 0.26}{t + 2} \quad (0 \le t \le 12)$$

where t is the age of the house in years. How fast is the formaldehyde level of the average house dropping when it is new? At the beginning of its fourth year?
Source: Bonneville Power Administration.

62. Population Growth A major corporation is building a 4325-acre complex of homes, offices, stores, schools, and churches in the rural community of Glen Cove. As a result of this development, the planners have estimated that Glen Cove's population (in thousands) t years from now will be given by

$$P(t) = \frac{25t^2 + 125t + 200}{t^2 + 5t + 40}$$

a. Find the rate at which Glen Cove's population is changing with respect to time.
b. What will be the population after 10 years? At what rate will the population be increasing when $t = 10$?

63. Optics The equation

$$\frac{1}{f} = \frac{1}{p} + \frac{1}{q}$$

sometimes called a **lens-maker's equation,** gives the relationship between the focal length f of a thin lens, the distance p of the object from the lens, and the distance q of its image from the lens. We can think of the eye as an optical system in which the ciliary muscle constantly adjusts the curvature of the cornea-lens system to focus the image on the retina. Assume that the distance from the cornea to the retina is 2.5 cm, as shown in the figure on the next page.

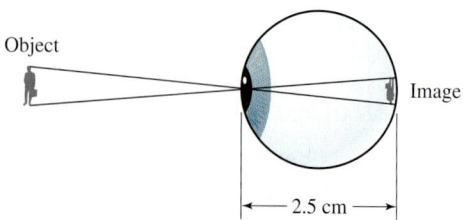

a. Find the focal length of the cornea-lens system if an object located 50 cm away is to be focused on the retina.

b. What is the rate of change of the focal length with respect to the distance of the object when the object is 50 cm away?

In Exercises 64–67, determine whether the statement is true or false. If it is true, explain why it is true. If it is false, give an example to show why it is false.

64. If f and g are differentiable, then
$$\frac{d}{dx}[f(x)g(x)] = f'(x)g'(x)$$

65. If f is differentiable, then
$$\frac{d}{dx}[xf(x)] = f(x) + xf'(x)$$

66. If f is differentiable, then
$$\frac{d}{dx}\left[\frac{f(x)}{x^2}\right] = \frac{f'(x)}{2x}$$

67. If f, g, and h are differentiable, then
$$\frac{d}{dx}\left[\frac{f(x)g(x)}{h(x)}\right] = \frac{f'(x)g(x)h(x) + f(x)g'(x)h(x) - f(x)g(x)h'(x)}{[h(x)]^2}$$

68. Extend the Product Rule for differentiation to the following case involving the product of three differentiable functions: Let $h(x) = u(x)v(x)w(x)$, and show that $h'(x) = u(x)v(x)w'(x) + u(x)v'(x)w(x) + u'(x)v(x)w(x)$.
Hint: Let $f(x) = u(x)v(x)$, $g(x) = w(x)$, and $h(x) = f(x)g(x)$, and apply the Product Rule to the function h.

69. Prove the Quotient Rule for differentiation (Rule 6).
Hint: Let $k(x) = f(x)/g(x)$, and verify the following steps:

a. $\dfrac{k(x+h) - k(x)}{h} = \dfrac{f(x+h)g(x) - f(x)g(x+h)}{hg(x+h)g(x)}$

b. By adding $[-f(x)g(x) + f(x)g(x)]$ to the numerator and simplifying, show that
$$\frac{k(x+h) - k(x)}{h} = \frac{1}{g(x+h)g(x)}$$
$$\times \left\{ \left[\frac{f(x+h) - f(x)}{h}\right] \cdot g(x) - \left[\frac{g(x+h) - g(x)}{h}\right] \cdot f(x) \right\}$$

c. $k'(x) = \lim\limits_{h \to 0} \dfrac{k(x+h) - k(x)}{h}$
$= \dfrac{g(x)f'(x) - f(x)g'(x)}{[g(x)]^2}$

3.2 Solutions to Self-Check Exercises

1. We use the Quotient Rule to obtain
$$f'(x) = \frac{(x^2 - 1)\dfrac{d}{dx}(2x+1) - (2x+1)\dfrac{d}{dx}(x^2-1)}{(x^2-1)^2}$$
$$= \frac{(x^2-1)(2) - (2x+1)(2x)}{(x^2-1)^2}$$
$$= \frac{2x^2 - 2 - 4x^2 - 2x}{(x^2-1)^2}$$
$$= \frac{-2x^2 - 2x - 2}{(x^2-1)^2}$$
$$= \frac{-2(x^2 + x + 1)}{(x^2-1)^2}$$

2. The slope of the tangent line to the graph of f at any point is given by
$$f'(x) = (x^2 + 1)\frac{d}{dx}(2x^3 - 3x^2 + 1)$$
$$+ (2x^3 - 3x^2 + 1)\frac{d}{dx}(x^2 + 1)$$
$$= (x^2 + 1)(6x^2 - 6x) + (2x^3 - 3x^2 + 1)(2x)$$

In particular, the slope of the tangent line to the graph of f when $x = 2$ is
$$f'(2) = (2^2 + 1)[6(2)^2 - 6(2)]$$
$$+ [2(2)^3 - 3(2)^2 + 1][2(2)]$$
$$= 60 + 20 = 80$$

Note that it is not necessary to simplify the expression for $f'(x)$, since we are required only to evaluate the expression at $x = 2$. We also conclude, from this result, that the function f is changing at the rate of 80 units/unit change in x when $x = 2$.

3. The rate at which the company's total sales are changing at any time t is given by

$$S'(t) = \frac{(1 + 0.4t^2)\frac{d}{dt}(0.3t^3) - (0.3t^3)\frac{d}{dt}(1 + 0.4t^2)}{(1 + 0.4t^2)^2}$$

$$= \frac{(1 + 0.4t^2)(0.9t^2) - (0.3t^3)(0.8t)}{(1 + 0.4t^2)^2}$$

Therefore, at the beginning of the second year of operation, Security Products' sales were increasing at the rate of

$$S'(1) = \frac{(1 + 0.4)(0.9) - (0.3)(0.8)}{(1 + 0.4)^2} \approx 0.52$$

or $520,000/year.

USING TECHNOLOGY

The Product and Quotient Rules

EXAMPLE 1 Let $f(x) = (2\sqrt{x} + 0.5x)(0.3x^3 + 2x - \frac{0.3}{x})$. Find $f'(0.2)$.

Solution Using the numerical derivative operation of a graphing utility, we find

$$f'(0.2) = 6.4797499802$$

See Figure T1.

```
nDeriv((2X^.5+.5
X)(.3X^3+2X-.3/X),
X,.2)
           6.4797499802
```

FIGURE T1
The TI-83/84 numerical derivative screen for computing $f'(0.2)$

APPLIED EXAMPLE 2 Importance of Time in Treating Heart Attacks According to the American Heart Association, the treatment benefit for heart attacks depends on the time until treatment and is described by the function

$$f(t) = \frac{0.44t^4 + 700}{0.1t^4 + 7} \quad (0 \le t \le 24)$$

where t is measured in hours and $f(t)$ is expressed as a percent.

a. Use a graphing utility to graph the function f using the viewing window $[0, 24] \times [0, 100]$.
b. Use a graphing utility to find the derivative of f when $t = 0$ and $t = 2$.
c. Interpret the results obtained in part (b).

Source: American Heart Association.

Solution

a. The graph of f is shown in Figure T2.
b. Using the numerical derivative operation of a graphing utility, we find

$$f'(0) \approx 0$$
$$f'(2) \approx -28.95402429$$

(see Figure T3).

FIGURE T2

```
nDeriv((.44X^4+7
00)/(.1X^4+7),X,
0)
              0
```
(a)

```
nDeriv((.44X^4+7
00)/(.1X^4+7),X,
2)
           -28.95402429
```
(b)

FIGURE T3
TI-83/84 numerical derivative screens (a) for computing $f'(0)$ and (b) for computing $f'(2)$

(continued)

c. The results of part (b) show that there is no drop in the treatment benefit when the heart attack is treated immediately. But the treatment benefit drops off at the rate of approximately 29% per hour when the time to treatment is 2 hours. Thus, it is extremely urgent that a patient suffering a heart attack receive medical attention as soon as possible.

TECHNOLOGY EXERCISES

In Exercises 1–6, use the numerical derivative operation to find the rate of change of $f(x)$ at the given value of x. Give your answer accurate to four decimal places.

1. $f(x) = (2x^2 + 1)(x^3 + 3x + 4); x = -0.5$
2. $f(x) = (\sqrt{x} + 1)(2x^2 + x - 3); x = 1.5$
3. $f(x) = \dfrac{\sqrt{x} - 1}{\sqrt{x} + 1}; x = 3$
4. $f(x) = \dfrac{\sqrt{x}(x^2 + 4)}{x^3 + 1}; x = 4$
5. $f(x) = \dfrac{\sqrt{x}(1 + x^{-1})}{x + 1}; x = 1$
6. $f(x) = \dfrac{x^2(2 + \sqrt{x})}{1 + \sqrt{x}}; x = 1$

7. **NEW CONSTRUCTION JOBS** The president of a major housing construction company claims that the number of construction jobs created in the next t months is given by

$$f(t) = 1.42\left(\dfrac{7t^2 + 140t + 700}{3t^2 + 80t + 550}\right)$$

where $f(t)$ is measured in millions of jobs per year. At what rate will construction jobs be created 1 year from now, assuming that her projection is correct?

8. **POPULATION GROWTH** A major corporation is building a 4325-acre complex of homes, offices, stores, schools, and churches in the rural community of Glen Cove. As a result of this development, the planners have estimated that Glen Cove's population (in thousands) t years from now will be given by

$$P(t) = \dfrac{25t^2 + 125t + 200}{t^2 + 5t + 40}$$

a. What will be the population 10 years from now?
b. At what rate will the population be increasing 10 years from now?

3.3 The Chain Rule

The population of Americans age 55 years and older as a percentage of the total population is approximated by the function

$$f(t) = 10.72(0.9t + 10)^{0.3} \qquad (0 \le t \le 20)$$

where t is measured in years with $t = 0$ corresponding to the year 2000 (Figure 6).

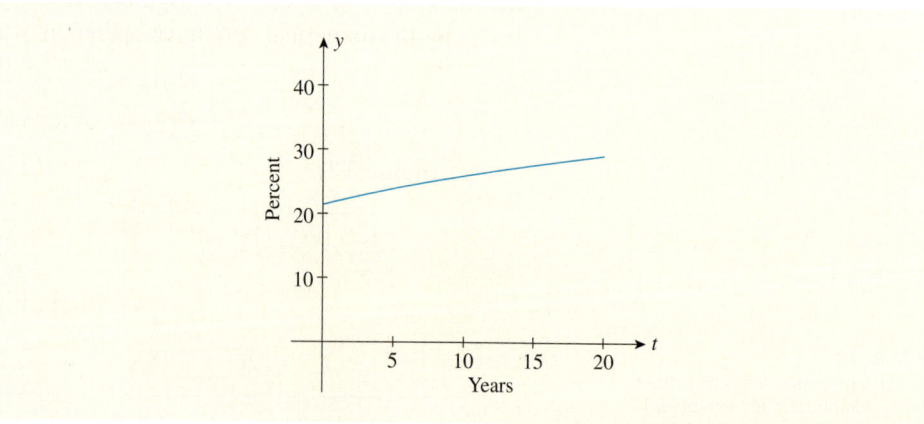

FIGURE 6
Population of Americans age 55 years and older

Source: U.S. Census Bureau.

How fast will the population age 55 years and older be increasing at the beginning of 2015? To answer this question, we have to evaluate $f'(15)$, where f' is the derivative of f. But the rules of differentiation that we have developed up to now will not help us find the derivative of f.

In this section, we will introduce another rule of differentiation called the **Chain Rule**. When used in conjunction with the rules of differentiation developed in the last two sections, the Chain Rule enables us to greatly enlarge the class of functions that we are able to differentiate. (In Exercise 72, page 191, we will use the Chain Rule to answer the question posed in the introductory example.)

The Chain Rule

Consider the function $h(x) = (x^2 + x + 1)^2$. If we were to compute $h'(x)$ using only the rules of differentiation from the previous sections, then our approach might be to expand $h(x)$. Thus,

$$h(x) = (x^2 + x + 1)^2 = (x^2 + x + 1)(x^2 + x + 1)$$
$$= x^4 + 2x^3 + 3x^2 + 2x + 1$$

from which we find

$$h'(x) = 4x^3 + 6x^2 + 6x + 2$$

But what about the function $H(x) = (x^2 + x + 1)^{100}$? The same technique may be used to find the derivative of the function H, but the amount of work involved in this case would be prodigious! Consider, also, the function $G(x) = \sqrt{x^2 + 1}$. For each of the two functions H and G, the rules of differentiation of the previous sections cannot be applied directly to compute the derivatives H' and G'.

Observe that both H and G are **composite functions;** that is, each is composed of, or built up from, simpler functions. For example, the function H is composed of the two simpler functions $f(x) = x^2 + x + 1$ and $g(x) = x^{100}$ as follows:

$$H(x) = g[f(x)] = [f(x)]^{100}$$
$$= (x^2 + x + 1)^{100}$$

In a similar manner, we see that the function G is composed of the two simpler functions $f(x) = x^2 + 1$ and $g(x) = \sqrt{x}$. Thus,

$$G(x) = g[f(x)] = \sqrt{f(x)}$$
$$= \sqrt{x^2 + 1}$$

As a first step toward finding the derivative h' of a composite function $h = g \circ f$ defined by $h(x) = g[f(x)]$, we write

$$u = f(x) \quad \text{and} \quad y = g[f(x)] = g(u)$$

The dependency of h on g and f is illustrated in Figure 7. Since u is a function of x, we may compute the derivative of u with respect to x, if f is a differentiable function, obtaining $du/dx = f'(x)$. Next, if g is a differentiable function of u, we may compute the derivative of g with respect to u, obtaining $dy/du = g'(u)$. Now, since the function h is composed of the function g and the function f, we might suspect that the rule $h'(x)$ for the derivative h' of h will be given by an expression that involves the rules for the derivatives of f and g. But how do we combine these derivatives to yield h'?

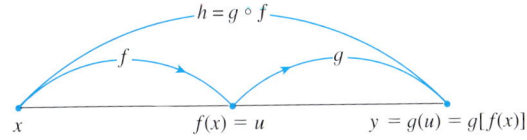

FIGURE 7
The composite function $h(x) = g[f(x)]$

This question can be answered by interpreting the derivative of each function as the rate of change of that function. For example, suppose $u = f(x)$ changes three times as fast as x—that is,

$$f'(x) = \frac{du}{dx} = 3$$

And suppose $y = g(u)$ changes twice as fast as u—that is,

$$g'(u) = \frac{dy}{du} = 2$$

Then we would expect $y = h(x)$ to change six times as fast as x—that is,

$$h'(x) = g'(u)f'(x) = (2)(3) = 6$$

or, equivalently,

$$\frac{dy}{dx} = \frac{dy}{du} \cdot \frac{du}{dx} = (2)(3) = 6$$

This observation suggests the following result, which we state without proof.

Rule 7: The Chain Rule
If $h(x) = g[f(x)]$, then

$$h'(x) = \frac{d}{dx} g[f(x)] = g'[f(x)]f'(x) \tag{1}$$

Equivalently, if we write $y = h(x) = g(u)$, where $u = f(x)$, then

$$\frac{dy}{dx} = \frac{dy}{du} \cdot \frac{du}{dx} \tag{2}$$

Notes

1. If we label the composite function h in the following manner:

$$h(x) = g[f(x)]$$

with the Inside function being $f(x)$ and the Outside function being g,

then $h'(x)$ is just the *derivative* of the "outside function" *evaluated at* the "inside function" times the *derivative* of the "inside function."

2. Equation (2) can be remembered by observing that if we "cancel" the du's, then

$$\frac{dy}{dx} = \frac{dy}{du} \cdot \frac{du}{dx} = \frac{dy}{dx}$$ ∎

The Chain Rule for Powers of Functions

Many composite functions have the special form $h(x) = g(f(x))$ where g is defined by the rule $g(x) = x^n$ (n, a real number)—that is,

$$h(x) = [f(x)]^n$$

In other words, the function h is given by the power of a function f. The functions

$$h(x) = (x^2 + x + 1)^2 \qquad H(x) = (x^2 + x + 1)^{100} \qquad G(x) = \sqrt{x^2 + 1}$$

discussed earlier are examples of this type of composite function. By using the following corollary of the Chain Rule, the General Power Rule, we can find the derivative of this type of function much more easily than by using the Chain Rule directly.

> **The General Power Rule**
> If the function f is differentiable and $h(x) = [f(x)]^n$ (n, a real number), then
> $$h'(x) = \frac{d}{dx}[f(x)]^n = n[f(x)]^{n-1}f'(x) \tag{3}$$

To see this, we observe that $h(x) = g(f(x))$ where $g(x) = x^n$, so by virtue of the Chain Rule, we have

$$\begin{aligned} h'(x) &= g'[f(x)]f'(x) \\ &= n[f(x)]^{n-1}f'(x) \end{aligned}$$

since $g'(x) = nx^{n-1}$.

EXAMPLE 1 Let $F(x) = (3x + 1)^2$.

a. Find $F'(x)$, using the General Power Rule.
b. Verify your result without the benefit of the Chain Rule or the General Power Rule.

Solution
a. Using the General Power Rule, we obtain

$$\begin{aligned} F'(x) &= 2(3x + 1)^1 \frac{d}{dx}(3x + 1) \\ &= 2(3x + 1)(3) \\ &= 6(3x + 1) \end{aligned}$$

b. We first expand $F(x)$. Thus,
$$F(x) = (3x + 1)^2 = 9x^2 + 6x + 1$$

Next, differentiating, we have
$$\begin{aligned} F'(x) &= \frac{d}{dx}(9x^2 + 6x + 1) \\ &= 18x + 6 \\ &= 6(3x + 1) \end{aligned}$$

as before.

EXAMPLE 2 Differentiate the function $G(x) = \sqrt{x^2 + 1}$.

Solution We rewrite the function $G(x)$ as
$$G(x) = (x^2 + 1)^{1/2}$$

and apply the General Power Rule, obtaining

$$\begin{aligned} G'(x) &= \frac{1}{2}(x^2 + 1)^{-1/2}\frac{d}{dx}(x^2 + 1) \\ &= \frac{1}{2}(x^2 + 1)^{-1/2} \cdot 2x = \frac{x}{\sqrt{x^2 + 1}} \end{aligned}$$

EXAMPLE 3 Differentiate the function $f(x) = x^2(2x + 3)^5$.

Solution Applying the Product Rule followed by the General Power Rule, we obtain

$$f'(x) = x^2 \frac{d}{dx}(2x + 3)^5 + (2x + 3)^5 \frac{d}{dx}(x^2)$$

$$= (x^2)5(2x + 3)^4 \cdot \frac{d}{dx}(2x + 3) + (2x + 3)^5(2x)$$

$$= 5x^2(2x + 3)^4(2) + 2x(2x + 3)^5$$

$$= 2x(2x + 3)^4(5x + 2x + 3) = 2x(7x + 3)(2x + 3)^4$$

EXAMPLE 4 Find $f'(x)$ if $f(x) = (2x^2 + 3)^4(3x - 1)^5$.

Solution Applying the Product Rule, we have

$$f'(x) = (2x^2 + 3)^4 \frac{d}{dx}(3x - 1)^5 + (3x - 1)^5 \frac{d}{dx}(2x^2 + 3)^4$$

Next, we apply the General Power Rule to each term, obtaining

$$f'(x) = (2x^2 + 3)^4 \cdot 5(3x - 1)^4 \frac{d}{dx}(3x - 1) + (3x - 1)^5 \cdot 4(2x^2 + 3)^3 \frac{d}{dx}(2x^2 + 3)$$

$$= 5(2x^2 + 3)^4(3x - 1)^4 \cdot 3 + 4(3x - 1)^5(2x^2 + 3)^3(4x)$$

Finally, observing that $(2x^2 + 3)^3(3x - 1)^4$ is common to both terms, we can factor and simplify as follows:

$$f'(x) = (2x^2 + 3)^3(3x - 1)^4[15(2x^2 + 3) + 16x(3x - 1)]$$

$$= (2x^2 + 3)^3(3x - 1)^4(30x^2 + 45 + 48x^2 - 16x)$$

$$= (2x^2 + 3)^3(3x - 1)^4(78x^2 - 16x + 45)$$

EXAMPLE 5 Find $f'(x)$ if $f(x) = \dfrac{1}{(4x^2 - 7)^2}$.

Solution Rewriting $f(x)$ and then applying the General Power Rule, we obtain

$$f'(x) = \frac{d}{dx}\left[\frac{1}{(4x^2 - 7)^2}\right] = \frac{d}{dx}(4x^2 - 7)^{-2}$$

$$= -2(4x^2 - 7)^{-3} \frac{d}{dx}(4x^2 - 7)$$

$$= -2(4x^2 - 7)^{-3}(8x) = -\frac{16x}{(4x^2 - 7)^3}$$

EXAMPLE 6 Find the slope of the tangent line to the graph of the function

$$f(x) = \left(\frac{2x + 1}{3x + 2}\right)^3$$

at the point $(0, \frac{1}{8})$.

Solution The slope of the tangent line to the graph of f at any point x is given by $f'(x)$. To compute $f'(x)$, we use the General Power Rule followed by the Quotient Rule, obtaining

$$f'(x) = 3\left(\frac{2x+1}{3x+2}\right)^2 \frac{d}{dx}\left(\frac{2x+1}{3x+2}\right)$$

$$= 3\left(\frac{2x+1}{3x+2}\right)^2 \left[\frac{(3x+2)(2) - (2x+1)(3)}{(3x+2)^2}\right] \qquad (x^2) \text{ See page 16.}$$

$$= 3\left(\frac{2x+1}{3x+2}\right)^2 \left[\frac{6x+4-6x-3}{(3x+2)^2}\right]$$

$$= \frac{3(2x+1)^2}{(3x+2)^4} \qquad \text{Combine like terms, and simplify.}$$

In particular, the slope of the tangent line to the graph of f at $(0, \frac{1}{8})$ is given by

$$f'(0) = \frac{3(0+1)^2}{(0+2)^4} = \frac{3}{16}$$

Exploring with TECHNOLOGY

Refer to Example 6.

1. Use a graphing utility to plot the graph of the function f, using the viewing window $[-2, 1] \times [-1, 2]$. Then draw the tangent line to the graph of f at the point $(0, \frac{1}{8})$.
2. For a better picture, repeat part 1 using the viewing window $[-1, 1] \times [-0.1, 0.3]$.
3. Use the numerical differentiation capability of the graphing utility to verify that the slope of the tangent line at $(0, \frac{1}{8})$ is $\frac{3}{16}$.

APPLIED EXAMPLE 7 Growth in a Health Club's Membership The membership of The Fitness Center, which opened a few years ago, is approximated by the function

$$N(t) = 100(64 + 4t)^{2/3} \qquad (0 \le t \le 52)$$

where $N(t)$ gives the number of members at the beginning of week t.

a. Find $N'(t)$.
b. How fast was the center's membership increasing initially ($t = 0$)?
c. How fast was the membership increasing at the beginning of the 40th week?
d. What was the membership when the center first opened? At the beginning of the 40th week?

Solution

a. Using the General Power Rule, we obtain

$$N'(t) = \frac{d}{dt}\left[100(64 + 4t)^{2/3}\right]$$

$$= 100 \frac{d}{dt}(64 + 4t)^{2/3}$$

$$= 100 \left(\frac{2}{3}\right)(64 + 4t)^{-1/3} \frac{d}{dt}(64 + 4t)$$

$$= \frac{200}{3}(64 + 4t)^{-1/3}(4)$$

$$= \frac{800}{3(64 + 4t)^{1/3}}$$

b. The rate at which the membership was increasing when the center first opened is given by

$$N'(0) = \frac{800}{3(64)^{1/3}} \approx 66.7$$

or approximately 67 people per week.

c. The rate at which the membership was increasing at the beginning of the 40th week is given by

$$N'(40) = \frac{800}{3(64 + 160)^{1/3}} \approx 43.9$$

or approximately 44 people per week.

d. The membership when the center first opened is given by

$$N(0) = 100(64)^{2/3} = 100(16)$$

or approximately 1600 people. The membership at the beginning of the 40th week is given by

$$N(40) = 100(64 + 160)^{2/3} \approx 3688.3$$

or approximately 3688 people.

Explore & Discuss

The profit P of a one-product software manufacturer depends on the number of units of its product sold. The manufacturer estimates that it will sell x units of its product per week. Suppose $P = g(x)$ and $x = f(t)$, where g and f are differentiable functions and t is measured in weeks.

1. Write an expression giving the rate of change of the profit with respect to the number of units sold.
2. Write an expression giving the rate of change of the number of units sold per week.
3. Write an expression giving the rate of change of the profit per week.

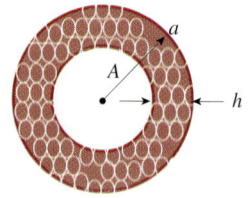

FIGURE 8
Cross section of the aorta

APPLIED EXAMPLE 8 *Arteriosclerosis* Arteriosclerosis begins during childhood when plaque (soft masses of fatty material) forms in the arterial walls, blocking the flow of blood through the arteries and leading to heart attacks, strokes, and gangrene. Suppose the idealized cross section of the aorta is circular with radius a cm and by year t, the thickness of the plaque (assume that it is uniform) is $h = g(t)$ cm (Figure 8). Then the area of the opening is given by $A = \pi(a - h)^2$ square centimeters (cm²).

Suppose the radius of an individual's artery is 1 cm ($a = 1$) and the thickness of the plaque in cm in year t is given by

$$h = g(t) = 1 - 0.01(10{,}000 - t^2)^{1/2} - 0.001t$$

Since the area of the arterial opening is given by

$$A = f(h) = \pi(1 - h)^2$$

the rate at which A is changing with respect to time is given by

$$\frac{dA}{dt} = \frac{dA}{dh} \cdot \frac{dh}{dt} = f'(h) \cdot g'(t) \qquad \text{By the Chain Rule}$$

$$= 2\pi(1 - h)(-1)\left[-0.01\left(\frac{1}{2}\right)(10{,}000 - t^2)^{-1/2}(-2t) - 0.001\right] \qquad \text{Use the Chain Rule thrice.}$$

$$= -2\pi(1 - h)\left[\frac{0.01t}{(10{,}000 - t^2)^{1/2}} - 0.001\right]$$

For example, when $t = 50$,

$$h = g(50) = 1 - 0.01(10{,}000 - 2500)^{1/2} - (0.001)(50) \approx 0.08397$$

so that

$$\frac{dA}{dt} = -2\pi(1 - 0.08397)\left[\frac{(0.01)(50)}{\sqrt{10{,}000 - (50)^2}} - 0.001\right] \approx -0.027$$

That is, the area of the arterial opening is decreasing at the rate of 0.03 cm² per year.

Explore & Discuss

Suppose the population P of a certain bacteria culture is given by $P = f(T)$, where T is the temperature of the medium. Further, suppose the temperature T is a function of time t in seconds—that is, $T = g(t)$. Give an interpretation of each of the following quantities:

1. $\dfrac{dP}{dT}$ 2. $\dfrac{dT}{dt}$ 3. $\dfrac{dP}{dt}$ 4. $(f \circ g)(t)$ 5. $f'(g(t))g'(t)$

3.3 Self-Check Exercises

1. Find the derivative of

$$f(x) = -\frac{1}{\sqrt{2x^2 - 1}}$$

2. Suppose the life expectancy at birth (in years) of a female in a certain country is described by the function

$$g(t) = 50.02(1 + 1.09t)^{0.1} \quad (0 \le t \le 150)$$

where t is measured in years, with $t = 0$ corresponding to the beginning of 1900.

a. What is the life expectancy at birth of a female born at the beginning of 1980? At the beginning of 2010?

b. How fast is the life expectancy at birth of a female born at any time t changing?

Solutions to Self-Check Exercises 3.3 can be found on page 193.

3.3 Concept Questions

1. In your own words, state the Chain Rule for differentiating the composite function $h(x) = g[f(x)]$.

2. In your own words, state the General Power Rule for differentiating the function $h(x) = [f(x)]^n$, where n is a real number.

3. If $f(t)$ gives the number of units of a certain product sold by a company after t days and $g(x)$ gives the revenue (in dollars) realized from the sale of x units of the company's products, what does $(g \circ f)'(t)$ describe?

4. Suppose $f(x)$ gives the air temperature in the gondola of a hot-air balloon when it is at an altitude of x ft from the ground and $g(t)$ gives the altitude of the balloon t min after lifting off from the ground. Find a function giving the rate of change of the air temperature in the gondola at time t.

3.3 Exercises

In Exercises 1–48, find the derivative of each function.

1. $f(x) = (2x - 1)^3$
2. $f(x) = (1 - x)^4$
3. $f(x) = (x^2 + 2)^5$
4. $f(t) = 2(t^3 - 1)^5$
5. $f(x) = (2x - x^2)^3$
6. $f(x) = 3(x^3 - x)^4$
7. $f(x) = (2x + 1)^{-2}$
8. $f(t) = \dfrac{1}{2}(2t^2 + t)^{-3}$
9. $f(x) = (x^2 - 4)^{5/2}$
10. $f(t) = (3t^2 - 2t + 1)^{3/2}$
11. $f(x) = \sqrt{3x - 2}$
12. $f(t) = \sqrt{3t^2 - t}$

13. $f(x) = \sqrt[3]{1 - x^2}$

14. $f(x) = \sqrt{2x^2 - 2x + 3}$

15. $f(x) = \dfrac{1}{(2x + 3)^3}$

16. $f(x) = \dfrac{2}{(x^2 - 1)^4}$

17. $f(t) = \dfrac{1}{\sqrt{2t - 4}}$

18. $f(x) = \dfrac{1}{\sqrt{2x^2 - 1}}$

19. $y = \dfrac{1}{(4x^4 + x)^{3/2}}$

20. $f(t) = \dfrac{4}{\sqrt[3]{2t^2 + t}}$

21. $f(x) = (3x^2 + 2x + 1)^{-2}$

22. $f(t) = (5t^3 + 2t^2 - t + 4)^{-3}$

23. $f(x) = (x^2 + 1)^3 - (x^3 + 1)^2$

24. $f(t) = (2t - 1)^4 + (2t + 1)^4$

25. $f(t) = (t^{-1} - t^{-2})^3$

26. $f(v) = (v^{-3} + 4v^{-2})^3$

27. $f(x) = \sqrt{x + 1} + \sqrt{x - 1}$

28. $f(u) = (2u + 1)^{3/2} + (u^2 - 1)^{-3/2}$

29. $f(x) = 2x^2(3 - 4x)^4$

30. $h(t) = t^2(3t + 4)^3$

31. $f(x) = (x - 1)^2(2x + 1)^4$

32. $g(u) = (1 + u^2)^5(1 - 2u^2)^8$

33. $f(x) = \left(\dfrac{x + 3}{x - 2}\right)^3$

34. $f(x) = \left(\dfrac{x + 1}{x - 1}\right)^5$

35. $s(t) = \left(\dfrac{t}{2t + 1}\right)^{3/2}$

36. $g(s) = \left(s^2 + \dfrac{1}{s}\right)^{3/2}$

37. $g(u) = \sqrt{\dfrac{u + 1}{3u + 2}}$

38. $g(x) = \sqrt{\dfrac{2x + 1}{2x - 1}}$

39. $f(x) = \dfrac{x^2}{(x^2 - 1)^4}$

40. $g(u) = \dfrac{2u^2}{(u^2 + u)^3}$

41. $h(x) = \dfrac{(3x^2 + 1)^3}{(x^2 - 1)^4}$

42. $g(t) = \dfrac{(2t - 1)^2}{(3t + 2)^4}$

43. $f(x) = \dfrac{\sqrt{2x + 1}}{x^2 - 1}$

44. $f(t) = \dfrac{4t^2}{\sqrt{2t^2 + 2t - 1}}$

45. $g(t) = \dfrac{\sqrt{t + 1}}{\sqrt{t^2 + 1}}$

46. $f(x) = \dfrac{\sqrt{x^2 + 1}}{\sqrt{x^2 - 1}}$

47. $f(x) = (3x + 1)^4(x^2 - x + 1)^3$

48. $g(t) = (2t + 3)^2(3t^2 - 1)^{-3}$

In Exercises 49–54, find $\dfrac{dy}{du}, \dfrac{du}{dx}$, and $\dfrac{dy}{dx}$.

49. $y = u^{4/3}$ and $u = 3x^2 - 1$

50. $y = \sqrt{u}$ and $u = 7x - 2x^2$

51. $y = u^{-2/3}$ and $u = 2x^3 - x + 1$

52. $y = 2u^2 + 1$ and $u = x^2 + 1$

53. $y = \sqrt{u} + \dfrac{1}{\sqrt{u}}$ and $u = x^3 - x$

54. $y = \dfrac{1}{u}$ and $u = \sqrt{x} + 1$

55. If $g(x) = f(2x + 1)$, what is $g'(x)$?

56. If $h(x) = f(-x^3)$, what is $h'(x)$?

57. Suppose $F(x) = g(f(x))$ and $f(2) = 3, f'(2) = -3$, $g(3) = 5$, and $g'(3) = 4$. Find $F'(2)$.

58. Suppose $h = f \circ g$. Find $h'(0)$ given that $f(0) = 6$, $f'(5) = -2$, $g(0) = 5$, and $g'(0) = 3$.

59. Suppose $F(x) = f(x^2 + 1)$. Find $F'(1)$ if $f'(2) = 3$.

60. Let $F(x) = f(f(x))$. Does it follow that $F'(x) = [f'(x)]^2$?
Hint: Let $f(x) = x^2$.

61. Suppose $h = g \circ f$. Does it follow that $h' = g' \circ f'$?
Hint: Let $f(x) = x$ and $g(x) = x^2$.

62. Suppose $h = f \circ g$. Show that $h' = (f' \circ g)g'$.

In Exercises 63–66, find an equation of the tangent line to the graph of the function at the given point.

63. $f(x) = (1 - x)(x^2 - 1)^2; (2, -9)$

64. $f(x) = \left(\dfrac{x + 1}{x - 1}\right)^2; (3, 4)$

65. $f(x) = x\sqrt{2x^2 + 7}; (3, 15)$

66. $f(x) = \dfrac{8}{\sqrt{x^2 + 6x}}; (2, 2)$

67. **TELEVISION VIEWING** The number of viewers of a television series introduced several years ago is approximated by the function

$$N(t) = (60 + 2t)^{2/3} \qquad (1 \leq t \leq 26)$$

where $N(t)$ (measured in millions) denotes the number of weekly viewers of the series in the tth week. Find the rate of increase of the weekly audience at the end of week 2 and at the end of week 12. How many viewers were there in week 2? In week 24?

68. **OUTSOURCING OF JOBS** According to a study conducted in 2003, the total number of U.S. jobs that are projected to leave the country by year t, where $t = 0$ corresponds to the beginning of 2000, is

$$N(t) = 0.0018425(t + 5)^{2.5} \qquad (0 \leq t \leq 15)$$

where $N(t)$ is measured in millions. How fast was the number of U.S. jobs that were outsourced changing in 2005 ($t = 5$)? In 2012 ($t = 12$)?

Source: Forrester Research.

69. **WORKING MOTHERS** The percentage of mothers who work outside the home and have children younger than age 6 is approximated by the function

$$P(t) = 33.55(t + 5)^{0.205} \qquad (0 \leq t \leq 21)$$

where t is measured in years, with $t = 0$ corresponding to the beginning of 1980. Compute $P'(t)$. At what rate was the percentage of these mothers changing at the beginning of 2000? What was the percentage of these mothers at the beginning of 2000?

Source: U.S. Bureau of Labor Statistics.

70. Selling Price of DVD Recorders The rise of digital music and the improvement to the DVD format are some of the reasons why the average selling price of stand-alone DVD recorders will drop in the coming years. The function

$$A(t) = \frac{699}{(t+1)^{0.94}} \qquad (0 \le t \le 5)$$

gives the projected average selling price (in dollars) of stand-alone DVD recorders in year t, where $t = 0$ corresponds to the beginning of 2002. How fast was the average selling price of stand-alone DVD recorders falling at the beginning of 2002? How fast was it falling at the beginning of 2006?

Source: Consumer Electronics Association.

71. Brain Cancer Survival Rate Glioblastoma is the most common and most deadly of brain tumors, and it kills most patients in a little over a year. The probability of survival for patients with a glioblastoma t years after diagnosis is approximated by the function

$$P(t) = \frac{100}{(1 + 0.14t)^{9.2}} \qquad (0 \le t \le 10)$$

where P is the percent of surviving patients.
a. Compute $P(0)$ and $P(1)$, and interpret your results.
b. Compute $P'(0)$ and $P'(1)$, and interpret your results.

Source: National Cancer Institute.

72. Aging Population The population of Americans age 55 and older as a percentage of the total population is approximated by the function

$$f(t) = 10.72(0.9t + 10)^{0.3} \qquad (0 \le t \le 20)$$

where t is measured in years, with $t = 0$ corresponding to the year 2000. At what rate was the percentage of Americans age 55 and older changing at the beginning of 2000? At what rate will the percentage of Americans age 55 and older be changing in 2015? What will the percentage of the population of Americans age 55 and older be in 2015?

Source: U.S. Census Bureau.

73. Concentration of Carbon Monoxide (CO) in the Air According to a joint study conducted by Oxnard's Environmental Management Department and a state government agency, the concentration of CO in the air due to automobile exhaust t years from now is given by

$$C(t) = 0.01(0.2t^2 + 4t + 64)^{2/3}$$

parts per million.
a. Find the rate at which the level of CO is changing with respect to time.
b. Find the rate at which the level of CO will be changing 5 years from now.

74. Continuing Education Enrollment The registrar of Kellogg University estimates that the total student enrollment in the Continuing Education division will be given by

$$N(t) = -\frac{20{,}000}{\sqrt{1 + 0.2t}} + 21{,}000$$

where $N(t)$ denotes the number of students enrolled in the division t years from now. Find an expression for $N'(t)$. How fast is the student enrollment increasing currently? How fast will it be increasing 5 years from now?

75. Air Pollution According to the South Coast Air Quality Management District, the level of nitrogen dioxide, a brown gas that impairs breathing, present in the atmosphere on a certain May day in downtown Los Angeles is approximated by

$$A(t) = 0.03t^3(t-7)^4 + 60.2 \qquad (0 \le t \le 7)$$

where $A(t)$ is measured in pollutant standard index and t is measured in hours, with $t = 0$ corresponding to 7 A.M.
a. Find $A'(t)$.
b. Find $A'(1)$, $A'(3)$, and $A'(4)$, and interpret your results.

Source: Los Angeles Times.

76. Effect of Luxury Tax on Consumption Government economists of a developing country determined that the purchase of imported perfume is related to a proposed "luxury tax" by the formula

$$N(x) = \sqrt{10{,}000 - 40x - 0.02x^2} \qquad (0 \le x \le 200)$$

where $N(x)$ measures the percentage of normal consumption of perfume when a "luxury tax" of $x\%$ is imposed on it. Find the rate of change of $N(x)$ for taxes of 10%, 100%, and 150%.

77. Pulse Rate of an Athlete The pulse rate (the number of heartbeats/minute) of a long-distance runner t sec after leaving the starting line is given by

$$P(t) = \frac{300\sqrt{\tfrac{1}{2}t^2 + 2t + 25}}{t + 25} \qquad (t \ge 0)$$

Compute $P'(t)$. How fast is the athlete's pulse rate increasing 10 sec, 60 sec, and 2 min into the run? What is her pulse rate 2 min into the run?

78. Thurstone Learning Model Psychologist L. L. Thurstone suggested the following relationship between learning time T and the length of a list n:

$$T = f(n) = An\sqrt{n - b}$$

where A and b are constants that depend on the person and the task.
a. Compute dT/dn, and interpret your result.
b. For a certain person and a certain task, suppose $A = 4$ and $b = 4$. Compute $f'(13)$ and $f'(29)$, and interpret your results.

79. Oil Spills In calm waters, the oil spilling from the ruptured hull of a grounded tanker spreads in all directions. Assuming that the area polluted is a circle and that its radius is increasing at a rate of 2 ft/sec, determine how fast the area is increasing when the radius of the circle is 40 ft.

80. **ARTERIOSCLEROSIS** Refer to Example 8, page 188. Suppose the radius of an individual's artery is 1 cm and the thickness of the plaque (in centimeters) t years from now is given by

$$h = g(t) = \frac{0.5t^2}{t^2 + 10} \quad (0 \le t \le 10)$$

How fast will the arterial opening be decreasing 5 years from now?

81. **TRAFFIC FLOW** Opened in the late 1950s, the Central Artery in downtown Boston was designed to move 75,000 vehicles a day. The number of vehicles moved per day is approximated by the function

$$x = f(t) = 6.25t^2 + 19.75t + 74.75 \quad (0 \le t \le 4.5)$$

where x is measured in thousands and t in decades, with $t = 0$ corresponding to the beginning of 1959. Suppose the average speed of traffic flow in mph is given by

$$S = g(x) = -0.00075x^2 + 67.5 \quad (75 \le x \le 350)$$

where x has the same meaning as before. What was the rate of change of the average speed of traffic flow at the beginning of 1999? What was the average speed of traffic flow at that time?
Hint: $S = g[f(t)]$

82. **HOTEL OCCUPANCY RATES** The occupancy rate of the all-suite Wonderland Hotel, located near an amusement park, is given by the function

$$r(t) = \frac{10}{81}t^3 - \frac{10}{3}t^2 + \frac{200}{9}t + 56.2 \quad (0 \le t \le 12)$$

where t is measured in months, with $t = 0$ corresponding to the beginning of January. Management has estimated that the monthly revenue (in thousands of dollars per month) is approximated by the function

$$R(r) = -\frac{3}{5000}r^3 + \frac{9}{50}r^2 \quad (0 \le r \le 100)$$

where r is the occupancy rate.
a. Find an expression that gives the rate of change of Wonderland's occupancy rate with respect to time.
b. Find an expression that gives the rate of change of Wonderland's monthly revenue with respect to the occupancy rate.
c. What is the rate of change of Wonderland's monthly revenue with respect to time at the beginning of January? At the beginning of July?
Hint: Use the Chain Rule to find $R'(r(0))r'(0)$ and $R'(r(6))r'(6)$.

83. **EFFECT OF HOUSING STARTS ON JOBS** The president of a major housing construction firm claims that the number of construction jobs created is given by

$$N(x) = 1.42x$$

where x denotes the number of housing starts. Suppose the number of housing starts in the next t months is expected to be

$$x(t) = \frac{7t^2 + 140t + 700}{3t^2 + 80t + 550}$$

million units per year. Find an expression that gives the rate at which the number of construction jobs will be created t months from now. At what rate will construction jobs be created 1 year from now?

84. **DEMAND FOR PCs** The quantity demanded per month, x, of a certain make of tablet PC is related to the average unit price, p (in dollars), of tablet PCs by the equation

$$x = f(p) = \frac{100}{9}\sqrt{810,000 - p^2}$$

It is estimated that t months from now, the average price of a tablet PC will be given by

$$p(t) = \frac{400}{1 + \frac{1}{8}\sqrt{t}} + 200 \quad (0 \le t \le 60)$$

dollars. Find the rate at which the quantity demanded per month of the tablet PCs will be changing 16 months from now.

85. **DEMAND FOR WATCHES** The demand equation for the Sicard sports watch is given by

$$x = f(p) = 10\sqrt{\frac{50 - p}{p}} \quad (0 < p \le 50)$$

where x (measured in units of a thousand) is the quantity demanded each week and p is the unit price in dollars. Find the rate of change of the quantity demanded of the sports watches with respect to the unit price when the unit price is $25.

86. **CRUISE SHIP BOOKINGS** The management of Cruise World, operators of Caribbean luxury cruises, expects that the percentage of young adults booking passage on their cruises in the years ahead will rise dramatically. They have constructed the following model, which gives the percentage of young adult passengers in year t:

$$p = f(t) = 50\left(\frac{t^2 + 2t + 4}{t^2 + 4t + 8}\right) \quad (0 \le t \le 5)$$

Young adults normally pick shorter cruises and generally spend less on their passage. The following model gives an approximation of the average amount of money R (in dollars) spent per passenger on a cruise when the percentage of young adults is p:

$$R(p) = 1000\left(\frac{p + 4}{p + 2}\right)$$

Find the rate at which the amount of money spent per passenger on a cruise will be changing 2 years from now.

In Exercises 87–90, determine whether the statement is true or false. If it is true, explain why it is true. If it is false, give an example to show why it is false.

87. If f and g are differentiable and $h = f \circ g$, then $h'(x) = f'[g(x)]g'(x)$.

88. If f is differentiable and c is a constant, then
$$\frac{d}{dx}[f(cx)] = cf'(cx)$$

89. If f is differentiable, then
$$\frac{d}{dx}\sqrt{f(x)} = \frac{f'(x)}{2\sqrt{f(x)}}$$

90. If f is differentiable, then
$$\frac{d}{dx}\left[f\left(\frac{1}{x}\right)\right] = f'\left(\frac{1}{x}\right)$$

91. In Section 3.1, we proved that
$$\frac{d}{dx}(x^n) = nx^{n-1}$$
for the special case when $n = 2$. Use the Chain Rule to show that
$$\frac{d}{dx}(x^{1/n}) = \frac{1}{n}x^{1/n-1}$$
for any nonzero integer n, assuming that $f(x) = x^{1/n}$ is differentiable.
Hint: Let $f(x) = x^{1/n}$, so that $[f(x)]^n = x$. Differentiate both sides with respect to x.

92. With the aid of Exercise 91, prove that
$$\frac{d}{dx}(x^r) = rx^{r-1}$$
for every rational number r.
Hint: Let $r = m/n$, where m and n are integers, with $n \neq 0$, and write $x^r = (x^m)^{1/n}$.

3.3 Solutions to Self-Check Exercises

1. Rewriting, we have
$$f(x) = -(2x^2 - 1)^{-1/2}$$
Using the General Power Rule, we find
$$f'(x) = -\frac{d}{dx}(2x^2 - 1)^{-1/2}$$
$$= -\left(-\frac{1}{2}\right)(2x^2 - 1)^{-3/2}\frac{d}{dx}(2x^2 - 1)$$
$$= \frac{1}{2}(2x^2 - 1)^{-3/2}(4x)$$
$$= \frac{2x}{(2x^2 - 1)^{3/2}}$$

2. a. The life expectancy at birth of a female born at the beginning of 1980 is given by
$$g(80) = 50.02[1 + 1.09(80)]^{0.1} \approx 78.29$$

or approximately 78 years. Similarly, the life expectancy at birth of a female born at the beginning of the year 2010 is given by
$$g(110) = 50.02[1 + 1.09(110)]^{0.1} \approx 80.80$$
or approximately 81 years.

b. The rate of change of the life expectancy at birth of a female born at any time t is given by $g'(t)$. Using the General Power Rule, we have
$$g'(t) = 50.02\frac{d}{dt}(1 + 1.09t)^{0.1}$$
$$= (50.02)(0.1)(1 + 1.09t)^{-0.9}\frac{d}{dt}(1 + 1.09t)$$
$$= (50.02)(0.1)(1.09)(1 + 1.09t)^{-0.9}$$
$$= 5.45218(1 + 1.09t)^{-0.9}$$
$$= \frac{5.45218}{(1 + 1.09t)^{0.9}}$$

USING TECHNOLOGY

Finding the Derivative of a Composite Function

EXAMPLE 1 Find the rate of change of $f(x) = \sqrt{x}(1 + 0.02x^2)^{3/2}$ when $x = 2.1$.

Solution Using the numerical derivative operation of a graphing utility, we find
$$f'(2.1) = 0.5821463392$$

(continued)

or approximately 0.58 unit per unit change in *x*. (See Figure T1.)

FIGURE T1
The TI-83/84 numerical derivative screen for computing $f'(2.1)$

APPLIED EXAMPLE 2 Amusement Park Attendance The management of AstroWorld ("The Amusement Park of the Future") estimates that the total number of visitors (in thousands) to the amusement park t hours after opening time at 9 A.M. is given by

$$N(t) = \frac{30t}{\sqrt{2 + t^2}}$$

What is the rate at which visitors are admitted to the amusement park at 10:30 A.M.?

Solution Using the numerical derivative operation of a graphing utility, we find

$$N'(1.5) \approx 6.8481$$

or approximately 6848 visitors per hour. (See Figure T2.)

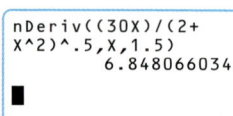

FIGURE T2
The TI-83/84 numerical derivative screen for computing $N'(1.5)$

TECHNOLOGY EXERCISES

In Exercises 1–6, use the numerical derivative operation to find the rate of change of *f* at the given value of *x*. Give your answer accurate to four decimal places.

1. $f(x) = \sqrt{x^2 - x^4}$; $x = 0.5$
2. $f(x) = x - \sqrt{1 - x^2}$; $x = 0.4$
3. $f(x) = x\sqrt{1 - x^2}$; $x = 0.2$
4. $f(x) = (x + \sqrt{x^2 + 4})^{3/2}$; $x = 1$
5. $f(x) = \dfrac{\sqrt{1 + x^2}}{x^3 + 2}$; $x = -1$
6. $f(x) = \dfrac{x^3}{1 + (1 + x^2)^{3/2}}$; $x = 3$

7. **TV Mobile Phones** The number of people watching TV on mobile phones (in millions) is approximated by

$$N(t) = 11.9\sqrt{1 + 0.91t} \quad (0 \le t \le 4)$$

where *t* is measured in years, with $t = 0$ corresponding to the beginning of 2007.

a. What was the rate of change of the number of people watching TV on mobile phones at the beginning of 2007?
b. What is the rate of change of the number of people watching TV on mobile phones expected to be at the beginning of 2011?

Source: IDC, U.S. forecast.

8. **Accumulation Years** Demographic studies pertaining to investors are of particular importance to financial institutions. People from their mid-40s to their mid-50s are in the prime investing years. The function

$$N(t) = 34.4(1 + 0.32125t)^{0.15} \quad (0 \le t \le 12)$$

gives the projected number of people in this age group in the United States (in millions) in year *t*, where $t = 0$ corresponds to the beginning of 1996.

a. How large was this segment of the population projected to be at the beginning of 2008?
b. How fast was this segment of the population growing at the beginning of 2008?

Source: U.S. Census Bureau.

3.4 Marginal Functions in Economics

Marginal analysis is the study of the rate of change of economic quantities. For example, an economist is not merely concerned with the value of an economy's gross domestic product (GDP) at a given time but is equally concerned with the rate at which it is growing or declining. In the same vein, a manufacturer is not only interested in the total cost corresponding to a certain level of production of a commodity but also is interested in the rate of change of the total cost with respect to the level of production, and so on. Let's begin with an example to explain the meaning of the adjective *marginal*, as used by economists.

Cost Functions

APPLIED EXAMPLE 1 Rate of Change of Cost Functions Suppose the total cost in dollars incurred each week by Polaraire for manufacturing x refrigerators is given by the total cost function

$$C(x) = 8000 + 200x - 0.2x^2 \quad (0 \leq x \leq 400)$$

a. What is the actual cost incurred for manufacturing the 251st refrigerator?
b. Find the rate of change of the total cost function with respect to x when $x = 250$.
c. Compare the results obtained in parts (a) and (b).

Solution

a. The actual cost incurred in producing the 251st refrigerator is the difference between the total cost incurred in producing the first 251 refrigerators and the total cost of producing the first 250 refrigerators:

$$\begin{aligned} C(251) - C(250) &= [8000 + 200(251) - 0.2(251)^2] \\ &\quad - [8000 + 200(250) - 0.2(250)^2] \\ &= 45{,}599.8 - 45{,}500 \\ &= 99.8 \end{aligned}$$

or $99.80.

b. The rate of change of the total cost function C with respect to x is given by the derivative of C—that is, $C'(x) = 200 - 0.4x$. Thus, when the level of production is 250 refrigerators, the rate of change of the total cost with respect to x is given by

$$\begin{aligned} C'(250) &= 200 - 0.4(250) \\ &= 100 \end{aligned}$$

or $100.

c. From the solution to part (a), we know that the actual cost for producing the 251st refrigerator is $99.80. This answer is very closely approximated by the answer to part (b), $100. To see why this is so, observe that the difference $C(251) - C(250)$ may be written in the form

$$\frac{C(251) - C(250)}{1} = \frac{C(250 + 1) - C(250)}{1} = \frac{C(250 + h) - C(250)}{h}$$

where $h = 1$. In other words, the difference $C(251) - C(250)$ is precisely the average rate of change of the total cost function C over the interval $[250, 251]$ or, equivalently, the slope of the secant line through the points $(250, 45{,}500)$

and (251, 45,599.8). However, the number $C'(250) = 100$ is the instantaneous rate of change of the total cost function C at $x = 250$ or, equivalently, the slope of the tangent line to the graph of C at $x = 250$.

Now when h is small, the average rate of change of the function C is a good approximation to the instantaneous rate of change of the function C, or, equivalently, the slope of the secant line through the points in question is a good approximation to the slope of the tangent line through the point in question. Therefore, we may expect

$$C(251) - C(250) = \frac{C(251) - C(250)}{1} \approx \frac{C(250 + h) - C(250)}{h} \quad (h \text{ small})$$

$$\approx \lim_{h \to 0} \frac{C(250 + h) - C(250)}{h} = C'(250)$$

which is precisely the case in this example.

The actual cost incurred in producing an additional unit of a certain commodity given that a plant is already at a certain level of operation is called the **marginal cost**. Knowing this cost is very important to management. As we saw in Example 1, the marginal cost is approximated by the rate of change of the total cost function evaluated at the appropriate point. For this reason, economists have defined the **marginal cost function** to be the derivative of the corresponding total cost function. In other words, if C is a total cost function, then the marginal cost function is defined to be its derivative C'. Thus, the adjective *marginal* is synonymous with *derivative of*.

APPLIED EXAMPLE 2 Marginal Cost Functions A subsidiary of Elektra Electronics manufactures a portable DVD player. Management determined that the daily total cost of producing these DVD players (in dollars) is given by

$$C(x) = 0.0001x^3 - 0.08x^2 + 40x + 5000$$

where x stands for the number of DVD players produced.

a. Find the marginal cost function.
b. What is the marginal cost when $x = 200, 300, 400,$ and 600?
c. Interpret your results.

Solution

a. The marginal cost function C' is given by the derivative of the total cost function C. Thus,

$$C'(x) = 0.0003x^2 - 0.16x + 40$$

b. The marginal cost when $x = 200, 300, 400,$ and 600 is given by

$$C'(200) = 0.0003(200)^2 - 0.16(200) + 40 = 20$$
$$C'(300) = 0.0003(300)^2 - 0.16(300) + 40 = 19$$
$$C'(400) = 0.0003(400)^2 - 0.16(400) + 40 = 24$$
$$C'(600) = 0.0003(600)^2 - 0.16(600) + 40 = 52$$

or \$20/unit, \$19/unit, \$24/unit, and \$52/unit, respectively.

c. From the results of part (b), we see that Elektra's actual cost for producing the 201st DVD player is approximately \$20. The actual cost incurred for producing one additional DVD player when the level of production is already 300

3.4 MARGINAL FUNCTIONS IN ECONOMICS   197

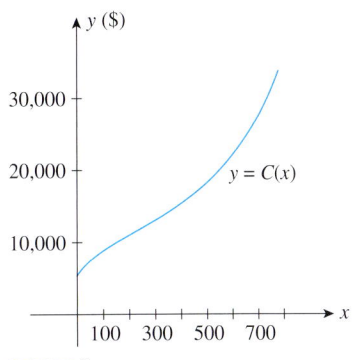

FIGURE 9
The cost of producing x DVD players is given by C(x).

players is approximately $19, and so on. Observe that when the level of production is already 600 units, the actual cost of producing one additional unit is approximately $52. The higher cost for producing this additional unit when the level of production is 600 units may be the result of several factors, among them excessive costs incurred because of overtime or higher maintenance, production breakdown caused by greater stress and strain on the equipment, and so on. The graph of the total cost function appears in Figure 9.

Average Cost Functions

Let's now introduce another marginal concept that is closely related to the marginal cost. Let $C(x)$ denote the total cost incurred in producing x units of a certain commodity. Then the **average cost** of producing x units of the commodity is obtained by dividing the total production cost by the number of units produced. This leads to the following definition:

> **Average Cost Function**
>
> Suppose $C(x)$ is a total cost function. Then the **average cost function,** denoted by $\overline{C}(x)$ (read "C bar of x"), is
>
> $$\frac{C(x)}{x} \tag{4}$$

The derivative $\overline{C}'(x)$ of the average cost function, called the **marginal average cost function,** measures the rate of change of the average cost function with respect to the number of units produced.

APPLIED EXAMPLE 3 Marginal Average Cost Functions The total cost of producing x units of a certain commodity is given by

$$C(x) = 400 + 20x$$

dollars.

a. Find the average cost function $\overline{C}$.
b. Find the marginal average cost function $\overline{C}'$.
c. What are the economic implications of your results?

Solution

a. The average cost function is given by

$$\overline{C}(x) = \frac{C(x)}{x} = \frac{400 + 20x}{x}$$

$$= 20 + \frac{400}{x}$$

b. The marginal average cost function is

$$\overline{C}'(x) = -\frac{400}{x^2}$$

c. Since the marginal average cost function is negative for all admissible values of x, the rate of change of the average cost function is negative for all $x > 0$;

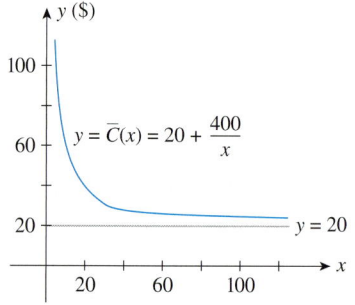

FIGURE 10
As the level of production increases, the average cost approaches $20.

that is, $\overline{C}(x)$ decreases as x increases. However, the graph of $\overline{C}$ always lies above the horizontal line $y = 20$, but it approaches the line, since

$$\lim_{x \to \infty} \overline{C}(x) = \lim_{x \to \infty} \left(20 + \frac{400}{x}\right) = 20$$

A sketch of the graph of the function $\overline{C}(x)$ appears in Figure 10. This result is fully expected if we consider the economic implications. Note that as the level of production increases, the fixed cost per unit of production, represented by the term $(400/x)$, drops steadily. The average cost approaches the constant unit cost of production, which is $20 in this case.

 APPLIED EXAMPLE 4 Marginal Average Cost Functions Once again consider the subsidiary of Elektra Electronics. The daily total cost for producing its portable DVD players is given by

$$C(x) = 0.0001x^3 - 0.08x^2 + 40x + 5000$$

dollars, where x stands for the number of DVD players produced (see Example 2).

a. Find the average cost function $\overline{C}$.
b. Find the marginal average cost function $\overline{C}'$. Compute $\overline{C}'(500)$.
c. Sketch the graph of the function $\overline{C}$ and interpret the results obtained in parts (a) and (b).

Solution

a. The average cost function is given by

$$\overline{C}(x) = \frac{C(x)}{x} = 0.0001x^2 - 0.08x + 40 + \frac{5000}{x}$$

b. The marginal average cost function is given by

$$\overline{C}'(x) = 0.0002x - 0.08 - \frac{5000}{x^2}$$

Also,

$$\overline{C}'(500) = 0.0002(500) - 0.08 - \frac{5000}{(500)^2} = 0$$

c. To sketch the graph of the function $\overline{C}$, observe that if x is a small positive number, then $\overline{C}(x) > 0$. Furthermore, $\overline{C}(x)$ becomes arbitrarily large as x approaches zero from the right, since the term $(5000/x)$ becomes arbitrarily large as x approaches zero. Next, the result $\overline{C}'(500) = 0$ obtained in part (b) tells us that the tangent line to the graph of the function $\overline{C}$ is horizontal at the point $(500, 35)$ on the graph. Finally, plotting the points on the graph corresponding to, say, $x = 100, 200, 300, \ldots, 900$, we obtain the sketch in Figure 11. As expected, the average cost drops as the level of production increases. But in this case, in contrast to the case in Example 3, the average cost reaches a minimum value of $35, corresponding to a production level of 500, and *increases* thereafter.

This phenomenon is typical in situations in which the marginal cost increases from some point on as production increases, as in Example 2. This situation is in contrast to that of Example 3, in which the marginal cost remains constant at any level of production.

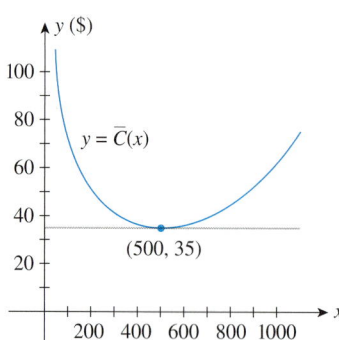

FIGURE 11
The average cost reaches a minimum of $35 when 500 DVD players are produced.

> **Exploring with TECHNOLOGY**
>
> Refer to Example 4.
>
> 1. Use a graphing utility to plot the graph of the average cost function
>
> $$\overline{C}(x) = 0.0001x^2 - 0.08x + 40 + \frac{5000}{x}$$
>
> using the viewing window $[0, 1000] \times [0, 100]$. Then, using ZOOM and TRACE, show that the lowest point on the graph of $\overline{C}$ is $(500, 35)$.
>
> 2. Draw the tangent line to the graph of $\overline{C}$ $(500, 35)$. What is its slope? Is this expected?
>
> 3. Plot the graph of the marginal average cost function
>
> $$\overline{C}'(x) = 0.0002x - 0.08 - \frac{5000}{x^2}$$
>
> using the viewing window $[0, 2000] \times [-1, 1]$. Then use ZOOM and TRACE to show that the zero of the function $\overline{C}'$ occurs at $x = 500$. Verify this result using the root-finding capability of your graphing utility. Is this result compatible with that obtained in part 2? Explain your answer.

Revenue Functions

Recall that a revenue function $R(x)$ gives the revenue realized by a company from the sale of x units of a certain commodity. If the company charges p dollars per unit, then

$$R(x) = px \qquad (5)$$

However, the price that a company can command for the product depends on the market in which the company operates. If the company is one of many—none of which is able to dictate the price of the commodity—then in this competitive market environment, the price is determined by market equilibrium (see Section 2.3). On the other hand, if the company is the sole supplier of the product, then under this monopolistic situation, it can manipulate the price of the commodity by controlling the supply. The unit selling price p of the commodity is related to the quantity x of the commodity demanded. This relationship between p and x is called a *demand equation* (see Section 2.3). Solving the demand equation for p in terms of x, we obtain the unit price function f. Thus,

$$p = f(x)$$

and the revenue function R is given by

$$R(x) = px = xf(x)$$

The **marginal revenue** gives the actual revenue realized from the sale of an additional unit of the commodity given that sales are already at a certain level. Following an argument parallel to that applied to the cost function in Example 1, you can convince yourself that the marginal revenue is approximated by $R'(x)$. Thus, we define the **marginal revenue function** to be $R'(x)$, where R is the revenue function. The derivative R' of the function R measures the rate of change of the revenue function.

APPLIED EXAMPLE 5 Marginal Revenue Functions Suppose the relationship between the unit price p in dollars and the quantity demanded x of the Acrosonic model F loudspeaker system is given by the equation

$$p = -0.02x + 400 \quad (0 \leq x \leq 20{,}000)$$

a. Find the revenue function R.
b. Find the marginal revenue function R'.
c. Compute $R'(2000)$, and interpret your result.

Solution

a. The revenue function R is given by

$$\begin{aligned} R(x) &= px \\ &= x(-0.02x + 400) \\ &= -0.02x^2 + 400x \quad (0 \leq x \leq 20{,}000) \end{aligned}$$

b. The marginal revenue function R' is given by

$$R'(x) = -0.04x + 400$$

c.
$$R'(2000) = -0.04(2000) + 400 = 320$$

Thus, the actual revenue to be realized from the sale of the 2001st loudspeaker system is approximately $320.

Profit Functions

Our final example of a marginal function involves the profit function. The profit function P is given by

$$P(x) = R(x) - C(x) \tag{6}$$

where R and C are the revenue and cost functions and x is the number of units of a commodity produced and sold. The **marginal profit function** $P'(x)$ measures the rate of change of the profit function P and provides us with a good approximation of the actual profit or loss realized from the sale of the $(x + 1)$st unit of the commodity (assuming that the xth unit has been sold).

APPLIED EXAMPLE 6 Marginal Profit Functions Refer to Example 5. Suppose the cost of producing x units of the Acrosonic model F loudspeaker is

$$C(x) = 100x + 200{,}000$$

dollars.
a. Find the profit function P.
b. Find the marginal profit function P'.
c. Compute $P'(2000)$, and interpret your result.
d. Sketch the graph of the profit function P.

Solution

a. From the solution to Example 5a, we have

$$R(x) = -0.02x^2 + 400x$$

Thus, the required profit function P is given by

$$\begin{aligned} P(x) &= R(x) - C(x) \\ &= (-0.02x^2 + 400x) - (100x + 200{,}000) \\ &= -0.02x^2 + 300x - 200{,}000 \end{aligned}$$

b. The marginal profit function P' is given by
$$P'(x) = -0.04x + 300$$

c.
$$P'(2000) = -0.04(2000) + 300 = 220$$

Thus, the actual profit realized from the sale of the 2001st loudspeaker system is approximately $220.

d. The graph of the profit function P appears in Figure 12.

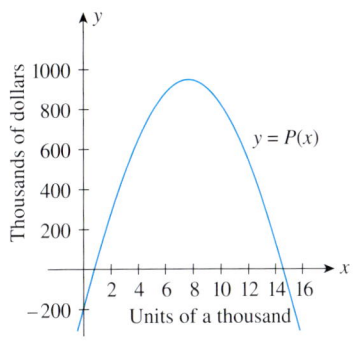

FIGURE 12
The total profit made when x loudspeakers are produced and sold is given by P(x).

Elasticity of Demand

Finally, let's use the marginal concepts introduced in this section to derive an important criterion that economists use to analyze a demand function: elasticity of demand. This measurement enables a businessperson to see how a small percentage change in the price of a commodity affects the percentage change in the quantity demanded of the commodity.

But first we need to introduce the notion of the *relative change* of the size of a quantity, which is given by

$$\frac{\text{Change in the size of the quantity}}{\text{Size of the quantity}}$$

For example, suppose that the mortgage rate increases from the current rate of 10% per year to a rate of 11% per year. Then the relative change in the mortgage rate is

$$\frac{1}{10} = 0.1 \quad \text{or} \quad 10\%$$

But if the current mortgage rate is 5% per year, then a change of 1% per year, to 6% per year, in the mortgage rate would yield a relative change in the mortgage rate of

$$\frac{1}{5} = 0.2 \quad \text{or} \quad 20\%$$

This example shows that the relative change sometimes conveys a better sense of what is going on than does a simple look at the change in the quantity itself.

The next notion we need is that of the *relative rate of change* of a quantity, which is given by

$$\frac{\text{Rate of change of the size of the quantity}}{\text{Size of the quantity}}$$

For example, suppose that the current mortgage rate of 6% per year is changing at the rate of 1.2% per year. Then the relative rate of change of the mortgage rate is

$$\frac{1.2}{6} = 0.2$$

or 20% per year.

Let's generalize this notion using the derivative of a function. Recall that the derivative f' of a function f with respect to x measures the rate of change of f at x with respect to x. Thus, we can write the relative rate of change of f with respect to x at x as

$$\frac{f'(x)}{f(x)} \quad \text{or} \quad \frac{100 f'(x)}{f(x)}\%$$

We are now in a position to discuss elasticity of demand. In what follows, it will be convenient to write the demand function f in the form $x = f(p)$; that is, we will

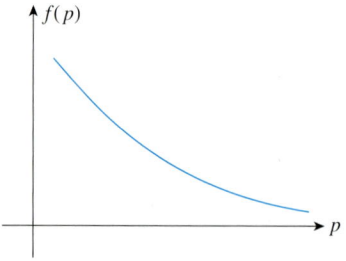

FIGURE 13
The graph of a demand function

think of the quantity demanded of a certain commodity as a function of its unit price. Since the quantity demanded of a commodity usually decreases as its unit price increases, the function f is typically a decreasing function of p (Figure 13).

Since the rate of change of the demand function with respect to p is $f'(p)$, we see that the relative rate of change of f with respect to p at p is

$$\frac{f'(p)}{f(p)} \quad \text{or} \quad \frac{100 f'(p)}{f(p)}\%$$

Next, we see that the relative rate of change of the price p of the commodity is

$$\frac{\frac{d}{dp}(p)}{p} = \frac{1}{p} \quad \text{or} \quad \frac{100}{p}\%$$

Therefore,

$$\frac{\text{Percentage rate of change of } f}{\text{Percentage rate of change of } p} = \frac{\frac{100 f'(p)}{f(p)}}{\frac{100}{p}} = \frac{p f'(p)}{f(p)}$$

Economists call the *negative* of this quantity the elasticity of demand.

Elasticity of Demand

If f is a differentiable demand function defined by $x = f(p)$, then the **elasticity of demand** at price p is given by

$$E(p) = -\frac{p f'(p)}{f(p)} \qquad (7)$$

Note It will be shown later (in Section 4.1) that if f is decreasing on an interval, then $f'(p) < 0$ for p in that interval. In light of this, we see that since both p and $f(p)$ are positive, the quantity $\frac{p f'(p)}{f(p)}$ is negative. Because economists would rather work with a positive value, the elasticity of demand $E(p)$ is defined to be the negative of this quantity. ∎

 APPLIED EXAMPLE 7 Elasticity of Demand Consider the demand equation

$$p = -0.02x + 400 \qquad (0 \leq x \leq 20{,}000)$$

which describes the relationship between the unit price in dollars and the quantity demanded x of the Acrosonic model F loudspeaker systems.

a. Find the elasticity of demand $E(p)$.
b. Compute $E(100)$, and interpret your result.
c. Compute $E(300)$, and interpret your result.

Solution

a. Solving the given demand equation for x in terms of p, we find

$$x = f(p) = -50p + 20{,}000$$

from which we see that
$$f'(p) = -50$$

Therefore,
$$E(p) = -\frac{pf'(p)}{f(p)} = -\frac{p(-50)}{-50p + 20{,}000}$$
$$= \frac{p}{400 - p}$$

b.
$$E(100) = \frac{100}{400 - 100} = \frac{1}{3}$$

which is the elasticity of demand when $p = 100$. To interpret this result, recall that $E(100)$ is the negative of the ratio of the percentage change in the quantity demanded to the percentage change in the unit price when $p = 100$. Therefore, our result tells us that when the unit price p is set at \$100 per speaker, an increase of 1% in the unit price will cause a decrease of approximately 0.33% in the quantity demanded.

c.
$$E(300) = \frac{300}{400 - 300} = 3$$

which is the elasticity of demand when $p = 300$. It tells us that when the unit price is set at \$300 per speaker, an increase of 1% in the unit price will cause a decrease of approximately 3% in the quantity demanded.

Economists often use the following terminology to describe demand in terms of elasticity.

Elasticity of Demand
The demand is said to be **elastic** if $E(p) > 1$.
The demand is said to be **unitary** if $E(p) = 1$.
The demand is said to be **inelastic** if $E(p) < 1$.

As an illustration, our computations in Example 7 revealed that demand for Acrosonic loudspeakers is elastic when $p = 300$ but inelastic when $p = 100$. These computations confirm that when demand is elastic, a small percentage change in the unit price will result in a greater percentage change in the quantity demanded; and when demand is inelastic, a small percentage change in the unit price will cause a smaller percentage change in the quantity demanded. Finally, when demand is unitary, a small percentage change in the unit price will result in the same percentage change in the quantity demanded.

We can describe the way revenue responds to changes in the unit price using the notion of elasticity. If the quantity demanded of a certain commodity is related to its unit price by the equation $x = f(p)$, then the revenue realized through the sale of x units of the commodity at a price of p dollars each is
$$R(p) = px = pf(p)$$

The rate of change of the revenue with respect to the unit price p is given by
$$R'(p) = f(p) + pf'(p)$$
$$= f(p)\left[1 + \frac{pf'(p)}{f(p)}\right]$$
$$= f(p)[1 - E(p)]$$

Now, suppose demand is elastic when the unit price is set at a dollars. Then $E(a) > 1$, and so $1 - E(a) < 0$. Since $f(p)$ is positive for all values of p, we see that

$$R'(a) = f(a)[1 - E(a)] < 0$$

and so $R(p)$ is decreasing at $p = a$. This implies that a small increase in the unit price when $p = a$ results in a decrease in the revenue, whereas a small decrease in the unit price will result in an increase in the revenue. Similarly, you can show that if the demand is inelastic when the unit price is set at a dollars, then a small increase in the unit price will cause the revenue to increase, and a small decrease in the unit price will cause the revenue to decrease. Finally, if the demand is unitary when the unit price is set at a dollars, then $E(a) = 1$ and $R'(a) = 0$. This implies that a small increase or decrease in the unit price will not result in a change in the revenue. The following statements summarize this discussion.

1. If the demand is elastic at p [$E(p) > 1$], then an increase in the unit price will cause the revenue to decrease, whereas a decrease in the unit price will cause the revenue to increase.
2. If the demand is inelastic at p [$E(p) < 1$], then an increase in the unit price will cause the revenue to increase, and a decrease in the unit price will cause the revenue to decrease.
3. If the demand is unitary at p [$E(p) = 1$], then an increase in the unit price will cause the revenue to stay about the same.

These results are illustrated in Figure 14.

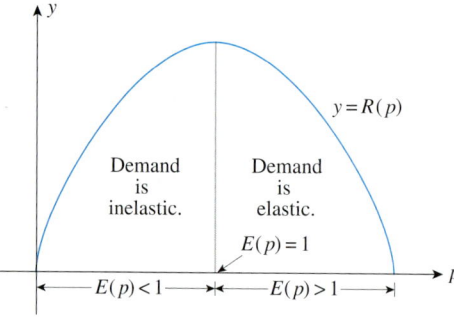

FIGURE 14
The revenue is increasing on an interval where the demand is inelastic, decreasing on an interval where the demand is elastic, and stationary at the point where the demand is unitary.

Note As an aid to remembering, note the following:

1. If demand is elastic, then the change in revenue and the change in the unit price move in opposite directions.
2. If demand is inelastic, then they move in the same direction.

 APPLIED EXAMPLE 8 Elasticity of Demand Refer to Example 7.

a. Is demand elastic, unitary, or inelastic when $p = 100$? When $p = 300$?
b. If the price is $100, will raising the unit price slightly cause the revenue to increase or decrease?

Solution

a. From the results of Example 7, we see that $E(100) = \frac{1}{3} < 1$ and $E(300) = 3 > 1$. We conclude accordingly that demand is inelastic when $p = 100$ and elastic when $p = 300$.
b. Since demand is inelastic when $p = 100$, raising the unit price slightly will cause the revenue to increase.

3.4 Self-Check Exercises

1. The weekly demand for Pulsar DVD recorders is given by the demand equation

$$p = -0.02x + 300 \quad (0 \le x \le 15{,}000)$$

where p denotes the wholesale unit price in dollars and x denotes the quantity demanded. The weekly total cost function associated with manufacturing these recorders is

$$C(x) = 0.000003x^3 - 0.04x^2 + 200x + 70{,}000$$

dollars.
a. Find the revenue function R and the profit function P.
b. Find the marginal cost function C', the marginal revenue function R', and the marginal profit function P'.
c. Find the marginal average cost function $\overline{C}'$.
d. Compute $C'(3000)$, $R'(3000)$, and $P'(3000)$ and interpret your results.

2. Refer to the preceding exercise. Determine whether the demand is elastic, unitary, or inelastic when $p = 100$ and when $p = 200$.

Solutions to Self-Check Exercises 3.4 can be found on page 208.

3.4 Concept Questions

1. Explain each term in your own words:
   a. Marginal cost function
   b. Average cost function
   c. Marginal average cost function
   d. Marginal revenue function
   e. Marginal profit function

2. a. Define the elasticity of demand.
   b. When is the elasticity of demand elastic? Unitary? Inelastic? Explain the meaning of each term.

3.4 Exercises

1. **PRODUCTION COSTS** The graph of a typical total cost function $C(x)$ associated with the manufacture of x units of a certain commodity is shown in the following figure.
   a. Explain why the function C is always increasing.
   b. As the level of production x increases, the cost per unit drops so that $C(x)$ increases but at a slower pace. However, a level of production is soon reached at which the cost per unit begins to increase dramatically (owing to a shortage of raw material, overtime, breakdown of machinery due to excessive stress and strain), so $C(x)$ continues to increase at a faster pace. Use the graph of C to find the approximate level of production x_0 where this occurs.

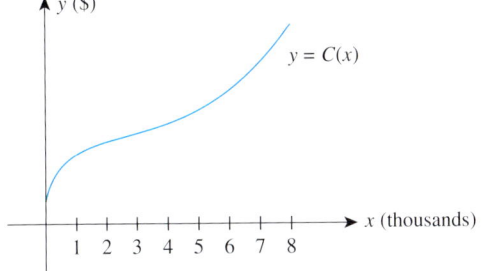

2. **PRODUCTION COSTS** The graph of a typical average cost function $A(x) = C(x)/x$, where $C(x)$ is a total cost function associated with the manufacture of x units of a certain commodity, is shown in the following figure.
   a. Explain in economic terms why $A(x)$ is large if x is small and why $A(x)$ is large if x is large.
   b. What is the significance of the numbers x_0 and y_0, the x- and y-coordinates of the lowest point on the graph of the function A?

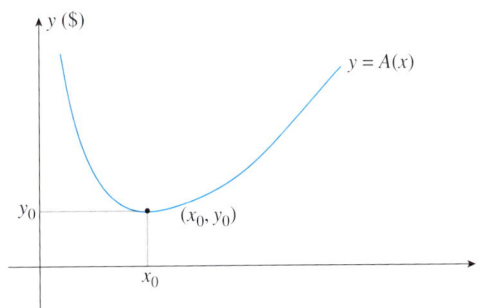

3. **MARGINAL COST** The total weekly cost (in dollars) incurred by Lincoln Records in pressing x compact discs is

$$C(x) = 2000 + 2x - 0.0001x^2 \quad (0 \le x \le 6000)$$

a. What is the actual cost incurred in producing the 1001st disc and the 2001st disc?
b. What is the marginal cost when $x = 1000$ and 2000?

4. **MARGINAL COST** A division of Ditton Industries manufactures the Futura model microwave oven. The daily cost (in dollars) of producing these microwave ovens is

$$C(x) = 0.0002x^3 - 0.06x^2 + 120x + 5000$$

where x stands for the number of units produced.

a. What is the actual cost incurred in manufacturing the 101st oven? The 201st oven? The 301st oven?
b. What is the marginal cost when $x = 100$, 200, and 300?

5. MARGINAL AVERAGE COST Custom Office makes a line of executive desks. It is estimated that the total cost for making x units of their Senior Executive model is

$$C(x) = 100x + 200,000$$

dollars/year.
a. Find the average cost function $\overline{C}$.
b. Find the marginal average cost function $\overline{C}'$.
c. What happens to $\overline{C}(x)$ when x is very large? Interpret your results.

6. MARGINAL AVERAGE COST The management of ThermoMaster Company, whose Mexican subsidiary manufactures an indoor–outdoor thermometer, has estimated that the total weekly cost (in dollars) for producing x thermometers is

$$C(x) = 5000 + 2x$$

a. Find the average cost function $\overline{C}$.
b. Find the marginal average cost function $\overline{C}'$.
c. Interpret your results.

7. Find the average cost function $\overline{C}$ and the marginal average cost function $\overline{C}'$ associated with the total cost function C of Exercise 3.

8. Find the average cost function $\overline{C}$ and the marginal average cost function $\overline{C}'$ associated with the total cost function C of Exercise 4.

9. MARGINAL REVENUE Williams Commuter Air Service realizes a monthly revenue of

$$R(x) = 8000x - 100x^2$$

dollars when the price charged per passenger is x dollars.
a. Find the marginal revenue R'.
b. Compute $R'(39)$, $R'(40)$, and $R'(41)$.
c. Based on the results of part (b), what price should the airline charge in order to maximize their revenue?

10. MARGINAL REVENUE The management of Acrosonic plans to market the ElectroStat, an electrostatic speaker system. The marketing department has determined that the demand function for these speakers is

$$p = -0.04x + 800 \quad (0 \leq x \leq 20,000)$$

where p denotes the speaker's unit price (in dollars) and x denotes the quantity demanded.
a. Find the revenue function R.
b. Find the marginal revenue function R'.
c. Compute $R'(5000)$, and interpret your results.

11. MARGINAL PROFIT Refer to Exercise 10. Acrosonic's production department estimates that the total cost (in dollars) incurred in manufacturing x ElectroStat speaker systems in the first year of production will be

$$C(x) = 200x + 300,000$$

a. Find the profit function P.
b. Find the marginal profit function P'.
c. Compute $P'(5000)$ and $P'(8000)$.
d. Sketch the graph of the profit function, and interpret your results.

12. MARGINAL PROFIT Lynbrook West, an apartment complex, has 100 two-bedroom units. The monthly profit (in dollars) realized from renting x apartments is

$$P(x) = -10x^2 + 1760x - 50,000$$

a. What is the actual profit realized from renting the 51st unit, assuming that 50 units have already been rented?
b. Compute the marginal profit when $x = 50$, and compare your results with that obtained in part (a).

13. MARGINAL COST, REVENUE, AND PROFIT The weekly demand for the Pulsar 25 color LED television is

$$p = 600 - 0.05x \quad (0 \leq x \leq 12,000)$$

where p denotes the wholesale unit price in dollars and x denotes the quantity demanded. The weekly total cost function associated with manufacturing the Pulsar 25 is given by

$$C(x) = 0.000002x^3 - 0.03x^2 + 400x + 80,000$$

where $C(x)$ denotes the total cost incurred in producing x sets.
a. Find the revenue function R and the profit function P.
b. Find the marginal cost function C', the marginal revenue function R', and the marginal profit function P'.
c. Compute $C'(2000)$, $R'(2000)$, and $P'(2000)$, and interpret your results.
d. Sketch the graphs of the functions C, R, and P, and interpret parts (b) and (c), using the graphs obtained.

14. MARGINAL COST, REVENUE, AND PROFIT Pulsar manufactures a series of 20-in. flat-tube digital televisions. The quantity x of these sets demanded each week is related to the wholesale unit price p by the equation

$$p = -0.006x + 180$$

The weekly total cost incurred by Pulsar for producing x sets is

$$C(x) = 0.000002x^3 - 0.02x^2 + 120x + 60,000$$

dollars. Answer the questions in Exercise 13 for these data.

15. MARGINAL AVERAGE COST Refer to Exercise 13.
a. Find the average cost function $\overline{C}$ associated with the total cost function C of Exercise 13.
b. What is the marginal average cost function $\overline{C}'$?
c. Compute $\overline{C}'(5000)$ and $\overline{C}'(10,000)$, and interpret your results.
d. Sketch the graph of $\overline{C}$.

16. MARGINAL AVERAGE COST Refer to Exercise 14.
a. Find the average cost function $\overline{C}$ associated with the total cost function C of Exercise 14.
b. What is the marginal average cost function $\overline{C}'$?
c. Compute $\overline{C}'(5000)$ and $\overline{C}'(10,000)$, and interpret your results.
d. Sketch the graph of $\overline{C}$.

3.4 MARGINAL FUNCTIONS IN ECONOMICS

17. MARGINAL REVENUE The quantity of Sicard sports watches demanded each month is related to the unit price by the equation

$$p = \frac{50}{0.01x^2 + 1} \quad (0 \le x \le 20)$$

where p is measured in thousands of dollars and x in units of a thousand.
a. Find the revenue function R.
b. Find the marginal revenue function R'.
c. Compute $R'(2)$, and interpret your result.

18. MARGINAL PROPENSITY TO CONSUME The consumption function of the U.S. economy from 1929 to 1941 is

$$C(x) = 0.712x + 95.05$$

where $C(x)$ is the personal consumption expenditure and x is the personal income, both measured in billions of dollars. Find the rate of change of consumption with respect to income, dC/dx. This quantity is called the *marginal propensity to consume*.

19. MARGINAL PROPENSITY TO CONSUME Refer to Exercise 18. Suppose a certain economy's consumption function is

$$C(x) = 0.873x^{1.1} + 20.34$$

where $C(x)$ and x are measured in billions of dollars. Find the marginal propensity to consume when $x = 10$.

20. MARGINAL PROPENSITY TO SAVE Suppose $C(x)$ measures an economy's personal consumption expenditure and x measures the personal income, both in billions of dollars. Then

$$S(x) = x - C(x) \quad \text{Income minus consumption}$$

measures the economy's savings corresponding to an income of x billion dollars. Show that

$$\frac{dS}{dx} = 1 - \frac{dC}{dx}$$

The quantity dS/dx is called the *marginal propensity to save*.

21. Refer to Exercise 20. For the consumption function of Exercise 18, find the marginal propensity to save.

22. Refer to Exercise 20. For the consumption function of Exercise 19, find the marginal propensity to save when $x = 10$.

For each demand equation in Exercises 23–28, compute the elasticity of demand and determine whether the demand is elastic, unitary, or inelastic at the indicated price.

23. $x = -\dfrac{5}{4}p + 20; \, p = 10$

24. $x = -\dfrac{3}{2}p + 9; \, p = 2$

25. $x + \dfrac{1}{3}p - 20 = 0; \, p = 30$

26. $0.4x + p - 20 = 0; \, p = 10$

27. $p = 169 - x^2; \, p = 29$ **28.** $p = 144 - x^2; \, p = 96$

29. ELASTICITY OF DEMAND The demand equation for the Roland portable hair dryer is given by

$$x = \frac{1}{5}(225 - p^2) \quad (0 \le p \le 15)$$

where x (measured in units of a hundred) is the quantity demanded per week and p is the unit price in dollars.
a. Is the demand elastic or inelastic when $p = 8$ and when $p = 10$?
b. When is the demand unitary?
   Hint: Solve $E(p) = 1$ for p.
c. If the unit price is lowered slightly from $10, will the revenue increase or decrease?
d. If the unit price is increased slightly from $8, will the revenue increase or decrease?

30. ELASTICITY OF DEMAND The management of Titan Tire Company has determined that the quantity demanded x of their Super Titan tires per week is related to the unit price p by the equation

$$x = \sqrt{144 - p} \quad (0 \le p \le 144)$$

where p is measured in dollars and x in units of a thousand.
a. Compute the elasticity of demand when $p = 63, 96$, and 108.
b. Interpret the results obtained in part (a).
c. Is the demand elastic, unitary, or inelastic when $p = 63$, 96, and 108?

31. ELASTICITY OF DEMAND The proprietor of Showplace, a video store, has estimated that the rental price p (in dollars) of prerecorded DVDs is related to the quantity x (in thousands) rented per day by the demand equation

$$x = \frac{2}{3}\sqrt{36 - p^2} \quad (0 \le p \le 6)$$

Currently, the rental price is $2/disc.
a. Is the demand elastic or inelastic at this rental price?
b. If the rental price is increased, will the revenue increase or decrease?

32. ELASTICITY OF DEMAND The quantity demanded each week x (in units of a hundred) of the Mikado digital camera is related to the unit price p (in dollars) by the demand equation

$$x = \sqrt{400 - 5p} \quad (0 \le p \le 80)$$

a. Is the demand elastic or inelastic when $p = 40$? When $p = 60$?
b. When is the demand unitary?
   Hint: Solve $E(p) = 1$ for p.
c. If the unit price is lowered slightly from $60, will the revenue increase or decrease?
d. If the unit price is increased slightly from $40, will the revenue increase or decrease?

33. ELASTICITY OF DEMAND The demand function for a certain make of exercise bicycle sold exclusively through cable television is

$$p = \sqrt{9 - 0.02x} \quad (0 \leq x \leq 450)$$

where p is the unit price in hundreds of dollars and x is the quantity demanded per week. Compute the elasticity of demand and determine the range of prices corresponding to inelastic, unitary, and elastic demand.
Hint: Solve the equation $E(p) = 1$.

34. ELASTICITY OF DEMAND The demand equation for the Sicard sports watch is given by

$$x = 10\sqrt{\frac{50 - p}{p}} \quad (0 < p \leq 50)$$

where x (measured in units of a thousand) is the quantity demanded per week and p is the unit price in dollars. Compute the elasticity of demand and determine the range of prices corresponding to inelastic, unitary, and elastic demand.

In Exercises 35 and 36, determine whether the statement is true or false. If it is true, explain why it is true. If it is false, give an example to show why it is false.

35. If C is a differentiable total cost function, then the marginal average cost function is

$$\overline{C}'(x) = \frac{xC'(x) - C(x)}{x^2}$$

36. If the marginal profit function is positive at $x = a$, then it makes sense to decrease the level of production.

3.4 Solutions to Self-Check Exercises

1. a. $R(x) = px$
$= x(-0.02x + 300)$
$= -0.02x^2 + 300x \quad (0 \leq x \leq 15,000)$
$P(x) = R(x) - C(x)$
$= -0.02x^2 + 300x$
$\quad - (0.000003x^3 - 0.04x^2 + 200x + 70,000)$
$= -0.000003x^3 + 0.02x^2 + 100x - 70,000$

b. $C'(x) = 0.000009x^2 - 0.08x + 200$
$R'(x) = -0.04x + 300$
$P'(x) = -0.000009x^2 + 0.04x + 100$

c. The average cost function is

$$\overline{C}(x) = \frac{C(x)}{x}$$
$$= \frac{0.000003x^3 - 0.04x^2 + 200x + 70,000}{x}$$
$$= 0.000003x^2 - 0.04x + 200 + \frac{70,000}{x}$$

Therefore, the marginal average cost function is

$$\overline{C}'(x) = 0.000006x - 0.04 - \frac{70,000}{x^2}$$

d. Using the results from part (b), we find
$C'(3000) = 0.000009(3000)^2 - 0.08(3000) + 200$
$= 41$

That is, when the level of production is already 3000 recorders, the actual cost of producing one additional recorder is approximately $41. Next,

$R'(3000) = -0.04(3000) + 300 = 180$

That is, the actual revenue to be realized from selling the 3001st recorder is approximately $180. Finally,

$P'(3000) = -0.000009(3000)^2 + 0.04(3000) + 100$
$= 139$

That is, the actual profit realized from selling the 3001st DVD recorder is approximately $139.

2. We first solve the given demand equation for x in terms of p, obtaining

$$x = f(p) = -50p + 15,000$$
$$f'(p) = -50$$

Therefore,

$$E(p) = -\frac{pf'(p)}{f(p)} = -\frac{p(-50)}{-50p + 15,000}$$
$$= \frac{p}{300 - p} \quad (0 \leq p < 300)$$

Next, we compute

$$E(100) = \frac{100}{300 - 100} = \frac{1}{2} < 1$$

and we conclude that demand is inelastic when $p = 100$. Also,

$$E(200) = \frac{200}{300 - 200} = 2 > 1$$

and we see that demand is elastic when $p = 200$.

3.5 Higher-Order Derivatives

Higher-Order Derivatives

The derivative f' of a function f is also a function. As such, the differentiability of f' may be considered. Thus, the function f' has a derivative f'' at a point x in the domain of f' if the limit of the quotient

$$\frac{f'(x+h) - f'(x)}{h}$$

exists as h approaches zero. In other words, it is the derivative of the first derivative.

The function f'' obtained in this manner is called the **second derivative of the function f,** just as the derivative f' of f is often called the first derivative of f. Continuing in this fashion, we are led to considering the third, fourth, and higher-order derivatives of f whenever they exist. Notations for the first, second, third, and, in general, nth derivatives of a function f at a point x are

$$f'(x), f''(x), f'''(x), \ldots, f^{(n)}(x)$$

or

$$D^1 f(x), D^2 f(x), D^3 f(x), \ldots, D^n f(x)$$

If f is written in the form $y = f(x)$, then the notations for its derivatives are

$$y', y'', y''', \ldots, y^{(n)}$$

$$\frac{dy}{dx}, \frac{d^2 y}{dx^2}, \frac{d^3 y}{dx^3}, \ldots, \frac{d^n y}{dx^n}$$

or

$$D^1 y, D^2 y, D^3 y, \ldots, D^n y$$

EXAMPLE 1 Find the derivatives of all orders of the polynomial function

$$f(x) = x^5 - 3x^4 + 4x^3 - 2x^2 + x - 8$$

Solution We have

$$f'(x) = 5x^4 - 12x^3 + 12x^2 - 4x + 1$$

$$f''(x) = \frac{d}{dx} f'(x) = 20x^3 - 36x^2 + 24x - 4$$

$$f'''(x) = \frac{d}{dx} f''(x) = 60x^2 - 72x + 24$$

$$f^{(4)}(x) = \frac{d}{dx} f'''(x) = 120x - 72$$

$$f^{(5)}(x) = \frac{d}{dx} f^{(4)}(x) = 120$$

and

$$f^{(n)}(x) = 0 \quad \text{for } n > 5$$

EXAMPLE 2 Find the third derivative of the function f defined by $y - x^{2/3}$. What is its domain?

Solution We have

$$y' = \frac{2}{3}x^{-1/3}$$

$$y'' = \left(\frac{2}{3}\right)\left(-\frac{1}{3}\right)x^{-4/3} = -\frac{2}{9}x^{-4/3}$$

so the required derivative is

$$y''' = \left(-\frac{2}{9}\right)\left(-\frac{4}{3}\right)x^{-7/3} = \frac{8}{27}x^{-7/3} = \frac{8}{27x^{7/3}}$$

The common domain of the functions f', f'', and f''' is the set of all real numbers except $x = 0$. The domain of $y = x^{2/3}$ is the set of all real numbers. The graph of the function $y = x^{2/3}$ appears in Figure 15.

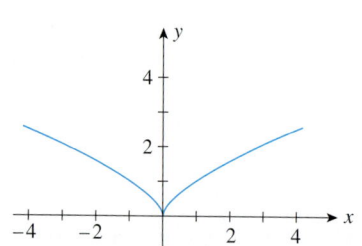

FIGURE 15
The graph of the function $y = x^{2/3}$

Note Always simplify an expression before differentiating it to obtain the next order derivative.

VIDEO **EXAMPLE 3** Find the second derivative of the function $y = (2x^2 + 3)^{3/2}$.

Solution We have, using the General Power Rule,

$$y' = \frac{3}{2}(2x^2 + 3)^{1/2}(4x) = 6x(2x^2 + 3)^{1/2}$$

Next, using the Product Rule and then the General Power Rule, we find

$$y'' = (6x) \cdot \frac{d}{dx}(2x^2 + 3)^{1/2} + \left[\frac{d}{dx}(6x)\right](2x^2 + 3)^{1/2}$$

$$= (6x)\left(\frac{1}{2}\right)(2x^2 + 3)^{-1/2}(4x) + 6(2x^2 + 3)^{1/2} \quad \text{(x²) See page 9.}$$

$$= 12x^2(2x^2 + 3)^{-1/2} + 6(2x^2 + 3)^{1/2}$$

$$= 6(2x^2 + 3)^{-1/2}[2x^2 + (2x^2 + 3)] \quad \text{Factor out } 6(2x^2 + 3)^{-1/2}.$$

$$= \frac{6(4x^2 + 3)}{\sqrt{2x^2 + 3}}$$

Just as the derivative of a function f at a point x measures the rate of change of the function f at that point, the second derivative of f (the derivative of f') measures the rate of change of the derivative f' of the function f. The third derivative of the function f, f''', measures the rate of change of f'', and so on.

In Chapter 4, we will discuss applications involving the geometric interpretation of the second derivative of a function. The following example gives an interpretation of the second derivative in a familiar role.

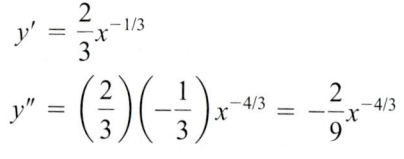 **APPLIED EXAMPLE 4** Acceleration of a Maglev Refer to the example on pages 97–98. The distance s (in feet) covered by a maglev moving along a straight track t seconds after starting from rest is given by the function $s = 4t^2$ ($0 \le t \le 10$). What is the maglev's acceleration at any time t?

Solution The velocity of the maglev t seconds from rest is given by

$$v = \frac{ds}{dt} = \frac{d}{dt}(4t^2) = 8t$$

The acceleration of the maglev t seconds from rest is given by the rate of change of the velocity of t—that is,

$$a = \frac{d}{dt}v = \frac{d}{dt}\left(\frac{ds}{dt}\right) = \frac{d^2s}{dt^2} = \frac{d}{dt}(8t) = 8$$

or 8 feet per second per second, normally abbreviated 8 ft/sec².

APPLIED EXAMPLE 5 Acceleration and Velocity of a Falling Object A ball is thrown straight up into the air from the roof of a building. The height of the ball as measured from the ground is given by

$$s = -16t^2 + 24t + 120$$

where s is measured in feet and t in seconds. Find the velocity and acceleration of the ball 3 seconds after it is thrown into the air.

Solution The velocity v and acceleration a of the ball at any time t are given by

$$v = \frac{ds}{dt} = \frac{d}{dt}(-16t^2 + 24t + 120) = -32t + 24$$

and

$$a = \frac{d^2s}{dt^2} = \frac{d}{dt}\left(\frac{ds}{dt}\right) = \frac{d}{dt}(-32t + 24) = -32$$

Therefore, the velocity of the ball 3 seconds after it is thrown into the air is

$$v = -32(3) + 24 = -72$$

That is, the ball is falling downward at a speed of 72 ft/sec. The acceleration of the ball is 32 ft/sec² downward at any time during the motion.

Another interpretation of the second derivative of a function—this time from the field of economics—follows. Suppose the consumer price index (CPI) of an economy between the years a and b is described by the function $I(t)$ ($a \leq t \leq b$) (Figure 16). Then the first derivative of I at $t = c$, $I'(c)$, where $a < c < b$, gives the rate of change of I at c. The quantity

$$\frac{I'(c)}{I(c)}$$

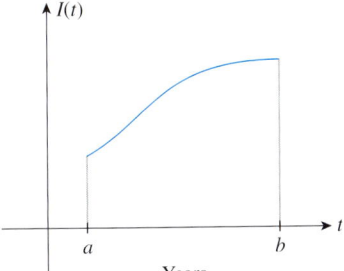

FIGURE 16
The CPI of a certain economy from year a to year b is given by $I(t)$.

measures the *inflation rate* of the economy at $t = c$. The second derivative of I at $t = c$, $I''(c)$, gives the rate of change of I' at $t = c$. Now, it is possible for $I'(t)$ to be positive and $I''(t)$ to be negative at $t = c$ (see Example 6). This tells us that at $t = c$ the economy is experiencing inflation (the CPI is increasing) but the rate at which the CPI is growing is in fact decreasing. This is precisely the situation described by an economist or a politician when she claims that "inflation is slowing." One may not jump to the conclusion from the aforementioned quote that prices of goods and services are about to drop!

APPLIED EXAMPLE 6 Inflation Rate of an Economy The function

$$I(t) = -0.2t^3 + 3t^2 + 100 \quad (0 \leq t \leq 9)$$

gives the CPI of an economy, where $t = 0$ corresponds to the beginning of 2004.

a. Find the inflation rate at the beginning of 2010 ($t = 6$).
b. Show that inflation was moderating at that time.

Solution

a. We find $I'(t) = -0.6t^2 + 6t$. Next, we compute

$$I'(6) = -0.6(6)^2 + 6(6) = 14.4 \quad \text{and} \quad I(6) = -0.2(6)^3 + 3(6)^2 + 100 = 164.8$$

from which we see that the inflation rate is

$$\frac{I'(6)}{I(6)} = \frac{14.4}{164.8} \approx 0.0874$$

or approximately 8.7%.

b. We find

$$I''(t) = \frac{d}{dt}(-0.6t^2 + 6t) = -1.2t + 6$$

Since

$$I''(6) = -1.2(6) + 6 = -1.2$$

we see that I' is indeed decreasing at $t = 6$, and we conclude that inflation was moderating at that time (Figure 17).

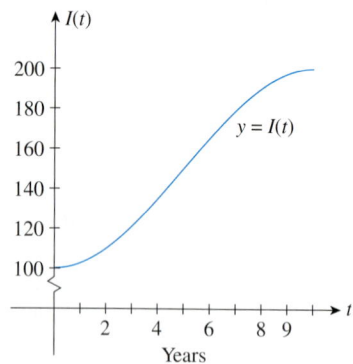

FIGURE 17
The CPI of an economy is given by $I(t)$.

3.5 Self-Check Exercises

1. Find the third derivative of
$$f(x) = 2x^5 - 3x^3 + x^2 - 6x + 10$$

2. Let
$$f(x) = \frac{1}{1+x}$$
Find $f'(x)$, $f''(x)$, and $f'''(x)$.

3. **CONSERVATION OF SPECIES** A certain species of turtles faces extinction because dealers collect truckloads of turtle eggs to be sold as aphrodisiacs. After severe conservation measures are implemented, it is hoped that the turtle population will grow according to the rule

$$N(t) = 2t^3 + 3t^2 - 4t + 1000 \quad (0 \leq t \leq 10)$$

where $N(t)$ denotes the population at the end of year t. Compute $N''(2)$ and $N''(8)$. What do your results tell you about the effectiveness of the program?

Solutions to Self-Check Exercises 3.5 can be found on page 214.

3.5 Concept Questions

1. a. What is the second derivative of a function f?
   b. How do you find the second derivative of a function f, assuming that it exists?

2. If $s = f(t)$ gives the position of an object moving on the coordinate line, what do $f'(t)$ and $f''(t)$ measure?

3. Suppose $f(t)$ measures the population of a country at time t. What can you say about the signs of $f'(t)$ and $f''(t)$ over the time interval (a, b)
   a. If the population is increasing at an increasing rate over (a, b)?
   b. If the population is increasing at a decreasing rate over (a, b)?
   c. If the population is decreasing at an increasing rate over (a, b)?
   d. If the population is decreasing at a decreasing rate over (a, b)?

4. Suppose $f(t)$ measures the population of a country at time t. What can you say about the signs of $f'(t)$ and $f''(t)$ over the time interval
   a. If the population is increasing at a constant rate over (a, b)?
   b. If the population is decreasing at a constant rate over (a, b)?
   c. If the population is constant over (a, b)?

3.5 Exercises

In Exercises 1–20, find the first and second derivatives of the function.

1. $f(x) = 4x^2 - 2x + 1$
2. $f(x) = -0.2x^2 + 0.3x + 4$
3. $f(x) = 2x^3 - 3x^2 + 1$
4. $g(x) = -3x^3 + 24x^2 + 6x - 64$
5. $h(t) = t^4 - 2t^3 + 6t^2 - 3t + 10$
6. $f(x) = x^5 - x^4 + x^3 - x^2 + x - 1$
7. $f(x) = (x^2 + 2)^5$
8. $g(t) = t^2(3t + 1)^4$
9. $g(t) = (2t^2 - 1)^2(3t^2)$
10. $h(x) = (x^2 + 1)^2(x - 1)$
11. $f(x) = (2x^2 + 2)^{7/2}$
12. $h(w) = (w^2 + 2w + 4)^{5/2}$
13. $f(x) = x(x^2 + 1)^2$
14. $g(u) = u(2u - 1)^3$
15. $f(x) = \dfrac{x}{2x + 1}$
16. $g(t) = \dfrac{t^2}{t - 1}$
17. $f(s) = \dfrac{s - 1}{s + 1}$
18. $f(u) = \dfrac{u}{u^2 + 1}$
19. $f(u) = \sqrt{4 - 3u}$
20. $f(x) = \sqrt{2x - 1}$

In Exercises 21–28, find the third derivative of the given function.

21. $f(x) = 3x^4 - 4x^3$
22. $f(x) = 3x^5 - 6x^4 + 2x^2 - 8x + 12$
23. $f(x) = \dfrac{1}{x}$
24. $f(x) = \dfrac{2}{x^2}$
25. $g(s) = \sqrt{3s - 2}$
26. $g(t) = \sqrt{2t + 3}$
27. $f(x) = (2x - 3)^4$
28. $g(t) = \left(\dfrac{1}{2}t^2 - 1\right)^5$

29. **ACCELERATION OF A FALLING OBJECT** During the construction of an office building, a hammer is accidentally dropped from a height of 256 ft. The distance (in feet) the hammer falls in t sec is $s = 16t^2$. What is the hammer's velocity when it strikes the ground? What is its acceleration?

30. **ACCELERATION OF A CAR** The distance s (in feet) covered by a car after t sec is given by

$$s = -t^3 + 8t^2 + 20t \quad (0 \le t \le 6)$$

Find a general expression for the car's acceleration at any time t ($0 \le t \le 6$). Show that the car is decelerating after $2\frac{2}{3}$ sec.

31. **CRIME RATES** The number of major crimes committed in Bronxville between 2005 and 2012 is approximated by the function

$$N(t) = -0.1t^3 + 1.5t^2 + 100 \quad (0 \le t \le 7)$$

where $N(t)$ denotes the number of crimes committed in year t, with $t = 0$ corresponding to 2005. Enraged by the dramatic increase in the crime rate, Bronxville's citizens, with the help of the local police, organized "Neighborhood Crime Watch" groups in early 2009 to combat this menace.
a. Verify that the crime rate was increasing from 2005 through 2012.
   Hint: Compute $N'(0), N'(1), \ldots, N'(7)$.
b. Show that the Neighborhood Crime Watch program was working by computing $N''(4), N''(5), N''(6)$, and $N''(7)$.

32. **GDP OF A DEVELOPING COUNTRY** A developing country's gross domestic product (GDP) from 2004 to 2012 is approximated by the function

$$G(t) = -0.2t^3 + 2.4t^2 + 60 \quad (0 \le t \le 8)$$

where $G(t)$ is measured in billions of dollars, with $t = 0$ corresponding to 2004.
a. Compute $G'(0), G'(1), \ldots, G'(8)$.
b. Compute $G''(0), G''(1), \ldots, G''(8)$.
c. Using the results obtained in parts (a) and (b), show that after a slow start, the GDP increases quickly and then cools off.

33. **DISABILITY BENEFITS** The number of persons age 18–64 years receiving disability benefits through Social Security, Supplemental Security Income, or both from 1990 through 2000 is approximated by the function

$$N(t) = 0.00037t^3 - 0.0242t^2 + 0.52t + 5.3 \quad (0 \le t \le 10)$$

where $N(t)$ is measured in units of a million and t is measured in years, with $t = 0$ corresponding to the beginning of 1990. Compute $N(8), N'(8)$, and $N''(8)$, and interpret your results.
Source: Social Security Administration.

34. **OBESITY IN AMERICA** The body mass index (BMI) measures body weight in relation to height. A BMI of 25 to 29.9 is considered overweight, a BMI of 30 or more is considered obese, and a BMI of 40 or more is morbidly obese. The percent of the U.S. population that is obese is approximated by the function

$$P(t) = 0.0004t^3 + 0.0036t^2 + 0.8t + 12 \quad (0 \le t \le 13)$$

where t is measured in years, with $t = 0$ corresponding to the beginning of 1991. Show that the rate of the rate of change of the percent of the U.S. population that is deemed obese was positive from 1991 to 2004. What does this mean?
Source: Centers for Disease Control and Prevention.

35. TEST FLIGHT OF A VTOL In a test flight of the McCord Terrier, McCord Aviation's experimental VTOL (vertical takeoff and landing) aircraft, it was determined that t sec after liftoff, when the craft was operated in the vertical takeoff mode, its altitude (in feet) was

$$h(t) = \frac{1}{16}t^4 - t^3 + 4t^2 \quad (0 \le t \le 8)$$

a. Find an expression for the craft's velocity at time t.
b. Find the craft's velocity when $t = 0$ (the initial velocity), $t = 4$, and $t = 8$.
c. Find an expression for the craft's acceleration at time t.
d. Find the craft's acceleration when $t = 0$, 4, and 8.
e. Find the craft's height when $t = 0$, 4, and 8.

36. AIR PURIFICATION During testing of a certain brand of air purifier, the amount of smoke remaining t min after the start of the test was

$$A(t) = -0.00006t^5 + 0.00468t^4 - 0.1316t^3 + 1.915t^2 - 17.63t + 100$$

percent of the original amount. Compute $A'(10)$ and $A''(10)$, and interpret your results.
Source: Consumer Reports.

37. AGING POPULATION The population of Americans age 55 and older as a percentage of the total population is approximated by the function

$$f(t) = 10.72(0.9t + 10)^{0.3} \quad (0 \le t \le 20)$$

where t is measured in years, with $t = 0$ corresponding to 2000. Compute $f''(10)$, and interpret your result.
Source: U.S. Census Bureau.

38. WORKING MOTHERS The percentage of mothers who work outside the home and have children younger than 6 years old is approximated by the function

$$P(t) = 33.55(t + 5)^{0.205} \quad (0 \le t \le 21)$$

where t is measured in years, with $t = 0$ corresponding to the beginning of 1980. Compute $P''(20)$, and interpret your result.
Source: U.S. Bureau of Labor Statistics.

In Exercises 39–43, determine whether the statement is true or false. If it is true, explain why it is true. If it is false, give an example to show why it is false.

39. If the second derivative of f exists at $x = a$, then $f''(a) = [f'(a)]^2$.

40. If $h = fg$ where f and g have second-order derivatives, then
$$h''(x) = f''(x)g(x) + 2f'(x)g'(x) + f(x)g''(x)$$

41. If f is a polynomial of degree n, then $f^{(n+1)}(x) = 0$.

42. Suppose $P(t)$ represents the population of bacteria at time t, and suppose $P'(t) > 0$ and $P''(t) < 0$; then the population is increasing at time t but at a decreasing rate.

43. If $h(x) = f(2x)$, then $h''(x) = 4f''(2x)$.

44. Let f be the function defined by the rule $f(x) = x^{7/3}$. Show that f has first- and second-order derivatives at all points x, and in particular at $x = 0$. Also show that the third derivative of f does *not* exist at $x = 0$.

45. Construct a function f that has derivatives of order up through and including n at a point a but fails to have the $(n + 1)$st derivative there.
Hint: See Exercise 44.

46. Show that a polynomial function has derivatives of all orders.
Hint: Let $P(x) = a_n x^n + a_{n-1} x^{n-1} + a_{n-2} x^{n-2} + \cdots + a_0$ be a polynomial of degree n, where n is a positive integer and $a_0, a_1, \ldots, a_n$ are constants with $a_n \ne 0$. Compute $P'(x)$, $P''(x)$, ...

3.5 Solutions to Self-Check Exercises

1. $f'(x) = 10x^4 - 9x^2 + 2x - 6$
$f''(x) = 40x^3 - 18x + 2$
$f'''(x) = 120x^2 - 18$

2. We write $f(x) = (1 + x)^{-1}$ and use the General Power Rule, obtaining

$$f'(x) = (-1)(1 + x)^{-2} \frac{d}{dx}(1 + x) = -(1 + x)^{-2}(1)$$

$$= -(1 + x)^{-2} = -\frac{1}{(1 + x)^2}$$

Continuing, we find

$$f''(x) = -(-2)(1 + x)^{-3}$$

$$= 2(1 + x)^{-3} = \frac{2}{(1 + x)^3}$$

$$f'''(x) = 2(-3)(1 + x)^{-4}$$

$$= -6(1 + x)^{-4}$$

$$= -\frac{6}{(1 + x)^4}$$

3. $N'(t) = 6t^2 + 6t - 4$ and $N''(t) = 12t + 6 = 6(2t + 1)$

Therefore, $N''(2) = 30$ and $N''(8) = 102$. The results of our computations reveal that at the end of year 2, the *rate* of growth of the turtle population is increasing at the rate of 30 turtles/year/year. At the end of year 8, the rate is increasing at the rate of 102 turtles/year/year. Clearly, the conservation measures are paying off handsomely.

3.5 HIGHER-ORDER DERIVATIVES

USING TECHNOLOGY

Finding the Second Derivative of a Function at a Given Point

Some graphing utilities have the capability of numerically computing the second derivative of a function at a point. If your graphing utility has this capability, use it to work through the examples and exercises of this section.

EXAMPLE 1 Use the (second) numerical derivative operation of a graphing utility to find the second derivative of $f(x) = \sqrt{x}$ when $x = 4$.

Solution Using the (second) numerical derivative operation, we find

$$f''(4) = \text{der2}(x\char`\^.5, x, 4) = -.03125$$

(Figure T1).

FIGURE T1
The TI-86 second derivative screen for computing $f''(4)$

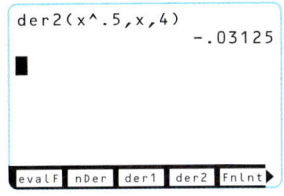

APPLIED EXAMPLE 2 Prevalence of Alzheimer's Patients The number of Alzheimer's patients in the United States is given by

$$f(t) = -0.02765t^4 + 0.3346t^3 - 1.1261t^2 + 1.7575t + 3.7745 \quad (0 \leq t \leq 5)$$

where $f(t)$ is measured in millions and t is measured in decades, with $t = 0$ corresponding to the beginning of 1990.

a. How fast is the number of Alzheimer's patients in the United States anticipated to be changing at the beginning of 2030?
b. How fast is the rate of change of the number of Alzheimer's patients in the United States anticipated to be changing at the beginning of 2030?
c. Plot the graph of f in the viewing window $[0, 5] \times [0, 12]$.

Source: Alzheimer's Association.

Solution

a. Using the numerical derivative operation of a graphing utility, we find that the number of Alzheimer's patients at the beginning of 2030 can be anticipated to be changing at the rate of

$$f'(4) = 1.7311$$

That is, the number is increasing at the rate of approximately 1.7 million patients per decade.

b. Using the (second) numerical derivative operation of a graphing utility, we find that

$$f''(4) = 0.4694$$

(Figure T2); that is, the rate of change of the number of Alzheimer's patients is increasing at the rate of approximately 0.5 million patients per decade per decade.

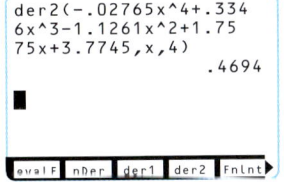

FIGURE T2
The TI-86 second derivative screen for computing $f''(4)$

(continued)

c. Figure T3 shows the graph.

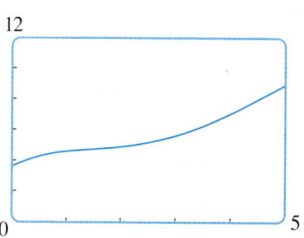

FIGURE T3
The graph of f in the viewing window
$[0, 5] \times [0, 12]$

TECHNOLOGY EXERCISES

In Exercises 1–8, find the value of the second derivative of f at the given value of x. Express your answer correct to four decimal places.

1. $f(x) = 2x^3 - 3x^2 + 1$; $x = -1$
2. $f(x) = 2.5x^5 - 3x^3 + 1.5x + 4$; $x = 2.1$
3. $f(x) = 2.1x^{3.1} - 4.2x^{1.7} + 4.2$; $x = 1.4$
4. $f(x) = 1.7x^{4.2} - 3.2x^{1.3} + 4.2x - 3.2$; $x = 2.2$
5. $f(x) = \dfrac{x^2 + 2x - 5}{x^3 + 1}$; $x = 2.1$
6. $f(x) = \dfrac{x^3 + x + 2}{2x^2 - 5x + 4}$; $x = 1.2$
7. $f(x) = \dfrac{x^{1/2} + 2x^{3/2} + 1}{2x^{1/2} + 3}$; $x = 0.5$
8. $f(x) = \dfrac{\sqrt{x} - 1}{2x + \sqrt{x} + 4}$; $x = 2.3$

9. **Rate of Bank Failures** The rate at which banks were failing between 1982 and 1988 is given by

$$f(t) = -0.063447t^4 - 1.953283t^3 + 14.632576t^2 - 6.684704t + 47.458874 \quad (0 \le t \le 6)$$

where $f(t)$ is measured in the number of banks per year and t is measured in years, with $t = 0$ corresponding to the beginning of 1982. Compute $f''(6)$, and interpret your results.
Source: Federal Deposit Insurance Corporation.

10. **Modeling With Data** The revenues (in billions of dollars) from cable advertisement for the years 1995 through 2000 follow:

Year	1995	1996	1997	1998	1999	2000
Revenue	5.1	6.6	8.1	9.4	11.1	13.7

a. Use **CubicReg** to find a third-degree polynomial regression model for the data. Let $t = 0$ correspond to 1995.
b. Plot the graph of the function f found in part (a), using the viewing window $[0, 6] \times [0, 14]$.
c. Compute $f''(5)$, and interpret your results.
Source: National Cable Television Association.

3.6 Implicit Differentiation and Related Rates

Differentiating Implicitly

Up to now, we have dealt with functions expressed in the form $y = f(x)$, that is, functions in which the dependent variable y is expressed *explicitly* in terms of the independent variable x. However, not all functions are expressed in this form. Consider, for example, the equation

$$x^2y + y - x^2 + 1 = 0 \quad (8)$$

This equation does express y *implicitly* as a function of x. In fact, solving (8) for y in terms of x, we obtain

$$(x^2 + 1)y = x^2 - 1 \quad \text{Implicit equation}$$

$$y = f(x) = \dfrac{x^2 - 1}{x^2 + 1} \quad \text{Explicit equation}$$

which gives an explicit representation of f.

Next, consider the equation

$$y^4 - y^3 - y + 2x^3 - x = 8$$

If certain restrictions are placed on x and y, this equation defines y as a function of x. But in this instance, we would be hard pressed to find y explicitly in terms of x. The following question arises naturally: How do we compute dy/dx in this case?

As it turns out, thanks to the Chain Rule, a method *does* exist for computing the derivative of a function directly from the implicit equation defining the function. This method is called **implicit differentiation** and is demonstrated in the next several examples.

EXAMPLE 1 Given the equation $y^2 = x$, find $\dfrac{dy}{dx}$.

Solution Differentiating both sides of the equation with respect to x, we obtain

$$\frac{d}{dx}(y^2) = \frac{d}{dx}(x)$$

To carry out the differentiation of the term $\dfrac{d}{dx}(y^2)$, we note that y (with suitable restrictions) is a function of x. Writing $y = f(x)$ to remind us of this fact, we find that

$$\begin{aligned}\frac{d}{dx}(y^2) &= \frac{d}{dx}[f(x)]^2 &&\text{Write } y = f(x).\\ &= 2f(x)f'(x) &&\text{Use the Chain Rule.}\\ &= 2y\frac{dy}{dx} &&\text{Replace } f(x) \text{ with } y.\end{aligned}$$

Therefore, the equation

$$\frac{d}{dx}(y^2) = \frac{d}{dx}(x)$$

is equivalent to

$$2y\frac{dy}{dx} = 1$$

Solving for $\dfrac{dy}{dx}$ yields

$$\frac{dy}{dx} = \frac{1}{2y}$$

Before considering other examples, let's summarize the important steps involved in implicit differentiation. (Here, we assume that dy/dx exists.)

> **Finding $\dfrac{dy}{dx}$ by Implicit Differentiation**
>
> 1. Differentiate both sides of the equation *with respect to x*. (Make sure that the derivative of any term involving y includes the factor dy/dx.)
> 2. Solve the resulting equation for dy/dx in terms of x and y.

EXAMPLE 2 Find $\dfrac{dy}{dx}$ given the equation

$$y^3 - y + 2x^3 - x = 8$$

Solution Differentiating both sides of the given equation with respect to x, we obtain

$$\frac{d}{dx}(y^3 - y + 2x^3 - x) = \frac{d}{dx}(8)$$

$$\frac{d}{dx}(y^3) - \frac{d}{dx}(y) + \frac{d}{dx}(2x^3) - \frac{d}{dx}(x) = 0$$

Now, recalling that y is a function of x, we apply the Chain Rule to the first two terms on the left. Thus,

$$3y^2 \frac{dy}{dx} - \frac{dy}{dx} + 6x^2 - 1 = 0$$

$$(3y^2 - 1)\frac{dy}{dx} = 1 - 6x^2$$

$$\frac{dy}{dx} = \frac{1 - 6x^2}{3y^2 - 1}$$

> **Explore & Discuss**
>
> Refer to Example 2. Suppose we think of the equation $y^3 - y + 2x^3 - x = 8$ as defining x implicitly as a function of y. Find dx/dy, and justify your method of solution.

EXAMPLE 3 Consider the equation $x^2 + y^2 = 4$.

  a. Find dy/dx by implicit differentiation.
  b. Find the slope of the tangent line to the graph of the function $y = f(x)$ at the point $(1, \sqrt{3})$.
  c. Find an equation of the tangent line of part (b).

Solution

a. Differentiating both sides of the equation with respect to x, we obtain

$$\frac{d}{dx}(x^2 + y^2) = \frac{d}{dx}(4)$$

$$\frac{d}{dx}(x^2) + \frac{d}{dx}(y^2) = 0$$

$$2x + 2y\frac{dy}{dx} = 0$$

$$2y\frac{dy}{dx} = -2x$$

$$\frac{dy}{dx} = -\frac{x}{y} \quad (y \neq 0)$$

b. The slope of the tangent line to the graph of the function at the point $(1, \sqrt{3})$ is given by

$$\left.\frac{dy}{dx}\right|_{(1, \sqrt{3})} = -\left.\frac{x}{y}\right|_{(1, \sqrt{3})} = -\frac{1}{\sqrt{3}}$$

(*Note:* This notation is read "dy/dx evaluated at the point $(1, \sqrt{3})$.")

c. An equation of the tangent line in question is found by using the point-slope form of the equation of a line with the slope $m = -1/\sqrt{3}$ and the point $(1, \sqrt{3})$. Thus,

$$y - \sqrt{3} = -\frac{1}{\sqrt{3}}(x - 1) \quad \text{(x²) See page 36.}$$

$$\sqrt{3}y - 3 = -x + 1$$

$$x + \sqrt{3}y - 4 = 0$$

A sketch of this tangent line is shown in Figure 18.

We can also solve the equation $x^2 + y^2 = 4$ explicitly for y in terms of x. If we do this, we obtain

$$y = \pm\sqrt{4 - x^2}$$

From this, we see that the equation $x^2 + y^2 = 4$ defines the two functions

$$y = f(x) = \sqrt{4 - x^2}$$
$$y = g(x) = -\sqrt{4 - x^2}$$

Since the point $(1, \sqrt{3})$ does not lie on the graph of $y = g(x)$, we conclude that

$$y = f(x) = \sqrt{4 - x^2}$$

is the required function. The graph of f is the upper semicircle shown in Figure 18.

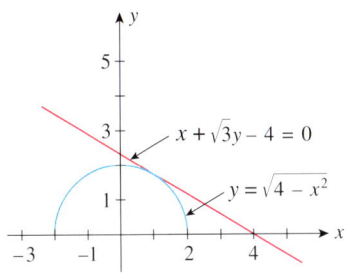

FIGURE 18
The line $x + \sqrt{3}y - 4 = 0$ is tangent to the graph of the function $y = f(x)$.

Note The notation

$$\left.\frac{dy}{dx}\right|_{(a,\,b)}$$

is used to denote the value of dy/dx at the point (a, b).

Explore & Discuss

Refer to Example 3. Yet another function defined implicitly by the equation $x^2 + y^2 = 4$ is the function

$$y = h(x) = \begin{cases} \sqrt{4 - x^2} & \text{if } -2 \leq x < 0 \\ -\sqrt{4 - x^2} & \text{if } 0 \leq x \leq 2 \end{cases}$$

1. Sketch the graph of h.
2. Show that $h'(x) = -x/y$ if $x \neq 0$ and $y \neq 0$.
3. Find an equation of the tangent line to the graph of h at the point $(1, -\sqrt{3})$.

To find dy/dx at a *specific point* (a, b), differentiate the given equation implicitly with respect to x and then replace x and y by a and b, respectively, *before* solving the equation for dy/dx. This often simplifies the amount of algebra involved.

EXAMPLE 4 Find $\dfrac{dy}{dx}$ given that x and y are related by the equation

$$x^2 y^3 + 6x^2 = y + 12$$

and that $y = 2$ when $x = 1$.

Solution Differentiating both sides of the given equation with respect to x, we obtain

$$\frac{d}{dx}(x^2 y^3) + \frac{d}{dx}(6x^2) = \frac{d}{dx}(y) + \frac{d}{dx}(12)$$

$$x^2 \cdot \frac{d}{dx}(y^3) + y^3 \cdot \frac{d}{dx}(x^2) + 12x = \frac{dy}{dx} \qquad \text{Use the Product Rule on } \frac{d}{dx}(x^2 y^3).$$

$$3x^2 y^2 \frac{dy}{dx} + 2xy^3 + 12x = \frac{dy}{dx}$$

Substituting $x = 1$ and $y = 2$ into this equation gives

$$3(1)^2(2)^2 \frac{dy}{dx} + 2(1)(2)^3 + 12(1) = \frac{dy}{dx}$$

$$12 \frac{dy}{dx} + 16 + 12 = \frac{dy}{dx}$$

and, solving for $\frac{dy}{dx}$,

$$\frac{dy}{dx} = -\frac{28}{11}$$

Note that it is not necessary to find an explicit expression for dy/dx.

Note In Examples 3 and 4, you can verify that the points at which we evaluated dy/dx actually lie on the curve in question by showing that the coordinates of the points satisfy the given equations.

EXAMPLE 5 Find $\frac{d^2y}{dx^2}$ if $xy - y^3 = 4$.

Solution Differentiating both sides of the given equation with respect to x, we obtain

$$\frac{d}{dx}(xy) - \frac{d}{dx}(y^3) = \frac{d}{dx}(4)$$

$$x\frac{dy}{dx} + y - 3y^2\frac{dy}{dx} = 0$$

$$y - (3y^2 - x)\frac{dy}{dx} = 0$$

or

$$\frac{dy}{dx} = \frac{y}{3y^2 - x}$$

Next, we differentiate both sides of the last equation with respect to x again, obtaining

$$\frac{d}{dx}\left(\frac{dy}{dx}\right) = \frac{d}{dx}\left(\frac{y}{3y^2 - x}\right)$$

$$\frac{d^2y}{dx^2} = \frac{(3y^2 - x)\frac{d}{dx}(y) - y\frac{d}{dx}(3y^2 - x)}{(3y^2 - x)^2}$$

$$= \frac{(3y^2 - x)\frac{dy}{dx} - y\left(6y\frac{dy}{dx} - 1\right)}{(3y^2 - x)^2} = \frac{y - (3y^2 + x)\frac{dy}{dx}}{(3y^2 - x)^2}$$

$$= \frac{y - (3y^2 + x)\left(\frac{y}{3y^2 - x}\right)}{(3y^2 - x)^2} = \frac{3y^3 - xy - 3y^3 - xy}{(3y^2 - x)^3}$$

$$= -\frac{2xy}{(3y^2 - x)^3}$$

APPLIED EXAMPLE 6 Production The chief economist of a country estimates that the output of the country is given by $Q = 10x^{3/4}y^{1/4}$, where x is the amount of money spent on labor and y is the amount spent on capital. Here x, y, and Q are measured in billions of dollars.

a. Find the output of the country if $625 billion is spent on labor and $81 billion is spent on capital.
b. Suppose that the output of the country is to be maintained at the level found in part (a). By how much should the amount spent on capital be changed if the amount spent on labor is to be increased by $1 billion?

Solution

a. Replacing x and y by 625 and 81, respectively, in the expression for Q, we see that the required output is

$$Q = 10(625)^{3/4}(81)^{1/4} = 3750$$

or $3750 billion.

b. We want to find the change in y per unit change in x when $x = 625$ and $y = 81$ given that $10x^{3/4}y^{1/4} = 3750$. But as you saw in Section 3.4, this quantity is approximated by dy/dx (evaluated at the specified values of x and y). To find dy/dx, we differentiate the equation

$$10x^{3/4}y^{1/4} = 3750$$

or, equivalently, the equation

$$x^{3/4}y^{1/4} = 375$$

implicitly. Thus,

$$\frac{d}{dx}(x^{3/4}y^{1/4}) = \frac{d}{dx}(375)$$

$$\frac{3}{4}x^{-1/4}y^{1/4} + x^{3/4}\left(\frac{1}{4}y^{-3/4}\frac{dy}{dx}\right) = 0$$

$$\frac{1}{4}x^{3/4}y^{-3/4}\frac{dy}{dx} = -\frac{3}{4}x^{-1/4}y^{1/4}$$

or

$$\frac{dy}{dx} = (4x^{-3/4}y^{3/4})\left(-\frac{3}{4}x^{-1/4}y^{1/4}\right) = -3x^{-1}y = -3\left(\frac{y}{x}\right)$$

Therefore, when $x = 625$ and $y = 81$, we find

$$\frac{dy}{dx} = -3\left(\frac{81}{625}\right) = -0.3888$$

So to keep the output at the constant level of $3750 billion, the amount spent on capital should decrease by $0.3888 billion if the amount spent on labor were to be increased by $1 billion.

Note The negative of dy/dx found in Example 6 is called the **marginal rate of technical substitution** (MRTS). In general, the MRTS measures the rate at which a producer is technically able to reduce one input (capital) for a unit increase of another input (labor) while maintaining a constant level of output. Thus, the MRTS for the country in Example 6 is $0.3888 billion per $1 billion.

Related Rates

Implicit differentiation is a useful technique for solving a class of problems known as **related-rates** problems. The following is a typical related-rates problem: Suppose x and y are two quantities that depend on a third quantity t and we know the relationship between x and y in the form of an equation. Can we find a relationship between dx/dt and dy/dt? In particular, if we know one of the rates of change at a specific value of t—say, dx/dt—can we find the other rate, dy/dt, at that value of t?

APPLIED EXAMPLE 7 Rate of Change of Housing Starts A study prepared for the National Association of Realtors estimates that the number of housing starts in the southwest, $N(t)$ (in units of a million), over the next 5 years is related to the mortgage rate $r(t)$ (percent per year) by the equation

$$9N^2 + r = 36$$

What is the rate of change of the number of housing starts with respect to time when the mortgage rate is 11% per year and is increasing at the rate of 1.5% per year?

Solution We are given that

$$r = 11 \quad \text{and} \quad \frac{dr}{dt} = 1.5$$

at a certain instant of time, and we are required to find dN/dt. First, by substituting $r = 11$ into the given equation, we find

$$9N^2 + 11 = 36$$
$$N^2 = \frac{25}{9}$$

or $N = 5/3$ (we reject the negative root). Next, differentiating the given equation implicitly on both sides with respect to t, we obtain

$$\frac{d}{dt}(9N^2) + \frac{d}{dt}(r) = \frac{d}{dt}(36)$$

$$18N\frac{dN}{dt} + \frac{dr}{dt} = 0 \qquad \text{Use the Chain Rule on the first term.}$$

Then, substituting $N = 5/3$ and $dr/dt = 1.5$ into this equation gives

$$18\left(\frac{5}{3}\right)\frac{dN}{dt} + 1.5 = 0$$

Solving this equation for dN/dt then gives

$$\frac{dN}{dt} = -\frac{1.5}{30} = -0.05$$

Thus, at the instant of time under consideration, the number of housing starts is decreasing at the rate of 50,000 units per year.

APPLIED EXAMPLE 8 Supply–Demand Texar Inc., a manufacturer of disk drives is willing to make x thousand IGB USB flash drives available in the marketplace each week when the wholesale price is \$$p$ per drive. It is known that the relationship between x and p is governed by the supply equation

$$x^2 - 3xp + p^2 = 5$$

How fast is the supply of drives changing when the price per drive is \$11, the quantity supplied is 4000 drives, and the wholesale price per drive is increasing at the rate of \$0.10 per drive each week?

3.6 IMPLICIT DIFFERENTIATION AND RELATED RATES

Solution We are given that

$$p = 11 \quad x = 4 \quad \frac{dp}{dt} = 0.1$$

at a certain instant of time, and we are required to find dx/dt. Differentiating the given equation on both sides with respect to t, we obtain

$$\frac{d}{dt}(x^2) - \frac{d}{dt}(3xp) + \frac{d}{dt}(p^2) = \frac{d}{dt}(5)$$

$$2x\frac{dx}{dt} - 3\left(p\frac{dx}{dt} + x\frac{dp}{dt}\right) + 2p\frac{dp}{dt} = 0 \qquad \text{Use the Product Rule on the second term.}$$

Substituting the given values of p, x, and dp/dt into the last equation, we have

$$2(4)\frac{dx}{dt} - 3\left[(11)\frac{dx}{dt} + 4(0.1)\right] + 2(11)(0.1) = 0$$

$$8\frac{dx}{dt} - 33\frac{dx}{dt} - 1.2 + 2.2 = 0$$

$$25\frac{dx}{dt} = 1$$

$$\frac{dx}{dt} = 0.04$$

Thus, at the instant of time under consideration, the supply of drives is increasing at the rate of (0.04)(1000), or 40, drives per week.

In certain related-rates problems, we need to formulate the problem mathematically before analyzing it. The following guidelines can be used to help solve problems of this type.

> **Solving Related-Rates Problems**
> 1. Assign a variable to each quantity. Draw a diagram if needed.
> 2. Write the *given* values of the variables and their rates of change with respect to t.
> 3. Find an equation giving the relationship between the variables.
> 4. Differentiate both sides of this equation implicitly with respect to t.
> 5. Replace the variables and their derivatives by the numerical data found in Step 2, and solve the equation for the required rate of change.

APPLIED EXAMPLE 9 Watching a Rocket Launch At a distance of 4000 feet from the launch site, a spectator is observing a rocket being launched. If the rocket lifts off vertically and is rising at a speed of 600 feet/second when it is at an altitude of 3000 feet, how fast is the distance between the rocket and the spectator changing at that instant?

Solution

Step 1 Let

y = altitude of the rocket

x = distance between the rocket and the spectator

at any time t (Figure 19).

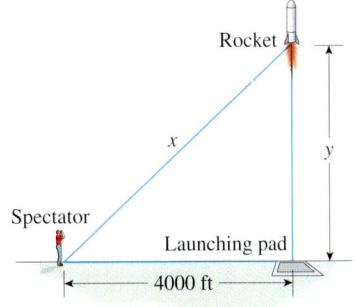

FIGURE 19
The rate at which x is changing with respect to time is related to the rate of change of y with respect to time.

Step 2 We are given that at a certain instant of time

$$y = 3000 \quad \text{and} \quad \frac{dy}{dt} = 600$$

and are asked to find dx/dt at that instant.

Step 3 Applying the Pythagorean Theorem to the right triangle in Figure 19, we find that

$$x^2 = y^2 + 4000^2$$

Therefore, when $y = 3000$,

$$x = \sqrt{3000^2 + 4000^2} = 5000$$

Step 4 Next, we differentiate the equation $x^2 = y^2 + 4000^2$ with respect to t, obtaining

$$2x \frac{dx}{dt} = 2y \frac{dy}{dt}$$

(Remember, both x and y are functions of t.)

Step 5 Substituting $x = 5000$, $y = 3000$, and $dy/dt = 600$, we find

$$2(5000) \frac{dx}{dt} = 2(3000)(600)$$

$$\frac{dx}{dt} = 360$$

Therefore, the distance between the rocket and the spectator is changing at a rate of 360 feet/second.

 Be sure that you do *not* replace the variables in the equation found in Step 3 by their numerical values before differentiating the equation.

EXAMPLE 10 A passenger ship and an oil tanker left port together sometime in the morning; the former headed north, and the latter headed east. At noon, the passenger ship was 40 miles from port and sailing at 30 mph, while the oil tanker was 30 miles from port and sailing at 20 mph. How fast was the distance between the two ships changing at that time?

Solution

Step 1 Let

$$x = \text{distance of the oil tanker from port}$$
$$y = \text{distance of the passenger ship from port}$$
$$z = \text{distance between the two ships}$$

See Figure 20.

Step 2 We are given that at noon,

$$x = 30 \quad y = 40 \quad \frac{dx}{dt} = 20 \quad \frac{dy}{dt} = 30$$

and we are required to find dz/dt at that time.

Step 3 Applying the Pythagorean Theorem to the right triangle in Figure 20, we find that

$$z^2 = x^2 + y^2 \tag{9}$$

In particular, when $x = 30$ and $y = 40$, we have

$$z^2 = 30^2 + 40^2 = 2500 \quad \text{or} \quad z = 50$$

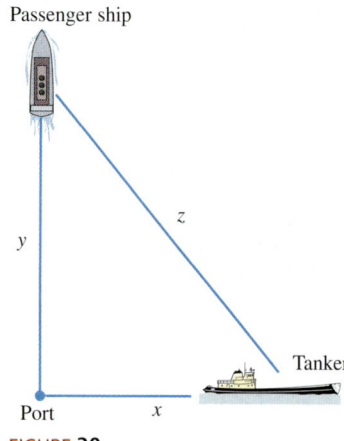

FIGURE 20
We want to find dz/dt, the rate at which the distance between the two ships is changing at a certain instant of time.

Step 4 Differentiating (9) implicitly with respect to t, we obtain

$$2z \frac{dz}{dt} = 2x \frac{dx}{dt} + 2y \frac{dy}{dt}$$

$$z \frac{dz}{dt} = x \frac{dx}{dt} + y \frac{dy}{dt}$$

Step 5 Finally, substituting $x = 30$, $y = 40$, $z = 50$, $dx/dt = 20$, and $dy/dt = 30$ into the last equation, we find

$$50 \frac{dz}{dt} = (30)(20) + (40)(30) \quad \text{and} \quad \frac{dz}{dt} = 36$$

Therefore, at noon on the day in question, the ships are moving apart at the rate of 36 mph.

3.6 Self-Check Exercises

1. Given the equation $x^3 + 3xy + y^3 = 4$, find dy/dx by implicit differentiation.

2. Find an equation of the tangent line to the graph of

$$16x^2 + 9y^2 = 144$$

at the point

$$\left(2, -\frac{4\sqrt{5}}{3}\right)$$

Solutions to Self-Check Exercises 3.6 can be found on page 228.

3.6 Concept Questions

1. **a.** Suppose the equation $F(x, y) = 0$ defines y as a function of x. Explain how implicit differentiation can be used to find dy/dx.
   b. What is the role of the Chain Rule in implicit differentiation?

2. Suppose the equation $xg(y) + yf(x) = 0$, where f and g are differentiable functions, defines y as a function of x. Find an expression for dy/dx.

3. In your own words, describe what a related-rates problem is.

4. Give the steps that you would use to solve a related-rates problem.

3.6 Exercises

In Exercises 1–8, find the derivative dy/dx (a) by solving each of the implicit equations for y explicitly in terms of x and (b) by differentiating each of the equations implicitly. Show that in each case, the results are equivalent.

1. $x + 2y = 5$
2. $3x + 4y = 6$
3. $xy = 1$
4. $xy - y - 1 = 0$
5. $x^3 - x^2 - xy = 4$
6. $x^2y - x^2 + y - 1 = 0$
7. $\dfrac{x}{y} - x^2 = 1$
8. $\dfrac{y}{x} - 2x^3 = 4$

In Exercises 9–30, find dy/dx by implicit differentiation.

9. $x^2 + y^2 = 16$
10. $2x^2 + y^2 = 16$
11. $x^2 - 2y^2 = 16$
12. $x^3 + y^3 + y - 4 = 0$
13. $x^2 - 2xy = 6$
14. $x^2 + 5xy + y^2 = 10$
15. $x^2y^2 - xy = 8$
16. $x^2y^3 - 2xy^2 = 5$
17. $x^{1/2} + y^{1/2} = 1$
18. $x^{1/3} + y^{1/3} = 1$
19. $\sqrt{x + y} = x$
20. $(2x + 3y)^{1/3} = x^2$
21. $\dfrac{1}{x^2} + \dfrac{1}{y^2} = 1$
22. $\dfrac{1}{x^3} + \dfrac{1}{y^3} = 5$
23. $\sqrt{xy} = x + y$
24. $\sqrt{xy} = 2x + y^2$
25. $\dfrac{x + y}{x - y} = 3x$
26. $\dfrac{x - y}{2x + 3y} = 2x$
27. $xy^{3/2} = x^2 + y^2$
28. $x^2y^{1/2} = x + 2y^3$
29. $(x + y)^3 + x^3 + y^3 = 0$
30. $(x + y^2)^{10} = x^2 + 25$

In Exercises 31–34, find an equation of the tangent line to the graph of the function *f* defined by the equation at the indicated point.

31. $4x^2 + 9y^2 = 36; (0, 2)$

32. $y^2 - x^2 = 16; (2, 2\sqrt{5})$

33. $x^2y^3 - y^2 + xy - 1 = 0; (1, 1)$

34. $(x - y - 1)^3 = x; (1, -1)$

In Exercises 35–38, find the second derivative d^2y/dx^2 of each of the functions defined implicitly by the equation.

35. $xy = 1$

36. $x^3 + y^3 = 28$

37. $y^2 - xy = 8$

38. $x^{1/3} + y^{1/3} = 1$

39. The volume of a right-circular cylinder of radius *r* and height *h* is $V = \pi r^2 h$. Suppose the radius and height of the cylinder are changing with respect to time *t*.
   a. Find a relationship between dV/dt, dr/dt, and dh/dt.
   b. At a certain instant of time, the radius and height of the cylinder are 2 and 6 in. and are increasing at the rate of 0.1 and 0.3 in./sec, respectively. How fast is the volume of the cylinder increasing?

40. A car leaves an intersection traveling west. Its position 4 sec later is 20 ft from the intersection. At the same time, another car leaves the same intersection heading north so that its position 4 sec later is 28 ft from the intersection. If the speeds of the cars at that instant of time are 9 ft/sec and 11 ft/sec, respectively, find the rate at which the distance between the two cars is changing.

41. PRICE-DEMAND Suppose the quantity demanded weekly of the Super Titan radial tires is related to its unit price by the equation

$$p + x^2 = 144$$

where *p* is measured in dollars and *x* is measured in units of a thousand. How fast is the quantity demanded weekly changing when $x = 9$, $p = 63$, and the price per tire is increasing at the rate of $2/week?

42. PRICE-SUPPLY Suppose the quantity *x* of Super Titan radial tires made available each week in the marketplace is related to the unit selling price by the equation

$$p - \frac{1}{2}x^2 = 48$$

where *x* is measured in units of a thousand and *p* is in dollars. How fast is the weekly supply of Super Titan radial tires being introduced into the marketplace changing when $x = 6$, $p = 66$, and the price per tire is decreasing at the rate of $3/week?

43. PRICE-DEMAND The demand equation for a certain brand of two-way headphones is

$$100x^2 + 9p^2 = 3600$$

where *x* represents the number (in thousands) of headphones demanded each week when the unit price is $*p*. How fast is the quantity demanded increasing when the unit price per headphone is $14 and the price is dropping at the rate of $0.15/headphone/week?
Hint: To find the value of *x* when $p = 14$, solve the equation $100x^2 + 9p^2 = 3600$ for *x* when $p = 14$.

44. EFFECT OF PRICE ON SUPPLY Suppose the wholesale price of a certain brand of medium-sized eggs *p* (in dollars per carton) is related to the weekly supply *x* (in thousands of cartons) by the equation

$$625p^2 - x^2 = 100$$

If 25,000 cartons of eggs are available at the beginning of a certain week and the price is falling at the rate of 2¢/carton/week, at what rate is the weekly supply falling?
Hint: To find the value of *p* when $x = 25$, solve the supply equation for *p* when $x = 25$.

45. SUPPLY-DEMAND Refer to Exercise 44. If 25,000 cartons of eggs are available at the beginning of a certain week and the weekly supply is falling at the rate of 1000 cartons/week, at what rate is the wholesale price changing?

46. ELASTICITY OF DEMAND The demand function for a certain make of ink-jet cartridge is

$$p = -0.01x^2 - 0.1x + 6$$

where *p* is the unit price in dollars and *x* is the quantity demanded each week, measured in units of a thousand. Compute the elasticity of demand and determine whether the demand is inelastic, unitary, or elastic when $x = 10$.

47. ELASTICITY OF DEMAND The demand function for a certain brand of compact disc is

$$p = -0.01x^2 - 0.2x + 8$$

where *p* is the wholesale unit price in dollars and *x* is the quantity demanded each week, measured in units of a thousand. Compute the elasticity of demand and determine whether the demand is inelastic, unitary, or elastic when $x = 15$.

48. PRODUCTION The manager of Dixie Furniture Company estimates that the daily output of her factory (in thousands of dollars) *Q* is given by

$$Q = 5x^{1/4}y^{3/4}$$

where *x* is the amount spent on labor and *y* is the amount spent on capital (both measured in thousands of dollars).
   a. Find the daily output of the factory if $16,000 per day is spent on labor and $81,000 per day is spent on capital.
   b. Suppose that the output of the factory is to be maintained at the level found in part (a). By how much should the amount spent on capital be changed if the amount on labor is increased by $1000?

49. **Production** Suppose that the output Q of a certain country is given by $Q = 20x^{3/5}y^{2/5}$ billion dollars if x billion dollars are spent on labor and y billion dollars are spent on capital.
    a. Find the output of the country if it spends $32 billion on labor and $243 billion on capital.
    b. Suppose that the output of the country is to be maintained at the level found in part (a). By how much should the amount spent on capital be changed if the amount spent on labor is increased by $1 billion? What is the MRTS?

50. The volume V of a cube with sides of length x in. is changing with respect to time. At a certain instant of time, the sides of the cube are 5 in. long and increasing at the rate of 0.1 in./sec. How fast is the volume of the cube changing at that instant of time?

51. **Oil Spills** In calm waters, the oil spilling from the ruptured hull of a grounded tanker spreads in all directions. Assuming that the area polluted is circular, determine how fast the area is increasing when the radius of the circle is 60 ft and is increasing at the rate of $\frac{1}{2}$ ft/sec?

52. Two ships leave the same port at noon. Ship A sails north at 15 mph, and Ship B sails east at 12 mph. How fast is the distance between them changing at 1 P.M.?

53. **Oil Spills** In calm waters, the oil spilling from the ruptured hull of a grounded tanker spreads in all directions. Assuming that the area polluted is circular, determine how fast the radius of the circle is changing when the area of the circle is 1600π ft^2 and increasing at the rate of 80π ft^2/sec.

54. A car leaves an intersection traveling east. Its position t sec later is given by $x = t^2 + t$ ft. At the same time, another car leaves the same intersection heading north, traveling $y = t^2 + 3t$ ft in t sec. Find the rate at which the distance between the two cars will be changing 5 sec later.

55. A car leaves an intersection traveling west. Its position 4 sec later is 20 ft from the intersection. At the same time, another car leaves the same intersection heading north so that its position t sec later is $t^2 + 2t$ ft from the intersection. If the speed of the first car 4 sec after leaving the intersection is 9 ft/sec, find the rate at which the distance between the two cars is changing at that instant of time.

56. At a distance of 50 ft from the pad, a man observes a helicopter taking off from a heliport. If the helicopter lifts off vertically and is rising at a speed of 44 ft/sec when it is at an altitude of 120 ft, how fast is the distance between the helicopter and the man changing at that instant?

57. A spectator watches a rowing race from the edge of a river bank. The lead boat is moving in a straight line that is 120 ft from the river bank. If the boat is moving at a constant speed of 20 ft/sec, how fast is the boat moving away from the spectator when it is 50 ft past her?

58. **Docking a Boat** A boat is pulled toward a dock by means of a rope wound on a drum that is located 4 ft above the bow of the boat. If the rope is being pulled in at the rate of 3 ft/sec, how fast is the boat approaching the dock when it is 25 ft from the dock?

59. Assume that a snowball is in the shape of a sphere. If the snowball melts at a rate that is proportional to its surface area, show that its radius decreases at a constant rate.
    Hint: Its volume is $V = (4/3)\pi r^3$, and its surface area is $S = 4\pi r^2$.

60. **Blowing Soap Bubbles** Carlos is blowing air into a soap bubble at the rate of 8 cm^3/sec. Assuming that the bubble is spherical, how fast is its radius changing at the instant of time when the radius is 10 cm? How fast is the surface area of the bubble changing at that instant of time?

61. **Coast Guard Patrol Search Mission** The pilot of a Coast Guard patrol aircraft on a search mission had just spotted a disabled fishing trawler and decided to go in for a closer look. Flying in a straight line at a constant altitude of 1000 ft and at a steady speed of 264 ft/sec, the aircraft passed directly over the trawler. How fast was the aircraft receding from the trawler when it was 1500 ft from the trawler?

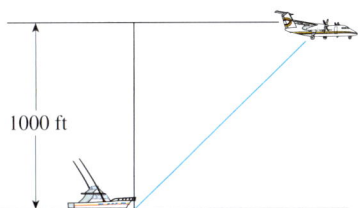

62. A coffee pot in the form of a circular cylinder of radius 4 in. is being filled with water flowing at a constant rate. If the water level is rising at the rate of 0.4 in./sec, what is the rate at which water is flowing into the coffee pot?

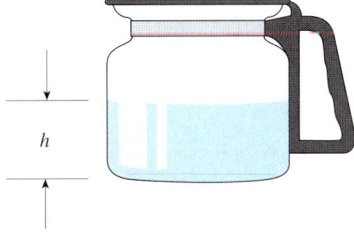

63. **Movement of a Shadow** A 6-ft tall man is walking away from a street light 18 ft high at a speed of 6 ft/sec. How fast is the tip of his shadow moving along the ground?

64. A 20-ft ladder leaning against a wall begins to slide. How fast is the top of the ladder sliding down the wall at the instant of time when the bottom of the ladder is 12 ft from the wall and sliding away from the wall at the rate of 5 ft/sec?

Hint: Refer to the accompanying figure. By the Pythagorean Theorem, $x^2 + y^2 = 400$. Find dy/dt when $x = 12$ and $dx/dt = 5$.

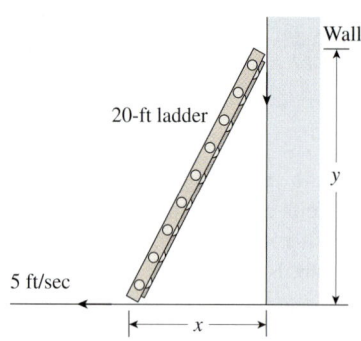

65. The base of a 13-ft ladder leaning against a wall begins to slide away from the wall. At the instant of time when the base is 12 ft from the wall, the base is moving at the rate of 8 ft/sec. How fast is the top of the ladder sliding down the wall at that instant of time?

Hint: Refer to the hint in Problem 64.

66. FLOW OF WATER FROM A TANK Water flows from a tank of constant cross-sectional area 50 ft² through an orifice of constant cross-sectional area 1.4 ft² located at the bottom of the tank (see the figure).

Initially, the height of the water in the tank was 20 ft, and its height t sec later is given by the equation

$$2\sqrt{h} + \frac{1}{25}t - 2\sqrt{20} = 0 \qquad (0 \le t \le 50\sqrt{20})$$

How fast was the height of the water decreasing when its height was 8 ft?

67. VOLUME OF A GAS In an adiabatic process (one in which no heat transfer takes place), the pressure P and volume V of an ideal gas such as oxygen satisfy the equation $P^5V^7 = C$, where C is a constant. Suppose that at a certain instant of time, the volume of the gas is 4 L, the pressure is 100 kPa, and the pressure is decreasing at the rate of 5 kPa/sec. Find the rate at which the volume is changing.

68. MASS OF A MOVING PARTICLE The mass m of a particle moving at a velocity v is related to its rest mass m_0 by the equation

$$m = \frac{m_0}{\sqrt{1 - \dfrac{v^2}{c^2}}}$$

where c (2.98 × 10⁸ m/sec) is the speed of light. Suppose an electron of mass 9.11×10^{-31} kg is being accelerated in a particle accelerator. When its velocity is 2.92×10^8 m/sec and its acceleration is 2.42×10^5 m/sec², how fast is the mass of the electron changing?

In Exercises 69–72, determine whether the statement is true or false. If it is true, explain why it is true. If it is false, give an example to show why it is false.

69. The equation $x^2 + y^2 + 1 = 0$ defines y as a function of x.

70. The function

$$f(x) = \begin{cases} \sqrt{1 - x^2} & \text{if } -1 \le x < 0 \\ -\sqrt{1 - x^2} & \text{if } 0 \le x \le 1 \end{cases}$$

may be defined implicitly by the equation $x^2 + y^2 = 1$.

71. If f and g are differentiable and $f(x)g(y) = 0$, then

$$\frac{dy}{dx} = -\frac{f'(x)g(y)}{f(x)g'(y)} \qquad f(x) \ne 0 \text{ and } g'(y) \ne 0$$

72. If f and g are differentiable and $f(x) + g(y) = 0$, then

$$\frac{dy}{dx} = -\frac{f'(x)}{g'(y)}$$

3.6 Solutions to Self-Check Exercises

1. Differentiating both sides of the equation with respect to x, we have

$$3x^2 + 3y + 3xy' + 3y^2y' = 0$$
$$(x^2 + y) + (x + y^2)y' = 0$$
$$(x + y^2)y' = -(x^2 + y)$$
$$y' = -\frac{x^2 + y}{x + y^2}$$

2. To find the slope of the tangent line to the graph of the function at any point, we differentiate the equation implicitly with respect to x, obtaining

$$32x + 18yy' = 0$$
$$y' = -\frac{16x}{9y}$$

In particular, the slope of the tangent line at $\left(2, -\dfrac{4\sqrt{5}}{3}\right)$ is

$$m = -\dfrac{16(2)}{9\left(-\dfrac{4\sqrt{5}}{3}\right)} = \dfrac{8}{3\sqrt{5}}$$

Using the point-slope form of the equation of a line, we find

$$y - \left(-\dfrac{4\sqrt{5}}{3}\right) = \dfrac{8}{3\sqrt{5}}(x - 2)$$

$$y = \dfrac{8\sqrt{5}}{15}x - \dfrac{36\sqrt{5}}{15} = \dfrac{8\sqrt{5}}{15}x - \dfrac{12\sqrt{5}}{5}$$

3.7 Differentials

The Millers are planning to buy a house in the near future and estimate that they will need a 30-year fixed-rate mortgage of $240,000. If the interest rate increases from the present rate of 5% per year to 5.4% per year between now and the time the Millers decide to secure the loan, approximately how much more per month will their mortgage be? (You will be asked to answer this question in Exercise 46, page 237.)

Questions such as this, in which one wishes to *estimate* the change in the dependent variable (monthly mortgage payment) corresponding to a small change in the independent variable (interest rate per year), occur in many real-life applications. For example:

- An economist would like to know how a small increase in a country's capital expenditure will affect the country's gross domestic output.
- A sociologist would like to know how a small increase in the amount of capital investment in a housing project will affect the crime rate.
- A businesswoman would like to know how raising a product's unit price by a small amount will affect her profit.
- A bacteriologist would like to know how a small increase in the amount of a bactericide will affect a population of bacteria.

To calculate these changes and estimate their effects, we use the **differential** of a function, a concept that will be introduced shortly.

Increments

Let x denote a variable quantity and suppose x changes from x_1 to x_2. This change in x is called the **increment in x** and is denoted by the symbol Δx (read "delta x"). Thus,

$$\Delta x = x_2 - x_1 \qquad \text{Final value } - \text{ initial value} \tag{10}$$

EXAMPLE 1 Find the increment in x as x changes (a) from 3 to 3.2 and (b) from 3 to 2.7.

Solution

a. Here, $x_1 = 3$ and $x_2 = 3.2$, so

$$\Delta x = x_2 - x_1 = 3.2 - 3 = 0.2$$

b. Here, $x_1 = 3$ and $x_2 = 2.7$. Therefore,

$$\Delta x = x_2 - x_1 = 2.7 - 3 = -0.3$$

Observe that Δx plays the same role that h played in Section 2.4.

Now, suppose two quantities, x and y, are related by an equation $y = f(x)$, where f is a function. If x changes from x to $x + \Delta x$, then the corresponding change in y is called the **increment in y.** It is denoted by Δy and is defined by

$$\Delta y = f(x + \Delta x) - f(x) \tag{11}$$

(see Figure 21).

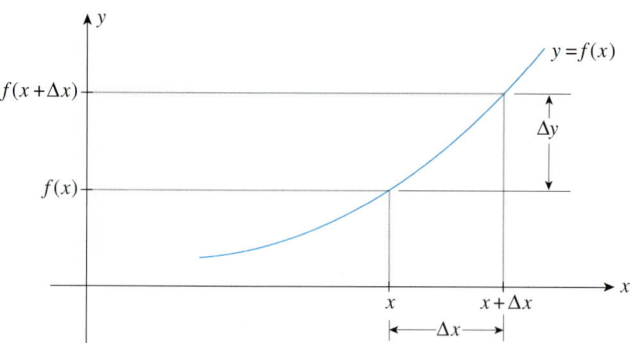

FIGURE 21
An increment of Δx in x induces an increment of $\Delta y = f(x + \Delta x) - f(x)$ in y.

VIDEO   **EXAMPLE 2** Let $y = x^3$. Find Δx and Δy when x changes (a) from 2 to 2.01 and (b) from 2 to 1.98.

Solution   Let $f(x) = x^3$.

a. Here, $\Delta x = 2.01 - 2 = 0.01$. Next,

$$\Delta y = f(x + \Delta x) - f(x) = f(2.01) - f(2)$$
$$= (2.01)^3 - 2^3 = 8.120601 - 8 = 0.120601$$

b. Here, $\Delta x = 1.98 - 2 = -0.02$. Next,

$$\Delta y = f(x + \Delta x) - f(x) = f(1.98) - f(2)$$
$$= (1.98)^3 - 2^3 = 7.762392 - 8 = -0.237608$$

Differentials

We can obtain a relatively quick and simple way of approximating Δy, the change in y due to a small change Δx, by examining the graph of the function f shown in Figure 22.

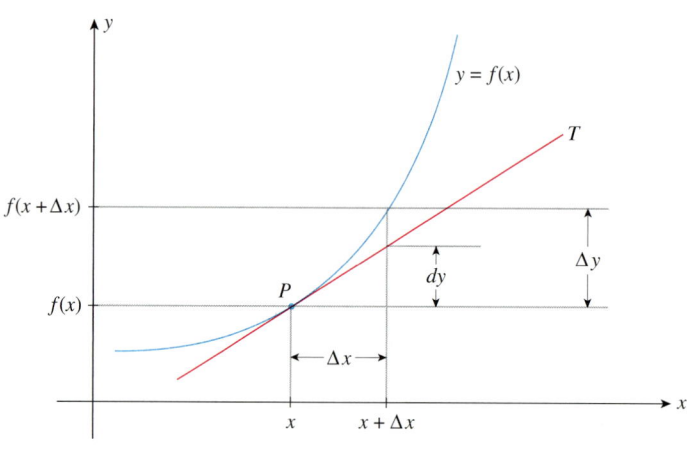

FIGURE 22
If Δx is small, dy is a good approximation of Δy.

Observe that near the point of tangency P, the tangent line T is close to the graph of f. Therefore, if Δx is small, then dy is a good approximation of Δy. We can find an expression for dy as follows. Notice that the slope of T is given by

$$\frac{dy}{\Delta x} \quad \text{Rise} \div \text{run}$$

However, the slope of T is given by $f'(x)$. Therefore, we have

$$\frac{dy}{\Delta x} = f'(x)$$

or $dy = f'(x)\,\Delta x$. Thus, we have the approximation

$$\Delta y \approx dy = f'(x)\,\Delta x$$

in terms of the derivative of f at x. The quantity dy is called the *differential of y*.

The Differential
Let $y = f(x)$ define a differentiable function of x. Then,

1. The **differential dx** of the independent variable x is $dx = \Delta x$.
2. The **differential dy** of the dependent variable y is

$$dy = f'(x)\,\Delta x = f'(x)\,dx \tag{12}$$

Notes

1. For the independent variable x: There is no difference between Δx and dx—both measure the change in x from x to $x + \Delta x$.
2. For the dependent variable y: Δy measures the *actual* change in y as x changes from x to $x + \Delta x$, whereas dy measures the *approximate* change in y corresponding to the same change in x.
3. The differential dy depends on both x and dx, but for fixed x, dy is a linear function of dx.

EXAMPLE 3 Let $y = x^3$.

a. Find the differential dy of y.
b. Use dy to approximate Δy when x changes from 2 to 2.01.
c. Use dy to approximate Δy when x changes from 2 to 1.98.
d. Compare the results of part (b) with those of Example 2.

Solution

a. Let $f(x) = x^3$. Then,

$$dy = f'(x)\,dx = 3x^2\,dx$$

b. Here, $x = 2$ and $dx = 2.01 - 2 = 0.01$. Therefore,

$$dy = 3x^2\,dx = 3(2)^2(0.01) = 0.12$$

c. Here, $x = 2$ and $dx = 1.98 - 2 = -0.02$. Therefore,

$$dy = 3x^2\,dx = 3(2)^2(-0.02) = -0.24$$

d. As you can see, both approximations 0.12 and -0.24 are quite close to the actual changes Δy obtained in Example 2: 0.120601 and -0.237608.

Observe how much easier it is to find an approximation to the exact change in a function with the help of the differential, rather than calculating the exact change in the function itself. In the following examples, we take advantage of this fact.

EXAMPLE 4 Approximate the value of $\sqrt{26.5}$ using differentials. Verify your result using the $\boxed{\sqrt{}}$ key on your calculator.

Solution Since we want to compute the square root of a number, let's consider the function $y = f(x) = \sqrt{x}$. Since 25 is the number nearest 26.5 whose square root is readily recognized, let's take $x = 25$. We want to know the change in y, Δy, as x changes from $x = 25$ to $x = 26.5$, an increase of $\Delta x = 1.5$ units. Using Equation (12), we find

$$\Delta y \approx dy = f'(x)\,\Delta x$$
$$= \left(\frac{1}{2\sqrt{x}}\bigg|_{x=25}\right) \cdot (1.5) = \left(\frac{1}{10}\right)(1.5) = 0.15$$

Therefore,
$$\sqrt{26.5} - \sqrt{25} = \Delta y \approx 0.15$$
$$\sqrt{26.5} \approx \sqrt{25} + 0.15 = 5.15$$

The exact value of $\sqrt{26.5}$, rounded off to five decimal places, is 5.14782. Thus, the error incurred in the approximation is 0.00218. ∎

APPLIED EXAMPLE 5 The Effect of Speed on Vehicular Operating Cost The total cost incurred in operating a certain type of truck on a 500-mile trip, traveling at an average speed of v mph, is estimated to be

$$C(v) = 125 + v + \frac{4500}{v}$$

dollars. Find the approximate change in the total operating cost when the average speed is increased from 55 mph to 58 mph.

Solution With $v = 55$ and $\Delta v = dv = 3$, we find

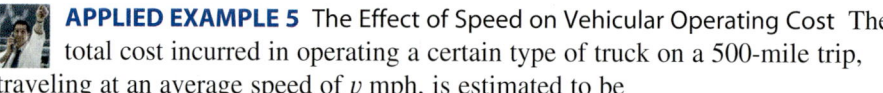

$$\Delta C \approx dC = C'(v)\,dv = \left(1 - \frac{4500}{v^2}\right)\bigg|_{v=55} \cdot (3)$$
$$= \left(1 - \frac{4500}{3025}\right)(3) \approx -1.46$$

so the total operating cost is found to decrease by \$1.46. This might explain why so many independent truckers often exceed the speed limit where it is 55 mph. ∎

APPLIED EXAMPLE 6 The Effect of Advertising on Sales The relationship between the amount of money x spent by Cannon Precision Instruments on advertising and Cannon's total sales $S(x)$ is given by the function

$$S(x) = -0.002x^3 + 0.6x^2 + x + 500 \qquad (0 \le x \le 200)$$

where x is measured in thousands of dollars. Use differentials to estimate the change in Cannon's total sales if advertising expenditures are increased from \$100,000 ($x = 100$) to \$105,000 ($x = 105$).

Solution The required change in sales is given by

$$\Delta S \approx dS = S'(100)\,dx$$
$$= -0.006x^2 + 1.2x + 1\bigg|_{x=100} \cdot (5) \qquad dx = 105 - 100 = 5$$
$$= (-60 + 120 + 1)(5) = 305$$

—that is, an increase of \$305,000. ∎

APPLIED EXAMPLE 7 The Rings of Neptune

a. A ring has an inner radius of r units and an outer radius of R units, where $(R - r)$ is small in comparison to r (Figure 23a). Use differentials to estimate the area of the ring.

b. Recent observations, including those of *Voyager I* and *II*, showed that Neptune's ring system is considerably more complex than had been believed. For one thing, it is made up of a large number of distinguishable rings rather than one continuous great ring as was previously thought (Figure 23b). The outermost ring, 1989N1R, has an inner radius of approximately 62,900 kilometers (measured from the center of the planet) and a radial width of approximately 50 kilometers. Using these data, estimate the area of the ring.

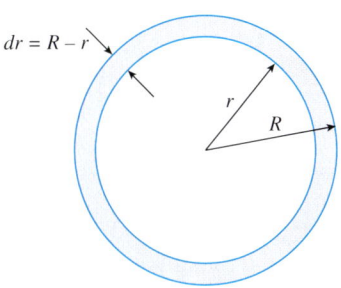

(a) The area of the ring is approximately equal to the circumference of the inner circle times the thickness.

(b) Neptune and its rings

FIGURE 23

Solution

a. Since the area of a circle of radius x is $A = f(x) = \pi x^2$, we find

$$\pi R^2 - \pi r^2 = f(R) - f(r)$$
$$= \Delta A \qquad \text{Remember, } \Delta A = \text{change in } f \text{ when } x \text{ changes from } x = r \text{ to } x = R.$$
$$\approx dA$$
$$= f'(r)\, dr$$

where $dr = R - r$. So we see that the area of the ring is approximately $2\pi r(R - r)$ square units. In words, the area of the ring is approximately equal to

Circumference of the inner circle $\times$ Thickness of the ring

b. Applying the results of part (a) with $r = 62{,}900$ and $dr = 50$, we find that the area of the ring is approximately $2\pi(62{,}900)(50)$, or $19{,}760{,}000$ square kilometers, which is roughly 4% of the earth's surface. ∎

Before looking at the next example, we need to familiarize ourselves with some terminology. If a quantity with exact value q is measured or calculated with an error of Δq, then the quantity $\Delta q/q$ is called the *relative error* in the measurement or calculation of q. If the quantity $\Delta q/q$ is expressed as a percentage, it is then called the *percentage error*. Because Δq is approximated by dq, we normally approximate the relative error $\Delta q/q$ by dq/q.

APPLIED EXAMPLE 8 Estimating Errors in Measurement Suppose the radius of a ball bearing is measured to be 0.5 inch, with a maximum error of ± 0.0002 inch. Then, the relative error in r is

$$\frac{dr}{r} = \frac{\pm 0.0002}{0.5} = \pm 0.0004$$

and the percentage error is $\pm 0.04\%$. ∎

APPLIED EXAMPLE 9 Estimating Errors in Measurement Suppose the side of a cube is measured with a maximum percentage error of 2%. Use differentials to estimate the maximum percentage error in the calculated volume of the cube.

Solution Suppose the side of the cube is x, so that its volume is

$$V = x^3$$

We are given that $\left|\dfrac{dx}{x}\right| \leq 0.02$. Now,

$$dV = 3x^2\, dx$$

and so

$$\frac{dV}{V} = \frac{3x^2\, dx}{x^3} = 3\,\frac{dx}{x}$$

Therefore,

$$\left|\frac{dV}{V}\right| = 3\left|\frac{dx}{x}\right| \leq 3(0.02) = 0.06$$

and we see that the maximum percentage error in the measurement of the volume of the cube is approximately 6%.

Finally, if at some point in reading this section you have a sense of déjà vu, do not be surprised, because the notion of the differential was first used in Section 3.4 (see Example 1). There, we took $\Delta x = 1$, since we were interested in finding the marginal cost when the level of production was increased from $x = 250$ to $x = 251$. If we had used differentials, we would have found

$$C(251) - C(250) \approx C'(250)\, dx$$

so taking $dx = \Delta x = 1$, we have $C(251) - C(250) \approx C'(250)$, which agrees with the result obtained in Example 1. Thus, in Section 3.4, we touched upon the notion of the differential, albeit in the special case in which $dx = 1$.

3.7 Self-Check Exercises

1. Find the differential of $f(x) = \sqrt{x} + 1$.

2. A certain country's government economists have determined that the demand equation for corn in that country is given by

$$p = f(x) = \frac{125}{x^2 + 1}$$

where p is expressed in dollars per bushel and x, the quantity demanded each year, is measured in billions of bushels.

The economists are forecasting a harvest of 6 billion bushels for the year. If the actual production of corn were 6.2 billion bushels for the year instead, what would be the approximate drop in the predicted price of corn per bushel?

Solutions to Self-Check Exercises 3.7 can be found on page 237.

3.7 Concept Questions

1. If $y = f(x)$, what is the differential of x? Write an expression for the differential dy.
2. Refer to the following figure.

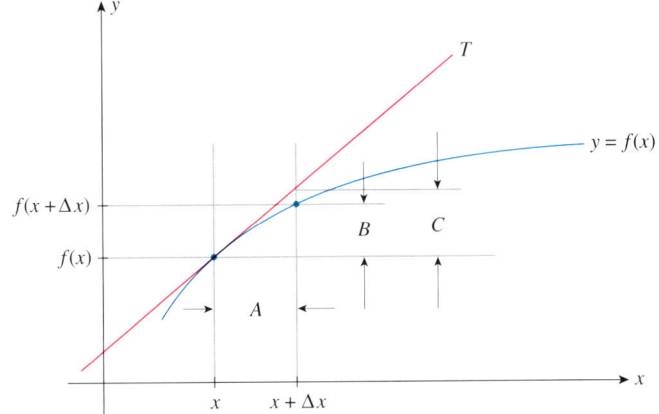

   a. Write A, B, and C in terms of Δx, Δy, and dy.
   b. Write $f'(x)$ (the slope of the tangent line T at x) in terms of Δx and dy.
   c. Observing that $B \approx C$, explain why $\Delta y \approx f'(x)\Delta x = f'(x)\,dx$.

3.7 Exercises

In Exercises 1–14, find the differential of the function.

1. $f(x) = 2x^2$
2. $f(x) = 3x^2 + 1$
3. $f(x) = x^3 - x$
4. $f(x) = 2x^3 + x$
5. $f(x) = \sqrt{x+1}$
6. $f(x) = \dfrac{3}{\sqrt{x}}$
7. $f(x) = 2x^{3/2} + x^{1/2}$
8. $f(x) = 3x^{5/6} + 7x^{2/3}$
9. $f(x) = x + \dfrac{2}{x}$
10. $f(x) = \dfrac{3}{x-1}$
11. $f(x) = \dfrac{x-1}{x^2+1}$
12. $f(x) = \dfrac{2x^2+1}{x+1}$
13. $f(x) = \sqrt{3x^2 - x}$
14. $f(x) = (2x^2 + 3)^{1/3}$

15. Let f be the function defined by
$$y = f(x) = x^2 - 1$$
   a. Find the differential of f.
   b. Use your result from part (a) to find the approximate change in y if x changes from 1 to 1.02.
   c. Find the actual change in y if x changes from 1 to 1.02, and compare your result with that obtained in part (b).

16. Let f be the function defined by
$$y = f(x) = 3x^2 - 2x + 6$$
   a. Find the differential of f.
   b. Use your result from part (a) to find the approximate change in y if x changes from 2 to 1.97.
   c. Find the actual change in y if x changes from 2 to 1.97, and compare your result with that obtained in part (b).

17. Let f be the function defined by
$$y = f(x) = \dfrac{1}{x}$$
   a. Find the differential of f.
   b. Use your result from part (a) to find the approximate change in y if x changes from -1 to -0.95.
   c. Find the actual change in y if x changes from -1 to -0.95, and compare your result with that obtained in part (b).

18. Let f be the function defined by
$$y = f(x) = \sqrt{2x+1}$$
   a. Find the differential of f.
   b. Use your result from part (a) to find the approximate change in y if x changes from 4 to 4.1.
   c. Find the actual change in y if x changes from 4 to 4.1, and compare your result with that obtained in part (b).

In Exercises 19–26, use differentials to approximate the quantity.

19. $\sqrt{10}$
20. $\sqrt{17}$
21. $\sqrt{49.5}$
22. $\sqrt{99.7}$
23. $\sqrt[3]{7.8}$
24. $\sqrt[4]{81.6}$
25. $\sqrt{0.089}$
26. $\sqrt[3]{0.00096}$

27. Use a differential to approximate $\sqrt{4.02} + \dfrac{1}{\sqrt{4.02}}$.

Hint: Let $f(x) = \sqrt{x} + \dfrac{1}{\sqrt{x}}$ and compute dy with $x = 4$ and $dx = 0.02$.

28. Use a differential to approximate $\dfrac{2(4.98)}{(4.98)^2 + 1}$.

Hint: Study the hint for Exercise 27.

29. ERROR ESTIMATION The length of each edge of a cube is 12 cm, with a possible error in measurement of 0.02 cm. Use differentials to estimate the error that might occur when the volume of the cube is calculated.

30. ESTIMATING THE AMOUNT OF PAINT REQUIRED A coat of paint of thickness 0.05 cm is to be applied uniformly to the faces of a cube of edge 30 cm. Use differentials to find the approximate amount of paint required for the job.

31. ERROR ESTIMATION A hemisphere-shaped dome of radius 60 ft is to be coated with a layer of rust-proofer before painting. Use differentials to estimate the amount of rust-proofer needed if the coat is to be 0.01 in. thick.

Hint: The volume of a hemisphere of radius r is $V = \frac{2}{3}\pi r^3$.

32. GROWTH OF A CANCEROUS TUMOR The volume of a spherical cancerous tumor is given by

$$V(r) = \frac{4}{3}\pi r^3$$

If the radius of a tumor is estimated at 1.1 cm, with a maximum error in measurement of 0.005 cm, determine the error that might occur when the volume of the tumor is calculated.

33. UNCLOGGING ARTERIES Research done in the 1930s by the French physiologist Jean Poiseuille showed that the resistance R of a blood vessel of length l and radius r is $R = kl/r^4$, where k is a constant. Suppose a dose of the drug TPA increases r by 10%. How will this affect the resistance R? Assume that l is constant.

34. GROSS DOMESTIC PRODUCT An economist has determined that a certain country's gross domestic product (GDP) is approximated by the function $f(x) = 640x^{1/5}$, where $f(x)$ is measured in billions of dollars and x is the capital outlay in billions of dollars. Use differentials to estimate the change in the country's GDP if the country's capital expenditure changes from $243 billion to $248 billion.

35. LEARNING CURVES The length of time (in seconds) a certain individual takes to learn a list of n items is approximated by

$$f(n) = 4n\sqrt{n-4}$$

Use differentials to approximate the additional time it takes the individual to learn the items on a list when n is increased from 85 to 90 items.

36. EFFECT OF ADVERTISING ON PROFITS The relationship between Cunningham Realty's quarterly profits, $P(x)$, and the amount of money x spent on advertising per quarter is described by the function

$$P(x) = -\frac{1}{8}x^2 + 7x + 30 \quad (0 \leq x \leq 50)$$

where both $P(x)$ and x are measured in thousands of dollars. Use differentials to estimate the increase in profits when advertising expenditure each quarter is increased from $24,000 to $26,000.

37. EFFECT OF MORTGAGE RATES ON HOUSING STARTS A study prepared for the National Association of Realtors estimates that the number of housing starts per year over the next 5 years will be

$$N(r) = \frac{7}{1 + 0.02r^2}$$

million units, where r (percent) is the mortgage rate. Use differentials to estimate the decrease in the number of annual housing starts when the mortgage rate is increased from 6% to 6.5%.

38. SUPPLY-PRICE The supply equation for a certain brand of radio is given by

$$p = s(x) = 0.3\sqrt{x} + 10$$

where x is the quantity supplied and p is the unit price in dollars. Use differentials to approximate the change in price when the quantity supplied is increased from 10,000 units to 10,500 units.

39. DEMAND-PRICE The demand function for the Sentinel smoke alarm is given by

$$p = d(x) = \frac{30}{0.02x^2 + 1}$$

where x is the quantity demanded (in units of a thousand) and p is the unit price in dollars. Use differentials to estimate the change in the price p when the quantity demanded changes from 5000 to 5500 units/week.

40. SURFACE AREA OF AN ANIMAL Animal physiologists use the formula

$$S = kW^{2/3}$$

to calculate an animal's surface area (in square meters) from its weight W (in kilograms), where k is a constant that depends on the animal under consideration. Suppose a physiologist calculates the surface area of a horse ($k = 0.1$). If the horse's weight is estimated at 300 kg, with a maximum error in measurement of 0.6 kg, determine the percentage error in the calculation of the horse's surface area.

41. FORECASTING PROFITS The management of Trappee and Sons forecast that they will sell 200,000 cases of their TexaPep hot sauce next year. Their annual profit is described by

$$P(x) = -0.000032x^3 + 6x - 100$$

thousand dollars, where x is measured in thousands of cases. If the maximum error in the forecast is 15%, determine the corresponding error in Trappee's profits.

42. FORECASTING COMMODITY PRICES A certain country's government economists have determined that the demand equation for soybeans in that country is given by

$$p = f(x) = \frac{55}{2x^2 + 1}$$

where p is expressed in dollars per bushel and x, the quantity demanded each year, is measured in billions of bushels. The economists are forecasting a harvest of 1.8 billion bushels for the year, with a maximum error of 15% in their forecast. Determine the corresponding maximum error in the predicted price per bushel of soybeans.

43. DEMAND FOR TIRES The management of Titan Tire Company has determined that the quantity demanded x of the Super Titan tires per week is related to the unit price p by the equation

$$x = \sqrt{144 - p} \quad (0 \le p \le 144)$$

where p is measured in dollars and x is measured in units of a thousand. Use differentials to find the approximate change in the quantity of the tires demanded per week if the unit price of the tires is increased from $108 per tire to $110 per tire.

44. AIR POLLUTION The amount of nitrogen dioxide, a brown gas that impairs breathing, present in the atmosphere on a certain May day in the city of Long Beach is approximated by

$$A(t) = \frac{136}{1 + 0.25(t - 4.5)^2} + 28 \quad (0 \le t \le 11)$$

where $A(t)$ is measured in pollutant standard index (PSI) and t is measured in hours, with $t = 0$ corresponding to 7 A.M. Use differentials to find the approximate change in the PSI from 8 A.M. to 8:05 A.M.

45. CRIME STUDIES A sociologist has found that the number of serious crimes in a certain city each year is described by the function

$$N(x) = \frac{500(400 + 20x)^{1/2}}{(5 + 0.2x)^2}$$

where x (in cents per dollar deposited) is the level of reinvestment in the area in conventional mortgages by the city's ten largest banks. Use differentials to estimate the change in the number of crimes if the level of reinvestment changes from 20¢/dollar deposited to 22¢/dollar deposited.

46. FINANCING A HOME The Millers are planning to buy a home in the near future and estimate that they will need a 30-year fixed-rate mortgage for $240,000. Their monthly payment P (in dollars) can be computed by using the formula

$$P = \frac{20{,}000r}{1 - (1 + \frac{r}{12})^{-360}}$$

where r is the interest rate per year.

a. Find the differential of P.
b. If the interest rate increases from the present rate of 5%/year to 5.2%/year between now and the time the Millers decide to secure the loan, approximately how much more will their monthly mortgage payment be? How much more will it be if the interest rate increases to 5.3%/year? To 5.4%/year? To 5.5%/year?

47. INVESTMENTS Lupé deposits a sum of $10,000 into an account that pays interest at the rate of r/year compounded monthly. Her investment at the end of 10 years is given by

$$A = 10{,}000\left(1 + \frac{r}{12}\right)^{120}$$

a. Find the differential of A.
b. Approximately how much more would Lupé's account be worth at the end of the term if her account paid 3.1%/year instead of 3%/year? 3.2%/year instead of 3%/year? 3.3%/year instead of 3%/year?

48. KEOGH ACCOUNTS Ian, who is self-employed, contributes $2000 a month into a Keogh retirement account earning interest at the rate of r/year compounded monthly. At the end of 25 years, his account will be worth

$$S = \frac{24{,}000[(1 + \frac{r}{12})^{300} - 1]}{r}$$

dollars.
a. Find the differential of S.
b. Approximately how much more would Ian's account be worth at the end of 25 years if his account earned 4.1%/year instead of 4%/year? 4.2%/year instead of 4%/year? 4.3%/year instead of 4%/year?

In Exercises 49 and 50, determine whether the statement is true or false. If it is true, explain why it is true. If it is false, give an example to show why it is false.

49. If $y = ax + b$ where a and b are constants, then $\Delta y = dy$.

50. If $A = f(x)$, then the percentage change in A is

$$\frac{100f'(x)}{f(x)} dx$$

3.7 Solutions to Self-Check Exercises

1. We find

$$f'(x) = \frac{1}{2}x^{-1/2} = \frac{1}{2\sqrt{x}}$$

Therefore, the required differential of f is

$$dy = \frac{1}{2\sqrt{x}} dx$$

2. We first compute the differential

$$dp = -\frac{250x}{(x^2 + 1)^2} dx$$

Next, using Equation (12) with $x = 6$ and $dx = 0.2$, we find

$$\Delta p \approx dp = -\frac{250(6)}{(36 + 1)^2}(0.2) \approx -0.22$$

or a drop in price of 22¢/bushel.

USING TECHNOLOGY

Finding the Differential of a Function

The calculation of the differential of f at a given value of x involves the evaluation of the derivative of f at that point and can be facilitated through the use of the numerical derivative function.

EXAMPLE 1 Use dy to approximate Δy if $y = x^2(2x^2 + x + 1)^{2/3}$ and x changes from 2 to 1.98.

Solution Let $f(x) = x^2(2x^2 + x + 1)^{2/3}$. Since $dx = 1.98 - 2 = -0.02$, we find the required approximation to be

$$dy = f'(2)(-0.02)$$

But using the numerical derivative operation, we find

$$f'(2) = 30.57581679$$

(see Figure T1). Thus,

$$dy = (-0.02)(30.57581679) = -0.6115163358$$

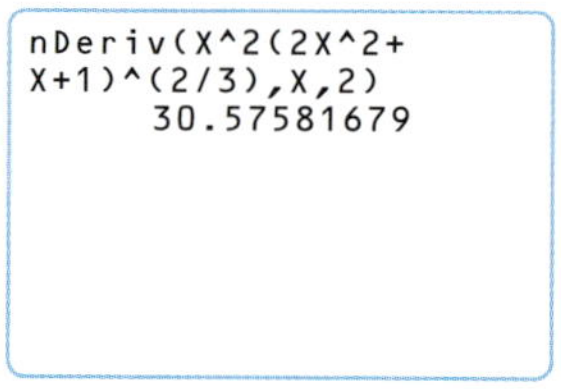

```
nDeriv(X^2(2X^2+
X+1)^(2/3),X,2)
              30.57581679
```

FIGURE T1
The TI-83/84 numerical derivative screen for computing $f'(2)$

APPLIED EXAMPLE 2 Financing a Home The Meyers are considering the purchase of a house in the near future and estimate that they will need a loan of $240,000. Based on a 30-year conventional mortgage with an annual interest rate of r, their monthly repayment will be

$$P = \frac{20{,}000r}{1 - \left(1 + \dfrac{r}{12}\right)^{-360}}$$

dollars. If the interest rate increases from 7% per year to 7.2% per year between now and the time the Meyers decide to secure the loan, approximately how much more will their monthly mortgage payment be?

Solution Let's write

$$P = f(r) = \frac{20{,}000r}{1 - \left(1 + \dfrac{r}{12}\right)^{-360}}$$

Then the increase in the mortgage payment will be approximately

$$dP = f'(0.07)\, dr = f'(0.07)(0.002) \quad \text{Since } dr = 0.072 - 0.07$$
$$\approx 32.2364 \quad \text{Use the numerical derivative operation.}$$

```
nDeriv((20000X)/
(1-(1+X/12)^-360
),X,.07)
        16118.19243
```

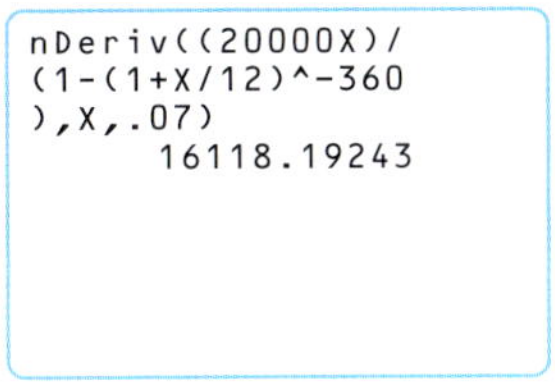

FIGURE T2
The TI-83/84 numerical derivative screen for computing $f'(0.07)$

or approximately $32.24 per month. (See Figure T2.)

TECHNOLOGY EXERCISES

In Exercises 1–6, use dy to approximate Δy for the function $y = f(x)$ when x changes from $x = a$ to $x = b$.

1. $f(x) = 0.21x^7 - 3.22x^4 + 5.43x^2 + 1.42x + 12.42$; $a = 3$, $b = 3.01$

2. $f(x) = \dfrac{0.2x^2 + 3.1}{1.2x + 1.3}$; $a = 2$, $b = 1.96$

3. $f(x) = \sqrt{2.2x^2 + 1.3x + 4}$; $a = 1$, $b = 1.03$

4. $f(x) = x\sqrt{2x^3 - x + 4}$; $a = 2$, $b = 1.98$

5. $f(x) = \dfrac{\sqrt{x^2 + 4}}{x - 1}$; $a = 4$, $b = 4.1$

6. $f(x) = 2.1x^2 + \dfrac{3}{\sqrt{x}} + 5$; $a = 3$, $b = 2.95$

7. **Calculating Mortgage Payments** Refer to Example 2. How much more will the Meyers' mortgage payment be each month if the interest rate increases from 7% to 7.3%/year? To 7.4%/year? To 7.5%/year?

8. **Estimating the Area of a Ring of Neptune** The ring 1989N2R of the planet Neptune has an inner radius of approximately 53,200 km (measured from the center of the planet) and a radial width of 15 km. Use differentials to estimate the area of the ring.

9. **Effect of Price Increase on Quantity Demanded** The quantity demanded each week of the Alpha Sports Watch, x (in thousands), is related to its unit price of p dollars by the equation

$$x = f(p) = 10\sqrt{\dfrac{50 - p}{p}} \quad (0 \le p \le 50)$$

Use differentials to find the decrease in the quantity of the watches demanded each week if the unit price is increased from $40 to $42.

10. **Period of a Communications Satellite** According to Kepler's Third Law of Planetary Motion, the period T (in days) of a satellite moving in a circular orbit d mi above the surface of the earth is given by

$$T = 0.0588\left(1 + \dfrac{d}{3959}\right)^{3/2}$$

Suppose a communications satellite that was moving in a circular orbit 22,000 mi above the earth's surface at one time has, because of friction, dropped down to a new orbit that is 21,500 mi above the earth's surface. Estimate the decrease in the period of the satellite to the nearest $\frac{1}{100}$th hr.

CHAPTER 3 Summary of Principal Formulas and Terms

FORMULAS

1. Derivative of a constant	$\dfrac{d}{dx}(c) = 0 \quad (c, \text{ a constant})$
2. Power Rule	$\dfrac{d}{dx}(x^n) = nx^{n-1}$
3. Constant Multiple Rule	$\dfrac{d}{dx}[cf(x)] = cf'(x)$
4. Sum Rule	$\dfrac{d}{dx}[f(x) \pm g(x)] = f'(x) \pm g'(x)$
5. Product Rule	$\dfrac{d}{dx}[f(x)g(x)] = f(x)g'(x) + g(x)f'(x)$

6. Quotient Rule	$\dfrac{d}{dx}\left[\dfrac{f(x)}{g(x)}\right] = \dfrac{g(x)f'(x) - f(x)g'(x)}{[g(x)]^2}$	
7. Chain Rule	$\dfrac{d}{dx}g(f(x)) = g'(f(x))f'(x)$	
8. General Power Rule	$\dfrac{d}{dx}[f(x)]^n = n[f(x)]^{n-1}f'(x)$	
9. Average cost function	$\overline{C}(x) = \dfrac{C(x)}{x}$	
10. Revenue function	$R(x) = px$	
11. Profit function	$P(x) = R(x) - C(x)$	
12. Elasticity of demand	$E(p) = -\dfrac{pf'(p)}{f(p)}$	
13. Differential of y	$dy = f'(x)\,dx$	

TERMS

marginal cost (196)
marginal cost function (196)
average cost (197)
marginal average cost function (197)
marginal revenue (199)
marginal revenue function (199)

marginal profit function (200)
elasticity of demand (202)
elastic demand (203)
unitary demand (203)
inelastic demand (203)
second derivative of f (209)

implicit differentiation (217)
marginal rate of technical substitution (221)
related rates (222)
differential (229)

CHAPTER 3 Concept Review Questions

Fill in the blanks.

1. **a.** If c is a constant, then $\dfrac{d}{dx}(c) = $ _____.

   b. The Power Rule states that if n is any real number, then $\dfrac{d}{dx}(x^n) = $ _____.

   c. The Constant Multiple Rule states that if c is a constant, then $\dfrac{d}{dx}[cf(x)] = $ _____.

   d. The Sum Rule states that $\dfrac{d}{dx}[f(x) \pm g(x)] = $ _____.

2. **a.** The Product Rule states that $\dfrac{d}{dx}[f(x)g(x)] = $ _____.

   b. The Quotient Rule states that $\dfrac{d}{dx}[f(x)/g(x)] = $ _____.

3. **a.** The Chain Rule states that if $h(x) = g[f(x)]$, then $h'(x) = $ _____.

   b. The General Power Rule states that if $h(x) = [f(x)]^n$, then $h'(x) = $ _____.

4. If C, R, P, and $\overline{C}$ denote the total cost function, the total revenue function, the profit function, and the average cost function, respectively, then C' denotes the _____ _____ function, R' denotes the _____ _____ function, P' denotes the _____ _____ function, and $\overline{C}'$ denotes the _____ _____ function.

5. **a.** If f is a differentiable demand function defined by $x = f(p)$, then the elasticity of demand at price p is given by $E(p) = $ _____.

   b. The demand is _____ if $E(p) > 1$; it is _____ if $E(p) = 1$; it is _____ if $E(p) < 1$.

6. Suppose a function $y = f(x)$ is defined implicitly by an equation in x and y. To find $\dfrac{dy}{dx}$, we differentiate _____ of the equation with respect to x and then solve the resulting equation for $\dfrac{dy}{dx}$. The derivative of a term involving y includes _____ as a factor.

7. In a related-rates problem, we are given a relationship between x and _____ that depends on a third variable t. Knowing the values of x, y, and $\dfrac{dx}{dt}$ at a, we want to find _____ at _____.

8. Let $y = f(t)$ and $x = g(t)$. If $x^2 + y^2 = 4$, then $\dfrac{dx}{dt} = $ _____.

   If $xy = 1$, then $\dfrac{dy}{dt} = $ _____.

9. **a.** If a variable quantity x changes from x_1 to x_2, then the increment in x is $\Delta x = $ _____.

   b. If $y = f(x)$ and x changes from x to $x + \Delta x$, then the increment in y is $\Delta y = $ _____.

10. If $y = f(x)$, where f is a differentiable function, then the differential dx of x is $dx = $ _____, where _____ is an increment in _____, and the differential dy of y is $dy = $ _____.

CHAPTER 3   Review Exercises

In Exercises 1–30, find the derivative of the function.

1. $f(x) = 3x^5 - 2x^4 + 3x^2 - 2x + 1$

2. $f(x) = 4x^6 + 2x^4 + 3x^2 - 2$

3. $g(x) = -2x^{-3} + 3x^{-1} + 2$

4. $f(t) = 2t^2 - 3t^3 - t^{-1/2}$

5. $g(t) = 2t^{-1/2} + 4t^{-3/2} + 2$

6. $h(x) = x^2 + \dfrac{2}{x}$

7. $f(t) = t + \dfrac{2}{t} + \dfrac{3}{t^2}$

8. $g(s) = 2s^2 - \dfrac{4}{s} + \dfrac{2}{\sqrt{s}}$

9. $h(x) = x^2 - \dfrac{2}{x^{3/2}}$

10. $f(x) = \dfrac{x+1}{2x-1}$

11. $g(t) = \dfrac{t^2}{2t^2+1}$

12. $h(t) = \dfrac{\sqrt{t}}{\sqrt{t}+1}$

13. $f(x) = \dfrac{\sqrt{x}-1}{\sqrt{x}+1}$

14. $f(t) = \dfrac{t}{2t^2+1}$

15. $f(x) = \dfrac{x^2(x^2+1)}{x^2-1}$

16. $f(x) = (2x^2 + x)^3$

17. $f(x) = (3x^3 - 2)^8$

18. $h(x) = (\sqrt{x} + 2)^5$

19. $f(t) = \sqrt{2t^2+1}$

20. $g(t) = \sqrt[3]{1 - 2t^3}$

21. $s(t) = (3t^2 - 2t + 5)^{-2}$

22. $f(x) = (2x^3 - 3x^2 + 1)^{-3/2}$

23. $h(x) = \left(x + \dfrac{1}{x}\right)^2$

24. $h(x) = \dfrac{1+x}{(2x^2+1)^2}$

25. $h(t) = (t^2 + t)^4(2t^2)$

26. $f(x) = (2x+1)^3(x^2+x)^2$

27. $g(x) = \sqrt{x}(x^2 - 1)^3$

28. $f(x) = \dfrac{x}{\sqrt{x^3+2}}$

29. $h(x) = \dfrac{\sqrt{3x+2}}{4x-3}$

30. $f(t) = \dfrac{\sqrt{2t+1}}{(t+1)^3}$

In Exercises 31–36, find the second derivative of the function.

31. $f(x) = 2x^4 - 3x^3 + 2x^2 + x + 4$

32. $g(x) = \sqrt{x} + \dfrac{1}{\sqrt{x}}$

33. $h(t) = \dfrac{t}{t^2+4}$

34. $f(x) = (x^3 + x + 1)^2$

35. $f(x) = \sqrt{2x^2+1}$

36. $f(t) = t(t^2+1)^3$

In Exercises 37–42, find dy/dx by implicit differentiation.

37. $6x^2 - 3y^2 = 9$

38. $2x^3 - 3xy = 4$

39. $y^3 + 3x^2 = 3y$

40. $x^2 + 2x^2y^2 + y^2 = 10$

41. $x^2 - 4xy - y^2 = 12$

42. $3x^2y - 4xy + x - 2y = 6$

43. Find the differential of $f(x) = x^2 + \dfrac{1}{x^2}$.

44. Find the differential of $f(x) = \dfrac{1}{\sqrt{x^3+1}}$.

45. Let f be the function defined by $f(x) = \sqrt{2x^2+4}$.
    a. Find the differential of f.
    b. Use your result from part (a) to find the approximate change in $y = f(x)$ if x changes from 4 to 4.1.
    c. Find the actual change in y if x changes from 4 to 4.1 and compare your result with that obtained in part (b).

46. Use a differential to approximate $\sqrt[3]{26.8}$.

47. Let $f(x) = 2x^3 - 3x^2 - 16x + 3$.
    a. Find the points on the graph of f at which the slope of the tangent line is equal to -4.
    b. Find the equation(s) of the tangent line(s) of part (a).

48. Let $f(x) = \tfrac{1}{3}x^3 + \tfrac{1}{2}x^2 - 4x + 1$.
    a. Find the points on the graph of f at which the slope of the tangent line is equal to -2.
    b. Find the equation(s) of the tangent line(s) of part (a).

49. Find an equation of the tangent line to the graph of $y = \sqrt{4 - x^2}$ at the point $(1, \sqrt{3})$.

50. Find an equation of the tangent line to the graph of $y = x(x+1)^5$ at the point $(1, 32)$.

51. Find the third derivative of the function
$$f(x) = \dfrac{1}{2x-1}$$
What is its domain?

52. **Elasticity of Demand** The demand equation for a certain product is $2x + 5p - 60 = 0$, where p is the unit price and x is the quantity demanded of the product. Find the elasticity of demand and determine whether the demand is elastic or inelastic, at the indicated prices.
    a. $p = 3$
    b. $p = 6$
    c. $p = 9$

53. ELASTICITY OF DEMAND The demand equation for a certain product is

$$x = \frac{25}{\sqrt{p}} - 1$$

where p is the unit price and x is the quantity demanded for the product. Compute the elasticity of demand and determine the range of prices corresponding to inelastic, unitary, and elastic demand.

54. ELASTICITY OF DEMAND The demand equation for a certain product is $x = 100 - 0.01p^2$.
a. Is the demand elastic, unitary, or inelastic when $p = 40$?
b. If the price is $40, will raising the price slightly cause the revenue to increase or decrease?

55. ELASTICITY OF DEMAND The demand equation for a certain product is

$$p = 9\sqrt[3]{1000 - x}$$

a. Is the demand elastic, unitary, or inelastic when $p = 60$?
b. If the price is $60, will raising the price slightly cause the revenue to increase or decrease?

56. GDP OF A COUNTRY The gross domestic product (GDP) of a certain country is

$$f(t) = 0.1t^3 + 0.5t^2 + 2t + 20 \quad (0 \leq t \leq 4)$$

billion dollars in year t, where t is measured in years with $t = 0$ corresponding to 2010.
a. What was the GDP of the country in 2013?
b. How fast was the GDP of the country changing in 2013?

57. CELL PHONES The percent of the U.S. population with cell phones is projected to be

$$P(t) = 24.4t^{0.34} \quad (1 \leq t \leq 10)$$

where t is measured in years, with $t = 1$ corresponding to the beginning of 1998.
a. What percentage of the U.S. population had cell phones by the beginning of 2006?
b. How fast was the percentage of the U.S. population with cell phones changing at the beginning of 2006?
Source: BancAmerica Robertson Stephens.

58. SALES OF CAMERAS The shipments of Lica digital single-lens reflex cameras (SLRs) are projected to be

$$N(t) = 6t^2 + 200t + 4\sqrt{t} + 20,000 \quad (0 \leq t \leq 4)$$

units t years from now.
a. How many Lica SLRs will be shipped after 2 years?
b. At what rate will the number of Lica SLRs shipped be changing after 2 years?

59. SALES OF DSPS The annual sales of digital signal processors (DSPs) in billions of dollars is approximated by

$$S(t) = 0.14t^2 + 0.68t + 3.1 \quad (0 \leq t \leq 6)$$

where t is measured in years, with $t = 0$ corresponding to the beginning of 1997.
a. What were the sales of DSPs at the beginning of 1997? What were the sales at the beginning of 2002?
b. How fast was the level of sales increasing at the beginning of 1997? How fast were sales increasing at the beginning of 2002?
Source: World Semiconductor Trade Statistics.

60. ADULT OBESITY In the United States, the percentage of adults (age 20–74) classified as obese held steady through the 1960s and 1970s at around 14% but began to rise rapidly during the 1980s and 1990s. This rise in adult obesity coincided with the period when an increasing number of Americans began eating more sugar and fats. The function

$$P(t) = 0.01484t^2 + 0.446t + 15 \quad (0 \leq t \leq 22)$$

gives the percentage of obese adults from 1978 ($t = 0$) through 2000 ($t = 22$).
a. What percentage of adults were obese in 1978? In 2000?
b. How fast was the percentage of obese adults increasing in 1980 ($t = 2$)? In 1998 ($t = 20$)?
Source: Journal of the American Medical Association.

61. POPULATION GROWTH The population of a certain suburb is expected to be

$$P(t) = 30 - \frac{20}{2t + 3} \quad (0 \leq t \leq 5)$$

thousand t years from now.
a. By how much will the population have grown after 3 years?
b. How fast is the population changing after 3 years?

62. BEST-SELLING NOVEL The number of copies of a best-selling novel sold t weeks after it was introduced is given by

$$N(t) = (4 + 5t)^{5/3} \quad (1 \leq t \leq 30)$$

where $N(t)$ is measured in thousands.
a. How many copies of the novel were sold after 12 weeks?
b. How fast were the sales of the novel changing after 12 weeks?

63. CABLE TV SUBSCRIBERS The number of subscribers to CNC Cable Television in the town of Randolph is approximated by the function

$$N(x) = 1000(1 + 2x)^{1/2} \quad (1 \leq x \leq 30)$$

where $N(x)$ denotes the number of subscribers to the service in the xth week. Find the rate of increase in the number of subscribers at the end of the 12th week.

64. COST OF WIRELESS PHONE CALLS As cellular phone usage continues to soar, the airtime costs have dropped. The average price per minute of use (in cents) is approximated by

$$f(t) = 31.88(1 + t)^{-0.45} \quad (0 \le t \le 6)$$

where t is measured in years and $t = 0$ corresponds to the beginning of 1998. Compute $f'(t)$. How fast was the average price/minute of use changing at the beginning of 2000? What was the average price/minute of use at the beginning of 2000?

Source: Cellular Telecommunications Industry Association.

65. **MALE LIFE EXPECTANCY** Suppose the life expectancy of a male at birth in a certain country is described by the function

$$f(t) = 46.9(1 + 1.09t)^{0.1} \quad (0 \le t \le 150)$$

where t is measured in years, with $t = 0$ corresponding to the beginning of 1900. How long can a male born at the beginning of 2000 in that country expect to live? What is the rate of change of the life expectancy of a male born in that country at the beginning of 2000?

66. **COST OF PRODUCING DVDS** The total weekly cost in dollars incurred by Herald Media Corp. in producing x DVDs is given by the total cost function

$$C(x) = 2500 + 2.2x \quad (0 \le x \le 8000)$$

a. What is the marginal cost when $x = 1000$ and 2000?
b. Find the average cost function $\overline{C}$ and the marginal average cost function $\overline{C}'$.
c. Using the results from part (b), show that the average cost incurred by Herald in producing a DVD approaches $2.20/disc when the level of production is high enough.

67. **SUPPLY FUNCTION** The supply function for a certain brand of satellite radios is given by

$$p = \frac{1}{10}x^{3/2} + 10 \quad (0 \le x \le 50)$$

where x is the quantity demanded (in thousands) if the unit price is $\$p$. Find $p'(40)$, and interpret your result.

68. **DEMAND FUNCTION** The demand for a certain brand of electric shavers is given by

$$p = 20\sqrt{-x^2 + 100} \quad (0 \le x \le 10)$$

where x (in thousands) is the quantity demanded if the unit price is $\$p$. Find $p'(6)$, and interpret your result.

69. **MARGINAL COST** The total daily cost (in dollars) incurred by Delta Electronics in producing x MP3 players is

$$C(x) = 0.0001x^3 - 0.02x^2 + 24x + 2000 \quad (0 \le x \le 500)$$

where x stands for the number of units produced.
a. What is the actual cost incurred in the manufacturing the 301st MP3 player, assuming that the 300th player was manufactured?
b. What is the marginal cost when $x = 300$?

70. **DEMAND FOR CORDLESS PHONES** The marketing department of Telecon has determined that the demand for their smartphones obeys the relationship

$$p = -0.02x + 600 \quad (0 \le x \le 30,000)$$

where p denotes the phone's unit price (in dollars) and x denotes the quantity demanded.
a. Find the revenue function R.
b. Find the marginal revenue function R'.
c. Compute $R'(10,000)$, and interpret your result.

71. **DEMAND FOR PHOTOCOPYING MACHINES** The weekly demand for the LectroCopy photocopying machine is given by the demand equation

$$p = 2000 - 0.04x \quad (0 \le x \le 50,000)$$

where p denotes the wholesale unit price in dollars and x denotes the quantity demanded. The weekly total cost function for manufacturing these copiers is given by

$$C(x) = 0.000002x^3 - 0.02x^2 + 1000x + 120,000$$

where $C(x)$ denotes the total cost incurred in producing x units.
a. Find the revenue function R, the profit function P, and the average cost function $\overline{C}$.
b. Find the marginal cost function C', the marginal revenue function R', the marginal profit function P', and the marginal average cost function $\overline{C}'$.
c. Compute $C'(3000)$, $R'(3000)$, and $P'(3000)$.
d. Compute $\overline{C}'(5000)$ and $\overline{C}'(8000)$, and interpret your results.

72. **MARGINAL AVERAGE COST** The Custom Office makes a line of executive desks. It is estimated that the total cost for making x units of the Junior Executive model is

$$C(x) = 80x + 150,000 \quad (0 \le x \le 20,000)$$

dollars/year.
a. Find the average cost function $\overline{C}$.
b. Find the marginal average cost function $\overline{C}'$.
c. What happens to $\overline{C}(x)$ when x is very large? Interpret your result.

73. **GDP OF A COUNTRY** The GDP of a country from the years 2006 to 2013 is approximated by the function

$$G(t) = -0.3t^3 + 1.2t^2 + 500 \quad (0 \le t \le 7)$$

where $G(t)$ is measured in billions of dollars and $t = 0$ corresponds to 2006. Find $G'(2)$ and $G''(2)$, and interpret your results.

74. **MOTION OF AN OBJECT** The position of an object moving along a straight line is given by

$$s = t\sqrt{2t^2 + 1} \quad (0 \le t \le 5)$$

where s is measured in feet and t in seconds. Find the velocity and acceleration of the object after 2 sec.

CHAPTER 3 Before Moving On...

1. Find the derivative of $f(x) = 2x^3 - 3x^{1/3} + 5x^{-2/3}$.
2. Differentiate $g(x) = x\sqrt{2x^2 - 1}$.
3. Find $\dfrac{dy}{dx}$ if $y = \dfrac{2x + 1}{x^2 + x + 1}$.
4. Find the first three derivatives of $f(x) = \dfrac{1}{\sqrt{x + 1}}$.
5. Find $\dfrac{dy}{dx}$ given that $xy^2 - x^2y + x^3 = 4$.
6. Let $y = x\sqrt{x^2 + 5}$.
   a. Find the differential of y.
   b. If x changes from $x = 2$ to $x = 2.01$, what is the approximate change in y?

4 APPLICATIONS OF THE DERIVATIVE

THIS CHAPTER FURTHER EXPLORES the power of the derivative as a tool to help analyze the properties of functions. The information obtained can then be used to accurately sketch graphs of functions. We also see how the derivative is used in solving a large class of optimization problems, including finding what level of production will yield a maximum profit for a company, finding what level of production will result in minimal cost to a company, finding the maximum height attained by a rocket, finding the maximum velocity at which air is expelled when a person coughs, and a host of other problems.

How many loudspeaker systems should the Acrosonic company produce to maximize its profit? In Example 4, page 301, you will see how the techniques of calculus can be used to help answer this question.

4.1 Applications of the First Derivative

Determining the Intervals Where a Function Is Increasing or Decreasing

According to a study by the U.S. Department of Energy and the Shell Development Company, a typical car's fuel economy as a function of its speed is described by the graph shown in Figure 1. Observe that the fuel economy $f(x)$ in miles per gallon (mpg) improves as x, the vehicle's speed in miles per hour (mph), increases from 0 to 42 and then drops as the speed increases beyond 42 mph. We use the terms *increasing* and *decreasing* to describe the behavior of a function as we move from left to right along its graph.

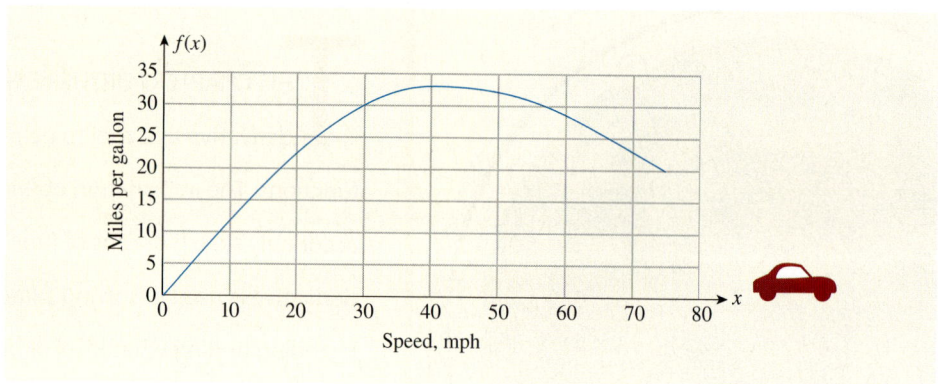

FIGURE 1
A typical car's fuel economy improves as the speed at which it is driven increases from 0 mph to 42 mph and drops at speeds greater than 42 mph.

Source: U.S. Department of Energy and Shell Development Co.

More precisely, we have the following definitions:

Increasing and Decreasing Functions

A function f is **increasing** on an interval (a, b) if for every two numbers x_1 and x_2 in (a, b), $f(x_1) < f(x_2)$ whenever $x_1 < x_2$ (Figure 2a).

A function f is **decreasing** on an interval (a, b) if for every two numbers x_1 and x_2 in (a, b), $f(x_1) > f(x_2)$ whenever $x_1 < x_2$ (Figure 2b).

(a) f is increasing on (a, b). (b) f is decreasing on (a, b).
FIGURE 2

We say that f is *increasing at a number* c if there exists an interval (a, b) containing c such that f is increasing on (a, b). Similarly, we say that f is *decreasing at a number* c if there exists an interval (a, b) containing c such that f is decreasing on (a, b).

Since the rate of change of a function at $x = c$ is given by the derivative of the function at that number, the derivative lends itself naturally to being a tool for determining the intervals where a differentiable function is increasing or decreasing.

Indeed, as we saw in Chapter 2, the derivative of a function at a number measures both the slope of the tangent line to the graph of the function at the point on the graph of f corresponding to that number and the rate of change of the function at that number. In fact, at a number where the derivative is positive, the slope of the tangent line to the graph is positive, and the function is increasing. At a number where the derivative is negative, the slope of the tangent line to the graph is negative, and the function is decreasing (Figure 3).

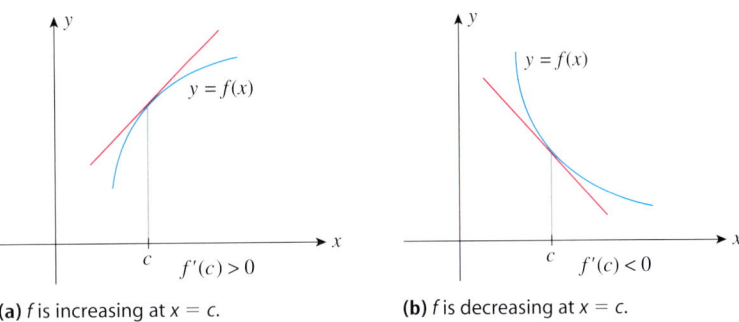

FIGURE 3

(a) f is increasing at $x = c$.

(b) f is decreasing at $x = c$.

These observations lead to the following important theorem, which we state without proof.

> **THEOREM 1**
> **a.** If $f'(x) > 0$ for every value of x in an interval (a, b), then f is increasing on (a, b).
> **b.** If $f'(x) < 0$ for every value of x in an interval (a, b), then f is decreasing on (a, b).
> **c.** If $f'(x) = 0$ for every value of x in an interval (a, b), then f is constant on (a, b).

EXAMPLE 1 Find the interval where the function $f(x) = x^2$ is increasing and the interval where it is decreasing.

Solution The derivative of $f(x) = x^2$ is $f'(x) = 2x$. Since

$$f'(x) = 2x > 0 \quad \text{if } x > 0 \quad \text{and} \quad f'(x) = 2x < 0 \quad \text{if } x < 0$$

f is increasing on the interval $(0, \infty)$ and decreasing on the interval $(-\infty, 0)$ (Figure 4).

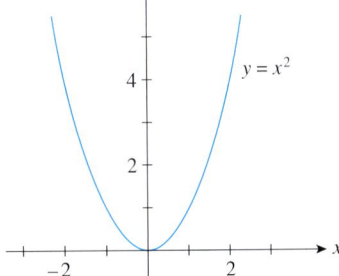

FIGURE 4
The graph of f falls on $(-\infty, 0)$ where $f'(x) < 0$ and rises on $(0, \infty)$ where $f'(x) > 0$.

Recall that the graph of a continuous function cannot have any breaks. As a consequence, a continuous function cannot change sign unless it equals zero for some value of x. (See Theorem 5, page 124.) This observation suggests the following procedure for determining the sign of the derivative f' of a function f and hence the intervals where the function f is increasing and where it is decreasing.

> **Determining the Intervals Where a Function Is Increasing or Decreasing**
> 1. Find all values of x for which $f'(x) = 0$ or f' is discontinuous, and identify the open intervals determined by these numbers.
> 2. Select a test number c in each interval found in step 1, and determine the sign of $f'(c)$ in that interval.
>   **a.** If $f'(c) > 0$, f is increasing on that interval.
>   **b.** If $f'(c) < 0$, f is decreasing on that interval.

Explore & Discuss

True or false? If f is continuous at c and f is increasing at c, then $f'(c) > 0$. Explain your answer.
Hint: Consider $f(x) = x^3$ and $c = 0$.

EXAMPLE 2 Determine the intervals where the function $f(x) = x^3 - 3x^2 - 24x + 32$ is increasing and where it is decreasing.

Solution

1. The derivative of f is

$$f'(x) = 3x^2 - 6x - 24 = 3(x + 2)(x - 4) \quad \text{(x²) See page 11.}$$

and it is continuous everywhere. The zeros of $f'(x)$ are $x = -2$ and $x = 4$, and these numbers divide the real line into the intervals $(-\infty, -2)$, $(-2, 4)$, and $(4, \infty)$.

2. To determine the sign of $f'(x)$ in the intervals $(-\infty, -2)$, $(-2, 4)$, and $(4, \infty)$, compute $f'(x)$ at a convenient test point in each interval. The results are shown in the following table:

Interval	Test Point c	$f'(c)$	Sign of $f'(x)$
$(-\infty, -2)$	-3	21	$+$
$(-2, 4)$	0	-24	$-$
$(4, \infty)$	5	21	$+$

(x²) See page 20.

Using these results, we obtain the sign diagram shown in Figure 5. We conclude that f is increasing on the intervals $(-\infty, -2)$ and $(4, \infty)$ and is decreasing on the interval $(-2, 4)$. Figure 6 shows the graph of f.

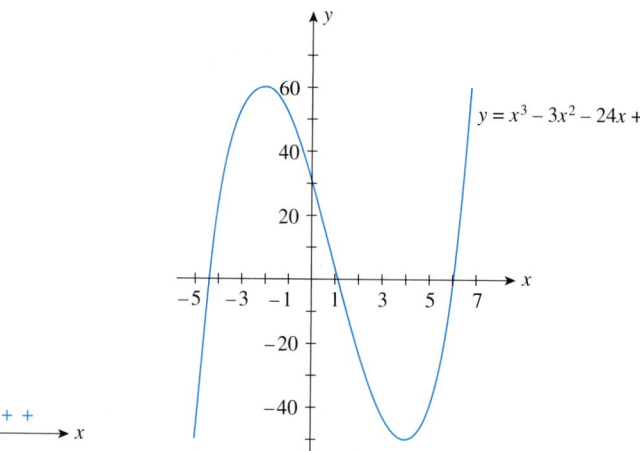

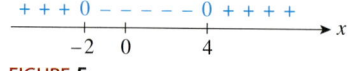

FIGURE 5
Sign diagram for f'

FIGURE 6
The graph of f rises on $(-\infty, -2)$, falls on $(-2, 4)$, and rises again on $(4, \infty)$.

Note We will learn how to sketch these graphs later. However, if you are familiar with the use of a graphing utility, you may go ahead and verify each graph.

Exploring with TECHNOLOGY

Refer to Example 2.

1. Use a graphing utility to plot the graphs of $f(x) = x^3 - 3x^2 - 24x + 32$ and its derivative function $f'(x) = 3x^2 - 6x - 24$ using the viewing window $[-10, 10] \times [-50, 70]$.

2. By looking at the graph of f', determine the intervals where $f'(x) > 0$ and the intervals where $f'(x) < 0$. Next, look at the graph of f and determine the intervals where it is increasing and the intervals where it is decreasing. Describe the relationship. Is it what you expected?

4.1 APPLICATIONS OF THE FIRST DERIVATIVE

EXAMPLE 3 Find the interval where the function $f(x) = x^{2/3}$ is increasing and the interval where it is decreasing.

Solution

1. The derivative of f is

$$f'(x) = \frac{2}{3}x^{-1/3} = \frac{2}{3x^{1/3}}$$

The function f' is not defined at $x = 0$, so f' is discontinuous there. It is continuous everywhere else. Furthermore, f' is not equal to zero anywhere. The number 0 divides the real line (the domain of f) into the intervals $(-\infty, 0)$ and $(0, \infty)$.

2. Pick a test point (say, $x = -1$) in the interval $(-\infty, 0)$, and compute

$$f'(-1) = -\frac{2}{3}$$

Since $f'(-1) < 0$, we know that $f'(x) < 0$ on $(-\infty, 0)$. Next, we pick a test point (say, $x = 1$) in the interval $(0, \infty)$ and compute

$$f'(1) = \frac{2}{3}$$

Since $f'(1) > 0$, we know that $f'(x) > 0$ on $(0, \infty)$. Figure 7 shows these results in the form of a sign diagram.

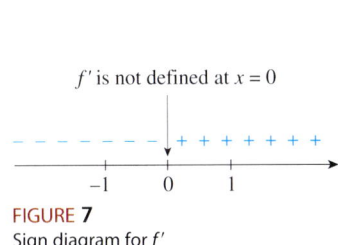

FIGURE 7
Sign diagram for f'

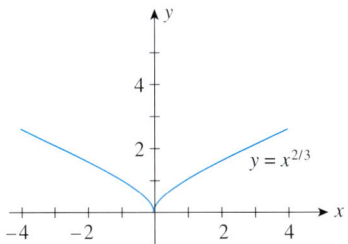

FIGURE 8
f decreases on $(-\infty, 0)$ and increases on $(0, \infty)$.

We conclude that f is decreasing on the interval $(-\infty, 0)$ and increasing on the interval $(0, \infty)$. The graph of f, shown in Figure 8, confirms these results.

EXAMPLE 4 Find the intervals where the function $f(x) = x + \frac{1}{x}$ is increasing and where it is decreasing.

Solution

1. The derivative of f is

$$f'(x) = 1 - \frac{1}{x^2} = \frac{x^2 - 1}{x^2} \quad \text{(x^2) See page 17.}$$

Since f' is not defined at $x = 0$, it is discontinuous there. Furthermore, $f'(x)$ is equal to zero when $x^2 - 1 = 0$ or $x = \pm 1$. Note that the value of f' is different from zero in the open intervals $(-\infty, -1)$, $(-1, 0)$, $(0, 1)$, and $(1, \infty)$.

2. To determine the sign of f' in each of these intervals, we compute $f'(x)$ at the test points $x = -2, -\frac{1}{2}, \frac{1}{2},$ and 2, respectively, obtaining $f'(-2) = \frac{3}{4}, f'(-\frac{1}{2}) = -3$,

$f'(\frac{1}{2}) = -3$, and $f'(2) = \frac{3}{4}$. From the sign diagram for f' (Figure 9), we conclude that f is increasing on $(-\infty, -1)$ and $(1, \infty)$ and decreasing on $(-1, 0)$ and $(0, 1)$.

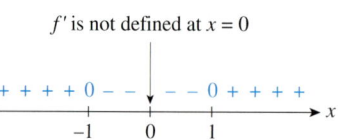

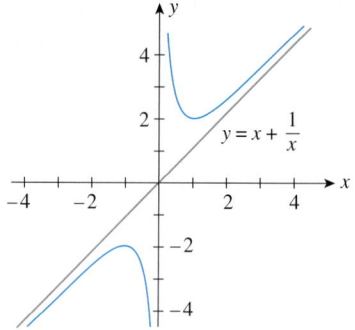

FIGURE 9
f' does not change sign as we move across $x = 0$.

FIGURE 10
The graph of f rises on $(-\infty, -1)$, falls on $(-1, 0)$ and $(0, 1)$, and rises again on $(1, \infty)$.

The graph of f appears in Figure 10. Note that f' does not change sign as we move across $x = 0$. (Compare this with Example 3.)

 Example 4 reminds us that we must *not* automatically conclude that the derivative f' must change sign when we move across a number where f' is discontinuous or a zero of f'.

Explore & Discuss

Consider the profit function P associated with a certain commodity defined by

$$P(x) = R(x) - C(x) \quad (x \geq 0)$$

where R is the revenue function, C is the total cost function, and x is the number of units of the product produced and sold.

1. Find an expression for $P'(x)$.
2. Find relationships in terms of the derivatives of R and C such that
   a. P is increasing at $x = a$.
   b. P is decreasing at $x = a$.
   c. P is neither increasing nor decreasing at $x = a$.
   Hint: Recall that the derivative of a function at $x = a$ measures the rate of change of the function at that number.
3. Explain the results of part 2 in economic terms.

Exploring with TECHNOLOGY

1. Use a graphing utility to sketch the graphs of $f(x) = x^3 - ax$ for $a = -2, -1, 0, 1,$ and 2, using the viewing window $[-2, 2] \times [-2, 2]$.
2. Use the results of part 1 to guess at the values of a such that f is increasing on $(-\infty, \infty)$.
3. Prove your conjecture analytically.

Relative Extrema

Besides helping us determine where the graph of a function is increasing and decreasing, the first derivative may be used to help us locate certain "high points" and "low points" on the graph of f. Knowing these points is invaluable in sketching the graphs of functions and solving optimization problems. These "high points" and "low points" correspond to the *relative (local) maxima* and *relative minima* of a function. They are so called because they are the highest or the lowest points when compared with points nearby.

The graph shown in Figure 11 gives the U.S. budget surplus (deficit) from 1996 ($t = 0$) to 2008 ($t = 12$). The relative maxima and the relative minima of the function f are indicated on the graph.

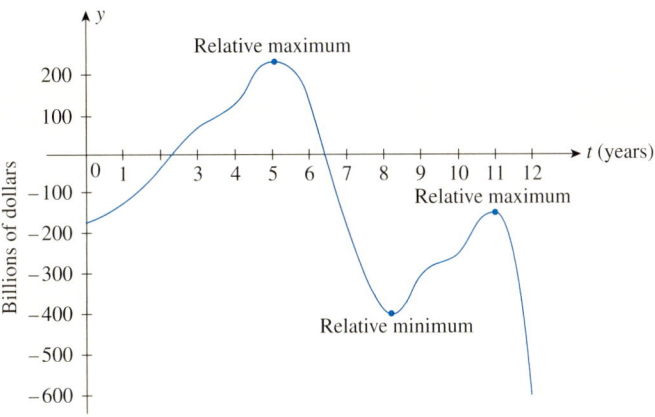

FIGURE 11
U.S. budget surplus (deficit) from 1996 to 2008

Source: Office of Management and Budget.

More generally, we have the following definition:

Relative Maximum

A function f has a **relative maximum** at $x = c$ if there exists an open interval (a, b) containing c such that $f(x) \leq f(c)$ for all x in (a, b).

Geometrically, this means that there is *some* interval containing $x = c$ such that no point on the graph of f with its x-coordinate in that interval can lie above the point $(c, f(c))$; that is, $f(c)$ is the largest value of $f(x)$ in some interval around $x = c$. Figure 12 depicts the graph of a function f that has a relative maximum at $x = x_1$ and another at $x = x_3$.

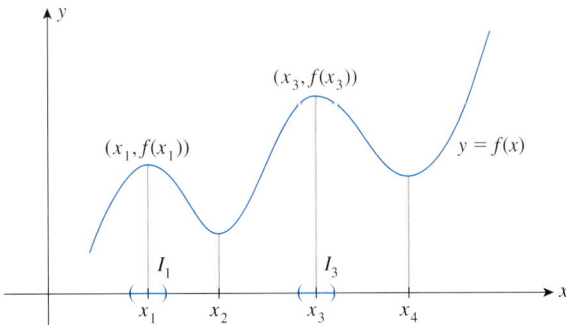

FIGURE 12
f has a relative maximum at $x = x_1$ and at $x = x_3$.

Observe that all the points on the graph of f with x-coordinates in the interval I_1 containing x_1 (shown in blue) lie on or below the point $(x_1, f(x_1))$. This is also true for the point $(x_3, f(x_3))$ and the interval I_3. Thus, even though there are points on the graph of f that are "higher" than the points $(x_1, f(x_1))$ and $(x_3, f(x_3))$, the latter points are "highest" relative to points in their respective neighborhoods (intervals). Points on the graph of a function f that are "highest" and "lowest" with respect to *all* points in the domain of f will be studied in Section 4.4.

The definition of the relative minimum of a function parallels that of the relative maximum of a function.

Relative Minimum

A function f has a **relative minimum** at $x = c$ if there exists an open interval (a, b) containing c such that $f(x) \geq f(c)$ for all x in (a, b).

The graph of the function f depicted in Figure 12 has a relative minimum at $x = x_2$ and another at $x = x_4$.

Finding the Relative Extrema

We refer to the relative maxima and relative minima of a function as the **relative extrema** of that function. As a first step in our quest to find the relative extrema of a function, we consider functions that have derivatives at such points. Suppose that f is a function that is differentiable on some interval (a, b) that contains a number c and that f has a relative maximum at $x = c$ (Figure 13a).

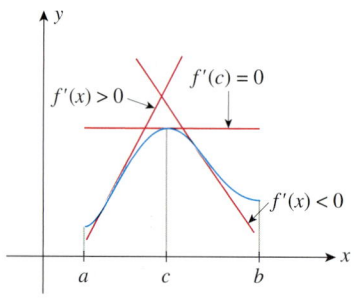 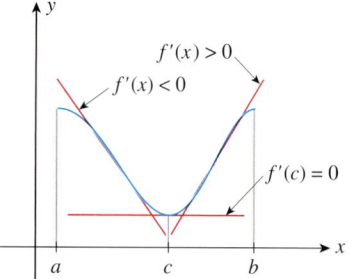

FIGURE 13

(a) f has a relative maximum at $x = c$. (b) f has a relative minimum at $x = c$.

Observe that the slope of the tangent line to the graph of f must change from positive to negative as we move across $x = c$ from left to right. Therefore, the tangent line to the graph of f at the point $(c, f(c))$ must be horizontal; that is, $f'(c) = 0$ (Figure 13a).

Using a similar argument, it may be shown that the derivative f' of a differentiable function f must also be equal to zero at $x = c$ if f has a relative minimum at $x = c$ (Figure 13b).

This analysis reveals an important characteristic of the relative extrema of a differentiable function f: At any number c where f has a relative extremum, $f'(c) = 0$.

 Before we develop a procedure for finding such numbers, a few words of caution are in order. First, this result tells us that if a differentiable function f has a relative extremum at a number $x = c$, then $f'(c) = 0$. The converse of this statement—if $f'(c) = 0$ at $x = c$, then f must have a relative extremum at that number—is *not* true. Consider, for example, the function $f(x) = x^3$. Here, $f'(x) = 3x^2$, so $f'(0) = 0$. However, f has neither a relative maximum nor a relative minimum at $x = 0$ (Figure 14).

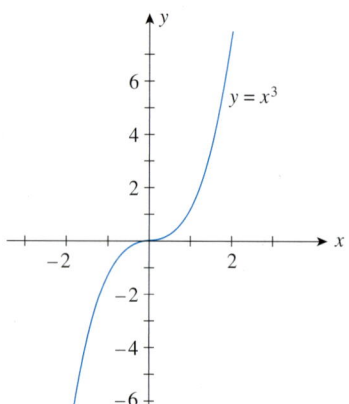

FIGURE 14
$f'(0) = 0$, but f does not have a relative extremum at $(0, 0)$.

Second, our result assumes that the function is differentiable and therefore has a derivative at a number that gives rise to a relative extremum. The functions $f(x) = |x|$ and $g(x) = x^{2/3}$ demonstrate that a relative extremum of a function may exist at a number at which the derivative does not exist. Both these functions fail to be differentiable at $x = 0$, but each has a relative minimum there. Figure 15 shows the graphs of these functions. Note that the slopes of the tangent lines change from negative to positive as we move across $x = 0$, just as in the case of a function that is differentiable at a value of x that gives rise to a relative minimum.

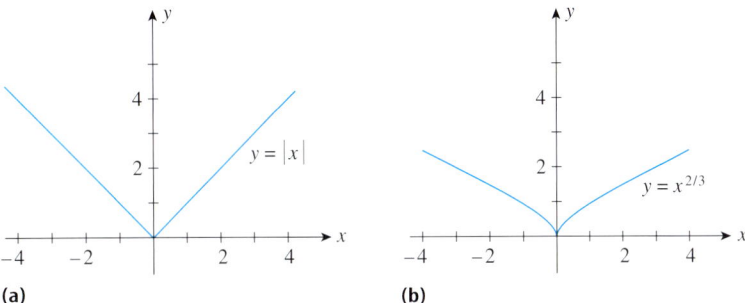

FIGURE 15
Each of these functions has a relative extremum at (0, 0), but the derivative does not exist there.

We refer to a number in the domain of f that *might* give rise to a relative extremum as a critical number.

Critical Number of f

A **critical number** of a function f is any number x in the domain of f such that $f'(x) = 0$ or $f'(x)$ does not exist.

Figure 16 depicts the graph of a function that has critical numbers at $x = a, b, c, d,$ and e. Observe that $f'(x) = 0$ at $x = a, b,$ and c. Next, since there is a corner at $x = d$, $f'(x)$ does not exist there. Finally, $f'(x)$ does not exist at $x = e$ because the tangent line there is vertical. Also, observe that the critical numbers $x = a, b,$ and d give rise to relative extrema of f, whereas the critical numbers $x = c$ and $x = e$ do not.

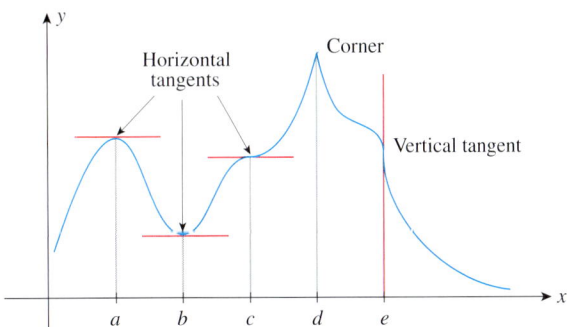

FIGURE 16
Critical numbers of f

Having defined what a critical number is, we can now state a formal procedure for finding the relative extrema of a continuous function that is differentiable everywhere except at isolated values of x. Incorporated into the procedure is the so-called **First Derivative Test,** which helps us determine whether a number gives rise to a relative maximum or a relative minimum of the function f.

The First Derivative Test

Procedure for Finding the Relative Extrema of a Continuous Function f

1. Determine the critical numbers of f.
2. Determine the sign of $f'(x)$ to the left and right of each critical number.
   a. If $f'(x)$ changes sign from *positive* to *negative* as we move across a critical number c, then f has a relative maximum at $x = c$.
   b. If $f'(x)$ changes sign from *negative* to *positive* as we move across a critical number c, then f has a relative minimum at $x = c$.
   c. If $f'(x)$ does not change sign as we move across a critical number c, then f does not have a relative extremum at $x = c$.

EXAMPLE 5 Find the relative maxima and relative minima of the function $f(x) = x^2$.

Solution The derivative of $f(x) = x^2$ is given by $f'(x) = 2x$. Setting $f'(x) = 0$ yields $x = 0$ as the only critical number of f. Since

$$f'(x) < 0 \quad \text{if } x < 0 \quad \text{and} \quad f'(x) > 0 \quad \text{if } x > 0$$

we see that $f'(x)$ changes sign from negative to positive as we move across the critical number 0. Thus, we conclude that $f(0) = 0$ is a relative minimum of f (Figure 17).

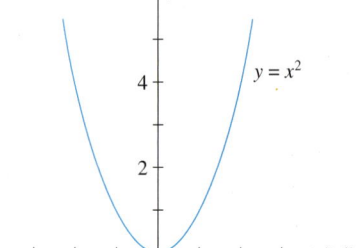

FIGURE 17
f has a relative minimum at $x = 0$.

EXAMPLE 6 Find the relative maxima and relative minima of the function $f(x) = x^{2/3}$ (see Example 3).

Solution The derivative of f is $f'(x) = \frac{2}{3}x^{-1/3}$. As was noted in Example 3, f' is not defined at $x = 0$, is continuous everywhere else, and is not equal to zero in its domain. Thus, $x = 0$ is the only critical number of the function f.

The sign diagram obtained in Example 3 is reproduced in Figure 18. We can see that the sign of $f'(x)$ changes from negative to positive as we move across $x = 0$ from left to right. Thus, an application of the First Derivative Test tells us that $f(0) = 0$ is a relative minimum of f (Figure 19).

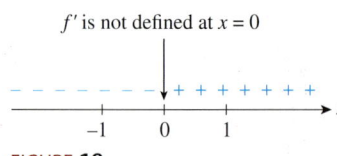

FIGURE 18
Sign diagram for f'

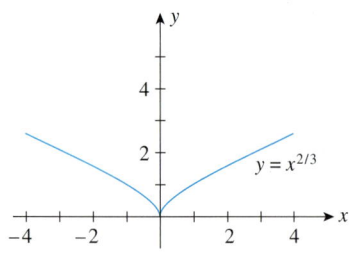

FIGURE 19
f has a relative minimum at $x = 0$.

Explore & Discuss

Recall that the average cost function $\overline{C}$ is defined by

$$\overline{C} = \frac{C(x)}{x}$$

where $C(x)$ is the total cost function and x is the number of units of a commodity manufactured (see Section 3.4).

1. Show that

$$\overline{C}'(x) = \frac{C'(x) - \overline{C}(x)}{x} \quad (x > 0)$$

2. Use the result of part 1 to conclude that $\overline{C}$ is decreasing for values of x at which $C'(x) < \overline{C}(x)$. Find similar conditions for which $\overline{C}$ is increasing and for which $\overline{C}$ is constant.

3. Explain the results of part 2 in economic terms.

4.1 APPLICATIONS OF THE FIRST DERIVATIVE

EXAMPLE 7 Find the relative maxima and relative minima of the function
$$f(x) = x^3 - 3x^2 - 24x + 32$$

Solution The derivative of f is
$$f'(x) = 3x^2 - 6x - 24 = 3(x + 2)(x - 4) \qquad (x^2) \text{ See page 11.}$$

and it is continuous everywhere. The zeros of $f'(x)$, $x = -2$ and $x = 4$, are the only critical numbers of the function f. The sign diagram for f' is shown in Figure 20. Examine the two critical numbers $x = -2$ and $x = 4$ for a relative extremum using the First Derivative Test and the sign diagram for f'.

1. *The critical number* -2: Since the function $f'(x)$ changes sign from positive to negative as we move across $x = -2$ from left to right, we conclude that a relative maximum of f occurs at $x = -2$. The value of $f(x)$ when $x = -2$ is
$$f(-2) = (-2)^3 - 3(-2)^2 - 24(-2) + 32 = 60$$

2. *The critical number* 4: $f'(x)$ changes sign from negative to positive as we move across $x = 4$ from left to right, so $f(4) = -48$ is a relative minimum of f. The graph of f appears in Figure 21.

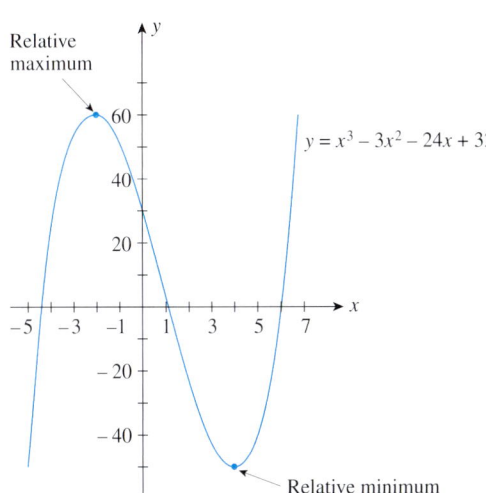

FIGURE 21
f has a relative maximum at $x = -2$ and a relative minimum at $x = 4$.

```
+ + + 0 - - - - - 0 + + + +
-----+------------+-------→ x
    -2    0       4
```
FIGURE 20
Sign diagram for f'

EXAMPLE 8 Find the relative maxima and the relative minima of the function
$$f(x) = x + \frac{1}{x}$$

Solution The derivative of f is
$$f'(x) = 1 - \frac{1}{x^2} = \frac{x^2 - 1}{x^2} = \frac{(x + 1)(x - 1)}{x^2}$$

Since f' is equal to zero at $x = -1$ and $x = 1$, these are critical numbers for the function f. Next, observe that f' is discontinuous at $x = 0$. However, because f is not defined at that number, $x = 0$ does not qualify as a critical number of f. Figure 22 shows the sign diagram for f'.

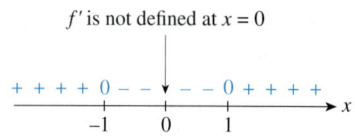

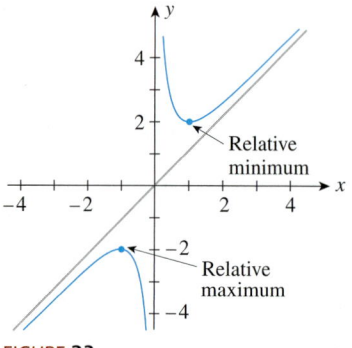

FIGURE 22
$x = 0$ is not a critical number because f is not defined at $x = 0$.

FIGURE 23
$f(x) = x + \dfrac{1}{x}$

Since $f'(x)$ changes sign from positive to negative as we move across $x = -1$ from left to right, the First Derivative Test implies that $f(-1) = -2$ is a relative maximum of the function f. Next, $f'(x)$ changes sign from negative to positive as we move across $x = 1$ from left to right, so $f(1) = 2$ is a relative minimum of the function f. The graph of f appears in Figure 23. Note that this function has a relative maximum that lies below its relative minimum. ■

Exploring with TECHNOLOGY

Refer to Example 8.

1. Use a graphing utility to plot the graphs of $f(x) = x + 1/x$ and its derivative function $f'(x) = 1 - 1/x^2$, using the viewing window $[-4, 4] \times [-8, 8]$.
2. By studying the graph of f', determine the critical numbers of f. Next, note the sign of $f'(x)$ immediately to the left and to the right of each critical number. What can you conclude about each critical number? Are your conclusions borne out by the graph of f?

 APPLIED EXAMPLE 9 Profit Functions The profit function of Acrosonic Company is given by

$$P(x) = -0.02x^2 + 300x - 200{,}000$$

dollars, where x is the number of Acrosonic model F loudspeaker systems produced. Find where the function P is increasing and where it is decreasing.

Solution The derivative P' of the function P is

$$P'(x) = -0.04x + 300 = -0.04(x - 7500)$$

Thus, $P'(x) = 0$ when $x = 7500$. Furthermore, $P'(x) > 0$ for x in the interval $(0, 7500)$, and $P'(x) < 0$ for x in the interval $(7500, \infty)$. This means that the profit function P is increasing on $(0, 7500)$ and decreasing on $(7500, \infty)$ (Figure 24). ■

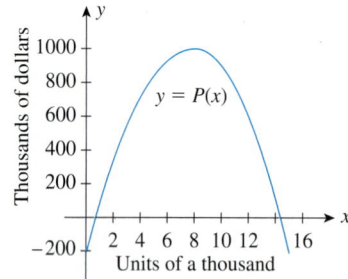

FIGURE 24
The profit function is increasing on $(0, 7500)$ and decreasing on $(7500, \infty)$.

 APPLIED EXAMPLE 10 Crime Rates The number of major crimes committed in the city of Bronxville from 2005 to 2012 is approximated by the function

$$N(t) = -0.1t^3 + 1.5t^2 + 100 \qquad (0 \leq t \leq 7)$$

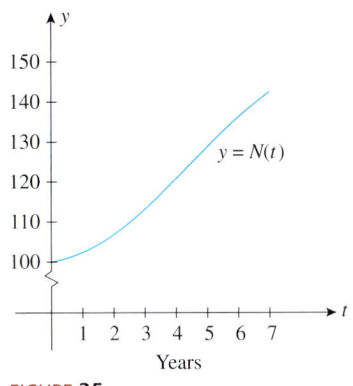

FIGURE 25
The number of crimes, $N(t)$, is increasing over the 7-year interval.

where $N(t)$ denotes the number of crimes committed in year t, with $t = 0$ corresponding to the beginning of 2005. Find where the function N is increasing and where it is decreasing.

Solution The derivative N' of the function N is

$$N'(t) = -0.3t^2 + 3t = -0.3t(t - 10)$$

Since $N'(t) > 0$ for t in the interval $(0, 7)$, the function N is increasing throughout that interval (Figure 25).

4.1 Self-Check Exercises

1. Find the intervals where the function

$$f(x) = \tfrac{2}{3}x^3 - x^2 - 12x + 3$$

is increasing and the intervals where it is decreasing.

2. Find the relative extrema of $f(x) = \dfrac{x^2}{1 - x^2}$.

Solutions to Self-Check Exercises 4.1 can be found on page 263.

4.1 Concept Questions

1. Explain each of the following:
   a. f is increasing on an interval I.
   b. f is decreasing on an interval I.

2. Describe a procedure for determining where a function is increasing and where it is decreasing.

3. Explain each term: (a) relative maximum and (b) relative minimum.

4. **a.** What is a critical number of a function f?
   b. Explain the role of critical numbers in determining the relative extrema of a function.

5. Describe the First Derivative Test, and describe a procedure for finding the relative extrema of a function.

4.1 Exercises

In Exercises 1–8, you are given the graph of a function f. Determine the intervals where f is increasing, constant, or decreasing.

1.
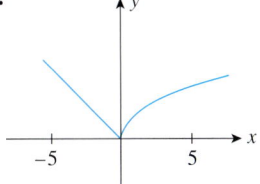

2.
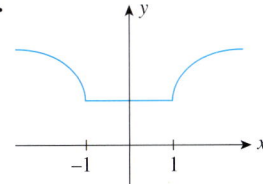

3.
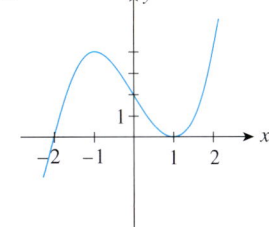

4.
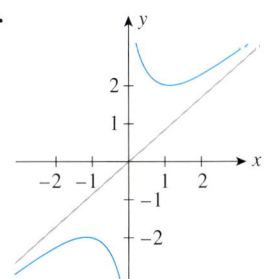

5.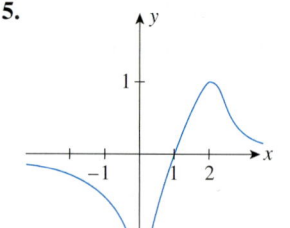
6.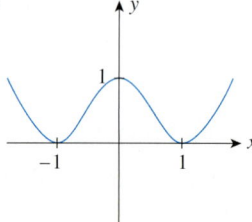
7.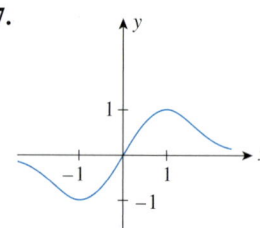
8.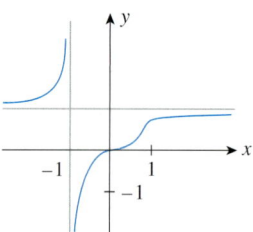

9. **THE BOSTON MARATHON** The graph of the function f shown in the accompanying figure gives the elevation of the part of the Boston Marathon course that includes the notorious Heartbreak Hill. Determine the intervals (stretches of the course) where the function f is increasing (the runner is laboring), where it is constant (the runner is taking a breather), and where it is decreasing (the runner is coasting).

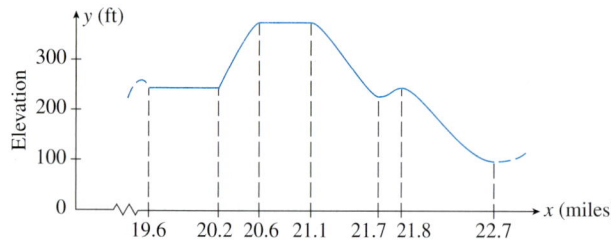

10. **AIRCRAFT STRUCTURAL INTEGRITY** Among the important factors in determining the structural integrity of an aircraft is its age. Advancing age makes planes more likely to crack. The graph of the function f, shown in the accompanying figure, is referred to as a "bathtub curve" in the airline industry. It gives the fleet damage rate (damage due to corrosion, accident, and metal fatigue) of a typical fleet of commercial aircraft as a function of the number of years of service.

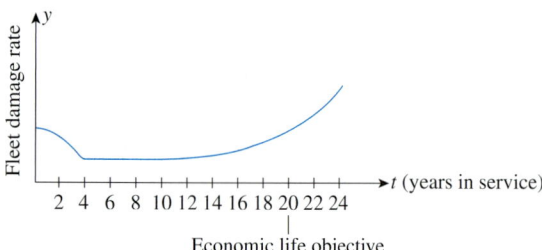

a. Determine the interval where f is decreasing. This corresponds to the time period when the fleet damage rate is dropping as problems are found and corrected during the initial "shakedown" period.

b. Determine the interval where f is constant. After the initial shakedown period, planes have few structural problems, and this is reflected by the fact that the function is constant on this interval.

c. Determine the interval where f is increasing. Beyond the time period mentioned in part (b), the function is increasing—reflecting an increase in structural defects due mainly to metal fatigue.

11. Refer to the following figure:

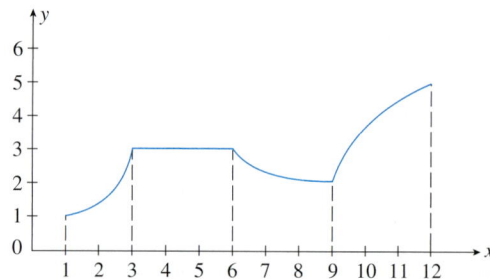

What is the sign of the following?
a. $f'(2)$
b. $f'(x)$ in the interval $(1, 3)$
c. $f'(4)$
d. $f'(x)$ in the interval $(3, 6)$
e. $f'(7)$
f. $f'(x)$ in the interval $(6, 9)$
g. $f'(x)$ in the interval $(9, 12)$

12. Refer to the following figure:

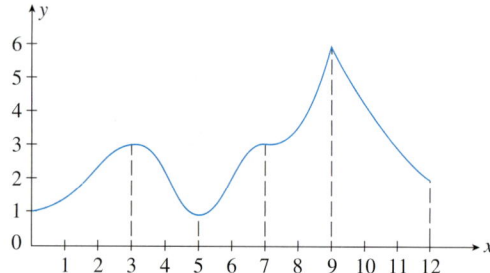

a. What are the critical numbers of f. Give reasons for your answers.
b. Draw the sign diagram for f'.
c. Find the relative extrema of f.

In Exercises 13–36, find the interval(s) where the function is increasing and the interval(s) where it is decreasing.

13. $f(x) = 3x + 5$
14. $f(x) = 4 - 5x$
15. $f(x) = x^2 - 3x$
16. $f(x) = 2x^2 + x + 1$
17. $g(x) = x - x^3$
18. $f(x) = x^3 - 3x^2$
19. $g(x) = x^3 + 3x^2 + 1$
20. $f(x) = x^3 - 3x + 4$
21. $f(x) = \frac{1}{3}x^3 - 3x^2 + 9x + 20$

22. $f(x) = \dfrac{2}{3}x^3 - 2x^2 - 6x - 2$

23. $h(x) = x^4 - 4x^3 + 10$ 24. $g(x) = x^4 - 2x^2 + 4$

25. $f(x) = \dfrac{1}{x-2}$ 26. $h(x) = \dfrac{1}{2x+3}$

27. $h(t) = \dfrac{t}{t-1}$ 28. $g(t) = \dfrac{2t}{t^2+1}$

29. $f(x) = x^{3/5}$ 30. $f(x) = x^{2/3} + 5$

31. $f(x) = \sqrt{x+1}$ 32. $f(x) = (x-5)^{2/3}$

33. $f(x) = \sqrt{16-x^2}$ 34. $g(x) = x\sqrt{x+1}$

35. $f(x) = \dfrac{1-x^2}{x}$ 36. $h(x) = \dfrac{x^2}{x-1}$

In Exercises 37–44, you are given the graph of a function f. Determine the relative maxima and relative minima, if any.

37. 38.

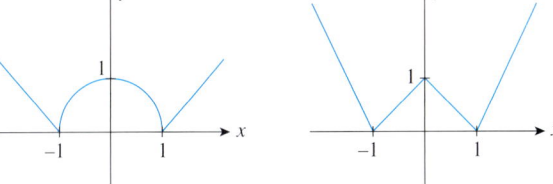

39. 40.

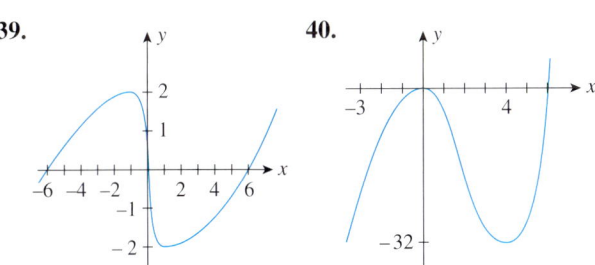

41. 42.

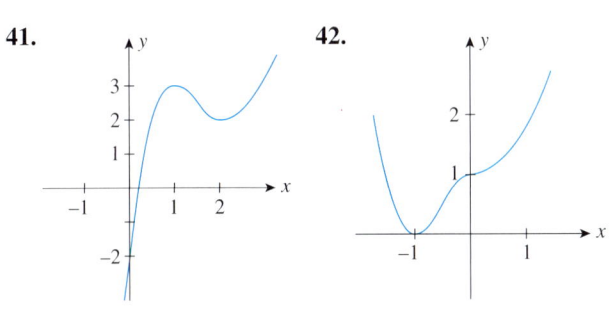

43.

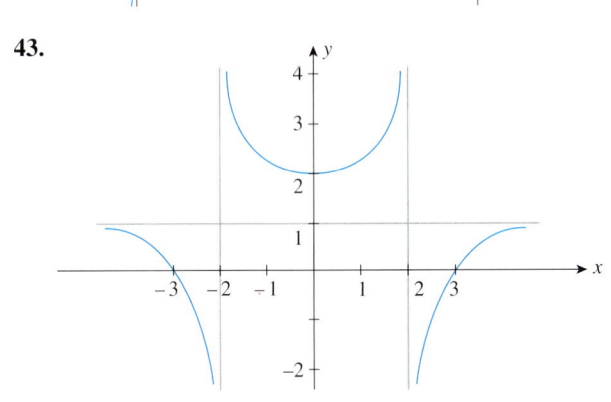

44.

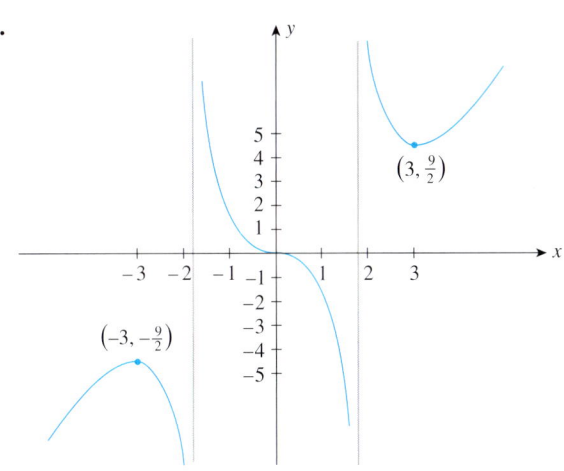

In Exercises 45–48, match the graph of the function with the graph of its derivative in (a)–(d).

(a) (b)

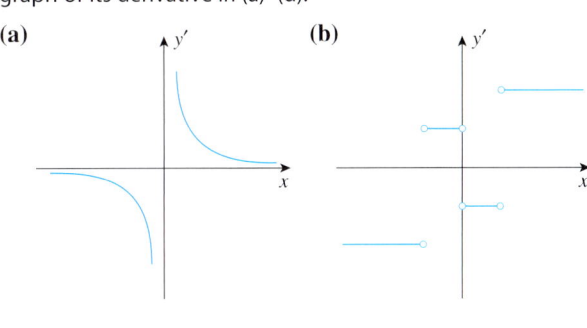

(c) (d)

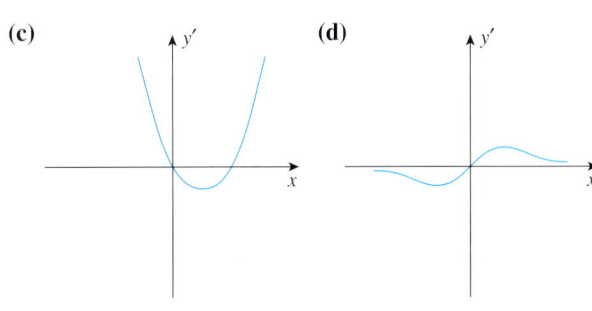

45. 46.

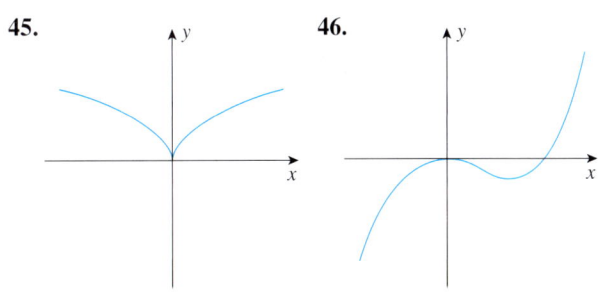

47. 48.

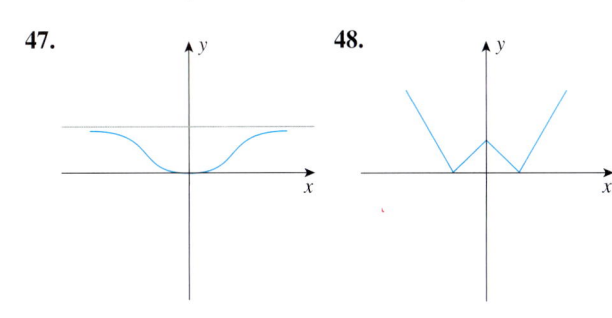

In Exercises 49–70, find the relative maxima and relative minima, if any, of each function.

49. $f(x) = x^2 - 4x$
50. $g(x) = x^2 + 3x + 8$
51. $h(t) = -t^2 + 6t + 6$
52. $f(x) = \frac{1}{2}x^2 - 2x + 4$
53. $f(x) = x^{5/3}$
54. $f(x) = x^{2/3} + 2$
55. $g(x) = x^3 - 3x^2 + 5$
56. $f(x) = x^3 - 3x + 6$
57. $f(x) = \frac{1}{2}x^4 - x^2$
58. $h(x) = \frac{1}{2}x^4 - 3x^2 + 4x - 8$
59. $F(x) = \frac{1}{3}x^3 - x^2 - 3x + 4$
60. $F(t) = 3t^5 - 20t^3 + 20$
61. $g(x) = x^4 - 4x^3 + 12$
62. $f(x) = 3x^4 - 2x^3 + 4$
63. $g(x) = \dfrac{x+1}{x}$
64. $h(x) = \dfrac{x}{x+1}$
65. $f(x) = x + \dfrac{9}{x} + 2$
66. $g(x) = 2x^2 + \dfrac{4000}{x} + 10$
67. $f(x) = \dfrac{x}{1+x^2}$
68. $g(x) = \dfrac{x}{x^2-1}$
69. $f(x) = (x-1)^{2/3}$
70. $g(x) = x\sqrt{x-4}$

71. A stone is thrown straight up from the roof of an 80-ft building. The distance (in feet) of the stone from the ground at any time t (in seconds) is given by

$$h(t) = -16t^2 + 64t + 80$$

When is the stone rising, and when is it falling? If the stone were to miss the building, when would it hit the ground? Sketch the graph of h.
Hint: The stone is on the ground when $h(t) = 0$.

72. **Profit Functions** The Mexican subsidiary of ThermoMaster manufactures an indoor–outdoor thermometer. Management estimates that the profit (in dollars) realizable by the company for the manufacture and sale of x units of thermometers each week is

$$P(x) = -0.001x^2 + 8x - 5000$$

Find the intervals where the profit function P is increasing and the intervals where P is decreasing.

73. **U.S. Cell Phone Subscribers** The number of U.S. cell phone subscribers (in millions) t years after 1989 is approximated by

$$N(t) = 0.63t^2 + 1.02t + 2.7 \qquad (0 \leq t \leq 20)$$

Show that N is an increasing function of t on the interval $(0, 20)$. What does your result tell you about the number of phone subscribers during the period under consideration?

74. **Growth of Managed Services** Almost half of companies let other firms manage some of their Web operations—a practice called Web hosting. Managed services—monitoring a customer's technology services—is the fastest-growing part of Web hosting. Managed services sales are approximated by the function

$$f(t) = 0.469t^2 + 0.758t + 0.44 \qquad (0 \leq t \leq 10)$$

where $f(t)$ is measured in billions of dollars and t is measured in years, with $t = 0$ corresponding to 1999.
a. Find the interval where f is increasing and the interval where f is decreasing.
b. What does your result tell you about sales in managed services from 1999 through 2009?
Source: International Data Corp.

75. **Flight of a Rocket** The height (in feet) attained by a rocket t sec into flight is given by the function

$$h(t) = -\frac{1}{3}t^3 + 16t^2 + 33t + 10 \qquad (t \geq 0)$$

When is the rocket rising, and when is it descending?

76. **Environment of Forests** Following the lead of the National Wildlife Federation, the Department of the Interior of a South American country began to record an index of environmental quality that measured progress and decline in the environmental quality of its forests. The index for the years 2000 through 2010 is approximated by the function

$$I(t) = \frac{1}{3}t^3 - \frac{5}{2}t^2 + 80 \qquad (0 \leq t \leq 10)$$

where $t = 0$ corresponds to 2000. Find the intervals where the function I is increasing and the intervals where it is decreasing. Interpret your results.

77. **Average Speed of a Highway Vehicle** The average speed of a vehicle on a stretch of Route 134 between 6 A.M. and 10 A.M. on a typical weekday is approximated by the function

$$f(t) = 20t - 40\sqrt{t} + 50 \qquad (0 \leq t \leq 4)$$

where $f(t)$ is measured in miles per hour and t is measured in hours, with $t = 0$ corresponding to 6 A.M. Find the interval where f is increasing and the interval where f is decreasing and interpret your results.

78. **Average Cost** The average cost (in dollars) incurred by Lincoln Records each week in pressing x compact discs is given by

$$\overline{C}(x) = -0.0001x + 2 + \dfrac{2000}{x} \qquad (0 < x \leq 6000)$$

Show that $\overline{C}(x)$ is always decreasing over the interval $(0, 6000)$.

79. Web Hosting Refer to Exercise 74. Sales in the Web-hosting industry are projected to grow in accordance with the function

$$f(t) = -0.05t^3 + 0.56t^2 + 5.47t + 7.5 \quad (0 \le t \le 10)$$

where $f(t)$ is measured in billions of dollars and t is measured in years, with $t = 0$ corresponding to 1999.
  a. Find the interval where f is increasing and the interval where f is decreasing.
    Hint: Use the quadratic formula.
  b. What does your result tell you about sales in the Web-hosting industry from 1999 through 2009?
Source: International Data Corp.

80. Medical School Applicants According to a study from the American Medical Association, the number of medical school applicants from academic year 1997–1998 ($t = 0$) through the academic year 2008–2009 is approximated by the function

$$N(t) = \begin{cases} 0.36t^2 - 3.10t + 41.2 & \text{if } 0 \le t < 9 \\ 42.46 & \text{if } 9 \le t \le 11 \end{cases}$$

Find the years when the number of medical school applicants was increasing, when it was decreasing, and when it was approximately constant.
Source: Journal of the American Medical Association.

81. Spending on Fiber-Optic Links U.S. telephone company spending on fiber-optic links to homes and businesses from 2001 to 2006 is approximated by

$$S(t) = -2.315t^3 + 34.325t^2 + 1.32t + 23 \quad (0 \le t \le 5)$$

billion dollars in year t, where t is measured in years with $t = 0$ corresponding to 2001. Show that $S'(t) > 0$ for all t in the interval $[0, 5]$. What conclusion can you draw from this result?
Hint: Use the quadratic formula.
Source: RHK Inc.

82. Projected Retirement Funds Based on data from the Central Provident Fund of a certain country (a government agency similar to the Social Security Administration), the estimated cash in the fund in 2005 is given by

$$A(t) = -96.6t^4 + 403.6t^3 + 660.9t^2 + 250 \quad (0 \le t \le 5)$$

where $A(t)$ is measured in billions of dollars and t is measured in decades, with $t = 0$ corresponding to 2005. Find the interval where A is increasing and the interval where A is decreasing, and interpret your results.
Hint: Use the quadratic formula.

83. Air Pollution According to the South Coast Air Quality Management District, the level of nitrogen dioxide, a brown gas that impairs breathing, present in the atmosphere on a certain May day in downtown Los Angeles is approximated by

$$A(t) = 0.03t^3(t - 7)^4 + 60.2 \quad (0 \le t \le 7)$$

where $A(t)$ is measured in pollutant standard index (PSI) and t is measured in hours, with $t = 0$ corresponding to 7 A.M. At what time of day is the air pollution increasing, and at what time is it decreasing?

84. Drug Concentration in the Blood The concentration (in milligrams per cubic centimeter) of a certain drug in a patient's body t hr after injection is given by

$$C(t) = \frac{t^2}{2t^3 + 1} \quad (0 \le t \le 4)$$

When is the concentration of the drug increasing, and when is it decreasing?

85. Small Car Market Share Owing in part to an aging population and the squeeze from carbon-cutting regulations, the percentage of small and lower-midsize vehicles is expected to increase in the near future. The function

$$f(t) = \frac{5.3\sqrt{t} - 300}{\sqrt{t} - 10} \quad (0 \le t \le 10)$$

gives the projected percentage of small and lower-midsize vehicles t years after 2005.
  a. What was the percentage of small and lower-midsize vehicles in 2005? What is the projected percentage of small and lower-midsize vehicles in 2015?
  b. Show that f is increasing on the interval $(0, 10)$, and interpret your results.
Source: J.D. Power Automotive.

86. Age of Drivers in Crash Fatalities The number of crash fatalities per 100,000 vehicle miles of travel in a certain year is approximated by the model

$$f(x) = \frac{15}{0.08333x^2 + 1.91667x + 1} \quad (0 \le x \le 11)$$

where x is the age of the driver in years, with $x = 0$ corresponding to age 16. Show that f is decreasing on $(0, 11)$, and interpret your result.
Source: National Highway Traffic Safety Administration.

87. Air Pollution The amount of nitrogen dioxide, a brown gas that impairs breathing, present in the atmosphere on a certain May day in the city of Long Beach is approximated by

$$A(t) = \frac{136}{1 + 0.25(t - 4.5)^2} + 28 \quad (0 \le t \le 11)$$

where $A(t)$ is measured in pollutant standard index (PSI) and t is measured in hours, with $t = 0$ corresponding to 7 A.M. Find the intervals where A is increasing and where A is decreasing, and interpret your results.
Source: Los Angeles Times.

88. Prison Overcrowding The 1980s saw a trend toward old-fashioned punitive deterrence as opposed to the more liberal penal policies and community-based corrections popular in the 1960s and early 1970s. As a result, prisons became more crowded, and the gap between the number of people in prison and the prison capacity widened. The number of prisoners (in thousands) in federal and state prisons is approximated by the function

$$N(t) = 3.5t^2 + 26.7t + 436.2 \quad (0 \leq t \leq 10)$$

where t is measured in years, with $t = 0$ corresponding to 1984. The number of inmates for which prisons were designed is given by

$$C(t) = 24.3t + 365 \quad (0 \leq t \leq 10)$$

where $C(t)$ is measured in thousands and t has the same meaning as before. Show that the gap between the number of prisoners and the number for which the prisons were designed widened throughout the decade from 1984 to 1994.

Hint: First, write a function G that gives the gap between the number of prisoners and the number for which the prisons were designed at any time t. Then show that $G'(t) > 0$ for all values of t in the interval $(0, 10)$.

Source: U.S. Department of Justice.

89. U.S. Nursing Shortage The demand for nurses between 2000 and 2015 is estimated to be

$$D(t) = 0.0007t^2 + 0.0265t + 2 \quad (0 \leq t \leq 15)$$

where $D(t)$ is measured in millions and $t = 0$ corresponds to the year 2000. The supply of nurses over the same time period is estimated to be

$$S(t) = -0.0014t^2 + 0.0326t + 1.9 \quad (0 \leq t \leq 15)$$

where $S(t)$ is also measured in millions.

a. Find an expression $G(t)$ giving the gap between the demand and supply of nurses over the period in question.
b. Find the interval where G is decreasing and where it is increasing. Interpret your result.
c. Find the relative extrema of G. Interpret your result.

Source: U.S. Department of Health and Human Services.

In Exercises 90–95, determine whether the statement is true or false. If it is true, explain why it is true. If it is false, give an example to show why it is false.

90. If f is decreasing on (a, b), then $f'(x) < 0$ for each x in (a, b).

91. If f and g are both increasing on (a, b), then $f + g$ is increasing on (a, b).

92. If f and g are both decreasing on (a, b), then $f - g$ is decreasing on (a, b).

93. If $f(x)$ and $g(x)$ are positive on (a, b) and both f and g are increasing on (a, b), then fg is increasing on (a, b).

94. If $f'(c) = 0$, then f has a relative maximum or a relative minimum at $x = c$.

95. If f has a relative minimum at $x = c$, then $f'(c) = 0$.

96. Using Theorem 1, verify that the linear function $f(x) = mx + b$ is (a) increasing everywhere if $m > 0$, (b) decreasing everywhere if $m < 0$, and (c) constant if $m = 0$.

97. Show that the function $f(x) = x^3 + x + 1$ has no relative extrema on $(-\infty, \infty)$.

98. Let $f(x) = -x^2 + ax + b$. Determine the constants a and b such that f has a relative maximum at $x = 2$ and the relative maximum value is 7.

99. Let $f(x) = ax^3 + 6x^2 + bx + 4$. Determine the constants a and b such that f has a relative minimum at $x = -1$ and a relative maximum at $x = 2$.

100. Let

$$f(x) = \begin{cases} -3x & \text{if } x < 0 \\ 2x + 4 & \text{if } x \geq 0 \end{cases}$$

a. Compute $f'(x)$, and show that it changes sign from negative to positive as we move across $x = 0$.
b. Show that f does not have a relative minimum at $x = 0$. Does this contradict the First Derivative Test? Explain your answer.

101. Let

$$f(x) = \begin{cases} -x^2 + 3 & \text{if } x \neq 0 \\ 2 & \text{if } x = 0 \end{cases}$$

a. Compute $f'(x)$, and show that it changes sign from positive to negative as we move across $x = 0$.
b. Show that f does not have a relative maximum at $x = 0$. Does this contradict the First Derivative Test? Explain your answer.

102. Let

$$f(x) = \begin{cases} \dfrac{1}{x^2} & \text{if } x > 0 \\ x^2 & \text{if } x \leq 0 \end{cases}$$

a. Compute $f'(x)$, and show that it does not change sign as we move across $x = 0$.
b. Show that f has a relative minimum at $x = 0$. Does this contradict the First Derivative Test? Explain your answer.

103. Show that the quadratic function

$$f(x) = ax^2 + bx + c \quad (a \neq 0)$$

has a relative extremum when $x = -b/2a$. Also, show that the relative extremum is a relative maximum if $a < 0$ and a relative minimum if $a > 0$.

104. Show that the cubic function

$$f(x) = ax^3 + bx^2 + cx + d \quad (a \neq 0)$$

has no relative extremum if and only if $b^2 - 3ac \leq 0$.

105. Refer to Example 6, page 124.
  a. Show that f is increasing on the interval $(0, 1)$.
  b. Show that $f(0) = -1$ and $f(1) = 1$, and use the result of part (a) together with the Intermediate Value Theorem to conclude that there is exactly one root of $f(x) = 0$ in $(0, 1)$.

106. Show that the function

$$f(x) = \frac{ax + b}{cx + d}$$

does not have a relative extremum if $ad - bc \neq 0$. What can you say about f if $ad - bc = 0$?

4.1 Solutions to Self-Check Exercises

1. The derivative of f is

$$f'(x) = 2x^2 - 2x - 12 = 2(x + 2)(x - 3)$$

and it is continuous everywhere. The zeros of $f'(x)$ are $x = -2$ and $x = 3$. The sign diagram of f' is shown in the accompanying figure. We conclude that f is increasing on the intervals $(-\infty, -2)$ and $(3, \infty)$ and decreasing on the interval $(-2, 3)$.

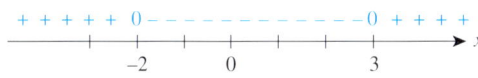

2. The derivative of f is

$$f'(x) = \frac{(1 - x^2)\dfrac{d}{dx}(x^2) - x^2 \dfrac{d}{dx}(1 - x^2)}{(1 - x^2)^2}$$

$$= \frac{(1 - x^2)(2x) - x^2(-2x)}{(1 - x^2)^2} = \frac{2x}{(1 - x^2)^2}$$

and it is continuous everywhere except at $x = \pm 1$. Since $f'(x)$ is equal to zero at $x = 0$, $x = 0$ is a critical number of f. Next, observe that $f'(x)$ is discontinuous at $x = \pm 1$, but since these numbers are not in the domain of f, they do not qualify as critical numbers of f. Finally, from the sign diagram of f' shown in the accompanying figure, we conclude that $f(0) = 0$ is a relative minimum of f.

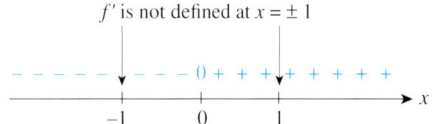

USING TECHNOLOGY

Using the First Derivative to Analyze a Function

A graphing utility is an effective tool for analyzing the properties of functions. This is especially true when we also bring into play the power of calculus, as the following examples show.

EXAMPLE 1 Let $f(x) = 2.4x^4 - 8.2x^3 + 2.7x^2 + 4x + 1$.

  a. Use a graphing utility to plot the graph of f.
  b. Find the intervals where f is increasing and the intervals where f is decreasing.
  c. Find the relative extrema of f.

Solution

  a. The graph of f in the viewing window $[-2, 4] \times [-10, 10]$ is shown in Figure T1.
  b. We compute

$$f'(x) = 9.6x^3 - 24.6x^2 + 5.4x + 4$$

and observe that f' is continuous everywhere, so the critical numbers of f occur at values of x where $f'(x) = 0$. To solve this last equation, observe that $f'(x)$ is

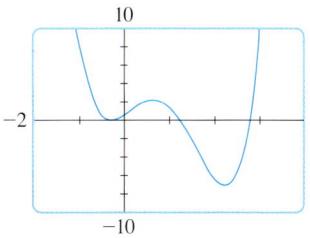

FIGURE T1
The graph of f in the viewing window $[-2, 4] \times [-10, 10]$

a *polynomial function* of degree 3. The easiest way to solve the polynomial equation

$$9.6x^3 - 24.6x^2 + 5.4x + 4 = 0$$

is to use the function on a graphing utility for solving polynomial equations. (Not all graphing utilities have this function.) You can also use TRACE and ZOOM, but this will not give the same accuracy without a much greater effort.

We find

$$x_1 \approx 2.22564943249 \qquad x_2 \approx 0.63272944121 \qquad x_3 \approx -0.295878873696$$

Referring to Figure T1, we conclude that f is decreasing on $(-\infty, -0.2959)$ and $(0.6327, 2.2256)$ (correct to four decimal places) and f is increasing on $(-0.2959, 0.6327)$ and $(2.2256, \infty)$.

c. Using the evaluation function of a graphing utility, we find the value of f at each of the critical numbers found in part (b). Upon referring to Figure T1 once again, we see that $f(x_3) \approx 0.2836$ and $f(x_1) \approx -8.2366$ are relative minimum values of f and $f(x_2) \approx 2.9194$ is a relative maximum value of f. ∎

Note The equation $f'(x) = 0$ in Example 1 is a polynomial equation, so it is easily solved by using the function for solving polynomial equations. We could also solve the equation using the function for finding the roots of equations, but that would require much more work. For equations that are *not* polynomial equations, however, our only choice is to use the function for finding the roots of equations. ∎

If the derivative of a function is difficult to compute or simplify and we do not require great precision in the solution, we can find the relative extrema of the function using a combination of ZOOM and TRACE. This technique, which does not require the use of the derivative of f, is illustrated in the following example.

EXAMPLE 2 Let $f(x) = x^{1/3}(x^2 + 1)^{-3/2} 3^{-x}$.

a. Use a graphing utility to plot the graph of f.*
b. Find the relative extrema of f.

Solution

a. The graph of f in the viewing window $[-4, 2] \times [-2, 1]$ is shown in Figure T2.
b. From the graph of f in Figure T2, we see that f has relative maxima when $x \approx -2$ and $x \approx 0.25$ and a relative minimum when $x \approx -0.75$. To obtain a better approximation of the first relative maximum, we zoom in with the cursor at approximately the point on the graph corresponding to $x \approx -2$. Then, using TRACE, we see that a relative maximum occurs when $x \approx -1.76$ with value $y \approx -1.01$. Similarly, we find the other relative maximum where $x \approx 0.20$ with value $y \approx 0.44$.

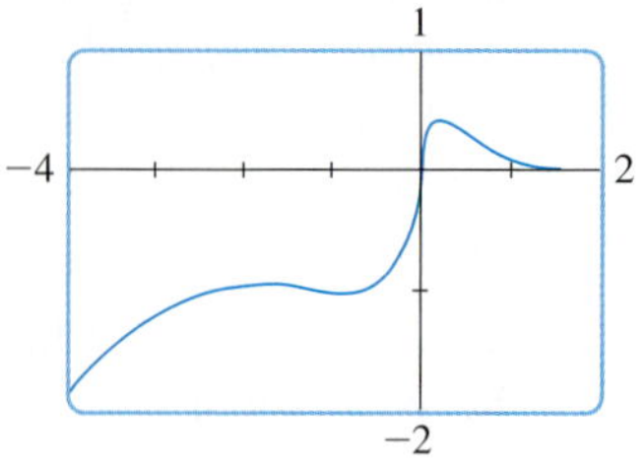

FIGURE T2
The graph of f in the viewing window $[-4, 2] \times [-2, 1]$

---

*Functions of the form $f(x) = 3^{-x}$ are called *exponential functions*, and we will study them in greater detail in Chapter 5.

Repeating the procedure, we find the relative minimum at $x \approx -0.86$ and $y \approx -1.07$.

You can also use the "minimum" and "maximum" functions of a graphing utility to find the relative extrema of the function. See the Web site for the procedure.

Finally, we comment that if you have access to a computer and software such as Derive, Maple, or Mathematica, then symbolic differentiation will yield the derivative $f'(x)$ of a differentiable function. This software will also solve the equation $f'(x) = 0$ with ease. Thus, the use of a computer will simplify even more greatly the analysis of functions.

TECHNOLOGY EXERCISES

In Exercises 1–4, find (a) the intervals where f is increasing and the intervals where f is decreasing and (b) the relative extrema of f. Express your answers accurate to four decimal places.

1. $f(x) = 3.4x^4 - 6.2x^3 + 1.8x^2 + 3x - 2$
2. $f(x) = 1.8x^4 - 9.1x^3 + 5x - 4$
3. $f(x) = 2x^5 - 5x^3 + 8x^2 - 3x + 2$
4. $f(x) = 3x^5 - 4x^2 + 3x - 1$

In Exercises 5–8, use the ZOOM and TRACE features to find (a) the intervals where f is increasing and the intervals where f is decreasing and (b) the relative extrema of f. Express your answers accurate to two decimal places.

5. $f(x) = (2x + 1)^{1/3}(x^2 + 1)^{-2/3}$
6. $f(x) = [x^2(x^3 - 1)]^{1/3} + \dfrac{1}{x}$
7. $f(x) = x - \sqrt{1 - x^2}$
8. $f(x) = \dfrac{\sqrt{x}(x^2 - 1)^2}{x - 2}$

9. **Manufacturing Capacity** Data show that the annual increase in manufacturing capacity between 1994 and 2000 is given by

$$f(t) = 0.009417t^3 - 0.426571t^2 + 2.74894t + 5.54$$
$$(0 \leq t \leq 6)$$

percent where t is measured in years, with $t = 0$ corresponding to the beginning of 1994.
   a. Plot the graph of f in the viewing window $[0, 6] \times [0, 11]$.
   b. Determine the interval where f is increasing and the interval where f is decreasing, and interpret your result.

   Source: Federal Reserve.

10. **Surgeries in Physicians' Offices** Driven by technological advances and financial pressures, the number of surgeries performed in physicians' offices nationwide has been increasing over the years. The function

$$f(t) = -0.00447t^3 + 0.09864t^2 + 0.05192t + 0.8$$
$$(0 \leq t \leq 15)$$

gives the number of surgeries (in millions) performed in physicians' offices in year t, with $t = 0$ corresponding to the beginning of 1986.
   a. Plot the graph of f in the viewing window $[0, 15] \times [0, 10]$.
   b. Prove that f is increasing on the interval $[0, 15]$.
      Hint: Show that f' is positive on the interval.

   Source: SMG Marketing Group.

11. **Air Pollution** The amount of nitrogen dioxide, a brown gas that impairs breathing, present in the atmosphere on a certain May day in the city of Long Beach, is approximated by

$$A(t) = \dfrac{136}{1 + 0.25(t - 4.5)^2} + 28 \qquad (0 \leq t \leq 11)$$

where $A(t)$ is measured in pollutant standard index (PSI) and t is measured in hours, with $t = 0$ corresponding to 7 A.M. When is the PSI increasing, and when is it decreasing? At what time is the PSI highest, and what is its value at that time?

12. **Modeling With Data** The following data gives the median age of women in the United States at first marriage from 1960 through 2000.

Decade	1960	1970	1980	1990	2000
Median Age	20.1	21.0	22.0	24.0	25.1

   a. Use **CubicReg** to find a third-degree polynomial regression model for these data. Let t be measured in decades, with $t = 0$ corresponding to 1960.
   b. Plot the graph of the polynomial function f found in part (a) using the viewing window $[0, 4] \times [18, 28]$.
   c. Where is f increasing? What does this tell us?
   d. Verify the result of part (c) analytically.

   Source: The National Marriage Project, Rutgers University.

4.2 Applications of the Second Derivative

Determining the Intervals of Concavity

Consider the graphs shown in Figure 26, which give the estimated population of the world and of the United States through the year 2000. Both graphs are rising, indicating that both the U.S. population and the world population continued to increase through the year 2000. But observe that the graph in Figure 26a opens upward, whereas the graph in Figure 26b opens downward. What is the significance of this? To answer this question, let's look at the slopes of the tangent lines to various points on each graph (Figure 27).

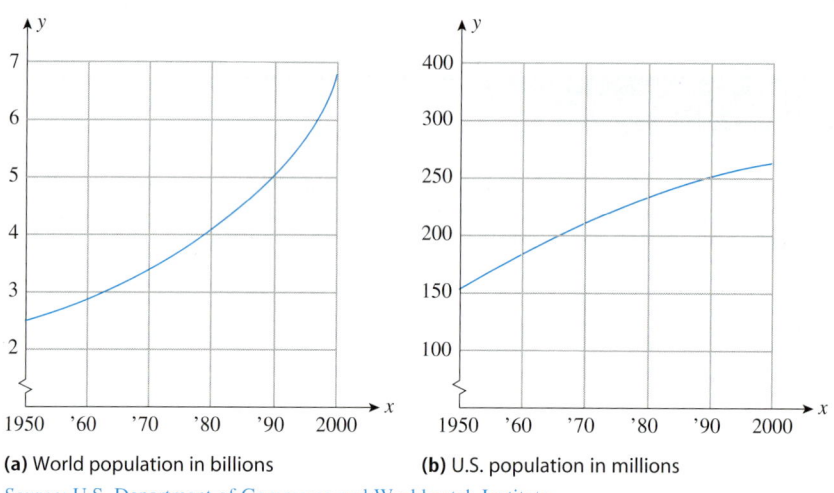

(a) World population in billions

(b) U.S. population in millions

Source: U.S. Department of Commerce and Worldwatch Institute.

FIGURE 26

In Figure 27a, we see that the slopes of the tangent lines to the graph are increasing as we move from left to right. Since the slope of the tangent line to the graph at a point on the graph measures the rate of change of the function at that point, we conclude that the world population not only was increasing through the year 2000 but also was increasing at an *increasing* pace. A similar analysis of Figure 27b reveals that the U.S. population was increasing, but at a *decreasing* pace.

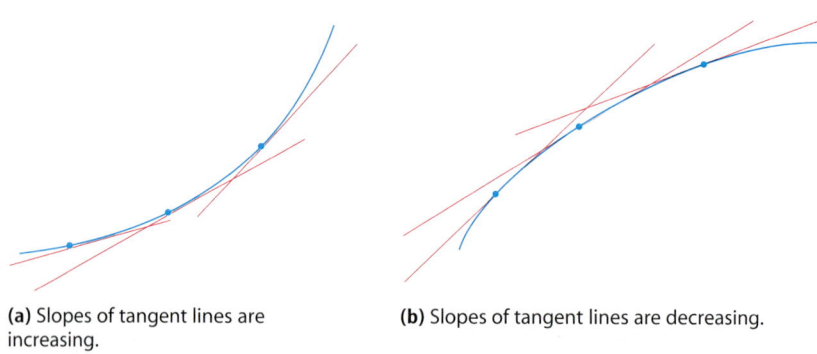

(a) Slopes of tangent lines are increasing.

(b) Slopes of tangent lines are decreasing.

FIGURE 27

The shape of a curve can be described by using the notion of concavity.

> **Concavity of a Function f**
>
> Let the function f be differentiable on an interval (a, b). Then,
>
> 1. The graph of f is **concave upward** on (a, b) if f' is increasing on (a, b).
> 2. The graph of f is **concave downward** on (a, b) if f' is decreasing on (a, b).

4.2 APPLICATIONS OF THE SECOND DERIVATIVE

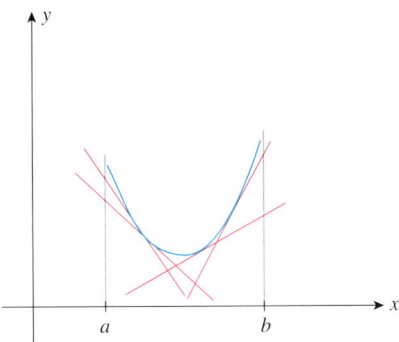
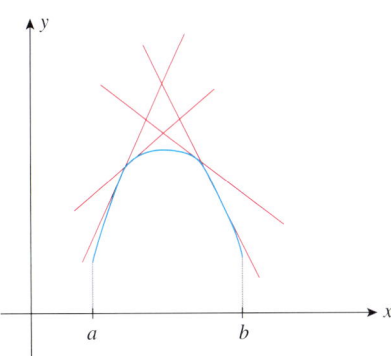

FIGURE 28

(a) The graph of f is concave upward on (a, b).

(b) The graph of f is concave downward on (a, b).

Geometrically, a curve is concave upward if it lies above its tangent lines (Figure 28a). Similarly, a curve is concave downward if it lies below its tangent lines (Figure 28b).

We also say that the graph of f is *concave upward at a number c* if there exists an interval (a, b) containing c on which the graph of f is concave upward. Similarly, we say that the graph of f is *concave downward at a number c* if there exists an interval (a, b) containing c on which the graph of f is concave downward.

If a function f has a second derivative f'', we can use f'' to determine the intervals of concavity of the graph of the function. Recall that $f''(x)$ measures the rate of change of the slope $f'(x)$ of the tangent line to the graph of f at the point $(x, f(x))$. Thus, if $f''(x) > 0$ on an interval (a, b), then the slopes of the tangent lines to the graph of f are increasing on (a, b), so the graph of f is concave upward on (a, b). Similarly, if $f''(x) < 0$ on (a, b), then the graph of f is concave downward on (a, b). These observations suggest the following theorem.

THEOREM 2
a. If $f''(x) > 0$ for every value of x in (a, b), then the graph of f is concave upward on (a, b).

b. If $f''(x) < 0$ for every value of x in (a, b), then the graph of f is concave downward on (a, b).

The following procedure, based on the conclusions of Theorem 2, may be used to determine the intervals of concavity of the graph of a function.

Determining the Intervals of Concavity of the Graph of f
1. Determine the values of x for which f'' is zero or where f'' is not defined, and identify the open intervals determined by these numbers.
2. Determine the sign of f'' in each interval found in Step 1. To do this, compute $f''(c)$, where c is any conveniently chosen test number in the interval.
   a. If $f''(c) > 0$, then the graph of f is concave upward on that interval.
   b. If $f''(c) < 0$, then the graph of f is concave downward on that interval.

EXAMPLE 1 Determine where the graph of the function $f(x) = x^3 - 3x^2 - 24x + 32$ is concave upward and where it is concave downward.

Solution Here,

$$f'(x) = 3x^2 - 6x - 24$$
$$f''(x) = 6x - 6 = 6(x - 1)$$

and f'' is defined everywhere. Setting $f''(x) = 0$ gives $x = 1$. The sign diagram of f'' appears in Figure 29. We conclude that the graph of f is concave downward on the interval $(-\infty, 1)$ and is concave upward on the interval $(1, \infty)$. Figure 30 shows the graph of f.

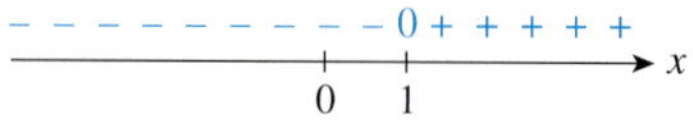

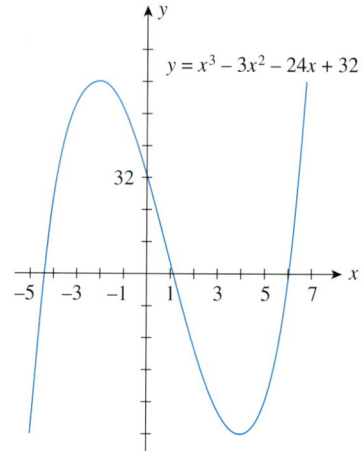

FIGURE 29
Sign diagram for f''

FIGURE 30
The graph of f is concave downward on $(-\infty, 1)$ and concave upward on $(1, \infty)$.

Exploring with TECHNOLOGY

Refer to Example 1.

1. Use a graphing utility to plot the graph of $f(x) = x^3 - 3x^2 - 24x + 32$ and its second derivative $f''(x) = 6x - 6$ using the viewing window $[-10, 10] \times [-80, 90]$.

2. By studying the graph of f'', determine the intervals where $f''(x) > 0$ and the intervals where $f''(x) < 0$. Next, look at the graph of f, and determine the intervals where the graph of f is concave upward and the intervals where it is concave downward. Are these observations what you might have expected?

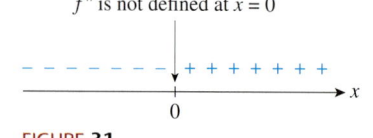

f'' is not defined at $x = 0$

FIGURE 31
The sign diagram for f''

EXAMPLE 2 Determine the intervals where the graph of the function $f(x) = x + \dfrac{1}{x}$ is concave upward and where it is concave downward.

Solution We have

$$f'(x) = 1 - \frac{1}{x^2}$$
$$f''(x) = \frac{2}{x^3}$$

We deduce from the sign diagram for f'' (Figure 31) that the graph of the function f is concave downward on the interval $(-\infty, 0)$ and concave upward on the interval $(0, \infty)$. The graph of f is sketched in Figure 32.

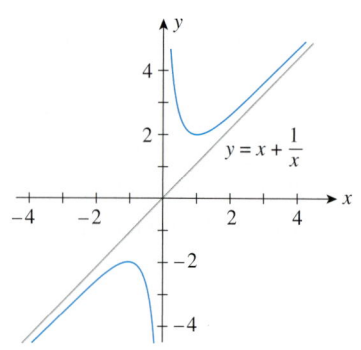

FIGURE 32
The graph of f is concave downward on $(-\infty, 0)$ and concave upward on $(0, \infty)$.

Inflection Points

Figure 33 shows the total sales S of a manufacturer of automobile air conditioners versus the amount of money x that the company spends on advertising its product.

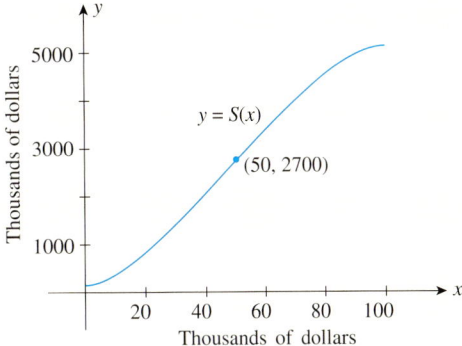

FIGURE 33
The graph of S has a point of inflection at (50, 2700).

Notice that the graph of the continuous function $y = S(x)$ changes concavity—from upward to downward—at the point (50, 2700). This point is called an *inflection point* of S. To understand the significance of this inflection point, observe that the total sales increase rather slowly at first, but as more money is spent on advertising, the total sales increase rapidly. This rapid increase reflects the effectiveness of the company's ads. However, a point is soon reached after which any additional advertising expenditure results in increased sales but at a slower rate of increase. This point, commonly known as the *point of diminishing returns,* is the point of inflection of the function S. We will return to this example later.

Let's now state formally the definition of an inflection point.

> ### Inflection Point
> A point on the graph of a continuous function f where the tangent line exists and where the concavity changes is called a **point of inflection** or an **inflection point.**

Observe that the graph of a function crosses its tangent line at a point of inflection (Figure 34).

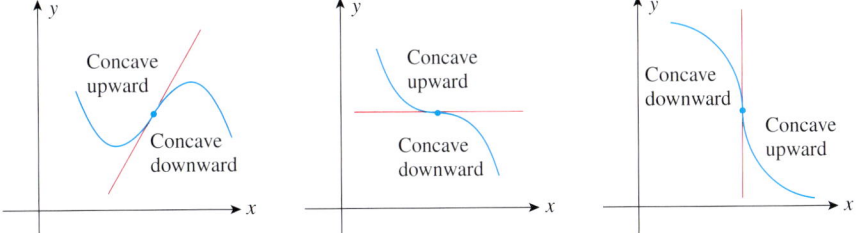

FIGURE 34
At each point of inflection, the graph of a function crosses its tangent line.

The following procedure may be used to find inflection points:

> ### Finding Inflection Points
> 1. Compute $f''(x)$.
> 2. Determine the numbers in the domain of f for which $f''(x) = 0$ or $f''(x)$ does not exist.
> 3. Determine the sign of $f''(x)$ to the left and right of each number c found in Step 2. If there is a change in the sign of $f''(x)$ as we move across $x = c$, then $(c, f(c))$ is an inflection point of f.

⚠ The numbers determined in Step 2 are only *candidates* for the inflection points of f. For example, you can easily verify that $f''(0) = 0$ if $f(x) = x^4$, but a sketch of the graph of f will show that $(0, 0)$ is *not* an inflection point.

EXAMPLE 3 Find the point of inflection of the function $f(x) = x^3$.

Solution

$$f'(x) = 3x^2$$
$$f''(x) = 6x$$

Observe that f'' is continuous everywhere and is zero if $x = 0$. The sign diagram of f'' is shown in Figure 35. From this diagram, we see that $f''(x)$ changes sign as we move across $x = 0$. Thus, the point $(0, 0)$ is an inflection point (Figure 36).

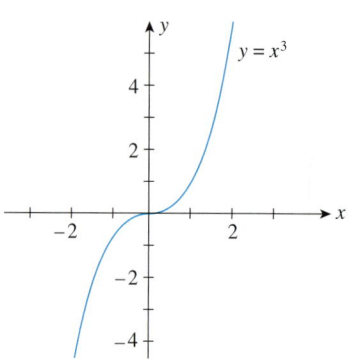

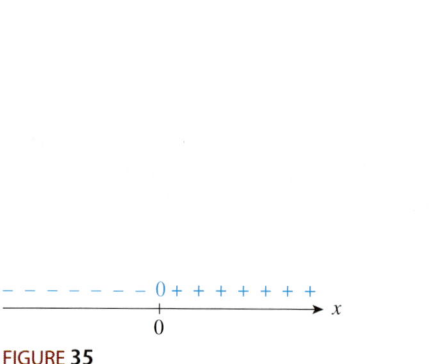

FIGURE 35
Sign diagram for f''

FIGURE 36
f has an inflection point at $(0, 0)$.

EXAMPLE 4 Determine the intervals where the graph of the function $f(x) = (x - 1)^{5/3}$ is concave upward and where it is concave downward, and find the inflection points of f.

Solution The first derivative of f is

$$f'(x) = \frac{5}{3}(x - 1)^{2/3}$$

and the second derivative of f is

$$f''(x) = \frac{10}{9}(x - 1)^{-1/3} = \frac{10}{9(x - 1)^{1/3}}$$

We see that f'' is not defined at $x = 1$. Furthermore, $f''(x)$ is not equal to zero anywhere. The sign diagram of f'' is shown in Figure 37. From the sign diagram, we see that the graph of f is concave downward on $(-\infty, 1)$ and concave upward on $(1, \infty)$. Next, since $x = 1$ does lie in the domain of f, our computations also reveal that the point $(1, 0)$ is an inflection point (Figure 38).

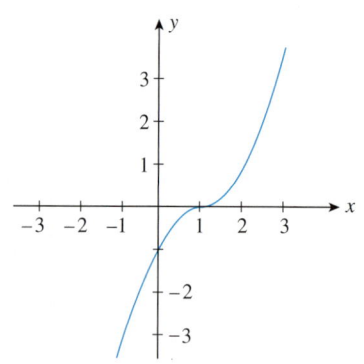

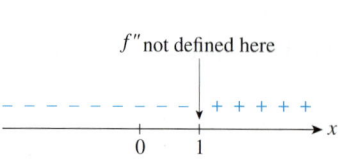

FIGURE 37
Sign diagram for f''

FIGURE 38
f has an inflection point at $(1, 0)$.

EXAMPLE 5 Determine the intervals where the graph of the function

$$f(x) = \frac{1}{x^2 + 1}$$

is concave upward and where it is concave downward, and find the inflection points of f.

Solution The first derivative of f is

$$f'(x) = \frac{d}{dx}(x^2 + 1)^{-1} = -2x(x^2 + 1)^{-2} \quad \text{Rewrite the original function, and use the General Power Rule.}$$

$$= -\frac{2x}{(x^2 + 1)^2}$$

Next, using the Quotient Rule, we find

$$f''(x) = \frac{(x^2 + 1)^2(-2) + (2x)2(x^2 + 1)(2x)}{(x^2 + 1)^4} \quad (x^2) \text{ See page 16.}$$

$$= \frac{(x^2 + 1)[-2(x^2 + 1) + 8x^2]}{(x^2 + 1)^4} = \frac{(x^2 + 1)(6x^2 - 2)}{(x^2 + 1)^4}$$

$$= \frac{2(3x^2 - 1)}{(x^2 + 1)^3} \quad \text{Cancel the common factors.}$$

Observe that f'' is continuous everywhere and is zero if

$$3x^2 - 1 = 0$$

$$x^2 = \frac{1}{3}$$

or $x = \pm\sqrt{3}/3$. The sign diagram for f'' is shown in Figure 39. From the sign diagram for f'', we see that the graph of f is concave upward on $(-\infty, -\sqrt{3}/3)$ and $(\sqrt{3}/3, \infty)$ and is concave downward on $(-\sqrt{3}/3, \sqrt{3}/3)$. Also, observe that $f''(x)$ changes sign as we move across the numbers $x = -\sqrt{3}/3$ and $x = \sqrt{3}/3$. Since

$$f\left(-\frac{\sqrt{3}}{3}\right) = \frac{1}{\frac{1}{3} + 1} = \frac{3}{4} \quad \text{and} \quad f\left(\frac{\sqrt{3}}{3}\right) = \frac{3}{4}$$

we see that the points $(-\sqrt{3}/3, 3/4)$ and $(\sqrt{3}/3, 3/4)$ are inflection points. The graph of f is shown in Figure 40.

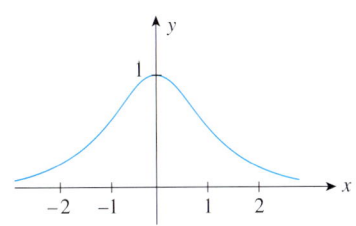

FIGURE 40

The graph of $f(x) = \dfrac{1}{x^2 + 1}$ is concave upward on $(-\infty, -\sqrt{3}/3)$ and $(\sqrt{3}/3, \infty)$ and is concave downward on $(-\sqrt{3}/3, \sqrt{3}/3)$.

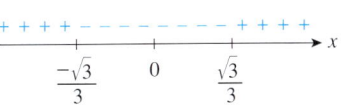

FIGURE 39
Sign diagram for f''

Explore & Discuss

1. Suppose $(c, f(c))$ is an inflection point of f. Can you conclude that f has no relative extremum at $x = c$? Explain your answer.

2. True or false: A polynomial function of degree 3 has exactly one inflection point.
   Hint: Study the function $f(x) = ax^3 + bx^2 + cx + d$ $(a \neq 0)$.

The next example uses an interpretation of the first and second derivatives to help us sketch the graph of a function.

EXAMPLE 6 Sketch the graph of a function having the following properties:

$$f(-1) = 4$$
$$f(0) = 2$$
$$f(1) = 0$$
$$f'(-1) = 0$$
$$f'(1) = 0$$
$$f'(x) > 0 \quad \text{on } (-\infty, -1) \text{ and } (1, \infty)$$
$$f'(x) < 0 \quad \text{on } (-1, 1)$$
$$f''(x) < 0 \quad \text{on } (-\infty, 0)$$
$$f''(x) > 0 \quad \text{on } (0, \infty)$$

Solution First, we plot the points $(-1, 4)$, $(0, 2)$, and $(1, 0)$ that lie on the graph of f. Since $f'(-1) = 0$ and $f'(1) = 0$, the tangent lines at the points $(-1, 4)$ and $(1, 0)$ are horizontal. Since $f'(x) > 0$ on $(-\infty, -1)$ and $f'(x) < 0$ on $(-1, 1)$, we see that f has a relative maximum at the point $(-1, 4)$. Also, $f'(x) < 0$ on $(-1, 1)$ and $f'(x) > 0$ on $(1, \infty)$ implies that f has a relative minimum at the point $(1, 0)$ (Figure 41a).

Since $f''(x) < 0$ on $(-\infty, 0)$ and $f''(x) > 0$ on $(0, \infty)$, we see that the point $(0, 2)$ is an inflection point. Finally, we complete the graph making use of the fact that f is increasing on $(-\infty, -1)$ and $(1, \infty)$, where it is given that $f'(x) > 0$, and f is decreasing on $(-1, 1)$, where $f'(x) < 0$. Also, make sure that the graph of f is concave downward on $(-\infty, 0)$ and concave upward on $(0, \infty)$ (Figure 41b).

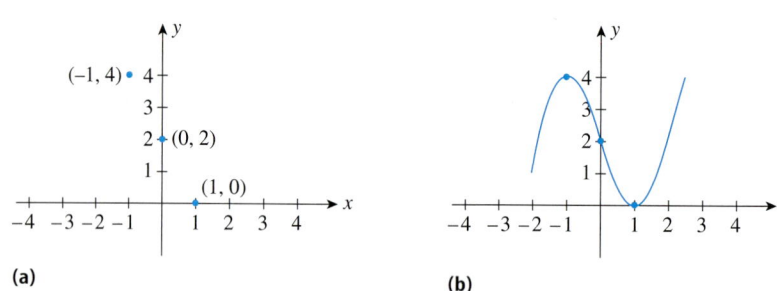

FIGURE 41   (a)   (b)

Examples 7 and 8 illustrate familiar interpretations of the significance of the inflection point of a function.

APPLIED EXAMPLE 7 Effect of Advertising on Sales The total sales S (in thousands of dollars) of Arctic Air Corporation, a manufacturer of automobile air conditioners, is related to the amount of money x (in thousands of dollars) the company spends on advertising its products by the formula

$$S(x) = -0.01x^3 + 1.5x^2 + 200 \quad (0 \leq x \leq 100)$$

Find the inflection point of the function S.

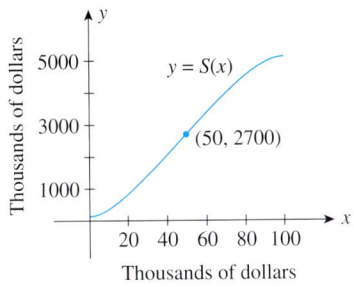

FIGURE 42
The graph of S has a point of inflection at (50, 2700).

Solution The first two derivatives of S are given by

$$S'(x) = -0.03x^2 + 3x$$
$$S''(x) = -0.06x + 3$$

Setting $S''(x) = 0$ gives $x = 50$. So $(50, S(50))$ is the only candidate for an inflection point of S. Moreover, since

$$S''(x) > 0 \quad \text{for } x < 50$$
$$S''(x) < 0 \quad \text{for } x > 50$$

the point (50, 2700) is an inflection point of the function S. The graph of S appears in Figure 42. Notice that this is the same graph as in Figure 33.

APPLIED EXAMPLE 8 Consumer Price Index An economy's consumer price index (CPI) is described by the function

$$I(t) = -0.2t^3 + 3t^2 + 100 \quad (0 \le t \le 10)$$

where $t = 0$ corresponds to the beginning of 2003. Find the point of inflection of the function I, and discuss its significance.

Solution The first two derivatives of I are given by

$$I'(t) = -0.6t^2 + 6t$$
$$I''(t) = -1.2t + 6 = -1.2(t - 5)$$

Setting $I''(t) = 0$ gives $t = 5$. So $(5, I(5))$ is the only candidate for an inflection point of I. Next, we observe that

$$I''(t) > 0 \quad \text{for } t < 5$$
$$I''(t) < 0 \quad \text{for } t > 5$$

so the point (5, 150) is an inflection point of I. The graph of I is sketched in Figure 43.

Since the second derivative of I measures the *rate of change* of the rate of change of the CPI, our computations reveal that the rate of inflation had in fact peaked at $t = 5$. Thus, relief actually began at the beginning of 2008.

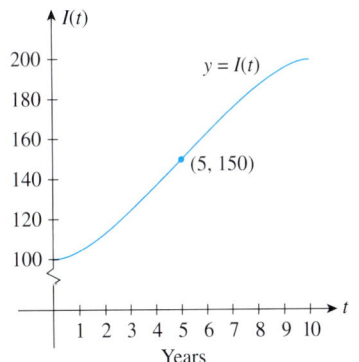

FIGURE 43
The graph of I has a point of inflection at (5, 150).

The Second Derivative Test

We now show how the second derivative f'' of a function f can be used to help us determine whether a critical number of f gives rise to a relative extremum of f. Figure 44a shows the graph of a function that has a relative maximum at $x = c$.

Observe that the graph of f is concave downward at that number. Similarly, Figure 44b shows that at a relative minimum of f, the graph is concave upward. But from our previous work, we know that the graph of f is concave downward at $x = c$ if $f''(c) < 0$ and the graph of f is concave upward at $x = c$ if $f''(c) > 0$. These observations suggest the following alternative procedure for determining whether a critical number of f gives rise to a relative extremum of f. This result is called the **Second Derivative Test** and is applicable when f'' exists.

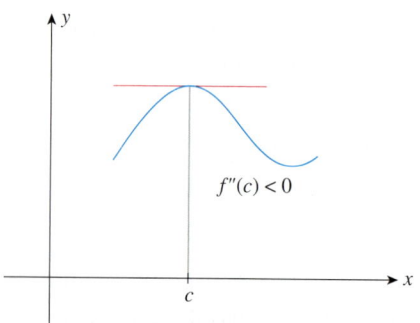

FIGURE 44

(a) f has a relative maximum at $x = c$.

(b) f has a relative minimum at $x = c$.

The Second Derivative Test
1. Compute $f'(x)$ and $f''(x)$.
2. Find all the critical numbers of f at which $f'(x) = 0$.
3. Compute $f''(c)$ for each such critical number c.
   a. If $f''(c) < 0$, then f has a relative maximum at c.
   b. If $f''(c) > 0$, then f has a relative minimum at c.
   c. If $f''(c) = 0$ or $f''(c)$ does not exist, the test fails; that is, it is inconclusive.

Note The Second Derivative Test does not yield a conclusion if $f''(c) = 0$ or if $f''(c)$ does not exist. In other words, $x = c$ may or may not give rise to a relative extremum (see Exercise 110, page 283). In such cases, you should revert to the First Derivative Test.

EXAMPLE 9 Determine the relative extrema of the function
$$f(x) = x^3 - 3x^2 - 24x + 32$$
using the Second Derivative Test. (See Example 7, Section 4.1.)

Solution We have
$$f'(x) = 3x^2 - 6x - 24 = 3(x + 2)(x - 4)$$
so $f'(x) = 0$ when $x = -2$ and $x = 4$, the critical numbers of f, as in Example 7, Section 4.1. Next, we compute
$$f''(x) = 6x - 6 = 6(x - 1)$$
Since
$$f''(-2) = 6(-2 - 1) = -18 < 0$$

the Second Derivative Test implies that $f(-2) = 60$ is a relative maximum of f. Also,
$$f''(4) = 6(4 - 1) = 18 > 0$$
and the Second Derivative Test implies that $f(4) = -48$ is a relative minimum of f, which confirms the results obtained earlier.

Explore & Discuss

Suppose a function f has the following properties:

1. $f''(x) > 0$ for all x in an interval (a, b).
2. There is a number c between a and b such that $f'(c) = 0$.

What special property can you ascribe to the point $(c, f(c))$? Answer the question if Property 1 is replaced by the property that $f''(x) < 0$ for all x in (a, b).

Comparing the First and Second Derivative Tests

Notice that both the First Derivative Test and the Second Derivative Test are used to classify the critical numbers of f. What are the pros and cons of the two tests? Since the Second Derivative Test is applicable only when f'' exists, it is less versatile than the First Derivative Test. For example, it cannot be used to locate the relative minimum $f(0) = 0$ of the function $f(x) = x^{2/3}$.

Furthermore, the Second Derivative Test is inconclusive when f'' is equal to zero at a critical number of f, whereas the First Derivative Test always yields a conclusion; that is, it tells us if f has a relative maximum, relative minimum, or neither. The Second Derivative Test is also inconvenient to use when f'' is difficult to compute. On the plus side, if f'' is computed easily, then we use the Second Derivative Test, since it involves just the evaluation of f'' at the critical number(s) of f. Also, the conclusions of the Second Derivative Test are important in theoretical work.

We close this section by summarizing the different roles played by the first derivative f' and the second derivative f'' of a function f in determining the properties of the graph of f. The first derivative f' tells us where f is increasing and where f is decreasing, whereas the second derivative f'' tells us where the graph of f is concave upward and where it is concave downward. These different properties of f are reflected by the signs of f' and f'' in the interval of interest. The following table shows the general characteristics of the function f for various possible combinations of the signs of f' and f'' in an interval (a, b).

Signs of f' and f''	Properties of f	General Shape of the Graph of f
$f'(x) > 0$ $f''(x) > 0$	f is increasing. The graph of f is concave upward.	
$f'(x) > 0$ $f''(x) < 0$	f is increasing. The graph of f is concave downward.	
$f'(x) < 0$ $f''(x) > 0$	f is decreasing. The graph of f is concave upward.	
$f'(x) < 0$ $f''(x) < 0$	f is decreasing. The graph of f is concave downward.	

4.2 Self-Check Exercises

1. Determine where the graph of the function $f(x) = 4x^3 - 3x^2 + 6$ is concave upward and where it is concave downward.

2. Using the Second Derivative Test, if applicable, find the relative extrema of the function $f(x) = 2x^3 - \frac{1}{2}x^2 - 12x - 10$.

3. A certain country's gross domestic product (GDP) (in millions of dollars) in year t is described by the function

$$G(t) = -2t^3 + 45t^2 + 20t + 6000 \quad (0 \le t \le 11)$$

where $t = 0$ corresponds to the beginning of 2002. Find the inflection point of the function G, and discuss its significance.

Solutions to Self-Check Exercises 4.2 can be found on page 283.

4.2 Concept Questions

1. Explain what it means for the graph of a function f to be (a) concave upward and (b) concave downward on an open interval I. Given that f has a second derivative on I (except at isolated numbers), how do you determine where the graph of f is concave upward and where it is concave downward?

2. What is an inflection point of a function f? How do you find the inflection point(s) of a function f whose rule is given?

3. State the Second Derivative Test. What are the pros and cons of using the First Derivative Test and the Second Derivative Test?

4.2 Exercises

In Exercises 1–8, you are given the graph of a function f. Determine the intervals where the graph of f is concave upward and where it is concave downward. Also, find all inflection points of f, if any.

1.
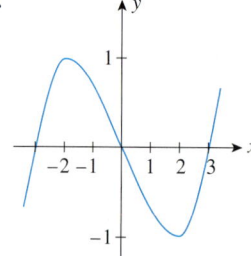

2.
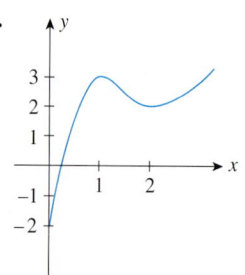

3.
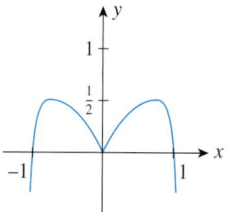

4.
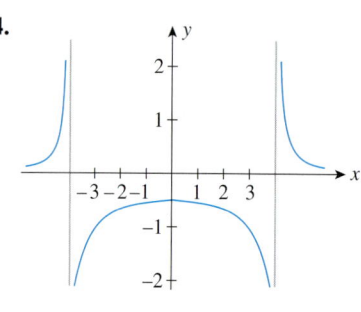

5.
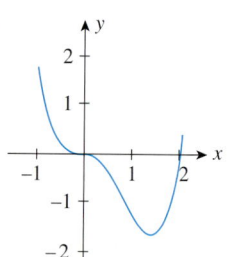

6.
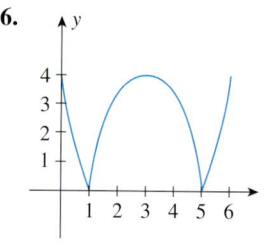

7.
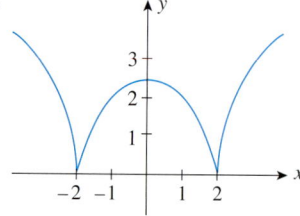

8.
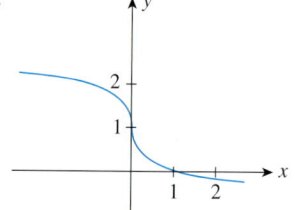

9. Refer to the graph of f shown in the following figure:

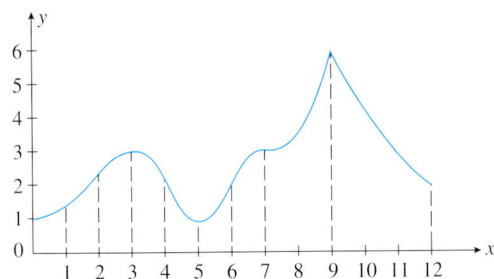

a. Find the intervals where f is concave upward and the intervals where f is concave downward.
b. Find the inflection points of f.

10. Refer to the figure for Exercise 9.
a. Explain how the Second Derivative Test can be used to show that the critical number 3 gives rise to a relative maximum of f and the critical number 5 gives rise to a relative minimum of f.
b. Explain why the Second Derivative Test cannot be used to show that the critical number 7 does not give rise to a relative extremum of f nor can it be used to show that the critical number 9 gives rise to a relative maximum of f.

In Exercises 11–14, determine which graph—(a), (b), or (c)—is the graph of the function f with the specified properties.

11. $f(2) = 1, f'(2) > 0$, and $f''(2) < 0$
(a)
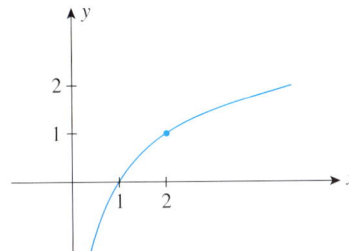

(b)
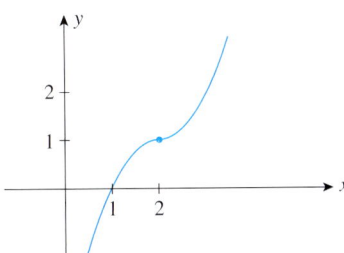

(c)
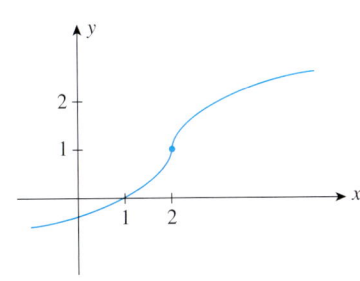

12. $f(1) = 2, f'(x) > 0$ on $(-\infty, 1)$ and $(1, \infty)$, and $f''(1) = 0$
(a)
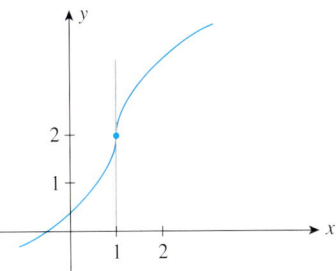

(b)
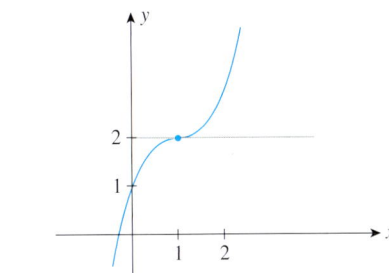

(c)
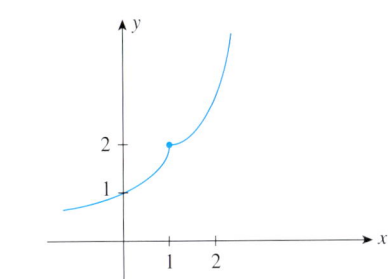

13. $f'(0)$ is undefined, f is decreasing on $(-\infty, 0)$, f is concave downward on $(0, 3)$, and f has an inflection point at $x = 3$.
(a)
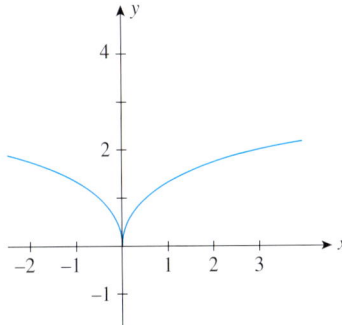

(b)
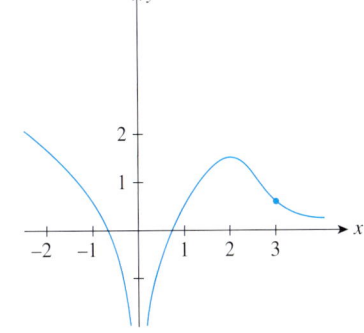

(c)

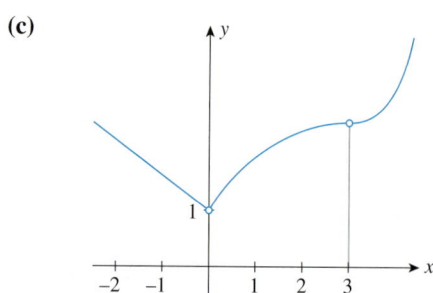

14. f is decreasing on $(-\infty, 2)$ and increasing on $(2, \infty)$, the graph of f is concave upward on $(1, \infty)$, and f has inflection points at $x = 0$ and $x = 1$.

(a)

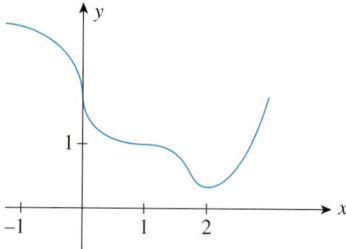

(b)

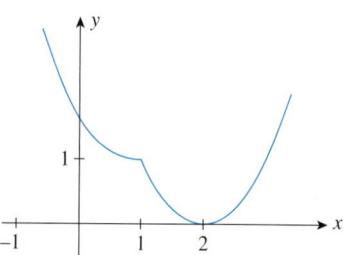

(c)

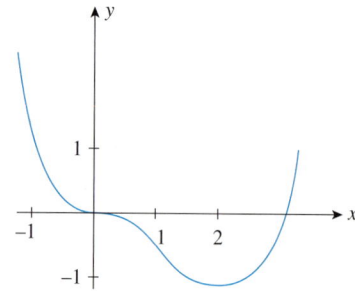

15. Effect of Advertising on Bank Deposits The following graphs were used by the CEO of the Madison Savings Bank to illustrate what effect a projected promotional campaign would have on its deposits over the next year. The functions D_1 and D_2 give the projected amount of money on deposit with the bank over the next 12 months with and without the proposed promotional campaign, respectively.

a. Determine the signs of $D_1'(t)$, $D_2'(t)$, $D_1''(t)$, and $D_2''(t)$ on the interval $(0, 12)$.

b. What can you conclude about the rate of change of the growth rate of the money on deposit with the bank with and without the proposed promotional campaign?

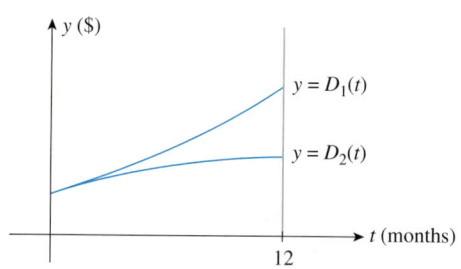

16. Motion of Cars Two cars start out side by side and travel along a straight road. The velocity of Car A is $f(t)$ ft/sec, and the velocity of Car B is $g(t)$ ft/sec over the interval $[0, t_2]$. Furthermore, suppose the graphs of f and g are as depicted in the accompanying figure.

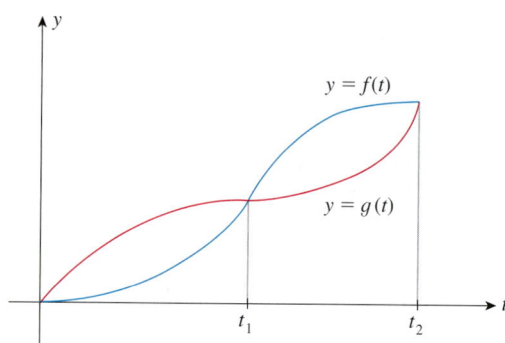

a. What can you say about the acceleration of Car A on the interval $(0, t_1)$? The acceleration of Car B on the interval $(0, t_1)$?

b. What can you say about the acceleration of Car A on the interval (t_1, t_2)? The acceleration of Car B over (t_1, t_2)?

c. What can you say about the acceleration of Car A at t_1? The acceleration of Car B at t_1?

d. At what time do both cars have the same velocity?

In Exercises 17–20, match the graphs (a), (b), (c), or (d) with the corresponding statement.

(a) 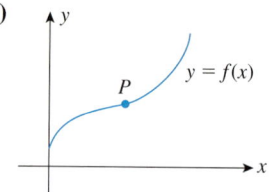   **(b)**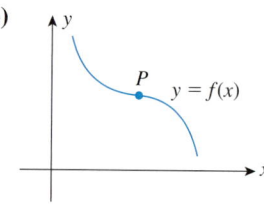

(c) 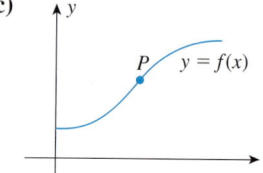   **(d)**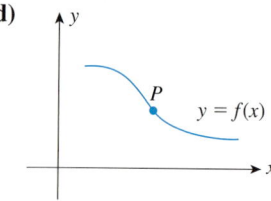

17. The function f is increasing most rapidly at P.

18. The function f is increasing least rapidly at P.

19. The function f is decreasing most rapidly at P.

20. The function f is decreasing least rapidly at P.

21. ASSEMBLY TIME OF A WORKER In the following graph, $N(t)$ gives the number of smartphones assembled by the average worker by the tth hr, where $t = 0$ corresponds to 8 A.M. and $0 \leq t \leq 4$. The point P is an inflection point.
  a. What can you say about the rate of change of the rate of change of the number of smartphones assembled by the average worker between 8 A.M. and 10 A.M.? Between 10 A.M. and 12 noon?
  b. At what time is the rate at which the smartphones are being assembled by the average worker greatest?

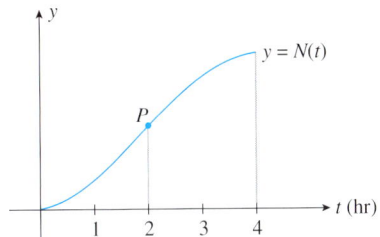

22. RUMORS OF A RUN ON A BANK The graph of the function f shows the total deposits with a bank t days after rumors abounded that there was a run on the bank due to heavy loan losses incurred by the bank.

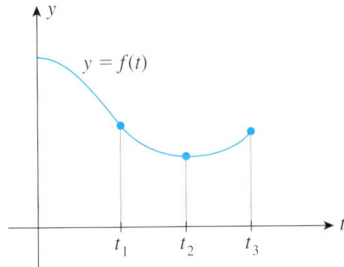

  a. Determine the signs of $f'(t)$ on the intervals $(0, t_2)$ and (t_2, t_3), and determine the signs of $f''(t)$ on the intervals $(0, t_1)$ and (t_1, t_3).
  b. Find where the inflection point(s) of f occur.
  c. Interpret the results of parts (a) and (b).

23. WATER POLLUTION When organic waste is dumped into a pond, the oxidation process that takes place reduces the pond's oxygen content. However, given time, nature will restore the oxygen content to its natural level. In the following graph, $P(t)$ gives the oxygen content (as a percent of its normal level) t days after organic waste has been dumped into the pond. Explain the significance of the inflection point Q.

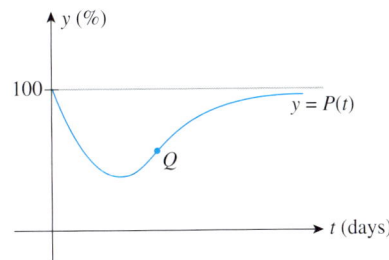

24. CASE-SHILLER HOME PRICE INDEX The following graph shows the change in the S&P/Case-Shiller Home Price Index based on a 20-city average from June 2001 ($t = \frac{1}{2}$) through June 2008 ($t = 7\frac{1}{2}$), adjusted for inflation.

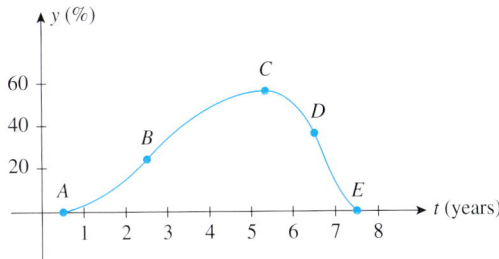

  a. What do the points $A(\frac{1}{2}, 0)$ and $E(7\frac{1}{2}, 0)$ tell you about the change in the Case-Shiller Home Price Index over the period under consideration?
  b. What does the point $C(5\frac{1}{3}, 56)$ tell you about the Case-Shiller Home Price Index?
  c. Give an interpretation of the inflection points $B(2\frac{1}{2}, 24)$ and $D(6\frac{1}{2}, 36)$.

Source: New York Times.

In Exercises 25–28, show that the graph of the function is concave upward wherever it is defined.

25. $f(x) = 4x^2 - 12x + 7$

26. $g(x) = x^4 + \frac{1}{2}x^2 + 6x + 10$

27. $f(x) = \dfrac{1}{x^4}$

28. $g(x) = -\sqrt{4 - x^2}$

In Exercises 29–48, determine where the graph of the function is concave upward and where it is concave downward.

29. $f(x) = 2x^2 - 3x + 4$

30. $g(x) = -x^2 + 3x + 4$

31. $f(x) = 1 - x^3$

32. $g(x) = x^3 - x$

33. $f(x) = x^4 - 6x^3 + 2x + 8$

34. $f(x) = 3x^4 - 6x^3 + x - 8$

35. $f(x) = x^{4/7}$

36. $f(x) = \sqrt[3]{x}$

37. $f(x) = \sqrt{4 - x}$

38. $g(x) = \sqrt{x - 2}$

39. $f(x) = \dfrac{1}{x - 2}$

40. $g(x) = \dfrac{x}{x + 1}$

41. $f(x) = \dfrac{1}{2 + x^2}$

42. $g(x) = \dfrac{x}{1 + x^2}$

43. $h(t) = \dfrac{t^2}{t - 1}$

44. $f(x) = \dfrac{x + 1}{x - 1}$

45. $g(x) = x + \dfrac{1}{x^2}$

46. $h(r) = -\dfrac{1}{(r - 2)^2}$

47. $g(t) = (2t - 4)^{1/3}$

48. $f(x) = (x - 2)^{2/3}$

In Exercises 49–60, find the inflection point(s), if any, of each function.

49. $f(x) = x^3 - 2$
50. $g(x) = x^3 - 6x$
51. $f(x) = 6x^3 - 18x^2 + 12x - 20$
52. $g(x) = 2x^3 - 3x^2 + 18x - 8$
53. $f(x) = 3x^4 - 4x^3 + 1$
54. $f(x) = x^4 - 2x^3 + 6$
55. $g(t) = \sqrt[3]{t}$
56. $f(x) = \sqrt[5]{x}$
57. $f(x) = (x - 1)^3 + 2$
58. $f(x) = (x - 2)^{4/3}$
59. $f(x) = \dfrac{2}{1 + x^2}$
60. $f(x) = 2 + \dfrac{3}{x}$

In Exercises 61–76, find the relative extrema, if any, of each function. Use the Second Derivative Test if applicable.

61. $f(x) = -x^2 + 2x + 4$
62. $g(x) = 2x^2 + 3x + 7$
63. $f(x) = 2x^3 + 1$
64. $g(x) = x^3 - 6x$
65. $f(x) = \dfrac{1}{3}x^3 - 2x^2 - 5x - 5$
66. $f(x) = 2x^3 + 3x^2 - 12x - 4$
67. $g(t) = t + \dfrac{9}{t}$
68. $f(t) = 2t + \dfrac{3}{t}$
69. $f(x) = \dfrac{x}{1 - x}$
70. $f(x) = \dfrac{2x}{x^2 + 1}$
71. $f(t) = t^2 - \dfrac{16}{t}$
72. $g(x) = x^2 + \dfrac{2}{x}$
73. $g(s) = \dfrac{s}{1 + s^2}$
74. $g(x) = \dfrac{1}{1 + x^2}$
75. $f(x) = \dfrac{x^4}{x - 1}$
76. $f(x) = \dfrac{x^2}{x^2 + 1}$

In Exercises 77–82, sketch the graph of a function having the given properties.

77. $f(2) = 4, f'(2) = 0, f''(x) < 0$ on $(-\infty, \infty)$

78. $f(2) = 2, f'(2) = 0, f'(x) > 0$ on $(-\infty, 2), f'(x) > 0$ on $(2, \infty), f''(x) < 0$ on $(-\infty, 2), f''(x) > 0$ on $(2, \infty)$

79. $f(-2) = 4, f(3) = -2, f'(-2) = 0, f'(3) = 0, f'(x) > 0$ on $(-\infty, -2)$ and $(3, \infty), f'(x) < 0$ on $(-2, 3)$, inflection point at $(1, 1)$

80. $f(0) = 0, f'(0)$ does not exist, $f''(x) < 0$ if $x \ne 0$

81. $f(0) = 1, f'(0) = 0, f(x) > 0$ on $(-\infty, \infty), f''(x) < 0$ on $(-\sqrt{2}/2, \sqrt{2}/2), f''(x) > 0$ on $(-\infty, -\sqrt{2}/2)$ and $(\sqrt{2}/2, \infty)$

82. f has domain $[-1, 1], f(-1) = -1, f(-\tfrac{1}{2}) = -2, f'(-\tfrac{1}{2}) = 0, f''(x) > 0$ on $(-1, 1)$

83. Demand for RNs The following graph gives the total number of help-wanted ads for RNs (registered nurses) in 22 cities over the last 12 months as a function of time t (t measured in months).

a. Explain why $N'(t)$ is positive on the interval $(0, 12)$.
b. Determine the signs of $N''(t)$ on the interval $(0, 6)$ and the interval $(6, 12)$.
c. Interpret the results of part (b).

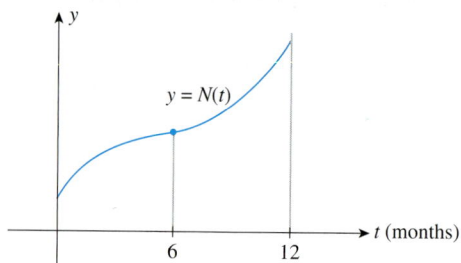

84. Effect of Budget Cuts on Drug-Related Crimes The graphs below were used by a police commissioner to illustrate what effect a budget cut would have on crime in the city. The number $N_1(t)$ gives the projected number of drug-related crimes in the next 12 months. The number $N_2(t)$ gives the projected number of drug-related crimes in the same time frame if next year's budget is cut.

a. Explain why $N_1'(t)$ and $N_2'(t)$ are both positive on the interval $(0, 12)$.
b. What are the signs of $N_1''(t)$ and $N_2''(t)$ on the interval $(0, 12)$?
c. Interpret the results of part (b).

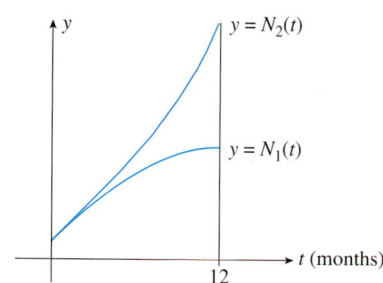

85. In the following figure, water is poured into the vase at a constant rate (in appropriate units), and the water level rises to a height of $f(t)$ units at time t as measured from the base of the vase. The graph of f follows. Explain the shape of the curve in terms of its concavity. What is the significance of the inflection point?

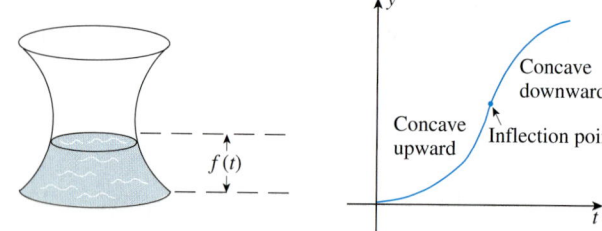

86. In the following figure, water is poured into an urn at a constant rate (in appropriate units), and the water level rises to a height of $f(t)$ units at time t as measured from the

base of the urn. Sketch the graph of f and explain its shape, indicating where it is concave upward and concave downward. Indicate the inflection point on the graph, and explain its significance.
Hint: Study Exercise 85.

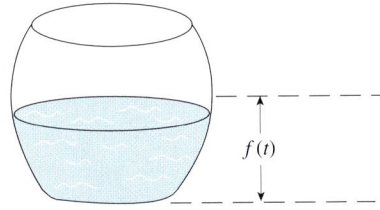

87. STATE CIGARETTE TAXES The average state cigarette tax per pack (in dollars) from 2001 through 2007 is approximated by the function

$$T(t) = 0.43t^{0.43} \quad (1 \leq t \leq 7)$$

where t is measured in years, with $t = 1$ corresponding to 2001.
a. Show that the average state cigarette tax per pack was increasing throughout the period in question.
b. What can you say about the rate at which the average state cigarette tax per pack was increasing over the period in question?
Source: Campaign for Tobacco-Free Kids.

88. GLOBAL WARMING The increase in carbon dioxide (CO_2) in the atmosphere is a major cause of global warming. Using data obtained by Charles David Keeling, professor at Scripps Institution of Oceanography, the average amount of CO_2 in the atmosphere from 1958 through 2010 is approximated by

$$A(t) = 0.012313t^2 + 0.7545t + 313.9 \quad (1 \leq t \leq 53)$$

where $A(t)$ is measured in parts per million volume (ppmv) and t in years, with $t = 1$ corresponding to 1958.
a. What can you say about the rate of change of the average amount of atmospheric CO_2 from 1958 through 2010?
b. What can you say about the rate of change of the rate of change of the average amount of atmospheric CO_2 from 1958 through 2010?
Source: Scripps Institution of Oceanography.

89. EFFECT OF SMOKING BANS The sales (in billions of dollars) in restaurants and bars in California from 1993 ($t = 0$) through 2000 ($t = 7$) are approximated by the function

$$S(t) = 0.195t^2 + 0.32t + 23.7 \quad (0 \leq t \leq 7)$$

a. Show that the sales in restaurants and bars continued to rise after smoking bans were implemented in restaurants in 1995 and in bars in 1998.
Hint: Show that S is increasing on the interval (2, 7).
b. What can you say about the rate at which the sales were rising after smoking bans were implemented?
Source: California Board of Equalization.

90. FLIGHT OF A ROCKET The altitude (in feet) of a rocket t sec into flight is given by

$$s = f(t) = -t^3 + 54t^2 + 480t + 6 \quad (t \geq 0)$$

Find the point of inflection of the function f, and interpret your result. What is the maximum velocity attained by the rocket?

91. ALTERNATIVE MINIMUM TAX Congress created the alternative minimum tax (AMT) in the late 1970s to ensure that wealthy people paid their fair share of taxes. But because of quirks in the law, even middle-income taxpayers have started to get hit with the tax. The AMT (in billions of dollars) projected to be collected by the IRS from 2001 through 2010 is

$$f(t) = 0.0117t^3 + 0.0037t^2 + 0.7563t + 4.1 \quad (0 \leq t \leq 9)$$

where t is measured in years, with $t = 0$ corresponding to 2001.
a. Show that f is increasing on the interval (0, 9). What does this result tell you about the projected amount of AMT paid over the years in question?
b. Show that f' is increasing on the interval (0, 9). What conclusion can you draw from this result concerning the rate of growth of the amount of the AMT collected over the years in question?
Source: U.S. Congress Joint Economic Committee.

92. AGE AT FIRST MARRIAGE According to a study conducted by Rutgers University, the median age of women in the U.S. at first marriage is approximated by the function

$$f(t) = -0.083t^3 + 0.6t^2 + 0.18t + 20.1 \quad (0 \leq t \leq 5)$$

where t is measured in decades and $t = 0$ corresponds to the beginning of 1960.
a. What was the median age of women at first marriage at the beginning of 1960? At the beginning of the year 2000?
b. When was the median age of women at first marriage changing most rapidly over the time period under consideration?
Source: The National Marriage Project, Rutgers University.

93. EFFECT OF ADVERTISING ON HOTEL REVENUE The total annual revenue R of the Miramar Resorts Hotel is related to the amount of money x the hotel spends on advertising its services by the function

$$R(x) = -0.003x^3 + 1.35x^2 + 2x + 8000 \quad (0 \leq x \leq 400)$$

where both R and x are measured in thousands of dollars.
a. Find the interval where the graph of R is concave upward and the interval where the graph of R is concave downward. What is the inflection point of the graph of R?
b. Would it be more beneficial for the hotel to increase its advertising budget slightly when the budget is $140,000 or when it is $160,000?

94. Forecasting Profits As a result of increasing energy costs, the growth rate of the profit of the 4-year old Venice Glassblowing Company has begun to decline. Venice's management, after consulting with energy experts, decides to implement certain energy-conservation measures aimed at cutting energy bills. The general manager reports that, according to his calculations, the growth rate of Venice's profit should be on the increase again within 4 years. If Venice's profit (in hundreds of dollars) t years from now is given by the function

$$P(t) = t^3 - 9t^2 + 40t + 50 \quad (0 \leq t \leq 8)$$

determine whether the general manager's forecast will be accurate.

Hint: Find the inflection point of the function P and study the concavity of the graph of P.

95. Outsourcing The amount (in billions of dollars) spent by the top 15 U.S. financial institutions on IT (information technology) offshore outsourcing is approximated by

$$A(t) = 0.92(t + 1)^{0.61} \quad (0 \leq t \leq 4)$$

where t is measured in years, with $t = 0$ corresponding to 2004.
a. Show that A is increasing on $(0, 4)$, and interpret your result.
b. Show that the graph of A is concave downward on $(0, 4)$. Interpret your result.
Source: Tower Group.

96. Public Transportation Budget Deficit According to the Massachusetts Bay Transportation Authority (MBTA), the projected cumulative MBTA budget deficit with the $160 million rescue package (in billions of dollars) is given by

$$D_1(t) = 0.0275t^2 + 0.081t + 0.07 \quad (0 \leq t \leq 3)$$

and the budget deficit without the rescue package is given by

$$D_2(t) = 0.035t^2 + 0.211t + 0.24 \quad (0 \leq t \leq 3)$$

Here t is measured in years, with $t = 0$ corresponding to 2011. Let $D = D_2 - D_1$.
a. Show that D is increasing on $(0, 3)$, and interpret your result.
b. Show that the graph of D is concave upward on $(0, 3)$, and interpret your result.
Source: MBTA Review.

97. Google's Revenue The revenue for Google from 2004 ($t = 0$) through 2008 ($t = 4$) is approximated by the function

$$R(t) = -0.2t^3 + 1.64t^2 + 1.31t + 3.2 \quad (0 \leq t \leq 4)$$

where $R(t)$ is measured in billions of dollars.
a. Find $R'(t)$ and $R''(t)$.
b. Show that $R'(t) > 0$ for all t in the interval $(0, 4)$ and interpret your result.
Hint: Use the quadratic formula.
c. Find the inflection point of R, and interpret your result.
Source: Google company report.

98. Surveillance Cameras Research reports indicate that surveillance cameras at major intersections dramatically reduce the number of drivers who barrel through red lights. The cameras automatically photograph vehicles that drive into intersections after the light turns red. Vehicle owners are then mailed citations instructing them to pay a fine or sign an affidavit that they weren't driving at the time. The function

$$N(t) = 6.08t^3 - 26.79t^2 + 53.06t + 69.5 \quad (0 \leq t \leq 4)$$

gives the number, $N(t)$, of U.S. communities using surveillance cameras at intersections in year t, with $t = 0$ corresponding to 2003.
a. Show that N is increasing on $(0, 4)$.
b. When was the number of communities using surveillance cameras at intersections increasing least rapidly? What is the rate of increase?
Source: Insurance Institute for Highway Safety.

99. Population Growth in Clark County Clark County in Nevada—dominated by greater Las Vegas—is one of the fastest-growing metropolitan areas in the United States. The population of the county from 1970 through 2010 is approximated by the function

$$P(t) = 44{,}560t^3 - 89{,}394t^2 + 234{,}633t + 273{,}288$$
$$(0 \leq t \leq 4)$$

where t is measured in decades, with $t = 0$ corresponding to the beginning of 1970.
a. Show that the population of Clark County was always increasing over the time period in question.
Hint: Show that $P'(t) > 0$ for all t in the interval $(0, 4)$.
b. Show that the population of Clark County was increasing at the slowest pace some time in August 1976.
Hint: Find the inflection point of P in the interval $(0, 4)$.
Source: U.S. Census Bureau.

100. Hiring Lobbyists Many public entities such as cities, counties, states, utilities, and Indian tribes are hiring firms to lobby Congress. One goal of such lobbying is to place earmarks—money directed at a specific project—into appropriation bills. The amount (in millions of dollars) spent by public entities on lobbying from 1998 through 2004 is given by

$$f(t) = -0.425t^3 + 3.6571t^2 + 4.018t + 43.7$$
$$(0 \leq t \leq 6)$$

where t is measured in years, with $t = 0$ corresponding to 1998.
a. Show that f is increasing on $(0, 6)$. What does this say about the spending by public entities on lobbying over the years in question?
b. Find the inflection point of f. What does your result tell you about the growth of spending by the public entities on lobbying?
Source: Center for Public Integrity.

101. Measles Deaths Measles is still a leading cause of vaccine-preventable death among children, but because of improvements in immunizations, measles deaths have dropped globally. The function

$$N(t) = -2.42t^3 + 24.5t^2 - 123.3t + 506 \quad (0 \leq t \leq 6)$$

gives the number of measles deaths (in thousands) in sub-Saharan Africa in year t, with $t = 0$ corresponding to the beginning of 1999.
  a. How many measles deaths were there in 1999? In 2005?
  b. Show that $N'(t) < 0$ on $(0, 6)$. What does this say about the number of measles deaths from 1999 through 2005?
  c. When was the number of measles deaths decreasing most rapidly? What was the rate of decline of measles deaths at that instant of time?

Source: Centers for Disease Control and Prevention and World Health Organization.

102. **Air Pollution** The level of ozone, an invisible gas that irritates and impairs breathing, present in the atmosphere on a certain May day in the city of Riverside was approximated by

$$A(t) = 1.0974t^3 - 0.0915t^4 \quad (0 \leq t \leq 11)$$

where $A(t)$ is measured in pollutant standard index (PSI) and t is measured in hours, with $t = 0$ corresponding to 7 A.M. Use the Second Derivative Test to show that the function A has a relative maximum at approximately $t = 9$. Interpret your results.

103. **Women's Soccer** Starting with the youth movement that took hold in the 1970s and buoyed by the success of the U.S. national women's team in international competition in recent years, girls and women have taken to soccer in ever-growing numbers. The function

$$N(t) = -0.9307t^3 + 74.04t^2 + 46.8667t + 3967 \quad (0 \leq t \leq 16)$$

gives the number of participants in women's soccer in year t, with $t = 0$ corresponding to the beginning of 1985.
  a. Verify that the number of participants per year in women's soccer had been increasing from 1985 through 2000.
  Hint: Use the quadratic formula.
  b. Show that the number of participants per year in women's soccer had been increasing at an increasing rate from 1985 through 2000.
  Hint: Show that the sign of N'' is positive on the interval in question.

Source: NCCA News.

104. **Dependency Ratio** The share of the world population that is over 60 years of age compared to the rest of the working population in the world is of concern to economists. An increasing dependency ratio means that there will be fewer workers to support an aging population. The dependency ratio over the next century is forecast to be

$$R(t) = 0.00731t^4 - 0.174t^3 + 1.528t^2 + 0.48t + 19.3 \quad (0 \leq t \leq 6)$$

in year t, where t is measured in decades with $t = 0$ corresponding to 2000.
  a. Show that the dependency ratio will be increasing at the fastest pace around 2052.
  Hint: Use the quadratic formula.
  b. What will the dependency ratio be at that time?

Source: International Institute for Applied Systems Analysis.

In Exercises 105–108, determine whether the statement is true or false. If it is true, explain why it is true. If it is false, give an example to show why it is false.

105. If the graph of f is concave upward on (a, b), then the graph of $-f$ is concave downward on (a, b).

106. If the graph of f is concave upward on (a, c) and concave downward on (c, b), where $a < c < b$, then f has an inflection point at $(c, f(c))$.

107. If c is a critical number of f where $a < c < b$ and $f''(x) < 0$ on (a, b), then f has a relative maximum at $x = c$.

108. A polynomial function of degree n $(n \geq 3)$ can have at most $(n - 2)$ inflection points.

109. Show that the graph of the quadratic function

$$f(x) = ax^2 + bx + c \quad (a \neq 0)$$

is concave upward if $a > 0$ and concave downward if $a < 0$. Thus, by examining the sign of the coefficient of x^2, one can tell immediately whether the parabola opens upward or downward.

110. Consider the functions $f(x) = x^3$, $g(x) = x^4$, and $h(x) = -x^4$.
  a. Show that $x = 0$ is a critical number of each of the functions f, g, and h.
  b. Show that the second derivative of each of the functions f, g, and h equals zero at $x = 0$.
  c. Show that f has neither a relative maximum nor a relative minimum at $x = 0$, that g has a relative minimum at $x = 0$, and that h has a relative maximum at $x = 0$.

4.2 Solutions to Self-Check Exercises

1. We first compute

$$f'(x) = 12x^2 - 6x$$
$$f''(x) = 24x - 6 = 6(4x - 1)$$

Observe that f'' is continuous everywhere and has a zero at $x = \frac{1}{4}$. The sign diagram of f'' is shown in the accompanying figure.

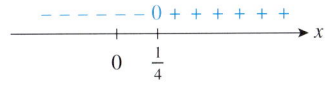

From the sign diagram for f'', we see that the graph of f is concave upward on $(\frac{1}{4}, \infty)$ and concave downward on $(-\infty, \frac{1}{4})$.

2. First, we find the critical numbers of f by solving the equation

$$f'(x) = 6x^2 - x - 12 = 0$$

That is,

$$(3x + 4)(2x - 3) = 0$$

giving $x = -\frac{4}{3}$ and $x = \frac{3}{2}$. Next, we compute

$$f''(x) = 12x - 1$$

Since

$$f''\left(-\frac{4}{3}\right) = 12\left(-\frac{4}{3}\right) - 1 = -17 < 0$$

the Second Derivative Test implies that $f(-\frac{4}{3}) = \frac{10}{27}$ is a relative maximum of f. Also,

$$f''\left(\frac{3}{2}\right) = 12\left(\frac{3}{2}\right) - 1 = 17 > 0$$

and we see that $f(\frac{3}{2}) = -\frac{179}{8}$ is a relative minimum.

3. We compute the second derivative of G. Thus,

$$G'(t) = -6t^2 + 90t + 20$$
$$G''(t) = -12t + 90$$

Now, G'' is continuous everywhere, and $G''(t) = 0$, when $t = \frac{15}{2}$, giving $t = \frac{15}{2}$ as the only candidate for an inflection point of G. Since $G''(t) > 0$ for $t < \frac{15}{2}$ and $G''(t) < 0$ for $t > \frac{15}{2}$, we see that $(\frac{15}{2}, \frac{15,675}{2})$ is an inflection point of G. The results of our computations tell us that the country's GDP was increasing most rapidly at the beginning of July 2009.

4.3 Curve Sketching

A Real-Life Example

As we have seen on numerous occasions, the graph of a function is a useful aid for visualizing the function's properties. From a practical point of view, the graph of a function also gives, at one glance, a complete summary of all the information captured by the function.

Consider, for example, the graph of the function giving the Dow-Jones Industrial Average (DJIA) on Black Monday, October 19, 1987 (Figure 45). Here, $t = 0$ corresponds to 9:30 A.M., when the market was open for business, and $t = 6.5$ corresponds to 4 P.M., the closing time. The following information may be gleaned from studying the graph.

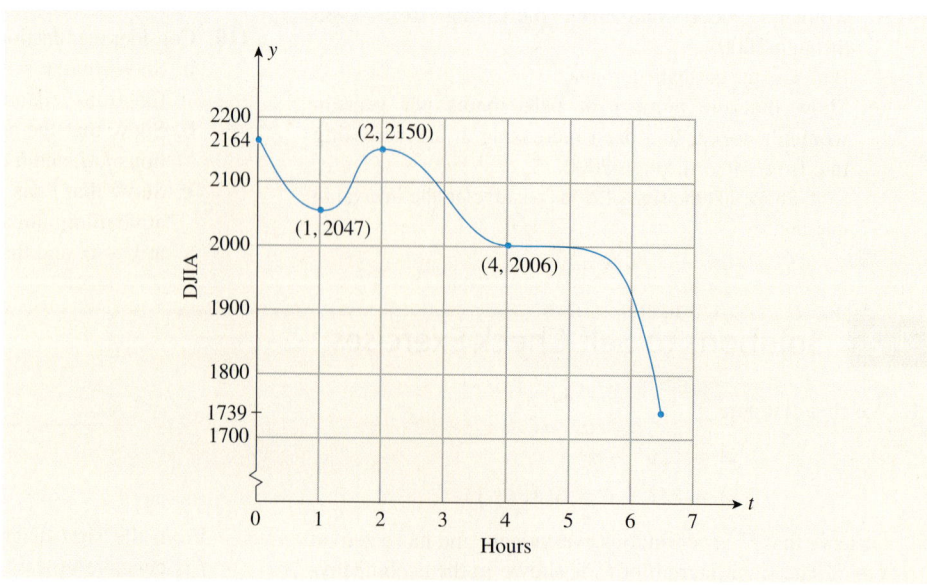

FIGURE 45
The Dow-Jones Industrial Average on Black Monday

Source: Wall Street Journal.

The graph is *falling* rapidly from $t = 0$ to $t = 1$, reflecting the sharp drop in the index in the first hour of trading. The point (1, 2047) is a *relative minimum* point of the function, and this turning point coincides with the start of an aborted recovery. The short-lived rally, represented by the portion of the graph that is *rising* on the interval (1, 2), quickly fizzled out at $t = 2$ (11:30 A.M.). The *relative maximum* point (2, 2150) marks the highest point of the recovery. The function is decreasing in the rest of the interval. The point (4, 2006) is an *inflection point* of the function; it shows that there was a temporary respite at $t = 4$ (1:30 P.M.). However, selling pressure continued unabated, and the DJIA continued to fall until the closing bell. Finally, the graph also shows that the index opened at the high of the day [$f(0) = 2164$ is the *absolute maximum* of the function] and closed at the low of the day [$f(\frac{13}{2}) = 1739$ is the *absolute minimum* of the function], a drop of 508 points from the previous close!*

Before we turn our attention to the actual task of sketching the graph of a function, let's look at some properties of graphs that will be helpful in this connection.

Vertical Asymptotes

Before going on, you might want to review the material on one-sided limits and the limit at infinity of a function (Sections 2.4 and 2.5).

Consider the graph of the function

$$f(x) = \frac{x+1}{x-1}$$

shown in Figure 46. Observe that $f(x)$ increases without bound (tends to infinity) as x approaches $x = 1$ from the right; that is,

$$\lim_{x \to 1^+} \frac{x+1}{x-1} = \infty$$

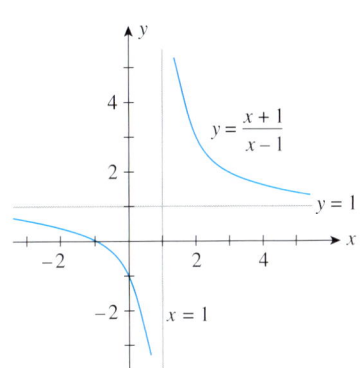

FIGURE 46
The graph of *f* has a vertical asymptote at $x = 1$.

You can verify this by taking a sequence of values of x approaching $x = 1$ from the right and looking at the corresponding values of $f(x)$.

Here is another way of looking at the situation: Observe that if x is a number that is a little larger than 1, then both $(x + 1)$ and $(x - 1)$ are positive, so $(x + 1)/(x - 1)$ is also positive. As x approaches $x = 1$, the numerator $(x + 1)$ approaches the number 2, but the denominator $(x - 1)$ approaches zero, so the quotient $(x + 1)/(x - 1)$ approaches infinity, as observed earlier. The line $x = 1$ is called a vertical asymptote of the graph of *f*.

For the function $f(x) = (x + 1)/(x - 1)$, you can show that

$$\lim_{x \to 1^-} \frac{x+1}{x-1} = -\infty$$

and this tells us how $f(x)$ approaches the asymptote $x = 1$ from the left.

More generally, we have the following definition:

> **Vertical Asymptote**
> The line $x = a$ is a **vertical asymptote** of the graph of a function *f* if
> $$\lim_{x \to a^+} f(x) = \infty \quad \text{or} \quad -\infty$$
> or
> $$\lim_{x \to a^-} f(x) = \infty \quad \text{or} \quad -\infty$$

---

*Absolute maxima and absolute minima of functions are covered in Section 4.4.

Note Although a vertical asymptote of a graph is not part of the graph, it serves as a useful aid for sketching the graph.

For rational functions

$$f(x) = \frac{P(x)}{Q(x)}$$

there is a simple criterion for determining whether the graph of f has any vertical asymptotes.

> **Finding Vertical Asymptotes of Rational Functions**
> Suppose f is a rational function
> $$f(x) = \frac{P(x)}{Q(x)}$$
> where P and Q are polynomial functions. Then, the line $x = a$ is a vertical asymptote of the graph of f if $Q(a) = 0$ but $P(a) \neq 0$.

For the function

$$f(x) = \frac{x+1}{x-1}$$

considered earlier, $P(x) = x + 1$ and $Q(x) = x - 1$. Observe that $Q(1) = 0$ but $P(1) = 2 \neq 0$, so $x = 1$ is a vertical asymptote of the graph of f.

EXAMPLE 1 Find the vertical asymptotes of the graph of the function

$$f(x) = \frac{x^2}{4 - x^2}$$

Solution The function f is a rational function with $P(x) = x^2$ and $Q(x) = 4 - x^2$. The zeros of Q are found by solving

$$4 - x^2 = 0$$

—that is,

$$(2 + x)(2 - x) = 0$$

giving $x = -2$ and $x = 2$. These are candidates for the vertical asymptotes of the graph of f. Examining $x = -2$, we compute $P(-2) = (-2)^2 = 4 \neq 0$, and we see that $x = -2$ is indeed a vertical asymptote of the graph of f. Similarly, we find $P(2) = 2^2 = 4 \neq 0$, so $x = 2$ is also a vertical asymptote of the graph of f. The graph of f sketched in Figure 47 confirms these results.

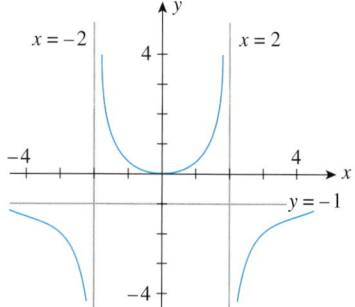

FIGURE 47
$x = -2$ and $x = 2$ are vertical asymptotes of the graph of f.

Recall that in order for the line $x = a$ to be a vertical asymptote of the graph of a rational function f, *only* the denominator of $f(x)$ must be equal to zero at $x = a$. If *both* $P(a)$ and $Q(a)$ are equal to zero, then $x = a$ need *not* be a vertical asymptote. For example, look at the function

$$f(x) = \frac{4(x^2 - 4)}{x - 2}$$

whose graph appears in Figure 32a in Chapter 2 (page 104).

Horizontal Asymptotes

Let's return to the function f defined by

$$f(x) = \frac{x+1}{x-1}$$

(Figure 48).

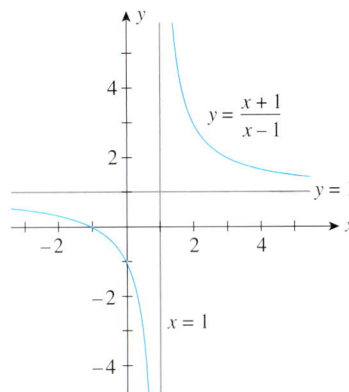

FIGURE 48
The graph of f has a horizontal asymptote at $y = 1$.

Observe that the graph of f approaches the horizontal line $y = 1$ as x approaches infinity, and in this case, the graph of f approaches $y = 1$ as x approaches minus infinity as well. The line $y = 1$ is called a horizontal asymptote of the graph of f. More generally, we have the following definition:

> **Horizontal Asymptote**
> The line $y = b$ is a **horizontal asymptote** of the graph of a function f if either
> $$\lim_{x \to \infty} f(x) = b \quad \text{or} \quad \lim_{x \to -\infty} f(x) = b$$

For the function

$$f(x) = \frac{x+1}{x-1}$$

we see that

$$\lim_{x \to \infty} \frac{x+1}{x-1} = \lim_{x \to \infty} \frac{1 + \frac{1}{x}}{1 - \frac{1}{x}} \quad \text{Divide numerator and denominator by } x.$$

$$= 1$$

Also,

$$\lim_{x \to -\infty} \frac{x+1}{x-1} = \lim_{x \to -\infty} \frac{1 + \frac{1}{x}}{1 - \frac{1}{x}}$$

$$= 1$$

In either case, we conclude that $y = 1$ is a horizontal asymptote of the graph of f, as observed earlier.

EXAMPLE 2 Find the horizontal asymptotes of the graph of the function

$$f(x) = \frac{x^2}{4 - x^2}$$

Solution We compute

$$\lim_{x \to \infty} \frac{x^2}{4 - x^2} = \lim_{x \to \infty} \frac{1}{\frac{4}{x^2} - 1} \quad \text{Divide numerator and denominator by } x^2.$$

$$= -1$$

so $y = -1$ is a horizontal asymptote, as before. (Similarly, $\lim_{x \to -\infty} f(x) = -1$ as well.) The graph of f sketched in Figure 49 confirms this result.

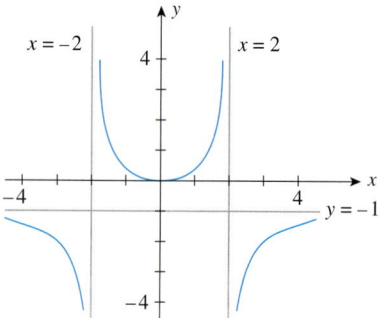

FIGURE 49
The graph of f has a horizontal asymptote at $y = -1$.

We next state an important property of polynomial functions.

> The graph of a polynomial function has no vertical or horizontal asymptotes.

To see this, note that a polynomial function $P(x)$ can be written as a rational function with denominator equal to 1. Thus,

$$P(x) = \frac{P(x)}{1}$$

Since the denominator is never equal to zero, P has no vertical asymptotes. Next, if P is a polynomial of degree greater than or equal to 1, then

$$\lim_{x \to \infty} P(x) \quad \text{and} \quad \lim_{x \to -\infty} P(x)$$

are either infinity or minus infinity; that is, they do not exist. Therefore, P has no horizontal asymptotes.

In the last two sections, we saw how the first and second derivatives of a function are used to reveal various properties of the graph of a function f. We now show how this information can be used to help us sketch the graph of f. We begin by giving a general procedure for curve sketching.

A Guide to Curve Sketching
1. Determine the domain of f.
2. Find the x- and y-intercepts of f.*

---

*The equation $f(x) = 0$ may be difficult to solve, in which case one may decide against finding the x-intercepts or to use technology, if available, for assistance.

3. Determine the behavior of f for large absolute values of x.
4. Find all horizontal and vertical asymptotes of the graph of f.
5. Determine the intervals where f is increasing and where f is decreasing.
6. Find the relative extrema of f.
7. Determine the concavity of the graph of f.
8. Find the inflection points of f.
9. Plot a few additional points to help further identify the shape of the graph of f and sketch the graph.

We now illustrate the techniques of curve sketching in the next two examples.

Two Step-by-Step Examples

EXAMPLE 3 Sketch the graph of the function
$$y = f(x) = x^3 - 6x^2 + 9x + 2$$

Solution Obtain the following information on the graph of f.

Step 1 The domain of f is the interval $(-\infty, \infty)$.

Step 2 By setting $x = 0$, we find that the y-intercept is 2. The x-intercept is found by setting $y = 0$, which in this case leads to a cubic equation. Since the solution is not readily found, we will not use this information.

Step 3 Since
$$\lim_{x \to -\infty} f(x) = \lim_{x \to -\infty} (x^3 - 6x^2 + 9x + 2) = -\infty$$
$$\lim_{x \to \infty} f(x) = \lim_{x \to \infty} (x^3 - 6x^2 + 9x + 2) = \infty$$
we see that f decreases without bound as x decreases without bound and that f increases without bound as x increases without bound.

Step 4 Since f is a polynomial function, there are no asymptotes.

Step 5
$$f'(x) = 3x^2 - 12x + 9 = 3(x^2 - 4x + 3)$$
$$= 3(x - 1)(x - 3)$$

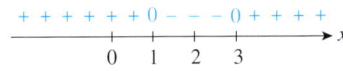

FIGURE 50
Sign diagram for f'

Setting $f'(x) = 0$ gives $x = 1$ or $x = 3$. The sign diagram for f' shows that f is increasing on the intervals $(-\infty, 1)$ and $(3, \infty)$ and decreasing on the interval $(1, 3)$ (Figure 50).

Step 6 From the results of Step 5, we see that $x = 1$ and $x = 3$ are critical numbers of f. Furthermore, f' changes sign from positive to negative as we move across $x = 1$, so a relative maximum of f occurs at $x = 1$. Similarly, we see that a relative minimum of f occurs at $x = 3$. Now,
$$f(1) = 1 - 6 + 9 + 2 = 6$$
$$f(3) = 3^3 - 6(3)^2 + 9(3) + 2 = 2$$
so $f(1) = 6$ is a relative maximum of f and $f(3) = 2$ is a relative minimum of f.

Step 7
$$f''(x) = 6x - 12 = 6(x - 2)$$
which is equal to zero when $x = 2$. The sign diagram of f'' shows that the graph of f is concave downward on the interval $(-\infty, 2)$ and concave upward on the interval $(2, \infty)$ (Figure 51).

FIGURE 51
Sign diagram for f''

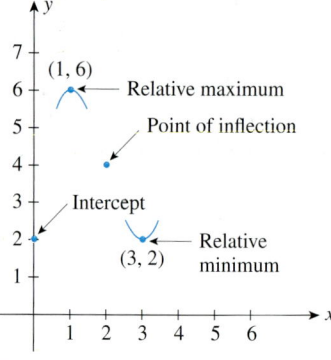

FIGURE 52
We first plot the intercept, the relative extrema, and the inflection point.

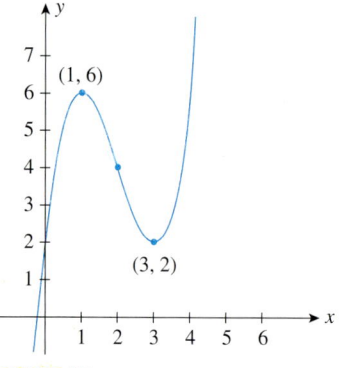

FIGURE 53
The graph of $y = x^3 - 6x^2 + 9x + 2$

Step 8   From the results of Step 7, we see that f'' changes sign as we move across $x = 2$. Next,
$$f(2) = 2^3 - 6(2)^2 + 9(2) + 2 = 4$$
so the required inflection point of f is $(2, 4)$.

Step 9   Summarizing, we have the following:

> Domain: $(-\infty, \infty)$
> Intercept: $(0, 2)$
> $\lim_{x \to -\infty} f(x);\ \lim_{x \to \infty} f(x):\ -\infty;\ \infty$
> Asymptotes: None
> Intervals where f is ↗ or ↘: ↗ on $(-\infty, 1)$ and $(3, \infty)$; ↘ on $(1, 3)$
> Relative extrema: Relative maximum at $(1, 6)$; relative minimum at $(3, 2)$
> Concavity: Downward on $(-\infty, 2)$; upward on $(2, \infty)$
> Point of inflection: $(2, 4)$

In general, it is a good idea to start graphing by plotting the intercept(s), relative extrema, and inflection point(s) (Figure 52). Then, using the rest of the information, we complete the graph of f, as sketched in Figure 53.

Explore & Discuss

The average price of gasoline at the pump over a 3-month period, during which there was a temporary shortage of oil, is described by the function f defined on the interval $[0, 3]$. During the first month, the price was increasing at an increasing rate. Starting with the second month, the good news was that the rate of increase was slowing down, although the price of gas was still increasing. This pattern continued until the end of the second month. The price of gas peaked at $t = 2$ and began to fall at an increasing rate until $t = 3$.

1. Describe the signs of $f'(t)$ and $f''(t)$ over each of the intervals $(0, 1)$, $(1, 2)$, and $(2, 3)$.
2. Make a sketch showing a plausible graph of f over $[0, 3]$.

EXAMPLE 4 Sketch the graph of the function
$$y = f(x) = \frac{x + 1}{x - 1}$$

Solution   Obtain the following information:

Step 1 f is undefined when $x = 1$, so the domain of f is the set of all real numbers other than $x = 1$.

Step 2   Setting $y = 0$ gives $x = -1$, the x-intercept of f. Next, setting $x = 0$ gives $y = -1$ as the y-intercept of f.

Step 3   Earlier, we found that
$$\lim_{x \to \infty} \frac{x + 1}{x - 1} = 1 \quad \text{and} \quad \lim_{x \to -\infty} \frac{x + 1}{x - 1} = 1$$

(see page 287). Consequently, we see that the graph of f approaches the line $y = 1$ as $|x|$ becomes arbitrarily large. For $x > 1$, $f(x) > 1$ and the graph of f approaches the line $y = 1$ from above. For $x < 1$, $f(x) < 1$, so the graph of f approaches the line $y = 1$ from below.

Step 4 The straight line $x = 1$ is a vertical asymptote of the graph of f. Also, from the results of Step 3, we conclude that $y = 1$ is a horizontal asymptote of the graph of f.

Step 5
$$f'(x) = \frac{(x-1)(1) - (x+1)(1)}{(x-1)^2} = -\frac{2}{(x-1)^2}$$

and is discontinuous at $x = 1$. The sign diagram of f' shows that $f'(x) < 0$ whenever it is defined. Thus, f is decreasing on the intervals $(-\infty, 1)$ and $(1, \infty)$ (Figure 54).

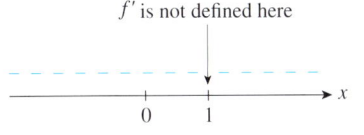

FIGURE 54
The sign diagram for f'

Step 6 From the results of Step 5, we see that there are no critical numbers of f, since $f'(x)$ is never equal to zero for any value of x in the domain of f.

Step 7
$$f''(x) = \frac{d}{dx}[-2(x-1)^{-2}] = 4(x-1)^{-3} = \frac{4}{(x-1)^3}$$

The sign diagram of f'' shows immediately that the graph of f is concave downward on the interval $(-\infty, 1)$ and concave upward on the interval $(1, \infty)$ (Figure 55).

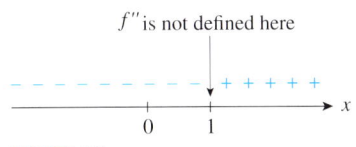

FIGURE 55
The sign diagram for f''

Step 8 From the results of Step 7, we see that there are no candidates for inflection points of f, since $f''(x)$ is never equal to zero for any value of x in the domain of f. Hence, f has no inflection points.

Step 9 Summarizing, we have the following:

> Domain: $(-\infty, 1) \cup (1, \infty)$
> Intercepts: $(-1, 0)$; $(0, -1)$
> $\lim_{x \to -\infty} f(x)$; $\lim_{x \to \infty} f(x)$: 1; 1
> Asymptotes: $x = 1$ is a vertical asymptote
> $\qquad\qquad\;\;$ $y = 1$ is a horizontal asymptote
> Intervals where f is ↗ or ↘: ↘ on $(-\infty, 1)$ and $(1, \infty)$
> Relative extrema: None
> Concavity: Downward on $(-\infty, 1)$; upward on $(1, \infty)$
> Points of inflection: None

The graph of f is sketched in Figure 56.

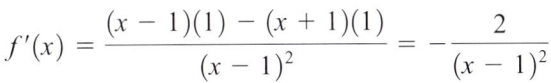

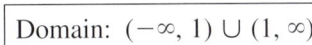

FIGURE 56
The graph of f has a horizontal asymptote at $y = 1$ and a vertical asymptote at $x = 1$.

4.3 Self-Check Exercises

1. Find the horizontal and vertical asymptotes of the graph of the function
$$f(x) = \frac{2x^2}{x^2 - 1}$$

2. Sketch the graph of the function
$$f(x) = \frac{2}{3}x^3 - 2x^2 - 6x + 4$$

Solutions to Self-Check Exercises 4.3 can be found on page 295.

4.3 Concept Questions

1. Explain the following terms in your own words:
   a. Vertical asymptote   b. Horizontal asymptote

2. a. How many vertical asymptotes can the graph of a function f have? Explain using graphs.
   b. How many horizontal asymptotes can the graph of a function f have? Explain using graphs.

3. How do you find the vertical asymptotes of the graph of a rational function?

4. Give a procedure for sketching the graph of a function.

4.3 Exercises

In Exercises 1–10, find the horizontal and vertical asymptotes of the graph of the function.

1.
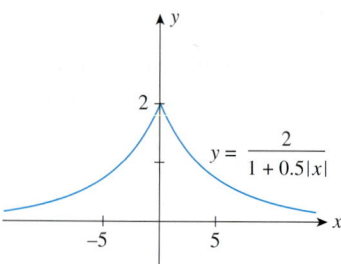

2.
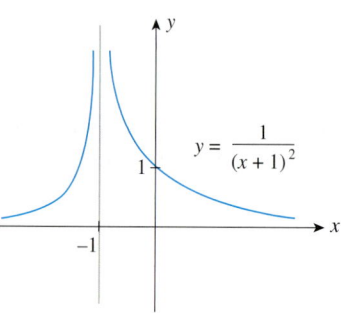

3.
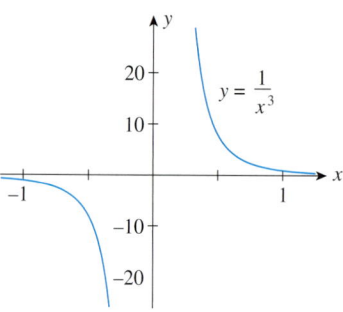

4.
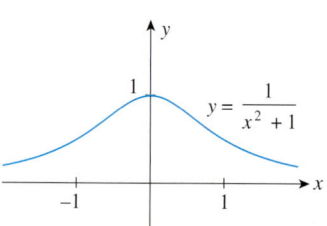

5.
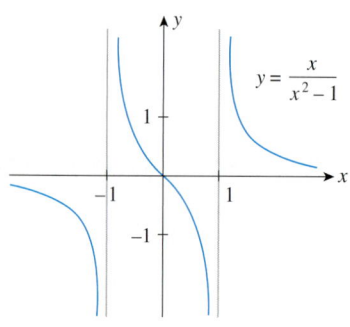

6.
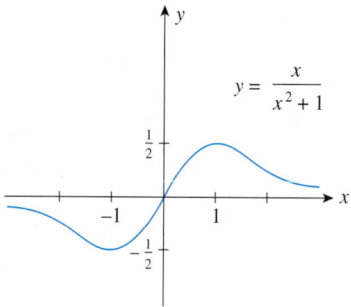

7.
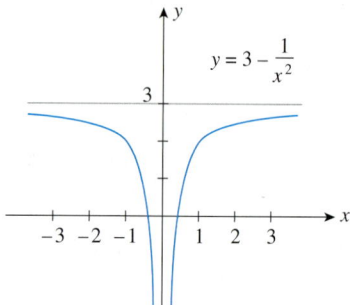

8.
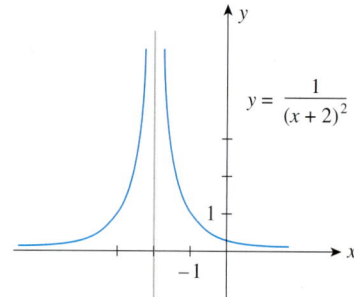

9.
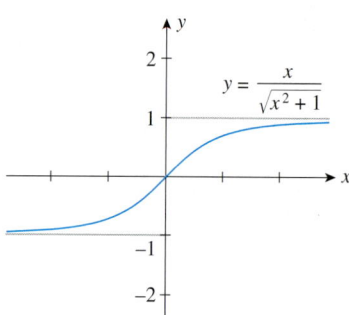

10.
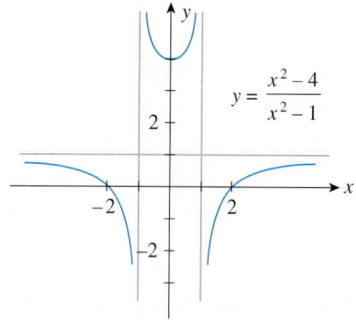

In Exercises 11–28, find the horizontal and vertical asymptotes of the graph of the function. (You need not sketch the graph.)

11. $f(x) = \dfrac{1}{x}$

12. $f(x) = \dfrac{1}{x+2}$

13. $f(x) = -\dfrac{2}{x^2}$

14. $g(x) = \dfrac{1}{1+2x^2}$

15. $f(x) = \dfrac{x-2}{x+2}$

16. $g(t) = \dfrac{t+1}{2t-1}$

17. $h(x) = x^3 - 3x^2 + x + 1$

18. $g(x) = 2x^3 + x^2 + 1$

19. $f(t) = \dfrac{t^2}{t^2 - 16}$

20. $g(x) = \dfrac{x^3}{x^2 - 4}$

21. $f(x) = \dfrac{3x}{x^2 - x - 6}$

22. $g(x) = \dfrac{2x}{x^2 + x - 2}$

23. $g(t) = 2 + \dfrac{5}{(t-2)^2}$

24. $f(x) = 1 + \dfrac{2}{x-3}$

25. $f(x) = \dfrac{x^2 - 2}{x^2 - 4}$

26. $h(x) = \dfrac{2 - x^2}{x^2 + x}$

27. $g(x) = \dfrac{x^3 - x}{x(x+1)}$

28. $f(x) = \dfrac{x^4 - x^2}{x(x-1)(x+2)}$

In Exercises 29 and 30, you are given the graphs of two functions f and g. One function is the derivative function of the other. Identify each of them.

29.

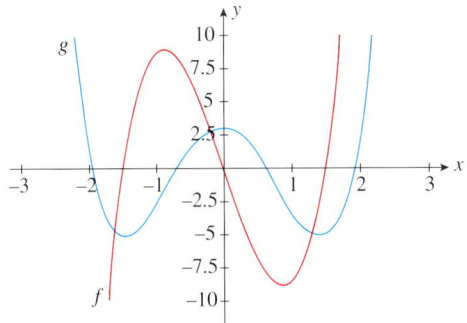

30.

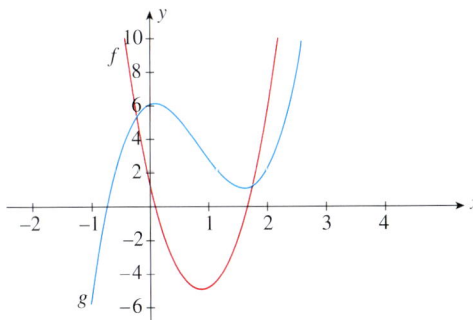

31. **Terminal Velocity** A skydiver leaps from the gondola of a hot-air balloon. As she free-falls, air resistance, which is proportional to her velocity, builds up to a point where it balances the force due to gravity. The resulting motion may be described in terms of her velocity as follows: Starting at rest (zero velocity), her velocity increases and approaches a constant velocity, called the *terminal velocity*. Sketch a graph of her velocity v versus time t.

32. **Spread of a Flu Epidemic** Initially, 10 students at a junior high school contracted influenza. The flu spread over time, and the total number of students who eventually contracted the flu approached but never exceeded 200. Let $P(t)$ denote the number of students who had contracted the flu after t days, where P is an appropriate function.
   a. Make a sketch of the graph of P. (Your answer will *not* be unique.)
   b. Where is the function increasing?
   c. Does the graph of P have a horizontal asymptote? If so, what is it?
   d. Discuss the concavity of the graph of P. Explain its significance.
   e. Is there an inflection point on the graph of P? If so, explain its significance.

In Exercises 33–36, use the information summarized in the table to sketch the graph of f.

33. $f(x) = x^3 - 3x^2 + 1$

Domain: $(-\infty, \infty)$
Intercept: y-intercept: 1
Asymptotes: None
Intervals where f is ↗ and ↘: ↗ on $(-\infty, 0)$ and $(2, \infty)$;
  ↘ on $(0, 2)$
Relative extrema: Rel. max. at $(0, 1)$; rel. min. at $(2, -3)$
Concavity: Downward on $(-\infty, 1)$; upward on $(1, \infty)$
Point of inflection: $(1, -1)$

34. $f(x) = \dfrac{1}{9}(x^4 - 4x^3)$

Domain: $(-\infty, \infty)$
Intercepts: x-intercepts: 0, 4; y-intercept: 0
Asymptotes: None
Intervals where f is ↗ and ↘: ↗ on $(3, \infty)$;
  ↘ on $(-\infty, 3)$
Relative extrema: Rel. min. at $(3, -3)$
Concavity: Downward on $(0, 2)$;
  upward on $(-\infty, 0)$ and $(2, \infty)$
Points of inflection: $(0, 0)$ and $(2, -\tfrac{16}{9})$

35. $f(x) = \dfrac{4x - 4}{x^2}$

Domain: $(-\infty, 0) \cup (0, \infty)$
Intercept: x-intercept: 1
Asymptotes: x-axis and y-axis
Intervals where f is ↗ and ↘: ↗ on $(0, 2)$;
  ↘ on $(-\infty, 0)$ and $(2, \infty)$
Relative extrema: Rel. max. at $(2, 1)$
Concavity: Downward on $(-\infty, 0)$ and $(0, 3)$
  upward on $(3, \infty)$
Point of inflection: $(3, \tfrac{8}{9})$

36. $f(x) = x - 3x^{1/3}$

Domain: $(-\infty, \infty)$
Intercepts: x-intercepts: $\pm 3\sqrt{3}, 0$; y-intercept: 0
Asymptotes: None
Intervals where f is ↗ and ↘: ↗ on $(-\infty, -1)$ and $(1, \infty)$; ↘ on $(-1, 1)$
Relative extrema: Rel. max. at $(-1, 2)$; rel. min. at $(1, -2)$
Concavity: Downward on $(-\infty, 0)$; upward on $(0, \infty)$
Point of inflection: $(0, 0)$

In Exercises 37–60, sketch the graph of the function, using the curve-sketching guide of this section.

37. $g(x) = 4 - 3x - 2x^3$ **38.** $f(x) = x^2 - 2x + 3$

39. $h(x) = x^3 - 3x + 1$ **40.** $f(x) = 2x^3 + 1$

41. $f(x) = -2x^3 + 3x^2 + 12x + 2$

42. $f(t) = 2t^3 - 15t^2 + 36t - 20$

43. $h(x) = \dfrac{3}{2}x^4 - 2x^3 - 6x^2 + 8$

44. $f(t) = 3t^4 + 4t^3$

45. $f(t) = \sqrt{t^2 - 4}$ **46.** $f(x) = \sqrt{x^2 + 5}$

47. $g(x) = \dfrac{1}{2}x - \sqrt{x}$ **48.** $f(x) = \sqrt[3]{x^2}$

49. $g(x) = \dfrac{2}{x-1}$ **50.** $f(x) = \dfrac{1}{x+1}$

51. $h(x) = \dfrac{x+2}{x-2}$ **52.** $g(x) = \dfrac{x}{x-1}$

53. $f(t) = \dfrac{t^2}{1+t^2}$ **54.** $g(x) = \dfrac{x}{x^2-4}$

55. $g(t) = -\dfrac{t^2-2}{t-1}$ **56.** $f(x) = \dfrac{x^2-9}{x^2-4}$

57. $g(t) = \dfrac{t^2}{t^2-1}$ **58.** $h(x) = \dfrac{1}{x^2-x-2}$

59. $h(x) = (x-1)^{2/3} + 1$ **60.** $g(x) = (x+2)^{3/2} + 1$

61. Cost of Removing Toxic Pollutants A city's main well was recently found to be contaminated with trichloroethylene (a cancer-causing chemical) as a result of an abandoned chemical dump that leached chemicals into the water. A proposal submitted to the city council indicated that the cost, measured in millions of dollars, of removing $x\%$ of the toxic pollutants is given by

$$C(x) = \dfrac{0.5x}{100-x}$$

a. Find the vertical asymptote of the graph of C.
b. Is it possible to remove 100% of the toxic pollutant from the water?

62. Average Cost of Producing DVDs The average cost per disc (in dollars) incurred by Herald Media Corporation in pressing x DVDs is given by the average cost function

$$\overline{C}(x) = 2.2 + \dfrac{2500}{x}$$

a. Find the horizontal asymptote of the graph of $\overline{C}$.
b. What is the limiting value of the average cost?

63. Concentration of a Drug in the Bloodstream The concentration (in milligrams per cubic centimeter) of a certain drug in a patient's bloodstream t hr after injection is given by

$$C(t) = \dfrac{0.2t}{t^2+1}$$

a. Find the horizontal asymptote of the graph of C.
b. Interpret your result.

64. Effect of Enzymes on Chemical Reactions Certain proteins, known as enzymes, serve as catalysts for chemical reactions in living things. In 1913, Leonor Michaelis and L. M. Menten discovered the following formula giving the initial speed V (in moles per liter per second) at which the reaction begins in terms of the amount of substrate x (the substance that is being acted upon, measured in moles per liter):

$$V = \dfrac{ax}{x+b}$$

where a and b are positive constants.
a. Find the horizontal asymptote of the graph of V.
b. What does the result of part (a) tell you about the initial speed at which the reaction begins, if the amount of substrate is very large?

65. GDP of a Developing Country A developing country's gross domestic product (GDP) from 2002 to 2010 is approximated by the function

$$G(t) = -0.2t^3 + 2.4t^2 + 60 \quad (0 \le t \le 8)$$

where $G(t)$ is measured in billions of dollars, and t is measured in years, with $t = 0$ corresponding to 2002. Sketch the graph of the function G, and interpret your results.

66. Crime Rate The number of major crimes per 100,000 committed in a city between 2003 and 2010 is approximated by the function

$$N(t) = -0.1t^3 + 1.5t^2 + 80 \quad (0 \le t \le 7)$$

where $N(t)$ denotes the number of crimes per 100,000 committed in year t, with $t = 0$ corresponding to 2003. Enraged by the dramatic increase in the crime rate, the citizens, with the help of the local police, organized Neighborhood Crime Watch groups in early 2007 to combat this menace. Sketch the graph of the function N', and interpret your results. Is the Neighborhood Crime Watch program working?

4.3 CURVE SKETCHING

67. Worker Efficiency An efficiency study showed that the total number of smartphones assembled by an average worker at Delphi Electronics t hr after starting work at 8 A.M. is given by

$$N(t) = -\frac{1}{2}t^3 + 3t^2 + 10t \qquad (0 \le t \le 4)$$

Sketch the graph of the function N, and interpret your results.

68. Concentration of a Drug in the Bloodstream The concentration (in milligrams per cubic centimeter) of a certain drug in a patient's bloodstream t hr after injection is given by

$$C(t) = \frac{0.2t}{t^2 + 1}$$

Sketch the graph of the function C, and interpret your results.

69. Box-Office Receipts The total worldwide box-office receipts for a long-running movie are approximated by the function

$$T(x) = \frac{120x^2}{x^2 + 4}$$

where $T(x)$ is measured in millions of dollars and x is the number of years since the movie's release. Sketch the graph of the function T, and interpret your results.

70. Oxygen Content of a Pond When organic waste is dumped into a pond, the oxidation process that takes place reduces the pond's oxygen content. However, given time, nature will restore the oxygen content to its natural level. Suppose the oxygen content t days after organic waste has been dumped into the pond is given by

$$f(t) = 100\left(\frac{t^2 - 4t + 4}{t^2 + 4}\right) \qquad (0 \le t < \infty)$$

percent of its normal level. Sketch the graph of the function f, and interpret your results.

71. Cost of Removing Toxic Pollutants Refer to Exercise 61. The cost, measured in millions of dollars, of removing $x\%$ of a toxic pollutant is given by

$$C(x) = \frac{0.5x}{100 - x}$$

Sketch the graph of the function C, and interpret your results.

72. Traffic Flow Analysis The speed of traffic flow in miles per hour on a stretch of Route 123 between 6 A.M. and 10 A.M. on a typical workday is approximated by the function

$$f(t) = 20t - 40\sqrt{t} + 52 \qquad (0 \le t \le 4)$$

where t is measured in hours, with $t = 0$ corresponding to 6 A.M. Sketch the graph of f, and interpret your results.

4.3 Solutions to Self-Check Exercises

1. Since

$$\lim_{x \to \infty} \frac{2x^2}{x^2 - 1} = \lim_{x \to \infty} \frac{2}{1 - \frac{1}{x^2}} \quad \text{Divide the numerator and denominator by } x^2.$$

$$= 2$$

we see that $y = 2$ is a horizontal asymptote. Next, since

$$x^2 - 1 = (x + 1)(x - 1) = 0$$

implies $x = -1$ or $x = 1$, these are candidates for the vertical asymptotes of the graph of f. Since the numerator of f is not equal to zero for $x = -1$ or $x = 1$, we conclude that $x = -1$ and $x = 1$ are vertical asymptotes of the graph of f.

2. We obtain the following information on the graph of f.
(1) The domain of f is the interval $(-\infty, \infty)$.
(2) By setting $x = 0$, we find that the y-intercept is 4.
(3) Since

$$\lim_{x \to -\infty} f(x) = \lim_{x \to -\infty} \left(\frac{2}{3}x^3 - 2x^2 - 6x + 4\right) = -\infty$$

$$\lim_{x \to \infty} f(x) = \lim_{x \to \infty} \left(\frac{2}{3}x^3 - 2x^2 - 6x + 4\right) = \infty$$

we see that $f(x)$ decreases without bound as x decreases without bound and that $f(x)$ increases without bound as x increases without bound.

(4) Since f is a polynomial function, the graph of f has no asymptotes.

(5) $\quad f'(x) = 2x^2 - 4x - 6 = 2(x^2 - 2x - 3)$
$\qquad\qquad = 2(x + 1)(x - 3)$

Setting $f'(x) = 0$ gives $x = -1$ or $x = 3$. The accompanying sign diagram for f' shows that f is increasing on the intervals $(-\infty, -1)$ and $(3, \infty)$ and decreasing on $(-1, 3)$.

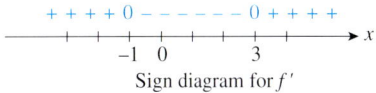

Sign diagram for f'

(6) From the results of Step 5, we see that $x = -1$ and $x = 3$ are critical numbers of f. Furthermore, the sign diagram of f' tells us that $x = -1$ gives rise to a relative maximum of f and $x = 3$ gives rise to a relative minimum of f. Now,

$$f(-1) = \frac{2}{3}(-1)^3 - 2(-1)^2 - 6(-1) + 4 = \frac{22}{3}$$

$$f(3) = \frac{2}{3}(3)^3 - 2(3)^2 - 6(3) + 4 = -14$$

so $f(-1) = \frac{22}{3}$ is a relative maximum of f and $f(3) = -14$ is a relative minimum of f.

(7) $$f''(x) = 4x - 4 = 4(x - 1)$$

which is equal to zero when $x = 1$. The accompanying sign diagram of f'' shows that the graph of f is concave downward on the interval $(-\infty, 1)$ and concave upward on the interval $(1, \infty)$.

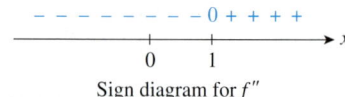

Sign diagram for f''

(8) From the results of Step 7, we see that $x = 1$ is the only candidate for an inflection point of f. Since $f''(x)$ changes sign as we move across the point $x = 1$ and

$$f(1) = \frac{2}{3}(1)^3 - 2(1)^2 - 6(1) + 4 = -\frac{10}{3}$$

we see that the required inflection point is $(1, -\frac{10}{3})$.

(9) Summarizing this information, we have the following:

Domain: $(-\infty, \infty)$
Intercept: $(0, 4)$
$\lim_{x \to -\infty} f(x); \lim_{x \to \infty} f(x): -\infty; \infty$
Asymptotes: None

Intervals where f is ↗ or ↘: ↗ on $(-\infty, -1)$ and $(3, \infty)$; ↘ on $(-1, 3)$
Relative extrema: Rel. max. at $(-1, \frac{22}{3})$; rel. min. at $(3, -14)$
Concavity: Downward on $(-\infty, 1)$; upward on $(1, \infty)$
Point of inflection: $(1, -\frac{10}{3})$

The graph of f is sketched in the accompanying figure.

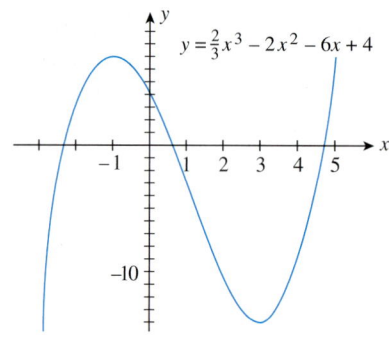

USING TECHNOLOGY

Analyzing the Properties of a Function

One of the main purposes of studying Section 4.3 is to see how the many concepts of calculus come together to paint a picture of a function. The techniques of graphing also play a very practical role. For example, using the techniques of graphing developed in Section 4.3, you can tell whether the graph of a function generated by a graphing utility is reasonably complete. Furthermore, these techniques can often reveal details that are missing from a graph.

EXAMPLE 1 Consider the function $f(x) = 2x^3 - 3.5x^2 + x - 10$. A plot of the graph of f in the standard viewing window is shown in Figure T1. Since the domain of f is the interval $(-\infty, \infty)$, we see that Figure T1 does not reveal the part of the graph to the left of the y-axis. This suggests that we enlarge the viewing window accordingly. Figure T2 shows the graph of f in the viewing window $[-10, 10] \times [-20, 10]$.

The behavior of f for large values of x

$$\lim_{x \to -\infty} f(x) = -\infty \quad \text{and} \quad \lim_{x \to \infty} f(x) = \infty$$

suggests that this viewing window has captured a sufficiently complete picture of f.

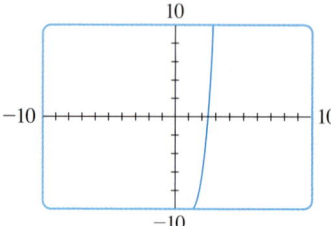

FIGURE T1
The graph of f in the standard viewing window

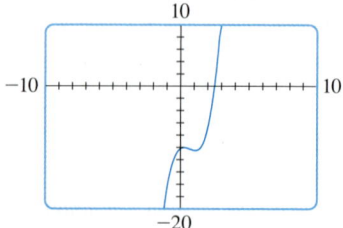

FIGURE T2
The graph of f in the viewing window $[-10, 10] \times [-20, 10]$

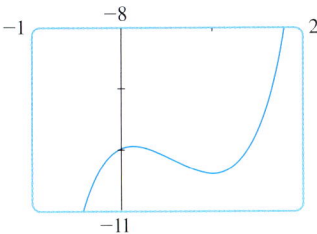

FIGURE T3
The graph of f in the viewing window $[-1, 2] \times [-11, -8]$

Next, an analysis of the first derivative of x,

$$f'(x) = 6x^2 - 7x + 1 = (6x - 1)(x - 1)$$

reveals that f has critical values at $x = \frac{1}{6}$ and $x = 1$. In fact, a sign diagram of f' shows that f has a relative maximum at $x = \frac{1}{6}$ and a relative minimum at $x = 1$, details that are not revealed in the graph of f shown in Figure T2. To examine this portion of the graph of f, we use, say, the viewing window $[-1, 2] \times [-11, -8]$. The resulting graph of f is shown in Figure T3, which certainly reveals the hitherto missing details! Thus, through an interaction of calculus and a graphing utility, we are able to obtain a good picture of the properties of f.

Finding x-Intercepts

As was noted in Section 4.3, it is not always easy to find the x-intercepts of the graph of a function. But this information is very important in applications. By using the function for solving polynomial equations or the function for finding the roots of an equation, we can solve the equation $f(x) = 0$ quite easily and hence yield the x-intercepts of the graph of a function.

EXAMPLE 2 Let $f(x) = x^3 - 3x^2 + x + 1.5$.

a. Use the function for solving polynomial equations on a graphing utility to find the x-intercepts of the graph of f.
b. Use the function for finding the roots of an equation on a graphing utility to find the x-intercepts of the graph of f.

Solution

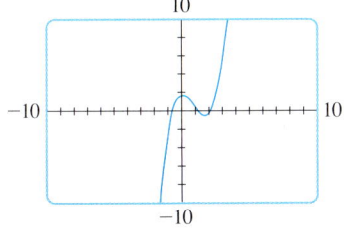

FIGURE T4
The graph of $f(x) = x^3 - 3x^2 + x + 1.5$

a. Observe that f is a polynomial function of degree 3, and so we may use the function for solving polynomial equations to solve the equation $x^3 - 3x^2 + x + 1.5 = 0$ [$f(x) = 0$]. We find that the solutions (x-intercepts) are

$$x_1 \approx -0.525687120865 \qquad x_2 \approx 1.2586520225 \qquad x_3 \approx 2.26703509836$$

b. Using the graph of f (Figure T4), we see that $x_1 \approx -0.5$, $x_2 \approx 1$, and $x_3 \approx 2$. Using the function for finding the roots of an equation on a graphing utility and these values of x as initial guesses, we find

$$x_1 \approx -0.5256871209 \qquad x_2 \approx 1.2586520225 \qquad x_3 \approx 2.2670350984$$

Note The function for solving polynomial equations on a graphing utility will solve a polynomial equation $f(x) = 0$, where f is a polynomial function. The function for finding the roots of an equation, however, will solve equations $f(x) = 0$ even if f is not a polynomial.

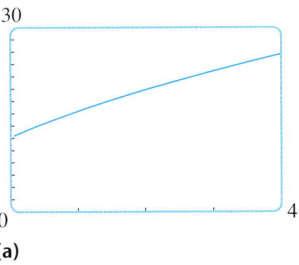

(a)

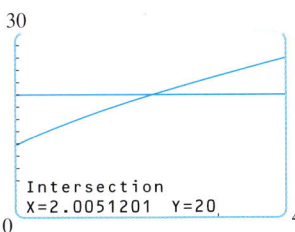

(b)

FIGURE T5
(a) The graph of N in the viewing window $[0, 4] \times [0, 30]$; (b) the graph showing the intersection of $y_1 = N(t)$ and $y_2 = 20$ on the TI 83/84.

 APPLIED EXAMPLE 3 TV Mobile Phones The number of people watching TV on mobile phones (in millions) is approximated by

$$N(t) = 11.9\sqrt{1 + 0.91t} \qquad (0 \leq t \leq 4)$$

where t is measured in years, with $t = 0$ corresponding to the beginning of 2007.

a. Use a graphing calculator to plot the graph of N.
b. Based on this model, when did the number of people watching TV on mobile phones first exceed 20 million?

Source: IDC, U.S. forecast.

Solution

a. The graph of N in the window $[0, 4] \times [0, 30]$ is shown in Figure T5a.
b. Using the function for finding the intersection of the graphs of $y_1 = N(t)$ and $y_2 = 20$, we find $t \approx 2.005$ (see Figure T5b). So the number of people watching TV on mobile phones first exceeded 20 million at the beginning of 2009.

(continued)

TECHNOLOGY EXERCISES

In Exercises 1–4, use the method of Example 1 to analyze the function. (*Note:* Your answers will *not* be unique.)

1. $f(x) = 4x^3 - 4x^2 + x + 10$
2. $f(x) = x^3 + 2x^2 + x - 12$
3. $f(x) = \frac{1}{2}x^4 + x^3 + \frac{1}{2}x^2 - 10$
4. $f(x) = 2.25x^4 - 4x^3 + 2x^2 + 2$

In Exercises 5–8, find the *x*-intercepts of the graph of *f*. Give your answers accurate to four decimal places.

5. $f(x) = 0.2x^3 - 1.2x^2 + 0.8x + 2.1$
6. $f(x) = -0.2x^4 + 0.8x^3 - 2.1x + 1.2$
7. $f(x) = 2x^2 - \sqrt{x+1} - 3$
8. $f(x) = x - \sqrt{1 - x^2}$

9. **AIR POLLUTION** The level of ozone, an invisible gas that irritates and impairs breathing, present in the atmosphere on a certain day in June in the city of Riverside is approximated by

$$S(t) = 1.0974t^3 - 0.0915t^4 \quad (0 \le t \le 11)$$

where $S(t)$ is measured in pollutant standard index (PSI) and *t* is measured in hours, with $t = 0$ corresponding to 7 A.M. Sketch the graph of *S*, and interpret your results.
Source: Los Angeles Times.

10. **FLIGHT PATH OF A PLANE** The function

$$f(x) = \begin{cases} 0 & \text{if } 0 \le x < 1 \\ -0.0411523x^3 + 0.679012x^2 \\ \quad -1.23457x + 0.596708 & \text{if } 1 \le x < 10 \\ 15 & \text{if } 10 \le x \le 11 \end{cases}$$

where both *x* and $f(x)$ are measured in units of 1000 ft, describes the flight path of a plane taking off from the origin and climbing to an altitude of 15,000 ft. Sketch the graph of *f* to visualize the trajectory of the plane.

4.4 Optimization I

Absolute Extrema

The graph of the function *f* in Figure 57 shows the average age of cars in use in the United States from the beginning of 1946 ($t = 0$) to the beginning of 2009 ($t = 63$). Observe that the highest average age of cars in use during this period is 9.3 years, whereas the lowest average age of cars in use during the same period is 5.5 years. The number 9.3, the largest value of $f(t)$ for all values of *t* in the interval [0, 63] (the domain of *f*), is called the *absolute maximum value of f* on that interval. The number 5.5, the smallest value of $f(t)$ for all values of *t* in [0, 63], is called the *absolute minimum value of f* on that interval. Notice, too, that the absolute maximum value of *f* is attained at the endpoint $t = 63$ of the interval, whereas the absolute minimum value of *f* is attained at the points $t = 12$ (corresponding to 1958) and $t = 23$ (corresponding to 1969) that lie within the interval (0, 63).

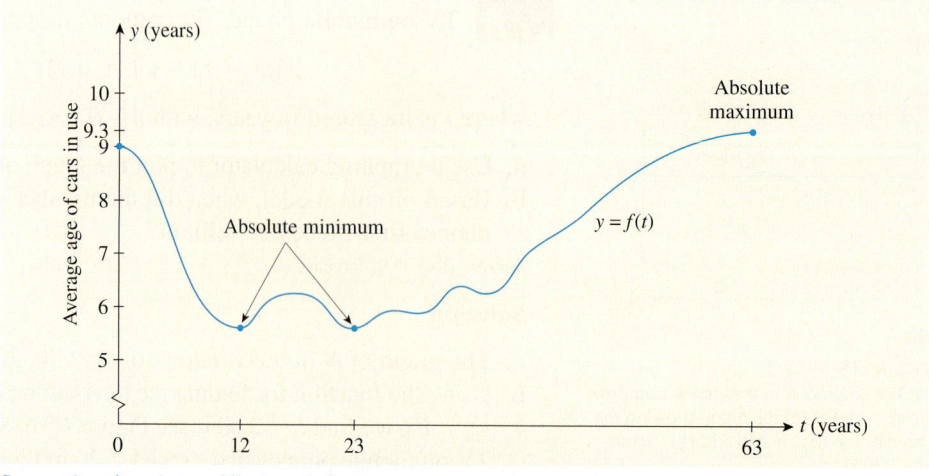

FIGURE 57
f(t) gives the average age of cars in use in year t, t in [0, 63].
Source: American Automobile Association.

(Incidentally, it is interesting to note that 1946 marked the first year of peace following World War II and the two years 1958 and 1969 marked the end of two periods of prosperity in recent U.S. history.)

A precise definition of the **absolute extrema** (absolute maximum or absolute minimum) of a function follows.

> **The Absolute Extrema of a Function f**
>
> If $f(x) \leq f(c)$ for all x in the domain of f, then $f(c)$ is called the **absolute maximum value** of f.
>
> If $f(x) \geq f(c)$ for all x in the domain of f, then $f(c)$ is called the **absolute minimum value** of f.

Figure 58 shows the graphs of several functions and gives the absolute maximum and absolute minimum of each function, if they exist.

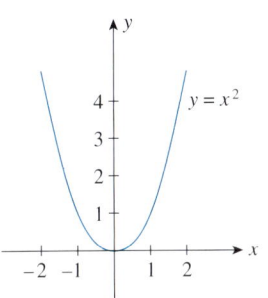
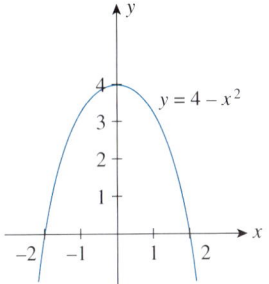
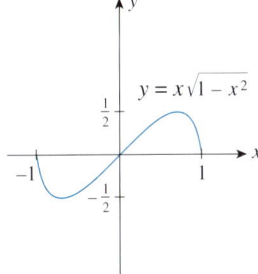
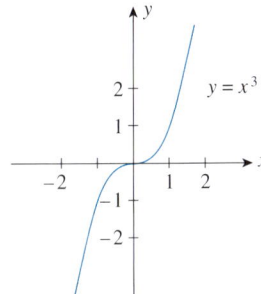

(a) $f(0) = 0$ is the absolute minimum of f; f has no absolute maximum.

(b) $f(0) = 4$ is the absolute maximum of f; f has no absolute minimum.

(c) $f(\sqrt{2}/2) = 1/2$ is the absolute maximum of f; $f(-\sqrt{2}/2) = -1/2$ is the absolute minimum of f.

(d) f has no absolute extrema.

FIGURE 58

Absolute Extrema on a Closed Interval

As the preceding examples show, a continuous function defined on an arbitrary interval does not always have an absolute maximum or an absolute minimum. But an important case arises often in practical applications in which both the absolute maximum and the absolute minimum of a function are guaranteed to exist. This occurs when a continuous function is defined on a *closed* interval. Let's state this important result in the form of a theorem, whose proof we will omit.

> **THEOREM 3**
> **The Extreme Value Theorem**
>
> If a function f is continuous on a closed interval $[a, b]$, then f has both an absolute maximum value and an absolute minimum value on $[a, b]$.

Observe that if an absolute extremum of a continuous function f occurs at a point in an open interval (a, b), then it must be a relative extremum of f, and hence its x-coordinate must be a critical number of f. Otherwise, the absolute extremum of f must occur at one or both of the endpoints of the interval $[a, b]$. A typical situation is illustrated in Figure 59.

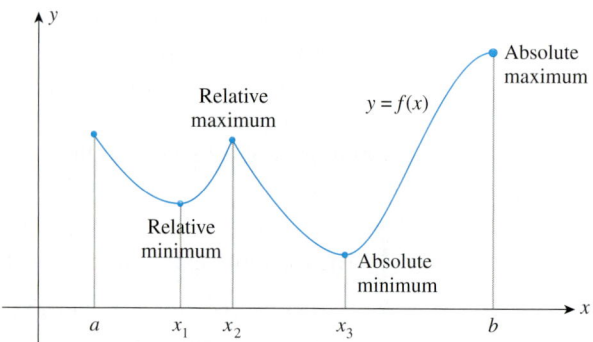

FIGURE 59
The relative minimum of f at x_3 is the absolute minimum of f. The right endpoint b of the interval $[a, b]$ gives rise to the absolute maximum value $f(b)$ of f.

Here, x_1, x_2, and x_3 are critical numbers of f. The absolute minimum of f occurs at x_3, which lies in the open interval (a, b) and is a critical number of f. The absolute maximum of f occurs at b, an endpoint. This observation suggests the following procedure for finding the absolute extrema of a continuous function on a closed interval.

> **Finding the Absolute Extrema of f on a Closed Interval**
> 1. Find the critical numbers of f that lie in (a, b).
> 2. Compute the value of f at each critical number found in Step 1 and compute $f(a)$ and $f(b)$.
> 3. The absolute maximum value and absolute minimum value of f will correspond to the largest and smallest numbers, respectively, found in Step 2.

EXAMPLE 1 Find the absolute extrema of the function $F(x) = x^2$ defined on the interval $[-1, 2]$.

Solution The function F is continuous on the closed interval $[-1, 2]$ and differentiable on the open interval $(-1, 2)$. The derivative of F is

$$F'(x) = 2x$$

so 0 is the only critical number of F. Next, evaluate $F(x)$ at $x = -1$, $x = 0$, and $x = 2$. Thus,

$$F(-1) = 1 \qquad F(0) = 0 \qquad F(2) = 4$$

It follows that 0 is the absolute minimum value of F and 4 is the absolute maximum value of F. The graph of F, in Figure 60, confirms our results.

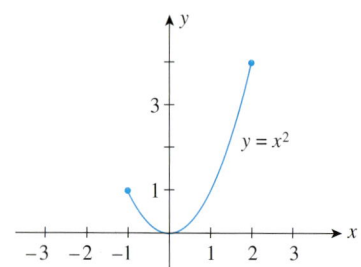

FIGURE 60
F has an absolute minimum value of 0 and an absolute maximum value of 4.

EXAMPLE 2 Find the absolute extrema of the function

$$f(x) = x^3 - 2x^2 - 4x + 4$$

defined on the interval $[0, 3]$.

Solution The function f is continuous on the closed interval $[0, 3]$ and differentiable on the open interval $(0, 3)$. The derivative of f is

$$f'(x) = 3x^2 - 4x - 4 = (3x + 2)(x - 2)$$

and it is equal to zero when $x = -\frac{2}{3}$ and $x = 2$. Since $x = -\frac{2}{3}$ lies outside the interval $[0, 3]$, it is dropped from further consideration, and $x = 2$ is seen to be the sole critical number of f. Next, we evaluate $f(x)$ at the critical number of f as well as the endpoints of f, obtaining

$$f(0) = 4 \qquad f(2) = -4 \qquad f(3) = 1$$

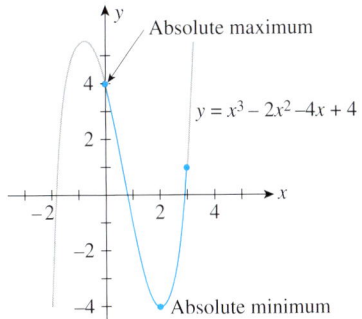

FIGURE 61
f has an absolute maximum value of 4 and an absolute minimum value of −4.

From these results, we conclude that −4 is the absolute minimum value of f and 4 is the absolute maximum value of f. The graph of f, which appears in Figure 61, confirms our results. Observe that the absolute maximum of f occurs at the endpoint $x = 0$ of the interval $[0, 3]$, while the absolute minimum of f occurs at $x = 2$, which lies in the interval $(0, 3)$.

Exploring with TECHNOLOGY

Let $f(x) = x^3 − 2x^2 − 4x + 4$. (This is the function of Example 2.)

1. Use a graphing utility to plot the graph of f, using the viewing window $[0, 3] \times [−5, 5]$. Use ZOOM and TRACE to find the absolute extrema of f on the interval $[0, 3]$ and thus verify the results obtained analytically in Example 2.

2. Plot the graph of f, using the viewing window $[−2, 1] \times [−5, 6]$. Use ZOOM and TRACE to find the absolute extrema of f on the interval $[−2, 1]$. Verify your results analytically.

EXAMPLE 3 Find the absolute maximum and absolute minimum values of the function $f(x) = x^{2/3}$ on the interval $[−1, 8]$.

Solution The derivative of f is

$$f'(x) = \frac{2}{3}x^{-1/3} = \frac{2}{3x^{1/3}}$$

Note that f' is not defined at $x = 0$, and does not equal zero for any x. Therefore, 0 is the only critical number of f. Evaluating $f(x)$ at $x = −1, 0,$ and 8, we obtain

$$f(−1) = 1 \qquad f(0) = 0 \qquad f(8) = 4$$

We conclude that the absolute minimum value of f is 0, attained at $x = 0$, and the absolute maximum value of f is 4, attained at $x = 8$ (Figure 62).

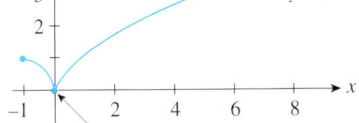

FIGURE 62
f has an absolute minimum value of $f(0) = 0$ and an absolute maximum value of $f(8) = 4$.

Many real-world applications call for finding the absolute maximum value or the absolute minimum value of a given function. For example, management is interested in finding what level of production will yield the maximum profit for a company; a farmer is interested in finding the right amount of fertilizer to maximize crop yield; a doctor is interested in finding the maximum concentration of a drug in a patient's body and the time at which it occurs; and an engineer is interested in finding the dimensions of a container with a specified shape and volume that can be constructed at a minimum cost.

APPLIED EXAMPLE 4 Maximizing Profits Acrosonic's total profit (in dollars) from manufacturing and selling x units of their model F loudspeaker systems is given by

$$P(x) = −0.02x^2 + 300x − 200{,}000 \qquad (0 \le x \le 20{,}000)$$

How many units of the loudspeaker system must Acrosonic produce to maximize its profits?

Solution To find the absolute maximum of P on $[0, 20{,}000]$, first find the critical points of P on the interval $(0, 20{,}000)$. To do this, compute

$$P'(x) = −0.04x + 300$$

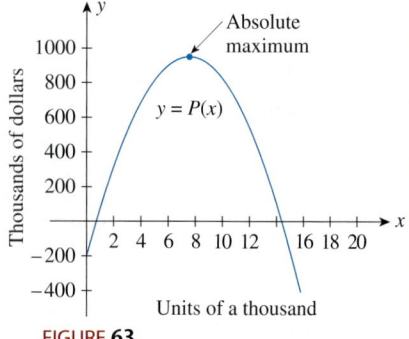

FIGURE 63
P has an absolute maximum at (7500, 925,000).

Solving the equation $P'(x) = 0$ gives $x = 7500$. Next, evaluate $P(x)$ at $x = 7500$ as well as the endpoints $x = 0$ and $x = 20{,}000$ of the interval $[0, 20{,}000]$, obtaining

$$P(0) = -200{,}000$$
$$P(7500) = 925{,}000$$
$$P(20{,}000) = -2{,}200{,}000$$

From these computations, we see that the absolute maximum value of the function P is 925,000. Thus, by producing 7500 units, Acrosonic will realize a maximum profit of $925,000. The graph of P is sketched in Figure 63. ∎

Explore & Discuss

Recall that the total profit function P is defined as $P(x) = R(x) - C(x)$, where R is the total revenue function, C is the total cost function, and x is the number of units of a product produced and sold. (Assume that all derivatives exist.)

1. Show that at the level of production x_0 that yields the maximum profit for the company, the following two conditions are satisfied:

$$R'(x_0) = C'(x_0) \quad \text{and} \quad R''(x_0) < C''(x_0)$$

2. Interpret the two conditions in part 1 in economic terms, and explain why they make sense.

APPLIED EXAMPLE 5 Trachea Contraction During a Cough When a person coughs, the trachea (windpipe) contracts, allowing air to be expelled at a maximum velocity. It can be shown that during a cough, the velocity v of airflow is given by the function

$$v = f(r) = kr^2(R - r)$$

where r is the trachea's radius (in centimeters) during a cough, R is the trachea's normal radius (in centimeters), and k is a positive constant that depends on the length of the trachea. Find the radius r for which the velocity of airflow is greatest.

Solution To find the absolute maximum of f on $[0, R]$, first find the critical numbers of f on the interval $(0, R)$. We compute

$$f'(r) = 2kr(R - r) - kr^2 \quad \text{Use the Product Rule.}$$
$$= -3kr^2 + 2kRr = kr(-3r + 2R)$$

Setting $f'(r) = 0$ gives $r = 0$ or $r = \frac{2}{3}R$, so $\frac{2}{3}R$ is the sole critical number of f ($r = 0$ is an endpoint). Evaluating $f(r)$ at $r = \frac{2}{3}R$, as well as at the endpoints $r = 0$ and $r = R$, we obtain

$$f(0) = 0$$
$$f\left(\frac{2}{3}R\right) = \frac{4k}{27}R^3$$
$$f(R) = 0$$

from which we deduce that the velocity of airflow is greatest when the radius of the contracted trachea is $\frac{2}{3}R$—that is, when the radius is contracted by approximately 33%. The graph of the function f is shown in Figure 64. ∎

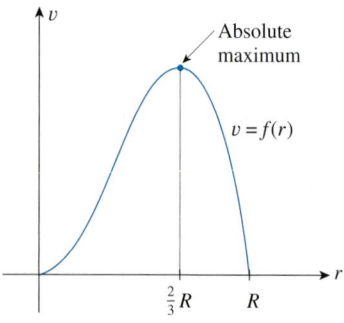

FIGURE 64
The velocity of airflow is greatest when the radius of the contracted trachea is $\frac{2}{3}R$.

Explore & Discuss

Prove that if a cost function $C(x)$ is concave upward $[C''(x) > 0]$, then the level of production that will result in the smallest average production cost occurs when

$$\overline{C}(x) = C'(x)$$

—that is, when the average cost $\overline{C}(x)$ is equal to the marginal cost $C'(x)$.

Hints:

1. Show that

$$\overline{C}'(x) = \frac{xC'(x) - C(x)}{x^2}$$

so the critical number of the function $\overline{C}$ occurs when

$$xC'(x) - C(x) = 0$$

2. Show that at a critical number of $\overline{C}$

$$\overline{C}''(x) = \frac{C''(x)}{x}$$

Use the Second Derivative Test to reach the desired conclusion.

 APPLIED EXAMPLE 6 Minimizing Average Cost The daily average cost function (in dollars per unit) of Elektra Electronics is given by

$$\overline{C}(x) = 0.0001x^2 - 0.08x + 40 + \frac{5000}{x} \quad (x > 0)$$

where x stands for the number of graphing calculators that Elektra produces. Show that a production level of 500 units per day results in a minimum average cost for the company.

Solution The domain of the function $\overline{C}$ is the interval $(0, \infty)$, which is not closed. To solve the problem, we resort to the graphical method. Using the techniques of graphing from the last section, we sketch the graph of $\overline{C}$ (Figure 65).
Now,

$$\overline{C}'(x) = 0.0002x - 0.08 - \frac{5000}{x^2}$$

Substituting the given value of x, 500, into $\overline{C}'(x)$ gives $\overline{C}'(500) = 0$, so 500 is a critical number of $\overline{C}$. Next,

$$\overline{C}''(x) = 0.0002 + \frac{10{,}000}{x^3}$$

Thus,

$$\overline{C}''(500) = 0.0002 + \frac{10{,}000}{(500)^3} > 0$$

and by the Second Derivative Test, a relative minimum of the function $\overline{C}$ occurs at 500. Furthermore, $\overline{C}''(x) > 0$ for $x > 0$, which implies that the graph of $\overline{C}$ is

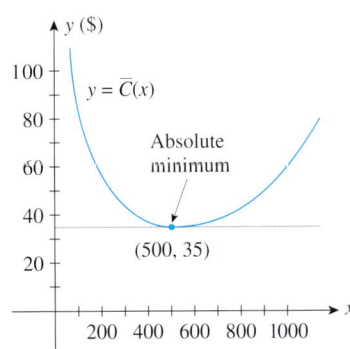

FIGURE 65
The minimum average cost is $35 per unit.

concave upward everywhere, so the relative minimum of $\overline{C}$ must be the absolute minimum of $\overline{C}$. The minimum average cost is given by

$$\overline{C}(500) = 0.0001(500)^2 - 0.08(500) + 40 + \frac{5000}{500}$$

$$= 35$$

or $35 per unit.

> ### Exploring with TECHNOLOGY
>
> Refer to the preceding Explore & Discuss and Example 6.
>
> 1. Using a graphing utility, plot the graphs of
>
> $$\overline{C}(x) = 0.0001x^2 - 0.08x + 40 + \frac{5000}{x}$$
>
> $$C'(x) = 0.0003x^2 - 0.16x + 40$$
>
> using the viewing window $[0, 1000] \times [0, 150]$.
>
> *Note:* $C(x) = 0.0001x^3 - 0.08x^2 + 40x + 5000$. (Why?)
>
> 2. Find the point of intersection of the graphs of $\overline{C}$ and C' and thus verify the assertion in the Explore & Discuss for the special case studied in Example 6.

APPLIED EXAMPLE 7 Flight of a Rocket The altitude (in feet) of a rocket t seconds into flight is given by

$$s = f(t) = -t^3 + 96t^2 + 5 \qquad (t \geq 0)$$

a. Find the maximum altitude attained by the rocket.
b. Find the maximum velocity attained by the rocket.

Solution

a. The maximum altitude attained by the rocket is given by the largest value of the function f in the closed interval $[0, T]$, where T denotes the time the rocket touches the earth. We know that such a number exists because the dominant term in the expression for the continuous function f is $-t^3$. So for t large enough, the value of $f(t)$ must change from positive to negative and, in particular, it must attain the value 0 for some T.

To find the absolute maximum of f, compute

$$f'(t) = -3t^2 + 192t$$

$$= -3t(t - 64) \qquad \text{(x^2) See page 11.}$$

and solve the equation $f'(t) = 0$, obtaining $t = 0$ and $t = 64$. Evaluating f at the critical number $t = 64$ and the endpoints of f, we have

$$f(0) = 5 \qquad f(64) = 131{,}077$$

and conclude, accordingly, that the absolute maximum value of f is 131,077. Thus, the maximum altitude of the rocket is 131,077 feet, attained 64 seconds into flight. The graph of f is sketched in Figure 66.

b. To find the maximum velocity attained by the rocket, find the largest value of the function that describes the rocket's velocity at any time t—namely,

$$v = f'(t) = -3t^2 + 192t \qquad (t \geq 0)$$

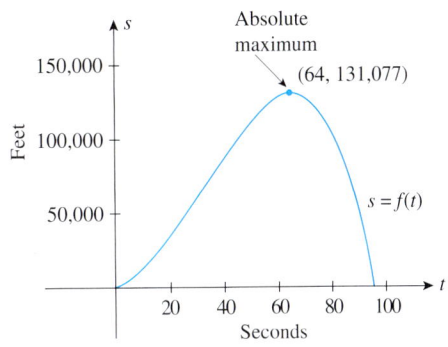

FIGURE 66
The maximum altitude of the rocket is 131,077 feet.

We find the critical number of v by setting $v' = 0$. But

$$v' = -6t + 192$$

and the critical number of v is 32. Since

$$v'' = -6 < 0$$

the Second Derivative Test implies that a relative maximum of v occurs at $t = 32$. Our computation has in fact clarified the property of the "velocity curve." Since $v'' < 0$ everywhere, the velocity curve is concave downward everywhere. With this observation, we assert that the relative maximum must in fact be the absolute maximum of v. The maximum velocity of the rocket is given by evaluating v at $t = 32$:

$$f'(32) = -3(32)^2 + 192(32)$$

or 3072 feet per second. The graph of the velocity function v is sketched in Figure 67.

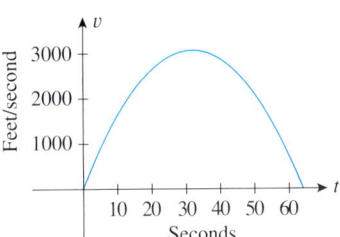

FIGURE 67
The maximum velocity of the rocket is 3072 feet per second.

4.4 Self-Check Exercises

1. Let $f(x) = x - 2\sqrt{x}$.
   a. Find the absolute extrema of f on the interval $[0, 9]$.
   b. Find the absolute extrema of f.

2. Find the absolute extrema of $f(x) = 3x^4 + 4x^3 + 1$ on $[-2, 1]$.

3. The operating rate (expressed as a percent) of factories, mines, and utilities in a certain region of the country on the tth day of 2010 is given by the function

$$f(t) = 80 + \frac{1200t}{t^2 + 40{,}000} \quad (0 \le t \le 250)$$

On which of the first 250 days of 2010 was the operating rate highest?

Solutions to Self-Check Exercises 4.4 can be found on page 310.

4.4 Concept Questions

1. Explain the following terms: (a) absolute maximum and (b) absolute minimum.

2. Describe the procedure for finding the absolute extrema of a continuous function on a closed interval.

4.4 Exercises

In Exercises 1–8, you are given the graph of a function f defined on the indicated interval. Find the absolute maximum and the absolute minimum of f, if they exist.

1.
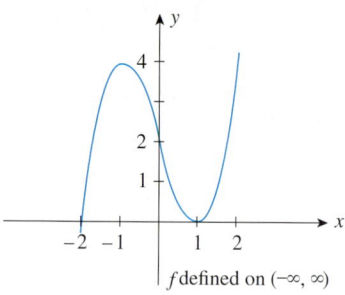
f defined on $(-\infty, \infty)$

2.
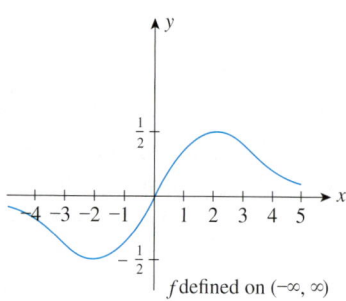
f defined on $(-\infty, \infty)$

3.
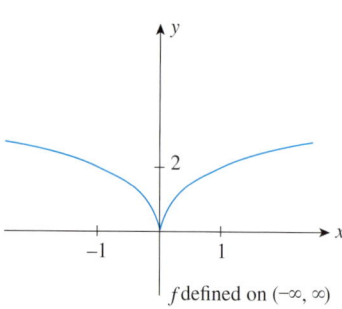
f defined on $(-\infty, \infty)$

4.
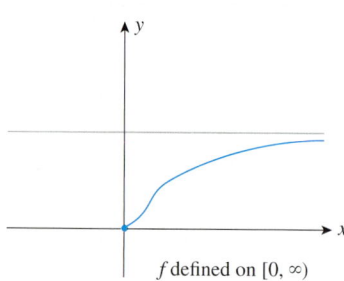
f defined on $[0, \infty)$

5.
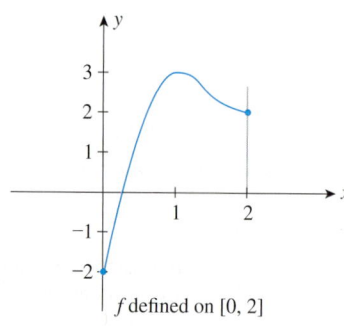
f defined on $[0, 2]$

6.
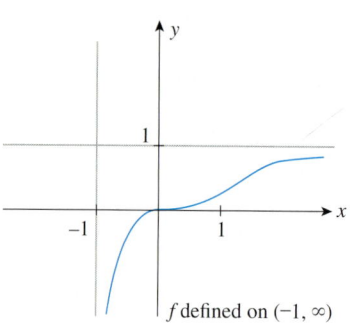
f defined on $(-1, \infty)$

7.
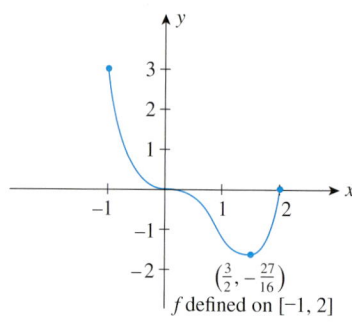
f defined on $[-1, 2]$

8.
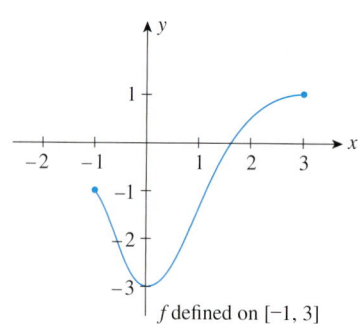
f defined on $[-1, 3]$

In Exercises 9–38, find the absolute maximum value and the absolute minimum value, if any, of each function.

9. $f(x) = 2x^2 + 3x - 4$
10. $g(x) = -x^2 + 4x + 3$
11. $h(x) = x^{1/3}$
12. $f(x) = x^{2/3}$
13. $f(x) = \dfrac{1}{1 + x^2}$
14. $f(x) = \dfrac{x}{1 + x^2}$
15. $f(x) = x^2 - 2x - 3$ on $[-2, 3]$
16. $g(x) = x^2 - 2x - 3$ on $[0, 4]$
17. $f(x) = -x^2 + 4x + 6$ on $[0, 5]$
18. $f(x) = -x^2 + 4x + 6$ on $[3, 6]$
19. $f(x) = x^3 + 3x^2 - 1$ on $[-3, 2]$
20. $g(x) = x^3 + 3x^2 - 1$ on $[-3, 1]$
21. $g(x) = 3x^4 + 4x^3$ on $[-2, 1]$
22. $f(x) = \dfrac{1}{2}x^4 - \dfrac{2}{3}x^3 - 2x^2 + 3$ on $[-2, 3]$

23. $f(x) = \dfrac{x+1}{x-1}$ on $[2, 4]$ 24. $g(t) = \dfrac{t}{t-1}$ on $[2, 4]$

25. $f(x) = 4x + \dfrac{1}{x}$ on $[1, 4]$

26. $f(x) = 9x - \dfrac{1}{x}$ on $[1, 3]$

27. $f(x) = \dfrac{1}{2}x^2 - 2\sqrt{x}$ on $[0, 3]$

28. $g(x) = \dfrac{1}{8}x^2 - 4\sqrt{x}$ on $[0, 9]$

29. $f(x) = \dfrac{1}{x}$ on $(0, \infty)$ 30. $g(x) = \dfrac{1}{x+1}$ on $(0, \infty)$

31. $f(x) = 3x^{2/3} - 2x$ on $[0, 3]$

32. $g(x) = x^2 + 2x^{2/3}$ on $[-2, 2]$

33. $f(x) = x^{2/3}(x^2 - 4)$ on $[-1, 2]$

34. $f(x) = x^{2/3}(x^2 - 4)$ on $[-1, 3]$

35. $f(x) = \dfrac{x}{x^2 + 2}$ on $[-1, 2]$

36. $f(x) = \dfrac{1}{x^2 + 2x + 5}$ on $[-2, 1]$

37. $f(x) = \dfrac{x}{\sqrt{x^2 + 1}}$ on $[-1, 1]$

38. $g(x) = x\sqrt{4 - x^2}$ on $[0, 2]$

39. A stone is thrown straight up from the roof of an 80-ft building. The height (in feet) of the stone at any time t (in seconds), measured from the ground, is given by

$$h(t) = -16t^2 + 64t + 80$$

What is the maximum height the stone reaches?

40. **Maximizing Profits** Lynbrook West, an apartment complex, has 100 two-bedroom units. The monthly profit (in dollars) realized from renting out x apartments is given by

$$P(x) = -10x^2 + 1760x - 50,000$$

To maximize the monthly rental profit, how many units should be rented out? What is the maximum monthly profit realizable?

41. **Seniors in the Workforce** The percentage of men age 65 years and older in the workforce from 1950 ($t = 0$) through 2000 ($t = 50$) is approximately

$$P(t) = 0.0135t^2 - 1.126t + 41.2 \quad (0 \leq t \leq 50)$$

Show that the percentage of men age 65 years and older in the workforce in the period of time under consideration was smallest around mid-September 1991. What is that percentage?

Source: U.S. Census Bureau

42. **Flight of a Rocket** The altitude (in feet) attained by a model rocket t sec into flight is given by the function

$$h(t) = -\dfrac{1}{3}t^3 + 4t^2 + 20t + 2 \quad (t \geq 0)$$

Find the maximum altitude attained by the rocket.

43. **Female Self-Employed Workforce** Data show that the number of nonfarm, full-time, self-employed women can be approximated by

$$N(t) = 0.81t - 1.14\sqrt{t} + 1.53 \quad (0 \leq t \leq 6)$$

where $N(t)$ is measured in millions and t is measured in 5-year intervals, with $t = 0$ corresponding to the beginning of 1963. Determine the absolute extrema of the function N on the interval $[0, 6]$. Interpret your results.

Source: U.S. Department of Labor

44. **Average Speed of a Vehicle** The average speed of a vehicle on a stretch of Route 134 between 6 A.M. and 10 A.M. on a typical weekday is approximated by the function

$$f(t) = 20t - 40\sqrt{t} + 50 \quad (0 \leq t \leq 4)$$

where $f(t)$ is measured in miles per hour and t is measured in hours, with $t = 0$ corresponding to 6 A.M. At what time of the morning commute is the traffic moving at the slowest rate? What is the average speed of a vehicle at that time?

45. **Maximizing Profits** The management of Trappee and Sons, producers of the famous TexaPep hot sauce, estimate that their profit (in dollars) from the daily production and sale of x cases (each case consisting of 24 bottles) of the hot sauce is given by

$$P(x) = -0.000002x^3 + 6x - 400$$

What is the largest possible profit Trappee can make in 1 day?

46. **Maximizing Profits** The quantity demanded each month of the Walter Serkin recording of Beethoven's *Moonlight Sonata*, manufactured by Phonola Record Industries, is related to the price per compact disc. The equation

$$p = -0.00042x + 6 \quad (0 \leq x \leq 12,000)$$

where p denotes the unit price in dollars and x is the number of discs demanded, relates the demand to the price. The total monthly cost (in dollars) for pressing and packaging x copies of this classical recording is given by

$$C(x) = 600 + 2x - 0.00002x^2 \quad (0 \leq x \leq 20,000)$$

To maximize its profits, how many copies should Phonola produce each month?
Hint: The revenue is $R(x) = px$, and the profit is $P(x) = R(x) - C(x)$.

47. **Maximizing Profit** A manufacturer of tennis rackets finds that the total cost $C(x)$ (in dollars) of manufacturing x rackets/day is given by $C(x) = 400 + 4x + 0.0001x^2$. Each racket can be sold at a price of p dollars, where p is related to x by the demand equation $p = 10 - 0.0004x$. If all rackets that are manufactured can be sold, find the daily level of production that will yield a maximum profit for the manufacturer.

48. MAXIMIZING PROFIT The weekly demand for the Pulsar 40-in. high-definition television is given by the demand equation

$$p = -0.05x + 600 \quad (0 \leq x \leq 12{,}000)$$

where p denotes the wholesale unit price in dollars and x denotes the quantity demanded. The weekly total cost function associated with manufacturing these sets is given by

$$C(x) = 0.000002x^3 - 0.03x^2 + 400x + 80{,}000$$

where $C(x)$ denotes the total cost incurred in producing x sets. Find the level of production that will yield a maximum profit for the manufacturer.
Hint: Use the quadratic formula.

49. MAXIMIZING PROFIT A division of Chapman Corporation manufactures a pager. The weekly fixed cost for the division is $20,000, and the variable cost for producing x pagers per week is

$$V(x) = 0.000001x^3 - 0.01x^2 + 50x$$

dollars. The company realizes a revenue of

$$R(x) = -0.02x^2 + 150x \quad (0 \leq x \leq 7500)$$

dollars from the sale of x pagers/week. Find the level of production that will yield a maximum profit for the manufacturer.
Hint: Use the quadratic formula.

50. MINIMIZING AVERAGE COST Suppose the total cost function for manufacturing a certain product is $C(x) = 0.2(0.01x^2 + 120)$ dollars, where x represents the number of units produced. Find the level of production that will minimize the average cost.

51. MINIMIZING PRODUCTION COSTS The total monthly cost (in dollars) incurred by Cannon Precision Instruments for manufacturing x units of the model M1 digital camera is given by the function

$$C(x) = 0.0025x^2 + 80x + 10{,}000$$

a. Find the average cost function $\overline{C}$.
b. Find the level of production that results in the smallest average production cost.
c. Find the level of production for which the average cost is equal to the marginal cost.
d. Compare the result of part (c) with that of part (b).

52. MINIMIZING PRODUCTION COSTS The daily total cost (in dollars) incurred by Trappee and Sons for producing x cases of TexaPep hot sauce is given by the function

$$C(x) = 0.000002x^3 + 5x + 400$$

Using this function, answer the questions posed in Exercise 51.

53. MINIMIZING AVERAGE COST Suppose that the total cost incurred in manufacturing x units of a certain product is given by $C(x)$, where C is a differentiable cost function. Show that the average cost is minimized at the level of production where the average cost is equal to the marginal cost.

54. Re-solve Exercise 51 using the result of Exercise 53.

55. MAXIMIZING REVENUE Suppose the quantity demanded per week of a certain dress is related to the unit price p by the demand equation $p = \sqrt{800 - x}$, where p is in dollars and x is the number of dresses made. To maximize the revenue, how many dresses should be made and sold each week?
Hint: $R(x) = px$.

56. MAXIMIZING REVENUE The quantity demanded each month of the Sicard sports watch is related to the unit price by the equation

$$p = \frac{50}{0.01x^2 + 1} \quad (0 \leq x \leq 20)$$

where p is measured in dollars and x is measured in units of a thousand. To yield a maximum revenue, how many watches must be sold?

57. OXYGEN CONTENT OF A POND When organic waste is dumped into a pond, the oxidation process that takes place reduces the pond's oxygen content. However, given time, nature will restore the oxygen content to its natural level. Suppose the oxygen content t days after organic waste has been dumped into the pond is given by

$$f(t) = 100\left(\frac{t^2 - 4t + 4}{t^2 + 4}\right) \quad (0 \leq t < \infty)$$

percent of its normal level.
a. When is the level of oxygen content lowest?
b. When is the rate of oxygen regeneration greatest?

58. AIR POLLUTION The amount of nitrogen dioxide, a brown gas that impairs breathing, present in the atmosphere on a certain May day in the city of Long Beach is approximated by

$$A(t) = \frac{136}{1 + 0.25(t - 4.5)^2} + 28 \quad (0 \leq t \leq 11)$$

where $A(t)$ is measured in pollutant standard index (PSI) and t is measured in hours, with $t = 0$ corresponding to 7 A.M. Determine the time of day when the pollution is at its highest level.

59. MAXIMIZING REVENUE The average revenue is defined as the function

$$\overline{R}(x) = \frac{R(x)}{x} \quad (x > 0)$$

Prove that if a revenue function $R(x)$ is concave downward $[R''(x) < 0]$, then the level of sales that will result in the largest average revenue occurs when $\overline{R}(x) = R'(x)$.

60. VELOCITY OF BLOOD According to a law discovered by the 19th-century physician Jean Louis Marie Poiseuille, the velocity (in centimeters per second) of blood r cm from the central axis of an artery is given by

$$v(r) = k(R^2 - r^2)$$

where k is a constant and R is the radius of the artery. Show that the velocity of blood is greatest along the central axis.

61. GDP OF A DEVELOPING COUNTRY A developing country's gross domestic product (GDP) from 2004 to 2012 is approximated by the function

$$G(t) = -0.2t^3 + 2.4t^2 + 60 \quad (0 \leq t \leq 8)$$

where $G(t)$ is measured in billions of dollars and $t = 0$ corresponds to 2004. Show that the growth rate of the country's GDP was maximal in 2008.

62. CRIME RATES The number of major crimes committed in the city of Bronxville between 2003 and 2010 is approximated by the function

$$N(t) = -0.1t^3 + 1.5t^2 + 100 \quad (0 \leq t \leq 7)$$

where $N(t)$ denotes the number of crimes committed in year t ($t = 0$ corresponds to 2003). Enraged by the dramatic increase in the crime rate, the citizens of Bronxville, with the help of the local police, organized Neighborhood Crime Watch groups in early 2007 to combat this menace. Show that the growth in the crime rate was maximal in 2008, giving credence to the claim that the Neighborhood Crime Watch program was working.

63. FOREIGN-BORN MEDICAL RESIDENTS The percentage of foreign-born medical residents in the United States from 1910 through 2000 is approximated by the function

$$P(t) = 0.04363t^3 - 0.267t^2 - 1.59t + 14.7 \quad (0 \leq t \leq 9)$$

where t is measured in decades, with $t = 0$ corresponding to 1910. Show that the percentage of foreign-born medical residents was lowest in early 1970.
Hint: Use the quadratic formula.
Source: Journal of the American Medical Association.

64. BRAIN GROWTH AND IQs In a study conducted at the National Institute of Mental Health, researchers followed the development of the cortex, the thinking part of the brain, in 307 children. Using repeated magnetic resonance imaging scans from childhood to the latter teens, they measured the thickness (in millimeters) of the cortex of children of age t years with the highest IQs—121 to 149. These data lead to the model

$$S(t) = 0.000989t^3 - 0.0486t^2 + 0.7116t + 1.46 \quad (5 \leq t \leq 19)$$

Show that the cortex of children with superior intelligence reaches maximum thickness around age 11 years.
Hint: Use the quadratic formula.
Source: Nature.

65. BRAIN GROWTH AND IQs Refer to Exercise 64. The researchers at the Institute also measured the thickness (also in millimeters) of the cortex of children of age t years who were of average intelligence. These data lead to the model

$$A(t) = -0.00005t^3 - 0.000826t^2 + 0.0153t + 4.55 \quad (5 \leq t \leq 19)$$

Show that the cortex of children with average intelligence reaches maximum thickness at age 6 years.
Source: Nature.

66. WORLD POPULATION The total world population is forecast to be

$$P(t) = 0.00074t^3 - 0.0704t^2 + 0.89t + 6.04 \quad (0 \leq t \leq 10)$$

in year t, where t is measured in decades, with $t = 0$ corresponding to 2000 and $P(t)$ is measured in billions.

a. Show that the world population is forecast to peak around 2071.
Hint: Use the quadratic formula.

b. At what number will the population peak?
Source: International Institute for Applied Systems Analysis.

67. VENTURE-CAPITAL INVESTMENT Venture-capital investment increased dramatically in the late 1990s but came to a screeching halt after the dot-com bust. The venture-capital investment (in billions of dollars) from 1995 ($t = 0$) through 2003 ($t = 8$) is approximated by the function

$$C(t) = \begin{cases} 0.6t^2 + 2.4t + 7.6 & \text{if } 0 \leq t < 3 \\ 3t^2 + 18.8t - 63.2 & \text{if } 3 \leq t < 5 \\ -3.3167t^3 + 80.1t^2 \\ \quad - 642.583t + 1730.8025 & \text{if } 5 \leq t < 8 \end{cases}$$

a. In what year did venture-capital investment peak over the period under consideration? What was the amount of that investment?

b. In what year was the venture-capital investment lowest over this period? What was the amount of that investment?
Hint: Find the absolute extrema of C on each of the closed intervals [0, 3], [3, 5], and [5, 8].
Sources: Venture One; Ernst & Young.

68. ENERGY EXPENDED BY A FISH It has been conjectured that a fish swimming a distance of L ft at a speed of v ft/sec relative to the water and against a current flowing at the rate of u ft/sec ($u < v$) expends a total energy given by

$$E(v) = \frac{aLv^3}{v - u}$$

where E is measured in foot-pounds (ft-lb) and a is a constant. Find the speed v at which the fish must swim in order to minimize the total energy expended. (*Note:* This result has been verified by biologists.)

69. REACTION TO A DRUG The strength of a human body's reaction R to a dosage D of a certain drug is given by

$$R = D^2 \left(\frac{k}{2} - \frac{D}{3} \right)$$

where k is a positive constant. Show that the maximum reaction is achieved if the dosage is k units.

70. Refer to Exercise 69. Show that the rate of change in the reaction R with respect to the dosage D is maximal if $D = k/2$.

71. Maximum Power Output Suppose the source of current in an electric circuit is a battery. Then the power output P (in watts) obtained if the circuit has a resistance of R ohms is given by

$$P = \frac{E^2 R}{(R + r)^2}$$

where E is the electromotive force in volts and r is the internal resistance of the battery in ohms. If E and r are constant, find the value of R that will result in the greatest power output. What is the maximum power output?

72. Velocity of a Wave In deep water, a wave of length L travels with a velocity

$$v = k\sqrt{\frac{L}{C} + \frac{C}{L}}$$

where k and C are positive constants. Find the length of the wave that has a minimum velocity.

73. Chemical Reaction In an autocatalytic chemical reaction, the product formed acts as a catalyst for the reaction. If Q is the amount of the original substrate present initially and x is the amount of catalyst formed, then the rate of change of the chemical reaction with respect to the amount of catalyst present in the reaction is

$$R(x) = kx(Q - x) \qquad (0 \leq x \leq Q)$$

where k is a constant. Show that the rate of the chemical reaction is greatest at the point when exactly half of the original substrate has been transformed.

74. A Mixture Problem A tank initially contains 10 gal of brine with 2 lb of salt. Brine with 1.5 lb of salt per gallon enters the tank at the rate of 3 gal/min, and the well-stirred mixture leaves the tank at the rate of 4 gal/min. It can be shown that the amount of salt in the tank after t minutes is x lb, where

$$x = f(t) = 1.5(10 - t) - 0.0013(10 - t)^4 \qquad (0 \leq t \leq 10)$$

What is the maximum amount of salt present in the tank at any time?

In Exercises 75–78, determine whether the statement is true or false. If it is true, explain why it is true. If it is false, give an example to show why it is false.

75. If f is defined on a closed interval $[a, b]$, then f has an absolute maximum value.

76. If f is continuous on an open interval (a, b), then f does not have an absolute minimum value.

77. If f is not continuous on the closed interval $[a, b]$, then f cannot have an absolute maximum value.

78. If $f''(x) < 0$ on (a, b) and $f'(c) = 0$, where $a < c < b$, then $f(c)$ is the absolute maximum value of f on $[a, b]$.

79. Let f be a constant function—that is, let $f(x) = c$, where c is some real number. Show that every number a gives rise to an absolute maximum and, at the same time, an absolute minimum of f.

80. Show that a polynomial function defined on the interval $(-\infty, \infty)$ cannot have both an absolute maximum and an absolute minimum unless it is a constant function.

81. One condition that must be satisfied before Theorem 3 (page 299) is applicable is that the function f must be continuous on the closed interval $[a, b]$. Define a function f on the closed interval $[-1, 1]$ by

$$f(x) = \begin{cases} \dfrac{1}{x} & \text{if } x \in [-1, 1] \quad (x \neq 0) \\ 0 & \text{if } x = 0 \end{cases}$$

a. Show that f is not continuous at $x = 0$.
b. Show that f does not attain an absolute maximum or an absolute minimum on the interval $[-1, 1]$.
c. Confirm your results by sketching the function f.

82. One condition that must be satisfied before Theorem 3 (page 299) is applicable is that the interval on which f is defined must be a closed interval $[a, b]$. Define a function f on the *open* interval $(-1, 1)$ by $f(x) = x$. Show that f does not attain an absolute maximum or an absolute minimum on the interval $(-1, 1)$.
Hint: What happens to $f(x)$ if x is close to but not equal to $x = -1$? If x is close to but not equal to $x = 1$?

4.4 Solutions to Self-Check Exercises

1. a. The function f is continuous in its domain and differentiable on the interval $(0, 9)$. The derivative of f is

$$f'(x) = 1 - x^{-1/2} = \frac{x^{1/2} - 1}{x^{1/2}}$$

and it is equal to zero when $x = 1$. Evaluating $f(x)$ at the endpoints $x = 0$ and $x = 9$ and at the critical number 1 of f, we have

$$f(0) = 0 \qquad f(1) = -1 \qquad f(9) = 3$$

From these results, we see that -1 is the absolute minimum value of f and 3 is the absolute maximum value of f.

b. In this case, the domain of f is the interval $[0, \infty)$, which is not closed. Therefore, we resort to the graphical method. Using the techniques of graphing, we sketch the graph of f in the accompanying figure.

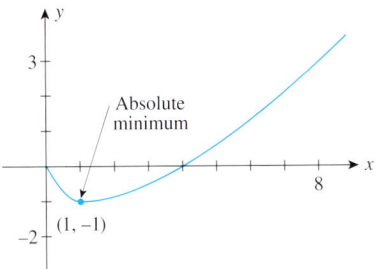

The graph of f shows that -1 is the absolute minimum value of f but f has no absolute maximum, since $f(x)$ increases without bound as x increases without bound.

2. The function f is continuous on the interval $[-2, 1]$. It is also differentiable on the open interval $(-2, 1)$. The derivative of f is

$$f'(x) = 12x^3 + 12x^2 = 12x^2(x + 1)$$

and it is continuous on $(-2, 1)$. Setting $f'(x) = 0$ gives -1 and 0 as critical numbers of f. Evaluating $f(x)$ at these critical numbers of f as well as at the endpoints of the interval $[-2, 1]$, we obtain

$$f(-2) = 17 \quad f(-1) = 0 \quad f(0) = 1 \quad f(1) = 8$$

From these results, we see that 0 is the absolute minimum value of f and 17 is the absolute maximum value of f.

3. The problem is solved by finding the absolute maximum of the function f on $[0, 250]$. Differentiating $f(t)$, we obtain

$$f'(t) = \frac{(t^2 + 40{,}000)(1200) - 1200t(2t)}{(t^2 + 40{,}000)^2}$$

$$= \frac{-1200(t^2 - 40{,}000)}{(t^2 + 40{,}000)^2}$$

Upon setting $f'(t) = 0$ and solving the resulting equation, we obtain $t = -200$ or 200. Since -200 lies outside the interval $[0, 250]$, we are interested only in the critical number 200 of f. Evaluating $f(t)$ at $t = 0$, $t = 200$, and $t = 250$, we find

$$f(0) = 80 \quad f(200) = 83 \quad f(250) \approx 82.93$$

We conclude that the operating rate was the highest on the 200th day of 2010—that is, a little past the middle of July 2010.

USING TECHNOLOGY

Finding the Absolute Extrema of a Function

Some graphing utilities have a function for finding the absolute maximum and the absolute minimum values of a continuous function on a closed interval. If your graphing utility has this capability, use it to work through the example and exercises of this section.

EXAMPLE 1 Let $f(x) = \dfrac{2x + 4}{(x^2 + 1)^{3/2}}$.

a. Use a graphing utility to plot the graph of f in the viewing window $[-3, 3] \times [-1, 5]$.
b. Find the absolute maximum and absolute minimum values of f on the interval $[-3, 3]$. Express your answers accurate to four decimal places.

Solution

a. The graph of f is shown in Figure T1.
b. Using the function on a graphing utility for finding the absolute minimum value of a continuous function on a closed interval, we find the absolute minimum value of f to be -0.0632. Similarly, using the function for finding the absolute maximum value, we find the absolute maximum value to be 4.1593.

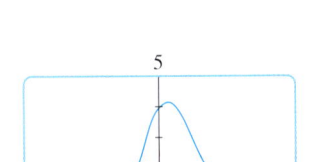

FIGURE T1
The graph of f in the viewing window $[-3, 3] \times [-1, 5]$.

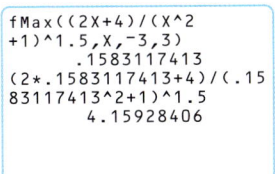

FIGURE T2
The TI-83/84 screen for Example 1

Note Some graphing utilities will enable you to find the absolute minimum and absolute maximum values of a continuous function on a closed interval without having to graph the function. For example, using **fMax** on the TI-83/84 will yield the x-coordinate of the absolute maximum of f. The absolute maximum value can then be found by evaluating f at that value of x. Figure T2 shows the work involved in finding the absolute maximum of the function of Example 1.

(continued)

TECHNOLOGY EXERCISES

In Exercises 1–6, find the absolute maximum and the absolute minimum values of f in the given interval using the method of Example 1. Express your answers accurate to four decimal places.

1. $f(x) = 3x^4 - 4.2x^3 + 6.1x - 2; [-2, 3]$
2. $f(x) = 2.1x^4 - 3.2x^3 + 4.1x^2 + 3x - 4; [-1, 2]$
3. $f(x) = \dfrac{2x^3 - 3x^2 + 1}{x^2 + 2x - 8}; [-3, 1]$
4. $f(x) = \sqrt{x}(x^3 - 4)^2; [0.5, 1]$
5. $f(x) = \dfrac{x^3 - 1}{x^2}; [1, 3]$
6. $f(x) = \dfrac{x^3 - x^2 + 1}{x - 2}; [1, 3]$

7. **BANK FAILURE** The Haven Trust Bank of Duluth, Ga., founded in 2000, quickly increased its risky commercial real estate portfolio, despite many red flags from regulators. The bank failed in December 2008. The amount of construction loans of the bank as a percentage of its capital is approximated by the function

$$f(t) = -5.92t^4 + 58.89t^3 - 165.75t^2 + 56.21t + 629$$
$$(0 \le t \le 5)$$

where $t = 0$ corresponds to the beginning of 2003.
a. Plot the graph of f using the viewing window $[0, 5] \times [0, 650]$.
b. Show that at no time during the period from the beginning of 2003 through the beginning of 2008 did the amount of construction loans of the bank as a percentage of its capital fall below 415%. Note: The maximum percentage recommended by regulators in 2008 was 100%.

Source: FDIC Office of Inspector General.

8. **CONSTRUCTION LOANS** Refer to Exercise 7. The amount of construction loans of peer banks as a percentage of capital from the beginning of 2003 ($t = 0$) through the beginning of 2008 is approximated by the function

$$g(t) = -0.656t^4 + 5.693t^3 - 16.798t^2 + 36.083t^2 + 51.9$$
$$(0 \le t \le 5)$$

a. When did the amount of construction loans of peer banks as a percentage of capital first exceed the maximum of 100% as recommended by regulators in 2006?
b. What was the highest amount of construction loans as a share of capital of peer banks over the period from the beginning of 2003 through the beginning of 2008?

Source: FDIC Office of Inspector General.

9. **SICKOUTS** In a sickout by pilots of American Airlines in February 1999, the number of canceled flights from February 6 ($t = 0$) through February 14 ($t = 8$) is approximated by the function

$$N(t) = 1.2576t^4 - 26.357t^3 + 127.98t^2 + 82.3t + 43$$
$$(0 \le t \le 8)$$

where t is measured in days. The sickout ended after the union was threatened with millions of dollars in fines.
a. Show that the number of canceled flights was increasing at the fastest rate on February 8.
b. Estimate the maximum number of canceled flights in a day during the sickout.

Source: Associated Press.

10. **MODELING WITH DATA** The following data give the average account balance (in thousands of dollars) of a 401(k) investor during a 6-year period.

Year	0	1	2	3	4	5	6
Account Balance	37.5	40.8	47.3	55.5	49.4	43	40

a. Use **QuartReg** to find a fourth-degree polynomial regression model for the data. Let $t = 0$ correspond to the beginning of the first year.
b. Plot the graph of the function found in part (a), using the viewing window $[0, 6] \times [0, 60]$.
c. When was the average account balance lowest in the period under consideration? When was it highest?
d. What were the lowest average account balance and the highest average account balance during the period under consideration?

Source: Investment Company Institute.

4.5 Optimization II

Section 4.4 outlined how to find the solution to certain optimization problems in which the objective function is given. In this section, we consider problems in which we are required to find first the appropriate function to be optimized. The following guidelines will be useful for solving these problems.

Guidelines for Solving Optimization Problems

1. Assign a letter to each variable mentioned in the problem. If appropriate, draw and label a figure.
2. Find an expression for the quantity to be optimized.
3. Use the conditions given in the problem to write the quantity to be optimized as a function f of *one* variable. Note any restrictions to be placed on the domain of f from physical considerations of the problem.
4. Optimize the function f over its domain using the methods of Section 4.4.

Note In carrying out Step 4, remember that if the function f to be optimized is continuous on a closed interval, then the absolute maximum and absolute minimum of f are, respectively, the largest and smallest values of $f(x)$ on the set composed of the critical numbers of f and the endpoints of the interval. If the domain of f is not a closed interval, then we resort to the graphical method.

Maximization Problems

APPLIED EXAMPLE 1 Fencing a Garden A man wishes to have a rectangular-shaped garden in his backyard. He has 50 feet of fencing with which to enclose his garden. Find the dimensions for the largest garden he can have if he uses all of the fencing.

Solution

Step 1 Let x and y denote the dimensions (in feet) of two adjacent sides of the garden (Figure 68), and let A denote its area. **(x^2) See pages 83–84.**

Step 2 The area of the garden
$$A = xy \qquad (1)$$
is the quantity to be maximized.

Step 3 The perimeter of the rectangle, $(2x + 2y)$ feet, must equal 50 feet. Therefore, we have the equation
$$2x + 2y = 50$$
Next, solving this equation for y in terms of x yields
$$y = 25 - x \qquad (2)$$
which, when substituted into Equation (1), gives
$$A = x(25 - x)$$
$$= -x^2 + 25x$$
(Remember, the function to be optimized must involve just one variable.) Since the sides of the rectangle must be nonnegative, we must have $x \geq 0$ and $y = 25 - x \geq 0$; that is, we must have $0 \leq x \leq 25$. Thus, the problem is reduced to that of finding the absolute maximum of $A = f(x) = -x^2 + 25x$ on the closed interval $[0, 25]$.

Step 4 Observe that f is continuous on $[0, 25]$, so the absolute maximum value of f must occur at the endpoint(s) of the interval or at the critical number(s) of f. The derivative of the function A is given by
$$A' = f'(x) = -2x + 25$$
Setting $A' = 0$ gives
$$-2x + 25 = 0$$

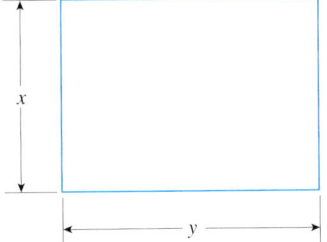

FIGURE 68
What is the maximum rectangular area that can be enclosed with 50 feet of fencing?

or 12.5, as the critical number of A. Next, we evaluate the function $A = f(x)$ at $x = 12.5$ and at the endpoints $x = 0$ and $x = 25$ of the interval $[0, 25]$, obtaining

$$f(0) = 0 \quad f(12.5) = 156.25 \quad f(25) = 0$$

We see that the absolute maximum value of the function f is 156.25. From Equation (2), we see that $y = 12.5$ when $x = 12.5$. Thus, the garden of maximum area (156.25 square feet) is a square with sides of length 12.5 feet.

APPLIED EXAMPLE 2 Packaging By cutting away identical squares from each corner of a rectangular piece of cardboard and folding up the resulting flaps, the cardboard may be turned into an open box. If the cardboard is 16 inches long and 10 inches wide, find the dimensions of the box that will yield the maximum volume.

Solution

Step 1   Let x denote the length (in inches) of one side of each of the identical squares to be cut out of the cardboard (Figure 69), and let V denote the volume of the resulting box.

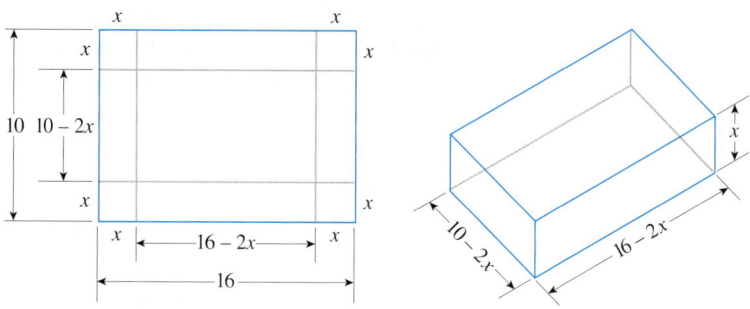

FIGURE 69
The dimensions of the open box are $(16 - 2x)$ inches by $(10 - 2x)$ inches by x inches.

Step 2   The dimensions of the box are $(16 - 2x)$ inches by $(10 - 2x)$ inches by x inches. Therefore, its volume (in cubic inches),

$$\begin{aligned} V &= (16 - 2x)(10 - 2x)x \\ &= 4(x^3 - 13x^2 + 40x) \quad \text{Expand the expression.} \end{aligned}$$

is the quantity to be maximized.

Step 3   Since each side of the box must be nonnegative, x must satisfy the inequalities $x \geq 0$, $16 - 2x \geq 0$, and $10 - 2x \geq 0$. This set of inequalities is equivalent to $0 \leq x \leq 5$. Thus, the problem at hand is equivalent to that of finding the absolute maximum of

$$V = f(x) = 4(x^3 - 13x^2 + 40x)$$

on the closed interval $[0, 5]$.

Step 4   Observe that f is continuous on $[0, 5]$, so the absolute maximum value of f must be attained at the endpoint(s) or at the critical number(s) of f.
Differentiating $f(x)$, we obtain

$$\begin{aligned} f'(x) &= 4(3x^2 - 26x + 40) \\ &= 4(3x - 20)(x - 2) \end{aligned}$$

Upon setting $f'(x) = 0$ and solving the resulting equation for x, we obtain $x = \frac{20}{3}$ or $x = 2$. Since $\frac{20}{3}$ lies outside the interval $[0, 5]$, it is no longer considered, and we are interested only in the critical number 2 of f. Next, evaluating $f(x)$ at $x = 0$, $x = 5$ (the endpoints of the interval $[0, 5]$), and $x = 2$, we obtain

$$f(0) = 0 \quad f(2) = 144 \quad f(5) = 0$$

Thus, the volume of the box is maximized by taking $x = 2$. The dimensions of the box are $12'' \times 6'' \times 2''$, and the volume is 144 cubic inches.

Exploring with TECHNOLOGY

Refer to Example 2.

1. Use a graphing utility to plot the graph of
$$f(x) = 4(x^3 - 13x^2 + 40x)$$
using the viewing window $[0, 5] \times [0, 150]$. Explain what happens to $f(x)$ as x increases from $x = 0$ to $x = 5$ and give a physical interpretation.

2. Using ZOOM and TRACE, find the absolute maximum of f on the interval $[0, 5]$ and thus verify the solution for Example 2 obtained analytically.

APPLIED EXAMPLE 3 Optimal Subway Fare A city's Metropolitan Transit Authority (MTA) operates a subway line for commuters from a certain suburb to the downtown metropolitan area. Currently, an average of 6000 passengers a day take the trains, paying a fare of $3.00 per ride. The board of the MTA, contemplating raising the fare to $3.50 per ride in order to generate a larger revenue, engages the services of a consulting firm. The firm's study reveals that for each $0.50 increase in fare, the ridership will be reduced by an average of 1000 passengers a day. Thus, the consulting firm recommends that MTA stick to the current fare of $3.00 per ride, which already yields a maximum revenue. Show that the consultants are correct.

Solution

Step 1 Let x denote the number of passengers per day, p denote the fare per ride, and R be MTA's revenue.   (x²) See pages 83–84.

Step 2 To find a relationship between x and p, observe that the given data imply that when $x = 6000$, $p = 3$, and when $x = 5000$, $p = 3.50$. Therefore, the points $(6000, 3)$ and $(5000, 3.50)$ lie on a straight line. (Why?) To find the linear relationship between p and x, use the point-slope form of the equation of a straight line. Now, the slope of the line is

$$m = \frac{3.50 - 3}{5000 - 6000} = -0.0005$$

Therefore, the required equation is

$$p - 3 = -0.0005(x - 6000)$$
$$= -0.0005x + 3$$
$$p = -0.0005x + 6$$

Therefore, the revenue

$$R = f(x) = xp = -0.0005x^2 + 6x \quad \text{Number of riders} \times \text{unit fare}$$

is the quantity to be maximized.

Step 3 Since both p and x must be nonnegative, we see that $0 \leq x \leq 12{,}000$, and the problem is that of finding the absolute maximum of the function f on the closed interval $[0, 12{,}000]$.

Step 4 Observe that f is continuous on $[0, 12{,}000]$. To find the critical number of R, we compute
$$f'(x) = -0.001x + 6$$

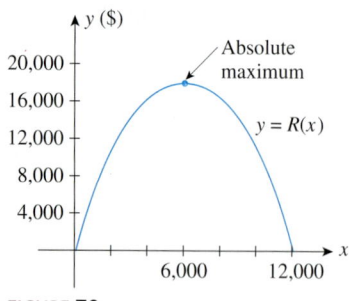

FIGURE 70
f has an absolute maximum of 18,000 when x = 6000.

and set it equal to zero, giving $x = 6000$. Evaluating the function f at $x = 6000$, as well as at the endpoints $x = 0$ and $x = 12{,}000$, yields

$$f(0) = 0$$
$$f(6000) = 18{,}000$$
$$f(12{,}000) = 0$$

We conclude that a maximum revenue of $18,000 per day is realized when the ridership is 6000 per day. The optimum price of the fare per ride is therefore $3.00, as recommended by the consultants. The graph of the revenue function R is shown in Figure 70.

Minimization Problems

APPLIED EXAMPLE 4 Packaging Betty Moore Company requires that its corned beef hash containers have a capacity of 54 cubic inches, have the shape of a right circular cylinder, and be made of aluminum. Determine the radius and height of the container that requires the least amount of metal.

Solution

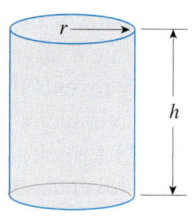

FIGURE 71
We want to minimize the amount of material used to construct the container.

Step 1   Let the radius and height of the container be r and h inches, respectively, and let S denote the surface area of the container (Figure 71).

Step 2   The amount of aluminum used to construct the container is given by the total surface area of the cylinder. Now, the area of the base and the top of the cylinder are each πr^2 square inches, and the area of the side is $2\pi rh$ square inches. Therefore,

$$S = 2\pi r^2 + 2\pi rh \qquad (3)$$

is the quantity to be minimized.

Step 3   The requirement that the volume of a container be 54 cubic inches implies that

$$\pi r^2 h = 54 \qquad (4)$$

Solving Equation (4) for h, we obtain

$$h = \frac{54}{\pi r^2} \qquad (5)$$

which, when substituted into (3), yields

$$S = 2\pi r^2 + 2\pi r \left(\frac{54}{\pi r^2} \right)$$
$$= 2\pi r^2 + \frac{108}{r}$$

Clearly, the radius r of the container must satisfy the inequality $r > 0$. The problem now is reduced to finding the absolute minimum of the function $S = f(r)$ on the interval $(0, \infty)$.

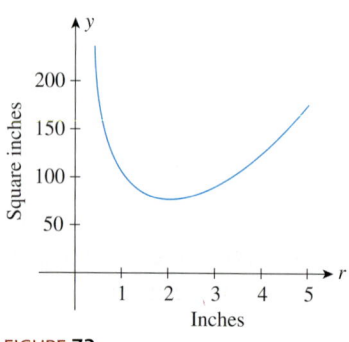

FIGURE 72
The total surface area of the right cylindrical container is graphed as a function of r.

Step 4   Using the curve-sketching techniques of Section 4.3, we obtain the graph of f in Figure 72.

To find the critical number of f, we compute

$$S' = 4\pi r - \frac{108}{r^2}$$

and solve the equation $S' = 0$ for r:

$$4\pi r - \frac{108}{r^2} = 0$$

$$4\pi r^3 - 108 = 0$$

$$r^3 = \frac{27}{\pi}$$

$$r = \frac{3}{\sqrt[3]{\pi}} \approx 2 \qquad (6)$$

Next, let's show that this value of r gives rise to the absolute minimum of f. To show this, we first compute

$$S'' = 4\pi + \frac{216}{r^3}$$

Since $S'' > 0$ for $r = 3/\sqrt[3]{\pi}$, the second derivative test implies that the value of r in Equation (6) gives rise to a relative minimum of f. Finally, this relative minimum of f is also the absolute minimum of f, since the graph of f is always concave upward ($S'' > 0$ for all $r > 0$). To find the height of the given container, we substitute the value of r given in Equation (6) into Equation (5). Thus,

$$h = \frac{54}{\pi r^2} = \frac{54}{\pi \left(\dfrac{3}{\pi^{1/3}}\right)^2}$$

$$= \frac{54\pi^{2/3}}{(\pi)9}$$

$$= \frac{6}{\pi^{1/3}} = \frac{6}{\sqrt[3]{\pi}}$$

$$= 2r$$

We conclude that the required container has a radius of approximately 2 inches and a height of approximately 4 inches, or twice the size of the radius.

An Inventory Problem

One problem faced by many companies is that of controlling the inventory of goods carried. Ideally, the manager must ensure that the company has sufficient stock to meet customer demand at all times. At the same time, she must make sure that this is accomplished without overstocking (incurring unnecessary storage costs) and also without having to place orders too frequently (incurring reordering costs).

APPLIED EXAMPLE 5 Inventory Control and Planning Dixie Import-Export is the sole agent for the Excalibur 250-cc motorcycle. Management estimates that the demand for these motorcycles is 10,000 per year and that they will sell at a uniform rate throughout the year. The cost incurred in ordering each shipment of motorcycles is $10,000, and the cost per year of storing each motorcycle is $200.

Dixie's management faces the following problem: Ordering too many motorcycles at one time ties up valuable storage space and increases the storage cost. On the other hand, placing orders too frequently increases the ordering costs. How large should each order be, and how often should orders be placed, to minimize ordering and storage costs?

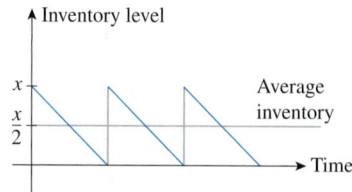

FIGURE 73
As each lot is depleted, the new lot arrives. The average inventory level is x/2 if x is the lot size.

Solution Let x denote the number of motorcycles in each order (the lot size). Then, assuming that each shipment arrives just as the previous shipment has been sold, the average number of motorcycles in storage during the year is $x/2$. You can see that this is the case by examining Figure 73. Thus, Dixie's storage cost for the year is given by $200(x/2)$, or $100x$ dollars.

Next, since the company requires 10,000 motorcycles for the year and since each order is for x motorcycles, the number of orders required is

$$\frac{10,000}{x}$$

This gives an ordering cost of

$$10,000\left(\frac{10,000}{x}\right) = \frac{100,000,000}{x}$$

dollars for the year. Thus, the total yearly cost incurred by Dixie, which includes the ordering and storage costs attributed to the sale of these motorcycles, is given by

$$C(x) = 100x + \frac{100,000,000}{x}$$

The problem is reduced to finding the absolute minimum of the function C on the interval $(0, 10,000]$. To accomplish this, we compute

$$C'(x) = 100 - \frac{100,000,000}{x^2}$$

Setting $C'(x) = 0$ and solving the resulting equation, we obtain $x = \pm 1000$. Since the number -1000 is outside the domain of the function C, it is rejected, leaving 1000 as the only critical number of C. Next, we find

$$C''(x) = \frac{200,000,000}{x^3}$$

Since $C''(1000) > 0$, the second derivative test implies that the critical number 1000 is a relative minimum of the function C (Figure 74). Also, since $C''(x) > 0$ for all x in $(0, 10,000)$, the graph of C is concave upward everywhere, so $x = 1000$ also gives the absolute minimum of C. Thus, to minimize the ordering and storage costs, Dixie should place 10,000/1000, or 10, orders a year, each for a shipment of 1000 motorcycles.

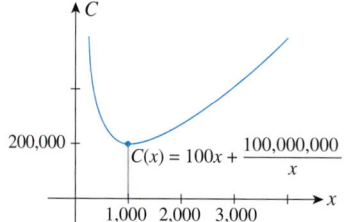

FIGURE 74
C has an absolute minimum at (1000, 200,000).

4.5 Self-Check Exercises

1. A man wishes to have an enclosed vegetable garden in his backyard. If the garden is to be a rectangular area of 300 ft², find the dimensions of the garden that will minimize the amount of fencing needed.

2. The demand for Super Titan tires is 1,000,000/year. The setup cost for each production run is $4000, and the manufacturing cost is $20/tire. The cost of storing each tire over the year is $2. Assuming uniformity of demand throughout the year and instantaneous production, determine how many tires should be manufactured per production run in order to keep the total cost to a minimum.

Solutions to Self-Check Exercises 4.5 can be found on page 323.

4.5 Concept Questions

1. If the domain of a function f is not a closed interval, how would you find the absolute extrema of f, if they exist?

2. Refer to Example 4 (page 316). In the solution given in the example, we solved for h in terms of r, resulting in a function of r, which we then optimized with respect to r. Write S in terms of h and re-solve the problem. Which choice is better?

4.5 Exercises

1. Find the dimensions of a rectangle with a perimeter of 100 ft that has the largest possible area.

2. Find the dimensions of a rectangle of area 144 sq ft that has the smallest possible perimeter.

3. **Enclosing the Largest Area** The owner of the Rancho Los Feliz has 3000 yd of fencing with which to enclose a rectangular piece of grazing land along the straight portion of a river. If fencing is not required along the river, what are the dimensions of the largest area that he can enclose? What is this area?

4. **Enclosing the Largest Area** Refer to Exercise 3. As an alternative plan, the owner of the Rancho Los Feliz might use the 3000 yd of fencing to enclose the rectangular piece of grazing land along the straight portion of the river and then subdivide it by means of a fence running parallel to the sides. Again, no fencing is required along the river. What are the dimensions of the largest area that can be enclosed? What is this area? (See the accompanying figure.)

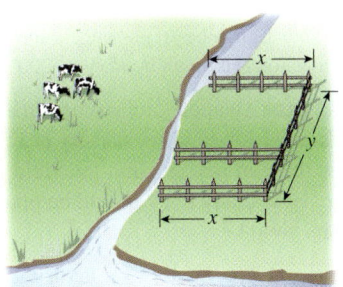

5. **Minimizing Construction Costs** The management of the UNICO department store has decided to enclose an 800-ft² area outside the building for displaying potted plants and flowers. One side will be formed by the external wall of the store, two sides will be constructed of pine boards, and the fourth side will be made of galvanized steel fencing. If the pine board fencing costs $6/running foot and the steel fencing costs $3/running foot, determine the dimensions of the enclosure that can be erected at minimum cost.

6. **Packaging** By cutting away identical squares from each corner of a rectangular piece of cardboard and folding up the resulting flaps, an open box may be made. If the cardboard is 15 in. long and 8 in. wide, find the dimensions of the box that will yield the maximum volume.

7. **Metal Fabrication** If an open box is made from a tin sheet 8 in. square by cutting out identical squares from each corner and bending up the resulting flaps, determine the dimensions of the largest box that can be made.

8. **Minimizing Packaging Costs** If an open box has a square base and a volume of 108 in.³ and is constructed from a tin sheet, find the dimensions of the box, assuming that a minimum amount of material is used in its construction.

9. **Minimizing Packaging Costs** What are the dimensions of a closed rectangular box that has a square cross section and a capacity of 128 in.³ and is constructed using the least amount of material?

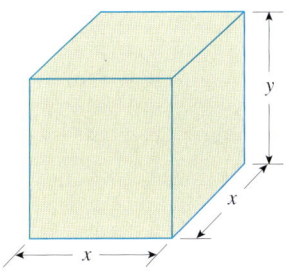

10. **MINIMIZING PACKAGING COSTS** A rectangular box is to have a square base and a volume of 20 ft^3. If the material for the base costs 30¢/square foot, the material for the sides costs 10¢/square foot, and the material for the top costs 20¢/square foot, determine the dimensions of the box that can be constructed at minimum cost. (Refer to the figure for Exercise 9.)

11. **PARCEL POST REGULATIONS** Postal regulations specify that a parcel sent by priority mail may have a combined length and girth of no more than 108 in. Find the dimensions of a rectangular package that has a square cross section and the largest volume that may be sent via priority mail. What is the volume of such a package?

    Hint: The length plus the girth is $4x + h$ (see the accompanying figure).

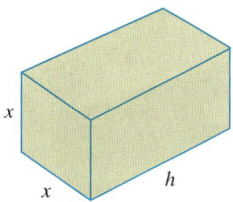

12. **BOOK DESIGN** A book designer has decided that the pages of a book should have 1-in. margins at the top and bottom and $\frac{1}{2}$-in. margins on the sides. She further stipulated that each page should have an area of 50 in.^{2} (see the accompanying figure). Determine the page dimensions that will result in the maximum printed area on the page.

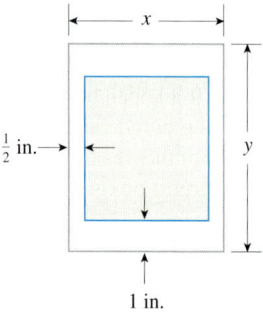

13. **PARCEL POST REGULATIONS** Postal regulations specify that a parcel sent by priority mail may have a combined length and girth of no more than 108 in. Find the dimensions of the cylindrical package of greatest volume that may be sent via priority mail. What is the volume of such a package? Compare with Exercise 11.

    Hint: The length plus the girth is $2\pi r + l$.

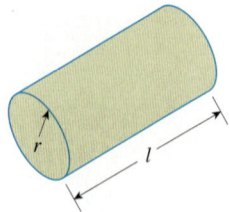

14. **MINIMIZING COSTS** For its beef stew, Betty Moore Company uses aluminum containers that have the form of right circular cylinders. Find the radius and height of a container if it has a capacity of 36 in.^{3} and is constructed using the least amount of metal.

15. **PRODUCT DESIGN** The cabinet that will enclose the Acrosonic model D loudspeaker system will be rectangular and will have an internal volume of 2.4 ft^3. For aesthetic reasons, the design team has decided that the height of the cabinet is to be 1.5 times its width. If the top, bottom, and sides of the cabinet are constructed of veneer costing 40¢/square foot and the front (ignore the cutouts in the baffle) and rear are constructed of particle board costing 20¢/square foot, what are the dimensions of the enclosure that can be constructed at a minimum cost?

16. **DESIGNING A NORMAN WINDOW** A Norman window has the shape of a rectangle surmounted by a semicircle (see the accompanying figure). If a Norman window is to have a perimeter of 28 ft, what should its dimensions be in order to allow the maximum amount of light through the window?

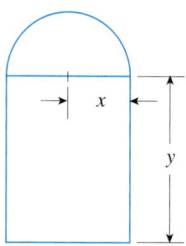

17. **OPTIMAL CHARTER-FLIGHT FARE** If exactly 200 people sign up for a charter flight, Leisure World Travel Agency charges $300/person. However, if more than 200 people sign up for the flight (assume that this is the case), then every fare is reduced by $1 times the number of passengers above 200. Determine how many passengers will result in a maximum revenue for the travel agency. What is the maximum revenue? What would be the fare per passenger in this case?

    Hint: Let x denote the number of passengers above 200. Show that the revenue function R is given by $R(x) = (200 + x)(300 - x)$.

18. **MAXIMIZING YIELD** An apple orchard has an average yield of 36 bushels of apples/tree if tree density is 22 trees/acre. For each unit increase in tree density, the yield decreases by 2 bushels/tree. How many trees should be planted in order to maximize the yield?

19. **CHARTER REVENUE** The owner of a luxury motor yacht that sails among the 4000 Greek islands charges $600/person/day if exactly 20 people sign up for the cruise. However, if more than 20 people sign up (up to the maximum capacity of 90) for the cruise, then every fare is reduced by $4 times the number of passengers above 20. Assuming that at least 20 people sign up for the cruise, determine how many passengers will result in the maximum revenue for the owner of the yacht. What is the maximum revenue? What would be the fare per passenger in this case?

20. **PROFIT OF A VINEYARD** Phillip, the proprietor of a vineyard, estimates that the first 10,000 bottles of wine produced this season will fetch a profit of $5/bottle. But if more than 10,000 bottles were produced, then the profit per bottle for

the entire lot would drop by $0.0002 for each additional bottle sold. Assuming that at least 10,000 bottles of wine are produced and sold, what is the maximum profit?

21. **OPTIMAL SPEED OF A TRUCK** A truck gets $600/x$ mpg when driven at a constant speed of x mph (between 50 and 70 mph). If the price of fuel is $3/gallon and the driver is paid $18/hour, at what speed between 50 and 70 mph is it most economical to drive?

22. **MINIMIZING COSTS** Suppose the cost incurred in operating a cruise ship for one hour is $a + bv^3$ dollars, where a and b are positive constants and v is the ship's speed in miles per hour. At what speed should the ship be operated between two ports to minimize the cost?

23. **STRENGTH OF A BEAM** A wooden beam has a rectangular cross section of height h in. and width w in. (see the accompanying figure). The strength S of the beam is directly proportional to its width and the square of its height. What are the dimensions of the cross section of the strongest beam that can be cut from a round log of diameter 24 in.?
Hint: $S = kh^2w$, where k is a constant of proportionality.

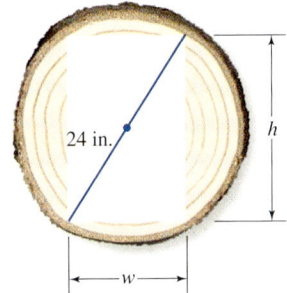

24. **DESIGNING A GRAIN SILO** A grain silo has the shape of a right circular cylinder surmounted by a hemisphere (see the accompanying figure). If the silo is to have a capacity of 504π ft³, find the radius and height of the silo that requires the least amount of material to construct.
Hint: The volume of the silo is $\pi r^2 h + \frac{2}{3}\pi r^3$, and the surface area (including the floor) is $\pi(3r^2 + 2rh)$.

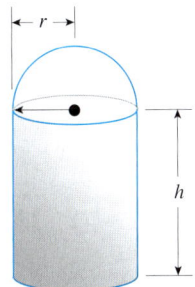

25. **MINIMIZING COST OF LAYING CABLE** In the following diagram, S represents the position of a power relay station located on a straight coast, and E shows the location of a marine biology experimental station on an island. A cable is to be laid connecting the relay station with the experimental station. If the cost of running the cable on land is $1.50/running foot and the cost of running the cable under water is $2.50/running foot, locate the point P that will result in a minimum cost (solve for x).

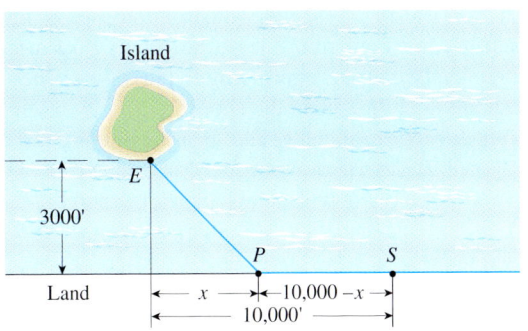

26. **STORING RADIOACTIVE WASTE** A cylindrical container for storing radioactive waste is to be constructed from lead and have a thickness of 6 in. (see the accompanying figure). If the volume of the outside cylinder is to be 16π ft³, find the radius and the height of the inside cylinder that will result in a container of maximum storage capacity.

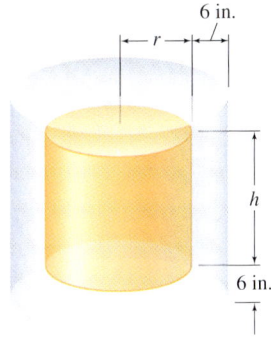

Hint: Show that the storage capacity (inside volume) is given by

$$V(r) = \pi r^2 \left[\frac{16}{(r + \frac{1}{2})^2} - 1 \right] \quad (0 \leq r \leq \tfrac{7}{2})$$

27. **FLIGHTS OF BIRDS** During daylight hours, some birds fly more slowly over water than over land because some of their energy is expended in overcoming the downdrafts of air over open bodies of water. Suppose a bird that flies at a constant speed of 4 mph over water and 6 mph over land starts its journey at the point E on an island and ends at its nest N on the shore of the mainland, as shown in the accompanying figure. Find the location of the point P that allows the bird to complete its journey in the minimum time (solve for x).

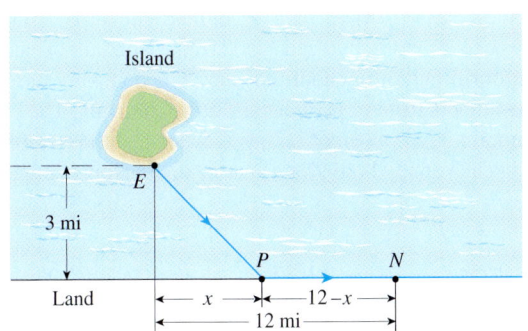

28. **Minimizing Travel Time** A woman is on a lake in a row boat located 1 mi from the closest point P of a straight shoreline (see the accompanying figure). She wishes to get to point Q, 10 mi along the shore from P, by rowing to a point R between P and Q and then jogging the rest of the distance. If she can row at a speed of 3 mph and walk at a speed of 4 mph, how should she pick the point R in order to get to Q as quickly as possible? How much time does she require?

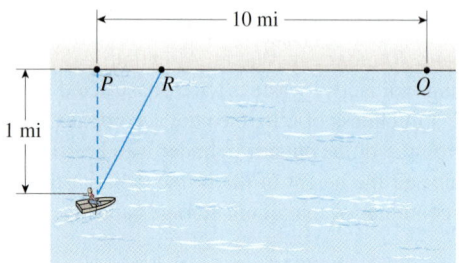

29. **Racetrack Design** The accompanying figure depicts a racetrack with ends that are semicircular in shape. The length of the track is 1760 ft ($\frac{1}{3}$ mi). Find l and r such that the area enclosed by the rectangular region of the racetrack is as large as possible. What is the area enclosed by the track in this case?

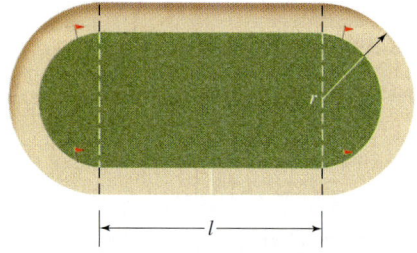

30. **Inventory Control and Planning** The demand for motorcycle tires imported by Dixie Import-Export is 40,000/year and may be assumed to be uniform throughout the year. The cost of ordering a shipment of tires is $400, and the cost of storing each tire for a year is $2. Determine how many tires should be in each shipment if the ordering and storage costs are to be minimized. (Assume that each shipment arrives just as the previous one has been sold.)

31. **Inventory Control and Planning** McDuff Preserves expects to bottle and sell 2,000,000 32-oz jars of jam at a uniform rate throughout the year. The company orders its containers from Consolidated Bottle Company. The cost of ordering a shipment of bottles is $200, and the cost of storing each empty bottle for a year is $0.40. How many orders should McDuff place per year, and how many bottles should be in each shipment if the ordering and storage costs are to be minimized? (Assume that each shipment of bottles is used up before the next shipment arrives.)

32. **Inventory Control and Planning** Neilsen Cookie Company sells its assorted butter cookies in containers that have a net content of 1 lb. The estimated demand for the cookies is 1,000,000 1-lb containers per year. The setup cost for each production run is $500, and the manufacturing cost is $0.50 for each container of cookies. The cost of storing each container of cookies over the year is $0.40. Assuming uniformity of demand throughout the year and instantaneous production, how many containers of cookies should Neilsen produce per production run in order to minimize the production cost?
    Hint: Following the method of Example 5, show that the total production cost is given by the function
    $$C(x) = \frac{500{,}000{,}000}{x} + 0.2x + 500{,}000$$
    Then minimize the function C on the interval $(0, 1{,}000{,}000)$.

33. **Inventory Control and Planning** A company expects to sell D units of a certain product per year. Sales are assumed to be at a steady rate with no shortages allowed. Each time an order for the product is placed, an ordering cost of K dollars is incurred. Each item costs p dollars, and the holding cost is h dollars per item per year.
    a. Show that the inventory cost (the combined ordering cost, purchasing cost, and holding cost) is
    $$C(x) = \frac{KD}{x} + pD + \frac{hx}{2} \quad (x > 0)$$
    where x is the order quantity (the number of items in each order).
    b. Use the result of part (a) to show that the inventory cost is minimized if
    $$x = \sqrt{\frac{2KD}{h}}$$
    This quantity is called the *economic order quantity* (EOQ).

34. **Inventory Control and Planning** Refer to Exercise 33. The Camera Store sells 960 Yamaha A35 digital cameras per year. Each time an order for cameras is placed with the manufacturer, an ordering cost of $10 is incurred. The store pays $80 for each camera, and the cost for holding a camera (mainly due to the opportunity cost incurred in tying up capital in inventory) is $12/year. Assume that the cameras sell at a uniform rate and no shortages are allowed.
    a. What is the EOQ?
    b. How many orders will be placed each year?
    c. What is the interval between orders?

4.5 Solutions to Self-Check Exercises

1. Let x and y (measured in feet) denote the length and width of the rectangular garden.

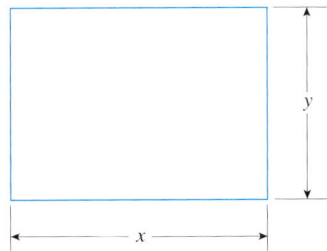

Since the area is to be 300 ft², we have

$$xy = 300$$

Next, the amount of fencing to be used is given by the perimeter, and this quantity is to be minimized. Thus, we want to minimize

$$2x + 2y$$

or, since $y = 300/x$ (obtained by solving for y in the first equation), we see that the expression to be minimized is

$$f(x) = 2x + 2\left(\frac{300}{x}\right)$$
$$= 2x + \frac{600}{x}$$

for positive values of x. Now,

$$f'(x) = 2 - \frac{600}{x^2}$$

Setting $f'(x) = 0$ yields $x = -\sqrt{300}$ or $x = \sqrt{300}$. We consider only the critical number $\sqrt{300}$, since $-\sqrt{300}$ lies outside the interval $(0, \infty)$. We then compute

$$f''(x) = \frac{1200}{x^3}$$

Since

$$f''(300) > 0$$

the second derivative test implies that a relative minimum of f occurs at $x = \sqrt{300}$. In fact, since $f''(x) > 0$ for all x in $(0, \infty)$, we conclude that $x = \sqrt{300}$ gives rise to the absolute minimum of f. The corresponding value of y, obtained by substituting this value of x into the equation $xy = 300$, is $y = \sqrt{300}$. Therefore, the required dimensions of the vegetable garden are approximately 17.3 ft × 17.3 ft.

2. Let x denote the number of tires in each production run. Then, the average number of tires in storage is $x/2$, so the storage cost incurred by the company is $2(x/2)$, or x dollars. Next, since the company needs to manufacture 1,000,000 tires for the year to meet the demand, the number of production runs is $1{,}000{,}000/x$. This gives setup costs amounting to

$$4000\left(\frac{1{,}000{,}000}{x}\right) = \frac{4{,}000{,}000{,}000}{x}$$

dollars for the year. The total manufacturing cost is $20,000,000. Thus, the total yearly cost incurred by the company is given by

$$C(x) = x + \frac{4{,}000{,}000{,}000}{x} + 20{,}000{,}000$$

Differentiating $C(x)$, we find

$$C'(x) = 1 - \frac{4{,}000{,}000{,}000}{x^2}$$

Setting $C'(x) = 0$ gives 63,246 as the critical number in the interval $(0, 1{,}000{,}000)$. Next, we find

$$C''(x) = \frac{8{,}000{,}000{,}000}{x^3}$$

Since $C''(x) > 0$ for all $x > 0$, we see that the graph of C is concave upward for all $x > 0$. Furthermore, $C''(63{,}246) > 0$ implies that $x = 63{,}246$ gives rise to a relative minimum of C (by the Second Derivative Test). Since the graph of C is always concave upward for $x > 0$, $x = 63{,}246$ gives the absolute minimum of C. Therefore, the company should manufacture 63,246 tires in each production run.

CHAPTER 4 Summary of Principal Terms

TERMS

increasing function (246)
decreasing function (246)
relative maximum (251)
relative minimum (252)
relative extrema (252)
critical number (253)

First Derivative Test (253)
concave upward (266)
concave downward (266)
point of inflection (269)
inflection point (269)
Second Derivative Test (274)

vertical asymptote (285)
horizontal asymptote (287)
absolute extrema (299)
absolute maximum value (299)
absolute minimum value (299)
Extreme Value Theorem (299)

CHAPTER 4 Concept Review Questions

Fill in the blanks.

1. a. A function f is increasing on an interval I if for any two numbers x_1 and x_2 in I, $x_1 < x_2$ implies that _____.
   b. A function f is decreasing on an interval I if for any two numbers x_1 and x_2 in I, $x_1 < x_2$ implies that _____.

2. a. If f is differentiable on an open interval (a, b) and $f'(x) > 0$ on (a, b), then f is _____ on (a, b).
   b. If f is differentiable on an open interval (a, b) and _____ on (a, b), then f is decreasing on (a, b).
   c. If $f'(x) = 0$ for each value of x in the interval (a, b), then f is _____ on (a, b).

3. a. A function f has a relative maximum at c if there exists an open interval (a, b) containing c such that _____ for all x in (a, b).
   b. A function f has a relative minimum at c if there exists an open interval (a, b) containing c such that _____ for all x in (a, b).

4. a. A critical number of a function f is any number in the _____ of f at which $f'(c)$ _____ or $f'(c)$ does not _____.
   b. If f has a relative extremum at c, then c must be a/an _____ _____ of f.
   c. If c is a critical number of f, then f may or may not have a/an _____ _____ at c.

5. a. The graph of a differentiable function f is concave upward on an interval I if _____ is increasing on I.
   b. If f has a second derivative on an open interval I and $f''(x)$ _____ on I, then the graph of f is concave upward on I.
   c. If the graph of a continuous function f has a tangent line at $P(c, f(c))$ and the graph of f changes _____ at P, then P is called an inflection point of f.
   d. Suppose f has a second derivative on an interval (a, b), containing a critical number c of f. If $f''(c) < 0$, then f has a/an _____ _____ at c. If $f''(c) = 0$, then f may or may not have a/an _____ _____ at c.

6. The line $x = a$ is a vertical asymptote of the graph f if at least one of the following is true: $\lim_{x \to a^+} f(x) =$ _____ or $\lim_{x \to a^-} f(x) =$ _____.

7. For a rational function $f(x) = \dfrac{P(x)}{Q(x)}$, the line $x = a$ is a vertical asymptote of the graph of f if $Q(a) =$ _____ but $P(a) \neq$ _____.

8. The line $y = b$ is a horizontal asymptote of the graph of a function f if either $\lim_{x \to \infty} f(x) =$ _____ or $\lim_{x \to -\infty} f(x) =$ _____.

9. a. A function f has an absolute maximum at c if _____ for all x in the domain D of f. The number $f(c)$ is called the _____ _____ _____ of f on D.
   b. A function f has a relative minimum at c if _____ for all values of x in some _____ _____ containing c.

10. The extreme value theorem states that if f is _____ on the closed interval $[a, b]$, then f has both a/an _____ maximum value and a/an _____ minimum value on $[a, b]$.

CHAPTER 4 Review Exercises

In Exercises 1–10, (a) find the intervals where the function f is increasing and where it is decreasing, (b) find the relative extrema of f, (c) find the intervals where the graph of f is concave upward and where it is concave downward, and (d) find the inflection points, if any, of f.

1. $f(x) = \dfrac{1}{3}x^3 - x^2 + x - 6$

2. $f(x) = (x - 2)^3$

3. $f(x) = x^4 - 2x^2$

4. $f(x) = x + \dfrac{4}{x}$

5. $f(x) = \dfrac{x^2}{x - 1}$

6. $f(x) = \sqrt{x - 1}$

7. $f(x) = (1 - x)^{1/3}$

8. $f(x) = x\sqrt{x - 1}$

9. $f(x) = \dfrac{2x}{x + 1}$

10. $f(x) = \dfrac{-1}{1 + x^2}$

In Exercises 11–18, use the curve-sketching guide on page 288 to sketch the graph of the function.

11. $f(x) = x^2 - 5x + 5$

12. $f(x) = -2x^2 - x + 1$

13. $g(x) = 2x^3 - 6x^2 + 6x + 1$

14. $g(x) = \dfrac{1}{3}x^3 - x^2 + x - 3$

15. $h(x) = x\sqrt{x - 2}$

16. $h(x) = \dfrac{2x}{1 + x^2}$

17. $f(x) = \dfrac{x - 2}{x + 2}$

18. $f(x) = x - \dfrac{1}{x}$

In Exercises 19–22, find the horizontal and vertical asymptotes of the graph of each function. Do not sketch the graph.

19. $f(x) = \dfrac{1}{2x + 3}$

20. $f(x) = \dfrac{2x}{x + 1}$

21. $f(x) = \dfrac{5x}{x^2 - 2x - 8}$

22. $f(x) = \dfrac{x^2 + x}{x(x - 1)}$

In Exercises 23–32, find the absolute maximum value and the absolute minimum value, if any, of the function.

23. $f(x) = 2x^2 + 3x - 2$

24. $g(x) = x^{2/3}$

25. $g(t) = \sqrt{25 - t^2}$

26. $f(x) = \frac{1}{3}x^3 - x^2 + x + 1$ on $[0, 2]$

27. $h(t) = t^3 - 6t^2$ on $[2, 5]$

28. $g(x) = \dfrac{x}{x^2 + 1}$ on $[0, 5]$

29. $f(x) = x - \dfrac{1}{x}$ on $[1, 3]$

30. $h(t) = 8t - \dfrac{1}{t^2}$ on $[1, 3]$

31. $f(s) = s\sqrt{1 - s^2}$ on $[-1, 1]$

32. $f(x) = \dfrac{x^2}{x - 1}$ on $[-1, 3]$

33. **Revenue of Two Bookstores** The graphs of R_1 and R_2 that follow show the revenue of a neighborhood bookstore and that of a branch of a national bookstore, three months after the opening of the latter ($t = 0$) until sometime later T.

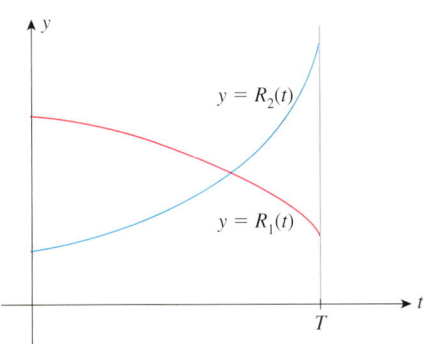

a. Find the signs of the first and second derivatives of R_1 and R_2 on the interval $(0, T)$.
b. Give an interpretation of the results obtained in part (a) in terms of the revenues of the two bookstores.

34. **Spread of a Rumor** Initially, a handful of students heard a rumor on campus. The rumor spread, and after t hr, the number of students who had heard it had grown to $N(t)$. The graph of the function N is shown in the following figure.

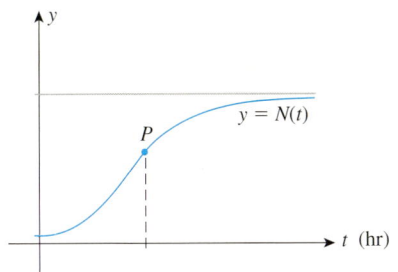

Describe the spread of the rumor in terms of the speed at which it was spread. In particular, explain the significance of the inflection point P of the graph of N.

35. **Maximizing Profits** Odyssey Travel Agency's monthly profit (in thousands of dollars) depends on the amount of money x (in thousands of dollars) spent on advertising each month according to the rule

$$P(x) = -x^2 + 8x + 20$$

To maximize its monthly profits, what should be Odyssey's monthly advertising budget?

36. **Sales of Camera Phones** Camera phones, virtually nonexistent ten years ago, are quickly gaining in popularity. The function

$$N(t) = 8.125t^2 + 24.625t + 18.375 \qquad (0 \le t \le 3)$$

gives the projected worldwide shipments of camera phones (in millions of units) in year t, with $t = 0$ corresponding to 2002.
a. Find $N'(t)$. What does this say about the sales of camera phones between 2002 and 2005?
b. Find $N''(t)$. What does this say about the rate of increase of the rate of sales of camera phones between 2002 and 2005?

Source: In-Stat/MDR.

37. **Effect of Advertising on Sales** The total sales S of Cannon Precision Instruments is related to the amount of money x that Cannon spends on advertising its products by the function

$$S(x) = -0.002x^3 + 0.6x^2 + x + 500 \qquad (0 \le x \le 200)$$

where S and x are measured in thousands of dollars. Find the inflection point of the function S, and discuss its significance.

38. **Elderly Workforce** The percentage of men 65 years and older in the workforce from 1970 through 2000 is approximated by the function

$$P(t) = 0.00093t^3 - 0.018t^2 - 0.51t + 25 \qquad (0 \le t \le 30)$$

where t is measured in years, with $t = 0$ corresponding to the beginning of 1970.
a. Find the interval where P is decreasing and the interval where P is increasing.
b. Interpret the results of part (a).

Source: U.S. Census Bureau.

39. **Cost of Producing Calculators** A subsidiary of Elektra Electronics manufactures graphing calculators. Management determines that the daily cost $C(x)$ (in dollars) of producing these calculators is

$$C(x) = 0.0001x^3 - 0.08x^2 + 40x + 5000$$

where x is the number of calculators produced. Find the inflection point of the function C, and interpret your result.

40. Sales of Mobile Processors The rising popularity of notebook computers is fueling the sales of mobile PC processors. In a study conducted in 2003, the sales of these chips (in billions of dollars) was projected to be

$$S(t) = 6.8(t + 1.03)^{0.49} \quad (0 \le t \le 4)$$

where t is measured in years, with $t = 0$ corresponding to 2003.
  a. Show that S is increasing on the interval $(0, 4)$, and interpret your result.
  b. Show that the graph of S is concave downward on the interval $(0, 4)$. Interpret your result.

Source: International Data Corp.

41. Small Car Market Share in Europe Owing in part to an aging population and the squeeze from carbon-cutting regulations, the percentage of small and lower-midsize vehicles in Europe is expected to increase in the near future. The function

$$f(t) = \frac{150\sqrt{t} + 766}{59 - \sqrt{t}} \quad (0 \le t \le 10)$$

gives the projected percentage of small and lower-midsize vehicles t years after 2005.
  a. What was the percentage of small and lower-midsize vehicles in 2005? What is the projected percentage of small and lower-midsize vehicles in 2015?
  b. Show that f is increasing on the interval $(0, 10)$, and interpret your results.

Source: J.D. Power Automotive.

42. Index of Environmental Quality The Department of the Interior of an African country began to record an index of environmental quality to measure progress or decline in the environmental quality of its wildlife. The index for the years 1998 through 2008 is approximated by the function

$$I(t) = \frac{50t^2 + 600}{t^2 + 10} \quad (0 \le t \le 10)$$

  a. Compute $I'(t)$ and show that $I(t)$ is decreasing on the interval $(0, 10)$.
  b. Compute $I''(t)$. Study the concavity of the graph of I.
  c. Sketch the graph of I.
  d. Interpret your results.

43. Maximizing Profits The weekly demand for DVDs manufactured by Herald Media Corporation is given by

$$p = -0.0005x^2 + 60$$

where p denotes the unit price in dollars and x denotes the quantity demanded. The weekly total cost function associated with producing these discs is given by

$$C(x) = -0.001x^2 + 18x + 4000$$

where $C(x)$ denotes the total cost (in dollars) incurred in pressing x discs. Find the production level that will yield a maximum profit for the manufacturer.
Hint: Use the quadratic formula.

44. Maximizing Profits The estimated monthly profit (in dollars) realizable by Cannon Precision Instruments for manufacturing and selling x units of its model M1 digital camera is

$$P(x) = -0.04x^2 + 240x - 10,000$$

To maximize its profits, how many cameras should Cannon produce each month?

45. Minimizing Average Cost The total monthly cost (in dollars) incurred by Carlota Music in manufacturing x units of its Professional Series guitars is given by the function

$$C(x) = 0.001x^2 + 100x + 4000$$

  a. Find the average cost function $\overline{C}$.
  b. Determine the production level that will result in the smallest average production cost.

46. Worker Efficiency The average worker at Wakefield Avionics will have assembled

$$N(t) = -2t^3 + 12t^2 + 2t \quad (0 \le t \le 4)$$

ready-to-fly radio-controlled model airplanes t hr into the 8 A.M. to 12 noon shift. At what time during this shift is the average worker performing at peak efficiency?

47. Senior Workforce The percentage of women 65 years and older in the workforce from 1970 through 2000 is approximated by the function

$$P(t) = -0.0002t^3 + 0.018t^2 - 0.36t + 10 \quad (0 \le t \le 30)$$

where t is measured in years, with $t = 0$ corresponding to the beginning of 1970.
  a. Find the interval where P is decreasing and the interval where P is increasing.
  b. Find the absolute minimum of P.
  c. Interpret the results of parts (a) and (b).

Source: U.S. Census Bureau.

48. Spread of a Contagious Disease The incidence (number of new cases per day) of a contagious disease spreading in a population of M people is given by

$$R(x) = kx(M - x)$$

where k is a positive constant and x denotes the number of people already infected. Show that the incidence R is greatest when half the population is infected.

49. Maximizing the Volume of a Box A box with an open top is to be constructed from a square piece of cardboard, 10 in. wide, by cutting out a square from each of the four corners and bending up the sides. What is the maximum volume of such a box?

50. Minimizing Construction Costs A man wishes to construct a cylindrical barrel with a capacity of 32π ft^3. The cost per square foot of the material for the side of the barrel is half that of the cost per square foot for the top and bottom. Help him find the dimensions of the barrel that can be constructed at a minimum cost in terms of material used.

51. Packaging You wish to construct a closed rectangular box that has a volume of 4 ft³. The length of the base of the box will be twice as long as its width. The material for the top and bottom of the box costs 30¢/square foot. The material for the sides of the box costs 20¢/square foot. Find the dimensions of the least expensive box that can be constructed.

52. Inventory Control and Planning Lehen Vinters imports a certain brand of beer. The demand, which may be assumed to be uniform, is 800,000 cases/year. The cost of ordering a shipment of beer is $500, and the cost of storing each case of beer for a year is $2. Determine how many cases of beer should be in each shipment if the ordering and storage costs are to be kept at a minimum. (Assume that each shipment of beer arrives just as the previous one has been sold.)

53. In what interval is the quadratic function

$$f(x) = ax^2 + bx + c \quad (a \neq 0)$$

increasing? In what interval is f decreasing?

54. Let $f(x) = x^2 + ax + b$. Determine the constants a and b such that f has a relative minimum at $x = 2$ and the relative minimum value is 7.

55. Find the values of c such that the graph of

$$f(x) = x^4 + 2x^3 + cx^2 + 2x + 2$$

is concave upward everywhere.

56. Suppose that the point $(a, f(a))$ is an inflection point of the graph of $y = f(x)$. Show that the number a gives rise to a relative extremum of the function f'.

57. Let

$$f(x) = \begin{cases} x^3 + 1 & \text{if } x \neq 0 \\ 2 & \text{if } x = 0 \end{cases}$$

a. Compute $f'(x)$, and show that it does not change sign as we move across $x = 0$.

b. Show that f has a relative maximum at $x = 0$. Does this contradict the First Derivative Test? Explain your answer.

CHAPTER 4 Before Moving On . . .

1. Find the interval(s) where $f(x) = \dfrac{x^2}{1-x}$ is increasing and where it is decreasing.

2. Find the relative maxima and relative minima, if any, of $f(x) = 2x^2 - 12x^{1/3}$.

3. Find the intervals where the graph of $f(x) = \frac{1}{3}x^3 - \frac{1}{4}x^2 - \frac{1}{2}x + 1$ is concave upward, the intervals where the graph of f is concave downward, and the inflection point(s) of f.

4. Sketch the graph of $f(x) = 2x^3 - 9x^2 + 12x - 1$.

5. Find the absolute maximum and absolute minimum values of $f(x) = 2x^3 + 3x^2 - 1$ on the interval $[-2, 3]$.

6. An open bucket in the form of a right circular cylinder is to be constructed with a capacity of 1 ft³. Find the radius and height of the cylinder if the amount of material used is minimal.

5 EXPONENTIAL AND LOGARITHMIC FUNCTIONS

How many cameras can a new employee at Eastman Optical assemble after completing the basic training program, and how many cameras can he assemble after being on the job for 6 months? In Example 5, page 385, you will see how to answer these questions.

THE EXPONENTIAL FUNCTION is, without doubt, the most important function in mathematics and its applications. After a brief introduction to the exponential function and its *inverse*, the logarithmic function, we learn how to differentiate such functions. This lays the foundation for exploring the many applications involving exponential functions. For example, we look at the role played by exponential functions in computing earned interest on a bank account and in studying the growth of a bacteria population in the laboratory, the rate at which radioactive matter decays, the rate at which a factory worker learns a certain process, and the rate at which a communicable disease is spread over time.

5.1 Exponential Functions

Exponential Functions and Their Graphs

Suppose you deposit a sum of $1000 in an account earning interest at the rate of 10% per year *compounded continuously* (the way most financial institutions compute interest). Then, the accumulated amount at the end of t years ($0 \leq t \leq 20$) is described by the function f, whose graph appears in Figure 1.* This function is called an *exponential function*. Observe that the graph of f rises rather slowly at first but very rapidly as time goes by. For purposes of comparison, we have also shown the graph of the function $y = g(t) = 1000(1 + 0.10t)$, giving the accumulated amount for the same principal ($1000) but earning *simple* interest at the rate of 10% per year. The moral of the story: It is never too early to save.

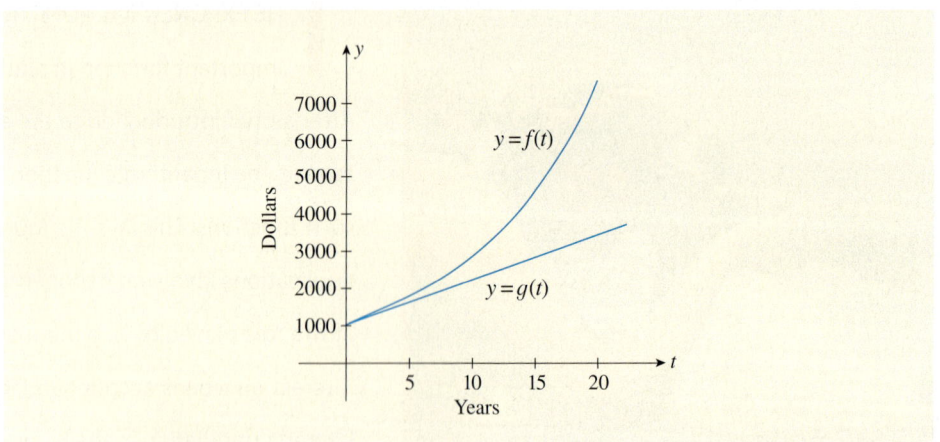

FIGURE 1
Under continuous compounding, a sum of money grows exponentially.

Exponential functions play an important role in many real-world applications, as you will see throughout this chapter.

Recall that whenever b is a positive number and n is any real number, the expression b^n is a real number. This enables us to define an exponential function as follows:

> **Exponential Function**
> The function defined by
> $$f(x) = b^x \quad (b > 0, b \neq 1)$$
> is called an **exponential function with base b and exponent x.** The domain of f is the set of all real numbers.

For example, the exponential function with base 2 is the function
$$f(x) = 2^x$$
with domain $(-\infty, \infty)$. The values of $f(x)$ for selected values of x follow:

$$f(3) = 2^3 = 8 \qquad f\left(\frac{3}{2}\right) = 2^{3/2} = 2 \cdot 2^{1/2} = 2\sqrt{2} \qquad f(0) = 2^0 = 1$$

$$f(-1) = 2^{-1} = \frac{1}{2} \qquad f\left(-\frac{2}{3}\right) = 2^{-2/3} = \frac{1}{2^{2/3}} = \frac{1}{\sqrt[3]{4}}$$

*We will derive the rule for f in Section 5.3.

5.1 EXPONENTIAL FUNCTIONS

Computations involving exponentials are facilitated by the laws of exponents. These laws were stated in Section 1.1, and you might want to review the material there. For convenience, however, we will restate these laws.

Laws of Exponents

Let a and b be positive numbers and let x and y be real numbers. Then,

1. $b^x \cdot b^y = b^{x+y}$
2. $\dfrac{b^x}{b^y} = b^{x-y}$
3. $(b^x)^y = b^{xy}$
4. $(ab)^x = a^x b^x$
5. $\left(\dfrac{a}{b}\right)^x = \dfrac{a^x}{b^x}$

The use of the laws of exponents is illustrated in the next two examples.

EXAMPLE 1

a. $16^{7/4} \cdot 16^{-1/2} = 16^{7/4 - 1/2} = 16^{5/4} = 2^5 = 32$ Law 1

b. $\dfrac{8^{5/3}}{8^{-1/3}} = 8^{5/3 - (-1/3)} = 8^2 = 64$ Law 2

c. $(64^{4/3})^{-1/2} = 64^{(4/3)(-1/2)} = 64^{-2/3}$

$\qquad = \dfrac{1}{64^{2/3}} = \dfrac{1}{(64^{1/3})^2} = \dfrac{1}{4^2} = \dfrac{1}{16}$ Law 3

d. $(16 \cdot 81)^{-1/4} = 16^{-1/4} \cdot 81^{-1/4} = \dfrac{1}{16^{1/4}} \cdot \dfrac{1}{81^{1/4}} = \dfrac{1}{2} \cdot \dfrac{1}{3} = \dfrac{1}{6}$ Law 4

e. $\left(\dfrac{3^{1/2}}{2^{1/3}}\right)^4 = \dfrac{3^{4/2}}{2^{4/3}} = \dfrac{9}{2^{4/3}}$ Law 5

EXAMPLE 2 Let $f(x) = 2^{2x-1}$. Find the value of x for which $f(x) = 16$.

Solution We want to solve the equation

$$2^{2x-1} = 16 = 2^4$$

But this equation holds if and only if

$$2x - 1 = 4 \qquad b^m = b^n \Rightarrow m = n$$

giving $x = \frac{5}{2}$.

Exponential functions play an important role in mathematical analysis. Because of their special characteristics, they are some of the most useful functions and are found in virtually every field in which mathematics is applied. To mention a few examples: Under ideal conditions, the number of bacteria present at any time t in a culture may be described by an exponential function of t; radioactive substances decay over time in accordance with an "exponential" law of decay; money left on fixed deposit and earning compound interest grows exponentially; and some of the most important distribution functions encountered in statistics are exponential.

Let's begin our investigation into the properties of exponential functions by studying their graphs.

EXAMPLE 3 Sketch the graph of the exponential function $y = 2^x$.

Solution First, as was discussed earlier, the domain of the exponential function $y = f(x) = 2^x$ is the set of real numbers. Next, putting $x = 0$ gives $y = 2^0 = 1$, the

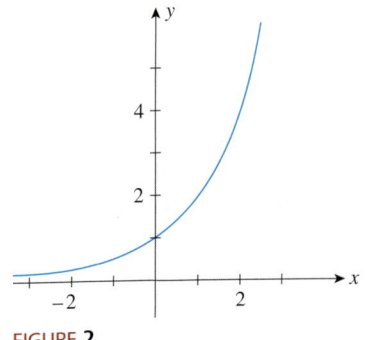

FIGURE 2
The graph of $y = 2^x$

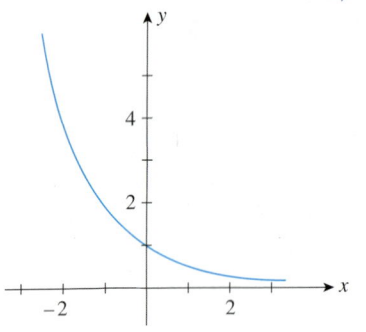

FIGURE 3
The graph of $y = \left(\frac{1}{2}\right)^x$

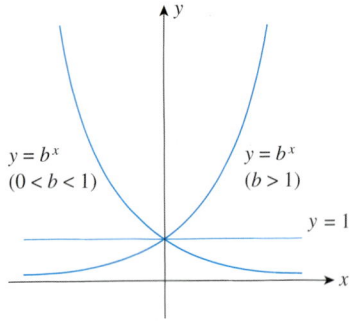

FIGURE 4
$y = b^x$ is an increasing function of x if $b > 1$, a constant function if $b = 1$, and a decreasing function if $0 < b < 1$.

y-intercept of f. There is no x-intercept, since there is no value of x for which $y = 0$. To find the range of f, consider the following table of values:

x	−5	−4	−3	−2	−1	0	1	2	3	4	5
y	$\frac{1}{32}$	$\frac{1}{16}$	$\frac{1}{8}$	$\frac{1}{4}$	$\frac{1}{2}$	1	2	4	8	16	32

We see from these computations that 2^x decreases and approaches zero as x decreases without bound and that 2^x increases without bound as x increases without bound. Thus, the range of f is the interval $(0, \infty)$—that is, the set of positive real numbers. Finally, we sketch the graph of $y = f(x) = 2^x$ in Figure 2.

EXAMPLE 4 Sketch the graph of the exponential function $y = (1/2)^x$.

Solution The domain of the exponential function $y = (1/2)^x$ is the set of all real numbers. The y-intercept is $(1/2)^0 = 1$; there is no x-intercept, since there is no value of x for which $y = 0$. From the following table of values

x	−5	−4	−3	−2	−1	0	1	2	3	4	5
y	32	16	8	4	2	1	$\frac{1}{2}$	$\frac{1}{4}$	$\frac{1}{8}$	$\frac{1}{16}$	$\frac{1}{32}$

we deduce that $(1/2)^x = 1/2^x$ increases without bound as x decreases without bound and that $(1/2)^x$ decreases and approaches zero as x increases without bound. Thus, the range of f is the interval $(0, \infty)$. The graph of $y = f(x) = (1/2)^x$ is sketched in Figure 3.

The functions $y = 2^x$ and $y = (1/2)^x$, whose graphs you studied in Examples 3 and 4, are special cases of the exponential function $y = f(x) = b^x$, obtained by setting $b = 2$ and $b = 1/2$, respectively. In general, the exponential function $y = b^x$ with $b > 1$ has a graph similar to that of $y = 2^x$, whereas the graph of $y = b^x$ for $0 < b < 1$ is similar to that of $y = (1/2)^x$ (Exercises 27 and 28 on page 334). When $b = 1$, the function $y = b^x$ reduces to the constant function $y = 1$. For comparison, the graphs of all three functions are sketched in Figure 4.

Properties of the Exponential Function

The exponential function $y = b^x$ ($b > 0$, $b \neq 1$) has the following properties:

1. Its domain is $(-\infty, \infty)$.
2. Its range is $(0, \infty)$.
3. Its graph passes through the point $(0, 1)$.
4. It is continuous on $(-\infty, \infty)$.
5. It is increasing on $(-\infty, \infty)$ if $b > 1$ and decreasing on $(-\infty, \infty)$ if $b < 1$.

The Base e

Exponential functions to the base e, where e is an irrational number whose value is $2.7182818\ldots$, play an important role in both theoretical and applied problems. It can be shown, although we will not do so here, that

$$e = \lim_{m \to \infty} \left(1 + \frac{1}{m}\right)^m \tag{1}$$

TABLE 1

m	$\left(1 + \dfrac{1}{m}\right)^m$
10	2.59374
100	2.70481
1000	2.71692
10,000	2.71815
100,000	2.71827
1,000,000	2.71828

However, you may convince yourself of the plausibility of this definition of the number e by examining Table 1, which can be constructed with the help of a calculator.

Exploring with TECHNOLOGY

To obtain a visual confirmation of the fact that the expression $(1 + 1/m)^m$ approaches the number $e = 2.71828\ldots$ as m increases without bound, plot the graph of $f(x) = (1 + 1/x)^x$ in a suitable viewing window and observe that $f(x)$ approaches $2.71828\ldots$ as x increases without bound. Use ZOOM and TRACE to find the value of $f(x)$ for large values of x.

VIDEO

EXAMPLE 5 Sketch the graph of the function $y = e^x$.

Solution Since $e > 1$, it follows from our previous discussion that the graph of $y = e^x$ is similar to the graph of $y = 2^x$ (see Figure 2). With the aid of a calculator, we obtain the following table:

x	-3	-2	-1	0	1	2	3
y	0.05	0.14	0.37	1	2.72	7.39	20.09

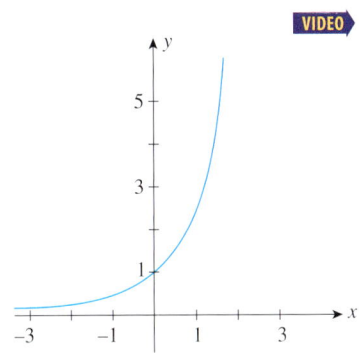

FIGURE 5
The graph of $y = e^x$

The graph of $y = e^x$ is sketched in Figure 5.

Next, we consider another exponential function to the base e that is closely related to the previous function and is particularly useful in constructing models that describe "exponential decay."

EXAMPLE 6 Sketch the graph of the function $y = e^{-x}$.

Solution Since $e > 1$, it follows that $0 < 1/e < 1$, so $f(x) = e^{-x} = 1/e^x = (1/e)^x$ is an exponential function with base less than 1. Therefore, it has a graph similar to that of the exponential function $y = (1/2)^x$. As before, we construct the following table of values of $y = e^{-x}$ for selected values of x:

x	-3	-2	-1	0	1	2	3
y	20.09	7.39	2.72	1	0.37	0.14	0.05

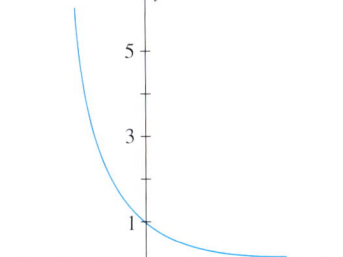

FIGURE 6
The graph of $y = e^{-x}$

Using this table, we sketch the graph of $y = e^{-x}$ in Figure 6.

5.1 Self-Check Exercises

1. Solve the equation $2^{2x+1} \cdot 2^{-3} = 2^{x-1}$.
2. Sketch the graph of $y = e^{0.4x}$.

Solutions to Self-Check Exercises 5.1 can be found on page 336.

5.1 Concept Questions

1. Define the exponential function f with base b and exponent x. What restrictions, if any, are placed on b?

2. For the exponential function $y = b^x$ ($b > 0$, $b \neq 1$), state (a) its domain and range, (b) its y-intercept, (c) where it is continuous, and (d) where it is increasing and where it is decreasing for the case $b > 1$ and the case $b < 1$.

5.1 Exercises

In Exercises 1–8, evaluate the expression.

1. a. $4^{-3} \cdot 4^5$ b. $3^{-3} \cdot 3^6$
2. a. $(2^{-1})^3$ b. $(3^{-2})^3$
3. a. $9(9)^{-1/2}$ b. $5(5)^{-1/2}$
4. a. $\left[\left(-\dfrac{1}{2}\right)^3\right]^{-2}$ b. $\left[\left(-\dfrac{1}{3}\right)^2\right]^{-3}$
5. a. $\dfrac{(-3)^4(-3)^5}{(-3)^8}$ b. $\dfrac{(2^{-4})(2^6)}{2^{-1}}$
6. a. $3^{1/4} \cdot 9^{-5/8}$ b. $2^{3/4} \cdot 4^{-3/2}$
7. a. $\dfrac{5^{3.3} \cdot 5^{-1.6}}{5^{-0.3}}$ b. $\dfrac{4^{2.7} \cdot 4^{-1.3}}{4^{-0.4}}$
8. a. $\left(\dfrac{1}{16}\right)^{-1/4}\left(\dfrac{27}{64}\right)^{-1/3}$ b. $\left(\dfrac{8}{27}\right)^{-1/3}\left(\dfrac{81}{256}\right)^{-1/4}$

In Exercises 9–16, simplify the expression.

9. a. $(64x^9)^{1/3}$ b. $(25x^3y^4)^{1/2}$
10. a. $(2x^3)(-4x^{-2})$ b. $(4x^{-2})(-3x^5)$
11. a. $\dfrac{6a^{-4}}{3a^{-3}}$ b. $\dfrac{4b^{-4}}{12b^{-6}}$
12. a. $y^{-3/2}y^{5/3}$ b. $x^{-3/5}x^{8/3}$
13. a. $(2x^3y^2)^3$ b. $(4x^2y^2z^3)^2$
14. a. $(x^{r/s})^{s/r}$ b. $(x^{-b/a})^{-a/b}$
15. a. $\dfrac{5^0}{(2^{-3}x^{-3}y^2)^2}$ b. $\dfrac{(x+y)(x-y)}{(x-y)^0}$
16. a. $\dfrac{(a^m \cdot a^{-n})^{-2}}{(a^{m+n})^2}$ b. $\left(\dfrac{x^{2n-2}y^{2n}}{x^{5n+1}y^{-n}}\right)^{1/3}$

In Exercises 17–26, solve the equation for x.

17. $6^{2x} = 6^6$
18. $5^{-x} = 5^3$
19. $3^{3x-4} = 3^5$
20. $10^{2x-1} = 10^{x+3}$
21. $(2.1)^{x+2} = (2.1)^5$
22. $(-1.3)^{x-2} = (-1.3)^{2x+1}$
23. $8^x = \left(\dfrac{1}{32}\right)^{x-2}$
24. $3^{x-x^2} = \dfrac{1}{9^x}$
25. $2^x - 12 \cdot 3^x + 27 = 0$
26. $2^{2x} - 4 \cdot 2^x + 4 = 0$

In Exercises 27–36, sketch the graphs of the given functions on the same axes.

27. $y = 2^x$, $y = 3^x$, and $y = 4^x$
28. $y = \left(\dfrac{1}{2}\right)^x$, $y = \left(\dfrac{1}{3}\right)^x$, and $y = \left(\dfrac{1}{4}\right)^x$
29. $y = 2^{-x}$, $y = 3^{-x}$, and $y = 4^{-x}$
30. $y = 4^{0.5x}$ and $y = 4^{-0.5x}$
31. $y = 4^{0.5x}$, $y = 4^x$, and $y = 4^{2x}$
32. $y = e^x$, $y = 2e^x$, and $y = 3e^x$
33. $y = e^{0.5x}$, $y = e^x$, and $y = e^{1.5x}$
34. $y = e^{-0.5x}$, $y = e^{-x}$, and $y = e^{-1.5x}$
35. $y = 0.5e^{-x}$, $y = e^{-x}$, and $y = 2e^{-x}$
36. $y = 1 - e^{-x}$ and $y = 1 - e^{-0.5x}$

37. A function f has the form $f(x) = Ae^{kx}$. Find f if it is known that $f(0) = 100$ and $f(1) = 120$.
Hint: $e^{kx} = (e^k)^x$

38. If $f(x) = Axe^{-kx}$, find $f(3)$ if $f(1) = 5$ and $f(2) = 7$.
Hint: $e^{kx} = (e^k)^x$

39. If
$$f(t) = \dfrac{1000}{1 + Be^{-kt}}$$
find $f(5)$ given that $f(0) = 20$ and $f(2) = 30$.
Hint: $e^{kx} = (e^k)^x$

40. **Tracking With GPS** Employers are increasingly turning to GPS (global positioning system) technology to keep track of their fleet vehicles. The number of automatic vehicle trackers installed on fleet vehicles in the United States is approximated by
$$N(t) = 0.6e^{0.17t} \quad (0 \leq t \leq 5)$$
where $N(t)$ is measured in millions and t is measured in years, with $t = 0$ corresponding to 2000.
a. What was the number of automatic vehicle trackers installed in the year 2000? How many were installed in 2005?
b. Sketch the graph of N.
Source: C. J. Driscoll Associates.

41. DISABILITY RATES Because of medical technology advances, the disability rates for people over 65 years old have been dropping rather dramatically. The function

$$R(t) = 26.3e^{-0.016t} \quad (0 \le t \le 18)$$

gives the disability rate $R(t)$, in percent, for people over age 65 from 1982 ($t = 0$) through 2000, where t is measured in years.
a. What was the disability rate in 1982? In 1986? In 1994? In 2000?
b. Sketch the graph of R.
Source: Frost and Sullivan.

42. MARRIED HOUSEHOLDS The percentage of families that were married households between 1970 and 2000 is approximately

$$P(t) = 86.9e^{-0.05t} \quad (0 \le t \le 3)$$

where t is measured in decades, with $t = 0$ corresponding to 1970.
a. What percentage of families were married households in 1970? In 1980? In 1990? In 2000?
b. Sketch the graph of P.
Source: U.S. Census Bureau.

43. GROWTH IN NUMBER OF WEBSITES According to a study conducted in 2000, the projected number of Web addresses (in billions) is approximated by the function

$$N(t) = 0.45e^{0.5696t} \quad (0 \le t \le 5)$$

where t is measured in years, with $t = 0$ corresponding to 1997.
a. Complete the following table by finding the number of Web addresses in each year:

Year	0	1	2	3	4	5
Number of Web Addresses (billions)						

b. Sketch the graph of N.

44. INTERNET USERS IN CHINA The number of Internet users in China is approximated by

$$N(t) = 94.5e^{0.2t} \quad (1 \le t \le 6)$$

where $N(t)$ is measured in millions and t is measured in years, with $t = 1$ corresponding to 2005.
a. How many Internet users were there in 2005? In 2006? In 2010?
b. Sketch the graph of N.
Source: C. E. Unterberg.

45. ALTERNATIVE MINIMUM TAX The alternative minimum tax was created in 1969 to prevent the very wealthy from using creative deductions and shelters to avoid having to pay anything to the Internal Revenue Service. But it has increasingly hit the middle class. The number of taxpayers subjected to an alternative minimum tax is projected to be

$$N(t) = \frac{35.5}{1 + 6.89e^{-0.8674t}} \quad (0 \le t \le 6)$$

where $N(t)$ is measured in millions and t is measured in years, with $t = 0$ corresponding to 2004. What was the projected number of taxpayers subjected to an alternative minimum tax in 2010?
Source: Brookings Institution.

46. ABSORPTION OF DRUGS The concentration of a drug in an organ at any time t (in seconds) is given by

$$C(t) = \begin{cases} 0.3t - 18(1 - e^{-t/60}) & \text{if } 0 \le t \le 20 \\ 18e^{-t/60} - 12e^{-(t-20)/60} & \text{if } t > 20 \end{cases}$$

where $C(t)$ is measured in grams per cubic centimeter (g/cm³).
a. What is the initial concentration of the drug in the organ?
b. What is the concentration of the drug in the organ after 10 sec?
c. What is the concentration of the drug in the organ after 30 sec?
d. What will be the concentration of the drug in the long run?
Hint: Evaluate $\lim_{t \to \infty} C(t)$.

47. ABSORPTION OF DRUGS The concentration of a drug in an organ at any time t (in seconds) is given by

$$x(t) = 0.08 + 0.12(1 - e^{-0.02t})$$

where $x(t)$ is measured in grams per cubic centimeter (g/cm³).
a. What is the initial concentration of the drug in the organ?
b. What is the concentration of the drug in the organ after 20 sec?
c. What will be the concentration of the drug in the organ in the long run?
Hint: Evaluate $\lim_{t \to \infty} x(t)$.
d. Sketch the graph of x.

48. ABSORPTION OF DRUGS Jane took 100 mg of a drug in the morning and another 100 mg of the same drug at the same time the following morning. The amount of the drug in her body t days after the first dose was taken is given by

$$A(t) = \begin{cases} 100e^{-1.4t} & \text{if } 0 \le t < 1 \\ 100(1 + e^{1.4})e^{-1.4t} & \text{if } t \ge 1 \end{cases}$$

a. What was the amount of drug in Jane's body immediately after taking the second dose? After 2 days? In the long run?
b. Sketch the graph of A.

In Exercises 49–52, determine whether the statement is true or false. If it is true, explain why it is true. If it is false, give an example to show why it is false.

49. $(x^2 + 1)^3 = x^6 + 1$ **50.** $e^{xy} = e^x e^y$

51. If $x < y$, then $e^x < e^y$.

52. If $0 < b < 1$ and $x < y$, then $b^x > b^y$.

5.1 Solutions to Self-Check Exercises

1. $2^{2x+1} \cdot 2^{-3} = 2^{x-1}$

    $\dfrac{2^{2x+1}}{2^{x-1}} \cdot 2^{-3} = 1$ Divide both sides by 2^{x-1}.

    $2^{(2x+1)-(x-1)-3} = 1$

    $2^{x-1} = 1$

    This is true if and only if $x - 1 = 0$ or $x = 1$.

2. We first construct the following table of values:

x	-3	-2	-1	0	1	2	3	4
$y = e^{0.4x}$	0.3	0.4	0.7	1	1.5	2.2	3.3	5

Next, we plot these points and join them by a smooth curve to obtain the graph of f shown in the accompanying figure.

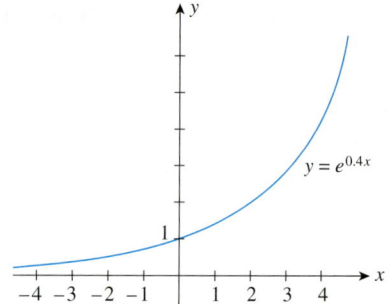

USING TECHNOLOGY

Although the proof is outside the scope of this book, it can be proved that an exponential function of the form $f(x) = b^x$, where $b > 1$, will ultimately grow faster than the power function $g(x) = x^n$ for *any* positive real number n. To give a visual demonstration of this result for the special case of the exponential function $f(x) = e^x$, we can use a graphing utility to plot the graphs of both f and g (for selected values of n) on the same set of axes in an appropriate viewing window and observe that the graph of f ultimately lies above that of g.

EXAMPLE 1 Use a graphing utility to plot the graphs of (a) $f(x) = e^x$ and $g(x) = x^3$ on the same set of axes in the viewing window $[0, 6] \times [0, 250]$ and (b) $f(x) = e^x$ and $g(x) = x^5$ in the viewing window $[0, 20] \times [0, 1,000,000]$.

Solution

a. The graphs of $f(x) = e^x$ and $g(x) = x^3$ in the viewing window $[0, 6] \times [0, 250]$ are shown in Figure T1a.

b. The graphs of $f(x) = e^x$ and $g(x) = x^5$ in the viewing window $[0, 20] \times [0, 1,000,000]$ are shown in Figure T1b.

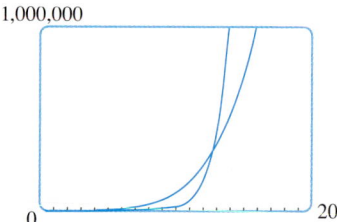

(a) The graphs of $f(x) = e^x$ and $g(x) = x^3$ in the viewing window $[0, 6] \times [0, 250]$

(b) The graphs of $f(x) = e^x$ and $g(x) = x^5$ in the viewing window $[0, 20] \times [0, 1,000,000]$

FIGURE T1

In the exercises that follow, you are asked to use a graphing utility to reveal the properties of exponential functions.

TECHNOLOGY EXERCISES

In Exercises 1 and 2, plot the graphs of the functions f and g on the same set of axes in the specified viewing window.

1. $f(x) = e^x$ and $g(x) = x^2$; $[0, 4] \times [0, 30]$
2. $f(x) = e^x$ and $g(x) = x^4$; $[0, 15] \times [0, 20{,}000]$

In Exercises 3 and 4, plot the graphs of the functions f and g on the same set of axes in an appropriate viewing window to demonstrate that f ultimately grows faster than g. (Note: Your answer will not be unique.)

3. $f(x) = 2^x$ and $g(x) = x^{2.5}$
4. $f(x) = 3^x$ and $g(x) = x^3$

5. Plot the graphs of $f(x) = 2^x$, $g(x) = 3^x$, and $h(x) = 4^x$ on the same set of axes in the viewing window $[0, 5] \times [0, 100]$. Comment on the relationship between the base b and the growth of the function $f(x) = b^x$.

6. Plot the graphs of $f(x) = (1/2)^x$, $g(x) = (1/3)^x$, and $h(x) = (1/4)^x$ on the same set of axes in the viewing window $[0, 4] \times [0, 1]$. Comment on the relationship between the base b and the growth of the function $f(x) = b^x$.

7. Plot the graphs of $f(x) = e^x$, $g(x) = 2e^x$, and $h(x) = 3e^x$ on the same set of axes in the viewing window $[-3, 3] \times [0, 10]$. Comment on the role played by the constant k in the graph of $f(x) = ke^x$.

8. Plot the graphs of $f(x) = -e^x$, $g(x) = -2e^x$, and $h(x) = -3e^x$ on the same set of axes in the viewing window $[-3, 3] \times [-10, 0]$. Comment on the role played by the constant k in the graph of $f(x) = ke^x$.

9. Plot the graphs of $f(x) = e^{0.5x}$, $g(x) = e^x$, and $h(x) = e^{1.5x}$ on the same set of axes in the viewing window $[-2, 2] \times [0, 4]$. Comment on the role played by the constant k in the graph of $f(x) = e^{kx}$.

10. Plot the graphs of $f(x) = e^{-0.5x}$, $g(x) = e^{-x}$, and $h(x) = e^{-1.5x}$ on the same set of axes in the viewing window $[-2, 2] \times [0, 4]$. Comment on the role played by the constant k in the graph of $f(x) = e^{kx}$.

11. **Absorption of Drugs** The concentration of a drug in an organ at any time t (in seconds) is given by

    $$x(t) = 0.08 + 0.12(1 - e^{-0.02t})$$

    where $x(t)$ is measured in grams per cubic centimeter (g/cm³).
    a. Plot the graph of the function x in the viewing window $[0, 200] \times [0, 0.2]$.
    b. What is the initial concentration of the drug in the organ?
    c. What is the concentration of the drug in the organ after 20 sec?
    d. What will be the concentration of the drug in the organ in the long run?
    Hint: Evaluate $\lim_{t \to \infty} x(t)$.

12. **Absorption of Drugs** Jane took 100 mg of a drug in the morning and another 100 mg of the same drug at the same time the following morning. The amount of the drug in her body t days after the first dosage was taken is given by

    $$A(t) = \begin{cases} 100e^{-1.4t} & \text{if } 0 \leq t < 1 \\ 100(1 + e^{1.4})e^{-1.4t} & \text{if } t \geq 1 \end{cases}$$

    a. Plot the graph of the function A in the viewing window $[0, 5] \times [0, 140]$.
    b. Verify the results of Exercise 48, page 335.

13. **Absorption of Drugs** The concentration of a drug in an organ at any time t (in seconds) is given by

    $$C(t) = \begin{cases} 0.3t - 18(1 - e^{-t/60}) & \text{if } 0 \leq t \leq 20 \\ 18e^{-t/60} - 12e^{-(t-20)/60} & \text{if } t > 20 \end{cases}$$

    where $C(t)$ is measured in grams per cubic centimeter (g/cm³).
    a. Plot the graph of the function C in the viewing window $[0, 120] \times [0, 1]$.
    b. How long after the drug is first introduced will it take for the concentration of the drug to reach a peak?
    c. How long after the concentration of the drug has peaked will it take for the concentration of the drug to fall back to 0.5 g/cm³?
    Hint: Plot the graphs of $y_1 = C(x)$ and $y_2 = 0.5$, and use the ISECT function of your graphing utility.

14. **Modeling With Data** The estimated number of Internet users in China (in millions) from 2005 through 2010 are shown in the following table:

Year	2005	2006	2007	2008	2009	2010
Number	116.1	141.9	169.0	209.0	258.1	314.8

    a. Use **ExpReg** to find an exponential regression model for the data. Let $t = 1$ correspond to 2005.
    Hint: $a^x = e^{x \ln a}$
    b. Plot the scatter diagram and the graph of the function f found in part (a).

5.2 Logarithmic Functions

Logarithms

You are already familiar with exponential equations of the form

$$b^y = x \quad (b > 0, b \neq 1)$$

where the variable x is expressed in terms of a real number b and a variable y. But what about solving this same equation for y? You may recall from your study of algebra that the number y is called the **logarithm of x to the base b** and is denoted by $\log_b x$. It is the power to which the base b must be raised to obtain the number x.

> **Logarithm of x to the Base b**
>
> $y = \log_b x \quad$ if and only if $\quad x = b^y \quad (b > 0, b \neq 1, \text{ and } x > 0)$

 Observe that the logarithm $\log_b x$ is defined only for positive values of x.

EXAMPLE 1

a. $\log_{10} 100 = 2$ since $100 = 10^2$
b. $\log_5 125 = 3$ since $125 = 5^3$
c. $\log_3 \dfrac{1}{27} = -3$ since $\dfrac{1}{27} = \dfrac{1}{3^3} = 3^{-3}$
d. $\log_{20} 20 = 1$ since $20 = 20^1$

EXAMPLE 2 Solve each of the following equations for x.

a. $\log_3 x = 4 \quad$ b. $\log_{16} 4 = x \quad$ c. $\log_x 8 = 3$

Solution

a. By definition, $\log_3 x = 4$ implies $x = 3^4 = 81$.
b. $\log_{16} 4 = x$ is equivalent to $4 = 16^x = (4^2)^x = 4^{2x}$, or $4^1 = 4^{2x}$, from which we deduce that

$$2x = 1 \qquad b^m = b^n \Rightarrow m = n$$
$$x = \frac{1}{2}$$

c. Referring once again to the definition, we see that the equation $\log_x 8 = 3$ is equivalent to

$$8 = 2^3 = x^3$$
$$x = 2 \qquad a^m = b^m \Rightarrow a = b$$

The two most widely used systems of logarithms are the system of **common logarithms,** which uses the number 10 as its base, and the system of **natural logarithms,** which uses the irrational number $e = 2.71828\ldots$ as its base. Also, it is standard practice to write **log** for $\log_{10}$ and **ln** for $\log_e$.

> **Logarithmic Notation**
>
> $\log x = \log_{10} x \qquad$ Common logarithm
>
> $\ln x = \log_e x \qquad$ Natural logarithm

The system of natural logarithms is widely used in theoretical work. Using natural logarithms rather than logarithms to other bases often leads to simpler expressions.

Laws of Logarithms

Computations involving logarithms are facilitated by the following **laws of logarithms.**

> **Laws of Logarithms**
> If m and n are positive numbers and $b > 0$, $b \neq 1$, then
> 1. $\log_b mn = \log_b m + \log_b n$
> 2. $\log_b \dfrac{m}{n} = \log_b m - \log_b n$
> 3. $\log_b m^n = n \log_b m$
> 4. $\log_b 1 = 0$
> 5. $\log_b b = 1$

 Do not confuse the expression $\log m/n$ (Law 2) with the expression $\log m/\log n$. For example,

$$\log \frac{100}{10} = \log 100 - \log 10 = 2 - 1 = 1 \neq \frac{\log 100}{\log 10} = \frac{2}{1} = 2$$

You will be asked to prove these laws in Exercises 62–64 on page 345. Their derivations are based on the definition of a logarithm and the corresponding laws of exponents. The following examples illustrate the properties of logarithms.

EXAMPLE 3

a. $\log(2 \cdot 3) = \log 2 + \log 3$ **b.** $\ln \dfrac{5}{3} = \ln 5 - \ln 3$

c. $\log \sqrt{7} = \log 7^{1/2} = \dfrac{1}{2} \log 7$ **d.** $\log_5 1 = 0$

e. $\log_{45} 45 = 1$

EXAMPLE 4 Given that $\log 2 \approx 0.3010$, $\log 3 \approx 0.4771$, and $\log 5 \approx 0.6990$, use the laws of logarithms to find

a. $\log 15$ **b.** $\log 7.5$ **c.** $\log 81$ **d.** $\log 50$

Solution

a. Note that $15 = 3 \cdot 5$, so by Law 1 for logarithms,

$$\log 15 = \log 3 \cdot 5$$
$$= \log 3 + \log 5$$
$$\approx 0.4771 + 0.6990$$
$$= 1.1761$$

b. Observing that $7.5 = 15/2 = (3 \cdot 5)/2$, we apply Laws 1 and 2, obtaining

$$\log 7.5 = \log \frac{(3)(5)}{2}$$
$$= \log 3 + \log 5 - \log 2$$
$$\approx 0.4771 + 0.6990 - 0.3010$$
$$= 0.8751$$

c. Since $81 = 3^4$, we apply Law 3 to obtain

$$\log 81 = \log 3^4$$
$$= 4 \log 3$$
$$\approx 4(0.4771)$$
$$= 1.9084$$

d. We write $50 = 5 \cdot 10$ and find

$$\log 50 = \log(5)(10)$$
$$= \log 5 + \log 10$$
$$\log 50 \approx 0.6990 + 1 \quad \text{Use Law 5}$$
$$= 1.6990$$

■

VIDEO **EXAMPLE 5** Expand and simplify the following expressions:

a. $\log_3 x^2 y^3$ **b.** $\log_2 \dfrac{x^2 + 1}{2^x}$ **c.** $\ln \dfrac{x^2 \sqrt{x^2 - 1}}{e^x}$

Solution

a. $\log_3 x^2 y^3 = \log_3 x^2 + \log_3 y^3$ Law 1
$ = 2 \log_3 x + 3 \log_3 y$ Law 3

b. $\log_2 \dfrac{x^2 + 1}{2^x} = \log_2(x^2 + 1) - \log_2 2^x$ Law 2
$\phantom{\log_2 \dfrac{x^2 + 1}{2^x}} = \log_2(x^2 + 1) - x \log_2 2$ Law 3
$\phantom{\log_2 \dfrac{x^2 + 1}{2^x}} = \log_2(x^2 + 1) - x$ Law 5

c. $\ln \dfrac{x^2 \sqrt{x^2 - 1}}{e^x} = \ln \dfrac{x^2(x^2 - 1)^{1/2}}{e^x}$ Rewrite
$\phantom{\ln \dfrac{x^2 \sqrt{x^2 - 1}}{e^x}} = \ln x^2 + \ln(x^2 - 1)^{1/2} - \ln e^x$ Laws 1 and 2
$\phantom{\ln \dfrac{x^2 \sqrt{x^2 - 1}}{e^x}} = 2 \ln x + \dfrac{1}{2} \ln(x^2 - 1) - x \ln e$ Law 3
$\phantom{\ln \dfrac{x^2 \sqrt{x^2 - 1}}{e^x}} = 2 \ln x + \dfrac{1}{2} \ln(x^2 - 1) - x$ Law 5

■

Logarithmic Functions and Their Graphs

The definition of a logarithm implies that if b and n are positive numbers and b is different from 1, then the expression $\log_b n$ is a real number. This enables us to define a logarithmic function as follows.

Logarithmic Function

The function defined by

$$f(x) = \log_b x \quad (b > 0, b \neq 1)$$

is called the **logarithmic function with base b.** The domain of f is the set of all positive numbers.

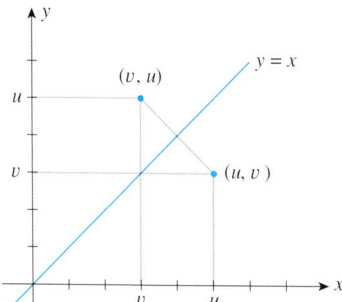

FIGURE 7
The points (u, v) and (v, u) are mirror reflections of each other.

One easy way to obtain the graph of the logarithmic function $y = \log_b x$ is to construct a table of values of the logarithm (base b). However, another method—and a more instructive one—is based on exploiting the intimate relationship between logarithmic and exponential functions.

If a point (u, v) lies on the graph of $y = \log_b x$, then

$$v = \log_b u$$

But we can also write this equation in exponential form as

$$u = b^v$$

So the point (v, u) also lies on the graph of the function $y = b^x$. Let's look at the relationship between the points (u, v) and (v, u) and the line $y = x$ (Figure 7). If we think of the line $y = x$ as a mirror, then the point (v, u) is the mirror reflection of the point (u, v). Similarly, the point (u, v) is the mirror reflection of the point (v, u). We can take advantage of this relationship to help us draw the graph of logarithmic functions. For example, if we wish to draw the graph of $y = \log_b x$, where $b > 1$, then we need only draw the mirror reflection of the graph of $y = b^x$ with respect to the line $y = x$ (Figure 8).

You may discover the following properties of the logarithmic function by taking the reflection of the graph of an appropriate exponential function (Exercises 33 and 34 on page 344).

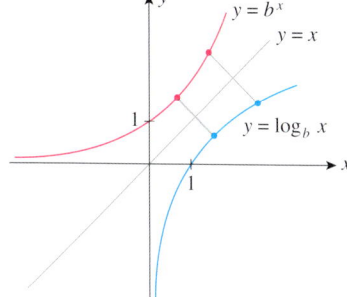

FIGURE 8
The graphs of $y = b^x$ and $y = \log_b x$ are mirror reflections of each other.

Properties of the Logarithmic Function

The logarithmic function $y = \log_b x$ ($b > 0, b \neq 1$) has the following properties:

1. Its domain is $(0, \infty)$.
2. Its range is $(-\infty, \infty)$.
3. Its graph passes through the point $(1, 0)$.
4. It is continuous on $(0, \infty)$.
5. It is increasing on $(0, \infty)$ if $b > 1$ and decreasing on $(0, \infty)$ if $b < 1$.

EXAMPLE 6 Sketch the graph of the function $y = \ln x$.

Solution We first sketch the graph of $y = e^x$. Then, the required graph is obtained by tracing the mirror reflection of the graph of $y = e^x$ with respect to the line $y = x$ (Figure 9). ∎

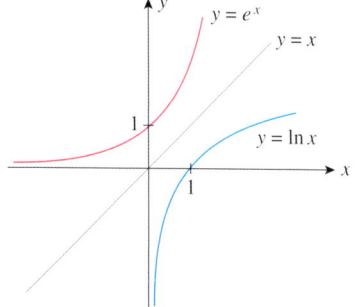

FIGURE 9
The graph of $y = \ln x$ is the mirror reflection of the graph of $y = e^x$.

Properties Relating the Exponential and Logarithmic Functions

We made use of the relationship that exists between the exponential function $f(x) = e^x$ and the logarithmic function $g(x) = \ln x$ when we sketched the graph of g in Example 6. This relationship is further described by the following properties, which are an immediate consequence of the definition of the logarithm of a number.

> **Properties Relating e^x and $\ln x$**
>
> $$e^{\ln x} = x \quad \text{(for } x > 0\text{)} \tag{2}$$
>
> $$\ln e^x = x \quad \text{(for any real number } x\text{)} \tag{3}$$

(Try to verify these properties.)

From Properties 2 and 3, we conclude that the composite function satisfies

$$(f \circ g)(x) = f[g(x)]$$
$$= e^{\ln x} = x \quad \text{(for all } x > 0\text{)}$$
$$(g \circ f)(x) = g[f(x)]$$
$$= \ln e^x = x \quad \text{(for all } x > 0\text{)}$$

Any two functions f and g that satisfy this relationship are said to be **inverses** of each other. Note that the function f undoes what the function g does, and vice versa, so the composition of the two functions in any order results in the identity function $F(x) = x$.*

The relationships expressed in Equations (2) and (3) are useful in solving equations that involve exponentials and logarithms.

Exploring with TECHNOLOGY

You can demonstrate the validity of Properties 2 and 3, which state that the exponential function $f(x) = e^x$ and the logarithmic function $g(x) = \ln x$ are inverses of each other, as follows:

1. Sketch the graph of $(f \circ g)(x) = e^{\ln x}$, using the viewing window $[0, 10] \times [0, 10]$. Interpret the result.
2. Sketch the graph of $(g \circ f)(x) = \ln e^x$, using the standard viewing window. Interpret the result.

VIDEO **EXAMPLE 7** Solve the equation $2e^{x+2} = 5$.

Solution We first divide both sides of the equation by 2 to obtain

$$e^{x+2} = \frac{5}{2} = 2.5$$

Next, taking the natural logarithm of each side of the equation and using Equation (3), we have

$$\ln e^{x+2} = \ln 2.5$$
$$x + 2 = \ln 2.5$$
$$x = -2 + \ln 2.5$$
$$\approx -1.08$$

Explore & Discuss

Consider the equation $y = y_0 b^{kx}$, where y_0 and k are positive constants and $b > 0$, $b \neq 1$. Suppose we want to express y in the form $y = y_0 e^{px}$. Use the laws of logarithms to show that $p = k \ln b$ and hence that $y = y_0 e^{(k \ln b)x}$ is an alternative form of $y = y_0 b^{kx}$ using the base e.

---
*For a more extensive treatment of inverse functions, see Appendix A.

EXAMPLE 8 Solve the equation $5 \ln x + 3 = 0$.

Solution Adding -3 to both sides of the equation leads to
$$5 \ln x = -3$$
$$\ln x = -\frac{3}{5} = -0.6$$

and so
$$e^{\ln x} = e^{-0.6}$$

Using Equation (2), we conclude that
$$x = e^{-0.6}$$
$$\approx 0.55$$

5.2 Self-Check Exercises

1. Sketch the graph of $y = 3^x$ and $y = \log_3 x$ on the same set of axes.

2. Solve the equation $3e^{x+1} - 2 = 4$.

Solutions to Self-Check Exercises 5.2 can be found on page 345.

5.2 Concept Questions

1. **a.** Define $y = \log_b x$.
   b. Define the logarithmic function f with base b. What restrictions, if any, are placed on b?

2. For the logarithmic function $y = \log_b x$ ($b > 0$, $b \neq 1$), state (a) its domain and range, (b) its x-intercept, (c) where it is continuous, and (d) where it is increasing and where it is decreasing for the case $b > 1$ and the case $b < 1$.

3. **a.** If $x > 0$, what is $e^{\ln x}$?
   b. If x is any real number, what is $\ln e^x$?

4. Let $f(x) = \ln x^2$ and $g(x) = 2 \ln x$. Are f and g identical?
   Hint: Look at their domains.

5.2 Exercises

In Exercises 1–10, express each equation in logarithmic form.

1. $2^6 = 64$
2. $3^5 = 243$
3. $4^{-2} = \frac{1}{16}$
4. $5^{-3} = \frac{1}{125}$
5. $\left(\frac{1}{3}\right)^1 = \frac{1}{3}$
6. $\left(\frac{1}{2}\right)^{-4} = 16$
7. $32^{4/5} = 16$
8. $81^{3/4} = 27$
9. $10^{-3} = 0.001$
10. $16^{-1/4} = 0.5$

In Exercises 11–16, given that $\log 3 \approx 0.4771$ and $\log 4 \approx 0.6021$, find the value of each logarithm.

11. $\log 12$
12. $\log \frac{3}{4}$
13. $\log 16$
14. $\log \sqrt{3}$
15. $\log 48$
16. $\log \frac{1}{300}$

In Exercises 17–20, write the expression as the logarithm of a single quantity.

17. $2 \ln a + 3 \ln b$
18. $\frac{1}{2} \ln x + 2 \ln y - 3 \ln z$
19. $\ln 3 + \frac{1}{2} \ln x + \ln y - \frac{1}{3} \ln z$
20. $\ln 2 + \frac{1}{2} \ln(x + 1) - 2 \ln(1 + \sqrt{x})$

In Exercises 21–28, use the laws of logarithms to expand and simplify the expression.

21. $\log x(x+1)^4$
22. $\log x(x^2+1)^{-1/2}$
23. $\log \dfrac{\sqrt{x+1}}{x^2+1}$
24. $\ln \dfrac{e^x}{1+e^x}$
25. $\ln xe^{-x^2}$
26. $\ln x(x+1)(x+2)$
27. $\ln \dfrac{x^{1/2}}{x^2\sqrt{1+x^2}}$
28. $\ln \dfrac{x^2}{\sqrt{x}(1+x)^2}$

In Exercises 29–32, sketch the graph of the equation.

29. $y = \log_3 x$
30. $y = \log_{1/3} x$
31. $y = \ln 2x$
32. $y = \ln \dfrac{1}{2}x$

In Exercises 33 and 34, sketch the graphs of the equations on the same coordinate axes.

33. $y = 2^x$ and $y = \log_2 x$
34. $y = e^{3x}$ and $y = \dfrac{1}{3}\ln x$

In Exercises 35–44, use logarithms to solve the equation for t.

35. $e^{0.4t} = 8$
36. $\dfrac{1}{3}e^{-3t} = 0.9$
37. $5e^{-2t} = 6$
38. $4e^{t-1} = 4$
39. $2e^{-0.2t} - 4 = 6$
40. $12 - e^{0.4t} = 3$
41. $\dfrac{50}{1+4e^{0.2t}} = 20$
42. $\dfrac{200}{1+3e^{-0.3t}} = 100$
43. $A = Be^{-t/2}$
44. $\dfrac{A}{1+Be^{t/2}} = C$

45. A function f has the form $f(x) = a + b \ln x$. Find f if it is known that $f(1) = 2$ and $f(2) = 4$.

46. **AVERAGE LIFE SPAN** One reason for the increase in human life span over the years has been the advances in medical technology. The average life span for American women from 1907 through 2007 is given by
$$W(t) = 49.9 + 17.1 \ln t \quad (1 \le t \le 6)$$
where $W(t)$ is measured in years and t is measured in 20-year intervals, with $t=1$ corresponding to 1907.
a. What was the average life expectancy for women in 1907?
b. If the trend continues, what will be the average life expectancy for women in 2027 ($t=7$)?
Source: American Association of Retired Persons (AARP).

47. **BLOOD PRESSURE** A normal child's systolic blood pressure may be approximated by the function
$$p(x) = m(\ln x) + b$$
where $p(x)$ is measured in millimeters of mercury, x is measured in pounds, and m and b are constants. Given that $m = 19.4$ and $b = 18$, determine the systolic blood pressure of a child who weighs 92 lb.

48. **MAGNITUDE OF EARTHQUAKES** On the Richter scale, the magnitude R of an earthquake is given by the formula
$$R = \log \dfrac{I}{I_0}$$
where I is the intensity of the earthquake being measured and I_0 is the standard reference intensity.
a. Express the intensity I of an earthquake of magnitude $R = 5$ in terms of the standard intensity I_0.
b. Express the intensity I of an earthquake of magnitude $R = 8$ in terms of the standard intensity I_0. How many times greater is the intensity of an earthquake of magnitude 8 than one of magnitude 5?
c. In modern times, the greatest loss of life attributable to an earthquake occurred in Haiti in 2010. Known as the Haiti earthquake, it registered 7.0 on the Richter scale. How does the intensity of this earthquake compare with the intensity of an earthquake of magnitude $R = 5$?

49. **SOUND INTENSITY** The relative loudness of a sound D of intensity I is measured in decibels (db), where
$$D = 10 \log \dfrac{I}{I_0}$$
and I_0 is the standard threshold of audibility.
a. Express the intensity I of a 30-db sound (the sound level of normal conversation) in terms of I_0.
b. Determine how many times greater the intensity of an 80-db sound (rock music) is than that of a 30-db sound.
c. Prolonged noise above 150 db causes permanent deafness. How does the intensity of a 150-db sound compare with the intensity of an 80-db sound?

50. **BAROMETRIC PRESSURE** Halley's Law states that the barometric pressure (in inches of mercury) at an altitude of x mi above sea level is approximated by the equation
$$p(x) = 29.92e^{-0.2x} \quad (x \ge 0)$$
If the barometric pressure as measured by a hot-air balloonist is 20 in. of mercury, what is the balloonist's altitude?

51. **HEIGHT OF TREES** The height (in feet) of a certain kind of tree is approximated by
$$h(t) = \dfrac{160}{1 + 240e^{-0.2t}}$$
where t is the age of the tree in years. Estimate the age of an 80-ft tree.

52. **NEWTON'S LAW OF COOLING** The temperature of a cup of coffee t min after it is poured is given by
$$T = 70 + 100e^{-0.0446t}$$
where T is measured in degrees Fahrenheit.
a. What was the temperature of the coffee when it was poured?
b. When will the coffee be cool enough to drink (say, 120°F)?

53. Lengths of Fish The length (in centimeters) of a typical Pacific halibut t years old is approximately

$$f(t) = 200(1 - 0.956e^{-0.18t})$$

Suppose a Pacific halibut caught by Mike measures 140 cm. What is its approximate age?

54. Absorption of Drugs The concentration of a drug in an organ t seconds after it has been administered is given by

$$x(t) = 0.08(1 - e^{-0.02t})$$

where $x(t)$ is measured in grams per cubic centimeter (g/cm³).
a. How long would it take for the concentration of the drug in the organ to reach 0.02 g/cm³?
b. How long would it take for the concentration of the drug in the organ to reach 0.04 g/cm³?

55. Absorption of Drugs The concentration of a drug in an organ t seconds after it has been administered is given by

$$x(t) = 0.08 + 0.12e^{-0.02t}$$

where $x(t)$ is measured in grams per cubic centimeter (g/cm³).
a. How long would it take for the concentration of the drug in the organ to reach 0.18 g/cm³?
b. How long would it take for the concentration of the drug in the organ to reach 0.16 g/cm³?

56. Forensic Science Forensic scientists use the following law to determine the time of death of accident or murder victims. If T denotes the temperature of a body t hr after death, then

$$T = T_0 + (T_1 - T_0)(0.97)^t$$

where T_0 is the air temperature and T_1 is the body temperature at the time of death. John Doe was found murdered at midnight in his house; the room temperature was 70°F, and his body temperature was 80°F when he was found. When was he killed? Assume that the normal body temperature is 98.6°F.

In Exercises 57–60, determine whether the statement is true or false. If it is true, explain why it is true. If it is false, give an example to show why it is false.

57. $(\ln x)^3 = 3 \ln x$ for all x in $(0, \infty)$.

58. $\ln a - \ln b = \ln(a - b)$ for all positive real numbers a and b.

59. The function $f(x) = \dfrac{1}{\ln x}$ is continuous on $(1, \infty)$.

60. The function $f(x) = \ln|x|$ is continuous for all $x \neq 0$.

61. a. Given that $2^x = e^{kx}$, find k.
b. Show that, in general, if b is a positive real number, then any equation of the form $y = b^x$ may be written in the form $y = e^{kx}$, for some real number k.

62. Use the definition of a logarithm to prove
a. $\log_b mn = \log_b m + \log_b n$
b. $\log_b \dfrac{m}{n} = \log_b m - \log_b n$

Hint: Let $\log_b m = p$ and $\log_b n = q$. Then, $b^p = m$ and $b^q = n$.

63. Use the definition of a logarithm to prove

$$\log_b m^n = n \log_b m$$

64. Use the definition of a logarithm to prove
a. $\log_b 1 = 0$
b. $\log_b b = 1$

5.2 Solutions to Self-Check Exercises

1. First, sketch the graph of $y = 3^x$ with the help of the following table of values:

x	-3	-2	-1	0	1	2	3
$y = 3^x$	$\frac{1}{27}$	$\frac{1}{9}$	$\frac{1}{3}$	1	3	9	27

Next, take the mirror reflection of this graph with respect to the line $y = x$ to obtain the graph of $y = \log_3 x$.

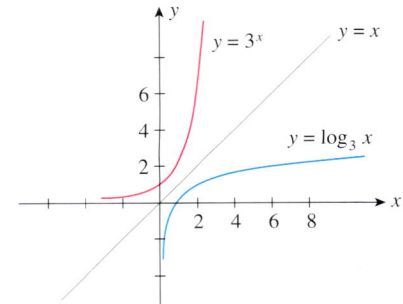

2. $3e^{x+1} - 2 = 4$

$3e^{x+1} = 6$

$e^{x+1} = 2$

$\ln e^{x+1} = \ln 2$ Take the logarithm of both sides.

$(x + 1)\ln e = \ln 2$ Law 3

$x + 1 = \ln 2$ Law 5

$x = \ln 2 - 1$

≈ -0.3069

5.3 Compound Interest

Compound Interest

Compound interest is a natural application of the exponential function to the business world. We begin by recalling that simple interest is interest that is computed only on the original principal. Thus, if I denotes the interest on a principal P (in dollars) at an interest rate of r per year for t years, then we have

$$I = Prt$$

The **accumulated amount** A, the sum of the principal and interest after t years, is given by

$$A = P + I = P + Prt$$
$$= P(1 + rt) \quad \text{Simple interest formula} \quad (4)$$

Frequently, interest earned is periodically added to the principal and thereafter earns interest itself at the same rate. This is called **compound interest.** To find a formula for the accumulated amount, let's consider a numerical example. Suppose $1000 (the principal) is deposited in a bank for a **term** of 3 years, earning interest at the rate of 8% per year (called the **nominal, or stated, rate**) compounded annually. Then, using Formula (4) with $P = 1000$, $r = 0.08$, and $t = 1$, we see that the accumulated amount at the end of the first year is

$$A_1 = P(1 + rt)$$
$$= 1000[1 + 0.08(1)] = 1000(1.08) = 1080$$

or $1080.

To find the accumulated amount A_2 at the end of the second year, we use Equation (4) once again, this time with $P = A_1$. (Remember, the principal *and* interest now earn interest over the second year.) We obtain

$$A_2 = P(1 + rt) = A_1(1 + rt)$$
$$= 1000[1 + 0.08(1)][1 + 0.08(1)]$$
$$= 1000(1 + 0.08)^2 = 1000(1.08)^2 = 1166.40$$

or $1166.40.

Finally, the accumulated amount A_3 at the end of the third year is found by using Equation (4) with $P = A_2$, giving

$$A_3 = P(1 + rt) = A_2(1 + rt)$$
$$= 1000[1 + 0.08(1)]^2[1 + 0.08(1)]$$
$$= 1000(1 + 0.08)^3 = 1000(1.08)^3 \approx 1259.71$$

or approximately $1259.71.

If you reexamine our calculations in this example, you will see that the accumulated amounts at the end of each year have the following form:

First year: $A_1 = 1000(1 + 0.08)$ or $A_1 = P(1 + r)$
Second year: $A_2 = 1000(1 + 0.08)^2$ or $A_2 = P(1 + r)^2$
Third year: $A_3 = 1000(1 + 0.08)^3$ or $A_3 = P(1 + r)^3$

These observations suggest the following general result: If P dollars are invested over a term of t years earning interest at the rate of r per year compounded annually, then the accumulated amount is

$$A = P(1 + r)^t \quad (5)$$

Formula (5) was derived under the assumption that interest was compounded *annually*. In practice, however, interest is usually compounded more than once a year. The interval of time between successive interest calculations is called the **conversion period.**

If interest at a nominal rate of r per year is compounded m times a year on a principal of P dollars, then the simple interest rate per conversion period is

$$i = \frac{r}{m} \quad \frac{\text{Annual interest rate}}{\text{Periods per year}}$$

For example, if the nominal interest rate is 8% per year ($r = 0.08$) and interest is compounded quarterly ($m = 4$), then

$$i = \frac{r}{m} = \frac{0.08}{4} = 0.02$$

or 2% per period.

To find a general formula for the accumulated amount when a principal of P dollars is deposited in a bank for a term of t years and earns interest at the (nominal) rate of r per year compounded m times per year, we proceed as before, using Formula (5) repeatedly with the interest rate $i = r/m$. We see that the accumulated amount at the end of each period is as follows:

First period: $A_1 = P(1 + i)$
Second period: $A_2 = A_1(1 + i) = [P(1 + i)](1 + i) = P(1 + i)^2$
Third period: $A_3 = A_2(1 + i) = [P(1 + i)^2](1 + i) = P(1 + i)^3$
$\vdots$
nth period: $A_n = A_{n-1}(1 + i) = [P(1 + i)^{n-1}](1 + i) = P(1 + i)^n$

But there are $n = mt$ periods in t years (number of conversion periods per year times the term). Therefore, the accumulated amount at the end of t years is given by

Compound Interest Formula

$$A = P\left(1 + \frac{r}{m}\right)^{mt} \tag{6}$$

where

A = Accumulated amount at the end of t years
P = Principal
r = Nominal interest rate per year
m = Number of conversion periods per year
t = Term (number of years)

EXAMPLE 1 Find the accumulated amount after 3 years if $1000 is invested at 8% per year compounded (a) annually, (b) semiannually, (c) quarterly, (d) monthly, and (e) daily.

Solution

a. Here, $P = 1000$, $r = 0.08$, $m = 1$, and $t = 3$, so Formula (6) gives

$$A = 1000(1 + 0.08)^3$$
$$\approx 1259.71$$

or $1259.71.

b. Here, $P = 1000$, $r = 0.08$, $m = 2$, and $t = 3$, so Formula (6) gives

$$A = 1000\left(1 + \frac{0.08}{2}\right)^{(2)(3)}$$
$$\approx 1265.32$$

or $1265.32.

c. In this case, $P = 1000$, $r = 0.08$, $m = 4$, and $t = 3$, so Formula (6) gives

$$A = 1000\left(1 + \frac{0.08}{4}\right)^{(4)(3)}$$
$$\approx 1268.24$$

or $1268.24

d. Here, $P = 1000$, $r = 0.08$, $m = 12$, and $t = 3$, so Formula (6) gives

$$A = 1000\left(1 + \frac{0.08}{12}\right)^{(12)(3)}$$
$$\approx 1270.24$$

or $1270.24.

e. Here, $P = 1000$, $r = 0.08$, $m = 365$, and $t = 3$, so Formula (6) gives

$$A = 1000\left(1 + \frac{0.08}{365}\right)^{(365)(3)}$$
$$\approx 1271.22$$

or $1271.22. These results are summarized in Table 2.

TABLE 2

Nominal Rate, r	Conversion Period	Term in Years	Initial Investment	Accumulated Amount
8%	Annual ($m = 1$)	3	$1000	$1259.71
8	Semiannual ($m = 2$)	3	1000	1265.32
8	Quarterly ($m = 4$)	3	1000	1268.24
8	Monthly ($m = 12$)	3	1000	1270.24
8	Daily ($m = 365$)	3	1000	1271.22

Effective Rate of Interest

In Example 1, we saw that the interest earned on an investment depends on the frequency with which the interest is compounded. Thus, the stated, or nominal, rate of 8% per year does not reflect the actual rate at which interest is earned. This suggests that we need to find a common basis for comparing interest rates. One way of comparing interest rates is provided by using the effective rate. The **effective rate** is the *simple interest rate that would produce the same accumulated amount in 1 year as the nominal rate compounded m times a year.* The effective rate is also called the **true rate.**

To derive a relation between the nominal interest rate, r per year compounded m times and its corresponding effective rate, r_{eff} per year, let's assume an initial investment of P dollars. Then, the accumulated amount after 1 year at a simple interest rate of r_{eff} per year is

$$A = P(1 + r_{\text{eff}})$$

Also, the accumulated amount after 1 year at an interest rate of r per year compounded m times a year is

$$A = P\left(1 + \frac{r}{m}\right)^m \quad \text{Since } t = 1$$

Equating the two expressions gives

$$P(1 + r_{\text{eff}}) = P\left(1 + \frac{r}{m}\right)^m$$

$$1 + r_{\text{eff}} = \left(1 + \frac{r}{m}\right)^m \quad \text{Divide both sides by } P.$$

or, upon solving for r_{eff}, we obtain the formula for computing the effective rate of interest:

Effective Rate of Interest Formula

$$r_{\text{eff}} = \left(1 + \frac{r}{m}\right)^m - 1 \tag{7}$$

where

r_{eff} = Effective rate of interest
r = Nominal interest rate per year
m = Number of conversion periods per year

EXAMPLE 2 Find the effective rate of interest corresponding to a nominal rate of 8% per year compounded (a) annually, (b) semiannually, (c) quarterly, (d) monthly, and (e) daily.

Solution

a. The effective rate of interest corresponding to a nominal rate of 8% per year compounded annually is of course given by 8% per year. This result is also confirmed by using Formula (7) with $r = 0.08$ and $m = 1$. Thus,

$$r_{\text{eff}} = (1 + 0.08) - 1 = 0.08$$

b. Let $r = 0.08$ and $m = 2$. Then, Formula (7) yields

$$r_{\text{eff}} = \left(1 + \frac{0.08}{2}\right)^2 - 1$$
$$= 0.0816$$

so the required effective rate is 8.16% per year.

c. Let $r = 0.08$ and $m = 4$. Then, Formula (7) yields

$$r_{\text{eff}} = \left(1 + \frac{0.08}{4}\right)^4 - 1$$
$$\approx 0.0824$$

so the corresponding effective rate in this case is 8.24% per year.

d. Let $r = 0.08$ and $m = 12$. Then, Formula (7) yields

$$r_{\text{eff}} = \left(1 + \frac{0.08}{12}\right)^{12} - 1$$
$$\approx 0.0830$$

so the corresponding effective rate in this case is 8.30% per year.

e. Let $r = 0.08$ and $m = 365$. Then, Formula (7) yields

$$r_{\text{eff}} = \left(1 + \frac{0.08}{365}\right)^{365} - 1$$
$$\approx 0.0833$$

so the corresponding effective rate in this case is 8.33% per year. ∎

Now, if the effective rate of interest r_{eff} is known, then the accumulated amount after t years on an investment of P dollars may be more readily computed by using the formula

$$A = P(1 + r_{\text{eff}})^t$$

The 1968 Truth in Lending Act passed by Congress requires that the effective rate of interest be disclosed in all contracts involving interest charges. The passage of this act has benefited consumers because they now have a common basis for comparing the various nominal rates quoted by different financial institutions. Furthermore, knowing the effective rate enables consumers to compute the actual charges involved in a transaction. The effective rate is also called the **annual percentage yield** (APY). Thus, if the effective rates of interest found in Example 2 were known, the accumulated values of Example 1, shown in Table 3, could have been readily found.

TABLE 3

Nominal Rate, r	Frequency of Interest Payment	Effective Rate	Initial Investment	Accumulated Amount after 3 Years	
8%	Annually	8%	$1000	$1000(1 + 0.08)^3$	= $1259.71
8	Semiannually	8.16	1000	$1000(1 + 0.0816)^3$	= 1265.32
8	Quarterly	8.243	1000	$1000(1 + 0.08243)^3$	= 1268.23
8	Monthly	8.300	1000	$1000(1 + 0.08300)^3$	= 1270.24
8	Daily	8.328	1000	$1000(1 + 0.08328)^3$	= 1271.22

Present Value

Let's return to the compound interest Formula (6), which expresses the accumulated amount at the end of t years when interest at the rate of r is compounded m times a year. The principal P in Formula (6) is often referred to as the **present value,** and the accumulated value A is called the **future value,** since it is realized at a future date. In certain instances, an investor may wish to determine how much money he or she should invest now, at a fixed rate of interest, to realize a certain sum at some future date. This problem may be solved by expressing P in terms of A. Thus, from Formula (6), we find

> **Present Value Formula for Compound Interest**
>
> $$P = A\left(1 + \frac{r}{m}\right)^{-mt} \qquad (8)$$

5.3 COMPOUND INTEREST

EXAMPLE 3 How much money should be deposited in a bank paying interest at the rate of 6% per year compounded monthly so that at the end of 3 years, the accumulated amount will be $20,000?

Solution Here, $A = 20{,}000$, $r = 0.06$, $m = 12$, and $t = 3$. Using Formula (8), we obtain

$$P = 20{,}000\left(1 + \frac{0.06}{12}\right)^{-(12)(3)}$$

$$\approx 16{,}713$$

or $16,713.

EXAMPLE 4 Find the present value of $49,158.60 due in 5 years at an interest rate of 10% per year compounded quarterly.

Solution Using Formula (8) with $A = 49{,}158.60$, $r = 0.1$, $m = 4$, and $t = 5$, we obtain

$$P = 49{,}158.6\left(1 + \frac{0.1}{4}\right)^{-(4)(5)} \approx 30{,}000$$

or $30,000.

Continuous Compounding of Interest

One question that arises naturally in the study of compound interest is: What happens to the accumulated amount over a fixed period of time if the interest is computed more and more frequently?

Intuition suggests that the more often interest is compounded, the larger the accumulated amount will be. This is confirmed by the results of Example 1, in which we found that the accumulated amounts did in fact increase when we increased the number of conversion periods per year.

This leads us to another question: Does the accumulated amount approach a limit when the interest is computed more and more frequently over a fixed period of time?

To answer this question, let's look again at the compound interest formula:

$$A = P\left(1 + \frac{r}{m}\right)^{mt} \qquad (9)$$

Recall that m is the number of conversion periods per year. So to find an answer to our question, we should let m get larger and larger (approach infinity) in Equation (9). But first we will rewrite this equation in the form

$$A = P\left[\left(1 + \frac{r}{m}\right)^{m}\right]^{t} \qquad \text{Since } b^{xy} = (b^x)^y$$

Now, letting $m \to \infty$, we find that

$$\lim_{m \to \infty}\left[P\left(1 + \frac{r}{m}\right)^{m}\right]^{t} = P\left[\lim_{m \to \infty}\left(1 + \frac{r}{m}\right)^{m}\right]^{t} \qquad \text{Why?}$$

Next, upon making the substitution $u = m/r$ and observing that $u \to \infty$ as $m \to \infty$, the foregoing expression reduces to

$$P\left[\lim_{u \to \infty}\left(1 + \frac{1}{u}\right)^{ur}\right]^{t} = P\left[\lim_{u \to \infty}\left(1 + \frac{1}{u}\right)^{u}\right]^{rt}$$

But

$$\lim_{u \to \infty}\left(1 + \frac{1}{u}\right)^{u} = e \qquad \text{Use Equation (1).}$$

so

$$\lim_{m\to\infty} P\left[\left(1 + \frac{r}{m}\right)^m\right]^t = Pe^{rt}$$

Our computations tell us that as the frequency with which interest is compounded increases without bound, the accumulated amount approaches Pe^{rt}. In this situation, we say that interest is *compounded continuously*. Let's summarize this important result.

Continuous Compound Interest Formula

$$A = Pe^{rt} \qquad (10)$$

where

$P = $ Principal

$r = $ Annual interest rate compounded continuously

$t = $ Time in years

$A = $ Accumulated amount at the end of t years

EXAMPLE 5 Find the accumulated amount after 3 years if $1000 is invested at 8% per year compounded (a) daily (assume a 365-day year) and (b) continuously.

Solution

a. Using Formula (6) with $P = 1000$, $r = 0.08$, $m = 365$, and $t = 3$, we find

$$A = 1000\left(1 + \frac{0.08}{365}\right)^{(365)(3)} \approx 1271.22$$

or $1271.22.

b. Here we use Formula (10) with $P = 1000$, $r = 0.08$, and $t = 3$, obtaining

$$A = 1000e^{(0.08)(3)}$$

$$\approx 1271.25$$

or $1271.25.

Observe that the accumulated amounts corresponding to interest compounded daily and interest compounded continuously differ by very little. The continuous compound interest formula is a very important tool in theoretical work in financial analysis.

Exploring with TECHNOLOGY

In the opening paragraph of Section 5.1, we pointed out that the accumulated amount of an account earning interest *compounded continuously* will eventually outgrow by far the accumulated amount of an account earning interest at the same nominal rate but earning simple interest. Illustrate this fact using the following example.

Suppose you deposit $1000 in Account I, earning interest at the rate of 10% per year compounded continuously so that the accumulated amount at the end of t years is $A_1(t) = 1000e^{0.1t}$. Suppose you also deposit $1000 in Account II, earning simple interest at the rate of 10% per year so that the accumulated amount at the end of t years is $A_2(t) = 1000(1 + 0.1t)$. Use a graphing utility to sketch the graphs of the functions A_1 and A_2 in the viewing window $[0, 20] \times [0, 10{,}000]$ to see the accumulated amounts $A_1(t)$ and $A_2(t)$ over a 20-year period.

If we solve Formula (10) for P, we obtain

$$P = Ae^{-rt} \tag{11}$$

which gives the present value in terms of the future (accumulated) value for the case of continuous compounding.

APPLIED EXAMPLE 6 Real Estate Investment Blakely Investment Company owns an office building located in the commercial district of a city. As a result of the continued success of an urban renewal program, local business is enjoying a miniboom. The market value of Blakely's property is

$$V(t) = 300{,}000 e^{\sqrt{t}/2}$$

where $V(t)$ is measured in dollars and t is the time in years from the present. If the expected rate of appreciation is 9% compounded continuously for the next 10 years, find an expression for the present value $P(t)$ of the market price of the property valid for the next 10 years. Compute $P(7)$, $P(8)$, and $P(9)$, and interpret your results.

Solution Using Formula (11) with $A = V(t)$ and $r = 0.09$, we find that the present value of the market price of the property t years from now is

$$P(t) = V(t) e^{-0.09t}$$
$$= 300{,}000 e^{-0.09t + \sqrt{t}/2} \qquad (0 \le t \le 10)$$

Letting $t = 7, 8,$ and 9, respectively, we find that

$$P(7) = 300{,}000 e^{-0.09(7) + \sqrt{7}/2} \approx 599{,}837 \quad \text{or} \quad \$599{,}837$$
$$P(8) = 300{,}000 e^{-0.09(8) + \sqrt{8}/2} \approx 600{,}640 \quad \text{or} \quad \$600{,}640$$
$$P(9) = 300{,}000 e^{-0.09(9) + \sqrt{9}/2} \approx 598{,}115 \quad \text{or} \quad \$598{,}115$$

From the results of these computations, we see that the present value of the property's market price seems to decrease after a certain period of growth. This suggests that there is an optimal time for the owners to sell. Later we will show that the highest present value of the property's market price is $600,779, which occurs at time $t = 7.72$ years.

Exploring with TECHNOLOGY

The effective rate of interest is given by

$$r_{\text{eff}} = \left(1 + \frac{r}{m}\right)^m - 1$$

where the number of conversion periods per year is m. In Exercise 48 on page 357 you will be asked to show that the effective rate of interest r_{eff} corresponding to a nominal interest rate r per year compounded continuously is given by

$$\hat{r}_{\text{eff}} = e^r - 1$$

To obtain a visual confirmation of this result, consider the special case in which $r = 0.1$ (10% per year).

1. Use a graphing utility to plot the graph of both

$$y_1 = \left(1 + \frac{0.1}{x}\right)^x - 1 \quad \text{and} \quad y_2 = e^{0.1} - 1$$

in the viewing window $[0, 3] \times [0, 0.12]$.

(continued)

2. Does your result seem to imply that

$$\left(1 + \frac{r}{m}\right)^m - 1$$

approaches

$$\hat{r}_{\text{eff}} = e^r - 1$$

as m increases without bound for the special case $r = 0.1$?

The next two examples show how logarithms can be used to solve problems involving compound interest.

EXAMPLE 7 How long will it take $10,000 to grow to $15,000 if the investment earns an interest rate of 12% per year compounded quarterly?

Solution Using Formula (6) with $A = 15{,}000$, $P = 10{,}000$, $r = 0.12$, and $m = 4$, we obtain

$$15{,}000 = 10{,}000\left(1 + \frac{0.12}{4}\right)^{4t}$$

$$(1.03)^{4t} = \frac{15{,}000}{10{,}000} = 1.5$$

Taking the logarithm on each side of the equation gives

$$\ln(1.03)^{4t} = \ln 1.5$$

$$4t \ln 1.03 = \ln 1.5 \qquad \log_b m^n = n \log_b m$$

$$4t = \frac{\ln 1.5}{\ln 1.03}$$

$$t = \frac{\ln 1.5}{4 \ln 1.03} \approx 3.43$$

So it will take approximately 3.4 years for the investment to grow from $10,000 to $15,000.

EXAMPLE 8 Find the interest rate needed for an investment of $10,000 to grow to an amount of $18,000 in 5 years if the interest is compounded monthly.

Solution Using Formula (6) with $A = 18{,}000$, $P = 10{,}000$, $m = 12$, and $t = 5$, we obtain

$$18{,}000 = 10{,}000\left(1 + \frac{r}{12}\right)^{12(5)}$$

Dividing both sides of the equation by 10,000 gives

$$\frac{18{,}000}{10{,}000} = \left(1 + \frac{r}{12}\right)^{60}$$

or, upon simplification,

$$\left(1 + \frac{r}{12}\right)^{60} = 1.8$$

Now, we take the logarithm on each side of the equation, obtaining

$$\ln\left(1 + \frac{r}{12}\right)^{60} = \ln 1.8$$

$$60 \ln\left(1 + \frac{r}{12}\right) = \ln 1.8$$

$$\ln\left(1 + \frac{r}{12}\right) = \frac{\ln 1.8}{60} \approx 0.009796$$

$$\left(1 + \frac{r}{12}\right) \approx e^{0.009796} \quad \text{By Property 2}$$

$$\approx 1.009844$$

and

$$\frac{r}{12} \approx 1.009844 - 1$$

$$r \approx 0.1181$$

or 11.81% per year.

5.3 Self-Check Exercises

1. Find the present value of $20,000 due in 3 years at an interest rate of 12%/year compounded monthly.

2. Glen is a retiree living on Social Security and the income from his investment. Currently, his $100,000 investment in a 1-year CD is yielding 6.6% interest compounded daily. If he reinvests the principal ($100,000) on the due date of the CD in another 1-year CD paying 5.2% interest compounded daily, find the net decrease in his yearly income from his investment.

3. a. What is the accumulated amount after 5 years if $10,000 is invested at 8%/year compounded continuously?
   b. Find the present value of $10,000 due in 5 years at an interest rate of 8%/year compounded continuously.

Solutions to Self-Check Exercises 5.3 can be found on page 358.

5.3 Concept Questions

1. a. What is the difference between simple interest and compound interest?
   b. State the simple interest formula and the compound interest formula.

2. a. What is the effective rate of interest?
   b. State the formula for computing the effective rate of interest.

3. What is the present value formula for compound interest?

4. State the continuous compound interest formula.

5.3 Exercises

In Exercises 1–4, find the accumulated amount A if the principal P is invested at an interest rate of r per year for t years.

1. $P = \$2500$, $r = 4\%$, $t = 10$, compounded semiannually

2. $P = \$12,000$, $r = 5\%$, $t = 10$, compounded quarterly

3. $P = \$150,000$, $r = 6\%$, $t = 4$, compounded monthly

4. $P = \$150,000$, $r = 7\%$, $t = 3$, compounded daily

In Exercises 5 and 6, find the effective rate corresponding to the given nominal rate.

5. a. 6%/year compounded semiannually
   b. 5%/year compounded quarterly

6. a. 4.5%/year compounded monthly
   b. 4.5%/year compounded daily

In Exercises 7 and 8, find the present value of $40,000 due in 4 years at the given rate of interest.

7. **a.** 5%/year compounded semiannually
   b. 5%/year compounded quarterly

8. **a.** 4%/year compounded monthly
   b. 7%/year compounded daily

9. Find the accumulated amount after 4 years if $5000 is invested at 5%/year compounded continuously.

10. An amount of $25,000 is deposited in a bank that pays interest at the rate of 4%/year, compounded annually. What is the total amount on deposit at the end of 6 years, assuming that there are no deposits or withdrawals during those 6 years? What is the interest earned in that period of time?

11. How much money should be deposited in a bank paying interest at the rate of 4%/year compounded daily (assume a 365-day year) so that at the end of 2 years the accumulated amount will be $10,000?

12. Jada deposited an amount of money in a bank 3 years ago. If the bank had been paying interest at the rate of 5%/year compounded daily (assume a 365-day year) and she has $15,000 on deposit today, what was her initial deposit?

13. How much money should Jack deposit in a bank paying interest at the rate of 6%/year compounded continuously so that at the end of 3 years the accumulated amount will be $20,000?

14. Diego deposited a certain sum of money in a bank 2 years ago. If the bank had been paying interest at the rate of 6% compounded continuously and he has $12,000 on deposit today, what was his initial deposit?

15. Find the interest rate needed for an investment of $5000 to grow to an amount of $7500 in 3 years if interest is compounded monthly.

16. Find the interest rate needed for an investment of $5000 to grow to an amount of $7500 in 3 years if interest is compounded quarterly.

17. Find the interest rate needed for an investment of $5000 to grow to an amount of $8000 in 4 years if interest is compounded semiannually.

18. Find the interest rate needed for an investment of $5000 to grow to an amount of $5500 in 6 months if interest is compounded monthly.

19. Find the interest rate needed for an investment of $2000 to double in 5 years if interest is compounded annually.

20. Find the interest rate needed for an investment of $2000 to triple in 5 years if interest is compounded monthly.

21. How long will it take $5000 to grow to $6500 if the investment earns interest at the rate of 6%/year compounded monthly?

22. How long will it take $12,000 to grow to $15,000 if the investment earns interest at the rate of 5%/year compounded monthly?

23. How long will it take an investment of $2000 to double if the investment earns interest at the rate of 6%/year compounded monthly?

24. How long will it take an investment of $5000 to triple if the investment earns interest at the rate of 4%/year compounded daily?

25. Find the interest rate needed for an investment of $5000 to grow to an amount of $6000 in 3 years if interest is compounded continuously.

26. Find the interest rate needed for an investment of $4000 to double in 5 years if interest is compounded continuously.

27. How long will it take an investment of $6000 to grow to $7000 if the investment earns interest at the rate of $7\frac{1}{2}$% compounded continuously?

28. How long will it take an investment of $8000 to double if the investment earns interest at the rate of 5% compounded continuously?

29. **Housing Prices** The Estradas are planning to buy a house 4 years from now. Housing experts in their area have estimated that the cost of a home will increase at a rate of 4%/year during that 4-year period. If this economic prediction holds true, how much can they expect to pay for a house that currently costs $180,000?

30. **Energy Consumption** A metropolitan utility company in a western city of the United States expects the consumption of electricity to increase by 8%/year during the next decade, owing mainly to an expected population increase. If consumption does increase at this rate, find the amount by which the utility company will have to increase its generating capacity to meet the area's needs at the end of the decade.

31. **Pension Funds** The managers of a pension fund have invested $1.5 million in U.S. government certificates of deposit (CDs) that pay interest at the rate of 4.5%/year compounded semiannually over a period of 10 years. At the end of this period, how much will the investment be worth?

32. **Savings Accounts** Bernie invested a sum of money 5 years ago in a savings account, which has since paid interest at the rate of 3%/year compounded quarterly. His investment is now worth $22,289.22. How much did he originally invest?

33. **Loan Consolidation** The proprietors of the Coachmen Inn secured two loans from the Union Bank: one for $8000 due in 3 years and one for $15,000 due in 6 years, both at an interest rate of 10%/year compounded semiannually. The bank agreed to allow the two loans to be consolidated into one loan payable in 5 years at the same interest rate. How much will the proprietors have to pay the bank at the end of 5 years?

34. **Tax-Deferred Annuities** Kate is in the 28% tax bracket and has $25,000 available for investment during her current tax year. Assume that she remains in the same tax bracket over

the next 10 years, and determine the accumulated amount of her investment if she puts the $25,000 into a
a. Tax-deferred annuity that pays 12%/year, tax deferred for 10 years.
b. Taxable instrument that pays 12%/year for 10 years.
   Hint: In this case the yield after taxes is 8.64%/year.

35. **REVENUE GROWTH OF A HOME THEATER BUSINESS** Maxwell started a home theater business in 2010. The revenue of his company for that year was $240,000. The revenue grew by 20% in 2011 and 30% in 2012. Maxwell projects that the revenue growth for his company in the next 3 years will be at least 25%/year. How much does Maxwell expect his minimum revenue to be for 2015?

36. **ONLINE RETAIL SALES** Online retail sales stood at $23.5 billion for 2000. For the next 2 years, they grew by 33.2% and 27.8% per year, respectively. For the next 6 years, online retail sales grew at the rate of approximately 30.5%, 19.9%, 24.3%, 14.0%, 17.6%, and 10.5% per year, respectively. What were the online sales for 2008?
   Source: Jupiter Research.

37. **PURCHASING POWER** The inflation rates in the U.S. economy in 2005, 2006, 2007, and 2008 were 3.4%, 3.2%, 2.9%, and 3.9%, respectively. What was the purchasing power of a dollar at the beginning of 2009 compared to that at the beginning of 2006?
   Source: U.S. Census Bureau.

38. **INVESTMENT RETURNS** Zoe purchased a house in 2002 for $160,000. In 2008, she sold the house and made a net profit of $56,000. Find the effective annual rate of return on her investment over the 6-year period.

39. **INVESTMENT RETURNS** Julio purchased 1000 shares of a certain stock for $25,250 (including commissions). He sold the shares 2 years later and received $32,100 after deducting commissions. Find the effective annual rate of return on his investment over the 2-year period.

40. **INVESTMENT OPTIONS** Investment A offers an 8% return compounded semiannually, and Investment B offers a 7.75% return compounded continuously. Which investment has a higher rate of return over a 4-year period?

41. **PRESENT VALUE** Find the present value of $59,673 due in 5 years at an interest rate of 6%/year compounded continuously.

42. **REAL ESTATE INVESTMENTS** A condominium complex was purchased by a group of private investors for $1.4 million and sold 6 years later for $3.6 million. Find the annual rate of return (compounded continuously) on their investment.

43. **SAVING FOR COLLEGE** Having received a large inheritance, a child's parents wish to establish a trust for the child's college education. If 7 years from now they need an estimated $70,000, how much should they set aside in trust now if they invest the money at 5.5% compounded quarterly? Continuously?

44. **EFFECT OF INFLATION ON SALARIES** Omar's current annual salary is $35,000. How much will he need to earn 10 years from now to retain his present purchasing power if the rate of inflation over that period is 6%/year? Assume that inflation is continuously compounded.

45. **PENSIONS** Eleni, who is now 50 years old, is employed by a firm that guarantees her a pension of $40,000/year at age 65. What is the present value of her first year's pension if inflation over the next 15 years is (a) 4%? (b) 6%? (c) 8%? Assume that inflation is continuously compounded.

46. **REAL ESTATE INVESTMENTS** An investor purchased a piece of waterfront property. Because of the development of a marina in the vicinity, the market value of the property is expected to increase according to the rule
$$V(t) = 80{,}000 e^{\sqrt{t}/2}$$
where $V(t)$ is measured in dollars and t is the time in years from the present. If the rate of appreciation is expected to be 9% compounded continuously for the next 8 years, find an expression for the present value $P(t)$ of the property's market price valid for the next 8 years. What is $P(t)$ expected to be in 4 years?

47. **REAL ESTATE INVESTMENTS** Tower Investments owns a shopping mall located just outside the city. The market value of the mall is
$$V(t) = 500{,}000 e^{0.5 t^{0.4}}$$
dollars, where t is measured in years from the present. The rate of appreciation of the mall is expected to be 8%/year compounded continuously for the next 8 years.
a. Find an expression for the present value $P(t)$ of the market price of the mall valid for the next 8 years.
b. Compute $P(4)$, $P(5)$, and $P(6)$, and interpret your result.

48. Show that the effective interest rate $\hat{r}_{\text{eff}}$ that corresponds to a nominal interest rate r per year compounded continuously is given by
$$\hat{r}_{\text{eff}} = e^r - 1$$
Hint: From Formula (7), we see that the effective rate $\hat{r}_{\text{eff}}$ corresponding to a nominal interest rate r per year compounded m times a year is given by
$$\hat{r}_{\text{eff}} = \left(1 + \frac{r}{m}\right)^m - 1$$
Let m tend to infinity in this expression.

49. Refer to Exercise 48. Find the effective interest rate that corresponds to a nominal rate of 10%/year compounded (a) quarterly, (b) monthly, and (c) continuously.

50. **INVESTMENT ANALYSIS** Refer to Exercise 47. Bank A pays interest on deposits at a 7% annual rate compounded quarterly, and Bank B pays interest on deposits at a $7\frac{1}{8}$% annual rate compounded continuously. Which bank has the higher effective rate of interest?

51. **INVESTMENT ANALYSIS** Find the nominal interest rate that, when compounded monthly, yields an effective interest rate of 8%/year.
   Hint: Use Equation (7).

52. INVESTMENT ANALYSIS Find the nominal interest rate that, when compounded continuously, yields an effective interest rate of 8%/year.
Hint: See Exercise 48.

53. ANNUITIES An annuity is a sequence of payments made at regular time intervals. The future value of an annuity of n payments of R dollars each paid at the end of each investment period into an account that earns an interest rate of i/period is

$$S = R\left[\frac{(1+i)^n - 1}{i}\right]$$

Determine

$$\lim_{i \to 0} R\left[\frac{(1+i)^n - 1}{i}\right]$$

and interpret your result.
Hint: Use the definition of the derivative.

5.3 Solutions to Self-Check Exercises

1. Using Formula (8) with $A = 20{,}000$, $r = 0.12$, $m = 12$, and $t = 3$, we find the required present value to be

$$P = 20{,}000\left(1 + \frac{0.12}{12}\right)^{-(12)(3)}$$

$$\approx 13{,}978.50$$

or $13,978.50.

2. The accumulated amount of Glen's current investment is found by using Formula (6) with $P = 100{,}000$, $r = 0.066$, and $m = 365$. Thus, the required accumulated amount is

$$A = 100{,}000\left(1 + \frac{0.066}{365}\right)^{365}$$

$$\approx 106{,}822.03$$

or $106,822.03. Next, we compute the accumulated amount of Glen's reinvestment. Once again, using Formula (6) with $P = 100{,}000$, $r = 0.052$, and $m = 365$, we find the required accumulated amount in this case to be

$$\bar{A} = 100{,}000\left(1 + \frac{0.052}{365}\right)^{365}$$

or approximately $105,337.18. Therefore, Glen can expect to experience a net decrease in yearly income of

$$106{,}822.03 - 105{,}337.18$$

or $1,484.85.

3. a. Using Formula (10) with $P = 10{,}000$, $r = 0.08$, and $t = 5$, we find that the required accumulated amount is given by

$$A = 10{,}000 e^{(0.08)(5)}$$

$$\approx 14{,}918.25$$

or $14,918.25.

b. Using Formula (11) with $A = 10{,}000$, $r = 0.08$, and $t = 5$, we see that the required present value is given by

$$P = 10{,}000 e^{-(0.08)(5)}$$

$$\approx 6703.20$$

or $6703.20.

USING TECHNOLOGY

Finding the Accumulated Amount of an Investment, the Effective Rate of Interest, and the Present Value of an Investment

Graphing Utility

Some graphing utilities have built-in routines for solving problems involving the mathematics of finance. For example, the TI-83/84 Solver function incorporates several functions that can be used to solve the problems that are encountered in this section. To access the TVM SOLVER on the TI-83 press 2ND, press FINANCE, and then select 1: TVM Solver. To access the TVM SOLVER on the TI-83 Plus and the TI-84, press APPS, press 1: Finance, and then select 1: TVM Solver.

5.3 COMPOUND INTEREST

EXAMPLE 1 Finding the Accumulated Amount of an Investment Find the accumulated amount after 10 years if $5000 is invested at a rate of 10% per year compounded monthly.

Solution Using the TI-83/84 TVM SOLVER with the following inputs,

$N = 120$ (10)(12)
$I\% = 10$
$PV = -5000$ We use a minus sign because an investment is an outflow.
$PMT = 0$
$FV = 0$
$P/Y = 12$ The number of payments each year
$C/Y = 12$ The number of conversion periods each year
PMT:END BEGIN

we obtain the display shown in Figure T1. We conclude that the required accumulated amount is $13,535.21.

```
N=120
I%=10
PV=-5000
PMT=0
■ FV=13535.20745
P/Y=12
C/Y=12
PMT:END  BEGIN
```

FIGURE T1
The TI-83/84 screen showing the future value (FV) of an investment

EXAMPLE 2 Finding the Effective Rate of Interest Find the effective rate of interest corresponding to a nominal rate of 10% per year compounded quarterly.

Solution Here we use the **Eff** function of the TI-83/84 calculator to obtain the result shown in Figure T2. The required effective rate is approximately 10.38% per year.

```
▶ Eff(10, 4)
       10.38128906
```

FIGURE T2
The TI-83/84 screen showing the effective rate of interest (Eff)

EXAMPLE 3 Finding the Present Value of an Investment Find the present value of $20,000 due in 5 years if the interest rate is 7.5% per year compounded daily.

Solution Using the TI-83/84 TVM SOLVER with the following inputs,

$N = 1825$ (5)(365)
$I\% = 7.5$
$PV = 0$
$PMT = 0$
$FV = 20000$
$P/Y = 365$ The number of payments each year
$C/Y = 365$ The number of conversions each year
PMT:END BEGIN

we obtain the display shown in Figure T3. We see that the required present value is approximately $13,746.32. Note that PV is negative because an investment is an outflow (money is paid out).

```
N=1825
I%=7.5
■ PV=-13746.3151
PMT=0
FV=20000
P/Y=365
C/Y=365
PMT:END  BEGIN
```

FIGURE T3
The TI-83/84 screen showing the present value (PV) of an investment

(continued)

TECHNOLOGY EXERCISES

1. Find the accumulated amount A if $5000 is invested at the interest rate of $5\frac{3}{8}$%/year compounded monthly for 3 years.

2. Find the accumulated amount A if $327.35 is invested at the interest rate of $5\frac{1}{3}$%/year compounded daily for 7 years.

3. Find the effective rate corresponding to $8\frac{2}{3}$%/year compounded quarterly.

4. Find the effective rate corresponding to $10\frac{5}{8}$%/year compounded monthly.

5. Find the present value of $38,000 due in 3 years at $8\frac{1}{4}$%/year compounded quarterly.

6. Find the present value of $150,000 due in 5 years at $9\frac{3}{8}$%/year compounded monthly.

5.4 Differentiation of Exponential Functions

The Derivative of the Exponential Function

To study the effects of budget deficit-reduction plans at different income levels, it is important to know the income distribution of American households. Based on data from the U.S. Census Bureau, the graph of f shown in Figure 10 gives the percentage of American households y as a function of their annual income x (in thousands of dollars) in 2010.

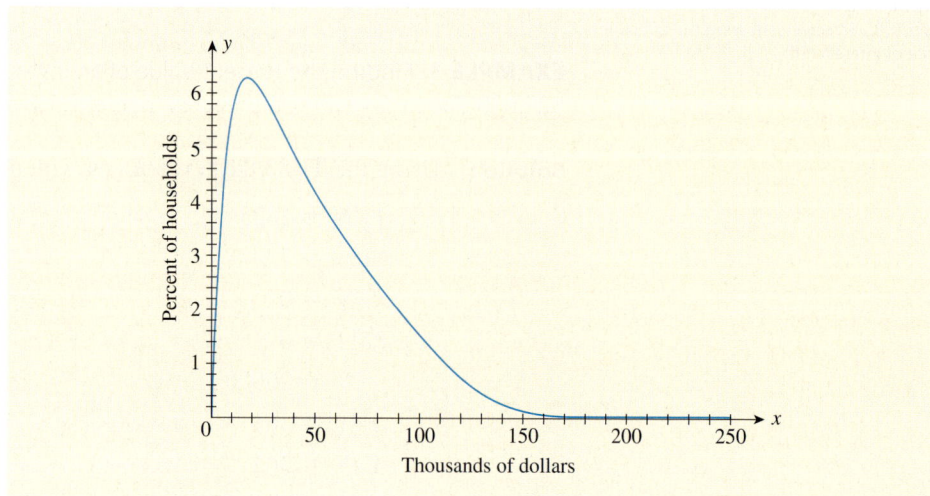

FIGURE 10
The graph of f shows the percentage of households versus their annual income.
Source: U.S. Census Bureau.

Observe that the graph of f rises very quickly and then tapers off. From the graph of f, you can see that the bulk of American households earned less than $150,000 per year. (We will refer to this model again in Using Technology at the end of this section.)

To analyze mathematical models involving exponential and logarithmic functions in greater detail, we need to develop rules for computing the derivative of these functions. We begin by looking at the rule for computing the derivative of the exponential function.

Rule 1: Derivative of the Exponential Function

$$\frac{d}{dx}(e^x) = e^x$$

Thus, the derivative of the exponential function with base e is equal to the function itself. To demonstrate the validity of this rule, we compute

$$f'(x) = \lim_{h \to 0} \frac{f(x+h) - f(x)}{h}$$

$$= \lim_{h \to 0} \frac{e^{x+h} - e^x}{h}$$

$$= \lim_{h \to 0} \frac{e^x(e^h - 1)}{h} \qquad \text{Write } e^{x+h} = e^x e^h \text{ and factor.}$$

$$= e^x \lim_{h \to 0} \frac{e^h - 1}{h} \qquad \text{Why?}$$

To evaluate

$$\lim_{h \to 0} \frac{e^h - 1}{h}$$

let's refer to Table 4, which is constructed with the aid of a calculator. From the table, we see that

$$\lim_{h \to 0} \frac{e^h - 1}{h} = 1$$

TABLE 4

h	$\frac{e^h - 1}{h}$
0.1	1.0517
0.01	1.0050
0.001	1.0005
−0.1	0.9516
−0.01	0.9950
−0.001	0.9995

(Although a rigorous proof of this fact is possible, it is beyond the scope of this book. Also see Example 1, Using Technology, page 370.) Using this result, we conclude that

$$f'(x) = e^x \cdot 1 = e^x$$

as we set out to show.

EXAMPLE 1 Find the derivative of each of the following functions:

a. $f(x) = x^2 e^x$ **b.** $g(t) = (e^t + 2)^{3/2}$

Solution

a. The Product Rule gives

$$f'(x) = \frac{d}{dx}(x^2 e^x) = x^2 \frac{d}{dx}(e^x) + e^x \frac{d}{dx}(x^2)$$

$$= x^2 e^x + e^x(2x) = xe^x(x + 2) \qquad (x^2) \text{ See page 10.}$$

b. Using the General Power Rule, we find

$$g'(t) = \frac{3}{2}(e^t + 2)^{1/2} \frac{d}{dt}(e^t + 2) = \frac{3}{2}(e^t + 2)^{1/2} e^t = \frac{3}{2} e^t (e^t + 2)^{1/2}$$

Exploring with TECHNOLOGY

Consider the exponential function $f(x) = b^x$ ($b > 0$, $b \neq 1$).

1. Use the definition of the derivative of a function to show that

$$f'(x) = b^x \cdot \lim_{h \to 0} \frac{b^h - 1}{h}$$

(continued)

2. Use the result of part 1 to show that

$$\frac{d}{dx}(2^x) = 2^x \cdot \lim_{h \to 0} \frac{2^h - 1}{h}$$

$$\frac{d}{dx}(3^x) = 3^x \cdot \lim_{h \to 0} \frac{3^h - 1}{h}$$

3. Use the technique in Using Technology, page 370, to show that (to two decimal places)

$$\lim_{h \to 0} \frac{2^h - 1}{h} = 0.69 \quad \text{and} \quad \lim_{h \to 0} \frac{3^h - 1}{h} = 1.10$$

4. Conclude from the results of parts 2 and 3 that

$$\frac{d}{dx}(2^x) \approx (0.69)2^x \quad \text{and} \quad \frac{d}{dx}(3^x) \approx (1.10)3^x$$

Thus,

$$\frac{d}{dx}(b^x) = k \cdot b^x$$

where k is an appropriate constant.

5. The results of part 4 suggest that, for convenience, we pick the base b, where $2 < b < 3$, so that $k = 1$. This value of b is $e \approx 2.718281828\ldots$. Thus,

$$\frac{d}{dx}(e^x) = e^x$$

This is why we prefer to work with the exponential function $f(x) = e^x$.

Applying the Chain Rule to Exponential Functions

To enlarge the class of exponential functions to be differentiated, we appeal to the Chain Rule to obtain the following rule for differentiating composite functions of the form $h(x) = e^{f(x)}$. An example of such a function is $h(x) = e^{x^2 - 2x}$. Here, $f(x) = x^2 - 2x$.

> **Rule 2: The Chain Rule for Exponential Functions**
> If $f(x)$ is a differentiable function, then
>
> $$\frac{d}{dx}(e^{f(x)}) = e^{f(x)} f'(x)$$

To see this, observe that if $h(x) = g[f(x)]$, where $g(x) = e^x$, then by virtue of the Chain Rule,

$$h'(x) = g'[f(x)]f'(x) = e^{f(x)} f'(x)$$

since $g'(x) = e^x$.

As an aid to remembering the Chain Rule for Exponential Functions, observe that it has the following form:

$$\frac{d}{dx}(e^{f(x)}) = e^{f(x)} \cdot \text{derivative of exponent}$$
$$\underbrace{\qquad\qquad\qquad}_{\text{Same}}$$

5.4 DIFFERENTIATION OF EXPONENTIAL FUNCTIONS

EXAMPLE 2 Find the derivative of each of the following functions:

a. $f(x) = e^{2x}$ **b.** $y = e^{-3x}$ **c.** $g(t) = e^{2t^2+t}$

Solution

a. $f'(x) = e^{2x} \dfrac{d}{dx}(2x) = e^{2x} \cdot 2 = 2e^{2x}$

b. $\dfrac{dy}{dx} = e^{-3x} \dfrac{d}{dx}(-3x) = -3e^{-3x}$

c. $g'(t) = e^{2t^2+t} \cdot \dfrac{d}{dt}(2t^2 + t) = (4t + 1)e^{2t^2+t}$

EXAMPLE 3 Differentiate the function $y = xe^{-2x}$.

Solution Using the Product Rule, followed by the Chain Rule, we find

$$\dfrac{dy}{dx} = x\dfrac{d}{dx}(e^{-2x}) + e^{-2x}\dfrac{d}{dx}(x)$$

$$= xe^{-2x}\dfrac{d}{dx}(-2x) + e^{-2x} \quad \text{Use the Chain Rule on the first term.}$$

$$= -2xe^{-2x} + e^{-2x}$$

$$= e^{-2x}(1 - 2x)$$

EXAMPLE 4 Differentiate the function $g(t) = \dfrac{e^t}{e^t + e^{-t}}$.

Solution Using the Quotient Rule, followed by the Chain Rule, we find

$$g'(t) = \dfrac{(e^t + e^{-t})\dfrac{d}{dt}(e^t) - e^t\dfrac{d}{dt}(e^t + e^{-t})}{(e^t + e^{-t})^2}$$

$$= \dfrac{(e^t + e^{-t})e^t - e^t(e^t - e^{-t})}{(e^t + e^{-t})^2} \quad (x^2) \text{ See page 9.}$$

$$= \dfrac{e^{2t} + 1 - e^{2t} + 1}{(e^t + e^{-t})^2} \quad e^0 = 1$$

$$= \dfrac{2}{(e^t + e^{-t})^2}$$

EXAMPLE 5 In Section 5.6, we will discuss some practical applications of the exponential function

$$Q(t) = Q_0 e^{kt}$$

where Q_0 and k are positive constants and $t \in [0, \infty)$. A quantity $Q(t)$ growing according to this law experiences exponential growth. Show that for a quantity $Q(t)$ experiencing exponential growth, the rate of growth of the quantity, $Q'(t)$, at any time t is directly proportional to the amount of the quantity present.

Solution Using the Chain Rule for Exponential Functions, we compute the derivative Q' of the function Q. Thus,

$$Q'(t) = Q_0 e^{kt} \frac{d}{dt}(kt)$$
$$= Q_0 e^{kt}(k)$$
$$= kQ_0 e^{kt}$$
$$= kQ(t) \qquad Q(t) = Q_0 e^{kt}$$

which is the desired conclusion.

EXAMPLE 6 Find the inflection points of the function $f(x) = e^{-x^2}$.

Solution The first derivative of f is

$$f'(x) = -2xe^{-x^2}$$

Differentiating $f'(x)$ with respect to x yields

$$f''(x) = (-2x)(-2xe^{-x^2}) - 2e^{-x^2}$$
$$= 2e^{-x^2}(2x^2 - 1)$$

Setting $f''(x) = 0$ gives

$$2e^{-x^2}(2x^2 - 1) = 0$$

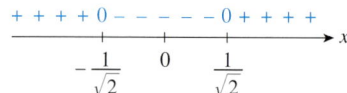

FIGURE 11
Sign diagram for f''

Since e^{-x^2} never equals zero for any real value of x, we see that $x = \pm 1/\sqrt{2}$ are the only candidates for inflection points of f. The sign diagram of f'', shown in Figure 11, tells us that both $x = -1/\sqrt{2}$ and $x = 1/\sqrt{2}$ give rise to inflection points of f.
Next,

$$f\left(-\frac{1}{\sqrt{2}}\right) = f\left(\frac{1}{\sqrt{2}}\right) = e^{-1/2}$$

and the inflection points of f are $(-1/\sqrt{2}, e^{-1/2})$ and $(1/\sqrt{2}, e^{-1/2})$. The graph of f appears in Figure 12.

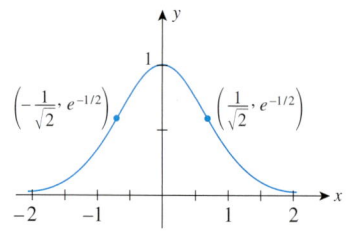

FIGURE 12
The graph of $y = e^{-x^2}$ has two inflection points.

Our final example involves finding the absolute maximum of an exponential function.

APPLIED EXAMPLE 7 Optimal Market Price Refer to Example 6, Section 5.3. The present value of the market price of the Blakely Office Building is given by

$$P(t) = 300{,}000 e^{-0.09t + \sqrt{t}/2} \qquad (0 \le t \le 10)$$

Find the optimal present value of the building's market price.

Solution To find the maximum value of P over $[0, 10]$, we compute

$$P'(t) = 300{,}000 e^{-0.09t + \sqrt{t}/2} \frac{d}{dt}\left(-0.09t + \frac{1}{2}t^{1/2}\right)$$
$$= 300{,}000 e^{-0.09t + \sqrt{t}/2}\left(-0.09 + \frac{1}{4}t^{-1/2}\right)$$

Setting $P'(t) = 0$ gives

$$-0.09 + \frac{1}{4t^{1/2}} = 0$$

since $e^{-0.09t+\sqrt{t}/2}$ is never zero for any value of t. Solving this equation, we find

$$\frac{1}{4t^{1/2}} = 0.09$$

$$t^{1/2} = \frac{1}{4(0.09)}$$

$$= \frac{1}{0.36}$$

$$t = \left(\frac{1}{0.36}\right)^2 \approx 7.72$$

the sole critical number of the function P. Finally, evaluating $P(t)$ at the critical number as well as at the endpoints of [0, 10], we have

t	0	7.72	10
$P(t)$	300,000	600,779	592,838

We conclude, accordingly, that the optimal present value of the property's market price is \$600,779 and that this will occur 7.72 years from now.

5.4 Self-Check Exercises

1. Let $f(x) = xe^{-x}$.
   a. Find the first and second derivatives of f.
   b. Find the relative extrema of f.
   c. Find the inflection points of f.

2. An industrial asset is being depreciated at a rate such that its book value t years from now will be

$$V(t) = 50{,}000e^{-0.4t}$$

dollars. How fast will the book value of the asset be changing 3 years from now?

Solutions to Self-Check Exercises 5.4 can be found on page 369.

5.4 Concept Questions

1. State the rule for differentiating (a) $f(x) = e^x$ and (b) $g(x) = e^{f(x)}$, where f is a differentiable function.

2. Let $f(x) = e^{kx}$.
   a. Compute $f'(x)$.
   b. Use the result to deduce the behavior of f for the case $k > 0$ and the case $k < 0$.

5.4 Exercises

In Exercises 1–28, find the derivative of the function.

1. $f(x) = e^{3x}$
2. $f(x) = 3e^x$
3. $g(t) = e^{-t}$
4. $f(x) = e^{-2x}$
5. $f(x) = e^x + x^2$
6. $f(x) = 2e^x - x^2$
7. $f(x) = x^3 e^x$
8. $f(u) = u^2 e^{-u}$
9. $f(x) = \dfrac{e^x}{x}$
10. $f(x) = \dfrac{x}{e^x}$
11. $f(x) = 3(e^x + e^{-x})$
12. $f(x) = \dfrac{e^x + e^{-x}}{2}$
13. $f(w) = \dfrac{e^w + 2}{e^w}$
14. $f(x) = \dfrac{e^x}{e^x + 1}$
15. $f(x) = 2e^{3x-1}$
16. $f(t) = 4e^{3t+2}$

17. $h(x) = e^{-x^2}$

18. $f(x) = e^{x^2-1}$

19. $f(x) = 3e^{-1/x}$

20. $f(x) = e^{1/(2x)}$

21. $f(x) = (e^x + 1)^{25}$

22. $f(x) = (4 - e^{-3x})^3$

23. $f(x) = e^{\sqrt{x}}$

24. $f(t) = -e^{-\sqrt{2t}}$

25. $f(x) = (x - 1)e^{3x+2}$

26. $f(s) = (s^2 + 1)e^{-s^2}$

27. $f(x) = \dfrac{e^x - 1}{e^x + 1}$

28. $g(t) = \dfrac{e^{-t}}{1 + t^2}$

In Exercises 29–32, find the second derivative of the function.

29. $f(x) = e^{-4x} + e^{3x}$

30. $f(t) = 3e^{-2t} - 5e^{-t}$

31. $f(x) = 2xe^{3x}$

32. $f(t) = t^2 e^{-2t}$

33. Find an equation of the tangent line to the graph of $y = e^{2x-3}$ at the point $(\frac{3}{2}, 1)$.

34. Find an equation of the tangent line to the graph of $y = e^{-x^2}$ at the point $(1, 1/e)$.

35. Determine the intervals where the function $f(x) = e^{-x^2/2}$ is increasing and where it is decreasing.

36. Determine the intervals where the function $f(x) = x^2 e^{-x}$ is increasing and where it is decreasing.

37. Determine the intervals of concavity for the graph of the function $f(x) = \dfrac{e^x - e^{-x}}{2}$.

38. Determine the intervals of concavity for the graph of the function $f(x) = xe^x$.

39. Find the inflection point of the function $f(x) = xe^{-2x}$.

40. Find the inflection point(s) of the function $f(x) = 2e^{-x^2}$.

41. Find the equations of the tangent lines to the graph of $f(x) = e^{-x^2}$ at its inflection points.

42. Find an equation of the tangent line to the graph of $f(x) = xe^{-x}$ at its inflection point.

In Exercises 43–46, find the absolute extrema of the function.

43. $f(x) = e^{-x^2}$ on $[-1, 1]$

44. $h(x) = e^{x^2-4}$ on $[-2, 2]$

45. $g(x) = (2x - 1)e^{-x}$ on $[0, \infty)$

46. $f(x) = xe^{-x^2}$ on $[0, 2]$

In Exercises 47–50, use the curve-sketching guidelines of Chapter 4, page 288, to sketch the graph of the function.

47. $f(t) = e^t - t$

48. $h(x) = \dfrac{e^x + e^{-x}}{2}$

49. $f(x) = 2 - e^{-x}$

50. $f(x) = \dfrac{3}{1 + e^{-x}}$

51. **Percentage of Population Relocating** On the basis of data obtained from the Census Bureau, the manager of Ply- mouth Van Lines estimates that the percent of the total population relocating in year t ($t = 0$ corresponds to the year 1960) may be approximated by the formula

$$P(t) = 20.6e^{-0.009t} \quad (0 \leq t \leq 35)$$

Compute $P'(10)$, $P'(20)$, and $P'(30)$, and interpret your results.

52. **Pharmaceutical Theft** Pharmaceutical theft has been rising rapidly in recent years. Experts believe that pharmaceuticals are the new "street gold." The value of stolen drugs (in millions of dollars per year) from 2007 through 2009 is approximated by the function

$$f(t) = 20.5e^{0.74t} \quad (1 \leq t \leq 3)$$

where t is the number of years since 2007. Find the rate of change of the value of stolen drugs at $t = 2$, and interpret your result.
Source: New York Times.

53. **Over-100 Population** On the basis of data obtained from the Census Bureau, the number of Americans over age 100 years is expected to be

$$P(t) = 0.07e^{0.54t} \quad (0 \leq t \leq 4)$$

where $P(t)$ is measured in millions and t is measured in decades, with $t = 0$ corresponding to the beginning of 2000.
a. What was the population of Americans over age 100 at the beginning of 2000? What will it be at the beginning of 2030?
b. How fast was the population of Americans over age 100 years changing at the beginning of 2000? How fast will it be changing at the beginning of 2030?
Source: U.S. Census Bureau.

54. **World Population Growth** After its fastest rate of growth ever during the 1980s and 1990s, the rate of growth of world population is expected to slow dramatically in the twenty-first century. The function

$$G(t) = 1.58e^{-0.213t}$$

gives the projected annual average percent population growth per decade in the tth decade, with $t = 1$ corresponding to 2000.
a. What will the projected annual average population growth rate be in 2020 ($t = 3$)?
b. How fast will the projected annual average population growth rate be changing in 2020?
Source: U.S. Census Bureau.

55. **Rate of Business Failures** The rate of business failure is highest in the first few years of the businesses' existence. According to a study of businesses that started in the second quarter of 1998, the percentage of companies still in business t years after the start of business is approximated by the function

$$f(t) = 93.1e^{-0.1626t} \quad (1 \leq t \leq 7)$$

Find the rate at which the percent of businesses in existence is dropping at $t = 1, 2, 3,$ and 4 years.
Source: Monthly Labor Review.

56. ENERGY CONSUMPTION OF APPLIANCES The average energy consumption of the typical refrigerator/freezer manufactured by York Industries is approximately

$$C(t) = 1486e^{-0.073t} + 500 \quad (0 \le t \le 20)$$

kilowatt-hours (kWh) per year, where t is measured in years, with $t = 0$ corresponding to 1972.
a. What was the average energy consumption of the York refrigerator/freezer at the beginning of 1972?
b. Prove that the average energy consumption of the York refrigerator/freezer is decreasing over the years in question.
c. All refrigerator/freezers manufactured as of January 1, 1990, must meet the 950-kWh/year maximum energy-consumption standard set by the National Appliance Conservation Act. Show that the York refrigerator/freezer satisfies this requirement.

57. SALES PROMOTION The Lady Bug, a women's clothing chain store, found that t days after the end of a sales promotion the volume of sales was given by

$$S(t) = 20{,}000(1 + e^{-0.5t}) \quad (0 \le t \le 5)$$

dollars.
a. Find the rate of change of The Lady Bug's sales volume when $t = 1$, $t = 2$, $t = 3$, and $t = 4$.
b. In how many days will the sales volume drop below $27,400?

58. BLOOD ALCOHOL LEVEL The percentage of alcohol in a person's bloodstream t hr after drinking 8 fluid oz of whiskey is given by

$$A(t) = 0.23te^{-0.4t} \quad (0 \le t \le 12)$$

a. What is the percentage of alcohol in a person's bloodstream after $\tfrac{1}{2}$ hr? After 8 hr?
b. How fast is the percentage of alcohol in a person's bloodstream changing after $\tfrac{1}{2}$ hr? After 8 hr?
Source: Encyclopedia Britannica.

59. POLIO IMMUNIZATION Polio, a once-feared killer, declined markedly in the United States in the 1950s after Jonas Salk developed the inactivated polio vaccine and mass immunization of children took place. The number of polio cases in the United States from the beginning of 1959 to the beginning of 1963 is approximated by the function

$$N(t) = 5.3e^{0.095t^2 - 0.85t} \quad (0 \le t \le 4)$$

where $N(t)$ gives the number of polio cases (in thousands) and t is measured in years, with $t = 0$ corresponding to the beginning of 1959.
a. Show that the function N is decreasing over the time interval under consideration.
b. How fast was the number of polio cases decreasing at the beginning of 1959? At the beginning of 1962? (*Comment:* Since the introduction of the oral vaccine developed by Dr. Albert B. Sabin in 1963, polio in the United States has, for all practical purposes, been eliminated.)

60. AUTISTIC BRAIN At birth, the autistic brain is similar in size to a healthy child's brain. Between birth and 2 years, it grows to be abnormally large, reaching its maximum size between 3 and 6 years of age. The percentage difference in size between the autistic brain and the normal brain to age 40 is approximated by

$$D(t) = 6.9te^{-0.24t} \quad (0 \le t \le 40)$$

where t is measured in years.
a. At what ages is the difference in the size between the autistic brain and the normal brain increasing? Decreasing?
b. At what age is the difference in the size between the autistic brain and the normal brain the greatest? What is the maximum difference?
c. At what age is the difference in the size between the autistic brain and the normal brain decreasing at the fastest rate?
d. Sketch the graph of D on the interval $[0, 40]$.
Source: Newsweek.

61. DEATH DUE TO STROKES Before 1950, little was known about strokes. By 1960, however, risk factors such as hypertension were identified. In recent years, CAT scans used as a diagnostic tool have helped to prevent strokes. As a result, the number of deaths due to strokes has fallen dramatically. The function

$$N(t) = 130.7e^{-0.1155t^2} + 50 \quad (0 \le t \le 6)$$

gives the number of deaths due to stroke per 100,000 people from 1950 through 2010, where t is measured in decades, with $t = 0$ corresponding to 1950.
a. How many deaths due to strokes per 100,000 people were there in 1950?
b. How fast was the number of deaths due to strokes per 100,000 people changing in 1950? In 1960? In 1970? In 1980?
c. When was the rate of decline in the number of deaths due to strokes per 100,000 people greatest?
d. How many deaths due to strokes per 100,000 people were there in 2010?
Source: American Heart Association, Centers for Disease Control, and National Institutes of Health.

62. ALZHEIMER'S DISEASE Alzheimer's disease can occur at any age, even as young as 40 years old, but its occurrence is much more common as the years go by. The frequency of occurrence of the disease (as a percentage) is given by

$$f(t) = 0.71e^{0.7t} \quad (1 \le t \le 5)$$

where t is measured in 5-year intervals, with $t = 1$ corresponding to an age of 70 years.
a. What is the frequency of occurrence of Alzheimer's disease for 70-year-old people? For 90-year-old people?
b. Show that f is increasing on the interval $(1, 5)$. Interpret your results.
c. Show that f is concave upward on the interval $(1, 5)$. Interpret your result.
Source: World Health Organization.

63. Marginal Revenue The relationship between the unit selling price p (in dollars) and the quantity demanded x (in pairs) of a certain brand of women's gloves are given by the demand equation

$$p = 100e^{-0.0001x} \quad (0 \leq x \leq 20{,}000)$$

a. Find the revenue function R.
  Hint: $R(x) = px$
b. Find the marginal revenue function R'.
c. What is the marginal revenue when $x = 10{,}000$?

64. Price of Perfume The monthly demand for a certain brand of perfume is given by the demand equation

$$p = 100e^{-0.0002x} + 150$$

where p denotes the retail unit price (in dollars) and x denotes the quantity (in 1-oz bottles) demanded.
a. Find the rate of change of the price per bottle when $x = 1000$ and when $x = 2000$.
b. What is the price per bottle when $x = 1000$? When $x = 2000$?

65. Price of Wine The monthly demand for a certain brand of table wine is given by the demand equation

$$p = 240\left(1 - \frac{3}{3 + e^{-0.0005x}}\right)$$

where p denotes the wholesale price per case (in dollars) and x denotes the number of cases demanded.
a. Find the rate of change of the price per case when $x = 1000$.
b. What is the price per case when $x = 1000$?

66. Spread of an Epidemic During a flu epidemic, the total number of students on a state university campus who had contracted influenza by the xth day was given by

$$N(x) = \frac{3000}{1 + 99e^{-x}} \quad (x \geq 0)$$

a. How many students had influenza initially?
b. Derive an expression for the rate at which the disease was being spread, and prove that the function N is increasing on the interval $(0, \infty)$.
c. Sketch the graph of N. What was the total number of students who contracted influenza during that particular epidemic?

67. Elasticity of Demand The quantity demanded each month x (in units of a hundred) of the Soundex model A alarm clock radio/CD player is related to the unit price p (in dollars) by the demand equation

$$x = 50e^{-0.02p} \quad (p > 0)$$

a. Find the elasticity of demand for the model A players.
b. Find the values of p for which the demand is inelastic, unitary, or elastic.
c. If the unit price of the player is decreased slightly from $40, will the revenue increase or decrease?
d. If the unit price of the player is increased slightly from $60, will the revenue increase or decrease?
  Hint: Refer to Section 3.4.

68. Elasticity of Demand Suppose that the demand equation for a certain commodity has the form $x = ae^{-bp}$, where a and b are positive constants.
a. Find the elasticity of demand $E(p)$.
b. Find the values of p for which the demand is inelastic, unitary, or elastic.

69. Weights of Children The Ehrenberg equation

$$W = 2.4e^{1.84h}$$

gives the relationship between the height h (in meters) and the average weight W (in kilograms) for children between 5 and 13 years of age.
a. What is the average weight of a 10-year-old child who stands 1.6 m tall?
b. Use differentials to estimate the change in the average weight of a 10-year-old child whose height increases from 1.6 m to 1.65 m.

70. Population Distribution The number of people living x mi from the center of town is given by

$$P(x) = 50{,}000(1 - e^{-0.01x^2}) \quad (0 < x < 25)$$

Use differentials to estimate the number of people living between 10 and 10.1 mi from the center of town.

71. Optimal Selling Time Refer to Exercise 46, page 357. The present value of a piece of waterfront property purchased by an investor is given by the function

$$P(t) = 80{,}000e^{\sqrt{t/2} - 0.09t} \quad (0 \leq t \leq 8)$$

Determine the optimal time (based on present value) for the investor to sell the property. What is the property's optimal present value?

72. Maximum Oil Production It has been estimated that the total production of oil from a certain oil well is given by

$$T(t) = -1000(t + 10)e^{-0.1t} + 10{,}000$$

thousand barrels t years after production has begun. Determine the year when the oil well will be producing at maximum capacity.

73. Blood Alcohol Level Refer to Exercise 58, page 367. At what time after drinking the alcohol is the percentage of alcohol in the person's bloodstream at its highest level? What is that level?

74. Price of a Commodity The price of a certain commodity in dollars per unit at time t (measured in weeks) is given by $p = 8 + 4e^{-2t} + te^{-2t}$.
a. What is the price of the commodity at $t = 0$?
b. How fast is the price of the commodity changing at $t = 0$?
c. Find the equilibrium price of the commodity.
  Hint: It's given by $\lim_{t \to \infty} p$. Also, use the fact that $\lim_{t \to \infty} te^{-2t} = 0$.

75. Thermometer Readiness A thermometer is moved from inside a house out to the deck. Its temperature t min after it has been moved is given by

$$T(t) = 30 + 40e^{-0.98t}$$

a. What is the temperature inside the house?

b. How fast is the reading on the thermometer changing 1 min after it has been taken out of the house?
c. What is the outdoor temperature?
Hint: Evaluate $\lim_{t \to \infty} T(t)$.

76. Chemical Reaction Two chemicals, A and B, interact to form a Chemical C. Suppose the amount (in grams) of Chemical C formed t min after the interaction begins is

$$A(t) = \frac{150(1 - e^{0.022662t})}{1 - 2.5e^{0.022662t}}$$

a. How fast is Chemical C being formed 1 min after the interaction first began?
b. How much Chemical C will there be eventually?
Hint: Evaluate $\lim_{t \to \infty} A(t)$.

77. Concentration of a Drug in the Bloodstream The concentration of a drug in the bloodstream t sec after injection into a muscle is given by

$$y = c(e^{-bt} - e^{-at}) \qquad (t \geq 0)$$

where a, b, and c are positive constants, with $a > b$.
a. Find the time at which the concentration is maximal.
b. Find the time at which the concentration of the drug in the bloodstream is decreasing most rapidly.

78. Absorption of Drugs A liquid carries a drug into an organ of volume V cm^3 at the rate of a cm^3/sec and leaves at the same rate. The concentration of the drug in the entering liquid is c g/cm^3. Letting $x(t)$ denote the concentration of the drug in the organ at any time t, we have $x(t) = c(1 - e^{-at/V})$.
a. Show that x is an increasing function on $(0, \infty)$.
b. Sketch the graph of x.

79. Absorption of Drugs Refer to Exercise 78. Suppose the maximum concentration of the drug in the organ must *not* exceed m g/cm^3, where $m < c$. Show that the liquid must not be allowed to enter the organ for a time longer than

$$T = \left(\frac{V}{a}\right) \ln\left(\frac{c}{c - m}\right)$$

minutes.

80. Absorption of Drugs Jane took 100 mg of a drug one morning and another 100 mg of the same drug at the same time the following morning. The amount of the drug in her body t days after the first dose was taken is given by

$$A(t) = \begin{cases} 100e^{-1.4t} & \text{if } 0 \leq t < 1 \\ 100(1 + e^{1.4})e^{-1.4t} & \text{if } t \geq 1 \end{cases}$$

a. How fast was the amount of drug in Jane's body changing after 12 hr ($t = \frac{1}{2}$)? After 2 days?
b. When was the amount of drug in Jane's body a maximum?
c. What was the maximum amount of drug in Jane's body?

81. Absorption of Drugs The concentration of a drug in an organ at any time t (in seconds) is given by

$$C(t) = \begin{cases} 0.3t - 18(1 - e^{-t/60}) & \text{if } 0 \leq t < 20 \\ 18e^{-t/60} - 12e^{-(t-20)/60} & \text{if } t > 20 \end{cases}$$

where $C(t)$ is measured in grams per cubic centimeter (g/cm^3).
a. How fast is the concentration of the drug in the organ changing after 10 sec?
b. How fast is the concentration of the drug in the organ changing after 30 sec?
c. When will the concentration of the drug in the organ reach a maximum?
d. What is the maximum drug concentration in the organ?

In Exercises 82–85, determine whether the statement is true or false. If it is true, explain why it is true. If it is false, give an example to show why it is false.

82. If $f(x) = 3^x$, then $f'(x) = x \cdot 3^{x-1}$.

83. If $f(x) = e^\pi$, then $f'(x) = e^\pi$.

84. If $f(x) = \pi^x$, then $f'(x) = \pi^x$.

85. If $x^2 + e^y = 10$, then $y' = \frac{-2x}{e^y}$.

5.4 Solutions to Self-Check Exercises

1. a. Using the Product Rule, we obtain

$$f'(x) = x \frac{d}{dx}(e^{-x}) + e^{-x}\frac{d}{dx}(x)$$
$$= -xe^{-x} + e^{-x} = (1 - x)e^{-x}$$

Using the Product Rule once again, we obtain

$$f''(x) = (1 - x)\frac{d}{dx}e^{-x} + e^{-x}\frac{d}{dx}(1 - x)$$
$$= (1 - x)(-e^{-x}) + e^{-x}(-1)$$
$$= -e^{-x} + xe^{-x} - e^{-x} = (x - 2)e^{-x}$$

b. Setting $f'(x) = 0$ gives

$$(1 - x)e^{-x} = 0$$

Since $e^{-x} \neq 0$, we see that $1 - x = 0$, and this gives 1 as the only critical number of f. The sign diagram of f' shown in the accompanying figure tells us that the point $(1, e^{-1})$ is a relative maximum of f.

c. Setting $f''(x) = 0$ gives $x - 2 = 0$, so $x = 2$ is a candidate for an inflection point of f. The sign diagram of f'' (see the accompanying figure) shows that $(2, 2e^{-2})$ is an inflection point of f.

2. The rate change of the book value of the asset t years from now is

$$V'(t) = 50{,}000\frac{d}{dt}e^{-0.4t}$$
$$= 50{,}000(-0.4)e^{-0.4t} = -20{,}000e^{-0.4t}$$

Therefore, 3 years from now, the book value of the asset will be changing at the rate of

$$V'(3) = -20{,}000e^{-0.4(3)} = -20{,}000e^{-1.2} \approx -6023.88$$

—that is, decreasing at the rate of approximately $6024/year.

USING TECHNOLOGY

EXAMPLE 1 At the beginning of Section 5.4, we demonstrated via a table of values of $(e^h - 1)/h$ for selected values of h the plausibility of the result

$$\lim_{h \to 0} \frac{e^h - 1}{h} = 1$$

To obtain a visual confirmation of this result, we plot the graph of

$$f(x) = \frac{e^x - 1}{x}$$

in the viewing window $[-1, 1] \times [0, 2]$ (Figure T1). From the graph of f, we see that $f(x)$ appears to approach 1 as x approaches 0.

The numerical derivative function of a graphing utility will yield the derivative of an exponential or logarithmic function for any value of x, just as it did for algebraic functions.*

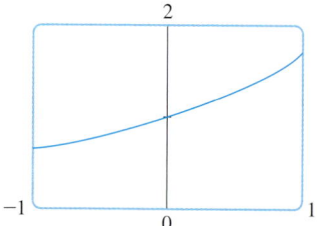

FIGURE T1
The graph of f in the viewing window $[-1, 1] \times [0, 2]$

*The rules for differentiating logarithmic functions will be covered in Section 5.5. However, the exercises given here can be done without using these rules.

TECHNOLOGY EXERCISES

In Exercises 1–6, use the numerical derivative operation to find the rate of change of $f(x)$ at the given value of x. Give your answer accurate to four decimal places.

1. $f(x) = x^3 e^{-1/x}$; $x = -1$
2. $f(x) = (\sqrt{x} + 1)^{3/2} e^{-x}$; $x = 0.5$
3. $f(x) = x^3 \sqrt{\ln x}$; $x = 2$
4. $f(x) = \dfrac{\sqrt{x} \ln x}{x + 1}$; $x = 3.2$
5. $f(x) = e^{-x} \ln(2x + 1)$; $x = 0.5$
6. $f(x) = \dfrac{e^{-\sqrt{x}}}{\ln(x^2 + 1)}$; $x = 1$

7. An Extinction Situation The number of saltwater crocodiles in a certain area of northern Australia is given by

$$P(t) = \frac{300e^{-0.024t}}{5e^{-0.024t} + 1}$$

a. How many crocodiles were in the population initially?
b. Show that $\lim\limits_{t \to \infty} P(t) = 0$.
c. Plot the graph of P in the viewing window $[0, 200] \times [0, 70]$.
(*Comment:* This phenomenon is referred to as an *extinction situation.*)

8. Income of American Households On the basis of government data, it is estimated that the percentage of American households y who earned x thousand dollars in 2010 is given by the equation

$$y = 1.168 x e^{-0.00000312 x^3 + 0.000659 x^2 - 0.0783 x} \quad (x > 0)$$

a. Plot the graph of the equation in the viewing window $[0, 150] \times [0, 2]$.
b. How fast is y changing with respect to x when $x = 10$? When $x = 50$? Interpret your results.

Source: U.S. Census Bureau.

9. **WORLD POPULATION GROWTH** On the basis of data obtained in a study, the world population (in billions) is approximated by the function

$$f(t) = \frac{12}{1 + 3.74914e^{-1.42804t}} \quad (0 \leq t \leq 4)$$

where t is measured in half centuries, with $t = 0$ corresponding to the beginning of 1950.
a. Plot the graph of f in the viewing window $[0, 5] \times [0, 14]$.
b. How fast was the world population expected to increase at the beginning of 2000?

Source: United Nations Population Division.

10. **LOAN AMORTIZATION** The Sotos plan to secure a loan of $160,000 to purchase a house. They are considering a conventional 30-year home mortgage at 9%/year on the unpaid balance. It can be shown that the Sotos will have an outstanding principal of

$$B(x) = \frac{160,000(1.0075^{360} - 1.0075^x)}{1.0075^{360} - 1}$$

dollars after making x monthly payments of $1287.40.
a. Plot the graph of $B(x)$, using the viewing window $[0, 360] \times [0, 160,000]$.
b. Compute $B(0)$ and $B'(0)$, and interpret your results; compute $B(180)$ and $B'(180)$, and interpret your results.

11. **INCREASE IN JUVENILE OFFENDERS** The number of youths aged 15 to 19 increased by 21% between 1994 and 2005, pushing up the crime rate. According to the National Council on Crime and Delinquency, the number of violent crime arrests of juveniles under age 18 in year t is given by

$$f(t) = -0.438t^2 + 9.002t + 107 \quad (0 \leq t \leq 13)$$

where $f(t)$ is measured in thousands and t in years, with $t = 0$ corresponding to 1989. According to the same source, if trends like inner-city drug use and wider availability of guns continue, then the number of violent crime arrests of juveniles under age 18 in year t is given by

$$g(t) = \begin{cases} -0.438t^2 + 9.002t + 107 & \text{if } 0 \leq t < 4 \\ 99.456e^{0.07824t} & \text{if } 4 \leq t \leq 13 \end{cases}$$

where $g(t)$ is measured in thousands and $t = 0$ corresponds to 1989.
a. Compute $f(11)$ and $g(11)$, and interpret your results.
b. Compute $f'(11)$ and $g'(11)$, and interpret your results.

Source: National Council on Crime and Delinquency.

12. **INCREASING CROP YIELDS** If left untreated on bean stems, aphids (small insects that suck plant juices) will multiply at an increasing rate during the summer months and reduce productivity and crop yield of cultivated crops. But if the aphids are treated in mid-June, the numbers decrease sharply to less than 100/bean stem, allowing for steep rises in crop yield. The function

$$F(t) = \begin{cases} 62e^{1.152t} & \text{if } 0 \leq t < 1.5 \\ 349e^{-1.324(t-1.5)} & \text{if } 1.5 \leq t \leq 3 \end{cases}$$

gives the number of aphids after treatment on a typical bean stem at time t, where t is measured in months, with $t = 0$ corresponding to the beginning of May.
a. How many aphids are there on a typical bean stem at the beginning of June ($t = 1$)? At the beginning of July ($t = 2$)?
b. How fast is the population of aphids changing at the beginning of June? At the beginning of July?

Source: The Random House Encyclopedia.

13. **WOMEN IN THE LABOR FORCE** On the basis of data from the U.S. Census Bureau, the chief economist of Manpower, Inc., constructed the following formula giving the percentage of the total female population in the civilian labor force, $P(t)$, at the beginning of the tth decade ($t = 0$ corresponds to the year 1900):

$$P(t) = \frac{74}{1 + 2.6e^{-0.166t + 0.04536t^2 - 0.0066t^3}} \quad (0 \leq t \leq 11)$$

Assume that this trend continued through 2010.
a. What percentage of the total female population was in the civilian labor force at the beginning of 2000?
b. What was the growth rate of the percentage of the total female population in the civilian labor force at the beginning of 2000?

Source: U.S. Census Bureau.

5.5 Differentiation of Logarithmic Functions

The Derivative of ln x

Let's now turn our attention to the differentiation of logarithmic functions.

Rule 3: Derivative of ln x

$$\frac{d}{dx} \ln |x| = \frac{1}{x} \quad (x \neq 0)$$

To derive Rule 3, suppose $x > 0$ and write $f(x) = \ln x$ in the equivalent form

$$x = e^{f(x)}$$

Differentiating both sides of the equation with respect to x, we find, using the Chain Rule,

$$1 = e^{f(x)} \cdot f'(x)$$

from which we see that

$$f'(x) = \frac{1}{e^{f(x)}}$$

or, since $e^{f(x)} = x$,

$$f'(x) = \frac{1}{x}$$

as we set out to show. You are asked to prove the rule for the case $x < 0$ in Exercise 92, page 379.

EXAMPLE 1 Find the derivative of each function:

a. $f(x) = x \ln x$ **b.** $g(x) = \dfrac{\ln x}{x}$

Solution

a. Using the Product Rule, we obtain

$$f'(x) = \frac{d}{dx}(x \ln x) = x \frac{d}{dx}(\ln x) + (\ln x)\frac{d}{dx}(x)$$

$$= x\left(\frac{1}{x}\right) + \ln x = 1 + \ln x$$

b. Using the Quotient Rule, we obtain

$$g'(x) = \frac{x \dfrac{d}{dx}(\ln x) - (\ln x)\dfrac{d}{dx}(x)}{x^2} = \frac{x\left(\dfrac{1}{x}\right) - \ln x}{x^2} = \frac{1 - \ln x}{x^2}$$

Explore & Discuss

You can derive the formula for the derivative of $f(x) = \ln x$ directly from the definition of the derivative, as follows.

1. Show that

$$f'(x) = \lim_{h \to 0} \frac{f(x+h) - f(x)}{h} = \lim_{h \to 0} \ln\left(1 + \frac{h}{x}\right)^{1/h}$$

2. Put $m = x/h$ and note that $m \to \infty$ as $h \to 0$. Then, $f'(x)$ can be written in the form

$$f'(x) = \lim_{m \to \infty} \ln\left(1 + \frac{1}{m}\right)^{m/x}$$

3. Finally, use both the fact that the natural logarithmic function is continuous and the definition of the number e to show that

$$f'(x) = \frac{1}{x} \ln\left[\lim_{m \to \infty}\left(1 + \frac{1}{m}\right)^m\right] = \frac{1}{x}$$

The Chain Rule and Logarithmic Functions

To enlarge the class of logarithmic functions to be differentiated, we appeal once again to the Chain Rule to obtain the following rule for differentiating composite functions of the form $h(x) = \ln f(x)$, where $f(x)$ is assumed to be a positive differentiable function.

> **Rule 4: Derivative of ln f(x)**
> If $f(x)$ is a differentiable function, then
> $$\frac{d}{dx}[\ln f(x)] = \frac{f'(x)}{f(x)} \quad (f(x) > 0)$$

To see this, observe that $h(x) = g[f(x)]$, where $g(x) = \ln x \; (x > 0)$. Since $g'(x) = 1/x$, we have, using the Chain Rule,

$$h'(x) = g'[f(x)]f'(x)$$
$$= \frac{1}{f(x)} f'(x) = \frac{f'(x)}{f(x)}$$

Observe that in the special case $f(x) = x$, $h(x) = \ln x$, so the derivative of h is, by Rule 3, given by $h'(x) = 1/x$.

EXAMPLE 2 Find the derivative of the function $f(x) = \ln(x^2 + 1)$.

Solution Using Rule 4, we see immediately that

$$f'(x) = \frac{\frac{d}{dx}(x^2 + 1)}{x^2 + 1} = \frac{2x}{x^2 + 1}$$

In differentiating functions involving logarithms, the rules of logarithms may be used to advantage, as shown in Examples 3 and 4.

EXAMPLE 3 Differentiate the function $y = \ln[(x^2 + 1)(x^3 + 2)^6]$.

Solution We first rewrite the given function using the properties of logarithms:

$$y = \ln[(x^2 + 1)(x^3 + 2)^6]$$
$$= \ln(x^2 + 1) + \ln(x^3 + 2)^6 \qquad \ln mn = \ln m + \ln n$$
$$= \ln(x^2 + 1) + 6\ln(x^3 + 2) \qquad \ln m^n = n \ln m$$

Differentiating and using Rule 4, we obtain

$$y' = \frac{\frac{d}{dx}(x^2 + 1)}{x^2 + 1} + \frac{6\frac{d}{dx}(x^3 + 2)}{x^3 + 2}$$
$$= \frac{2x}{x^2 + 1} + \frac{6(3x^2)}{x^3 + 2} = \frac{2x}{x^2 + 1} + \frac{18x^2}{x^3 + 2}$$

> **Exploring with TECHNOLOGY**
>
> Use a graphing utility to plot the graphs of $f(x) = \ln x$; its first derivative function, $f'(x) = 1/x$; and its second derivative function $f''(x) = -1/x^2$, using the same viewing window $[0, 4] \times [-3, 3]$.
>
> 1. Describe the properties of the graph of f revealed by studying the graph of $f'(x)$. What can you say about the rate of increase of f?
> 2. Describe the properties of the graph of f revealed by studying the graph of $f''(x)$. What can you say about the concavity of f?

EXAMPLE 4 Find the derivative of the function $g(t) = \ln(t^2 e^{-t^2})$.

Solution Here again, to save a lot of work, we first simplify the given expression using the properties of logarithms. We have

$$g(t) = \ln(t^2 e^{-t^2})$$
$$= \ln t^2 + \ln e^{-t^2} \quad \text{ln } mn = \ln m + \ln n$$
$$= 2 \ln t - t^2 \quad \text{ln } m^n = n \ln m \text{ and } \ln e = 1$$

Therefore,

$$g'(t) = \frac{2}{t} - 2t = \frac{2(1-t^2)}{t}$$

Logarithmic Differentiation

As we saw in Examples 3 and 4, the task of finding the derivative of a given function can sometimes be made easier by first applying the laws of logarithms to simplify the function. We now illustrate a process called **logarithmic differentiation,** which not only simplifies the calculation of the derivatives of certain functions but also enables us to compute the derivatives of functions that we could not otherwise differentiate using the techniques developed thus far.

EXAMPLE 5 Differentiate $y = x(x + 1)(x^2 + 1)$, using logarithmic differentiation.

Solution First, we take the natural logarithm on both sides of the given equation, obtaining

$$\ln y = \ln[x(x+1)(x^2+1)]$$

Next, we use the properties of logarithms to rewrite the right-hand side of this equation, obtaining

$$\ln y = \ln x + \ln(x+1) + \ln(x^2+1)$$

If we differentiate both sides of this equation, we have

$$\frac{d}{dx} \ln y = \frac{d}{dx}[\ln x + \ln(x+1) + \ln(x^2+1)]$$
$$= \frac{1}{x} + \frac{1}{x+1} + \frac{2x}{x^2+1} \quad \text{Use Rule 4.}$$

To evaluate the expression on the left-hand side, note that y is a function of x. Therefore, writing $y = f(x)$ to remind us of this fact, we have

$$\frac{d}{dx} \ln y = \frac{d}{dx} \ln[f(x)] \quad \text{Write } y = f(x).$$

$$= \frac{f'(x)}{f(x)} \quad \text{Use Rule 4.}$$

$$= \frac{y'}{y} \quad \text{Return to using } y \text{ instead of } f(x).$$

Therefore, we have

$$\frac{y'}{y} = \frac{1}{x} + \frac{1}{x+1} + \frac{2x}{x^2+1}$$

Finally, solving for y', we have

$$y' = y\left(\frac{1}{x} + \frac{1}{x+1} + \frac{2x}{x^2+1}\right)$$

$$= x(x+1)(x^2+1)\left(\frac{1}{x} + \frac{1}{x+1} + \frac{2x}{x^2+1}\right)$$

Before considering other examples, let's summarize the important steps involved in logarithmic differentiation.

Finding $\dfrac{dy}{dx}$ by Logarithmic Differentiation

1. Take the natural logarithm on both sides of the equation, and use the properties of logarithms to write any "complicated expression" as a sum of simpler terms.
2. Differentiate both sides of the equation with respect to x.
3. Solve the resulting equation for $\dfrac{dy}{dx}$.

EXAMPLE 6 Differentiate $y = x^2(x-1)(x^2+4)^3$.

Solution Taking the natural logarithm on both sides of the given equation and using the laws of logarithms, we obtain

$$\ln y = \ln[x^2(x-1)(x^2+4)^3]$$
$$= \ln x^2 + \ln(x-1) + \ln(x^2+4)^3$$
$$= 2\ln x + \ln(x-1) + 3\ln(x^2+4)$$

Differentiating both sides of the equation with respect to x, we have

$$\frac{d}{dx} \ln y = \frac{y'}{y} = \frac{2}{x} + \frac{1}{x-1} + 3 \cdot \frac{2x}{x^2+4}$$

Finally, solving for y', we have

$$y' = y\left(\frac{2}{x} + \frac{1}{x-1} + \frac{6x}{x^2+4}\right)$$

$$= x^2(x-1)(x^2+4)^3\left(\frac{2}{x} + \frac{1}{x-1} + \frac{6x}{x^2+4}\right)$$

Recall from Section 3.4 that the relative rate of change of a differentiable function Q of x is $Q'(x)/Q(x)$. In view of Rule 4, we see that the relative rate of change of Q at x can also be obtained by finding the derivative of $\ln Q$. We exploit this fact in Example 7.

APPLIED EXAMPLE 7 Population Growth The population of a town t months after the opening of an auto assembly plant in the surrounding area is given by the function

$$P(t) = 18000e^{-(\ln 9)e^{-0.1t}}$$

What is the relative rate of growth of the population 6 months after the opening of the auto assembly plant?

Solution We could find the required relative rate by computing $P'(t)/P(t)$ directly. Alternatively, we can proceed as follows:

$$\ln P(t) = \ln 18000e^{-(\ln 9)e^{-0.1t}}$$
$$= \ln 18000 + \ln e^{-(\ln 9)e^{-0.1t}}$$
$$= \ln 18000 - (\ln 9)e^{-0.1t} \qquad \ln e^x = x$$

So

$$\frac{P'(t)}{P(t)} = \frac{d}{dt}[\ln P(t)] = \frac{d}{dt}\ln 18000 - \frac{d}{dt}(\ln 9)e^{-0.1t}$$
$$= 0 - (\ln 9)(-0.1)e^{-0.1t} = (0.1)(\ln 9)e^{-0.1t}$$

Therefore,

$$\left.\frac{P'(t)}{P(t)}\right|_{t=6} = (0.1)(\ln 9)e^{-(0.1)(6)} \approx 0.121$$

This tells us that 6 months after the opening of the auto assembly plant, the relative rate of growth of the population is approximately 12.1% per month.

5.5 Self-Check Exercises

1. Find an equation of the tangent line to the graph of $f(x) = x \ln(2x + 3)$ at the point $(-1, 0)$.

2. Use logarithmic differentiation to compute y', given $y = (2x + 1)^3(3x + 4)^5$.

Solutions to Self-Check Exercises 5.5 can be found on page 380.

5.5 Concept Questions

1. State the rule for differentiating (a) $f(x) = \ln|x|$ ($x \neq 0$), and (b) $g(x) = \ln f(x)$ [$f(x) > 0$], where f is a differentiable function.

2. Explain the technique of logarithmic differentiation.

5.5 Exercises

In Exercises 1–34, find the derivative of the function.

1. $f(x) = 5 \ln x$
2. $f(x) = \ln 5x$
3. $f(x) = \ln(x + 1)$
4. $g(x) = \ln(2x + 1)$
5. $f(x) = \ln x^8$
6. $h(t) = 2 \ln t^5$
7. $f(x) = \ln \sqrt{x}$
8. $f(x) = \ln(\sqrt{x} + 1)$
9. $f(x) = \ln \dfrac{1}{x^2}$
10. $f(x) = \ln \dfrac{1}{2x^3}$
11. $f(x) = \ln(4x^2 - 5x + 3)$
12. $f(x) = \ln(3x^2 - 2x + 1)$
13. $f(x) = \ln \dfrac{2x}{x + 1}$
14. $f(x) = \ln \dfrac{x + 1}{x - 1}$
15. $f(x) = x^2 \ln x$
16. $f(x) = 3x^2 \ln 2x$
17. $f(x) = \dfrac{2 \ln x}{x}$
18. $f(x) = \dfrac{3 \ln x}{x^2}$
19. $f(u) = \ln(u - 2)^3$
20. $f(x) = \ln(x^3 - 3)^4$
21. $f(x) = \sqrt{\ln x}$
22. $f(x) = \sqrt{\ln x + x}$
23. $f(x) = (\ln x)^2$
24. $f(x) = 2(\ln x)^{3/2}$
25. $f(x) = \ln(x^3 + 1)$
26. $f(x) = \ln \sqrt{x^2 - 4}$
27. $f(x) = e^x \ln x$
28. $f(x) = e^x \ln \sqrt{x + 3}$
29. $f(t) = e^{2t} \ln(t + 1)$
30. $g(t) = t^2 \ln(e^{2t} + 1)$
31. $f(x) = \dfrac{\ln x}{x^2}$
32. $g(t) = \dfrac{t}{\ln t}$
33. $f(x) = \ln(\ln x)$
34. $g(x) = \ln(e^x + \ln x)$

In Exercises 35–40, find the second derivative of the function.

35. $f(x) = \ln 2x$
36. $f(x) = \ln(x + 5)$
37. $f(x) = \ln(x^2 + 2)$
38. $f(x) = (\ln x)^2$
39. $f(x) = x^2 \ln x$
40. $g(x) = e^{2x} \ln x$

In Exercises 41–50, use logarithmic differentiation to find the derivative of the function.

41. $y = (x + 1)^2 (x + 2)^3$
42. $y = (3x + 2)^4 (5x - 1)^2$
43. $y = (x - 1)^2 (x + 1)^3 (x + 3)^4$
44. $y = \sqrt{3x + 5}(2x - 3)^4$
45. $y = \dfrac{(2x^2 - 1)^5}{\sqrt{x + 1}}$
46. $y = \dfrac{\sqrt{4 + 3x^2}}{\sqrt[3]{x^2 + 1}}$
47. $y = 3^x$
48. $y = x^{x+2}$
49. $y = (x^2 + 1)^x$
50. $y = x^{\ln x}$

In Exercises 51 and 52, use implicit differentiation to find dy/dx.

51. $\ln y - x \ln x = -1$
52. $\ln xy - y^2 = 5$

53. Find an equation of the tangent line to the graph of $y = x \ln x$ at the point $(1, 0)$.

54. Find an equation of the tangent line to the graph of $y = \ln x^2$ at the point $(2, \ln 4)$.

55. Determine the intervals where the function $f(x) = \ln x^2$ is increasing and where it is decreasing.

56. Determine the intervals where the function $f(x) = \dfrac{\ln x}{x}$ is increasing and where it is decreasing.

57. Determine the intervals of concavity for the graph of the function $f(x) = x^2 + \ln x^2$.

58. Determine the intervals of concavity for the graph of the function $f(x) = \dfrac{\ln x}{x}$.

59. Find the inflection points of the function $f(x) = \ln(x^2 + 1)$.

60. Find the inflection points of the function $f(x) = x^2 \ln x$.

61. Find an equation of the tangent line to the graph of $f(x) = x^2 + 2 \ln x$ at its inflection point.

62. Find an equation of the tangent line to the graph of $f(x) = e^{x/2} \ln x$ at its inflection point.
    Hint: Show that $(1, 0)$ is the only inflection point of f.

63. Find the absolute extrema of the function $f(x) = x - \ln x$ on $[\tfrac{1}{2}, 3]$.

64. Find the absolute extrema of the function $g(x) = \dfrac{x}{\ln x}$ on $[2, 5]$.

65. **STRAIN ON VERTEBRAE** The strain (percentage of compression) on the lumbar vertebral disks in an adult human as a function of the load x (in kilograms) is given by

$$f(x) = 7.2956 \ln(0.0645012 x^{0.95} + 1)$$

What is the rate of change of the strain with respect to the load when the load is 100 kg? When the load is 500 kg?
Source: Benedek and Villars, *Physics with Illustrative Examples from Medicine and Biology.*

66. **HEIGHTS OF CHILDREN** For children between the ages of 5 and 13 years, the Ehrenberg equation

$$\ln W = \ln 2.4 + 1.84h$$

gives the relationship between the weight W (in kilograms) and the height h (in meters) of the child. Use differentials to estimate the change in the weight of a child who grows from 1 m to 1.1 m.

67. Yahoo! in Europe Yahoo! is putting more emphasis on Western Europe, where the number of online households is expected to grow steadily. In a study conducted in 2004, the number of online households (in millions) in Western Europe was projected to be

$$N(t) = 34.68 + 23.88 \ln(1.05t + 5.3) \quad (0 \le t \le 2)$$

where $t = 0$ corresponds to the beginning of 2004.
a. What was the projected number of online households in Western Europe at the beginning of 2005?
b. How fast was the projected number of online households in Western Europe increasing at the beginning of 2005?

Source: Jupiter Research.

68. Depreciation of Equipment For assets such as machines, whose market values drop rapidly in the early years of usage, businesses often use the double declining–balance method. In practice, a business firm normally employs the double declining–balance method for depreciating such assets for a certain number of years and then switches over to the linear method (see Exercise 43, page 88). The double declining–balance formula is

$$V(n) = C\left(1 - \frac{2}{N}\right)^n$$

where C is the initial value of the asset in dollars, $V(n)$ denotes the book value of the assets at the end of n years, and N is the number of years over which the asset is depreciated.
a. Find $V'(n)$.
   Hint: Use logarithmic differentiation.
b. What is the relative rate of change of $V(n)$?
   Hint: Find $[V'(n)]/[V(n)]$. See Section 3.5.

69. Depreciation of Equipment Refer to Exercise 68. A tractor purchased at a cost of $60,000 is to be depreciated by the double declining–balance method over 10 years.
a. What is the book value of the tractor at the end of 2 years?
b. What is the relative rate of change of the book value of the tractor at the end of 2 years?

70. Online Buyers The number of online buyers in Western Europe grew steadily over the past decade. The function

$$P(t) = 28.5 + 14.42 \ln t \quad (1 \le t \le 7)$$

gives the number of online buyers as a percentage of the total population, where t is measured in years, with $t = 1$ corresponding to 2001.
a. What was the percentage of online buyers in 2001 ($t = 1$)? How fast was it changing in 2001?
b. What was the percentage of online buyers in 2006 ($t = 6$)? How fast was it changing in 2006?

Source: Jupiter Research.

71. Average Life Span One reason for the increase in the life span over the years has been the advances in medical technology. The average life span for American women from 1907 through 2007 is given by

$$W(t) = 49.9 + 17.1 \ln t \quad (1 \le t \le 6)$$

where $W(t)$ is measured in years and t is measured in 20-year intervals, with $t = 1$ corresponding to the beginning of 1907.
a. Show that W is increasing on $(1, 6)$.
b. What can you say about the concavity of the graph of W on the interval $(1, 6)$?

72. Marginal Revenue The demand function for the Viking Boat's 34-ft *Sundancer* yacht is

$$p = 200 - 0.01x \ln x$$

where x denotes the number of yachts and p is the price per yacht in hundreds of dollars.
a. Find the revenue and the marginal revenue function for this model of yacht.
b. Use the result of part (a) to estimate the revenue to be realized from the sale of the 500th 34-ft *Sundancer* yacht.

73. Maximizing Profit The manager of Seko, an information technology (IT) consulting company, estimates that the annual profit of the company, in millions of dollars, is given by

$$P(x) = 2 \ln(2x + 1) + 2x - x^2 - 0.3$$

where x is the number of IT consultants (in hundreds) in its employ. Find the number of consultants the firm should hire so that its profit is maximized. What is the maximum profit?

74. Lambert's Law of Absorption Lambert's law of absorption states that the light intensity $I(x)$ (in calories per square centimeter per second) at a depth of x m as measured from the surface of a material is given by $I = I_0 a^x$, where I_0 and a are positive constants.
a. Find the rate of change of the light intensity with respect to x at a depth of x m from the surface of the material.
b. Using the result of part (a), conclude that the rate of change $I'(x)$ at a depth of x m is proportional to $I(x)$. What is the constant of proportion?

75. Population Growth The population of a town t months after the establishment of a biotech research center nearby is given by

$$P(t) = \frac{40 + 80e^{0.06t}}{20 + e^{0.06t}}$$

where $P(t)$ is measured in thousands. Find the relative rate of growth of the population 5 years after the establishment of the biotech research center.

76. Gompertz Function The Gompertz function defined by

$$P(t) = Le^{-\ln(L/P_0)e^{-ct}}$$

provides us with a model for population growth. Here, L is an upper bound for the population (called the carrying capacity for the environment), c is a positive constant, and P_0 is the population at $t = 0$. Show that the relative rate of change of P is

$$c \ln \frac{L}{P_0} e^{-ct}$$

77. ABSORPTION OF LIGHT When light passes through a window glass, some of it is absorbed. It can be shown that if $r\%$ of the light is absorbed by a glass of thickness w, then the percentage of light absorbed by a piece of glass of thickness nw is

$$A(n) = 100\left[1 - \left(1 - \frac{r}{100}\right)^n\right] \quad (0 \le r \le 100)$$

a. Show that A is an increasing function of n on $(0, \infty)$ if $0 < r < 100$.
Hint: Use logarithmic differentiation.
b. Sketch the graph of A for the special case in which $r = 10$.
c. Evaluate $\lim_{n \to \infty} A(n)$ and interpret your result.

78. MAGNITUDE OF EARTHQUAKES On the Richter scale, the magnitude R of an earthquake is given by the formula

$$R = \log \frac{I}{I_0}$$

where I is the intensity of the earthquake being measured and I_0 is the standard reference intensity.
a. What is the magnitude of an earthquake that has intensity 1 million times that of I_0?
b. Suppose an earthquake is measured with a magnitude of 6 on the Richter scale with an error of at most 2%. Use differentials to find the error in the intensity of the earthquake.
Hint: Observe that $I = I_0 10^R$, and use logarithmic differentiation.

79. WEBER–FECHNER LAW The Weber–Fechner Law

$$R = k \ln \frac{S}{S_0}$$

where k is a positive constant, describes the relationship between a stimulus S and the resulting response R. Here, S_0, a positive constant, is the threshold level.
a. Show that $R = 0$ if the stimulus is at the threshold level S_0.
b. The derivative dR/dS is the *sensitivity* corresponding to the stimulus level S and measures the capability to detect small changes in the stimulus level. Show that dR/dS is inversely proportional to S, and interpret your result.

80. PREDATOR–PREY MODEL The relationship between the number of rabbits $y(t)$ and the number of foxes $x(t)$ at any time t is given by

$$-C \ln y + Dy = A \ln x - Bx + E$$

where A, B, C, D, and E are constants. This relationship is based on a model by Lotka (1880–1949) and Volterra (1860–1940) for analyzing the ecological balance between two species of animals, one of which is a prey and the other a predator. Use implicit differentiation to find the relationship between the rate of change of the rabbit population in terms of the rate of change of the fox population.

81. RATE OF A CATALYTIC CHEMICAL REACTION A catalyst is a substance that either accelerates a chemical reaction or is necessary for the reaction to occur. Suppose an enzyme E (a catalyst) combines with a substrate S (a reacting chemical) to form an intermediate product X that then produces a product P and releases the enzyme. If initially there are x_0 moles/liter of S and there is no P, then based on the theory of Michaelis and Menten, the concentration of P, $p(t)$, after t hr is given by the equation

$$Vt = p - k \ln\left(1 - \frac{p}{x_0}\right)$$

where the constant V is the maximum possible speed of the reaction and the constant k is called the **Michaelis constant** for the reaction. Find the rate of change of the formation of the product P in this reaction.

In Exercises 82 and 83, use the guidelines on page 288 to sketch the graph of the given function.

82. $f(x) = \ln(x - 1)$

83. $f(x) = 2x - \ln x$

84. DERIVATIVE OF b^x
a. Let $f(x) = b^x$ ($b > 0$, $b \ne 1$). Show that $f'(x) = (\ln b)b^x$.
b. Use the result of part (a) to find the derivative of $f(x) = 3^x$.

85. DERIVATIVE OF $\log_b x$
a. Let $f(x) = \log_b x$ ($b > 0$, $b \ne 1$). Use the result of Exercise 84 to show that $f'(x) = \dfrac{1}{\ln b} \cdot \dfrac{1}{x}$.
b. Use the result of part (a) to find the derivative of $f(x) = \log_{10} x$.

In Exercises 86–89, use the results of Exercises 84 and 85 to find the derivative of the given function.

86. $f(x) = x^3 2^x$

87. $g(x) = \dfrac{10^x}{x + 1}$

88. $h(x) = x^2 \log_{10} x$

89. $f(x) = 3^{x^2} + \log_2(x^2 + 1)$

In Exercises 90 and 91, determine whether the statement is true or false. If it is true, explain why it is true. If it is false, give an example to show why it is false.

90. If $f(x) = \ln 5$, then $f'(x) = \frac{1}{5}$.

91. If $f(x) = \ln a^x$, then $f'(x) = \ln a$.

92. Prove that $\dfrac{d}{dx} \ln|x| = \dfrac{1}{x}$ ($x \ne 0$) for the case $x < 0$.

93. Use the definition of the derivative to show that

$$\lim_{x \to 0} \frac{\ln(x + 1)}{x} = 1$$

5.5 Solutions to Self-Check Exercises

1. The slope of the tangent line to the graph of f at any point $(x, f(x))$ lying on the graph of f is given by $f'(x)$. Using the Product Rule, we find

$$\begin{aligned} f'(x) &= \frac{d}{dx}[x \ln(2x+3)] \\ &= x \frac{d}{dx} \ln(2x+3) + \ln(2x+3) \cdot \frac{d}{dx}(x) \\ &= x\left(\frac{2}{2x+3}\right) + \ln(2x+3) \cdot 1 \\ &= \frac{2x}{2x+3} + \ln(2x+3) \end{aligned}$$

In particular, the slope of the tangent line to the graph of f at the point $(-1, 0)$ is

$$f'(-1) = \frac{-2}{-2+3} + \ln 1 = -2$$

Therefore, using the point-slope form of the equation of a line, we see that a required equation is

$$\begin{aligned} y - 0 &= -2(x+1) \\ y &= -2x - 2 \end{aligned}$$

2. Taking the logarithm on both sides of the equation gives

$$\begin{aligned} \ln y &= \ln[(2x+1)^3(3x+4)^5] \\ &= \ln(2x+1)^3 + \ln(3x+4)^5 \\ &= 3\ln(2x+1) + 5\ln(3x+4) \end{aligned}$$

Differentiating both sides of the equation with respect to x, keeping in mind that y is a function of x, we obtain

$$\begin{aligned} \frac{d}{dx}(\ln y) = \frac{y'}{y} &= 3 \cdot \frac{2}{2x+1} + 5 \cdot \frac{3}{3x+4} \\ &= 3\left(\frac{2}{2x+1} + \frac{5}{3x+4}\right) \end{aligned}$$

and

$$y' = 3(2x+1)^3(3x+4)^5\left(\frac{2}{2x+1} + \frac{5}{3x+4}\right)$$

5.6 Exponential Functions as Mathematical Models

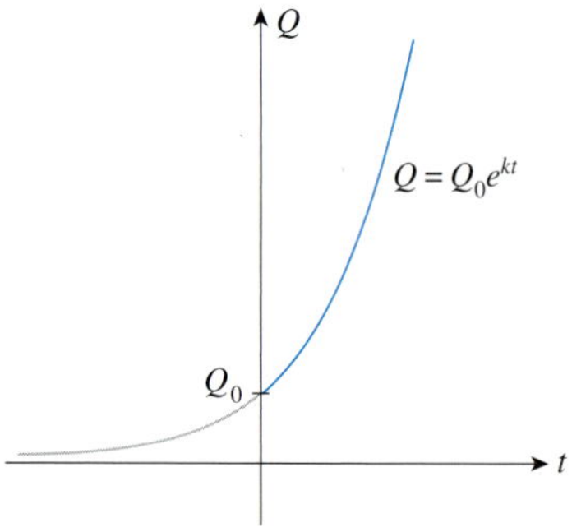

FIGURE 13
Exponential growth

Exponential Growth

Many problems arising from practical situations can be described mathematically in terms of exponential functions or functions closely related to the exponential function. In this section, we look at some applications involving exponential functions from the fields of the life and social sciences.

In Section 5.1, we saw that the exponential function $f(x) = b^x$ is an increasing function when $b > 1$. In particular, the function $f(x) = e^x$ has this property. Suppose that $Q(t)$ represents a quantity at time t, then one may deduce that the function $Q(t) = Q_0e^{kt}$, where Q_0 and k are positive constants, has the following properties:

1. $Q(0) = Q_0$
2. $Q(t)$ increases "rapidly" without bound as t increases without bound (Figure 13).

Property 1 follows from the computation

$$Q(0) = Q_0e^0 = Q_0$$

Next, to study the rate of change of the function $Q(t)$, we differentiate it with respect to t, obtaining

$$\begin{aligned} Q'(t) &= \frac{d}{dt}(Q_0e^{kt}) \\ &= Q_0\frac{d}{dt}(e^{kt}) \\ &= kQ_0e^{kt} \\ &= kQ(t) \end{aligned} \tag{12}$$

PORTFOLIO Carol A. Reeb, Ph.D.

TITLE   Research Associate
INSTITUTION   Hopkins Marine Station, Stanford University

Historically, the world's oceans were thought to provide an unlimited source of inexpensive seafood. However, in a world in which the human population now exceeds six billion people, overfishing has pushed one third of all marine fishery stocks toward a state of collapse.

As a fishery geneticist at Hopkins Marine Station, I study commercially harvested marine populations and use exponential models in my work. The equation for determining the size of a population that grows or declines exponentially is $x_t = x_0 e^{rt}$, where x_0 is the initial population, t is time, and r is the growth or decay constant (positive for growth, negative for decay).

This equation can be used to estimate the population in the past as well as in the future. We know that the demand for seafood increased as the human population grew, eventually causing fish populations to decline. Because genetic diversity is linked to population size, the exponential function is useful to model change in fishery populations and their gene pools over time.

Interestingly, exponential functions can also be used to model the increase in the market value of seafood in the United States over the past 60 years. In general, the price of seafood has increased exponentially, although the price did stabilize briefly in 1995.

Although exponential curves are important to my work, they are not always the best fit. Exponential curves are best applied across short time frames when environments or markets are unlimited. Over longer periods, the logistic growth function is more suitable. In my research, selecting the most accurate model requires examining many possibilities.

Michel Le Tallec; (inset) © Rich Carey/Shutterstock.com

Since $Q(t) > 0$ (because Q_0 is assumed to be positive) and $k > 0$, we see that $Q'(t) > 0$, so $Q(t)$ is an increasing function of t. Our computation has in fact shed more light on an important property of the function $Q(t)$. Equation (12) says that the rate of increase of the function $Q(t)$ is proportional to the amount $Q(t)$ of the quantity present at time t. The implication is that as $Q(t)$ increases, so does the *rate of increase* of $Q(t)$, resulting in a very rapid increase in $Q(t)$ as t increases without bound.

Thus, the exponential function

$$Q(t) = Q_0 e^{kt} \qquad (0 \le t < \infty) \tag{13}$$

provides us with a mathematical model of a quantity $Q(t)$ that is initially present in the amount of $Q(0) = Q_0$ and whose rate of growth at any time t is directly proportional to the amount of the quantity present at time t. Such a quantity is said to exhibit unrestricted **exponential growth**, and the constant k of proportionality is called the **growth constant**. Interest earned on a fixed deposit when compounded continuously exhibits exponential growth. Other examples of unrestricted exponential growth follow.

APPLIED EXAMPLE 1 Growth of Bacteria Under ideal laboratory conditions, the number of bacteria in a culture grows in accordance with the law $Q(t) = Q_0 e^{kt}$, where Q_0 denotes the number of bacteria initially present in the culture, k is a constant determined by the strain of bacteria under consideration and other factors, and t is the elapsed time measured in hours. Suppose 10,000 bacteria are present initially in the culture and 60,000 present 2 hours later.

a. How many bacteria will there be in the culture at the end of 4 hours?
b. What is the rate of growth of the population after 4 hours?

Solution

a. We are given that $Q(0) = Q_0 = 10{,}000$, so $Q(t) = 10{,}000e^{kt}$. Next, the fact that 60,000 bacteria are present 2 hours later translates into $Q(2) = 60{,}000$. Thus,

$$60{,}000 = 10{,}000e^{2k}$$
$$e^{2k} = 6$$

Taking the natural logarithm on both sides of the equation, we obtain

$$\ln e^{2k} = \ln 6$$
$$2k = \ln 6 \quad \text{Since } \ln e = 1$$
$$k = \frac{\ln 6}{2}$$
$$k \approx 0.8959$$

Thus, the number of bacteria present at any time t is given by

$$Q(t) \approx 10{,}000 e^{0.8959t}$$

In particular, the number of bacteria present in the culture at the end of 4 hours is given by

$$Q(4) \approx 10{,}000 e^{0.8959(4)}$$
$$\approx 360{,}000$$

b. The rate of growth of the bacteria population at any time t is given by

$$Q'(t) = kQ(t)$$

Thus, using the result from part (a), we find that the rate at which the population is growing at the end of 4 hours is

$$Q'(4) = kQ(4)$$
$$\approx (0.8959)(360{,}000)$$
$$\approx 322{,}500$$

or approximately 322,500 bacteria per hour.

Exponential Decay

In contrast to exponential growth, a quantity exhibits **exponential decay** if it decreases at a rate that is directly proportional to its size. Such a quantity may be described by the exponential function

$$Q(t) = Q_0 e^{-kt} \quad (0 \leq t < \infty) \tag{14}$$

where the positive constant Q_0 measures the amount present initially ($t = 0$) and k is some suitable positive number, called the **decay constant**. The choice of this number is determined by the nature of the substance under consideration and other factors. The graph of this function is sketched in Figure 14.

To verify the properties ascribed to the function $Q(t)$, we simply compute

$$Q(0) = Q_0 e^0 = Q_0$$
$$Q'(t) = \frac{d}{dt}(Q_0 e^{-kt})$$
$$= Q_0 \frac{d}{dt}(e^{-kt})$$
$$= -kQ_0 e^{-kt} = -kQ(t)$$

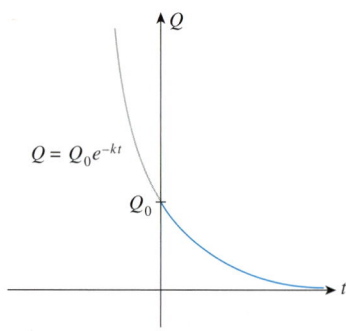

FIGURE 14
Exponential decay

APPLIED EXAMPLE 2 Radioactive Decay Radioactive substances decay exponentially. For example, the amount of radium present at any time t obeys the law $Q(t) = Q_0 e^{-kt}$, where Q_0 is the initial amount present and k is a specific positive constant. The **half-life of a radioactive substance** is the time required for a given amount to be reduced by one-half. Now, it is known that the half-life of radium is approximately 1600 years. Suppose initially there are 200 milligrams of pure radium. Find the amount left after t years. What is the amount left after 800 years?

Solution The initial amount of radium present is 200 milligrams, so $Q(0) = Q_0 = 200$. Thus, $Q(t) = 200 e^{-kt}$. Next, the datum concerning the half-life of radium implies that $Q(1600) = 100$, and this gives

$$100 = 200 e^{-1600k}$$

$$e^{-1600k} = \frac{1}{2}$$

Taking the natural logarithm on both sides of this equation yields

$$-1600k \ln e = \ln \frac{1}{2}$$

$$-1600k = \ln \frac{1}{2} \quad \text{\small ln } e = 1$$

$$k = -\frac{1}{1600} \ln \left(\frac{1}{2}\right) \approx 0.0004332$$

Therefore, the amount of radium left after t years is

$$Q(t) = 200 e^{-0.0004332 t}$$

In particular, the amount of radium left after 800 years is

$$Q(800) = 200 e^{-0.0004332(800)} \approx 141.42$$

or approximately 141 milligrams.

APPLIED EXAMPLE 3 Carbon-14 Decay Carbon 14, a radioactive isotope of carbon, has a half-life of 5730 years. What is its decay constant?

Solution We have $Q(t) = Q_0 e^{-kt}$. Since the half-life of the element is 5730 years, half of the substance is left at the end of that period; that is,

$$Q(5730) = Q_0 e^{-5730k} = \frac{1}{2} Q_0$$

$$e^{-5730k} = \frac{1}{2}$$

Taking the natural logarithm on both sides of this equation, we have

$$\ln e^{-5730k} = \ln \frac{1}{2}$$

$$-5730k = -0.693147$$

$$k \approx 0.000121$$

Carbon-14 dating is a well-known method used by anthropologists to establish the age of animal and plant fossils. This method assumes that the proportion of carbon 14 (C-14) present in the atmosphere has remained fairly constant over the past 50,000 years.

Professor Willard Libby, recipient of the Nobel Prize in chemistry in 1960, proposed this theory.

The amount of C-14 in the tissues of a living plant or animal is fairly constant. However, when an organism dies, it stops absorbing new quantities of C-14, and the amount of C-14 in the remains diminishes because of the natural decay of the radioactive substance. Therefore, the approximate age of a plant or animal fossil can be determined by measuring the amount of C-14 present in the remains.

APPLIED EXAMPLE 4 Carbon-14 Dating A skull from an archeological site has one tenth the amount of C-14 that it originally contained. Determine the approximate age of the skull.

Solution Here,

$$Q(t) = Q_0 e^{-kt}$$
$$= Q_0 e^{-0.000121t}$$

where Q_0 is the amount of C-14 present originally and k, the decay constant, is equal to 0.000121 (see Example 3). Since $Q(t) = (1/10)Q_0$, we have

$$\frac{1}{10} Q_0 = Q_0 e^{-0.000121t}$$

$$\ln \frac{1}{10} = -0.000121t \quad \text{Take the natural logarithm on both sides.}$$

$$t = \frac{\ln \frac{1}{10}}{-0.000121}$$

$$\approx 19{,}030$$

or approximately 19,030 years.

Learning Curves

The next example shows how the exponential function may be applied to describe certain types of learning processes. Consider the function

$$Q(t) = C - Ae^{-kt}$$

where C, A, and k are positive constants. To sketch the graph of the function Q, observe that its y-intercept is given by $Q(0) = C - A$. Next, we compute

$$Q'(t) = kAe^{-kt}$$

Since both k and A are positive, we see that $Q'(t) > 0$ for all values of t. Thus, $Q(t)$ is an increasing function of t. Also,

$$\lim_{t \to \infty} Q(t) = \lim_{t \to \infty} (C - Ae^{-kt})$$
$$= \lim_{t \to \infty} C - \lim_{t \to \infty} Ae^{-kt}$$
$$= C$$

so $y = C$ is a horizontal asymptote of Q. Thus, $Q(t)$ increases and approaches the number C as t increases without bound. The graph of the function Q is shown in Figure 15, where that part of the graph corresponding to the negative values of t is drawn with a gray line since, in practice, one normally restricts the domain of the function to the interval $[0, \infty)$.

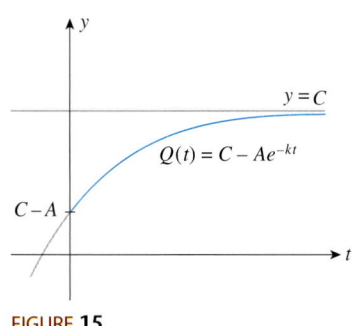

FIGURE 15
A learning curve

Observe that starting at $t = 0$, $Q(t)$ increases rather rapidly but then the rate of increase slows down considerably after a while. To see this, we compute

$$\lim_{t \to \infty} Q'(t) = \lim_{t \to \infty} kAe^{-kt} = 0$$

This behavior of the graph of the function Q closely resembles the learning pattern experienced by workers engaged in highly repetitive work. For example, the productivity of an assembly-line worker increases very rapidly in the early stages of the training period. This productivity increase is a direct result of the worker's training and accumulated experience. But the rate of increase of productivity slows as time goes by, and the worker's productivity level approaches some fixed level due to the limitations of the worker and the machine. Because of this characteristic, the graph of the function $Q(t) = C - Ae^{-kt}$ is often called a **learning curve**.

APPLIED EXAMPLE 5 Assembly Time The Camera Division of Eastman Optical produces a 35-mm single-lens reflex camera. Eastman's training department determines that after completing the basic training program, a new, previously inexperienced employee will be able to assemble

$$Q(t) = 50 - 30e^{-0.5t}$$

model F cameras per day t months after the employee starts work on the assembly line.

a. How many model F cameras can a new employee assemble per day after basic training?
b. How many model F cameras can an employee with 1 month of experience assemble per day? An employee with 2 months of experience? An employee with 6 months of experience?
c. How many model F cameras can the average experienced employee assemble per day?

Solution

a. The number of model F cameras a new employee can assemble is given by

$$Q(0) = 50 - 30 = 20$$

b. The number of model F cameras that an employee with 1 month of experience, 2 months of experience, and 6 months of experience can assemble per day is given by

$$Q(1) = 50 - 30e^{-0.5} \approx 31.80$$
$$Q(2) = 50 - 30e^{-1} \approx 38.96$$
$$Q(6) = 50 - 30e^{-3} \approx 48.51$$

or approximately 32, 39, and 49, respectively.

c. As t increases without bound, $Q(t)$ approaches 50. Hence, the average experienced employee can ultimately be expected to assemble 50 model F cameras per day.

Other applications of the learning curve are found in models that describe the dissemination of information about a product or the velocity of an object dropped into a viscous medium.

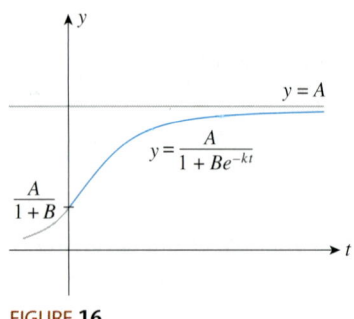

FIGURE 16
A logistic curve

Logistic Growth Functions

Our last example of an application of exponential functions to the description of natural phenomena involves the **logistic** (also called the **S-shaped**, or **sigmoidal**) **curve,** which is the graph of the function

$$Q(t) = \frac{A}{1 + Be^{-kt}}$$

where A, B, and k are positive constants. The function Q is called a **logistic growth function**. The graph of the function Q is sketched in Figure 16.

Observe that $Q(t)$ increases slowly at first but more rapidly as t increases. In fact, for small positive values of t, the logistic curve resembles an exponential growth curve. However, the *rate of growth* of $Q(t)$ decreases quite rapidly as t increases and $Q(t)$ approaches the number A as t increases without bound.

Thus, the logistic curve exhibits both the property of rapid growth of the exponential growth curve as well as the "saturation" property of the learning curve. Because of these characteristics, the logistic curve serves as a suitable mathematical model for describing many natural phenomena. For example, if a small number of rabbits were introduced to a tiny island in the South Pacific, the rabbit population might be expected to grow very rapidly at first, but the growth rate would decrease quickly as overcrowding, scarcity of food, and other environmental factors affected it. The population would eventually stabilize at a level compatible with the life-support capacity of the environment. This level, given by A, is called the *carrying capacity* of the environment. Models describing the spread of rumors and epidemics are other examples of the application of the logistic curve.

APPLIED EXAMPLE 6 Spread of Flu The number of soldiers at Fort MacArthur who contracted influenza after t days during a flu epidemic is approximated by the exponential model

$$Q(t) = \frac{5000}{1 + 1249e^{-kt}}$$

If 40 soldiers contracted the flu by day 7, find how many soldiers contracted the flu by day 15.

Solution The given information implies that

$$Q(7) = \frac{5000}{1 + 1249e^{-7k}} = 40$$

Thus,

$$40(1 + 1249e^{-7k}) = 5000$$
$$1 + 1249e^{-7k} = \frac{5000}{40} = 125$$
$$e^{-7k} = \frac{124}{1249}$$
$$-7k = \ln\frac{124}{1249}$$
$$k = -\frac{\ln\frac{124}{1249}}{7} \approx 0.33$$

Therefore, the number of soldiers who contracted the flu after t days is given by

$$Q(t) = \frac{5000}{1 + 1249e^{-0.33t}}$$

In particular, the number of soldiers who contracted the flu by day 15 is given by

$$Q(15) = \frac{5000}{1 + 1249e^{-15(0.33)}}$$
$$\approx 508$$

or approximately 508 soldiers.

Exploring with TECHNOLOGY

Refer to Example 6.

1. Use a graphing utility to plot the graph of the function Q, using the viewing window $[0, 40] \times [0, 5000]$.
2. Find how long it takes for the first 1000 soldiers to contract the flu.
   Hint: Plot the graphs of $y_1 = Q(t)$ and $y_2 = 1000$, and find the point of intersection of the two graphs.

5.6 Self-Check Exercise

Suppose the population (in millions) of a country at any time t grows in accordance with the rule

$$P = \left(P_0 + \frac{I}{k}\right)e^{kt} - \frac{I}{k}$$

where P denotes the population at any time t, k is a constant reflecting the natural growth rate of the population, I is a constant giving the (constant) rate of immigration into the country, and P_0 is the total population of the country at time $t = 0$. The population of the United States in 1980 ($t = 0$) was 226.5 million. If the natural growth rate is 0.8% annually ($k = 0.008$) and net immigration is allowed at the rate of half a million people per year ($I = 0.5$), what is the projected population of the United States in 2015?

The solution to Self-Check Exercise 5.6 can be found on page 391.

5.6 Concept Questions

1. Give the model for unrestricted exponential growth and the model for exponential decay. What effect does the magnitude of the growth (decay) constant have on the growth (decay) of a quantity?

2. What is the half-life of a radioactive substance?

3. What is the logistic growth function? What are its characteristics?

5.6 Exercises

1. **EXPONENTIAL GROWTH** Given that a quantity $Q(t)$ is described by the exponential growth function

$$Q(t) = 300e^{0.02t}$$

where t is measured in minutes, answer the following questions:

a. What is the growth constant?
b. What quantity is present initially?
c. Complete the following table of values:

t	0	10	20	100	1000
Q					

2. **EXPONENTIAL DECAY** Given that a quantity $Q(t)$ exhibiting exponential decay is described by the function

$$Q(t) = 2000e^{-0.06t}$$

where t is measured in years, answer the following questions:
   a. What is the decay constant?
   b. What quantity is present initially?
   c. Complete the following table of values:

t	0	5	10	20	100
Q					

3. **GROWTH OF BACTERIA** The growth rate of the bacterium *Escherichia coli*, a common bacterium found in the human intestine, is proportional to its size. Under ideal laboratory conditions, when this bacterium is grown in a nutrient broth medium, the number of cells in a culture doubles approximately every 20 min.
   a. If the initial cell population is 100, determine the function $Q(t)$ that expresses the exponential growth of the number of cells of this bacterium as a function of time t (in minutes).
   b. How long will it take for a colony of 100 cells to increase to a population of 1 million?
   c. If the initial cell population were 1000, how would this alter our model?

4. **WORLD POPULATION** The world population at the beginning of 1990 was 5.3 billion. Assume that the population continues to grow at the rate of approximately 2%/year and find the function $Q(t)$ that expresses the world population (in billions) as a function of time t (in years), with $t = 0$ corresponding to the beginning of 1990.
   a. Using this function, complete the following table of values and sketch the graph of the function Q.

Year	1990	1995	2000	2005
World Population				

Year	2010	2015	2020	2025
World Population				

   b. Find the estimated rate of growth in 2010.

5. **WORLD POPULATION** Refer to Exercise 4.
   a. If the world population continues to grow at the rate of approximately 2%/year, find the length of time t_0 required for the world population to triple in size.
   b. Using the time t_0 found in part (a), what would be the world population if the growth rate were reduced to 1.8%/year?

6. **RESALE VALUE** Garland Mills purchased a certain piece of machinery 3 years ago for $500,000. Its present resale value is $320,000. Assuming that the machine's resale value decreases exponentially, what will it be 4 years from now?

7. **ATMOSPHERIC PRESSURE** If the temperature is constant, then the atmospheric pressure P (in pounds per square inch) varies with the altitude above sea level h in accordance with the law

$$P = p_0 e^{-kh}$$

where p_0 is the atmospheric pressure at sea level and k is a constant. If the atmospheric pressure is 15 lb/in.^{2} at sea level and 12.5 lb/in.^{2} at 4000 ft, find the atmospheric pressure at an altitude of 12,000 ft. How fast is the atmospheric pressure changing with respect to altitude at an altitude of 12,000 ft?

8. **RADIOACTIVE DECAY** The radioactive element polonium decays according to the law

$$Q(t) = Q_0 \cdot 2^{-(t/140)}$$

where Q_0 is the initial amount and the time t is measured in days. If the amount of polonium left after 280 days is 20 mg, what was the initial amount present?

9. **RADIOACTIVE DECAY** Phosphorus 32 (P-32) has a half-life of 14.2 days. If 100 g of this substance are present initially, find the amount present after t days. What amount will be left after 7.1 days? How fast is P-32 decaying when $t = 7.1$?

10. **NUCLEAR FALLOUT** Strontium 90 (Sr-90), a radioactive isotope of strontium, is present in the fallout resulting from nuclear explosions. It is especially hazardous to animal life, including humans, because, upon ingestion of contaminated food, it is absorbed into the bone structure. Its half-life is 27 years. If the amount of Sr-90 in a certain area is found to be four times the "safe" level, find how much time must elapse before the safe level is reached.

11. **CARBON-14 DATING** Wood deposits recovered from an archeological site contain 20% of the C-14 they originally contained. How long ago did the tree from which the wood was obtained die?

12. **CARBON-14 DATING** The skeletal remains of the so-called Pittsburgh Man, unearthed in Pennsylvania, had lost 82% of the C-14 they originally contained. Determine the approximate age of the bones.

13. **LEARNING CURVES** The American Court Reporting Institute finds that the average student taking Advanced Machine Shorthand, an intensive 20-week course, progresses according to the function

$$Q(t) = 120(1 - e^{-0.05t}) + 60 \quad (0 \le t \le 20)$$

where $Q(t)$ measures the number of words (per minute) of dictation that the student can take in machine shorthand after t weeks in the course. Sketch the graph of the function Q and answer the following questions:
   a. What is the beginning shorthand speed for the average student in this course?
   b. What shorthand speed does the average student attain halfway through the course?
   c. How many words per minute can the average student take after completing this course?

14. PEOPLE LIVING WITH HIV On the basis of data compiled by WHO, the number of people living with HIV (human immunodeficiency virus) worldwide from 1985 through 2006 is approximated by

$$N(t) = \frac{39.88}{1 + 18.94e^{-0.2957t}} \quad (0 \leq t \leq 21)$$

where $N(t)$ is measured in millions and t in years, with $t = 0$ corresponding to the beginning of 1985.
a. How many people were living with HIV worldwide at the beginning of 1985? At the beginning of 2005?
b. Assuming that the trend continued, how many people were living with HIV worldwide at the beginning of 2008?

Source: World Health Organization.

15. FEDERAL DEBT According to data obtained from the CBO, the total federal debt (in trillions of dollars) from 2001 through 2006 is given by

$$f(t) = 5.37e^{0.078t} \quad (1 \leq t \leq 6)$$

where t is measured in years, with $t = 1$ corresponding to 2001.
a. What was the total federal debt in 2001? In 2006?
b. How fast was the total federal debt increasing in 2001? In 2006?

Source: Congressional Budget Office.

16. EFFECT OF ADVERTISING ON SALES Metro Department Store found that t weeks after the end of a sales promotion the volume of sales was given by

$$S(t) = B + Ae^{-kt} \quad (0 \leq t \leq 4)$$

where $B = 50{,}000$ and is equal to the average weekly volume of sales before the promotion. The sales volumes at the end of the first and third weeks were $83,515 and $65,055, respectively. Assume that the sales volume is decreasing exponentially.
a. Find the decay constant k.
b. Find the sales volume at the end of the fourth week.
c. How fast is the sales volume dropping at the end of the fourth week?

17. DEMAND FOR COMPUTERS Universal Instruments found that the monthly demand for its new line of Galaxy Home Computers t months after placing the line on the market was given by

$$D(t) = 2000 - 1500e^{-0.05t} \quad (t > 0)$$

Graph this function and answer the following questions:
a. What is the demand after 1 month? After 1 year? After 2 years? After 5 years?
b. At what level is the demand expected to stabilize?
c. Find the rate of growth of the demand after the tenth month.

18. RELIABILITY OF COMPUTER CHIPS The percentage of a certain brand of computer chips that will fail after t years of use is estimated to be

$$P(t) = 100(1 - e^{-0.1t})$$

a. What percentage of this brand of computer chips are expected to be usable after 3 years?
b. Evaluate $\lim_{t \to \infty} P(t)$. Did you expect this result?

19. LENGTHS OF FISH The length (in centimeters) of a typical Pacific halibut t years old is approximately

$$f(t) = 200(1 - 0.956e^{-0.18t})$$

a. What is the length of a typical 5-year-old Pacific halibut?
b. How fast is the length of a typical 5-year-old Pacific halibut increasing?
c. What is the maximum length a typical Pacific halibut can attain?

20. SPREAD OF AN EPIDEMIC During a flu epidemic, the number of children in the Woodbridge Community School System who contracted influenza after t days was given by

$$Q(t) = \frac{1000}{1 + 199e^{-0.8t}}$$

a. How many children were stricken by the flu after the first day?
b. How many children had the flu after 10 days?
c. How many children eventually contracted the disease?

21. LAY TEACHERS AT ROMAN CATHOLIC SCHOOLS The change from religious to lay teachers at Roman Catholic schools has been attributed partly to the decline in the number of women and men entering religious orders. The percentage of teachers who are lay teachers is given by

$$f(t) = \frac{98}{1 + 2.77e^{-t}} \quad (0 \leq t \leq 4)$$

where t is measured in decades, with $t = 0$ corresponding to the beginning of 1960.
a. What percentage of teachers were lay teachers at the beginning of 1990?
b. How fast was the percentage of lay teachers changing at the beginning of 1990?
c. Find the year when the percentage of lay teachers was increasing most rapidly.

Sources: National Catholic Education Association and the U.S. Department of Education.

22. GROWTH OF A FRUIT FLY POPULATION On the basis of data collected during an experiment, a biologist found that the growth of a fruit fly (*Drosophila*) with a limited food supply could be approximated by the exponential model

$$N(t) = \frac{400}{1 + 39e^{-0.16t}}$$

where t denotes the number of days since the beginning of the experiment.
a. What was the initial fruit fly population in the experiment?
b. What was the maximum fruit fly population that could be expected under this laboratory condition?
c. What was the population of the fruit fly colony on the 20th day?
d. How fast was the population changing on the 20th day?

23. **DEMOGRAPHICS** The number of citizens aged 45–64 years is approximated by

$$P(t) = \frac{197.9}{1 + 3.274e^{-0.0361t}} \quad (0 \le t \le 20)$$

where $P(t)$ is measured in millions and t is measured in years, with $t = 0$ corresponding to the beginning of 1990. People belonging to this age group are the targets of insurance companies that want to sell them annuities. What is the expected population of citizens aged 45–64 years in 2010?

Source: K. G. Securities.

24. **POPULATION GROWTH IN THE TWENTY-FIRST CENTURY** The U.S. population is approximated by the function

$$P(t) = \frac{616.5}{1 + 4.02e^{-0.5t}}$$

where $P(t)$ is measured in millions of people and t is measured in 30-year intervals, with $t = 0$ corresponding to 1930. What is the expected population of the United States in 2020 ($t = 3$)?

25. **DISSEMINATION OF INFORMATION** Three hundred students attended the dedication ceremony of a new building on a college campus. The president of the traditionally female college announced a new expansion program, which included plans to make the college coeducational. The number of students who learned of the new program t hr later is given by the function

$$f(t) = \frac{3000}{1 + Be^{-kt}}$$

If 600 students on campus had heard about the new program 2 hr after the ceremony, how many students had heard about the policy after 4 hr? How fast was the news spreading 4 hr after the ceremony?

26. **PRICE OF A COMMODITY** The unit price of a certain commodity is given by

$$p = f(t) = 6 + 4e^{-2t}$$

where p is measured in dollars and t is measured in months.
a. Show that f is decreasing on $(0, \infty)$.
b. Show that the graph of f is concave upward on $(0, \infty)$.
c. Evaluate $\lim_{t \to \infty} f(t)$. (*Note:* This value is called the *equilibrium price* of the commodity, and in this case, we have *price stability*.)
d. Sketch the graph of f.

27. **CHEMICAL MIXTURES** Two chemicals react to form a compound. Suppose the amount of the compound formed in time t (in hours) is given by

$$x(t) = \frac{15[1 - (\frac{2}{3})^{3t}]}{1 - \frac{1}{4}(\frac{2}{3})^{3t}}$$

where $x(t)$ is measured in pounds. How many pounds of the compound are formed eventually?
Hint: You need to evaluate $\lim_{t \to \infty} x(t)$.

28. **VON BERTALANFFY GROWTH FUNCTION** The length (in centimeters) of a common commercial fish is approximated by the von Bertalanffy growth function

$$f(t) = a(1 - be^{-kt})$$

where a, b, and k are positive constants.
a. Show that f is increasing on the interval $(0, \infty)$.
b. Show that the graph of f is concave downward on $(0, \infty)$.
c. Show that $\lim_{t \to \infty} f(t) = a$.
d. Use the results of parts (a)–(c) to sketch the graph of f.

29. **ABSORPTION OF DRUGS** The concentration of a drug in grams per cubic centimeter (g/cm³) t min after it has been injected into the bloodstream is given by

$$C(t) = \frac{k}{b - a}(e^{-at} - e^{-bt})$$

where a, b, and k are positive constants, with $b > a$.
a. At what time is the concentration of the drug the greatest?
b. What will be the concentration of the drug in the long run?

30. **CONCENTRATION OF GLUCOSE IN THE BLOODSTREAM** A glucose solution is administered intravenously into the bloodstream at a constant rate of r mg/hr. As the glucose is being administered, it is converted into other substances and removed from the bloodstream. Suppose the concentration of the glucose solution at time t is given by

$$C(t) = \frac{r}{k} - \left[\left(\frac{r}{k}\right) - C_0\right]e^{-kt}$$

where C_0 is the concentration at time $t = 0$ and k is a positive constant. Assuming that $C_0 < r/k$, evaluate $\lim_{t \to \infty} C(t)$.
a. What does your result say about the concentration of the glucose solution in the long run?
b. Show that the function C is increasing on $(0, \infty)$.
c. Show that the graph of C is concave downward on $(0, \infty)$.
d. Sketch the graph of the function C.

31. **RADIOACTIVE DECAY** A radioactive substance decays according to the formula

$$Q(t) = Q_0 e^{-kt}$$

where $Q(t)$ denotes the amount of the substance present at time t (measured in years), Q_0 denotes the amount of the substance present initially, and k (a positive constant) is the decay constant.
a. Show that half-life of the substance is $\bar{t} = (\ln 2)/k$.
b. Suppose a radioactive substance decays according to the formula

$$Q(t) = 20e^{-0.0001238t}$$

How long will it take for the substance to decay to half the original amount?

32. LOGISTIC GROWTH FUNCTION Consider the logistic growth function

$$Q(t) = \frac{A}{1 + Be^{-kt}}$$

where A, B, and k are positive constants.
  a. Show that Q satisfies the equation

$$Q'(t) = kQ\left(1 - \frac{Q}{A}\right)$$

  b. Show that $Q(t)$ is increasing on $(0, \infty)$.

33. a. Use the results of Exercise 32 to show that the graph of Q has an inflection point when $Q = A/2$ and that this occurs when $t = (\ln B)/k$.
  b. Interpret your results.

34. LOGISTIC GROWTH FUNCTION Consider the logistic growth function

$$Q(t) = \frac{A}{1 + Be^{-kt}}$$

Suppose the population is Q_1 when $t = t_1$ and Q_2 when $t = t_2$. Show that the value of k is

$$k = \frac{1}{t_2 - t_1} \ln\left[\frac{Q_2(A - Q_1)}{Q_1(A - Q_2)}\right]$$

35. LOGISTIC GROWTH FUNCTION The carrying capacity of a colony of fruit flies (*Drosophila*) is 600. The population of fruit flies after 14 days is 76, and the population after 21 days is 167. What is the value of the growth constant k?
Hint: Use the result of Exercise 34.

36. GOMPERTZ GROWTH CURVE Consider the function

$$Q(t) = Ce^{-Ae^{-kt}}$$

where $Q(t)$ is the size of a quantity at time t, and A, C, and k are positive constants. The graph of this function, called the *Gompertz growth curve*, is used by biologists to describe restricted population growth.
  a. Show that the function Q is always increasing.
  b. Find the time t at which the growth rate $Q'(t)$ is increasing most rapidly.
  Hint: Find the inflection point of Q.
  c. Show that $\lim_{t \to \infty} Q(t) = C$, and interpret your result.

5.6 Solution to Self-Check Exercise

We are given that $P_0 = 226.5$, $k = 0.008$, and $I = 0.5$. So

$$P = \left(226.5 + \frac{0.5}{0.008}\right)e^{0.008t} - \frac{0.5}{0.008}$$

$$= 289e^{0.008t} - 62.5$$

Therefore, the expected population in 2015 is given by

$$P(35) = 289e^{0.28} - 62.5$$

$$\approx 319.9$$

or approximately 319.9 million.

USING TECHNOLOGY

Analyzing Mathematical Models

We can use a graphing utility to analyze the mathematical models encountered in this section.

APPLIED EXAMPLE 1 Internet-Gaming Sales The estimated growth in global Internet-gaming revenue (in billions of dollars), as predicted by industry analysts, is given in the following table:

Year	2001	2002	2003	2004	2005	2006	2007	2008	2009	2010
Revenue	3.1	3.9	5.6	8.0	11.8	15.2	18.2	20.4	22.7	24.5

  a. Use **Logistic** to find a regression model for the data. Let $t = 0$ correspond to 2001.
  b. Plot the scatter diagram and the graph of the function f found in part (a) using the viewing window $[0, 9] \times [0, 35]$.
  c. How fast was the revenue from global gaming on the Internet changing in 2001? In 2007?

Source: Christiansen Capital/Advisors.

(continued)

Solution

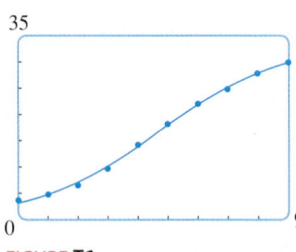

FIGURE T1
The graph of f in the viewing window [0, 9] × [0, 35]

a. Using **Logistic** we find

$$f(t) = \frac{27.11}{1 + 9.64e^{-0.49t}} \quad (0 \le t \le 9)$$

b. The scatter plot for the data, and the graph of f in the viewing window $[0, 9] \times [0, 35]$, are shown in Figure T1.

c. Using the numerical derivative operation, we find $f'(0) \approx 1.13$ and $f'(6) \approx 2.971$. We conclude that the revenue was increasing at the rate of $1.13 billion/year in 2001, and at the rate of $2.97 billion/year in 2007.

TECHNOLOGY EXERCISES

1. ONLINE BANKING In a study prepared in 2000, the percentage of households using online banking was projected to be

$$f(t) = 1.5e^{0.78t} \quad (0 \le t \le 4)$$

where t is measured in years, with $t = 0$ corresponding to the beginning of 2000.
a. Plot the graph of f, using the viewing window $[0, 4] \times [0, 40]$.
b. How fast was the projected percentage of households using online banking changing at the beginning of 2003?
c. How fast was the rate of increase of the projected percentage of households using online banking changing at the beginning of 2003?
Hint: We want $f''(3)$. Why?
Source: Online Banking Report.

2. NEWTON'S LAW OF COOLING The temperature of a cup of coffee t min after it is poured is given by

$$T = 70 + 100e^{-0.0446t}$$

where T is measured in degrees Fahrenheit.
a. Plot the graph of T, using the viewing window $[0, 30] \times [0, 200]$.
b. When will the coffee be cool enough to drink (say, 120°)?
Hint: Use the ISECT function.

3. AIR TRAVEL Air travel has been rising dramatically in the past 30 years. In a study conducted in 2000, the FAA projected further exponential growth for air travel through 2010. The function

$$f(t) = 666e^{0.0413t} \quad (0 \le t \le 10)$$

gives the number of passengers (in millions) in year t, with $t = 0$ corresponding to 2000.
a. Plot the graph of f, using the viewing window $[0, 10] \times [0, 1000]$.
b. How many air passengers were there in 2000? What was the projected number of air passengers for 2008?

c. What was the rate of change of the number of air passengers in 2008?
Source: Federal Aviation Administration.

4. COMPUTER GAME SALES The total number of Starr Communication's newest game, Laser Beams, sold t months after its release is given by

$$N(t) = -20(t + 20)e^{-0.05t} + 400$$

thousand units.
a. Plot the graph of N, using the viewing window $[0, 500] \times [0, 500]$.
b. Use the result of part (a) to find $\lim_{t \to \infty} N(t)$, and interpret this result.

5. POPULATION GROWTH IN THE TWENTY-FIRST CENTURY The U.S. population is approximated by the function

$$P(t) = \frac{616.5}{1 + 4.02e^{-0.5t}}$$

where $P(t)$ is measured in millions of people and t is measured in 30-year intervals, with $t = 0$ corresponding to 1930.
a. Plot the graph of P, using the viewing window $[0, 4] \times [0, 650]$.
b. What is the projected population of the United States in 2020 ($t = 3$)?
c. What is the projected rate of growth of the U.S. population in 2020?

6. TIME RATE OF GROWTH OF A TUMOR The rate at which a tumor grows, with respect to time, is given by

$$R = Ax \ln \frac{B}{x} \quad (\text{for } 0 < x < B)$$

where A and B are positive constants and x is the radius of the tumor.
a. Plot the graph of R for the case $A = B = 10$.
b. Find the radius of the tumor when the tumor is growing most rapidly with respect to time for the case $A = B = 10$.

7. **ABSORPTION OF DRUGS** The concentration of a drug in an organ at any time t (in seconds) is given by

$$C(t) = \begin{cases} 0.3t - 18(1 - e^{-t/60}) & \text{if } 0 \leq t \leq 20 \\ 18e^{-t/60} - 12e^{-(t-20)/60} & \text{if } t > 20 \end{cases}$$

where $C(t)$ is measured in grams per cubic centimeter (g/cm^3).
a. Plot the graph of C, using the viewing window $[0, 120] \times [0, 1]$.
b. What is the initial concentration of the drug in the organ?
c. What is the concentration of the drug in the organ after 10 sec?
d. What is the concentration of the drug in the organ after 30 sec?
e. What will be the concentration of the drug in the long run?

8. **ANNUITIES** At the time of retirement, Christine expects to have a sum of $500,000 in her retirement account. Assuming that the account pays interest at the rate of 5%/year compounded continuously, her accountant pointed out to her that if she made withdrawals amounting to x dollars per year ($x > 25{,}000$), then the time required to deplete her savings would be T years, where

$$T = f(x) = 20 \ln\left(\frac{x}{x - 25{,}000}\right) \quad (x > 25{,}000)$$

a. Plot the graph of f, using the viewing window $[25{,}000, 50{,}000] \times [0, 100]$.
b. How much should Christine plan to withdraw from her retirement account each year if she wants it to last for 25 years?
c. Evaluate $\lim_{x \to 25{,}000^+} f(x)$. Is the result expected? Explain.
d. Evaluate $\lim_{x \to \infty} f(x)$. Is the result expected? Explain.

9. **HOUSEHOLDS WITH MICROWAVES** The number of households with microwave ovens increased greatly in the 1980s and 1990s. The percentage of households with microwave ovens from 1981 through 1999 is given by

$$f(t) = \frac{87}{1 + 4.209e^{-0.3727t}} \quad (0 \leq t \leq 18)$$

where t is measured in years, with $t = 0$ corresponding to the beginning of 1981.
a. Use a graphing utility to plot the graph of f on the interval $[0, 18]$.
b. What percentage of households owned microwave ovens at the beginning of 1984? At the beginning of 1994?
c. At what rate was the ownership of microwave ovens increasing at the beginning of 1984? At the beginning of 1994?
d. At what time was the increase in ownership of microwave ovens greatest?

Source: Energy Information Agency.

10. **MODELING WITH DATA** The snowfall accumulation at Logan Airport (in inches), t hr after the beginning of a 33-hr snowstorm in Boston on a certain day, follows:

Hour	0	3	6	9	12	15	18	21	24	27	30	33
Inches	0.1	0.4	3.6	6.5	9.1	14.4	19.5	22	23.6	24.8	26.6	27

Here, $t = 0$ corresponds to noon of February 6.
a. Use **Logistic** to find a regression model for the data.
b. Plot the scatter diagram and the graph of the function f found in part (a), using the viewing window $[0, 33] \times [0, 30]$.
c. How fast was the snowfall accumulating at midnight on February 6? At noon on February 7?
d. At what time during the storm was the snowfall accumulating at the greatest rate? What was the rate of accumulation?

Source: Boston Globe.

11. **WORLDWIDE PC SHIPMENTS** Estimated worldwide PC shipments (in millions of units) from 2005 through 2009 are given in the following table:

Year	2005	2006	2007	2008	2009
PCs	207.1	226.2	252.9	283.3	302.4

a. Use **Logistic** to find a regression model for the data. Let $t = 0$ correspond to 2005.
b. Plot the graph of the function f found in part (a), using the viewing window $[0, 4] \times [200, 300]$.
c. How fast were the worldwide PC shipments increasing in 2006? In 2008?

Source: International Data Corporation.

12. **FEDERAL DEBT** According to data obtained from the CBO, the total federal debt (in trillions of dollars) from 2001 through 2006 is given in the following table:

Year	2001	2002	2003	2004	2005	2006
Debt	5.81	6.23	6.78	7.40	7.93	8.51

a. Use **ExpReg** to find a regression model for the data. Let $t = 1$ correspond to 2001.
b. Plot the graph of the function f found in part (a), using the viewing window $[1, 6] \times [4, 10]$.

Source: Congressional Budget Office.

CHAPTER 5 Summary of Principal Formulas and Terms

FORMULAS

1.	Exponential function with base b	$y = b^x$
2.	The number e	$e = \lim_{m \to \infty} \left(1 + \dfrac{1}{m}\right)^m = 2.71828\ldots$
3.	Exponential function with base e	$y = e^x$
4.	Logarithmic function with base b	$y = \log_b x$
5.	Logarithmic function with base e	$y = \ln x$
6.	Inverse properties of $\ln x$ and e^x	$\ln e^x = x$ and $e^{\ln x} = x$
7.	Compound interest (accumulated amount)	$A = P\left(1 + \dfrac{r}{m}\right)^{mt}$
8.	Effective rate of interest	$r_{\text{eff}} = \left(1 + \dfrac{r}{m}\right)^m - 1$
9.	Compound interest (present value)	$P = A\left(1 + \dfrac{r}{m}\right)^{-mt}$
10.	Continuous compound interest	$A = Pe^{rt}$
11.	Derivative of the exponential function	$\dfrac{d}{dx}(e^x) = e^x$
12.	Chain Rule for Exponential Functions	$\dfrac{d}{dx}(e^u) = e^u \dfrac{du}{dx}$
13.	Derivative of the logarithmic function	$\dfrac{d}{dx}\ln|x| = \dfrac{1}{x}$
14.	Chain Rule for Logarithmic Functions	$\dfrac{d}{dx}\ln|u| = \dfrac{1}{u}\dfrac{du}{dx}$

TERMS

common logarithm (338)
natural logarithm (338)
compound interest (346)
logarithmic differentiation (374)

exponential growth (381)
growth constant (381)
exponential decay (382)
decay constant (382)

half-life of a radioactive substance (383)
logistic growth function (386)

CHAPTER 5 Concept Review Questions

Fill in the blanks.

1. The function $f(x) = x^b$ (b, a real number) is called a/an _____ function, whereas the function $g(x) = b^x$, where $b > $ _____ and $b \neq $ _____, is called a/an _____ function.

2. **a.** The domain of the function $y = 3^x$ is _____, and its range is _____.
   b. The graph of the function $y = 0.3^x$ passes through the point _____ and is decreasing on _____.

3. **a.** If $b > 0$ and $b \neq 1$, then the logarithmic function $y = \log_b x$ has domain _____ and range _____; its graph passes through the point _____.
   b. The graph of $y = \log_b x$ is decreasing if b _____ and increasing if b _____.

4. **a.** If $x > 0$, then $e^{\ln x} = $ _____.
   b. If x is any real number, then $\ln e^x = $ _____.

5. In the compound interest formula $A = P\left(1 + \frac{r}{m}\right)^{mt}$, A stands for the _____ _____, P stands for the _____, r stands for the _____ _____ _____ per year, m stands for the _____ _____ _____ per year, and t stands for the _____.

6. The effective rate r_{eff} is related to the nominal interest rate r per year and the number of conversion periods per year by $r_{\text{eff}} = $ _____.

7. If interest earned at the rate of r per year is compounded continuously over t years, then a principal of P dollars will have an accumulated value of $A = $ _____ dollars.

8. **a.** If $g(x) = e^{f(x)}$, where f is a differentiable function, then $g'(x) = $ _____.
   b. If $g(x) = \ln f(x)$, where $f(x) > 0$ is differentiable, then $g'(x) = $ _____.

9. a. In the unrestricted exponential growth model $Q = Q_0 e^{kt}$, Q_0 represents the quantity present _____, and k is called the _____ constant.
  b. In the exponential decay model $Q = Q_0 e^{-kt}$, k is called the _____ constant.
  c. The half-life of a radioactive substance is the _____ required for a substance to decay to _____ of its original amount.

10. a. For the model $Q(t) = C - Ae^{-kt}$ describing a learning curve, $y = C$ is a/an _____ _____ of the graph of Q. The value of $Q(t)$ never exceeds _____.
  b. For the logistic growth model $Q(t) = \dfrac{A}{1 + Be^{-kt}}$, $y = A$ is a/an _____ _____ of the graph of Q. If the quantity $Q(t)$ is initially smaller than A, then $Q(t)$ will eventually approach _____ as t increases; the number A represents the life-support capacity of the environment and is called the _____ _____ of the environment.

CHAPTER 5 Review Exercises

1. Sketch on the same set of coordinate axes the graphs of the exponential functions defined by the equations.
  a. $y = 2^{-x}$
  b. $y = \left(\dfrac{1}{2}\right)^x$

In Exercises 2 and 3, express each equation in logarithmic form.

2. $\left(\dfrac{2}{3}\right)^{-3} = \dfrac{27}{8}$
3. $16^{-3/4} = 0.125$

In Exercises 4 and 5, solve each equation for x.

4. $\log_4(2x + 1) = 2$

5. $\ln(x - 1) + \ln 4 = \ln(2x + 4) - \ln 2$

In Exercises 6–8, given that $\ln 2 = x$, $\ln 3 = y$, and $\ln 5 = z$, express each of the given logarithmic values in terms of x, y, and z.

6. $\ln 30$ **7.** $\ln 3.6$ **8.** $\ln 75$

9. Sketch the graph of the function $y = \log_2(x + 3)$.

10. Sketch the graph of the function $y = \log_3(x + 1)$.

11. A sum of $10,000 is deposited in a bank. Find the amount on deposit after 2 years if the bank pays interest at the rate of 6%/year compounded (a) daily (assume a 365-day year) and (b) continuously.

12. Find the interest rate needed for an investment of $10,000 to grow to an amount of $12,000 in 3 years if interest is compounded quarterly.

13. How long will it take for an investment of $10,000 to grow to $15,000 if the investment earns interest at the rate of 6%/year compounded quarterly?

14. Find the nominal interest rate that yields an effective interest rate of 8%/year compounded quarterly.

In Exercises 15–32, find the derivative of the function.

15. $f(x) = xe^{2x}$
16. $f(t) = \sqrt{t}e^t + t$
17. $g(t) = \sqrt{t}e^{-2t}$
18. $g(x) = e^x\sqrt{1 + x^2}$
19. $y = \dfrac{e^{2x}}{1 + e^{-2x}}$
20. $f(x) = e^{2x^2 - 1}$
21. $f(x) = xe^{-x^2}$
22. $g(x) = (1 + e^{2x})^{3/2}$
23. $f(x) = x^2 e^x + e^x$
24. $g(t) = t \ln t$
25. $f(x) = \ln(e^{x^2} + 1)$
26. $f(x) = \dfrac{x}{\ln x}$
27. $f(x) = \dfrac{\ln x}{x + 1}$
28. $y = (x + 1)e^x$
29. $y = \ln(e^{4x} + 3)$
30. $f(r) = \dfrac{re^r}{1 + r^2}$
31. $f(x) = \dfrac{\ln x}{1 + e^x}$
32. $g(x) = \dfrac{e^{x^2}}{1 + \ln x}$

33. Find the second derivative of the function $y = \ln(3x + 1)$.

34. Find the second derivative of the function $y = x \ln x$.

35. Find $h'(0)$ if $h(x) = g(f(x))$, $g(x) = x + \dfrac{1}{x}$, and $f(x) = e^x$.

36. Find $h'(1)$ if $h(x) = g(f(x))$, $g(x) = \dfrac{x + 1}{x - 1}$, and $f(x) = \ln x$.

37. Use logarithmic differentiation to find the derivative of $f(x) = (2x^3 + 1)(x^2 + 2)^3$.

38. Use logarithmic differentiation to find the derivative of
$$f(x) = \dfrac{x(x^2 - 2)^2}{x - 1}$$

39. Find an equation of the tangent line to the graph of $y = e^{-2x}$ at the point $(1, e^{-2})$.

40. Find an equation of the tangent line to the graph of $y = xe^{-x}$ at the point $(1, e^{-1})$.

41. Sketch the graph of the function $f(x) = xe^{-2x}$.

42. Sketch the graph of the function $f(x) = x^2 - \ln x$.

43. Find the absolute extrema of the function $f(t) = te^{-t}$.

44. Find the absolute extrema of the function
$$g(t) = \dfrac{\ln t}{t}$$
on $[1, 2]$.

45. INVESTMENT RETURN A hotel was purchased by a conglomerate for $4.5 million and sold 5 years later for $8.2 million. Find the annual rate of return (compounded continuously).

46. Find the present value of $30,000 due in 5 years at an interest rate of 5%/year compounded monthly.

47. Find the present value of $119,346 due in 4 years at an interest rate of 6%/year compounded continuously.

48. CONSUMER PRICE INDEX At an annual inflation rate of 7.5%, how long will it take the Consumer Price Index (CPI) to double?

49. GROWTH OF BACTERIA A culture of bacteria that initially contained 2000 bacteria has a count of 18,000 bacteria after 2 hr.
  a. Determine the function $Q(t)$ that expresses the exponential growth of the number of cells of this bacterium as a function of time t (in minutes).
  b. Find the number of bacteria present after 4 hr.

50. RADIOACTIVE DECAY The radioactive element radium has a half-life of 1600 years. What is its decay constant?

51. DEMAND FOR DVD PLAYERS VCA Television found that the monthly demand for its new line of DVD players t months after placing the players on the market is given by

$$D(t) = 4000 - 3000e^{-0.06t} \quad (t \geq 0)$$

Graph this function and answer the following questions:
  a. What was the demand after 1 month? After 1 year? After 2 years?
  b. At what level is the demand expected to stabilize?

52. RADIOACTIVITY The mass of a radioactive isotope at time t (in years) is $M(t) = 200e^{-0.14t}$ g. What is the mass of the isotope initially? How fast is the mass of the isotope changing 2 years later?

53. OIL USED TO FUEL PRODUCTIVITY A study on worldwide oil use was prepared for a major oil company. The study predicted that the amount of oil used to fuel productivity in a certain country is given by

$$f(t) = 1.5 + 1.8te^{-1.2t} \quad (0 \leq t \leq 4)$$

where $f(t)$ denotes the number of barrels per $1000 of economic output and t is measured in decades ($t = 0$ corresponds to 1965). Compute $f'(0), f'(1), f'(2)$, and $f'(3)$ and interpret your results.

54. PRICE OF A COMMODITY The price of a certain commodity in dollars per unit at time t (measured in weeks) is given by $p = 18 - 3e^{-2t} - 6e^{-t/3}$.
  a. What is the price of the commodity at $t = 0$?
  b. How fast is the price of the commodity changing at $t = 0$?
  c. Find the equilibrium price of the commodity.
  Hint: It is given by $\lim_{t \to \infty} p$.

55. FLU EPIDEMIC During a flu epidemic, the number of students at a certain university who contracted influenza after t days could be approximated by the exponential model

$$Q(t) = \frac{3000}{1 + 499e^{-kt}}$$

If 90 students contracted the flu by day 10, how many students contracted the flu by day 20?

56. U.S. INFANT MORTALITY RATE The U.S. infant mortality rate (per 1000 live births) is approximated by the function

$$N(t) = 12.5e^{-0.0294t} \quad (0 \leq t \leq 21)$$

where t is measured in years, with $t = 0$ corresponding to 1980.
  a. What was the mortality rate in 1980? In 1990? In 2000?
  b. Sketch the graph of N.

Source: U.S. Department of Health and Human Services.

57. MAXIMIZING REVENUE The unit selling price p (in dollars) and the quantity demanded x (in pairs) of a certain brand of men's socks is given by the demand equation

$$p = 20e^{-0.0002x} \quad (0 \leq x \leq 10,000)$$

How many pairs of socks must be sold to yield a maximum revenue? What will the maximum revenue be?

58. ABSORPTION OF DRUGS The concentration of a drug in an organ at any time t (in seconds) is given by

$$x(t) = 0.08(1 - e^{-0.02t})$$

where $x(t)$ is measured in grams per cubic centimeter (g/cm³).
  a. What is the initial concentration of the drug in the organ?
  b. What is the concentration of the drug in the organ after 30 sec?
  c. What will be the concentration of the drug in the organ in the long run?
  d. Sketch the graph of x.

CHAPTER 5 Before Moving On . . .

1. Solve the equation $\dfrac{100}{1 + 2e^{0.3t}} = 40$ for t.

2. Find the accumulated amount after 4 years if $3000 is invested at 8%/year compounded weekly.

3. Find the slope of the tangent line to the graph of $f(x) = e^{\sqrt{x}}$.

4. Find the rate at which $y = x \ln(x^2 + 1)$ is changing at $x = 1$.

5. Find the second derivative of $y = e^{2x} \ln 3x$.

6. The temperature of a cup of coffee at time t (in minutes) is

$$T(t) = 70 + ce^{-kt}$$

Initially, the temperature of the coffee was 200°F. Three minutes later, it was 180°. When will the temperature of the coffee be 150°F?

8 CALCULUS OF SEVERAL VARIABLES

What should the dimensions of the new swimming pool be? It will be built in an elliptical area located in the rear of the promenade deck. Subject to this constraint, what are the dimensions of the largest pool that can be built? See Example 5, page 587, to see how to solve this problem.

UP TO NOW, we have dealt with functions involving one variable. However, many situations involve functions of two or more variables. For example, the Consumer Price Index (CPI) compiled by the Bureau of Labor Statistics depends on the price of more than 95,000 consumer items. To study such relationships, we need the notion of a function of several variables, the first topic in this chapter. Next, generalizing the concept of the derivative of a function of one variable, we study the *partial derivatives* of a function of two or more variables. Using partial derivatives, we study the rate of change of a function with respect to one variable while holding all other variables constant. We then learn how to find the extremum values of a function of several variables. As an application of optimization theory, we learn how to find an equation of the straight line that "best" fits a set of data points scattered about a straight line. Finally, we generalize the notion of the integral to the case involving a function of two variables.

8.1 Functions of Several Variables

Up to now, our study of calculus has been restricted to functions of one variable. In many practical situations, however, the formulation of a problem results in a mathematical model that involves a function of two or more variables. For example, suppose Ace Novelty determines that the profits are $6, $5, and $4 for three types of souvenirs it produces. Let x, y, and z denote the number of Type A, Type B, and Type C souvenirs to be made; then the company's profit is given by

$$P = 6x + 5y + 4z$$

and P is a function of the three variables, x, y, and z.

Functions of Two Variables

Although this chapter deals with real-valued functions of several variables, most of our definitions and results are stated in terms of a function of two variables. One reason for adopting this approach, as you will soon see, is that there is a geometric interpretation for this special case, which serves as an important visual aid. We can then draw upon the experience gained from studying the two-variable case to help us understand the concepts and results connected with the more general case, which, by and large, is just a simple extension of the lower-dimensional case.

> **A Function of Two Variables**
>
> A real-valued **function of two variables** f consists of
>
> 1. A set A of ordered pairs of real numbers (x, y) called the **domain** of the function.
> 2. A rule that associates with each ordered pair in the domain of f one and only one real number, denoted by $z = f(x, y)$.

The variables x and y are called **independent variables,** and the variable z, which is dependent on the values of x and y, is referred to as a **dependent variable.**

As in the case of a real-valued function of one real variable, the number $z = f(x, y)$ is called the **value of f** at the point (x, y). And, unless specified, the domain of the function f will be taken to be the largest possible set for which the rule defining f is meaningful.

EXAMPLE 1 Let f be the function defined by

$$f(x, y) = x + xy + y^2 + 2$$

Compute $f(0, 0)$, $f(1, 2)$, and $f(2, 1)$.

Solution We have

$$f(0, 0) = 0 + (0)(0) + 0^2 + 2 = 2$$
$$f(1, 2) = 1 + (1)(2) + 2^2 + 2 = 9$$
$$f(2, 1) = 2 + (2)(1) + 1^2 + 2 = 7$$

The domain of a function of two variables $f(x, y)$ is a set of ordered pairs of real numbers and may therefore be viewed as a subset of the xy-plane.

EXAMPLE 2 Find the domain of each function.

a. $f(x, y) = x^2 + y^2$ **b.** $g(x, y) = \dfrac{2}{x - y}$ **c.** $h(x, y) = \sqrt{1 - x^2 - y^2}$

Solution

a. $f(x, y)$ is defined for all real values of x and y, so the domain of the function f is the set of all points (x, y) in the xy-plane.

b. $g(x, y)$ is defined for all $x \neq y$, so the domain of the function g is the set of all points in the xy-plane except those lying on the line $y = x$ (Figure 1a).

c. We require that $1 - x^2 - y^2 \geq 0$ or $x^2 + y^2 \leq 1$, which is just the set of all points (x, y) lying on and inside the circle of radius 1 with center at the origin (Figure 1b).

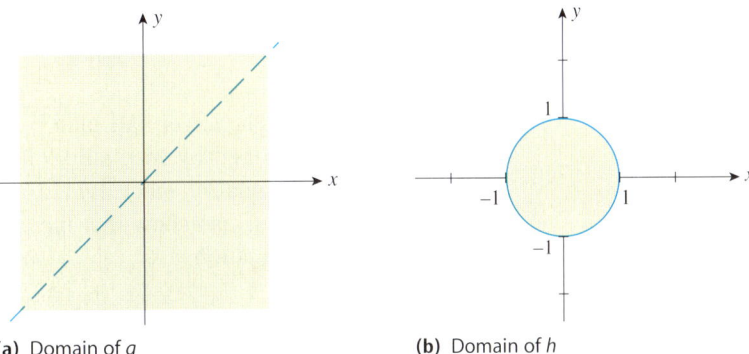

FIGURE 1 (a) Domain of g (b) Domain of h

APPLIED EXAMPLE 3 Revenue Functions Acrosonic manufactures a bookshelf loudspeaker system that may be bought fully assembled or in a kit. The demand equations that relate the unit prices, p and q, to the quantities demanded weekly, x and y, of the assembled and kit versions of the loudspeaker systems are given by

$$p = 300 - \frac{1}{4}x - \frac{1}{8}y \quad \text{and} \quad q = 240 - \frac{1}{8}x - \frac{3}{8}y$$

a. What is the weekly total revenue function $R(x, y)$?
b. What is the domain of the function R?

Solution

a. The weekly revenue realizable from the sale of x units of the assembled speaker systems at p dollars per unit is given by xp dollars. Similarly, the weekly revenue realizable from the sale of y units of the kits at q dollars per unit is given by yq dollars. Therefore, the weekly total revenue function R is given by

$$R(x, y) = xp + yq$$
$$= x\left(300 - \frac{1}{4}x - \frac{1}{8}y\right) + y\left(240 - \frac{1}{8}x - \frac{3}{8}y\right) \quad \text{(x²) See page 8.}$$
$$= -\frac{1}{4}x^2 - \frac{3}{8}y^2 - \frac{1}{4}xy + 300x + 240y$$

b. To find the domain of the function R, let's observe that the quantities x, y, p, and q must be nonnegative. This observation leads to the following system of linear inequalities:

$$300 - \frac{1}{4}x - \frac{1}{8}y \geq 0 \quad (1)$$

$$240 - \frac{1}{8}x - \frac{3}{8}y \geq 0 \quad (2)$$

$$x \geq 0 \quad (3)$$

$$y \geq 0 \quad (4)$$

This system of linear inequalities defines a region D in the xy-plane that is the domain of R. To sketch D, we first draw the line defined by the equation

$$300 - \frac{1}{4}x - \frac{1}{8}y = 0$$

obtained from Inequality (1) by replacing it by an equality (see Figure 2). Observe that this line divides the plane into two half-planes. To find the half-plane determined by Inequality (1), we pick a point lying in one of the half-planes. For simplicity, we pick the origin, (0, 0). This *test point* satisfies Inequality (1), since

$$300 - \frac{1}{4}(0) - \frac{1}{8}(0) \geq 0$$

This shows that the *lower* half-plane (the half-plane containing the test point) is the half-plane described by Inequality (1). Note that the line itself is included in the graph of the inequality, since equality is also allowed in Inequality (1).

Similarly, we can show that Inequality (2) defines the lower half-plane determined by the equation

$$240 - \frac{1}{8}x - \frac{3}{8}y = 0$$

obtained from replacing Inequality (2) by an equation. Once again, the line itself is included in the graph.

Finally, Inequalities (3) and (4) together define the first quadrant with the positive x- and y-axis. Therefore, the region D is obtained by taking the intersection of the two half-planes in the first quadrant.

The domain D of the function R is sketched in Figure 2.

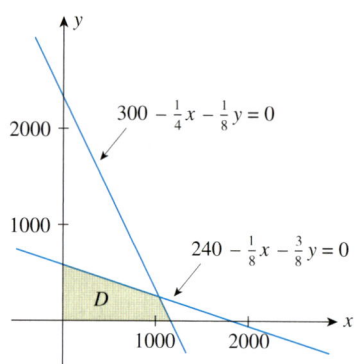

FIGURE 2
The domain of $R(x, y)$

Explore & Discuss

Suppose the total profit of a two-product company is given by $P(x, y)$, where x denotes the number of units of the first product produced and sold and y denotes the number of units of the second product produced and sold. Fix $x = a$, where a is a positive number such that (a, y) is in the domain of P. Describe and give an economic interpretation of the function $f(y) = P(a, y)$. Next, fix $y = b$, where b is a positive number such that (x, b) is in the domain of P. Describe and give an economic interpretation of the function $g(x) = P(x, b)$.

APPLIED EXAMPLE 4 Home Mortgage Payments The monthly payment that amortizes a loan of A dollars in t years when the interest rate is r per year compounded monthly is given by

$$P = f(A, r, t) = \frac{Ar}{12\left[1 - \left(1 + \frac{r}{12}\right)^{-12t}\right]}$$

Find the monthly payment for a home mortgage of \$270,000 to be amortized over 30 years when the interest rate is 6% per year, compounded monthly.

Solution Letting $A = 270{,}000$, $r = 0.06$, and $t = 30$, we find the required monthly payment to be

$$P = f(270{,}000, 0.06, 30) = \frac{270{,}000(0.06)}{12\left[1 - \left(1 + \frac{0.06}{12}\right)^{-360}\right]}$$
$$\approx 1618.79$$

or approximately \$1618.79.

Graphs of Functions of Two Variables

To graph a function of two variables, we need a three-dimensional coordinate system. This is readily constructed by adding a third axis to the plane Cartesian coordinate system in such a way that the three resulting axes are mutually perpendicular and intersect at O. Observe that, by construction, the zeros of the three number scales coincide at the origin of the **three-dimensional Cartesian coordinate system** (Figure 3).

A point in three-dimensional space can now be represented uniquely in this coordinate system by an **ordered triple** of numbers (x, y, z), and, conversely, every ordered triple of real numbers (x, y, z) represents a point in three-dimensional space (Figure 4a). For example, the points $A(2, 3, 4)$, $B(1, -2, -2)$, $C(2, 4, 0)$, and $D(0, 0, 4)$ are shown in Figure 4b.

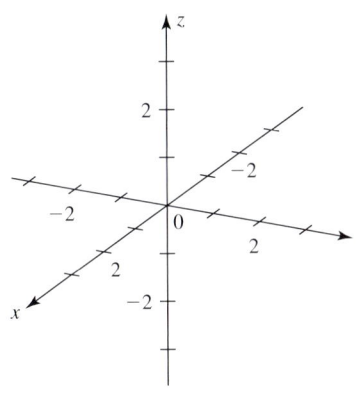

FIGURE 3
The three-dimensional Cartesian coordinate system

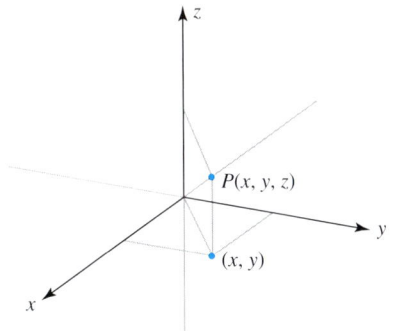

(a) A point in three-dimensional space

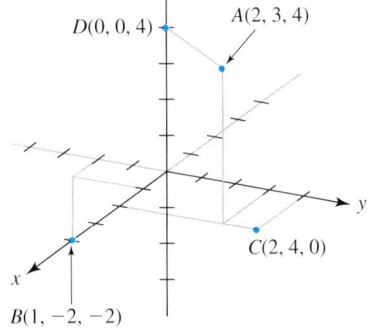

(b) Some sample points in three-dimensional space

FIGURE 4

Now, if $f(x, y)$ is a function of two variables x and y, the domain of f is a subset of the xy-plane. If we denote $f(x, y)$ by z, then the totality of all points (x, y, z), that is, $(x, y, f(x, y))$, makes up the **graph** of the function f and is, except for certain degenerate cases, a surface in three-dimensional space (Figure 5).

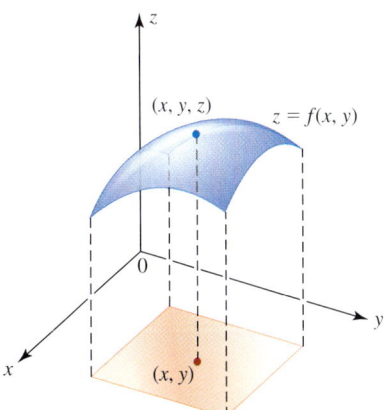

FIGURE 5
The graph of a function in three-dimensional space

In interpreting the graph of a function $f(x, y)$, one often thinks of the value $z = f(x, y)$ of the function at the point (x, y) as the "height" of the point (x, y, z) on the graph of f. If $f(x, y) > 0$, then the point (x, y, z) is $f(x, y)$ units above the xy-plane; if $f(x, y) < 0$, then the point (x, y, z) is $|f(x, y)|$ units below the xy-plane.

In general, it is quite difficult to draw the graph of a function of two variables. But techniques have been developed that enable us to generate such graphs with a minimum of effort, using a computer. Figure 6 shows the computer-generated graphs of some functions of two variables.

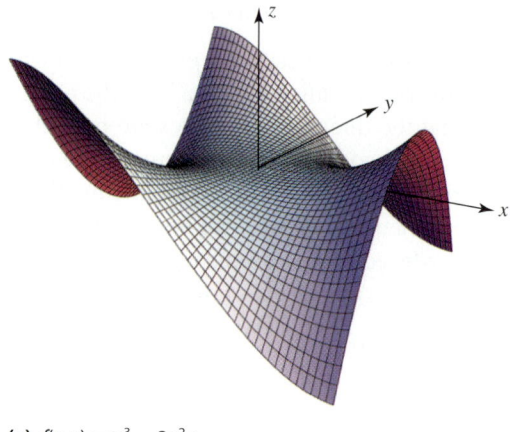
(a) $f(x, y) = x^3 - 3y^2 x$

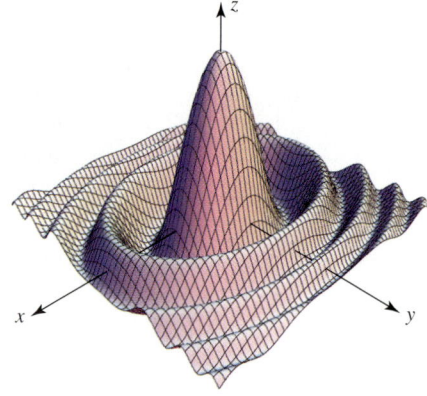
(b) $f(x, y) = \dfrac{\cos(x^2 + 2y^2)}{1 + x^2 + 2y^2}$

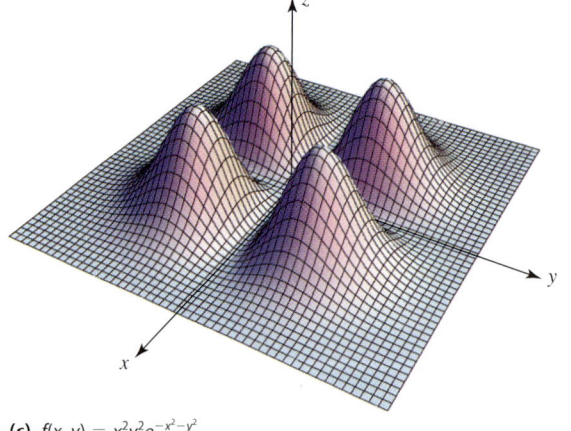
(c) $f(x, y) = x^2 y^2 e^{-x^2 - y^2}$

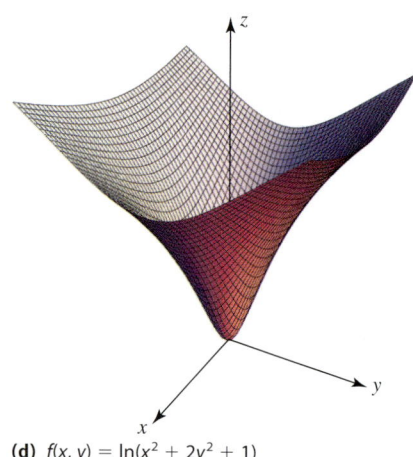
(d) $f(x, y) = \ln(x^2 + 2y^2 + 1)$

FIGURE 6
Four computer-generated graphs of functions of two variables

Level Curves

We can visualize the graph of a function of two variables by using *level curves*. To define the level curve of a function f of two variables, let $z = f(x, y)$ and consider the trace of f in the plane $z = k$ (k, a constant), as shown in Figure 7a. If we project this trace onto the xy-plane, we obtain a curve C with equation $f(x, y) = k$, called a *level curve* of f (Figure 7b).

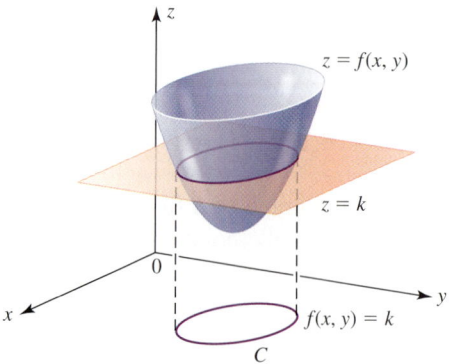
(a) The level curve C with equation $f(x, y) = k$ is the projection of the trace of f in the plane $z = k$ onto the xy-plane.

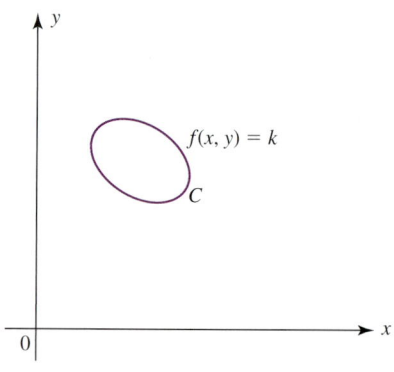
(b) The level curve C

FIGURE 7

Level Curves

The **level curves** of a function f of two variables are the curves with equations $f(x, y) = k$, where k is a constant in the range of f.

Notice that the level curve with equation $f(x, y) = k$ is the set of all points in the domain of f corresponding to the points on the surface $z = f(x, y)$ having the same height or depth k. By drawing the level curves corresponding to several admissible values of k, we obtain a *contour map*. The map enables us to visualize the surface represented by the graph of $z = f(x, y)$: We simply lift or depress the level curve to see the "cross sections" of the surface. Figure 8a shows a hill, and Figure 8b shows a contour map associated with that hill.

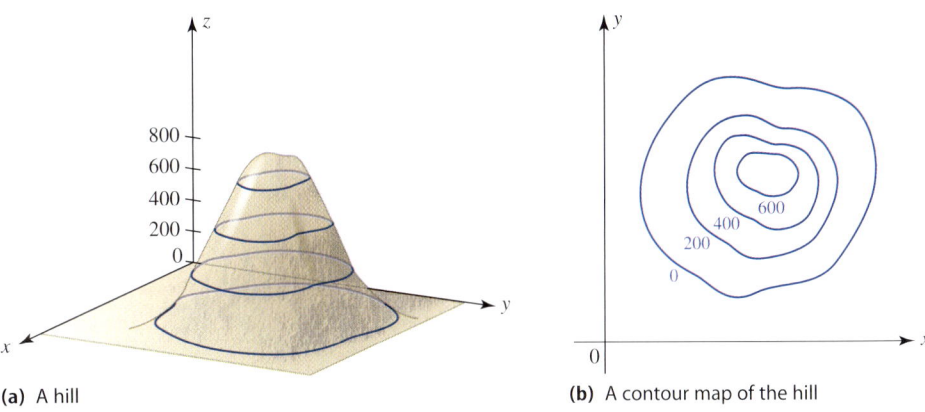

FIGURE 8   (a) A hill   (b) A contour map of the hill

EXAMPLE 5 Sketch a contour map for the function $f(x, y) = x^2 + y^2$.

Solution The level curves are the graphs of the equation $x^2 + y^2 = k$ for nonnegative numbers k. Taking $k = 0, 1, 4, 9,$ and 16, for example, we obtain

$$k = 0: \quad x^2 + y^2 = 0$$
$$k = 1: \quad x^2 + y^2 = 1$$
$$k = 4: \quad x^2 + y^2 = 4 = 2^2$$
$$k = 9: \quad x^2 + y^2 = 9 = 3^2$$
$$k = 16: \quad x^2 + y^2 = 16 = 4^2$$

The five level curves are concentric circles with center at the origin and radius given by $r = 0, 1, 2, 3,$ and 4, respectively (Figure 9a). A sketch of the graph of $f(x, y) = x^2 + y^2$ is included for your reference in Figure 9b.

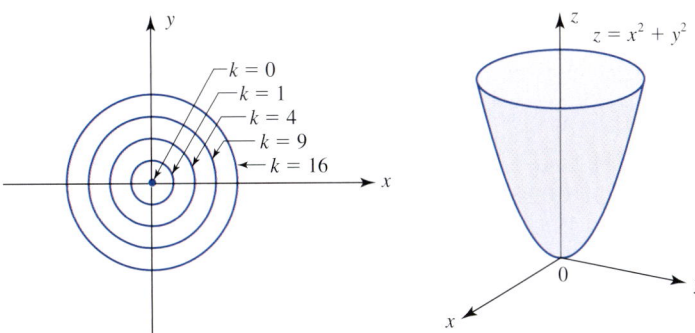

FIGURE 9   (a) Contour map for $f(x, y) = x^2 + y^2$ (b) The graph of $f(x, y) = x^2 + y^2$

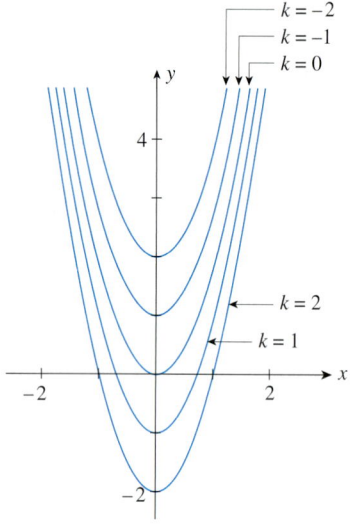

FIGURE 10
Level curves for $f(x, y) = 2x^2 - y$

EXAMPLE 6 Sketch the level curves for the function $f(x, y) = 2x^2 - y$ corresponding to $z = -2, -1, 0, 1,$ and 2.

Solution The level curves are the graphs of the equation $2x^2 - y = k$ or $y = 2x^2 - k$ for $k = -2, -1, 0, 1,$ and 2. The required level curves are shown in Figure 10. ∎

Level curves of functions of two variables are found in many practical applications. For example, if $f(x, y)$ denotes the temperature at a location within the continental United States with longitude x and latitude y at a certain point in time, then the temperature at the point (x, y) is given by the "height" of the surface, represented by $z = f(x, y)$. In this situation, the level curve $f(x, y) = k$ is a curve superimposed on a map of the United States, connecting points having the same temperature at a given time (Figure 11). These level curves are called **isotherms.**

Similarly, if $f(x, y)$ gives the barometric pressure at the location (x, y), then the level curves of the function f are called **isobars,** lines connecting points having the same barometric pressure at a given time.

As a final example, suppose $P(x, y, z)$ is a function of three variables $x, y,$ and z that gives the profit realized when $x, y,$ and z units of three products, $A, B,$ and C, respectively, are produced and sold. Then, the equation $P(x, y, z) = k$, where k is a constant, represents a surface in three-dimensional space called a **level surface** of P. In this situation, the level surface represented by $P(x, y, z) = k$ represents the product mix that results in a profit of exactly k dollars. Such a level surface is called an **isoprofit surface.**

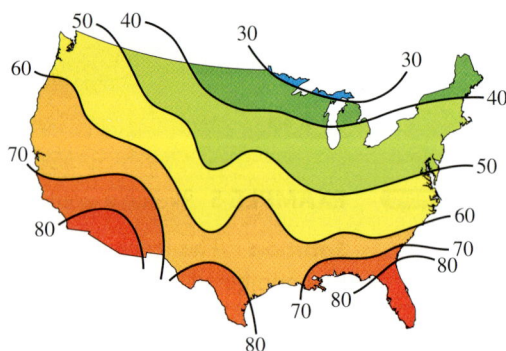

FIGURE 11
Isotherms: curves connecting points that have the same temperature

8.1 Self-Check Exercises

1. Let $f(x, y) = x^2 - 3xy + \sqrt{x + y}$. Compute $f(1, 3)$ and $f(-1, 1)$. Is the point $(-1, 0)$ in the domain of f?

2. Find the domain of $f(x, y) = \dfrac{1}{x} + \dfrac{1}{x - y} - e^{x+y}$.

3. Odyssey Travel Agency has a monthly advertising budget of $20,000. Odyssey's management estimates that if they spend x dollars on newspaper advertising and y dollars on television advertising, then the monthly revenue will be

$$f(x, y) = 30x^{1/4}y^{3/4}$$

dollars. What will be the monthly revenue if Odyssey spends $5000/month on newspaper ads and $15,000/month on television ads? If Odyssey spends $4000/month on newspaper ads and $16,000/month on television ads?

Solutions to Self-Check Exercises 8.1 can be found on page 544.

8.1 Concept Questions

1. What is a function of two variables? Give an example of a function of two variables and state its rule of definition and domain.

2. If f is a function of two variables, what can you say about the relationship between $f(a, b)$ and $f(c, d)$, if (a, b) is in the domain of f and $c = a$ and $d = b$?

3. Define (a) the graph of $f(x, y)$ and (b) a level curve of f.

4. Suppose f is a function of two variables and let $P = f(x, y)$ denote the profit realized when x units of Product A and y units of Product B are produced and sold. Give an interpretation of the level curve of f, defined by the equation $f(x, y) = k$, where k is a positive constant.

8.1 Exercises

1. Let $f(x, y) = 2x + 3y - 4$. Compute $f(0, 0)$, $f(1, 0)$, $f(0, 1)$, $f(1, 2)$, and $f(2, -1)$.

2. Let $g(x, y) = 2x^2 - y^2$. Compute $g(1, 2)$, $g(2, 1)$, $g(1, 1)$, $g(-1, 1)$, and $g(2, -1)$.

3. Let $f(x, y) = x^2 + 2xy - x + 3$. Compute $f(1, 2)$, $f(2, 1)$, $f(-1, 2)$, and $f(2, -1)$.

4. Let $h(x, y) = (x + y)/(x - y)$. Compute $h(0, 1)$, $h(-1, 1)$, $h(2, 1)$, and $h(\pi, -\pi)$.

5. Let $g(s, t) = 3s\sqrt{t} + t\sqrt{s} + 2$. Compute $g(1, 2)$, $g(2, 1)$, $g(0, 4)$, and $g(4, 9)$.

6. Let $f(x, y) = xye^{x^2+y^2}$. Compute $f(0, 0)$, $f(0, 1)$, $f(1, 1)$, and $f(-1, -1)$.

7. Let $h(s, t) = s \ln t - t \ln s$. Compute $h(1, e)$, $h(e, 1)$, and $h(e, e)$.

8. Let $f(u, v) = (u^2 + v^2)e^{uv^2}$. Compute $f(0, 1)$, $f(-1, -1)$, $f(a, b)$, and $f(b, a)$.

9. Let $g(r, s, t) = re^{s/t}$. Compute $g(1, 1, 1)$, $g(1, 0, 1)$, and $g(-1, -1, -1)$.

10. Let $g(u, v, w) = (ue^{vw} + ve^{uw} + we^{uv})/(u^2 + v^2 + w^2)$. Compute $g(1, 2, 3)$ and $g(3, 2, 1)$.

In Exercises 11–18, find the domain of the function.

11. $f(x, y) = 2x + 3y$
12. $g(x, y, z) = x^2 + y^2 + z^2$
13. $h(u, v) = \dfrac{uv}{u - v}$
14. $f(s, t) = \sqrt{s^2 + t^2}$
15. $g(r, s) = \sqrt{rs}$
16. $f(x, y) = e^{-xy}$
17. $h(x, y) = \ln(x + y - 5)$
18. $h(u, v) = \sqrt{4 - u^2 - v^2}$

In Exercises 19–24, sketch the level curves of the function corresponding to each value of z.

19. $f(x, y) = 2x + 3y$; $z = -2, -1, 0, 1, 2$
20. $f(x, y) = -x^2 + y$; $z = -2, -1, 0, 1, 2$
21. $f(x, y) = 2x^2 + y$; $z = -2, -1, 0, 1, 2$
22. $f(x, y) = xy$; $z = -4, -2, 2, 4$
23. $f(x, y) = \sqrt{16 - x^2 - y^2}$; $z = 0, 1, 2, 3, 4$
24. $f(x, y) = e^x - y$; $z = -2, -1, 0, 1, 2$

25. Find an equation of the level curve of $f(x, y) = \sqrt{x^2 + y^2}$ that contains the point $(3, 4)$.

26. Find an equation of the level surface of $f(x, y, z) = 2x^2 + 3y^2 - z$ that contains the point $(-1, 2, -3)$.

In Exercises 27 and 28, match the graph of the surface with one of the contour maps labeled (a) and (b).

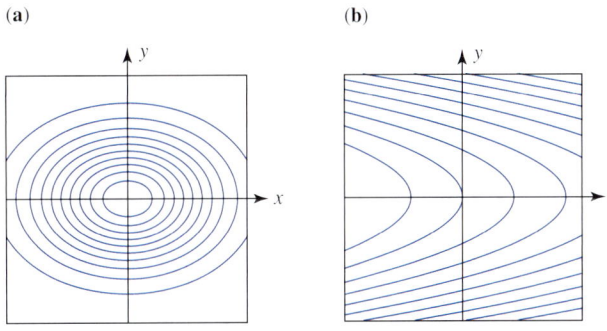

(a) (b)

27. $f(x, y) = x + y^2$
28. $f(x, y) = e^{1-2x^2-4y^2}$

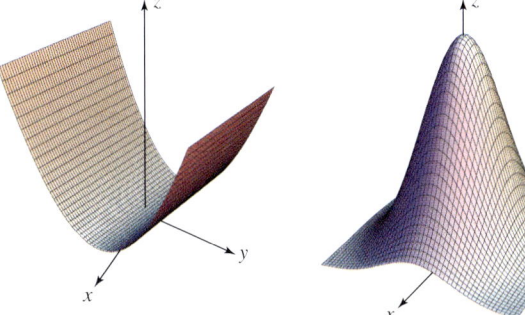

29. Can two level curves of a function f of two variables x and y intersect? Explain.

30. A *level set* of f is the set $S = \{(x, y) \mid f(x, y) = k\}$, where k is in the range of f. Let

$$f(x, y) = \begin{cases} 0 & \text{if } x^2 + y^2 < 1 \\ x^2 + y^2 - 1 & \text{if } x^2 + y^2 \geq 1 \end{cases}$$

Sketch the level set of f for $k = 0$ and 3.

31. The volume of a cylindrical tank of radius r and height h is given by

$$V = f(r, h) = \pi r^2 h$$

Find the volume of a cylindrical tank of radius 1.5 ft and height 4 ft.

32. IQs The IQ (intelligence quotient) of a person whose mental age is m years and whose chronological age is c years is defined as

$$f(m, c) = \frac{100m}{c}$$

What is the IQ of a 9-year-old child who has a mental age of 13.5 years?

33. Price-to-Earnings Ratio The current P/E ratio (price-to-earnings ratio) of a stock is defined as

$$R = \frac{P}{E}$$

where P is the current market price of a single share of the stock and E is the earnings per share for the most recent 12-month period.
a. What is the domain of the function R?
b. The earnings per share of IBM Corporation for 2011 were $13.09, and its price per share on March 19, 2012 was $205.56. What was the P/E ratio of IBM at that time?
Source: IBM Corporate Annual Report.

34. Current Dividend Yield The current dividend yield for common stock is calculated using the formula

$$Y = \frac{D}{P}$$

where D is the most recent full-year dividend and P is the current share price (both measured in dollars).
a. What is the domain of the function Y?
b. The annualized dividend of IBM Corporation for the year 2011 was $3, and its price per share was $205.56 on March 19, 2012. What was the current dividend yield for the common stock of IBM at that time?
Source: IBM Corporation Annual Report.

35. Body Mass Index The body mass index (BMI) is used to identify, evaluate, and treat overweight and obese adults. The BMI value for an adult of weight w (in kilograms) and height h (in meters) is defined to be

$$M = f(w, h) = \frac{w}{h^2}$$

According to federal guidelines, an adult is overweight if he or she has a BMI value greater than 25 but less than 30 and is "obese" if the value is greater than or equal to 30.
a. What is the BMI of an adult who weighs in at 80 kg and stands 1.8 m tall?
b. What is the maximum weight for an adult of height 1.8 m, who is not classified as overweight or obese?

36. Poiseuille's Law Poiseuille's Law states that the resistance R of blood flowing in a blood vessel of length l and radius r is given by

$$R = f(l, r) = \frac{kl}{r^4}$$

where k is a constant that depends on the viscosity of blood. What is the resistance, in terms of k, of blood flowing through an arteriole 4 cm long and of radius 0.1 cm?

37. Revenue Functions Country Workshop manufactures both finished and unfinished furniture for the home. The estimated quantities demanded each week of its rolltop desks in the finished and unfinished versions are x and y units when the corresponding unit prices are

$$p = 200 - \frac{1}{5}x - \frac{1}{10}y$$

$$q = 160 - \frac{1}{10}x - \frac{1}{4}y$$

dollars, respectively.
a. What is the weekly total revenue function $R(x, y)$?
b. Find the domain of the function R.

38. For the total revenue function $R(x, y)$ of Exercise 37, compute $R(100, 60)$ and $R(60, 100)$. Interpret your results.

39. Revenue Functions Weston Publishing publishes a deluxe edition and a standard edition of its English language dictionary. Weston's management estimates that the number of deluxe editions demanded is x copies/day and the number of standard editions demanded is y copies/day when the unit prices are

$$p = 20 - 0.005x - 0.001y$$

$$q = 15 - 0.001x - 0.003y$$

dollars, respectively.
a. Find the daily total revenue function $R(x, y)$.
b. Find the domain of the function R.

40. For the total revenue function $R(x, y)$ of Exercise 39, compute $R(300, 200)$ and $R(200, 300)$. Interpret your results.

41. Volume of a Gas The volume of a certain mass of gas is related to its pressure and temperature by the formula

$$V = \frac{30.9T}{P}$$

where the volume V is measured in liters, the temperature T is measured in kelvins (obtained by adding 273 to the Celsius temperature), and the pressure P is measured in millimeters of mercury pressure.
a. Find the domain of the function V.
b. Calculate the volume of the gas at standard temperature and pressure—that is, when $T = 273$ K and $P = 760$ mm of mercury.

42. Surface Area of a Human Body An empirical formula by E. F. Dubois relates the surface area S of a human body (in square meters) to its weight W (in kilograms) and its height H (in centimeters). The formula, given by

$$S = 0.007184 W^{0.425} H^{0.725}$$

is used by physiologists in metabolism studies.
a. Find the domain of the function S.
b. What is the surface area of a human body that weighs 70 kg and has a height of 178 cm?

43. ESTIMATING THE WEIGHT OF A TROUT A formula for estimating the weight of a trout (from measurements) is

$$W = \frac{LG^2}{800}$$

where L is its length and G is its girth (the distance around the body of the fish at it largest point), both measured in inches. The weight of the fish W is in pounds.
a. What is the domain of the function W?
b. Sue caught a trout and measured its length to be 20 in. and its girth to be 12 in. What is its approximate weight?

44. ESTIMATING THE WEIGHT OF A FISH A formula for estimating the weight of a trout (from measurements) is

$$W = \frac{LG^2}{800}$$

where L is its length and G is its girth (the distance around the body of the fish at its largest point), both measured in inches. The weight of the fish W is in pounds. A trout caught by Ashley is 20% longer and has a girth that is 10% shorter than the one caught by Jane. Whose catch is heavier? By how much does the weight of the trout caught by Ashley differ from the weight of the one caught by Jane?

45. PRODUCTION FUNCTION Suppose the output of a certain country is given by

$$f(x, y) = 100x^{3/5}y^{2/5}$$

billion dollars if x billion dollars are spent on labor and y billion dollars are spent on capital. Find the output if the country spent $32 billion on labor and $243 billion on capital.

46. PRODUCTION FUNCTION Economists have found that the output of a finished product, $f(x, y)$, is sometimes described by the function

$$f(x, y) = ax^b y^{1-b}$$

where x stands for the amount of money expended on labor, y stands for the amount expended on capital, and a and b are positive constants with $0 < b < 1$.
a. If p is a positive number, show that $f(px, py) = pf(x, y)$.
b. Use the result of part (a) to show that if the amount of money expended for labor and capital are both increased by $r\%$, then the output is also increased by $r\%$.

47. ARSON FOR PROFIT A study of arson for profit was conducted by a team of paid civilian experts and police detectives appointed by the mayor of a large city. It was found that the number of suspicious fires in that city in 2010 was very closely related to the concentration of tenants in the city's public housing and to the level of reinvestment in the area in conventional mortgages by the ten largest banks. In fact, the number of fires was closely approximated by the formula

$$N(x, y) = \frac{100(1000 + 0.03x^2 y)^{1/2}}{(5 + 0.2y)^2}$$

$$(0 \leq x \leq 150;\ 5 \leq y \leq 35)$$

where x denotes the number of persons/census tract and y denotes the level of reinvestment in the area in cents/dollar deposited. Using this formula, estimate the total number of suspicious fires in the districts of the city where the concentration of public housing tenants was 100/census tract and the level of reinvestment was 20 cents/dollar deposited.

48. CONTINUOUSLY COMPOUNDED INTEREST If a principal of P dollars is deposited in an account earning interest at the rate of r/year compounded continuously, then the accumulated amount at the end of t years is given by

$$A = f(P, r, t) = Pe^{rt}$$

dollars. Find the accumulated amount at the end of 3 years if a sum of $10,000 is deposited in an account earning interest at the rate of 6%/year.

49. HOME MORTGAGES The monthly payment that amortizes a loan of A dollars in t years when the interest rate is r per year, compounded monthly, is given by

$$P = f(A, r, t) = \frac{Ar}{12[1 - (1 + \frac{r}{12})^{-12t}]}$$

a. What is the monthly payment for a home mortgage of $300,000 that will be amortized over 30 years with an interest rate of 4%/year? An interest rate of 6%/year?
b. Find the monthly payment for a home mortgage of $300,000 that will be amortized over 20 years with an interest rate of 6%/year.

50 HOME MORTGAGES Suppose a home buyer secures a bank loan of A dollars to purchase a house. If the interest rate charged is r/year compounded monthly and the loan is to be amortized in t years, then the principal repayment at the end of i months is given by

$$B = f(A, r, t, i)$$
$$= A \left[\frac{(1 + \frac{r}{12})^i - 1}{(1 + \frac{r}{12})^{12t} - 1} \right] \quad (0 \leq i \leq 12t)$$

Suppose the Blakelys borrow a sum of $280,000 from a bank to help finance the purchase of a house and the bank charges interest at a rate of 6%/year. If the Blakelys agree to repay the loan in equal installments over 30 years, how much will they owe the bank after the 60th payment (5 years)? The 240th payment (20 years)?

51. WILSON LOT-SIZE FORMULA The Wilson lot-size formula in economics states that the optimal quantity Q of goods for a store to order is given by

$$Q = f(C, N, h) = \sqrt{\frac{2CN}{h}}$$

where C is the cost of placing an order, N is the number of items the store sells per week, and h is the weekly holding cost for each item. Find the most economical quantity of 10-speed bicycles to order if it costs the store $20 to place an order, $5 to hold a bicycle for a week, and the store expects to sell 40 bicycles per week.

52. Wind Power The power output (in watts) of a certain brand of wind turbine generators is estimated to be

$$P = f(R, V) = 0.772R^2V^3$$

where R is the radius (in meters) of a rotor blade and V is the wind speed (in meters per second). Estimate the power output of a model of these generators if its radius is 30 m and the wind speed is 16 m/s.

53. International America's Cup Class Drafted by an international committee in 1989, the rules for the new International America's Cup Class (IACC) include a formula that governs the basic yacht dimensions. The formula

$$f(L, S, D) \leq 42$$

where

$$f(L, S, D) = \frac{L + 1.25S^{1/2} - 9.80D^{1/3}}{0.388}$$

balances the rated length L (in meters), the rated sail area S (in square meters), and the displacement D (in cubic meters). All changes in the basic dimensions are trade-offs. For example, if you want to pick up speed by increasing the sail area, you must pay for it by decreasing the length or increasing the displacement, both of which slow down the boat. Show that Yacht A of rated length 20.95 m, rated sail area 277.3 m², and displacement 17.56 m³ and the longer and heavier Yacht B with $L = 21.87$, $S = 311.78$, and $D = 22.48$ both satisfy the formula.

54. Force Generated by a Centrifuge A centrifuge is a machine designed for the specific purpose of subjecting materials to a sustained centrifugal force. The actual amount of centrifugal force, F, expressed in dynes (1 gram of force = 980 dynes) is given by

$$F = f(M, S, R) = \frac{\pi^2 S^2 M R}{900}$$

where S is in revolutions per minute (rpm), M is in grams, and R is in centimeters. Show that an object revolving at the rate of 600 rpm in a circle with radius of 10 cm generates a centrifugal force that is approximately 40 times gravity.

55. Ideal Gas Law According to the *ideal gas law*, the volume V of an ideal gas is related to its pressure P and temperature T by the formula

$$V = \frac{kT}{P}$$

where k is a positive constant. Describe the level curves of V and give a physical interpretation of your result.

In Exercises 56–60, determine whether the statement is true or false. If it is true, explain why it is true. If it is false, explain why, or give an example to show why it is false.

56. If h is a function of x and y, then there are functions f and g of one variable such that

$$h(x, y) = f(x) + g(y)$$

57. If f is a function of x and y and a is a real number, then

$$f(ax, ay) = af(x, y)$$

58. The domain of $f(x, y) = 1/(x^2 - y^2)$ is $\{(x, y) | y \neq x\}$.

59. Every point on the level curve $f(x, y) = c$ corresponds to a point on the graph of f that is c units above the xy-plane if $c > 0$ and $|c|$ units below the xy-plane if $c < 0$.

60. f is a function of x and y if and only if for any two points $P_1(x_1, y_1)$ and $P_2(x_2, y_2)$ in the domain of f, $f(x_1, y_1) = f(x_2, y_2)$ implies that $P_1(x_1, y_1) = P_2(x_2, y_2)$.

8.1 Solutions to Self-Check Exercises

1. $f(1, 3) = 1^2 - 3(1)(3) + \sqrt{1 + 3} = -6$
$f(-1, 1) = (-1)^2 - 3(-1)(1) + \sqrt{-1 + 1} = 4$

The point $(-1, 0)$ is not in the domain of f because the term $\sqrt{x + y}$ is not defined when $x = -1$ and $y = 0$. In fact, the domain of f consists of all real values of x and y that satisfy the inequality $x + y \geq 0$, the shaded half-plane shown in the accompanying figure.

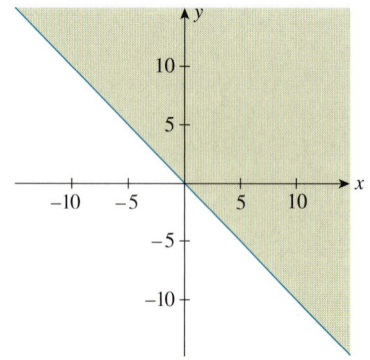

2. Since division by zero is not permitted, we see that $x \neq 0$ and $x - y \neq 0$. Therefore, the domain of f is the set of all points in the xy-plane except for the y-axis ($x = 0$) and the straight line $x = y$.

3. If Odyssey spends $5000/month on newspaper ads ($x = 5000$) and $15,000/month on television ads ($y = 15,000$), then its monthly revenue will be given by

$$f(5000, 15{,}000) = 30(5000)^{1/4}(15{,}000)^{3/4}$$
$$\approx 341{,}926.06$$

or approximately $341,926. If the agency spends $4000/month on newspaper ads and $16,000/month on television ads, then its monthly revenue will be given by

$$f(4000, 16{,}000) = 30(4000)^{1/4}(16{,}000)^{3/4}$$
$$\approx 339{,}411.26$$

or approximately $339,411.

8.2 Partial Derivatives

Partial Derivatives

For a function $f(x)$ of one variable x, there is no ambiguity when we speak about the rate of change of $f(x)$ with respect to x, since x must be constrained to move along the x-axis. The situation becomes more complicated, however, when we study the rate of change of a function of two or more variables. For example, the domain D of a function of two variables $f(x, y)$ is a subset of the plane, so if (a, b) is any point in the domain of f, there are infinitely many directions from which one can approach the point (a, b) (Figure 12). We may therefore ask for the rate of change of f at (a, b) along any of these directions.

However, we will not deal with this general problem. Instead, we will restrict ourselves to studying the rate of change of the function $f(x, y)$ at a point (a, b) in each of two *preferred directions*—namely, the direction parallel to the x-axis and the direction parallel to the y-axis. Let $y = b$, where b is a constant, so that $f(x, b)$ is a function of the one variable x. Since the equation $z = f(x, y)$ is the equation of a surface, the equation $z = f(x, b)$ is the equation of the curve C on the surface formed by the intersection of the surface and the plane $y = b$ (Figure 13).

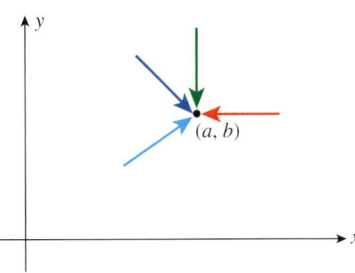

FIGURE 12
We can approach a point in the plane from infinitely many directions.

Because $f(x, b)$ is a function of one variable x, we may compute the derivative of f with respect to x at $x = a$. This derivative, obtained by keeping the variable y fixed at b and differentiating the resulting function $f(x, b)$ with respect to x, is called the **first partial derivative of f with respect to x** at (a, b), written

$$\frac{\partial z}{\partial x}(a, b) \quad \text{or} \quad \frac{\partial f}{\partial x}(a, b) \quad \text{or} \quad f_x(a, b)$$

Thus,

$$\frac{\partial z}{\partial x}(a, b) = \frac{\partial f}{\partial x}(a, b) = f_x(a, b) = \lim_{h \to 0} \frac{f(a + h, b) - f(a, b)}{h}$$

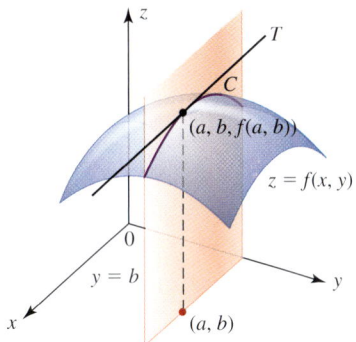

FIGURE 13
The curve C is formed by the intersection of the plane $y = b$ with the surface $z = f(x, y)$.

provided that the limit exists. The first partial derivative of f with respect to x at (a, b) measures both the slope of the tangent line T to the curve C and the rate of change of the function f in the x-direction when $x = a$ and $y = b$. We also write

$$\left.\frac{\partial f}{\partial x}\right|_{(a, b)} = f_x(a, b)$$

Similarly, we define the **first partial derivative of f with respect to y** at (a, b), written

$$\frac{\partial z}{\partial y}(a, b) \quad \text{or} \quad \frac{\partial f}{\partial y}(a, b) \quad \text{or} \quad f_y(a, b)$$

as the derivative obtained by keeping the variable x fixed at a and differentiating the resulting function $f(a, y)$ with respect to y. That is,

$$\frac{\partial z}{\partial y}(a, b) = \frac{\partial f}{\partial y}(a, b) = f_y(a, b)$$

$$= \lim_{k \to 0} \frac{f(a, b + k) - f(a, b)}{k}$$

if the limit exists. The first partial derivative of f with respect to y at (a, b) measures both the slope of the tangent line T to the curve C, obtained by holding x constant (Figure 14), and the rate of change of the function f in the y-direction when $x = a$ and $y = b$. We write

$$\left.\frac{\partial f}{\partial y}\right|_{(a, b)} = f_y(a, b)$$

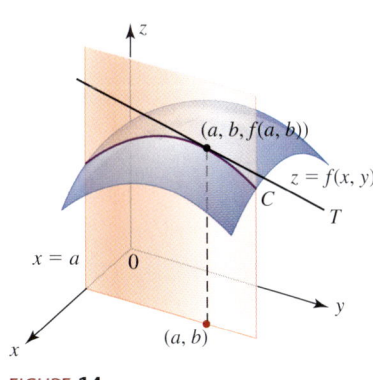

FIGURE 14
The first partial derivative of f with respect to y at (a, b) measures the slope of the tangent line T to the curve C with x held constant.

PORTFOLIO

Karthik Ramachandran

TITLE   Principal Software Engineer
INSTITUTION   Iron Mountain

Iron Mountain is the world's largest provider of document management solutions. Although the company has its roots in paper document storage, it has become the leader in digital document solutions as well. In addition to providing online backup and digital archiving services, Iron Mountain offers various digital document management tools.

As a principal software engineer at Iron Mountain, I help to develop eDiscovery products and services designed specifically for lawyers. When lawyers search for information relating to a corporate lawsuit, for example, it can be very difficult for them to find the information they need if they are searching through terabytes of unorganized data. To address this challenge, Iron Mountain has developed a variety of techniques to organize data automatically based on the statistical analysis of words and phrases contained within documents and email messages.

Although we use statistical analysis to initiate the organization of data, we apply calculus to refine our products and services. When word sets are very large, it is important to focus on particular words and phrases that convey vital information. At Iron Mountain, we use optimization techniques involving partial derivatives to detect the most informative words and phrases within documents and allot them an apportioned weight in the organization process.

The methods outlined above allow us to organize data into folders—or clusters—of related documents. Ultimately, this enables us to improve upon traditional search capabilities by grouping search results into concept folders, minimizing cost and reducing the time needed for lawyers to find relevant documents.

With the aid of mathematical analysis, we provide robust eDiscovery products and services that radically improve legal workflow. As the amount of electronically stored information grows, our application of statistics and calculus allows us to remain at the forefront of our field.

Greg Neustaetter/Iron Mountain; (inset) © istockphoto.com/Jacob Wackerhausen

Before looking at some examples, let's summarize these definitions.

First Partial Derivatives of f(x, y)

Suppose $f(x, y)$ is a function of the two variables x and y. Then the **first partial derivative of f** with respect to x at the point (x, y) is

$$\frac{\partial f}{\partial x} = \lim_{h \to 0} \frac{f(x + h, y) - f(x, y)}{h}$$

provided that the limit exists. The first partial derivative of f with respect to y at the point (x, y) is

$$\frac{\partial f}{\partial y} = \lim_{k \to 0} \frac{f(x, y + k) - f(x, y)}{k}$$

provided that the limit exists.

VIDEO **EXAMPLE 1** Find the partial derivatives $\dfrac{\partial f}{\partial x}$ and $\dfrac{\partial f}{\partial y}$ of the function

$$f(x, y) = x^2 - xy^2 + y^3$$

What is the rate of change of the function f in the x-direction at the point $(1, 2)$?
What is the rate of change of the function f in the y-direction at the point $(1, 2)$?

Solution To compute $\dfrac{\partial f}{\partial x}$, think of the variable y as a constant and differentiate the resulting function of x with respect to x. Let's write

$$f(x, y) = x^2 - xy^2 + y^3$$

where the variable y to be treated as a constant is shown in color. Then,

$$\frac{\partial f}{\partial x} = 2x - y^2$$

To compute $\dfrac{\partial f}{\partial y}$, think of the variable x as being fixed—that is, as a constant—and differentiate the resulting function of y with respect to y. In this case,

$$f(x, y) = x^2 - xy^2 + y^3$$

so

$$\frac{\partial f}{\partial y} = -2xy + 3y^2$$

The rate of change of the function f in the x-direction at the point $(1, 2)$ is given by

$$f_x(1, 2) = \left.\frac{\partial f}{\partial x}\right|_{(1,\,2)} = 2(1) - 2^2 = -2$$

That is, f decreases 2 units for each unit increase in the x-direction, y being kept constant ($y = 2$). The rate of change of the function f in the y-direction at the point $(1, 2)$ is given by

$$f_y(1, 2) = \left.\frac{\partial f}{\partial y}\right|_{(1,\,2)} = -2(1)(2) + 3(2)^2 = 8$$

That is, f increases 8 units for each unit increase in the y-direction, x being kept constant ($x = 1$).

Explore & Discuss

Refer to the Explore & Discuss on page 536. Suppose management has decided that the projected sales of the first product is a units. Describe how you might help management decide how many units of the second product the company should produce and sell in order to maximize the company's total profit. Justify your method to management. Suppose, however, that management feels that b units of the second product should be manufactured and sold. How would you help management decide how many units of the first product to manufacture in order to maximize the company's total profit?

EXAMPLE 2 Compute the first partial derivatives of each function.

a. $f(x, y) = \dfrac{xy}{x^2 + y^2}$ **b.** $g(s, t) = (s^2 - st + t^2)^5$

c. $h(u, v) = e^{u^2 - v^2}$ **d.** $f(x, y) = \ln(x^2 + 2y^2)$

Solution

a. To compute $\dfrac{\partial f}{\partial x}$, think of the variable y as a constant. Thus,

$$f(x, y) = \dfrac{xy}{x^2 + y^2}$$

Then using the quotient rule, we have

$$\dfrac{\partial f}{\partial x} = \dfrac{(x^2 + y^2)y - xy(2x)}{(x^2 + y^2)^2} = \dfrac{x^2y + y^3 - 2x^2y}{(x^2 + y^2)^2}$$

$$= \dfrac{y(y^2 - x^2)}{(x^2 + y^2)^2}$$

upon simplification and factorization. To compute $\dfrac{\partial f}{\partial y}$, think of the variable x as a constant. Thus,

$$f(x, y) = \dfrac{xy}{x^2 + y^2}$$

Then using the quotient rule once again, we obtain

$$\dfrac{\partial f}{\partial y} = \dfrac{(x^2 + y^2)x - xy(2y)}{(x^2 + y^2)^2} = \dfrac{x^3 + xy^2 - 2xy^2}{(x^2 + y^2)^2}$$

$$= \dfrac{x(x^2 - y^2)}{(x^2 + y^2)^2}$$

b. To compute $\dfrac{\partial g}{\partial s}$, we treat the variable t as if it were a constant. Thus,

$$g(s, t) = (s^2 - st + t^2)^5$$

Using the General Power Rule, we find

$$\dfrac{\partial g}{\partial s} = 5(s^2 - st + t^2)^4 \cdot (2s - t)$$

$$= 5(2s - t)(s^2 - st + t^2)^4$$

To compute $\dfrac{\partial g}{\partial t}$, we treat the variable s as if it were a constant. Thus,

$$g(s, t) = (s^2 - st + t^2)^5$$

$$\dfrac{\partial g}{\partial t} = 5(s^2 - st + t^2)^4(-s + 2t)$$

$$= 5(2t - s)(s^2 - st + t^2)^4$$

c. To compute $\dfrac{\partial h}{\partial u}$, think of the variable v as a constant. Thus,

$$h(u, v) = e^{u^2 - v^2}$$

Using the Chain Rule for Exponential Functions, we have

$$\dfrac{\partial h}{\partial u} = e^{u^2 - v^2} \cdot 2u = 2ue^{u^2 - v^2}$$

Next, we treat the variable u as if it were a constant,

$$h(u, v) = e^{u^2 - v^2}$$

and we obtain

$$\frac{\partial h}{\partial v} = e^{u^2-v^2} \cdot (-2v) = -2ve^{u^2-v^2}$$

d. To compute $\dfrac{\partial f}{\partial x}$, think of the variable y as a constant. Thus,

$$f(x, y) = \ln(x^2 + 2y^2)$$

so the Chain Rule for Logarithmic Functions gives

$$\frac{\partial f}{\partial x} = \frac{2x}{x^2 + 2y^2}$$

Next, treating the variable x as if it were a constant, we find

$$f(x, y) = \ln(x^2 + 2y^2)$$
$$\frac{\partial f}{\partial y} = \frac{4y}{x^2 + 2y^2}$$

To compute the partial derivative of a function of several variables with respect to one variable—say, x—we think of the other variables as if they were constants and differentiate the resulting function with respect to x.

Explore & Discuss

1. Let (a, b) be a point in the domain of $f(x, y)$. Put $g(x) = f(x, b)$ and suppose that g is differentiable at $x = a$. Explain why you can find $f_x(a, b)$ by computing $g'(a)$. How would you go about calculating $f_y(a, b)$ using a similar technique? Give a geometric interpretation of these processes.

2. Let $f(x, y) = x^2y^3 - 3x^2y + 2$. Use the method of Problem 1 to find $f_x(1, 2)$ and $f_y(1, 2)$.

EXAMPLE 3 Compute the first partial derivatives of the function

$$w = f(x, y, z) = xyz - xe^{yz} + x \ln y$$

Solution Here we have a function of three variables, x, y, and z, and we are required to compute

$$\frac{\partial f}{\partial x}, \frac{\partial f}{\partial y}, \frac{\partial f}{\partial z}$$

To compute f_x, we think of the other two variables, y and z, as fixed, and we differentiate the resulting function of x with respect to x, thereby obtaining

$$f_x = yz - e^{yz} + \ln y$$

To compute f_y, we think of the other two variables, x and z, as constants, and we differentiate the resulting function of y with respect to y. We then obtain

$$f_y = xz - xze^{yz} + \frac{x}{y}$$

Finally, to compute f_z, we treat the variables x and y as constants and differentiate the function f with respect to z, obtaining

$$f_z = xy - xye^{yz}$$

> **Exploring with TECHNOLOGY**
>
> Refer to the Explore & Discuss on page 549. Let
>
> $$f(x, y) = \frac{e^{\sqrt{xy}}}{(1 + xy^2)^{3/2}}$$
>
> 1. Compute $g(x) = f(x, 1)$, and use a graphing utility to plot the graph of g in the viewing window $[0, 2] \times [0, 2]$.
> 2. Use the differentiation operation of your graphing utility to find $g'(1)$ and hence $f_x(1, 1)$.
> 3. Compute $h(y) = f(1, y)$, and use a graphing utility to plot the graph of h in the viewing window $[0, 2] \times [0, 2]$.
> 4. Use the differentiation operation of your graphing utility to find $h'(1)$ and hence $f_y(1, 1)$.

The Cobb–Douglas Production Function

For an economic interpretation of the first partial derivatives of a function of two variables, let's turn our attention to the function

$$f(x, y) = ax^b y^{1-b} \tag{5}$$

where a and b are positive constants with $0 < b < 1$. This function is called the **Cobb–Douglas production function**. Here, x stands for the amount of money expended for labor, y stands for the cost of capital equipment (buildings, machinery, and other tools of production), and the function f measures the output of the finished product (in suitable units) and is called, accordingly, the production function.

The partial derivative f_x is called the **marginal productivity of labor**. It measures the rate of change of production with respect to the amount of money expended for labor, with the level of capital expenditure held constant. Similarly, the partial derivative f_y, called the **marginal productivity of capital**, measures the rate of change of production with respect to the amount expended on capital, with the level of labor expenditure held fixed.

APPLIED EXAMPLE 4 Marginal Productivity A certain country's production in the early years following World War II is described by the function

$$f(x, y) = 30x^{2/3} y^{1/3}$$

units, when x units of labor and y units of capital were used.

a. Compute f_x and f_y.
b. What is the marginal productivity of labor and the marginal productivity of capital when the amounts expended on labor and capital are 125 units and 27 units, respectively?
c. Should the government have encouraged capital investment rather than increasing expenditure on labor to increase the country's productivity?

Solution

a. $f_x = 30 \cdot \dfrac{2}{3} x^{-1/3} y^{1/3} = 20 \left(\dfrac{y}{x}\right)^{1/3}$

$f_y = 30 x^{2/3} \cdot \dfrac{1}{3} y^{-2/3} = 10 \left(\dfrac{x}{y}\right)^{2/3}$

b. The required marginal productivity of labor is given by

$$f_x(125, 27) = 20 \left(\frac{27}{125}\right)^{1/3} = 20 \left(\frac{3}{5}\right)$$

or 12 units per unit increase in labor expenditure (capital expenditure is held constant at 27 units). The required marginal productivity of capital is given by

$$f_y(125, 27) = 10\left(\frac{125}{27}\right)^{2/3} = 10\left(\frac{25}{9}\right)$$

or $27\frac{7}{9}$ units per unit increase in capital expenditure (labor outlay is held constant at 125 units).

c. From the results of part (b), we see that a unit increase in capital expenditure resulted in a much faster increase in productivity than a unit increase in labor expenditure would have. Therefore, the government should have encouraged increased spending on capital rather than on labor during the early years of reconstruction.

Substitute and Complementary Commodities

For another application of the first partial derivatives of a function of two variables in the field of economics, let's consider the relative demands of two commodities. We say that the two commodities are **substitute** (competitive) **commodities** if a decrease in the demand for one results in an increase in the demand for the other. Examples of substitute commodities are coffee and tea. Conversely, two commodities are referred to as **complementary commodities** if a decrease in the demand for one results in a decrease in the demand for the other as well. Examples of complementary commodities are automobiles and tires.

We now derive a criterion for determining whether two commodities A and B are substitute or complementary. Suppose the demand equations that relate the quantities demanded, x and y, to the unit prices, p and q, of the two commodities are given by

$$x = f(p, q) \quad \text{and} \quad y = g(p, q)$$

Let's consider the partial derivative $\partial f/\partial p$. Since f is the demand function for Commodity A, we see that, for fixed q, f is typically a decreasing function of p—that is, $\partial f/\partial p < 0$. Now, if the two commodities were substitute commodities, then the quantity demanded of Commodity B would increase with respect to p—that is, $\partial g/\partial p > 0$. A similar argument with p fixed shows that if A and B are substitute commodities, then $\partial f/\partial q > 0$. Thus, the two Commodities A and B are substitute commodities if

$$\frac{\partial f}{\partial q} > 0 \quad \text{and} \quad \frac{\partial g}{\partial p} > 0$$

Similarly, A and B are complementary commodities if

$$\frac{\partial f}{\partial q} < 0 \quad \text{and} \quad \frac{\partial g}{\partial p} < 0$$

Substitute and Complementary Commodities

Two commodities A and B are substitute commodities if

$$\frac{\partial f}{\partial q} > 0 \quad \text{and} \quad \frac{\partial g}{\partial p} > 0 \tag{6}$$

Two commodities A and B are complementary commodities if

$$\frac{\partial f}{\partial q} < 0 \quad \text{and} \quad \frac{\partial g}{\partial p} < 0 \tag{7}$$

APPLIED EXAMPLE 5 Substitute and Complementary Commodities
Suppose that the daily demand for butter is given by

$$x = f(p, q) = \frac{3q}{1 + p^2}$$

and the daily demand for margarine is given by

$$y = g(p, q) = \frac{2p}{1 + \sqrt{q}} \qquad (p > 0, q > 0)$$

where p and q denote the prices per pound (in dollars) of butter and margarine, respectively, and x and y are measured in millions of pounds. Determine whether these two commodities are substitute, complementary, or neither.

Solution We compute

$$\frac{\partial f}{\partial q} = \frac{3}{1 + p^2} \quad \text{and} \quad \frac{\partial g}{\partial p} = \frac{2}{1 + \sqrt{q}}$$

Since

$$\frac{\partial f}{\partial q} > 0 \quad \text{and} \quad \frac{\partial g}{\partial p} > 0$$

for all values of $p > 0$ and $q > 0$, we conclude that butter and margarine are substitute commodities.

Second-Order Partial Derivatives

The first partial derivatives $f_x(x, y)$ and $f_y(x, y)$ of a function $f(x, y)$ of the two variables x and y are also functions of x and y. As such, we may differentiate each of the functions f_x and f_y to obtain the **second-order partial derivatives of f** (Figure 15). Thus, differentiating the function f_x with respect to x leads to the second partial derivative

$$f_{xx} = \frac{\partial^2 f}{\partial x^2} = \frac{\partial}{\partial x}(f_x)$$

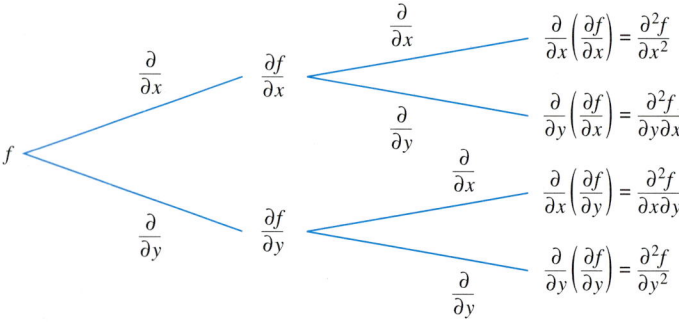

FIGURE 15
A schematic showing the four second-order partial derivatives of f

However, differentiation of f_x with respect to y leads to the second partial derivative

$$f_{xy} = \frac{\partial^2 f}{\partial y \partial x} = \frac{\partial}{\partial y}(f_x)$$

Similarly, differentiation of the function f_y with respect to x and with respect to y leads to

$$f_{yx} = \frac{\partial^2 f}{\partial x \partial y} = \frac{\partial}{\partial x}(f_y)$$

$$f_{yy} = \frac{\partial^2 f}{\partial y^2} = \frac{\partial}{\partial y}(f_y)$$

respectively. Although it is not always true that $f_{xy} = f_{yx}$, they are equal, however, if both f_{xy} and f_{yx} are continuous. We might add that this is the case in most practical applications.

EXAMPLE 6 Find the second-order partial derivatives of the function

$$f(x, y) = x^3 - 3x^2y + 3xy^2 + y^2$$

Solution The first partial derivatives of f are

$$f_x = \frac{\partial}{\partial x}(x^3 - 3x^2y + 3xy^2 + y^2)$$
$$= 3x^2 - 6xy + 3y^2$$
$$f_y = \frac{\partial}{\partial y}(x^3 - 3x^2y + 3xy^2 + y^2)$$
$$= -3x^2 + 6xy + 2y$$

Therefore,

$$f_{xx} = \frac{\partial}{\partial x}(f_x) = \frac{\partial}{\partial x}(3x^2 - 6xy + 3y^2)$$
$$= 6x - 6y = 6(x - y)$$
$$f_{xy} = \frac{\partial}{\partial y}(f_x) = \frac{\partial}{\partial y}(3x^2 - 6xy + 3y^2)$$
$$= -6x + 6y = 6(y - x)$$
$$f_{yx} = \frac{\partial}{\partial x}(f_y) = \frac{\partial}{\partial x}(-3x^2 + 6xy + 2y)$$
$$= -6x + 6y = 6(y - x)$$
$$f_{yy} = \frac{\partial}{\partial y}(f_y) = \frac{\partial}{\partial y}(-3x^2 + 6xy + 2y)$$
$$= 6x + 2$$

Note that $f_{xy} = f_{yx}$ everywhere.

EXAMPLE 7 Find the second-order partial derivatives of the function

$$f(x, y) = e^{xy^2}$$

Solution We have

$$f_x = \frac{\partial}{\partial x}(e^{xy^2})$$
$$= y^2 e^{xy^2}$$
$$f_y = \frac{\partial}{\partial y}(e^{xy^2})$$
$$= 2xy e^{xy^2}$$

so the required second-order partial derivatives of f are

$$f_{xx} = \frac{\partial}{\partial x}(f_x) = \frac{\partial}{\partial x}(y^2 e^{xy^2})$$
$$= y^4 e^{xy^2}$$

$$f_{xy} = \frac{\partial}{\partial y}(f_x) = \frac{\partial}{\partial y}(y^2 e^{xy^2})$$
$$= 2y e^{xy^2} + 2xy^3 e^{xy^2} \qquad \text{(x^2) See page 9.}$$
$$= 2y e^{xy^2}(1 + xy^2)$$

$$f_{yx} = \frac{\partial}{\partial x}(f_y) = \frac{\partial}{\partial x}(2xy e^{xy^2})$$
$$= 2y e^{xy^2} + 2xy^3 e^{xy^2}$$
$$= 2y e^{xy^2}(1 + xy^2)$$

$$f_{yy} = \frac{\partial}{\partial y}(f_y) = \frac{\partial}{\partial y}(2xy e^{xy^2})$$
$$= 2x e^{xy^2} + (2xy)(2xy) e^{xy^2}$$
$$= 2x e^{xy^2}(1 + 2xy^2)$$

Note that $f_{xy} = f_{yx}$ everywhere.

8.2 Self-Check Exercises

1. Compute the first partial derivatives of $f(x, y) = x^3 - 2xy^2 + y^2 - 8$.

2. Find the first partial derivatives of $f(x, y) = x \ln y + y e^x - x^2$ at $(0, 1)$ and interpret your results.

3. Find the second-order partial derivatives of the function of Self-Check Exercise 1.

4. A certain country's production is described by the function
$$f(x, y) = 60 x^{1/3} y^{2/3}$$
when x units of labor and y units of capital are used.

  a. What are the marginal productivity of labor and the marginal productivity of capital when the amounts expended on labor and capital are 125 units and 8 units, respectively?

  b. Should the government encourage capital investment rather than increased expenditure on labor at this time to increase the country's productivity?

Solutions to Self-Check Exercises 8.2 can be found on page 558.

8.2 Concept Questions

1. a. What is the partial derivative of $f(x, y)$ with respect to x at (a, b)?
   b. Give a geometric interpretation of $f_x(a, b)$ and a practical interpretation of $f_x(a, b)$.

2. a. What are substitute commodities and complementary commodities? Give an example of each.
   b. Suppose $x = f(p, q)$ and $y = g(p, q)$ are demand functions for two commodities A and B, respectively. Give conditions for determining whether A and B are substitute or complementary commodities.

3. List all second-order partial derivatives of a function of two variables.

4. How many second-order partial derivatives are there for a function of three variables f, assuming all such derivatives exist? List all of them.

8.2 Exercises

1. Let $f(x, y) = x^2 + 2y^2$.
   a. Find $f_x(2, 1)$ and $f_y(2, 1)$.
   b. Interpret the numbers in part (a) as slopes.
   c. Interpret the numbers in part (a) as rates of change.

2. Let $f(x, y) = 9 - x^2 + xy - 2y^2$.
   a. Find $f_x(1, 2)$ and $f_y(1, 2)$.
   b. Interpret the numbers in part (a) as slopes.
   c. Interpret the numbers in part (a) as rates of change.

In Exercises 3–24, find the first partial derivatives of the function.

3. $f(x, y) = 2x + 3y + 5$
4. $f(x, y) = 2xy$
5. $g(x, y) = 2x^2 + 4y + 1$
6. $f(x, y) = 1 + x^2 + y^2$
7. $f(x, y) = \dfrac{2y}{x^2}$
8. $f(x, y) = \dfrac{x}{1 + y}$
9. $g(u, v) = \dfrac{u - v}{u + v}$
10. $f(x, y) = \dfrac{x^2 - y^2}{x^2 + y^2}$
11. $f(s, t) = (s^2 - st + t^2)^3$
12. $g(s, t) = s^2 t + st^{-3}$
13. $f(x, y) = (x^2 + y^2)^{2/3}$
14. $f(x, y) = x\sqrt{1 + y^2}$
15. $f(x, y) = e^{xy+1}$
16. $f(x, y) = (e^x + e^y)^5$
17. $f(x, y) = x \ln y + y \ln x$
18. $f(x, y) = x^2 e^{y^2}$
19. $g(u, v) = e^u \ln v$
20. $f(x, y) = \dfrac{e^{xy}}{x + y}$
21. $f(x, y, z) = xyz + xy^2 + yz^2 + zx^2$
22. $g(u, v, w) = \dfrac{2uvw}{u^2 + v^2 + w^2}$
23. $h(r, s, t) = e^{rst}$
24. $f(x, y, z) = xe^{y/z}$

In Exercises 25–34, evaluate the first partial derivatives of the function at the given point.

25. $f(x, y) = x^2 y + xy^2;\ (1, 2)$
26. $f(x, y) = x^2 + xy + y^2 + 2x - y;\ (-1, 2)$
27. $f(x, y) = x\sqrt{y} + y^2;\ (2, 1)$
28. $g(x, y) = \sqrt{x^2 + y^2};\ (3, 4)$
29. $f(x, y) = \dfrac{x}{y};\ (1, 2)$
30. $f(x, y) = \dfrac{x + y}{x - y};\ (1, -2)$
31. $f(x, y) = e^{xy};\ (1, 1)$
32. $f(x, y) = e^x \ln y;\ (0, e)$
33. $f(x, y, z) = x^2 yz^3;\ (1, 0, 2)$
34. $f(x, y, z) = x^2 y^2 + z^2;\ (1, 1, 2)$

In Exercises 35–42, find the second-order partial derivatives of the function. In each case, show that the mixed partial derivatives f_{xy} and f_{yx} are equal.

35. $f(x, y) = x^2 y + xy^3$
36. $f(x, y) = x^3 + x^2 y + x + 4$
37. $f(x, y) = x^2 - 2xy + 2y^2 + x - 2y$
38. $f(x, y) = x^3 + x^2 y^2 + y^3 + x + y$
39. $f(x, y) = \sqrt{x^2 + y^2}$
40. $f(x, y) = x\sqrt{y} + y\sqrt{x}$
41. $f(x, y) = e^{-x/y}$
42. $f(x, y) = \ln(1 + x^2 y^2)$

43. **PRODUCTIVITY OF A COUNTRY** The productivity of a South American country is given by the function
$$f(x, y) = 20x^{3/4} y^{1/4}$$
when x units of labor and y units of capital are used.
a. What are the marginal productivity of labor and the marginal productivity of capital when the amounts expended on labor and capital are 256 units and 16 units, respectively?
b. Should the government encourage capital investment rather than increased expenditure on labor at this time to increase the country's productivity?

44. **PRODUCTIVITY OF A COUNTRY** The productivity of a country in Western Europe is given by the function
$$f(x, y) = 40x^{4/5} y^{1/5}$$
when x units of labor and y units of capital are used.
a. What are the marginal productivity of labor and the marginal productivity of capital when the amounts expended on labor and capital are 32 units and 243 units, respectively?
b. Should the government encourage capital investment rather than increased expenditure on labor at this time to increase the country's productivity?

45. **LAND PRICES** The rectangular region R shown in the following figure represents a city's financial district. The price of land within the district is approximated by the function
$$p(x, y) = 200 - 10\left(x - \dfrac{1}{2}\right)^2 - 15(y - 1)^2$$
where $p(x, y)$ is the price of land at the point (x, y) in dollars per square foot and x and y are measured in miles. Compute
$$\dfrac{\partial p}{\partial x}(0, 1) \quad \text{and} \quad \dfrac{\partial p}{\partial y}(0, 1)$$
and interpret your results.

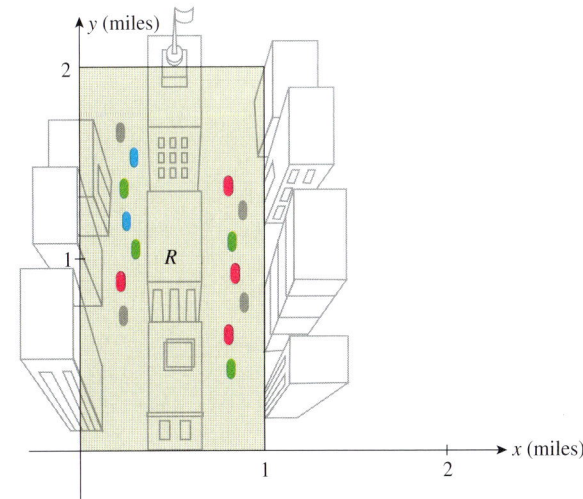

46. COMPLEMENTARY AND SUBSTITUTE COMMODITIES In a survey conducted by a video magazine, it was determined that the demand equation for DVD players is given by

$$x = f(p, q) = 10{,}000 - 10p + 0.2q^2$$

and the demand equation for Blu-ray players is given by

$$y = g(p, q) = 5000 + 0.8p^2 - 20q$$

where p and q denote the unit prices (in dollars) for the DVD and Blu-ray players, respectively, and x and y denote the number of DVD and Blu-ray players demanded per week. Determine whether these two products are substitute, complementary, or neither.

47. COMPLEMENTARY AND SUBSTITUTE COMMODITIES In a survey, it was determined that the demand equation for DVD players is given by

$$x = f(p, q) = 10{,}000 - 10p - e^{0.5q}$$

and the demand equation for blank DVD discs is given by

$$y = g(p, q) = 50{,}000 - 4000q - 10p$$

where p and q denote the unit prices, respectively, and x and y denote the number of DVD players and the number of blank DVD discs demanded each week. Determine whether these two products are substitute, complementary, or neither.

48. COMPLEMENTARY AND SUBSTITUTE COMMODITIES Refer to Exercise 37, Exercises 8.1. Show that the finished and unfinished home furniture manufactured by Country Workshop are substitute commodities.

Hint: Solve the system of equations for x and y in terms of p and q.

49. REVENUE FUNCTIONS The total weekly revenue (in dollars) of Country Workshop associated with manufacturing and selling their rolltop desks is given by the function

$$R(x, y) = -0.2x^2 - 0.25y^2 - 0.2xy + 200x + 160y$$

where x denotes the number of finished units and y denotes the number of unfinished units manufactured and sold each week. Compute $\partial R/\partial x$ and $\partial R/\partial y$ when $x = 300$ and $y = 250$. Interpret your results.

50. PROFIT FUNCTIONS The monthly profit (in dollars) of Bond and Barker Department Store depends on the level of inventory x (in thousands of dollars) and the floor space y (in thousands of square feet) available for display of the merchandise, as given by the equation

$$P(x, y) = -0.02x^2 - 15y^2 + xy \\ + 39x + 25y - 20{,}000$$

Compute $\partial P/\partial x$ and $\partial P/\partial y$ when $x = 4000$ and $y = 150$. Interpret your results. Repeat with $x = 5000$ and $y = 150$.

For Exercises 51–53, let $x = f(p, q)$ be the demand equation for the commodities A and B, where p and q are the respective unit prices. The **elasticity of demand for A** is

$$E_p = -\frac{p \dfrac{\partial x}{\partial p}}{x}$$

and the **cross elasticity of demand for A with respect to q** is

$$E_q = -\frac{q \dfrac{\partial x}{\partial q}}{x}$$

(see Section 3.4).

51. ELASTICITY OF DEMAND The demand equation for Product A is

$$x = 400 - 8p^2 + 0.4q$$

where x is the quantity demanded of Product A, and p and q are the respective unit prices of Product A and a related Product B. Compute E_p and E_q when $p = 6$ and $q = 40$, and interpret your results.

52. ELASTICITY OF DEMAND Suppose that the daily demand for butter is given by

$$x = f(p, q) = \frac{3q}{1 + p^2}$$

where p and q denote the prices per pound (in dollars) of butter and margarine, respectively, and x is measured in millions of pounds. Compute E_p and E_q where $p = 5$ and $q = 4$. Interpret your results.

53. ELASTICITY OF DEMAND Suppose that the daily demand for margarine is given by

$$x = g(p, q) = \frac{2q}{1 + \sqrt{p}} \quad (p, q > 0)$$

where p and q denote the prices per pound (in dollars) of margarine and butter, respectively, and x is measured in millions of pounds. Compute E_p and E_q where $p = 4$ and $q = 5$. Interpret your results.

54. WIND CHILL FACTOR The wind chill temperature is the temperature that you would feel in still air that is the same as the actual temperature when the presence of wind is taken into consideration. The following table gives the wind chill temperature $T = f(t, s)$ in degrees Fahrenheit in terms of the actual air temperature t in degrees Fahrenheit and the wind speed s in mph.

			Wind speed (mph)				
	s \ t	10	15	20	25	30	35	40
Actual air temperature (°F)	30	21.2	19.0	17.4	16.0	14.9	13.9	13.0
	32	23.7	21.6	20.0	18.7	17.6	16.6	15.8
	34	26.2	24.2	22.6	21.4	20.3	19.4	18.6
	36	28.7	26.7	25.2	24.0	23.0	22.2	21.4
	38	31.2	29.3	27.9	26.7	25.7	24.9	24.2
	40	33.6	31.8	30.5	29.4	28.5	27.7	26.9

a. Estimate the rate of change of the wind chill temperature T with respect to the actual air temperature when the wind speed is constant at 25 mph and the actual air temperature is 34°F.

Hint: Show that it is given by

$$\frac{\partial T}{\partial t}(34, 25) \approx \frac{f(36, 25) - f(34, 25)}{2}$$

b. Estimate the rate of change of the wind chill temperature T with respect to the wind speed when the actual air temperature is constant at 34°F and the wind speed is 25 mph.

Source: National Weather Service.

55. WIND CHILL FACTOR A formula used by meteorologists to calculate the wind chill temperature (the temperature that you feel in still air that is the same as the actual temperature when the presence of wind is taken into consideration) is

$$T = f(t, s) = 35.74 + 0.6215t - 35.75s^{0.16} + 0.4275ts^{0.16}$$
$$(s \geq 1)$$

where t is the actual air temperature in degrees Fahrenheit and s is the wind speed in mph.
a. What is the wind chill temperature when the actual air temperature is 32°F and the wind speed is 20 mph?
b. If the temperature is 32°F, by how much approximately will the wind chill temperature change if the wind speed increases from 20 mph to 21 mph?

56. ARSON STUDY A study of arson for profit conducted for a certain city found that the number of suspicious fires is approximated by the formula

$$N(x, y) = \frac{120\sqrt{1000 + 0.03x^2 y}}{(5 + 0.2y)^2}$$
$$(0 \leq x \leq 150, 5 \leq y \leq 35)$$

where x denotes the number of persons per census tract and y denotes the level of reinvestment in conventional mortgages by the city's ten largest banks measured in cents per dollars deposited. Compute $\partial N/\partial x$ and $\partial N/\partial y$ when $x = 100$ and $y = 20$, and interpret your results.

57. ENGINE EFFICIENCY The efficiency of an internal combustion engine is given by

$$E = \left(1 - \frac{v}{V}\right)^{0.4}$$

where V and v are the respective maximum and minimum volumes of air in each cylinder.
a. Show that $\partial E/\partial V > 0$ and interpret your result.
b. Show that $\partial E/\partial v < 0$ and interpret your result.

58. SURFACE AREA OF A HUMAN BODY The formula

$$S = 0.007184 W^{0.425} H^{0.725}$$

gives the surface area S of a human body (in square meters) in terms of its weight W (in kilograms) and its height H (in centimeters). Compute $\partial S/\partial W$ and $\partial S/\partial H$ when $W = 70$ kg and $H = 180$ cm. Interpret your results.

59. VOLUME OF A GAS The volume V (in liters) of a certain mass of gas is related to its pressure P (in millimeters of mercury) and its temperature T (in kelvins) by the law

$$V = \frac{30.9T}{P}$$

Compute $\partial V/\partial T$ and $\partial V/\partial P$ when $T = 300$ and $P = 800$. Interpret your results.

60. According to the *ideal gas law*, the volume V (in liters) of an ideal gas is related to its pressure P (in pascals) and temperature T (in kelvins) by the formula

$$V = \frac{kT}{P}$$

where k is a constant. Show that

$$\frac{\partial V}{\partial T} \cdot \frac{\partial T}{\partial P} \cdot \frac{\partial P}{\partial V} = -1$$

61. KINETIC ENERGY OF A BODY The kinetic energy K of a body of mass m and velocity v is given by

$$K = \frac{1}{2} mv^2$$

Show that $\dfrac{\partial K}{\partial m} \cdot \dfrac{\partial^2 K}{\partial v^2} = K$.

62. COBB–DOUGLAS PRODUCTION FUNCTION Show that the Cobb–Douglas production function $P = kx^{\alpha} y^{1-\alpha}$, where $0 < \alpha < 1$, satisfies the equation

$$x \frac{\partial P}{\partial x} + y \frac{\partial P}{\partial y} = P$$

This equation is called Euler's equation.

In Exercises 63–68, determine whether the statement is true or false. If it is true, explain why it is true. If it is false, give an example to show why it is false.

63. If $f_x(x, y)$ is defined at (a, b), then $f_y(x, y)$ must also be defined at (a, b).

64. If $f(x, y)$ is continuous at (a, b), then both $f_x(a, b)$ and $f_y(a, b)$ exist.

65. If $f_x(x, y) = 0$ and $f_y(x, y) = 0$ for all x and y, then f must be a constant function.

66. If $f_x(a, b) < 0$, then f is decreasing with respect to x near (a, b).

67. If $f_{xy}(x, y)$ and $f_{yx}(x, y)$ are both continuous for all values of x and y, then $f_{xy} = f_{yx}$ for all values of x and y.

68. If both f_{xy} and f_{yx} are defined at (a, b), then f_{xx} and f_{yy} must be defined at (a, b).

8.2 Solutions to Self-Check Exercises

1. $f_x = \dfrac{\partial f}{\partial x} = 3x^2 - 2y^2$

$f_y = \dfrac{\partial f}{\partial y} = -2x(2y) + 2y$
$= 2y(1 - 2x)$

2. $f_x = \ln y + ye^x - 2x;\ f_y = \dfrac{x}{y} + e^x$

In particular,

$f_x(0, 1) = \ln 1 + 1e^0 - 2(0) = 1$

$f_y(0, 1) = \dfrac{0}{1} + e^0 = 1$

The results tell us that at the point $(0, 1)$, $f(x, y)$ increases 1 unit for each unit increase in the x-direction, y being kept constant; $f(x, y)$ also increases 1 unit for each unit increase in the y-direction, x being kept constant.

3. From the results of Self-Check Exercise 1,

$f_x = 3x^2 - 2y^2$

Therefore,

$f_{xx} = \dfrac{\partial}{\partial x}(3x^2 - 2y^2) = 6x$

$f_{xy} = \dfrac{\partial}{\partial y}(3x^2 - 2y^2) = -4y$

Also, from the results of Self-Check Exercise 1,

$f_y = 2y(1 - 2x)$

Thus,

$f_{yx} = \dfrac{\partial}{\partial x}[2y(1 - 2x)] = -4y$

$f_{yy} = \dfrac{\partial}{\partial y}[2y(1 - 2x)] = 2(1 - 2x)$

4. a. The marginal productivity of labor when the amounts expended on labor and capital are x and y units, respectively, is given by

$f_x(x, y) = 60\left(\dfrac{1}{3}x^{-2/3}\right)y^{2/3} = 20\left(\dfrac{y}{x}\right)^{2/3}$

In particular, the required marginal productivity of labor is given by

$f_x(125, 8) = 20\left(\dfrac{8}{125}\right)^{2/3} = 20\left(\dfrac{4}{25}\right)$

or 3.2 units/unit increase in labor expenditure, capital expenditure being held constant at 8 units. Next, we compute

$f_y(x, y) = 60x^{1/3}\left(\dfrac{2}{3}y^{-1/3}\right) = 40\left(\dfrac{x}{y}\right)^{1/3}$

and deduce that the required marginal productivity of capital is given by

$f_y(125, 8) = 40\left(\dfrac{125}{8}\right)^{1/3} = 40\left(\dfrac{5}{2}\right)$

or 100 units/unit increase in capital expenditure, labor expenditure being held constant at 125 units.

b. The results of part (a) tell us that the government should encourage increased spending on capital rather than on labor.

USING TECHNOLOGY

Finding Partial Derivatives at a Given Point

Suppose $f(x, y)$ is a function of two variables and we wish to compute

$$f_x(a, b) = \dfrac{\partial f}{\partial x}\bigg|_{(a, b)}$$

Recall that in computing $\partial f/\partial x$, we think of y as being fixed. But in this situation, we are evaluating $\partial f/\partial x$ at (a, b). Therefore, we set y equal to b. Doing this leads to the function g of one variable, x, defined by

$$g(x) = f(x, b)$$

It follows from the definition of the partial derivative that

$$f_x(a, b) = g'(a)$$

Thus, the value of the partial derivative $\partial f/\partial x$ at a given point (a, b) can be found by evaluating the derivative of a function of one variable. In particular, the latter can be found by using the numerical derivative operation of a graphing utility. We find $f_y(a, b)$ in a similar manner.

EXAMPLE 1 Let $f(x, y) = (1 + xy^2)^{3/2} e^{x^2 y}$. Find (a) $f_x(1, 2)$ and (b) $f_y(1, 2)$.

Solution

a. Define $g(x) = f(x, 2) = (1 + 4x)^{3/2} e^{2x^2}$. Using the numerical derivative operation to find $g'(1)$, we obtain

$$f_x(1, 2) = g'(1) \approx 429.585835$$

b. Define $h(y) = f(1, y) = (1 + y^2)^{3/2} e^y$. Using the numerical derivative operation to find $h'(2)$, we obtain

$$f_y(1, 2) = h'(2) \approx 181.7468642$$

TECHNOLOGY EXERCISES

For each of the following functions, compute

$$\frac{\partial f}{\partial x} \quad \text{and} \quad \frac{\partial f}{\partial y}$$

at the given point:

1. $f(x, y) = \sqrt{x}(2 + xy^2)^{1/3}$; $(1, 2)$
2. $f(x, y) = \sqrt{xy}(1 + 2xy)^{2/3}$; $(1, 4)$
3. $f(x, y) = \dfrac{x + y^2}{1 + x^2 y}$; $(1, 2)$
4. $f(x, y) = \dfrac{xy^2}{(\sqrt{x} + \sqrt{y})^2}$; $(4, 1)$
5. $f(x, y) = e^{-xy^2}(x + y)^{1/3}$; $(1, 1)$
6. $f(x, y) = \dfrac{\ln(\sqrt{x} + y^2)}{x^2 + y^2}$; $(4, 1)$

8.3 Maxima and Minima of Functions of Several Variables

Maxima and Minima

In Chapter 4, we saw that the solution of a problem often reduces to finding the extreme values of a function of one variable. In practice, however, situations also arise in which a problem is solved by finding the absolute maximum or absolute minimum value of a function of two or more variables.

For example, suppose Scandi Company manufactures computer desks in both assembled and unassembled versions. Its profit P is therefore a function of the number of assembled units, x, and the number of unassembled units, y, manufactured and sold per week; that is, $P = f(x, y)$. A question of paramount importance to the manufacturer is: How many assembled and unassembled desks should the company manufacture per week to maximize its weekly profit? Mathematically, the problem is solved by finding the values of x and y that will make $f(x, y)$ a maximum.

In this section, we will focus our attention on finding the extrema of a function of two variables. As in the case of a function of one variable, we distinguish between the relative (or local) extrema and the absolute extrema of a function of two variables.

Relative Extrema of a Function of Two Variables

Let f be a function defined on a region R containing the point (a, b). Then f has a **relative maximum** at (a, b) if $f(x, y) \leq f(a, b)$ for all points (x, y) that are sufficiently close to (a, b). The number $f(a, b)$ is called a **relative maximum value**. Similarly, f has a **relative minimum** at (a, b), with **relative minimum value** $f(a, b)$, if $f(x, y) \geq f(a, b)$ for all points (x, y) that are sufficiently close to (a, b).

Loosely speaking, f has a relative maximum at (a, b) if the point $(a, b, f(a, b))$ is the highest point on the graph of f when compared with all nearby points. A similar interpretation holds for a relative minimum.

If the inequalities in this last definition hold for *all* points (x, y) in the domain of f, then f has an **absolute maximum** (or **absolute minimum**) at (a, b) with **absolute maximum value** (or **absolute minimum value**) $f(a, b)$. Figure 16 shows the graph of a function with relative maxima at (a, b) and (e, g) and a relative minimum at (c, d). The absolute maximum of f occurs at (e, g), and the absolute minimum of f occurs at (h, i).

Observe that in the case of a function of one variable, a relative extremum (relative maximum or relative minimum) may or may not be an absolute extremum.

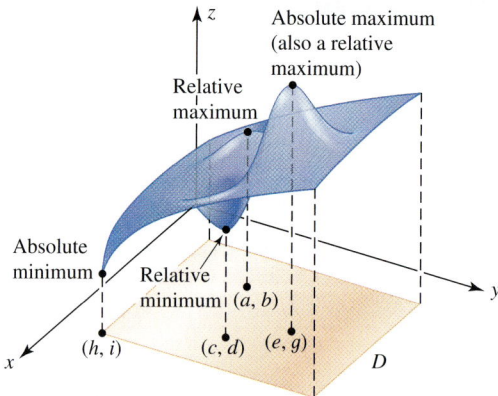

FIGURE 16
The relative and absolute extrema of the function f over the domain D

Now let's turn our attention to the study of relative extrema of a function. Suppose that a differentiable function $f(x, y)$ of two variables has a relative maximum (relative minimum) at a point (a, b) in the domain of f. From Figure 17, it is clear that at the point (a, b) the slopes of the "tangent lines" to the surface in any direction must be zero. In particular, this implies that both

$$\frac{\partial f}{\partial x}(a, b) \quad \text{and} \quad \frac{\partial f}{\partial y}(a, b)$$

must be zero.

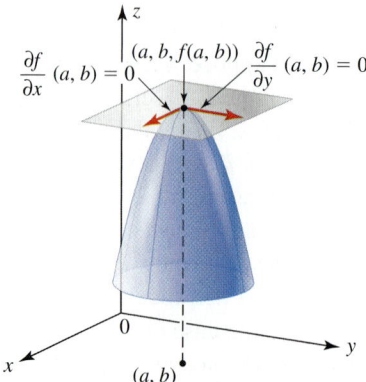 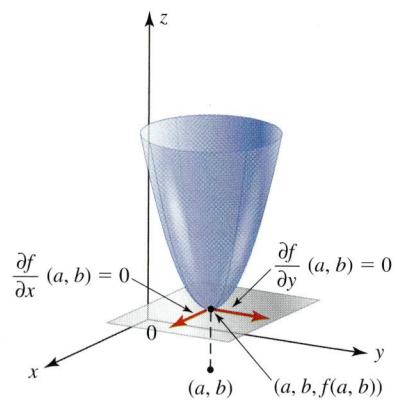

FIGURE 17

(a) f has a relative maximum at (a, b).

(b) f has a relative minimum at (a, b).

Lest we be tempted to jump to the conclusion that a differentiable function f satisfying both the conditions

$$\frac{\partial f}{\partial x}(a, b) = 0 \quad \text{and} \quad \frac{\partial f}{\partial y}(a, b) = 0$$

at a point (a, b) must have a relative extremum at the point (a, b), let's examine the graph of the function f depicted in Figure 18. Here, both

$$\frac{\partial f}{\partial x}(a, b) = 0 \quad \text{and} \quad \frac{\partial f}{\partial y}(a, b) = 0$$

but f has neither a relative maximum nor a relative minimum at the point (a, b) because some nearby points are higher and some are lower than the point $(a, b, f(a, b))$. The point $(a, b, f(a, b))$ is called a **saddle point.**

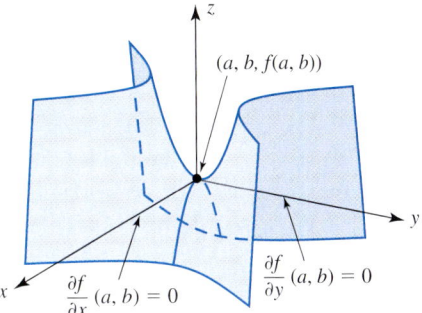

FIGURE 18
The point $(a, b, f(a, b))$ is called a saddle point.

Finally, an examination of the graph of the function f depicted in Figure 19 should convince you that f has a relative maximum at the point (a, b). But both $\partial f/\partial x$ and $\partial f/\partial y$ fail to be defined at (a, b).

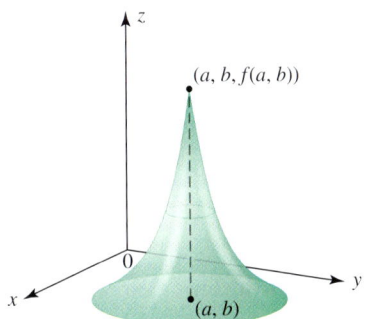

FIGURE 19
f has a relative maximum at (a, b), but neither $\partial f/\partial x$ nor $\partial f/\partial y$ exists at (a, b).

To summarize, a function f of two variables can have a relative extremum only at a point (a, b) in its domain where $\partial f/\partial x$ and $\partial f/\partial y$ both exist and are equal to zero at (a, b) or at least one of the partial derivatives does not exist. As in the case of one variable, we refer to a point in the domain of f that *may* give rise to a relative extremum as a critical point. The precise definition follows.

Critical Point of f

A **critical point** of f is a point (a, b) in the domain of f such that both

$$\frac{\partial f}{\partial x}(a, b) = 0 \quad \text{and} \quad \frac{\partial f}{\partial y}(a, b) = 0$$

or at least one of the partial derivatives does not exist.

To determine the nature of a critical point of a function $f(x, y)$ of two variables, we use the second partial derivatives of f. The resulting test, which helps us to classify these points, is called the **second derivative test** and is incorporated in the following procedure for finding and classifying the relative extrema of f.

Determining Relative Extrema

1. Find the critical points of $f(x, y)$ by solving the system of simultaneous equations
$$f_x(x, y) = 0$$
$$f_y(x, y) = 0$$

2. The second derivative test: Let
$$D(x, y) = f_{xx}(x, y)f_{yy}(x, y) - f_{xy}^2(x, y)$$
Then
   a. $D(a, b) > 0$ and $f_{xx}(a, b) < 0$ implies that $f(x, y)$ has a **relative maximum** at the point (a, b).
   b. $D(a, b) > 0$ and $f_{xx}(a, b) > 0$ implies that $f(x, y)$ has a **relative minimum** at the point (a, b).
   c. $D(a, b) < 0$ implies that $f(x, y)$ has neither a relative maximum nor a relative minimum at the point (a, b). The point $(a, b, f(a, b))$ is called a **saddle point**.
   d. $D(a, b) = 0$ implies that the test is inconclusive, so some other technique must be used to solve the problem.

Note We can replace $f_{xx}(a, b)$ by $f_{yy}(a, b)$ in 2a and 2b because $D(a, b) > 0$ implies that $f_{xx}(a, b)$ and $f_{yy}(a, b)$ must have the same sign. ■

EXAMPLE 1 Find the relative extrema of the function
$$f(x, y) = x^2 + y^2$$

Solution We have
$$f_x(x, y) = 2x$$
$$f_y(x, y) = 2y$$

To find the critical point(s) of f, we set $f_x(x, y) = 0$ and $f_y(x, y) = 0$ and solve the resulting system of simultaneous equations $2x = 0$ and $2y = 0$. We obtain $x = 0$, $y = 0$, or $(0, 0)$, as the sole critical point of f. Next, we apply the second derivative test to determine the nature of the critical point $(0, 0)$. We compute
$$f_{xx}(x, y) = 2 \qquad f_{xy}(x, y) = 0 \qquad f_{yy}(x, y) = 2$$
and
$$D(x, y) = f_{xx}(x, y)f_{yy}(x, y) - f_{xy}^2(x, y) = (2)(2) - 0 = 4$$

In particular, $D(0, 0) = 4$. Since $D(0, 0) > 0$ and $f_{xx}(0, 0) = 2 > 0$, we conclude that $f(x, y)$ has a relative minimum at the point $(0, 0)$. The relative minimum value, 0, also happens to be the absolute minimum of f. The graph of the function f, shown in Figure 20, confirms these results. ■

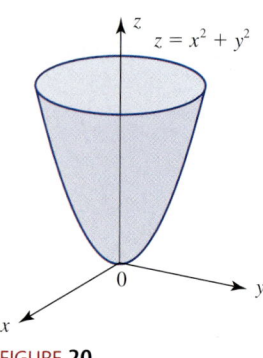

FIGURE 20
The graph of $f(x, y) = x^2 + y^2$

8.3 MAXIMA AND MINIMA OF FUNCTIONS OF SEVERAL VARIABLES

EXAMPLE 2 Find the relative extrema of the function
$$f(x, y) = 3x^2 - 4xy + 4y^2 - 4x + 8y + 4$$

Solution We have
$$f_x(x, y) = 6x - 4y - 4$$
$$f_y(x, y) = -4x + 8y + 8$$

To find the critical points of f, we set $f_x(x, y) = 0$ and $f_y(x, y) = 0$ and solve the resulting system of simultaneous equations
$$6x - 4y = 4$$
$$-4x + 8y = -8$$

Multiplying the first equation by 2 and the second equation by 3, we obtain the equivalent system
$$12x - 8y = 8$$
$$-12x + 24y = -24$$

Adding the two equations gives $16y = -16$, or $y = -1$. We substitute this value for y into either equation in the system to get $x = 0$. Thus, the only critical point of f is the point $(0, -1)$. Next, we apply the second derivative test to determine whether the point $(0, -1)$ gives rise to a relative extremum of f. We compute
$$f_{xx}(x, y) = 6 \quad f_{xy}(x, y) = -4 \quad f_{yy}(x, y) = 8$$
and
$$D(x, y) = f_{xx}(x, y)f_{yy}(x, y) - f_{xy}^2(x, y) = (6)(8) - (-4)^2 = 32$$

Since $D(0, -1) = 32 > 0$ and $f_{xx}(0, -1) = 6 > 0$, we conclude that $f(x, y)$ has a relative minimum at the point $(0, -1)$. The value of $f(x, y)$ at the point $(0, -1)$ is given by
$$f(0, -1) = 3(0)^2 - 4(0)(-1) + 4(-1)^2 - 4(0) + 8(-1) + 4 = 0$$

Explore & Discuss

Suppose $f(x, y)$ has a relative extremum (relative maximum or relative minimum) at a point (a, b). Let $g(x) = f(x, b)$ and $h(y) = f(a, y)$. Assuming that f and g are differentiable, explain why $g'(a) = 0$ and $h'(b) = 0$. Explain why these results are equivalent to the conditions $f_x(a, b) = 0$ and $f_y(a, b) = 0$.

VIDEO **EXAMPLE 3** Find the relative extrema of the function
$$f(x, y) = 4y^3 + x^2 - 12y^2 - 36y + 2$$

Solution To find the critical points of f, we set $f_x = 0$ and $f_y = 0$ simultaneously, obtaining
$$f_x(x, y) = 2x = 0$$
$$f_y(x, y) = 12y^2 - 24y - 36 = 0$$

The first equation implies that $x = 0$. The second equation implies that
$$y^2 - 2y - 3 = 0$$
$$(y + 1)(y - 3) = 0$$

—that is, $y = -1$ or 3. Therefore, there are two critical points of the function f: $(0, -1)$ and $(0, 3)$.

Next, we apply the second derivative test to determine the nature of each of the two critical points. We compute
$$f_{xx}(x, y) = 2 \quad f_{xy}(x, y) = 0 \quad f_{yy}(x, y) = 24y - 24 = 24(y - 1)$$

Therefore,
$$D(x, y) = f_{xx}(x, y)f_{yy}(x, y) - f_{xy}^2(x, y) = 48(y - 1)$$

For the point $(0, -1)$,
$$D(0, -1) = 48(-1 - 1) = -96 < 0$$

Explore & Discuss

1. Refer to the second derivative test. Can the condition $f_{xx}(a, b) < 0$ in part 2a be replaced by the condition $f_{yy}(a, b) < 0$? Explain your answer. How about the condition $f_{xx}(a, b) > 0$ in part 2b?

2. Let $f(x, y) = x^4 + y^4$.
   a. Show that $(0, 0)$ is a critical point of f and that $D(0, 0) = 0$.
   b. Explain why f has a relative (in fact, an absolute) minimum at $(0, 0)$. Does this contradict the second derivative test? Explain your answer.

Since $D(0, -1) < 0$, we conclude that the point $(0, -1)$ gives a saddle point of f. For the point $(0, 3)$,

$$D(0, 3) = 48(3 - 1) = 96 > 0$$

Since $D(0, 3) > 0$ and $f_{xx}(0, 3) > 0$, we conclude that the function f has a relative minimum at the point $(0, 3)$. Furthermore, since

$$f(0, 3) = 4(3)^3 + (0)^2 - 12(3)^2 - 36(3) + 2$$
$$= -106$$

we see that the relative minimum value of f is -106.

As in the case of a practical optimization problem involving a function of one variable, the solution to an optimization problem involving a function of several variables calls for finding the *absolute* extremum of the function. Determining the absolute extremum of a function of several variables is more difficult than merely finding the relative extrema of the function. However, in many situations, the absolute extremum of a function actually coincides with the largest relative extremum of the function that occurs in the interior of its domain. We assume that the problems considered here belong to this category. Furthermore, the existence of the absolute extremum (solution) of a practical problem is often deduced from the geometric or practical nature of the problem.

APPLIED EXAMPLE 4 Maximizing Profits The total weekly revenue (in dollars) that Acrosonic realizes in producing and selling its bookshelf loudspeaker systems is given by

$$R(x, y) = -\frac{1}{4}x^2 - \frac{3}{8}y^2 - \frac{1}{4}xy + 300x + 240y$$

where x denotes the number of fully assembled units and y denotes the number of kits produced and sold each week. The total weekly cost attributable to the production of these loudspeakers is

$$C(x, y) = 180x + 140y + 5000$$

dollars, where x and y have the same meaning as before. Determine how many assembled units and how many kits Acrosonic should produce per week to maximize its profit. What is the maximum profit?

Solution The contribution to Acrosonic's weekly profit stemming from the production and sale of the bookshelf loudspeaker systems is given by

$$P(x, y) = R(x, y) - C(x, y)$$
$$= \left(-\frac{1}{4}x^2 - \frac{3}{8}y^2 - \frac{1}{4}xy + 300x + 240y\right) - (180x + 140y + 5000)$$
$$= -\frac{1}{4}x^2 - \frac{3}{8}y^2 - \frac{1}{4}xy + 120x + 100y - 5000$$

To find the relative maximum of the profit function $P(x, y)$, we first locate the critical point(s) of P. Setting $P_x(x, y)$ and $P_y(x, y)$ equal to zero, we obtain

$$P_x = -\frac{1}{2}x - \frac{1}{4}y + 120 = 0$$

$$P_y = -\frac{3}{4}y - \frac{1}{4}x + 100 = 0$$

Solving the first of these equations for y yields

$$y = -2x + 480$$

which, upon substitution into the second equation, yields

$$-\frac{3}{4}(-2x + 480) - \frac{1}{4}x + 100 = 0$$
$$6x - 1440 - x + 400 = 0$$
$$x = 208$$

We substitute this value of x into the equation $y = -2x + 480$ to get

$$y = 64$$

Therefore, the function P has the sole critical point $(208, 64)$. To show that the point $(208, 64)$ is a solution to our problem, we use the second derivative test. We compute

$$P_{xx} = -\frac{1}{2} \qquad P_{xy} = -\frac{1}{4} \qquad P_{yy} = -\frac{3}{4}$$

So,

$$D(x, y) = \left(-\frac{1}{2}\right)\left(-\frac{3}{4}\right) - \left(-\frac{1}{4}\right)^2 = \frac{3}{8} - \frac{1}{16} = \frac{5}{16}$$

Since $D(208, 64) > 0$ and $P_{xx}(208, 64) < 0$, the point $(208, 64)$ yields a relative maximum of P. It can be shown that this relative maximum is also the absolute maximum of P. We conclude that Acrosonic can maximize its weekly profit by manufacturing 208 assembled units and 64 kits of their bookshelf loudspeaker systems. The maximum weekly profit realizable from the production and sale of these loudspeaker systems is given by

$$P(208, 64) = -\frac{1}{4}(208)^2 - \frac{3}{8}(64)^2 - \frac{1}{4}(208)(64)$$
$$+ 120(208) + 100(64) - 5000$$
$$= 10{,}680$$

or \$10,680.

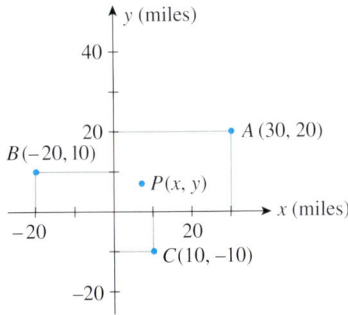

FIGURE 21
Locating a site for a television relay station

APPLIED EXAMPLE 5 Locating a Television Relay Station Site A television relay station will serve Towns A, B, and C, whose relative locations are shown in Figure 21. Determine a site for the location of the station if the sum of the squares of the distances from each town to the site is minimized.

Solution Suppose the required site is located at the point $P(x, y)$. With the aid of the distance formula, we find that the square of the distance from Town A to the site is

$$(x - 30)^2 + (y - 20)^2 \qquad (x^2) \text{ See page 26.}$$

The respective distances from Towns B and C to the site are found in a similar manner, so the sum of the squares of the distances from each town to the site is given by

$$f(x, y) = (x - 30)^2 + (y - 20)^2 + (x + 20)^2$$
$$+ (y - 10)^2 + (x - 10)^2 + (y + 10)^2$$

To find the relative minimum of $f(x, y)$, we first find the critical point(s) of f. Using the Chain Rule to find $f_x(x, y)$ and $f_y(x, y)$ and setting each equal to zero, we obtain

$$f_x = 2(x - 30) + 2(x + 20) + 2(x - 10) = 6x - 40 = 0$$
$$f_y = 2(y - 20) + 2(y - 10) + 2(y + 10) = 6y - 40 = 0$$

from which we deduce that $(\frac{20}{3}, \frac{20}{3})$ is the sole critical point of f. Since

$$f_{xx} = 6 \quad f_{xy} = 0 \quad f_{yy} = 6$$

we have

$$D(x, y) = f_{xx} f_{yy} - f_{xy}^2 = (6)(6) - 0 = 36$$

Since $D(\frac{20}{3}, \frac{20}{3}) > 0$ and $f_{xx}(\frac{20}{3}, \frac{20}{3}) > 0$, we conclude that the point $(\frac{20}{3}, \frac{20}{3})$ yields a relative minimum of f. Thus, the required site has coordinates $x = \frac{20}{3}$ and $y = \frac{20}{3}$.

8.3 Self-Check Exercises

1. Let $f(x, y) = 2x^2 + 3y^2 - 4xy + 4x - 2y + 3$.
   a. Find the critical point of f.
   b. Use the second derivative test to classify the nature of the critical point.
   c. Find the relative extremum of f, if it exists.

2. Robertson Controls manufactures two basic models of setback thermostats: a standard mechanical thermostat and a deluxe electronic thermostat. Robertson's monthly revenue (in hundreds of dollars) is

$$R(x, y) = -\frac{1}{8}x^2 - \frac{1}{2}y^2 - \frac{1}{4}xy + 20x + 60y$$

where x (in units of a hundred) denotes the number of mechanical thermostats manufactured each month and y (in units of a hundred) denotes the number of electronic thermostats manufactured each month. The total monthly cost incurred in producing these thermostats is

$$C(x, y) = 7x + 20y + 280$$

hundred dollars. Find how many thermostats of each model Robertson should manufacture each month to maximize its profits. What is the maximum profit?

Solutions to Self-Check Exercises 8.3 can be found on page 569.

8.3 Concept Questions

1. Explain the terms (a) relative maximum of a function $f(x, y)$ and (b) absolute maximum of a function $f(x, y)$.

2. a. What is a critical point of a function $f(x, y)$?
   b. Explain the role of a critical point in determining the relative extrema of a function of two variables.

3. Explain how the second derivative test is used to determine the relative extrema of a function of two variables.

4. In (a)–(d), suppose that (a, b) is a critical point of f. Are the given conditions sufficient for you to determine whether f has a relative maximum, a relative minimum, or a saddle point at (a, b)? Explain.
   a. $f_{xx}(a, b) = -2, f_{xy}(a, b) = 3, f_{yy}(a, b) = -5$
   b. $f_{xx}(a, b) = 3, f_{xy}(a, b) = 3, f_{yy}(a, b) = 2$
   c. $f_{xx}(a, b) = 1, f_{xy}(a, b) = 2, f_{yy}(a, b) = 4$
   d. $f_{xx}(a, b) = 2, f_{xy}(a, b) = 2, f_{yy}(a, b) = 4$

8.3 Exercises

In Exercises 1–20, find the critical point(s) of the function. Then use the second derivative test to classify the nature of each point, if possible. Finally, determine the relative extrema of the function.

1. $f(x, y) = 1 - 2x^2 - 3y^2$
2. $f(x, y) = x^2 - xy + y^2 + 1$
3. $f(x, y) = x^2 - y^2 - 2x + 4y + 1$
4. $f(x, y) = 2x^2 + y^2 - 4x + 6y + 3$
5. $f(x, y) = x^2 + 2xy + 2y^2 - 4x + 8y - 1$
6. $f(x, y) = x^2 - 4xy + 2y^2 + 4x + 8y - 1$
7. $f(x, y) = 2x^3 + y^2 - 9x^2 - 4y + 12x - 2$
8. $f(x, y) = 2x^3 + y^2 - 6x^2 - 4y + 12x - 2$
9. $f(x, y) = x^3 + y^2 - 2xy + 7x - 8y + 4$
10. $f(x, y) = 2y^3 - 3y^2 - 12y + 2x^2 - 6x + 2$
11. $f(x, y) = x^3 - 3xy + y^3 - 2$
12. $f(x, y) = x^3 - 2xy + y^2 + 5$
13. $f(x, y) = xy + \frac{4}{x} + \frac{2}{y}$
14. $f(x, y) = \frac{x}{y^2} + xy$
15. $f(x, y) = x^2 - e^{y^2}$
16. $f(x, y) = e^{x^2-y^2}$
17. $f(x, y) = e^{x^2+y^2}$
18. $f(x, y) = e^{xy}$
19. $f(x, y) = \ln(1 + x^2 + y^2)$
20. $f(x, y) = xy + \ln x + 2y^2$

21. **MAXIMIZING PROFIT** The total weekly revenue (in dollars) of the Country Workshop realized in manufacturing and selling its rolltop desks is given by

$$R(x, y) = -0.2x^2 - 0.25y^2 - 0.2xy + 200x + 160y$$

where x denotes the number of finished units and y denotes the number of unfinished units manufactured and sold each week. The total weekly cost attributable to the manufacture of these desks is given by

$$C(x, y) = 100x + 70y + 4000$$

dollars. Determine how many finished units and how many unfinished units the company should manufacture each week to maximize its profit. What is the maximum profit realizable?

22. **MAXIMIZING PROFIT** The total daily revenue (in dollars) that Weston Publishing realizes in publishing and selling its English-language dictionaries is given by

$$R(x, y) = -0.005x^2 - 0.003y^2 - 0.002xy + 20x + 15y$$

where x denotes the number of deluxe copies and y denotes the number of standard copies published and sold daily. The total daily cost of publishing these dictionaries is given by

$$C(x, y) = 6x + 3y + 200$$

dollars. Determine how many deluxe copies and how many standard copies Weston should publish each day to maximize its profits. What is the maximum profit realizable?

23. **MAXIMUM PRICE** The rectangular region R shown in the accompanying figure represents the financial district of a city. The price of land within the district is approximated by the function

$$p(x, y) = 200 - 10\left(x - \frac{1}{2}\right)^2 - 15(y - 1)^2$$

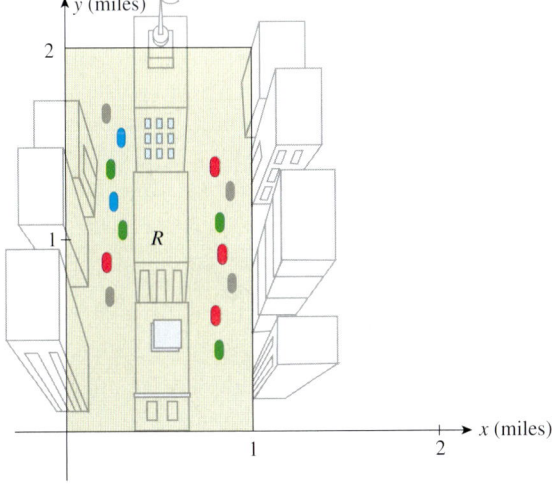

where $p(x, y)$ is the price of land at the point (x, y) in dollars/square foot and x and y are measured in miles. At what point within the financial district is the price of land highest?

24. **MAXIMIZING PROFIT** C&G Imports imports two brands of white wine, one from Germany and the other from Italy. The German wine costs $4/bottle, and the Italian wine costs $3/bottle. It has been estimated that if the German wine sells for p dollars/bottle and the Italian wine sells for for q dollars/bottle, then

$$2000 - 150p + 100q$$

bottles of the German wine and

$$1000 + 80p - 120q$$

bottles of the Italian wine will be sold each week. Determine the unit price for each brand that will allow C&G to realize the largest possible weekly profit.

25. **MAXIMIZING REVENUE** The management of Cal Supermarkets has determined that the quantity demanded per week of their 90% lean ground sirloin, x, and the quantity demanded per week of their 80% ground beef, y (both measured in pounds), are related to their unit prices p and q (in dollars), respectively, by the equations

$$x = 6400 - 400p - 200q \quad \text{and} \quad y = 5600 - 200p - 400q$$

a. What is the total revenue function $R(p, q)$?
   Hint: $R(p, q) = xp + yq$
b. What price should Cal Supermarkets charge for each product to maximize its weekly revenue? How many pounds of each product will then be sold? What is the maximum revenue?

26. **MAXIMIZING PROFIT** Johnson's Household Products has a division that produces two sizes of bar soap. The demand equations that relate the prices p and q (in dollars/hundred bars), to the quantities demanded, x and y (in units of a hundred), of the 3.5-oz size bar soap and the 5-oz bath size bar soap are given by

$$p = 80 - 0.01x - 0.005y \quad \text{and} \quad q = 60 - 0.005x - 0.015y$$

The fixed cost attributed to the division is $10,000/week, and the cost for producing 100 3.5-oz size bars and 100 5-oz bath size bars is $8 and $12, respectively.
a. What is the weekly profit function $P(x, y)$?
b. How many of the 3.5-oz size bars and how many of the 5-oz bath size bars should the division produce per week to maximize its profit? What is the maximum weekly profit?

27. **MAXIMIZING PROFIT** Johnson's Household Products has a division that produces two types of toothpaste: a regular toothpaste and a whitening tooth paste. The demand equations that relate the prices, p and q (in dollars/thousand units), to the quantities demanded weekly, x and y (in units of a thousand), of the regular toothpaste and the whitening toothpaste are given by

$$p = 3000 - 20x - 10y \quad \text{and} \quad q = 4000 - 10x - 30y$$

respectively. The fixed cost attributed to the division is $20,000/week, and the cost for producing 1000 tubes of regular and 1000 tubes of whitening toothpaste is $400 and $500, respectively.
  a. What is the weekly total revenue function $R(x, y)$?
  b. What is the weekly total cost function $C(x, y)$?
  c. What is the weekly profit function $P(x, y)$?
  d. How many tubes of regular and whitening toothpaste should be produced weekly to maximize the division's profit? What is the maximum weekly profit?

28. **Determining the Optimal Site** An auxiliary electric power station will serve three communities, A, B, and C, whose relative locations are shown in the accompanying figure. Determine where the power station should be located if the sum of the squares of the distances from each community to the site is minimized.

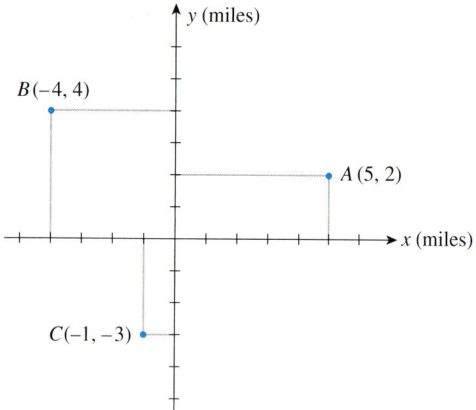

29. **Locating a Radio Station** The following figure shows the locations of three neighboring communities. The operators of a newly proposed radio station have decided that the site $P(x, y)$ for the station should be chosen so that the sum of the squares of the distances from the site to each community is minimized. Find the location of the proposed radio station.

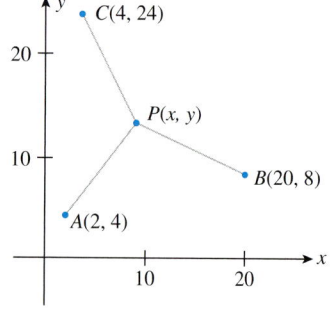

30. **Parcel Post Regulations** Postal regulations specify that a parcel sent by parcel post may have a combined length and girth of no more than 130 in. Find the dimensions of a cylindrical package of greatest volume that can be sent through the mail. What is the volume of such a package?
Hint: The length plus the girth is $2\pi r + l$.

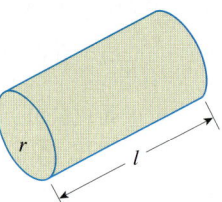

31. **Packaging** An open rectangular box having a volume of 108 in.^{3} is to be constructed from a tin sheet. Find the dimensions of such a box if the amount of material used in its construction is to be minimal.
Hint: Let the dimensions of the box be x in. by y in. by z in. Then $xyz = 108$, and the amount of material used is given by $S = xy + 2yz + 2xz$. Show that
$$S = f(x, y) = xy + \frac{216}{x} + \frac{216}{y}$$
Minimize $f(x, y)$.

32. **Packaging** An open rectangular box having a surface area of 300 in.^{2} is to be constructed from a tin sheet. Find the dimensions of the box if the volume of the box is to be as large as possible. What is the maximum volume?
Hint: Let the dimensions of the box be $x \times y \times z$ (see the figure that follows). Then the surface area is $xy + 2xz + 2yz$, and its volume is xyz.

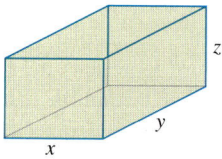

33. **Packaging** Postal regulations specify that the combined length and girth of a parcel sent by parcel post may not exceed 130 in. Find the dimensions of the rectangular package that would have the greatest possible volume under these regulations.
Hint: Let the dimensions of the box be x in. by y in. by z in. (see the figure below). Then $2x + 2z + y = 130$, and the volume $V = xyz$. Show that
$$V = f(x, z) = 130xz - 2x^2z - 2xz^2$$
Maximize $f(x, z)$.

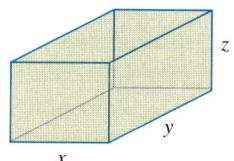

34. Minimizing Heating and Cooling Costs A building in the shape of a rectangular box is to have a volume of 12,000 ft³ (see the figure). It is estimated that the annual heating and cooling costs will be $2/square foot for the top, $4/square foot for the front and back, and $3/square foot for the sides. Find the dimensions of the building that will result in a minimal annual heating and cooling cost. What is the minimal annual heating and cooling cost?

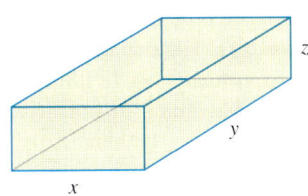

35. Packaging An open box having a volume of 48 in.³ is to be constructed. If the box is to include a partition that is parallel to a side of the box, as shown in the figure, and the amount of material used is to be minimal, what should be the dimensions of the box?

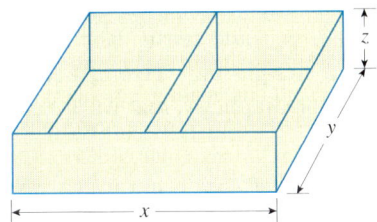

In Exercises 36–42, determine whether the statement is true or false. If it is true, explain why it is true. If it is false, give an example to show why it is false.

36. If $f_x(a, b) = 0$ and $f_y(a, b) = 0$, then f must have a relative extremum at (a, b).

37. If (a, b) is a critical point of f and both the conditions $f_{xx}(a, b) < 0$ and $f_{yy}(a, b) < 0$ hold, then f has a relative maximum at (a, b).

38. If $f(x, y)$ has a relative maximum at (a, b), then $f_x(a, b) = 0$ and $f_y(a, b) = 0$.

39. Let $h(x, y) = f(x) + g(y)$. If $f(x) > 0$ and $g(y) < 0$, then h cannot have a relative maximum or a relative minimum at any point.

40. If $f(x, y)$ satisfies $f_{xx}(a, b) \neq 0, f_{xy}(a, b) = 0, f_{yy}(a, b) \neq 0$, and $f_{xx}(a, b) + f_{yy}(a, b) = 0$ at the critical point (a, b) of f, then f cannot have a relative extremum at (a, b).

41. Suppose $h(x, y) = f(x) + g(y)$, where f and g have continuous second derivatives near a and b, respectively. If a is a critical number of f, b is a critical number of g, and $f''(a)g''(b) > 0$, then h has a relative extremum at (a, b).

42. If f_{xx} and f_{yy} have opposite signs at a critical point (a, b) of f, then f has a saddle point at (a, b).

8.3 Solutions to Self-Check Exercises

1. a. To find the critical point(s) of f, we solve the system of equations

$$f_x = 4x - 4y + 4 = 0$$
$$f_y = -4x + 6y - 2 = 0$$

obtaining $x = -2$ and $y = -1$. Thus, the only critical point of f is the point $(-2, -1)$.

b. We have $f_{xx} = 4, f_{xy} = -4$, and $f_{yy} = 6$, so

$$D(x, y) = f_{xx}f_{yy} - f_{xy}^2$$
$$= (4)(6) - (-4)^2 = 8$$

Since $D(-2, -1) > 0$ and $f_{xx}(-2, -1) > 0$, we conclude that f has a relative minimum at the point $(-2, -1)$.

c. The relative minimum value of $f(x, y)$ at the point $(-2, -1)$ is

$$f(-2, -1) = 2(-2)^2 + 3(-1)^2 - 4(-2)(-1)$$
$$+ 4(-2) - 2(-1) + 3$$
$$= 0$$

2. Robertson's monthly profit is

$$P(x, y) = R(x, y) - C(x, y)$$
$$= \left(-\frac{1}{8}x^2 - \frac{1}{2}y^2 - \frac{1}{4}xy + 20x + 60y\right) - (7x + 20y + 280)$$
$$= -\frac{1}{8}x^2 - \frac{1}{2}y^2 - \frac{1}{4}xy + 13x + 40y - 280$$

The critical point of P is found by solving the system

$$P_x = -\frac{1}{4}x - \frac{1}{4}y + 13 = 0$$

$$P_y = -\frac{1}{4}x - y + 40 = 0$$

giving $x = 16$ and $y = 36$. Thus, $(16, 36)$ is the critical point of P. Next,

$$P_{xx} = -\frac{1}{4} \quad P_{xy} = -\frac{1}{4} \quad P_{yy} = -1$$

and

$$D(x, y) = P_{xx} P_{yy} - P_{xy}^2$$
$$= \left(-\frac{1}{4}\right)(-1) - \left(-\frac{1}{4}\right)^2 = \frac{3}{16}$$

Since $D(16, 36) > 0$ and $P_{xx}(16, 36) < 0$, the point $(16, 36)$ yields a relative maximum of P. We conclude that the monthly profit is maximized by manufacturing 1600 mechanical and 3600 electronic setback thermostats each month. The maximum monthly profit realizable is

$$P(16, 36) = -\frac{1}{8}(16)^2 - \frac{1}{2}(36)^2 - \frac{1}{4}(16)(36)$$
$$+ 13(16) + 40(36) - 280$$
$$= 544$$

or $54,400.

8.4 The Method of Least Squares

The Method of Least Squares

In Section 1.4, Example 10, we saw how a linear equation can be used to approximate the sales trend for a local sporting goods store. As we saw there, one use of a **trend line** is to predict a store's future sales. Recall that we obtained the line by requiring that it pass through two data points, the rationale being that such a line seems to *fit* the data reasonably well.

In this section, we describe a general method, known as the **method of least squares**, for determining a straight line that, in some sense, *best* fits a set of data points when the points are scattered about a straight line. To illustrate the principle behind the method of least squares, suppose, for simplicity, that we are given five data points,

$$P_1(x_1, y_1), \quad P_2(x_2, y_2), \quad P_3(x_3, y_3), \quad P_4(x_4, y_4), \quad P_5(x_5, y_5)$$

that describe the relationship between the two variables x and y. By plotting these data points, we obtain a graph called a **scatter diagram** (Figure 22).

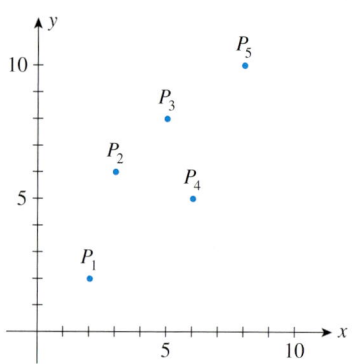

FIGURE 22
A scatter diagram

If we try to fit a straight line to these data points, the line will miss the first, second, third, fourth, and fifth data points by the amounts d_1, d_2, d_3, d_4, and d_5, respectively (Figure 23).

The **principle of least squares** states that the straight line L that fits the data points best is the one chosen by requiring that the sum of the squares of d_1, d_2, d_3, d_4, d_5—that is,

$$d_1^2 + d_2^2 + d_3^2 + d_4^2 + d_5^2$$

be made as small as possible. If we think of the amount d_1 as the error made when the value y_1 is approximated by the corresponding value of y lying on the straight line L, d_2 as the error made when the value y_2 is approximated by the corresponding value of y, and so on, then it can be seen that the least-squares criterion calls for minimizing the sum of the squares of the errors. The line L obtained in this manner is called the **least-squares line**, or **regression line**.

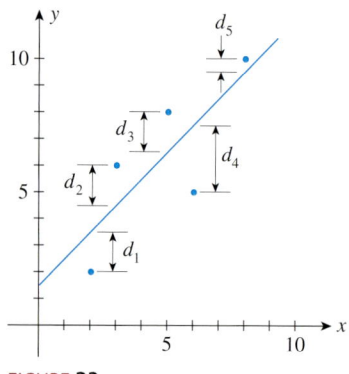

FIGURE 23
The approximating line misses the points by the amounts $d_1, d_2, \ldots, d_5$, respectively.

To find a method for computing the regression line L, suppose L has representation $y = f(x) = mx + b$, where m and b are to be determined. Observe that

$$d_1^2 + d_2^2 + d_3^2 + d_4^2 + d_5^2$$
$$= [f(x_1) - y_1]^2 + [f(x_2) - y_2]^2 + [f(x_3) - y_3]^2$$
$$+ [f(x_4) - y_4]^2 + [f(x_5) - y_5]^2$$
$$= (mx_1 + b - y_1)^2 + (mx_2 + b - y_2)^2 + (mx_3 + b - y_3)^2$$
$$+ (mx_4 + b - y_4)^2 + (mx_5 + b - y_5)^2$$

and may be viewed as a function of the two variables m and b. Thus, the least-squares criterion is equivalent to minimizing the function

$$f(m, b) = (mx_1 + b - y_1)^2 + (mx_2 + b - y_2)^2 + (mx_3 + b - y_3)^2 \\ + (mx_4 + b - y_4)^2 + (mx_5 + b - y_5)^2$$

with respect to m and b. To find the minimum of $f(m, b)$, we use the Chain Rule and compute

$$\begin{aligned}\frac{\partial f}{\partial m} &= 2(mx_1 + b - y_1)x_1 + 2(mx_2 + b - y_2)x_2 + 2(mx_3 + b - y_3)x_3 \\ &\quad + 2(mx_4 + b - y_4)x_4 + 2(mx_5 + b - y_5)x_5 \\ &= 2[mx_1^2 + bx_1 - x_1y_1 + mx_2^2 + bx_2 - x_2y_2 + mx_3^2 + bx_3 - x_3y_3 \\ &\quad + mx_4^2 + bx_4 - x_4y_4 + mx_5^2 + bx_5 - x_5y_5] \\ &= 2[(x_1^2 + x_2^2 + x_3^2 + x_4^2 + x_5^2)m + (x_1 + x_2 + x_3 + x_4 + x_5)b \\ &\quad - (x_1y_1 + x_2y_2 + x_3y_3 + x_4y_4 + x_5y_5)]\end{aligned}$$

and

$$\begin{aligned}\frac{\partial f}{\partial b} &= 2(mx_1 + b - y_1) + 2(mx_2 + b - y_2) + 2(mx_3 + b - y_3) \\ &\quad + 2(mx_4 + b - y_4) + 2(mx_5 + b - y_5) \\ &= 2[(x_1 + x_2 + x_3 + x_4 + x_5)m + 5b - (y_1 + y_2 + y_3 + y_4 + y_5)]\end{aligned}$$

Setting

$$\frac{\partial f}{\partial m} = 0 \quad \text{and} \quad \frac{\partial f}{\partial b} = 0$$

gives

$$(x_1^2 + x_2^2 + x_3^2 + x_4^2 + x_5^2)m + (x_1 + x_2 + x_3 + x_4 + x_5)b \\ = x_1y_1 + x_2y_2 + x_3y_3 + x_4y_4 + x_5y_5$$

and

$$(x_1 + x_2 + x_3 + x_4 + x_5)m + 5b = y_1 + y_2 + y_3 + y_4 + y_5$$

Solving these two simultaneous equations for m and b then leads to an equation $y = mx + b$ of a straight line.

Before looking at an example, we state a more general result whose derivation is identical to the special case involving the five data points just discussed.

The Method of Least Squares

Suppose we are given n data points:

$$P_1(x_1, y_1), \quad P_2(x_2, y_2), \quad P_3(x_3, y_3), \ldots, P_n(x_n, y_n)$$

Then the least-squares (regression) line for the data is given by the linear equation

$$y = f(x) = mx + b$$

(continued)

where the constants m and b satisfy the equations

$$(x_1^2 + x_2^2 + x_3^2 + \cdots + x_n^2)m + (x_1 + x_2 + x_3 + \cdots + x_n)b$$
$$= x_1 y_1 + x_2 y_2 + x_3 y_3 + \cdots + x_n y_n \qquad (8)$$

and

$$(x_1 + x_2 + x_3 + \cdots + x_n)m + nb$$
$$= y_1 + y_2 + y_3 + \cdots + y_n \qquad (9)$$

simultaneously. Equations (8) and (9) are called **normal equations**.

VIDEO **EXAMPLE 1** Find an equation of the least-squares line for the data

$$P_1(1, 1), \quad P_2(2, 3), \quad P_3(3, 4), \quad P_4(4, 3), \quad P_5(5, 6)$$

Solution Here, we have $n = 5$ and

$$\begin{array}{lllll} x_1 = 1 & x_2 = 2 & x_3 = 3 & x_4 = 4 & x_5 = 5 \\ y_1 = 1 & y_2 = 3 & y_3 = 4 & y_4 = 3 & y_5 = 6 \end{array}$$

so Equation (8) becomes

$$(1 + 4 + 9 + 16 + 25)m + (1 + 2 + 3 + 4 + 5)b = 1 + 6 + 12 + 12 + 30$$

or

$$55m + 15b = 61 \qquad (10)$$

and Equation (9) becomes

$$(1 + 2 + 3 + 4 + 5)m + 5b = 1 + 3 + 4 + 3 + 6$$

or

$$15m + 5b = 17 \qquad (11)$$

Solving Equation (11) for b gives

$$b = -3m + \frac{17}{5} \qquad (12)$$

which, upon substitution into Equation (10), gives

$$55m + 15\left(-3m + \frac{17}{5}\right) = 61$$
$$55m - 45m + 51 = 61$$
$$10m = 10$$
$$m = 1$$

Substituting this value of m into Equation (12) gives

$$b = -3 + \frac{17}{5} = \frac{2}{5} = 0.4$$

Therefore, the required equation of the least-squares line is

$$y = x + 0.4$$

The scatter diagram and the regression line are shown in Figure 24.

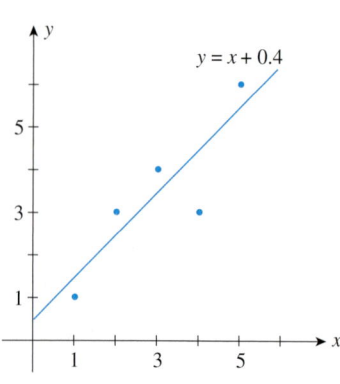

FIGURE 24
The scatter diagram and the least-squares line $y = x + 0.4$

APPLIED EXAMPLE 2 A Firm's Advertising Expense and Profit

The proprietor of Leisure Travel Service compiled the following data relating the firm's annual profit to its annual advertising expenditure (both measured in thousands of dollars).

Annual Advertising Expenditure, x	12	14	17	21	26	30
Annual Profit, y	60	70	90	100	100	120

a. Determine an equation of the least-squares line for these data.
b. Draw a scatter diagram and the least-squares line for these data.
c. Use the result obtained in part (a) to predict Leisure Travel's annual profit if the annual advertising budget is $20,000.

Solution

a. The calculations required for obtaining the normal equations may be summarized as follows:

	x	y	x^2	xy
	12	60	144	720
	14	70	196	980
	17	90	289	1,530
	21	100	441	2,100
	26	100	676	2,600
	30	120	900	3,600
Sum	120	540	2,646	11,530

The normal equations are

$$6b + 120m = 540 \quad (13)$$
$$120b + 2646m = 11{,}530 \quad (14)$$

Solving Equation (13) for b gives

$$b = -20m + 90 \quad (15)$$

which, upon substitution into Equation (14), gives

$$120(-20m + 90) + 2646m = 11{,}530$$
$$-2400m + 10{,}800 + 2646m = 11{,}530$$
$$246m = 730$$
$$m \approx 2.97$$

Substituting this value of m into Equation (15) gives

$$b \approx -20(2.97) + 90 = 30.6$$

Therefore, the required equation of the least-squares line is given by

$$y = f(x) \approx 2.97x + 30.6$$

b. The scatter diagram and the least-squares line are shown in Figure 25.

c. Leisure Travel's predicted annual profit corresponding to an annual budget of $20,000 is given by

$$f(20) \approx 2.97(20) + 30.6$$
$$= 90$$

or $90,000.

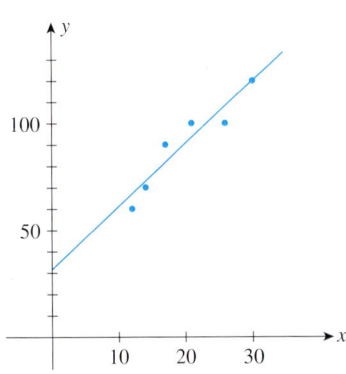

FIGURE 25
The scatter diagram and the least-squares line $y = 2.97x + 30.6$

APPLIED EXAMPLE 3 Maximizing Profit A market research study conducted for Century Communications provided the following data based on the projected monthly sales x (in thousands) of Century's DVD version of a box-office hit adventure movie with a proposed wholesale unit price of p dollars.

p	38	36	34.5	30	28.5
x	2.2	5.4	7.0	11.5	14.6

a. Find the demand equation if the demand curve is the least-squares line for these data.

b. The total monthly cost function associated with producing and distributing the DVD movies is given by

$$C(x) = 4x + 25$$

where x denotes the number of discs (in thousands) produced and sold and $C(x)$ is in thousands of dollars. Determine the unit wholesale price that will maximize Century's monthly profit.

Solution

a. The calculations required for obtaining the normal equations may be summarized as follows:

	x	p	x^2	xp
	2.2	38	4.84	83.6
	5.4	36	29.16	194.4
	7.0	34.5	49	241.5
	11.5	30	132.25	345
	14.6	28.5	213.16	416.1
Sum	40.7	167	428.41	1280.6

The normal equations are

$$5b + 40.7m = 167$$
$$40.7b + 428.41m = 1280.6$$

Solving this system of linear equations simultaneously, we find that

$$m \approx -0.81 \quad \text{and} \quad b \approx 40.00$$

Therefore, the required equation of the least-squares line is given by

$$p = f(x) \approx -0.81x + 40.00$$

which is the required demand equation, provided $0 \le x \le 49.38$.

b. The total revenue function in this case is given by

$$R(x) = xp \approx -0.81x^2 + 40.00x$$

and since the total cost function is

$$C(x) = 4x + 25$$

we see that the profit function is

$$P(x) \approx -0.81x^2 + 40.00x - (4x + 25)$$
$$= -0.81x^2 + 36.00x - 25$$

To find the absolute maximum of $P(x)$ over the closed interval $[0, 49.38]$, we compute

$$P'(x) \approx -1.62x + 36.00$$

Setting $P'(x) = 0$ gives $x \approx 22.22$ as the only critical point of P. Finally, from the table

x	0	22.22	49.38
$P(x)$	-25	375.00	-222.41

we see that the optimal wholesale price is

$$p = -0.81(22.22) + 40.00 = 22.00$$

or $22 per disc.

8.4 Self-Check Exercises

1. Find an equation of the least-squares line for the data

   $P_1(0, 3)$, $P_2(2, 6.5)$, $P_3(4, 10)$, $P_4(6, 16)$, $P_5(7, 16.5)$

2. **GLOBAL BOX-OFFICE RECEIPTS** Global box-office ticket sales have been growing steadily over the years, reflecting the rapid growth in overseas markets, particularly in China. The sales (in billions of dollars) from 2007 through 2011 are summarized in the following table.

Year	2007	2008	2009	2010	2011
Sales, y	26.1	27.2	28.9	31.1	32.6

   a. Find an equation of the least-squares line for these data. (Let $x = 1$ represent 2007.)
   b. Use the result of part (a) to predict the global ticket sales for 2014, assuming that the trend continues.

   Source: Motion Picture Association of America.

   Solutions to Self-Check Exercises 8.4 can be found on page 579.

8.4 Concept Questions

1. Explain the terms (a) *scatter diagram* and (b) *least-squares line*.

2. Explain the *principle of least squares* in your own words.

3. In the **method of least squares,** we are required to minimize the sum of the *squares* of the errors. Comment on replacing this criterion by (a) minimizing the sum of the errors, and (b) minimizing the sum of the absolute values of the errors.

4. Given the array $(x_1, y_1), (x_2, y_2), \ldots, (x_n, y_n)$, the mean or average of the array is the point $(\bar{x}, \bar{y})$, where

   $$\bar{x} = \frac{1}{n}(x_1 + x_2 + \cdots + x_n)$$

   and

   $$\bar{y} = \frac{1}{n}(y_1 + y_2 + \cdots + y_n)$$

   Show that the least-squares line for the array passes through $(\bar{x}, \bar{y})$.

8.4 Exercises

In Exercises 1–6, (a) find an equation of the least-squares line for the data and (b) draw a scatter diagram for the data and graph the least-squares line.

1.

x	1	2	3	4
y	4	6	8	11

2.

x	1	3	5	7	9
y	9	8	6	3	2

3.

x	1	2	3	4	4	6
y	4.5	5	3	2	3.5	1

4.

x	1	1	2	3	4	4	5
y	2	3	3	3.5	3.5	4	5

5. $P_1(1, 3)$, $P_2(2, 5)$, $P_3(3, 5)$, $P_4(4, 7)$, $P_5(5, 8)$

6. $P_1(1, 8)$, $P_2(2, 6)$, $P_3(5, 6)$, $P_4(7, 4)$, $P_5(10, 1)$

7. COLLEGE ADMISSIONS The following data, compiled by the admissions office at Faber College during the past 5 years, relate the number of college brochures and follow-up letters (x) sent to a preselected list of high school juniors who had taken the PSAT and the number of completed applications (y) received from these students (both measured in units of 1000):

x	4	4.5	5	5.5	6
y	0.5	0.6	0.8	0.9	1.2

a. Determine the equation of the least-squares line for these data.
b. Draw a scatter diagram and the least-squares line for these data.
c. Use the result obtained in part (a) to predict the number of completed applications that might be expected if 6400 brochures and follow-up letters are sent out during the next year.

8. BOUNCED-CHECK CHARGES Overdraft fees have become an important piece of a bank's total fee income. The following table gives the bank revenue from overdraft fees (in billions of dollars) from 2004 through 2009. Here, $x = 4$ corresponds to 2004:

Year, x	4	5	6	7	8	9
Revenue, y	27.5	29	31	34	36	38

a. Find an equation of the least-squares line for these data.
b. Use the result of part (a) to estimate the average rate of increase in overdraft fees over the period under consideration.
c. Assuming that the trend continued, what was the revenue from overdraft fees in 2011?

Source: New York Times.

9. SAT VERBAL SCORES The following data, compiled by the superintendent of schools in a large metropolitan area, shows the average SAT verbal scores of high school seniors during the 5 years since the district implemented the "back-to-basics" program:

Year, x	1	2	3	4	5
Average Score, y	436	438	428	430	426

a. Determine the equation of the least-squares line for these data.
b. Draw a scatter diagram and the least-squares line for these data.
c. Use the result obtained in part (a) to predict the average SAT verbal score of high school seniors 2 years from now ($x = 7$).

10. NET SALES The management of Kaldor, a manufacturer of electric motors, submitted the following data in the annual report to its stockholders. The table shows the net sales (in millions of dollars) during the 5 years that have elapsed since the new management team took over. (The first year the firm operated under the new management corresponds to the time period $x = 1$, and the four subsequent years correspond to $x = 2, 3, 4, 5$.)

Year, x	1	2	3	4	5
Net Sales, y	426	437	460	473	477

a. Determine the equation of the least-squares line for these data.
b. Draw a scatter diagram and the least-squares line for these data.
c. Use the result obtained in part (a) to predict the net sales for the upcoming year.

11. MASS TRANSIT SUBSIDIES The following table gives the projected state subsidies (in millions of dollars) to the MBTA over a 5-year period ($x = 1$ corresponds to the first year):

Year, x	1	2	3	4	5
Subsidy, y	20	24	26	28	32

a. Find an equation of the least-squares line for these data.
b. Use the result of part (a) to estimate the state subsidy to the MBTA for the eighth year ($x = 8$).

Source: Massachusetts Bay Transportation Authority.

12. INFORMATION SECURITY SOFTWARE SALES As online attacks persist, spending on information security software continues to rise. The following table gives the forecast for the worldwide sales (in billions of dollars) of information security software through 2007 ($x = 0$ corresponds to 2002):

Year, x	0	1	2	3	4	5
Spending, y	6.8	8.3	9.8	11.3	12.8	14.9

a. Find an equation of the least-squares line for these data.
b. Use the result of part (a) to estimate the spending on information security software in 2010, assuming the trend continued.

Source: International Data Corporation.

13. HEALTH-CARE SPENDING The projected spending on home care and durable medical equipment (in billions of dollars) through 2016 ($x = 0$ corresponds to 2004) is given in the following table:

Year, x	0	2	4	6	8	10	12
Spending, y	60	74	90	106	118	128	150

a. Find an equation of the least-squares line for these data.
b. Use the result of part (a) to give the approximate projected spending on home care and durable medical equipment in 2015.
c. Use the result of part (a) to estimate the projected rate of change of the spending on home care and durable medical equipment for the period from 2004 through 2016.

Source: National Association of Home Care and Hospice.

14. FACEBOOK USERS End-of-year data for the number of Facebook users (in millions) from 2008 ($x = 0$) through 2011 are given in the following table.

Year	2008	2009	2010	2011
Number, y	154.5	381.8	654.5	845.0

a. Find an equation of the least-squares line for these data.
b. Use the result of part (a) to project the number of Facebook users at the end of 2015, assuming that the trend continues.

Source: Company Reports.

15. E-BOOK AUDIENCE The number of adults (in millions) using e-book devices is expected to climb in the years ahead. The projected number of e-book readers in the United States from 2011 ($x = 0$) through 2015 is given in the following table.

Year	2011	2012	2013	2014	2015
Number, y	25.3	33.4	39.5	50.0	59.6

a. Find an equation of the least squares line for these data.
b. Use the result of part (a) to estimate the projected average rate of growth of the number of e-book readers between 2011 and 2015.

Source: Forrester Research, Inc.

16. GROWTH OF CREDIT UNIONS Credit union membership is on the rise. The following table gives the number (in millions) of credit union members from 2003 ($x = 0$) through 2011 in two-year intervals.

Year	2003	2005	2007	2009	2011
Number, y	82.0	84.7	86.8	89.7	91.8

a. Find an equation of the least-squares line for these data.
b. Use the result of part (a) to estimate the number of credit union members in 2013 ($x = 5$).

Source: National Credit Union Association.

17. GLOBAL DEFENSE SPENDING The following table gives the projected global defense spending (in trillions of dollars) from the beginning of 2008 ($t = 0$) through 2015 ($t = 7$):

Year, x	0	1	2	3	4	5	6	7
Sales, y	1.38	1.44	1.49	1.56	1.61	1.67	1.74	1.78

a. Find an equation of the least-squares line for these data.
b. Use the result of part (a) to estimate the rate of change in the projected global defense spending from 2008 through 2015.
c. Assuming that the trend continues, what will the global spending on defense be in 2018?

Source: Homeland Security Research.

18. WORLDWIDE CONSULTING SPENDING The following table gives the projected worldwide consulting spending (in billions of dollars) from 2005 through 2009. Here, $x = 5$ corresponds to 2005.

Year, x	5	6	7	8	9
Spending, y	254	279	300	320	345

a. Find an equation of the least-squares line for these data.
b. Use the results of part (a) to estimate the average rate of increase of worldwide consulting spending over the period under consideration.
c. Use the results of part (a) to estimate the amount of spending in 2010, assuming that the trend continued.

Source: Kennedy Information.

19. REVENUE OF MOODY'S CORPORATION Moody's Corporation is the holding company for Moody's Investors Service, which has a 40% share in the world credit-rating market. According to company reports, the projected total revenue (in billions of dollars) of the company is as follows:

Year	2004	2005	2006	2007	2008
Revenue, y	1.42	1.73	1.98	2.32	2.65

a. Letting $x = 4$ denote 2004, find an equation of the least-squares line for these data.
b. Use the results of part (a) to estimate the rate of change of the revenue of the company for the period in question.
c. Use the result of part (a) to estimate the total revenue of the company in 2010, assuming that the trend continued.

Source: Company reports.

20. U.S. ONLINE BANKING HOUSEHOLDS The following table gives the projected U.S. online banking households as a percentage of all U.S. banking households from 2001 ($x = 1$) through 2007 ($x = 7$):

Year, x	1	2	3	4	5	6	7
Percentage of Households, y	21.2	26.7	32.2	37.7	43.2	48.7	54.2

a. Find an equation of the least-squares line for these data.
b. Use the result of part (a) to estimate the percentage of U.S. online banking households in 2010.

Source: Jupiter Research.

21. U.S. OUTDOOR ADVERTISING U.S. outdoor advertising expenditure (in billions of dollars) from 2011 through 2015 is given in the following table:

Year	2011	2012	2013	2014	2015
Expenditure, y	6.4	6.8	7.1	7.4	7.6

a. Letting $x = 1$ denote 2011, find an equation of the least-squares line for these data.
b. Use the result of part (a) to estimate the rate of change of the advertising expenditures for the period in question.

Source: eMarketer.

22. ONLINE SALES OF USED AUTOS The amount (in millions of dollars) of used autos projected to be sold online in the United States is expected to grow in accordance with the figures given in the following table:

Year, x	0	1	2	3	4
Sales, y	12.9	13.9	14.65	15.25	15.85

(Here, $x = 0$ corresponds to 2011.)

a. Find an equation of the least-squares line for these data.
b. Use the result of part (a) to estimate the sales of used autos online in 2016, assuming that the predicted trend continued through that year.

Source: Edmunds.com.

23. **SOCIAL SECURITY WAGE BASE** The Social Security (FICA) wage base (in thousands of dollars) from 2004 to 2009 is given in the following table:

Year	2004	2005	2006	2007	2008	2009
Wage Base, y	87.9	90	94.2	97.5	102.6	106.8

a. Find an equation of the least-squares line for these data. (Let $x = 1$ represent 2004.)
b. Use your result of part (a) to estimate the FICA wage base in 2012.

Source: The World Almanac.

24. **MARKET FOR DRUGS** Because of new, lower standards, experts in a study conducted in early 2000 projected a rise in the market for cholesterol-reducing drugs. The U.S. market (in billions of dollars) for such drugs from 1999 through 2004 is given in the following table:

Year	1999	2000	2001	2002	2003	2004
Market, y	12.07	14.07	16.21	18.28	20.00	21.72

a. Find an equation of the least-squares line for these data. (Let $x = 0$ represent 1999.)
b. Use the result of part (a) to estimate the U.S. market for cholesterol-reducing drugs in 2008, assuming that the trend continued.

Source: S. G. Cowen.

25. **MALE LIFE EXPECTANCY AT 65** The projections of male life expectancy at age 65 years in the United States are summarized in the following table ($x = 0$ corresponds to 2000):

Year, x	0	10	20	30	40	50
Years beyond 65, y	15.9	16.8	17.6	18.5	19.3	20.3

a. Find an equation of the least-squares line for these data.
b. Use the result of (a) to estimate the life expectancy at 65 of a male in 2040. How does this result compare with the given data for that year?
c. Use the result of (a) to estimate the life expectancy at 65 of a male in 2030.

Source: U.S. Census Bureau.

26. **SATELLITE TV SUBSCRIBERS** The number of satellite and telecommunications subscribers continues to grow over the years. The following table gives the number of subscribers (in millions) from 2006 ($x = 0$) through 2010.

Year	2006	2007	2008	2009	2010
Number, y	29.4	32.2	34.8	37.7	40.4

a. Find an equation of the least-squares line for these data.
b. Use the result of part (a) to estimate the average rate of growth of the number of subscribers between 2006 and 2010.

Source: SNL Ragan.

27. **CORN USED IN U.S. ETHANOL PRODUCTION** The amount of corn used in the United States for the production of ethanol is expected to rise steadily as the demand for plant-based fuels continues to increase. The following table gives the projected amount of corn (in billions of bushels) used for ethanol production from 2005 through 2010 ($x = 1$ corresponds to 2005):

Year, x	1	2	3	4	5	6
Amount, y	1.4	1.6	1.8	2.1	2.3	2.5

a. Find an equation of the least-squares line for these data.
b. Use the result of part (a) to estimate the amount of corn that was used for the production of ethanol in 2011 if the trend continues.

Source: U.S. Department of Agriculture.

28. **OPERATIONS MANAGEMENT CONSULTING SPENDING** The following table gives the projected operations management consulting spending (in billions of dollars) from 2005 through 2010. Here, $x = 5$ corresponds to 2005.

Year, x	5	6	7	8	9	10
Spending, y	40	43.2	47.4	50.5	53.7	56.8

a. Find an equation of the least-squares line for these data.
b. Use the results of part (a) to estimate the average rate of change of operations management consulting spending from 2005 through 2010.
c. Use the results of part (a) to estimate the amount of spending on operations management consulting in 2011, assuming that the trend continues.

Source: Kennedy Information.

In Exercises 29–32, determine whether the statement is true or false. If it is true, explain why it is true. If it is false, give an example to show why it is false.

29. The least-squares line must pass through at least one of the data points.

30. The sum of the squares of the errors incurred in approximating n data points using the least-squares linear function is zero if and only if the n data points lie on a nonvertical straight line.

31. If the data consist of two distinct points, then the least-squares line is just the line that passes through the two points.

32. A data point lies on the least-squares line if and only if the vertical distance between the point and the line is equal to zero.

8.4 Solutions to Self-Check Exercises

1. We first construct the table:

x	y	x^2	xy
0	3	0	0
2	6.5	4	13
4	10	16	40
6	16	36	96
7	16.5	49	115.5
Sum 19	52	105	264.5

The normal equations are

$$5b + 19m = 52$$
$$19b + 105m = 264.5$$

Solving the first equation for b gives

$$b = -3.8m + 10.4$$

which, upon substitution into the second equation, gives

$$19(-3.8m + 10.4) + 105m = 264.5$$
$$-72.2m + 197.6 + 105m = 264.5$$
$$32.8m = 66.9$$
$$m \approx 2.04$$

Substituting this value of m into the expression for b found earlier gives

$$b \approx -3.8(2.04) + 10.4 \approx 2.65$$

Therefore, the required least-squares line has the equation given by

$$y = 2.04x + 2.65$$

2. a. We summarize the calculations as follows:

	x	y	x^2	xy
	1	26.1	1	26.1
	2	27.2	4	54.4
	3	28.9	9	86.7
	4	31.1	16	124.4
	5	32.6	25	163.0
Sum	15	145.9	55	454.6

The normal equations are

$$5b + 15m = 145.9$$
$$15b + 55m = 454.6$$

Solving this system, we find $m = 1.69$ and $b = 24.11$. Therefore, the required equation is

$$y = f(x) = 1.69x + 24.11$$

b. The predicted global sales for 2014 are given by

$$f(8) = 1.69(8) + 24.11$$
$$= 37.63$$

or approximately \$37.6 billion.

USING TECHNOLOGY

Finding an Equation of a Least-Squares Line

A graphing utility is especially useful in calculating an equation of the least-squares line for a set of data. We simply enter the given data in the form of lists into the calculator and then use the linear regression function to obtain the coefficients of the required equation.

EXAMPLE 1 Find an equation of the least-squares line for the following data:

x	1.1	2.3	3.2	4.6	5.8	6.7	8.0
y	−5.8	−5.1	−4.8	−4.4	−3.7	−3.2	−2.5

Solution First, we enter the data as follows:

$x_1 = 1.1$ $y_1 = -5.8$ $x_2 = 2.3$ $y_2 = -5.1$ $x_3 = 3.2$
$y_3 = -4.8$ $x_4 = 4.6$ $y_4 = -4.4$ $x_5 = 5.8$ $y_5 = -3.7$
$x_6 = 6.7$ $y_6 = -3.2$ $x_7 = 8$ $y_7 = -2.5$

Then, using the linear regression function from the statistics menu, we find

$a = -6.29996900666$ $b = 0.460560979389$ corr. $= 0.994488871079$ $n = 7$

(continued)

Therefore, an equation of the least-squares line ($y = a + bx$) is
$$y = -6.30 + 0.46x$$
The correlation coefficient of 0.99449 attests to the excellent fit of the regression line.

 APPLIED EXAMPLE 2 Demand for Electricity According to Pacific Gas and Electric, the nation's largest utility company, the demand for electricity from 1990 through 2000 was as follows:

t	0	2	4	6	8	10
y	333	917	1500	2117	2667	3292

Here, $t = 0$ corresponds to 1990, and y gives the amount of electricity demanded in year t, measured in megawatts. Find an equation of the least-squares line for these data.

Source: Pacific Gas and Electric.

Solution First, we enter the data as

$$x_1 = 0 \quad y_1 = 333 \quad x_2 = 2 \quad y_2 = 917$$
$$x_3 = 4 \quad y_3 = 1500 \quad x_4 = 6 \quad y_4 = 2117$$
$$y_5 = 8 \quad x_5 = 2667 \quad x_6 = 10 \quad y_6 = 3292$$

Then, using the linear regression function from the statistics menu, we find

$$a = 328.476190476 \quad \text{and} \quad b = 295.171428571$$

Therefore, an equation of the least-squares line is

$$y = 328 + 295t$$

TECHNOLOGY EXERCISES

In Exercises 1–4, find an equation of the least-squares line for the data.

1.

x	2.1	3.4	4.7	5.6	6.8	7.2
y	8.8	12.1	14.8	16.9	19.8	21.1

2.

x	1.1	2.4	3.2	4.7	5.6	7.2
y	−0.5	1.2	2.4	4.4	5.7	8.1

3.

x	−2.1	−1.1	0.1	1.4	2.5	4.2	5.1
y	6.2	4.7	3.5	1.9	0.4	−1.4	−2.5

4.

x	−1.12	0.1	1.24	2.76	4.21	6.82
y	7.61	4.9	2.74	−0.47	−3.51	−8.94

5. STARBUCKS' ANNUAL SALES According to company reports, Starbucks' annual sales (in billions of dollars) for 2001 through 2006 are the following ($x = 0$ corresponds to 2001):

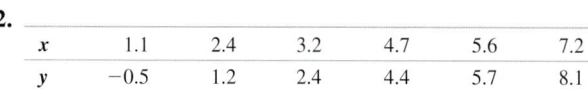

Year, x	0	1	2	3	4	5
Sales, y	2.65	3.29	4.08	5.29	6.37	7.79

a. Find an equation of the least-squares line for these data.
b. Use the result of part (a) to project Starbucks' sales for 2011, assuming that the trend continued.

Source: Company reports.

6. SALES OF GPS EQUIPMENT The annual sales (in billions of dollars) of global positioning system (GPS) equipment from 2000 through 2006 follow. Here, $x = 0$ corresponds to 2000.

Year, x	0	1	2	3	4	5	6
Sales, y	7.9	9.6	11.5	13.3	15.2	16	18.8

a. Find an equation of the least-squares line for these data.
b. Use the equation found in part (a) to estimate the annual sales of GPS equipment for 2010.

Source: ABI Research.

7. Waste Generation
The amount of waste (in millions of tons per year) generated in the United States from 1960 to 1990 was as follows:

Year	1960	1965	1970	1975
Amount, y	81	100	120	124

Year	1980	1985	1990
Amount, y	140	152	164

a. Find an equation of the least-squares line for these data. (Let x be in units of 5 and let $x = 1$ represent 1960.)
b. Use the result of part (a) to estimate the amount of waste generated in 2000, assuming that the trend continued.
Source: Council on Environmental Quality.

8. Online Travel
More and more travelers are purchasing their tickets online. According to industry projections, the U.S. online travel revenues (in billions of dollars) from 2001 through 2005 are the following ($t = 0$ corresponds to 2001):

Year, t	0	1	2	3	4
Revenue, y	16.3	21.0	25.0	28.8	32.7

a. Find an equation of the least-squares line for these data.
b. Use the result of part (a) to estimate the U.S. online travel revenue for 2008, assuming that the trend continued.
Source: Forrester Research, Inc.

9. Market for Drugs
Because of new, lower standards, experts in a study conducted in early 2000 projected a rise in the market for cholesterol-reducing drugs. The following table gives the U.S. market (in billions of dollars) for such drugs from 1999 through 2004:

Year	1999	2000	2001	2002	2003	2004
Market, y	12.07	14.07	16.21	18.28	20	21.72

a. Find an equation of the least-squares line for these data. (Let $x = 0$ represent 1999.)
b. Use the result of part (a) to estimate the U.S. market for cholesterol-reducing drugs in 2008, assuming that the trend continued.
Source: S. G. Cowen.

10. Outpatient Visits
With an aging population, the demand for health care, as measured by outpatient visits, is steadily growing. The number of outpatient visits (in millions) from 1991 through 2001 is recorded in the following table ($x = 0$ corresponds to 1991):

Year, x	0	1	2	3	4	5	6	7	8	9	10
Number of Visits, y	320	340	362	380	416	440	444	470	495	520	530

a. Find an equation of the least-squares line for these data.
b. Use the result of part (a) to estimate the number of outpatient visits in 2006, assuming that the trend continued.
Source: PriceWaterhouse Coopers.

8.5 Constrained Maxima and Minima and the Method of Lagrange Multipliers

Constrained Relative Extrema

In Section 8.3, we studied the problem of determining the relative extrema of a function $f(x, y)$ without placing any restrictions on the independent variables x and y—except, of course, that the point (x, y) lie in the domain of f. Such a relative extremum of a function f is referred to as an **unconstrained relative extremum** of f. However, in many practical optimization problems, we must maximize or minimize a function in which the independent variables are subjected to certain further constraints.

In this section, we discuss a powerful method for determining the relative extrema of a function $f(x, y)$ whose independent variables x and y are required to satisfy one or more constraints of the form $g(x, y) = 0$. Such a relative extremum of a function f is called a **constrained relative extremum** of f. We can see the difference between an unconstrained extremum of a function $f(x, y)$ of two variables and a constrained extremum of f, where the independent variables x and y are subjected to a constraint of the form $g(x, y) = 0$, by considering the geometry of the two cases. Figure 26a depicts the graph of a function $f(x, y)$ that has an unconstrained relative minimum at the point $(0, 0)$. However, when the independent variables x and y are subjected to an equality constraint of the form $g(x, y) = 0$, the points (x, y, z) that satisfy both $z = f(x, y)$ and the constraint equation $g(x, y) = 0$ lie on a curve C. Therefore, the constrained relative minimum of f must also lie on C (Figure 26b).

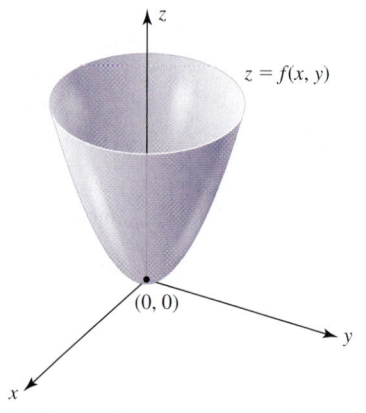
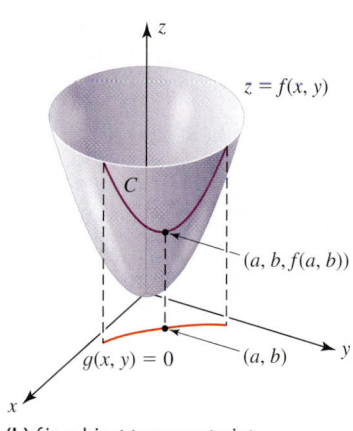

FIGURE 26
The function *f* has an unconstrained minimum value of 0, but it has a constrained minimum value of *f*(*a*, *b*) when subjected to the constraint *g*(*x*, *y*) = 0.

(a) *f* is not subject to any constraints.

(b) *f* is subject to a constraint.

Our first example involves an equality constraint $g(x, y) = 0$ in which we solve for the variable y explicitly in terms of x. In this case we may apply the technique used in Chapter 4 to find the relative extrema of a function of one variable.

EXAMPLE 1 Find the relative minimum of the function

$$f(x, y) = 2x^2 + y^2$$

subject to the constraint $g(x, y) = x + y - 1 = 0$.

Solution Solving the constraint equation for y explicitly in terms of x, we obtain $y = -x + 1$. Substituting this value of y into the function $f(x, y) = 2x^2 + y^2$ results in a function of x,

$$h(x) = 2x^2 + (-x + 1)^2 = 3x^2 - 2x + 1$$

The function h describes the curve C lying on the graph of f on which the constrained relative minimum of f occurs. To find this point, use the technique developed in Chapter 4 to determine the relative extrema of a function of one variable:

$$h'(x) = 6x - 2 = 2(3x - 1)$$

Setting $h'(x) = 0$ gives $x = \frac{1}{3}$ as the sole critical number of the function h. Next, we find

$$h''(x) = 6$$

and, in particular,

$$h''\left(\frac{1}{3}\right) = 6 > 0$$

Therefore, by the second derivative test, the point $x = \frac{1}{3}$ gives rise to a relative minimum of h. Substitute this value of x into the constraint equation $x + y - 1 = 0$ to get $y = \frac{2}{3}$. Thus, the point $(\frac{1}{3}, \frac{2}{3})$ gives rise to the required constrained relative minimum of f. Since

$$f\left(\frac{1}{3}, \frac{2}{3}\right) = 2\left(\frac{1}{3}\right)^2 + \left(\frac{2}{3}\right)^2 = \frac{2}{3}$$

the required constrained relative minimum value of f is $\frac{2}{3}$ at the point $(\frac{1}{3}, \frac{2}{3})$. It may be shown that $\frac{2}{3}$ is in fact a constrained absolute minimum value of f (Figure 27).

8.5 CONSTRAINED MAXIMA AND MINIMA AND THE METHOD OF LAGRANGE MULTIPLIERS

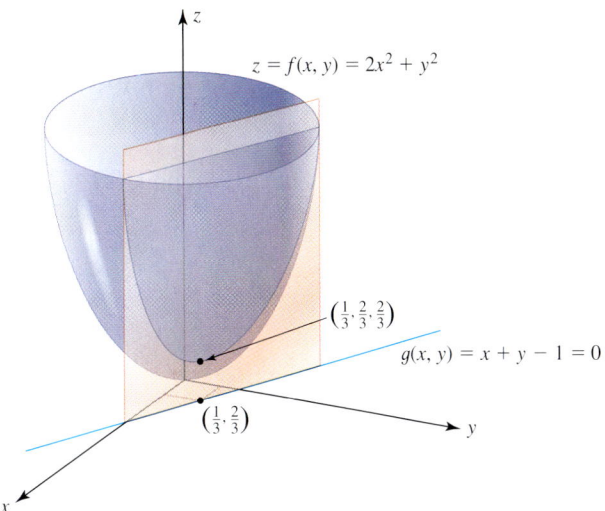

FIGURE 27
f has a constrained absolute minimum of $\frac{2}{3}$ at $(\frac{1}{3}, \frac{2}{3})$.

The Method of Lagrange Multipliers

The major drawback of the technique used in Example 1 is that it relies on our ability to solve the constraint equation $g(x, y) = 0$ for y explicitly in terms of x. This is not always an easy task. Moreover, even when we can solve the constraint equation $g(x, y) = 0$ for y explicitly in terms of x, the resulting function of one variable that is to be optimized may turn out to be unnecessarily complicated. Fortunately, an easier method exists. This method, called the **method of Lagrange multipliers** (Joseph Lagrange, 1736–1813), is as follows:

The Method of Lagrange Multipliers

To find the relative extrema of the function $f(x, y)$ subject to the constraint $g(x, y) = 0$ (assuming that these extreme values exist),

1. Form an auxiliary function

$$F(x, y, \lambda) = f(x, y) + \lambda g(x, y)$$

called the Lagrangian function (the variable λ is called the Lagrange multiplier).

2. Solve the system that consists of the equations

$$F_x = 0 \quad F_y = 0 \quad F_\lambda = 0$$

for all values of x, y, and λ.

3. The solutions found in step 2 are candidates for the extrema of f.

Let's re-solve Example 1 using the method of Lagrange multipliers.

EXAMPLE 2 Using the method of Lagrange multipliers, find the relative minimum of the function

$$f(x, y) = 2x^2 + y^2$$

subject to the constraint $x + y = 1$.

Solution Write the constraint equation $x + y = 1$ in the form $g(x, y) = x + y - 1 = 0$. Then, form the Lagrangian function

$$F(x, y, \lambda) = f(x, y) + \lambda g(x, y)$$
$$= 2x^2 + y^2 + \lambda(x + y - 1)$$

To find the critical point(s) of the function F, solve the system composed of the equations

$$F_x = 4x + \lambda = 0$$
$$F_y = 2y + \lambda = 0$$
$$F_\lambda = x + y - 1 = 0$$

Solving the first and second equations in this system for x and y in terms of λ, we obtain

$$x = -\frac{1}{4}\lambda \quad \text{and} \quad y = -\frac{1}{2}\lambda$$

which, upon substitution into the third equation, yields

$$-\frac{1}{4}\lambda - \frac{1}{2}\lambda - 1 = 0 \quad \text{or} \quad \lambda = -\frac{4}{3}$$

Therefore, $x = \frac{1}{3}$ and $y = \frac{2}{3}$, and $(\frac{1}{3}, \frac{2}{3})$ affords a constrained minimum of the function f, in agreement with the result obtained earlier.

Note A disadvantage of the method of Lagrange multipliers is that there is no test analogous to the second derivative test mentioned in Section 8.3 for determining whether a critical point of a function of two or more variables leads to a relative maximum or relative minimum of the function. Here we have to rely on the geometric or physical nature of the problem to help us draw the necessary conclusions (see Example 2).

The method of Lagrange multipliers may be used to solve a problem involving a function of three or more variables, as illustrated in the next example.

EXAMPLE 3 Use the method of Lagrange multipliers to find the minimum of the function

$$f(x, y, z) = 2xy + 6yz + 8xz$$

subject to the constraint

$$xyz = 12{,}000$$

(*Note:* The existence of the minimum is suggested by the geometry of the problem.)

Solution Write the constraint equation $xyz = 12{,}000$ in the form $g(x, y, z) = xyz - 12{,}000 = 0$. Then the Lagrangian function is

$$F(x, y, z, \lambda) = f(x, y, z) + \lambda g(x, y, z)$$
$$= 2xy + 6yz + 8xz + \lambda(xyz - 12{,}000)$$

To find the critical point(s) of the function F, we solve the system composed of the equations

$$F_x = 2y + 8z + \lambda yz = 0$$
$$F_y = 2x + 6z + \lambda xz = 0$$
$$F_z = 6y + 8x + \lambda xy = 0$$
$$F_\lambda = xyz - 12{,}000 = 0$$

Solving the first three equations of the system for λ in terms of x, y, and z, we have

$$\lambda = -\frac{2y + 8z}{yz}$$

$$\lambda = -\frac{2x + 6z}{xz}$$

$$\lambda = -\frac{6y + 8x}{xy}$$

Equating the first two expressions for λ leads to

$$\frac{2y + 8z}{yz} = \frac{2x + 6z}{xz}$$

$$2xy + 8xz = 2xy + 6yz$$

$$x = \frac{3}{4}y$$

Next, equating the second and third expressions for λ above yields

$$\frac{2x + 6z}{xz} = \frac{6y + 8x}{xy}$$

$$2xy + 6yz = 6yz + 8xz$$

$$z = \frac{1}{4}y$$

Finally, substituting these values of x and z into the equation $xyz - 12{,}000 = 0$, the fourth equation of the first system of equations, we have

$$\left(\frac{3}{4}y\right)(y)\left(\frac{1}{4}y\right) - 12{,}000 = 0$$

$$y^3 = \frac{(12{,}000)(4)(4)}{3} = 64{,}000$$

$$y = 40$$

The corresponding values of x and z are given by $x = \frac{3}{4}(40) = 30$ and $z = \frac{1}{4}(40) = 10$. Therefore, we see that the point $(30, 40, 10)$ gives the constrained minimum of f. The minimum value is

$$f(30, 40, 10) = 2(30)(40) + 6(40)(10) + 8(30)(10) = 7200$$

APPLIED EXAMPLE 4 Maximizing Profit Refer to Example 4, Section 8.3. The total weekly profit (in dollars) that Acrosonic realized in producing and selling its bookshelf loudspeaker systems is given by the profit function

$$P(x, y) = -\frac{1}{4}x^2 - \frac{3}{8}y^2 - \frac{1}{4}xy + 120x + 100y - 5000$$

where x denotes the number of fully assembled units and y denotes the number of kits produced and sold per week. Acrosonic's management decides that production of these loudspeaker systems should be restricted to a total of exactly 230 units each week. Under this condition, how many fully assembled units and how many kits should be produced each week to maximize Acrosonic's weekly profit?

Solution The problem is equivalent to the problem of maximizing the function

$$P(x, y) = -\frac{1}{4}x^2 - \frac{3}{8}y^2 - \frac{1}{4}xy + 120x + 100y - 5000$$

subject to the constraint

$$g(x, y) = x + y - 230 = 0$$

The Lagrangian function is

$$F(x, y, \lambda) = P(x, y) + \lambda g(x, y)$$
$$= -\frac{1}{4}x^2 - \frac{3}{8}y^2 - \frac{1}{4}xy + 120x + 100y$$
$$- 5000 + \lambda(x + y - 230)$$

To find the critical point(s) of F, solve the following system of equations:

$$F_x = -\frac{1}{2}x - \frac{1}{4}y + 120 + \lambda = 0$$

$$F_y = -\frac{3}{4}y - \frac{1}{4}x + 100 + \lambda = 0$$

$$F_\lambda = x + y - 230 = 0$$

Solving the first equation of this system for λ, we obtain

$$\lambda = \frac{1}{2}x + \frac{1}{4}y - 120$$

which, upon substitution into the second equation, yields

$$-\frac{3}{4}y - \frac{1}{4}x + 100 + \frac{1}{2}x + \frac{1}{4}y - 120 = 0$$

$$-\frac{1}{2}y + \frac{1}{4}x - 20 = 0$$

Solving the last equation for y gives

$$y = \frac{1}{2}x - 40$$

Substituting this value of y into the third equation of the system, we have

$$x + \frac{1}{2}x - 40 - 230 = 0$$

$$x = 180$$

The corresponding value of y is $\frac{1}{2}(180) - 40$, or 50. Thus, the required constrained relative maximum of P occurs at the point (180, 50). Again, we can show that the point (180, 50) in fact yields a constrained absolute maximum for P. Thus, Acrosonic's profit is maximized by producing 180 assembled and 50 kit versions of their bookshelf loudspeaker systems. The maximum weekly profit realizable is given by

$$P(180, 50) = -\frac{1}{4}(180)^2 - \frac{3}{8}(50)^2 - \frac{1}{4}(180)(50)$$
$$+ 120(180) + 100(50) - 5000$$
$$= 10{,}312.5$$

or $10,312.50.

8.5 CONSTRAINED MAXIMA AND MINIMA AND THE METHOD OF LAGRANGE MULTIPLIERS

FIGURE 28
A rectangular-shaped pool will be built in the elliptical-shaped poolside area.

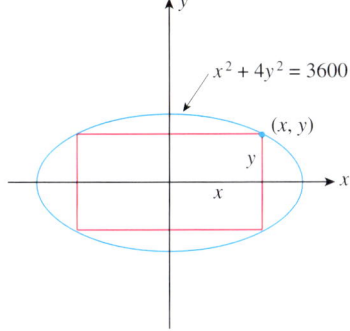

FIGURE 29
We want to find the largest rectangle that can be inscribed in the ellipse described by $x^2 + 4y^2 = 3600$.

APPLIED EXAMPLE 5 Designing a Cruise-Ship Pool The operators of the *Viking Princess*, a luxury cruise liner, are contemplating the addition of another swimming pool to the ship. The chief engineer has suggested that an area in the form of an ellipse located in the rear of the promenade deck would be suitable for this purpose. This location would provide a poolside area with sufficient space for passenger movement and placement of deck chairs (Figure 28). It has been determined that the shape of the ellipse may be described by the equation $x^2 + 4y^2 = 3600$, where x and y are measured in feet. *Viking*'s operators would like to know the dimensions of the rectangular pool with the largest possible area that would meet these requirements.

Solution To solve this problem, we need to find the rectangle of largest area that can be inscribed in the ellipse with equation $x^2 + 4y^2 = 3600$. Letting the sides of the rectangle be $2x$ and $2y$ feet, we see that the area of the rectangle is $A = 4xy$ (Figure 29). Furthermore, the point (x, y) must be constrained to lie on the ellipse, so it satisfies the equation $x^2 + 4y^2 = 3600$. Thus, the problem is equivalent to the problem of maximizing the function

$$f(x, y) = 4xy$$

subject to the constraint $g(x, y) = x^2 + 4y^2 - 3600 = 0$. The Lagrangian function is

$$\begin{aligned} F(x, y, \lambda) &= f(x, y) + \lambda g(x, y) \\ &= 4xy + \lambda(x^2 + 4y^2 - 3600) \end{aligned}$$

To find the critical point(s) of F, we solve the following system of equations:

$$\begin{aligned} F_x &= 4y + 2\lambda x = 0 \\ F_y &= 4x + 8\lambda y = 0 \\ F_\lambda &= x^2 + 4y^2 - 3600 = 0 \end{aligned}$$

Solving the first equation of this system for λ, we obtain

$$\lambda = -\frac{2y}{x}$$

which, upon substitution into the second equation, yields

$$4x + 8\left(-\frac{2y}{x}\right)y = 0 \quad \text{or} \quad x^2 - 4y^2 = 0$$

—that is, $x = \pm 2y$. Substituting these values of x into the third equation of the system, we have

$$4y^2 + 4y^2 - 3600 = 0$$

or, upon solving $y = \pm\sqrt{450} = \pm 15\sqrt{2}$. The corresponding values of x are $\pm 30\sqrt{2}$. Because both x and y must be nonnegative, we have $x = 30\sqrt{2}$ and $y = 15\sqrt{2}$. Thus, the dimensions of the pool with maximum area are $30\sqrt{2}$ feet × $60\sqrt{2}$ feet, or approximately 42 feet × 85 feet.

APPLIED EXAMPLE 6 Cobb–Douglas Production Function Suppose x units of labor and y units of capital are required to produce

$$f(x, y) = 100x^{3/4}y^{1/4}$$

units of a certain product (recall that this is a Cobb–Douglas production function). If each unit of labor costs $200 and each unit of capital costs $300 and a total of $60,000 is available for production, determine how many units of labor and how many units of capital should be used in order to maximize production.

Solution The total cost of x units of labor at \$200 per unit and y units of capital at \$300 per unit is equal to $200x + 300y$ dollars. But \$60,000 is budgeted for production, so $200x + 300y = 60,000$, which we rewrite as

$$g(x, y) = 200x + 300y - 60,000 = 0$$

To maximize $f(x, y) = 100x^{3/4}y^{1/4}$ subject to the constraint $g(x, y) = 0$, we form the Lagrangian function

$$\begin{aligned} F(x, y, \lambda) &= f(x, y) + \lambda g(x, y) \\ &= 100x^{3/4}y^{1/4} + \lambda(200x + 300y - 60,000) \end{aligned}$$

To find the critical point(s) of F, we solve the following system of equations:

$$\begin{aligned} F_x &= 75x^{-1/4}y^{1/4} + 200\lambda = 0 \\ F_y &= 25x^{3/4}y^{-3/4} + 300\lambda = 0 \\ F_\lambda &= 200x + 300y - 60,000 = 0 \end{aligned}$$

Solving the first equation for λ, we have

$$\lambda = -\frac{75x^{-1/4}y^{1/4}}{200} = -\frac{3}{8}\left(\frac{y}{x}\right)^{1/4}$$

which, when substituted into the second equation, yields

$$25\left(\frac{x}{y}\right)^{3/4} + 300\left(-\frac{3}{8}\right)\left(\frac{y}{x}\right)^{1/4} = 0$$

Multiplying the last equation by $\left(\frac{x}{y}\right)^{1/4}$ then gives

$$25\left(\frac{x}{y}\right) - \frac{900}{8} = 0$$

$$x = \left(\frac{900}{8}\right)\left(\frac{1}{25}\right)y = \frac{9}{2}y$$

Substituting this value of x into the third equation of the first system of equations, we have

$$200\left(\frac{9}{2}y\right) + 300y - 60,000 = 0$$

from which we deduce that $y = 50$. Hence, $x = 225$. Thus, maximum production is achieved when 225 units of labor and 50 units of capital are used. ∎

When used in the context of Example 6, the negative of the Lagrange multiplier λ is called the **marginal productivity of money.** That is, if one additional dollar is available for production, then approximately $-\lambda$ additional units of a product can be produced. Here,

$$\lambda = -\frac{3}{8}\left(\frac{y}{x}\right)^{1/4} = -\frac{3}{8}\left(\frac{50}{225}\right)^{1/4} \approx -0.257$$

so in this case, the marginal productivity of money is 0.257. For example, if \$65,000 is available for production instead of the originally budgeted figure of \$60,000, then the maximum production may be boosted from the original

$$f(225, 50) = 100(225)^{3/4}(50)^{1/4}$$

or 15,448 units, to approximately

$$15,448 + 5000(0.257)$$

or 16,733 units.

8.5 Self-Check Exercises

1. Use the method of Lagrange multipliers to find the relative maximum of the function

$$f(x, y) = -2x^2 - y^2$$

subject to the constraint $3x + 4y = 12$.

2. The total monthly profit of Robertson Controls in manufacturing and selling x hundred of its standard mechanical setback thermostats and y hundred of its deluxe electronic setback thermostats each month is given by the total profit function

$$P(x, y) = -\frac{1}{8}x^2 - \frac{1}{2}y^2 - \frac{1}{4}xy + 13x + 40y - 280$$

where P is in hundreds of dollars. If the production of setback thermostats is to be restricted to a total of exactly 4000/month, how many of each model should Robertson manufacture in order to maximize its monthly profits? What is the maximum monthly profit?

Solutions to Self-Check Exercises 8.5 can be found on page 591.

8.5 Concept Questions

1. What is a constrained relative extremum of a function f?

2. Explain how the method of Lagrange multipliers is used to find the relative extrema of $f(x, y)$ subject to $g(x, y) = 0$.

8.5 Exercises

In Exercises 1–16, use the method of Lagrange multipliers to optimize the function subject to the given constraint.

1. Minimize the function $f(x, y) = x^2 + 3y^2$ subject to the constraint $x + y - 1 = 0$.

2. Minimize the function $f(x, y) = x^2 + y^2 - xy$ subject to the constraint $x + 2y - 14 = 0$.

3. Maximize the function $f(x, y) = 2x + 3y - x^2 - y^2$ subject to the constraint $x + 2y = 9$.

4. Maximize the function $f(x, y) = 16 - x^2 - y^2$ subject to the constraint $x + y - 6 = 0$.

5. Minimize the function $f(x, y) = x^2 + 4y^2$ subject to the constraint $xy = 1$.

6. Minimize the function $f(x, y) = xy$ subject to the constraint $x^2 + 4y^2 = 4$.

7. Maximize the function $f(x, y) = x + 5y - 2xy - x^2 - 2y^2$ subject to the constraint $2x + y = 4$.

8. Maximize the function $f(x, y) = xy$ subject to the constraint $2x + 3y - 6 = 0$.

9. Maximize the function $f(x, y) = xy^2$ subject to the constraint $9x^2 + y^2 = 9$.

10. Minimize the function $f(x, y) = \sqrt{y^2 - x^2}$ subject to the constraint $x + 2y - 5 = 0$.

11. Find the maximum and minimum values of the function $f(x, y) = xy$ subject to the constraint $x^2 + y^2 = 16$.

12. Find the maximum and minimum values of the function $f(x, y) = e^{xy}$ subject to the constraint $x^2 + y^2 = 8$.

13. Find the maximum and minimum values of the function $f(x, y) = xy^2$ subject to the constraint $x^2 + y^2 = 1$.

14. Maximize the function $f(x, y, z) = xyz$ subject to the constraint $2x + 2y + z = 84$.

15. Minimize the function $f(x, y, z) = x^2 + y^2 + z^2$ subject to the constraint $3x + 2y + z = 6$.

16. Find the maximum value of the function $f(x, y, z) = x + 2y - 3z$ subject to the constraint $z = 4x^2 + y^2$.

17. **Maximizing Profit** The total weekly profit (in dollars) realized by Country Workshop in manufacturing and selling its rolltop desks is given by the profit function

$$P(x, y) = -0.2x^2 - 0.25y^2 - 0.2xy + 100x + 90y - 4000$$

where x denotes the number of finished units and y denotes the number of unfinished units manufactured and sold each week. The company's management decides to restrict the manufacture of these desks to a total of exactly 200 units/week. How many finished and how many unfinished units should be manufactured each week to maximize the company's weekly profit?

18. **Maximizing Profit** The total daily profit (in dollars) realized by Weston Publishing in publishing and selling its dictionaries is given by the profit function

$$P(x, y) = -0.005x^2 - 0.003y^2 - 0.002xy + 14x + 12y - 200$$

where x stands for the number of deluxe editions and y denotes the number of standard editions sold daily. Weston's management decides that publication of these dictionaries should be restricted to a total of exactly 400 copies/day. How many deluxe copies and how many standard copies should be published each day to maximize Weston's daily profit?

19. MINIMIZING CONSTRUCTION COSTS The management of UNICO Department Store decides to enclose an 800-ft² area outside their building to display potted plants. The enclosed area will be a rectangle, one side of which is provided by the external wall of the store. Two sides of the enclosure will be made of pine board, and the fourth side will be made of galvanized steel fencing material. If the pine board fencing costs $6/running foot and the steel fencing costs $3/running foot, determine the dimensions of the enclosure that will cost the least to erect.

20. PACKAGING Find the dimensions of an open rectangular box of maximum volume and having an area of 48 ft² that can be constructed from a piece of cardboard. What is the volume of the box?

21. PACKAGING Find the dimensions of an open rectangular box of maximum volume and having an area of 12 ft² that can be constructed from a piece of cardboard. What is the volume of the box?

22. MAXIMIZING PROFIT The Ace Novelty company produces two souvenirs: Type A and Type B. The number of Type A souvenirs, x, and the number of Type B souvenirs, y, that the company can produce weekly are related by the equation $2x^2 + y - 3 = 0$, where x and y are measured in units of a thousand. The profits for a Type A souvenir and a Type B souvenir are $4 and $2, respectively. How many of each type of souvenirs should the company produce to maximize its profit?

23. POSTAL REGULATIONS Find the dimensions of a rectangular package having the greatest possible volume and satisfying the postal regulation that specifies that the combined length and girth of an express mail or priority mail package may not exceed 108 in.

24. PARCEL POST REGULATIONS Postal regulations specify that a parcel sent by parcel post may have a combined length and girth of no more than 130 in. Find the dimensions of the cylindrical package of greatest volume that may be sent through the mail. What is the volume of such a package?
Hint: The length plus the girth is $2\pi r + l$, and the volume is $\pi r^2 l$.

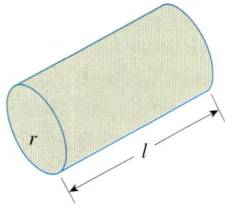

25. MINIMIZING CONTAINER COSTS The Betty Moore Company requires that its corned beef hash containers have a capacity of 64 in.³, be right circular cylinders, and be made of a tin alloy. Find the radius and height of the least expensive container that can be made if the metal for the side and bottom costs 4¢/in.² and the metal for the pull-off lid costs 2¢/in.².
Hint: Let the radius and height of the container be r and h in., respectively. Then, the volume of the container is $\pi r^2 h = 64$, and the cost is given by $C(r, h) = 8\pi rh + 6\pi r^2$.

26. MINIMIZING CONSTRUCTION COSTS An open rectangular box is to be constructed from material that costs $3/ft² for the bottom and $1/ft² for its sides. Find the dimensions of the box of greatest volume that can be constructed for $36.

27. MINIMIZING CONSTRUCTION COSTS A closed rectangular box having a volume of 4 ft³ is to be constructed. If the material for the sides costs $1.00/ft² and the material for the top and bottom costs $1.50/ft², find the dimensions of the box that can be constructed with minimum cost.

28. MINIMIZING CONSTRUCTION COSTS An open rectangular box is to have a volume of 12 ft³. If the material for its base costs three times as much (per square foot) as the material for its sides, what are the dimensions of the box that can be constructed at the minimum cost?

29. MINIMIZING CONSTRUCTION COSTS A rectangular box is to have a volume of 16 ft³. If the material for its base costs twice as much (per square foot) as the material for its top and sides, find the dimensions of the box that can be constructed at the minimum cost.

30. MAXIMIZING SALES Ross–Simons Company has a monthly advertising budget of $60,000. Their marketing department estimates that if they spend x dollars on newspaper advertising and y dollars on television advertising, then the monthly sales will be given by

$$z = f(x, y) = 90x^{1/4}y^{3/4}$$

dollars. Determine how much money Ross–Simons should spend on newspaper ads and on television ads each month to maximize its monthly sales.

31. MAXIMIZING PRODUCTION John Mills—the proprietor of Mills Engine Company, a manufacturer of model airplane engines—finds that it takes x units of labor and y units of capital to produce

$$f(x, y) = 100x^{3/4}y^{1/4}$$

units of the product. If a unit of labor costs $100, a unit of capital costs $200, and $200,000 is budgeted for production, determine how many units should be expended on labor and how many units should be expended on capital in order to maximize production.

32. Use the method of Lagrange multipliers to solve Exercise 34, Exercises 8.3.

In Exercises 33–36, determine whether the statement is true or false. If it is true, explain why it is true. If it is false, give an example to show why it is false.

33. If (a, b) gives rise to a (constrained) relative extremum of f subject to the constraint $g(x, y) = 0$, then (a, b) also gives rise to the unconstrained relative extremum of f.

34. If (a, b) gives rise to a (constrained) relative extremum of f subject to the constraint $g(x, y) = 0$, then $f_x(a, b) = 0$ and $f_y(a, b) = 0$, simultaneously.

35. Suppose f and g have continuous first partial derivatives in some region D in the plane. If f has an extremum at a point (a, b) subject to the constraint $g(x, y) = c$, then there exists a constant λ such that

$$f_x(a, b) = -\lambda g_x(a, b)$$
$$f_y(a, b) = -\lambda g_y(a, b) \quad \text{and} \quad g(a, b) = 0$$

36. If f is defined everywhere, then the constrained maximum (minimum) of f, if it exists, is always smaller (larger) than the unconstrained maximum (minimum).

8.5 Solutions to Self-Check Exercises

1. Write the constraint equation in the form $g(x, y) = 3x + 4y - 12 = 0$. Then, the Lagrangian function is

$$F(x, y, \lambda) = -2x^2 - y^2 + \lambda(3x + 4y - 12)$$

To find the critical point(s) of F, we solve the system

$$F_x = -4x + 3\lambda = 0$$
$$F_y = -2y + 4\lambda = 0$$
$$F_\lambda = 3x + 4y - 12 = 0$$

Solving the first two equations for x and y in terms of λ, we find $x = \frac{3}{4}\lambda$ and $y = 2\lambda$. Substituting these values of x and y into the third equation of the system yields

$$3\left(\frac{3}{4}\lambda\right) + 4(2\lambda) - 12 = 0$$

or $\lambda = \frac{48}{41}$. Therefore, $x = \left(\frac{3}{4}\right)\left(\frac{48}{41}\right) = \frac{36}{41}$ and $y = 2\left(\frac{48}{41}\right) = \frac{96}{41}$, and we see that the point $\left(\frac{36}{41}, \frac{96}{41}\right)$ gives the constrained maximum of f. The maximum value is

$$f\left(\frac{36}{41}, \frac{96}{41}\right) = -2\left(\frac{36}{41}\right)^2 - \left(\frac{96}{41}\right)^2$$

$$= -\frac{11{,}808}{1681} = -\frac{288}{41}$$

2. We want to maximize

$$P(x, y) = -\frac{1}{8}x^2 - \frac{1}{2}y^2 - \frac{1}{4}xy + 13x + 40y - 280$$

subject to the constraint $g(x, y) = x + y - 40 = 0$.

The Lagrangian function is

$$F(x, y, \lambda) = P(x, y) + \lambda g(x, y)$$

$$= -\frac{1}{8}x^2 - \frac{1}{2}y^2 - \frac{1}{4}xy + 13x$$

$$+ 40y - 280 + \lambda(x + y - 40)$$

To find the critical points of F, solve the following system of equations:

$$F_x = -\frac{1}{4}x - \frac{1}{4}y + 13 + \lambda = 0$$

$$F_y = -\frac{1}{4}x - y + 40 + \lambda = 0$$

$$F_\lambda = x + y - 40 = 0$$

Subtracting the first equation from the second gives

$$-\frac{3}{4}y + 27 = 0 \quad \text{or} \quad y = 36$$

Substituting this value of y into the third equation yields $x = 4$. Therefore, to maximize its monthly profits, Robertson should manufacture 400 standard and 3600 deluxe thermostats. The maximum monthly profit is given by

$$P(4, 36) = -\frac{1}{8}(4)^2 - \frac{1}{2}(36)^2 - \frac{1}{4}(4)(36)$$

$$+ 13(4) + 40(36) - 280$$

$$= 526$$

or $52{,}600$.

8.6 Total Differentials

Increments

Recall that if f is a function of one variable defined by $y = f(x)$, then the *increment* in y is defined to be

$$\Delta y = f(x + \Delta x) - f(x)$$

where Δx is an increment in x (Figure 30a). The increment of a function of two or more variables is defined in an analogous manner. For example, if z is a function of two variables defined by $z = f(x, y)$, then the **increment** in z is

$$\Delta z = f(x + \Delta x, y + \Delta y) - f(x, y) \tag{16}$$

where Δx and Δy are the increments in the independent variables x and y, respectively. (See Figure 30b.)

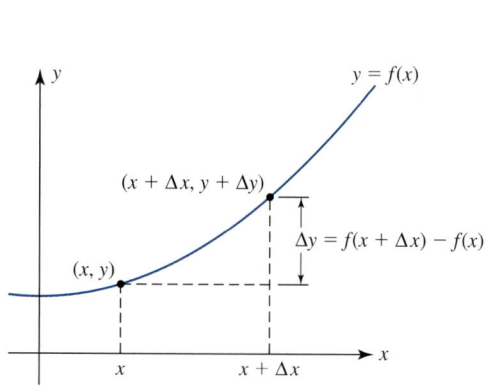
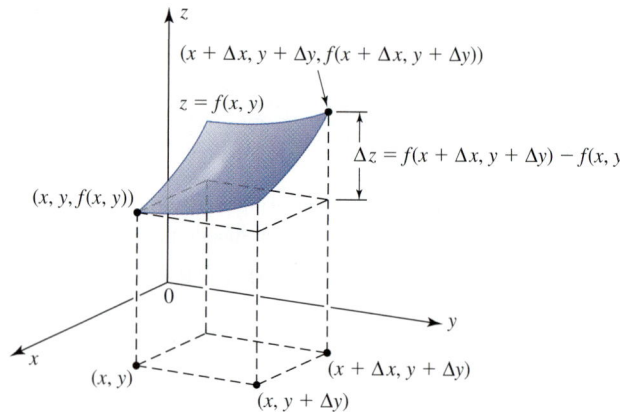

FIGURE 30
(a) The increment Δy is the change in y as x changes from x to $x + \Delta x$.
(b) The increment Δz is the change in z as x changes from x to $x + \Delta x$ and y changes from y to $y + \Delta y$.

EXAMPLE 1 Let $z = f(x, y) = 2x^2 - xy$. Find Δz. Then use your result to find the change in z if (x, y) changes from $(1, 1)$ to $(0.98, 1.03)$.

Solution Using (16), we obtain

$$\begin{aligned}\Delta z &= f(x + \Delta x, y + \Delta y) - f(x, y) \\ &= [2(x + \Delta x)^2 - (x + \Delta x)(y + \Delta y)] - (2x^2 - xy) \\ &= 2x^2 + 4x\Delta x + 2(\Delta x)^2 - xy - x\Delta y - y\Delta x - \Delta x\Delta y - 2x^2 + xy \\ &= (4x - y)\Delta x - x\Delta y + 2(\Delta x)^2 - \Delta x\Delta y\end{aligned}$$

Next, to find the increment in z if (x, y) changes from $(1, 1)$ to $(0.98, 1.03)$, we note that $\Delta x = 0.98 - 1 = -0.02$ and $\Delta y = 1.03 - 1 = 0.03$. Therefore, using the result obtained earlier with $x = 1$, $y = 1$, $\Delta x = -0.02$, and $\Delta y = 0.03$, we obtain

$$\begin{aligned}\Delta z &= [4(1) - 1](-0.02) - (1)(0.03) + 2(-0.02)^2 - (-0.02)(0.03) \\ &= -0.0886\end{aligned}$$

You can verify the correctness of this result by calculating the quantity $f(0.98, 1.03) - f(1, 1)$. ∎

The Total Differential

Recall from Section 3.7 that if f is a function of one variable defined by $y = f(x)$, then the differential of f at x is defined by

$$dy = f'(x)\, dx$$

where $dx = \Delta x$ is the differential in x. Furthermore, we saw that

$$\Delta y \approx dy$$

if Δx is small (Figure 31a).

8.6 TOTAL DIFFERENTIALS

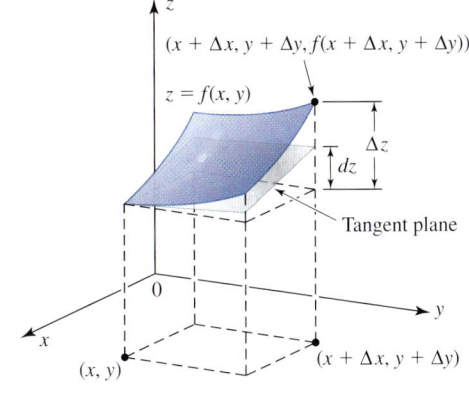

FIGURE 31
(a) Relationship between dy and Δy
(b) Relationship between dz and Δz. The tangent plane is the analog of tangent line T in the one-variable case.

The concept of the differential extends readily to a function of two or more variables.

> **Total Differential**
> Let $z = f(x, y)$ define a differentiable function of x and y.
>
> 1. The **differentials** of the independent variables x and y are $dx = \Delta x$ and $dy = \Delta y$.
> 2. The **differential** of the dependent variable z is
>
> $$dz = \frac{\partial f}{\partial x} dx + \frac{\partial f}{\partial y} dy \qquad (17)$$

Thus, analogous to the one-variable case, the total differential of z is a linear function of dx and dy. Furthermore, it provides us with an approximation of the exact change in z,

$$\Delta z = f(x + \Delta x, y + \Delta y) - f(x, y)$$

corresponding to a net change Δx in x from x to $x + \Delta x$ and a net change Δy in y from y to $y + \Delta y$; that is,

$$\Delta z \approx dz = \frac{\partial f}{\partial x}(x, y)\, dx + \frac{\partial f}{\partial y}(x, y)\, dy \qquad (18)$$

provided that $\Delta x = dx$ and $\Delta y = dy$ are sufficiently small.

VIDEO **EXAMPLE 2** Let $z = 2x^2y + y^3$.

a. Find the differential dz of z.
b. Find the approximate change in z when x changes from $x = 1$ to $x = 1.01$ and y changes from $y = 2$ to $y = 1.98$.
c. Find the actual change in z when x changes from $x = 1$ to $x = 1.01$ and y changes from $y = 2$ to $y = 1.98$. Compare the result with that obtained in part (b).

Solution

a. Let $f(x, y) = 2x^2y + y^3$. Then the required differential is

$$dz = \frac{\partial f}{\partial x} dx + \frac{\partial f}{\partial y} dy = 4xy\, dx + (2x^2 + 3y^2)\, dy$$

b. Here $x = 1$, $y = 2$, and $dx = 1.01 - 1 = 0.01$ and $dy = 1.98 - 2 = -0.02$. Therefore,

$$\Delta z \approx dz = 4(1)(2)(0.01) + [2(1) + 3(4)](-0.02) = -0.20$$

c. The actual change in z is given by

$$\begin{aligned}\Delta z &= f(1.01, 1.98) - f(1, 2) \\ &= [2(1.01)^2(1.98) + (1.98)^3] - [2(1)^2(2) + (2)^3] \\ &= 11.801988 - 12 \\ &\approx -0.1980\end{aligned}$$

We see that $\Delta z \approx dz$, as expected.

APPLIED EXAMPLE 3 Approximating Changes in Revenue The weekly total revenue of Acrosonic Company resulting from the production and sales of x fully assembled bookshelf loudspeaker systems and y kit versions of the same loudspeaker system is

$$R(x, y) = -\frac{1}{4}x^2 - \frac{3}{8}y^2 - \frac{1}{4}xy + 300x + 240y$$

dollars. Determine the approximate change in Acrosonic's weekly total revenue when the level of production is increased from 200 assembled units and 60 kits per week to 206 assembled units and 64 kits per week.

Solution The approximate change in the weekly total revenue is given by the total differential R at $x = 200$ and $y = 60$, $dx = 206 - 200 = 6$ and $dy = 64 - 60 = 4$, that is, by

$$\begin{aligned} dR &= \frac{\partial R}{\partial x} dx + \frac{\partial R}{\partial y} dy \bigg|_{\substack{x=200, y=60 \\ dx=6, dy=4}} \\ &= \left(-\frac{1}{2}x - \frac{1}{4}y + 300\right)\bigg|_{(200, 60)} \cdot (6) \\ &\quad + \left(-\frac{3}{4}y - \frac{1}{4}x + 240\right)\bigg|_{(200, 60)} \cdot (4) \\ &= (-100 - 15 + 300)6 + (-45 - 50 + 240)4 \\ &= 1690 \end{aligned}$$

or $1690.

APPLIED EXAMPLE 4 Cobb–Douglas Production Function The production for a certain country in the early years following World War II is described by the function

$$f(x, y) = 30x^{2/3}y^{1/3}$$

units, when x units of labor and y units of capital were utilized. Find the approximate change in output if the amount expended on labor had been decreased from 125 units to 123 units and the amount expended on capital had been increased from 27 to 29 units. Is your result as expected given the result of Example 4c, Section 8.2?

Solution The approximate change in output is given by the total differential of f at $x = 125$, $y = 27$, $dx = 123 - 125 = -2$, and $dy = 29 - 27 = 2$, that is, by

$$df = \frac{\partial f}{\partial x} dx + \frac{\partial f}{\partial y} dy \bigg|_{\substack{x=125,\, y=27 \\ dx=-2,\, dy=2}}$$

$$= 20x^{-1/3}y^{1/3}\bigg|_{(125,27)} \cdot (-2) + 10x^{2/3}y^{-2/3}\bigg|_{(125,27)} \cdot (2)$$

$$= 20\left(\frac{27}{125}\right)^{1/3}(-2) + 10\left(\frac{125}{27}\right)^{2/3}(2)$$

$$= -20\left(\frac{3}{5}\right)(2) + 10\left(\frac{25}{9}\right)(2) = \frac{284}{9}$$

or $31\tfrac{5}{9}$ units. This result is fully compatible with the result of Example 4c, Section 8.2, in which the recommendation was to encourage increased spending on capital rather than on labor.

If f is a function of the three variables x, y, and z, then the total differential of $w = f(x, y, z)$ is defined to be

$$dw = \frac{\partial f}{\partial x} dx + \frac{\partial f}{\partial y} dy + \frac{\partial f}{\partial z} dz$$

where $dx = \Delta x$, $dy = \Delta y$, and $dz = \Delta z$ are the actual changes in the independent variables x, y, and z as x changes from $x = a$ to $x = a + \Delta x$, y changes from $y = b$ to $y = b + \Delta y$, and z changes from $z = c$ to $z = c + \Delta z$, respectively.

APPLIED EXAMPLE 5 Error Analysis   Find the maximum percentage error in calculating the volume of a rectangular box if an error of at most 1% is made in measuring the length, width, and height of the box.

Solution   Let x, y, and z denote the length, width, and height, respectively, of the rectangular box. Then the volume of the box is given by $V = f(x, y, z) = xyz$ cubic units. Now suppose that the true dimensions of the rectangular box are a, b, and c units, respectively. Since the error committed in measuring the length, width, and height of the box is at most 1%, we have

$$|\Delta x| = |x - a| \le 0.01a$$
$$|\Delta y| = |y - b| \le 0.01b$$
$$|\Delta z| = |z - c| \le 0.01c$$

Therefore, the maximum error in calculating the volume of the box is

$$|\Delta V| \approx |dV| = \left| \frac{\partial f}{\partial x} dx + \frac{\partial f}{\partial y} dy + \frac{\partial f}{\partial z} dz \right|_{x=a,\, y=b,\, z=c}$$

$$= |yz\, dx + xz\, dy + xy\, dz|_{x=a,\, y=b,\, z=c}$$

$$= |bc\, dx + ac\, dy + ab\, dz|$$
$$\le bc|dx| + ac|dy| + ab|dz|$$
$$\le bc(0.01a) + ac(0.01b) + ab(0.01c)$$
$$= (0.03)abc$$

Since the actual volume of the box is abc cubic units, we see that the maximum percentage error in calculating its volume is

$$\frac{|\Delta V|}{V}\bigg|_{(a,b,c)} \approx \frac{(0.03)abc}{abc} = 0.03$$

—that is, approximately 3%.

Explore & Discuss

Refer to Example 5, in which we found the maximum percentage error in calculating the volume of the rectangular box to be *approximately* 3%. What is the precise maximum percentage error?

8.6 Self-Check Exercise

Let f be a function defined by $z = f(x, y) = 3xy^2 - 4y$. Find the total differential of f at $(-1, 3)$. Then find the approximate change in z when x changes from $x = -1$ to $x = -0.98$ and y changes from $y = 3$ to $y = 3.01$.

The solution to Self-Check Exercise 8.6 can be found on page 598.

8.6 Concept Questions

1. If $z = f(x, y)$, what is the differential of x? The differential of y? What is the total differential of z?

2. Let $z = f(x, y)$. What is the relationship between the actual change, Δz, when x changes from x to $x + \Delta x$ and y changes from y to $y + \Delta y$, and the total differential, dz, of f at (x, y)?

8.6 Exercises

1. Let $z = 2x^2 + 3y^2$, and suppose that (x, y) changes from $(2, -1)$ to $(2.01, -0.98)$.
   a. Compute Δz.
   b. Compute dz.
   c. Compare the values of Δz and dz.

2. Let $z = x^2 - 2xy + 3y^2$, and suppose that (x, y) changes from $(2, 1)$ to $(1.97, 1.02)$.
   a. Compute Δz.
   b. Compute dz.
   c. Compare the values of Δz and dz.

In Exercises 3–18, find the total differential of the function.

3. $f(x, y) = 2x^2 - 3xy + 4x$
4. $f(x, y) = xy^3 - x^2y^2$
5. $f(x, y) = \sqrt{x^2 + y^2}$
6. $f(x, y) = (x + 3y^2)^{1/3}$
7. $f(x, y) = \dfrac{5y}{x - y}$
8. $f(x, y) = \dfrac{x + y}{x - y}$
9. $f(x, y) = 2x^5 - ye^{-3x}$
10. $f(x, y) = xye^{x+y}$
11. $f(x, y) = x^2e^y + y \ln x$
12. $f(x, y) = \ln(x^2 + y^2)$
13. $f(x, y, z) = xy^2z^3$
14. $f(x, y, z) = x\sqrt{y} + y\sqrt{z}$
15. $f(x, y, z) = \dfrac{x}{y + z}$
16. $f(x, y, z) = \dfrac{x + y}{y + z}$
17. $f(x, y, z) = xyz + xe^{yz}$
18. $f(x, y, z) = \sqrt{e^x + e^y + ze^{xy}}$

In Exercises 19–30, find the approximate change in z when the point (x, y) changes from (x_0, y_0) to (x_1, y_1).

19. $f(x, y) = 4x^2 - xy$; from $(1, 2)$ to $(1.01, 2.02)$
20. $f(x, y) = 2x^2 - 2x^3y^2 - y^3$; from $(-1, 2)$ to $(-0.98, 2.01)$
21. $f(x, y) = x^{2/3}y^{1/2}$; from $(8, 9)$ to $(7.97, 9.03)$
22. $f(x, y) = \sqrt{x^2 + y^2}$; from $(1, 3)$ to $(1.03, 3.03)$
23. $f(x, y) = \dfrac{x}{x - y}$; from $(-3, -2)$ to $(-3.02, -1.98)$
24. $f(x, y) = \dfrac{x - y}{x + y}$; from $(-3, -2)$ to $(-3.02, -1.98)$
25. $f(x, y) = 2xe^{-y}$; from $(4, 0)$ to $(4.03, 0.03)$
26. $f(x, y) = \sqrt{x}e^y$; from $(1, 1)$ to $(1.01, 0.98)$
27. $f(x, y) = xe^{xy} - y^2$; from $(-1, 0)$ to $(-0.97, 0.03)$
28. $f(x, y) = xe^{-y} + ye^{-x}$; from $(1, 1)$ to $(1.01, 0.90)$
29. $f(x, y) = x \ln x + y \ln x$; from $(2, 3)$ to $(1.98, 2.89)$
30. $f(x, y) = \ln(xy)^{1/2}$; from $(5, 10)$ to $(5.05, 9.95)$

31. **Effect of Inventory and Floor Space on Profit** The monthly profit (in dollars) of Bond and Barker Department Store depends on the level of inventory x (in thousands of dollars) and the floor space y (in thousands of square feet) available for display of the merchandise, as given by

$$P(x, y) = -0.02x^2 - 15y^2 + xy + 39x + 25y - 20{,}000$$

Currently, the level of inventory is \$4,000,000 ($x = 4000$), and the floor space is 150,000 ft^2 ($y = 150$). Find the anticipated change in monthly profit if management increases the level of inventory by \$500,000 and decreases the floor space for display of merchandise by 10,000 ft^2.

32. **Effect of Production on Profit** The Country Workshop's total weekly profit (in dollars) realized in manufacturing and selling its rolltop desks is given by

$$P(x, y) = -0.2x^2 - 0.25y^2 - 0.2xy + 100x + 90y - 4000$$

where x stands for the number of finished units and y denotes the number of unfinished units manufactured and sold per week. Currently, the weekly output is 190 finished units and 105 unfinished units. Determine the approximate change in the total weekly profit if the sole proprietor of the Country Workshop decides to increase the number of finished units to 200/week and decrease the number of unfinished units to 100/week.

33. REVENUE OF A TRAVEL AGENCY The Odyssey Travel Agency's monthly revenue (in thousands of dollars) depends on the amount of money x (in thousands) spent on advertising per month and the number of agents y in its employ in accordance with the rule

$$R(x, y) = -x^2 - 0.5y^2 + xy + 8x + 3y + 20$$

Currently, the amount of money spent on advertising is $10,000 per month, and there are 15 agents in the agency's employ. Estimate the change in revenue resulting from an increase of $1000 per month in advertising expenditure and a decrease of 1 agent.

34. EFFECT OF CAPITAL AND LABOR ON PRODUCTIVITY The production of a South American country is given by the function

$$f(x, y) = 20x^{3/4}y^{1/4}$$

when x units of labor and y units of capital are utilized. Find the approximate change in output if the amount expended on labor is decreased from 256 to 254 units and the amount expended on capital is increased from 16 to 18 units.

35. EFFECT OF CAPITAL AND LABOR ON PRODUCTIVITY The productivity of a certain country is given by the function

$$f(x, y) = 30x^{4/5}y^{1/5}$$

when x units of labor and y units of capital are utilized. What is the approximate change in the number of units produced if the amount expended on labor is decreased from 243 to 240 units and the amount expended on capital is increased from 32 to 35 units?

36. MARGINAL PRODUCTIVITY OF MONEY Refer to Applied Example 6, page 587. Suppose x units of labor and y units of capital are required to produce

$$f(x, y) = 100x^{3/4}y^{1/4}$$

units of a certain product. Suppose that each unit of labor costs $200 and each unit of capital costs $300 and a total of $60,000 is available for production. Show that at the maximum level of production, the availability of one additional dollar for production can result in the production of $-\lambda$ additional units of the product, where λ is the Lagrange multiplier. (*Note:* The negative of λ is called the **marginal productivity of labor.**)

37. SUSPENSION BRIDGE CABLES The supports of a cable of a suspension bridge are at the same level and at a distance of L ft apart. The supports are a feet higher than the lowest point of the cable (see the figure). If the weight of the cable is negligible and the bridge has a uniform weight of W lb/ft, then the tension (in pounds) in the cable at its lowest point is given by

$$H = \frac{WL^2}{8a}$$

If W, L, and a are measured with possible maximum errors of 1%, 2%, and 2%, respectively, determine the maximum percentage error in calculating H.

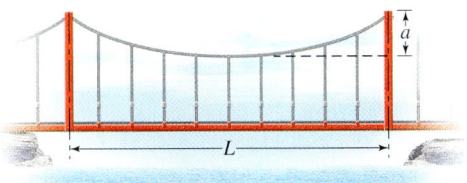

38. PRICE–EARNINGS RATIO The price-to-earnings ratio (P/E ratio) of a stock is given by

$$R(x, y) = \frac{x}{y}$$

where x denotes the price per share of the stock and y denotes the earnings per share. Estimate the change in the P/E ratio of a stock if its price increases from $60/share to $62/share while its earnings decrease from $4/share to $3.80/share.

39. ERROR IN CALCULATING THE SURFACE AREA OF A HUMAN The formula

$$S = 0.007184W^{0.425}H^{0.725}$$

gives the surface area S of a human body (in square meters) in terms of its weight W in kilograms and its height H in centimeters. If an error of 1% is made in measuring the weight of a person and an error of 2% is made in measuring the height, what is the percentage error in the measurement of the person's surface area?

40. ERROR IN CALCULATING THE VOLUME OF A STORAGE TANK A storage tank has the shape of a right circular cylinder. Suppose that the radius and height of the tank are measured at 1.5 ft and 5 ft, respectively, with a possible error of 0.05 ft and 0.1 ft, respectively. Use differentials to estimate the error in calculating the capacity of the tank.

41. ERROR IN CALCULATING THE VOLUME OF A CYLINDER The radius and height of a right circular cylinder are measured with a maximum error of 0.1 cm in each measurement. Approximate the maximum error in calculating the volume of the cylinder if the measured dimensions $r = 8$ cm and $h = 20$ cm are used.

42. The dimensions of a closed rectangular box are measured as 30 in., 40 in., and 60 in., with a maximum error of 0.2 in. in each measurement. Use differentials to estimate the maximum error in calculating the volume of the box.

43. Use differentials to estimate the maximum error in calculating the surface area of the box of Exercise 42.

44. Effect of Capital and Labor on Productivity The production of a certain company is given by the function

$$f(x, y) = 50x^{1/3}y^{2/3}$$

when x units of labor and y units of capital are utilized. Find the approximate percentage change in the production of the company if labor is increased by 2% and capital is increased by 1%.

45. The pressure P (in pascals), the volume V (in liters), and the temperature T (in kelvins) of an ideal gas are related by the equation $PV = 8.314T$. Use differentials to find the approximate change in the pressure of the gas if its volume increases from 20 L to 20.2 L and its temperature decreases from 300 K to 295 K.

46. Specific Gravity The specific gravity of an object with density greater than that of water can be determined by using the formula

$$S = \frac{A}{A - W}$$

where A and W are the weights of the object in air and in water, respectively. If the measurements of an object are $A = 2.2$ lb and $W = 1.8$ lb with maximum errors of 0.02 lb and 0.04 lb, respectively, find the approximate maximum error in calculating S.

47. Error in Calculating Total Resistance The total resistance R of three resistors with resistance R_1, R_2, and R_3, connected in parallel, is given by the relationship

$$\frac{1}{R} = \frac{1}{R_1} + \frac{1}{R_2} + \frac{1}{R_3}$$

If R_1, R_2, and R_3 are measured at 100, 200, and 300 ohms, respectively, with a maximum error of 1% in each measurement, find the approximate maximum error in the calculated value of R.

48. Error in Measuring Arterial Blood Flow The flow of blood through an arteriole in cubic centimeters per second is given by

$$V = \frac{\pi p r^4}{8kl}$$

where l is the length (in cm) of the arteriole, r is its radius (in cm), p is the difference in pressure between the two ends of the arteriole (in dyne/cm^2), and k is the viscosity of blood (in dyne-sec/cm^2). Find the approximate percentage change in the flow of blood if an error of 2% is made in measuring the length of the arteriole and an error of 1% is made in measuring its radius. Assume that p and k are constant.

8.6 Solutions to Self-Check Exercises

We find

$$\frac{\partial f}{\partial x} = 3y^2 \quad \text{and} \quad \frac{\partial f}{\partial y} = 6xy - 4$$

so that

$$\frac{\partial f}{\partial x}(-1, 3) = 3(3)^2 = 27$$

and

$$\frac{\partial f}{\partial y}(-1, 3) = 6(-1)(3) - 4 = -22$$

Therefore, the total differential is

$$dz = \frac{\partial f}{\partial x}(-1, 3)\, dx + \frac{\partial f}{\partial y}(-1, 3)\, dy$$
$$= 27\, dx - 22\, dy$$

Now $dx = -0.98 - (-1) = 0.02$ and $dy = 3.01 - 3 = 0.01$, so the approximate change in z is

$$dz = 27(0.02) - 22(0.01) = 0.32$$

8.7 Double Integrals

A Geometric Interpretation of the Double Integral

To introduce the notion of the integral of a function of two variables, let's first recall the definition of the definite integral of a continuous function of one variable $y = f(x)$ over the interval $[a, b]$. We first divide the interval $[a, b]$ into n subintervals, each of equal length, by the points $x_0 = a < x_1 < x_2 < \cdots < x_n = b$ and define the **Riemann sum** by

$$S_n = f(p_1)h + f(p_2)h + \cdots + f(p_n)h$$

where $h = (b - a)/n$ and p_i is an arbitrary point in the interval $[x_{i-1}, x_i]$. The definite integral of f over $[a, b]$ is then defined as the limit of the Riemann sum S_n as n tends to infinity, whenever it exists. Furthermore, recall that when f is a nonnegative continuous function on $[a, b]$, then the ith term of the Riemann sum, $f(p_i)h$, is an approximation (by the area of a rectangle) of the area under that part of the graph of $y = f(x)$ between $x = x_{i-1}$ and $x = x_i$, so that the Riemann sum S_n provides us with an approximation of the area under the curve $y = f(x)$ from $x = a$ to $x = b$. The integral

$$\int_a^b f(x)\,dx = \lim_{n \to \infty} S_n$$

gives the *actual* area under the curve from $x = a$ to $x = b$.

Now suppose $f(x, y)$ is a continuous function of two variables defined over a region R. For simplicity, we assume for the moment that R is a rectangular region in the plane (Figure 32). Let's construct a Riemann sum for this function over the rectangle R by following a procedure that parallels the case for a function of one variable over an interval I. We begin by observing that the analog of a *partition* in the two-dimensional case is a rectangular **grid** composed of mn rectangles, each of length h and width k, as a result of partitioning the side of the rectangle R of length $(b - a)$ into m segments and the side of length $(d - c)$ into n segments. By construction,

$$h = \frac{b - a}{m} \quad \text{and} \quad k = \frac{d - c}{n}$$

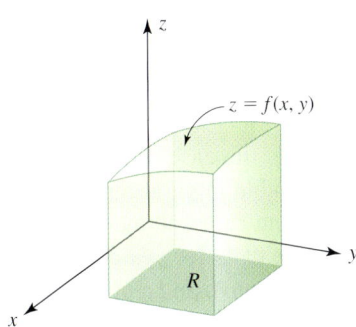

FIGURE 32
$f(x, y)$ is a function defined over a rectangular region R.

A sample grid with $m = 5$ and $n = 4$ is shown in Figure 33.

Let's label the rectangles $R_1, R_2, R_3, \ldots, R_{mn}$. If (x_i, y_i) is *any* point in R_i ($1 \le i \le mn$), then the **Riemann sum of $f(x, y)$ over the region R** is defined as

$$S(m, n) = f(x_1, y_1)hk + f(x_2, y_2)hk + \cdots + f(x_{mn}, y_{mn})hk$$

If the limit of $S(m, n)$ exists as both m and n tend to infinity, we call this limit the value of the **double integral of $f(x, y)$ over the region R** and denote it by

$$\iint_R f(x, y)\,dA$$

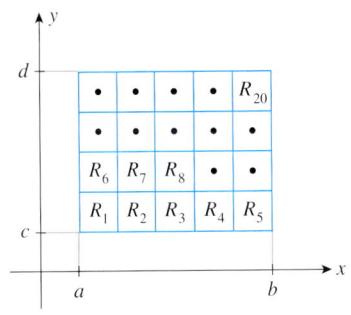

FIGURE 33
Grid with $m = 5$ and $n = 4$.

If $f(x, y)$ is a nonnegative function, then it defines a solid S bounded above by the graph of f and below by the rectangular region R. Furthermore, the solid S is the union of the mn solids bounded above by the graph of f and below by the mn rectangular regions corresponding to the partition of R (Figure 34). The volume of

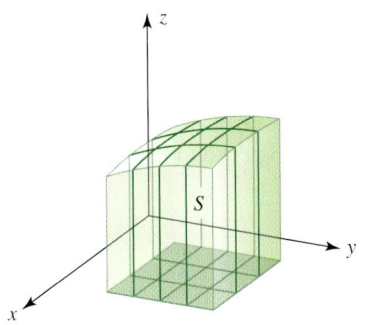
(a) The solid S is the union of mn solids (shown here with $m = 3$ and $n = 4$).

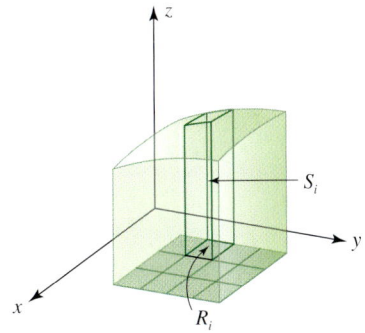
(b) A typical solid S_i is bounded above by the graph of f and lies above R_i.

FIGURE 34

a typical solid S_i can be approximated by a parallelepiped with base R_i and height $f(x_i, y_i)$ (Figure 35).

Therefore, the Riemann sum $S(m, n)$ gives us an approximation of the volume of the solid bounded above by the surface $z = f(x, y)$ and below by the plane region R. As both m and n tend to infinity, the Riemann sum $S(m, n)$ approaches the *actual* volume under the solid.

Evaluating a Double Integral Over a Rectangular Region

Let's turn our attention to the evaluation of the double integral

$$\iint_R f(x, y)\, dA$$

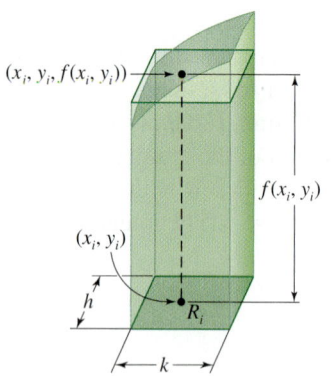

FIGURE 35
The volume of S_i is approximated by the parallelepiped with base R_i and height $f(x_i, y_i)$.

where R is the rectangular region shown in Figure 32. As in the case of the definite integral of a function of one variable, it turns out that the double integral can be evaluated without our having to first find an appropriate Riemann sum and then take the limit of that sum. Instead, as we will now see, the technique calls for evaluating two single integrals—the so-called **iterated integrals**—in succession, using a process that might be called "antipartial differentiation." The technique is described in the following result, which we state without proof.

Let R be the rectangle defined by the inequalities $a \leq x \leq b$ and $c \leq y \leq d$ (see Figure 33). Then

$$\iint_R f(x, y)\, dA = \int_c^d \left[\int_a^b f(x, y)\, dx \right] dy \tag{19}$$

Explore & Discuss

Using a geometric interpretation, evaluate

$$\iint_R \sqrt{4 - x^2 - y^2}\, dA$$

where $R = \{(x, y)\,|\, x^2 + y^2 \leq 4\}$.

where the iterated integrals on the right-hand side are evaluated as follows. We first compute the integral

$$\int_a^b f(x, y)\, dx$$

by treating y as if it were a constant and integrating the resulting function of x with respect to x (dx reminds us that we are integrating with respect to x). In this manner we obtain a value for the integral that may contain the variable y. Thus,

$$\int_a^b f(x, y)\, dx = g(y)$$

for some function g. Substituting this value into Equation (19) gives

$$\int_c^d g(y)\, dy$$

which may be integrated in the usual manner.

VIDEO **EXAMPLE 1** Evaluate $\iint_R f(x, y)\, dA$, where $f(x, y) = x + 2y$ and R is the rectangle defined by $1 \leq x \leq 4$ and $1 \leq y \leq 2$.

Solution Using Equation (19), we find

$$\iint_R f(x, y)\, dA = \int_1^2 \left[\int_1^4 (x + 2y)\, dx \right] dy$$

To compute

$$\int_1^4 (x + 2y)\, dx$$

we treat y as if it were a constant (remember that dx reminds us that we are integrating with respect to x). We obtain

$$\int_1^4 (x + 2y)\, dx = \frac{1}{2}x^2 + 2xy \Big|_{x=1}^{x=4}$$

$$= \left[\frac{1}{2}(16) + 2(4)y\right] - \left[\frac{1}{2}(1) + 2(1)y\right]$$

$$= \frac{15}{2} + 6y$$

Thus,

$$\iint_R f(x,y)\, dA = \int_1^2 \left(\frac{15}{2} + 6y\right) dy = \left(\frac{15}{2}y + 3y^2\right)\Big|_1^2$$

$$= (15 + 12) - \left(\frac{15}{2} + 3\right) = 16\tfrac{1}{2}$$

Evaluating a Double Integral Over a Plane Region

Up to now, we have assumed that the region over which a double integral is to be evaluated is rectangular. In fact, however, it is possible to compute the double integral of functions over rather arbitrary regions. The next theorem, which we state without proof, expands the number of types of regions over which we may integrate.

THEOREM 1

a. Suppose $g_1(x)$ and $g_2(x)$ are continuous functions on $[a, b]$ and the region R is defined by $R = \{(x, y)\mid g_1(x) \le y \le g_2(x); a \le x \le b\}$. Then

$$\iint_R f(x, y)\, dA = \int_a^b \left[\int_{g_1(x)}^{g_2(x)} f(x, y)\, dy\right] dx \qquad (20)$$

(Figure 36a).

b. Suppose $h_1(y)$ and $h_2(y)$ are continuous functions on $[c, d]$ and the region R is defined by $R = \{(x, y)\mid h_1(y) \le x \le h_2(y); c \le y \le d\}$. Then

$$\iint_R f(x, y)\, dA = \int_c^d \left[\int_{h_1(y)}^{h_2(y)} f(x, y)\, dx\right] dy \qquad (21)$$

(Figure 36b).

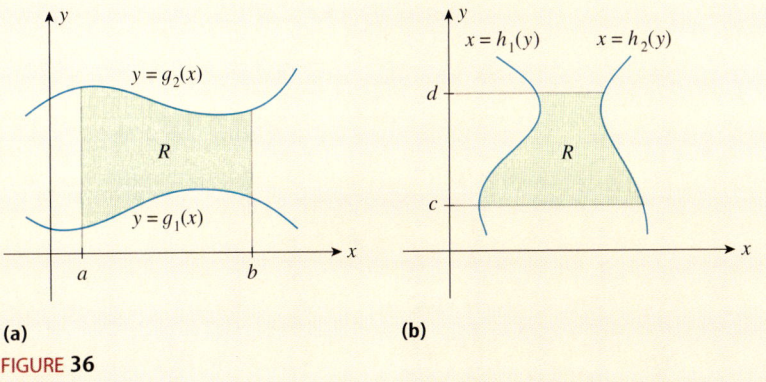

(a)   (b)
FIGURE 36

Notes

1. Observe that in Equation (20), the lower and upper limits of integration with respect to y are given by $y = g_1(x)$ and $y = g_2(x)$. This is to be expected, since for a fixed value of x lying between $x = a$ and $x = b$, y runs between the lower curve

defined by $y = g_1(x)$ and the upper curve defined by $y = g_2(x)$ (see Figure 36a). Observe, too, that in the special case when $g_1(x) = c$ and $g_2(x) = d$, the region R is rectangular, and Equation (20) reduces to Equation (19).

2. For a fixed value of y, x runs between $x = h_1(y)$ and $x = h_2(y)$, giving the indicated limits of integration with respect to x in Equation (21) (see Figure 36b).

3. Note that the two curves in Figure 36b are not graphs of functions of x (use the Vertical Line Test), but they are graphs of functions of y. It is this observation that justifies the approach leading to Equation (21).

We now look at several examples.

EXAMPLE 2 Evaluate $\int_R \int f(x, y)\, dA$ given that $f(x, y) = x^2 + y^2$ and R is the region bounded by the graphs of $g_1(x) = x$ and $g_2(x) = 2x$ for $0 \leq x \leq 2$.

Solution The region under consideration is shown in Figure 37. Using Equation (20), we find

$$\int_R \int f(x, y)\, dA = \int_0^2 \left[\int_x^{2x} (x^2 + y^2)\, dy \right] dx$$

$$= \int_0^2 \left[\left(x^2 y + \frac{1}{3} y^3 \right) \Big|_x^{2x} \right] dx$$

$$= \int_0^2 \left[\left(2x^3 + \frac{8}{3} x^3 \right) - \left(x^3 + \frac{1}{3} x^3 \right) \right] dx$$

$$= \int_0^2 \frac{10}{3} x^3\, dx = \frac{5}{6} x^4 \Big|_0^2 = 13 \tfrac{1}{3}$$

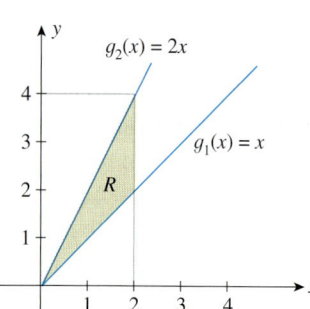

FIGURE 37
R is the region bounded by $g_1(x) = x$ and $g_2(x) = 2x$ for $0 \leq x \leq 2$.

EXAMPLE 3 Evaluate $\int_R \int f(x, y)\, dA$, where $f(x, y) = xe^y$ and R is the plane region bounded by the graphs of $y = x^2$ and $y = x$.

Solution The region in question is shown in Figure 38. The points of intersection of the two curves are found by solving the equation $x^2 = x$, giving $x = 0$ and $x = 1$. Using Equation (20), we find

$$\int_R \int f(x, y)\, dA = \int_0^1 \left(\int_{x^2}^x xe^y\, dy \right) dx = \int_0^1 \left(xe^y \Big|_{x^2}^x \right) dx$$

$$= \int_0^1 (xe^x - xe^{x^2})\, dx = \int_0^1 xe^x\, dx - \int_0^1 xe^{x^2}\, dx$$

and, integrating the first integral on the right-hand side by parts, we have

$$\int_R \int f(x, y)\, dA = \left[(x - 1)e^x - \frac{1}{2} e^{x^2} \right] \Big|_0^1$$

$$= -\frac{1}{2} e - \left(-1 - \frac{1}{2} \right) = \frac{1}{2}(3 - e)$$

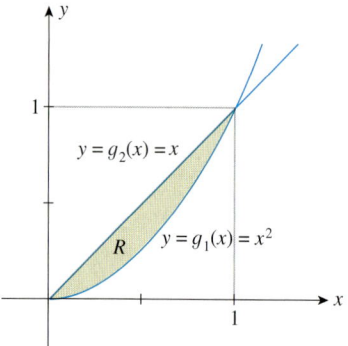

FIGURE 38
R is the region bounded by $y = x^2$ and $y = x$.

Explore & Discuss

Refer to Example 3.

1. You can also view the region R as an example of the region shown in Figure 36b. Doing so, find the functions h_1 and h_2 and the numbers c and d.

2. Find an expression for $\iint_R f(x, y)\, dA$ in terms of iterated integrals using Equation (21).
3. Evaluate the iterated integrals of part 2 and hence verify the result of Example 3.
   Hint: Integrate by parts twice.
4. Does viewing the region R in two different ways make a difference?

The next example not only illustrates the use of Equation (21) but also shows that it may be the only viable way to evaluate the given double integral.

EXAMPLE 4 Evaluate

$$\iint_R x e^{y^2}\, dA$$

where R is the plane region bounded by the y-axis, $x = 0$, the horizontal line $y = 4$, and the graph of $y = x^2$.

Solution The region R is shown in Figure 39. The point of intersection of the line $y = 4$ and the graph of $y = x^2$ is found by solving the equation $x^2 = 4$, giving $x = 2$ and the required point $(2, 4)$. Using Equation (20) with $y = g_1(x) = x^2$ and $y = g_2(x) = 4$ leads to

$$\iint_R x e^{y^2}\, dA = \int_0^2 \left[\int_{x^2}^4 x e^{y^2}\, dy \right] dx$$

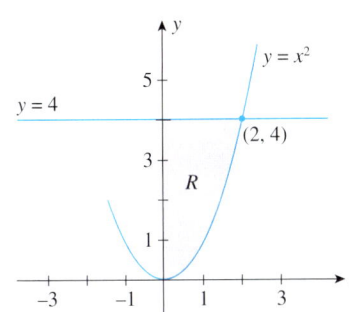

FIGURE 39
R is the region bounded by the y-axis, $x = 0$, $y = 4$, and $y = x^2$.

Now evaluation of the integral

$$\int_{x^2}^4 x e^{y^2}\, dy = x \int_{x^2}^4 e^{y^2}\, dy$$

calls for finding the antiderivative of the integrand e^{y^2} in terms of elementary functions, a task that, as was pointed out in Section 7.3, cannot be done. Let's begin afresh and attempt to make use of Equation (21).

Since the equation $y = x^2$ is equivalent to the equation $x = \sqrt{y}$, which clearly expresses x as a function of y, we may write, with $x = h_1(y) = 0$ and $h_2(y) = \sqrt{y}$,

$$\iint_R x e^{y^2}\, dA = \int_0^4 \left[\int_0^{\sqrt{y}} x e^{y^2}\, dx \right] dy = \int_0^4 \left[\frac{1}{2} x^2 e^{y^2} \right]_0^{\sqrt{y}} dy$$

$$= \int_0^4 \frac{1}{2} y e^{y^2}\, dy = \frac{1}{4} e^{y^2} \Big|_0^4 = \frac{1}{4}(e^{16} - 1)$$

8.7 Self-Check Exercise

Evaluate $\iint_R (x + y)\, dA$, where R is the region bounded by the graphs of $g_1(x) = x$ and $g_2(x) = x^{1/3}$ in the first quadrant.

The solution to Self-Check Exercise 8.7 can be found on page 605.

8.7 Concept Questions

1. What is an iterated integral? How is $\int_R \int f(x, y)\, dA$ evaluated in terms of iterated integrals, where R is the rectangular region defined by $a \leq x \leq b$ and $c \leq y \leq d$?

2. Suppose g_1 and g_2 are continuous on the interval $[a, b]$ and $R = \{(x, y) \mid g_1(x) \leq y \leq g_2(x),\ a \leq x \leq b\}$, what is $\int_R \int f(x, y)\, dA$, where f is a continuous function defined on R?

3. Suppose h_1 and h_2 are continuous on the interval $[c, d]$ and $R = \{(x, y) \mid h_1(y) \leq x \leq h_2(y),\ c \leq y \leq d\}$, what is $\int_R \int f(x, y)\, dA$, where f is a continuous function defined on R?

8.7 Exercises

In Exercises 1–25, evaluate the double integral

$$\int_R \int f(x, y)\, dA$$

for the function $f(x, y)$ and the region R.

1. $f(x, y) = y + 2x$; R is the rectangle defined by $1 \leq x \leq 2$ and $0 \leq y \leq 1$.

2. $f(x, y) = x + 2y$; R is the rectangle defined by $-1 \leq x \leq 2$ and $0 \leq y \leq 2$.

3. $f(x, y) = xy^2$; R is the rectangle defined by $-1 \leq x \leq 1$ and $0 \leq y \leq 1$.

4. $f(x, y) = 12xy^2 + 8y^3$; R is the rectangle defined by $0 \leq x \leq 1$ and $0 \leq y \leq 2$.

5. $f(x, y) = \dfrac{x}{y}$; R is the rectangle defined by $-1 \leq x \leq 2$ and $1 \leq y \leq e^3$.

6. $f(x, y) = \dfrac{xy}{1 + y^2}$; R is the rectangle defined by $-2 \leq x \leq 2$ and $0 \leq y \leq 1$.

7. $f(x, y) = 4xe^{2x^2 + y}$; R is the rectangle defined by $0 \leq x \leq 1$ and $-2 \leq y \leq 0$.

8. $f(x, y) = \dfrac{y}{x^2} e^{y/x}$; R is the rectangle defined by $1 \leq x \leq 2$ and $0 \leq y \leq 1$.

9. $f(x, y) = \ln y$; R is the rectangle defined by $0 \leq x \leq 1$ and $1 \leq y \leq e$.

10. $f(x, y) = \dfrac{\ln y}{x}$; R is the rectangle defined by $1 \leq x \leq e^2$ and $1 \leq y \leq e$.

11. $f(x, y) = x + 2y$; R is bounded by the lines $x = 1$, $y = 0$, and $y = x$.

12. $f(x, y) = xy$; R is bounded by the lines $x = 1$, $y = 0$ and $y = x$.

13. $f(x, y) = 2x + 4y$; R is bounded by $x = 1$, $x = 3$, $y = 0$, and $y = x + 1$.

14. $f(x, y) = 2 - y$; R is bounded by $x = -1$, $x = 1 - y$, $y = 0$, and $y = 2$.

15. $f(x, y) = x + y$; R is bounded by $x = 0$, $x = \sqrt{y}$, and $y = 4$.

16. $f(x, y) = x^2 y^2$; R is bounded by $x = 0$, $x = 1$, $y = x^2$, and $y = x^3$.

17. $f(x, y) = y$; R is bounded by $x = 0$, $x = \sqrt{4 - y^2}$, and $y = 0$.

18. $f(x, y) = \dfrac{y}{x^3 + 2}$; R is bounded by the lines $x = 1$, $y = 0$, and $y = x$.

19. $f(x, y) = 2xe^y$; R is bounded by the lines $x = 1$, $y = 0$, and $y = x$.

20. $f(x, y) = 2x$; R is bounded by $x = e^{2y}$, $x = y$, $y = 0$, and $y = 1$.

21. $f(x, y) = ye^x$; R is bounded by $y = \sqrt{x}$ and $y = x$.

22. $f(x, y) = xe^{-y^2}$; R is bounded by $x = 0$, $y = x^2$, and $y = 4$.

23. $f(x, y) = e^{y^2}$; R is bounded by $x = 0$, $y = 2x$, and $y = 2$.

24. $f(x, y) = y$; R is bounded by $y = \ln x$, $x = e$, and $y = 0$.

25. $f(x, y) = ye^{x^3}$; R is bounded by $x = \dfrac{y}{2}$, $x = 1$, and $y = 0$.

In Exercises 26–27, determine whether the statement is true or false. If it is true, explain why it is true. If it is false, give an example to show why it is false.

26. If $h(x, y) = f(x)g(y)$, where f is continuous on $[a, b]$ and g is continuous on $[c, d]$, then $\int_R \int h(x, y)\, dA = [\int_a^b f(x)\, dx][\int_c^d g(y)\, dy]$, where $R = \{(x, y) \mid a \leq x \leq b;\ c \leq y \leq d\}$.

27. If $\int_{R_1} \int f(x, y)\, dA$ exists, where $R_1 = \{(x, y) \mid a \leq x \leq b;\ c \leq y \leq d\}$, then $\int_{R_2} \int f(x, y)\, dA$ exists, where $R_2 = \{(x, y) \mid c \leq x \leq d;\ a \leq y \leq b\}$.

8.7 Solution to Self-Check Exercise

The region R is shown in the accompanying figure. The points of intersection of the two curves are found by solving the equation $x = x^{1/3}$, giving $x = 0$ and $x = 1$. Using Equation (20), we find

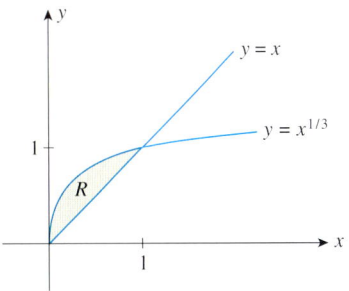

$$\iint_R (x+y)\, dA = \int_0^1 \left[\int_x^{x^{1/3}} (x+y)\, dy \right] dx$$

$$= \int_0^1 \left[xy + \frac{1}{2} y^2 \right]_x^{x^{1/3}} dx$$

$$= \int_0^1 \left[\left(x^{4/3} + \frac{1}{2} x^{2/3} \right) - \left(x^2 + \frac{1}{2} x^2 \right) \right] dx$$

$$= \int_0^1 \left(x^{4/3} + \frac{1}{2} x^{2/3} - \frac{3}{2} x^2 \right) dx$$

$$= \frac{3}{7} x^{7/3} + \frac{3}{10} x^{5/3} - \frac{1}{2} x^3 \Big|_0^1$$

$$= \frac{3}{7} + \frac{3}{10} - \frac{1}{2} = \frac{8}{35}$$

8.8 Applications of Double Integrals

In this section we will discuss applications involving the double integral.

Finding the Volume of a Solid by Double Integrals

As we saw earlier, the double integral

$$\iint_R f(x, y)\, dA$$

gives the volume of the solid bounded by the graph of $f(x, y)$ over the region R.

> **The Volume of a Solid Under a Surface**
> Let R be a region in the xy-plane and let f be continuous and nonnegative on R. Then, the **volume of a solid under a surface** bounded above by $z = f(x, y)$ and below by R is given by
> $$V = \iint_R f(x, y)\, dA$$

EXAMPLE 1 Find the volume of the solid bounded above by the plane $z = f(x, y) = y$ and below by the plane region R defined by $y = \sqrt{1 - x^2}$ ($0 \le x \le 1$).

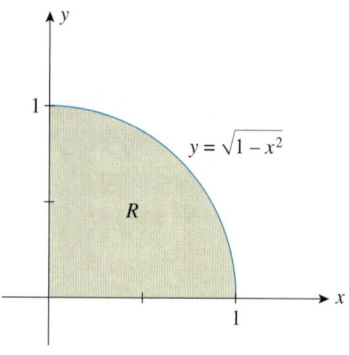

FIGURE 40
The plane region R defined by $y = \sqrt{1 - x^2}$ $(0 \leq x \leq 1)$

Solution The region R is sketched in Figure 40. Observe that $f(x, y) = y \geq 0$ for $(x, y) \in R$. Therefore, the required volume is given by

$$\iint_R y \, dA = \int_0^1 \left(\int_0^{\sqrt{1-x^2}} y \, dy \right) dx = \int_0^1 \left(\frac{1}{2} y^2 \Big|_0^{\sqrt{1-x^2}} \right) dx$$

$$= \int_0^1 \frac{1}{2} (1 - x^2) \, dx = \frac{1}{2} \left(x - \frac{1}{3} x^3 \right) \Big|_0^1 = \frac{1}{3}$$

The solid is shown in Figure 41. Note that it is not necessary to make a sketch of the solid in order to compute its volume.

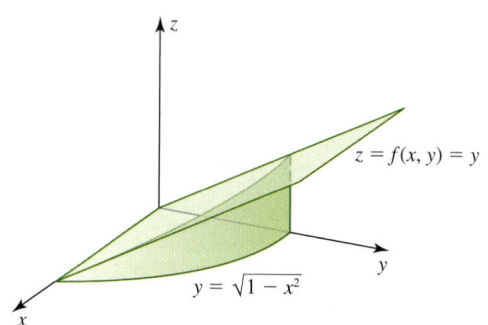

FIGURE 41
The solid bounded above by the plane $z = y$ and below by the plane region defined by $y = \sqrt{1 - x^2}$ $(0 \leq x \leq 1)$

Population of a City

Suppose the plane region R represents a certain district of a city and $f(x, y)$ gives the population density (the number of people per square mile) at any point (x, y) in R. Enclose the set R by a rectangle, and construct a grid for it in the usual manner. In any rectangular region of the grid that has no point in common with R, set $f(x_i, y_i)hk = 0$ (Figure 42). Then, corresponding to any grid covering the set R, the general term of the Riemann sum $f(x_i, y_i)hk$ (population density times area) gives the number of people living in that part of the city corresponding to the rectangular region R_i. Therefore, the Riemann sum gives an approximation of the number of people living in the district represented by R and, in the limit, the double integral

$$\iint_R f(x, y) \, dA$$

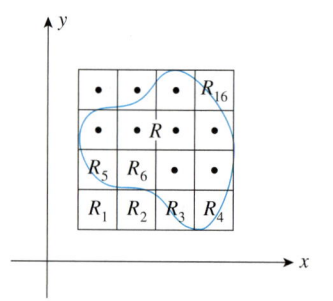

FIGURE 42
The rectangular region R representing a certain district of a city is enclosed by a rectangular grid.

gives the actual number of people living in the district under consideration.

 APPLIED EXAMPLE 2 Population Density The population density (number of people per square mile) of a certain city is described by the function

$$f(x, y) = 10{,}000 e^{-0.2|x| - 0.1|y|}$$

where the origin $(0, 0)$ gives the location of the city hall. What is the population inside the rectangular area described by

$$R = \{(x, y) \mid -10 \leq x \leq 10; \, -5 \leq y \leq 5\}$$

if x and y are in miles? (See Figure 43.)

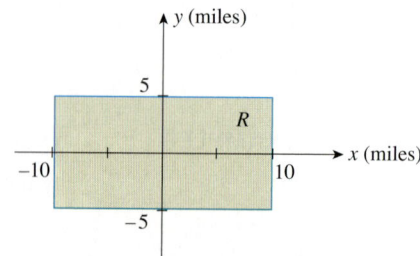

FIGURE 43
The rectangular region R represents a certain district of a city.

Solution By symmetry, it suffices to compute the population in the first quadrant. (Why?) Then, upon observing that in this quadrant

$$f(x, y) = 10{,}000e^{-0.2x-0.1y} = 10{,}000e^{-0.2x}e^{-0.1y}$$

we see that the population in R is given by

$$\iint_R f(x, y)\, dA = 4\int_0^{10}\left(\int_0^5 10{,}000e^{-0.2x}e^{-0.1y}\, dy\right) dx$$

$$= 4\int_0^{10}\left(-100{,}000e^{-0.2x}e^{-0.1y}\Big|_0^5\right) dx$$

$$= 400{,}000(1 - e^{-0.5})\int_0^{10} e^{-0.2x}\, dx$$

$$= 2{,}000{,}000(1 - e^{-0.5})(1 - e^{-2})$$

or approximately 680,438.

Explore & Discuss

1. Consider the improper double integral $\iint_D f(x, y)\, dA$ of the continuous function f of two variables defined over the plane region

$$D = \{(x, y)\mid 0 \le x < \infty;\ 0 \le y < \infty\}$$

Using the definition of improper integrals of functions of one variable (Section 7.4), explain why it makes sense to define

$$\iint_D f(x, y)\, dA = \lim_{N\to\infty}\int_0^N \left[\lim_{M\to\infty}\int_0^M f(x, y)\, dx\right] dy$$

$$= \lim_{M\to\infty}\int_0^M \left[\lim_{N\to\infty}\int_0^N f(x, y)\, dy\right] dx$$

provided that the limits exist.

2. Refer to Example 2. Assuming that the population density of the city is described by

$$f(x, y) = 10{,}000e^{-0.2|x|-0.1|y|}$$

for $-\infty < x < \infty$ and $-\infty < y < \infty$, show that the population outside the rectangular region

$$R = \{(x, y)\mid -10 < x < 10;\ -5 < y \le 5\}$$

of Example 2 is given by

$$4\iint_D f(x, y)\, dx\, dy - 680{,}438$$

(Recall that 680,438 is the approximate population inside R.)

3. Use the results of parts 1 and 2 to determine the population of the city outside the rectangular area R. (Assume that there are no other major cities nearby.)

Average Value of a Function

In Section 6.5, we showed that the average value of a continuous function $f(x)$ over an interval $[a, b]$ is given by

$$\frac{1}{b-a}\int_a^b f(x)\, dx$$

That is, the average value of a function over $[a, b]$ is the integral of f over $[a, b]$ divided by the length of the interval. An analogous result holds for a function of two variables $f(x, y)$ over a plane region R. To see this, we enclose R by a rectangle and construct a rectangular grid. Let (x_i, y_i) be any point in the rectangle R_i of area hk. Now, the average value of the mn numbers $f(x_1, y_1), f(x_2, y_2), \ldots, f(x_{mn}, y_{mn})$ is given by

$$\frac{f(x_1, y_1) + f(x_2, y_2) + \cdots + f(x_{mn}, y_{mn})}{mn}$$

which can also be written as

$$\frac{hk}{hk}\left[\frac{f(x_1, y_1) + f(x_2, y_2) + \cdots + f(x_{mn}, y_{mn})}{mn}\right]$$

$$= \frac{1}{(mn)hk}[f(x_1, y_1) + f(x_2, y_2) + \cdots + f(x_{mn}, y_{mn})]hk$$

Now the area of R is approximated by the sum of the mn rectangles (*omitting* those having no points in common with R), each of area hk. Note that this is the denominator of the previous expression. Therefore, taking the limit as m and n both tend to infinity, we obtain the following formula for the *average value of $f(x, y)$ over R.*

Average Value of f(x, y) over the Region R

If f is integrable over the plane region R, then its average value over R is given by

$$\frac{\iint_R f(x, y)\, dA}{\text{Area of } R} \quad \text{or} \quad \frac{\iint_R f(x, y)\, dA}{\iint_R dA} \tag{22}$$

Note If we let $f(x, y) = 1$ for all (x, y) in R, then

$$\iint_R f(x, y)\, dA = \iint_R dA = \text{Area of } R \qquad \blacksquare$$

EXAMPLE 3 Find the average value of the function $f(x, y) = xy$ over the plane region R bounded by the graph of $y = e^x$ and the lines $x = 0$ and $x = 1$.

Solution The region R is shown in Figure 44. The area of the region R is given by

$$\int_0^1 \left(\int_0^{e^x} dy\right) dx = \int_0^1 \left(y\Big|_0^{e^x}\right) dx$$

$$= \int_0^1 e^x\, dx$$

$$= e^x\Big|_0^1$$

$$= e - 1$$

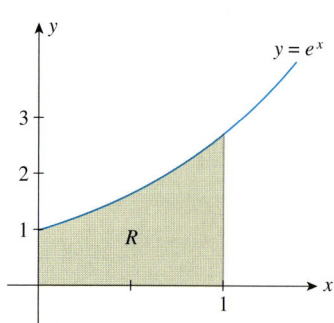

FIGURE 44
The plane region R defined by $y = e^x$ $(0 \le x \le 1)$

We would obtain the same result had we viewed the area of this region as the area of the region under the curve $y = e^x$ from $x = 0$ to $x = 1$. Next, we compute

$$\iint_R f(x, y)\, dA = \int_0^1 \left(\int_0^{e^x} xy\, dy \right) dx$$

$$= \int_0^1 \left(\frac{1}{2} xy^2 \Big|_0^{e^x} \right) dx$$

$$= \int_0^1 \frac{1}{2} xe^{2x}\, dx$$

$$= \frac{1}{4} xe^{2x} - \frac{1}{8} e^{2x} \Big|_0^1 \qquad \text{Integrate by parts.}$$

$$= \left(\frac{1}{4} e^2 - \frac{1}{8} e^2 \right) + \frac{1}{8}$$

$$= \frac{1}{8}(e^2 + 1)$$

Therefore, the required average value is given by

$$\frac{\iint_R f(x, y)\, dA}{\iint_R dA} = \frac{\frac{1}{8}(e^2 + 1)}{e - 1} = \frac{e^2 + 1}{8(e - 1)}$$

APPLIED EXAMPLE 4 Population Density (Refer to Example 2.) The population density of a certain city (number of people per square mile) is described by the function

$$f(x, y) = 10{,}000 e^{-0.2|x| - 0.1|y|}$$

where the origin gives the location of the city hall. What is the average population density inside the rectangular area described by

$$R = \{(x, y) \mid -10 \leq x \leq 10;\ -5 \leq y \leq 5\}$$

where x and y are measured in miles?

Solution From the results of Example 2, we know that

$$\iint_R f(x, y)\, dA \approx 680{,}438$$

From Figure 44, we see that the area of the plane rectangular region R is $(20)(10)$, or 200, square miles. Therefore, the average population density inside R is

$$\frac{\iint_R f(x, y)\, dA}{\iint_R dA} = \frac{680{,}438}{200} = 3402.19$$

or approximately 3402 people per square mile.

8.8 Self-Check Exercise

The population density of a coastal town located on an island is described by the function

$$f(x, y) = \frac{5000xe^y}{1 + 2x^2} \quad (0 \le x \le 4; -2 \le y \le 0)$$

where x and y are measured in miles (see the accompanying figure).

What is the population inside the rectangular area defined by $R = \{(x, y) \mid 0 \le x \le 4; -2 \le y \le 0\}$? What is the average population density in the area?

The solution to Self-Check Exercise 8.8 can be found on page 612.

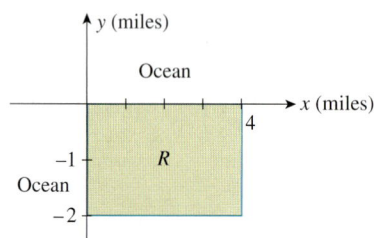

8.8 Concept Questions

1. Give the formula for finding the volume of the solid bounded above by the graph of $z = f(x, y)$ and below by the region R in the xy-plane.
2. What is the average value of $f(x, y)$ over the region R?
3. Suppose a plane region R represents a certain district of a city, $f(x, y)$ gives the population density at any point (x, y) in R, and the set R is enclosed by a rectangle.

   a. Explain how a Riemann sum can be used to approximate the number of people living in the district represented by R.
   b. What does the general term of the Riemann sum $f(x_i, y_i)hk$ represent?
   c. What does $\int_R \int f(x, y) \, dA$ represent?

8.8 Exercises

In Exercises 1–8, use a double integral to find the volume of the solid shown in the figure.

1.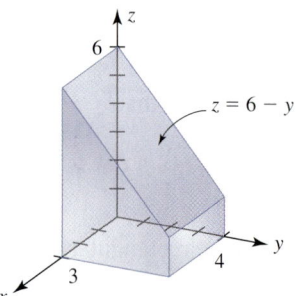

2.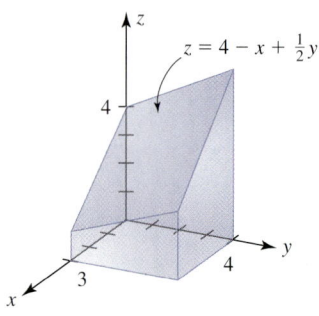

3.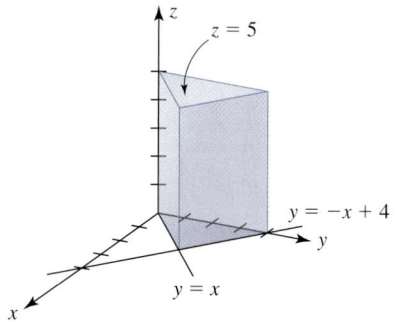

4.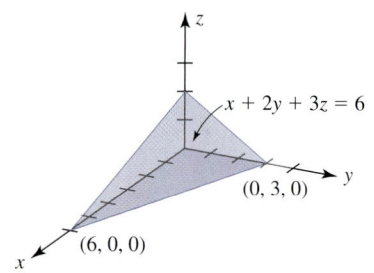

5.

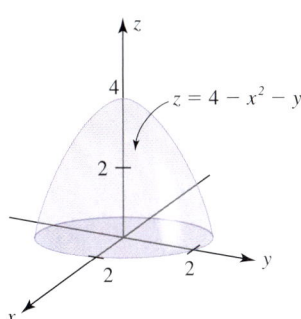

6.

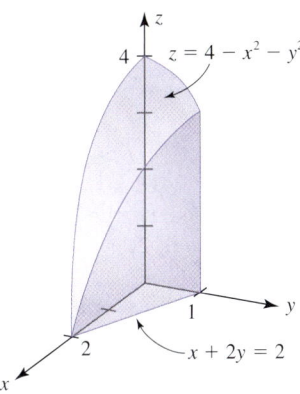

7.

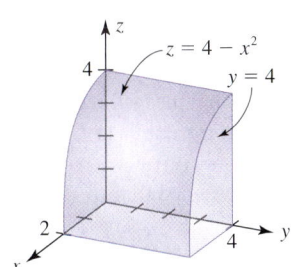

8.

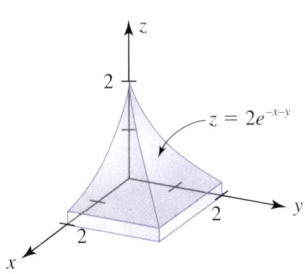

In Exercises 9–16, find the volume of the solid bounded above by the surface $z = f(x, y)$ and below by the plane region R.

9. $f(x, y) = 4 - 2x - y$; $R = \{(x, y) \mid 0 \leq x \leq 1; 0 \leq y \leq 2\}$

10. $f(x, y) = 2x + y$; R is the triangle bounded by $y = 2x$, $y = 0$, and $x = 2$.

11. $f(x, y) = x^2 + y^2$; R is the rectangle with vertices $(0, 0)$, $(1, 0)$, $(1, 2)$, and $(0, 2)$.

12. $f(x, y) = e^{x+2y}$; R is the triangle with vertices $(0, 0)$, $(1, 0)$, and $(0, 1)$.

13. $f(x, y) = 2xe^y$; R is the triangle bounded by $y = x$, $y = 2$, and $x = 0$.

14. $f(x, y) = \dfrac{2y}{1 + x^2}$; R is the region bounded by $y = \sqrt{x}$, $y = 0$, and $x = 4$.

15. $f(x, y) = 2x^2y$; R is the region bounded by the graphs of $y = x$ and $y = x^2$.

16. $f(x, y) = x$; R is the region in the first quadrant bounded by the semicircle $y = \sqrt{16 - x^2}$, the x-axis, and the y-axis.

In Exercises 17–22, find the average value of the function $f(x, y)$ over the plane region R.

17. $f(x, y) = 6x^2y^3$; $R = \{(x, y) \mid 0 \leq x \leq 2; 0 \leq y \leq 3\}$

18. $f(x, y) = x + 2y$; R is the triangle with vertices $(0, 0)$, $(1, 0)$, and $(1, 1)$.

19. $f(x, y) = xy$; R is the triangle bounded by $y = x$, $y = 2 - x$, and $y = 0$.

20. $f(x, y) = e^{-x^2}$; R is the triangle with vertices $(0, 0)$, $(1, 0)$, and $(1, 1)$.

21. $f(x, y) = xe^y$; R is the triangle with vertices $(0, 0)$, $(1, 0)$, and $(1, 1)$.

22. $f(x, y) = \ln x$; R is the region bounded by the graphs of $y = 2x$ and $y = 0$ from $x = 1$ to $x = 3$.
Hint: Use integration by parts.

23. POPULATION DENSITY The population density (number of people per square mile) of a coastal town is described by the function

$$f(x, y) = \dfrac{10{,}000 e^y}{1 + 0.5|x|} \quad (-10 \leq x \leq 10; -4 \leq y \leq 0)$$

where x and y are measured in miles (see the accompanying figure). Find the population inside the rectangular area described by

$$R = \{(x, y) \mid -5 \leq x \leq 5; -2 \leq y \leq 0\}$$

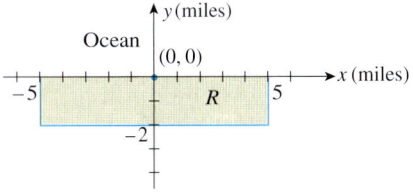

24. AVERAGE POPULATION DENSITY Refer to Exercise 23. Find the average population density inside the rectangular area R.

25. **POPULATION DENSITY** The population density (number of people per square mile) of a certain city is given by the function

$$f(x, y) = \frac{50,000|xy|}{(x^2 + 20)(y^2 + 36)}$$

where the origin (0, 0) gives the location of the government center. Find the population inside the rectangular area described by

$$R = \{(x, y) | -15 \leq x \leq 15; -20 \leq y \leq 20\}$$

26. **AVERAGE PROFIT** The Country Workshop's total weekly profit (in dollars) realized in manufacturing and selling its rolltop desks is given by the profit function

$$P(x, y) = -0.2x^2 - 0.25y^2 - 0.2xy + 100x + 90y - 4000$$

where x stands for the number of finished units and y stands for the number of unfinished units manufactured and sold each week. Find the average weekly profit if the number of finished units manufactured and sold varies between 180 and 200 and the number of unfinished units varies between 100 and 120/week.

27. **AVERAGE PRICE OF LAND** The rectangular region R shown in the accompanying figure represents a city's financial district. The price of land in the district is approximated by the function

$$p(x, y) = 200 - 10\left(x - \frac{1}{2}\right)^2 - 15(y - 1)^2$$

where $p(x, y)$ is the price of land at the point (x, y) in dollars/square foot and x and y are measured in miles. What is the average price of land per square foot in the district?

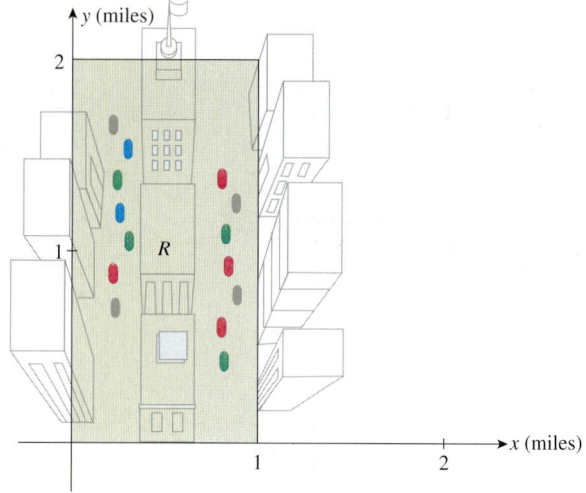

In Exercises 28–29, determine whether the statement is true or false. If it is true, explain why it is true. If it is false, give an example to show why it is false.

28. Let R be a region in the xy-plane and let f and g be continuous functions on R that satisfy the condition $f(x, y) \leq g(x, y)$ for all (x, y) in R. Then $\int_R \int [g(x, y) - f(x, y)] \, dA$ gives the volume of the solid bounded above by the surface $z = g(x, y)$ and below by the surface $z = f(x, y)$.

29. Suppose that f is nonnegative and integrable over the plane region R. Then the average value of f over R can be thought of as the (constant) height of the cylinder with base R and volume that is exactly equal to the volume of the solid under the graph of $z = f(x, y)$. (*Note:* The cylinder referred to here has sides perpendicular to R.)

8.8 Solution to Self-Check Exercise

The population in R is given by

$$\iint_R f(x, y) \, dA = \int_0^4 \left[\int_{-2}^0 \frac{5000xe^y}{1 + 2x^2} \, dy\right] dx$$

$$= \int_0^4 \left[\frac{5000xe^y}{1 + 2x^2}\bigg|_{-2}^0\right] dx$$

$$= 5000(1 - e^{-2}) \int_0^4 \frac{x}{1 + 2x^2} \, dx$$

$$= 5000(1 - e^{-2}) \left[\frac{1}{4} \ln(1 + 2x^2)\right]_0^4$$

$$= 5000(1 - e^{-2})\left(\frac{1}{4}\right) \ln 33$$

or approximately 3779 people. The average population density inside R is

$$\frac{\iint_R f(x, y) \, dA}{\iint_R dA} = \frac{3779}{(2)(4)}$$

or approximately 472 people/square mile.

CHAPTER 8 Summary of Principal Terms

TERMS

function of two variables (534)
domain (534)
three-dimensional Cartesian coordinate system (537)
level curve (539)
first partial derivatives of f (546)
Cobb–Douglas production function (550)
marginal productivity of labor (550)
marginal productivity of capital (550)
substitute commodity (551)
complementary commodity (551)

second-order partial derivative of f (552)
relative maximum (560)
relative maximum value (560)
relative minimum (560)
relative minimum value (560)
absolute maximum (560)
absolute minimum (560)
absolute maximum value (560)
absolute minimum value (560)
saddle point (561)
critical point (561)
second derivative test (562)

method of least squares (570)
scatter diagram (570)
least-squares line (regression line) (570)
normal equation (572)
constrained relative extremum (581)
method of Lagrange multipliers (583)
increment (592)
total differential (593)
Riemann sum (598)
double integral (599)
volume of the solid under a surface (605)

CHAPTER 8 Concept Review Questions

Fill in the blanks.

1. The domain of a function f of two variables is a subset of the _____-plane. The rule of f associates with each _____ _____ in the domain of f one and only one _____ _____, denoted by $z =$ _____.

2. If the function f has rule $z = f(x, y)$, then x and y are called _____ variables, and z is a/an _____ variable. The number z is also called the _____ of f at (x, y).

3. The graph of a function f of two variables is the set of all points (x, y, z), where _____, and (x, y) is the domain of _____. The graph of a function of two variables is a/an _____ in three-dimensional space.

4. The trace of the graph of $f(x, y)$ in the plane $z = k$ is the curve with equation _____ lying in the plane $z = k$. The projection of the trace of f in the plane $z = k$ onto the xy-plane is called the _____ _____ of f. The contour map associated with f is obtained by drawing the _____ _____ of f corresponding to several admissible values of _____.

5. The partial derivative $\partial f/\partial x$ of f at (x, y) can be found by thinking of y as a/an _____ in the expression for f and differentiating this expression with respect to _____ as if it were a function of x alone.

6. The number $f_x(a, b)$ measures the _____ of the tangent line to the curve C obtained by the intersection of the graph of f and the plane $y = b$ at the point _____. It also measures the rate of change of f with respect to _____ at the point (a, b) with y held fixed with value _____.

7. A function $f(x, y)$ has a relative maximum at (a, b) if $f(x, y)$ _____ $f(a, b)$ for all points (x, y) that are sufficiently close to _____. The absolute maximum value of $f(x, y)$ is the number $f(a, b)$ such that $f(x, y)$ _____ $f(a, b)$ for all (x, y) in the _____ of f.

8. A critical point of $f(x, y)$ is a point (a, b) in the _____ of f such that _____ _____ _____ or at least one of the partial derivatives of f does not _____. A critical point of f is a/an _____ for a relative extremum of f.

9. By plotting the points associated with a set of data we obtain the _____ diagram for the data. The least-squares line is the line obtained by _____ the sum of the squares of the errors made when the y-values of the data points are approximated by the corresponding y-values of the _____ line. The least-squares equation is found by solving the _____ equations.

10. The method of Lagrange multipliers solves the problem of finding the relative extrema of a function $f(x, y)$ subject to the constraint _____. We first form the Lagrangian function $F(x, y, \lambda) =$ _____. Then we solve the system consisting of the three equations _____, _____, and _____ for x, y, and λ. These solutions give the critical points that give rise to the relative _____ of f.

11. For a differentiable function $z = f(x, y)$, the differentials of the _____ variables x and y are _____ and _____, and the differential of the dependent variable z is $dz =$ _____.

12. If $f(x, y)$ is continuous and nonnegative over a region R in the xy-plane and $\int_R \int f(x, y)\, dA$ exists, then the double integral gives the _____ of the _____ bounded by the graph of $f(x, y)$ over the region R.

13. The integral $\int_R \int f(x, y)\, dA$ is evaluated by using _____ integrals. For example, $\int_R \int (2x + y^2)\, dA$, where $R = \{(x, y) \mid 0 \le x \le 1;\ 3 \le y \le 5\}$ is equal to $\int_0^1 \int_3^5 (2x + y^2)\, dy\, dx$ or the (iterated) integral _____.

CHAPTER 8 Review Exercises

1. Let $f(x, y) = \dfrac{xy}{x^2 + y^2}$. Compute $f(0, 1), f(1, 0)$, and $f(1, 1)$. Does $f(0, 0)$ exist?

2. Let $f(x, y) = \dfrac{xe^y}{1 + \ln xy}$. Compute $f(1, 1)$, $f(1, 2)$, and $f(2, 1)$. Does $f(1, 0)$ exist?

3. Let $h(x, y, z) = xye^z + \dfrac{x}{y}$. Compute $h(1, 1, 0)$, $h(-1, 1, 1)$, and $h(1, -1, 1)$.

4. Find the domain of the function $f(u, v) = \dfrac{\sqrt{u}}{u - v}$.

5. Find the domain of the function $f(x, y) = \dfrac{x - y}{x + y}$.

6. Find the domain of the function $f(x, y) = x\sqrt{y} + y\sqrt{1 - x}$.

7. Find the domain of the function
$$f(x, y, z) = \dfrac{xy\sqrt{z}}{(1 - x)(1 - y)(1 - z)}$$

In Exercises 8–11, sketch the level curves of the function corresponding to each value of z.

8. $z = f(x, y) = 2x + 3y$; $z = -2, -1, 0, 1, 2$
9. $z = f(x, y) = y - x^2$; $z = -2, -1, 0, 1, 2$
10. $z = f(x, y) = \sqrt{x^2 + y^2}$; $z = 0, 1, 2, 3, 4$
11. $z = f(x, y) = e^{xy}$; $z = 1, 2, 3$

In Exercises 12–21, compute the first partial derivatives of the function.

12. $f(x, y) = x^2y^3 + 3xy^2 + \dfrac{x}{y}$
13. $f(x, y) = x\sqrt{y} + y\sqrt{x}$ 14. $f(u, v) = \sqrt{uv^2 - 2u}$
15. $f(x, y) = \dfrac{x - y}{y + 2x}$ 16. $g(x, y) = \dfrac{xy}{x^2 + y^2}$
17. $h(x, y) = (2xy + 3y^2)^5$ 18. $f(x, y) = (xe^y + 1)^{1/2}$
19. $f(x, y) = (x^2 + y^2)e^{x^2+y^2}$
20. $f(x, y) = \ln(1 + 2x^2 + 4y^4)$
21. $f(x, y) = \ln\left(1 + \dfrac{x^2}{y^2}\right)$

In Exercises 22–27, compute the second-order partial derivatives of the function.

22. $f(x, y) = x^3 - 2x^2y + y^2 + x - 2y$
23. $f(x, y) = x^4 + 2x^2y^2 - y^4$
24. $f(x, y) = (2x^2 + 3y^2)^3$ 25. $g(x, y) = \dfrac{x}{x + y^2}$
26. $g(x, y) = e^{x^2+y^2}$ 27. $h(s, t) = \ln\left(\dfrac{s}{t}\right)$

28. Let $f(x, y, z) = x^3y^2z + xy^2z + 3xy - 4z$. Compute $f_x(1, 1, 0)$, $f_y(1, 1, 0)$, and $f_z(1, 1, 0)$, and interpret your results.

In Exercises 29–34, find the critical point(s) of the functions. Then use the second derivative test to classify the nature of each of these points, if possible. Finally, determine the relative extrema of each function.

29. $f(x, y) = 2x^2 + y^2 - 8x - 6y + 4$
30. $f(x, y) = x^2 + 3xy + y^2 - 10x - 20y + 12$
31. $f(x, y) = x^3 - 3xy + y^2$
32. $f(x, y) = x^3 + y^2 - 4xy + 17x - 10y + 8$
33. $f(x, y) = e^{2x^2+y^2}$
34. $f(x, y) = \ln(x^2 + y^2 - 2x - 2y + 4)$

In Exercises 35–38, use the method of Lagrange multipliers to optimize the function subject to the given constraints.

35. Maximize the function $f(x, y) = -3x^2 - y^2 + 2xy$ subject to the constraint $2x + y = 4$.

36. Minimize the function $f(x, y) = 2x^2 + 3y^2 - 6xy + 4x - 9y + 10$ subject to the constraint $x + y = 1$.

37. Find the maximum and minimum values of the function $f(x, y) = 2x - 3y + 1$ subject to the constraint $2x^2 + 3y^2 - 125 = 0$.

38. Find the maximum and minimum values of the function $f(x, y) = e^{x-y}$ subject to the constraint $x^2 + y^2 = 1$.

In Exercises 39 and 40, find the total differential of each function at the given point.

39. $f(x, y) = (x^2 + y^4)^{3/2}$; $(3, 2)$
40. $f(x, y) = xe^{x-y} + x \ln y$; $(1, 1)$

In Exercises 41 and 42, find the total approximate change in z when the point (x, y) changes from (x_0, y_0) to (x_1, y_1).

41. $f(x, y) = 2x^2y^3 + 3y^2x^2 - 2xy$; from $(1, -1)$ to $(1.02, -0.98)$

42. $f(x, y) = 4x^{3/4}y^{1/4}$; from $(16, 81)$ to $(17, 80)$

In Exercises 43–46, evaluate the double integrals.

43. $f(x, y) = 3x - 2y$; R is the rectangle defined by $2 \le x \le 4$ and $-1 \le y \le 2$.

44. $f(x, y) = e^{-x-2y}$; R is the rectangle defined by $0 \le x \le 2$ and $0 \le y \le 1$.

45. $f(x, y) = 2x^2y$; R is bounded by the curves $y = x^2$ and $y = x^3$ in the first quadrant.

46. $f(x, y) = \dfrac{y}{x}$, R is bounded by the lines $x = 2$, $y = 1$, and $y = x$.

In Exercises 47 and 48, find the volume of the solid bounded above by the surface $z = f(x, y)$ and below by the plane region R.

47. $f(x, y) = 4x^2 + y^2$; $R = \{0 \leq x \leq 2; 0 \leq y \leq 1\}$

48. $f(x, y) = x + y$; R is the region bounded by $y = x^2$, $y = 4x$, and $y = 4$.

49. Find the average value of the function
$$f(x, y) = xy + 1$$
over the plane region R bounded by $y = x^2$ and $y = 2x$.

50. IQs The IQ (intelligence quotient) of a person whose chronological age is c and whose mental age is m is defined as
$$I(c, m) = \dfrac{100m}{c}$$
Describe the level curves of I. Sketch the level curves corresponding to $I = 90, 100, 120, 180$. Interpret your results.

51. Revenue Functions A division of Ditton Industries makes a 16-speed and a 10-speed electric blender. The company's management estimates that x units of the 16-speed model and y units of the 10-speed model are demanded daily when the unit prices are
$$p = 80 - 0.02x - 0.1y$$
$$q = 60 - 0.1x - 0.05y$$
dollars, respectively.
a. Find the daily total revenue function $R(x, y)$.
b. Find the domain of the function R.
c. Compute $R(100, 300)$, and interpret your result.

52. Demand for CD Players In a survey conducted by *Home Entertainment* magazine, it was determined that the demand equation for CD players is given by
$$x = f(p, q) = 900 - 9p - e^{0.4q}$$
whereas the demand equation for audio CDs is given by
$$y = g(p, q) = 20{,}000 - 3000q - 4p$$
where p and q denote the unit prices (in dollars) for the CD players and audio CDs, respectively, and x and y denote the number of CD players and audio CDs demanded per week. Determine whether these two products are substitute, complementary, or neither.

53. Estimating Changes in Profit The total daily profit function (in dollars) of Weston Publishing Company realized in publishing and selling its English language dictionaries is given by
$$P(x, y) = -0.0005x^2 - 0.003y^2 - 0.002xy + 14x + 12y - 200$$
where x denotes the number of deluxe copies and y denotes the number of standard copies published and sold daily. Currently, the number of deluxe and standard copies of the dictionaries published and sold daily are 1000 and 1700, respectively. Determine the approximate daily change in the total daily profit if the number of deluxe copies is increased to 1050 and the number of standard copies is decreased to 1650 per day.

54. Average Daily TV-Viewing Time The following data were compiled by the Bureau of Television Advertising in a large metropolitan area, giving the average daily TV-viewing time per household in that area over the years 2004 to 2012.

Year	2004	2006
Daily Viewing Time, y	6 hr 9 min	6 hr 30 min

Year	2008	2010	2012
Daily Viewing Time, y	6 hr 36 min	7 hr	7 hr 16 min

a. Find the least-squares line for these data. (Let $x = 1$ represent 2004.)
b. Estimate the average daily TV-viewing time per household in 2014.

55. Female Life Expectancy at 65 The projections of female life expectancy at age 65 years in the United States are summarized in the following table ($x = 0$ corresponds to 2000):

Year, x	0	10	20	30	40	50
Years beyond 65, y	19.5	20.0	20.6	21.2	21.8	22.4

a. Find an equation of the least-squares line for these data.
b. Use the result of (a) to estimate life expectancy at 65 of a female in 2040. How does this result compare with the given data for that year?
c. Use the result of (a) to estimate the life expectancy at 65 of a female in 2030.

Source: U.S. Census Bureau.

56. Mobile Phone Use in China The number of mobile phone users (in millions) in China from 2007 ($x = 0$) through 2011 is given in the following table:

Year	2007	2008	2009	2010	2011
Number, y	547.2	638.9	750.1	861.2	929.8

a. Find an equation of the least squares line for these data.
b. Use the result of part (a) to estimate the number of mobile phone users in China in 2014, assuming that the trend continues.

Source: HIS iSuppli.

57. Maximizing Revenue Odyssey Travel Agency's monthly revenue depends on the amount of money x (in thousands of dollars) spent on advertising per month and the number of agents y in its employ in accordance with the rule
$$R(x, y) = -x^2 - 0.5y^2 + xy + 8x + 3y + 20$$
Determine the amount of money the agency should spend per month and the number of agents it should employ to maximize its monthly revenue.

58. **Minimizing Fencing Costs** The owner of the Rancho Grande wants to enclose a rectangular piece of grazing land along the straight portion of a river and then subdivide it using a fence running parallel to the sides. No fencing is required along the river. If the material for the sides costs $3/running yard and the material for the divider costs $2/running yard, what will be the dimensions of a 303,750 yd² pasture if the cost of fencing is kept to a minimum?

59. **Maximizing Revenue** The annual profit (in hundreds of dollars) of Apex Travel Agency is given by

$$P(x, y) = -\frac{3}{2}x^2 - 2y^2 + xy + 38x + 178y - 1500$$

where x denotes the number of agents it has on its payroll and y denotes the number of advertisements it places on television. How many agents and how many advertisements should Apex place on television to maximize its annual profit? What is the maximum annual profit?

60. **Maximizing the Weights of Fish** A pond is stacked with x bass and y trout (in hundreds). The average weights of bass and trout after 1 year are $(8 - x - \frac{1}{2}y)$ lb and $(11 - \frac{1}{2}x - 2y)$ lb, respectively. Find the number of each species of fish in the pond that will make the total weight of fish a maximum.

61. **Cobb–Douglas Production Functions** The production of Q units of a commodity is related to the amount of labor x and the amount of capital y (in suitable units) expended by the equation

$$Q = f(x, y) = x^{3/4} y^{1/4}$$

If an expenditure of 100 units is available for production, how should it be apportioned between labor and capital so that Q is maximized?

Hint: Use the method of Lagrange multipliers to maximize the function Q subject to the constraint $x + y = 100$.

62. **Cobb–Douglas Production Function** Show that the Cobb–Douglas production function $P = kx^a y^{1-a}$, where $0 < a < 1$, satisfies the equation

$$x\frac{\partial P}{\partial x} + y\frac{\partial P}{\partial y} = P$$

CHAPTER 8 Before Moving On . . .

1. Find the domain of

$$f(x, y) = \frac{\sqrt{x} + \sqrt{y}}{(1 - x)(2 - y)}$$

2. Find the first- and second-order partial derivatives of $f(x, y) = x^2 y + e^{xy}$.

3. Find the relative extrema, if any, of $f(x, y) = 2x^3 + 2y^3 - 6xy - 5$.

4. Find an equation of the least-squares line for the following data:

x	0	1	2	3	5
y	2.9	5.1	6.8	8.8	13.2

5. Use the method of Lagrange multipliers to find the minimum of $f(x, y) = 3x^2 + 3y^2 + 1$ subject to $x + y = 1$.

6. Let $z = 2x^2 - xy$.
   a. Find the differential dz.
   b. Compute the value of dz if (x, y) changes from $(1, 1)$ to $(0.98, 1.03)$.

7. Evaluate $\int_R \int (1 - xy)\, dA$, where R is the region bounded by $x = 0$, $x = 1$, $y = x$, and $y = x^2$.

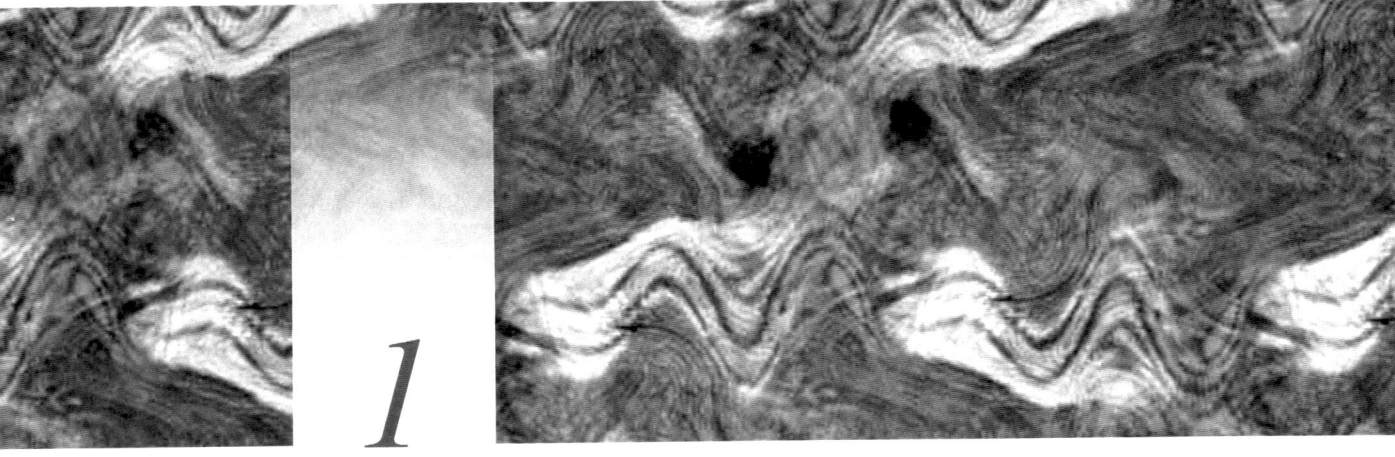

1

An Introduction to Regression Analysis

Advances in technology including computers, scanners, and telecommunications equipment have buried present-day managers under a mountain of data. Although the purpose of these data is to assist managers in the decision-making process, corporate executives who face the task of juggling data on many variables may find themselves at a loss when attempting to make sense of such information. The decision-making process is further complicated by the dynamic elements in the business environment and the complex interrelationships among these elements.

This text has been prepared to give managers (and future managers) tools for examining possible relationships between two or more variables. For example, sales and advertising are two variables commonly thought to be related. When a soft drink company increases advertising expenditures by paying entertainers or professional athletes millions of dollars to do its advertisements, it expects this outlay to increase sales. In general, it is comforting to have some evidence that past increases in advertising expenditures indeed led to increased sales.

Another example is the relationship between the selling price of a house and its square footage. When a new house is listed for sale, how should the price be determined? Is a 4000-square-foot house worth twice as much as a 2000-square-foot house? What other factors might be involved in the pricing of houses, and how should these factors be included in determining the price?

In a study of absenteeism at a large manufacturing plant, management may feel that several variables have an impact. These variables might include job complexity, base pay, the number of years a worker has been with the plant, and the age of that worker. If absenteeism can cost the company tens of thousands of dollars, then the importance of identifying its associated factors becomes clear.

Perhaps the most important analytic tool for examining the relationships between two or more variables is regression analysis. *Regression analysis* is a statistical technique for developing an equation that describes the relationship between two or more variables. One variable is specified to be the *dependent variable,* or the variable to be explained. The other one or more variables are called the *independent* or *explanatory variables.* Using the previous examples, the soft drink firm would identify sales as the dependent variable and advertising expenditures as the explanatory variable. The real estate firm would choose selling price as the dependent variable and size of the house as the explanatory variable.

There are several reasons business researchers might want to know how certain variables are related. The soft drink firm may want to know how much advertising is necessary to achieve a certain level of sales. An equation expressing the relationship between sales and advertising is useful in answering this question. For the real estate firm, the relationship might be used in assigning prices to houses coming onto the market. To try to lower the absenteeism rate, the management of the manufacturing firm wants to know which variables are most highly related to absenteeism. Reasons for wanting to develop an equation relating two or more variables can be classified as follows: (a) to describe the relationship, (b) for control purposes (what value of the explanatory variable is needed to produce a certain level of the dependent variable), or (c) for prediction.

Much statistical analysis is a multistage process of trial and error. A good deal of exploratory work must be done to select appropriate variables for study and to determine the relationships between or among them. A variety of statistical tests and other procedures must be performed and sound judgments made in order to arrive at satisfactory choices of dependent and explanatory variables. The emphasis in this text is on this multistage process rather than on computations or an in-depth study of the theory behind the techniques presented. In this sense, the text is directed at the applied researcher or the consumer of statistics.

Except for a few preparatory examples, it is assumed that the reader has a computer available to perform the actual computations. The use of statistical software frees the user to concentrate on the multistage "model-building" process.

Three software packages, Microsoft® Excel 2000, MINITAB™ (Version 14), and SAS® (Version 8.2), are discussed in this text. MINITAB and SAS are included because they are widely used as teaching tools in universities and are also used in industry. Excel is included because it is the prevalent spreadsheet in businesses throughout the world. In a business environment, not all managers have access to a statistical package, but nearly all have access to a spreadsheet package, and that package is usually Excel. Knowing how to perform statistical routines with Excel enables the manager who lacks a statistical package to still conduct some statistical analyses. The output from Excel, MINITAB, and SAS is fairly standard and is described in the text and illustrated in at least one example. Most examples and exercises use generic output, but the content of the generic output is very similar to that found in most statistical software. Many of the exercises are intended to be done with the aid of a computer, and most statistical software packages or spreadsheets could be used for this purpose. Certain options available in the software discussed in this text may not be present in other packages, but this should not create a problem in completing the exercises.

One note of caution at this point: Excel is a spreadsheet, not created specifically for statistical analysis. It can be useful for many analyses, but it cannot take the place of a true statistical package such as MINITAB or SAS. If the reader is involved in a substantial amount of data analysis, I recommend using a statistical package rather than a spreadsheet. This text contains many examples of regression analysis where Excel, MINITAB and SAS are used on the same data set. Unless specifically noted in the example, Excel produces the same answers as MINITAB and SAS. There are situations where Excel may fail to produce correct answers, however. The data sets where this occurs are likely to be those that impose a severe computational burden on Excel. The reader is cautioned to make sure the results of any analysis make sense. Results that are contrary to intuition should be questioned and verified with a statistical package. Also some procedures that are useful in more advanced analyses are not available in Excel. These require the use of a package specifically designed for statistical analysis.

Data sets for all exercises are available on the CD that accompanies this text. Data sets come in a variety of formats, including ASCII, EViews®, JMP®, MINITAB, SAS, SPSS and Stata® files, and also as Excel spreadsheets. In each problem where data sets are provided, the file names required to read the data are given.

A section called Using the Computer is included at the end of each chapter. It presents the procedures used in Excel, MINITAB, and SAS to produce the statistical analyses discussed in the chapter. Appendix C provides a brief, general discussion of the use of Excel, MINITAB, and SAS. However, this book is not intended to provide full information on the use of these software packages. For further information on Excel, MINITAB, and SAS, the interested reader should consult one of the following references:

Berk, K., and Carey, P. *Data Analysis with Microsoft® Excel: Updated for Office XP*. Pacific Grove, CA: Duxbury Press, 2004.

Freund, R., and Littell, R. *SAS® System for Regression* (3rd ed.). Cary, NC: SAS Institute, 2000.

McKenzie, J., and Goldman, R. *The Student Edition of MINITAB™ for Windows 95/NT™, Release 12* (4th ed.). Reading, MA.: Addison-Wesley, 1999.

Middleton, M. *Data Analysis Using Microsoft® Excel: Updated for Microsoft Office XP*. Pacific Grove, CA: Duxbury Press, 2004.

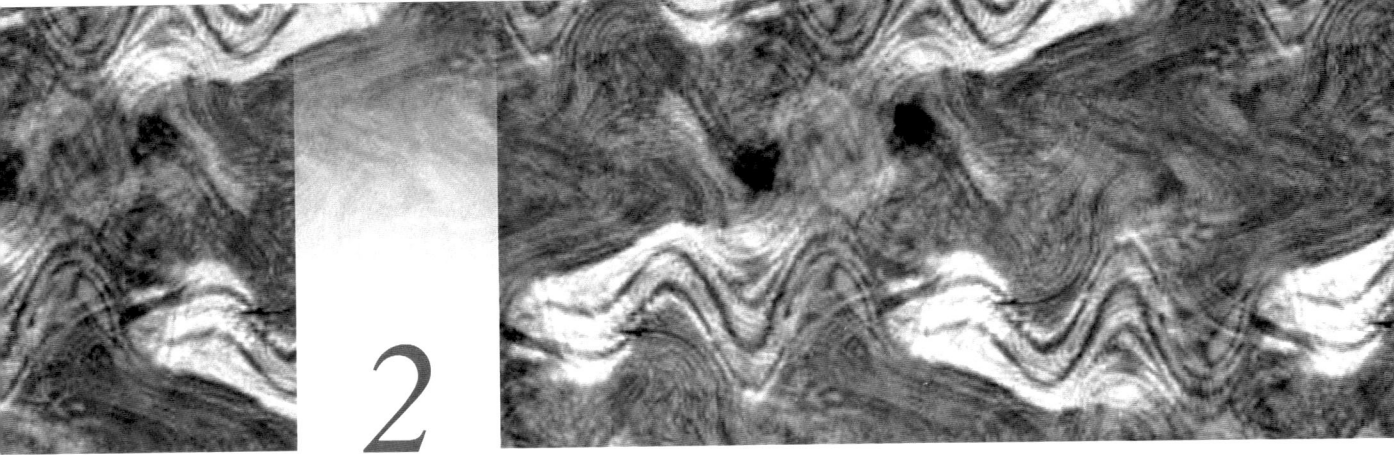

2

Review of Basic Statistical Concepts

2.1 INTRODUCTION

This chapter summarizes and reviews many of the basic statistical concepts taught in an introductory statistics course. For the most part, introductory courses in statistics deal with three main areas of interest: descriptive statistics, probability, and statistical inference.

Typically, statistics is used to study a particular population. A *population*, for purposes of this text, may be defined as the collection of all items of interest to a researcher. The researcher may want to study the sales figures for firms in a particular industry, the rates of return on public utility firms, or the lifetimes of a new brand of automobile tires. But because of time limitations, cost, or the destructive nature of testing, it is not always possible to examine all elements in a population. Instead, a subset of the population, called a *sample*, is chosen, and the characteristic of interest is determined for the items in the sample.

Descriptive statistics is that area of statistics that summarizes the information contained in a sample. This summary may be achieved by condensing the information and presenting it in tabular form. For example, frequency distributions are one way to summarize data in a table. Graphical methods of summarizing data also may be used. The types of graphs discussed in introductory statistics courses include histograms, pie charts, bar charts, and scatterplots.

Data also may be summarized by numerical values. For example, to describe the center of a data set, the mean or median is often used. To describe variability, the variance, standard deviation, or interquartile range might be used. Each of the numerical values is a single number computed from the data that describe a certain characteristic of a sample.

Describing the information contained in a sample is only a first step for most statistical studies. If the study of a population's characteristics is the goal, then the researcher wants to use the information obtained from the sample to make statements about the population. The process of generalizing from characteristics of a sample to those of a population is called *statistical inference*. The bridge leading from descriptive measures computed for a sample to inferences made about population characteristics is the field of probability.

Statistical sampling is an additional topic discussed in introductory statistics. By choosing the elements of a sample in a particular manner, objective evaluations can be made of the quality of the inferences concerning population characteristics. Without proper choice of a sample, there is no way to evaluate these inferences objectively. Thus, the manner in which the sample is chosen is important.

The most common type of sampling procedure discussed in introductory statistics is simple random sampling. Suppose a sample of n items is desired. To qualify as a *simple random sample* (SRS), the items in the sample are selected so that each possible sample of size n is equally likely to be chosen. In other words, each possible sample has an equal probability of being the one actually chosen for study. This is one of the pieces of the bridge that links descriptive statistics and statistical inference. Another piece of the bridge is a description of the behavior of certain numerical summaries that are computed as descriptive statistics.

Any numerical summary computed from a sample is called a *statistic*. A researcher often computes a single statistic from one sample chosen from the population of interest and uses the numerical value of this statistic to make a statement about the value of some population characteristic. For example, suppose a particular brand of tires is to be studied to determine their average life. If the average life is known, the tire company might use this information to establish a warranty for its tires. A SRS of n tires is chosen, and each tire is tested to determine its individual lifetime. Then the sample average lifetime is computed. This sample average can be used as an estimate of the population average lifetime of these tires.

The statistic computed, however, is the sample average lifetime for one particular sample of tires chosen. If a different set of n tires had been chosen, a different sample average would have resulted because of individual variation in the tire-lifetimes. Thus, the sample means themselves vary depending on which set of n tires is chosen as the sample. If this variation in the sample means was without any pattern, then there is no way to relate the value of the sample mean obtained to the unknown value of the population mean. Fortunately, the behavior of the sample means (and other statistics) from random samples is not without a pattern. The behavior of statistics is described by a concept called a *sampling distribution*. Probability enters the picture because sampling distributions are simply probability distributions. Through knowledge of the sampling distribution of a statistic, procedures can be developed to objectively evaluate the quality of sample statistics used to approximate population characteristics.

In this chapter, many of these concepts are reviewed, including descriptive statistics, random variables and probability distributions, sampling distributions, and statistical inference. Because most or all of these topics are covered in an introductory course in statistics, the coverage here is brief.

For detailed discussions of introductory statistics, the interested reader is referred to such texts as

Albright, S., Winston, W., and Zappe, C. *Data Analysis for Managers with Microsoft® Excel* (2nd ed.). Pacific Grove, CA: Duxbury Press, 2004.

Keller, G., and Warrack, B. *Statistics for Management and Economics* (6th ed.). Pacific Grove, CA: Duxbury Press, 2003.

Mendenhall, W., Beaver, R., and Beaver, B. *A Brief Course in Business Statistics* (2nd ed.). Pacific Grove, CA: Duxbury Press, 2001.

Weiers, R. *Introduction to Business Statistics* (4th ed.). Pacific Grove, CA: Duxbury Press, 2002.

2.2 DESCRIPTIVE STATISTICS

Table 2.1 shows the 5-year returns as of July 1, 2002, for a random sample of 83 mutual funds. Examining the 83 numbers in this list provides little useful information. Just looking at a list of numbers is confusing even when the sample size is only 83. For larger samples, the confusion becomes even greater.

The field of descriptive statistics provides ways to summarize the information in a data set. Summaries can be tabular, graphical, or numerical. One common tabular method of summarizing data is the frequency distribution. A *frequency distribution* is a table that is used to summarize quantitative data. The frequency distribution is set up by defining *bins* or *classes* that contain the data values. An examination of the returns in Table 2.1 shows that the largest 5-year rate of return is 17.0% and the smallest is –7.8%. We want to make sure that we include all the data in our frequency distribution, so the bins of the frequency distribution must begin at or below the smallest value and end at or above the largest. One example of how we might set up the frequency distribution is as follows: Start the first bin at –8.0%, end the last bin at 20.0%, and use a total of seven bins. The resulting frequency distribution is shown in Figure 2.1.

The bins are constructed so that there is no confusion about where a data value should go. Each bin excludes the lower limit, but includes the upper limit. A 5-year return of 4% belongs in the third bin; a 16% return belongs in the sixth bin. Also note that each of the bins has the same width—4%. Two guidelines for constructing an effective frequency distribution are (1) make sure each data value belongs in a

FIGURE 2.1
Frequency Distribution for 5-Year Rates of Return.

Five-Year Rates of Return	Number of Funds
Greater than –8% but less than or equal to –4%	5
Greater than –4% but less than or equal to 0%	6
Greater than 0% but less than or equal to 4%	17
Greater than 4% but less than or equal to 8%	34
Greater than 8% but less than or equal to 12%	12
Greater than 12% but less than or equal to 16%	8
Greater than 16% but less than or equal to 20%	1

TABLE 2.1 Five-Year Rates of Return for Mutual Funds

Name of Fund	Five-Year Return	Name of Fund	Five-Year Return	Name of Fund	Five-Year Return
AAL Capital Growth A	5.7	Fifth Third Quality Growth A	2.7	T. Rowe Price Blue Chip Growth	3.4
Aim Constellation A	1.0	First American Small Cap Value A	6.6	T. Rowe Price Mid-Cap Value	11.6
Aim Weingarten A	−4.0	FPA Paramount	−5.0	Principal MidCap A	5.4
Alpine US Real Estate Equity Y	9.6	Franklin Small Mid Cap Growth A	4.2	Prudential US Emerging Growth A	6.3
American AAdvantage Balanced Plan	5.2	Gabelli Small Cap Growth	9.8	Putnam Classic Equity A	0.9
American Century Income & Growth	5.0	Goldman Sachs Capital Growth A	4.3	Putnam New Value A	5.4
Ariel Appreciation	14.2	J. Hancock Large Cap Equity A	4.0	Rainier Core Equity	5.0
AXA Rosenberg US Small Cap	9.5	Hartford Growth Opportunities L	4.3	RS Diversified Growth A	14.0
AXP Small Companies Index A	7.1	ICAP Equity	5.7	Salomon Brothers Investors Value A	6.3
Berger Growth	−5.0	Invesco Health Sciences	6.4	Scudder Capital Growth AARP	0.7
BlackRock Small Cap Growth A	−1.5	Janus Core Equity	10.6	Security Equity A	0.2
Calamos Convertible A	10.6	Janus Twenty	4.3	Sit Mid Cap Growth	−0.9
CG Capital Markets Large Cap Value	3.9	Liberty Fund A	2.4	Smith Barney Large Cap Value A	1.7
Columbia Balanced	4.5	Lord Abbett Developing Growth A	2.9	State Street Research Aurora A	15.9
Davis Financial A	7.7	Mairs & Power Growth	10.8	Strong Growth	4.1
Deutsche Flag Value Builder A	5.6	Merrill Lynch Balanced Capital D	3.0	Third Avenue Value	8.9
Dreyfus Growth & Value Emerging Leaders	10.1	MFS Emerging Growth A	−4.4	Turner Small Cap Value	14.4
Dreyfus Premier Third Century Z	0.4	MFS Value A	10.5	US Global Investors World Precious Minerals	−7.8
Evergreen Fund A	−3.2	Morgan Stanley Global Utilities B	5.9	Van Kampen Comstock A	10.4
Evergreen Small Cap Value A	9.4	Morgan Stanley Total Return B	2.7	Vanguard Capital Opportunity	15.6
Federated Equity Income A	0.0	Nations Convertible Securities A	7.9	Vanguard LifeStrategy Conservative Growth	6.0
Fidelity Blue Chip Growth	2.5	Needham Growth	17.0	Vanguard Small Cap Index	5.2
Fidelity Equity-Income II	5.1	Nuveen Large Cap Value A	4.8	Vanguard Utilities Income	4.6
Fidelity New Millennium	14.5	One Group Equity Income A	3.2	Waddell & Reed Continental	4.1
Fidelity Advisor Balanced T	1.9	Oppenheimer Cap Appreciation A	6.5	Wasatch Small Cap Growth	15.2
Fidelity Advisor Technology T	−1.1	Oppenheimer Total Return A	4.1	WM Growth & Income A	4.9
Fidelity Select Air Transportation	14.1	PBHG Growth	−3.9		
Fidelity Select Health Care	7.3	Pimco Capital Appreciation A	6.3		
		Pioneer Fund A	6.6		

Reprinted by permission from the September issue of *Kiplinger's Personal Finance*. Copyright © 2002 The Kiplinger Washington Editors, Inc.

unique bin and (2) if possible, make each bin width the same. Intervals covering the range of the data are constructed, and the number of observations in each interval is then tabulated and recorded.

If the proportion or percentage of items in each class is noted rather than the number, the table is referred to as a *relative frequency distribution*. It is also possible to construct a *cumulative frequency distribution* in which the number of items at or below each class limit is noted.

A *histogram* is a graphical representation of a frequency distribution. The horizontal axis of the graph is marked off into classes or bins over the full range of the data, and the vertical axis represents the number or proportion (relative frequency) of observations in each of the classes. The bin limits for the horizontal axis are the limits established in the frequency distribution. Rectangles (bars) are drawn over the bin limits, with the area of the bar proportional to the frequency in that particular bin. If the bin limits are all the same width, the height of the bars can be equal to the frequency in each bin. If the bin limits differ in width, adjustments must be made. It is recommended that bin widths be made the same whenever possible. The adjustments for unequal bin widths are not discussed here.

From the frequency distribution or the histogram, one can obtain a quick picture of certain characteristics of the data. For example, the center of the data and how much variability is present can be observed. The data have been summarized so that these characteristics are more obvious. When the frequency distribution in Figure 2.1 was constructed, we arbitrarily decided to use seven bins. In general, you do not want to have too few bins because the data will be oversummarized and it will be hard to see patterns in it. Also, too many bins make the frequency distribution confusing and difficult to read. Various rules have been suggested concerning the appropriate number of bins. A good guideline is to use between 5 and 20 bins. As a rough idea of the number of bins to try, start with the square root of the number of observations. We have 83 5-year rates of return. The square root of 83 is between 9 and 10, so we might start with 10 as a possible number of bins. There is no right or wrong number of bins. However, there are better and worse numbers—too few or too many bins are not good—and there are better choices than others (my choices are better than yours, for example, because I wrote this book), but constructing a frequency distribution is in large part a matter of preference. It may take two or three tries to get the table the way you believe is most helpful in representing the data or clearest for presentation purposes.

Most statistical software packages and spreadsheets provide various tabular and graphical methods of summarizing data. Figure 2.2a shows the frequency distribution constructed by Excel for the mutual-fund-return data. This is the distribution without any modifications and would obviously be modified before use in a presentation. Such a modification might appear as in Figure 2.2b. Figures 2.3 and 2.4 show histograms constructed using Excel and MINITAB, respectively. I chose to use ten bins for the Excel frequency distribution and histogram, starting the first bin at −8% and giving each bin a width of 2.5%. The MINITAB histogram was allowed to choose its own limits.[1] Note that the numbers shown in the bin column of the

FIGURE 2.2A Excel Frequency Distribution for 5-Year Rates of Return for Mutual Funds.

bins	Frequency
−5.5	1
−3	6
−0.5	3
2	8
4.5	18
7	22
9.5	7
12	9
14.5	5
17	4

---

[1] Excel can be allowed to choose its own bin limits as well. However, I find the limits chosen by Excel are often (actually, always) not to my liking. I prefer to set up the bin limits myself to make them easier to work with. This process is discussed in the Using the Computer section at the end of this chapter.

FIGURE 2.2B Modified Excel Frequency Distribution for 5-Year Rates of Return for Mutual Funds.

Five-Year Returns	Number
Greater than -8% but less than or equal to -5.5%	1
Greater than -5.5% but less than or equal to -3%	6
Greater than -3% but less than or equal to -0.5%	3
Greater than -0.5% but less than or equal to 2%	8
Greater than 2% but less than or equal to 4.5%	18
Greater than 4.5% but less than or equal to 7%	22
Greater than 7% but less than or equal to 9.5%	7
Greater than 9.5% but less than or equal to 12%	9
Greater than 12% but less than or equal to 14.5%	5
Greater than 14.5% but less than or equal to 17%	4

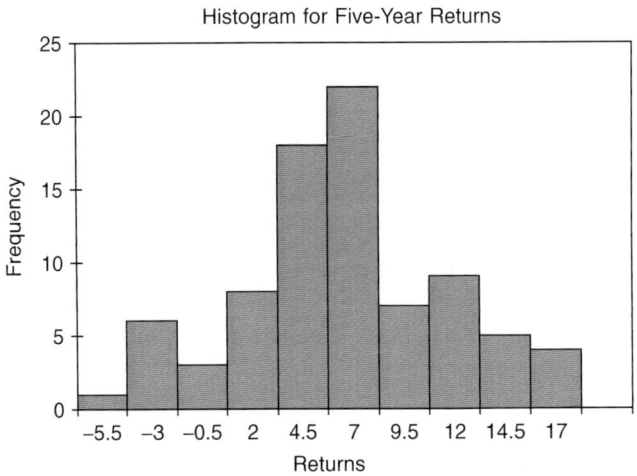

FIGURE 2.3 Excel Histogram for 5-Year Rates of Return for Mutual Funds.

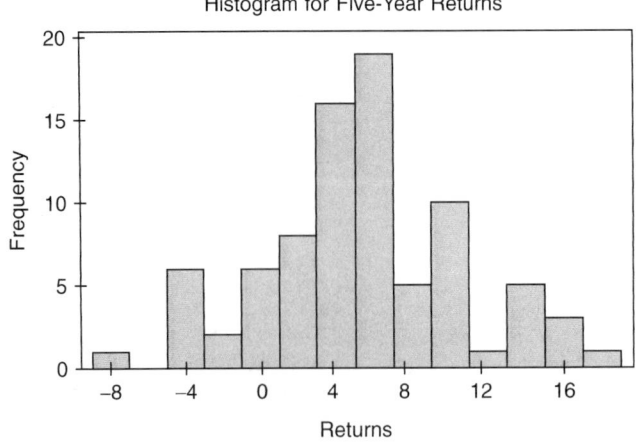

FIGURE 2.4 MINITAB Histogram for 5-Year Rates of Return for Mutual Funds.

Excel frequency distribution are the inclusive upper bin limits. Excel includes a final bin that is represented by the label "More" (indicating values more than the last bin limit shown). When constructing a frequency distribution or histogram in Excel, I prefer not to use the More class but to use numerical bin limits instead. I have eliminated the More class in the frequency distribution shown. On the Excel histogram, the numbers shown on the horizontal axis also represent the upper limits of the bin under which they are printed. Even though they are printed in the middle of the bin, they do not represent midpoints, but upper limits.

On the MINITAB histogram, the numbers shown on the horizontal axis are the midpoints of the bin intervals. For example, the first bin has a midpoint of −8. The bin limits are −9 and −7. The next bin has no fund returns in it, but would have a midpoint of −6 and bin limits of −7 and −5. Each bin limit is halfway between two of the midpoints. MINITAB includes the value of the lower limit in the bin, but excludes the upper limit.

The idea of a graph such as a histogram is to summarize the data so that the viewer can get a quick picture of what is going on but without masking too much of the information. Using too few classes on a histogram oversummarizes the data, whereas using too many does not summarize the data sufficiently. In either case, a histogram that is difficult to read is produced. Again, there is no right or wrong number of bins to use when constructing a histogram.

The histograms show that the returns for these funds vary from a low of about −8.0% (the lower boundary of the first interval is −9.0% on the MINITAB histogram and −8.0% on the Excel histogram) to a high of around 17.0% (17.0% is the upper boundary of the last interval on the Excel histogram; 18.5% is the upper limit of the last interval on the MINITAB histogram). Very high returns and very low returns are rare. Most returns cluster toward the center of the histogram (we might identify the center as roughly around 5%).

Numerical summaries are single numbers computed from a sample to describe some characteristic of the data set. Some common numerical summaries are sample mean, sample median, sample variance, and sample standard deviation. The *sample mean* and *sample median* are measures of the center, or central tendency, of the data. The sample mean is the average of all the observations in the data set:

$$\bar{y} = \frac{\sum_{i=1}^{n} y_i}{n}$$

where y_i represents the *i*th observation in the data set. (See Appendix A for an explanation of summation notation.) The sample median is the midpoint of the data after they have been ordered. If n, the number of observations, is even, then the median is the average of the two middle observations after the observations have been ordered from smallest to largest. If n is odd, the unique middle value in the ordered data set is the median.

The *sample variance* and *sample standard deviation* are measures of variability. The sample variance is computed as

$$s^2 = \frac{\sum_{i=1}^{n}(y_i - \bar{y})^2}{n - 1}$$

This is the average squared distance of each data point, y_i, from the center of the data, $\bar{y}$. The divisor $n - 1$ is used, rather than n, to provide an unbiased estimator (one that neither consistently overestimates nor underestimates the true parameter) of the population variance. Because s^2 expresses variability in squared units, an intuitively more appealing measure is the sample standard deviation, s, which is simply the square root of s^2. Although many other numerical summaries exist, they are not discussed in this review.

Figure 2.5 shows the results of using MINITAB to compute several descriptive measures for the mutual-fund-return data, including the sample mean, sample median, sample variance, and sample standard deviation. Figure 2.6 shows similar results using Excel.

FIGURE 2.5 MINITAB Numerical Summaries for the 5-Year Rates of Return.[a]

Descriptive Statistics: Five-Year Return

Variable	N	N*	Mean	SE Mean	StDev	Minimum	Q1	Median	Q3	Maximum
5yr ret	83	0	5.371	0.574	5.229	-7.800	2.700	5.100	8.900	17.000

[a]The summaries computed include the number of observations (N), the number of missing observations (N*), mean, standard error of the mean (SE Mean), sample standard deviation (StDev), minimum, first quartile (Q1), median, third quartile (Q3), and maximum.

FIGURE 2.6 Excel Numerical Summaries for the 5-Year Rates of Return.[a]

Five-Year Return	
Mean	5.37
Standard Error	0.57
Median	5.10
Mode	4.30
Standard Deviation	5.23
Sample Variance	27.35
Kurtosis	0.07
Skewness	0.02
Range	24.80
Minimum	-7.80
Maximum	17.00
Sum	445.80
Count	83.00

[a]The summaries computed include the mean, standard error of the mean (Standard Error), median, mode, sample standard deviation (Standard Deviation), sample variance, kurtosis, skewness, range, minimum, maximum, the sum of all the data values (Sum), and the number of observations (Count).

EXERCISES

1. **Highway Mileages.** The highway mileages of 147 cars are in a file named CARS2. These cars are 2003 models. Find the mean, standard deviation, and median for the mileages. Construct a histogram of the data.
(Data from *Road & Track: The New Cars.* October 2002. Copyright 2002 by Hachette Filipacchi Media, Inc. Used with permission.)

2. **Graduation Rates.** The National Collegiate Athletic Association (NCAA) is concerned with the graduation rate of student athletes. Part of an effort to increase graduation rates for student athletes involved implementing Proposition 48, beginning with the 1986–1987 school year. Proposition 48 mandated that student athletes

obtain a 700 SAT or a 15 ACT test score to be eligible to play.

Graduation rates are provided in a file named GRADRATE2. The file contains graduation rates (percentages) for several groups of students.

   all students entering freshman classes in 1983–1984 through 1985–1986 (AS83)
   student athletes entering in 1983–1984 through 1985–1986 (SA83)
   all students entering freshman classes in 1986–1987 (AS86)
   student athletes entering in 1986–1987 (SA86)
   all students entering freshman classes in 1987–1988 (AS87)
   student athletes entering in 1987–1988 (SA87)
   all students entering freshman classes in 1988–1989 (AS88)
   student athletes entering in 1988–1989 (SA88)
   all students entering freshman classes in 1989–1990 (AS89)
   student athletes entering in 1989–1990 (SA89)
   all students entering freshman classes in 1990–1991 (AS90)
   student athletes entering in 1990-1991 (SA90)
   all students entering freshman classes in 1991–1992 (AS91)
   student athletes entering in 1991–1992 (SA91)
   all students entering freshman classes in 1993–1994 (AS93)
   student athletes entering in 1993–1994 (SA93)
   all students entering freshman classes in 1994–1995 (AS94)
   student athletes entering in 1994–1995 (SA94)

All Division I schools with complete data for all years are represented. [The data were obtained from the Fort Worth Star-Telegram (July 2, 1993; May 20, 1993; July 1, 1994; June 30, 1995; June 28, 1996; June 27, 1997; and November 9, 1998 issues)] and from the NCAA web site for the remaining years. Note that the 1983–1984 through 1985–1986 data provide graduation rates prior to the implementation of Proposition 48. All other years provide graduation rates after its implementation. (Note that rates were not available for the 1992 groups.)
(*Source*: Data used courtesy of the *Fort Worth Star-Telegram* and the NCAA.)

a. Examine the groups by finding the mean and median graduation rate for each. Construct a histogram for each set of graduation rates.

b. U.S. District Judge Ronald Buckwalter invalidated the NCAA's academic eligibility standards for incoming freshmen athletes on March 8, 1999. The NCAA still believes in the standards and in the positive effect of Proposition 48 on graduation rates. To help support its position, the NCAA has asked for your input concerning the effect of Proposition 48 on graduation rates. Using any graphical or numerical summaries to support your position, what would you report to the NCAA based on the data you have been given?

2.3 DISCRETE RANDOM VARIABLES AND PROBABILITY DISTRIBUTIONS

A *random variable* can be defined as a rule that assigns a number to every possible outcome of an experiment. A *discrete random variable* is one with a definite distance between each of its possible values. For example, consider the toss of a coin. The two possible outcomes are head (H) and tail (T). A random variable of interest could be defined as

$$X = \text{number of heads on a single coin toss}$$

Then X assigns the number 1 to the outcome H and the number 0 to outcome T. As another example, suppose two cards are randomly drawn without replacement from a deck of 52 cards. Let

$$Y = \text{number of kings on two draws}$$

Then Y assigns the number 0, 1, or 2 to each possible outcome of the experiment.

In each of these examples, the outcome of the experiment is determined by chance. Probabilities can be assigned to the outcomes of the experiment and thus to the values of the random variables. A table listing the values of a random variable and the probabilities associated with each value is called a *probability distribution* for the random variable.

For the coin toss, the probability distribution of X is

x	$P(x)$
0	1/2
1	1/2

Here, the notation $P(x)$ means "the probability that the random variable X has the value x" or $P(x) = P(X = x)$. The function $P(x)$ is called the *probability mass function* (pmf) of X.

For the card-drawing experiment, the probability distribution of Y is

y	$P(y)$
0	188/221
1	32/221
2	1/221

Note that probabilities must satisfy the following conditions:

1. They must be between 0 and 1. $0 \leq P(x) \leq 1$
2. They must sum to 1. $\sum P(x) = 1$

When we discussed a sample of observations drawn from a population in the previous section, certain characteristics were of interest, primarily center and variability. Numerical summaries were used to measure these characteristics. Describing the center and the variation in a probability distribution also is often useful. The measures most commonly used to do this are the mean and variance (or standard deviation) of the random variable.

As an example, consider two random variables X and Y, representing the profit from two different investments. Suppose the two probability distributions have been set up as follows:

x	$P(x)$	y	$P(y)$
−2000	.05	0	.40
−1000	.10	1000	.20
1000	.10	2000	.20
2000	.25	3000	.10
5000	.50	4000	.10

If only one of the investments can be chosen, some methods to compare the two would be useful. As can be seen, the chances of a loss are greater for investment X than for investment Y, although the chances for a large profit are also greater for investment X.

One way to compare the investments might be to use the expected value, or mean, of the random variables representing the outcomes of the investments. The *expected value* of a discrete random variable X is defined as the sum of each value of X times the probability associated with that value:

$$E(X) = \mu_X = \sum xP(x)$$

The subscript X on μ_X often is dropped if it is clear which random variable is being discussed. The computation of the expected values of X and Y is shown in Figure 2.7. The expected value of X is greater than the expected value of Y. Thus, on the basis of maximizing expected values, investment X would be chosen.

The expected value of a random variable deserves some additional explanation. Consider again the coin-toss experiment with X equal to the "number of heads" and probability distribution:

x	$P(x)$
0	1/2
1	1/2

Computing the expected value of X gives $E(x) = 1/2$.

Obviously, if a coin is tossed once, either the outcome 0 (tail) or 1 (head) will appear. The expected value of X represents the average obtained over a large number of trials. If the coin is tossed a large number of times and 0s were recorded for tails and 1s for heads, then the average of these 0s and 1s will be close to one-half. The same interpretation can be made for the case of the investments. The expected outcomes represent the averages obtained over a large number of trials rather than the outcome of a single trial. Thus, in the long run, investment X will provide a higher average profit than investment Y.

There are, of course, other criteria for choosing between investments than simply maximizing the expected returns. A measure of each investment's risk also might be important. The variability of the outcomes is sometimes used as a measure of risk. One measure of a random variable's variation is the *variance*, defined for a discrete random variable, X, as

$$Var(X) = \sigma_X^2 = \sum (x - \mu)^2 P(x)$$

To compute $Var(X)$, the mean is subtracted from each possible value of X, and the differences are squared and then multiplied times the probability of the associated value of X. The resulting sum is the variance, which represents an average squared distance of each value of X to the center of the probability distribution. Note that no division is used in computing this "average." The division used to compute a sample variance has been replaced by the weighting of each outcome by its

FIGURE 2.7 Computation of $E(X)$ and $E(Y)$.

x	$P(x)$	$xP(x)$	y	$P(y)$	$yP(y)$
−2000	0.05	−100	0	0.40	0
−1000	0.10	−100	1000	0.20	200
1000	0.10	100	2000	0.20	400
2000	0.25	500	3000	0.10	300
5000	0.50	2500	4000	0.10	400
	$E(X) = \sum xP(x) = 2900$			$E(Y) = \sum yP(y) = 1300$	

x	$P(x)$	$xP(x)$	$x - \mu_X$	$(x - \mu_X)^2$	$(x - \mu_X)^2 P(x)$
−2000	0.05	−100	−4900	24,010,000	1,200,500
−1000	0.10	−100	−3900	15,210,000	1,521,000
1000	0.10	100	−1900	3,610,000	361,000
2000	0.25	500	−900	810,000	202,500
5000	0.50	2500	2100	4,410,000	2,205,000
		$\mu_X = 2900$			$\sigma_X^2 = 5,490,000$

y	$P(y)$	$yP(y)$	$y - \mu_Y$	$(y - \mu_Y)^2$	$(y - \mu_Y)^2 P(y)$
0	0.40	0	−1300	1,690,000	676,000
1000	0.20	200	−300	90,000	18,000
2000	0.20	400	700	490,000	98,000
3000	0.10	300	1700	2,890,000	289,000
4000	0.10	400	2700	7,290,000	729,000
		$\mu_Y = 1300$			$\sigma_Y^2 = 1,810,000$

FIGURE 2.8 Computation of and σ_X^2 and σ_Y^2.

probability. An alternative formula for computing the variance of a discrete random variable is

$$\sigma_X^2 = \sum x^2 P(x) - \mu^2$$

This formula is sometimes preferred when doing computations on a calculator. Both formulas provide the same answer. The variances of X and Y are computed in Figure 2.8. The variances are

$$\sigma_X^2 = 5{,}490{,}000 \quad \text{and} \quad \sigma_Y^2 = 1{,}810{,}000$$

Obviously, investment X is more variable than investment Y. The variances are somewhat difficult to interpret, however, because they measure variability in squared units (squared dollars for the investments). To return to the original units of the problem, the square root of the variance, called the *standard deviation*, may be used:

$$\sigma_X = 2343.07 \quad \text{and} \quad \sigma_Y = 1345.36$$

The standard deviations are expressed in the original units of the problem (dollars for the investments).

All of the random variables discussed so far have been discrete random variables. A *continuous random variable* is one whose values are measured on a continuous scale. It is measured over a range of values with all numbers within that range as possible values (at least in theory). Examples of quantities that might be represented by continuous random variables are temperature, gas mileage, and stock prices. In the next section, a very useful continuous random variable in statistics, the normal random variable, is introduced.

EXERCISES

3. Consider the roll of a single die. Construct the probability distribution of the random variable $X =$ number of dots showing on the die. Find the expected value and standard deviation of X. How would you interpret the number obtained for the expected value?

4. Let X be a random variable defined as

   X = 1 if an even number of dots appears on the roll of a single die

   X = 0 if an odd number of dots appears on the roll of a single die

   Construct the probability distribution of X. Find the expected value and standard deviation of X. How would you interpret the number obtained for the expected value?

5. Consider the roll of two dice. Let X be a random variable representing the sum of the number of dots appearing on each of the dice. The probabilities of each possible value of X are as follows:

x	P(x)
2	1/36
3	2/36
4	3/36
5	4/36
6	5/36
7	6/36
8	5/36
9	4/36
10	3/36
11	2/36
12	1/36

   Determine the expected value and standard deviation of X.

6. The game of craps deals with rolling a pair of fair dice. In one version of the game, a field bet is a one-roll bet based on the outcome of the pair of dice. For every $1 bet, you lose $1 if the sum is 5, 6, 7, or 8; you win $1 if the sum is 3, 4, 9, 10, or 11; or you win $2 if the sum is 2 or 12.

   a. Using the probability distribution in Exercise 5, construct the probability distribution of the different outcomes available in a field bet.

   b. Determine the expected value of this probability distribution. How would you interpret this number?

7. A computer shop builds PCs from shipments of parts it receives from various suppliers. The number of defective hard drives per shipment is to be modeled as a random variable X. The random variable is assumed to have the following distribution:

x	P(x)
0	0.55
1	0.15
2	0.10
3	0.10
4	0.05
5	0.05

   a. What is the expected number of defective hard drives per shipment?

   b. If each defective drive costs the company $100 in rework costs, what is the expected rework cost per shipment?

   c. What is the probability that a shipment has more than two defective hard drives?

2.4 THE NORMAL DISTRIBUTION

A continuous random variable is a random variable that can take any value over a given range. An example that is important in statistical inference is the normal random variable. The probability distribution of the normal random variable, called the *normal distribution*, is often depicted as a bell-shaped symmetric curve, as shown in Figure 2.9. The normal distribution is centered at the mean, μ. Variation in the distribution is described by the variance, σ^2, or standard deviation, σ.

Figure 2.10 shows two normal distributions with different means but equal standard deviations, and Figure 2.11 shows two distributions with the same mean but different standard deviations. The location of the distribution is determined by the mean; the spread of the distribution (how compressed or spread out it appears) is determined by the standard deviation.

For a continuous distribution such as the normal distribution, the probability that the random variable takes on a value within a certain range can be determined by computing the area under the curve that defines the probability distribution between the limits of the range. To determine the probability that a normal random

18 Review of Basic Statistical Concepts

FIGURE 2.9 The Normal Distribution with Mean μ and Standard Deviation σ.

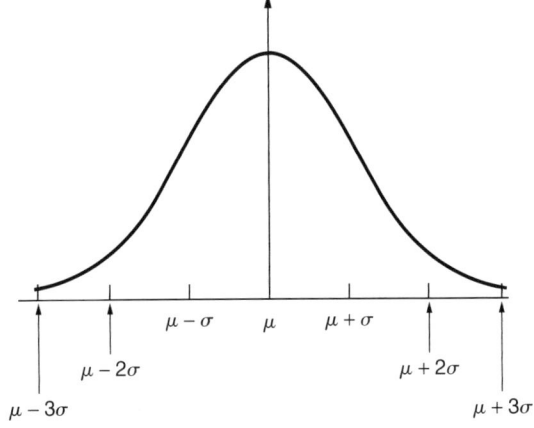

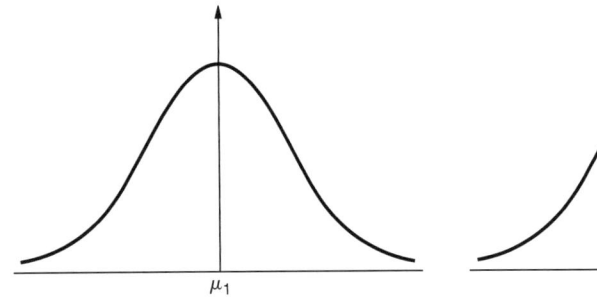
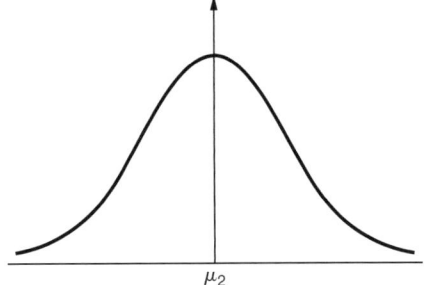

FIGURE 2.10 Normal Distributions with Equal Standard Deviations but Different Means.

FIGURE 2.11 Normal Distributions with Equal Means but Different Standard Deviations.

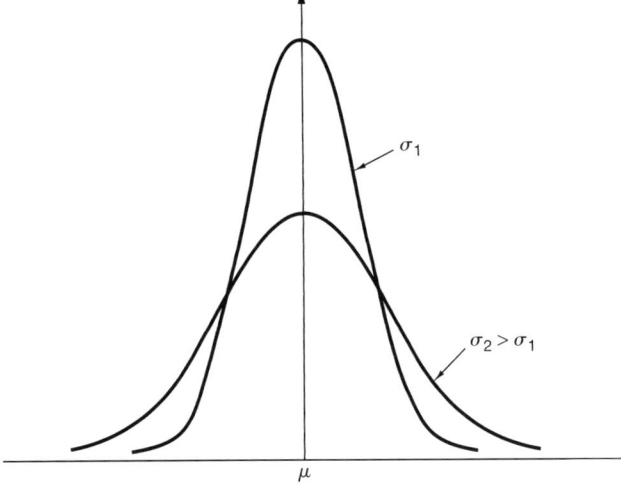

variable is between 0 and 2, the area under the normal curve between these values must be computed. This computation is a fairly difficult task if done from scratch. Fortunately, a table of certain areas or probabilities under the normal curve is available to simplify these computations considerably.

Table B.1 in Appendix B lists probabilities between certain values of the standard normal distribution. The standard normal distribution has a mean of $\mu = 0$ and a standard deviation of $\sigma = 1$. Throughout this text, the standard normal random variable is denoted by the letter Z. This table is set up to show the probability that the normal random variable is between 0 and some number z written

$$P(0 \leq Z \leq z)$$

The z numbers are given by the values in the far left-hand column of the table to one decimal place. A second decimal place is provided by using the values in the top row of the table. For example, to find the probability that Z is between 0 and 1, the value 1.0 is located in the left-hand column and the probability is read from the .00 column of the table:

$$P(0 \leq Z \leq 1) = 0.3413$$

This area is illustrated by the shaded region in Figure 2.12.

Similarly, the probability between 0 and 2.3 is

$$P(0 \leq Z \leq 2.3) = 0.4893$$

To compute the probability between 0 and 1.96, first find 1.9 in the left-hand column. The probability is then read from the .06 column of the table as

$$P(0 \leq Z \leq 1.96) = 0.4750$$

Because the standard normal curve has a mean of 0, the numbers to the right of the mean are positive, as illustrated in the examples thus far. The numbers to the left of the mean are negative. How is the table used to find the probability that Z is between, say, -1.0 and 1.0? There are no negative z values in the table. But the fact that the curve is symmetric can be used to determine the probabilities for numbers to the left of the mean.

The probability between 0 and 1.0 has been determined to be 0.3413. Because the curve is symmetric, the half of the curve to the left of the mean is a mirror image of the half to the right. Thus, in an interval between 0 and -1.0, there is exactly the same probability as in the interval between 0 and 1.0 because these regions are mirror images of each other. So,

$$P(-1.0 \leq Z \leq 1.0) = 0.3413 + 0.3413 = 0.6826$$

This probability is illustrated in Figure 2.13.

Now, consider finding the following probability

$$P(Z \geq 1.7)$$

The area between 0 and 1.7 can be found from Table B.1 in Appendix B as

$$P(0 \leq Z \leq 1.7) = 0.4554$$

FIGURE 2.12 Area or Probability Under the Standard Normal Curve Between 0 and 1.0.

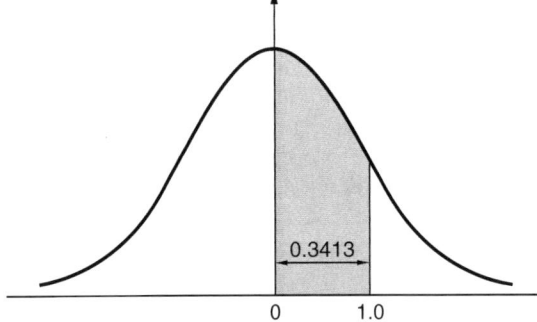

FIGURE 2.13 Area or Probability Under the Standard Normal Curve Between −1.0 and 1.0.

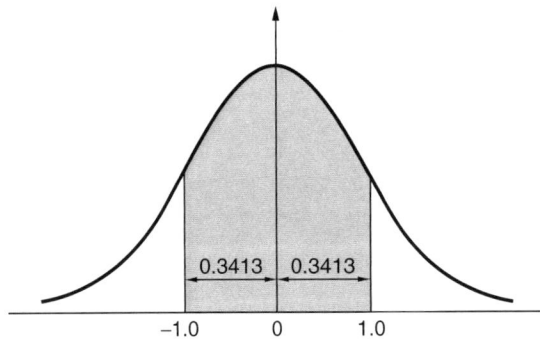

The desired area is to the right of 1.7, however. Here, we use the facts that the total area under the curve must be 1.0 and that the curve is symmetric. The total area under the standard normal curve must be 1.0 because this area represents probability, and probability must sum to 1. Because the curve is symmetric, the area to the right of the mean (0) must be 0.5. In Figure 2.14, if the unshaded area between 0 and 1.7 is subtracted from the total area to the right of 0, the remainder is the area in the shaded region:

$$P(Z \geq 1.7) = 0.5 - 0.4554 = 0.0446$$

The probabilities in the standard normal table also can be used to find probabilities for normal distributions other than the standard normal distribution. For example, suppose X is a normal random variable with mean $\mu = 10$ and standard deviation $\sigma = 2$. Find the probability that X is between 10 and 12: $P(10 \leq X \leq 12)$. The standard normal table cannot be used to find this probability as it is currently stated. But the problem can be solved by translating it into *standardized units*, or units of standard deviation away from the mean. Referring to Figure 2.15, first recognize that 10 is the mean of the normal distribution represented by X. Because $\sigma = 2$, 12 is 1 standard deviation above the mean ($10 + 2 = 12$). Then, in standardized units, the problem becomes

$$P(10 \leq X \leq 12) = P(0 \leq Z \leq 1)$$

FIGURE 2.14 Area or Probability Under the Standard Normal Curve Between 0 and 1.7 and Above 1.7.

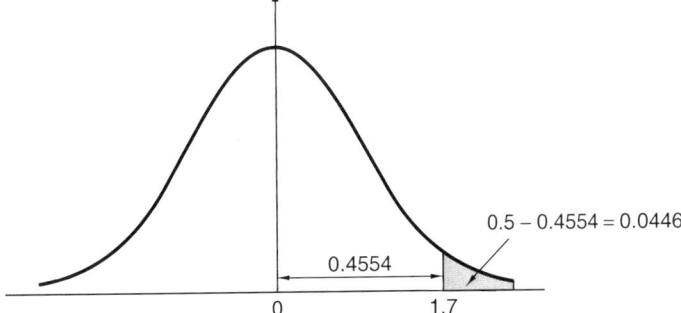

FIGURE 2.15 Finding $P(10 \leq X \leq 12)$ When X Is a Normal Random Variable with $\mu = 10$ and $\sigma = 2$.

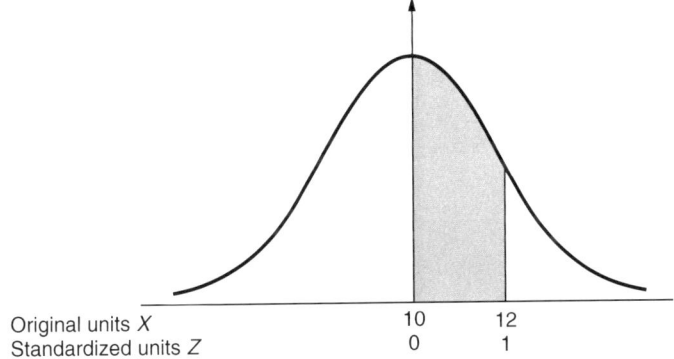

The area between 10 and 12 under the normal curve with $\mu = 10$ and $\sigma = 2$ is the same as the area between 0 and 1 under the standard normal curve. By translating the original units into standardized units, any probability can be determined from the standard normal table. The general transformation is given by the formula

$$Z = \frac{X - \mu_X}{\sigma_X}$$

To translate a number, X, into standardized units, Z, simply subtract the mean and divide by the standard deviation. The following examples should help further illustrate.

EXAMPLE 2.1 Suppose X is a normal random variable with $\mu = 50$ and $\sigma = 10$. What is $P(30 \leq X \leq 60)$?
Answer:

$$P(30 \leq X \leq 60) = P\left(\frac{30 - 50}{10} \leq Z \leq \frac{60 - 50}{10}\right) = P(-2 \leq Z \leq 1)$$
$$= 0.4772 + 0.3413 = 0.8185$$

The solution is illustrated in Figure 2.16.

FIGURE 2.16

Finding $P(30 \leq X \leq 60)$ When X Is a Normal Random Variable with $\mu = 50$ and $\sigma = 10$.

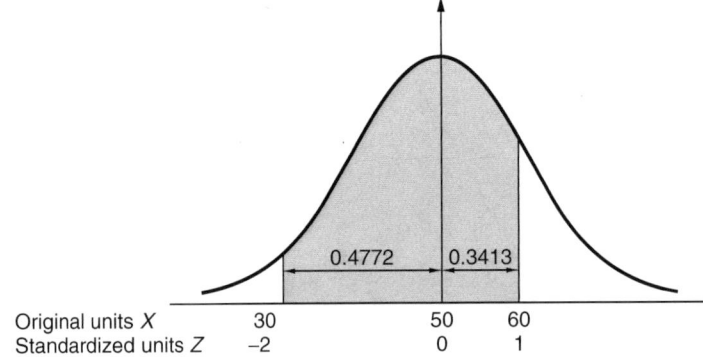

EXAMPLE 2.2 A large retail firm has accounts receivable that are assumed to be normally distributed with mean $\mu = \$281$ and standard deviation $\sigma = \$35$.

1. What is the proportion of accounts with balances greater than $316?
Answer:

$$P(X > 316) = P\left(Z > \frac{316 - 281}{35}\right) = P(Z > 1) = 0.5 - 0.3413 = 0.1587$$

Thus, 0.1587 or 15.87% of all accounts have balances greater than $316. The solution is illustrated in Figure 2.17.

2. Above what value do 13.57% of all account balances lie?
Answer: Figure 2.18 illustrates the problem to be solved. Find an account balance, k, such that the probability above k is 0.1357. This means the probability between the mean and k must be $0.5 - 0.1357 = 0.3643$. Looking up 0.3643 in the standard normal table, k has a z value of 1.1, so k is 1.1 standard deviations above the mean:

$$k = \mu + z\sigma = 281 + (1.1)(35) = \$319.50$$

FIGURE 2.17

Finding $P(X > 316)$ When X Is a Normal Random Variable with $\mu = 281$ and $\sigma = 35$.

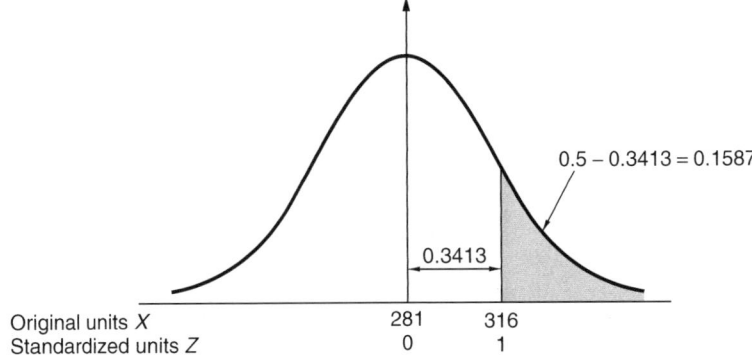

FIGURE 2.18
Above What Value Do 13.57% of All Account Balances Lie When $\mu = 281$ and $\sigma = 35$?

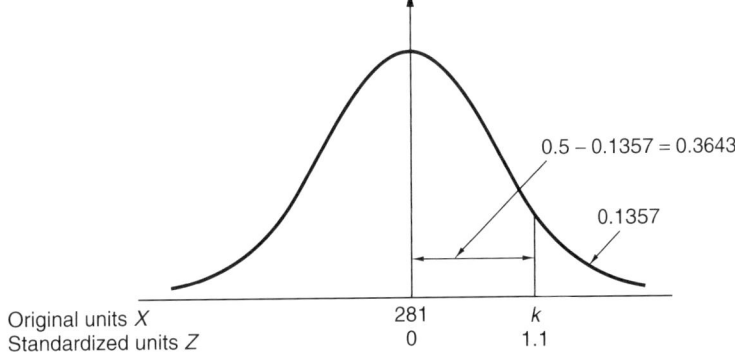

EXERCISES

8. Calculate the following probabilities using the standard normal distribution:
   a. $P(0.0 \le Z \le 1.2)$
   b. $P(-0.9 \le Z \le 0.0)$
   c. $P(0.0 \le Z \le 1.45)$
   d. $P(0.3 \le Z \le 1.56)$
   e. $P(-2.03 \le Z \le -1.71)$
   f. $P(-0.02 \le Z \le 3.54)$
   g. $P(Z \ge 2.50)$
   h. $P(Z \le 1.66)$
   i. $P(Z \ge 5)$
   j. $P(Z \ge -6)$

9. All applicants at a large university are required to take a special entrance exam before they are admitted. The exam scores are known to be normally distributed with a mean of 800 and a standard deviation of 100. Applicants must score 700 or more on the exam before they are admitted.
   a. What proportion of all applicants taking the exam is granted admission?
   b. What proportion of all applicants will score 1000 or higher on the exam?
   c. For the coming academic year, 2500 applicants have registered to take the exam. How many do we expect to be qualified for admission to the university?

10. A manufacturer produces bearings, but because of variability in the production process, not all of the bearings have the same diameter. The diameters have a normal distribution with a mean of 1.2 centimeters (cm) and a standard deviation of 0.01 cm. The manufacturer has determined that diameters in the range of 1.18 to 1.22 cm are acceptable. What proportion of all bearings falls in the acceptable range?

11. Find the value of Z from the standard normal table such that the probability between Z and –Z is
    a. 0.9544
    b. 0.9010
    c. 0.9802
    d. 0.9902

12. A large manufacturing plant uses lightbulbs with lifetimes that are normally distributed with a mean of 1000 hours and a standard deviation of 50 hours. To minimize the number of bulbs that burn out during operating hours, all bulbs are replaced at once. How often should the bulbs be replaced so that no more than 1% burn out between replacement periods?

13. A company that produces an expensive stereo component is considering offering a warranty on the component. Suppose the population of lifetimes of the components is a normal distribution with a mean of 84 months and a standard deviation of 7 months. If the company wants no more than 2% of the components to wear out before they reach the warranty date, what number of months should be used for the warranty?

14. The average amount of time that students use computers at a university computer center is 36 minutes with a standard deviation of 5 minutes. The times are known to be normally distributed. Around 10,000 uses are recorded each week in the computer center. The computer center administrative committee has decided that if more than 2000 uses of longer than 40 minutes at each sitting are recorded weekly, some new terminals must be purchased to meet usage needs. Should the computer center purchase the new computers?

2.5 POPULATIONS, SAMPLES, AND SAMPLING DISTRIBUTIONS

A *population* is a group to be studied and may consist of people, households, firms, automobile tires, and so on. A *sample* is a subset of a population. In other words, a sample is a group of items chosen from the population. Statistics is concerned with the use of sample information to make generalizations or inferences about a *population*.

Typically, the study of every item in a population is not feasible. It may be too time-consuming or too expensive to examine every item. As an alternative, a few items are chosen from the population. From the information provided by this sample, we hope to make reliable generalizations about characteristics of the population.

In this section, we assume that the items of the sample are randomly chosen from the population. By choosing the sample in this way, it is possible to objectively evaluate the quality of the generalizations made about the population characteristics of interest.

As discussed in Section 2.1, many possible random samples can be chosen from a particular population. In practice, typically only one such sample is chosen and examined. (In some applications, such as statistical quality control, repeated samples may be used.) To understand the processes that govern how inferences should be made, it is necessary to imagine all possible random samples of a given sample size n chosen from a particular population. Suppose the characteristic of interest for this population is the mean, μ. To estimate the population mean, the statistic most often used is the sample mean, $\bar{y}$; each possible random sample has an associated value of $\bar{y}$. Thus, the sample mean acts just like a random variable: It assigns a number (the value of the sample mean for each sample) to each of the possible outcomes of an experiment. The experiment, in this instance, is the process of choosing samples of size n from the population. Because the samples are chosen randomly, each one has an equal probability of being chosen. Thus, each value of $\bar{y}$ has a probability associated with it.

Because the sample mean, $\bar{y}$, can be viewed as a random variable, it has a probability distribution. This probability distribution is called the *sampling distribution of the sample mean*. In this section, some of the properties of the sampling distribution of the sample mean are reviewed. These are discussed in more detail in most introductory statistics courses. The sampling distribution of the sample mean is important because it allows us to make the link between population characteristics and sample values that make it possible to assess the quality of our inferences.

First, suppose the population of interest has a mean μ and variance σ^2. The mean of the sampling distribution of $\bar{y}$, written $\mu_{\bar{y}}$, is equal to the population mean:

$$\mu_{\bar{y}} = \mu$$

The variance of the sampling distribution of $\bar{y}$, written $\sigma_{\bar{y}}^2$, is

$$\sigma_{\bar{y}}^2 = \frac{\sigma^2}{n}$$

The standard deviation of the sampling distribution of $\bar{y}$, $\sigma_{\bar{y}}$, is the square root of the variance:

$$\sigma_{\bar{y}} = \frac{\sigma}{\sqrt{n}}$$

Thus, if all possible sample means for samples of size n could be collected, the average of the sample means would be the same as the average of all the individual population values. The sample mean values, however, would be less spread out than the individual population values because $\sigma_{\bar{y}}$ is always less than σ.

If the original population from which the samples were drawn is a normal distribution, then the sampling distribution of the sample mean is also a normal distribution for any sample size n. Thus, if μ and σ are known and the population to be sampled is normal, probability statements could be made about the sample mean, $\bar{y}$. Consider the following example.

EXAMPLE 2.3 In a certain manufacturing process, the diameter of a part produced is 40 centimeters (cm) on average, although it varies somewhat from part to part. This variation is thought to be well represented by a normal distribution with a standard deviation of 0.2 cm. If a random sample of 16 parts is chosen, what is the probability that the average diameter of the 16 parts is greater than 40.1 cm?

Answer: Because the population is normal, the sampling distribution of sample means also is normal. The mean of the sampling distribution is $\mu_{\bar{y}} = 40$ and the standard deviation is:

$$\sigma_{\bar{y}} = \frac{0.2}{\sqrt{16}} = 0.05$$

Thus,

$$P(\bar{y} > 40) = P\left(Z > \frac{40.1 - 40}{0.05}\right) = 0.5 - 0.4772 = 0.0228$$

Knowledge of the sampling distribution provides information about how sample means from a particular population should behave. But what if the population does not have a normal distribution, or what if the actual distribution of the population is unknown? In this case, there is an important result in statistics called the *central limit theorem* (CLT) that states

As long as the sample size is large, the sampling distribution of the sample mean is approximately normal, regardless of the population distribution.

The CLT states that probabilities still can be computed concerning sample means, even though the population does not have a normal distribution, as long as

EXAMPLE 2.4

A cereal manufacturer claims that boxes of its cereal weigh 20 ounces (oz) on average with a population standard deviation of 0.5 oz. The manufacturer does not know whether the population distribution is normal. A random sample of 100 boxes is selected. What is the probability that the sample mean is between 19.9 and 20.1 oz?

Answer: Because the sample size is large ($n = 100$), the sampling distribution is approximately normal even though the population distribution may be nonnormal. The mean and standard deviation of the sampling distribution are

$$\mu_{\bar{y}} = 20 \quad \text{and} \quad \sigma_{\bar{y}} = \frac{0.5}{\sqrt{100}} = 0.05$$

Thus,

$$P(19.9 \leq \bar{y} \leq 20.1) = P\left(\frac{19.9 - 20}{0.05} \leq Z \leq \frac{20.1 - 20}{0.05}\right) = P(-2 \leq Z \leq 2)$$

$$= 0.4772 + 0.4772 = 0.9544$$

EXERCISES

15. The daily receipts of a fast-food franchise are normally distributed with a mean of $2200 per day and a standard deviation of $50. A random sample of 25 days' receipts is chosen for an audit.

   a. What is the probability that the sample mean is larger than $2220?

   b. What is the probability that the sample mean differs from the true population mean by more than ±$10?

16. The accounts receivable of a large department store are normally distributed with a mean of $250 and a standard deviation of $80. If a random sample of 225 accounts is chosen, what is the probability that the mean of the sample is between $232 and $268?

17. When a certain manufacturing process is correctly adjusted, the length of a machine part produced is a random variable with a mean of 200 cm and a standard deviation of 0.1 cm. The individual measurements are normally distributed.

   a. What is the probability that an individual part is longer than 200.2 cm?

   b. Suppose a sample of 25 parts is chosen randomly. What is the probability that the mean of the sample is bigger than 200.2 cm?

   c. Is it correct to use the normal distribution to compute the probability in part b? Why or why not?

18. Suppose we have a large population of houses in a community. The average annual heating expense for each house is $400 with a population standard deviation of $25. A random sample of 25 houses had a sample mean of $380 and a sample standard deviation of $35.

   a. What is the mean of the sampling distribution of the sample mean for samples of size 25 chosen from the population of all houses?

   b. What is the standard deviation of the sampling distribution of the sample mean for samples of size 25 chosen from the population of all houses?

   c. Is it safe to assume that the sampling distribution of the sample mean for samples of size 25 is normally distributed? Why or why not?

19. The average time to complete a certain production-line task is assumed to be normally distributed with a standard deviation of 5 minutes (min). A random sample of 16 workers' times is selected to estimate the average time taken to complete the task. What is the probability that the sample mean is within ±1 min of the population's true mean time?

20. The speed of automobiles on I-20 west of Fort Worth, Texas, is being investigated by the Texas Department of Public Safety (DPS). If the average speed of cars on the highway exceeds 80 miles per hour (mph), the DPS plans to add more patrol cars to the area. To decide what to do, it takes a random sample of 150 cars and finds the sample average speed to be 80.5 mph. Assuming that the population standard deviation is 8 mph, should the DPS add patrol cars to the area?

21. Suppose past evidence shows that the lifetimes of hard drives from a certain production line have a population standard deviation of 700 hours. But a modification has been made in the material used to manufacture the hard drives. The manufacturer wants to know if the average lifetime of the modified hard drives is longer than the previous average lifetime. It is believed that the modification will not affect the population standard deviation of the lifetimes, but it may affect the average life. The previous average lifetime was 3250 hours. A random sample of 50 hard drives with the modification is taken and the drives are tested. The sample average lifetime for the drives from the new process is found to be 3575 hours. Should the manufacturer conclude that the new process produces hard drives with longer average lifetimes?

2.6 ESTIMATING A POPULATION MEAN

Two types of estimates can be constructed for any population parameter: point estimates and interval estimates. *Point estimates* are single numbers used as an estimate of a parameter. To estimate the population mean μ, the sample mean $\bar{y}$ typically is used. The sample mean is a point estimate.

An *interval estimate* is a range of values used as an estimate of a population parameter. The width of the interval provides a sense of the accuracy of the point estimate. The interval tells us the likely values of the population mean.

Assuming the population standard deviation, σ, is known, a confidence interval for the population mean, μ, can be constructed as

$$\left(\bar{y} - z_{\alpha/2}\frac{\sigma}{\sqrt{n}},\; \bar{y} + z_{\alpha/2}\frac{\sigma}{\sqrt{n}}\right)$$

where $z_{\alpha/2}$ is a standard normal value chosen so that the probability above $z_{\alpha/2}$ is $\alpha/2$. Thus, between $z_{\alpha/2}$ and $-z_{\alpha/2}$, there is a probability of $1 - \alpha$ (see Figure 2.19). For this reason, the confidence interval written in its general form is referred to as a $(1 - \alpha)100\%$ (read "one minus alpha times 100 percent") confidence interval estimate of μ. By replacing $z_{\alpha/2}$ by the appropriate standard normal value, the desired level of confidence can be achieved. For example, to achieve 95% confidence, use $z_{0.025} = 1.96$ because the probability under the standard normal curve between -1.96 and 1.96 is 0.95. Note that lowercase z's are used here to represent the specific values chosen from the standard normal distribution as opposed to uppercase Z's, which represent the standard normal random variable.

The term "95% confidence interval" means that if repeated samples of size n are taken from the same population, and a confidence interval is constructed in the manner just described for each sample, 95% of those intervals contain the

FIGURE 2.19
Choosing z Values for a $(1 - \alpha)100\%$ Confidence Interval.

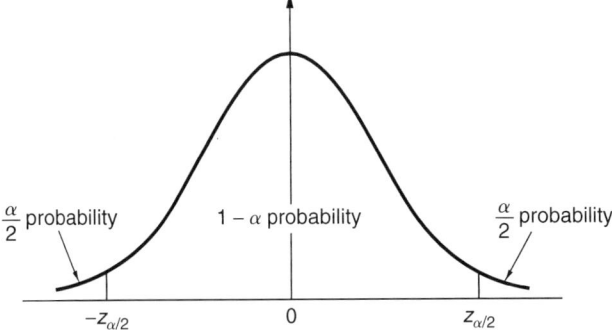

population mean. The use of the interval, assuming that σ is known, is justified for any sample size if the population is normal. When the population is normal, the sampling distribution of sample means is also normal, which is the basis for the construction of the interval. If the distribution of the population is unknown or if it is known to be nonnormal, the interval still can be used as long as the sample size is large (generally $n \geq 30$) because the CLT guarantees that the sampling distribution of sample means is approximately normal.

EXAMPLE 2.5 Managers of Newman-Markups Department Store want a 90% confidence interval estimate of the current average balance of charge customers. With a random sample of 100 accounts, a sample mean of $245, and a population standard deviation of $45, what is the 90% interval estimate of the true average balance?
Answer: The 90% confidence interval estimate of the true average balance is

$$\left(245 - 1.65\frac{45}{\sqrt{100}}, 245 + 1.645\frac{45}{\sqrt{100}}\right)$$

or ($237.58, $252.43)

In Example 2.5, the population standard deviation σ was assumed to be known. In most instances, however, σ is unknown. In this case, σ can be estimated by the sample standard deviation:

$$s = \sqrt{\frac{\sum_{i=1}^{n}(y_i - \bar{y})^2}{n - 1}}$$

Replacing σ by s in the previous interval and $z_{\alpha/2}$ by $t_{\alpha/2, n-1}$ gives

$$\left(\bar{y} - t_{\alpha/2, n-1}\frac{s}{\sqrt{n}}, \bar{y} + t_{\alpha/2, n-1}\frac{s}{\sqrt{n}}\right)$$

Changing $z_{\alpha/2}$ to $t_{\alpha/2, n-1}$ reflects the fact that s, an estimator of σ, is being used to construct a confidence interval estimate for μ. The value of $t_{\alpha/2, n-1}$ is chosen from the t table (Table B.2) in Appendix B. The t value chosen depends on the number of *degrees of freedom* (df) and on the confidence level desired. The number of degrees of freedom for estimating μ is $n - 1$. These values are listed on the left-hand side of the t table. The confidence levels are reflected through the upper-tail areas at the top of the table. Note that the 0.025 column is used for a 95% level of confidence ($\alpha/2 = 0.025$).

The shape of the t distribution depends on the number of degrees of freedom. The t distribution has fatter tails than the normal distribution and, thus, has greater probability in its tails. The t value for a given level of confidence is therefore larger than the standard normal value, producing wider confidence intervals (less precise estimates) because s rather than σ is used to construct the interval estimate. Also, as the number of degrees of freedom increases, the t distribution begins to look more like the normal distribution. In the last row of the t table, which is the ∞ degree of freedom row, the z values corresponding to the given upper-tail areas are shown. This indicates that, for a very large sample, the t and Z distributions are identical.

When the number of degrees of freedom is 30 or more, the t values are sometimes replaced by z values because there is little difference between the two in these cases. Although using t values is more correct in all cases when σ is unknown, replacing the t values by z values for large degrees of freedom is often adopted simply for convenience. Note that Excel, MINITAB, and SAS continue to use t values in certain applications when σ is unknown, regardless of the sample size. Also, the estimated standard deviation of the sampling distribution $(s/\sqrt{n})$ is often referred to as the *standard error of the mean*.

In small samples ($n < 30$), the population should be normal, or close to a normal distribution, before the interval is used. If the population is nonnormal and small samples are used, nonparametric methods may be appropriate. These methods are discussed in many introductory statistics texts (for example, see Chapter 17 of *Statistics for Management and Economics* by Keller and Warrack[2]). When the sample size is large, the assumption of a normal population is not necessary. The flowchart in Figure 2.20 describes the interval choices when σ is unknown. The σ known case is not shown because it rarely occurs in practice.

EXAMPLE 2.6 A manufacturer wants to estimate the average life of an expensive electrical component. Because the test to be used destroys the component, a small sample is desired. The lifetimes in hours of five randomly selected components are

$$92, 110, 115, 103, 98$$

Find a point estimate and a 95% confidence interval estimate of the population average lifetime of the components. The population of lifetimes is assumed to be normal. Answer: Because σ is unknown, the interval to be used is

$$\left(\bar{y} - t_{0.025,4}\frac{s}{\sqrt{n}}, \bar{y} + t_{0.025,4}\frac{s}{\sqrt{n}}\right)$$

---

[2] See References for complete publication information.

FIGURE 2.20
Estimating μ Assuming σ is Unknown.

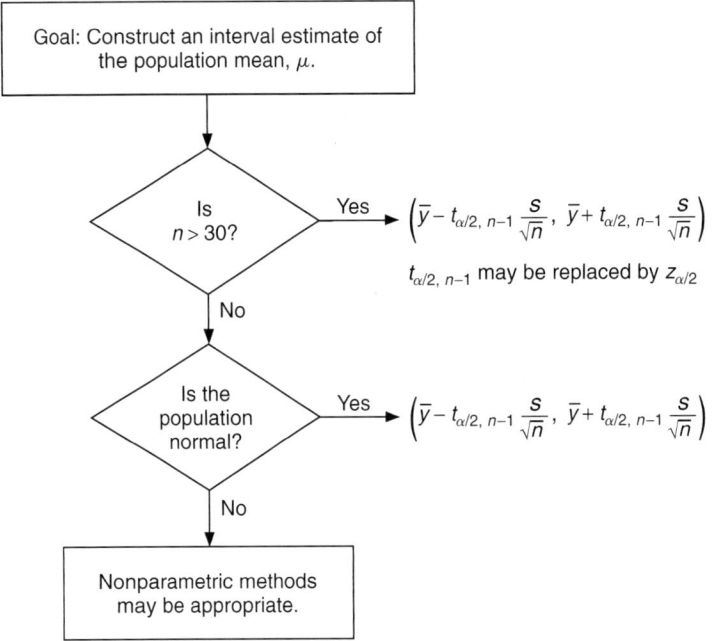

FIGURE 2.21 Using MINITAB to Construct Confidence Intervals.[a]

```
One-Sample T: hours
Variable       N      Mean     StDev    SE Mean         90% CI
hours          5    103.600    9.182     4.106     (94.846, 112.354)

One-Sample T: hours
Variable       N      Mean     StDev    SE Mean         95% CI
hours          5    103.600    9.182     4.106     (92.200, 115.000)

One-Sample T: hours
Variable       N      Mean     StDev    SE Mean         99% CI
hours          5    103.600    9.182     4.106     (84.695, 122.505)
```

[a]The quantities shown include the sample size (N), the sample mean (Mean), the sample standard deviation (StDev), the standard error of the mean (SE Mean) and the confidence interval for the requested level of confidence.

The quantities needed to construct the interval are

$$\bar{y} = \frac{\sum_{i=1}^{n} y_i}{n} = 103.6$$

$$s = \sqrt{\frac{\sum_{i=1}^{n}(y_i - \bar{y})^2}{n-1}} = 9.18$$

$$t_{0.025,4} = 2.776$$

The point estimate of μ is $\bar{y} = 103.6$. The interval estimate of μ is

$$\left[103.6 - 2.776\left(\frac{9.18}{\sqrt{5}}\right), 103.6 + 2.776\left(\frac{9.18}{\sqrt{5}}\right)\right] \quad \text{or} \quad (92.2, 115.0)$$

Computer packages such as Excel, MINITAB, and SAS also can be used to construct confidence intervals. Figure 2.21 shows the MINITAB output for requests for 90%, 95%, and 99% confidence intervals using the data from Example 2.6. There is more information on such procedures in the Using the Computer section at the end of this chapter.

EXERCISES

22. A local department store wants to determine the average age of the adults in its existing marketing area to help target its advertising. A random sample of 400 adults is selected. The sample mean age is found to be 35 years with a sample standard deviation of 5 years. Construct a 95% confidence interval estimate of the population average age of the adults in the area.

23. The management of a manufacturing plant is studying the number of times employees in a large population of workers are absent. A random sample of 25 employees is chosen, and the average number of annual absences per employee in the sample is found to be six. The sample standard deviation is 0.6. Assuming the population of absences is normally distributed, construct a 99% confidence interval estimate of the population average number of absences.

24. A quality control inspector is concerned with the average amount of weight that can be held by a type of steel beam. A random sample of five beams is tested with the following amounts of weight added before the beams begin to show stress (in thousands of pounds):

9, 11, 10, 10, 8

Assuming that the population of weights is normally distributed, construct a 95% confidence interval estimate of the population average weight that can be held.

25. Refer to the data in Exercise 1. Highway mileages of 147 cars are in a file named CARS2. Assume these cars represent a random sample of all new cars produced in 2003. Find a 95% confidence interval estimate for the population mean miles per gallon.

26. The July 1, 2002, one-year returns for a random sample of 83 mutual funds are available in a data file named ONERET2. Find a 95% confidence interval estimate for the population mean rate of return.
(*Source*: The data are used by permission from the September issue of *Kiplinger's's Personal Finance*. Copyright © 2002 The Kiplinger Washington Editors, Inc.)

2.7 HYPOTHESIS TESTS ABOUT A POPULATION MEAN

Section 2.6 discussed estimation of a population mean. Estimation was the first of our two main topics of statistical inference. In this section, we discuss the second topic: hypothesis tests. Again, the population mean is used to demonstrate tests of hypotheses.

The following definitions are useful in testing hypotheses:

Null Hypothesis, H_0:	The null hypothesis states a hypothesis to be tested.
Alternative Hypothesis, H_a:	The alternative hypothesis includes possible values of the population parameter not included in the null hypothesis.
Test Statistic:	A number computed from sample information.
Decision Rule:	A rule used in conjunction with the test statistic to determine whether or not the null hypothesis should be rejected.

In setting up a hypothesis test, the null hypothesis is initially assumed to be true. Under this assumption, a decision rule is constructed based on the sampling distribution of the test statistic. The decision rule states a range of values for the test statistic that are plausible if H_0 is true and a range of values for the test statistic that seem implausible if H_0 is true. Depending on the test statistic value, the statistical decision made is either "reject H_0" (the test statistic falls in the implausible range) or "do not reject H_0" (the test statistic falls in the plausible range).

Because the decision is based on sample information, it is not possible to be certain that the correct decision has been made. A statistical decision does not prove or disprove the null hypothesis with certainty, although it does present support for one of the two hypotheses. In a business environment, decisions are typically made on the basis of limited information. Hypothesis-testing results provide support for alternative possible courses of action based on such limited (sample) information.

Two types of errors are possible in hypothesis testing; these are illustrated in Figure 2.22. On the left-hand side of the figure are the two possible states of nature: Either H_0 is true or H_0 is false. The statistical decisions "do not reject H_0" and "reject H_0" are listed at the top of the figure. If H_0 is true and the sample information says do not reject H_0, a correct decision has been made. Also, if H_0 is false and the sample information says to reject H_0, the decision is correct. If H_0 is true, however, and the sample information says to reject H_0, the decision is incorrect. Rejecting the null hypothesis when it is true is called a *Type I error*. A *Type II error* occurs if the null hypothesis is actually false but the sample information says do not reject H_0.

FIGURE 2.22
The Risks of Hypothesis Testing.

		Decision	
		Do Not Reject H_0	Reject H_0
State of Nature	H_0 True	Correct Decision	Type I Error
	H_0 False	Type II Error	Correct Decision

Note that the decision made is always stated with reference to the null hypothesis: Either reject H_0 or do not reject H_0. Also note that because only two possibilities are considered (either H_0 or H_a), rejecting H_0 implies agreement with H_a.

Note that the phrase "do not reject H_0" (or "fail to reject H_0") rather than "accept H_0" is used. When the data suggest that we should not reject H_0, this is not proof that H_0 is true. This is a statistical decision and may imply simply that we do not have enough evidence to reject H_0. The expression "accept H_0" is viewed as too strong in suggesting that H_0 has been proven true. "Do not reject H_0" emphasizes that our result is not proof that the null hypothesis is true. Our sample may simply not provide enough evidence to reject it.

When testing any hypothesis, it is desirable to keep the chances of an error occurring as small as possible. Typically, when setting up the test, a desired level for the probability of a Type I error is established. By specifying a small probability, control can be exercised over the chances of making such an error. The probability of a Type I error is called the *level of significance* (or *significance level*) of the test and is denoted α.

To illustrate, suppose the following hypotheses are to be tested:

$$H_0: \mu = 10$$
$$H_a: \mu \neq 10$$

Also assume that the population standard deviation is known to be 2. This assumption will be relaxed later. The sample to be drawn consists of 100 items, and the desired level of significance is $\alpha = 0.05$, or a 5% chance of making a Type I error.

The hypotheses have now been set up and a level of significance chosen. The next step is to establish the decision rule for the test. To do this, consider the sampling distribution of sample means as shown in Figure 2.23. If the null hypothesis is true ($\mu = 10$), then a region can be determined in which 95% of all sample means will fall. The upper bound of this region is denoted C_1 and the lower bound C_2. Above and below these bounds, there is a combined probability of only 0.05 of obtaining a sample mean. The decision rule for the test can be set up as

Reject H_0 if $\bar{y} > C_1$ or $\bar{y} < C_2$
Do not reject H_0 if $C_2 \leq \bar{y} \leq C_1$

FIGURE 2.23
Sampling Distribution of Sample Means Assuming $H_0: \mu = 10$ Is True.

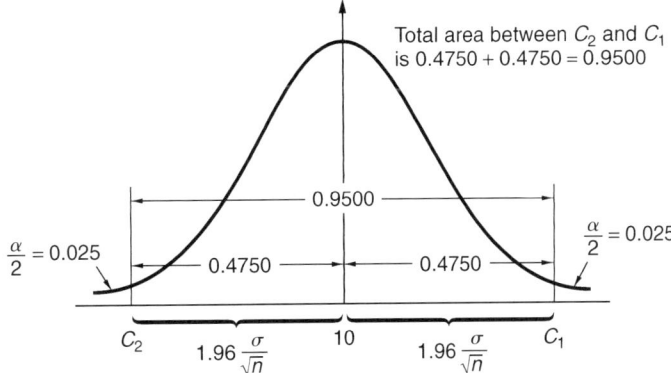

If the null hypothesis is true, 95% of all sample means will fall between C_1 and C_2. So there is only a 5% chance of obtaining a sample mean that falls in the rejection region. In other words, there is a 5% chance of rejecting H_0 if H_0 is true. Thus, the desired level of significance for the test has been achieved.

The critical values C_1 and C_2 in the decision rule can be determined because the sampling distribution is normal (as guaranteed by the CLT because the sample size is large). Between C_1 and C_2, we want a probability of 0.95; the associated z value to produce this probability is 1.96. The upper critical value is

$$C_1 = 10 + 1.96 \frac{2}{\sqrt{100}} = 10.392$$

and the lower critical value is

$$C_1 = 10 - 1.96 \frac{2}{\sqrt{100}} = 9.608$$

The decision rule becomes

Reject H_0 if $\bar{y} > 10.392$ or $\bar{y} < 9.608$

Do not reject H_0 if $9.608 \leq \bar{y} \leq 10.392$

Now suppose a sample of size 100 is randomly selected, and the sample mean computed is $\bar{y} = 11.2$. Based on this value of the sample mean, the null hypothesis is not supported. A sample mean of 11.2 is too extreme to believe that the individual values which produced it came from a population with mean $\mu = 10$. So the null hypothesis is rejected. Note that the null hypothesis has not been proved false. There is simply contradictory evidence, and so a statistical decision to reject was made. The alternative hypothesis, $\mu \neq 10$, seems more plausible given the evidence obtained.

The previous hypothesis-testing problem was set up in terms of the sampling distribution of the sample mean. An alternative and more typical way of performing hypothesis tests is with a standardized test statistic. The basic philosophy and structure of the test are the same. The only difference is that the test statistic is standardized and compared directly with the z value. For example, the standardized test statistic for testing hypotheses about the population mean is

$$z = \frac{\bar{y} - \mu_0}{\sigma/\sqrt{n}}$$

where μ_0 is the hypothesized value of the population mean. For the previous example, the standardized decision rule is

Reject H_0 if $\dfrac{\bar{y} - \mu_0}{\sigma/\sqrt{n}} > 1.96$ or $\dfrac{\bar{y} - \mu_0}{\sigma/\sqrt{n}} < -1.96$

Do not reject H_0 if $-1.96 \leq \dfrac{\bar{y} - \mu_0}{\sigma/\sqrt{n}} \leq 1.96$

The standardized test statistic value is

$$\frac{\bar{y} - \mu_0}{\sigma/\sqrt{n}} = \frac{11.2 - 10}{2/\sqrt{100}} = 6$$

Because the test statistic value falls in the rejection region ($6 > 1.96$), the null hypothesis is rejected.

Whether the standardized or unstandardized form of the test is used, the decision made (reject H_0 or do not reject H_0) will always be the same. Throughout this text, the standardized form is used unless otherwise noted.

If σ is unknown, the same adjustment is used as with confidence intervals. The unknown population standard deviation, σ, is replaced by an estimate, s, and the z value is replaced by the t value with $n - 1$ degrees of freedom.

The hypothesis structure previously discussed is called a *two-tailed test* because rejection occurs in both the upper and lower tails of the sampling distribution. Two other hypothesis structures need to be considered: *upper-tailed* and *lower-tailed tests*. Both of these involve rejection in only one tail of the sampling distribution and are therefore referred to as *one-tailed tests*. The hypothesis structures, test statistic, and decision rules for the case when σ is unknown are shown in Figure 2.24. The σ known case is omitted because it is rarely encountered in practice.

As was the case for confidence intervals, the t tests are constructed with the assumption that the sampling distribution is normally distributed. The population should be normal (or nearly so) before tests are used with small samples ($n < 30$), or nonparametric methods may be appropriate. Because the CLT guarantees that the sampling distribution is close to normal when n is large, the assumption of a normal population is unnecessary in cases with large sample sizes.

The following examples help illustrate hypothesis-testing techniques.

Hypotheses	Test Statistic	Decision Rules
$H_0: \mu = \mu_0$	$t = \dfrac{\bar{x} - \mu_0}{s/\sqrt{n}}$	Reject H_0 if $t > t_{\alpha/2, n-1}$ or if $t < -t_{\alpha/2, n-1}$
$H_a: \mu \neq \mu_0$		Do not reject H_0 if $-t_{\alpha/2, n-1} \leq t \leq t_{\alpha/2, n-1}$
$H_0: \mu \leq \mu_0$	$t = \dfrac{\bar{x} - \mu_0}{s/\sqrt{n}}$	Reject H_0 if $t > t_{\alpha, n-1}$
$H_a: \mu > \mu_0$		Do not reject H_0 if $t \leq t_{\alpha, n-1}$
$H_0: \mu \geq \mu_0$	$t = \dfrac{\bar{x} - \mu_0}{s/\sqrt{n}}$	Reject H_0 if $t < -t_{\alpha, n-1}$
$H_a: \mu < \mu_0$		Do not reject H_0 if $t \geq -t_{\alpha, n-1}$

FIGURE 2.24 Hypotheses, Test Statistics, and Decision Rules for Testing Hypotheses About Population Means (σ unknown)

EXAMPLE 2.7

Consider again the manufacturer of electrical components in Example 2.6. Suppose the manufacturer wishes to test whether the population average life of the components is 110 hours or more. If it is less than 110 hours, the components do not meet specifications, and the production process must be adjusted to increase the average life of the components.

As in Example 2.6, five components are randomly selected, and the lifetimes in hours for these components are determined to be

$$92, 110, 115, 103, 98$$

Assume the population of lifetimes is known to be normally distributed. The hypotheses to be tested are

$$H_0: \mu \geq 110$$
$$H_a: \mu < 110$$

Using a 5% level of significance, the decision rule for the test is

Reject H_0 if $t < -t_{0.05,4} = -2.132$

Do not reject H_0 if $t \geq -t_{0.05,4} = -2.132$

The sample mean and standard deviation are

$$\bar{y} = 103.6, s = 9.18$$

and the standardized test statistic, t, is

$$t = \frac{\bar{y} - \mu_0}{\sigma/\sqrt{n}} = \frac{103.6 - 110}{9.18/\sqrt{5}} = -1.56$$

so we do not reject the null hypothesis. Note that the sample mean of 103.6 hours is below the desired average lifetime of 110 hours. However, it is possible that the population mean could be 110 hours and our sample will produce a sample mean of 103.6 hours. We have not proved the null hypothesis is true, but there is not enough evidence in our small sample to reject it.

EXAMPLE 2.8

A company that manufactures rulers wants to ensure that the average length of its rulers is 12 inches (for obvious reasons). From each production run, a random sample of 25 rulers is selected and their lengths determined by very accurate measuring instruments. On one particular run, the average length of the 25 rulers is determined to be 12.02 inches, with a sample standard deviation of 0.02 inch.

Using a 1% level of significance, is the average length of the rulers produced by this manufacturer equal to 12 inches?

Answer: The hypotheses to be tested are

$$H_0: \mu = 12$$
$$H_a: \mu \neq 12$$

Using a 1% level of significance, the decision rule for the test is

Reject H_0 if $t > t_{0.005, 24} = 2.797$ or if $t < -t_{0.005, 24} = -2.797$
Do not reject H_0 if $-2.797 \leq t \leq 2.797$

From the sample information, the standardized test statistic is

$$t = \frac{\bar{y} - \mu_0}{\sigma/\sqrt{n}} = \frac{12.02 - 12}{0.02/\sqrt{25}} = 5.0$$

resulting in a decision to reject the null hypothesis. This decision suggests that the average length of rulers is not 12 inches. Some adjustment to the production process is necessary.

Most statistical software packages perform tests of hypotheses. Instead of reporting a decision to reject or fail to reject the null hypothesis, however, the output often includes a number called a *p* value. By comparing the *p* value to the level of significance, α, an alternative decision rule can be constructed:

Reject H_0 if *p* value $< \alpha$
Do not reject H_0 if *p* value $\geq \alpha$

The *p* value is the computed area under the sampling distribution at or beyond the value of the standardized test statistic. That is, it is the probability of observing a *t* value or *z* value as extreme as, or more extreme than, the sample test statistic. For example, suppose the hypotheses to be tested are

$$H_0: \mu \leq 10$$
$$H_a: \mu > 10$$

A sample size of 100 is to be used and σ is known to be 10 (for illustrative purposes, we assume σ is known; the process is similar when σ is unknown). From the random sample of 100 items, a sample mean of $\bar{y} = 12$ is obtained. The standardized test statistic is

$$z = \frac{12 - 10}{10/\sqrt{100}} = 2.0$$

In Figure 2.25, the standard normal distribution is shown, and the position of the test statistic has been located. The *p* value for this test is

$$p \text{ value} = P(Z \geq 2.0) = 0.5 - 0.4772 = 0.0228$$

FIGURE 2.25
Computation of p Value.

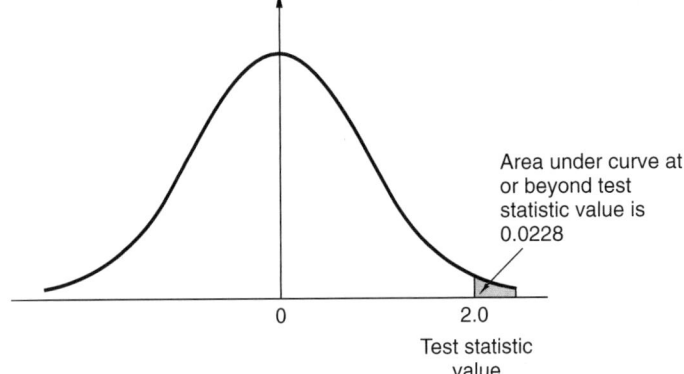

If we require a 5% level of significance, our decision is to reject the null hypothesis because the p value of 0.0228 is less than 0.05.

Regardless of which approach to hypothesis testing is chosen (unstandardized or standardized test statistic or p value), the decision made is identical for a given level of significance.

The way p values are computed differs depending on whether the test is an upper-tailed, lower-tailed, or two-tailed test. In all three cases, the first step is to compute the standardized test statistic. For an upper-tailed test (as in the previous illustration), compute the probability to the right of the standardized test statistic. For a lower-tailed test, compute the probability to the left of the standardized test statistic. For a two-tailed test, compute the probability in the tail area closest to the standardized test statistic, and then multiply this area by 2. By computing p values in this manner, the following decision rule can always be used:

Reject H_0 if p value $< \alpha$

Do not reject H_0 if p value $\geq \alpha$

EXAMPLE 2.9 Figure 2.26 shows the MINITAB output for testing the hypothesis discussed in Example 2.7. The output shows $\bar{y} = 103.600$, $s = 9.182$, $s/\sqrt{n} = 4.106$, $t = -1.56$, and the p value $= 0.097$. Using the p value decision rule, note that the null hypothesis is not rejected for levels of significance of 1% or 5%, but is rejected at

FIGURE 2.26
MINITAB Output for Test of Hypothesis in Example 2.9.

One-Sample T: hours

Test of mu = 110 vs < 110

Variable	N	Mean	StDev	SE Mean	95% Upper Bound	T	P
hours	5	103.600	9.182	4.106	112.354	-1.56	0.097

10%. Reporting the p value is somewhat more informative in a case such as this than simply reporting a decision to reject or fail to reject the null hypothesis.[3]

EXERCISES

27. If a null hypothesis is rejected at the 5% level of significance, what decision would have been made at the 10% level? Why?

28. To investigate an alleged unfair trade practice, the Federal Trade Commission (FTC) takes a random sample of sixteen "5-ounce" candy bars from a large shipment. The mean of the sample weights is 4.85 ounces and the sample standard deviation is 0.1 ounce. Test the hypotheses

$$H_0: \mu \geq 5$$
$$H_a: \mu < 5$$

at the 5% level of significance. Assume the population of candy bar weights is approximately normally distributed. Based on the results of the test, does the FTC have grounds to proceed against the manufacturer for the unfair practice of short-weight selling? State the decision rule, the test statistic, and your decision.

29. A quality inspector is interested in the time spent replacing defective parts in one of the company's products. The average time spent should be at most 20 minutes (min) per day according to company standards. The following hypotheses are set up to examine whether the standards are being met:

$$H_0: \mu \leq 20$$
$$H_a: \mu > 20$$

where μ represents the population average time spent replacing defective parts. To conduct the test, a random sample of 16 employees is chosen. The average time spent replacing defective parts for the sample was 20.5 min, with a sample standard deviation of 4 min. Perform the test at a 5% level of significance. Assume the population of service times is approximately normally distributed. State the decision rule, the test statistic, and your decision. Are company standards being met?

30. Refer to the data in Exercise 1. Highway mileages of 147 cars are in a file named CARS2. Assume these cars represent a random sample of all new cars produced in 2003. The corporate average fuel economy (CAFE) standards set by the government require that average fuel economy for cars be more than 27.5 miles per gallon. To examine whether the CAFE standard is being met, the following hypotheses are tested:

$$H_0: \mu \leq 27.5$$
$$H_a: \mu > 27.5$$

where μ is the population average gas mileage for all 2003 cars. Use a 5% level of significance. State the decision rule, the test statistic, and your decision. What is your conclusion regarding the average mileage of 2003 cars?

31. Refer to the data in Exercise 26. The July 1, 2002, one-year returns for a random sample of 83 mutual funds are available in a file named ONERET2. The return for the S&P 500 stock index for the same one-year time period was −18.0%. Test to see if there is evidence that the average one-year return for the population of funds is more than the return for the S&P 500 stock index. Use a 5% level of significance. State the hypotheses to be tested, the decision rule, the test statistic, and your decision. What is your conclusion regarding the average one-year return for mutual funds?

---

[3] MINITAB also reports a 95% upper bound of 112.354 for the population mean. This number is the upper limit of a one-sided confidence interval for the mean. We are 95% sure that the population mean is less than 112.354. One-sided confidence intervals were not discussed in this text but would be consistent with the use of a one-sided test.

2.8 ESTIMATING THE DIFFERENCE BETWEEN TWO POPULATION MEANS

The comparison of two separate populations often is of more concern than the estimation of the parameters of a single population as discussed in Section 2.6. In many cases, the comparison is between the means of the two populations. In this section, point and interval estimates of the difference between two population means are discussed.

Throughout this section, we assume that two populations with parameters μ_1, σ_1 and μ_2, σ_2 are being studied. Random samples are drawn independently from the two populations. Summary statistics from the two samples are:

	Sample 1	Sample 2
Sample size	n_1	n_2
Sample mean	$\bar{y}_1$	$\bar{y}_2$
Sample standard deviation	s_1	s_2

A point estimate of the difference between the two population means, $\mu_1 - \mu_2$, is given by the difference between the two sample means, $\bar{y}_1 - \bar{y}_2$.

If σ_1 and σ_2 are known, the standard deviation of the sampling distribution of $\bar{y}_1 - \bar{y}_2$ is written as

$$\sqrt{\frac{\sigma_1^2}{n_1} + \frac{\sigma_2^2}{n_2}}$$

In addition, the sampling distribution of $\bar{y}_1 - \bar{y}_2$ can be shown to be normally distributed if each of the populations is normal. The sampling distribution is approximately normal if both sample sizes are large ($n_1 \geq 30$ and $n_2 \geq 30$) even if the populations are not normal. A $(1 - \alpha)100\%$ confidence interval for $\mu_1 - \mu_2$ would be

$$\bar{y}_1 - \bar{y}_2 \pm z_{\alpha/2} \sqrt{\frac{\sigma_1^2}{n_1} + \frac{\sigma_2^2}{n_2}}$$

As in the previous situations we discussed, it is unlikely that the population variances, σ_1^2 and σ_2^2, are known in practice. They must be estimated by the sample variances, s_1^2 and s_2^2. These estimates are substituted into the formula for the standard deviation of the sampling distribution of $\bar{y}_1 - \bar{y}_2$, which results in

$$\sqrt{\frac{s_1^2}{n_1} + \frac{s_2^2}{n_2}}$$

An approximate $(1 - \alpha)100\%$ confidence interval for $\mu_1 - \mu_2$ is then

$$\bar{y}_1 - \bar{y}_2 \pm t_{\alpha/2,\Delta} \sqrt{\frac{s_1^2}{n_1} + \frac{s_2^2}{n_2}}$$

The interval is approximate because the population variances may differ. When $\sigma_1^2 \neq \sigma_2^2$, an exact interval cannot be constructed. This interval is referred to as the *approximate* interval.

The approximate degrees of freedom for the sampling distribution of $\bar{y}_1 - \bar{y}_2$ are given by

$$\Delta = \frac{(s_1^2/n_1 + s_2^2/n_2)^2}{[(s_1^2/n_1)^2/(n_1 - 1)] + [(s_2^2/n_2)^2/(n_2 - 1)]}$$

If the population variances can be assumed equal, $\sigma_1^2 = \sigma_2^2$, then an exact interval can be constructed. Because $\sigma_1^2 = \sigma_2^2$, it is no longer necessary to provide separate estimates of the two variances. The information in both samples can be combined, or pooled, to estimate the common variance. The pooled estimator of the population variance is

$$s_p^2 = \frac{(n_1 - 1)s_1^2 + (n_2 - 1)s_2^2}{n_1 + n_2 - 2}$$

The $(1 - \alpha)100\%$ confidence interval is given by

$$\bar{y}_1 - \bar{y}_2 \pm t_{\alpha/2, n_1+n_2-2} \sqrt{s_p^2\left(\frac{1}{n_1} + \frac{1}{n_2}\right)}$$

This interval is referred to as the *exact* interval.

Which of these intervals, exact or approximate, should be used in practice? The answer depends on what we know about the population variances, σ_1^2 and σ_2^2. If σ_1^2 and σ_2^2 are known to be equal, choose the exact interval. If σ_1^2 and σ_2^2 are known to be unequal, choose the approximate interval. But what if we have no information on whether or not the variances are equal? In this case, current research recommends using the approximate interval.[4]

EXAMPLE 2.10 Table 2.2, panel A, lists the 5-year returns for a random sample of 51 load mutual funds. (Load funds require the payment of an up-front sales charge to invest in the fund.) Table 2.2, panel B, shows the returns over the same 5-year period for a random sample of 32 no-load funds (no up-front sales charge is required). Construct a 95% confidence interval estimate of the difference between the population average 5-year returns for no-load and load funds.

---

[4] We could test for equality of the variances and choose an approach based on the test result, but this procedure has been shown to be less powerful than simply using the approximate interval. See, for example, "Homogeneity of Variance in the Two-Sample Means Test" by Moser and Stevens in *The American Statistician* 46(1992): 19–21.

TABLE 2.2 5-Year Rates of Return

Panel A: Load Funds		Panel A: Load Funds (cont.)		Panel B: No-Load Funds	
Name of Fund	5-yr Return	Name of Fund	5-yr Return	Name of Fund	5-yr Return
Dreyfus Growth & Value Emerging Leaders	10.1	Aim Weingarten A	−4	American AAdvantage Balanced Plan	5.2
Vanguard Capital Opportunity	15.6	Deutsche Flag Value Builder A	5.6	American Century Income & Growth	5
Fidelity New Millennium	14.5	Federated Equity Income A	0	Ariel Appreciation	14.2
Fidelity Select Air Transportation	14.1	First American Small Cap Value A	6.6	AXA Rosenberg US Small Cap	9.5
Fidelity Select Health Care	7.3	Goldman Sachs Capital Growth A	4.3	Berger Growth	−5
Fidelity Advisor Balanced T	1.9	Pimco Capital Appreciation A	6.3	CG Capital Markets Large Cap Value	3.9
Fidelity Advisor Technology T	−1.1	WM Growth & Income A	4.9	Columbia Balanced	4.5
AAL Capital Growth A	5.7	AXP Small Companies Index A	7.1	Dreyfus Premier Third Century Z	0.4
MFS Emerging Growth A	−4.4	Evergreen Fund A	−3.2	Fidelity Blue Chip Growth	2.5
BlackRock Small Cap Growth A	−1.5	Evergreen Small Cap Value A	9.4	Fidelity Equity-Income II	5.1
Fifth Third Quality Growth A	2.7	Franklin Small Mid Cap Growth A	4.2	Gabelli Small Cap Growth	9.8
Calamos Convertible A	10.6	Liberty Fund A	2.4	ICAP Equity	5.7
Davis Financial A	7.7	Lord Abbott Developing Growth A	2.9	Invesco Health Sciences	6.4
Hartford Growth Opportunities L	4.3	MFS Value A	10.5	Janus Core Equity	10.6
Principal MidCap A	5.4	Nations Convertible Securities A	7.9	Janus Twenty	4.3
Alpine US Real Estate Equity Y	9.6	Nuveen Large Cap Value A	4.8	Mairs & Power Growth	10.8
J. Hancock Large Cap Equity A	4	Oppenheimer Cap Appreciation A	6.5	Needham Growth	17
Morgan Stanley Global Utilities B	5.9	Oppenheimer Total Return A	4.1	PBHG Growth	−3.9
Morgan Stanley Total Return B	2.7	Pioneer Fund A	6.6	T. Rowe Price Blue Chip Growth	3.4
Prudential US Emerging Growth A	6.3	Putnam Classic Equity A	0.9	T. Rowe Price Mid-Cap Value	11.6
Smith Barney Large Cap Value A	1.7	Putnam New Value A	5.4	Rainier Core Equity	5
FPA Paramount	−5	Salomon Brothers Investors Value A	6.3	RS Diversified Growth A	14
Merrill Lynch Balanced Capital D	3	Security Equity A	0.2	Scudder Capital Growth AARP	0.7
One Group Equity Income A	3.2	State Street Research Aurora A	15.9	Sit Mid Cap Growth	−0.9
Aim Constellation A	1	Van Kampen Comstock A	10.4	Strong Growth	4.1
		Waddell & Reed Continental	4.1	Third Avenue Value	8.9
				Turner Small Cap Value	14.4
				US Global Investors World Precious Minerals	−7.8
				Vanguard LifeStrategy Conservative Growth	6
				Vanguard Small Cap Index	5.2
				Vanguard Utilities Income	4.6
				Wasatch Small Cap Growth	15.2

Reprinted by permission from the September issue of *Kiplinger's Personal Finance*. Copyright © 2002 The Kiplinger Washington Editors, Inc.

The MINITAB output shown in Figure 2.27 can be used to obtain the confidence interval. As will be discussed in Section 2.9, this output also can be used to test hypotheses about the difference between the two population means. In panel A of Figure 2.27, the results correspond to the approximate interval (assuming $\sigma_1^2 \neq \sigma_2^2$). Panel B shows the results for the exact interval (assuming $\sigma_1^2 = \sigma_2^2$).

Which interval is appropriate in this problem? Since we really have no information on the population variances, the approximate interval is the better choice. In this case, there is little difference between the two intervals, but sizable differences

FIGURE 2.27
MINITAB Output for
Examples 2.10 and 2.11.

Panel A: Assumes Variances Are Unequal

```
Two-sample T for FIVEYRRET

LOAD    N    Mean    StDev    SE Mean
0       32   5.95    5.88     1.0
1       51   5.01    4.80     0.67

Difference = mu (0) - mu (1)
Estimate for difference:  0.942157
95% CI for difference:  (-1.538493, 3.422807)
T-Test of difference = 0 (vs not =): T-Value = 0.76   P-Value = 0.450   DF = 56
```

Panel B: Assumes Variances Are Equal

```
Two-sample T for FIVEYRRET

LOAD    N    Mean    StDev    SE Mean
0       32   5.95    5.88     1.0
1       51   5.01    4.80     0.67

Difference = mu (0) - mu (1)
Estimate for difference:  0.942157
95% CI for difference:  (-1.409585, 3.293899)
T-Test of difference = 0 (vs not =): T-Value = 0.80   P-Value = 0.428   DF = 81
Both use Pooled StDev = 5.2411
```

can occur that can produce misleading conclusions if the exact interval is used in an inappropriate situation. The approximate interval provides a conservative result when the population variances are equal but also provides protection against the case when the variances are not equal.

The interval produced by the approximate method is (−1.54%, 3.42%). This result suggests that we can be 95% confident that the difference in population average 5-year returns for load and no-load funds ($\mu_{NoLoad} - \mu_{Load}$) is between −1.54% and 3.42%. What does this result suggest to you regarding the two types of funds?

EXERCISES

32. A graduate school of business is interested in estimating the difference between mean GMAT scores for applicants with and without work experience. Independent random samples of 50 applicants with and 50 applicants without work experience are chosen. The following results were obtained:

	With Work Experience	Without Work Experience
Sample size	50	50
Sample mean	545	510
Sample standard deviation	104	95

Construct a 95% confidence interval estimate of the difference between the mean GMAT scores for the two groups.

33. Two suppliers are being considered by a manufacturer. Independent random samples of ten parts from shipments from each supplier are selected, and the lifetime in hours for each part is determined for each sample. Use the following information to construct a 98% confidence interval estimate of the difference in the population average lifetimes. Assume that the population variances are equal and the populations are normally distributed.

	Supplier 1	Supplier 2
Sample size	10.0	10.0
Sample mean	15.0	11.0
Sample standard deviation	1.5	1.0

34. To help validate a new employee-rating form, a company administers it to independent random samples of employees in two different divisions. The following information is obtained from the scores on the forms:

	Division 1	Division 2
Sample size	15.0	15.0
Sample mean	82.0	78.0
Sample standard deviation	3.0	2.5

Use the information to construct a 95% confidence interval estimate of the difference in mean scores between the two divisions. Assume that the population variances are equal and the populations are normally distributed.

35. The one-year returns for a random sample of 51 load mutual funds and 32 no-load funds were obtained. The returns are in a file named RETURNS2. Construct a 95% confidence interval estimate of the difference between the population mean returns.

These data are arranged in the file in two columns. The first column contains the returns for the funds and the second column indicates to which sample (load = 1, no-load = 0) each value in column 1 belongs (in the Excel spreadsheet, the returns are in two separate columns).

(*Source*: Used with permission from the September issue of *Kiplinger's Personal Finance*. Copyright © 2002. The Kiplinger Washington Editors, Inc.)

36. The file named PRIVATE2 contains the graduation rates for 195 schools and a variable coded 1 for private schools and 0 for public schools. Construct a 99% confidence interval estimate for the difference between population average graduation rates for public and private schools.

These data are arranged in the file in two columns. The first column contains the graduation rates and the second column contains the private school variable (in the Excel spreadsheet, the graduation rates are in two separate columns for private and public schools).

(*Source*: Used by permission from the November and December 2003 issues of *Kiplinger's Personal Finance*. Copyright © 2003 The Kiplinger Washington Editors, Inc. Visit our website at www.kiplingers.com for further information).

2.9 HYPOTHESIS TESTS ABOUT THE DIFFERENCE BETWEEN TWO POPULATION MEANS

We may be interested in testing hypotheses about the difference between two population means rather than estimating that difference. The most common hypotheses tested in comparing two populations are

$$H_0: \mu_1 = \mu_2$$

$$H_a: \mu_1 \neq \mu_2$$

The null hypothesis states that the means of the two populations are equal, whereas the alternate states that the two population means differ. These hypotheses can be restated in terms of the difference between two means as

$$H_0: \mu_1 - \mu_2 = 0$$
$$H_a: \mu_1 - \mu_2 \neq 0$$

The decision rule for the test is

Reject H_0 if $t > t_{\alpha/2, df}$ or $t < -t_{\alpha/2, df}$
Do not reject H_0 if $-t_{\alpha/2, df} \leq t \leq t_{\alpha/2, df}$

The construction of the test statistic, t, depends on whether the population variances can be assumed equal. If $\sigma_1^2 = \sigma_2^2$, then

$$t = \frac{\bar{y}_1 - \bar{y}_2}{\sqrt{s_p^2 \left(\frac{1}{n_1} + \frac{1}{n_2} \right)}}$$

and the critical value, $t_{\alpha/2, df}$, is chosen with df $= n_1 + n_2 - 2$ degrees of freedom. The pooled estimate of the population variance, s_p^2, is used in computing the standard deviation of the sampling distribution.

If $\sigma_1^2 \neq \sigma_2^2$, then

$$t = \frac{\bar{y}_1 - \bar{y}_2}{\sqrt{\left(\frac{s_1^2}{n_1} + \frac{s_2^2}{n_2} \right)}}$$

and the approximate critical value is chosen with

$$\Delta = \frac{(s_1^2/n_1 + s_2^2/n_2)^2}{[(s_1^2/n_1)^2/(n_1 - 1)] + [(s_2^2/n_2)^2/(n_2 - 1)]}$$

degrees of freedom.

The justification for using two different standard errors in constructing the test statistics is the same as that for constructing the confidence intervals in the previous section.

Figure 2.28 shows the three possible hypothesis structures, the test statistic to be used, and the decision rules for the case when $\sigma_1^2 = \sigma_2^2$. Figure 2.29 presents similar information for $\sigma_1^2 \neq \sigma_2^2$.

EXAMPLE 2.11 Consider again the 5-year return data for the random sample of mutual funds examined in Example 2.10 of Section 2.8 ( ).

Let μ_L represent the population mean 5-year return for load funds and μ_N represent the population mean 5-year return for no-load funds. Then the hypotheses

$$H_0: \mu_N - \mu_L = 0$$
$$H_a: \mu_N - \mu_L \neq 0$$

Hypotheses	Test Statistic	Decision Rules
$H_0: \mu_1 - \mu_2 = 0$	$t = \dfrac{\bar{y}_1 - \bar{y}_2}{\sqrt{s_p^2\left(\dfrac{1}{n_1} + \dfrac{1}{n_2}\right)}}$	Reject H_0 if $t > t_{\alpha/2,\, n_1+n_2-2}$ or if $t < -t_{\alpha/2,\, n_1+n_2-2}$
$H_a: \mu_1 - \mu_2 \neq 0$		Do not reject H_0 if $-t_{\alpha/2,\, n_1+n_2-2} \leq t \leq t_{\alpha/2,\, n_1+n_2-2}$
$H_0: \mu_1 - \mu_2 \leq 0$	$t = \dfrac{\bar{y}_1 - \bar{y}_2}{\sqrt{s_p^2\left(\dfrac{1}{n_1} + \dfrac{1}{n_2}\right)}}$	Reject H_0 if $t > t_{\alpha,\, n_1+n_2-2}$
$H_a: \mu_1 - \mu_2 > 0$		Do not reject H_0 if $t \leq t_{\alpha,\, n_1+n_2-2}$
$H_0: \mu_1 - \mu_2 \geq 0$	$t = \dfrac{\bar{y}_1 - \bar{y}_2}{\sqrt{s_p^2\left(\dfrac{1}{n_1} + \dfrac{1}{n_2}\right)}}$	Reject H_0 if $t < -t_{\alpha,\, n_1+n_2-2}$
$H_a: \mu_1 - \mu_2 < 0$		Do not reject H_0 if $t \geq -t_{\alpha,\, n_1+n_2-2}$

FIGURE 2.28 Hypotheses, Test Statistics, and Decision Rules for Testing Hypotheses About Differences Between Population Means When $\sigma_1^2 = \sigma_2^2$.

Hypotheses	Test Statistic	Decision Rules
$H_0: \mu_1 - \mu_2 = 0$	$t = \dfrac{\bar{y}_1 - \bar{y}_2}{\sqrt{\left(\dfrac{s_1^2}{n_1} + \dfrac{s_2^2}{n_2}\right)}}$	Reject H_0 if $t > t_{\alpha/2,\Delta}$ or if $t < -t_{\alpha/2,\Delta}$
$H_a: \mu_1 - \mu_2 \neq 0$		Do not reject H_0 if $-t_{\alpha/2,\Delta} \leq t \leq t_{\alpha/2,\Delta}$
$H_0: \mu_1 - \mu_2 \leq 0$	$t = \dfrac{\bar{y}_1 - \bar{y}_2}{\sqrt{\left(\dfrac{s_1^2}{n_1} + \dfrac{s_2^2}{n_2}\right)}}$	Reject H_0 if $t > t_{\alpha,\Delta}$
$H_a: \mu_1 - \mu_2 > 0$		Do not reject H_0 if $t \leq t_{\alpha,\Delta}$
$H_0: \mu_1 - \mu_2 \geq 0$	$t = \dfrac{\bar{y}_1 - \bar{y}_2}{\sqrt{\left(\dfrac{s_1^2}{n_1} + \dfrac{s_2^2}{n_2}\right)}}$	Reject H_0 if $t < -t_{\alpha,\Delta}$
$H_a: \mu_1 - \mu_2 < 0$		Do not reject H_0 if $t \geq -t_{\alpha,\Delta}$

FIGURE 2.29 Hypotheses, Test Statistics, and Decision Rules for Testing Hypotheses About Differences Between Population Means When $\sigma_1^2 \neq \sigma_2^2$.

can be tested to determine if there is a difference between the average returns for these two groups.

Figure 2.29 shows the MINITAB output for testing the hypotheses. Figure 2.30 shows the Excel output and Figure 2.31 shows the SAS output. Assuming that the population variances are unequal (or that we have no information about the variances), the panel A output is appropriate in MINITAB and Excel. In the SAS output, the two versions of the test are referred to as the Pooled Method and the Satterthwaite Method. Both results are printed whenever a two-sample comparison is requested.

2.9 Hypothesis Tests About The Difference Between Two Population Means

FIGURE 2.30 Excel Output for Examples 2.10 and 2.11.

Panel A: Assumes Variances Are Unequal

t-Test: Two-Sample Assuming Unequal Variances

	NOLOAD5	LOAD5
Mean	5.95	5.01
Variance	34.61	23.04
Observations	32.00	51.00
Hypothesized Mean Difference	0.00	
df	56.00	
t Stat	0.76	
P(T<=t) one-tail	0.22	
t Critical one-tail	1.67	
P(T<=t) two-tail	0.45	
t Critical two-tail	2.00	

Panel B: Assumes Variances Are Equal

t-Test: Two-Sample Assuming Equal Variances

	NOLOAD5	LOAD5
Mean	5.95	5.01
Variance	34.61	23.04
Observations	32.00	51.00
Pooled Variance	27.47	
Hypothesized Mean Difference	0.00	
df	81.00	
t Stat	0.80	
P(T<=t) one-tail	0.21	
t Critical one-tail	1.66	
P(T<=t) two-tail	0.43	
t Critical two-tail	1.99	

The TTEST Procedure

Statistics

Variable	type	N	Lower CL Mean	Mean	Upper CL Mean	Lower CL Std Dev	Std Dev	Upper CL Std Dev	Std Err
FIVEYR	0	32	3.8289	5.95	8.0711	4.7166	5.8833	7.8217	1.04
FIVEYR	1	51	3.6578	5.0078	6.3579	4.0163	4.8001	5.9669	0.6721
FIVEYR	Diff (1-2)		-1.41	0.9422	3.2939	4.5435	5.2411	6.1938	1.182

T-Tests

Variable	Method	Variances	DF	t Value	Pr>|t|
FIVEYR	Pooled	Equal	81	0.80	0.4277
FIVEYR	Satterthwaite	Unequal	56.2	0.76	0.4499

Equality of Variances

Variable	Method	Num DF	Den DF	F Value	Pr>F
FIVEYR	Folded F	31	50	1.50	0.1963

FIGURE 2.31 SAS Output for Examples 2.10 and 2.11.

Based on the test results, the null hypothesis should not be rejected. The t statistic or the p value can be used to reach this decision. If the t statistic is used, the decision rule is

Reject H_0 if $t > 1.96$ or $t < -1.96$

Do not reject H_0 if $-1.96 \leq t \leq 1.96$

There are 56 degrees of freedom for this test, so the z value of 1.96 was used as the critical value. Note that Excel provides the t value for 56 degrees of freedom for a two-tailed test: $t = 2.00$. This value could be used (and is in fact preferred) rather than the z value if it is available. The test statistic value is $t = 0.76$.

If the p value is used, the decision rule is

Reject H_0 if p value < 0.05

Do not reject H_0 if p value ≥ 0.05

where the p value is 0.45. As always, both procedures lead to the same decision.

The statistical decision is do not reject the null hypothesis, so we conclude that there is no difference between the average 5-year returns for load and no-load mutual funds.

EXERCISES

37. Consider again Exercise 32 in Section 2.8. Two independent random samples of applicants to business schools who had and did not have work experience were chosen. Each sample contained 50 applicants. To determine whether there is a difference in the population average test scores, the following hypotheses should be tested:

$$H_0: \mu_1 - \mu_2 = 0$$
$$H_a: \mu_1 - \mu_2 \neq 0$$

Use a 5% level of significance. State the decision rule, the test statistic, and your decision. What implication do these test results have for admissions officers in MBA programs?

38. Use the information in Exercise 33. Suppose that the manufacturer currently uses supplier 2. A change to supplier 1 will be made only if the average lifetime of parts for supplier 1 is greater than the average for supplier 2. Using a 1% level of significance, conduct the appropriate test. State the hypotheses to be tested, the decision rule, the test statistic, and your decision. Assume that the population variances are equal and the populations are normally distributed. On the basis of the test result, which supplier will the manufacturer choose?

39. Use the information in Exercise 34. Is there a difference in population mean rating scores for the two divisions? State the hypotheses to be tested, the decision rule, the test statistic, and your decision. Assume that the population variances are equal and the populations are normally distributed. Use a 5% level of significance.

40. The 1-year returns for a random sample of 51 load mutual funds and 32 no-load funds were obtained. Test to see if there is any difference between the population average 1-year returns for load and no-load funds. Use a 5% level of significance. State the decision rule, the test statistic, and your decision. Is there a difference in the population averages? Based on the test results, what conclusions do you draw concerning investment in load versus no-load funds? The returns are in a file named RETURNS2. See Exercise 35 for a further description of the data in the file.

41. Let $\mu_0 =$ the population average graduation rate for public schools and $\mu_1 =$ the population average graduation rate for private schools. Suppose that a claim is made that private schools have higher graduation rates, on average, than public schools. Examine the claim by testing the following hypotheses:

$$H_0: \mu_0 - \mu_1 \geq 0$$

$$H_a: \mu_0 - \mu_1 < 0$$

Use a 5% level of significance. State the decision rule, the test statistic, and your decision. Is the claim supported? The file named PRIVATE2 contains the graduation rates. See Exercise 36 for a further description of the data in the file.

ADDITIONAL EXERCISES

42. A university wants to examine starting monthly salaries for its finance and marketing graduates. Independent random samples of 12 finance graduates and 12 marketing graduates are selected from the files of last year's graduates. The following starting salaries are obtained from these people:

Finance	Marketing
1850	1675
2150	1275
1700	1800
1500	2100
2200	2200
1650	2250
2100	1950
2140	1850
1790	2000
1650	1800
2300	2100
2000	2150

Use these data to determine the following:

a. Find the sample mean starting salaries for finance graduates and marketing graduates (separately).

b. Construct a histogram for starting salaries in both finance and marketing.

c. Construct a 95% confidence interval estimate of the population mean starting salary for finance majors. Do the same for marketing majors. Assume that the populations of starting salaries for both groups are normally distributed.

The data are available in the file named SALARY2. The salary data are in column one. The second column indicates to which sample each column one value belongs: 1 = finance; 0 = marketing (in the Excel spreadsheet, the salary data are in two separate columns.)

43. Consider again the finance and marketing starting salaries in Exercise 42.

a. Conduct a test to determine whether the population mean starting salaries for the two majors are equal. State the hypotheses to be tested, the decision rule, the test statistic, and your decision. Use a 5% level of significance.

b. What conclusion can be drawn from the result in part a?

44. A telemarketing firm is considering two different sales approaches for selling magazine subscriptions over the phone. Two independent random samples of 20 salespeople each are selected to use either approach 1 or 2 for the sample period and for the same number of household contacts. The sales data (number of subscriptions sold per period) are given here.

Approach 1	Approach 2
12	8
15	10
28	24
14	14
18	10
10	20
15	20
20	15
5	0
4	7
12	10
10	16
24	17
16	20
13	12
14	4
18	12
22	18
6	3
12	16

a. Is there a difference in average sales produced by the two approaches? Assume that the populations are normally distributed. State the hypotheses to be tested, the decision rule, the test statistic, and your decision. Use a 10% level of significance.

b. What does the result in part a suggest to a sales division manager?

The data are available in the file named SALES2. The sales data are in the first column; the number in the second column indicates whether the sales value belongs to approach 1 or approach 2 (in the Excel spreadsheet, the sales data are in two separate columns.)

45. The file named HARRIS2 contains 1977 annual starting salary data for 93 employees of Harris Bank of Chicago. The column of data denoted MALE indicates whether the employees were MALE(1) or FEMALE(0) (in the Excel spreadsheet, the salary data are in two separate columns.) Let μ_0 = average starting salary for females and μ_1 = average starting salary for males.

   a. Set up and test hypotheses to determine whether there is evidence of wage discrimination for the Harris Bank employees. Use a 5% level of significance. Set up the hypotheses assuming that discrimination is represented by an average wage for females that is less than the average wage for males.

   b. What implications do your test results have for Harris Bank?

   c. Are there other factors that might need to be considered in this analysis? If so, state them and why you believe they are important.

   (Source: These data were obtained from D. Schafer, "Measurement-Error Diagnostics and the Sex Discrimination Problem," *Journal of Business and Economic Statistics*. Copyright 1987 by the American Statistical Association. Used with permission. All rights reserved.)

46. Can expert stock analysts pick stocks that perform better than stocks chosen at random? Or better than a stock market index? There are no definitive answers to these questions, but people have lots of fun trying to find out. For example, in Fort Worth, TX, we have Rusty, a 1700-pound steer. At the start of each calendar year, Rusty is pitted against several of the state's top stock analysts. The analysts pick their portfolio of stocks and Rusty picks his. Rusty makes his picks in a special corral in Sundance Square in downtown Fort Worth with rectangles representing local companies. He lets the chips (as they say) fall where they may and his stocks are chosen accordingly. The results for 1997: Rusty's stocks gained 62.87% and the experts' stocks gained 37.09% (the S&P 500 gained 31% in 1997). For 1997–2003, Rusty came out ahead four times to three for the experts.

In another comparison, the *Wall Street Journal* formed a panel of four experts each month and compared the performance of their four stock picks to four stocks chosen by throwing darts at stock tables. A 6-month holding period was used for the comparison. Table 2.3 shows the results of the first five contests along with the return for the Dow Jones Industrial Average for each time period. The complete data set is available in the file named DARTS2. The three columns of data represent returns for the pros, darts, and the Dow Jones Industrial Average, respectively.

   a. Assume that the returns for the experts' portfolios and the dartboard portfolios represent two independent random samples of results. Test to see if there is a difference in the average returns for the experts and the average of the randomly chosen stocks.

   b. Perform the same test to compare the experts' average performance with the average performance of the Dow Jones Industrial Average.

   c. Based on the results of the tests in parts a and b, what do you conclude about the stock-picking ability of the experts?

   d. What problems might there be in comparing performance in this manner? What other comparisons might be useful in deciding whether expert stock analysts can pick stocks that perform better than stocks chosen at random or better than a stock market index?

47. The following is a relative frequency distribution that appeared in the *Wall Street Journal* (1/31/96 issue, reprinted courtesy of *The Wall Street Journal*). Use the distribution to help answer the questions that follow.

TABLE 2.3 Returns for Experts (PROS), Dartboard Portfolio (DARTS), and Dow Jones Industrial Index (DJIA) for First Five *Wall Street Journal* Darts Contests

Contest Period	PROS	DARTS	DJIA
1 January–June 1990	12.7	0.0	2.5
2 February–July 1990	26.4	1.8	11.5
3 March–August 1990	2.5	–14.3	–2.3
4 April–September 1990	–20.0	–7.2	–9.2
5 May–October 1990	–37.8	–16.3	–8.5

The Going Rate

Price distribution of homes sold through the Multiple Listing Service in Texas for the year ended October 1995.

Price of Home	% of Homes Sold
$29,999 or less	5.3
$30,000–39,999	5.1
$40,000–49,999	7.8
$50,000–59,999	10.0
$60,000–69,999	10.6
$70,000–79,999	9.9
$80,000–89,999	8.9
$90,000–99,999	6.4
$100,000–119,999	9.5
$120,000–139,999	7.4
$140,000–159,999	4.9
$160,000–179,999	3.5
$180,000–199,999	2.4
$200,000–299,999	5.2
$300,000–399,999	1.6
$400,000–499,999	0.7
$500,000 and more	0.8

a. What percentage of homes sold for less than $50,000?

b. What percentage of homes sold for $120,000 or more but less than $400,000?

c. Can you determine what percentage of homes sold for more than $130,000?

d. In what range of prices is the median price of the homes sold?

48. Our company produces metal parts that must have holes of a certain diameter punched for later use. As long as the center of each hole is within 1.5 centimeters (cm) of a particular spot on the metal part, the part will be acceptable to the buyer. The distance from the center of the hole punched to the desired center is called the error (errors can be positive or negative depending on their direction). Figure 2.32 shows a MINITAB histogram of the errors for a sample of the last 50 parts to come off the punch line. Managers have decided that as long as no more than 2 parts in a sample of 50 are in error by more than 1.5 cm, the punch machine is operating acceptably and no adjustments will be made.

a. How many of the parts from the sample of 50 have an error of 1.5 cm or more?

b. What percentage of parts in this sample has an error of 1.5 cm or more?

c. Based on the histogram, what is management's decision?

49. Figure 2.33 is a MINITAB time-series plot of the errors for the 50 parts examined in Exercise 48. The errors are plotted in the order that the parts were handled by the punch machine. Based on the histogram in Exercise 48, management decided that the punch machine was operating correctly and needed no adjustment. After examining the time-series plot in addition to the histogram, do you agree or disagree with management's conclusion? Justify your answer.

50. The owner of a construction company makes bids on contracts for jobs. The company describes the probability distribution of X = "the number of jobs it is awarded per year" as shown:

x	p(x)
2	0.05
3	0.15
4	0.20
5	0.35
6	0.25

FIGURE 2.32 Histogram of Punch Errors.

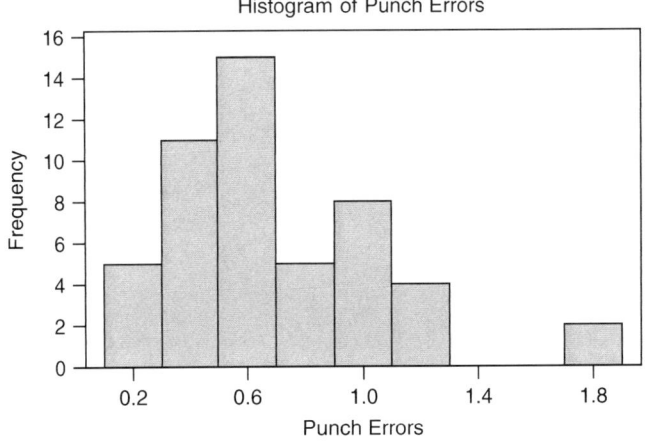

FIGURE 2.33 Time-Series Plot of Punch Errors.

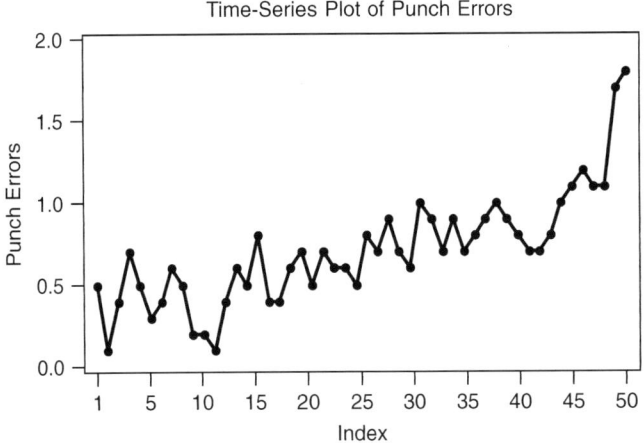

a. What is the expected or mean number of contracts that will be awarded per year?

b. Find the probability that more than three contracts will be awarded in a given year.

c. What is the most likely number of contracts that will be awarded in a given year?

51. An insurance company insures homes in Ft. Worth, TX. Roof damage due to hailstorms is always a persistent problem in this area. Suppose we simplify the range of damages in a given year to include just four possible values. We define a discrete random variable, X, to represent the dollar loss per year due to hail and assign the probabilities shown in the following distribution. For example, the probability is 0.9 that there will be no loss; it is 0.02 that there will be a $1000 loss, etc. Note that these losses are per household.

x	$p(x)$
0	0.90
1000	0.02
2000	0.04
3000	0.04

If the insurance company only wants to break even in the long run (dumb company), what should it charge annually for a policy for each household? Assume that the policy insures only against hail damage.

52. The filling machine for gallons of milk to be used by school districts in Texas can be set so that it discharges an average of μ ounces of milk per gallon. It is impossible to put exactly 1 gallon of milk in each gallon container due to natural variability in the dispensing machine. The amount of milk discharged is known to have a normal distribution with standard deviation equal to 0.4 ounce. When school officials discover that containers regularly have less than 1 gallon of milk in them, the state gets very upset. So we want to find a setting for μ so that only 1% of the containers of milk will have less than 1 gallon (128 ounces). What value should be used for μ?

53. A computer shop builds PCs from shipments of parts it receives from various suppliers. The number of defective hard drives per shipment is to be modeled as a random variable X. The random variable is assumed to have the following distribution:

x	$P(x)$
0	0.55
1	0.15
2	0.10
3	0.10
4	0.05
5	0.05

  a. What is the expected number of defective hard drives per shipment?

  b. What is the probability that a shipment will have more than two defective hard drives?

  c. What is the probability that a shipment will have four or fewer defective hard drives?

  d. What is the probability that a shipment will have more than two but less than five defective hard drives?

  e. What is the most likely number of defectives in a shipment?

54. Proctor and Gamble (P&G) has developed a new brand of toothpaste and plan to begin marketing the new toothpaste next month. From test market studies, P&G estimates that demand for the new toothpaste will average 200,000 tubes nationally with a standard deviation of 15,000 tubes during the first year. The *break-even point* for the toothpaste is the amount of sales necessary for P&G to begin making a profit. Management estimates that P&G will need to sell 180,000 tubes to break even in the first year. Assume that demand for the new toothpaste follows a normal distribution with mean 200,000 and standard deviation 15,000. What is the probability that P&G will sell at or above the break-even point during the first year?

55. High-Tech, Inc. produces an electronic component, GS-7, that has an average life span of 1000 hours. The life span is normally distributed with a standard deviation of 25 hours. The company is considering how long a warranty to place on the component. If it wants to replace no more than 10% of the components for free, how many hours should it guarantee the components would last under the warranty?

56. A bank reports that the population of its demand deposit balances has a population mean of $7500 and a population standard deviation of $1000. An auditor refuses to certify the bank's claim and takes a random sample of 100 account balances. He will certify the bank's report only if the sample mean is no more than $125 above or below the bank's stated population mean.

  a. Assuming the bank's reported figures are correct, what is the probability that the auditor will certify the bank's report?

  b. What would be the probability that the auditor will certify the bank's report if the reported population mean were $9000 rather than $7500 (assuming the standard deviation is the same)?

57. Researchers at American Airlines and Sabre have developed a system to coordinate scheduling, yield management, and pricing decisions (*ORMS Today*, August 2000, pp. 36–44). In one of their examples they set up the following scenario: "Consider the case of a single flight leg with one fare class. For this example, we assume that the passenger demand follows a normal distribution with a mean of 125 passengers and a coefficient of variation of 0.3. The capacity of the leg is assumed to be 150 seats." The coefficient of variation is defined as the standard deviation divided by the mean. What is the probability that demand will exceed the flight capacity?

58. A city engineer recorded the number of vehicles passing through a certain intersection for each day of last year. One objective of this study was to determine the percentage of days that more than 425 vehicles used the intersection. Suppose the mean for the data was 375 vehicles per day and the standard deviation was 25 vehicles. What can you say about the percentage of days that more than 425 vehicles used the intersection? Assume the distribution of the data is a normal distribution.

59. Our company manufactures a certain electronic component for use in the new F-22 fighter being built by Lockheed. We want to test the reliability of the components, and we think testing a random sample of 50 components will be sufficient. The population standard deviation of the lives of the components is known to be 10 hours. If we use a random sample of 50 components, what is the probability that our sample mean will be within (plus or minus) 2 hours of the true population average lifetime of the components?

60. **Process Capability.** Specification limits for a particular characteristic of a product indicate the values at which the product will operate properly. Specification limits are set by engineering and are not determined by the data. The data must be examined to determine whether the product is within specifications. If the product is within specification limits, the process generating the product is often referred to as capable. For example, assume we have a product with an upper specification limit of USL = 4.25 and a lower specification limit of LSL = 3.0. As long as our product falls within these limits our buyers will be happy. We now investigate the process used to manufacture the product and find that the process mean is 4 and the process standard deviation is 0.2. Assuming the process is normally distributed, what percentage of items will fall outside the specification limits? Should we make changes in the process? If so, what would you suggest?

61. By law, a manufacturer of a food product is required to list Food and Drug Administration (FDA) estimates of the contents of the packaged product. Suppose the FDA wants to estimate the mean sugar content (by weight) in 16-ounce boxes of "Disney Chocolate Mud and Bugs" cereal. The FDA randomly selects 200 boxes of Disney Chocolate Mud and Bugs, measures the sugar content in each, and computes a 95% confidence interval estimate of the average sugar content to be (3.2, 4.5).

The manufacturer plans to use the interval to claim that 95% of all boxes of Disney Chocolate Mud and Bugs cereal have sugar content weights between 3.2 and 4.5 ounces. Is this a correct interpretation of the interval? Justify your answer.

62. Suppose a sample of $n = 100$ items is randomly chosen. The following hypotheses are to be tested:

$$H_0: \mu \leq 10$$
$$H_a: \mu > 10$$

A p value of 0.0409 for the test is obtained. If the population standard deviation is 5, what would the value of the sample mean have to be in order to achieve the stated p value?

USING THE COMPUTER

The Using the Computer section in each chapter describes how to perform the computer analyses in the chapter using Excel, MINITAB, and SAS. For further detail on Excel, MINITAB, and SAS, see Appendix C.

EXCEL

Descriptive Statistics

TOOLS: DATA ANALYSIS: HISTOGRAM

Click Tools, then Data Analysis, and then Histogram. Histogram creates a frequency distribution and a histogram. In the histogram dialog box (see Figure 2.34), fill in the input range of the variable to be graphed. In the language of worksheets, the range of the data indicates the cells in which the data are contained. For example, if you have

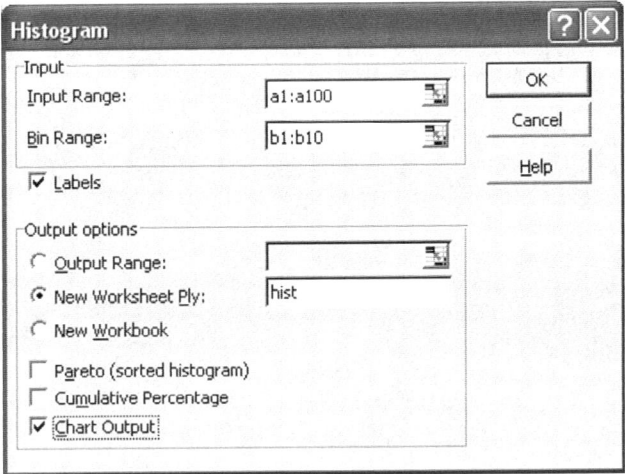

FIGURE 2.34 Excel Histogram Dialog Box.

100 observations to be included in your frequency distribution/histogram and you have typed these 100 numbers into cells A1 through A100, you specify the range as A1:A100 in the Input Range box.

You can specify the bins you want Excel to use. If you enter nothing in this box, Excel picks the bin limits for your frequency distribution/histogram. To specify the bins for Excel to use, just put the numbers you want to use as bin limits in a column. Then indicate the range of this column in the Bin Range box.

If the first row of your column of data contains a label, check the Label box. The next three options determine where you want the frequency distribution to appear. If you click Output Range, you can tell Excel a specific spot in the spreadsheet to put your frequency distribution and histogram. For example, if you check the Output Range button and put C3 in this box, the frequency distribution starts in cell C3. Check the New Worksheet Ply button and Excel puts your frequency distribution/histogram on a new ply in this same workbook. You can name the worksheet ply if you want. Check New Workbook and Excel puts your frequency distribution/histogram in a completely new workbook.

The next three options determine exactly what kind of output you want from Excel. If you want only a frequency distribution, do not check any of these options. The Pareto option constructs a histogram with the bins arranged from biggest to smallest. This option is often not very useful when quantitative data are used. It is more useful when qualitative data are used and the order of importance of certain categories is of interest. A Pareto chart is often used in quality control situations. For example, if customer complaints are being monitored and the most frequent complaints are of primary interest, a Pareto chart emphasizes those complaints by listing them first. The Cumulative Percentage option constructs a histogram but superimposes a line on the histogram that represents the cumulative percentage (sometimes called an *ogive*). Chart Output requests Excel to construct a histogram in addition to a frequency distribution.

Once you have the options set as you want, click OK.

TOOLS: DATA ANALYSIS: DESCRIPTIVE STATISTICS

Click Tools, then Data Analysis, and then Descriptive Statistics. This procedure generates a variety of descriptive statistics (see Figure 2.6). In the Descriptive Statistics dialog box (see Figure 2.35), fill in the input range for the variable for which descriptive statistics are desired. Typically, this variable is in a column, but the option is available to indicate whether it is in a column or row. Choose the output option and click Summary statistics. Click Confidence Level for Mean to produce a 95% error bound. (The 95% error bound is the default value and can be set at any desired level). Click OK.

FIGURE 2.35 Excel Descriptive Statistics Dialog Box.

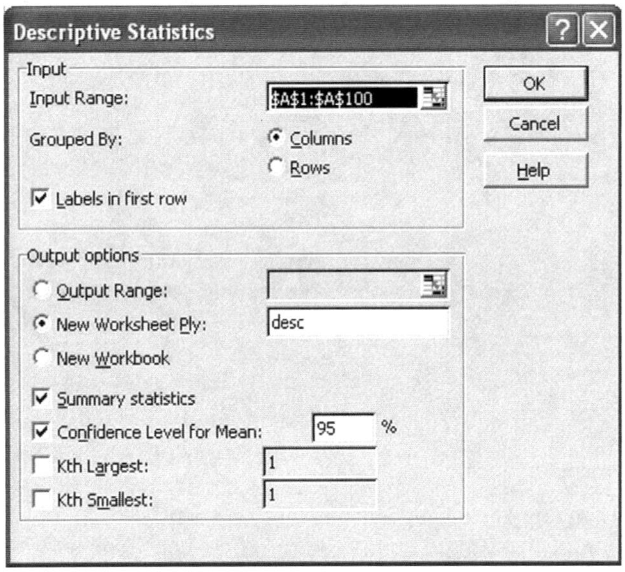

Confidence Interval Estimate of μ

The descriptive statistics option on the Data Analysis toolpack produces a value for the sample mean and a 95% (or any other desired) error bound, which can be used to construct a confidence interval. See Descriptive Statistics and Figure 2.35 for information on this procedure. The confidence interval can be constructed as sample mean ± error bound.

Alternatively, a formula can be constructed to compute the upper and lower confidence interval bounds. For example, suppose we have data in cells A1 through A100 and want to construct a 95% confidence interval. The formulas for the upper (UCL) and lower (LCL) confidence interval limits are

$$\text{UCL} = \text{average(A1:A100)} + \text{tinv(0.05,99)} * \text{stdev(A1:A100)}/\text{sqrt(100)}$$

$$\text{LCL} = \text{average(A1:A100)} - \text{tinv(0.05,99)} * \text{stdev(A1:A100)}/\text{sqrt(100)}$$

Note that tinv(0.05,99)*stdev(A1:A100)/sqrt(100) is the 95% error bound produced by the descriptive statistics procedure. The function tinv(0.05,99) returns the t value that puts a combined probability of 0.05 in the upper and lower tails of the t distribution with 99 degrees of freedom.

Hypotheses Tests About μ

Suppose we want to test the hypotheses $H_0: \mu = k$ versus $H_a: \mu \neq k$, where k represents the hypothesized value in the problem. Let's say the data are in cells A1 through A100. The t statistic for this test can be constructed using the formula

$$= (\text{average(A1:A100)} - k)/(\text{stdev(A1:A100)}/\text{sqrt(100)})$$

This value can be compared to the appropriate t critical value chosen with 99 degrees of freedom (or a z value since we have lots of degrees of freedom here) using a two-tailed decision rule.

To compute the p value associated with this test statistic requires use of the tdist function. Suppose we build the formula for the t statistic in cell C10. Then the p value is computed as

$$= \text{tdist(abs(C10),99,2)} \text{ or}$$

$$= \text{tdist(test statistic, degrees of freedom, number of tails)}$$

Note that the number of tails is 2 because this is a two-tailed test. You have to make sure that the numeric value provided to the tdist function is positive so the absolute value function is applied to C10.

For a one-tailed test, the test statistic is computed in exactly the same way, but a one-tailed decision rule is used. The computation of the p value is a little trickier.

Upper-tailed test: If the test statistic in C10 is positive, use = tdist(abs(C10),99,1)
   If the test statistic in C10 is negative, use = 1 − tdist(abs(C10),99,1)

Lower-tailed test: If the test statistic in C10 is positive, use = 1 − tdist(abs(C10),99,1)
   If the test statistic in C10 is negative, use = tdist(abs(C10),99,1)

Hypotheses Tests About $\mu_1 - \mu_2$
Confidence Interval Estimate of $\mu_1 - \mu_2$

To construct confidence intervals or test hypotheses about the difference between two population means, use either t-Test: Two-Sample Assuming Equal Variances or t-Test: Two-Sample Assuming Unequal Variances, as shown in Figure 2.36. The t-Test: Paired Two Sample for Means option is used when samples are matched rather than independent. The procedure for matched-sample tests was not discussed in this book. The z-Test: Two Sample for Means option is used when the population standard deviations are known (a fairly unlikely situation). The dialog boxes for the test procedures assuming either equal or unequal variances are identical. The dialog box from the unequal variance option is shown in Figure 2.37. Fill in the Variable Range 1 and 2 boxes with the ranges of the two independent samples. The Hypothesized Mean Difference is typically set to zero. This is the case discussed in this text, although tests for nonzero differences can be performed if these make sense. The Alpha level should be adjusted to the level desired for your test. This is because Excel prints out the critical value to compare with the standardized test statistic and needs to know the level of significance to do this. Specify the output option and click OK.

FIGURE 2.36 Excel Data Analysis Dialog Box with Two-Sample Test Options.

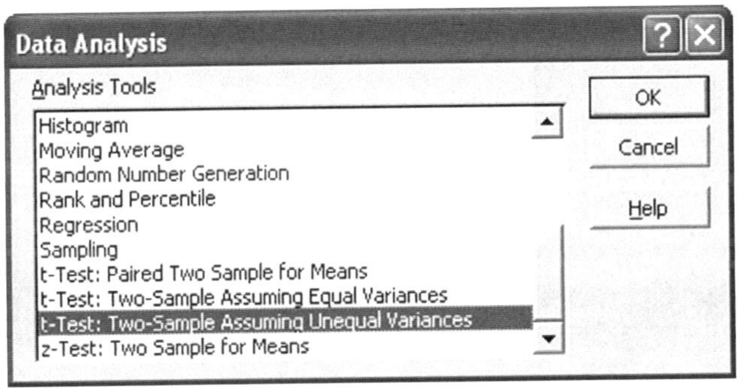

FIGURE 2.37 Excel Two-Sample Test Dialog Box with Unequal Variances Option.

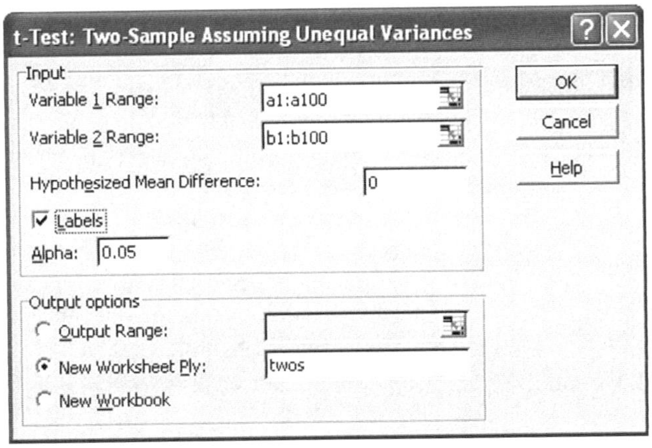

MINITAB
Descriptive Statistics

GRAPH: HISTOGRAM

Creates a histogram. Click Graph, then Histogram. Choose the type of histogram desired (See Figure 2.38). In the histogram dialog box (see Figure 2.39), fill in the variables to be graphed and click OK. There are a variety of options available, which are described in the MINITAB Help facility.

FIGURE 2.38 Choose Type of Histogram Desired in MINITAB.

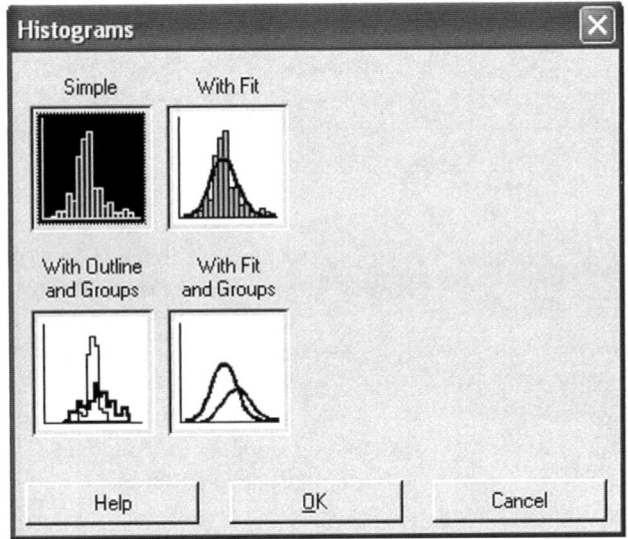

FIGURE 2.39 MINITAB Histogram Dialog Box.

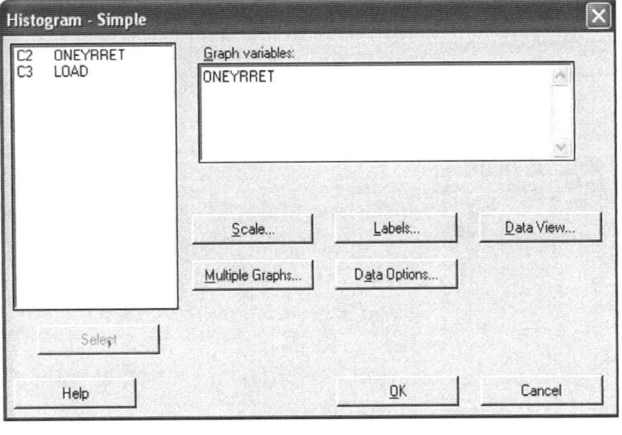

STAT: BASIC STATISTICS: DISPLAY DESCRIPTIVE STATISTICS

Click on Stat, then on Basic Statistics, then on Display Descriptive Statistics. This procedure generates a variety of descriptive statistics (see Figure 2.5). In the Display Descriptive Statistics dialog box (see Figure 2.40), fill in the variables for which descriptive statistics are desired and click OK.

Confidence Interval Estimate of μ
STAT: BASIC STATISTICS: 1-SAMPLE t

Click Stat, then Basic Statistics, then 1-Sample t. This procedure can be used to construct a confidence interval estimate of the population mean, μ. Fill in the name of the variable in the 1-Sample t dialog box and then click on Options (See Figures 2.41 and 2.42). Indicate the confidence level desired (the default value is 95%) and choose the not equal alternative for a two-sided confidence interval. One-sided intervals can also be constructed but were not discussed in this text. Choosing the alternative less than will provide an upper bound on the population mean, and the alternative greater than will provide a lower bound.

Hypotheses Tests About μ
STAT: BASIC STATISTICS: 1-SAMPLE t

Click Stat, then Basic Statistics, and then 1-Sample t. This procedure can be used to produce the output necessary to test the hypotheses $H_0: \mu = k$ versus $H_a: \mu \neq k$, where k represents the hypothesized value. The Options window is used to specify the type of hypothesis structure desired: "less than" for a lower-tailed test, "greater than" for an upper-tailed test, and "not equal" for a two-tailed (see Figures 2.43 and 2.44).

Confidence Interval Estimate of $\mu_1 - \mu_2$
STAT: BASIC STATISTICS: 2-SAMPLE T

Click Stat, then Basic Statistics, and then 2-Sample t. This procedure is used to construct confidence interval estimates of the difference between two population means, $\mu_1 - \mu_2$. Sample data from two different populations with means μ_1 and μ_2 are assumed to be available. These data can be in either one or two columns. The 2-Sample t dialog box is illustrated in Figure 2.45. If the data from both

FIGURE 2.40 MINITAB Descriptive Statistics Dialog Box.

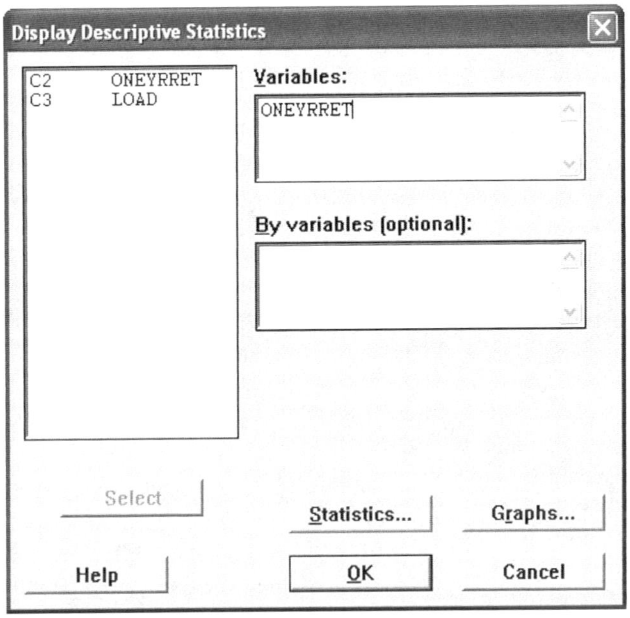

FIGURE 2.41 MINITAB 1-Sample t Dialog Box for Confidence Interval.

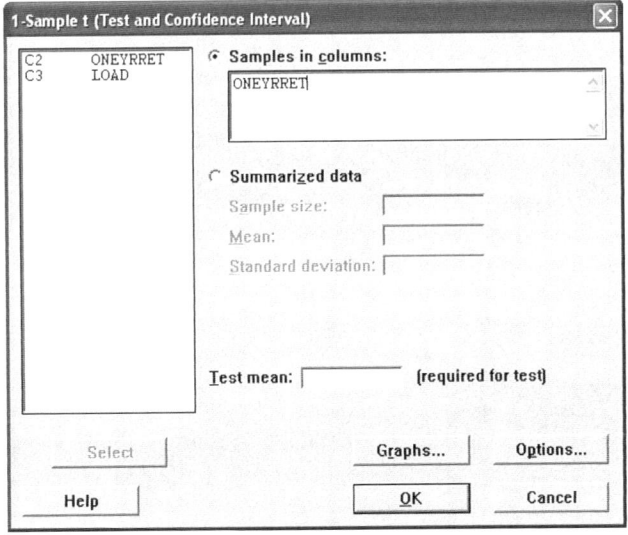

samples are contained in one column, check the "Samples in one column" button. If you use this option, a second column must contain numbers indicating from which sample each data point came (number the items from the first sample 0 and the items from the second sample 1, for example). If the data from each sample are in a different column, check the "Samples in different columns" button. Then indicate the two columns containing your data. The choice between these two options is based purely on how your data are arranged.

The assumption that the population variances (or standard deviations) are equal can be included by checking the "Assume equal variances" box in the dialog box. The Options window allows the user to choose a level of confidence and the hypothesized difference (Test difference). The default value for the hypothesized difference is zero because this is what is appropriate for a confidence interval for the difference between the two means. A two-sided confidence interval would be created if the Alternative is "not equal. " For a one-sided interval use either "less than" for an upper bound on the difference or "greater than" for a lower bound. See Figure 2.46.

Hypotheses Tests About $\mu_1 - \mu_2$

STAT: BASIC STATISTICS: 2-SAMPLE t

Click Stat, then Basic Statistics, and then 2-Sample *t*. This procedure is used to produce the output necessary to test $H_0: \mu_1 - \mu_2 = 0$ versus $H_a: \mu_1 - \mu_2 \neq 0$ (see Figure 2.45). The same dialog box is used for constructing a confidence interval for the difference between two means and testing hypotheses about the difference between two means. See the previous section for a description of how data should be arranged and how the equal variance option is handled.

The Options window allows the user to choose the type of hypothesis structure desired ("less than" for a lower-tailed test, "greater than" for an upper-tailed test, and "not equal" for a two-tailed test) and the hypothesized difference (Test difference).

FIGURE 2.42 MINITAB 1-Sample t Options Window Set for a Two-Sided Confidence Interval.

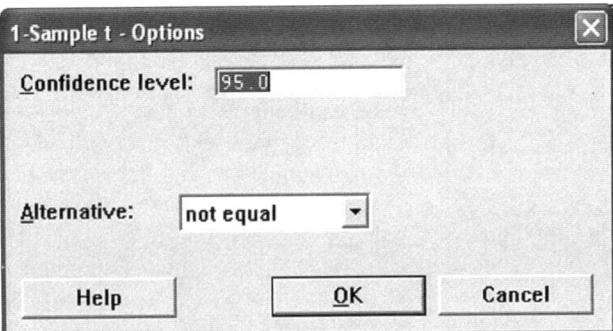

FIGURE 2.43 MINITAB 1-Sample t Dialog Box for Hypothesis Test.

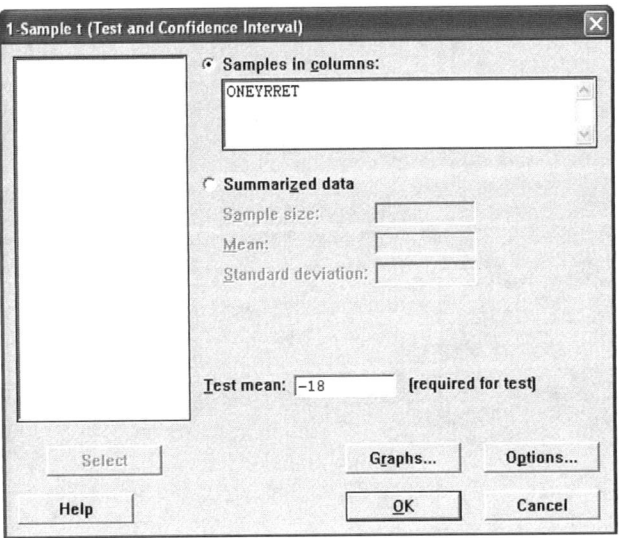

FIGURE 2.44 MINITAB 1-Sample t Options Window Set for a Lower-Tail Test.

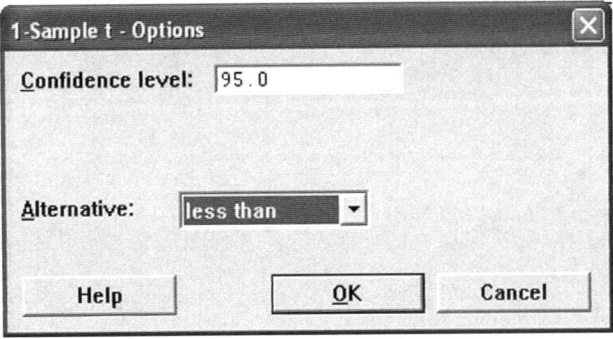

The default value for the hypothesized difference is zero because this is what is typically used (and is the only case discussed in this text). See Figure 2.46.

SAS

Hypothesis Tests About $\mu_1 - \mu_2$

PROC TTEST;
CLASS V1;
VAR V2;

Produces output to conduct a test of the difference between two population means. V1 is a variable used to separate the observations of the variable V2 into two groups. The two group means can be called μ_1 and μ_2. The t statistic produced can be used to conduct either a one- or two-tailed test. The p values in the Prob |T| column are designed specifically for two-tailed tests. Two versions of the test statistic are produced: the variances-equal version and the variances-unequal version. The appropriate statistic depends on the assumptions deemed correct for the population variances.

FIGURE 2.45 MINITAB 2-Sample *t* Dialog Box.

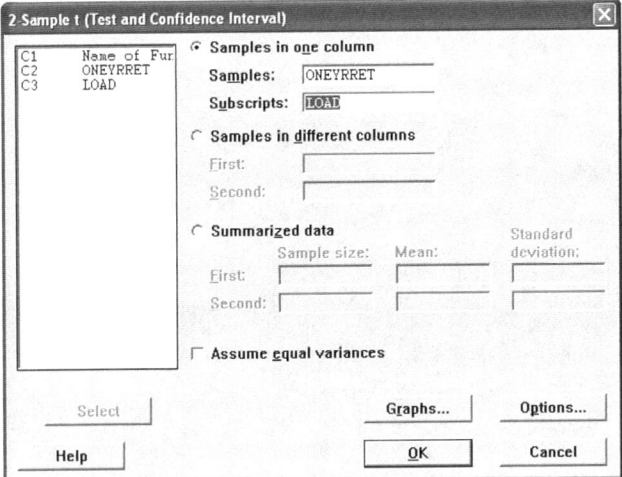

FIGURE 2.46 MINITAB 2-Sample *t* Options Window.

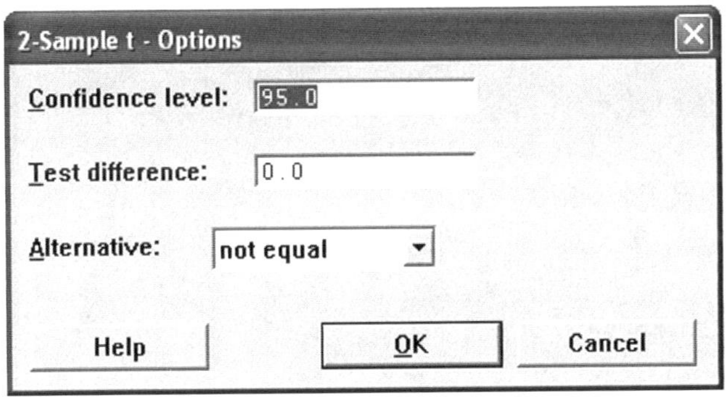

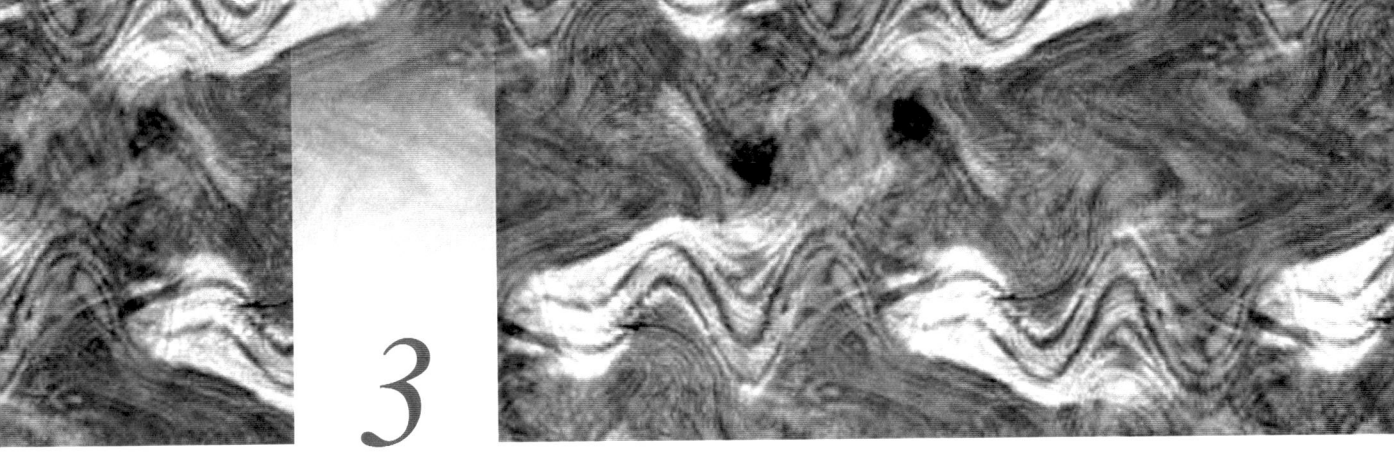

3 Simple Regression Analysis

3.1 USING SIMPLE REGRESSION TO DESCRIBE A LINEAR RELATIONSHIP

Regression analysis is a statistical technique used to describe relationships among variables. The simplest case to examine is one in which a variable y, referred to as the *dependent* variable, may be related to one variable x, called an *independent* or *explanatory* variable. If the relationship between y and x is believed linear, then the equation expressing this relationship is the equation for a line:

$$y = b_0 + b_1 x$$

If a graph of all the (x, y) pairs is constructed, then b_0 represents the *y intercept*, the point where the line crosses the vertical (y) axis, and b_1 represents the *slope* of the line.

Consider the data shown in Table 3.1. A graph of the (x, y) pairs would appear as shown in Figure 3.1. Regression analysis is not needed to obtain the equation expressing the relationship between these two variables. In equation form,

$$y = 1 + 2x$$

This is an *exact* or *deterministic* linear relationship. Exact linear relationships are sometimes encountered in business environments. For example, from accounting:

$$\text{assets} = \text{liabilities} + \text{owner equity}$$

$$\text{total costs} = \text{fixed costs} + \text{variable costs}$$

64 Simple Regression Analysis

TABLE 3.1 Example Data: Exact Relationship

x	1	2	3	4	5	6
y	3	5	7	9	11	13

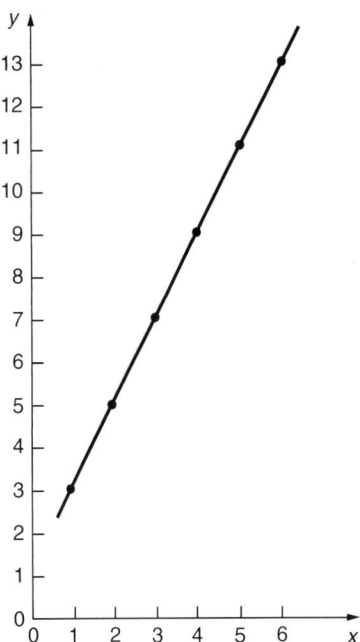

FIGURE 3.1 Graph of an Exact Linear Relationship (data in Table 3.1).

Other exact relationships may be encountered in various science courses (for example, physics or chemistry). In the social sciences (for example, psychology or sociology) and in business and economics, exact linear relationships are the exception rather than the rule. Data encountered in a business environment are more likely to appear as in Table 3.2. These data graph as shown in Figure 3.2.

It appears that x and y may be linearly related, but it is not an exact relationship. Still, it may be of use to describe the relationship in equation form. This can be done by drawing what appears to be the "best-fitting" line through the points and guessing what the values of b_0 and b_1 are for this line. This has been done in Figure 3.2. For the line drawn, a good guess might be the following equation:

$$\hat{y} = -1 + 2.5x$$

The notation $\hat{y}$ is used here to indicate that we do not expect this to be an exact relationship. Given a value of x, the equation will produce an estimate of y, which is denoted $\hat{y}$.

TABLE 3.2 Example Data: Not an Exact Relationship

x	1	2	3	4	5	6
y	3	2	8	8	11	13

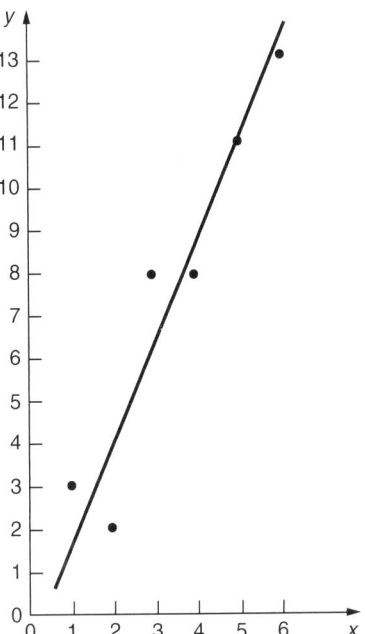

FIGURE 3.2 Graph of a Relationship That Is Not Exact (data in Table 3.2).

The drawbacks to this method of fitting the line should be clear. For example, if the (x, y) pairs graphed in Figure 3.2 were given to two people, each would probably guess different values for the intercept and slope of the best-fitting line. Furthermore, there is no way to assess who would be more correct. To make line fitting more precise, a definition of what it means for a line to be the "best" is needed. The criterion for a best-fitting line that we will use might be called the "minimum sum of squared errors" criterion or, as it is more commonly known, the least-squares criterion.

In Figure 3.3, the (x, y) pairs from Table 3.2 have been plotted and an arbitrary line drawn through the points. Consider the pair of values denoted (x^*, y^*). The actual y value is indicated as y^*; the value predicted to be associated with x^* if the line shown were used is indicated as $\hat{y}^*$. The difference between the actual y value and the predicted y value at the point x^* is called a *residual* and represents the "error" involved. This error is denoted $y^* - \hat{y}^*$. If the line is to fit the data points as accurately as possible, these errors should be minimized. This should be done not just for the single point (x^*, y^*), but for all the points on the graph. There are several possible ways to approach this task.

1. Use the line that minimizes the sum of the errors, $\sum_{i=1}^{n}(y_i - \hat{y}_i)$. The problem with this approach is that for any line that passes through the point $(\bar{x}, \bar{y})$, $\sum_{i=1}^{n}(y_i - \hat{y}_i) = 0$, so there are an infinite number of lines satisfying this criterion, some of which obviously do not fit the data well. For example, in Figure 3.4, lines A and B have both been constructed so that $\sum_{i=1}^{n}(y_i - \hat{y}_i) = 0$. But line A obviously fits the data better than line B; that is, it keeps the distances $y^* - \hat{y}^*$ small.

66 Simple Regression Analysis

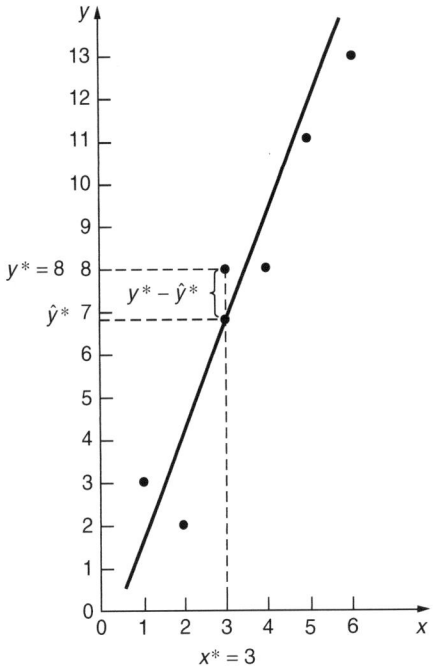

FIGURE 3.3
Motivation for the Least-Squares Regression Line.

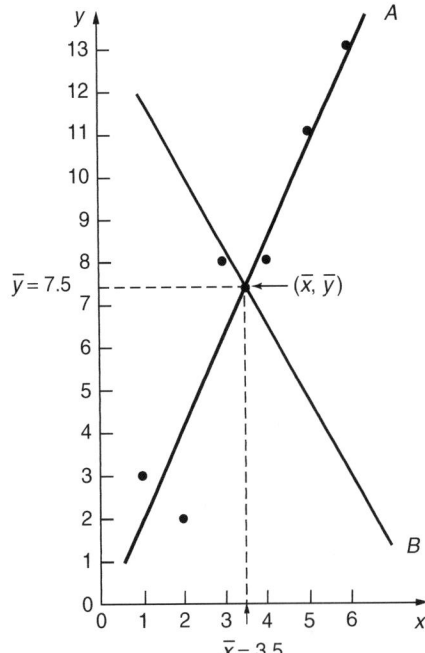

FIGURE 3.4 Lines *A* and *B* Both Satisfy the Criterion
$$\sum_{i=1}^{n}(y_i - \hat{y}_i) = 0.$$

As mentioned previously, any line that passes through the point represented by the means of x and y ($\bar{x}, \bar{y}$) has errors that sum to zero. The line passes through the points in such a way that positive and negative errors cancel each other out. Because a criterion is desired that makes the errors small regardless of whether they are positive or negative, some method of removing negative signs is required. One such method uses absolute values of the errors; another squares the errors. Each method provides another possible criterion.

2. Use the line that minimizes the sum of the absolute values of errors, $\sum_{i=1}^{n}|y_i - \hat{y}_i|$. This is called the *minimum sum of absolute errors criterion*. The resulting line is called the *least absolute value* (LAV) regression line. Although use of this criterion is gaining popularity in many situations, it is not the one that we use in this text. Finding the line that satisfies the minimum sum of absolute errors criterion requires solving a fairly complex problem by a technique called *linear programming*. This is a difficult problem by hand, and the LAV procedure is not readily available in most statistical software packages. Furthermore, there may be no unique LAV regression line.

3. Use the line that minimizes the sum of the squared errors, $\sum_{i=1}^{n}(y_i - \hat{y}_i)^2$. This is called the *least-squares criterion*, and the resulting line is called the *least-squares regression line*. Applying the least-squares (LS) criterion results in a unique least-squares regression line. Its advantages over the LAV line include computational simplicity and the wide availability of statistical packages that contain easily implemented least-squares regression routines.

Now that a criterion has been established, the next question is: Can convenient computational formulas for the values of b_0 and b_1 that minimize

$$\sum_{i=1}^{n}(y_i - \hat{y}_i)^2 \tag{3.1}$$

be developed? The answer is "yes" and the resulting equations are

$$b_1 = \frac{\sum_{i=1}^{n}(x_i - \bar{x})(y_i - \bar{y})}{\sum_{i=1}^{n}(x_i - \bar{x})^2} \tag{3.2}$$

$$b_0 = \bar{y} - b_1\bar{x} \tag{3.3}$$

A computationally simpler form of Equation (3.2) is

$$b_1 = \frac{\sum_{i=1}^{n}x_iy_i - \frac{1}{n}\sum_{i=1}^{n}x_i\sum_{i=1}^{n}y_i}{\sum_{i=1}^{n}x_i^2 - \frac{1}{n}\left(\sum_{i=1}^{n}x_i\right)^2} \tag{3.4}$$

EXAMPLE 3.1 As an example of the use of these formulas, consider again the data in Table 3.2. The intermediate computations necessary for finding b_0 and b_1 are shown in Figure 3.5. The slope, b_1, can now be computed using the formula in Equation (3.4):

$$b_1 = \frac{196 - \frac{1}{6}(21)(45)}{91 - \frac{1}{6}(21)^2} = \frac{38.5}{17.5} = 2.2$$

The intercept, b_0, is computed as in Equation (3.3):

$$b_0 = 7.5 - 2.2(3.5) = -0.2$$

because

$$\bar{x} = \frac{21}{6} = 3.5 \text{ and } \bar{y} = \frac{45}{6} = 7.5$$

The least-squares regression line for these data is

$$\hat{y} = -0.2 + 2.2x$$

There is no longer any guesswork associated with computing the best-fitting line once a criterion has been stated that defines "best." Using the criterion of minimum sum of squared errors, the regression line we computed provides the best description of the relationship between the variables x and y. Any other values used for b_0 and b_1 result in larger sum of squared errors. For example, Figure 3.6(a) shows the computation of the sum of squared errors for the original "guessed" line $\hat{y} = -1 + 2.5x$, and Figure 3.6(b) shows the same computation for the least-squares line $\hat{y} = -0.2 + 2.2x$. As expected, the sum of squared errors for the line in (a) is larger than that for the least-squares line.

FIGURE 3.5
Computations for Finding b_0 and b_1.

i	x_i	y_i	$x_i y_i$	x_i^2
1	1	3	3	1
2	2	2	4	4
3	3	8	24	9
4	4	8	32	16
5	5	11	55	25
6	6	13	78	36
Sums	21	45	196	91

FIGURE 3.6

(a) Computation of Sum of Squared Errors for: $\hat{y} = -1 + 2.5x$

x	y	$\hat{y}$	$y - \hat{y}$	$(y - \hat{y})^2$
1	3	1.5	1.5	2.25
2	2	4.0	-2.0	4.00
3	8	6.5	1.5	2.25
4	8	9.0	-1.0	1.00
5	11	11.5	-0.5	0.25
6	13	14.0	-1.0	1.00

$$\sum_{i=1}^{6}(y_i - \hat{y}_i)^2 = 10.75$$

(b) Computation of Sum of Squared Errors for: $\hat{y} = -0.2 + 2.2x$

x	y	$\hat{y}$	$y - \hat{y}$	$(y - \hat{y})^2$
1	3	2.0	1.0	1.00
2	2	4.2	-2.2	4.84
3	8	6.4	1.6	2.56
4	8	8.6	-0.6	0.36
5	11	10.8	0.2	0.04
6	13	13.0	0.0	0.00

$$\sum_{i=1}^{6}(y_i - \hat{y}_i)^2 = 8.80$$

EXERCISES

Exercises 1 and 2 should be done by hand.

1. **Flexible Budgeting.** A budget is an expression of management's expectations and goals concerning future revenues and costs. To increase their effectiveness, many budgets are flexible, including allowances for the effect of variation in uncontrolled variables. For example, the costs and revenues of many production plants are greatly affected by the number of units produced by the plant during the budget period, and this may be beyond a plant manager's control. Standard cost-accounting procedures can be used to adjust the direct-cost parts of the budget for the level of production, but it is often more difficult to handle overhead. In many cases, statistical methods are used to estimate the relationship between overhead (y) and the level of production (x) using historical data. As a simple example, consider the historical data for a certain plant:

   Production
   (in 10,000)
   units: 5   6   7   8   9   10   11

   Overhead
   costs
   (in $1000): 12   11.5   14   15   15.4   15.3   17.5

   a. Construct a scatterplot of y versus x.
   b. Find the least-squares regression line relating overhead costs to production.
   c. Graph the regression line on the scatterplot.

2. **Central Company.** The Central Company manufactures a certain specialty item once a month in a batch production run. The number of items produced in each run varies from month to month as demand fluctuates. The company is interested in the relationship between the size of the production run (x) and the number of hours of labor (y) required for the run. The company has collected the following data for the ten most recent runs:

   Number
   of items: 40   30   70   90   50   60   70   40   80   70

   Labor
   (hours):   83   60   138   180   97   118   140   75   159   144

a. Construct a scatterplot of y versus x.
b. Find the least-squares line relating hours of labor to number of items produced.
c. Graph the regression line on the scatterplot.

3.2 EXAMPLES OF REGRESSION AS A DESCRIPTIVE TECHNIQUE

EXAMPLE 3.2 **Pricing Communications Nodes** In recent years the growth of data communications networks has been amazing. The convenience and capabilities afforded by such networks are appealing to businesses with locations scattered throughout the United States and the world. Using networks allows centralization of an information system with access through personal computers at remote locations.

The cost of adding a new communications node at a location not currently included in the network was of concern for a major Fort Worth manufacturing company. To try to predict the price of new communications nodes, data were obtained on a sample of existing nodes. The installation cost and the number of ports available for access in each existing node were readily available information. These data are shown in Table 3.3 and a scatterplot of cost (y = COST) versus number of ports (x = NUMPORTS) is shown in Figure 3.7. (See the file COMNODE3 on the CD.)

Using a statistical package, the equation relating the price of the new communications node to the number of access ports to be included at the node was computed to be

$$\text{COST} = 16{,}594 + 650\,\text{NUMPORTS}$$

This equation could be used to help predict the cost of installing new communications nodes based on the number of access ports to be included.

TABLE 3.3 Cost and Number of Ports for Communications Nodes Example*

COST	NUMPORTS
52,388	68
51,761	52
50,221	44
36,095	32
27,500	16
57,088	56
54,475	56
33,969	28
31,309	24
23,444	24
24,269	12
53,479	52
33,543	20
33,056	24

*These data have been modified as requested by the company to provide confidentiality.

FIGURE 3.7
Scatterplot of Cost Versus Number of Ports for the Communications Nodes Example.

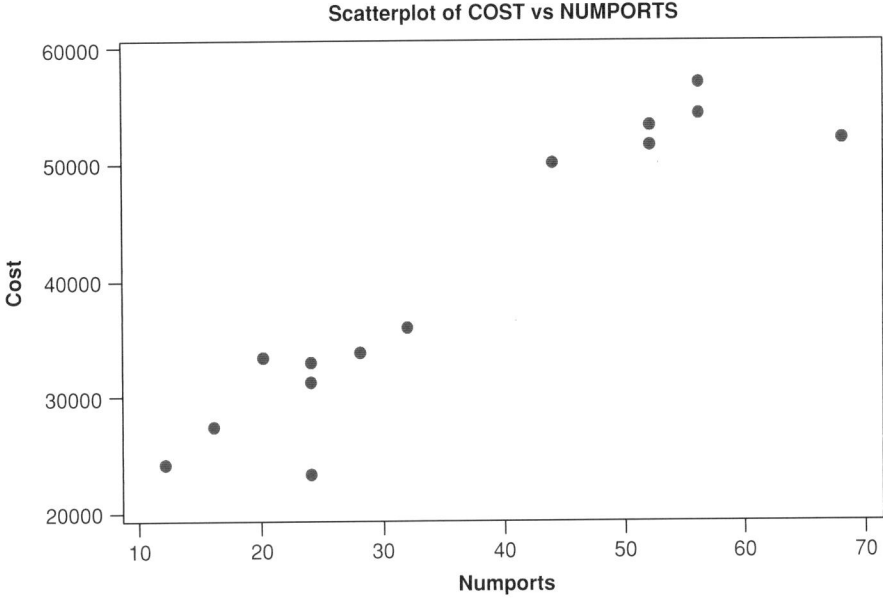

 **EXAMPLE 3.3** **Estimating Residential Real Estate Values** The Tarrant County Appraisal District must appraise properties for the entire county. The appraisal district uses data such as square footage of the individual houses as well as location, depreciation, and physical condition of an entire neighborhood to derive individual appraisal values on each house. This avoids labor-intensive inspections each year.

Regression can be used to establish the weight assigned to various factors used in assessing values. For example, Table 3.4 shows the value and size in square feet for a sample of 100 Tarrant County homes (these data are from 1990). A scatterplot of value (y = VALUE) versus size (x = SIZE) is shown in Figure 3.8. (See the file REALEST3 on the CD.)

Using a statistical package, the regression equation relating value to size can be determined as

$$\text{VALUE} = -50{,}035 + 72.8\,\text{SIZE}$$

If size were the only factor thought to be of importance in determining value, this equation could be used by the appraisal district. But obviously, other factors need to be considered. Developing an equation that includes more than one important factor (explanatory variable) is discussed in Chapter 4.

EXAMPLE 3.4 **Forecasting Housing Starts** Forecasts of various economic measures are important to the U.S. government and to various industries throughout the United States. The construction industry is concerned with the number of housing starts in a given year. Accurate forecasts can help with plans for expansion or cutbacks within the industry.

TABLE 3.4 VALUE and SIZE for Residential Real Estate Value Example

VALUE	SIZE	VALUE	SIZE	VALUE	SIZE
23,974	1442	12,001	783	21,536	1404
24,087	1426	37,650	1874	24,147	1676
16,781	1632	27,930	1242	17,867	1131
29,061	910	16,066	772	21,583	1397
37,982	972	20,411	908	15,482	888
29,433	912	23,672	1155	24,857	1448
33,624	1400	24,215	1004	17,716	1022
27,032	1087	22,020	958	224,182	2251
28,653	1139	52,863	1828	182,012	1126
33,075	1386	41,822	1146	201,597	2617
17,474	756	45,104	1368	49,683	966
33,852	1044	28,154	1392	60,647	1469
29,046	1032	20,943	1058	49,024	1322
20,715	720	17,851	1375	52,092	1509
19,461	734	16,616	648	55,645	1724
21,377	720	38,752	1313	51,919	1559
52,881	1635	44,377	1780	55,174	2133
43,889	1381	43,566	1148	48,760	1233
45,134	1372	38,950	1363	45,906	1323
47,655	1349	44,633	1262	52,013	1733
53,088	1599	12,372	840	56,612	1357
38,923	1171	12,148	840	69,197	1234
57,870	1966	19,852	839	84,416	1434
30,489	1504	20,012	852	60,962	1384
29,207	1296	20,314	852	47,359	995
44,919	1356	22,814	974	56,302	1372
48,090	1553	24,696	1135	88,285	1774
40,521	1142	23,443	1170	91,862	1903
43,403	1268	35,904	960	242,690	3581
38,112	1008	21,799	1052	296,251	4343
27,710	1120	28,212	1296	107,132	1861
27,621	960	27,553	1282	77,797	1542
22,258	920	15,826	916		
29,064	1259	18,660	864		

Table 3.5 shows data on the number of housing starts for the years 1963 to 2002. Also shown are data on home mortgage rates for new home purchases (U.S. average) for the same years. (See the file HSTARTS3 on the CD.) These data were obtained from the Department of Commerce and the Federal Housing Finance Board. A scatterplot of housing starts (y = STARTS) versus mortgage rates (x = RATES) is shown in Figure 3.9. Note that the relationship appears to be considerably "weaker" than in the other scatterplots presented in this section. Intuitively, we might expect the relationship between housing starts and mortgage rates to be a strong one. But from the data, this does not appear to be the case. Perhaps there are other variables that might be more strongly related to housing starts that could be used to provide accurate forecasts for future years. From viewing the scatterplot, mortgage rates alone do not appear to be particularly helpful.

Cross-sectional data are gathered on a number of different individual units at approximately the same point in time. Examples 3.2 and 3.3 use cross-sectional

FIGURE 3.8
Scatterplot of VALUE Versus SIZE for Residential Real Estate Value Example.

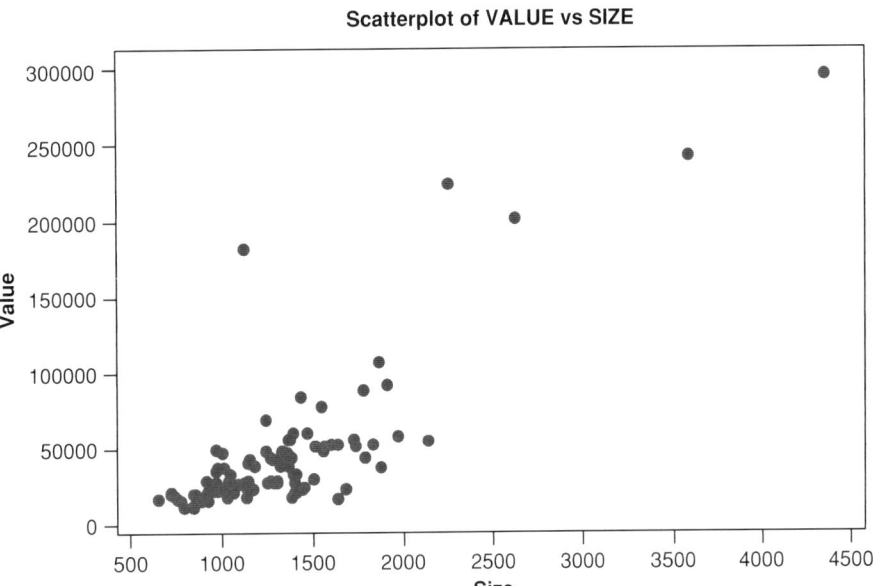

TABLE 3.5 Annual Housing Starts and Mortgage Rates for 1963–2002

Year	STARTS	RATES	Year	STARTS	RATES
1963	1,603.2	5.89	1983	1,703.0	12.57
1964	1,528.8	5.83	1984	1,749.5	12.38
1965	1,472.8	5.81	1985	1,741.8	11.55
1966	1,164.9	6.25	1986	1,805.4	10.17
1967	1,291.6	6.46	1987	1,620.5	9.31
1968	1,507.6	6.97	1988	1,488.1	9.19
1969	1,466.8	7.81	1989	1,376.1	10.13
1970	1,433.6	8.45	1990	1,192.7	10.05
1971	2,052.2	7.74	1991	1,013.9	9.32
1972	2,356.6	7.60	1992	1,199.7	8.24
1973	2,045.3	7.96	1993	1,287.6	7.20
1974	1,337.7	8.92	1994	1,457.0	7.49
1975	1,160.4	9.00	1995	1,354.1	7.87
1976	1,537.5	9.00	1996	1,476.8	7.80
1977	1,987.1	9.02	1997	1,474.0	7.71
1978	2,020.3	9.56	1998	1,616.9	7.07
1979	1,745.1	10.78	1999	1,640.9	7.04
1980	1,292.2	12.66	2000	1,568.7	7.52
1981	1,084.2	14.70	2001	1,602.7	7.00
1982	1,062.2	15.14	2002	1,704.9	6.43

data. The data examined in Example 3.4 are called *time-series data* because they are gathered over a time sequence (years in this case).

Most of the techniques discussed in this and subsequent chapters can be applied to either time-series or cross-sectional data. There are certain special techniques available when working with time-series data that may be helpful in developing forecasts. These techniques will be discussed throughout subsequent chapters where

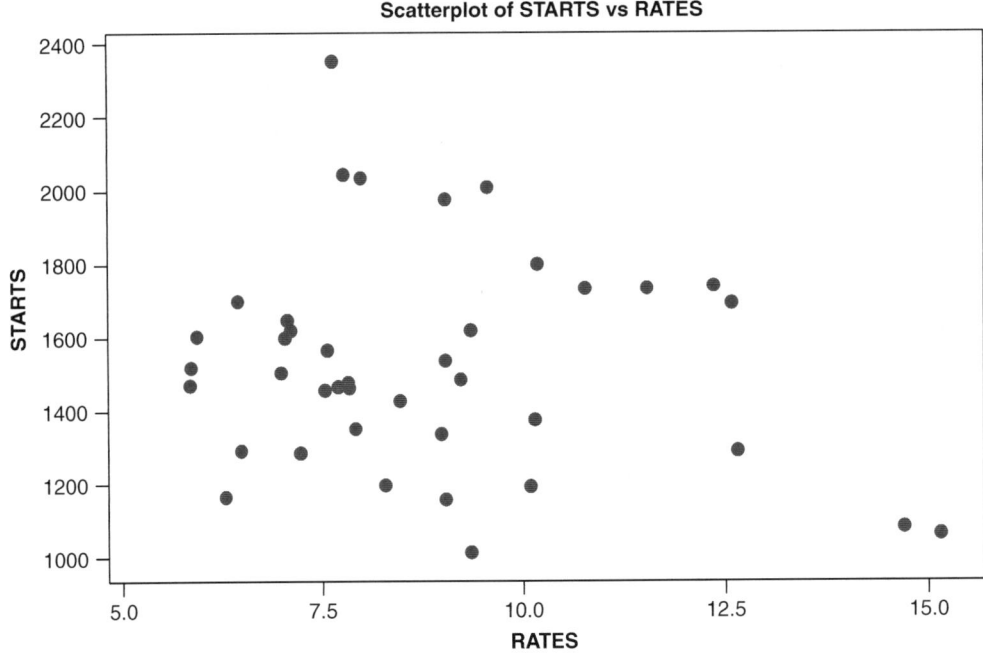

FIGURE 3.9 Scatterplot of Housing Starts Versus Mortgage Rates for Housing Starts Example.

appropriate. When special techniques apply to only one type of data, this will be mentioned.

EXERCISES

3. For each of the data sets discussed in this section, use a computer to access the data, construct a scatterplot of y versus x, and produce the regression output relating y to x.

   a. **Estimating Residential Real Estate Values:** The file name is REALEST3.

   b. **Pricing Communications Nodes:** The file name is COMNODE3.

   c. **Forecasting Housing Starts:** The file name is HSTARTS3.

3.3 INFERENCES FROM A SIMPLE REGRESSION ANALYSIS

3.3.1 ASSUMPTIONS CONCERNING THE POPULATION REGRESSION LINE

Thus far, regression analysis has been viewed as a way to describe the relationship between two variables. The regression equation obtained can be viewed in this manner simply as a descriptive statistic. However, the power of the technique of least-squares regression is not in its use as a descriptive measure for one particular sample, but in its ability to draw inferences or generalizations about the relationship for the entire population of values for the variables x and y.

To draw inferences from a sample regression equation, we must make some assumptions about how x and y are related in the population. These initial assumptions describe an "ideal" situation. Later, each of these assumptions is relaxed and we demonstrate modifications to the basic least-squares approach that provide a model that is still suitable for statistical inference.

Assume that the relationship between the variables x and y is represented by a population regression line. The equation of this line is written as

$$\mu_{y|x} = \beta_0 + \beta_1 x \tag{3.5}$$

where $\mu_{y|x}$ is the *conditional mean* of y given a value of x, β_0 is the y intercept for the population regression line, and β_1 is the slope of the population regression line. Examples of possible relationships are shown in Figure 3.10.

The use of $\mu_{y|x}$ requires some additional explanation. Suppose the y variable represents the cost of installing a new communications node as discussed in Example 3.3, and x represents the number of access ports (NUMPORTS) to be included. It is possible that these two variables are related. Now consider all possible communications nodes with 30 access ports. If the costs were known, the average value for all communications nodes with 30 access ports could be calculated. This is the conditional mean of y given $x = 30$:

$$\mu_{y|x=30}$$

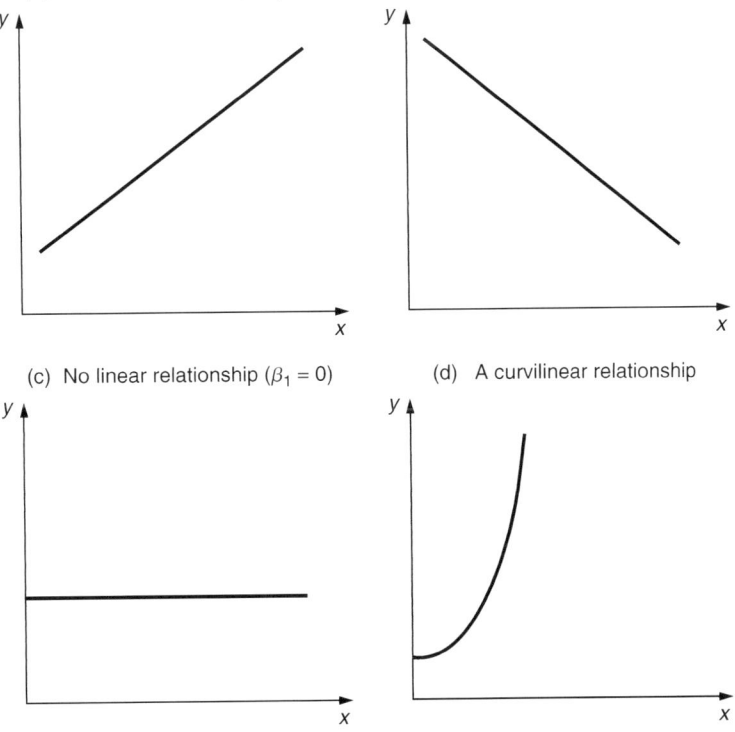

FIGURE 3.10
Examples of Possible Population Regression Lines.

Suppose this computation could be done for a number of x values and the resulting conditional means plotted as in Figure 3.11. The population regression line is the line passing through the conditional means. The relationship between y and x is linear if all conditional means lie on a straight line (or nearly so).

For a given number of ports (say, 30) costs vary; that is, not every communications node has a cost equal to the mean of y given $x = 30$. The actual cost is distributed around the point $\mu_{y|x=30}$ or around the regression line. Thus, in a sample of communications nodes with 30 ports, the costs are expected to differ from points on the population regression line (see Figure 3.12).

Because of this variation of the y values around the regression line, it is convenient to rewrite the equation representing an individual response as

$$y_i = \beta_0 + \beta_1 x_i + e_i \qquad (3.6)$$

where β_0 and β_1 have the same interpretation as they did in Equation (3.5). The term e_i represents the difference between the true cost for communications node i and the conditional mean of all costs for nodes with that number of ports:

$$e_i = \text{COST}_i - \mu_{y|x} = y_i - (\beta_0 + \beta_1 x_i)$$

The e_i are called *disturbances*. These disturbances keep the relationship from being an exact one. If the e_i were all equal to zero, then there would be an exact linear relationship between COST and NUMPORTS. The effects of all factors other than the x variable that influence COST are included in the disturbances. To allow statistical inference from a sample to the population, some assumptions about the population regression line are necessary.

FIGURE 3.11 The Population Regression Line Passes Through the Conditional Means.

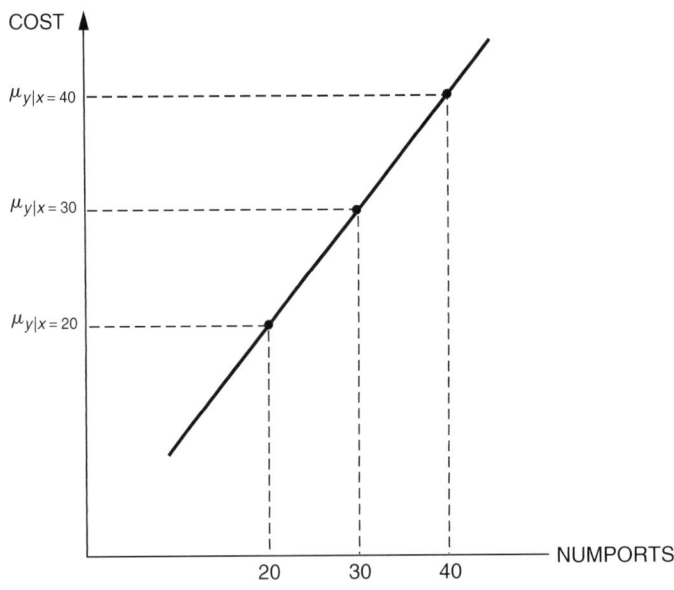

FIGURE 3.12
Distribution of Costs Around the Regression Line.

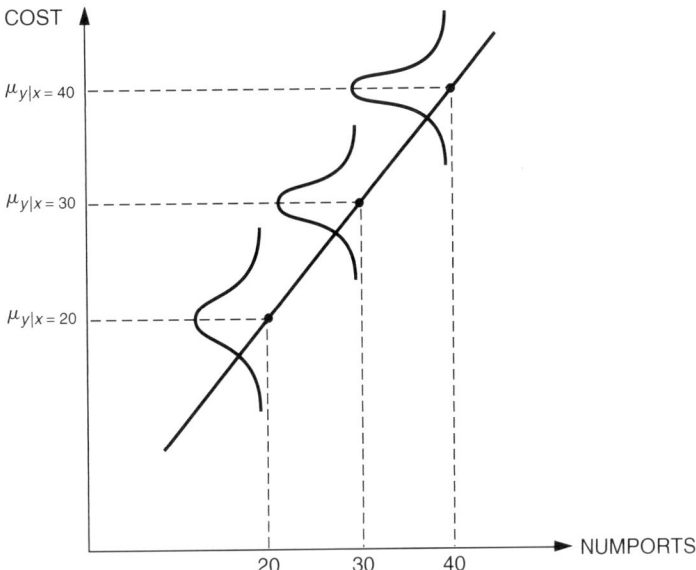

1. The expected value of the disturbances is zero: $E(e_i) = 0$. This implies that the regression line passes through the conditional means of the x variable. For our purposes, we interpret this assumption as: The population regression equation is linear in the explanatory variable.[1]

2. The variance of each e_i is equal to σ_e^2. Referring to Figure 3.12, this assumption means that each of the distributions along the regression line has the same variance regardless of the value of x.

3. The e_i are normally distributed.

4. The e_i are independent. This is an assumption that is most important when data are gathered over time. When the data are cross-sectional (that is, gathered at the same point in time for different individual units), this is typically not an assumption of concern.

These assumptions allow inferences to be made about the population regression line from a sample regression line. The first inferences considered will be those made about β_0 and β_1, the intercept and slope, respectively, of the population regression line.

The previous assumptions define an ideal case for linear regression. In Chapters 3 and 4, we examine regression procedures designed for this ideal case. In Chapter 6, we examine how violations of each of the assumptions might be detected and how corrections for these violations can be made.

---

[1] Our assumption here is that the population regression equation is linear in the x variable. In Chapter 5, we relax this assumption and find that we can fit curves by allowing equations that are not linear in the x variables. Throughout this text, however, we always assume that the equations are linear in the parameters. This means that equations such as $y = \beta_0 \beta_1^2 x + e$, for example, are not considered. These types of equations are beyond the scope of this text.

3.3.2 INFERENCES ABOUT β_0 AND β_1

The point estimates of β_0 and β_1 were previously justified by saying that b_0 and b_1 minimize the sum of squared errors for the sample. With the assumptions made concerning the random disturbances of the model, additional justification for the use of b_0 and b_1 can be made by stating certain properties these estimators possess. To fully discuss these properties, some characteristics of the sampling distributions of b_0 and b_1 must first be established.

Recall that a statistic is any value calculated from a sample. Thus, b_0 and b_1 are statistics. Because statistics are random variables, they have probability distributions called *sampling distributions*. Some characteristics of the sampling distributions of b_0 and b_1 are given here.

Sampling Distribution of b_0

1. $E(b_0) = \beta_0$ (3.7)

2. $Var(b_0) = \sigma_e^2 \left(\dfrac{1}{n} + \dfrac{\bar{x}^2}{\sum_{i=1}^{n}(x_i - \bar{x})^2} \right) = \sigma_e^2 \left(\dfrac{1}{n} + \dfrac{\bar{x}^2}{(n-1)s_x^2} \right)$ (3.8)

where $s_x^2 = \sum_{i=1}^{n}(x_i - \bar{x})^2/(n-1)$

3. The sampling distribution of b_0 is normally distributed.

Sampling Distribution of b_1

1. $E(b_1) = \beta_1$ (3.9)

2. $Var(b_1) = \dfrac{\sigma_e^2}{\sum_{i=1}^{n}(x_i - \bar{x})^2} = \dfrac{\sigma_e^2}{(n-1)s_x^2}$ (3.10)

3. The sampling distribution of b_1 is normally distributed.

The sampling distributions of both b_0 and b_1 are centered at the true parameter values of β_0 and β_1, respectively. Because the means of the sampling distributions are equal to the parameter values to be estimated, b_0 and b_1 are called unbiased estimators of β_0 and β_1 [see Figure 3.13(a)]. The variances of the sampling distributions are given in Equations (3.8) and (3.10). The standard deviations of the sampling distributions are obtained by taking the square roots of the variances:

$$\sigma_{b_0} = \sigma_e \sqrt{\dfrac{1}{n} + \dfrac{\bar{x}^2}{(n-1)s_x^2}} \tag{3.11}$$

$$\sigma_{b_1} = \sigma_e \sqrt{\dfrac{1}{(n-1)s_x^2}} \tag{3.12}$$

If the disturbances are normally distributed, the sampling distributions of b_0 and b_1 are also normally distributed regardless of the sample size.

FIGURE 3.13
Properties of b_0 and b_1 as Estimators of β_0 and β_1 (illustrated for b_1).

(a) *Unbiased Estimators:* The mean of the sampling distribution is equal to the population parameter being estimated.

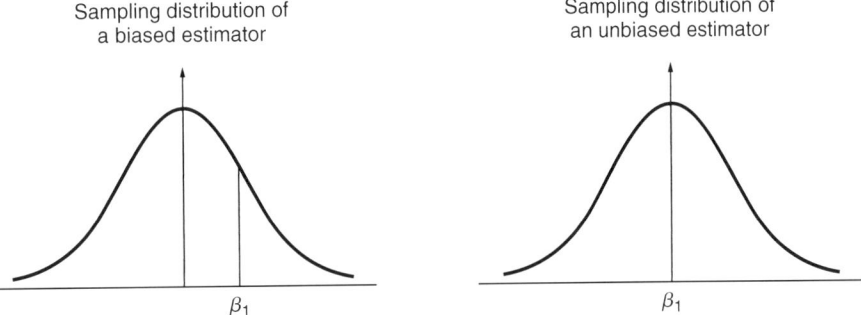

(b) *Consistent Estimators:* As n increases, the probability that the estimator will be close to the true parameter increases.

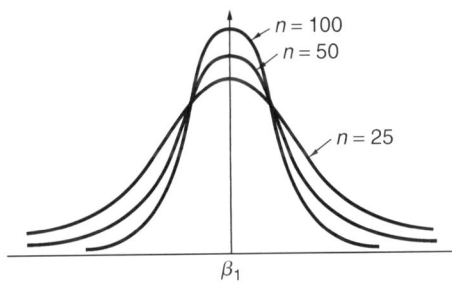

(c) *Minimum Variance Estimators:* The variance of b_1 is smaller than the variance of any other linear unbiased estimator of β_1, say b_1^*.

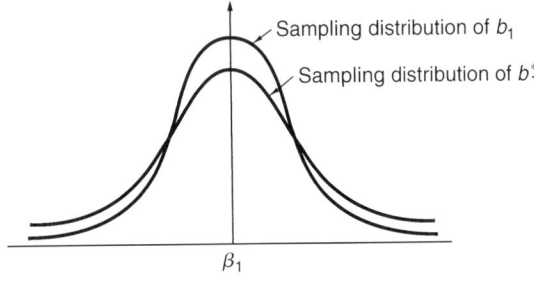

The estimators, b_0 and b_1, possess certain other properties that make them desirable as estimators of β_0 and β_1. Although these properties do not have a direct bearing on the work in this text, they are stated here for completeness. Each is a consistent estimator of its population counterpart. Using b_1 as an example, this means that as sample size increases, the probability increases that b_1 is "close" to β_1. Another way to view this property is by considering the standard deviation of the sampling distribution of b_1. As the sample size increases, the standard deviation of the sampling distribution decreases (as it did for the sample mean when viewed as an estimator of the population mean in Chapter 2). When this happens, the probability under the curve representing the sampling distribution becomes more concentrated near the center, β_1, and less concentrated in the extreme tails [see Figure 3.13(b)].

A final property of b_0 and b_1 can be illustrated by considering all other possible estimators of β_0 and β_1 that are unbiased. The standard deviations of the sampling distributions of b_0 and b_1 are smaller than those of any of the other unbiased estimators.

This minimum variance property can be restated by saying that b_0 and b_1 have smaller sampling errors than any other unbiased estimator. Of course, this says nothing about estimators that are biased. It does specify that the least-squares estimators are best within a certain class of estimators [see Figure 3.13(c)].

Using the properties of the sampling distributions of b_0 and b_1, inferences about the population parameters β_0 and β_1 can be made. This is analogous to what was done in Chapter 2 when properties of the sampling distribution of the sample mean, $\bar{y}$, were used to make inferences about the population mean, μ.

First, however, an estimate of one other unknown parameter in the regression model is needed: an estimate of σ_e^2, the variance around the regression line. This estimate of σ_e^2 is given by

$$s_e^2 = \frac{\sum_{i=1}^{n}(y_i - \hat{y}_i)^2}{n - 2} = \frac{SSE}{n - 2} = MSE$$

where $\hat{y}_i = b_0 + b_1 x_i$. The term s_e^2 represents an estimate of the variance around the regression line. Recall that in Figure 3.3, $y_i - \hat{y}_i$ represented the distance from the ith sample y value to the point on the regression line associated with the ith sample x value. The sum of the squares of these distances (*SSE*) is divided by the degrees of freedom ($n - 2$) and is used as an estimator of the variance around the regression line. *MSE* stands for mean square error. A *mean square* is any sum of squares divided by its degrees of freedom. In any regression problem, the number of degrees of freedom associated with *SSE* is n (the sample size) minus the number of regression coefficients to be estimated. In a simple regression, there are two regression coefficients, β_0 and β_1, so there are $n - 2$ degrees of freedom.

The square root of s_e^2, denoted s_e, is an estimate of the standard deviation around the regression line. It is often referred to as the *standard error of the regression.*

Now it is possible to discuss inferences about the population regression coefficients. Point estimates of β_0 and β_1 are simply the least-squares estimates b_0 and b_1. Point estimates are single numbers. Often it is more desirable to state a range of values in which the parameter is thought to lie rather than a single number. These ranges are called *confidence interval* estimates. They are less precise than point estimates because they cover a range of values rather than a single number. There is a trade-off, however, between the precision of an estimate and confidence in it.

Think back to Chapter 2 and the construction of estimates for the population mean, μ. A point estimate of μ was given by the sample mean, say, $\bar{y} = 40$. This is a very precise estimate, but the population mean will not be equal to 40. This value was obtained from one of a large number of possible samples, each of which has a different sample mean. The chances of one of these sample means being exactly equal to the population mean is so small that a 0% level of confidence is assigned to the possibility that μ is exactly 40. When constructing a confidence interval estimate of μ, however, the situation changes. For example, using the fact that the sampling distribution of $\bar{y}$ is normal (or approximately normal), 95% error bounds could be determined using a value from the standard normal table times the standard deviation of the sampling distribution. Putting these numbers into interval form, a 95% confidence interval for the population mean is

$$\left(\bar{y} - 1.96\frac{\sigma}{\sqrt{n}}, \bar{y} + 1.96\frac{\sigma}{\sqrt{n}}\right)$$

Although the interval estimate is less precise, confidence that the true population mean falls between the interval limits is considerably increased. In constructing interval estimates, a high level of confidence in the estimates is desired (90%, 95%, or 99% are commonly used), but the interval also should be precise enough to be practically useful. Telling the boss, "I'm 90% confident that our average monthly sales will be between $10,000 and $15,000" is probably better than "I'm 100% confident that our average monthly sales will be between $0 and $100,000." These same considerations in constructing interval estimates of the population mean also apply to interval estimates of population regression coefficients.

To construct a confidence interval estimate for β_1, the slope of the regression line, an estimate of the standard deviation of the sampling distribution is needed. This estimate is obtained by substituting s_e for σ_e in Equation (3.12):

$$s_{b_1} = s_e\sqrt{\frac{1}{(n-1)s_x^2}}$$

When sample sizes are small (say, $n \leq 30$), the t distribution is used to construct the interval estimate. A $(1 - \alpha)$ 100% confidence interval for β_1 is given by

$$\left(b_1 - t_{\alpha/2, n-2}\, s_{b_1},\, b_1 + t_{\alpha/2, n-2}\, s_{b_1}\right)$$

The value $t_{\alpha/2, n-2}$ is a number chosen from the t table to ensure the appropriate level of confidence. For example, for a 90% confidence interval estimate, $\alpha = 0.10$, so that $(1 - \alpha)$ 100% = 90%, and a t value with $\alpha/2 = 0.05$ probability in each tail of the t distribution with $n - 2$ degrees of freedom is used.

For β_0, the $(1 - \alpha)$ 100% confidence interval is

$$\left(b_0 - t_{\alpha/2, n-2}\, s_{b_0},\, b_0 + t_{\alpha/2, n-2}\, s_{b_0}\right)$$

where

$$s_{b_0} = s_e\sqrt{\frac{1}{n} + \frac{\bar{x}^2}{(n-1)s_x^2}}$$

The estimated standard deviations of the sampling distribution of b_0 and b_1 are sometimes referred to as *standard errors of the coefficients*. Thus, s_{b_0} is the standard error of b_0, and s_{b_1} is the standard error of b_1.

Hypothesis tests about β_0 and β_1 also can be performed. The most common hypothesis test in simple regression is

$$H_0: \beta_1 = 0$$
$$H_a: \beta_1 \neq 0$$

where H_0 represents the null hypothesis and H_a is the alternative hypothesis. The null hypothesis states that the slope of the population regression line is zero. This means that there is no linear relationship between y and x and that knowledge of x does not help explain the variation in y. The alternative hypothesis states that the slope of the population regression line is not equal to zero; that is, x and y are linearly related. Knowledge of the value of x does provide information concerning the associated value of y.

To test this hypothesis, a t statistic is used:

$$t = \frac{b_1}{s_{b_1}}$$

If the null hypothesis is true, then the t statistic has a t distribution with $n - 2$ degrees of freedom, and it should be small in absolute value. If the null hypothesis is false, then the t statistic should be large in absolute value.

To decide whether or not to reject the null hypothesis, a level of significance, α, must first be chosen. The level of significance is the probability of a Type I error; that is, α is equal to the probability of rejecting the null hypothesis if the null hypothesis is really true. Typical values for α are 0.01, 0.05, and 0.10. The decision rule for the test can be stated as:

Reject H_0 if $t > t_{\alpha/2, n-2}$ or $t < -t_{\alpha/2, n-2}$
Do not reject H_0 if $-t_{\alpha/2, n-2} \leq t \leq t_{\alpha/2, n-2}$

The value $t_{\alpha/2, n-2}$ is called a *critical value* and is chosen from the t table to ensure that the test is performed with the stated level of significance. A t value with probability $\alpha/2$ in each tail of the t distribution with $n - 2$ degrees of freedom is used.

Although the test of the null hypothesis $H_0: \beta_1 = 0$ is the most common and important test in simple regression analysis, tests of whether β_1 is equal to any value are possible. The general hypotheses can be stated as

$$H_0: \beta_1 = \beta_1^*$$
$$H_a: \beta_1 \neq \beta_1^*$$

where β_1^* is any number chosen as the hypothesized value. The decision rule is

Reject H_0 if $t > t_{\alpha/2, n-2}$ or $t < -t_{\alpha/2, n-2}$
Do not reject H_0 if $-t_{\alpha/2, n-2} \leq t \leq t_{\alpha/2, n-2}$

and the test statistic is

$$t = \frac{b_1 - \beta_1^*}{s_{b_1}}$$

When the null hypothesis is true, t should be small in absolute value because b_1 (the estimate of β_1) should be close to β_1^*, making the numerator, $b_1 - \beta_1^*$, close to zero. When the null hypothesis is false, b_1 should be different in value from the hypothesized value, β_1^*, and the difference $b_1 - \beta_1^*$ should be large in absolute value, resulting in a large absolute value for the t statistic.

Tests for hypotheses about β_0 proceed in a similar fashion. To test

$$H_0: \beta_0 = \beta_0^*$$
$$H_a: \beta_0 \ne \beta_0^*$$

the test statistic is

$$t = \frac{b_0 - \beta_0^*}{s_{b_0}}$$

where β_0^* is any hypothesized value. The decision rule for the test is

Reject H_0 if $t > t_{\alpha/2, n-2}$ or $t < -t_{\alpha/2, n-2}$
Do not reject H_0 if $-t_{\alpha/2, n-2} \le t \le t_{\alpha/2, n-2}$

Note that tests about β_0 do not provide information about the existence of a relationship between x and y. Testing whether the slope coefficient is equal to zero tells you if there is a relationship between x and y; testing whether the intercept is equal to zero does not.

An alternative method of reporting hypothesis-testing results also is available in many software packages. Consider again the hypotheses

$$H_0: \beta_1 = 0$$
$$H_a: \beta_1 \ne 0$$

The test statistic used is

$$t = \frac{b_1}{s_{b_1}}$$

Some computerized regression software routines perform this computation and then report the p value associated with the computed test statistic. The p value is the probability of obtaining a value of t at least as extreme as the actual computed value if the null hypothesis is true. Suppose a simple regression analysis is performed on a sample of 25 observations and the computed test statistic value is 2.50. The p value for the two-tailed test of $H_0: \beta_1 = 0$ is p-value $= P(t > 2.5 \text{ or } t < -2.5) = 0.02$

from the t table because there is a probability of 0.01 above 2.5 and 0.01 below -2.5 in the t distribution with $n - 2 = 23$ degrees of freedom.

The p value can be viewed as the minimum level of significance, α, that can be chosen for the test that results in rejection of the null hypothesis. Thus, a decision rule using p values can be stated as:

Reject H_0 if p value $< \alpha$

Do not reject H_0 if p value $\geq \alpha$

For a given level of significance, the same decision results regardless of which test procedure is used.

Using the previous example, we would reject H_0 if $\alpha = 0.05$, but would not reject H_0 if $\alpha = 0.01$. Reporting p values associated with hypothesis tests provides additional information beyond simply reporting whether the null hypothesis was rejected or not at a single level of significance. With a p value, readers can make their own decisions about the strength of the relationship by comparing the p value to any desired significance level.

Figure 3.14(a) shows the structure of the initial portion of the MINITAB output for a regression analysis. The estimated regression coefficients b_0 and b_1 are given along with the standard errors of the coefficients, s_{b_0} and s_{b_1}, and the t ratios for testing either H_0: $\beta_0 = 0$ or H_0: $\beta_1 = 0$. The last column reports the p values associated with the two-tailed test of the hypotheses H_0: $\beta_0 = 0$ and H_0: $\beta_1 = 0$.

Figure 3.14(b) shows the structure of the equivalent Excel output. Again, the coefficients, standard errors, t ratios, and p values are reported. Excel also prints out

(a) MINITAB

Predictor	Coef	SE Coef	T	P
Constant	b_0	s_{b_0}	b_0/s_{b_0}	p-value
x1 variable name	b_1	s_{b_1}	b_1/s_{b_1}	p-value

(b) Excel

	Coefficients	Standard Error	t stat	p value	Lower 95%	Upper 95%
Intercept	b_0	s_{b_0}	b_0/s_{b_0}	p-value	$b_0 - t_{\alpha/2, n-2} s_{b_0}$	$b_0 + t_{\alpha/2, n-2} s_{b_0}$
x1 variable name	b_1	s_{b_1}	b_1/s_{b_1}	p-value	$b_1 - t_{\alpha/2, n-2} s_{b_1}$	$b_1 + t_{\alpha/2, n-2} s_{b_1}$

(c) SAS

Parameter Estimates

Variable	DF	Parameter Estimate	Standard Error	t Value	Pr > |t|
Intercept	1	b_0	s_{b_0}	b_0/s_{b_0}	p-value
Variable name	1	b_1	s_{b_1}	b_1/s_{b_1}	p-value

FIGURE 3.14 Illustration of MINITAB, Excel and SAS Regression Outputs.

FIGURE 3.15
Hypothesis Structures and Their Associated Decision Rules for Hypotheses About β_0 and β_1.

Hypotheses		Decision Rules
$H_0: \beta_1 = \beta_1^*$	or $H_0: \beta_0 = \beta_0^*$	Reject H_0 if $t > t_{\alpha/2, n-2}$ or $t < -t_{\alpha/2, n-2}$
$H_a: \beta_1 \neq \beta_1^*$	$H_a: \beta_0 \neq \beta_0^*$	Do not reject H_0 if $-t_{\alpha/2, n-2} \leq t \leq t_{\alpha/2, n-2}$
$H_0: \beta_1 \geq \beta_1^*$	or $H_0: \beta_0 \geq \beta_0^*$	Reject H_0 if $t < -t_{\alpha, n-2}$
$H_a: \beta_1 < \beta_1^*$	$H_a: \beta_0 < \beta_0^*$	Do not reject H_0 if $t \geq -t_{\alpha, n-2}$
$H_0: \beta_1 \leq \beta_1^*$	or $H_0: \beta_0 \leq \beta_0^*$	Reject H_0 if $t > t_{\alpha, n-2}$
$H_a: \beta_1 > \beta_1^*$	$H_a: \beta_0 > \beta_0^*$	Do not reject H_0 if $t \leq t_{\alpha, n-2}$

the 95% confidence interval estimates for both β_0 and β_1. In addition, Excel allows the user to request confidence intervals for levels of confidence other than 95%.

Figure 3.14(c) shows the structure of the equivalent SAS output. The DF column indicates that 1 degree of freedom is lost for each parameter estimated. The remaining columns show the coefficients, standard errors, t ratios, and p values.

The p values for Excel, MINITAB, and SAS represent the appropriate values only for a two-tailed test. Note that the t ratios can be used for performing either one- or two-tailed tests as long as the hypothesized value is zero. The one-tailed tests simply require an adjustment in the decision rules. These are shown in Figure 3.15 for the more general hypothesis structures. Example 3.6 illustrates the use of the regression output.

EXAMPLE 3.5 Consider the data in Table 3.2 and the computations required to obtain the least-squares estimates in Figure 3.5. To compute s_{b_0} and s_{b_1}, the following quantities are needed:

$$n = 6$$

$$\bar{x}^2 = \left(\frac{\sum_{i=1}^{n} x_i}{n}\right)^2 = \left(\frac{21}{6}\right)^2 = 12.25$$

and

$$(n-1)s_x^2 = \sum_{i=1}^{n} x_i^2 - \frac{1}{n}\left(\sum_{i=1}^{n} x_i\right)^2 = 91 - \frac{1}{6}(21)^2 = 17.5$$

which can be determined using the sums obtained in Figure 3.5. In addition, the standard error of the regression must be computed:

$$s_e = \sqrt{\frac{\sum_{i=1}^{n}(y_i - \hat{y}_i)^2}{n-2}} = \sqrt{\frac{8.8}{4}} = 1.48$$

86 Simple Regression Analysis

The error sum of squares, $\sum_{i=1}^{n}(y_i - \hat{y}_i)^2$, was computed in Figure 3.6(b). Using this information,

$$s_{b_0} = s_e\sqrt{\frac{1}{n} + \frac{\bar{x}^2}{(n-1)s_x^2}} = 1.48\sqrt{\frac{1}{6} + \frac{12.25}{17.5}} = 1.38$$

and

$$s_{b_1} = s_e\sqrt{\frac{1}{(n-1)s_x^2}} = 1.48\sqrt{\frac{1}{17.5}} = 0$$

The 95% confidence interval estimates for β_0 and β_1 now can be constructed.

For β_0: $[-0.2 - 2.776(1.38), -0.2 + 2.776(1.38)]$ or $(-4.03, 3.63)$
For β_1: $[2.2 - 2.776(0.35), 2.2 + 2.776(0.35)]$ or $(1.23, 3.17)$

EXAMPLE 3.6 **Pricing Communications Nodes (continued)** Table 3.3 shows the cost of installing a sample of communications nodes for a large manufacturing firm whose headquarters is based in Fort Worth, Texas, with branches throughout the United States. The number of access ports at each of the sampled nodes is also shown. The administrator of the network wants to develop an equation that is helpful in pricing the installation of new communications nodes on the network.

Figure 3.7 shows the scatterplot of cost versus number of ports. The MINITAB, Excel, and SAS regression results are shown in Figures 3.16, 3.17, and 3.18, respectively. Use the outputs to answer the following questions:

1. What is the sample regression equation relating NUMPORTS to COST?
   Answer: COST = 16,594 + 650NUMPORTS

FIGURE 3.16
MINITAB Regression for Communications Node Example.

```
The regression equation is
COST = 16594 + 650 NUMPORTS

Predictor       Coef      SE Coef         T         P
Constant       16594         2687      6.18     0.000
NUMPORTS      650.17        66.91      9.72     0.000

S = 4306.91    R-Sq = 88.7%    R-Sq(adj) = 87.8%

Analysis of Variance

Source           DF            SS           MS         F         P
Regression        1    1751268376   1751268376     94.41     0.000
Residual Error   12     222594146     18549512

Total            13    1973862522

Unusual Observations

Obs    NUMPORTS         COST        Fit      SE Fit    Residual   St Resid
  1        68.0        52388      60805        2414       -8417     -2.36R
 10        24.0        23444      32198        1414       -8754     -2.15R

R denotes an observation with a large standardized residual.
```

3.3 Inferences from a Simple Regression Analysis

SUMMARY OUTPUT

Regression Statistics
Multiple R	0.942
R Square	0.887
Adjusted R Square	0.878
Standard Error	4306.914
Observations	14.000

ANOVA

	df	SS	MS	F	Significance F
Regression	1	1751268375.709	1751268375.709	94.410	0.000
Residual	12	222594145.791	18549512.149		
Total	13	1973862521.500			

	Coefficients	Standard Error	t Stat	P-value	Lower 95%	Upper 95%
Intercept	16593.647	2687.050	6.175	0.000	10739.068	22448.226
NUMPORTS	650.169	66.914	9.717	0.000	504.376	795.962

FIGURE 3.17 Excel Regression for Communications Node Example.

FIGURE 3.18 SAS Regression for Communications Node Example.

The REG Procedure
Model: MODEL1
Dependent Variable: COST

Analysis of Variance

Source	DF	Sum of Squares	Mean Square	F Value	Pr > F
Model	1	1751268376	1751268376	94.41	<.0001
Error	12	222594146	18549512		
Corrected Total	13	1973862522			

Root MSE	4306.91446	R-Square	0.8872
Dependent Mean	40186	Adj R-Sq	0.8778
Coeff Var	10.71758		

Parameter Estimates

Variable	DF	Parameter Estimate	Standard Error	t Value	Pr > |t|
Intercept	1	16594	2687.05000	6.18	<.0001
NUMPORTS	1	650.16917	66.91389	9.72	<.0001

2. Is there sufficient evidence to conclude that a linear relationship exists between COST and NUMPORTS?

Answer: To answer this question, the hypotheses

$$H_0: \beta_1 = 0$$

$$H_a: \beta_1 \neq 0$$

should be tested. Assuming that we want to use a 5% level of significance, here are two options to conduct the test.

(a) Using the standardized test statistic:

Decision Rule:	Reject H_0 if $t > 2.179$ or $t < -2.179$
	Do not reject H_0 if $-2.179 \leq t \leq 2.179$
Test Statistic:	$t = 9.72$ from either Figure 3.16 (MINITAB), 3.17 (Excel), or 3.18 (SAS)
Decision:	Reject H_0.
Conclusion:	There is sufficient evidence to conclude that a linear relationship between COST and NUMPORTS does exist.

(b) Using the p value:

Decision Rule:	Reject H_0 if p value < 0.05
	Do not reject H_0 if p value ≥ 0.05
Test Statistic:	p value $= 0.000$
Decision:	Reject H_0.

3. Find a 95% confidence interval estimate of β_1.

Answer: $650.17 \pm (2.179)(66.91)$ or 650.17 ± 145.80

Our point estimate of the change in cost, on average, for each additional port is $650.17. A 95% error bound for this estimate is $145.80. We can now construct a range of values that can be used as an estimate of the average change in cost for each additional port. The range is 650.17 ± 145.80 or $504.37 to $795.97. We can be 95% confident that the true average change falls within this range. The upper and lower confidence limits provide an idea of the accuracy of our estimate of the cost of each additional port. (Note that Excel prints out this interval).

4. Test whether there is a direct (positive) relationship between COST and NUMPORTS.

Answer: To answer this question, the hypotheses

$$H_0: \beta_1 \leq 0$$
$$H_a: \beta_1 > 0$$

should be tested. The null hypothesis states that the slope of the population regression line is either zero (no relationship) or negative (an inverse relationship). The alternate hypothesis states that the slope of the population regression line is positive (a direct relationship). Choosing a 5% level of significance:

Decision Rule:	Reject H_0 if $t > 1.782$
	Do not reject H_0 if $t \leq 1.782$
Test statistic:	$t = 9.72$ from either Figure 3.16 (MINITAB), 3.17 (Excel), or 3.18 (SAS)
Decision:	Reject H_0
Conclusion:	There is sufficient evidence to conclude that a direct (positive) linear relationship between COST and NUMPORTS does exist.

If the null hypothesis had not been rejected, the conclusion would be that either there is an inverse relationship or there is no relationship. The relationship is not direct, but failure to reject does not imply that no linear relationship exists as in the case of the two-tailed test. Care must be taken in interpreting the results of one-tailed tests.

Also note that the p values printed out by MINITAB, Excel, and SAS are not intended to be used for this test. The printed p values are computed for a two-tailed test.

5. A claim is made that each new access port adds at least $1000 to the installation cost of a communications node. To examine this claim, we test the hypotheses

$$H_0: \beta_1 \geq 1000$$

$$H_a: \beta_1 < 1000$$

using a 5% level of significance.

Answer:

Decision rule: Reject H_0 if $t < -1.782$
Do not reject H_0 if $t \geq -1.782$

Test statistic: $t = \dfrac{b_1 - \beta_1^*}{s_{b_1}} = \dfrac{650.17 - 1000}{66.91} = -5.23$

Decision: Reject H_0.

Conclusion: The slope of the line is not 1000 or more. In practical terms, the evidence does not support the claim that each port adds at least $1000 to the cost of installing a new communications node.

6. The intercept in this problem might be interpreted as a measure of the fixed cost of installing a new communications node. The slope would be a measure of the variable cost, that is, the cost of installation per port. If we view the intercept as a measure of the fixed cost, then it would be of interest to find a confidence interval estimate of this amount. Find a 95% confidence interval estimate of β_0.

Answer: $16{,}594 \pm 2.179(2687)$ or $16{,}594 \pm 5854.97$

7. The following test has no *practical* significance in this problem and is performed merely to illustrate the test for the intercept provided on most computer output. However, it might be of interest to test whether the intercept is equal to some value other than zero. If the intercept represents the fixed cost of installing a communications node, there might be a situation when we would want to know if the fixed cost would be more than $15,000, for example. In that case, we would construct the standardized test statistic in a similar fashion to Problem 5 in this example. To test whether the intercept is zero, as in this problem, the test statistic is provided on the computer output.

Test the hypotheses

$$H_0: \beta_0 = 0$$
$$H_a: \beta_0 \neq 0$$

using a 5% level of significance.

Answer:

(a) Using the standardized test statistic:

Decision rule:	Reject H_0 if $t > 2.179$ or $t < -2.179$
	Do not reject H_0 if $-2.179 \leq t \leq 2.179$
Test statistic:	$t = 6.18$
Decision:	Reject H_0
Conclusion:	The population intercept is not equal to zero.

(b) Using the p value:

Decision Rule:	Reject H_0 if p value < 0.05
	Do not reject H_0 if p value ≥ 0.05
Test statistic:	p value $= 0.000$
Decision:	Reject H_0.
Conclusion:	The population intercept is not equal to zero.

Note that rejection of the null hypothesis $H_0: \beta_0 = 0$ does not indicate that x and y are related. It merely makes a statement about the intercept of the population regression line.

EXERCISES

Exercises 4 and 5 should be done by hand.

4. **Flexible Budgeting (continued)** Refer to Exercise 1.

   a. Test the hypotheses $H_0: \beta_1 = 0$ versus $H_a: \beta_1 \neq 0$ at the 5% level of significance. State the decision rule, the test statistic value, and your decision.

   b. From the result in part a, are production and overhead costs linearly related?

   c. Test the hypotheses $H_0: \beta_1 = 1$ versus $H_a: \beta_1 \neq 1$ at the 5% level of significance. State the decision rule, the test statistic value, and your decision.

   d. From the result in part c, what can be concluded?

5. **Central Company (continued)** Refer to Exercise 2.

   a. Test the hypotheses $H_0: \beta_1 = 0$ versus $H_a: \beta_1 \neq 0$ at the 5% level of significance. State the decision rule, the test statistic value, and your decision.

   b. From the result in part a, are hours of labor and number of items linearly related?

   c. Test the hypotheses $H_a: \beta_0 = 0$ versus $H_a: \beta_0 \neq 0$ at the 5% level of significance. State the decision rule, the test statistic value, and your decision.

   d. From the result in part c, what can be concluded?

6. **Dividends** A random sample of 42 firms was chosen from the S&P 500 firms listed in the Spring 2003 Special Issue of *Business Week* (The Business Week Fifty Best Performers). The dividend yield (DIVYIELD) and the 2002 earnings

per share (EPS) were recorded for these 42 firms. These data are in a file named DIV3.

Using dividend yield as the dependent variable and EPS as the independent variable, a regression was run. Use the results to answer the questions. The scatterplot of DIVYIELD and EPS is shown in Figure 3.19. The regression results are shown in Figure 3.20.

a. What is the sample regression equation relating dividends to EPS?

b. Is there a linear relationship between dividend yield and EPS? Use $\alpha = 0.05$. State the hypotheses to be tested, the decision rule, the test statistic, and your decision.

c. What conclusion can be drawn from the test result?

d. Construct a 95% confidence interval estimate of β_1.

e. Construct a 95% confidence interval estimate of β_0.

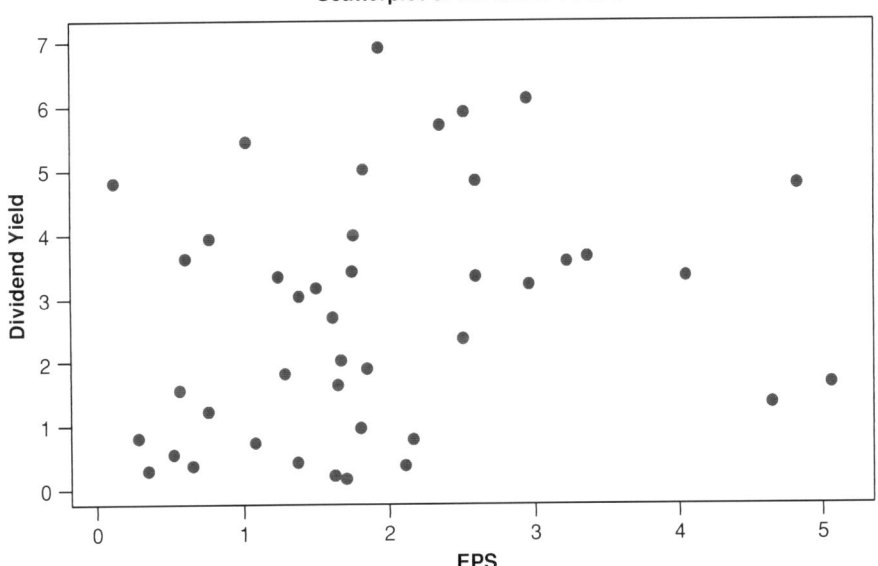

FIGURE 3.19
Scatterplot for Dividends Exercise.

FIGURE 3.20
Regression Results for Dividends Exercise.

Variable	Coefficient	Std Dev	T Stat	P Value
Intercept	2.0336	0.5405	3.76	0.001
EPS	0.3740	0.2395	1.56	0.126

Standard Error = 1.84975   R-Sq = 5.7%   R-Sq(adj) = 3.4%

Analysis of Variance

Source	DF	Sum of Squares	Mean Square	F Stat	P Value
Regression	1	8.345	8.345	2.44	0.126
Error	40	136.864	3.422		
Total	41	145.208			

7. Sales/Advertising

The vice-president of marketing for a large firm is concerned about the effect of advertising on sales of the firm's major product. To investigate the relationship between advertising and sales, data on the two variables were gathered from a random sample of 20 sales districts. These data are available in a file named SALESAD3. (Sales and advertising are both expressed in hundreds of dollars. For example, 4250 represents $425,000). The scatterplot is in Figure 3.21. The results for the regression of sales (SALES) on advertising (ADV) are shown in Figure 3.22. Using the information given, answer the following questions:

a. Is there a linear relationship between sales and advertising? Use $\alpha = 0.05$. State the hypotheses to be tested, the decision rule, the test statistic, and your decision.

b. What implications does this test result have for the firm?

c. What is the sample regression equation relating sales to advertising?

d. If ADV increases by $1000, what would be the resulting change in our prediction of SALES?

e. Construct a 90% confidence interval estimate of β_0.

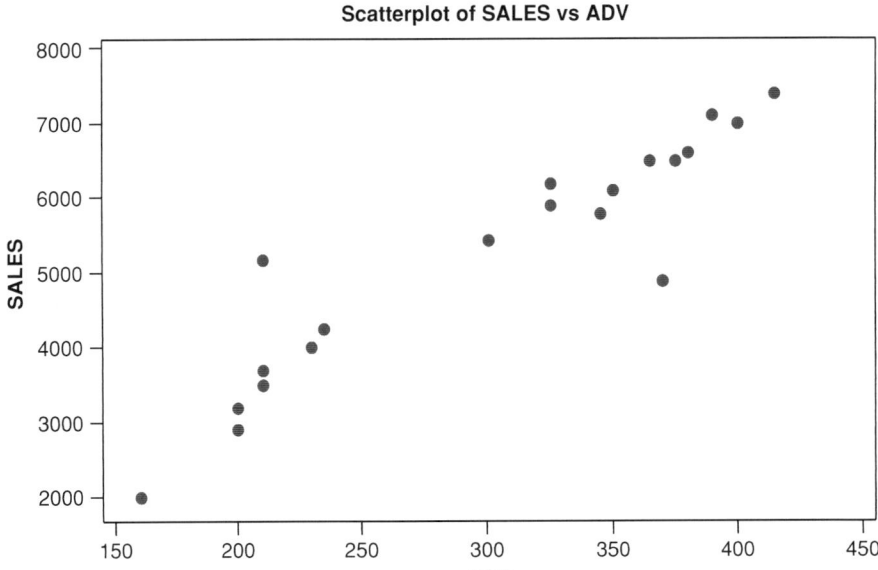

FIGURE 3.21 Scatterplot for Sales and Advertising Exercise.

FIGURE 3.22 Regression Results for Sales and Advertising Exercise.

Variable	Coefficient	Std Dev	T Stat	P Value
Intercept	-57.281	509.750	-0.11	0.912
ADV	17.570	1.642	10.70	0.000

Standard Error = 594.808   R-Sq = 86.4%   R-Sq(adj) = 85.7

Analysis of Variance

Source	DF	Sum of Squares	Mean Square	F Stat	P Value
Regression	1	40523671	40523671	114.54	0.000
Error	18	6368342	353797		
Total	19	46892014			

f. Construct a 95% confidence interval estimate of β_1.

g. Test the hypotheses

$$H_0: \beta_1 = 20$$
$$H_a: \beta_1 \neq 20$$

using a 5% level of significance. State the decision rule, the test statistic, and your decision.

h. What conclusion can be drawn from the result of the test in part g?

3.4 ASSESSING THE FIT OF THE REGRESSION LINE

3.4.1 THE ANOVA TABLE

In Example 3.6 (communications nodes), the goal might be to obtain the best possible prediction of the cost of a new node to be installed. Using a sample of n previously installed nodes, the sample mean of the n costs could be computed and used to predict the cost of any future node. But additional information on the number of access ports at each communications node might be used to obtain more accurate predictions of cost. This improvement in accuracy can be measured in terms of how much better the predictions are using the regression line instead of simply the mean of the y variable. If there is a significant improvement in prediction accuracy, then it is worthwhile to utilize the additional information.

In Figure 3.23, suppose that x^* represents the number of access ports at a particular communications node and y^* is the true cost of that node. If $\bar{y}$ is used to predict the cost of this node, then the prediction error is $y^* - \bar{y}$. But if the regression equation is used, the error is $y^* - \hat{y}$ thus reducing the error by $\hat{y} - \bar{y}$.

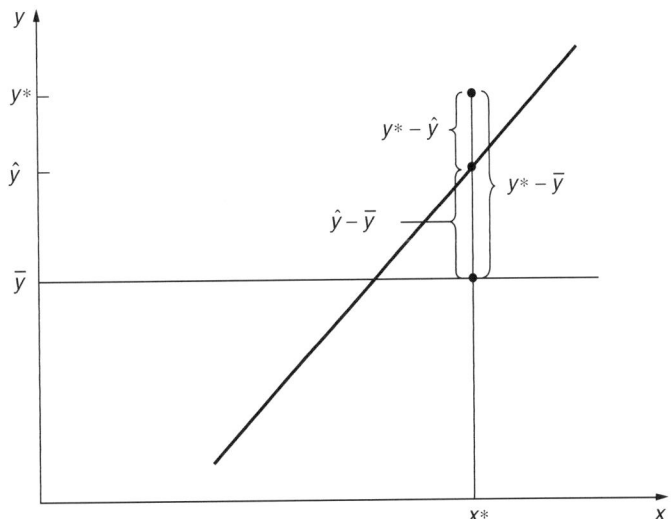

FIGURE 3.23
Partitioning the Variation in y.

94 Simple Regression Analysis

Note that

$$y^* - \bar{y} = (y^* - \hat{y}) + (\hat{y} - \bar{y}) \qquad (3.13)$$

In words, the error in using the sample mean as our predictor of y^*, $(y^* - \bar{y})$, is equal to the error produced by using the regression line $(y^* - \hat{y})$, plus the improvement over using the mean $(\hat{y} - \bar{y})$. Squaring both sides of Equation (3.13) gives

$$(y^* - \bar{y})^2 = [(y^* - \hat{y}) + (\hat{y} - \bar{y})]^2$$

or expanding the right-hand side,

$$(y^* - \bar{y})^2 = (y^* - \hat{y})^2 + 2(\hat{y} - \bar{y})(y^* - \hat{y}) + (\hat{y} - \bar{y})^2 \qquad (3.14)$$

The terms on either side of Equation (3.14) can be summed for all the individuals in the sample to obtain

$$\sum_{i=1}^{n}(y_i - \bar{y})^2 = \sum_{i=1}^{n}(y_i - \hat{y}_i)^2 + 2\sum_{i=1}^{n}(\hat{y}_i - \bar{y})(y_i - \hat{y}_i) + \sum_{i=1}^{n}(\hat{y}_i - \bar{y})^2$$

The term $\sum_{i=1}^{n}(\hat{y}_i - \bar{y})(y_i - \hat{y}_i)$ can be shown to equal zero so that

$$\sum_{i=1}^{n}(y_i - \bar{y})^2 = \sum_{i=1}^{n}(y_i - \hat{y}_i)^2 + \sum_{i=1}^{n}(\hat{y}_i - \bar{y})^2 \qquad (3.15)$$

Each term in Equation (3.15) is a sum of squares and has a special interpretation in regression analysis. The term on the left-hand side of the equality,

$$SST = \sum_{i=1}^{n}(y_i - \bar{y})^2$$

is called the *total sum of squares (SST)*. This is the numerator of the fraction used to compute the sample variance of y and is interpreted as the total variation in y.

On the right-hand side of the equation are two additional sums of squares:

$$SSE = \sum_{i=1}^{n}(y_i - \hat{y}_i)^2$$

called the *error sum of squares (SSE)*, and

$$SSR = \sum_{i=1}^{n}(\hat{y}_i - \bar{y})^2$$

called the *regression sum of squares (SSR)*.

The terms in SSE, $y_i - \hat{y}_i$, are the prediction errors for the sample—that is, the differences between the true y values and the values predicted by using the regression line. SSE is often referred to as the "unexplained" sum of squares or as a measure of "unexplained" variation in y.

The terms in SSR, $\hat{y}_i - \bar{y}$, are measures of the improvement in using the regression line rather than the sample mean to predict. SSR, which is referred to as the "explained" sum of squares or as a measure of "explained" variation in y. The "explaining" is done through the use of the variable x.

The terms *explained* and *unexplained* should be interpreted with some caution. "Explained" does not necessarily mean that x causes y. It means that the variation in y around the regression line is smaller than the variation around the sample mean, $\bar{y}$. This caveat is explored in more detail in Section 3.7.

Figure 3.24 shows how MINITAB, Excel, and SAS report certain of the quantities discussed in this section. The tables shown are called *analysis of variance* (ANOVA) tables. This name refers to the partitioning of the total variation in the dependent variable into the regression and error sums of squares. These tables are typical of ANOVA tables presented in most statistical packages.

In the MINITAB ANOVA table in Figure 3.24(a), the three sources of variation in y are denoted Regression, Residual Error, and Total. In the Excel ANOVA table in Figure 3.24(b), the name Residual rather than Residual Error is used. In Figure 3.24(c), SAS uses the terms Model and Error in place of Regression and Residual Error or Residual. Total variation is referred to as Corrected Total. The quantities SSR, SSE, and SST are found in the SS column in both MINITAB and Excel and in the Sum of Squares column in SAS. In addition, the DF columns report the degrees of freedom associated with each of these sums of squares. SSR has 1 degree of freedom, SSE has $n - 2$ degrees of freedom, and SST has $n - 1$ degrees of freedom. Just as $SSR + SSE = SST$, it is also true that the regression and residual degrees of freedom always add up to the total degrees of freedom:

$$1 + (n - 2) = n - 1$$

FIGURE 3.24
Analysis of Variance Tables from MINITAB, Excel, and SAS Regressions.

(a) MINITAB

Source	DF	SS	MS	F	P
Regression	1	SSR	MSR = SSR/1	F = MSR/MSE	p value
Residual Error	n – 2	SSE	MSE = SSE/(n – 2)		
Total	n – 1	SST			

(b) Excel

	df	SS	MS	F	Significance F
Regression	1	SSR	MSR = SSR/1	F = MSR/MSE	p value
Residual	n – 2	SSE	MSE = SSE/(n – 2)		
Total	n – 1	SST			

(c) SAS

Analysis of Variance

Source	DF	Sum of Squares	Mean Square	F Value	Pr > F
Model	1	SSR	MSR = SSR/1	F = MSR/MSE	p value
Error	n – 2	SSE	MSE = SSE/(n – 2)		
Corrected Total	n – 1	SST			

Mean squares are shown in the MS column in MINITAB and Excel and in the Mean Square column in SAS. The mean squares are sums of squares divided by their degrees of freedom:

$$MSR = \frac{SSR}{1}$$

$$MSE = \frac{SSE}{n-2}$$

MSR is referred to as the *mean square due to regression,* and *MSE* is the *mean square due to error* (or more simply, "mean square regression" and "mean square error"). Note that *MSR* always equals *SSR* because the divisor is 1. This holds true in the case of simple regression, but will differ when multiple regression is discussed in the next chapter. *MSE* was used earlier in this chapter and was denoted s_e^2 to represent an estimate of the variance around the regression line. The square root of *MSE*, denoted s_e, was called the standard deviation around the regression line or the standard error of the regression.

The remaining columns in the ANOVA tables and further uses of *MSR* and *MSE* are discussed in the next sections.

3.4.2 THE COEFFICIENT OF DETERMINATION AND THE CORRELATION COEFFICIENT

In an exact or deterministic relationship, $SSR = SST$ and $SSE = 0$. A line could be drawn that passed through every sample point. But this is not the case in most practical business situations. A measure of how well the regression line fits the data is needed. In other words, "What proportion of the total variation has been explained?" This measure is provided through a statistic called the *coefficient of determination,* denoted R^2 (this is read "R squared"):

$$R^2 = \frac{SSR}{SST}$$

R^2 is computed by dividing the explained sum of squares by the total sum of squares. The result is the proportion of variation in *y* explained by the regression. R^2 falls between 0 and 1. The closer to 1 the value of R^2 is, the better the "fit" of the regression line to the data. An alternative formula for computing R^2 is to compute the proportion of variation unexplained by the regression and subtract this proportion from 1:

$$R^2 = 1 - \frac{SSE}{SST}$$

Most computerized regression routines report R^2, and some also report another quantity called the *correlation coefficient,* which is the square root of R^2 with an appropriate sign attached:

$$R = \pm\sqrt{R^2}$$

Note that this relationship holds in the case of simple regression, but not for multiple regression (discussed in Chapter 4). The sign of R is positive if the relationship is direct ($b_1 > 0$, or an upward-sloping line) and negative if the relationship is inverse ($b_1 < 0$, or a downward-sloping line). R ranges between -1 and 1.

Note that R^2 is referred to as a measure of fit of the regression line. It is not interpreted as a measure of the predictive quality of the regression equation even though the ANOVA decomposition was motivated using the concept of improved predictions. R^2 generally overstates the regression equation's predictive ability. This fact is discussed further in Section 3.5, and alternative measures of the regression's predictive ability are suggested. R^2 will continue to be referred to as a measure of fit.

3.4.3 THE F STATISTIC

An additional measure of how well the regression line fits the data is provided by the F statistic, which tests whether the equation $\hat{y} = b_0 + b_1 x$ provides a better fit to the data than the equation $\hat{y} = \bar{y}$. The F statistic is computed as

$$F = \frac{MSR}{MSE}$$

MSR is the mean square due to the regression, or the regression sum of squares divided by its degrees of freedom. MSE is the mean square due to error, or the error sum of squares divided by its degrees of freedom, so $MSE = SSE/(n-2)$. If the regression line fits the data well (that is, if the variation around the regression line is small relative to the variation around the sample mean, $\bar{y}$), then MSR should be large relative to MSE. If the regression line does not fit well, then MSR is small relative to MSE. Thus, large values of F support the use of the regression line, whereas small values suggest that x is of little use in explaining the variation in y.

To formalize the use of the F statistic, consider again the hypotheses

$$H_0: \beta_1 = 0$$
$$H_a: \beta_1 \neq 0$$

The F statistic can be used to perform this test. The decision rule is

Reject H_0 if $F > F(\alpha; 1, n-2)$

Do not reject H_0 if $F \leq F(\alpha; 1, n-2)$

where $F(\alpha; 1, n-2)$ is a critical value chosen from the F table for level of significance α. The F statistic has degrees of freedom associated with both the numerator and denominator sums of squares used in its computation. For a simple regression, there is 1 numerator degree of freedom and $n-2$ denominator degrees of freedom. These are the degrees of freedom associated with SSR and SSE, respectively. F tables are provided in Appendix B.

If the null hypothesis $H_0: \beta_1 = 0$ is rejected, then the conclusion is that x and y are linearly related. In other words, the line $\hat{y} = b_0 + b_1 x$ provides a better fit to the data than $\hat{y} = \bar{y}$.

The hypotheses tested by the F statistic also can be tested using the t test previously discussed. The decision made using either test is exactly the same. This is because the F statistic is equal to the square of the t statistic. Also, the $F(\alpha; 1, n-2)$ critical value is the square of the $t_{\alpha/2, n-2}$ critical value:

$$F = \frac{MSR}{MSE} = t^2$$

and

$$F(\alpha; 1, n-2) = t^2_{\alpha/2, n-2}$$

Because the two test procedures yield exactly the same decision, it does not matter which is used when testing $H_0: \beta_1 = 0$ versus $H_a: \beta_1 \neq 0$ (when testing $H_0: \beta_1 \leq 0$ versus $H_a: \beta_1 > 0$, or $H_0: \beta_1 \geq 0$ versus $H_a: \beta_1 < 0$ or any tests where $\beta_1^* \neq 0$, the t test should be used). The importance of the F statistic in multiple regression, however, makes it necessary to learn how to use this test. When there are two or more explanatory variables, the F test can be used to test hypotheses that cannot be tested using the t test.

Figure 3.25 shows the additional statistics provided by MINITAB [3.25(a)], Excel [3.25(b)], and SAS [3.25(c)]. In MINITAB, these statistics appear as shown directly above the ANOVA table. In Excel, the statistics appear as shown as the first entries in the regression output (called Regression Statistics). In SAS, the statistics appear directly below the ANOVA table. Figure 3.26 shows a representation of the complete MINITAB, Excel, and SAS outputs. The bracketed sections are discussed in later chapters.

In the MINITAB output, s is the standard deviation around the regression line or standard error of the regression. This was denoted s_e in the text. Also, R-Sq is the R^2 value. The Excel output labels s_e as *Standard Error* and R^2 as *R Square*.

FIGURE 3.25 Additional Statistics Provided on MINITAB, Excel, and SAS Regression Outputs.

(a) MINITAB

$S = s_e \qquad R\text{-}Sq = R^2 \qquad R\text{-}Sq(adj) = R^2_{adj}$

(b) Excel

Regression Statistics
- Multiple R $\quad = R$
- R Square $\quad = R^2$
- Adjusted R Square $\quad = R^2_{adj}$
- Standard Error $\quad = s_e$
- Observations $\quad = n$

(c) SAS

- Root MSE $\quad s_e \qquad$ R-Square $\quad R^2$
- Dependent Mean $\quad \bar{y} \qquad$ Adj R-Sq $\quad R^2_{adj}$
- Coeff Var $\quad \frac{s_e}{\bar{y}} \times (100)$

FIGURE 3.26
Complete Regression Outputs for MINITAB, Excel, and SAS.

(a) MINITAB

The regression equation is
$y = b_0 + b_1 x$

Predictor	Coef	SE Coef	T	P
Constant	b_0	s_{b_0}	b_0/s_{b_0}	p-value
x1 variable name	b_1	s_{b_1}	b_1/s_{b_1}	p-value

$S = s_e$ R-Sq $= R^2$ $\left[\text{R-Sq(adj)} = R^2_{adj}\right]^*$

Analysis of Variance

Source	DF	SS	MS	F	P
Regression	1	SSR	$MSR = SSR/1$	$F = MSR/MSE$	p value
Residual Error	$n-2$	SSE	$MSE = SSE/(n-2)$		
Total	$n-1$	SST			

$\left[\begin{array}{l}\text{Unusual Observations}\\ \\ \text{Obs} \qquad\qquad \text{X} \qquad\qquad \text{Y} \qquad\qquad \text{Fit} \qquad\qquad \text{SE Fit} \qquad\qquad \text{Residual} \qquad \text{St Resid}\\ \text{Obs.No.} \quad \text{Value of X} \quad \text{Value of } y \quad\quad \hat{y} \qquad\qquad\quad s_m \qquad\qquad\quad y-\hat{y} \qquad\qquad \text{—}\\ \\ \text{R denotes an observation with a large standardized residual.}\\ \text{X denotes an observation whose X value gives it large influence.}\end{array}\right]^*$

(b) Excel

Regression Statistics

Multiple R	$= R$
R Square	$= R^2$
Adjusted R Square	$= R^2_{adj}$
Standard Error	$= s_e$
Observations	$= n$

ANOVA

	df	SS	MS	F	Significance F
Regression	1	SSR	$MSR = SSR/1$	$F = MSR/MSE$	p value
Residual	$n-2$	SSE	$MSE = SSE/(n-2)$		
Total	$n-1$	SST			

	Coefficients	Standard Error	t stat	P-value	Lower 95%	Upper 95%
Intercept	b_0	s_{b_0}	b_0/s_{b_0}	p value	$b_0 - t_{\alpha/2,n-2}s_{b_0}$	$b_0 + t_{\alpha/2,n-2}s_{b_0}$
x1 variable name	b_1	s_{b_1}	b_1/s_{b_1}	p value	$b_1 - t_{\alpha/2,n-2}s_{b_1}$	$b_1 + t_{\alpha/2,n-2}s_{b_1}$

FIGURE 3.26
(Continued)

(c) SAS

Analysis of Variance

Source	DF	Sum of Squares	Mean Square	F Value	Pr > F
Model	1	SSR	$MSR = SSR/1$	$F = MSR/MSE$	p value
Error	$n-2$	SSE	$MSE = SSE/(n-2)$		
Corrected Total	$n-1$	SST			

Root MSE	s_e	R-Square	R^2	
Dependent Mean	$\bar{y}$	Adj R-Sq	R^2_{adj}	
Coeff Var	$\dfrac{s_e}{\bar{y}} \times (100)$			

Parameter Estimates

Variable	DF	Parameter Estimate	Standard Error	t Value	Pr > |t|
Intercept	1	b_0	s_{b_0}	b_0/s_{b_0}	p value
Variable name	1	b_1	s_{b_1}	b_1/s_{b_1}	p value

*Bracketed sections will be discussed in later chapters.

The SAS output calls the standard error *Root MSE* and refers to R^2 as *R-Square*. In addition, Excel shows the *Multiple R*, which is the positive square root of R^2, and the number of observations *(Observations)*. All three outputs also show the *Adjusted R Square* [*R-Sq(adj)* in MINITAB and *Adj R-Sq* in SAS]. This quantity is discussed in Chapter 4. SAS also includes a number it calls the coefficient of variation *(Coeff Var)* which is computed by dividing the standard error of the regression by the mean of the *y* values and multiplying by 100. *Coeff Var* expresses the standard error of the regression in units of the mean of the dependent variable. The multiplication by 100 is merely a rescaling. *Coeff Var* is a number representing variability around the regression line and expresses this variation in unitless values. Thus, the coefficient of variation for two different regressions could be compared more readily than the standard errors because the influence of the units of the data has been removed.

EXAMPLE 3.7 To compute the R^2 for the data in Table 3.2, the quantities *SSE* and *SST* must be computed.

$$R^2 = 1 - \frac{SSE}{SST} = 1 - \frac{\sum_{i=1}^{n}(y_i - \hat{y}_i)^2}{\sum_{i=1}^{n}(y_i - \bar{y})^2}$$

SSE was computed in Figure 3.6 as $SSE = 8.8$. SST can be computed by the formula

$$\sum_{i=1}^{n} y_i^2 - \frac{1}{n}\left(\sum_{i=1}^{n} y_i\right)^2 = 431 - \frac{1}{6}(45)^2 = 93.5$$

The coefficient of determination or R^2 is

$$1 - \frac{8.8}{93.5} = 0.91$$

so 91% of the variation in y has been explained by the regression.

Note that the formula

$$R^2 = \frac{SSR}{SST}$$

could have been used here. But because SSE already had been computed and SSR had not, the alternative formula was used. If it were desired to compute SSR, this could be done by recalling that $SSR = SST - SSE = 93.5 - 8.8 = 84.7$.

The F statistic is computed as

$$F = \frac{MSR}{MSE} = \frac{SSR/1}{SSE/(n-2)} = \frac{84.7}{8.8/4} = 38.5$$

The hypotheses

$$H_0: \beta_1 = 0$$
$$H_a: \beta_1 \neq 0$$

can be tested using the F statistic. Using a 5% level of significance, the decision rule is

Reject H_0 if $F > F(0.05; 1,4) = 7.71$
Do not reject H_0 if $F \leq F(0.05; 1,4) = 7.71$

The test statistic was computed as $F = 38.5$, which results in a decision to reject H_0. In this case, the conclusion is that β_1 is not equal to zero and that the two variables x and y are linearly related.

EXAMPLE 3.8 **Pricing Communications Nodes (continued)** Refer to Example 3.6 to complete the following problems using the regression output in Figure 3.16 (MINITAB), 3.17 (Excel), or 3.18 (SAS).

1. What percentage of the variation in COST is explained by the regression?

    Answer: Using the R^2 value, 88.7% of the variation in COST has been explained by the regression.

2. Use the F test and a 5% level of significance to test the hypotheses

$$H_0: \beta_1 = 0$$
$$H_a: \beta_1 \neq 0$$

Answer:

(a) Using the standardized test statistic:

Decision rule: Reject H_0 if $F > 4.75$
                Do not reject H_0 if $F \leq 4.75$
Test Statistic: $F = 94.41$
Decision: Reject H_0
Conclusion: There is evidence to conclude that COST and NUMPORTS are linearly related.

(b) Using the p value:

Decision rule: Reject H_0 if p value < 0.05
                Do not reject H_0 if p value ≥ 0.05
Test Statistic: p value $= 0.000$
Decision: Reject H_0

EXERCISES

Exercises 8 and 9 should be done by hand.

8. **Flexible Budgeting (continued)** Refer to Exercise 1.
   a. Compute the coefficient of determination (R^2) for the regression of overhead costs on production.
   b. What percentage of the variation in overhead costs has been explained by the regression?
   c. Use the F test to test the hypotheses $H_0: \beta_1 = 0$ versus $H_a: \beta_1 \neq 0$ at the 5% level of significance. Be sure to state the decision rule, the test statistic value, and your decision.
   d. From the result in part c, are production and overhead costs linearly related?

9. **Central Company (continued)** Refer to Exercise 2.
   a. Compute the coefficient of determination (R^2) for the regression of the number of labor hours on number of items produced.
   b. What percentage of the variation in hours of labor has been explained by the regression?
   c. Use the F test to test the hypotheses $H_0: \beta_1 = 0$ versus $H_a: \beta_1 \neq 0$ at the 5% level of significance. Be sure to state the decision rule, the test statistic value, and your decision.
   d. From the result in part c, are hours of labor and number of items produced linearly related?

10. **Dividends (continued)** Use the regression results in Figure 3.20 to help answer the questions.
    a. What percentage of the variation in dividend yield has been explained by the regression?
    b. Use the F test to test the hypotheses $H_0: \beta_1 = 0$ versus $H_a: \beta_1 \neq 0$ at the 5% level of significance. Be sure to state the decision rule, the test statistic value, and your decision.

11. **Sales/Advertising (continued)** Use the regression results in Figure 3.22 to help answer the questions.
    a. What percentage of the variation in sales has been explained by the regression?
    b. Use the F test to test the hypotheses $H_0: \beta_1 = 0$ versus $H_a: \beta_1 \neq 0$ at the 5% level of significance. Be sure to state the decision rule, the test statistic value, and your decision.

3.5 PREDICTION OR FORECASTING WITH A SIMPLE LINEAR REGRESSION EQUATION

One of the possible goals for fitting a regression line to data is to be able to use the regression equation to predict or forecast values of the dependent variable y. Given that a value of x has been observed, what is the best prediction of the response value, y? To discuss how to best predict y and how to make inferences using predictions based on a random sample, two cases that may arise in practice are considered.

3.5.1 ESTIMATING THE CONDITIONAL MEAN OF y GIVEN x

In Example 3.6, suppose that the network administrator wants to consider all possible nodes with 40 communications ports. The question to be answered is, "What will the cost be, on average, for nodes with 40 ports?"

The average cost of all nodes with 40 communications ports is to be estimated. Thus, an estimate of the conditional mean cost given $x = 40$, or an estimate of $\mu_{y|x=40}$, is required (see Figure 3.27). If a relationship between y and x does exist, the best estimate of this point on the population regression line is given by

$$\hat{y}_m = b_0 + b_1 x_m$$

where b_0 and b_1 are the least-squares estimates of β_0 and β_1, x_m is the number of ports for which an estimate is desired, and $\hat{y}_m$ is the estimate of the conditional mean cost. In this case, $\hat{y}_m$ represents the estimate of the point on the regression line corresponding to (or conditional on) $x = x_m$. Thus, it is the estimate of a population mean. The variance of this estimate can be shown to equal

$$\sigma_m^2 = \sigma_e^2 \left(\frac{1}{n} + \frac{(x_m - \bar{x})^2}{(n-1)s_x^2} \right) \tag{3.16}$$

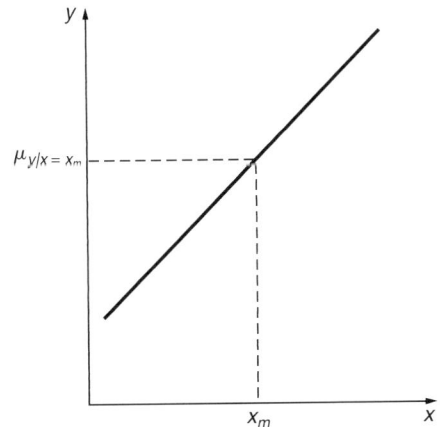

FIGURE 3.27
Estimating a Conditional Mean $\mu_{y|x=x_m}$.

Because σ_e^2 is unknown, s_e^2 is substituted to obtain an estimate of σ_m^2:

$$s_m^2 = s_e^2 \left(\frac{1}{n} + \frac{(x_m - \bar{x})^2}{(n-1)s_x^2} \right) \qquad (3.17)$$

The standard deviation or standard error of the estimate, s_m, is simply the square root of s_m^2.

The standard error of the estimate of the point on the regression line is affected by the distance of the value of x_m from the sample mean $\bar{x}$. The closer the value of x_m to the mean of the sample x values, the closer the term $(x_m - \bar{x})^2$ is to zero. If $(x_m = \bar{x})$, the term $(x_m - \bar{x})^2/[(n-1)s_x^2]$ equals zero, and the standard error equals $s_e/\sqrt{n}$. Thus, the closer the value x_m is to the sample mean, the smaller the standard error is or the more accurate the estimate is expected to be.

The reason is illustrated in Figure 3.28. Because all least-squares lines pass through the point $(\bar{x}, \bar{y})$, two least-squares lines have been drawn intersecting at the point $(\bar{x}, \bar{y})$. The two lines could represent least-squares lines fitted to two independent random samples taken from the same population. It is assumed that the two samples have means $\bar{x}$ and $\bar{y}$. Even though both lines pass through a common point, the slopes of the two estimated lines could be quite different, as illustrated. Thus, there is more certainty as to the value of a point on the regression line near the value $\bar{x}$ than at the extreme values of x. When estimating a point on the regression line for an extreme value of x, the greater uncertainty is reflected through a larger standard error.

It was previously stated that when the term $(x_m - \bar{x})^2/[(n-1)s_x^2]$ is zero, the standard error is $s_m = s_e/\sqrt{n}$. This happens when $(x_m = \bar{x})$. The quantity $s_e/\sqrt{n}$ is very much like the standard error associated with the sample mean $\bar{y}$, when it is used to estimate the (unconditional) population mean of the y values. A population mean is being estimated in the case of regression, so this is to be expected. The difference

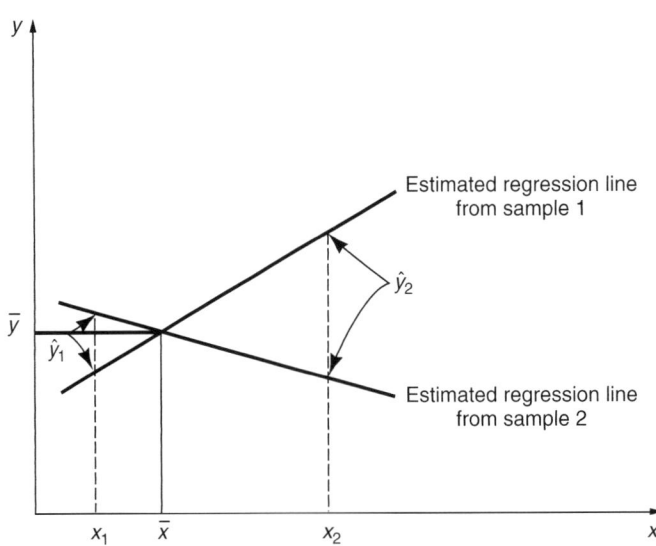

FIGURE 3.28 Effect on $\hat{y}_m$ of Variation in b_1 from Sample to Sample.

is that the mean in regression, $\mu_{y|x}$, is conditional on the value of x, rather than being unconditional.

Confidence intervals can be constructed for estimates of a conditional mean using

$$(\hat{y}_m - t_{\alpha/2, n-2} s_m, \hat{y}_m + t_{\alpha/2, n-2} s_m) \quad (3.18)$$

where $\hat{y}_m$ is the point estimate and $t_{\alpha/2, n-2}$ is chosen in the usual fashion from the t distribution with $n - 2$ degrees of freedom.

Hypothesis tests also can be conducted. To test

$$H_0: \mu_{y|x_m} = \mu^*_{y|x_m}$$
$$H_a: \mu_{y|x_m} \neq \mu^*_{y|x_m}$$

where $\mu^*_{y|x_m}$ is a hypothesized value for the point on the population regression line, the decision rule is

Reject H_0 if $t > t_{\alpha/2, n-2}$ or $t < -t_{\alpha/2, n-2}$

Do not reject H_0 if $-t_{\alpha/2, n-2} \leq t \leq t_{\alpha/2, n-2}$

The test statistic, t, is computed as

$$t = \frac{\hat{y}_m - \mu^*_{y|x_m}}{s_m}$$

and it has a t distribution with $n - 2$ degrees of freedom when H_0 is true.

One-tailed tests also can be performed provided the usual modifications are made in constructing the decision rule.

3.5.2 PREDICTING AN INDIVIDUAL VALUE OF y GIVEN x

Now suppose the network administrator is interested in a single communications node in a plant in Kansas City, Missouri, which will have 40 access ports. Predict the cost of installation for this particular node. With $x_p = 40$ access ports, the best prediction of the cost of this node is

$$\hat{y}_p = b_0 + b_1 x_p$$

which is exactly the same number that would be used to estimate the average cost for all nodes with 40 access ports. One can do no better in predicting cost for an individual node than to use the estimate of average cost for all nodes with the same number of access ports. This is because there is no additional information used in the regression that distinguishes this one node from all the others (see Figure 3.29).

The prediction for an individual value, however, is not as accurate as the estimate of a population mean for all individuals in a certain category. The variance of the prediction for an individual is

$$\sigma_p^2 = \sigma_e^2 \left(1 + \frac{1}{n} + \frac{(x_m - \bar{x})^2}{(n-1)s_x^2} \right) \quad (3.19)$$

FIGURE 3.29
Predicting an Individual y Value.

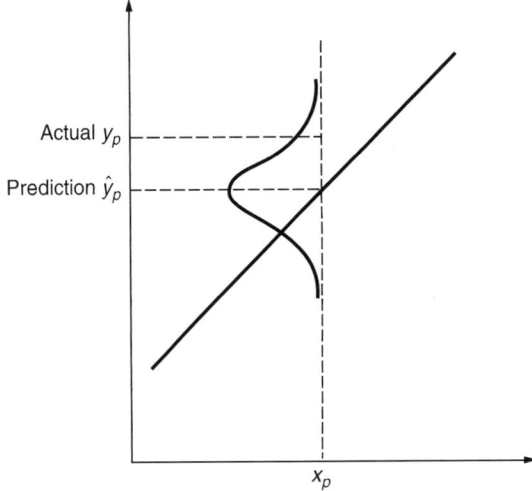

which can be estimated by replacing σ_e^2 by s_e^2:

$$s_p^2 = s_e^2\left(1 + \frac{1}{n} + \frac{(x_m - \bar{x})^2}{(n-1)s_x^2}\right) \qquad (3.20)$$

To compare s_p^2 to the variance of the estimate of a conditional mean, write the prediction variance as

$$s_p^2 = s_e^2 + s_m^2$$

The variance of the prediction for an individual value is equal to the variance from estimating the point on the regression line for $x = x_p$, s_m^2, plus the estimate of the variation of the individual y values around the regression line, s_e^2. Even if the exact position of $\mu_{y|x_p}$ were known, y_p still would not be known. The individual y values are distributed around $\mu_{y|x_p}$ with standard deviation σ_e. Because $\mu_{y|x_p}$ is actually unknown, there is uncertainty associated with the estimation of this value (reflected in s_m or s_m^2) plus the uncertainty in predicting an individual value (reflected in s_e or s_e^2).

Interval estimation of y_p is accomplished by constructing prediction intervals. The term *prediction interval* is used rather than confidence interval because a population parameter is not being estimated in this case; instead, the response or performance of a single individual in the population is being predicted.

A $(1 - \alpha)100\%$ prediction interval for y_p is

$$(\hat{y}_p - t_{\alpha/2,n-2}s_p, \hat{y}_p + t_{\alpha/2,n-2}s_p) \qquad (3.21)$$

where $\hat{y}_p$ is the predicted value and $t_{\alpha/2,n-2}$ is chosen from the t distribution with $n - 2$ degrees of freedom.

3.5 Prediction or Forecasting with a Simple Linear Regression Equation

FIGURE 3.30
Confidence Interval Limits Versus Prediction Interval Limits.

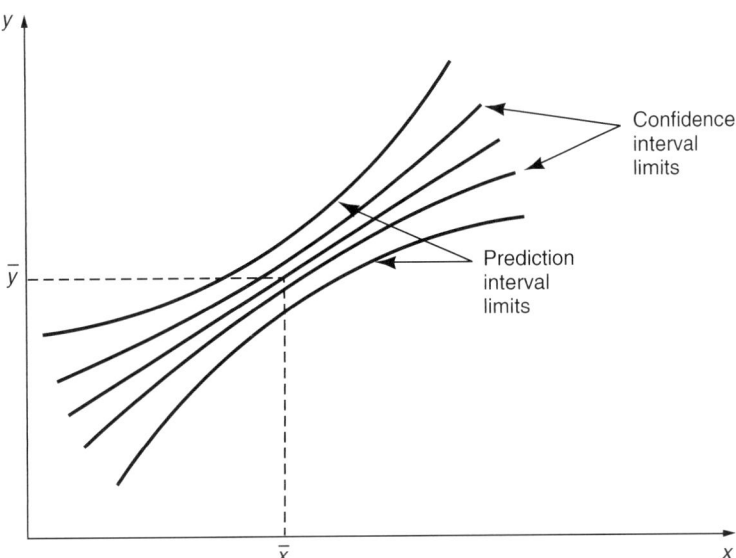

Figure 3.30 illustrates the difference between the confidence interval estimate of $\mu_{y|x}$ and the prediction interval for an individual. Both the confidence interval in Equation (3.18) and the prediction interval in Equation (3.21) are narrower (more precise) near $x = \bar{x}$ and wider at the extreme values of x. The prediction interval is always wider than the confidence interval because of the added uncertainty involved in predicting an individual response.

3.5.3 ASSESSING QUALITY OF PREDICTION

As noted in Section 3.4, the R^2 of the regression is a measure of the fit of the regression to the sample data. It is not generally considered an adequate measure of the regression equation's ability to estimate $\mu_{y|x}$ or to predict new responses. The standard R^2 overestimates the quality of future (or out-of-sample) predictions.

Two possible means of assessing prediction quality are presented in this section. The first is called *data splitting*. In this method, the data set is partitioned into two groups. One group of n_1 data points is used to estimate or fit possible equations used for forecasting. The second group of n_2 data points, called a *holdout sample* or *validation sample*, is used to assess predictive ability of the models estimated using the fitting sample. Any models that are considered possible candidates are estimated using the fitting sample. Predictions, $\hat{y}_i$, are then computed for these models using the explanatory variable values in the validation sample. For each candidate model, prediction errors, $y_i - \hat{y}_i$, are computed for all n_2 observations in the validation sample. A measure of forecast accuracy based on the forecast errors can then be computed. For example, the *mean square deviation (MSD)*

$$MSD = \frac{\sum_{i=1}^{n_2} (y_i - \hat{y}_i)^2}{n_2}$$

the *mean absolute deviation (MSD)*

$$MAD = \frac{\sum_{i=1}^{n_2} |y_i - \hat{y}_i|^2}{n_2}$$

and the *mean absolute percentage error (MAPE)*

$$MAPE = \frac{\sum_{i=1}^{n_2} \left(\frac{|y_i - \hat{y}_i|}{|y_i|}\right)}{n_2}$$

are three commonly computed measures.

Models with smaller *MSD*, *MAD*, or *MAPE* are better for prediction purposes. The advantage of using this approach is that the models are tested on data that were not used in the fitting or model estimation process. This provides an independent assessment of the models' predictive ability.

After an appropriate model has been chosen, the entire data set can be used to estimate the model parameters. This model is then used to produce future predictions.

A second means of assessing prediction quality is to use the *PRESS* statistic. *PRESS* stands for prediction sum of squares and is defined as

$$PRESS = \sum_{i=1}^{n} (y_i - \hat{y}_{i,-1})^2$$

In this formula, $\hat{y}_{i,-1}$ represents the prediction obtained from a model estimated with one of the sample observations deleted. If there are n observations in the sample, there are n different predictions, $\hat{y}_{i,-1}$. The prediction $\hat{y}_{i,-1}$ is obtained by evaluating the regression equation at x_i, but the data point (x_i, y_i) is not used in obtaining the estimated regression equation. Thus, as with the use of a validation sample, predictions are obtained from data that are not used to fit the model.

The quantities $y_i - \hat{y}_{i,-1}$ often are called *PRESS* residuals because they are similar to the actual regression residuals, $y_i - \hat{y}_i$. The prediction sum of squares also is similar to the error sum of squares, *SSE*. This suggests construction of an R^2-like statistic that might be called the prediction R^2:

$$R^2_{PRED} = 1 - \frac{PRESS}{SST}$$

Larger values of R^2_{PRED} (or smaller values of *PRESS*) suggest models of greater predictive ability.

EXAMPLE 3.9   **1.** Refer to the data in Table 3.2. Find an estimate of the conditional mean of y when $x = 6$ and find the standard deviation of this estimate.

Answer: Using the least-squares regression equation $\hat{y} = -0.2 + 2.2x$, an estimate of the point on the regression line when $x = 6$ is $\hat{y} = -0.2 + 2.2(6) = 13$. The standard deviation of the estimate of the point on the regression line is

$$s_m = s_e \sqrt{\frac{1}{n} + \frac{(x_m - \bar{x})^2}{(n-1)s_x^2}} = 1.48 \sqrt{\frac{1}{6} + \frac{(6 - 3.5)^2}{17.5}} = 1.07$$

[For computation of s_e, $\bar{x}$, and $(n - 1)s_x^2$, see Example 3.5.]

2. Find a prediction of the y value when $x = 6$ and find the standard deviation of the prediction.

Answer: Using the least-squares regression equation $\hat{y} = -0.2 + 2.2x$, the prediction of y when $x = 6$ is $\hat{y} = -0.2 + 2.2(6) = 13$. The standard deviation of the prediction is

$$s_p = s_e \sqrt{1 + \frac{1}{n} + \frac{(x_p - \bar{x})^2}{(n-1)s_x^2}} = 1.48 \sqrt{1 + \frac{1}{6} + \frac{(6 - 3.5)^2}{17.5}} = 1.83$$

3. Find a 95% confidence interval and 95% prediction interval when $x = 6$.

Answer: The 95% confidence and prediction intervals are, respectively,

$$[13 - 2.776(1.07), 13 + 2.776(1.07)] \text{ or } (10.03, 15.97)$$

and

$$[13 - 2.776(1.83), 13 + 2.776(1.83)] \text{ or } (7.92, 18.08)$$

EXAMPLE 3.10 **Pricing Communication Nodes (continued)**

1. On average, how much do we expect communication nodes to cost if there are to be 40 access ports?

Answer: Figure 3.31 shows the MINITAB regression output using the option to request a prediction with 40 as the value of the x variable. Figure 3.32 shows similar output for SAS. Both outputs show the predicted value (*Fit* in MINITAB and *Predicted Value* in SAS), the standard error of the estimate of the point on the regression line, s_m (*SE Fit* in MINITAB and *Std Error Mean Predict* in SAS), a 95% confidence interval for the estimate of the conditional mean (*95% CI* in MINITAB and *95% CL Mean* in SAS), and a 95% prediction interval for each prediction (*95% PI* in MINITAB and *95% CL Predict* in SAS). Note that the SAS output provides this information for the observations used to estimate the equation as well as the observations for which predictions are desired.

A point estimate of cost for all nodes with 40 access ports is $42,600. A 95% confidence interval estimate is given by ($40,035, $45,166). Thus, we can say with 95% confidence that the average cost of all nodes with 40 access ports is expected to be between $40,035 and $45,166.

110 Simple Regression Analysis

FIGURE 3.31
MINITAB Regression Prediction Output for Example 3.10.

```
Predicted Values for New Observations

New
Obs      Fit     SE Fit      95% CI           95% PI
 1     42600      1178    (40035, 45166)   (32872, 52329)


Values of Predictors for New Observations

New
Obs    NUMPORTS
 1       40.0
```

```
                              Output Statistics

       Dep Var   Predicted    Std Error
Obs     COST      Value     Mean Predict   95% CL Mean      95% CL Predict       Residual

  1     52388     60805        2414      55545    66065    50047    71563        -8417
  2     51761     50402        1559      47006    53799    40423    60382         1359
  3     50221     45201        1262      42452    47950    35423    54979         5020
  4     36095     37399        1186      34814    39984    27666    47132        -1304
  5     27500     26996        1780      23119    30874    16843    37150       503.6461
  6     57088     53003        1751      49189    56818    42873    63133         4085
  7     54475     53003        1751      49189    56818    42873    63133         1472
  8     33969     34798        1278      32015    37582    25010    44587      -829.3840
  9     31309     32198        1414      29116    35280    22321    42075      -888.7073
 10     23444     32198        1414      29116    35280    22321    42075        -8754
 11     24269     24396        1991      20057    28735    14057    34734      -126.6772
 12     53479     50402        1559      47006    53799    40423    60382         3077
 13     33543     29597        1585      26143    33051    19598    39596         3946
 14     33056     32198        1414      29116    35280    22321    42075       858.2927
 15       .       42600        1178      40035    45166    32872    52329           .

               Sum of Residuals                          0
               Sum of Squared Residuals          222594146
               Predicted Residual SS (PRESS)     345066019
```

FIGURE 3.32 SAS Regression Prediction Output for Example 3.10.

2. For an individual node with 40 access ports, find a prediction of cost.

Answer: The point prediction is again $42,600. A 95% prediction interval for the individual node is ($32,872, $52,329). Note that the prediction interval is considerably wider than the confidence interval, reflecting the additional uncertainty of predicting for an individual as opposed to estimating an average.

EXERCISES

Exercises 12 and 13 should be done by hand.

12. **Flexible Budgeting (continued)**, Refer to Exercise 1.
    a. Find a point estimate of the overhead costs, on average, for production runs of 80,000 units.
    b. Find a 95% confidence interval estimate of overhead costs, on average, for production runs of 80,000 units.
    c. Find a point prediction of the overhead costs for a single production run of 80,000 units.
    d. Find a 95% prediction interval for overhead costs for a single production run of 80,000 units.
    e. State why the prediction interval is wider than the confidence interval.

13. **Central Company (continued)**, Refer to Exercise 2.
    a. Find a point estimate for the number of hours of labor required, on average, when 60 units are produced.
    b. Find a 95% confidence interval estimate of hours of labor required, on average, when 60 units are produced.
    c. Find a point prediction of the number of hours of labor required for one run producing 60 units.
    d. Find a 95% prediction interval for the number of hours of labor required for one run producing 60 units.

14. **Dividends (continued)**, Consider the dividend-yield problem in Exercise 6 and the associated computer results in Figure 3.20. An analyst wants an estimate of dividend yield for all firms with earnings per share of $3. Does the equation developed provide a more accurate estimate than simply using the sample mean dividend yield for all 42 firms examined? State why or why not.

15. **Sales/Advertising (continued)**, Use the results in Figure 3.33 to help solve these problems. These results were obtained requesting a prediction with x = 200, 250, 300 and 350, respectively (representing $20,000, $25,000, $30,000 and $35,000).
    a. Find an estimate of average sales for all sales districts with advertising expenditures of $25,000. Find a point estimate and a 95% confidence interval estimate.
    b. Predict sales for individual districts having advertising expenditures of $20,000, $25,000, $30,000, and $35,000. Find point predictions as well as 95% prediction intervals.

FIGURE 3.33
Prediction Output for Exercise 15.

Predicted Value	SE Fit	95% Confidence Int.	95% Prediction Int.
3457	211	(3013, 3900)	(2131, 4783)
4335	156	(4007, 4663)	(3043, 5627)
5214	133	(4934, 5493)	(3933, 6494)
6092	157	(5763, 6421)	(4800, 7384)

3.6 FITTING A LINEAR TREND TO TIME-SERIES DATA

Data gathered on individuals at the same point in time are called *cross-sectional data*. *Time-series data* are data gathered on a single individual (person, firm, and so on) over a sequence of time periods, which may be days, weeks, months, quarters, years, or virtually any other measure of time. In a given problem, however, it is assumed that the data are gathered over only one interval of time (daily and weekly data are not combined, for example).

When dealing with time-series data, the primary goal often is to be able to produce forecasts of the dependent variable for future time periods. Two separate approaches to this problem can be identified. On the one hand, a researcher may identify variables that are related to the dependent variable in a causal manner and use these in developing a *causal regression model.* For example, when trying to forecast sales for a particular product, causal variables might include advertising expenditures and competitors' market share. Changes in these variables are felt to produce or cause changes in sales. Thus, the term *causal regression model* is used.

The researcher may, on the other hand, identify patterns of movement in past values of the dependent variable and extrapolate these patterns into the future using an *extrapolative regression model.* An extrapolative model uses explanatory variables, although they are not related to the dependent variable in a causal manner. They simply describe the past movements of the dependent variable so that these movements can be extended into future time periods. Variables that represent trend and seasonal components often are included in extrapolative models.

Both causal and extrapolative models have their benefits and drawbacks. Causal models require the identification of variables that are related to the dependent variable in a causal manner. Then data must be gathered on these explanatory variables to use the model. Furthermore, when forecasting for future time periods, the values of the explanatory variables in these periods must be known. In extrapolative models, only past values of the dependent variable are required, and thus variable selection and data gathering are simpler processes.

Whether a causal or extrapolative model performs better is determined to some extent by how far into the future the forecast refers. Forecasts often are classified as short-term (0 to 3 months), medium-term (3 months to 2 years), or long-term (2 years and longer). (Note that these cutoffs to classify forecast horizons are somewhat arbitrary and may not apply to all situations). Extrapolative models tend to perform well in the short term, but they can be reasonably accurate for medium-term forecasts. Often, extrapolative models can produce more accurate forecasts than causal models in the short term. Causal models often outperform extrapolative models when long-term forecasts are desired. In addition to being just as effective as causal models in the short term and often in the medium term, extrapolative models tend to be easier to develop and use.

The success of extrapolative models depends on the stability of the behavior of the time series. If past time-series patterns are expected to continue into the future, then an extrapolative model should be relatively successful in making accurate forecasts. If these past patterns are altered for some reason, and future movements differ in general from past movements, then extrapolative models do not perform well. Thus, an assumption when using an extrapolative model for forecasting is that past patterns of data movement are reflective of future patterns.

Causal models can respond, to some extent, to more drastic changes in patterns. Changes in the explanatory variables caused by changes in economic or market conditions should produce relatively accurate forecasts of changes in the dependent variable. Of course, this assumes that the changes in the explanatory variables will be known for future time periods. In addition, if the model itself changes (that is, if the way the variables are related changes in the future), the causal model is not capable of making accurate forecasts.

3.6 Fitting a Linear Trend to Time-Series Data

In this section, the use of a *linear trend model* for time-series data is examined. The linear trend model is a type of extrapolative model that may be useful in certain time-series applications. In subsequent chapters, other techniques useful in building extrapolative time-series models are examined.

A *trend* in time-series data is a tendency for the series to move upward or downward over many time periods. This movement may follow a straight line or a curvilinear pattern. Regression analysis can be used to model certain trends and to extrapolate these trends into future time periods.

The simplest form of a trend over time is a linear trend. The linear trend model can be written

$$y_i = \beta_0 + \beta_1 t + e_i$$

The explanatory variable simply indicates the time period ($x_i = t$). Usually, the variable t is constructed by using the integers, 1, 2, 3, . . . to indicate the time period.

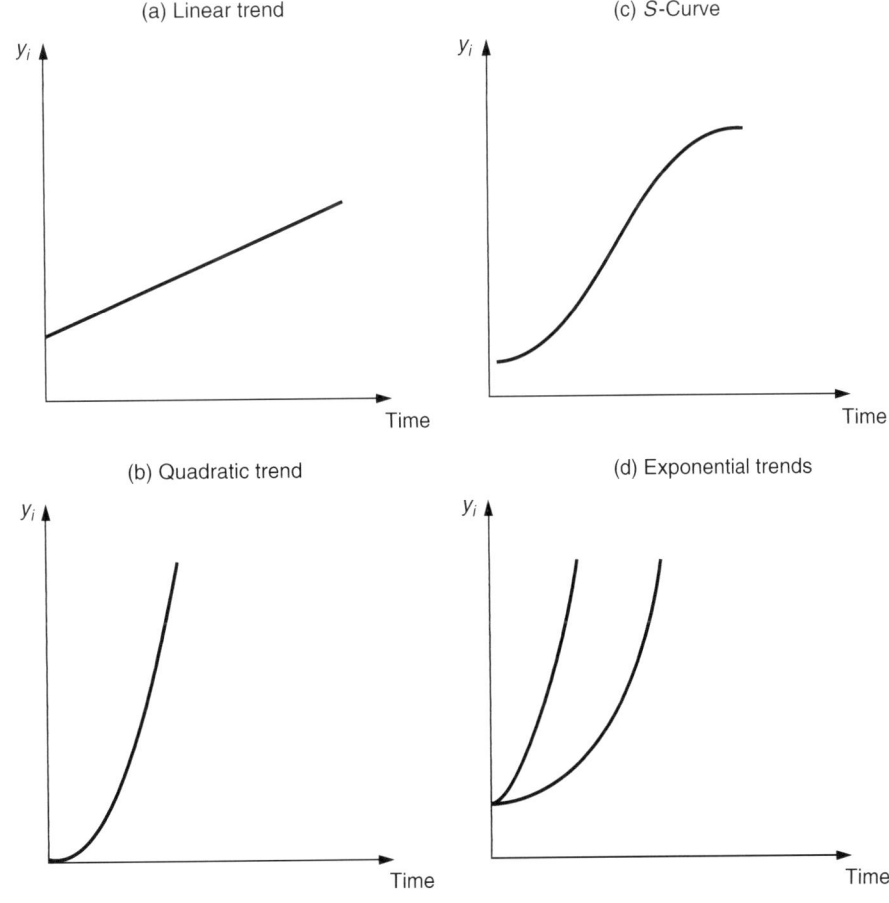

FIGURE 3.34
Examples of Types of Trends.

114 Simple Regression Analysis

This is preferred to using the actual years (1980, 1981, 1982, . . .) because it reduces computational problems.

Forecasts are simple to compute when the linear trend model is used. Simply insert the appropriate value for the time period to be forecast into the regression equation. The time period T forecast can be written as

$$\hat{y}_T = b_0 + b_1 T$$

Many other types of trends can be modeled using regression. Some examples, including the linear trend, are shown in Figure 3.34. Note that the other types of trends are represented by curves. Equations to represent curvilinear trends are discussed in Chapter 5.

EXAMPLE 3.11   **ABX Company Sales** The ABX Company sells winter sports merchandise including skis, ice skates, sleds, and so on. Quarterly sales (in thousands of dollars) for the ABX Company are shown in Table 3.6. The time period represented starts in the first quarter of 1994 and ends in the fourth quarter of 2003. (See the file ABXSALES3 on the CD.)

A time-series plot of the sales figures is shown in Figure 3.35. The time-series plot suggests a strong linear trend in the sales figures. A regression with a linear trend variable (labeled TIME) was estimated and the regression results are shown in Figure 3.36. The linear trend model estimated is

$$y_i = \beta_0 + \beta_1 t + e_i$$

TABLE 3.6 Data for ABX Company Sales Example

Year.Qtr	SALES	TIME	Year.Qtr	SALES	TIME
1994.1	221.0	1	1999.1	260.5	21
1994.2	203.5	2	1999.2	244.0	22
1994.3	190.0	3	1999.3	256.0	23
1994.4	225.5	4	1999.4	276.5	24
1995.1	223.0	5	2000.1	291.0	25
1995.2	190.0	6	2000.2	255.5	26
1995.3	206.0	7	2000.3	244.0	27
1995.4	226.5	8	2000.4	291.0	28
1996.1	236.0	9	2001.1	296.0	29
1996.2	214.0	10	2001.2	260.0	30
1996.3	210.5	11	2001.3	271.5	31
1996.4	237.0	12	2001.4	299.5	32
1997.1	245.5	13	2002.1	297.0	33
1997.2	201.0	14	2002.2	271.0	34
1997.3	230.0	15	2002.3	270.0	35
1997.4	254.5	16	2002.4	300.0	36
1998.1	257.0	17	2003.1	306.5	37
1998.2	238.0	18	2003.2	283.5	38
1998.3	228.0	19	2003.3	283.5	39
1998.4	255.0	20	2003.4	307.5	40

FIGURE 3.35
Time-Series Plot of ABX Company Sales.

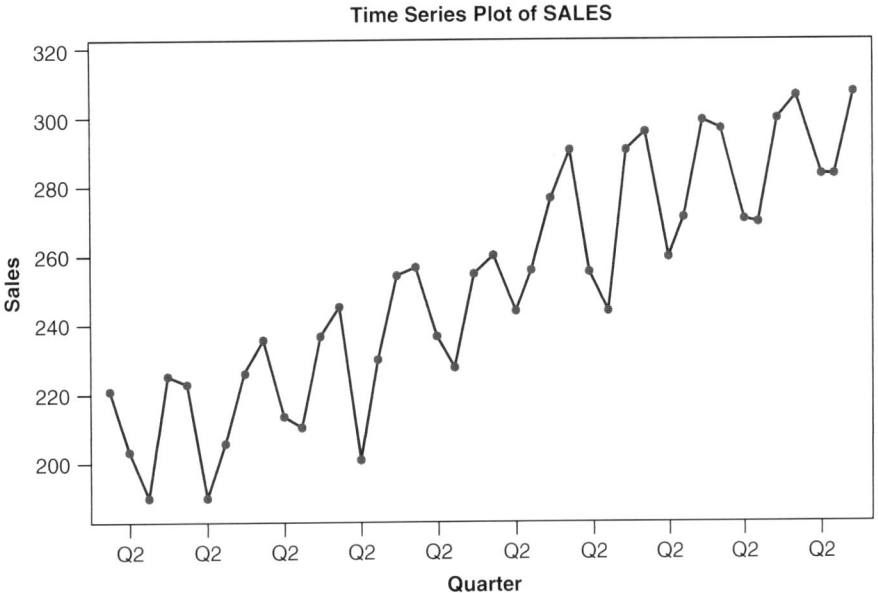

FIGURE 3.36
Regression Results for ABX Company Sales Example.

```
Variable     Coefficient    Std Dev    T Stat    P Value

Intercept    199.017        5.128      38.81     0.000
TIME         2.556          0.218      11.73     0.000

Standard Error = 15.9126   R-Sq = 78.3%    R-Sq(adj) = 77.8%

Analysis of Variance

Source       DF    Sum of Squares    Mean Square    F Stat    P Value

Regression   1     34818             34818          137.50    0.000
Error        38    9622              253
Total        39    44440
```

To test whether the linear trend component is useful in explaining the variation in sales, the following hypotheses should be tested:

$$H_0: \beta_1 = 0$$
$$H_a: \beta_1 \neq 0$$

Using a 5% level of significance, the decision rule is

Reject H_0 if $t > 1.96$ or $t < -1.96$

Do not reject H_0 if $-1.96 \leq t \leq 1.96$

The z value of 1.96 is used as a critical value because the number of degrees of freedom is large (38).

When using time-series data for forecasting, it is generally true that prediction intervals are more appropriate than confidence intervals for representing the uncertainty in predictions for future time periods. Figures 3.37 and 3.38 show the additional MINITAB and SAS output obtained when forecasts for the next four time periods are requested. Both outputs show the predicted value (*Fit* in MINITAB and *Predicted Value* in SAS), the standard error of the estimate of the point on the regression line (*SE Fit* in MINITAB and *Std Error Mean Predict* in SAS), a 95% confidence interval for the estimate of the conditional mean (*95% CI* in MINITAB and *95% CL Mean* in SAS) and a 95% prediction interval for each prediction (*95% PI* in MINITAB and *95% CL Predict* in SAS). Note that the SAS output provides this information for the observations used to estimate the equation as well as the observations for which predictions are desired.

Using the estimated linear trend equation, the point forecasts are determined by substituting values of the trend variable for the appropriate time period into the equation:

$$2004.1 \text{ sales} = 199.017 + 2.5559(41) = 303.81$$
$$2004.2 \text{ sales} = 199.017 + 2.5559(42) = 306.36$$
$$2004.3 \text{ sales} = 199.017 + 2.5559(43) = 308.92$$
$$2004.4 \text{ sales} = 199.017 + 2.5559(44) = 311.48$$

The prediction intervals in the outputs would be used as our interval predictions of sales in each of the four quarters. Thus, our interval prediction for sales in the first quarter of 2004 is $269,960 to $337,650.

If you look again at the time-series plot of sales in Figure 3.35, you may notice a pattern other than the trend. Note that the sales figures for the first and fourth quarters tend to be higher than the figures for the second and third quarters. This systematic variation among time periods from year to year is called *seasonal variation*. In Chapter 7, methods to account for seasonal variation are discussed.

FIGURE 3.37 MINITAB Output Showing the Next Four Periods' Forecasts.

```
Predicted Values for New Observations

New
Obs        Fit   SE Fit       95% CI            95% PI
  1     303.81     5.13  (293.43, 314.19)  (269.96, 337.65)
  2     306.36     5.32  (295.60, 317.13)  (272.40, 340.33)
  3     308.92     5.51  (297.76, 320.08)  (274.83, 343.01)
  4     311.48     5.71  (299.92, 323.03)  (277.25, 345.70)


Values of Predictors for New Observations

New
Obs      TIME
  1      41.0
  2      42.0
  3      43.0
  4      44.0
```

3.6 Fitting a Linear Trend to Time-Series Data 117

Obs	Dep Var sales	Predicted Value	Std Error Mean Predict	95% CL Mean		95% CL Predict		Residual
1	221.0000	201.5732	4.9391	191.5745	211.5719	167.8436	235.3027	19.4268
2	203.5000	204.1290	4.7528	194.5074	213.7507	170.5094	237.7487	-0.6290
3	190.0000	206.6849	4.5694	197.4347	215.9351	173.1696	240.2002	-16.6849
4	225.5000	209.2408	4.3891	200.3555	218.1260	175.8244	242.6571	16.2592
5	223.0000	211.7966	4.2123	203.2692	220.3241	178.4736	245.1196	11.2034
6	190.0000	214.3525	4.0396	206.1747	222.5303	181.1172	247.5878	-24.3525
7	206.0000	216.9083	3.8715	209.0709	224.7458	183.7552	250.0615	-10.9083
8	226.5000	219.4642	3.7085	211.9567	226.9718	186.3875	252.5409	7.0358
9	236.0000	222.0201	3.5515	214.8305	229.2097	189.0141	255.0261	13.9799
10	214.0000	224.5759	3.4012	217.6906	231.4612	191.6349	257.5170	-10.5759
11	210.5000	227.1318	3.2585	220.5353	233.7283	194.2499	260.0137	-16.6318
12	237.0000	229.6877	3.1245	223.3624	236.0129	196.8591	262.5162	7.3123
13	245.5000	232.2435	3.0004	226.1695	238.3176	199.4624	265.0246	13.2565
14	201.0000	234.7994	2.8875	228.9540	240.6448	202.0599	267.5389	-33.7994
15	230.0000	237.3553	2.7870	231.7133	242.9972	204.6514	270.0591	-7.3553
16	254.5000	239.9111	2.7004	234.4444	245.3778	207.2371	272.5851	14.5889
17	257.0000	242.4670	2.6291	237.1446	247.7894	209.8168	275.1172	14.5330
18	238.0000	245.0228	2.5743	239.8114	250.2343	212.3906	277.6551	-7.0228
19	228.0000	247.5787	2.5372	242.4425	252.7149	214.9584	280.1991	-19.5787
20	255.0000	250.1346	2.5184	245.0364	255.2327	217.5202	282.7490	4.8654
21	260.5000	252.6904	2.5184	247.5923	257.7886	220.0760	285.3048	7.8096
22	244.0000	255.2463	2.5372	250.1101	260.3825	222.6259	287.8666	-11.2463
23	256.0000	257.8022	2.5743	252.5907	263.0136	225.1699	290.4344	-1.8022
24	276.5000	260.3580	2.6291	255.0356	265.6804	227.7078	293.0082	16.1420
25	291.0000	262.9139	2.7004	257.4472	268.3806	230.2399	295.5879	28.0861
26	255.5000	265.4697	2.7870	259.8278	271.1117	232.7659	298.1736	-9.9697
27	244.0000	268.0256	2.8875	262.1802	273.8710	235.2861	300.7651	-24.0256
28	291.0000	270.5815	3.0004	264.5074	276.6555	237.8004	303.3626	20.4185
29	296.0000	273.1373	3.1245	266.8121	279.4626	240.3088	305.9659	22.8627
30	260.0000	275.6932	3.2585	269.0967	282.2897	242.8113	308.5751	-15.6932
31	271.5000	278.2491	3.4012	271.3638	285.1344	245.3080	311.1901	-6.7491
32	299.5000	280.8049	3.5515	273.6153	287.9945	247.7989	313.8109	18.6951
33	297.0000	283.3608	3.7085	275.8532	290.8683	250.2841	316.4375	13.6392
34	271.0000	285.9167	3.8715	278.0792	293.7541	252.7635	319.0698	-14.9167
35	270.0000	288.4725	4.0396	280.2947	296.6503	255.2372	321.7078	-18.4725
36	300.0000	291.0284	4.2123	282.5009	299.5558	257.7054	324.3514	8.9716
37	306.5000	293.5842	4.3891	284.6990	302.4695	260.1679	327.0006	12.9158
38	283.5000	296.1401	4.5694	286.8899	305.3903	262.6248	329.6554	-12.6401
39	283.5000	298.6960	4.7528	289.0743	308.3176	265.0763	332.3156	-15.1960
40	307.5000	301.2518	4.9391	291.2531	311.2505	267.5223	334.9814	6.2482
41	.	303.8077	5.1279	293.4269	314.1885	269.9629	337.6525	.
42	.	306.3636	5.3189	295.5961	317.1310	272.3982	340.3289	.
43	.	308.9194	5.5119	297.7612	320.0776	274.8282	343.0107	.
44	.	311.4753	5.7067	299.9227	323.0278	277.2530	345.6976	.

Sum of Residuals 0
Sum of Squared Residuals 9622.06053
Predicted Residual SS (PRESS) 10620

FIGURE 3.38 SAS Output Showing the Next Four Periods' Forecasts.

EXERCISES

16. Fort Worth Water Department. In 1990, the city of Fort Worth, Texas, conducted a study examining the level of water purity. One aspect helpful in maintaining water purity is monitoring the quality of water at storm drains that pour into the Trinity River. This river supplies drinking water for Fort Worth. Even though water from the river is filtered later, preventing contaminants from entering the river from storm drains is helpful in maintaining purity. Five of the variables the city monitored to test purity of water entering the river from storm drains are:

ODOR: determined by a sensory test

COLOR: no color is best—determined by comparison to a standard water sample

SCUM: floatable solids

hydrocarbon SHEEN: hydrocarbon (oil) sheen on surface of water

sewage BACTERIA: filamentous sewage bacteria

These are monthly data from January 1986 through December 1989. In all cases, lower numbers are better.

These data are in a file named WATER3 on the CD.

Your job is to use time-series plots and linear trend regression to examine the performance of the city's water department in improving the quality of storm drain water entering the Trinity River. Which of the variables show a significant decrease? Are there areas where the city might concentrate its efforts to achieve future improvements? Use a 5% level of significance in any tests.

3.7 SOME CAUTIONS IN INTERPRETING REGRESSION RESULTS

3.7.1 ASSOCIATION VERSUS CAUSALITY

A common mistake made when using regression analysis is to assume that a strong fit (high R^2) of a regression of y on x automatically means that "x causes y." This is not necessarily true. Some alternative explanations for the good fit include:

1. The reverse is true; y causes x. Linear regression computations pay no attention to the direction of causality. If x and y are highly correlated, a high R^2 value results even if the causal order of the variables is reversed.

2. There may be a third variable related to both x and y. It may be that neither x causes y nor y causes x. Both variables may be related to some third common cause. As an example, consider the price and gasoline mileage of automobiles. These two variables are inversely related. As mileage rises, price goes down (on average). But it is not the rise in mileage that "causes" the price to drop. A third variable, size of car, may be influencing both of the other two variables. As size increases, price increases and mileage drops. There are a variety of interesting examples in this category. For example, the mortality rate in countries is inversely related to the number of televisions. As the number of televisions increases, mortality rate decreases. I don't think this is a causal relationship.

To infer that x causes y requires that additional conditions be satisfied. A high R^2 for a regression of y on x might be considered supporting evidence for causality, but on its own, this is not enough to ensure that x causes y.

Note that the absence of causality is not necessarily a drawback in regression analysis. An equation showing a relationship between x and y can be important and useful even if it is recognized that x does not cause y.

3.7.2 FORECASTING OUTSIDE THE RANGE OF THE EXPLANATORY VARIABLE

When using an estimated regression equation to construct estimates of $\mu_{y|x}$ or to predict individual values of the dependent variable, some caution must be used if forecasts are outside the range of the x variable. Consider the communications nodes example. The explanatory variable was NUMPORTS, the number of access ports. The sample values ranged from 12 to 68. The estimated regression model can be expected to be reliable over this range of the x variable. If, however, a node is to be installed with 100 ports, there is some question as to how reliable the model will be. The relationship that holds over the range from 12 to 68 may differ from the relationship outside this range. Estimates of $\mu_{y|x}$ or predictions outside the range of the x variable require some caution for this reason.

There are often occasions where forecasts outside the range of the x variable must be made. One common example is when time-series data are used and forecasts for future time periods are desired. It may be that the values of the explanatory variables in future time periods are outside the range observed in the past, as, for example, when the linear trend model is used. In such cases, it must be recognized that the quality of the forecasts depends on whether the estimated relationship still holds for values of the explanatory variables that are outside the observed range.

EXERCISES

17. Sales/Advertising (continued). Use the results in Figure 3.22 to help answer the following questions.

  a. Find a point estimate of average sales for all sales districts with advertising expenditures of $60,000. Are there any cautions that should be exercised regarding this estimate?

  b. A district sales manager examines the model developed. The manager points out that $0 advertising expenditure results in sales of –$5700, which is impossible. She suggests that this means the model is of no use. Do you agree or disagree with her assessment? Explain why.

ADDITIONAL EXERCISES

18. Indicate whether the following statements are true or false:

  a. If the hypothesis $H_0: \beta_1 = 0$ is rejected, then it can be safely concluded that x causes y.

  b. Suppose a regression of y on x is run and the t statistic for testing $H_0: \beta_1 = 0$ versus $H_a: \beta_1 \neq 0$ has a p value of 0.0295 associated with it. Using a 5% level of significance, the null hypothesis should be rejected.

  c. If the correlation between y and x is 0.9, then the R^2 value for a regression of y on x is 90%.

  d. As long as the R^2 value is high for an estimated regression equation, it is safe to use the equation to predict for any value of x.

  e. If the R^2 value for a regression of y on x is 75%, then the R^2 value for a regression of x on y is also 75%.

19. Suppose a regression analysis provides the following results:

$$b_0 = 1, \quad b_1 = 2, \quad s_{b_0} = 0.05,$$
$$s_{b_1} = 0.25, \quad SST = 117.2873, \quad SSE = 30.0$$

and $n = 24$. Use this information to solve the following problems.

  a. Test the hypotheses

$$H_0: \beta_1 = 0$$
$$H_a: \beta_1 \neq 0$$

using a 5% level of significance. State the decision rule, the test statistic, and your decision. Use a t test.

    b. Perform the same test as in part a using an F test. Use a 10% level of significance.

    c. Compute the R^2 for the regression.

20. Suppose a regression analysis provides the following results:

$$b_0 = 4.0, \quad b_1 = 10.0, \quad s_{b_0} = 1.0,$$
$$s_{b_1} = 4.0, \quad SST = 67.36, \quad SSE = 50.0$$

and $n = 20$. Use this information to solve the following problems:

    a. Test the hypotheses

$$H_0: \beta_0 = 0$$
$$H_a: \beta_0 \neq 0$$

using a 5% level of significance. State the decision rule, the test statistic, and your decision.

    b. Test the hypotheses

$$H_0: \beta_1 \leq 0$$
$$H_a: \beta_1 > 0$$

using a 5% level of significance. State the decision rule, the test statistic, and your decision. What conclusion can be drawn from the test result?

    c. Compute the R^2 for the regression.

21. Fill in the missing blanks on the following ANOVA table:

ANOVA Source	DF	SS	MS	F
Regression	1		1000	
Error (Residual)		800		
Total	81			

22. Fill in the missing blanks on the following ANOVA table:

ANOVA Source	DF	SS	MS	F
Regression			100	10
Error (Residual)	40			
Total				

23. **Salary/Education.** Data on beginning salary ($y = SALARY$) and years of education ($x = EDUC$) for 93 employees of Harris Bank Chicago in 1977 are provided in a data file named SALED3 on the CD. These data were obtained from an article by Daniel W. Schafer, "Measurement-Error Diagnostics and the Sex Discrimination Problem," *Journal of Business and Economic Statistics*, 5: 529–537, 1987. (Copyright 1987 by the American Statistical Association. Used with permission. All rights reserved.)

The scatterplot of salary verses education is shown in Figure 3.39. The regression results are shown in Figure 3.40. Use the results to answer the following questions:

    a. Is there a linear relationship between salary and education? State the hypotheses to be tested, the decision rule, the test statistic, and your decision. Use a 10% level of significance.

    b. What percentage of the variation in salary has been explained by the regression?

    c. For an individual with 12 years of education, find a point prediction of beginning salary.

    d. For all individuals with 12 years of education, find a point estimate of the conditional mean beginning salary.

    e. What other factors, in addition to education, might be useful in helping to estimate beginning salary?

24. **Cost Estimation.** The file COSTEST3 on the CD contains data on production runs at a manufacturing plant. There are two columns of data:

    $y = $ COST is the total cost of the production run.

    $x = $ NUMBER is the number of items produced during that run.

Run the regression using COST as the dependent variable and NUMBER as the independent variable and use the result to help answer the following questions:

    a. What is the estimated regression equation relating y to x?

    b. What percentage of the variation in y has been explained by the regression?

    c. Are y and x linearly related? Conduct a hypothesis test to answer this question and use a 5% level of significance. State the hypotheses to be tested, the decision rule, the test statistic,

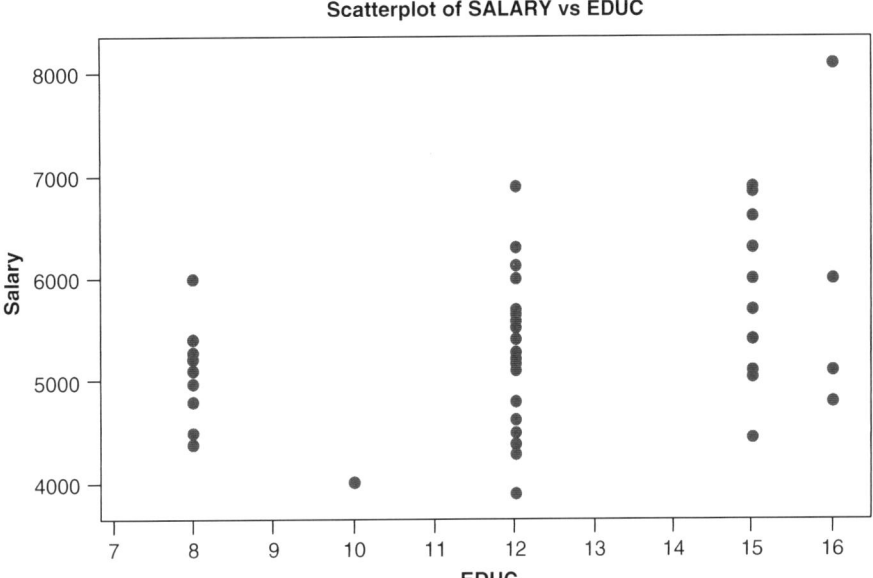

FIGURE 3.39
Scatterplot for Salary and Education Exercise.

FIGURE 3.40
Regression Results for Salary and Education Exercise.

```
Variable      Coefficient    Std Dev    T Stat    P Value

Intercept     3818.6         377.4      10.12     0.000
EDUC          128.1          29.7       4.31      0.000

Standard Error = 650.112    R-Sq = 17.0%    R-Sq(adj) = 16.1%

Analysis of Variance

Source        DF    Sum of Squares    Mean Square    F Stat    P Value

Regression    1     7862534           7862534        18.60     0.000
Error         91    38460756          422646
Total         92    46323290
```

and your decision. What conclusion can be drawn from the result of the test?

d. Estimate the fixed cost involved in the production process. Find a point estimate and a 95% confidence interval estimate.

e. Estimate the variable cost involved in the production process. Find a point estimate and a 95% confidence interval estimate.

25. Income/Consumption. The following data are annual disposable income and total annual consumption for 12 families selected at random from a large metropolitan area. Regard annual disposable income as the explanatory variable and total annual consumption as the dependent variable. From the regression of y on x, answer the questions that follow. These data are in a file named INCONS3 on the CD.

Annual Disposable Income ($)	Total Annual Consumption ($)
INC	CONS
16,000	14,000
30,000	24,545
43,000	36,776
70,000	63,254

Annual Disposable Income ($)	Total Annual Consumption ($)
56,000	40,176
50,000	49,548
16,000	16,000
26,000	22,386
14,000	16,032
12,000	12,000
24,000	20,768
30,000	34,780

a. What is the estimated regression equation relating y to x?

b. What percentage of the variation in y has been explained by the regression?

c. Construct a 90% confidence interval estimate of β_1.

d. Use a t test to test the hypotheses $H_0: \beta_1 = 0$ versus $H_a: \beta_1 \neq 0$ at the 5% level of significance. State the decision rule, the test statistic, and your decision. What conclusion can be drawn from the result of the test?

e. Use an F test to test the hypotheses $H_0: \beta_1 = 0$ versus $H_a: \beta_1 \neq 0$ at the 5% level of significance. State the decision rule, the test statistic, and your decision.

f. Can the F test be used to test the hypotheses $H_0: \beta_1 \leq 0$ versus $H_a: \beta_1 > 0$?

g. Test the hypotheses $H_0: \beta_1 = 1$ versus $H_a: \beta_1 \neq 1$ at the 5% level of significance. State the decision rule, the test statistic, and your decision. What conclusion can be drawn from the result of the test?

26. **Apex Corporation.** The Apex Corporation produces corrugated paper. It has collected monthly data from January 2001 through March 2003 on the following two variables:

   y, total manufacturing cost per month (in thousands of dollars) (COST)

   x, total machine hours used per month (MACHINE)

The data are shown in Table 3.7 and are available in a file named APEX3 on the CD. Perform any analyses necessary to answer the following questions:

a. What is the estimated regression equation relating y to x?

b. What percentage of the variation in y has been explained by the regression?

c. Are y and x linearly related? Conduct a hypothesis test to answer this question and use a 5% level of significance. State the hypotheses to be tested, the decision rule, the test statistic, and your decision. What conclusion can be drawn from the result of the test?

d. Use the equation developed to estimate the average manufacturing cost in a month with 350 machine hours. Find a point estimate and a 95% confidence interval estimate. How reliable do you believe this forecast might be?

e. Use the equation developed to estimate the average manufacturing cost in a month with 550 machine hours. Find a point estimate and a 95% confidence interval estimate. How reliable do you believe this forecast might be?

27. **New Construction.** Our construction firm is interested in forecasting new construction in the United States for the years 2002 and 2003. We have data in billions of dollars for the years 1991 through 2001 from the Department of Commerce. These data are in the file NEWCON3 on the CD and are shown in Table 3.8.

a. Fit a linear trend to these data. What is the resulting regression equation?

b. What percentage of the variation in y has been explained by the regression?

c. Based on your answer in part b and on any other regression results you obtain, how well does the equation fit the data? Does a good fit ensure that forecasts for future years will be accurate?

d. Use the equation developed to predict new construction in both 2002 and 2003. Find a point prediction and a 95% prediction interval.

e. How reliable do you believe the forecast in part d might be? What factors might influence this accuracy?

28. **U.S. Population.** The data file USPOP3 on the CD contains the population of the United States for the years 1930 through 1999. Fit a linear trend to these data.

a. What is the resulting regression equation?

b. What percentage of the variation in y has been explained by the regression?

TABLE 3.7 Data for APEX Exercise

Date	COST	MACHINE	Date	COST	MACHINE
1/01	1102	218	3/02	1287	259
2/01	1008	199	4/02	1451	286
3/01	1227	249	5/02	1828	389
4/01	1395	277	6/02	1903	404
5/01	1710	363	7/02	1997	430
6/01	1881	399	8/02	1363	271
7/01	1924	411	9/02	1421	286
8/01	1246	248	10/02	1543	317
9/01	1255	259	11/02	1774	376
10/01	1314	266	12/02	1929	415
11/01	1557	334	1/03	1317	260
12/01	1887	401	2/03	1302	255
1/02	1204	238	3/03	1388	281
2/02	1211	246			

Source: These data were created by Professor Roger L. Wright, RLW Analytics, Inc., and are used (with modification) with his permission.

TABLE 3.8 Data for New Construction Exercise

YEAR	NEWCON
1991	432.6
1992	463.7
1993	491
1994	539.2
1995	557.8
1996	615.9
1997	653.4
1998	705.7
1999	765.9
2000	820.3
2001	842.5

    c. Based on your answer in part b and on any other regression results you obtain, how well does the equation fit the data? Does a good fit ensure that forecasts for future years will be accurate?

    d. Use the equation developed to predict the U.S. population in the years 2000 and 2001. Find a point prediction and a 95% prediction interval.

    e. How reliable do you believe the forecast in part d might be? What factors might influence this accuracy?

29. Wheat Exports. The relationship between exchange rates and agricultural exports is of interest to agricultural economists. One such export of interest is wheat. The following data

    y, U.S. wheat export shipments (SHIPMENT)

    x, the real index of weighted-average exchange rates for the U.S. dollar (EXCHRATE)

are available in a file named WHEAT3 on the CD. These time-series data were observed monthly from January 1974 through March 1985. Perform any analyses necessary to answer the following questions:

    a. What is the estimated regression equation relating y to x?

    b. Are y and x linearly related? Conduct a hypothesis test to answer this question and use a 5% level of significance. State the hypotheses to be tested, the decision rule, the test statistic, and your decision. What conclusion can be drawn from the result of the test?

c. What percentage of the variation in y has been explained by the regression?

d. Construct a 95% confidence interval estimate of β_1.

(*Source:* Data are from D. A. Bessler and R. A. Babubla, "Forecasting Wheat Exports: Do Exchange Rates Really Matter?" *Journal of Business and Economic Statistics*, 5, 1987, pp. 397–406. Copyright 1987 by the American Statistical Association. Used with permission. All rights reserved.)

30. **Major League Baseball Salaries.** The owners of Major League Baseball (MLB) teams are concerned with rising salaries (as are owners of all professional sports teams). Table 3.9 provides the average salary (AVESAL) of the 30 MLB teams for the 2002 season. Also provided is the number of wins (WINS) for each team during the 2002 season. Is there evidence that teams with higher total payrolls tend to be more successful? Justify your answer. These data are available in a file named BBALL3 on the CD.

31. **Computing Beta.** In finance class you will discuss (or have discussed) the use of simple regression to estimate the relationship between the return on a stock and the market return. This relationship can be written as

$$y = \beta_0 + \beta_1 x + e$$

where y = return on the stock and x = the return on the market. The slope coefficient, β_1, is called the *beta coefficient* and is used to measure how responsive a stock's price is to movements in the market. The beta coefficient is used as a measure of a firm's systematic risk. In the file named BETA3 on the CD, the return on the stock of three companies is provided: Dell, Sabre, and Wal-Mart. Also provided is the return on the market (This is the value-weighted return computed by CRSP, the Center for Research on Security Prices). Five year's of monthly returns (January 1998 through December 2002) are used so there are a total of 60 observations for each company. Run the regression using the firm's return as the dependent variable and the market return as the independent variable for each of the three companies. Use the three regression results to answer the following questions:

a. What are the beta coefficients for each of the three companies?

b. Is there a relationship between the firm return and the market return for each of these three companies? Be sure to state the hypotheses to be tested, the decision rule, the test statistic, and your decision. Use a 5% level of significance.

c. The beta coefficient measures a security's responsiveness to movements in the market. For example, a beta of 2 would mean that a 1% increase (decrease) in the market return would result in, on average, a 2% increase (decrease) in the security's return. A beta of 1 would mean that movements in the market were matched, on average, by movements in the security's return. For each of the companies in the data file, test to see if the beta coefficient is equal to one or not. Be sure to state the decision rule, the test statistic, and your decision. Use a 5% level of significance.

d. Test to see if Dell's beta coefficient is greater than 1. Be sure to state the decision rule, the test statistic, and your decision. Use a 5% level of significance.

e. Test to see if Wal-Mart's beta coefficient is less than 1. Be sure to state the decision rule, the test statistic, and your decision. Use a 5% level of significance.

32. **Major League Baseball Wins.** What factor is most important in building a winning baseball team? Some might argue for a high batting average. Or it might be a team that hits for power as measured by the number of home runs. On the other hand, many believe that it is quality pitching as measured by the earned run average of the team's pitchers. The file MLB3 on the CD contains data on the following variables for the 30 major league baseball teams during the 2002 season:

   WINS = number of games won
   HR = number of home runs hit
   BA = average batting average
   ERA = earned run average

Using WINS as the dependent variable, use scatterplots and regression to investigate the relationship of the other three variables to WINS. Which of the three possible explanatory

TABLE 3.9 Data for Major League Baseball Salaries Exercise

Team	WINS	AVESAL	Team	WINS	AVESAL
Anaheim Angels	99	2160054	Atlanta Braves	101	3166233
Baltimore Orioles	67	1855318	Chicago Cubs	67	2528398
Boston Red Sox	93	3633457	Cincinnati Reds	78	1658363
Chicago White Sox	81	1791286	Colorado Rockies	73	1848858
Cleveland Indians	74	2106591	Florida Marlins	79	1506567
Detroit Tigers	55	1562847	Houston Astros	84	2449680
Kansas City Royals	62	1832594	Los Angeles Dodgers	92	3396961
Minnesota Twins	94	1430068	Milwaukee Brewers	56	1338991
New York Yankees	103	4902777	Montreal Expos	83	1497309
Oakland Athletics	103	1746264	New York Mets	75	3192482
Seattle Mariners	93	3337435	Philadelphia Phillies	80	2086812
Tampa Bay Devil Rays	55	1131474	Pittsburgh Pirates	72	1370088
Texas Rangers	72	3123803	St. Louis Cardinals	97	2998072
Toronto Blue Jays	78	1868356	San Diego Padres	66	1292744
Arizona Diamondbacks	98	3199608	San Francisco Giants	95	3030571

Source: Reprinted courtesy of the *Fort Worth Star-Telegram*.

variables exhibits the strongest relationship to WINS? What might this suggest to managers of major league baseball teams?
(*Source*: Data courtesy of the *Fort Worth Star-Telegram*.)

33. Work Orders. During the construction phase of a nuclear plant, the number of corrective work orders open should gradually decline until reaching a steady state that would be present during the operational phase. The Nuclear Regulatory Commission has licensing requirements that the number of work orders open at licensing and for operational plants be less than 1000. (This was, of course, back in the days when nuclear plants were still being constructed in the United States.) This number is set to provide a goal indicating operational readiness. The number of work orders for a consecutive 120-working-day period during the construction phase of a nuclear plant are available in a file named WKORDER3 on the CD.

As a consultant to the plant, you have been asked to estimate how many days it will take to reach the operational level of 1000 work orders. In determining the number of days, state any assumptions you make and any caveats that might be in order.

34. Fanfare. Fanfare International, Inc., designs, distributes, and markets ceiling fans and lighting fixtures. The company's product line includes 120 basic models of ceiling fans and 138 compatible fan light kits and table lamps. These products are marketed to over 1000 lighting showrooms and electrical wholesalers that supply the remodeling and new construction markets. The product line is distributed by a sales organization of 58 independent sales representatives.

In the summer of 1994, Fanfare decided it needed to develop forecasts of future sales to help determine future salesforce needs, capital expenditures, and so on. The file named FAN3 on the CD contains monthly sales data and data on three additional variables for the period July 1990 through May 1994. The variables are defined as follows

SALES = total monthly sales in thousands of dollars

ADEX = advertising expense in thousands of dollars

MTGRATE = mortgage rate for 30-year loans (%)

HSSTARTS = housing starts in thousands of units

The data file contains the four variables as shown, plus columns for year and month. The sales data have been transformed to provide confidentiality.

As a consultant to Fanfare, your job is to find the best single variable to forecast future sales. Try each of the three variables in a simple regression and decide which is the best to create a forecasting model for Fanfare. Justify your choice. What problems do you see with using each of the three possible variables to help forecast sales?

35. **College Graduation Rates.** *Kiplinger's Personal Finance* provides information on the best public and private college values. Some of the variables included in this issue are as follows. All are based on the most recent available data.

GRADRATE4	the percentage of students who earned a bachelor's degree in four years (expressed as a percentage)
ADMISRATE	admission rate expressed as a percentage
SFACRATIO	student faculty ratio
AVGDEBT	average debt at graduation

This information is included in a file on the CD named COLLEGE3 for 195 schools. Only schools with complete information on all the categories listed are included. Using the graduation rate as the dependent variable, examine simple regressions using the independent variables provided. Which variable appears to do the best job of explaining graduation rate? How might you go about determining which of the possible variables used in a simple regression provides the best equation for predicting graduation rates?

(*Source*: Used by permission from the November and December 2003 issues of *Kiplinger's Personal Finance*. Copyright © 2003 The Kiplinger Washington Editors, Inc. Visit our website at www.kiplingers.com for further information.)

36. **Retail Furniture Sales.** The file FURNSALES3 on the CD contains monthly sales data (in millions of dollars) for retail furniture stores from January 1992 through December 2002. The data file contains a column with the year, the month (coded 1 = Jan through 12 = Dec) and SALES.

   a. Use a linear trend model to forecast sales for each month of 2003.

   b. Are there patterns present in this time series besides the trend in the data? If so, what are they?

37. **Cubs Attendance.** The Chicago Cubs baseball organization is interested in examining the relationship between attendance and the number of wins during the season. One possible hypothesized model is

   $$\text{ATTENDANCE} = \beta_0 + \beta_1 \text{WINS} + e$$

   They plan to use the equation to forecast future attendance and have annual data from 1972 through 1999. These data are shown in Table 3.10 and are in the file named CUBSWIN3 on the CD.

   a. Find the regression equation using ATTENDANCE as the dependent variable and WINS as the explanatory variable.

TABLE 3.10 Cubs Attendance and Wins

YEAR	ATTENDANCE	WINS	YEAR	ATTENDANCE	WINS
1972	1299163	85	1986	1859102	70
1973	1351705	77	1987	2035130	76
1974	1015378	66	1988	2089034	77
1975	1034819	75	1989	2491942	93
1976	1026217	75	1990	2243791	77
1977	1439834	81	1991	2314250	77
1978	1525311	79	1992	2126720	78
1979	1648587	80	1993	2653763	84
1980	1206776	64	1994	1845208	49
1981	565637	38	1995	1918265	73
1982	1249278	73	1996	2219110	76
1983	1479717	71	1997	2190368	68
1984	2107655	96	1998	2623194	90
1985	2161534	77	1999	2813854	67

b. The Cubs organization wants to use the regression to forecast attendance next season: If we win 110 games next year, that means our forecast for attendance will be almost 2,600,000. With Sammy Sosa back, and if Kerry Wood is healthy, I think we've got a good chance at 110 wins. Do you see any problems with using the forecast for attendance specifically when the number of wins is 110?

USING THE COMPUTER

The Using the Computer section in each chapter describes how to perform the computer analyses in the chapter using Excel, MINITAB, and SAS. For further detail on Excel, MINITAB, and SAS, see Appendix C.

EXCEL

Plotting Data

FIGURE 3.41 Chart Wizard Dialog Box Showing Types of Scatterplots.

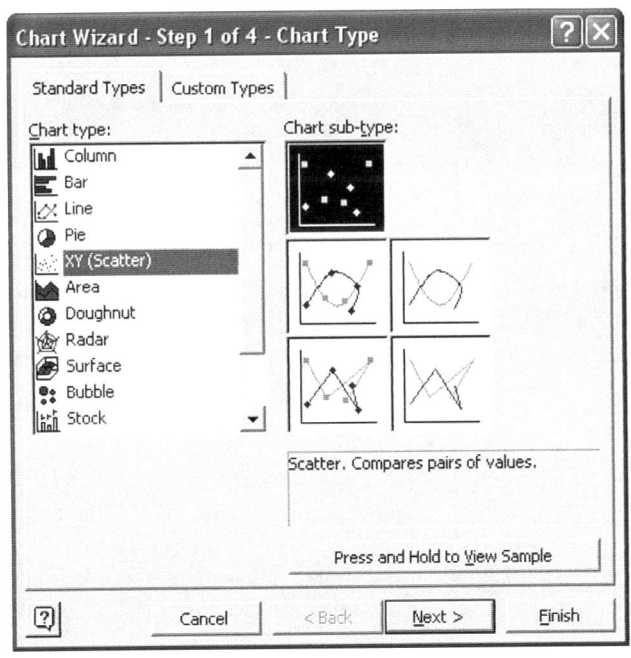

Use the Chart Wizard to create a scatterplot (or XY plot). Click the Chart Wizard button. A window opens showing chart types. Click on XY (Scatter), pick the type of scatterplot you want (see Figure 3.41), click Next> and follow the directions to create the scatterplot. To create a time-series plot, use the Line Plot feature in the Chart Wizard.

Regression

TOOLS: DATA ANALYSIS: REGRESSION

To perform a simple regression in Excel, use the Regression procedure on the Data Analysis menu. The Regression dialog box is shown in Figure 3.42. Fill in the Input Y Range and Input X Range with the cells containing the Y variable and X variable, respectively. Click the Labels box if your variables have labels in the first row. You can request an alternate level of confidence (instead of 95%) for confidence intervals for the regression coefficients by clicking the Confidence Level box and filling in the desired level. Choose the desired Output option and click OK.

Creating a Trend Variable

To put the numbers 1 through n in column B (for example), type 1 in B1, 2 in B2, then select these two cells, put the cursor on the rectangle at the bottom right-hand corner of cell B2, and drag through cell n.

128 Simple Regression Analysis

FIGURE 3.42 Excel Regression Dialog Box.

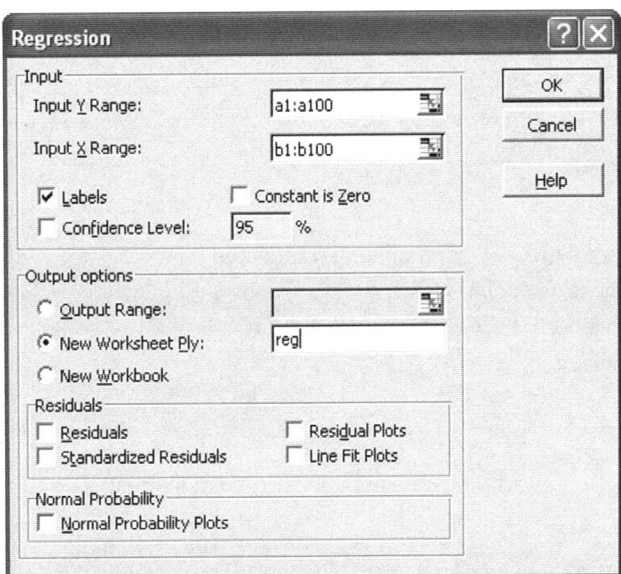

FIGURE 3.43 MINITAB Request for Type of Scatterplot Desired.

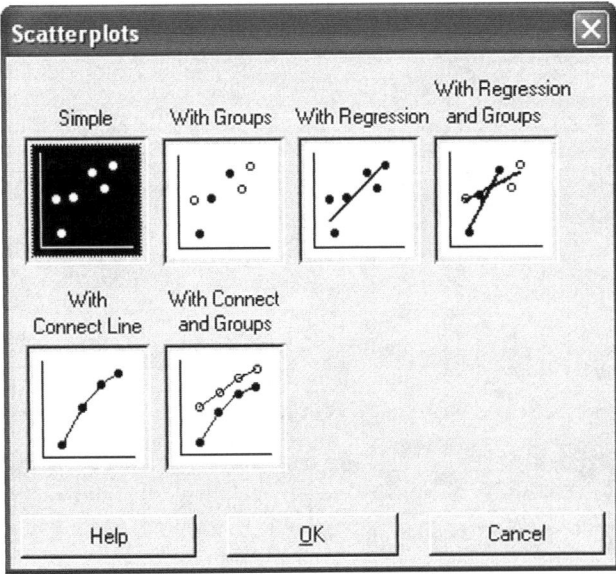

MINITAB

Plotting Data

GRAPH: SCATTERPLOT

Click Scatterplot and the dialog box in Figure 3.43 allows you to choose the type of plot you want. Click OK and the next dialog box (Figure 3.44) asks you indicate which columns represent the Y and X variables. When you click OK, MINITAB will plot the Y-data on the vertical axis and the X-data on the horizontal axis to create a scatterplot. There are a variety of other options available.

GRAPH: TIME-SERIES PLOT

Click Time Series Plot and the dialog box in Figure 3.45 allows you to choose the type of plot you want. Click OK and the next dialog box (Figure 3.46) asks you to indicate which column represents the data to be plotted. When you click OK, MINITAB plots the data values on the vertical axis versus a time indicator on the horizontal axis. MINITAB assumes the data are entered in a column with the most recent time period as the last entry in the column. There are a variety of other options available. For example, the use of month, quarter, and so on is chosen using the Time/Scale option.

Regression

STAT: REGRESSION: REGRESSION

See the Regression dialog box in Figure 3.47. Fill in the response (Y) and predictor (X) variable and click OK. A variety of options are available. For example, click Options and the PRESS, and predicted R^2 can be requested by checking the appropriate box. Other options are discussed in later chapters.

FIGURE 3.44 MINITAB Scatterplot Dialog Box.

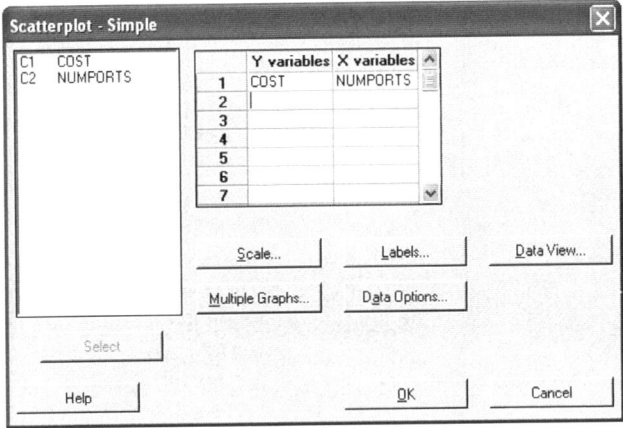

FIGURE 3.45 MINITAB Request for Type of Time-Series Plot Desired.

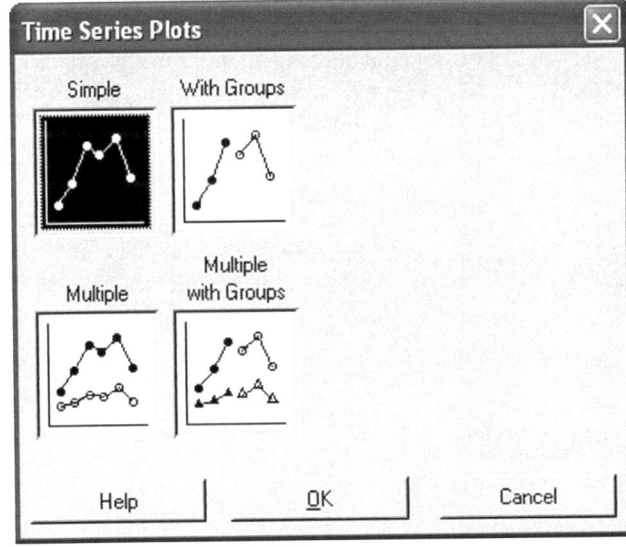

Forecasting with the Regression Equation

See the Regression–Options dialog box in Figure 3.48. Click OPTIONS in the Regression dialog box (Figure 3.47) to get to this screen. To generate forecasts and appropriate intervals, put the value of the x variable for which a forecast is desired in the line labeled Prediction intervals for new observations. If you want forecasts for several values, type those values in a column starting in row one and indicate that column in the Prediction intervals for new observations line.

Creating a Trend Variable

CALC: MAKE PATTERNED DATA: SIMPLE SET OF NUMBERS

See the Simple Set of Numbers dialog box in Figure 3.49. Type in the column number for the trend variable. Then enter a first value of 1 and a last value of n (the number of time periods) in steps of 1. Make sure "List each value" and "List the whole sequence" are set at 1. Then click OK.

SAS

Plotting Data

Plots in SAS are generated using the following command sequence:

```
PROC GPLOT;
PLOT COST*NUMPORTS;
```

or

```
PROC PLOT;
PLOT COST*NUMPORTS;
```

The variable to be plotted on the vertical axis (COST) is listed first, with the variable to be plotted on the horizontal axis (NUMPORTS) second. PLOT produces character plots. GPLOT produces high-resolution plots.

130 Simple Regression Analysis

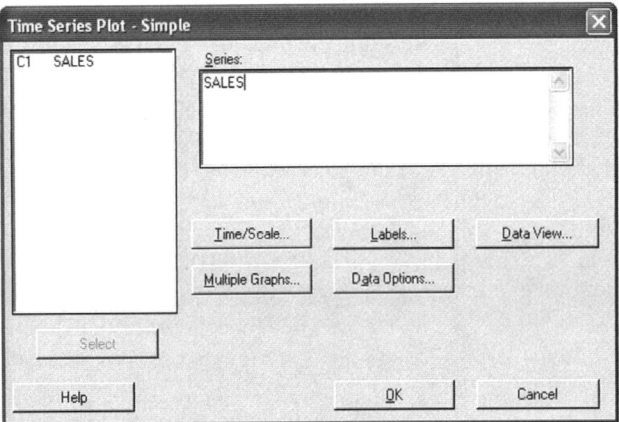

FIGURE 3.46 MINITAB Time-Series Plot Dialog Box.

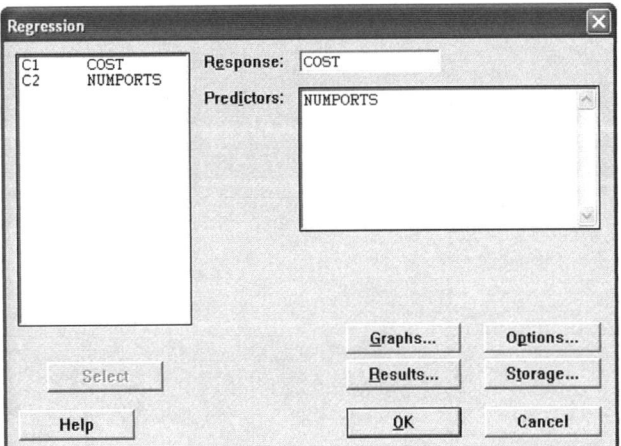

FIGURE 3.47 MINITAB Regression Dialog Box.

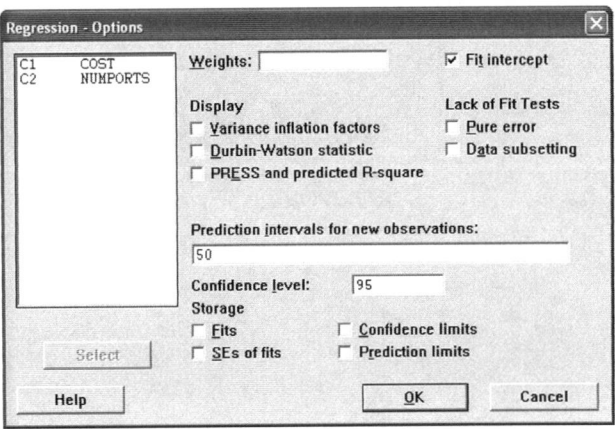

FIGURE 3.48 MINITAB Regression-Options Dialog Box.

Regression
The following command sequence produces a regression with COST as the dependent variable and NUMPORTS as the independent variable:
PROC REG;
MODEL COST=NUMPORTS;

Forecasting with the Regression Equation
Forecasts in SAS are generated using an "appended" data set. To the values of the independent variable in the original data set, add the values for which predictions are desired. Then add to the values of the dependent variable the SAS symbol for missing data, a period, because we do not know those values. Now rerun the regression as follows:
PROC REG;
MODEL COST=NUMPORTS/P CLM CLI;

The option P requests forecasts (or predicted values), CLM requests upper and lower confidence interval limits for the estimate of the conditional mean, and CLI requests upper and lower prediction interval limits for an individual prediction.

Creating a Trend Variable
In the data input phase in SAS, use the command
TREND=_N_;

to create a trend variable. The command TREND=_N_ sets the variable TREND

FIGURE 3.49 MINITAB Dialog Box to Create a Trend Variable.

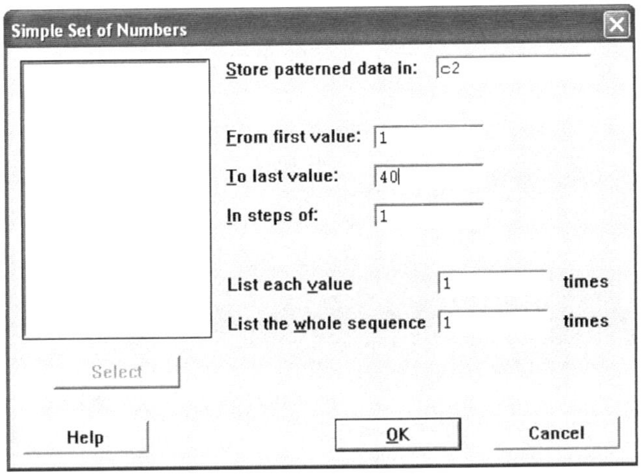

equal to the integers 1 through N, where N is the total number of observations in the data set. To do a time-series plot of the variable SALES, use the commands

```
PROC GPLOT;
PLOT SALES*TREND;
```

To fit the linear trend model, use the commands

```
PROC REG;
MODEL SALES=TREND;
```

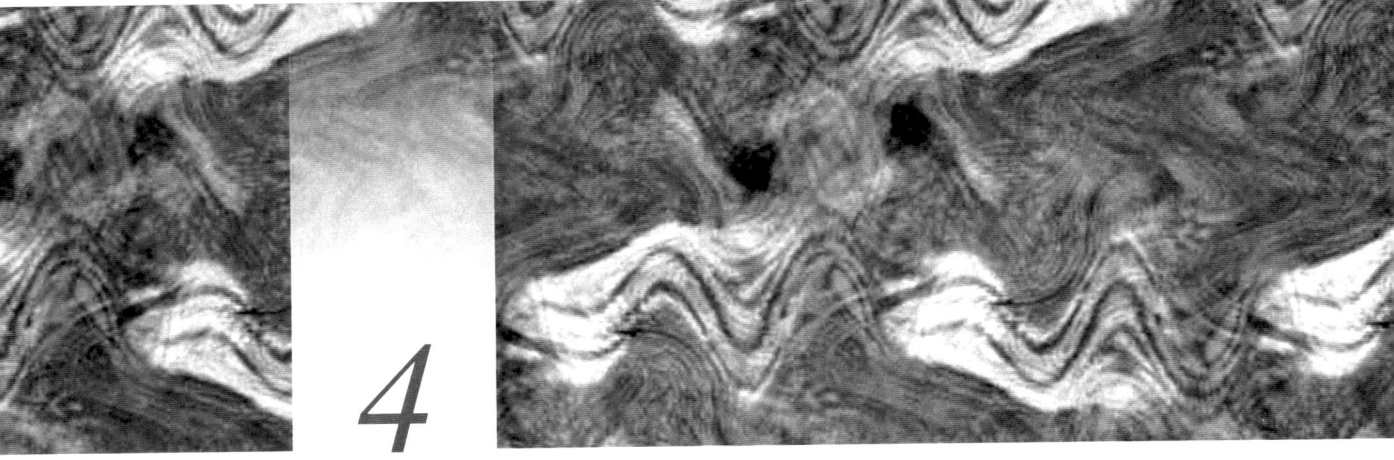

4 Multiple Regression Analysis

4.1 USING MULTIPLE REGRESSION TO DESCRIBE A LINEAR RELATIONSHIP

In Chapter 3, the method of least squares was used to develop the equation of a line that best described the relationship between a dependent variable y and an explanatory variable x. In business and economic applications, however, there may be more than one explanatory variable that is useful in explaining variation in the dependent variable y or obtaining better predictions of y. An equation of the form

$$\hat{y} = b_0 + b_1 x_1 + b_2 x_2$$

where x_1 and x_2 are the explanatory variables and b_1 and b_2 are estimates of the population regression coefficients may be desired. The relationship is still "linear"; each term on the right-hand side of the equation is additive, and the regression coefficients do not enter the equation in a nonlinear manner (such as $b_1^2 x_1$). The graph of the relationship is no longer a line, however, because there are three variables involved.

Graphing the equation thus requires the use of three dimensions rather than two, and the equation graphs as a plane passing through the three-dimensional space. Figure 4.1 shows how this graph might appear. The x_1 axis and y axis are drawn as before; the x_2 axis can be thought of as moving toward you to imitate the three-dimensional space. Because of the difficulty of drawing graphs in more than two dimensions on paper, the usefulness of graphical methods such as scatterplots is somewhat limited.

FIGURE 4.1 Graph Showing Regression "Plane."

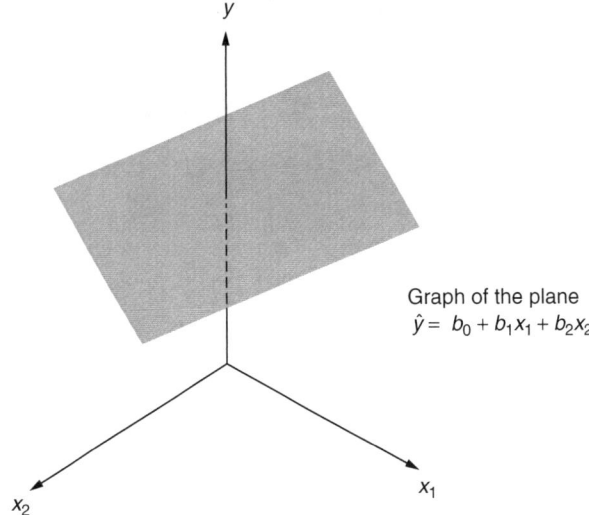

Graph of the plane
$\hat{y} = b_0 + b_1 x_1 + b_2 x_2$

Still, when two or more explanatory variables are involved, two-dimensional scatterplots between the dependent variable and each explanatory variable can provide an initial indication of the relationships present. The relationship involving more than one explanatory variable may differ, however, from that involving each explanatory variable individually. The least-squares method can still be used to develop regression equations involving more than one explanatory variable. These equations are referred to as *multiple regression equations*. As discussed, the equations no longer graph as lines, but the terms *linear regression* and even regression *line* (when perhaps regression *surface* might be more appropriate) still are used.

As the number of explanatory variables increases, the formulas for computing the estimates of the regression coefficients become increasingly complex. The availability of computerized regression routines precludes the need for hand computation of the estimates. The equations for the coefficient estimates when there are two or more explanatory variables are not presented in this text. There is a convenient method for writing the equations for the least-squares estimates for any number of explanatory variables, but it requires using matrices and matrix algebra. Because this text attempts to avoid as much mathematical detail as possible and concentrate on the use of computer regression output, the matrix presentation has been avoided; however, Appendix D does contain a brief introduction to the topic. A more advanced treatment of multiple regression that utilizes the matrix presentation is found, for example, in *Classical and Modern Regression with Applications* by R. Myers and in *Regression Analysis: Concepts and Applications* by F. Graybill and H. Iyer.[1]

---

[1] See References for complete publication information.

The concepts involved in producing least-squares coefficient estimates for a multiple regression equation are very similar to those for simple regression. An equation that "best" describes the relationship between a dependent variable y and K explanatory variables $x_1, x_2, \ldots, x_K$ can be written

$$\hat{y} = b_0 + b_1 x_1 + b_2 x_2 + \cdots + b_K x_K$$

where $b_0, b_1, b_2, \ldots, b_K$ are the least-squares coefficients. The case $K = 1$ is simple regression. The criterion for "best" is the same as it was for a simple regression; the difference between the true values of y and the values predicted by the multiple regression equation, $\hat{y}$, should be as small as possible. As before, this is accomplished by choosing $b_0, b_1, b_2, \ldots, b_K$ so that the sum of squares of the differences between the y and $\hat{y}$ values, $\sum_{i=1}^{n}(y_i - \hat{y}_i)^2$, is minimized. The optimizing values, $b_0, b_1, b_2, \ldots, b_K$ are the least-squares coefficients printed out by regression routines such as those available in Excel, MINITAB, and SAS.

EXAMPLE 4.1 **Meddicorp Sales** Meddicorp Company sells medical supplies to hospitals, clinics, and doctors' offices. The company currently markets in three regions of the United States: the South, the West, and the Midwest. These regions are each divided into many smaller sales territories. Data for Meddicorp is contained in the MEDDICORP4 file on the CD.

Meddicorp's management is concerned with the effectiveness of a new bonus program. This program is overseen by regional sales managers and provides bonuses to salespeople based on performance. Management wants to know if the bonuses paid in 2003 were related to sales. (Obviously, if there is a relationship here, the managers expect it to be a direct—positive—one.) In determining whether this relationship exists, they also want to take into account the effects of advertising. The variables to be used in the study include:

- y, Meddicorp's sales (in thousands of dollars) in each territory for 2003 (SALES)
- x_1, the amount Meddicorp spent on advertising in each territory (in hundreds of dollars) in 2003 (ADV)
- x_2, the total amount of bonuses paid in each territory (in hundreds of dollars) in 2003 (BONUS)

Data for a random sample of 25 of Meddicorp's sales territories are shown in Table 4.1.

Figures 4.2 and 4.3 show the scatterplots of SALES versus ADV and BONUS, respectively. Figures 4.4, 4.5, and 4.6, respectively, show the MINITAB, Excel, and SAS regression output obtained relating SALES(y) to ADV(x_1) and BONUS(x_2). These outputs provide the multiple regression equation, which is used in this example, as well as additional information that will be used in later examples.

After rounding, the multiple regression equation describing the relationship between sales and the two explanatory variables may be written

$$\hat{y} = -516.4 + 2.47 x_1 + 1.86 x_2$$

or

$$\text{SALES} = -516.4 + 2.47 \text{ADV} + 1.86 \text{BONUS}$$

TABLE 4.1 Data for Meddicorp Example

Territory	SALES (in thousand $)	ADV (in hundred $)	BONUS (in hundred $)
1	963.50	374.27	230.98
2	893.00	408.50	236.28
3	1057.25	414.31	271.57
4	1183.25	448.42	291.20
5	1419.50	517.88	282.17
6	1547.75	637.60	321.16
7	1580.00	635.72	294.32
8	1071.50	446.86	305.69
9	1078.25	489.59	238.41
10	1122.50	500.56	271.38
11	1304.75	484.18	332.64
12	1552.25	618.07	261.80
13	1040.00	453.39	235.63
14	1045.25	440.86	249.68
15	1102.25	487.79	232.99
16	1225.25	537.67	272.20
17	1508.00	612.21	266.64
18	1564.25	601.46	277.44
19	1634.75	585.10	312.25
20	1159.25	524.56	292.87
21	1202.75	535.17	268.27
22	1294.25	486.03	309.85
23	1467.50	540.17	291.03
24	1583.75	583.85	289.29
25	1124.75	499.15	272.55

FIGURE 4.2
Scatterplot of SALES versus ADV.

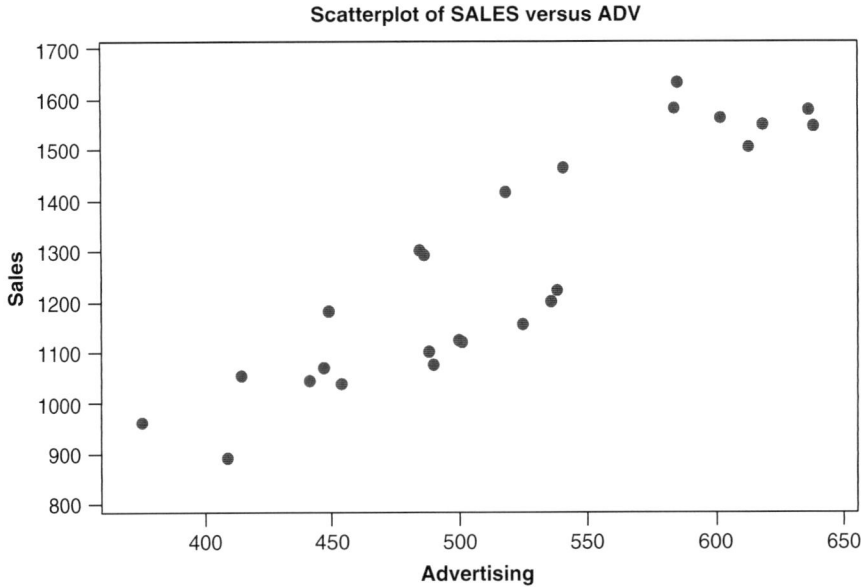

FIGURE 4.3 Scatterplot of SALES versus BONUS.

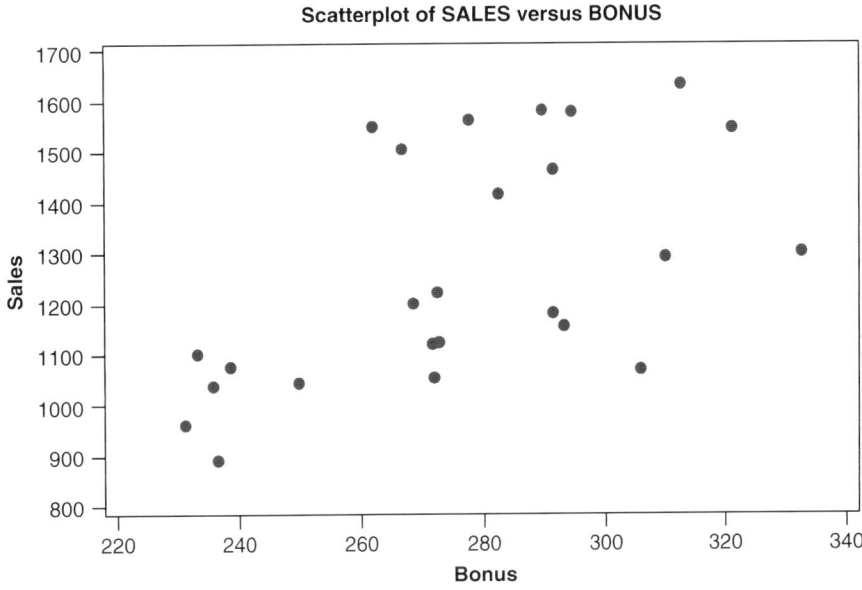

FIGURE 4.4 MINITAB Regression of SALES on ADV and BONUS for Meddicorp Example.

```
The regression equation is
SALES = -516 + 2.47 ADV + 1.86 BONUS

Predictor      Coef     SE Coef        T        P
Constant     -516.4       189.9    -2.72    0.013
ADV          2.4732      0.2753     8.98    0.000
BONUS        1.8562      0.7157     2.59    0.017

S = 90.7485      R-Sq = 85.5%     R-Sq(adj) = 84.2%

Analysis of Variance

Source           DF          SS         MS        F        P
Regression        2     1067797     533899    64.83    0.000
Residual Error   22      181176       8235
Total            24     1248974

Source           DF      Seq SS
ADV               1     1012408
BONUS             1       55389
```

This equation can be interpreted as providing an estimate of mean sales for a given level of advertising and bonus payment. Moreover, if advertising is held fixed, the equation shows that mean sales tends to rise by $1860 (1.86 thousands of dollars) for each unit increase in BONUS. Also, if bonus payment is held fixed, it shows that mean sales tends to rise by $2470 (2.47 thousands of dollars) for

138 Multiple Regression Analysis

SUMMARY OUTPUT

Regression Statistics
Multiple R	0.925
R Square	0.855
Adjusted R Square	0.842
Standard Error	90.749
Observations	25.000

ANOVA

	df	SS	MS	F	Significance F
Regression	2	1067797.321	533898.660	64.831	0.000
Residual	22	181176.419	8235.292		
Total	24	1248973.740			

	Coefficients	Standard Error	t Stat	P-value	Lower 95%	Upper 95%
Intercept	-516.444	189.876	-2.720	0.013	-910.223	-122.666
ADV	2.473	0.275	8.983	0.000	1.902	3.044
BONUS	1.856	0.716	2.593	0.017	0.372	3.341

FIGURE 4.5 Excel Regression of SALES on ADV and BONUS for Meddicorp Example.

FIGURE 4.6 SAS Regression of SALES on ADV and BONUS for Meddicorp Example.

The REG Procedure
Model: MODEL1
Dependent Variable: SALES

Analysis of Variance

Source	DF	Sum of Squares	Mean Square	F Value	Pr > F
Model	2	1067797	533899	64.83	<.0001
Error	22	181176	8235.29179		
Corrected Total	24	1248974			

Root MSE	90.74851	R-Square	0.8549	
Dependent Mean	1269.02000	Adj R-Sq	0.8418	
Coeff Var	7.15107			

Parameter Estimates

Variable	DF	Parameter Estimate	Standard Error	t Value	Pr > |t|
Intercept	1	-516.44428	189.87570	-2.72	0.0125
ADV	1	2.47318	0.27531	8.98	<.0001
BONUS	1	1.85618	0.71573	2.59	0.0166

each unit increase in ADV. (Note that a "unit" increase in either BONUS or ADV represents a $100 increase.) Clearly, such information provides a useful summary of the data.

4.2 INFERENCES FROM A MULTIPLE REGRESSION ANALYSIS

4.2.1 ASSUMPTIONS CONCERNING THE POPULATION REGRESSION LINE

In general, a population regression equation involving K explanatory variables can be written as

$$\mu_{y|x_1, x_2, \ldots, x_K} = \beta_0 + \beta_1 x_1 + \beta_2 x_2 + \cdots + \beta_K x_K$$

This equation says that the conditional mean of y given $x_1, x_2, \ldots, x_K$ is a point on the regression surface described by the terms on the right-hand side of the equation.

An alternative way of writing the relationship is

$$y_i = \beta_0 + \beta_1 x_{1i} + \beta_2 x_{2i} + \cdots + \beta_K x_{Ki} + e_i$$

where i denotes the ith observation and e_i is a random error or disturbance. Thus, y_i is related to the explanatory variables $x_{1i}, x_{2i}, \ldots, x_{Ki}$, although the relationship is not an exact one. The random error e_i shows that, given the same values for $x_{1i}, x_{2i}, \ldots, x_{Ki}$, each point y_i will not be exactly on the regression surface. Rather, the individual y_i values are distributed around the regression surface in the manner discussed for a simple regression line in Chapter 3. The following assumptions about the e_i are made:

1. The expected value of the disturbances is zero: $E(e_i) = 0$. This implies that the regression line passes through the conditional means of the y variable for each set of x variables. For our purposes, we interpret this assumption as: The population regression equation is linear in the explanatory variables.[2]

2. The variance of each e_i is equal to σ_e^2.

3. The e_i are normally distributed.

4. The e_i are independent. This is an assumption that is most important when data are gathered over time. When the data are cross-sectional (that is, gathered at the same point in time for different individual units), this is typically not an assumption of concern.

These assumptions allow inferences to be made about the population multiple regression line from a sample multiple regression line. The first inferences to be considered are those made about the individual population regression coefficients, $\beta_1, \beta_2, \ldots, \beta_K$. The effects of violations of the assumptions are considered in Chapter 6. In this chapter, each assumption is assumed to hold so that an ideal situation exists for the use of least-squares inference procedures.

4.2.2 INFERENCES ABOUT THE POPULATION REGRESSION COEFFICIENTS

This section considers estimates of the population regression coefficients and tests of hypotheses about the population coefficients. Much of the information required

---

[2] As pointed out in Chapter 3, the assumption here is that the population regression equation is linear in the x variables. In Chapter 5, we relax this assumption and find that we can fit curves by allowing equations that are not linear in the x variables. Throughout this text, however, we always assume that the equations are linear in the parameters. This means that equations such as $y = \beta_0 + \beta_1^2 x_1 + e$, are not considered. These types of equations are beyond the scope of this text.

to construct estimates and perform tests of hypotheses can be found in standard multiple regression output. For example, Figure 4.7 shows, in general, what information is provided by the multiple regression output for MINITAB, Excel, and SAS.

The least-squares estimates $b_0, b_1, b_2, \ldots, b_K$ are unbiased estimators of the corresponding population regression coefficients. A $(1 - \alpha)100\%$ confidence interval estimate of the population regression coefficient, β_k, is

$$b_k \pm t_{\alpha/2, n-K-1} s_{b_k}$$

Here, k refers to the kth regression coefficient, $k = 0, 1, \ldots, K$. The value $t_{\alpha/2, n-K-1}$ is a number chosen from the t table to ensure the appropriate level of confidence, and s_{b_k} is the standard deviation of the sampling distribution of b_k. The number of degrees of freedom used in determining the t value is $n - (K + 1)$, where $K + 1$ is the number of regression coefficients to be estimated (K coefficients corresponding to the K explanatory variables and one intercept or constant). Note that $n - (K + 1) = n - K - 1$ and is written in this manner throughout the text.

Hypothesis tests about the individual β_k also can be performed. The general form of two-tailed hypotheses about the individual β_k is as follows:

$$H_0: \beta_k = \beta_k^*$$
$$H_a: \beta_k \neq \beta_k^*$$

where β_k^* is any number chosen as the hypothesized value of the kth regression coefficient.

(a) MINITAB

The regression equation is
$y = b_0 + b_1 x_1 + b_2 x_2 + \cdots + b_K x_K$

Predictor	Coef	SE Coef	T	P
Constant	b_0	s_{b_0}	b_0/s_{b_0}	p value
x1 variable name	b_1	s_{b_1}	b_1/s_{b_1}	p value
x2 variable name	b_2	s_{b_2}	b_2/s_{b_2}	p value
.	.	.	.	.
.	.	.	.	.
.	.	.	.	.
xK variable name	b_K	s_{b_K}	b_K/s_{b_K}	p value

$S = s_e$ R-Sq $= R^2$ R-Sq(adj) $= R^2_{adj}$

Analysis of Variance

Source	DF	SS	MS	F	P
Regression	K	SSR	$MSR = SSR/K$	$F = MSR/MSE$	p value
Residual Error	$n - K - 1$	SSE	$MSE = SSE/(n - K - 1)$		
Total	$n - 1$	SST			

Unusual Observations

Obs.	X1	Y	Fit	SE Fit	Residual	St Resid
Obs.No.	Value of X1	Value of y	$\hat{y}$	s_m	$y - \hat{y}$	—

R denotes an observation with a large standardized residual
X denotes an observation whose X value gives it large influence

FIGURE 4.7 Illustration of MINITAB, Excel, and SAS Multiple Regression Output.

(b) Excel

Regression Statistics
Multiple R $\quad = R$
R Square $\quad = R^2$
Adjusted R Square $\quad = R^2_{adj}$
Standard Error $\quad = s_e$
Observations $\quad = n$

ANOVA	df	SS	MS	F	Significance F
Regression	K	SSR	$MSR = SSR/K$	$F = MSR/MSE$	p value
Residual	$n - K - 1$	SSE	$MSE = SSE/(n - K - 1)$		
Total	$n - 1$	SST			

	Coefficients	Standard Error	t stat	P-value	Lower 95%	Upper 95%
Intercept	b_0	s_{b_0}	b_0/s_{b_0}	p value	$b_0 - t_{\alpha/2, n-K-1} s_{b_0}$	$b_0 + t_{\alpha/2, n-K-1} s_{b_0}$
x1 variable name	b_1	s_{b_1}	b_1/s_{b_1}	p value	$b_1 - t_{\alpha/2, n-K-1} s_{b_1}$	$b_1 + t_{\alpha/2, n-K-1} s_{b_1}$
x2 variable name	b_2	s_{b_2}	b_2/s_{b_2}	p value	$b_2 - t_{\alpha/2, n-K-1} s_{b_2}$	$b_2 + t_{\alpha/2, n-K-1} s_{b_2}$
.	.	.	.	.	.	.
.	.	.	.	.	.	.
xK variable name	b_K	s_{b_K}	b_K/s_{b_K}	p value	$b_K - t_{\alpha/2, n-K-1} s_{b_K}$	$b_K + t_{\alpha/2, n-K-1} s_{b_K}$

(c) SAS

Analysis of Variance

Source	DF	Sum of Squares	Mean Square	F Value	Pr > F
Model	K	SSR	$MSR = SSR/K$	$F = MSR/MSE$	p value
Error	$n - K - 1$	SSE	$MSE = SSE/(n - K - 1)$		
Corrected Total	$n - 1$	SST			

Root MSE	s_e	R-Square	R^2
Dependent Mean	$\bar{y}$	Adj R-Sq	R^2_{adj}
Coeff Var	$\dfrac{s_e}{\bar{y}} \times (100)$		

Parameter Estimates

Variable	DF	Parameter Estimate	Standard Error	t Value	Pr > |t|
Intercept	1	b_0	s_{b_0}	b_0/s_{b_0}	p value
x1 variable name	1	b_1	s_{b_1}	b_1/s_{b_1}	p value
x2 variable name	1	b_2	s_{b_2}	b_2/s_{b_2}	p value
.	.	.	.	.	.
.	.	.	.	.	.
xK variable name	1	b_K	s_{b_K}	b_K/s_{b_K}	p value

FIGURE 4.7 (*Continued*).

The decision rule for this test is

Reject H_0 if $t > t_{\alpha/2, n-K-1}$ or $t < -t_{\alpha/2, n-K-1}$
Do not reject H_0 if $-t_{\alpha/2, n-K-1} \leq t \leq t_{\alpha/2, n-K-1}$

where α is the probability of a Type I error.

The standardized test statistic is

$$t = \frac{b_k - \beta_k^*}{s_{b_k}}$$

When the null hypothesis is true, the standardized test statistic, t, should be small in absolute value because the estimate, b_k, is close to the hypothesized value β_k^*, making the numerator $b_k - \beta_k^*$ close to zero. When the null hypothesis is false, the difference between b_k and β_k^* is large in absolute value, leading to a large absolute value of the test statistic and resulting in the decision to reject H_0.

The most common hypothesis test encountered in multiple regression analysis is

$$H_0: \beta_k = 0$$
$$H_a: \beta_k \neq 0$$

as in simple regression. This test is typically most important when β_k refers to the coefficient of the explanatory variable x_k rather than the intercept.

If the null hypothesis $H_0: \beta_k = 0$ is not rejected, then the conclusion is that, once the effects of all other variables in the multiple regression are included, x_k is not linearly related to y. In other words, adding x_k to the regression equation is of no help in explaining any additional variation in y left unexplained by the other explanatory variables.

On the other hand, if the null hypothesis is rejected, then there is evidence that y and x_k are linearly related and that x_k does help explain some of the variation in y not accounted for by the other explanatory variables.

Figure 4.7 shows that the test statistic for testing $H_0: \beta_k = 0$ is printed out on the regression output. The test statistic is

$$t = \frac{b_k}{s_{b_k}}$$

and is found in the column labeled "T" for MINITAB, the column labeled "t stat" for Excel, and the column labeled "t Value" for SAS. Also note that the p values for testing whether each population regression coefficient is equal to zero are found in the column labeled "P" in MINITAB, the column labeled "P-value" in Excel, and "Pr > |t|" in SAS.

EXAMPLE 4.2 **Meddicorp (Continued)** Refer again to the MINITAB output in Figure 4.4, the Excel output in Figure 4.5, or the SAS output in Figure 4.6.

1. Use the regression output to test the following hypotheses:
$$H_0: \beta_1 = 0$$
$$H_a: \beta_1 \neq 0$$

where β_1 is the coefficient of ADV. Use a 5% level of significance. What conclusion can be drawn from the result of the test?

Answer 1: Using the standardized test statistic:

Decision Rule: Reject H_0 if $t > 2.074$ or $t < -2.074$
Do not reject H_0 if $-2.074 \leq t \leq 2.074$

Note: The t value with 22 degrees of freedom is 2.074 for a two-tailed test with a 5% level of significance.

Test Statistic: 8.98
Decision: Reject H_0
Conclusion: ADV is related to SALES (even when the effect of BONUS is taken into account).

Answer 2: Using the p value:

Decision Rule: Reject H_0 if p value < 0.05
Do not reject H_0 if p value ≥ 0.05

Test Statistic: p value $= 0.000$
Decision: Reject H_0

2. Use the regression output to test the following hypotheses:

$$H_0: \beta_2 = 0$$
$$H_a: \beta_2 \neq 0$$

where β_2 is the coefficient of BONUS. Use a 5% level of significance. What conclusion can be drawn from the result of the test?

Answer 1: Using the standardized test statistic:

Decision Rule: Reject H_0 if $t > 2.074$ or $t < -2.074$
Do not reject H_0 if $-2.074 \leq t \leq 2.074$.

Note: The t value with 22 degrees of freedom is 2.074 for a two-tailed test with a 5% level of significance.

Test Statistic: 2.59
Decision: Reject H_0
Conclusion: BONUS is related to SALES (even when the effect of ADV is taken into account).

Answer 2: Using the p value:

Decision Rule: Reject H_0 if p value < 0.05
Do not reject H_0 if p value ≥ 0.05

Test Statistic: p value $= 0.017$
Decision: Reject H_0

3. Use the regression output to test the following hypotheses:

$$H_0: \beta_2 \leq 0$$
$$H_a: \beta_2 > 0$$

where β_2 is the coefficient of BONUS. Use a 5% level of significance. What conclusion can be drawn from the result of the test?

Answer: Decision Rule: Reject H_0 if $t > 1.717$
Do not reject H_0 if $t \leq 1.717$

Note: The t value with 22 degrees of freedom is 1.717 for a one-tailed test with a 5% level of significance.

Test Statistic: $t = 2.59$
Decision: Reject H_0
Conclusion: BONUS is directly related to SALES (even when the effect of ADV is taken into account).

4. Construct a 95% confidence interval estimate of β_1, the coefficient of ADV.

Answer: $(2.4732 - (2.074 * 0.2753), 2.4732 + (2.074 * 0.2753))$ or $(1.9022, 3.0442)$

4.3 ASSESSING THE FIT OF THE REGRESSION LINE

4.3.1 THE ANOVA TABLE, THE COEFFICIENT OF DETERMINATION, AND THE MULTIPLE CORRELATION COEFFICIENT

As with simple regression, the variation in the dependent variable y in a multiple regression can be written as follows:

$$SST = SSE + SSR$$

The total variation in y is given by the total sum of squares:

$$SST = \sum_{i=1}^{n}(y_i - \bar{y})^2$$

The error sum of squares represents the variation in y left "unexplained" by the regression:

$$SSE = \sum_{i=1}^{n}(y_i - \hat{y}_i)^2$$

The regression sum of squares represents the variation in y "explained" by the regression:

$$SSR = \sum_{i=1}^{n}(\hat{y}_i - \bar{y})^2$$

In SSE and SSR, the $\hat{y}_i$ values are the predicted or fitted values from the multiple regression equation. These three sums of squares can be interpreted as in a simple regression context.

Figure 4.7 shows the format of the MINITAB, Excel, and SAS multiple regression output. Each of the three sums of squares is listed in the analysis of variance

(ANOVA) table as shown in the output. Also listed in the ANOVA table are the numbers of degrees of freedom associated with each of the sums of squares. For SSR, the number of degrees of freedom is equal to the number of explanatory variables, K. For SSE, the number of degrees of freedom is $n - K - 1$. As in the simple regression ANOVA table, the mean squares are also shown. These are computed by dividing the sums of squares by the appropriate number of degrees of freedom.

SST, SSR, and SSE can be used to evaluate how well the regression equation is explaining the variation in y. One measure of the goodness of fit of the regression is the coefficient of determination, R^2. The R^2 value, as for a simple regression, is computed by dividing SSR by SST:

$$R^2 = \frac{SSR}{SST}$$

Thus, R^2 represents the proportion of the variation in y explained by the regression. As before, R^2 ranges between 0 and 1. The closer to 1 the value of R^2 is, the better the fit of the regression equation to the data. If R^2 is multiplied by 100, it represents the percentage of the variation in y explained by the regression.

Although R^2 has a nice interpretation, there is a drawback to its use in multiple regression. As more explanatory variables are added to the regression model, the value of R^2 will never decrease, even if the additional variables are explaining an insignificant proportion of the variation in y. The addition of these unnecessary explanatory variables is not desirable. An alternative measure of the goodness of fit that is useful in multiple regression is R^2 adjusted for degrees of freedom (or simply, adjusted R^2). Recall that another way of writing R^2 is as

$$R^2 = 1 - \frac{SSE}{SST}$$

SSE/SST can be interpreted as the unexplained proportion of the total variation in y. Because the addition of explanatory variables to the model causes SSE to decrease, R^2 gets increasingly closer to 1. This happens even if the added explanatory variables have little significant relationship to y.

The adjusted R^2 does not suffer from this limitation. The adjusted R^2 is denoted as R^2_{adj}. It is computed by

$$R^2_{adj} = 1 - \frac{SSE/(n - K - 1)}{SST/(n - 1)}$$

Note that the sums of squares have been divided by (adjusted by) their degrees of freedom before they are used in computing R^2_{adj}. Now, suppose an explanatory variable is added to the regression model that produces only a very small decrease in SSE. The divisor, $n - K - 1$, also decreases because the number of explanatory variables, K, has been incremented by 1. It is possible that $SSE/(n - K - 1)$ may increase if the decrease in SSE from the addition of an explanatory variable is very small, because there is also a decrease in the size of the divisor. Thus, R^2_{adj} may decrease when the added explanatory variable adds little to the ability of the model to explain the variation in y. It is also possible that negative R^2_{adj} values may occur. This is not a mistake, but a result of a model that fits the data very poorly. (In

MINITAB, when negative R_{adj}^2 values occur, they are printed as 0.0. Excel and SAS will print the actual value.)

R_{adj}^2 no longer represents the proportion of variation in y explained by the regression, but it can be useful when comparing two regressions with different numbers of explanatory variables (say, a two-variable model with a three- or more variable model). A decrease in R_{adj}^2 from the addition of one or more explanatory variables signals that the added variable(s) was of little importance in the regression equation. R_{adj}^2 is purely a descriptive measure, however. The t test discussed previously can be used to compare two regressions that differ by just one variable. Comparing two regressions that differ by more than one variable will be discussed in more detail in Section 4.4.

MINITAB, SAS, and Excel print out the R^2 value for the regression and the value of R_{adj}^2. For regression routines that do not print out R_{adj}^2, it can be computed using the equation

$$R_{adj}^2 = 1 - \frac{SSE/(n - K - 1)}{SST/(n - 1)}$$

or by using the relationship

$$R_{adj}^2 = 1 + \frac{(n - 1)}{(n - K - 1)}(R^2 - 1)$$

Excel also prints a measure called the multiple correlation coefficient, R, which is the positive square root of R^2. The multiple correlation coefficient is equal to the simple correlation between the predicted y values, $\hat{y}$, and the true y values. Thus, it represents a measure of how closely associated the true values of y are with the points on the regression line. R^2 may be a preferable measure of goodness of fit because of its interpretation as percentage of variance explained.

4.3.2 THE F STATISTIC

Another measure of how well the multiple regression equation fits the data is the F statistic:

$$F = \frac{MSR}{MSE}$$

MSR is the mean square due to the regression, or the regression sum of squares divided by its degrees of freedom:

$$MSR = \frac{SSR}{K}$$

Note that SSR has K degrees of freedom, where K is the number of explanatory variables in the model. MSE is the mean square due to error, or the error sum of squares divided by its degrees of freedom:

$$MSE = \frac{SSE}{n - K - 1}$$

SSE has $n - K - 1$ degrees of freedom.

The F statistic is used to test the hypotheses

$$H_0: \beta_1 = \beta_2 = \cdots = \beta_K = 0$$
H_a: At least one coefficient is not equal to zero

The decision rule for the test is:

Reject H_0 if $F > F(a; K, n - K - 1)$
Do not reject H_0 if $F \leq F(a; K, n - K - 1)$

where $F(a; K, n - K - 1)$ is a value chosen from the F table for the appropriate level of significance, a. The critical value depends on the number of degrees of freedom associated with the numerator of the F statistic, K, and the number of degrees of freedom associated with the denominator, $n - K - 1$.

Failing to reject the null hypothesis implies that the explanatory variables in the regression equation are of little or no use in explaining the variation in the dependent variable, y. Rejection of the null hypothesis implies that *at least one* of the explanatory variables helps explain the variation in y. Rejection does not mean that all the population regression coefficients are different from zero (although this *may* be the case). Rejection does mean that the regression equation is useful, however.

If the hypothesis that all the population regression coefficients are zero is rejected, the t test discussed previously can be used to determine which of the individual variables contribute significantly to the model's ability to explain the variation in y. If the null hypothesis is not rejected, there is no need to perform the individual t tests. The F test can be thought of as a global test designed to assess the overall fit of the regression.

The information necessary to perform F tests is typically included in computer regression output in the ANOVA table. The Excel, SAS, and MINITAB outputs provide the computed value of the F statistic and the p value associated with the statistic (as shown in Figure 4.7).

EXAMPLE 4.3 **Meddicorp (Continued)** Refer again to the MINITAB output in Figure 4.4, the Excel output in Figure 4.5, or the SAS output in Figure 4.6.

1. What percentage of the variation in sales has been explained by the regression?

Answer: 0.855, or 85.5%

2. What is the adjusted R^2?

Answer: 0.842

3. Conduct the F test for overall fit of the regression. Use a 5% level of significance.

$H_0: \beta_1 = \beta_2 = 0$
H_a: At least one coefficient is not equal to zero

Answer 1: Using the standardized test statistic:
   Decision Rule: Reject H_0 if $F > F(0.05;2,22) = 3.44$
                    Do not reject H_0 if $F \leq F(0.05;2,22) = 3.44$
   Test Statistic: $F = 64.83$
   Decision: Reject H_0

Answer 2: Using the p value:
   Decision Rule: Reject H_0 if p value < 0.05
                    Do not reject H_0 if p value ≥ 0.05
   Test Statistic: p value $= 0.000$
   Decision: Reject H_0

4. What conclusion can be drawn from the result of the F test for overall fit?

   Answer: At least one of the coefficients (β_1, β_2) is not equal to zero. In other words, at least one of the variables (x_1, x_2) is important in explaining the variation in y.

EXERCISES

1. **Cost Control.** Ms. Karen Ainsworth is an employee of a well-known accounting firm's management services division. She is currently on a consulting assignment to the Apex Corporation, a firm that produces corrugated paper for use in making boxes and other packing materials. Apex called in consulting help to improve its cost control program, and Ms. Ainsworth is analyzing manufacturing costs to understand more fully the important influences on these costs. She has assembled monthly data on a group of variables, and she is using regression analysis to help assess how these variables are related to total manufacturing cost. The variables Ms. Ainsworth has selected to study, the data for which are contained in the file COST4 on the CD, are

    y, total manufacturing cost per month in thousands of dollars (COST)

    x_1, total production of paper per month in tons (PAPER)

    x_2, total machine hours used per month (MACHINE)

    x_3, total variable overhead costs per month in thousands of dollars (OVERHEAD)

    x_4, total direct labor hours used each month (LABOR)

    The data shown in Table 4.2 refer to the period January 2001 through March 2003. Ms. Ainsworth wants to use a cost function developed by means of regression analysis that initially includes all four of the explanatory variables. Use the regression results in Figure 4.8 to help answer the following questions:

    a. What is the equation that is determined using all four explanatory variables?

    b. Conduct the F test for overall fit of the regression. State the hypotheses to be tested, the decision rule, the test statistic, and your decision. Use a 5% level of significance. What conclusion can be drawn from the result of the test?

    c. In the cost accounting literature, the sample regression coefficient corresponding to x_k is regarded as an estimate of the true marginal cost of output associated with the variable x_k. Find a point estimate of the true marginal cost associated with total machine hours per month. Also, find a 95% confidence interval estimate of the true marginal cost associated with total machine hours.

    d. Test the hypothesis that the true marginal cost of output associated with total production of paper is 1.0. Use a 5% level of significance and a two-tailed test procedure. State the hypotheses to be tested, the decision rule, the test statistic, and your decision. What conclusion can be drawn from the result of the test?

    e. What percentage of the variation in y has been explained by the regression?

TABLE 4.2 Data for Cost Control Exercise

COST	PAPER	MACHINE	OVERHEAD	LABOR
1102	550	218	112	325
1008	502	199	99	301
1227	616	249	126	376
1395	701	277	143	419
1710	838	363	191	682
1881	919	399	210	751
1924	939	411	216	813
1246	622	248	124	371
1255	626	259	127	383
1314	659	266	135	402
1557	740	334	181	546
1887	901	401	216	655
1204	610	238	117	351
1211	598	246	124	370
1287	646	259	127	387
1451	732	286	155	433
1828	891	389	208	878
1903	932	404	216	660
1997	964	430	233	694
1363	680	271	129	405
1421	723	286	146	426
1543	784	317	158	478
1774	841	376	199	601
1929	922	415	228	679
1317	647	260	126	378
1302	656	255	117	380
1388	704	281	142	429

Source: These data were created by Dr. Roger L. Wright, RLW Analytics, Inc., Sonoma, CA, and are used (with modification) with his permission.

FIGURE 4.8 Regression Results for Cost Control Exercise.

Variable	Coefficient	Std Dev	T Stat	P Value
Intercept	51.72	21.70	2.38	0.026
PAPER	0.95	0.12	7.90	0.000
MACHINE	2.47	0.47	5.31	0.000
OVERHEAD	0.05	0.53	0.09	0.927
LABOR	-0.05	0.04	-1.26	0.223

Standard Error = 11.0756   R-Sq = 99.9%   R-Sq(adj) = 99.9%

Analysis of Variance

Source	DF	Sum of Squares	Mean Square	F Stat	P Value
Regression	4	2271423	567856	4629.17	0.000
Error	22	2699	123		
Total	26	2274122			

FIGURE 4.9 Regression Results for Salaries Exercise.

```
Variable      Coefficient    Std Dev       T Stat      P Value

Intercept     3179.5         383.4         8.29        0.000
EDUC          139.6          27.7          5.04        0.000
EXPER         1.5            0.7           2.13        0.036
TIME          20.6           6.2           3.35        0.001

Standard Error = 602.728    R-Sq = 30.2%    R-Sq(adj) = 27.9%

Analysis of Variance
Source        DF       Sum of Squares    Mean Square    F Stat    P Value

Regression    3        13991247          4663749        12.84     0.000
Error         89       32332043          363281
Total         92       46323290
```

f. What is the adjusted R^2 for this regression?

g. Based on the regression equation, what actions might be taken to control costs?

2. **Salaries.** The file on the CD named HARRIS4 contains values of the following four variables for 93 employees of Harris Bank Chicago in 1977:

   y, beginning salary in dollars (SALARY)
   x_1, years of schooling at the time of hire (EDUC)
   x_2, number of months of previous work experience (EXPER)
   x_3, number of months after January 1, 1969, that the individual was hired (TIME)

The regression results for the regression of SALARY on the three explanatory variables are shown in Figure 4.9. Use the results to help answer the following questions:

a. What is the estimated regression equation relating SALARY to EDUC, EXPER, and TIME?

b. Conduct the F test for overall fit of the regression. Use a 5% level of significance. State the hypotheses to be tested, the decision rule, the test statistic, and your decision. What conclusion can be drawn from the result of the test?

c. Is education linearly related to beginning salary (after taking into account the effect of experience and time)? Perform the hypothesis test necessary to answer this question. State the hypotheses to be tested, the decision rule, the test statistic, and your decision. Use a 5% level of significance.

d. What percentage of the variation in salary has been explained by the regression?

4.4 COMPARING TWO REGRESSION MODELS

4.4.1 FULL AND REDUCED MODEL COMPARISONS USING SEPARATE REGRESSIONS

Thus far, two types of hypothesis tests for multiple regression models have been considered:

1. A test of the overall fit of the regression:

$$H_0: \beta_1 = \beta_2 = \cdots = \beta_K = 0$$

H_a: At least one coefficient is not equal to zero

2. A test of the significance of each individual regression coefficient:

$$H_0: \beta_k = 0$$
$$H_a: \beta_k \neq 0$$

In multiple regression models, it also may be useful to test whether subsets of coefficients are equal to zero. In this section, a *partial F test* to test whether any subset of coefficients in a multiple regression equals zero is considered.

To set up this hypothesis test, consider the following regression model:

$$y = \beta_0 + \beta_1 x_1 + \beta_2 x_2 + \cdots + \beta_L x_L + \beta_{L+1} x_{L+1} + \cdots + \beta_K x_K + e$$

Testing whether the variables $x_{L+1}, \ldots, x_K$ are useful in explaining any variation in y after taking account of the variation already explained by $x_1, \ldots, x_L$ can be viewed as a comparison of two regression models to determine whether it is worthwhile to include the additional variables. The two models for comparison are called the *full* and *reduced* models.

Full Model

$$y = \beta_0 + \beta_1 x_1 + \beta_2 x_2 + \cdots + \beta_L x_L + \beta_{L+1} x_{L+1} + \cdots + \beta_K x_K + e$$

This is called the full model because all K explanatory variables of interest are included.

Reduced Model

$$y = \beta_0 + \beta_1 x_1 + \beta_2 x_2 + \cdots + \beta_L x_L + e$$

This is called the reduced model because the variables $x_{L+1}, \ldots, x_K$ have been removed.

The question to be answered is, "Is the full model significantly better than the reduced model at explaining the variation in y?" This question can be formalized by setting up the following null and alternative hypotheses:

$H_0: \beta_{L+1} = \cdots = \beta_K = 0$

$H_a:$ At least one of the coefficients $\beta_{L+1}, \ldots, \beta_K$ is not equal to zero

If the null hypothesis is not rejected, choose the reduced model; if the null hypothesis is rejected, at least one of $x_{L+1}, \ldots, x_K$ is contributing to the explanation of the variation in y, and the full model is chosen as superior to the reduced.

To test the hypotheses (that is, to compare the full and reduced models), an F statistic is used. The F statistic can be written:

$$F = \frac{(SSE_R - SSE_F)/(K - L)}{SSE_F/(n - K - 1)}$$

where the subscript F stands for full model and the subscript R stands for reduced model.

Now consider what is being computed in the F statistic. If the full and reduced models are estimated, the regression output includes the error sum of squares for each of these regressions. In the F statistic, SSE_F refers to the error sum of squares from the full model output using all K explanatory variables. SSE_R refers to the error sum of squares from the reduced model output using only L explanatory variables. Recall that the error sum of squares represents the variation in y unexplained by the regression. Also, the reduced model error sum of squares can never be less than the full model error sum of squares, so the difference $SSE_R - SSE_F$ is always greater than or equal to zero. This difference represents the additional amount of the variation in y explained by adding $x_{L+1}, \ldots, x_K$ to the regression model. This measure of improvement is then divided by the number of additional variables to be added to the model, $K - L$. The numerator thus represents the additional variation in y explained per additional variable used. Note that the numerator degrees of freedom, $K - L$, is equal to the number of coefficients included in the null hypothesis or, equivalently, to the difference in the number of explanatory variables in the full and reduced models.

The mean square error for the full regression model is used in the denominator:

$$MSE_F = \frac{SSE_F}{n - K - 1}$$

If the measure of improvement is large relative to the mean square error for the full model, then the F statistic is large. If the improvement measure is small relative to MSE_F, then the value of the F statistic is small. The decision rule for the test is

Reject H_0 if $F > F(\alpha; K - L, n - K - 1)$
Do not reject H_0 if $F \leq F(\alpha; K - L, n - K - 1)$

Here, α is the probability of a Type I error, and $F(\alpha; K - L, n - K - 1)$ is a value chosen from the F table for level of significance α, $K - L$ numerator degrees of freedom, and $n - K - 1$ denominator degrees of freedom. This test is referred to as a partial F test and can be performed with any statistical package by running both the full and reduced model regressions.

EXAMPLE 4.4 **Meddicorp (continued)** Management of Meddicorp believes that, in addition to advertising and bonus, two other explanatory variables may be important in explaining the variation in sales. These variables are

x_3, market share currently held by Meddicorp in each territory (MKTSHR)

x_4, largest competitor's sales in each territory (COMPET)

These two additional variables are shown in Table 4.3 for each territory and in the file MEDDICORP4 on the CD.

The regression results for the regression of SALES on ADV, BONUS, MKTSHR, and COMPET are shown in Figure 4.10. This is the full model output. The hypothesized population regression model is

$$y = \beta_0 + \beta_1 x_1 + \beta_2 x_2 + \beta_3 x_3 + \beta_4 x_4 + e$$

Consider the test of the hypotheses

$H_0: \beta_3 = \beta_4 = 0$

$H_a:$ At least one of the coefficients (β_3, β_4) is not equal to zero

TABLE 4.3 Additional Data for Meddicorp Example

Territory	MKTSHR (percentage)	COMPET (in thousand $)
1	33	202.22
2	29	252.77
3	34	293.22
4	24	202.22
5	32	303.33
6	29	353.88
7	28	374.11
8	31	404.44
9	20	394.33
10	30	303.33
11	25	333.66
12	34	353.88
13	42	262.88
14	28	333.66
15	28	232.55
16	30	273.00
17	29	323.55
18	32	404.44
19	36	283.11
20	34	222.44
21	31	283.11
22	32	242.66
23	28	333.66
24	27	313.44
25	26	374.11

FIGURE 4.10 Results for the Regression of SALES on ADV, BONUS, MKTSHR, and COMPET.

Variable	Coefficient	Std Dev	T Stat	P Value
Intercept	-593.537	259.196	-2.29	0.033
ADV	2.513	0.314	8.00	0.000
BONUS	1.906	0.742	2.57	0.018
MKTSHR	2.651	4.636	0.57	0.574
COMPET	-0.122	0.372	-0.32	0.749

Standard Error = 93.7697   R-Sq = 85.9%   R-Sq(adj) = 83.1%

Analysis of Variance

Source	DF	Sum of Squares	Mean Square	F Stat	P Value
Regression	4	1073119	268280	30.51	0.000
Error	20	175855	8793		
Total	24	1248974			

The reduced model output is in Figure 4.4 (MINITAB), 4.5 (Excel), or 4.6 (SAS). The F statistic can be computed as

$$F = \frac{(181{,}176 - 175{,}855)/2}{175{,}855/20} = 0.303$$

(Note that $K = 4$ and $L = 2$, so $K - L = 2$.) If a 5% level of significance is used, the decision rule is

Reject H_0 if $F > 3.49$

Do not reject H_0 if $F \leq 3.49$

where 3.49 is the 5% F critical value with 2 numerator and 20 denominator degrees of freedom.

We do not reject H_0 and so the conclusion is that both coefficients β_3 and β_4 are equal to zero. Thus, the variables x_3 and x_4 are not useful in explaining any of the remaining variation in y.

SAS allows the user to request that the partial F test be conducted for any group of x variables desired (see Using the Computer at the end of this chapter). Figure 4.11 shows the result of requesting a partial F test for the variables MKTSHR and COMPET. The value of the F statistic is shown as 0.3. A p value of 0.7422 is also provided. Using the p value decision rule, the null hypothesis would not be rejected at reasonable levels of significance.

4.4.2 FULL AND REDUCED MODEL COMPARISONS USING CONDITIONAL SUMS OF SQUARES[3]

Another way to view partial F tests is through the use of conditional or sequential sums of squares. For a regression model with two explanatory variables,

$$\hat{y} = b_0 + b_1 x_1 + b_2 x_2$$

the standard ANOVA table appears as in Figure 4.12(a). In Figure 4.12(b), an alternative ANOVA table is presented. In this figure, the regression sum of squares has been decomposed into two parts. The first, $SSR(x_1)$, is the sum of squares explained by x_1 if it were the only explanatory variable. The second $SSR(x_2|x_1)$, is called a *conditional* or *sequential sum of squares*. It represents the sum of squares explained by x_2 in addition to that explained by x_1. That is, given that x_1 has explained a certain amount of variation in y, $SSR(x_2|x_1)$ shows how much of the remaining variation x_2 explains. Note that

$$SSR = SSR(x_1) + SSR(x_2|x_1)$$

Here, SSR is the variation explained by both x_1 and x_2.

To test the hypotheses

$$H_0: \beta_2 = 0$$
$$H_a: \beta_2 \neq 0$$

FIGURE 4.11 SAS Output Showing the Partial F Test for the MKTSHR and COMPET Variables.

Test 1 Results for Dependent Variable SALES

Source	DF	Mean Square	F Value	Pr > F
Numerator	2	2660.61069	0.30	0.7422
Denominator	20	8792.75990		

---

[3] Optional section.

FIGURE 4.12 ANOVA Tables for Two Explanatory Variable Regressions.

(a) Standard ANOVA Table

Source of Variation	DF	SS
Regression	2	SSR
Error	$n-3$	SSE
Total	$n-1$	SST

(b) ANOVA with Conditional Sums of Squares Explained by Each Explanatory Variable

Source of Variation	DF	SS
Regression		
x_1	1	$SSR(x_1)$
$x_2\|x_1$	1	$SSR(x_2\|x_1)$
Error	$n-3$	SSE
Total	$n-1$	SST

an F statistic can be constructed using the conditional sum of squares.

$$F = \frac{SSR(x_2|x_1)/1}{SSE/(n-3)}$$

The numerator of F is the conditional sum of squares for x_2, given that x_1 is in the model, divided by its degrees of freedom. The conditional sum of squares has 1 degree of freedom because it represents the sum of squares explained by only one variable. The denominator is the error sum of squares for the full model, the model with both x_1 and x_2, divided by its degrees of freedom. If x_2 explains little of the additional unexplained variation in y, then $SSR(x_2|x_1)$ is small as is the F statistic. The more variation explained by x_2, the bigger the F statistic is. The decision rule for the test is

Reject H_0 if $F > F(\alpha; 1, n-3)$

Do not reject H_0 if $F \leq F(\alpha; 1, n-3)$

where $F(\alpha; 1, n-3)$ is chosen from an F table for level of significance α, 1 is the numerator degree of freedom, and $n-3$ is the denominator degrees of freedom.

Of course, this hypothesis could be tested with a two-tailed t test as described in Section 4.2 because it involves only one coefficient. It can be shown that an F statistic with 1 numerator degree of freedom is equal to the square of a t statistic and that the F critical value equals the square of a t critical value for appropriately chosen levels of significance and degrees of freedom, denoted df:

$$F(\alpha; 1, df) = t^2_{\alpha/2, df}$$

Therefore, the decision made is the same regardless of which test is used. Since t statistics routinely appear on regression output, the t test is typically used when testing hypotheses about individual coefficients. The t test has additional advantages over the F test. The t test can be used to perform one-tailed hypothesis tests, whereas the F is restricted to the two-tailed test. It is also easier to test whether a coefficient is equal to some value other than zero using a t test than it is using an F test.

The F test gains its advantage when testing whether a subset of coefficients are all equal to zero—for example, to test

$H_0: \beta_{L+1} = \cdots = \beta_K = 0$

H_a: At least one of the coefficients $\beta_{L+1}, \ldots, \beta_K$ is not equal to zero

for the general model presented earlier. In this case, the t test cannot be used. Even performing individual t tests on each coefficient may not provide as much information as performing the F test on the coefficients as a group.

To test whether $\beta_{L+1}, \ldots, \beta_K$ are all zero, the following F statistic is used:

$$F = \frac{SSR(x_{L+1}, \ldots, x_K | x_1, x_2, \ldots, x_L)/(K - L)}{SSE/(n - K - 1)}$$

$SSR(x_{L+1}, \ldots, x_K | x_1, x_2, \ldots, x_L)$ is the additional variation in y explained by $x_{L+1}, \ldots, x_K$, given that $x_1, \ldots, x_L$ are already in the model. The number of degrees of freedom associated with this conditional sum of squares is $K - L$, the number of coefficients to be included in the test. SSE is the error sum of squares from the model with all the variables included and is divided by its degrees of freedom, $n - K - 1$. The conditional sum of squares can be computed as

$$SSR(x_{L+1}, \ldots, x_K | x_1, x_2, \ldots, x_L) = SSR(x_{L+1} | x_1, x_2, \ldots, x_L) \\ + SSR(x_{L+2} | x_1, x_2, \ldots, x_{L+1}) + \ldots + SSR(x_K | x_1, x_2, \ldots, x_{K-1})$$

The regression output from certain statistical packages contains the necessary information to compute the conditional sums of squares. For example, in MINITAB, an additional sum of squares breakdown is provided as in Figure 4.13. The table provides the conditional sums of squares for each of the variables individually, given that the previous variables are in the model. By adding these sums of squares for appropriate individual variables, the conditional sum of squares for x_{L+1} through x_K is obtained. In the MINITAB output, the conditional sums of squares are denoted SEQ SS for sequential sums of squares. This term is used to indicate that the sums of squares explained by each of the x variables when entered sequentially are represented.

The order in which the variables enter the regression is very important when using the conditional sums of squares to construct a partial F statistic. In MINITAB, for example, the explanatory variables whose coefficients are included in the null hypothesis must be the last ones in the list of variables in the regression dialog box (see Using the Computer at the end of this chapter). This ensures that the conditional sums of squares are computed appropriately for the hypothesis to be tested.

A more extensive look at the use of these conditional sums of square is provided in the following example.

EXAMPLE 4.5 **Meddicorp (continued)** Consider again the problem posed in Example 4.4. In addition to BONUS and ADV, Meddicorp wants to consider the possibility that MKTSHR and COMPET are important in explaining the variation in sales. The

4.4 Comparing Two Regression Models

FIGURE 4.13
Breakdown of SSR into Its Conditional Components as Provided by MINITAB Regression Output.

SOURCE	DF	SEQ SS
x_1	1	$SSR(x_1)$
x_2	1	$SSR(x_2|x_1)$
.	.	.
.	.	.
.	.	.
x_{L-1}	1	$SSR(x_{L-1}|x_1, \ldots, x_{L-2})$
x_L	1	$SSR(x_L|x_1, \ldots, x_{L-1})$
x_{L+1}	1	$SSR(x_{L+1}|x_1, \ldots, x_L)$
.	.	.
.	.	.
.	.	.
x_K	1	$SSR(x_K|x_1, \ldots, x_{K-1})$

MINITAB regression of SALES on ADV, BONUS, MKTSHR, and COMPET is shown in Figure 4.14.

The hypothesized population regression model is

$$y = \beta_0 + \beta_1 x_1 + \beta_2 x_2 + \beta_3 x_3 + \beta_4 x_4 + e$$

Consider the test of the hypotheses

$H_0: \beta_3 = \beta_4 = 0$

H_a: At least one of the coefficients (β_3, β_4) is not equal to zero

The MINITAB regression output gives the conditional sums of squares explained by each variable. The regression sum of squares for the full regression is $SSR = 1{,}073{,}119$. The conditional sums of squares are as follows:

$$SSR(x_1) = 1{,}012{,}408$$
$$SSR(x_2|x_1) = 55{,}389$$
$$SSR(x_3|x_1, x_2) = 4394$$
$$SSR(x_4|x_1, x_2, x_3) = 927$$

The decision rule for the test is

Reject H_0 if $F > 3.49$

Do not reject H_0 if $F \leq 3.49$

where 3.49 is the F critical value using a 5% level of significance with 2 and 20 degrees of freedom.

The test statistic, F, is

$$F = \frac{SSR(x_3, x_4|x_1, x_2)/2}{SSE/20} = \frac{[SSR(x_4|x_1, x_2, x_3) + SSR(x_3|x_1, x_2)]/2}{SSE/20}$$

$$= \frac{(927 + 4394)/2}{175{,}855/20} = 0.303$$

FIGURE 4.14
MINITAB Regression of SALES on ADV, BONUS, MKTSHR, and COMPET.

```
The regression equation is
SALES = - 594 + 2.51 ADV + 1.91 BONUS + 2.65 MKTSHR - 0.121 COMPET

Predictor            Coef      SE Coef         T         P
Constant           -593.5        259.2     -2.29     0.033
ADV                2.5131       0.3143      8.00     0.000
BONUS              1.9059       0.7424      2.57     0.018
MKTSHR              2.651        4.636      0.57     0.574
COMPET            -0.1207       0.3718     -0.32     0.749

S = 93.7697    R-Sq = 85.9%    R-Sq(adj) = 83.1%

Analysis of Variance
Source           DF         SS         MS         F         P
Regression        4    1073119     268280     30.51     0.000
Residual Error   20     175855       8793
Total            24    1248974

Source           DF     Seq SS
ADV               1    1012408
BONUS             1      55389
MKTSHR            1       4394
COMPET            1        927

Unusual Observations
Obs  ADV   SALES    Fit   SE Fit  Residual   St Resid
 20  525  1159.3  1346.2    39.4    -187.0     -2.20R
R denotes an observation with a large standardized residual.
```

The null hypothesis cannot be rejected. The variables x_3 and x_4 do not significantly improve the model's ability to explain sales.

EXERCISES

3. **Cost Control (continued).** Consider again the cost data from Exercise 4.1 and the regression results in Figure 4.8. Consider this output to be for the full model:

$$y = \beta_0 + \beta_1 x_1 + \beta_2 x_2 + \beta_3 x_3 + \beta_4 x_4 + e$$

where y, x_1, x_2, x_3, and x_4 were defined in the first exercise.

Now consider the reduced model:

$$y = \beta_0 + \beta_1 x_1 + \beta_2 x_2 + e$$

Conduct the test to compare these two models. State the hypotheses to be tested, the decision rule, the test statistic, and your decision. What conclusion can be drawn from the result of the test? The regression results for the reduced model can be found in Figure 4.15. Use a 5% level of significance.

4. **Salaries (continued).** Consider again the salary data from Exercise 4.2 and the regression results in Figure 4.9. Consider this output to be for the full model:

$$y = \beta_0 + \beta_1 x_1 + \beta_2 x_2 + \beta_3 x_3 + e$$

where y, x_1, x_2, and x_3 were defined in Exercise 4.2.

FIGURE 4.15
Regression Results for the Reduced Model in the Cost Control Exercise.

```
Variable     Coefficient    Std Dev      T Stat      P Value
Intercept    59.4318        19.6388      3.03        0.006
PAPER         0.9489         0.1101      8.62        0.000
MACHINE       2.3864         0.2101     11.36        0.000

Standard Error = 10.9835   R-Sq = 99.9%   R-Sq(adj) = 99.9%

Analysis of Variance
Source         DF    Sum of Squares   Mean Square   F Stat    P Value

Regression      2        2271227        1135613     9413.48   0.000
Error          24           2895            121
Total          26        2274122
```

FIGURE 4.16
Regression Results for the Reduced Model in the Salaries Exercise.

```
Variable     Coefficient    Std Dev      T Stat      P Value

Intercept    3818.56        377.44      10.12        0.000
EDUC          128.09         29.70       4.31        0.000

Standard Error = 650.112   R-Sq = 17.0%   R-Sq(adj) = 16.1%

Analysis of Variance
Source         DF    Sum of Squares   Mean Square   F Stat    P Value

Regression      1        7862534        7862534     18.60     0.000
Error          91       38460756         422646
Total          92       46323290
```

Now consider the reduced model:

$$y = \beta_0 + \beta_1 x_1 + e$$

Conduct the test to compare these two models. State the hypotheses to be tested, the decision rule, the test statistic, and your decision. What conclusion can be drawn from the result of the test? The regression results for the reduced model can be found in Figure 4.16. Use a 5% level of significance.

4.5 PREDICTION WITH A MULTIPLE REGRESSION EQUATION

As with simple regression, one of the possible goals of fitting a multiple regression equation is using it to predict values of the dependent variable. The two cases considered here are the same as in simple regression.

4.5.1 ESTIMATING THE CONDITIONAL MEAN OF Y GIVEN $X_1, X_2, \ldots, X_K$

In this case, the goal is to estimate the point on the regression surface for specific values of the explanatory variables. For example, in the Meddicorp example (Example 4.1), consider the population regression equation

$$\mu_{y|x_1, x_2} = \beta_0 + \beta_1 x_1 + \beta_2 x_2$$

where x_1 is ADV and x_2 is BONUS. The estimated regression equation from Figure 4.4 (MINITAB), 4.5(Excel), or 4.6(SAS) is

$$\hat{y} = -516.4 + 2.47x_1 + 1.86x_2$$

A point estimate of the conditional mean of y given x_1 and x_2 can be written as

$$\hat{y}_m = b_0 + b_1 x_1 + b_2 x_2$$

In the Meddicorp problem, the point estimate of the conditional mean of y given $x_1 = 500$ and $x_2 = 250$ is

$$\hat{y}_m = -516.4 + 2.47(500) + 1.86(250) = 1183.6$$

This is an estimate of average sales for *all* territories with advertising 500 and bonus 250. Confidence interval estimates can also be constructed. The formula for s_m, the standard deviation of $\hat{y}_m$, is omitted here due to its complexity.

Figure 4.17 shows the MINITAB output for this example, including the predicted value (*Fit*), the standard error of the estimate of the point on the regression line, s_m, (*SE Fit*), a 95% confidence interval for the estimate of the conditional mean (*95% CI*), and a 95% prediction interval for each prediction (*95% PI*). (The difference in $\hat{y}_m$ computed by MINITAB and by hand is due to rounding. The MINITAB forecasts are more accurate and therefore are preferred.)

4.5.2 PREDICTING AN INDIVIDUAL VALUE OF Y GIVEN $X_1, X_2, \ldots, X_K$

Write the population regression equation for a single individual as

$$y_i = \beta_0 + \beta_1 x_{1i} + \beta_2 x_{2i} + e_i$$

where e_i is the random disturbance. Denote the predicted value of y for an individual as $\hat{y}_p$. To predict the value of a dependent variable for a single individual, the point on the regression surface is used:

$$\hat{y}_p = b_0 + b_1 x_1 + b_2 x_2$$

As in simple regression, the point estimate of $\mu_{y|x_1, x_2, \ldots, x_K}$ and the point prediction for an individual are the same. However, the standard error of the prediction, s_p, is larger than the standard error of the forecast, s_m. Thus, the prediction interval is wider than the confidence interval, reflecting the greater uncertainty in predicting for individuals than in estimating a conditional mean.

FIGURE 4.17 Prediction in the Meddicorp Example Using MINITAB.

```
Predicted Values for New Observations
New
Obs        Fit      SE Fit      95% CI              95% PI
 1       1184.2      25.2    (1131.8, 1236.6)    (988.8, 1379.5)

Values of Predictors for New Observations
New
Obs      ADV       BONUS
 1       500        250
```

MINITAB and SAS produce prediction intervals when requested. The forecast standard error, s_m, also is printed in both outputs. If the prediction standard error is desired, it can be computed using the relationship $s_p^2 = s_m^2 + s_e^2$, where s_e^2 is the *MSE* of the regression.

4.6 MULTICOLLINEARITY: A POTENTIAL PROBLEM IN MULTIPLE REGRESSION

4.6.1 CONSEQUENCES OF MULTICOLLINEARITY

For a regression of y on K explanatory variables $x_1, x_2, \ldots, x_K$, it is hoped that the explanatory variables are highly correlated with the dependent variable. A relationship is sought that explains a large portion of the variation in y. At the same time, however, it is not desirable for strong relationships to exist among the explanatory variables. When explanatory variables are correlated with one another, the problem of *multicollinearity* is said to exist. How serious the problem is depends on the degree of multicollinearity. Low correlations among the explanatory variables generally do not result in serious deterioration of the quality of the least-squares estimates. But high correlations may result in highly unstable least-squares estimates of the regression coefficients.

The presence of a high degree of multicollinearity among the explanatory variables results in the following problems:

1. The standard deviations of the regression coefficients are disproportionately large. As a result, the t values computed to test whether the population regression coefficients are zero are small. The null hypothesis that the coefficients are zero may not be rejected even when the associated variable is important in explaining variation in y.

2. The regression coefficient estimates are unstable. Because of the high standard errors, reliable estimates of the regression coefficients are difficult to obtain. Signs of the coefficients may be the opposite of what is intuitively reasonable. Dropping one variable from the regression or adding a variable causes large changes in the estimates of the coefficients of other variables.

4.6.2 DETECTING MULTICOLLINEARITY

Numerous ways have been suggested in the literature to help detect multicollinearity. These are listed here with some recommendations on their usefulness:

1. **Pairwise Correlations:** Compute the pairwise correlations between the explanatory variables. Because multicollinearity exists when explanatory variables are highly correlated, these correlations should help identify any highly correlated pairs of variables. One rule of thumb suggested by some researchers is that multicollinearity may be a serious problem if any pairwise correlation is bigger than 0.5.

Limitations: There are two limitations to this approach. First, the correlation cutoff of 0.5 is somewhat arbitrary and not always effective in identifying serious pairwise multicollinearity problems. Second, only relationships between two explanatory variables can be investigated. For example, if there are three explanatory variables in the model, x_1, x_2, and x_3, the pairwise correlations can be computed between x_1 and x_2, x_1 and x_3, and x_2 and x_3. But the relationships resulting in the multicollinearity may be more complex than simple pairwise correlations. The variable x_1 may not be highly correlated with x_2 or x_3 individually, but may be highly correlated with some linear combination of the two variables. That is, x_1 may be highly correlated with $a_1 x_2 + a_2 x_3$.

Another suggested rule of thumb is that multicollinearity may be a serious problem if any of the pairwise correlations among the x variables is larger than the largest of the correlations between the y variable and the x variables. Although this rule does not suffer from an arbitrary cutoff point (such as 0.5), it does suffer from the same limitations concerning more complex relationships among the x variables.

2. **Large F, small t:** An indication of multicollinearity is a large overall F statistic but small t statistics. As mentioned, multicollinearity results in large standard deviations of the regression coefficients and small t ratios. Thus, the test for whether the individual regression coefficients are equal to zero may fail to reject the null hypotheses H_0: $\beta_k = 0$ even when the variables included in the regression are important in explaining the variation in y. The overall F statistic is typically not affected by the multicollinearity, however. If the variables are important, the F statistic should be large, indicating a good overall fit even if the t statistics appear to be saying that none of the variables is important.

Limitations: This method of detecting multicollinearity is not always effective because multicollinearity may result in some, but not all, of the t values being small. The question of whether the variable is unimportant or whether it just appears so because of multicollinearity cannot be answered by looking at the output. Although this approach may be helpful in pointing out that multicollinearity exists in some instances, it does not provide any information on which of the explanatory variables are highly correlated with others.

3. **Variance Inflation Factors (VIFs):** Let $x_1, x_2, \ldots, x_K$ be the K explanatory variables in a regression. Perform the regression of x_j on the remaining $K - 1$ explanatory variables and call the coefficient of determination from this regression R_j^2. The VIF for the variable x_j is

$$VIF_j = \frac{1}{1 - R_j^2}$$

A variance inflation factor can be computed for each explanatory variable. It is a measure of the strength of the relationship between each explanatory variable and all other explanatory variables in the regression. Thus, pairwise correlations are taken into account as well as more complex relationships with two or more of the other variables. The value R_j^2 measures the strength of the relationship between x_j and the other $K - 1$ explanatory variables. If there is no relationship (an ideal case), then $R_j^2 = 0.0$ and $VIF_j = 1/(1 - 0) = 1$. As R_j^2 increases, VIF_j increases also. For example, if $R_j^2 = 0.9$, then $VIF_j = 1/(1 - 0.9) = 10$; if $R_j^2 = 0.99$, then $VIF_j = 1/(1 - 0.99) = 100$. Large values of VIF_j suggest that x_j may be highly related to other explanatory variables and, thus, multicollinearity may be a problem. How large the VIFs must be to suggest a serious problem with multicollinearity is not completely clear. Some suggested guidelines are as follows:

a. Any individual VIF_j larger than 10 indicates that multicollinearity may be influencing the least-squares estimates of the regression coefficients.

b. If the average of the VIF_j, $\overline{VIF} = \sum_{j=1}^{K} VIF_j/K$, is considerably larger than 1, then serious problems may exist. $\overline{VIF}$ indicates how many times larger the error sum of squares for the regression is due to multicollinearity than it is if the variables are uncorrelated.

c. VIFs also need to be evaluated relative to the overall fit of the model. Freund and Wilson[4] note that whenever the VIFs are less than $1/(1 - R^2)$ where R^2 is the coefficient of determination for the model with all x variables included, multicollinearity is not strong enough to affect the coefficient estimates. In this case the independent variables are more strongly related to the y variable than they are to each other.

4.6.3 CORRECTION FOR MULTICOLLINEARITY

One obvious solution to the multicollinearity problem is to remove those variables that are highly correlated with others and thus eliminate the problem. One obvious drawback of this approach is that no information is obtained on the omitted variables. In addition, the omission of one variable causes changes in the estimates of the regression coefficients of variables left in the equation. It is important for the researcher to make a careful selection of potential explanatory variables to include in the regression. If it is found that some of these variables result in multicollinearity problems, it may be necessary to exclude certain of the variables in favor of others. This is where the insight and expertise of the researcher come into play in developing the most useful regression equation.

In certain cases, adding more data can break the pattern of multicollinearity. But this solution is not always possible, especially in many business and economics situations. It also does not always work even when it is possible.

When multicollinearity is present, it affects the regression coefficient estimates in the ways noted earlier. However, it does not affect the ability to obtain a good fit of the regression (high R^2). Nor does it affect the quality of forecasts or predictions from the regression (as long as the pattern of multicollinearity continues for those observations where forecasts are desired). Thus, if the regression model is to be used strictly for forecasting, corrections may be unnecessary. Even when developing a

---

[4] Freund, R. J., and Wilson, W. J., *Regression Analysis: Statistical Modeling of a Response Variable*, San Diego: Academic Press, 1998.

model for forecasting, however, it is often desirable to test whether individual variables are contributing significantly to the explanatory power of the model. In this case, multicollinearity remains a problem.

Finally, several other statistical procedures have been proposed as possible remedial measures with multicollinearity. These include ridge regression and principal components regression. These techniques are beyond the scope of this text. The interested reader is referred to Raymond H. Myers, *Classical and Modern Regression with Applications*, pp. 243–263.

4.7 LAGGED VARIABLES AS EXPLANATORY VARIABLES IN TIME-SERIES REGRESSION

When using time-series data, it is possible to relate values of the dependent variable in the current time period to explanatory variable values in the current time period. For example, sales for a firm in the current month can be related to advertising expenditures in the current month. It may be, however, that sales in the current month are not affected as much by advertising expenditures in the current month as by advertising expenditures from the previous month or from 2 months ago. This fact can be incorporated into a time-series regression. To illustrate, let y_i represent sales in time period i, x_i represent advertising expenditures in time period i, x_{i-1} represent advertising expenditures in time period $i - 1$, and so on. Then a possible model for sales could be written

$$y_i = \beta_0 + \beta_1 x_i + \beta_2 x_{i-1} + \beta_3 x_{i-2} + e_i$$

Here the variable sales is modeled as a function of advertising expenditures in the current month and the 2 previous months.

The variables x_{i-1} and x_{i-2} are called *lagged variables*. Any lags felt to be appropriate may be used. Here the one- and two-period lags are used. Some caution must be exercised, however. Because lagged variables are likely to be highly correlated with each other, including several such variables may result in multicollinearity problems discussed in the previous section.

When lagged variables are used, a certain number of data points in the initial time periods are lost. This is illustrated in Table 4.4. Note that no value can be computed for x_{i-1} in time period 1. No prior time period exists from which to take this value. For the same reason, no value for the first or second time period can be computed for x_{i-2}. These time periods have to be omitted from the analysis, reducing the effective sample size from eight to six in the example.

Lagged values of the dependent variable also can be used as explanatory variables. Consider again the sales example. Now, however, assume that no information on advertising is available. Sales in the current month are modeled simply as a function of sales in the previous month:

$$y_i = \beta_0 + \beta_1 y_{i-1} + e_i$$

The data are illustrated in Table 4.5. Of course, further lags can be used if desired. One observation is lost for each lag.

TABLE 4.4 Creation of Lagged Values of the Explanatory Variables

i	x_i	x_{i-1}	x_{i-2}
1	4	*	*
2	7	4	*
3	8	7	4
4	10	8	7
5	11	10	8
6	9	11	10
7	15	9	11
8	16	15	9

*Indicates a missing value.

TABLE 4.5 Creation of Lagged Values of the Dependent Variable

i	y_i	y_{i-1}
1	22	*
2	24	22
3	27	24
4	35	27
5	38	35
6	42	38
7	47	42
8	50	47

*Indicates a missing value.

Note that the model with the lagged value of the dependent variable as an explanatory variable may be viewed as an extrapolative time-series model (introduced in Chapter 3). We are using only past information from the series itself to help describe the behavior of the series and to forecast future values.

It should be noted, however, that regressions that include lagged dependent variable values are still sometimes interpreted as causal relationships. For example, the one-period lagged value of sales might be included along with the current and lagged values of advertising:

$$y_i = \beta_0 + \beta_1 y_{i-1} + \beta_2 x_i + \beta_3 x_{i-1} + e_i$$

If it is believed that last month's sales might help to generate new sales in the current month or to maintain sales, then this model could be justified as causal.

In the economics literature, a model with lagged values of the dependent variable (and possible lagged values of other explanatory variables) might be called an adaptive expectations model or a partial adjustment model (see, for example, G. Judge et al., *The Theory and Practice of Econometrics,* pp. 379–380). For alternatives to regression analysis useful in analyzing time-series data, see B. Bowerman and R. O'Connel, *Forecasting and Time Series: An Applied Approach.*[5]

---

[5] See References for complete publication information.

EXAMPLE 4.6 **Unemployment Rate** The file on the CD named UNEMP4 contains monthly unemployment rates from January 1983 until December 2002. (These data were obtained from the web site www.economagic.com and were obtained from the St. Louis Federal Reserve Bank.) The data have been seasonally adjusted.

In Figure 4.18, a time-series plot of unemployment rate is shown. The horizontal axis in this plot is an index that numbers each month from 1 to 240.

In Figure 4.19, the regression of unemployment rate (UNEMP) on the one-period lagged unemployment rate is shown. The one-period lagged variable has been called UNEMPL1. Note that in any analyses involving UNEMPL1, there is one less observation than the total time-series length because missing cases are not used.

The regression of the monthly unemployment rate on the previous month's rate obviously produces a good fit. The t value for UMEMPL1 is 134.43 (p value = 0.000), resulting in rejection of the hypothesis H_0: $\beta_1 = 0$. Also, 98.7% of the variation in unemployment rate has been explained by the regression.

Now consider adding a two-period lagged variable, UNEMPL2, to the equation. The regression of UNEMP on UNEMPL1 and UNEMPL2 is shown in Figure 4.20. The regression model can now be written

$$y_i = \beta_0 + \beta_1 y_{i-1} + \beta_2 y_{i-2} + e_i$$

To test whether the two-period lag is of any importance in the model, the following hypotheses should be tested:

$$H_0: \beta_2 = 0$$
$$H_a: \beta_2 \neq 0$$

The test statistic value is 1.23 (p value = 0.218). At a 5% level of significance, the decision rule for the test is

Reject H_0 if $t > 1.96$ or $t < -1.96$

Do not reject H_0 if $-1.96 \leq t \leq 1.96$

The z value of 1.96 is used since a large number of degrees of freedom are available (235). The null hypothesis is not rejected, suggesting that the two-period lagged variable is not useful in explaining any of the additional variation in unemployment rates. A three-period lagged variable was also tried, but it, too, was found to be of no additional help in the regression.

Obviously, this process of lagging the dependent variable can be continued for additional lags if desired. However, the use of continued lagged variables as explanatory variables is questionable in this problem. With an R^2 of 98.7% for the one-period lag model, it is obvious that not much room for improvement is available. Even if an additional variable were found to be significant, it still might be best for the sake of simplicity to use the one-period lagged variable model.

Also, if additional lags are examined, problems can arise. For example, the lagged variables are highly correlated among themselves, which could result in

4.7 Lagged Variables as Explanatory Variables in Time-Series Regression 167

FIGURE 4.18 Time-Series Plot of Unemployment Rates.

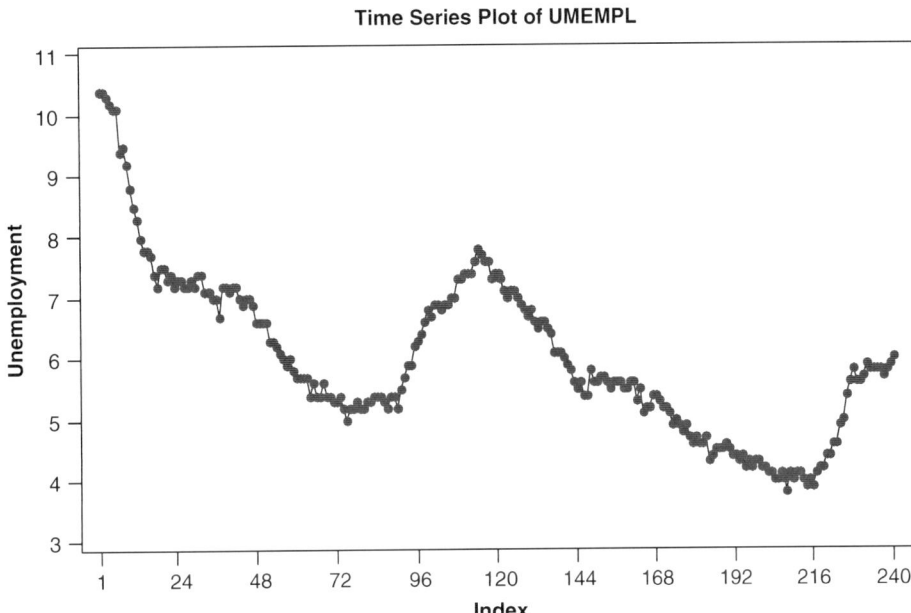

FIGURE 4.19 Regression of Unemployment on One-Period Lagged Unemployment.

Variable	Coefficient	Std Dev	T Stat	P Value
Intercept	0.153195	0.044597	3.43	0.001
UNEMPL1	0.971495	0.007227	134.43	0.000

Standard Error = 0.151529   R-Sq = 98.7%   R-Sq(adj) = 98.7%

Analysis of Variance

Source	DF	Sum of Squares	Mean Square	F Stat	P Value
Regression	1	414.92	414.92	18070.47	0.000
Error	237	5.44	0.02		
Total	238	420.36			

multicollinearity problems. Also, for each additional lag used, one data point is lost. Since there were originally 240 monthly observations on unemployment, this loss is not a substantial part of the data set. However, in smaller data sets, the loss could be significant.

FIGURE 4.20
Regression of Unemployment on One- and Two-Period Lagged Unemployment.

Variable	Coefficient	Std Dev	T Stat	P Value
Intercept	0.16764	0.04565	3.67	0.000
UNEMPL1	0.89032	0.06497	13.70	0.000
UNEMPL2	0.07842	0.06353	1.23	0.218

Standard Error = 0.151381   R-Sq = 98.7%   R-Sq(adj) = 98.6%

Analysis of Variance

Source	DF	Sum of Squares	Mean Square	F Stat	P Value
Regression	2	395.55	197.77	8630.30	0.000
Error	235	5.39	0.02		
Total	237	400.93			

EXERCISES

5. Mortgage Rates. The file on the CD named MRATES4 contains monthly 30-year conventional mortgage rates from January 1985 to December 2002. (These figures are found on the web site www.economagic.com and are obtained from the Federal Home Mortgage Corporation).

Figure 4.21 provides a time-series plot of the rates. The following model is to be used to forecast mortgage rates:

$$y_i = \beta_0 + \beta_1 y_{i-1} + e_i$$

where y_i is the mortgage rate at time i and y_{i-1} is the rate at time $i - 1$. The regression results are shown in Figure 4.22. Note that RATEL1 in the output represents the one-period lagged variable, y_{i-1}. Use the output to answer the following questions:

a. What is the estimated regression equation?

b. Is there a relationship between current and previous period mortgage rates? State the hypotheses to be tested, the decision rule, the test statistic, and your decision. Use a 5% level of significance.

FIGURE 4.21
Time-Series Plot of Mortgage Rates.

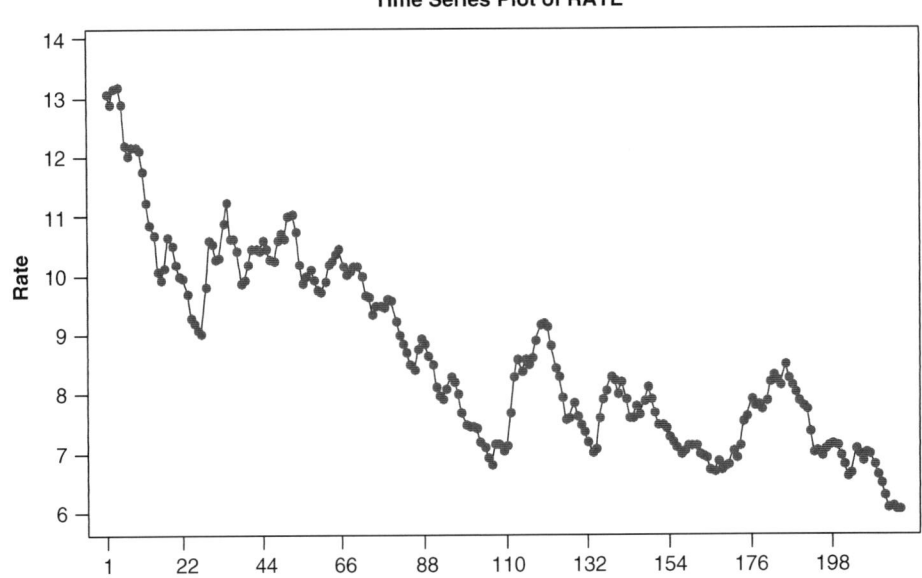

FIGURE 4.22
Regression Results for Mortgage Rates Exercise.

```
Variable        Coefficient       Std Dev         T Stat      P Value

Intercept       0.16077           0.08854         1.82        0.071
RATEL1          0.97774           0.01002         97.60       0.000


Standard Error = 0.235473    R-Sq = 97.8%    R-Sq(adj) = 97.8%

Analysis of Variance

Source          DF       Sum of Squares    Mean Square    F Stat      P Value
Regression      1        528.13            528.13         9524.82     0.000
Error           213      11.81             0.06
Total           214      539.94
```

c. What percentage of the variation in mortgage rates has been explained by the regression?

d. Use the estimated equation to produce a forecast of the mortgage rate in January 2003. Find out what the actual rate was and compare it to the forecast. How well did the equation do? Repeat this process for the remainder of 2003. Discuss any difficulties you encounter in forecasting more than 1 month ahead.

e. Test whether the intercept of the equation is equal to zero. State the hypotheses to be tested, the decision rule, the test statistic, and your decision. Use a 5% level of significance.

f. Test whether the slope of the equation is equal to 1. State the hypotheses to be tested, the decision rule, the test statistic, and your decision. Use a 5% level of significance.

ADDITIONAL EXERCISES

6. Fill in the missing blanks on the following ANOVA table:

ANOVA Source	DF	SS	MS	F
Regression	2		250	
Error (Residual)		400		
Total	82			

7. Fill in the missing blanks on the following ANOVA table:

ANOVA Source	DF	SS	MS	F
Regression		300	100	10
Error (Residual)	27			
Total				

8. **Wheat Exports.** The relationship between exchange rates, prices, and agricultural exports is of interest to agricultural economists. One such export of interest is wheat. The file named WHEAT4 on the CD contains data on the following variables:

y, U.S. wheat export shipments (SHIPMENT)
x_1, the real index of weighted-average exchange rates of the U.S. dollar (EXCHRATE)
x_2, the per-bushel real price of no. 1 red winter wheat (PRICE)

The dependent variable is U.S. wheat export shipments. The explanatory variables are exchange rate and price. The data are observed monthly from January 1974 through March 1985. The regression results are shown in Figure 4.23. Use the output to help answer the following questions:

a. What is the estimated regression equation relating SHIPMENT to EXCHRATE and PRICE?

b. Test the overall fit of the regression. State the hypotheses to be tested, the decision rule, the test statistic, and your decision. Use a 5% level of significance. What conclusion can be drawn from the result of the test?

c. After taking account of the effect of PRICE, are SHIPMENT and EXCHRATE related? Conduct a

FIGURE 4.23
Regression Results for Wheat Export Exercise.

Variable	Coefficient	Std Dev	T Stat	P Value
Intercept	3361.932	633.194	5.31	0.000
EXCHRATE	1.869	4.223	0.44	0.659
PRICE	-2413.837	846.480	-2.85	0.005

Standard Error = 798.260   R-Sq = 8.8%   R-Sq(adj) = 7.4%

Analysis of Variance

Source	DF	Sum of Squares	Mean Square	F Stat	P Value
Regression	2	8117338	4058669	6.37	0.002
Error	132	84112922	637219		
Total	134	92230260			

hypothesis test to answer this question and use a 5% level of significance. State the hypotheses to be tested, the decision rule, the test statistic, and your decision. What conclusion can be drawn from the result of the test?

   d. What percentage of the variation in the dependent variable has been explained by the regression?

   e. Construct a 95% confidence interval estimate for the population regression coefficient of PRICE.

   f. What is the value of the R^2 adjusted for degrees of freedom? What, if any, is the advantage of this number over the coefficient of determination?
   (*Source*: Data are from D. A. Bessler and R. A. Babubla, "Forecasting Wheat Exports: Do Exchange Rates Really Matter?" *Journal of Business and Economic Statistics*, 5, 1987, pp. 397–406. Copyright 1987 by the American Statistical Association. Used with permission. All rights reserved.)

9. **Mortgage Rates.** The regression in Figure 4.24 is an attempt to develop an equation to forecast mortgage rates. The explanatory variables include prime rate and the one through six-period lagged values of prime rate. The forecaster initially believed that prime rate in some of the past time periods and possibly the current time period may have an effect on mortgage rates. After examining the regression, the forecaster notes the small *t* statistics (large *p* values) for all the variables included in the regression and concludes that the regression is basically worthless: None of the variables included—current or lagged—are helpful in predicting mortgage rates. Do you agree or disagree? Justify your position.

10. **Dividends.** A random sample of 42 firms was chosen from the S&P 500 firms listed in the Spring 2003 Special Issue of *Business Week* (The Business Week Fifty Best Performers). The indicated dividend yield (DIVYIELD), the earnings per share (EPS), and the stock price (PRICE) were recorded for these 42 firms. These data are available on the CD in a file named DIV4. Run a regression using DIVYIELD as the dependent variable and EPS and PRICE as the independent variables. Use the output to answer the following questions:

   a. What is the sample regression equation relating DIVYIELD to PRICE and EPS?

   b. What percentage of the variation of DIVYIELD has been explained by the regression?

   c. Test the overall fit of the regression. Use a 10% level of significance. State the hypotheses to be tested, the decision rule, the test statistic, and your decision.

   d. What conclusion can be drawn from the test result?

   e. Is it necessary to test each coefficient individually to see if either PRICE or EPS is related to DIVYIELD? Why or why not?
   (*Source*: Copyright 2003, *Business Week*. Visit us at our Web site at www.businessweek.com for additional information.)

11. **Fuel Consumption.** The data file FUELCON4 on the CD contains the following variables for all 50 states plus the District of Columbia:

FIGURE 4.24
Mortgage Rate and Prime Rate Regression.

Variable	Coefficient	Std Dev	T Stat	P Value
Intercept	5.3188	0.5203	10.22	0.000
PRIMERT	0.6266	0.5047	1.24	0.218
PRIMEM1	-0.1148	0.8281	-0.14	0.890
PRIMEM2	-0.2286	0.8286	-0.28	0.783
PRIMEM3	-0.3137	0.8295	-0.38	0.706
PRIMEM4	-0.2468	0.8286	-0.30	0.767
PRIMEM5	-0.0368	0.8262	-0.04	0.965
PRIMEM6	0.6845	0.4947	1.38	0.170

Standard Error = 0.7587   R-Sq = 29.7%   R-Sq(adj) = 23.7%

Analysis of Variance

Source	DF	Sum of Squares	Mean Square	F Stat	P Value
Regression	7	19.9288	2.8470	4.95	0.000
Error	82	47.1977	0.5756		
Total	89	67.1265			

FUELCON: Per capita fuel consumption in gallons

DRIVERS: The ratio of licensed drivers to private and commercial motor vehicles registered

HWYMILES: The number of miles of federally funded highways

GASTAX: The tax per gallon of gasoline in cents

INCOME: The average household income in dollars

Run the regression with FUELCON as the dependent variable and the other four variables as independent variables. Use the output to help answer the following questions:

a. What is the estimated regression equation?

b. Test the overall fit of the regression. Use a 5% level of significance. Be sure to state your decision rule, test statistic value, and your decision. What conclusion can you draw from the result of the test?

c. What percentage of the variation in FUELCON has been explained by the regression?

d. Are there any variables that appear to be unnecessary in the regression? Justify your answer.

e. Are there variables that have been omitted from the regression that you believe might be useful in explaining more of the variation in FUELCON? If so, what variables?

(*Source*: Data are from the U.S. Department of Transportation, Federal Highway Administration. Most data were found in the online publication Highway Statistics 2001 at www.fhwa.dot.gov/ohim/hs01/index.htm)

12. Pricing Communications Nodes. The cost of adding a new communications node at a location not currently included on the network was of concern to a major Fort Worth manufacturing company. To try to predict the price of new communications nodes, data were obtained on a sample of existing nodes. The installation cost (COST) and the number of ports (NUMPORTS) available for access in each existing node were readily available. Data on two additional characteristics of communications nodes were also obtained: bandwidth (BANDWIDTH) and port speed (PORTSPEED). These data are shown in Table 4.6 and are available on the CD in a file named COMNODE4.

The network administrator wants to develop a method of estimating the cost of new nodes in a quick and fairly accurate manner. You have been asked to help in this project. Using the data available, develop an equation to help in the

TABLE 4.6 Data for Communications Nodes Exercise

Cost	Number of Ports	Bandwidth	Port Speed
52,388	68	58	653
51,761	52	179	499
50,221	44	123	422
36,095	32	38	307
27,500	16	29	154
57,088	56	141	538
54,475	56	141	538
33,969	28	48	269
31,309	24	29	230
23,444	24	10	230
24,269	12	56	115
53,479	52	131	499
33,543	20	38	192
33,056	24	29	230

pricing of new communications nodes. Justify your choice of equation.

Start with all three explanatory variables in the equation. Do you encounter any problems when you estimate this equation? Request variance inflation factors for the variables included in the regression. What do the VIFs tell you?

13. **Prime Rate.** The file named PRIME4 on the CD contains monthly prime rates for the time period from January 1988 through December 2002. (These data are from the web site www.economagic.com and are from the Federal Reserve Bank of St. Louis.) Develop an extrapolative model to forecast the prime rate for each month in 2003.

    Find the actual rates for each month in 2003 and compare them to your forecasts. How well did your model do? (How will you measure the accuracy of your forecasts?)

14. **Graduation Rates.** *Kiplinger's Personal Finance* provides information on the best public and private college values. Some of the variables included in this issue are as follows. All are based on the most recent available data.

GRADRATE4	the percentage of students who earned a bachelor's degree in four years (expressed as a percentage)
ADMISRATE	admission rate expressed as a percentage
SFACRATIO	student faculty ratio
AVGDEBT	average debt at graduation

    This information is included in a file named COLLEGE4 on the CD for 195 schools. Only schools with complete information on all the categories listed are included. Using the graduation rate as the dependent variable, use the techniques discussed in this chapter to develop an equation to explain four year graduation rates.

    (*Source*: Used by permission from the November and December 2003 issues of *Kiplinger's Personal Finance*. Copyright © 2003 The Kiplinger Washington Editors, Inc. Visit our website at www.kiplingers.com for further information).

15. **Absenteeism.** The ABX Company is interested in conducting a study of the factors that affect absenteeism among its production employees. After reviewing the literature on absenteeism and interviewing several production supervisors and a number of employees, the researcher in charge of the project defined the variables shown in Figure 4.25. Then a sample of 77 employees was randomly selected, and the data contained in the file on the CD named ABSENT4 was collected. The dependent variable is absenteeism. The other variables are considered possible explanatory variables.

    Use the procedures discussed in Chapters 3 and 4 to identify factors that may be related to absenteeism. Write down your final model and justify your choice of variables in the model. Check to see if your choice of variables and the coefficient estimates make intuitive sense. How much variation in absenteeism has been explained? What does this tell you? Does your model give you some sense of which

Additional Exercises   173

FIGURE 4.25
Absenteeism Study Variables.

	Variable	Description
1.	Absenteeism (ABSENT):	The number of distinct occasions that the worker was absent during 2003. Each occasion consists of one or more consecutive days of absence.
2.	Job Complexity (COMPLX):	An index ranging from 0 to 100.
3.	Base Pay (PAY):	Base hourly pay rate in dollars.
4.	Seniority (SENIOR):	Number of complete years with the company on December 31, 2003.
5.	Age (AGE):	Employee's age on December 31, 2003.
6.	Dependents (DEPEND):	Determined by employee response to the question: "How many individuals other than yourself depend on you for most of their financial support?"

Source: These data were created by Dr. Roger L. Wright, RLW Analytics, Inc., Sonoma, CA, and are used (with modification) with his permission.

employees might be absent most often? If so, which ones? What might be done to reduce absenteeism?

16. **Fanfare.** Fanfare International, Inc. designs, distributes, and markets ceiling fans and lighting fixtures. The company's product line includes 120 basic models of ceiling fans and 138 compatible fan light kits and table lamps. These products are marketed to over 1000 lighting showrooms and electrical wholesalers that supply the remodeling and new construction markets. The product line is distributed by a sales organization of 58 independent sales representatives.

In the summer of 1994, Fanfare decided it needed to develop forecasts of future sales to help determine future sales force needs, capital expenditures, and so on. The data file named FAN4 on the CD contains data on the following variables:

SALES	= total monthly sales in thousands of dollars
ADEX	= advertising expense in thousands of dollars
MTGRATE	= mortgage rate for 30-year loans (%)
HSSTARTS	= housing starts in thousands of units

The data are monthly and cover the period from July 1990 through May 1994. (*Note:* These data have been modified as requested by the company to provide confidentiality.)

As a consultant to Fanfare, your job is to find a causal regression model to forecast future sales. Use the techniques discussed in Chapters 3 and 4 to help you decide which variables you should include in the equation and which should be omitted. Justify your choices. How well do you believe the equation you developed will do at forecasting future sales? What additional analyses might you use to examine forecasting ability?

Now use the techniques discussed in Chapters 3 and 4 to build an extrapolative model to forecast sales. Generate forecasts from both the causal model and the extrapolative model. What are the benefits and drawbacks of each of the models? How could you compare the forecasting ability of the two models?

17. **Major League Baseball.** What factor is most important in building a winning baseball team? Some might argue for a high batting average. Or it might be a team that hits for power as measured by the number of home runs. On the other hand, many believe that it is quality pitching as measured by the earned run average of the team's pitchers. The file MLB4 on the CD contains data on the following variables for the 30 major league baseball teams during the 2002 season:

WINS	= number of games won for each team
HR	= number of home runs hit by each team
BA	= average batting average for each team
ERA	= earned run average for each team

Using WINS as the dependent variable, use scatterplots and regression to investigate the relationship of the other variables to WINS. Use the variables to build a multiple regression

174 Multiple Regression Analysis

model to explain WINS. Interpret what your model tells you about a successful baseball team.
(*Source*: Courtesy of the *Fort Worth Star-Telegram*.)

18. **NBA.** The following data were obtained from the *Fort Worth Star-Telegram* and refer to the 2002–2003 National Basketball Association (NBA) season. The data are included in a file on the CD named NBA4 for all 29 NBA teams:

> wins (WINS)
> field goals attempted (FGA)
> field goals made (FGM)
> field goals attempted for opponents (FGAOP)
> field goals made for opponents (FGMOP)
> three-point field goals attempted (TFGA)
> three-point field goals made (TFGM)
> three-point field goals attempted for opponents (TFGAOP)
> three-point field goals made for opponents (TFGMOP)
> offensive rebounds (OFFREB)
> total rebounds (TOTREB)
> offensive rebounds for opponents (OFFREBOP)
> total rebounds for opponents (TOTREBOP)
> assists (ASST)
> assists for opponents (ASSTOP)
> steals (STL)
> steals for opponents (STLOP)
> blocked shots (BLK)
> blocked shots for opponents (BLKOP)

The dependent variable is the number of wins (WINS) for the season. The other variables are to be considered possible explanatory variables.

You have been hired by your favorite NBA team to try and determine what factors might be important in helping to achieve a winning season. Using the available data, determine which combination of the variables provides the best explanation of what makes a winning team.

Write a report with your results. Your report should consist of a letter/executive summary of your results for team management and a technical section with a description and justification of your regression equation. In the technical section you will want to discuss aspects such as your regression equation, the choice of variables, the strength of the relationship, and the practical usefulness of the results. What do your results tell you about winning teams?
(*Source*: Courtesy of the *Fort Worth Star-Telegram*.)

19. **Multicollinearity.** Consider the following regression model:

$$y_i = \beta_0 + \beta_1 x_{1i} + \beta_2 x_{2i} + \beta_3 x_{3i} + \beta_4 x_{4i} + e_i$$

Multicollinearity is suspected to exist among the four explanatory variables used in this regression. The analyst using the regression computes pairwise correlations between the four explanatory variables and the dependent variable. The following correlation matrix results:

	y	x_1	x_2	x_3	x_4
x_1	0.4	1.0			
x_2	0.3	0.2	1.0		
x_3	0.6	0.3	0.4	1.0	
x_4	0.7	0.3	0.5	0.3	1.0

Based on these correlations, the analyst concludes that there will be no problems with multicollinearity. Do these correlations provide sufficient evidence to conclude that multicollinearity will not be a problem? Justify your answer.

USING THE COMPUTER

The Using the Computer section in each chapter describes how to perform the computer analyses in the chapter using Excel, MINITAB, and SAS. For further detail on Excel, MINITAB, and SAS, see Appendix C.

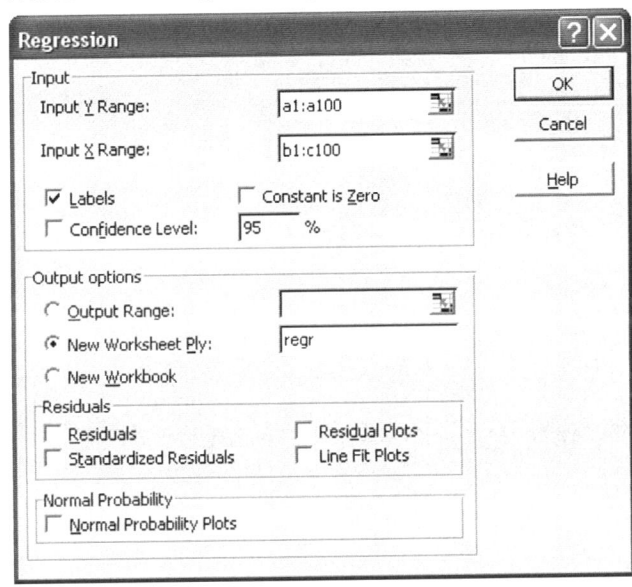

FIGURE 4.26 Excel Regression Dialog Box.

EXCEL

Multiple Regression

TOOLS: DATA ANALYSIS: REGRESSION

Figure 4.26 shows the Excel Regression dialog box. Regression is accessed in Excel by clicking on Tools and then Data Analysis. The Regression option is chosen from the Data Analysis menu. Put the range of the y variable in "Input Y Range." Put the range of the x variables in "Input X Range." Note that all x variables to be used in a multiple regression must be in adjacent columns. To accommodate this restriction, variables often must be moved around.

Click "Labels" if the variables have labels in the first row. Typically, "Constant is Zero" is not an option that is used. This option forces the constant or y intercept in a regression to be zero. It is seldom a good idea to use this option and can make interpretation of the regression results difficult.

Excel produces 95% confidence interval estimates of the population regression coefficients by default. If another level is required, click the "Confidence Level" box and insert the desired level.

Click the output option desired. The Residuals and Normal Probability options are discussed in Chapter 6.

Variance Inflation Factors

Excel does not provide options for automatically producing the variance inflation factors. Although these could be computed by creating formulas, this option is not discussed in this text.

A regression add-in is available that does compute a number of additional statistics. See the Excel section in Appendix C for more information on the add-in.

Creating a Lagged Variable

One way to create a lagged variable in Excel is simply to copy the necessary portion of the column to be lagged and paste it in the appropriate position. Figure 4.27 shows an example. Once the column is copied, the initial values with no matches are not used when running any regressions.

FIGURE 4.27 Creating a Lagged Variable with Excel.

	A	B	C	D	E
1	YEAR	MONTH	UNEMP	UNEMPL1	UNEMPL2
2	1983	1	10.4		
3	1983	2	10.4	10.4	
4	1983	3	10.3	10.4	10.4
5	1983	4	10.2	10.3	10.4
6	1983	5	10.1	10.2	10.3
7	1983	6	10.1	10.1	10.2
8	1983	7	9.4	10.1	10.1
9	1983	8	9.5	9.4	10.1
10	1983	9	9.2	9.5	9.4
11	1983	10	8.8	9.2	9.5
12	1983	11	8.5	8.8	9.2
13	1983	12	8.3	8.5	8.8
14	1984	1	8	8.3	8.5
15	1984	2	7.8	8	8.3
16	1984	3	7.8	7.8	8
17	1984	4	7.7	7.8	7.8
18	1984	5	7.4	7.7	7.8
19	1984	6	7.2	7.4	7.7
20	1984	7	7.5	7.2	7.4
21	1984	8	7.5	7.5	7.2
22	1984	9	7.3	7.5	7.5
23	1984	10	7.4	7.3	7.5
24	1984	11	7.2	7.4	7.3
25	1984	12	7.3	7.2	7.4
26	1985	1	7.3	7.3	7.2
27	1985	2	7.2	7.3	7.3
28	1985	3	7.2	7.2	7.3
29	1985	4	7.3	7.2	7.2
30	1985	5	7.2	7.3	7.2
31	1985	6	7.4	7.2	7.3

MINITAB

Multiple Regression

STAT:REGRESSION:REGRESSION

To do a multiple regression in MINITAB, click the Stat menu and the Regression option. Then click the regression option on the subsequent menu. The Regression dialog box is shown in Figure 4.28. Fill in the Response variable and the Predictor variables. Then click OK.

Forecasting with a Multiple Regression Equation

STAT: REGRESSION: REGRESSION: OPTIONS

Click on Stat, then choose Regression from the Stat menu, and choose Regression again from the next menu. In the Regression dialog box (see Figure 4.28), click Options. The Regression–Options dialog box is shown in Figure 4.29. Fill in "Prediction intervals for new observations" with the values of the independent variables for which predictions are desired. These values can be single numbers or columns with numbers. The number of entries on this line must be the same as the number of independent variables in the regression equation—one entry for each variable.

Variance Inflation Factors

STAT: REGRESSION: REGRESSION: OPTIONS

Click Stat, then choose Regression from the Stat menu, and choose Regression again from the next menu. In the Regression dialog box (see Figure 4.28), click Options. The Regression–Options dialog box is shown in Figure 4.29. Click the box for variance inflation factors.

FIGURE 4.28 MINITAB Regression Dialog Box.

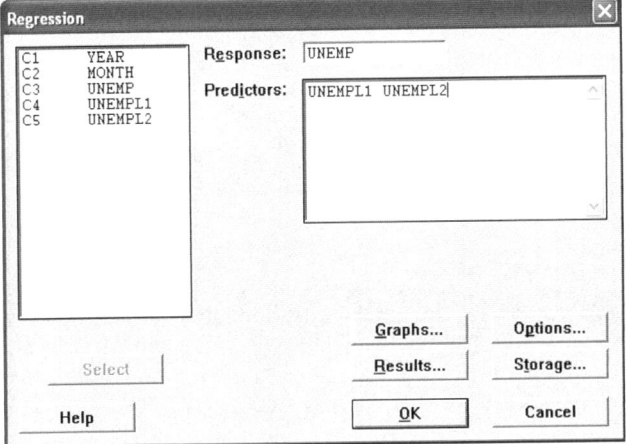

FIGURE 4.29 MINITAB Regression—Options Dialog Box.

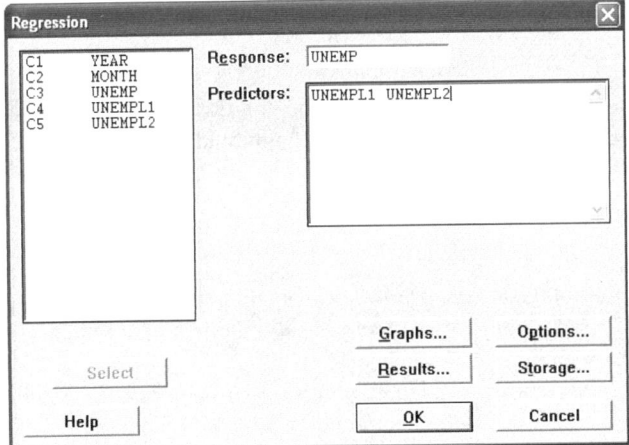

Creating a Lagged Variable
STAT: TIME-SERIES:LAG

To create a lagged variable in MINITAB, click on the Stat menu and then on Time Series. Choose Lag from the available options. Figure 4.30 shows the Lag dialog box. In "Series," put the column of the variable to be lagged. Put the location of the lagged variable in "Store lags in." In the "Lag" box, put the order of the lag desired: 1 for a one-period lag, 2 for a two-period lag, and so on.

SAS

Multiple Regression
The following command sequence produces a regression with dependent variable SALES and independent variables ADV and BONUS:

PROC REG;
MODEL SALES=ADV BONUS;

Partial F Tests in Multiple Regression
The TEST command in SAS produces the partial F statistic for testing whether several coefficients are equal to zero. The following command sequence illustrates the use of the TEST command:

PROC REG;
MODEL SALES=ADV BONUS MKTSHR COMPET;
TEST MKTSHR, COMPET;

This sequence produces the F statistic to test whether the coefficients of MKTSHR and COMPET are equal to zero. When the SAS TEST command is used, the explanatory variables in the MODEL command do not have to be listed in any particular order. For example, the following command sequence produces the same test result as the previous sequence:

178 Multiple Regression Analysis

FIGURE 4.30 MINITAB Lag Dialog Box.

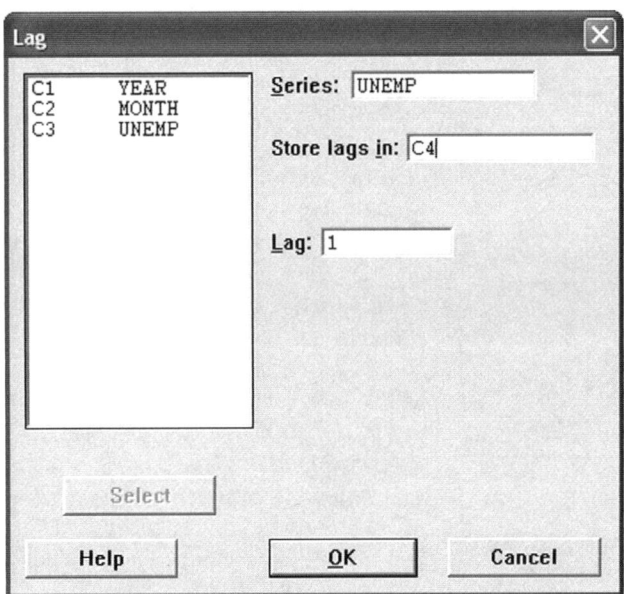

```
PROC REG;
MODEL SALES=ADV MKTSHR BONUS COMPET;
TEST MKTSHR, COMPET;
```

Forecasting with the Multiple Regression Equation

Forecasts in SAS for multiple regression are generated using an "appended" data set just as with simple regression. To the values of the independent variables in the original data set add the values for which predictions are desired. Then to the values of the dependent variable add the SAS symbol for mission data, which is a period, because we do not know those values. Now run the regression with the following options:

```
PROC REG;
MODEL SALES = ADV BONUS/P CLM CLI;
```

The option P requests forecasts (or predicted values), CLM requests upper and lower confidence interval limits for the estimate of the conditional mean, and CLI requests upper and lower prediction interval limits for an individual prediction.

Variance Inflation Factors

```
PROC REG;
MODEL SALES = ADV BONUS/VIF;
```

requests that variance inflation factors be printed.

Creating a Lagged Variable

In the data input phase in SAS, use the LAG_ command to create lagged variables. For example, to create a one-period lagged variable for the unemployment variable in Example 4.6, use

UNEMPL1 = LAG1(UNEMP);

For a two-period lagged variable, use

UNEMPL2 = LAG2(UNEMP);

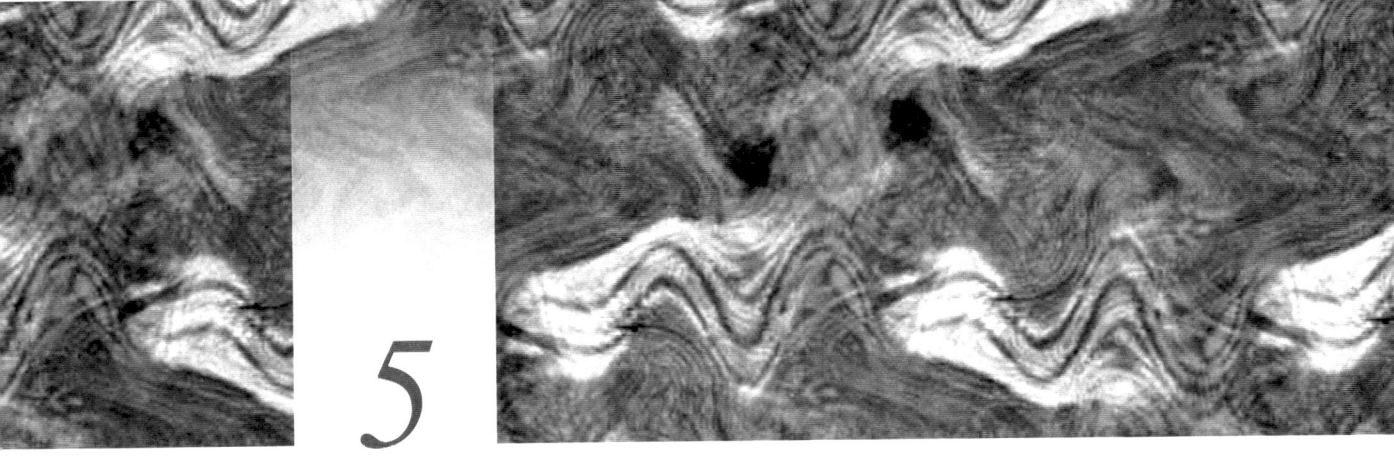

5
Fitting Curves to Data

5.1 INTRODUCTION

In Chapter 4, the multiple linear regression model was presented as

$$y_i = \beta_0 + \beta_1 x_{1i} + \beta_2 x_{2i} + \cdots + \beta_K x_{Ki} + e_i \tag{5.1}$$

There we assumed that the true relationship was linear in the x variables. In this chapter, we find that this assumption need not be true to fit a regression equation to the data. We can fit *curvilinear* as well as linear relationships. This is accomplished through transformations of the variables in the model. The equation $y_i = \beta_0 + \beta_1 x_i + e_i$ represents a straight-line relationship between y and x. But the equation $y_i = \beta_0 + \beta_1 x_i + \beta_2 x_i^2 + e_i$ represents a curve (a parabola). The same x variable is involved in the equation; the fitting of the curve is accomplished through the transformation of the x variable to x^2. There are many possible transformations that produce some type of curvilinear relationship. The most commonly used transformations in business and economic applications are discussed in this chapter.

When a curvilinear relationship is suspected, the appropriate transformation of the variables to produce the best-fitting curve for the data is not always obvious. The variables y and x are related in some curvilinear fashion, but there are many equations that describe curvilinear relationships. The idea behind the use of any equation of this sort is to transform the variables in such a way that a linear relationship is achieved. If x and y are related in a curvilinear fashion, then perhaps x^2 and y have a linear relationship.

In this text, the following four commonly used corrections are considered:

1. polynomial regression
2. reciprocal transformation of the x variable
3. log transformation of the x variable
4. log transformation of both the x and y variables

In this chapter, we suggest some ways to assess whether a good choice of transformations was used to fit a curve to the data. In Chapter 6, some additional methods of assessing the choice are considered. In some texts, the choice of linear or curvilinear model and the type of curvilinear model to be used is called selection of *functional form* of the model.

5.2 FITTING CURVILINEAR RELATIONSHIPS

5.2.1 POLYNOMIAL REGRESSION

A common correction when the linearity assumption is violated is to add powers of the explanatory variable that is viewed as the curvilinear component of the model. This type of model is called a *polynomial regression*. The *order* of the model is the highest power used for the explanatory variable. For example, a second-order polynomial regression in one variable is written as

$$y_i = \beta_0 + \beta_1 x_i + \beta_2 x_i^2 + e_i$$

Higher-order polynomial models may be developed by adding higher powers of x. A Kth-order polynomial regression model in one variable, x, is written as

$$y_i = \beta_0 + \beta_1 x_i + \beta_2 x_i^2 + \cdots + \beta_K x_i^K + e_i$$

In practice, the second-order model is often sufficient to describe curvilinear relationships encountered.

EXAMPLE 5.1 **Telemarketing** A company that provides transportation services uses a telemarketing division to help sell its services. The division manager is interested in the time spent on the phone by the telemarketers in the division. Data on the number of months of employment and the number of calls placed per day (an average for 20 working days) is recorded for 20 employees. These data are shown in Table 5.1 and in the file TELEMARK5 on the CD.

The average number of calls for all 20 employees is 28.95. The division manager, however, suspects that there may be a relationship between time on the job and number of calls. As time on the job increases, the employee becomes more familiar with the calling system and the correct procedures to use on the phone and also begins to acquire more regular clients. Thus, the longer the time on the job, the greater the number of calls per day. The scatterplot of CALLS (y) versus MONTHS (x) is shown in Figure 5.1. Looking at this scatterplot helps to verify that the relationship

TABLE 5.1 Data for Telemarketing Example

MONTHS	CALLS
10	18
10	19
11	22
14	23
15	25
17	28
18	29
20	29
20	31
21	31
22	33
22	32
24	31
25	32
25	32
25	33
25	31
28	33
29	33
30	34

FIGURE 5.1
Scatterplot of CALLS versus MONTHS.

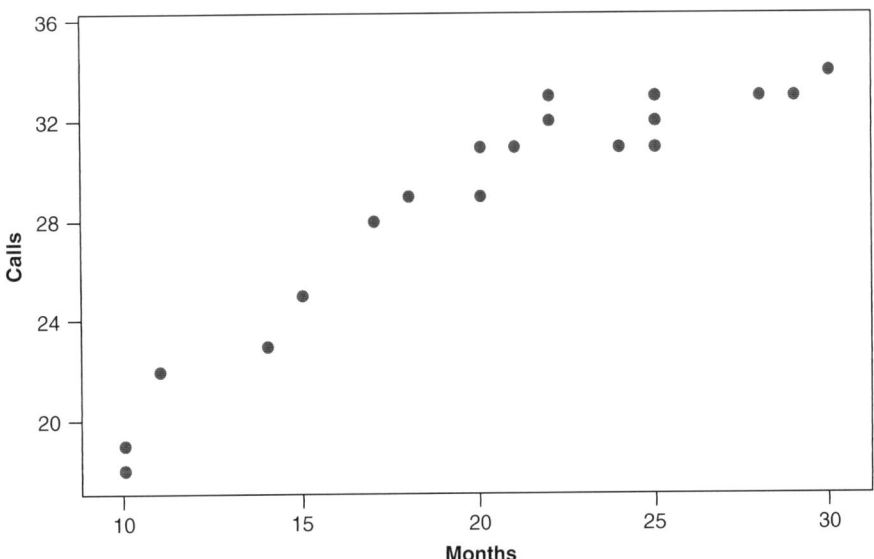

may not be linear. As the number of months on the job increases, the number of calls also increases. But the rate of increase begins to slow over time, thus resulting in a pattern that may be better modeled by a curve than a straight line.

In Figure 5.2, the plot of y versus x has been reproduced with a curve drawn in that approximates the relationship between the two variables.

FIGURE 5.2
Scatterplot of CALLS versus MONTHS with Curve Drawn to Represent the Relationship.

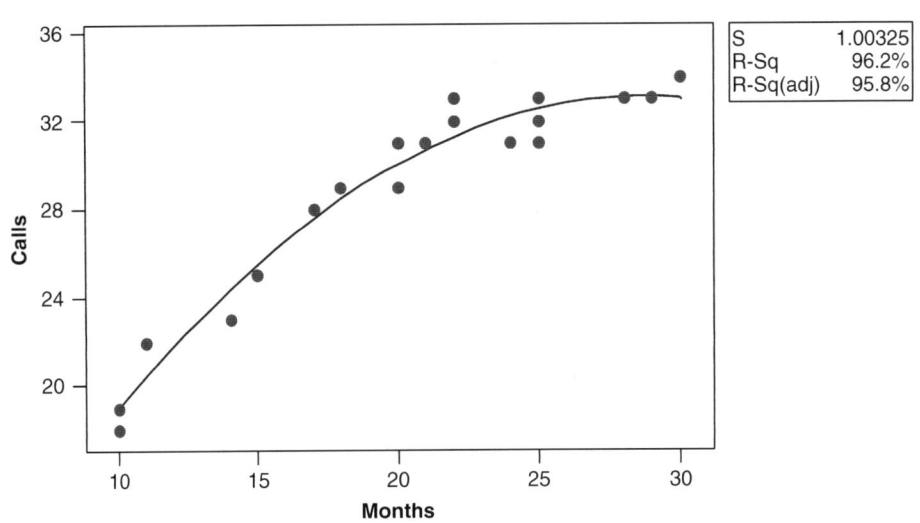

To model the curvilinear relationship in the telemarketing data, a second-order polynomial regression can be tried. The model can be written

$$\text{CALLS} = \beta_0 + \beta_1 \text{MONTHS} + \beta_2 \text{XSQR} + e$$

where XSQR is a variable created by squaring each value of the MONTHS variable.

For comparison purposes, the linear regression using CALLS as the dependent variable and MONTHS as the explanatory variable is shown in Figure 5.3. Figure 5.4 shows the regression estimates of the second-order model. The estimated regression is

$$\text{CALLS} = -0.14 + 2.31\,\text{MONTHS} - 0.04\,\text{XSQR}$$

The primary check that should be made at this point to determine whether the second-order model is preferred to the original linear model is to test whether the coefficient of the second-order term is significantly different from zero.

To determine whether the x^2 variable has significantly improved the fit of the regression, the following hypotheses can be tested:

$$H_0: \beta_2 = 0$$
$$H_a: \beta_2 \neq 0$$

where β_2 is the coefficient of x^2. The t test discussed in Chapter 4 can be used to conduct the test. For $\alpha = 0.05$, the decision rule is

Reject H_0 if $t > 2.11$ or $t < -2.11$

Do not reject H_0 if $-2.11 \leq t \leq 2.11$

FIGURE 5.3 Regression Results for Telemarketing Example Using Only MONTHS as an Explanatory Variable.

Variable	Coefficient	Std Dev	T Stat	P Value
Intercept	13.6708	1.4270	9.58	0.000
MONTHS	0.7435	0.0667	11.15	0.000

Standard Error = 1.78737   R-Sq = 87.4%   R-Sq(adj) = 86.7%

Analysis of Variance

Source	DF	Sum of Squares	Mean Square	F Stat	P Value
Regression	1	397.45	397.45	124.41	0.000
Error	18	57.50	3.19		
Total	19	454.95			

FIGURE 5.4 Regression Results for Telemarketing Example with Second-Order Term Added.

Variable	Coefficient	Std Dev	T Stat	P Value
Intercept	-0.14047	2.32226	-0.06	0.952
MONTHS	2.31020	0.25012	9.24	0.000
XSQR	-0.04012	0.00633	-6.33	0.000

Standard Error = 1.00325   R-Sq = 96.2%   R-Sq(adj)=95.8%

Analysis of Variance

Source	DF	Sum of Squares	Mean Square	F Stat	P Value
Regression	2	437.84	218.92	217.50	0.000
Error	17	17.11	1.01		
Total	19	454.95			

The test statistic value is $t = -6.33$. The null hypothesis is rejected. The x^2 term adds significantly to the ability of the regression to explain the variation in y. Thus, the term should remain in the equation. Note that the p value could also have been used to conduct this test ($p$value $= 0.000 < 0.05$, so reject H_0).

Once a decision is made to keep the second-order term in the model, the lower-order term is typically kept in the model regardless of the t test result on its coefficient. There are good statistical reasons for keeping lower-order terms in a polynomial regression when the higher-order terms are judged important (see "A Property of Well-Formulated Polynomial Regression Models," by J. L. Peixoto).[1] Other indicators that the regression has been improved by adding the x^2 term include the reduction in the standard error of the regression from 1.787 to 1.003 and the increase in adjusted R^2 from 86.7% to 95.8%.

---

[1] See References for complete publication information.

In our example, the second-order model is an improvement over the first-order model. Higher-order terms could be added to the model to see whether additional improvements are possible. Figure 5.5 shows the estimates of a third-order model. The explanatory variables are MONTHS, XSQR, and $X^\wedge 3$, the cube (third power) of the MONTHS variable. Note that the coefficient of the $X^\wedge 3$ variable is not significant at the 0.05 level, suggesting that the addition of this term is of little additional help in explaining the variation in CALLS. The third-order term is unnecessary in the model. Note that second-order term is not significant either in this model. This is a result of multicollinearity. The explanatory variables are highly correlated and this causes least-squares to have difficulty determining which of the variables (the second-order or the third-order term) is the important one. We will opt for the simpler second-order model in this case.

Table 5.2 summarizes the different measures that may be useful in determining the best model to use. The p values suggest that the second-order term is useful, but the third-order term is not. The R^2 increases from 87.4% for the linear model to 96.2% for the second-order model. The increase for the third-order model is very small, however. If R^2_{adj} is used, there is actually a decrease from the second-order to the third-order model. This is further verification that the third-order term is unnecessary. This is also reflected in the standard error, which decreases from 1.787 for the linear model to 1.003 for the second-order model but increases to 1.020 for the third-order model. (Recall that we want increases in R^2 and R^2_{adj} but decreases in the standard error.)

One caution should be observed in using higher-order polynomial regression models. Correlations between powers of a variable can result in *multicollinearity* problems, as discussed in Chapter 4. This problem was encountered in the telemarketing example when a third-order model was estimated. To reduce the possibility of computational difficulties, the use of explanatory variables that have been centered often is recommended. For example, instead of using the explanatory variables, x, x^2, x^3, use instead

$$x - \bar{x}, (x - \bar{x})^2, \text{ and } (x - \bar{x})^3$$

where $\bar{x}$ is the sample mean of the variable values. Using the centered variables helps avoid multicollinearity problems in polynomial regressions to some extent.

FIGURE 5.5 Regression Results for Telemarketing Example with Second-Order and Third-Order Terms Added.

Variable	Coefficient	Std Dev	T Stat	P Value
Intercept	-5.58003	8.38720	-0.67	0.515
MONTHS	3.25806	1.42526	2.29	0.036
XSQR	-0.09075	0.07518	-1.21	0.245
X^3	0.00085	0.00125	0.68	0.509

Standard Error = 1.01967   R-Sq = 96.3%   R-Sq(adj) = 95.7%

Analysis of Variance

Source	DF	Sum of Squares	Mean Square	F Stat	P Value
Regression	3	438.31	146.10	140.52	0.000
Error	16	16.64	1.04		
Total	19	454.95			

5.2 Fitting Curvilinear Relationships 185

TABLE 5.2 Summary Measures for Linear, Second-Order, and Third-Order Models for Telemarketing Example

Model	p Value for Highest-Order Term	R^2	R^2_{adj}	s_e
Linear Model	0.000	87.4%	86.7%	1.787
Second-Order Model	**0.000**	96.2%	**95.8%**	**1.003**
Third-Order Model	0.509	96.3%	95.7%	1.020

Choosing which curvilinear model to use in a particular case is not always a simple matter. In the telemarketing example, I chose to use a second-order polynomial regression as my starting point. Why not use the logarithm of the x variable instead? This is another of the transformations to be discussed in this chapter. Familiarity with the look of certain curves can be helpful in choosing the right curvilinear model. The curve shown in Figure 5.2 is similar to a parabola (or half of a parabola, at any rate), and this led me to make the second-order model my first choice (since the second-order equation is the equation of a parabola). If you are not sure about what type of transformation is best, you can always try different ones and check the summary measures used in the example to help make the choice. Chapter 6 also presents some additional methods of assessing the validity of a curvilinear model and choosing the best transformation.

5.2.2 RECIPROCAL TRANSFORMATION OF THE X VARIABLE

Other transformations may produce a linear relationship. A fairly common example is the reciprocal transformation:

$$y_i = \beta_0 + \beta_1\left(\frac{1}{x_i}\right) + e_i$$

In this equation, x and y are inversely related, but the inverse relationship is not a linear one. (Note that this transformation is not defined when $x = 0$.)

EXAMPLE 5.2 **MPG Versus HP** A scatterplot of a possible curvilinear inverse relationship is shown in Figure 5.6. The variables are HWYMPG (y), which is the number of miles per gallon obtained by a car in highway driving, and HP (x), the horsepower of the car. This information is available for 147 cars listed in the *Road and Track* October 2002 issue and in the file MPGHP5 on the CD.

As HP increases, the mileage decreases, as would be expected. However, it appears that the rate of decrease may be slower as the cars get more powerful. The regression results for the linear regression of HWYMPG on HP are shown in Figure 5.7. Can we find a curvilinear model that better describes the relationship between these two variables? The scatterplot suggests that the following curvilinear inverse relationship might be appropriate:

$$\text{HWYMPG} = \beta_0 + \beta_1\left(\frac{1}{\text{HP}}\right) + e$$

186 Fitting Curves to Data

FIGURE 5.6 Scatterplot of HWYMPG Versus HP.

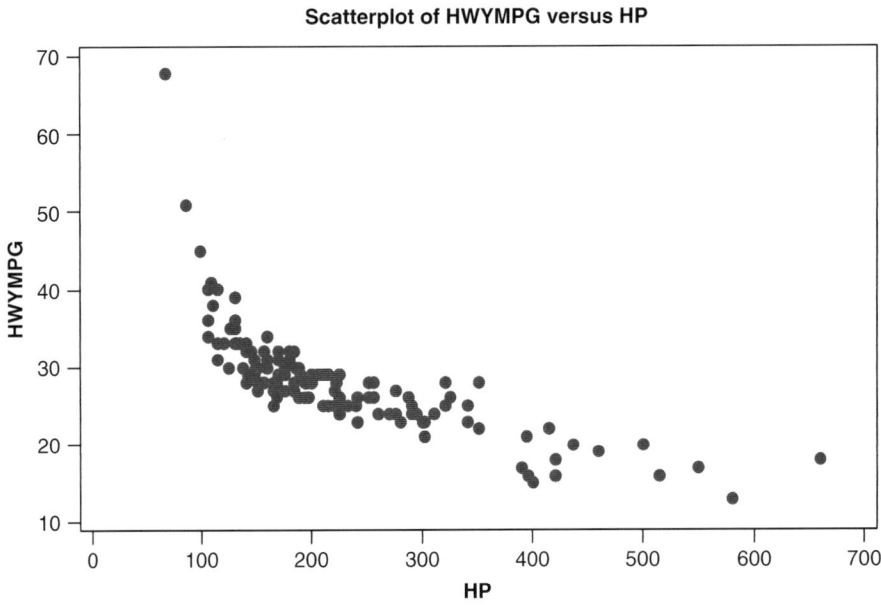

FIGURE 5.7 Results for Regression of HWYMPG on HP.

```
Variable        Coefficient      Std Dev        T Stat         P Value

Intercept       38.730875        0.803300       48.21          0.000
HP              -0.047732        0.003274       -14.58         0.000

Standard Error = 4.17503     R-Sq = 59.4%     R-Sq(adj) = 59.2%

Analysis of Variance

Source          DF     Sum of Squares    Mean Square    F Stat      P Value
Regression      1      3705.2            3705.2         212.57      0.000
Error           145    2527.5            17.4
Total           146    6232.7
```

An inverse relationship is one where y decreases as x increases. In a linear model, an inverse relationship results in a negative slope coefficient. If a relationship is expected to be inverse but curvilinear, then the reciprocal transformation of the x variable is often useful in representing this relationship.

Figure 5.8 shows the scatterplot of HWYMPG versus the transformed explanatory variable 1/HP (named HPINV).[2] This scatterplot appears linear. The regression results for the regression of HWYMPG on HPINV are shown in Figure 5.9.

---

[2] A graphical method to see if a transformation might be effective in modeling a curvilinear relationship is to plot the dependent variable versus the transformed x variable, as was done in this example. If the resulting graph looks linear, the transformation likely gives a good result. Note that this was not done in the first example using the second-order model. Because this model has two terms (x and x^2), examining the transformation graphically is more difficult.

5.2 Fitting Curvilinear Relationships 187

FIGURE 5.8
Scatterplot of HWYMPG versus HPINV = 1/HP.

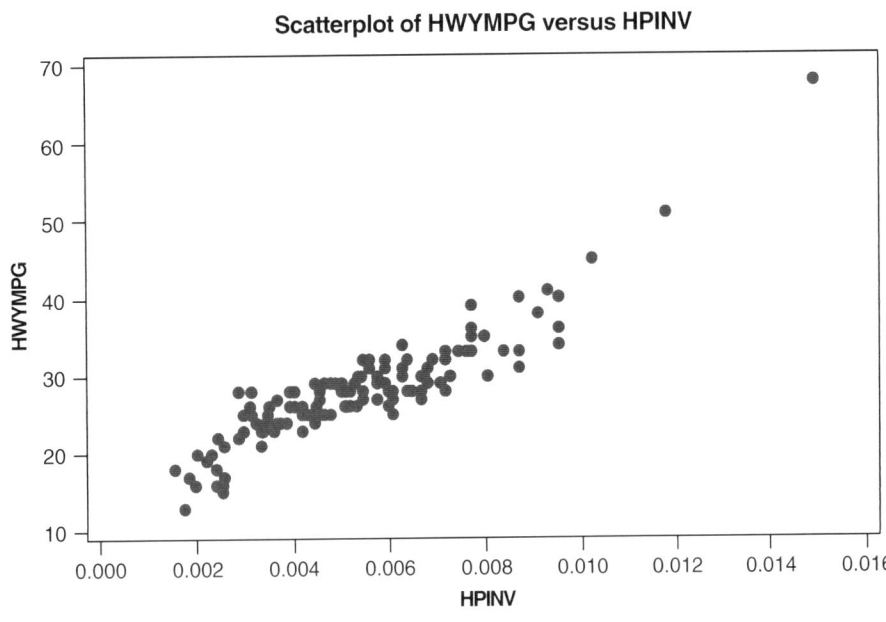

FIGURE 5.9
Results for Regression of HWYMPG on HPINV = 1/HP.

```
Variable         Coefficient     Std Dev       T Stat        P Value

Intercept          13.6310       0.6493        20.99         0.000
HPINV            2692.4675      11.7526        24.09         0.000

Standard Error = 2.93107    R-Sq = 80.0%    R-Sq(adj) = 79.9%

Analysis of Variance

Source           DF     Sum of Squares    Mean Square    F Stat     P Value
Regression        1          4987.0          4987.0      580.48     0.000
Error           145          1245.7             8.6
Total           146          6232.7
```

The R^2 value has increased from 59.4% for the linear model to 80.0% for the model with the transformed x variable. The standard error of the regression has decreased from 4.17503 to 2.93107. Both of these facts support the use of the curvilinear model. Table 5.3 summarizes the statistics for the two models.

TABLE 5.3 Summary Measures for the Linear Model and the Model Using the Reciprocal of HP for MPG Versus HP Example

Model	p Value for Highest-Order Term	R^2	R^2_{adj}	s_e
Linear Model	0.000	59.4%	59.2%	4.17503
Reciprocal Model	0.000	**80.0%**	**79.9%**	**2.93107**

5.2.3 LOG TRANSFORMATION OF THE X VARIABLE

Another useful curvilinear equation is

$$y_i = \beta_0 + \beta_1 \ln(x_i) + e_i$$

where $\ln(x)$ is the natural logarithm of x. It is assumed here that the x values are positive, because $\ln(x)$ is not defined for $x \leq 0$.

EXAMPLE 5.3 **Fuel Consumption** Table 5.4 shows the fuel consumption (FUELCON) in gallons per capita for each of the 50 states and Washington, DC. (*Source*: U.S. Department of Transportation, Federal Highway Administration; see the file FUELCON5 on the CD). The following variables are also shown: the population of the state (POP), the area of the state in square miles (AREA), and the population

TABLE 5.4 Data for Fuel Consumption Example

State	FUELCON	POP	AREA	DENSITY
Alabama	547.92	4,486,508	50750	88.4041
Alaska	440.38	643,786	570374	1.128709
Arizona	456.90	5,456,453	113642	48.0144
Arkansas	530.08	2,710,079	52075	52.04184
California	426.21	35,116,033	155973	225.1417
Colorado	474.78	4,506,542	103729	43.44534
Connecticut	432.44	3,460,503	4845	714.2421
Delaware	492.97	807,385	1955	412.9847
Florida	461.55	16,713,149	53997	309.52
Georgia	564.82	8,560,310	57919	147.798
Hawaii	336.97	1,244,898	6423	193.8188
Idaho	484.83	1,341,131	82751	16.20683
Illinois	406.99	12,600,620	55593	226.6584
Indiana	524.01	6,159,068	35870	171.7053
Iowa	532.39	2,936,760	55875	52.55946
Kansas	483.31	2,715,884	81823	33.19218
Kentucky	532.77	4,092,891	39732	103.0125
Louisiana	513.80	4,482,646	43566	102.8932
Maine	472.68	1,294,464	30865	41.93954
Maryland	463.46	5,458,137	9775	558.3772
Massachusetts	436.57	6,427,801	7838	820.0818
Michigan	504.95	10,050,446	56809	176.9164
Minnesota	532.52	5,019,720	79617	63.04834
Mississippi	541.06	2,871,782	46914	61.21375
Missouri	549.16	5,672,579	68898	82.333
Montana	549.35	909,453	145556	6.248131
Nebraska	503.10	1,729,180	76878	22.49252
Nevada	448.81	2,173,491	109806	19.79392
New Hampshire	541.67	1,275,056	8969	142.1626
New Jersey	465.52	8,590,300	7419	1157.878
New Mexico	504.77	1,855,059	121364	15.28508
New York	296.44	19,157,532	47224	405.6736
North Carolina	510.05	8,320,146	48718	170.7818
North Dakota	580.32	634,110	68994	9.190799

TABLE 5.4 (Continued)

State	FUELCON	POP	AREA	DENSITY
Ohio	458.31	11,421,267	40953	278.8872
Oklahoma	523.89	3,493,714	68679	50.87019
Oregon	439.09	3,521,515	96002	36.68168
Pennsylvania	417.36	12,335,091	44820	275.214
Rhode Island	382.82	1,069,725	1045	1023.66
South Carolina	557.53	4,107,183	30111	136.4014
South Dakota	577.84	761,063	75896	10.02771
Tennessee	506.30	5,797,289	41219	140.646
Texas	502.17	21,779,893	261914	83.15666
Utah	430.53	2,316,256	82168	28.18927
Vermont	555.78	616,592	9249	66.6658
Virginia	529.52	7,293,542	39598	184.1897
Washington	446.63	6,068,996	66581	91.15207
West Virginia	466.31	1,801,873	24087	74.80687
Wisconsin	466.08	5,441,196	54314	100.1804
Wyoming	715.55	498,703	97105	5.135709
Washington D.C.	289.99	570898	61	9358.984

density (DENSITY) defined as population/area. The object is to develop a regression equation to predict fuel consumption based on the population density. FUELCON is the dependent variable and DENSITY is the explanatory variable. The scatterplot of FUELCON versus DENSITY is shown in Figure 5.10. Looking at the scatterplot of FUELCON versus DENSITY, it is clear that this is not a linear

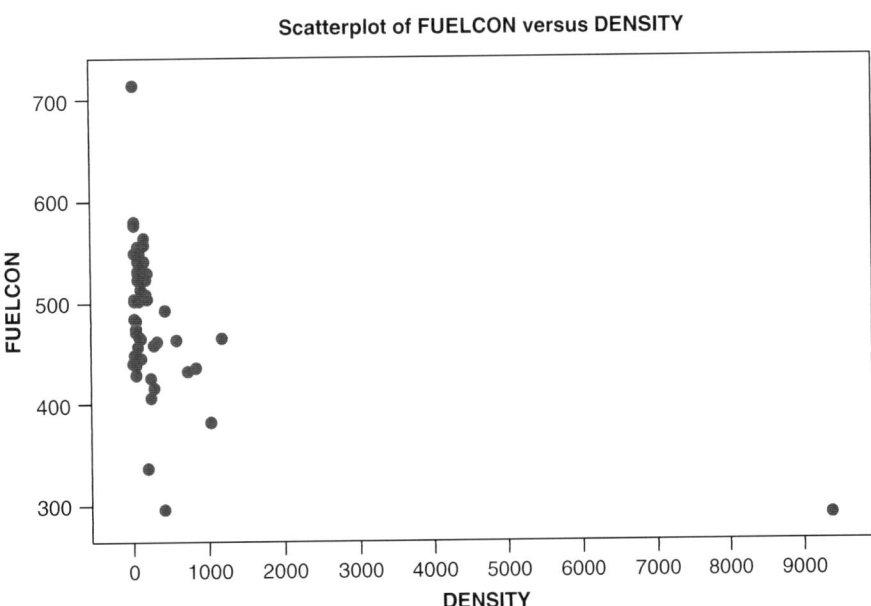

FIGURE 5.10 Scatterplot of FUELCON versus DENSITY.

relationship. One thing to note about this plot is how the values spread out on the x axis. At the left-hand side of the x axis, the values are clumped together. Moving from left to right, the values become progressively more spread out. This suggests the use of a log transformation of DENSITY. The log transformation puts values on a different scale that compresses large distances so that they are more comparable to smaller distances. Table 5.5 shows the effect of applying the log to the base 10 to a series of numbers. (If we let q represent the log to the base 10 of a number x, then q is defined as the value that makes the following equation true: $10^q = x$. If we use log to the base 2, then the defining equation becomes $2^q = x$.) Note that the values of x in the table are successively more and more spread out; the distances between the values are becoming greater. The $\log_{10}(x)$ values do not exhibit this tendency. The log transformation evens out the successively larger distances between the values.

The scatterplot of FUELCON versus the log of DENSITY (LOGDENS) is shown in Figure 5.11. The natural log of DENSITY is used. The natural log uses the number called e (e is approximately equal to 2.718) as its base. It is common in business and economic applications to use natural logarithms, although the base used is usually not important. The relationship in Figure 5.11 appears to be linear.

TABLE 5.5 Effect of Applying Log to the Base 10 to a Set of Numbers

x	10	100	1000	10000
$\log_{10}(x)$	1	2	3	4

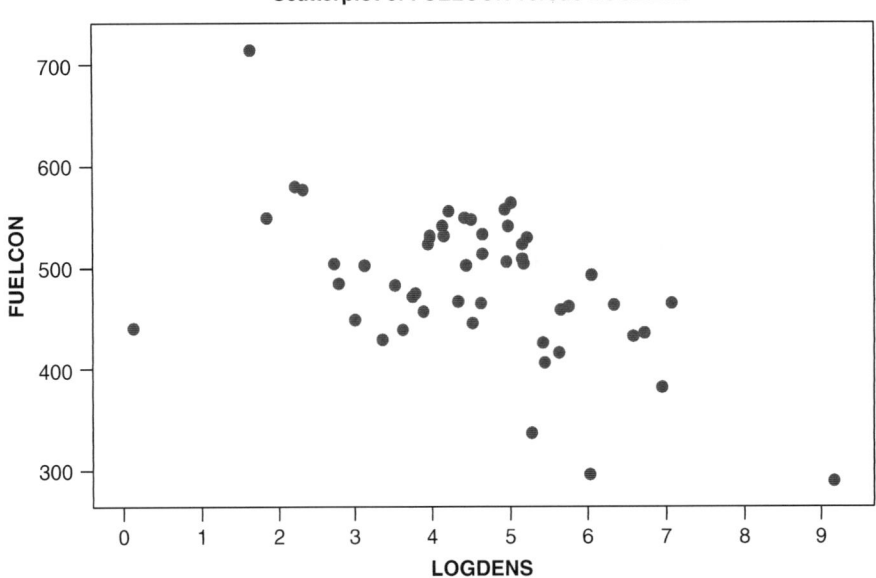

FIGURE 5.11 Scatterplot of FUELCON versus LOGDENS.

Fitting a line to the values in Figure 5.11 makes much more sense than trying to fit a line to the values in Figure 5.10. The log transformation is a good choice.

Figure 5.12 shows the linear regression results for the regression of FUELCON on DENSITY. These are used for comparison purposes. The regression results using the natural log of DENSITY as the explanatory variable are shown in Figure 5.13. The regression results indicate that using LOGDENS as the explanatory variable produces a better model fit than the regression using DENSITY. Table 5.6 provides summary statistics for the two models.

FIGURE 5.12 Regression of FUELCON on DENSITY.

Variable	Coefficient	Std Dev	T Stat	P Value
Intercept	495.628	9.481	52.28	0.000
DENSITY	-0.025	0.007	-3.56	0.001

Standard Error = 65.1675   R-Sq = 20.6%   R-Sq(adj) = 19.0%

Analysis of Variance

Source	DF	Sum of Squares	Mean Square	F Stat	P Value
Regression	1	53961	53961	12.71	0.001
Error	49	208093	4247		
Total	50	262054			

FIGURE 5.13 Regression of FUELCON on LOGDENS.

Variable	Coefficient	Std Dev	T Stat	P Value
Intercept	597.19	26.96	22.15	0.000
LOGDENS	-24.53	5.65	-4.34	0.000

Standard Error = 62.1561   R-Sq = 27.8%   R-Sq(adj) = 26.3%

Analysis of Variance

Source	DF	Sum of Squares	Mean Square	F Stat	P Value
Regression	1	72748	72748	18.83	0.000
Error	49	189306	3863		
Total	50	262054			

TABLE 5.6 Summary Measures for the Linear Model and the Model Using LOGDENS for Fuel Consumption Example

Model	p Value for Highest-Order Term	R^2	R^2_{adj}	S_e
Linear Model	0.001	20.6%	19.0%	65.17
Log x Model	0.000	**27.8%**	**26.3%**	**62.16**

5.2.4 LOG TRANSFORMATION OF BOTH THE X AND Y VARIABLES

It is also possible to transform the y variable in attempting to achieve a linear relationship. The natural logarithm of y is often used as the dependent variable with the natural logarithm of x as the explanatory variable:

$$\ln(y_i) = \beta_0 + \beta_1 \ln(x_i) + e_i$$

Some caution must be exercised if this model is chosen. First, all x and y values must be positive for the natural log transformation to be defined. Second, because $\ln(y)$ is used as the dependent variable, it becomes more difficult to compare this regression to any model using y as the dependent variable. The R^2 values of the two regressions cannot be compared, for example, because two different units of measurement are used for the dependent variable. (This applies as well to adjusted R^2 and the standard error.) Thus, increases in R^2 when the natural logarithm transformation is applied to y do not necessarily suggest an improved model. (Note that transformations of the explanatory variables do not create this type of problem. It is only when the y variable is transformed that comparison becomes more difficult.)

EXAMPLE 5.4   **Imports and GDP**   The gross domestic product (GDP) and dollar amount of total imports (IMPORTS), both in billions of dollars for 25 countries are shown in Table 5.7

TABLE 5.7 Data for Imports and GDP Example

Country	Imports	GDP
Argentina	20.300	391.000
Australia	68.000	528.000
Bolivia	1.500	21.400
Brazil	57.700	1340.000
Canada	229.000	923.000
Cuba	4.800	25.900
Denmark	47.900	155.500
Egypt	164.000	258.000
Finland	31.800	136.200
France	303.700	1540.000
Greece	31.400	201.100
Haiti	0.978	12.000
India	53.800	2660.000
Israel	30.800	122.000
Jamaica	3.100	9.800
Japan	292.100	3550.000
Liberia	0.170	3.600
Malaysia	76.900	200.000
Mauritius	2.000	12.900
Netherlands	201.100	434.000
Nigeria	13.700	105.900
Panama	6.700	16.900
Samoa	0.900	0.618
United Kingdom	330.100	1520.000
United States	1148.000	10,082.000

and contained in the file IMPGDP5 on the CD. These data were obtained from *The World Fact Book 2002* at www.odci.gov/cia/publications/factbook/index.html. The objective is to find an equation showing the relationship between IMPORTS (y) and GDP (x). The scatterplot of IMPORTS versus GDP in Figure 5.14 shows that this is not a linear relationship. One thing to note about this plot is how the values spread out on the x and y axes. At the left-hand side of the x axis and the bottom of the y axis, the values are clumped together. Moving from left to right on the x axis, the values become more spread out. The same thing happens when moving up the y axis —the values become progressively more spread out. This suggests the use of a log transformation for both the x and y variables. The motivation for using the log transformation is the same as in the Fuel Consumption example, but the transformation needs to be applied to both the x and y variables.

Figure 5.15 shows the scatterplot of the natural logarithm of imports (LOGIMP) versus the natural logarithm of GDP (LOGGDP). The relationship appears much closer to linear than in Figure 5.14. Figure 5.16 shows the results for the regression of LOGIMP on LOGGDP. The regression of IMPORTS on GDP is not shown for comparison purposes as in previous examples. As noted, since the dependent variable has been transformed, the usual comparisons are not valid. In Chapter 6, we find alternative methods for judging which of the functional forms of the model appears better. At this stage, the scatterplot strongly supports the use of the log transformation.

It is important to keep in mind that the type of transformation to correct for curvilinearity is not always obvious. If y and x are related in a curvilinear manner, the goal is to transform the variables in some manner to achieve a linear

FIGURE 5.14
Scatterplot of IMPORTS versus GDP.

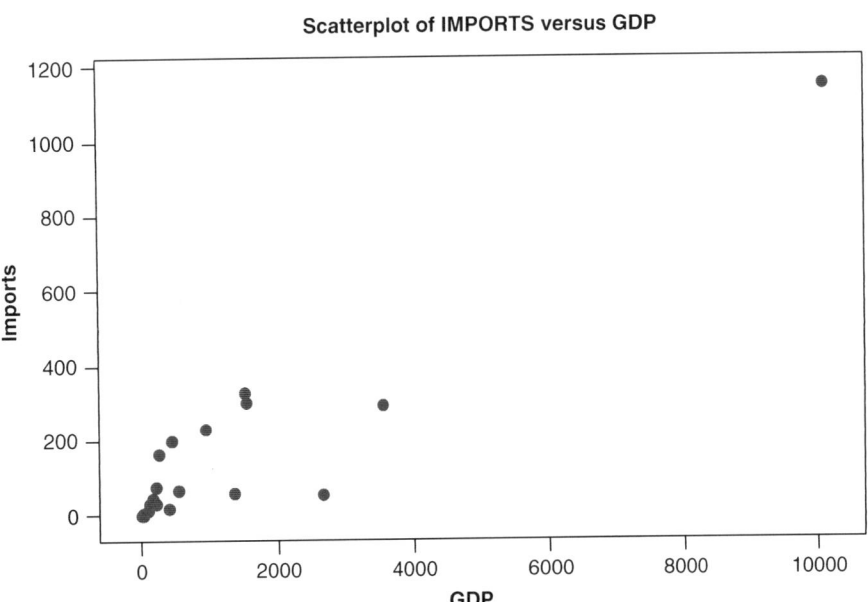

194 Fitting Curves to Data

FIGURE 5.15
Scatterplot of Log IMPORTS versus Log GDP.

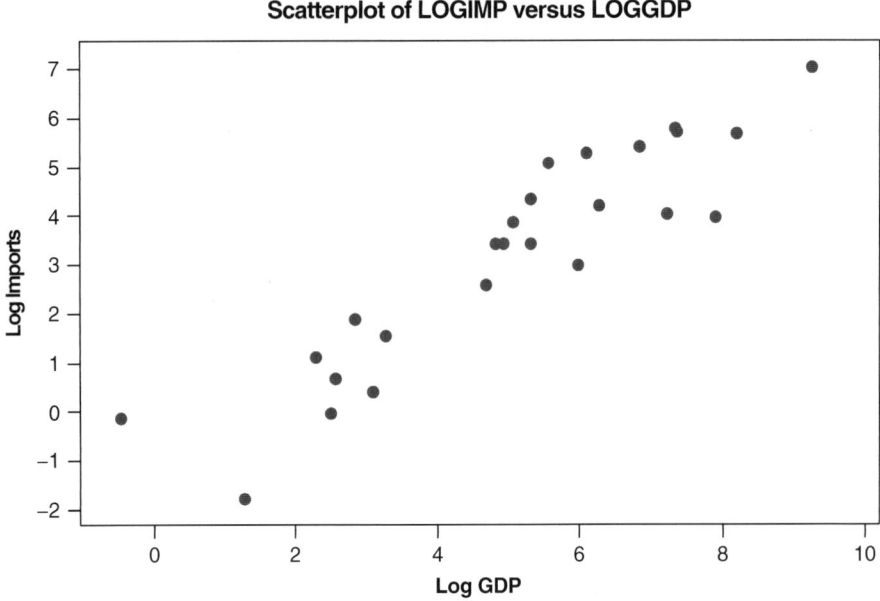

FIGURE 5.16
Regression of Log IMPORTS on Log GDP.

Variable	Coefficient	Std Dev	T Stat	P Value
Intercept	-1.1275	0.4346	-2.59	0.016
LOGGDP	0.8670	0.0788	11.01	0.000

Standard Error = 0.914202   R-Sq = 84.0%   R-Sq(adj) = 83.4%

Analysis of Variance

Source	DF	Sum of Squares	Mean Square	F Stat	P Value
Regression	1	101.26	101.26	121.15	0.000
Error	23	19.22	0.84		
Total	24	120.48			

relationship. Different transformations may be tried (including transformations not discussed in this section) and the one that appears to do the best job chosen. There may be theoretical results as well that support the use of certain transformations in certain cases. As always, subject matter expertise is important in any analysis.

In deciding what type of transformation to use, look at the scatterplot showing the relationship between y and x. This may help identify the form of the relationship between the two variables. In Chapter 6, we discuss other methods of recognizing when a linear model is not the best choice, when a curvilinear model may be more appropriate, and which curvilinear model provides an improvement.

5.2.5 FITTING CURVILINEAR TRENDS

In Chapter 3, the linear trend model was presented:

$$y_i = \beta_0 + \beta_1 t + e_i$$

where t is simply a variable indicating time sequence, $t = 1, 2, \ldots, n$. Just as curvilinear patterns can be observed with regard to x variables as discussed in this chapter, so can curvilinear trends occur. It is possible to model certain curvilinear trends using regression. This is done in a manner very similar to the fitting of curves to data just discussed. A few basic curvilinear trend models are presented here.

A *quadratic trend* equation can be written

$$y_i = \beta_0 + \beta_1 t + \beta_2 t^2 + e_i$$

Examples of the linear and quadratic trends are shown in Figures 5.17(a) and 5.17(b), respectively.

An equation for a curve called an *S-curve* is given by

$$y_i = \exp\left(\beta_0 + \beta_1\left(\frac{1}{t}\right) + e_i\right)$$

where exp denotes the exponential operator: the value $e = 2.718$ (approximately) is raised to the power

$$\beta_0 + \beta_1\left(\frac{1}{t}\right) + e_i$$

The S-curve is shown in Figure 5.17(c). This type of trend might be used to model demand for certain products over their lifetime. Demand is slow initially until the product becomes better known. Then demand picks up until a saturation point is reached. At that time, demand levels off.

The S-curve equation cannot be estimated directly using least-squares. By taking natural logarithms of both sides of the equation, however, a new equation is obtained that can be estimated. Because $\ln(\exp(x)) = x$ for any x, taking natural logarithms of both sides of the equation produces

$$\ln(y_i) = \beta_0 + \beta_1\left(\frac{1}{t}\right) + e_i$$

Regressing $\ln(y_i)$ on $1/t$ produces estimates of β_0 and β_1. When forecasting with this model, care should be taken. For example, write $y'_i = \ln(y_i)$ and $t' = \dfrac{1}{t}$, and write the estimated regression equation as

$$\hat{y}'_i = b_0 + b_1 t'$$

The estimated equation provides the forecast for time period T:

$$\hat{y}'_T = b_0 + b_1\left(\frac{1}{T}\right)$$

Note that this is a forecast of y'_T or the natural logarithm of y_T. To obtain a forecast of the original dependent variable, y_T, the conversion back to the original units from logarithmic units must be made:

$$\exp(\hat{y}'_T) = \hat{y}_T$$

Exponential trends also are used in time-series applications. The equation for an exponential trend is

$$y_i = \exp(\beta_0 + \beta_1 t + e_i)$$

Again, to estimate β_0 and β_1, the equation is transformed using natural logarithms. Writing $y'_i = \ln(y_i)$, the transformed equation is

$$y'_i = \beta_0 + \beta_1 t + e_i$$

Regressing $y'_i = \ln(y_i)$ on t produces estimates of β_0 and β_1. As with the S-curve, exercise caution when using this equation for forecasting. The natural logarithm of $\hat{y}'_i$ must be transformed back to its original units to obtain the desired forecast. Examples of exponential trends are shown in Figure 5.17(d).

FIGURE 5.17
Examples of Types of Trends.

(a) Linear trend
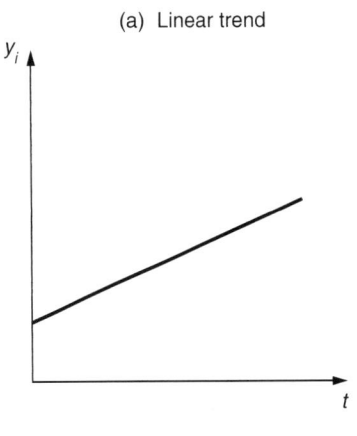

(c) S-Curve
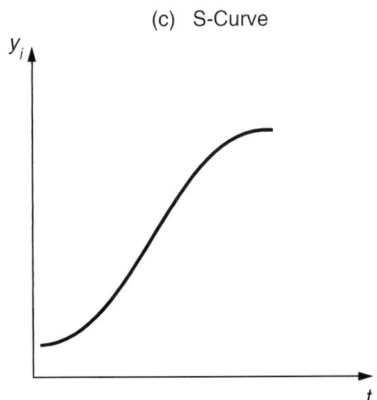

(b) Quadratic trend
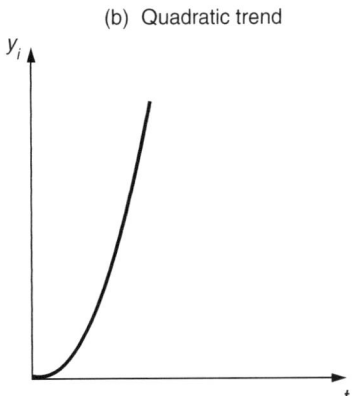

(d) Exponential trend $\beta_0 = 0.0$
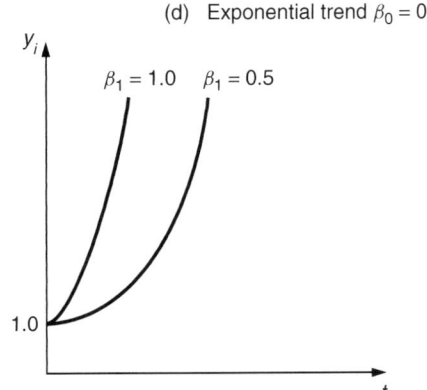

EXERCISES

1. **Research and Development.** A company is interested in the relationship between profit on a number of projects and two explanatory variables. These variables are the expenditure on research and development for the project (RD) and a measure of risk assigned at the outset of the project (RISK). The file RD5 on the CD and Table 5.8 show the data on the three variables PROFIT, RISK, and RD. PROFIT is measured in thousands of dollars and RD is measured in hundreds of dollars. The scatterplots of PROFIT versus RISK and PROFIT versus RD are shown in Figures 5.18 and 5.19, respectively. The regression of PROFIT on the two explanatory variables RISK and RD is shown in Figure 5.20.

   Figure 5.21 shows the results of a regression using PROFIT as the dependent variable with RISK, RD, and RDSQR (the square of the RD variable) as explanatory variables. Choose the model you prefer for PROFIT and provide a justification for your choice.

TABLE 5.8 Data for Research and Development Exercise

RD	RISK	PROFIT
132.580	8.5	396
81.928	7.5	130
145.992	10.0	508
90.020	8.0	172
114.408	7.0	256
53.704	7.5	32
76.244	7.0	102
71.680	8.0	102
151.592	9.5	536
74.816	7.5	102
108.752	6.0	214
92.372	8.5	200
92.260	7.0	158
60.732	6.5	32
78.120	7.5	116
90.000	5.5	120
105.532	9.0	270
111.832	8.0	270

2. **Piston Corporation (Part A).** Reginald Jackson was employed as a cost accountant by the Piston Corporation, a medium-size auto parts company located in the outskirts of Detroit. Kelly Jones, the controller for Piston, decided that she needed an assistant. Jackson was selected to fill that position. As part of his training program, Jackson was sent to night school to study quantitative applications in cost accounting.

   Because the Piston Corporation's products were replacement parts, its sales were, fortunately, not as volatile as the new car market's. Piston had

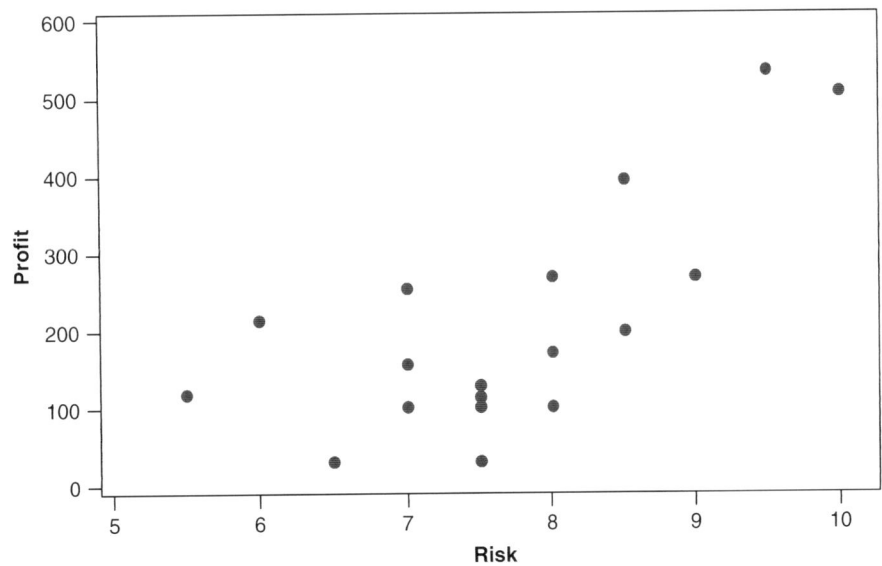

FIGURE 5.18
Scatterplot of PROFIT versus RISK for Research and Development Exercise.

198 Fitting Curves to Data

FIGURE 5.19
Scatterplot of PROFIT versus RD for Research and Development Exercise.

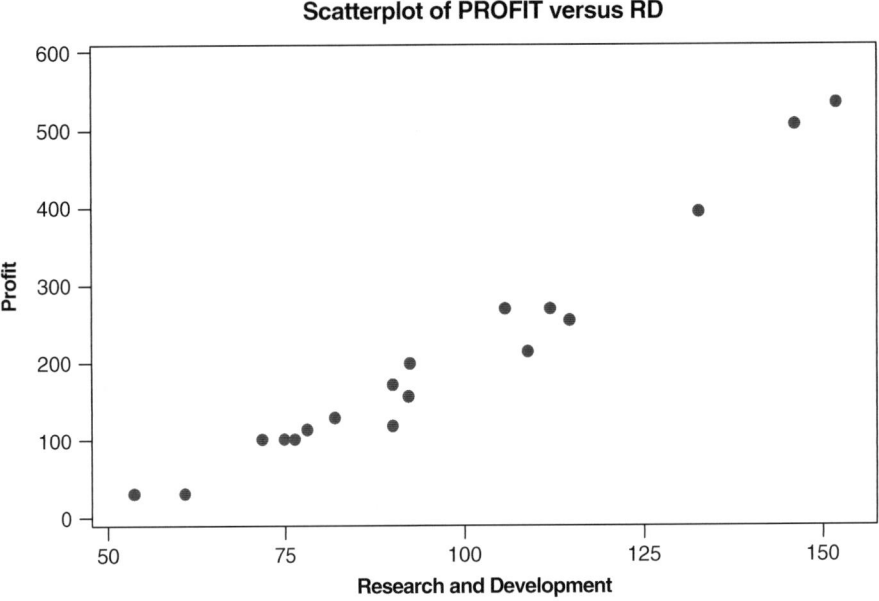

FIGURE 5.20
Regression of PROFIT on RISK and RD for Research and Development Exercise.

Variable	Coefficient	Std Dev	T Stat	P Value
Intercept	-453.1763	23.5061	-19.28	0.000
RISK	29.3090	3.6686	7.99	0.000
RD	4.5100	0.1538	29.33	0.000

Standard Error = 14.3420   R-Sq = 99.2%   R-Sq(adj) = 99.0%

Analysis of Variance

Source	DF	Sum of Squares	Mean Square	F Stat	P Value
Regression	2	361639	180820	879.08	0.000
Error	15	3085	206		
Total	17	364724			

FIGURE 5.21
Regression of PROFIT on RISK and First- and Second-Order Terms for RD for Research and Development Exercise.

Variable	Coefficient	Std Dev	T Stat	P Value
Intercept	-245.369584	14.811115	-16.57	0.000
RISK	23.249237	0.988413	23.52	0.000
RD	1.014314	0.232378	4.36	0.001
RDSQR	0.017567	0.001152	15.25	0.000

Standard Error = 3.53798   R-Sq = 100.0%   R-Sq(adj)=99.9%

Analysis of Variance

Source	DF	Sum of Squares	Mean Square	F Stat	P Value
Regression	3	364549	121516	9707.86	0.000
Error	14	175	13		
Total	17	364724			

experienced a rather stable growth in sales in recent years and had been required to increase its capacity regularly. It appeared to be time for another expansion, but with an uncertain stock market prevailing and uncertainty concerning interest rates, Jones was worried about obtaining funds at a reasonable cost. On the other hand, Piston's production manager had been complaining, more than usual, about various personnel, material handling, and scheduling bottlenecks that arose from the high level of output demanded of his present facilities.

The executive officers had been asked by Piston's directors to formulate a proposal for expansion and price adjustments. Jones asked Jackson what he could determine statistically about the effect of inflation and the level of production on unit costs.

By looking at old budgets, Jackson was able to obtain quarterly data on manufacturing costs per unit, production level (a percentage of the total capacity), and the index of direct material and direct labor costs for a five-year period (These data are in a file named PISTON5 on the CD). He immediately went to the computer and ran a regression of unit cost on production volume (PROD) and the cost index (INDEX). He began to wonder about the validity of modeling unit costs as a linear function of production level and the cost index.

The scatterplots of the dependent variable (COST) versus each explanatory variable are shown in Figures 5.22 and 5.23. The regression results are shown in Figure 5.24.

After spending several days trying to improve his model, Jackson had several new solutions but was still unsure which one was best. That night after his quantitative accounting class, he asked his professor for advice. The professor suggested that Jackson first derive a theoretically plausible solution and then see if the data satisfied this relationship.

Jackson knew that the basic relationship with which he was dealing was:

total cost = variable cost per unit × volume + fixed cost

He also theorized that variable cost per unit was composed of a constant multiple of the index of direct materials and labor (x_2) and that fixed cost was simply the current capacity of the company times some constant.

Jackson realized that the basic equation could be rewritten as

total cost = $\beta_1$capacity + $\beta_2 x_2$volume

FIGURE 5.22
Scatterplot of COST versus PROD.

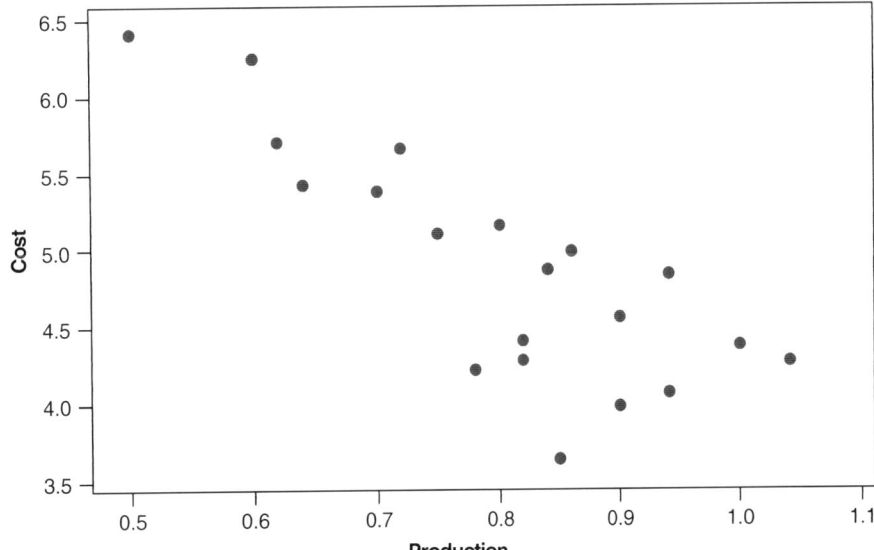

FIGURE 5.23
Scatterplot of COST versus INDEX.

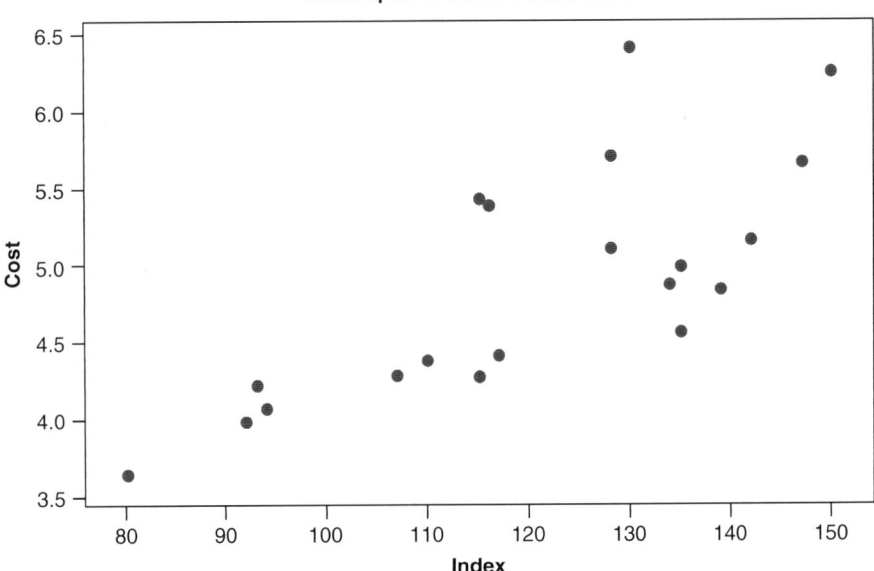

FIGURE 5.24
Regression of COST on PROD and INDEX.

```
Variable        Coefficient     Std Dev      T Stat     P Value

Intercept        5.1829         0.5364        9.66       0.000
PROD            -3.4482         0.3961       -8.70       0.000
INDEX            0.0205         0.0028        7.33       0.000


Standard Error = 0.227957     R-Sq = 91.9%   R-Sq(adj) = 91.0%

Analysis of Variance

Source         DF     Sum of Squares      Mean Square    F Stat    P Value

Regression      2        10.0586             5.0293       96.78     0.000
Error          17         0.8834             0.0520
Total          19        10.9420
```

Then it dawned on him that he actually wanted cost per unit. The preceding equation could be divided by volume to get

total cost/volume = $\beta_1$capacity/volume + $\beta_2 x_2$

Capacity/volume, however, is simply the reciprocal of production level, so the new equation becomes

$$y = \beta_1\left(\frac{1}{x_1}\right) + \beta_2 x_2$$

where y is total cost/volume (or cost per unit), x_1 is production level and x_2 is cost index. Allowing for random error and allowing the equation to have an intercept term produces:

$$y = \beta_0 + \beta_1\left(\frac{1}{x_1}\right) + \beta_2 x_2 + e$$

Try Jackson's new model. How does this model compare with the original regression? In answering this question, use the R^2 values, the standard error of the regression, and any other information you feel might be useful in the comparisons. Which model do you prefer?

3. **Piston Corporation (Part B).** Use the model developed in Problem 2 (Piston Part A) to answer the following questions:

   a. A three-point rise in the cost index will cause what change in unit costs (assuming production level remains constant)?

   b. What is the marginal unit cost of a rise in production volume from 0.94 to 0.95 of capacity (marginal cost implies all other variables remain constant)?

   c. If forecasts of production level of 0.87 and cost index of 120 are obtained, find a prediction of the manufacturing cost per unit (use a point prediction).

   d. Construct a 95% prediction interval for manufacturing cost per unit under the conditions described in part c.

4. **Computer Repair.** A computer repair service is examining the time taken on service calls to repair computers. Data are obtained for 30 service calls. The data are in a file named COMPREP5 on the CD. Information obtained includes:

   x_1 = number of machines to be repaired (NUMBER)
   x_2 = years of experience of service person (EXPER)
   y = time taken (in minutes) to provide service (TIME)

   Develop a polynomial regression model to predict average time on the service calls using EXPER and NUMBER as explanatory variables. Justify your model choice including transformations of any variables.

5. **Criminal Justice Expenditures.** The file named CRIMSPN5 on the CD contains the following data for each of the 50 states:

   total expenditures on a state's criminal justice system (in millions of dollars) (EXPEND)
   total number of police employed in the state (POLICE)

   State governments must try to project spending in many areas. Expenditure on the criminal justice system is one area of continually rising cost. Your job is to build a model that can be used to forecast spending on a state's criminal justice system.

   Once your model is complete, predict expenditures for a state that plans to hire 10,000 police personnel. Find a point prediction and a 95% prediction interval.
   (*Source*: These data were obtained from the U.S. Department of Criminal Justice web site and are for the year 1999.)

6. **Predicting Movie Grosses.** The file named MOVIES5 on the CD contains data on movies released in the United States during the calendar year 1998. The two variables in this file are

   TDOMGROSS, the total domestic gross revenue
   WEEKEND, first weekend gross

   We would like to find an equation to predict the total domestic gross revenue of movies based on their first weekend gross. People in the movie industry watch the first weekend gross revenues closely and use them to help make decisions about advertising, distribution, etc. We need to formalize the relationship between total domestic gross revenue and first weekend gross. Find an equation that represents this relationship. Use the scatterplot to guide in choosing the best model.
   (*Source*: These data are discussed in the article "Predicting Movie Grosses: Winners and Losers, Blockbusters and Sleepers," by Jeffrey S. Simonoff and Ilana R. Sparrow. Chance, Vol. 13, 2000, 15—24, and were obtained from Dr. Simonoff's web site.)

7. **Kentucky Derby.** On the first Saturday in May, the granddaddy of horse races—the Kentucky Derby—is run at Churchill Downs in Louisville, Kentucky. The amount of money bet, in millions of dollars, on this race for the 66-year period from 1927 through 1992 is in the file named DERBY5 on the CD.
   (*Source*: "How the Betting Went," Louisville *Courier-Journal*, May 3, 1992. Copyright 1992, Louisville *Courier-Journal*. Used with permission.).

   Build an extrapolative model for the amount bet and provide a justification for the model. Use linear and/or nonlinear trends to build the model. Once you have chosen your preferred model, use it to forecast the amount bet in 1993 and 1994.

8. **Mileage and Weight.** The variables CITYMPG (y), which is the number of miles per gallon obtained by a car in city driving, and WEIGHT (x), the weight in pounds of the car, are in a file named MPGWT5 on the CD. This information is available for 147 cars listed in the *Road and Track* October 2002 issue.

   Fit the linear regression using CITYMPG as the dependent variable and WEIGHT as the independent variable.

Examine a scatterplot of these two variables. Can you find a curvilinear model that better describes the relationship between these two variables? If so, what is the regression equation that describes this relationship? Justify your choice of equation.

USING THE COMPUTER

The Using the Computer section in each chapter describes how to perform the computer analyses in the chapter using Excel, MINITAB, and SAS. For further detail on Excel, MINITAB, and SAS, see Appendix C.

EXCEL

Variable Transformations

Variable transformations in Excel are accomplished through the use of formulas. Consider the screen shown in Figure 5.25. We want to create a new column containing the square of the numbers in column A. To do this, we create a formula in cell C2 to multiply the value in cell A2 by itself (=A2*A2). Then place the cursor on the lower right-hand corner of cell C2 and drag this cell to the last entry in column A (see Figure 5.26). Other transformations are created in a similar manner. If you are

FIGURE 5.25 Excel Screen Showing Formula Creation for Variable Transformation.

	A	B	C
1	MONTHS	CALLS	
2	10	18	=a2*a2
3	10	19	
4	11	22	
5	14	23	
6	15	25	
7	17	28	
8	18	29	
9	20	29	
10	20	31	
11	21	31	
12	22	33	
13	22	32	
14	24	31	
15	25	32	
16	25	32	
17	25	33	
18	25	31	
19	28	33	
20	29	33	
21	30	34	

FIGURE 5.26 Excel Screen Showing Formula Creation for Variable Transformation.

	MONTHS	CALLS	
2	10	18	100
3	10	19	100
4	11	22	121
5	14	23	196
6	15	25	225
7	17	28	289
8	18	29	324
9	20	29	400
10	20	31	400
11	21	31	441
12	22	33	484
13	22	32	484
14	24	31	576
15	25	32	625
16	25	32	625
17	25	33	625
18	25	31	625
19	28	33	784
20	29	33	841
21	30	34	900

FIGURE 5.27 Excel Screen Showing How to Access Functions.

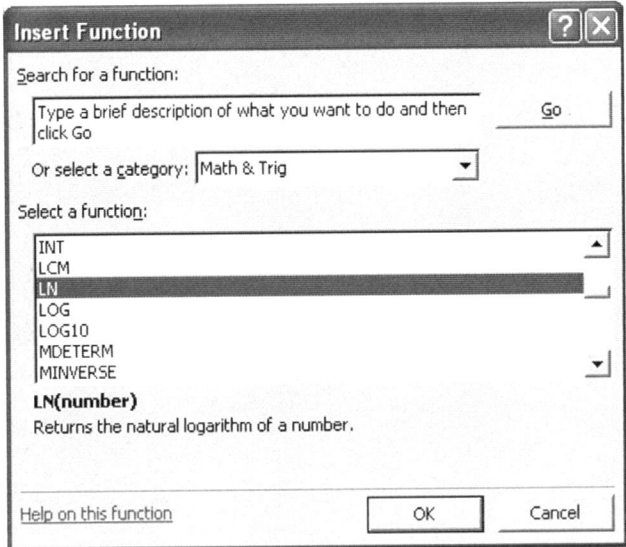

FIGURE 5.28 MINITAB Dialog Box for Calculator.

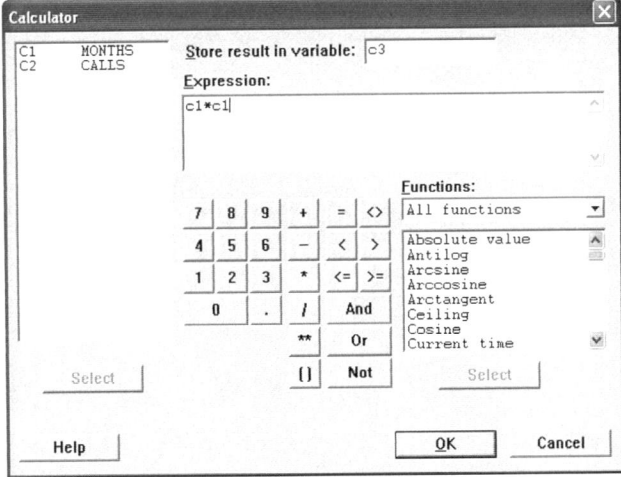

not sure of the form of a certain transformation, click the f_x button. Figure 5.27 shows an example using the natural log function. Most functions used to transform variables can be found in the Math & Trig category.

MINITAB

Variable Transformations

CALC: CALCULATOR

The CALCULATOR dialog box is shown in Figure 5.28. Any variable transformations can be performed using the calculator. As shown, the calculator is set up to multiply the numbers in column 1 by themselves (C1*C1) and place the result in C3. Thus, C3 will contain the square of the numbers in C1. Other transformations can be performed in similar fashion. Placing LOGE(C1) in the Expression box produces the natural logarithm of C1. Placing SQRT(C1) in the expression box produces the square root of C1. If you are not sure of the form of the function (LOGE, SQRT, and so on), just double click the desired function to the right of the keyboard and the appropriate expression appears in the Expression box.

SAS

Variable Transformations

In SAS, variable transformations are performed during the data input phase. Here are some typical examples:

Create the square of the variable MONTHS and call it XSQR:

```
XSQR = MONTHS**2;
```

Create the natural log of the variable MONTHS and call it LOGMONTH:

```
LOGMONTH = LOG(MONTH);
```

These transformed variables can then be used in PROC REG, PROC PLOT, and so on.

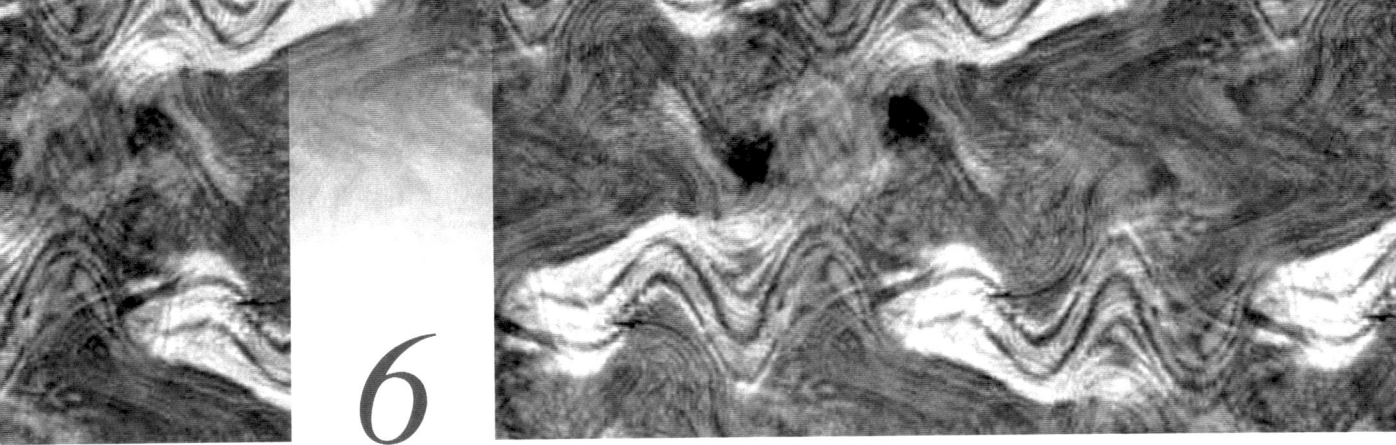

6
Assessing the Assumptions of the Regression Model

6.1 INTRODUCTION

In Chapter 4, the multiple linear regression model was presented as

$$y_i = \beta_0 + \beta_1 x_{1i} + \beta_2 x_{2i} + \cdots + \beta_K x_{Ki} + e_i \tag{6.1}$$

Certain assumptions were made concerning the disturbances, e_i, of this model. The e_i represent the differences between the true values of the dependent variable and the corresponding points on the population regression line. Because the true disturbances cannot be observed, they are modeled as realizations of a random variable about which certain assumptions are made. Under a set of ideal assumptions, the method of least squares provides the best possible estimates of the population regression coefficients. Certain assumptions are necessary for inference procedures (confidence interval estimates and hypothesis tests) to perform as desired. In this chapter, we consider the problems with estimation and inference that may arise if any of these assumptions are violated. Methods of assessing the validity of the assumptions also are discussed. Graphical procedures such as scatterplots and residual plots may be used to examine certain assumptions, and statistical tests are available for a more formal examination. Finally, we discuss appropriate techniques to correct for violated assumptions.

6.2 ASSUMPTIONS OF THE MULTIPLE LINEAR REGRESSION MODEL

The "ideal" conditions for estimation and inference in the multiple regression model are as follows:

a. The expected value of the disturbances is zero: $E(e_i) = 0$. This implies that the regression line passes through the conditional means of the y variable. For our purposes, we interpret this assumption as: The relationship is linear in the explanatory variables.

b. The disturbances have constant variance σ_e^2.

c. The disturbances are normally distributed.

d. The disturbances are independent.

The effects of violations of each of these assumptions on the least-squares estimates of the regression coefficients are examined in subsequent sections. Methods of assessing the validity of the assumptions are discussed and possible corrections for violations are offered. Because many of the methods of assessing assumption validity depend on the use of the residuals (the sample counterpart of the disturbances), the next section is devoted to a brief discussion of the computation and properties of the residuals.

6.3 THE REGRESSION RESIDUALS

The regression equation estimated from the sample data may be written

$$\hat{y}_i = b_0 + b_1 x_{1i} + b_2 x_{2i} + \cdots + b_K x_{Ki} \tag{6.2}$$

By substituting in the sample values for each explanatory variable, the predicted or fitted y value for each data point in the sample is obtained. The fitted y values are denoted as $\hat{y}_i$. The y values for each point in the sample are also available and are referred to as y_i. The differences between the true and fitted y values for the points in the sample are called the *residuals*. The residuals are denoted by $\hat{e}_i$:

$$\hat{e}_i = y_i - \hat{y}_i$$

They represent the distance that each dependent variable value is from the estimated regression line or the portion of the variation in y that cannot be "explained" with the data available. Because these "sample disturbances" approximate the population disturbances, they can be used to examine assumptions concerning the population disturbances.

After estimating a sample regression equation, it is highly recommended that some sort of analysis be conducted to assess the model assumptions. No regression analysis can be considered complete without such further examination. The residuals can be used to conduct such analyses through graphical techniques called *residual plots*. Often, violations of assumptions can be detected through the use of residual plots in combination with scatterplots without the use of statistical tests.

The use of graphical techniques, however, is not an exact science. It might, in fact, be considered an "art." It takes some experience at examining plots to become adept at determining which, if any, assumptions may be violated. Several examples are presented later to illustrate this art and to aid in mastering residual analysis.

First, consider some properties of the residuals.

Property 1: The average of the residuals is equal to zero. This property holds regardless of whether the assumptions are true or not and is a direct result of the way the least-squares method works. Least squares "forces" the mean of the residuals to be zero when it chooses the estimates of the regression coefficients.

Property 2: If assumptions a, b, and d of Section 6.2 are true, then the residuals should be randomly distributed about their mean (zero). There should be no systematic pattern in a residual plot.

Property 3: If assumptions a, b, and d are true and the disturbances are also normally distributed (assumption c), then the residuals should look like random numbers chosen from a normal distribution.

The residuals can be thought of as representing the variation in y that cannot be explained using the proposed regression model. Think of the process we are following as building a model for the data. We can write DATA = MODEL + ERROR. We have some DATA that we want to explain. We build a MODEL that we believe helps to explain patterns in the data. Any patterns in the DATA not included in the MODEL are accounted for in the ERROR term. These errors are represented by the residuals. Thus, if an assumption is violated, an indication of this violation appears as some type of pattern in the residuals. Identification of such patterns is a first step in correcting for the violation.

In a residual analysis, it is suggested that the following plots be used:

1. Plot the residuals versus each explanatory variable.
2. Plot the residuals versus the predicted or fitted values.
3. If the data are measured over time, plot the residuals versus some variable representing the time sequence.

As is shown in subsequent sections, each of these three types of plots plays a part in identifying violations of the basic assumptions.

If no assumptions are violated, then the residuals should be randomly distributed around their mean of zero and should look like numbers drawn randomly from a normal distribution. Figure 6.1 shows how a residual plot might appear when assumptions a through d are all true. There is no pattern visible in the scatter of residuals. For comparison, Figure 6.2 shows a residual plot with an obvious pattern to the residuals. Compare this to the random scatter of the residuals in Figure 6.1. A residual plot such as Figure 6.2 indicates that some assumption has been violated. (There are other patterns that could suggest violations, as is seen throughout this chapter.)

In most regression software packages (Excel, SAS, and MINITAB included), residual plots are easily constructed after a regression analysis has been performed.

FIGURE 6.1
Residual Plot Assuming No Violation of Assumptions a Through d of Section 6.2.

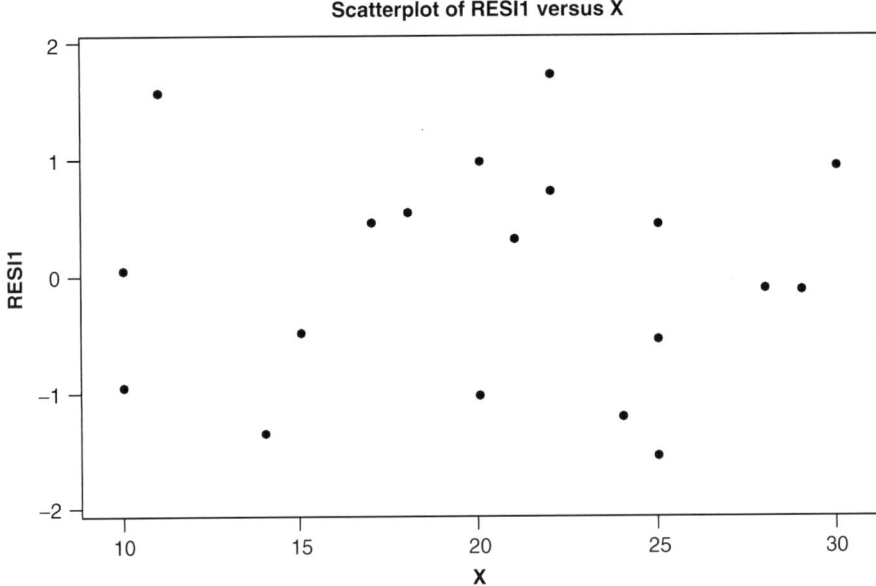

FIGURE 6.2
Residual Plot Indicating That at Least One of Assumptions a Through d of Section 6.2 Has Been Violated.

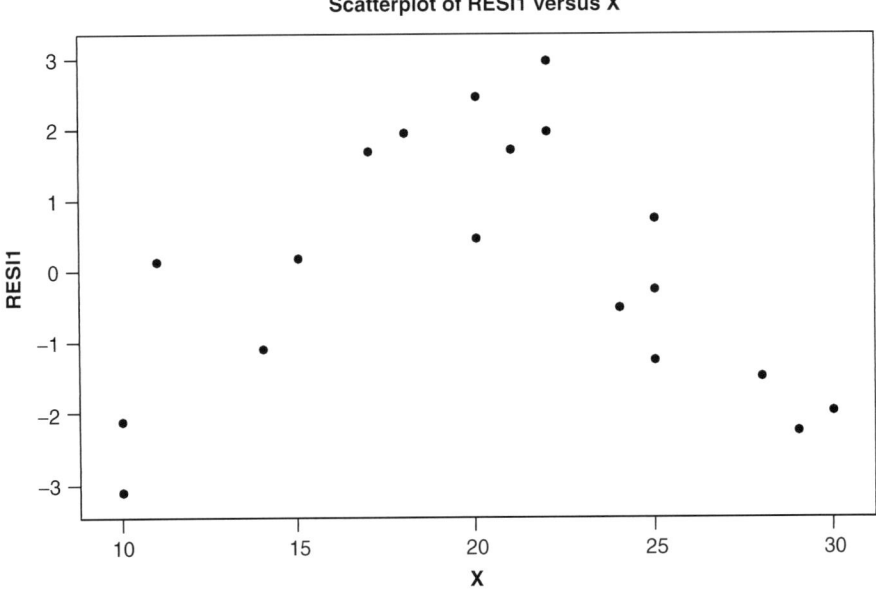

These plots may be constructed using the actual residuals, $\hat{e}_i$, or the standardized residuals. The *standardized residuals* are simply the residuals divided by their standard deviation. There is very little difference in the way residual plots with actual residuals or those with standardized residuals are used. To illustrate the difference in the two types of plots, compare Figure 6.1, a plot of actual residuals, to Figure 6.3, a plot of the same residuals after standardization. The residuals plotted in Figure 6.2

FIGURE 6.3 Plot of the Residuals from Figure 6.1 After Standardizing.

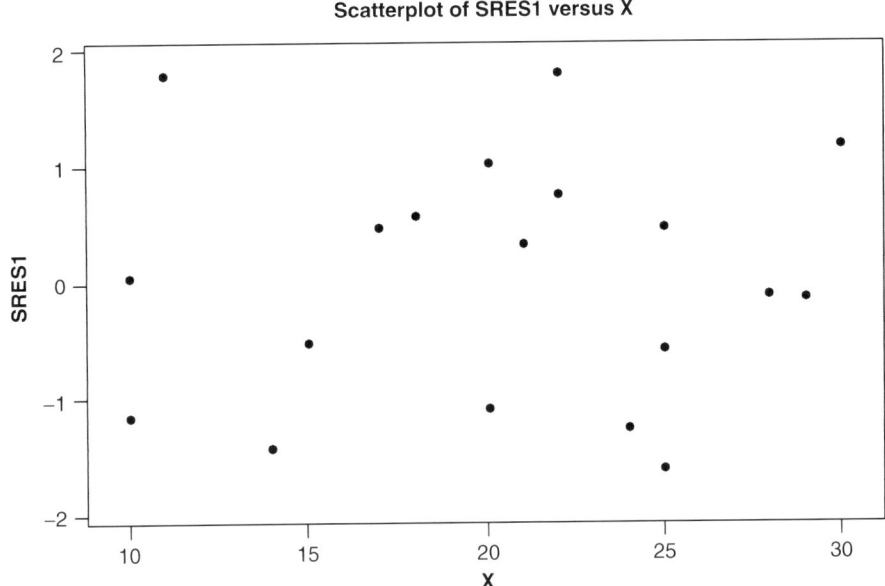

also have been standardized and plotted in Figure 6.4. The patterns in the actual and standardized plots are identical; only the scale has been changed. One advantage of using the standardized plots becomes more evident when the assumption of normality is discussed in Section 6.6. In this text, the residual plots shown are standardized plots unless otherwise indicated.

FIGURE 6.4 Plot of the Residuals from Figure 6.2 After Standardizing.

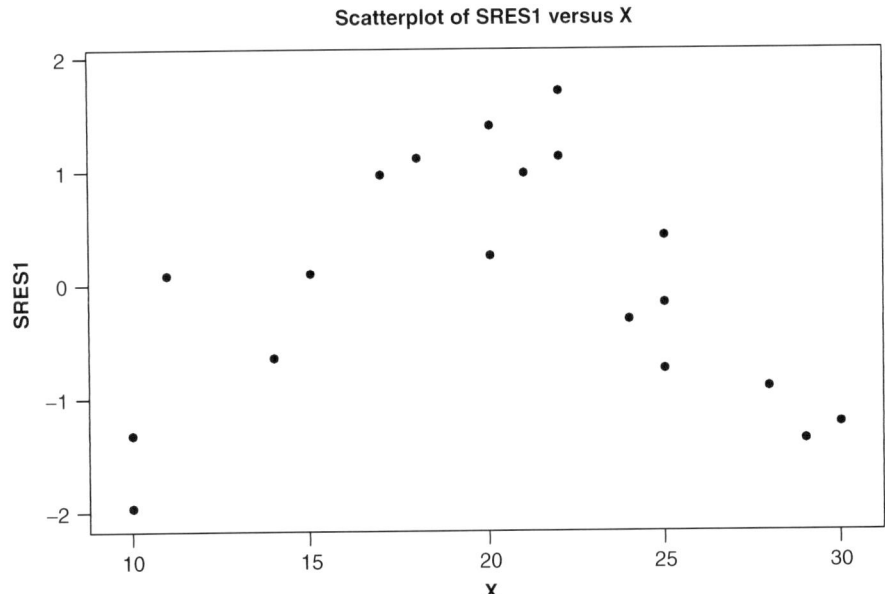

6.4 ASSESSING THE ASSUMPTION THAT THE RELATIONSHIP IS LINEAR

6.4.1 USING PLOTS TO ASSESS THE LINEARITY ASSUMPTION

The first assumption given in Section 6.2 was that the regression was linear in the explanatory variables. In Chapter 5, we saw that we can fit curvilinear as well as linear relationships using regression. In that chapter, we assumed that we could tell when a curvilinear relationship was needed simply by looking at the scatterplot. The scatterplots of y versus each of the explanatory variables may give an indication of whether the linearity assumption is an appropriate one, but this is not always the case. After performing a regression, this assumption can be checked visually through residual plots. Small deviations from linearity that are not evident in the scatterplots may show up clearly in the residual plots. The following example illustrates a violation of the linearity assumption.

EXAMPLE 6.1 **Telemarketing** Consider again the telemarketing data from Example 5.1 (see the TELEMARK6 file on the CD). A company that provides transportation services uses a telemarketing division to help sell its services. The division manager is interested in the time spent on the phone by the telemarketers in the division. Data on the number of months of employment and the number of calls placed per day (an average for 20 working days) are recorded for 20 employees. These data are shown in Table 6.1.

The average number of calls for all 20 employees is 28.95. The division manager suspects, however, that there may be a relationship between time on the job and

TABLE 6.1 Data for Telemarketing Example

MONTHS	CALLS
10	18
10	19
11	22
14	23
15	25
17	28
18	29
20	29
20	31
21	31
22	33
22	32
24	31
25	32
25	32
25	33
25	31
28	33
29	33
30	34

number of calls. As time on the job increases, the employee becomes more familiar with the calling system and the correct procedures to use on the phone and also begins to acquire more regular clients. Thus, the longer the time on the job, the greater the number of calls per day. The scatterplot of CALLS versus MONTHS is in Figure 6.5 and the regression relating CALLS to MONTHS is shown in Figure 6.6.

Plots of the standardized residuals versus the fitted values and the explanatory variable MONTHS are shown in Figures 6.7 and 6.8, respectively. The standardized residuals have been labeled SRES1 in the plots. The fitted values are labeled FITS1.

A systematic pattern can be observed in both of the residual plots. The standardized residuals plot in a curvilinear pattern, suggesting a curvilinear component may be omitted from the equation expressing the relationship between CALLS and MONTHS. The plots of the standardized residuals versus the fitted values and versus MONTHS show identical patterns in this case. The plot versus the fitted values

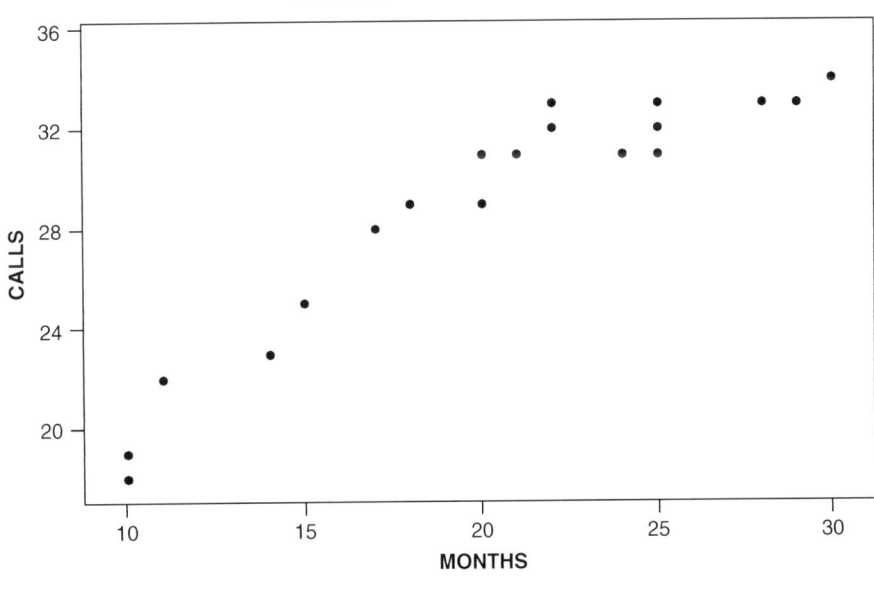

FIGURE 6.5
Scatterplot of CALLS versus MONTHS for Telemarketing Example.

FIGURE 6.6
Regression Results for Telemarketing Example.

Variable	Coefficient	Std Dev	T Stat	P Value
Intercept	13.6708	1.4270	9.58	0.000
MONTHS	0.7435	0.0667	11.15	0.000

Standard Error = 1.78737   R-Sq = 87.4%   R-Sq(adj) = 86.7%

Analysis of Variance

Source	DF	Sum of Squares	Mean Square	F Stat	P Value
Regression	1	397.45	397.45	124.41	0.000
Error	18	57.50	3.19		
Total	19	454.95			

212 Assessing the Assumptions of the Regression Model

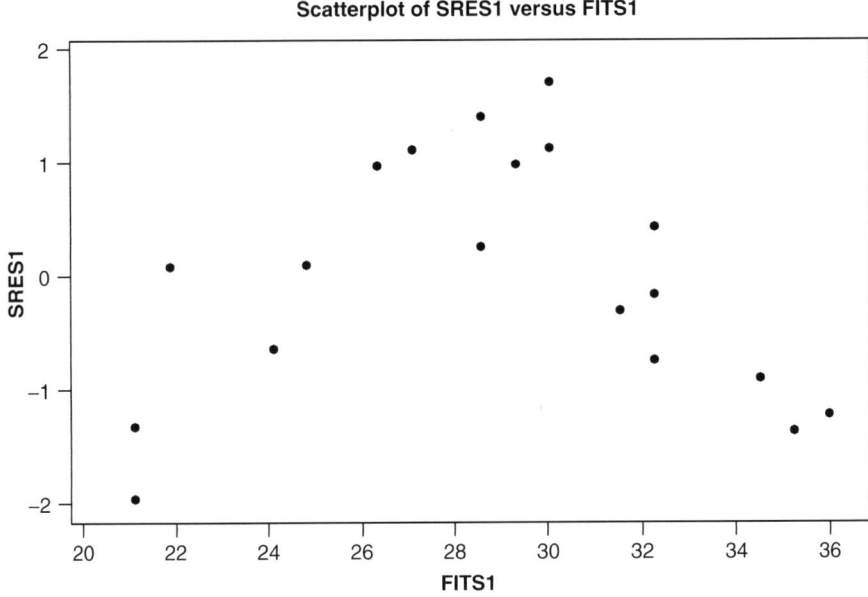

FIGURE 6.7 Plot of Standardized Residuals versus Fitted Values for Telemarketing Example.

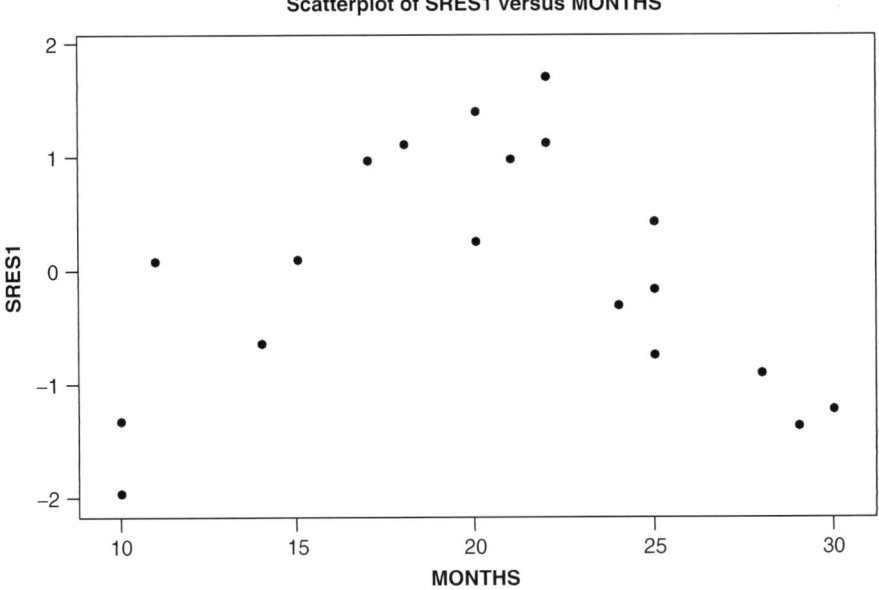

FIGURE 6.8 Plot of Standardized Residuals versus Explanatory Variable MONTHS for Telemarketing Example.

may differ from the plot versus one of the explanatory variables, especially in a multiple regression. The fitted values combine the effects of all the explanatory variables used in the regression. In a multiple regression, the plot of the standardized residuals versus the fitted values provides an overall picture, while the plots of the standardized residuals versus each explanatory variable may help identify any violations specifically related to an individual explanatory variable.

The systematic pattern observed in the residual plots suggests a violation of the linearity assumption. Looking at the scatterplot in Figure 6.5 also helps verify that the relationship may not be linear. As the number of months on the job increases, the number of calls also increases. But the rate of increase begins to slow over time, thus resulting in a pattern that may be better modeled by a curve than a straight line.

Note that the residual plots were used to determine whether the linearity assumption had been violated, although this violation might have been suspected from looking at only the scatterplot of CALLS versus MONTHS. In many cases, the violation of an assumption is not obvious from a scatterplot. The residual plot, however, is intended to magnify the consequences of any possible violation. Thus, the residual plot should be depended on to identify the violation.

6.4.2 TESTS FOR LACK OF FIT[1]

MINITAB provides two tests to determine whether a curvilinear model might fit the data better than a linear model. These tests are referred to as *tests for lack of fit*. The first test is called the *pure error lack-of-fit test*. To perform this test, the error sum of squares is decomposed into two parts: the pure error component and the lack-of-fit component. These two components are used to construct an F statistic to test the hypotheses

H_0: The relationship is linear

H_a: The relationship is not linear

If H_0 is not rejected, the linear regression model is appropriate. If H_0 is rejected, the linear model does not fit the data well, and some other function may provide a better fit, although the test does not specify what that function is.

To conduct the F test, the decision rule is

Reject H_0 if $F > F(\alpha; c - K - 1, n - c)$

Do not reject H_0 if $F \leq F(\alpha; c - K - 1, n - c)$

where K is the number of explanatory variables and n is the sample size. The value c requires some additional explanation.

The pure error lack-of-fit test requires that there be repeated observations (replications) for at least one level of the x variables. In the telemarketing data in Table 6.1, there are replicates for $x = 10, 20, 22,$ and 25. The value c is the number of distinct levels of x. In the telemarketing example, there are 14 levels of (or distinct values of) the explanatory variable. The decision rule to perform this test on the telemarketing data is:

Reject H_0 if $F > F(\alpha; 12, 6)$

Do not reject H_0 if $F \leq F(\alpha; 12, 6)$

---

[1] This section refers specifically to MINITAB output, but similar tests could be available with other software.

(because $c = 14$, $K = 1$, and $n = 20$). The results are shown in Figure 6.9. From the output, the F statistic value is seen to be 5.25. If a 5% level of significance is used, the critical value for the test is $F(0.05; 12, 6) = 4.00$, and the decision is to reject H_0 and conclude that a curvilinear model may fit the data better than the linear model. Also, the p value (0.026) can be used in the usual manner to perform this test.

Note that this test cannot be performed unless there are replicates for at least one level of x. MINITAB does provide another test for lack of fit that does not require replicates. The *data subsetting test* actually involves a series of tests, and the results of several of these tests may be printed out on the output. For example, in Figure 6.10, results of tests examining curvilinearity in the variable MONTHS, lack of fit at the outer x values, and overall lack of fit are reported. These results are reported in terms of the p values, so the p value decision rule can be applied.

Reject H_0 if p value $< \alpha$

Do not reject H_0 if p value $\geq \alpha$

For $\alpha = 0.05$, the test result indicates possible curvature in the variable MONTHS and an overall lack of fit.

6.4.3 CORRECTIONS FOR VIOLATIONS OF THE LINEARITY ASSUMPTION

When the linearity assumption is violated, the appropriate correction is not always obvious. The violation of this assumption implies that y and x are related in some curvilinear fashion, but there are many equations that describe curvilinear relationships. The idea behind the use of any equation of this sort is to transform the variables in such a way that a linear relationship is achieved. If x and y are related in a curvilinear fashion, then perhaps x^2 and y have a linear relationship.

The violation of the linearity assumption was originally noted in the residual plots. If we have corrected for the violation, we should not see the same patterns in

FIGURE 6.9 MINITAB Output Showing Pure Error Lack-of-Fit Test for Telemarketing Example.

```
The regression equation is
CALLS = 13.7 + 0.744    MONTHS


Predictor        Coef   SE Coef        T        P
Constant       13.671     1.427     9.58    0.000
MONTHS        0.74351   0.06666    11.15    0.000


S = 1.78737    R-Sq = 87.4%    R-Sq(adj) = 86.7%


Analysis of Variance

Source           DF       SS       MS        F        P
Regression        1   397.45   397.45   124.41    0.000
Residual Error   18    57.50     3.19
  Lack of Fit    12    52.50     4.38     5.25    0.026
  Pure Error      6     5.00     0.83
Total            19   454.95


10 rows with no replicates
```

FIGURE 6.10
MINITAB Output Showing Data Subsetting Test for Telemarketing Example.

```
The regression equation is
CALLS = 13.7 + 0.744 MONTHS

Predictor      Coef   SE Coef      T      P
Constant     13.671     1.427   9.58  0.000
MONTHS      0.74351   0.06666  11.15  0.000

S = 1.78737    R-Sq = 87.4%   R-Sq(adj) = 86.7%

Analysis of Variance
Source           DF       SS       MS       F      P
Regression        1   397.45   397.45  124.41  0.000
Residual Error   18    57.50     3.19
Total            19   454.95

Lack of fit test
Possible curvature in variable MONTHS    (P-Value = 0.0001)

Possible lack of fit at outer X-values (P-Value = 0.097)
Overall lack of fit test is significant at P = 0.000
```

the residual plots from the corrected model. The residuals from a properly corrected model should be randomly scattered.

In Chapter 5, the following four commonly used corrections were considered:

1. polynomial regression
2. reciprocal transformation of the x variable
3. log transformation of the x variable
4. log transformation of both the x and y variables

After trying one of these transformations, check the new residual plots to see if the violation was effectively corrected. If not, try one of the other corrections. Refer to Chapter 5 for examples of the use of curvilinear models. The four corrections just listed are described in greater detail in Chapter 5 as well.

EXAMPLE 6.2 **Telemarketing (continued)** In Chapter 5, a second-order polynomial regression was used to model the telemarketing data. The model can be written

$$\text{CALLS} = \beta_0 + \beta_1 \text{MONTHS} + \beta_2 \text{XSQR} + e$$

where XSQR is a variable created by squaring each value of the MONTHS variable. Figure 6.11 shows the regression estimates of the second-order model. The estimated regression is

$$\text{CALLS} = -0.14 + 2.31 \text{MONTHS} - 0.04 \text{XSQR}$$

Two checks should be made at this point to determine whether the second-order model is preferred to the original linear model: (a) test to see whether the coefficient

FIGURE 6.11 Regression Results for Telemarketing Example with Second-Order Term Added.

```
Variable        Coefficient     Std Dev         T Stat          P Value

Intercept       -0.14047        2.32226         -0.06           0.952
MONTHS           2.31020        0.25012          9.24           0.000
XSQR            -0.04012        0.00633         -6.33           0.000

Standard Error = 1.00325        R-Sq = 96.2%    R-Sq(adj) = 95.8%

Analysis of Variance
Source          DF      Sum of Squares          Mean Square         F Stat      P Value

Regression       2          437.84                218.92            217.50      0.000
Error           17           17.11                  1.01
Total           19          454.95
```

of the second-order term is significantly different from zero and (b) check to see whether the new residual plots indicate an improved model. The t stat or the p value for the coefficient of XSQR can be used to verify that the coefficient is significant. Next, we should examine the new residual plots.

The test result on the second-order term by itself is not sufficient evidence to judge this model to be adequate. The goal in adding the second-order term was to correct for the curvilinear patterns noted in the original residual plots. To see if this has been accomplished, the residual plots from the new equation must be examined. The residual plots of the standardized residuals versus the fitted values, the MONTHS variable, and the XSQR variable are shown in Figures 6.12, 6.13, and 6.14, respectively.

FIGURE 6.12 Plot of Standardized Residuals versus Fitted Values for Second-Order Model.

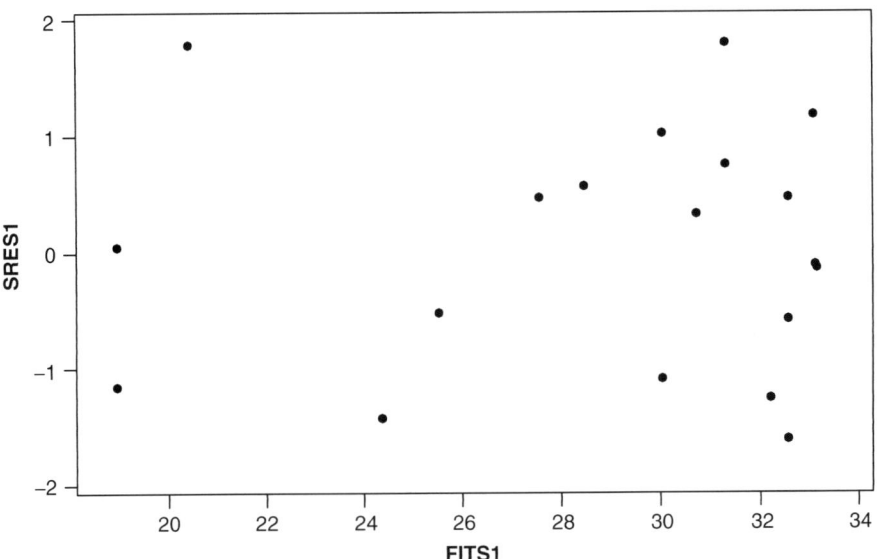

FIGURE 6.13 Plot of Standardized Residuals versus Explanatory Variable MONTHS for Second-Order Model.

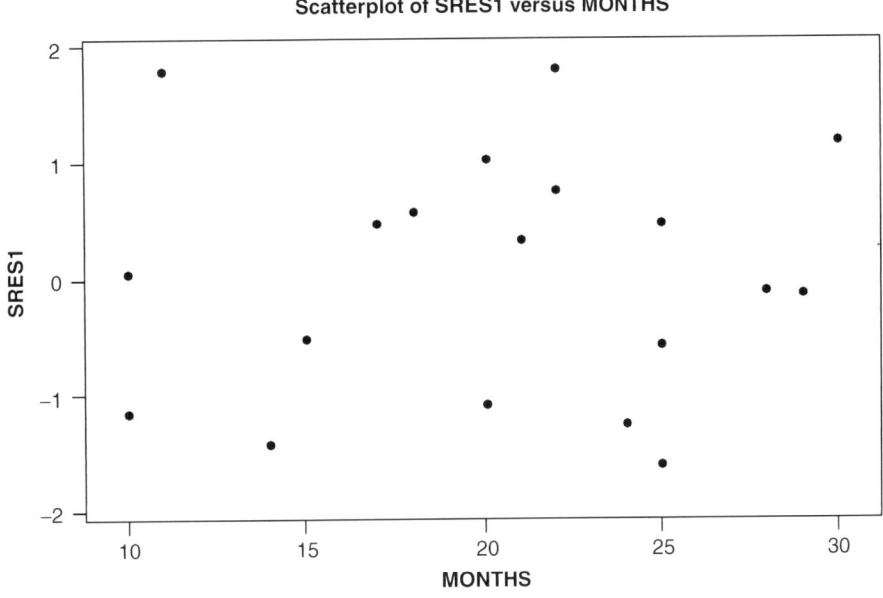

FIGURE 6.14 Plot of Standardized Residuals versus Explanatory Variable XSQR for Second-Order Model.

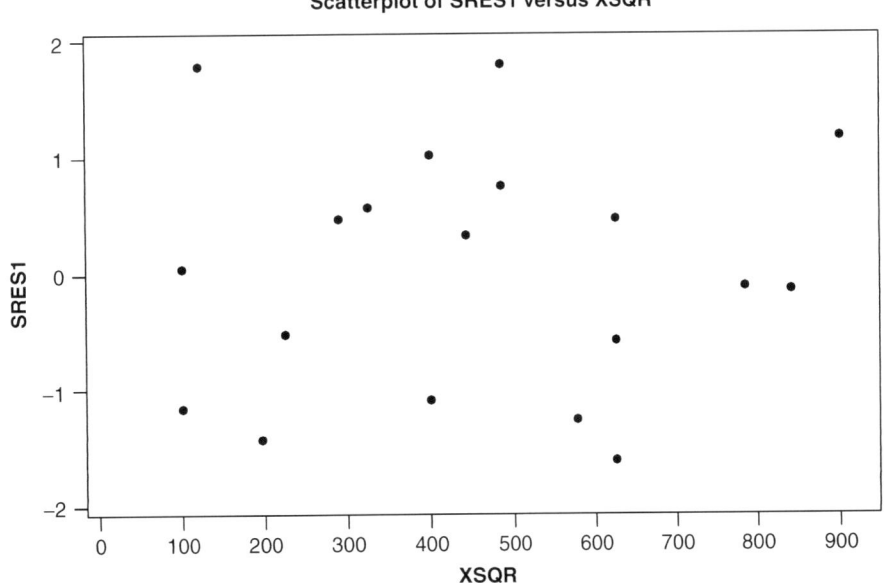

Looking at the residual plots, no distinct patterns can be seen. Contrast this with the obvious patterns of Figures 6.7 and 6.8. The addition of the x^2 variable appears to have corrected for the curvilinearity. The second-order model is an improvement over the first-order model, and the regression assumptions appear to be satisfied.

Higher-order terms could be added to the model, but there appears to be little justification in doing so from looking at the second-order model regression output and residual plots.

Other indicators that the regression has been improved by adding the x^2 term (as discussed in Chapter 5) include the reduction in the standard error of the regression from 1.787 to 1.003 and the increase in adjusted R^2 from 86.7% to 95.8%.

When curvilinear patterns appear in residuals plots, it is typically a sign that a linear model has been fit when a curvilinear model is more appropriate (or that an incorrect curvilinear model was fit). When this happens, a choice of the type of curvilinear model must be made. Some of the more common types of curvilinear models were discussed in Chapter 5. That discussion is not repeated again here in Chapter 6. The reader should be sure to review these models. When one of the models is chosen as a possible improvement, be sure to recheck the residual plots. A random scatter in the residual plots indicates that the correct model was fit to the data. If patterns in the residual plots persist, try a different correction.

EXERCISES

1. Parabola. Consider the following data:

y	16	4	1	9	1	25	16	4	0	9	25
x	−4	−2	1	3	−1	−5	4	2	0	−3	5

Regard x as the explanatory variable and y as the dependent variable. Figure 6.15 shows the scatterplot of y versus x. Figure 6.16 shows the regression output. These data are in a file named PARA6 on the CD.

a. Examine the scatterplot of y versus x. Does there appear to be a relationship between y and x?

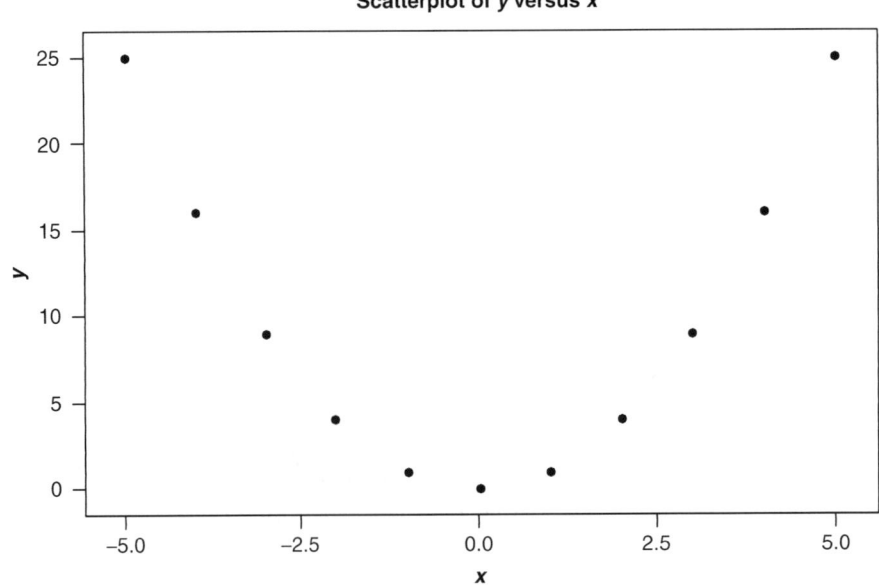

FIGURE 6.15
Scatterplot of y Versus x for Parabola Exercise.

b. What is the estimated linear regression equation relating y to x?

c. Test the hypothesis H_0: $\beta_1 = 0$ against the alternate H_a: $\beta_1 \neq 0$ at the 1% level of significance. What conclusion can be drawn from the result of the test?

d. Despite the outcome of the test in part c, does there appear to be a "strong" or "weak" association between x and y? Express this association in the form of an equation.

2. Research and Development. A company is interested in the relationship between profit (PROFIT) on a number of projects and two explanatory variables. These variables are the expenditure on research and development for the project (RD) and a measure of risk assigned at the outset of the project (RISK). PROFIT is measured in thousands of dollars and RD is measured in hundreds of dollars. The scatterplots of PROFIT versus RISK and PROFIT versus RD are shown in Figures 6.17 and 6.18, respectively. The regression results are in

FIGURE 6.16
Regression of y on x for Parabola Exercise.

```
Variable        Coefficient     Std Dev       T Stat      P Value
Intercept       10.000          2.944         3.40        0.008
X               -0.000          0.931         -0.00       1.000

Standard Error = 9.76388        R-Sq = 0.0%   R-Sq(adj) = 0.0%

Analysis of Variance
Source          DF     Sum of Squares    Mean Square     F Stat    P Value

Regression      1      0.00              0.00            0.00      1.000
Error           9      858.00            95.33
Total           10     858.00
```

FIGURE 6.17
Scatterplot of PROFIT versus RISK for Research and Development Exercise.

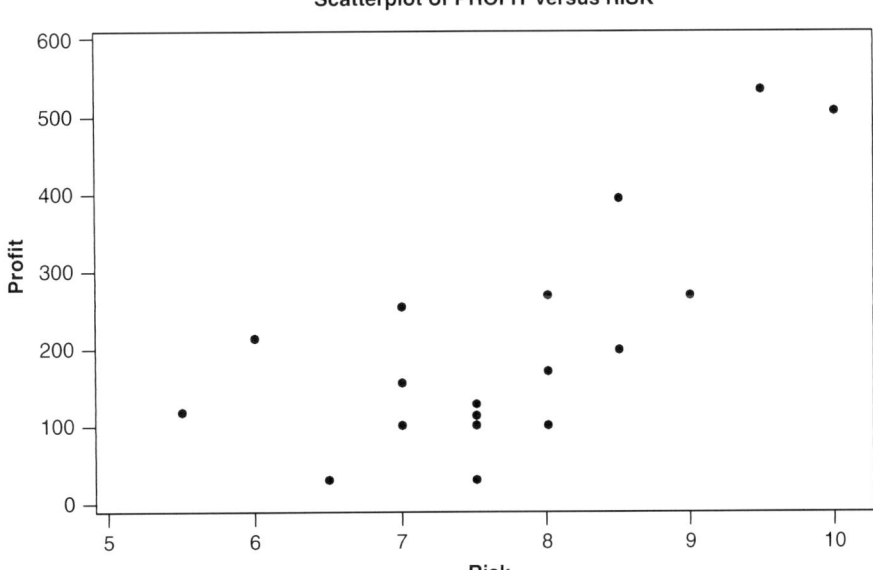

Scatterplot of PROFIT versus RISK

FIGURE 6.18
Scatterplot of PROFIT versus RD for Research and Development Exercise.

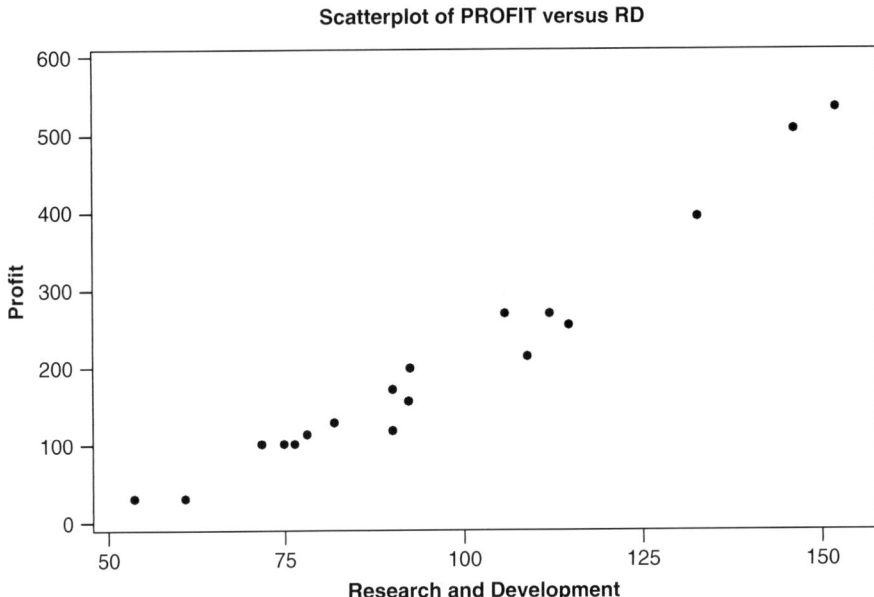

FIGURE 6.19
Regression Results for Research and Development Exercise.

```
Variable      Coefficient    Std Dev      T Stat      P Value

Intercept     -453.1763      23.5061      -19.28      0.000
RISK            29.3090       3.6686        7.99      0.000
RD               4.5100       0.1538       29.33      0.000


Standard Error = 14.3420    R-Sq = 99.2%  R-Sq(adj) = 99.0%

Analysis of Variance
Source         DF     Sum of Squares    Mean Square    F Stat    P Value

Regression      2         361639          180820       879.08    0.000
Error          15           3085             206
Total          17         364724
```

Figure 6.19. The residual plots of the standardized residuals versus the fitted values, RISK, and RD are shown in Figures 6.20, 6.21, and 6.22, respectively.

Using any of the given outputs, does the linearity assumption appear to be violated? Justify your answer. If you answered yes, state how the violation might be corrected. Then try your correction using a computer regression routine. Does your model appear to be an improvement over the original model? Justify your answer.

These data are available in a file named RD6 on the CD.

6.4 Assessing the Assumption That the Relationship is Linear 221

FIGURE 6.20
Plot of Standardized
Residuals versus
Fitted Values for
Research and
Development
Exercise.

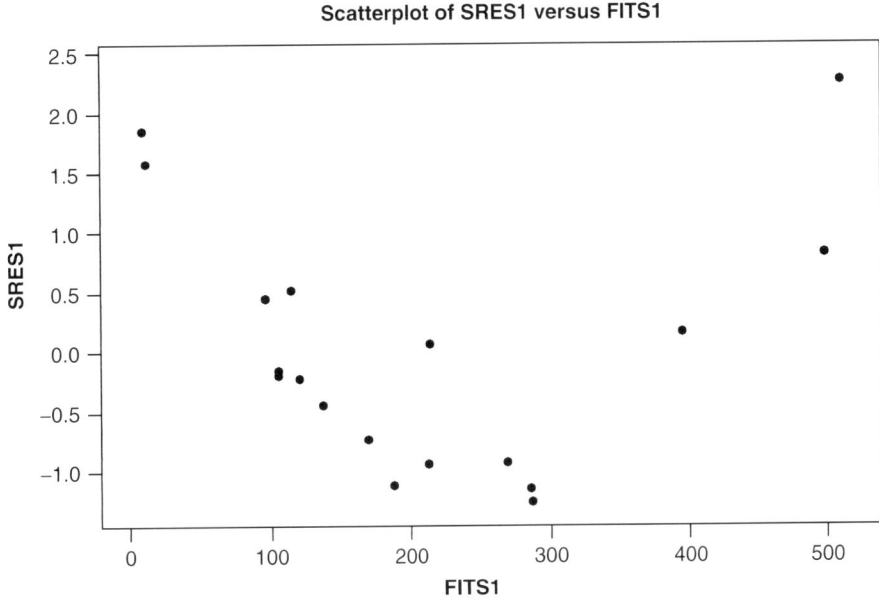

FIGURE 6.21
Plot of Standardized
Residuals versus
RISK for Research
and Development
Exercise.

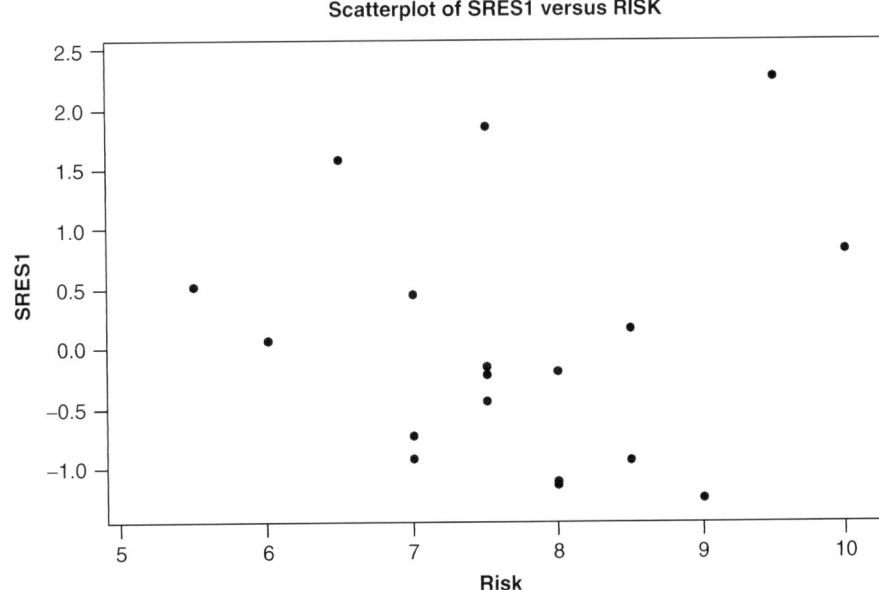

FIGURE 6.22
Plot of Standardized Residuals versus RD for Research and Development Exercise.

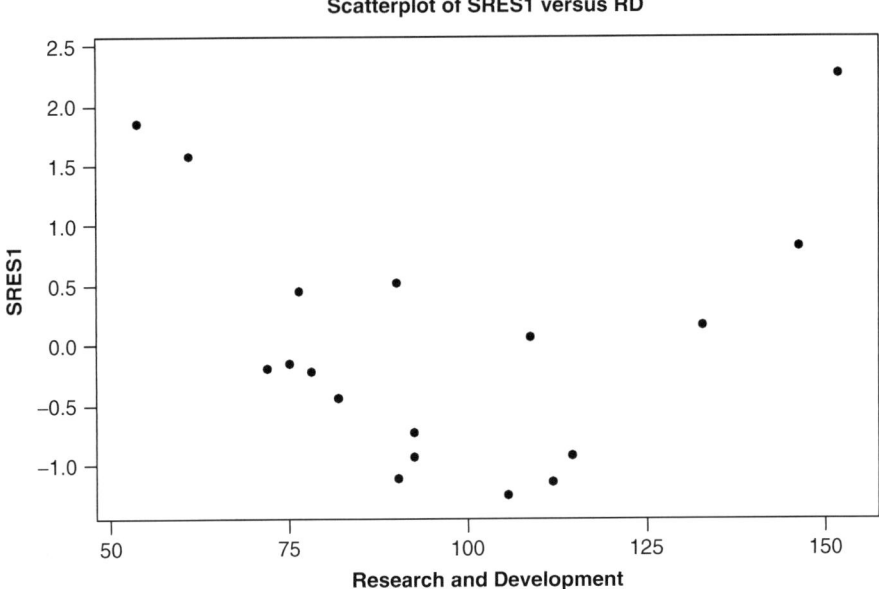

6.5 ASSESSING THE ASSUMPTION THAT THE VARIANCE AROUND THE REGRESSION LINE IS CONSTANT

6.5.1 USING PLOTS TO ASSESS THE ASSUMPTION OF CONSTANT VARIANCE

Assumption b of Section 6.2 states that the disturbances in the population regression equation, e_i, have constant variance σ_e^2. In a residual plot of $\hat{e}_i$ versus an explanatory variable x, the residuals should appear scattered randomly about the zero line with no differences in the amount of variation in the residuals regardless of the value of x. If there appears to be a difference in variation (for example, if the residuals are more spread out for large values of x than for small values), then the assumption of constant variance may be violated. In a residual plot, nonconstant variance is often identified by a "cone-shaped" pattern, as shown in Figure 6.23. Again, the violation is indicated by a systematic pattern in the residuals. In many texts, the term *heteroskedasticity* is used in place of "nonconstant variance." Example 6.3 illustrates the use of plots to assess the constant variance assumption.

EXAMPLE 6.3 **FOC Sales** Techcore is a high-tech company located in Fort Worth, Texas. The company produces a part called a fibre-optic connector (FOC) and wants to generate reasonably accurate but simple forecasts of the sales of FOCs over time. The company has weekly sales data for the past 265 weeks (in the file on the CD named FOC6). (The data have been disguised to provide confidentiality.) The time-series plot of FOC sales is shown in Figure 6.24. The regression of FOC sales using a

FIGURE 6.23
Cone-Shaped Pattern in Residual Plot Suggesting Nonconstant Variance.

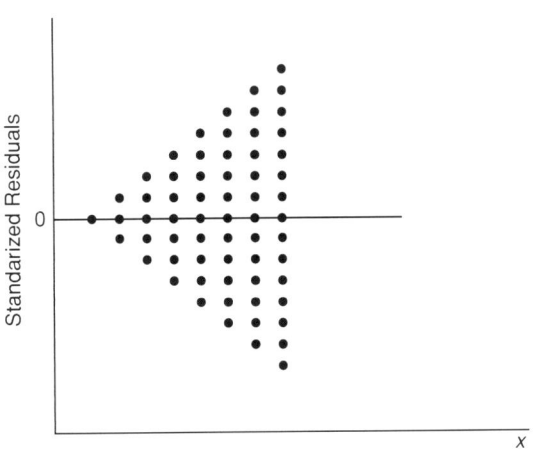

FIGURE 6.24
Time-Series Plot of FOC Sales.

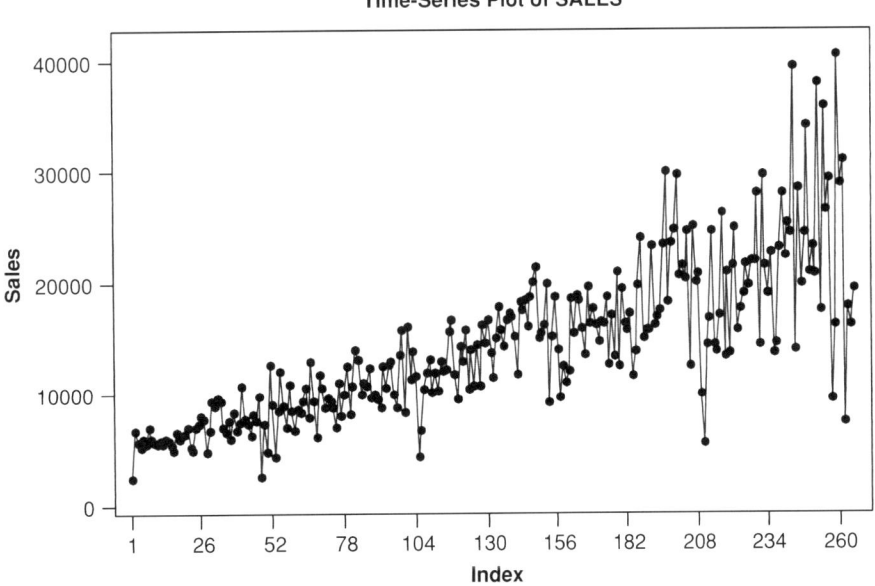

linear trend variable as the independent variable is shown in Figure 6.25. The plot of the standardized residuals versus the fitted values is shown in Figure 6.26. The cone-shaped pattern of residuals is obvious. Note that the variability of the residuals increases over time. This pattern also can be seen in the time-series plot of FOC sales. Such violations of assumptions, however, are typically magnified in the residual plots, as in this case.

FIGURE 6.25 Regression Results for FOC Sales Example.

```
Variable         Coefficient       Std Dev         T Stat         P Value

Intercept        4703.77           12.47           9.18           0.000
TIME             72.46             3.34            21.69          0.000


Standard Error = 4159.39        R-Sq = 64.2%   R-Sq(adj) = 64.0%

Analysis of Variance
Source          DF      Sum of Squares        Mean Square    F Stat    P Value

Regression      1       8142042913            8142042913     470.62    0.000
Error           263     4550042800            17300543
Total           264     12692085712
```

FIGURE 6.26 Plot of Standardized Residuals versus Fitted Values for FOC Sales Example.

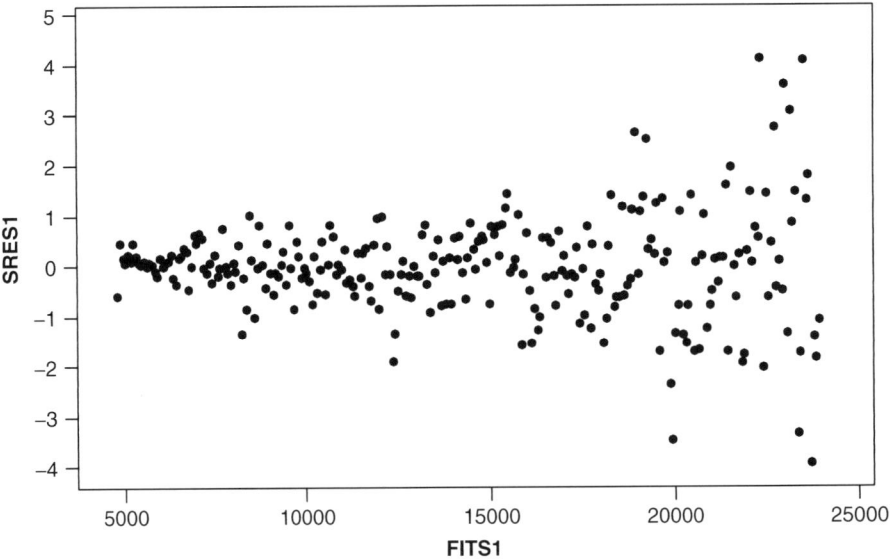

When the disturbance variance is not constant, the use of the least-squares method has two major drawbacks:

1. The estimates of the regression coefficients are no longer minimum variance estimates.
2. The estimates of the standard errors of the coefficients are biased.

The first drawback suggests that estimates of the coefficients with smaller sampling variability may exist. Because of the second drawback, hypothesis tests about the population regression parameters may provide misleading results.

6.5.2 A TEST FOR NONCONSTANT VARIANCE

Several tests are available for nonconstant variance, although a study by Griffiths and Surekha ("A Monte Carlo Evaluation of the Power of Some Tests for Heteroscedasticity"[2]) demonstrated that a test developed by J. Szroeter tends to be better at detecting nonconstant variance. The hypotheses to be tested are

H_0: Variance is constant

H_a: Variance is not constant

Szroeter's test statistic is

$$Q = \left(\frac{6n}{n^2 - 1}\right)^{1/2} \left(h - \frac{n+1}{2}\right)$$

where n is the sample size,

$$h = \frac{\sum_{i=1}^{n} i\hat{e}_i^2}{\sum_{i=1}^{n} \hat{e}_i^2}$$

and $\hat{e}_i$ is the residual from the ith observation in the regression equation. The decision rule for the test is

Reject H_0 if $Q > z_\alpha$

Do not reject H_0 if $Q \leq z_\alpha$

where α is the level of significance for the test and z_α is chosen from the standard normal table with upper-tail area α.

Szroeter's test assumes that all the observations can be arranged in order of increasing variance. Typically, it is assumed that the variance increases as the value of one of the explanatory variables increases. Thus, the data need to be arranged according to the values of this explanatory variable. As a simple example, suppose the values of x and y in a simple regression are as follows:

x	3	2	7	9	4
y	6	4	16	15	8

Arranging the values of x in ascending order and maintaining the associated values of y results in the following arrangement of the data:

x	2	3	4	7	9
y	4	6	8	16	15

After reordering the data in this way, a regression is run and the residuals, $\hat{e}_i$, are saved and used to compute h. The value for h is then substituted into the equation to compute the test statistic Q.

---

[2] See References for complete publication information.

6.5.3 CORRECTIONS FOR NONCONSTANT VARIANCE

There are a number of possible corrections for nonconstant variance. All require a transformation of the dependent variable. This often makes comparison of the new regression equation with the old equation difficult. Some commonly used corrections for nonconstant variance are as follows:

1. In place of the dependent variable y, use the natural logarithm of y, $\ln(y)$. The natural logarithms of the y values are less variable than the original y values and may stabilize the variance. For example, consider the following numbers:

y	1	2	5	10	50
$\ln(y)$	0	0.69	1.61	2.30	3.91

   Note the difference in variation between the original y values and their natural logarithms. The natural logarithms are more equally spaced than are the original values. Note also that the natural logarithm transformation is only defined for positive numbers.

   The natural logarithm transformation is the appropriate transformation when the error standard deviation is proportional to the mean of the dependent variable.

2. In place of the dependent variable y, use the square root of y, $\sqrt{y}$. The square roots of the y values are less variable than the original y values and may stabilize the variance. Note that the square root transformation is not defined for negative numbers.

   The square root transformation is appropriate when the dependent variable is a count variable that follows a Poisson distribution.

3. There is a general class of transformations, called Box-Cox transformations, that can be used to transform the y variable when attempting to stabilize the variance. Box-Cox transformations can be defined as follows:

   Use y^p in place of y where $0 \leq p \leq 1$

   Thus, powers of the y variable are used in place of the original variable. The square root transformation defined in correction 2 is a special case of the Box-Cox transformation because $p = 0.5$ in that case. Note that the extreme values of p for this type of transformation are 0 and 1. When $p = 1$, the original y values are used. When $p = 0$, we define the transformation to be the natural log transformation discussed under correction 1. When $p = 1$, no compression of the data are necessary. When $p = 0$, we use the log transformation and obtain maximum compression of the data. Powers between 0 and 1 provide flexibility in the amount of compression obtained.

   (For more on the use of Box-Cox transformations, the interested reader is referred to Neter, Wasserman, and Kutner, *Applied Linear Statistical Models*, pp. 394–400.)

4. If the disturbance variance is thought to be proportional to some function of one of the x variables, the values of that variable can be used to stabilize the variance. For example, if

$$\sigma^2_{e_i} = \sigma^2 x_i^2$$

is thought to express the relationship between the variance at each observation i and the associated value of the x variable, then dividing each variable in the regression by x stabilizes the variance. If the original equation is

$$y_i = \beta_0 + \beta_1 x_i + e_i$$

then, after dividing through by x_i, the transformed model becomes

$$\frac{y_i}{x_i} = \beta_0\left(\frac{1}{x_i}\right) + \beta_1 + e'_i$$

where $e'_i = e_i/x_i$ is a new disturbance with constant variance. Note that the roles played by β_0 and β_1 have been reversed in the transformed equation. The transformation is not defined when x_i is zero.

EXAMPLE 6.4 **FOC Sales (continued)** The regression output for the regression of the natural log of sales (LOGSALES) on TIME is shown in Figure 6.27, and the residual plot of the standardized residuals versus the fitted values is in Figure 6.28. In the residual plot, the cone-shaped pattern has been greatly reduced. Using the log transformation stabilized the variance, so it is now relatively constant for all values of x. Caution should be exercised in interpreting and using the regression output, however. The output in Figure 6.27 shows results concerning the relationship of the log of sales to TIME. All information from the output must be interpreted in light of this fact. Thus, 68.7% of the variation in the log of sales has been explained. This value is not directly comparable to the R^2 from the regression of y on x in Figure 6.25 (64.2%). No conclusions can be drawn concerning which regression does "better" based on the R^2 because two different measures of the dependent variable are used.

Also keep in mind that if the equation from Figure 6.27 is used for forecasting, natural logs of the y values are forecasted, not the y values themselves. For example, what is the forecast of sales in week 300? Using the estimated regression equation

$$\text{LOGSALES} = 8.74 + 0.00537(300) = 10.351$$

FIGURE 6.27
Regression of LOGSALES on TIME.

Variable	Coefficient	Std Dev	T Stat	P Value
Intercept	8.7390327	0.0342976	254.80	0.000
TIME	0.0053746	0.0002235	24.04	0.000

Standard Error = 0.278373   R-Sq = 68.7%   R-Sq(adj) = 68.6%

Analysis of Variance

Source	DF	Sum of Squares	Mean Square	F Stat	P Value
Regression	1	44.797	44.797	578.09	0.000
Error	263	20.380	0.077		
Total	264	65.177			

FIGURE 6.28
Plot of Standardized Residuals versus Fitted Values for Transformed Model.

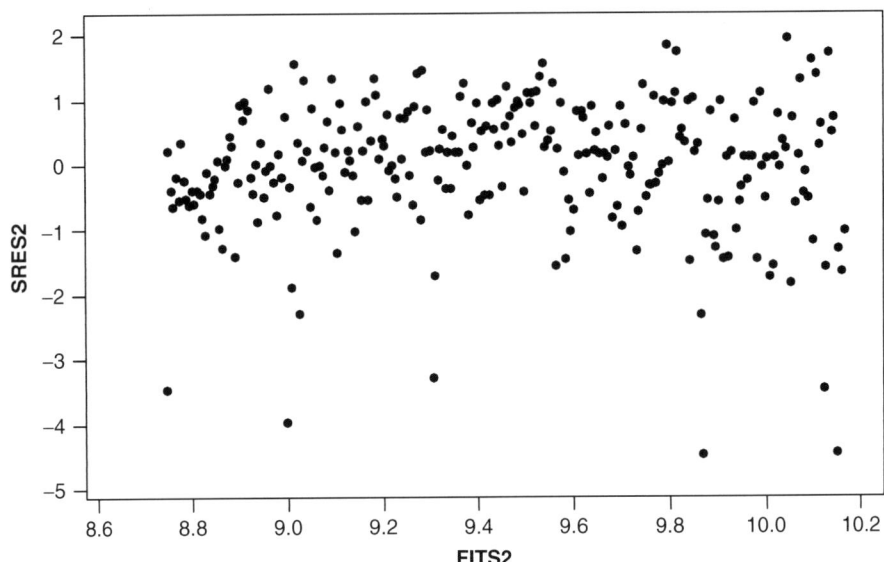

The forecast value for LOGSALES is 10.351. The resulting forecast value for y must be computed as

$$SALES = e^{10.351} = 31,288$$

When deciding whether the transformation has improved our results, comparing the R^2 is not always dependable, as noted. The residual plots can be used to help in this decision. If the pattern suggesting a violation in the original residual plots is no longer present in the plots from a transformed model, then the transformed model is preferable. If there is little or no improvement in the residual plots, then another transformation should be tried and the resulting residual plots examined to see whether they indicate an improved model. Choosing the correct transformation to produce an adequate model is thus an iterative process that may take several tries. And when all else fails, consult your neighborhood statistician!

EXERCISES

3. S&P 500 Index Prices. Data for the S&P 500 Stock Index for the time period January 1974 through December 2002 are in the file named SP5006 on the CD. These data were obtained from the web site *www.economagic.com*. The objective is to build a regression model relating the current price to the price in the previous month. The regression equation can be written as

$$y_i = \beta_0 + \beta_1 y_{i-1} + e_i$$

where y_i represents the S&P 500 index price in time period i.

Figure 6.29 is the output for the regression of the current S&P 500 index price on the previous month's price. Figure 6.30 is the plot of the standardized residuals versus the fitted values.

FIGURE 6.29
Regression of Current S&P 500 Index Price on Previous Month's Price.

Variable	Coefficient	Std Dev	T Stat	P Value
Intercept	8.806181	6.236058	1.41	0.159
S&P LAG1	0.998917	0.003629	275.27	0.000

Standard Error = 85.4760   R-Sq = 99.5%   R-Sq(adj) = 99.5%

Analysis of Variance

Source	DF	Sum of Squares	Mean Square	F Stat	P Value
Regression	1	553606990	553606990	75772.72	0.000
Error	345	2520622	7306		
Total	346	556127613			

FIGURE 6.30
Plot of Standardized Residuals versus Fitted Values for S&P 500 Index Exercise.

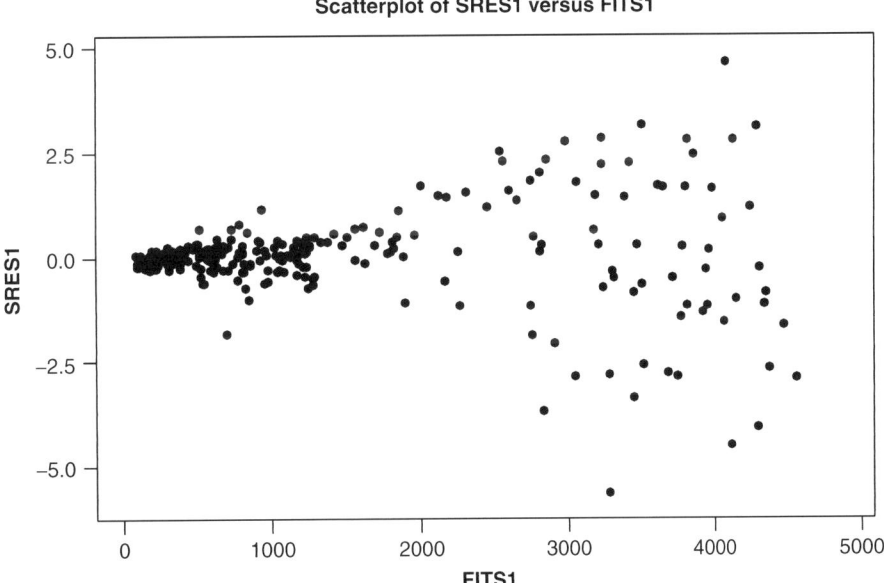

Is there evidence that the constant variance assumption has been violated? Justify your answer. Suggest a correction for the violation of the constant variance assumption for this example. Try the correction using a regression routine. Does the correction appear to have eliminated the problem of nonconstant variance? State why or why not.

6.6 ASSESSING THE ASSUMPTION THAT THE DISTURBANCES ARE NORMALLY DISTRIBUTED

6.6.1 USING PLOTS TO ASSESS THE ASSUMPTION OF NORMALITY

Residual plots of the standardized residuals versus the fitted values can be used to assess graphically whether the sample residuals have come from a normally distributed population. For normally distributed data, about 68% of the standardized residuals should be between −1 and +1, about 95% should be between −2 and +2, and about 99% should be between −3 and +3. Normal probability plots also can be a useful graphical technique in assessing normality.

EXAMPLE 6.5   **Communications Nodes (continued)**   Figure 6.31 shows the regression of cost (COST) on the number of ports (NUMPORTS) and the bandwidth (BANDWIDTH) for the communications nodes data discussed in several examples and exercises in Chapters 3 and 4. (The data can be found in the file named COMNODES6 on the CD.) The plot of the standardized residuals versus the fitted values is shown in Figure 6.32. A printout of the standardized residuals is shown in Table 6.2.

FIGURE 6.31 Regression of COST on NUMPORTS and BANDWIDTH.

```
Variable         Coefficient      Std Dev      T Stat       P Value
Intercept          17086.75       1865.41        9.16         0.000
NUMPORTS             469.03         66.98        7.00         0.000
BANDWIDTH             81.07         21.65        3.74         0.003

Standard Error = 2982.52     R-Sq = 95.0%    R-Sq(adj) = 94.1%

Analysis of Variance

Source          DF    Sum of Squares      Mean Square    F Stat    P Value

Regression       2        1876012662        938006331    105.45     0.000
Error           11          97849860          8895442
Total           13        1973862522
```

FIGURE 6.32 Plot of Standardized Residuals versus Fitted Values.

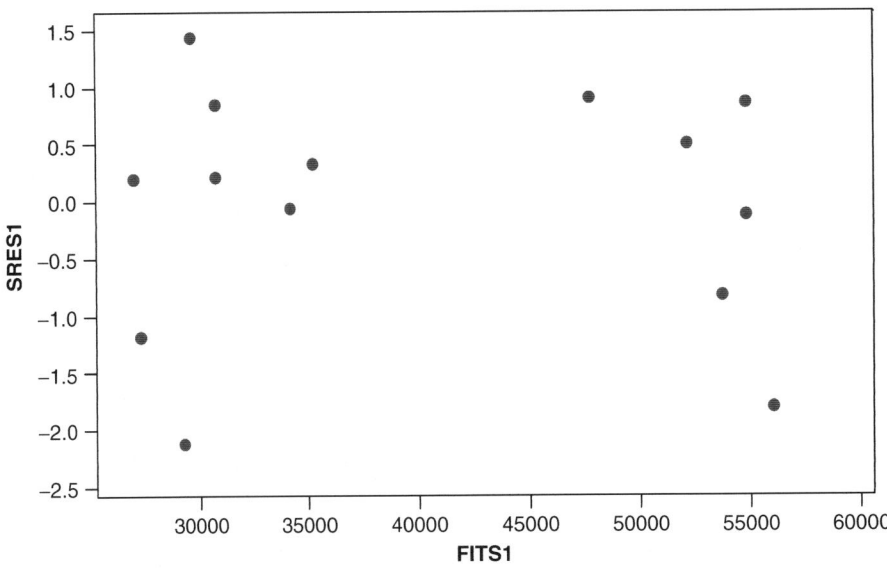

6.6 Assessing the Assumption That the Disturbances Are Normally Distributed 231

TABLE 6.2 List of Standardized Residuals for Regression of COST on NUMPORTS and BANDWIDTH

Observation	Standardized Residuals
1	−0.82144
2	−1.80142
3	0.91087
4	0.32777
5	0.20567
6	0.85903
7	−0.11480
8	−0.04969
9	0.22068
10	−2.11672
11	−1.17904
12	0.50409
13	1.44088
14	0.84722

As can be seen from the plots and the printout, 13 of the 14 standardized residuals (93%) are between ±2. So there are about the number of residuals we would expect to see if they came from a normal distribution. From these plots, we conclude that the normality assumption seems reasonable.

When examining the normality assumption, concern should be placed on relatively large standardized residuals. Thus, an excessive number of residuals outside the ±2 limit or the ±3 limit might cause concern about this assumption. But remember to expect some values outside these limits, especially in large data sets.

The MINITAB regression output is shown in Figure 6.33. Note that MINITAB flags any observations with standardized residuals that are greater than or equal to 2 in absolute value. These values are shown with an R next to the value of the standardized residual in the table of Unusual Observations. Often, it is the observations with large standardized residuals with which we are concerned. This is why MINITAB takes the time to flag these observations. This does not mean that there is anything wrong with these data values or that they should be deleted. It simply means that these observations may be different from the others in our data set for some reason and may deserve special attention.

Figure 6.34 shows the normal probability plot for the residuals produced by MINITAB. In this plot, the observed residuals (horizontal axis) are plotted against the "normal scores." The normal scores can be thought of as the values expected if a sample of the same size as the one used (14 in this case) was selected from a normal distribution. The vertical scale shows the cumulative percentage (probability) at or below the normal scores rather than the normal scores themselves. The normal scores and the percentages are computed by MINITAB.

When the plot of the normal scores (cumulative probabilities) and the data is approximately a straight line, the normality assumption appears reasonable. The

FIGURE 6.33 MINITAB Regression of COST on NUMPORTS and BANDWIDTH.

```
The regression equation is
COST = 17086 + 469 NUMPORTS + 81.1 BANDWIDTH

Predictor       Coef      SE Coef         T         P
Constant       17086         1865      9.16     0.000
NUMPORTS      469.03        66.98      7.00     0.000
BANDWIDTH      81.07        21.65      3.74     0.003

S = 2982.52      R-Sq = 95.0%     R-Sq(adj) = 94.1%

Analysis of Variance

Source           DF           SS           MS          F         P
Regression        2   1876012662    938006331     105.45     0.000
Residual Error   11     97849860      8895442
Total            13   1973862522

Source          DF       Seq SS
NUMPORTS         1   1751268376
BANDWIDTH        1    124744286

Unusual Observations

Obs   NUMPORTS     COST      Fit    SE Fit   Residual   St Resid
  1       68.0    52388    53682      2532      -1294      -0.82 X
 10       24.0    23444    29153      1273      -5709      -2.12R

R denotes an observation with a large standardized residual.
X denotes an observation whose X value gives it large influence.
```

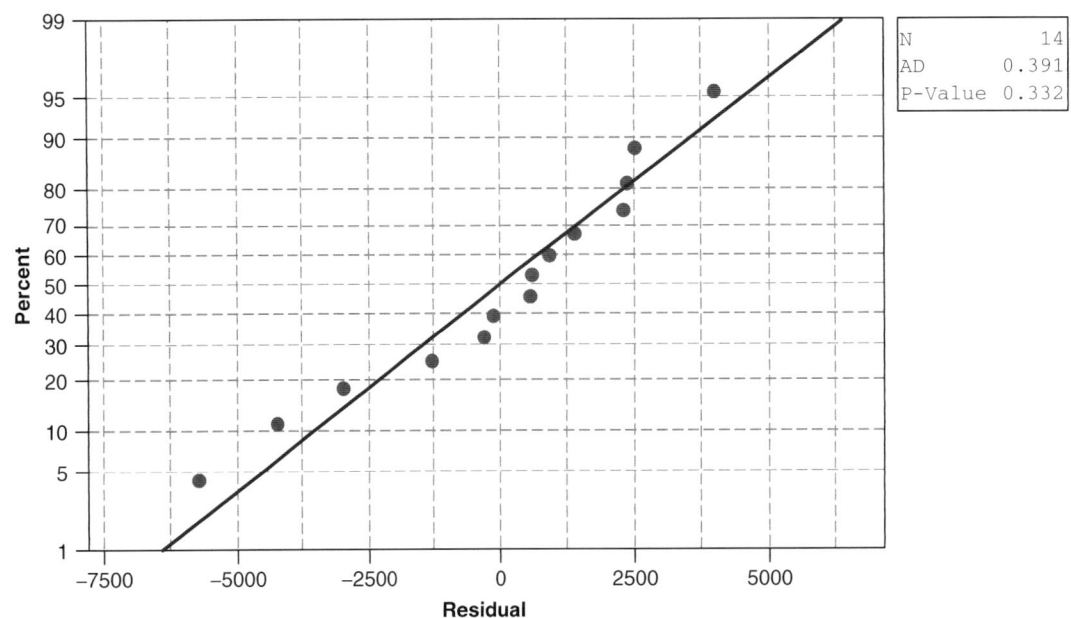

FIGURE 6.34 MINITAB Normal Probability Plot and Test for Normality.

normal scores are numbers we expect to see from a sample from a normal distribution, so for the two to plot on a straight line, the two sets of data have to be similar. Thus, we reason that the data must also have come from a normal distribution. If the data did not come from a normal distribution, the plot will show curvature. (Note that whether the normal probability plot is linear or not has nothing to do with whether the relationship between y and x is linear. We are not assessing whether the relationship between the original variables is linear, but whether the disturbances come from a normal distribution.)

In this example, the plot of the normal scores versus the data is nearly linear (MINITAB has drawn in the line). This suggests that the standardized residuals could have come from a normal distribution. Thus, the normality assumption is supported.

6.6.2 TEST FOR NORMALITY

The plot of the residuals versus the normal scores in a normal probability plot should be approximately linear if the disturbances are normal. The "straightness" of the line in the normal probability plot can be measured by computing a variety of test statistics including the Kolmogorov–Smirnov (KS) statistic, the Anderson–Darling (AD) statistic, and the Ryan–Joiner (RJ) statistic. Regardless of the test statistic used, the hypotheses can be stated as

H_0: Disturbances are normal

H_a: Disturbances are nonnormal

In most statistical packages a p value will be printed for the appropriate test statistic (KS, AD, or RJ). The decision rule is

Reject H_0 if p value $< \alpha$

Do not reject H_0 if p value $\geq \alpha$

where α is the level of significance for the test.

EXAMPLE 6.6 **Communications Nodes (continued)**   In the MINITAB normal probability plot in Figure 6.34 the value of the Anderson–Darling (AD) statistic is AD = 0.391. The associated p value is 0.332. Using a 5% level of significance, the decision rule for the normality test is

Reject H_0 if p value < 0.05

Do not reject H_0 if p value ≥ 0.05

With a reported p value of 0.332, H_0 cannot be rejected. The conclusion is that the disturbances are normally distributed.

EXAMPLE 6.7 **Saving and Loan (S&L) Rate of Return**   In "Return, Risk, and Cost of Equity for Stock S&L Firms: Theory and Empirical Results," Lee and Lynge discuss methods of estimating the cost of equity capital for S&L associations. One aspect of their

234 Assessing the Assumptions of the Regression Model

analysis included an examination of the relationship between the rate of return of the S&L stocks (y) and two measures of the risk of the stocks: the beta coefficient (x_1), which is a measure of nondiversifiable risk, and the standard deviation of the security returns (x_2), which measures total risk. The data for their sample of 35 S&Ls is shown in Table 6.3 and is contained in the file SL6 on the CD. Scatterplots of y versus x_1 and y versus x_2 are shown in Figures 6.35 and 6.36, respectively. The regression output is in Figure 6.37. Figures 6.38 through 6.40 show the residual plots of the standardized residuals versus the fitted values, x_1 and x_2, respectively. These plots highlight the presence of one standardized residual that falls well above the +3 limit. The normal probability plot and results of the normal probability plot

TABLE 6.3 Data for S&L Rate of Return Example

Name	Time Period	RETURN	BETA	SIGMA
1. H.F. Ahnanson	1/78–12/82	2.29	1.2862	14.2896
2. Alamo Savings Bank	9/78–12/82	0.34	0.6254	10.9786
3. American Federal S&L	12/80–12/82	2.57	1.1706	9.7917
4. American S&L	1/78–12/82	2.91	1.6328	17.7219
5. Bell National Ind.	1/78–12/82	3.50	1.2492	18.4450
6. Beverly Hills S&L	11/78–12/82	0.47	1.1363	12.4691
7. Broadview Financial Ind.	1/78–12/82	−0.28	1.3585	12.3396
8. Buckeye Financial Ind.	10/80–12/82	0.40	1.5415	14.7000
9. Citizens Savings Financial	6/80–12/82	2.42	2.1457	18.9970
10. City Federal Bank	6/80–12/82	5.48	2.2701	18.0840
11. Danney S&L	1/78–12/82	1.67	1.4527	14.2785
12. Far West Financial	1/78–12/82	1.01	1.4532	13.8673
13. Financial Corp. of America	1/78–12/82	6.06	1.8826	18.1800
14. Financial Corp. of Santa Barbara	1/78–12/82	0.48	1.4493	14.7792
15. Financial Federation	1/78–12/82	0.96	1.5590	14.9088
16. First Charter Financial	1/78–12/82	1.24	1.2274	12.3504
17. First City Federal Ind.	5/81–12/82	6.39	2.2567	22.0455
18. First Financial S&L	1/81–12/82	2.39	1.7003	12.8343
19. First Lincoln Financial Bank	1/78–12/82	0.30	2.2226	1.8750
20. First Western Financial Bank	1/78–12/82	2.09	1.6535	16.5737
21. Freedom S&L Bank	5/80–12/82	2.46	1.3616	14.2680
22. Gibraltar Financial	1/78–12/82	2.10	1.9851	17.8500
23. Golden West Financial Corp.	1/78–12/82	2.76	1.4311	14.1036
24. Great Western Financial	1/78–12/82	2.06	1.3448	12.4630
25. Guarantee Financial Ind.	1/78–12/82	1.42	1.4560	13.9728
26. Homestead Financial Bank	1/78–12/82	4.12	1.5543	17.2628
27. Imperial Corp. of America	1/78–12/82	1.82	1.7280	15.2880
28. Land of Lincoln Ind.	12/79–12/82	1.59	1.3389	10.3032
29. Mercury Saving	1/78–12/82	13.05	1.2973	13.3110
30. Naples Federal	2/80–12/82	3.04	1.0945	10.5792
31. Palmetto Federal S&L	11/79–12/82	3.72	1.2051	12.9456
32. Prudential Federal S&L	1/78–12/82	0.75	1.0756	11.5200
33. Texas Federal Bank	7/81–12/82	1.00	1.9157	16.6000
34. Transohio Financial	1/78–12/82	−3.35	1.4456	11.7705
35. Western Financial Corp.	1/78–12/82	2.26	1.9128	16.8370

Source: Data are from Lee and Lynge, "Return, Risk, and Cost of Equity for Stock S&L Firms: Theory and Empirical Results," *Journal of the American Real Estate and Urban Economic Association,* 1985. Copyright © 1985 School of Business, Indiana University. Reprinted by permission.

FIGURE 6.35
Scatterplot of RETURN versus BETA for S&L Rate of Return Example.

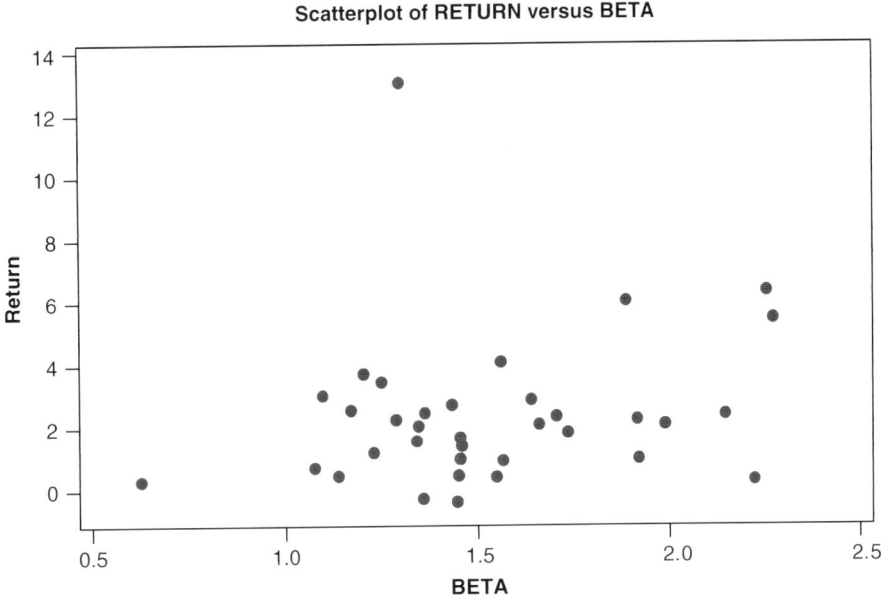

FIGURE 6.36
Scatterplot of RETURN versus SIGMA for S&L Rate of Return Example.

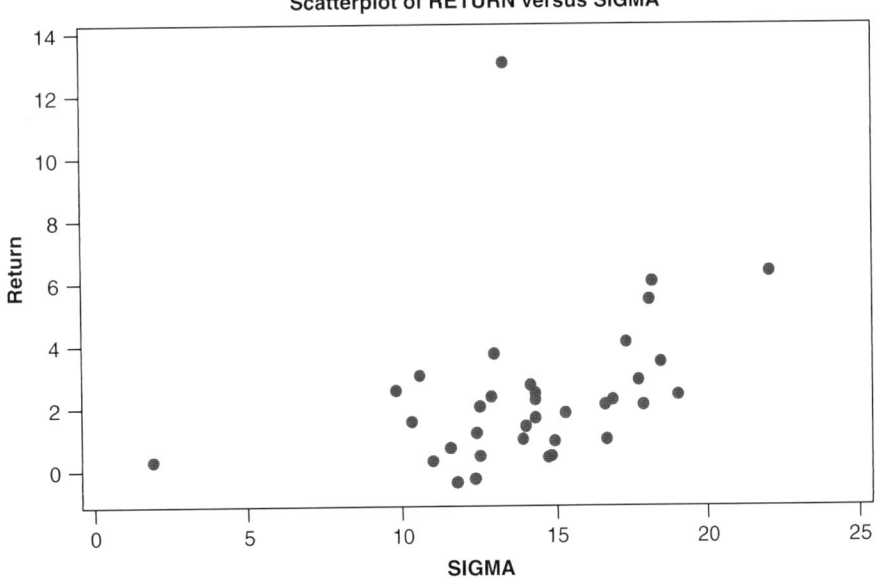

test for normality are shown in Figure 6.41. To perform the test for normal disturbances at the 5% level of significance, the decision rule is

Reject H_0 if $p$value < 0.05

Do not reject H_0 if $p$value ≥ 0.05

236 Assessing the Assumptions of the Regression Model

FIGURE 6.37
Regression Results for S&L Rate of Return Example.

Variable	Coefficient	Std Dev	T Stat	P Value
Intercept	-1.330	2.012	-0.66	0.513
BETA	0.300	1.198	0.25	0.804
SIGMA	0.231	0.126	1.84	0.075

Standard Error = 2.37707   R-Sq = 12.5%   R-Sq(adj) = 7.0%

Analysis of Variance

Source	DF	Sum of Squares	Mean Square	F	P
Regression	2	25.808	12.904	2.28	0.118
Error	32	180.815	5.650		
Total	34	206.624			

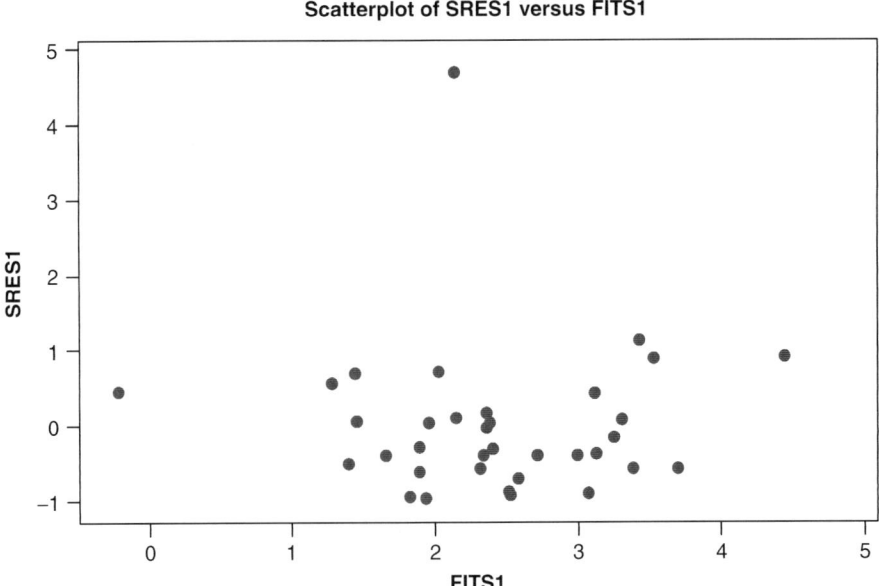

FIGURE 6.38
Plot of the Standardized Residuals versus the Fitted Values for S&L Rate of Return Example.

The p value computed by MINITAB is less than 0.005, so the null hypothesis is rejected. The disturbances do not appear to be normally distributed.

6.6.3 CORRECTIONS FOR NONNORMALITY

The assumption of normally distributed disturbances is not necessary to use least-squares estimation to produce an estimated regression equation. However, for making inferences with small samples, it is necessary. In large samples, the assumption is not as important because the sampling distribution of the estimators of the regression coefficients is still approximately normal. Recall that when estimating a population mean, μ, the cutoff point for a large sample is $n = 30$. When $n \geq 30$, the

FIGURE 6.39
Plot of the Standardized Residuals versus the Explanatory Variable BETA for S&L Rate of Return Example.

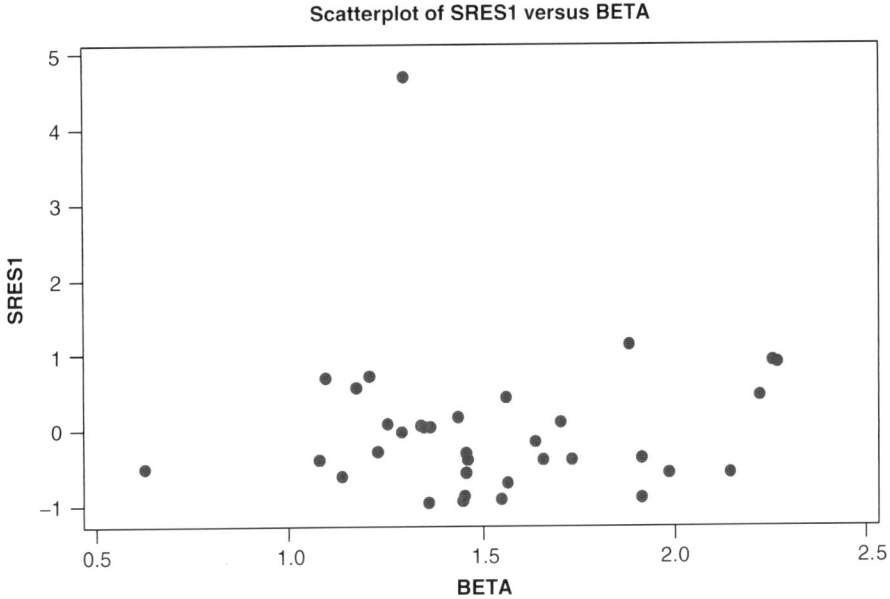

FIGURE 6.40
Plot of the Standardized Residuals versus the Explanatory Variable SIGMA for S&L Rate of Return Example.

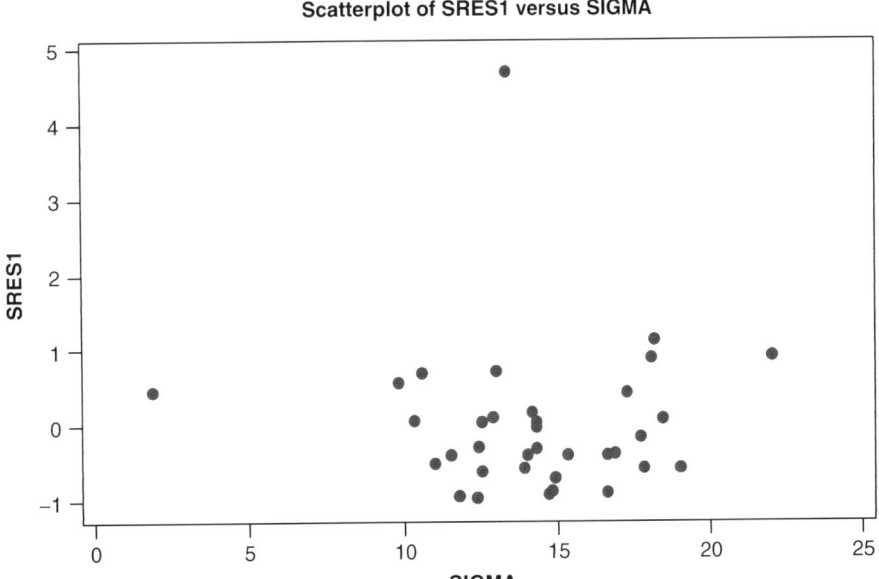

central limit theorem guarantees that the sampling distribution of the sample mean is approximately normal. A similar theorem operates in the regression context for the sampling distribution of the regression coefficients, b_k. The cutoff for a large sample may differ, however, because several coefficients may be estimated in a multiple regression context. It is uncertain exactly how many observations ensure

238 Assessing the Assumptions of the Regression Model

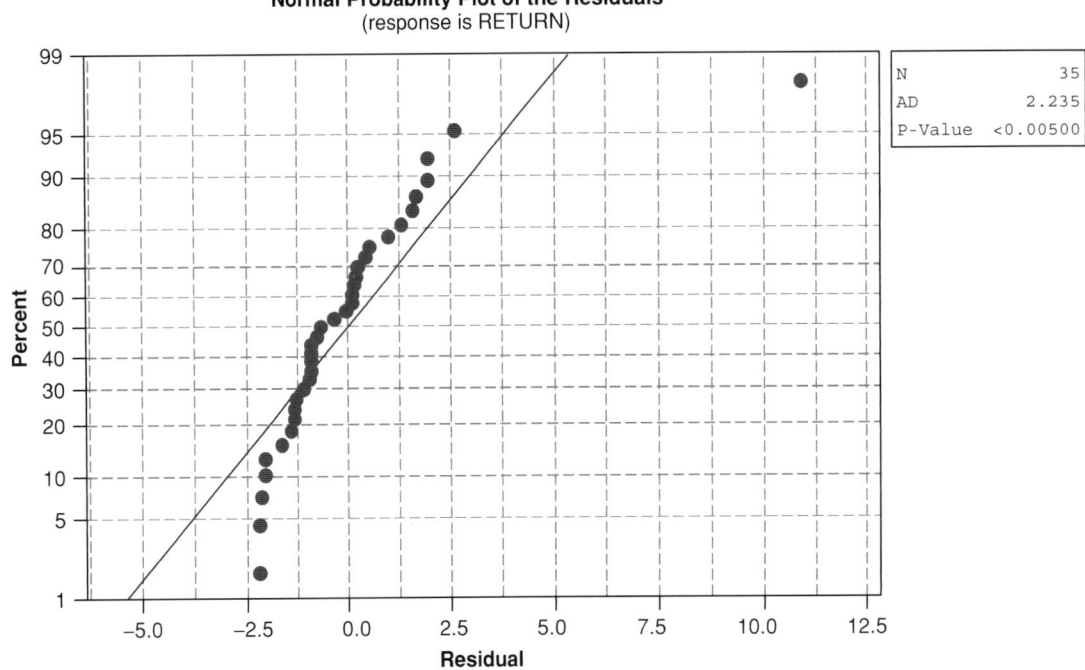

FIGURE 6.41 MINITAB Normal Probability Plot and Test for Normality for S&L Rate of Return Example.

normality of the sampling distributions in the multiple regression context. If the assumption of normal disturbances does not hold, additional observations are necessary for each additional explanatory variable. For a simple regression, 30 observations with 10 to 20 additional observations for *each* additional explanatory variable are commonly suggested.

When the normality assumption is violated and the sample size is too small to ensure normality of the sampling distributions, there are a variety of possible corrections. One type of correction is to transform the dependent variable using a Box-Cox transformation. (This was also a correction for nonconstant variance). Box-Cox transformations can be defined as follows:

$$\text{Use } y^p \text{ in place of } y \text{ where } 0 \leq p \leq 1$$

Thus, powers of the y variable are used in place of the original variable. Note that the extreme values of p for this type of transformation are 0 and 1. When $p = 1$, the original y values are used. When $p = 0$, we define the transformation to be the natural log transformation. When $p = 1$ is used, no correction is necessary; the disturbances are normal. The natural log transformation would provide the maximum amount of correction to the residuals to try to achieve normality. Powers of y between 0 and 1 can be used to adjust the amount of correction necessary. (For more on the use of Box-Cox transformations, the interested reader is referred to Neter, Wasserman, and Kutner, *Applied Linear Statistical Models*, pp. 394–400.)

When considering the normality assumption, be sure to correct for other violations before worrying about normality. A violation of the linearity or the constant variance assumption can introduce outliers (discussed in Section 6.7) into a data set that make the normality assumption appear to be violated also. Choosing the correct model by correcting for nonlinearity or nonconstant variance may eliminate the outliers, however. So check for violations of the linearity and constant variance assumptions before being too concerned with the normality assumption.

In some cases, the primary reason for the nonnormality of the disturbances may be the presence of one or a few data points that are much different from the remaining observations in the data set. Even in large samples, it is important to recognize such unusual observations because their presence may drastically alter results. In such instances, the sampling distributions of the estimated regression coefficients should not be assumed normal, even if the sample size is large. These cases, and some possible corrections, are discussed in the next section.

6.7 INFLUENTIAL OBSERVATIONS

6.7.1 INTRODUCTION

The method of least-squares estimation chooses the regression coefficient estimates so that the error sum of squares, *SSE*, is a minimum. In doing this, the sum of squared distances from the true y values, y_i, to the points on the regression line or surface, $\hat{y}_i$, are minimized. Least squares thus tries to avoid any large distances from y_i to $\hat{y}_i$. As shown in Figure 6.42, this can have an effect on the placement of the regression line. In Figure 6.42(a), the points are all clustered near the regression line. In Figure 6.42(b), one of the points has been moved so that its y value is much different from the y values of the remaining sample points. The effect of moving this one point on the placement of the estimated regression line is also shown. Note that the regression line has been pulled toward the point that was moved to the extreme position. This is a result of the requirement of the least-squares method to minimize the error sum of squares. Squaring the residuals gives proportionately more weight to extreme points in the error sum of squares. The least-squares regression line is often drawn toward such extreme points.

When a sample data point has a y value that is much different from the y values of the other points in the sample, it is called an *outlier*. Outliers can be either good or bad. They can provide information concerning the behavior of the process being studied that would be unavailable otherwise. In this sense, the presence of the outlier could be viewed as positive. On the other hand, the presence of an outlier can at times produce confusing results and mask important information that could otherwise be obtained from the regression. In either case, it is important to recognize an outlier when it is present. Outlier detection techniques are discussed in this section.

Now consider Figure 6.43. In both Figure 6.43(a) and 6.43(b), the same five initial points are used as in Figure 6.42(a). One additional point is added to each figure but is placed in a different position. The positioning of this point is seen to have a large effect on the estimated regression line. In Figure 6.43(a), the slope of the line appears to have changed little from its value with only the five original

FIGURE 6.42(a)
Scatterplot with Regression Line for Data Shown.

Data

x	1	2	3	4	5
y	1.1	2.2	2.4	2.8	3.3

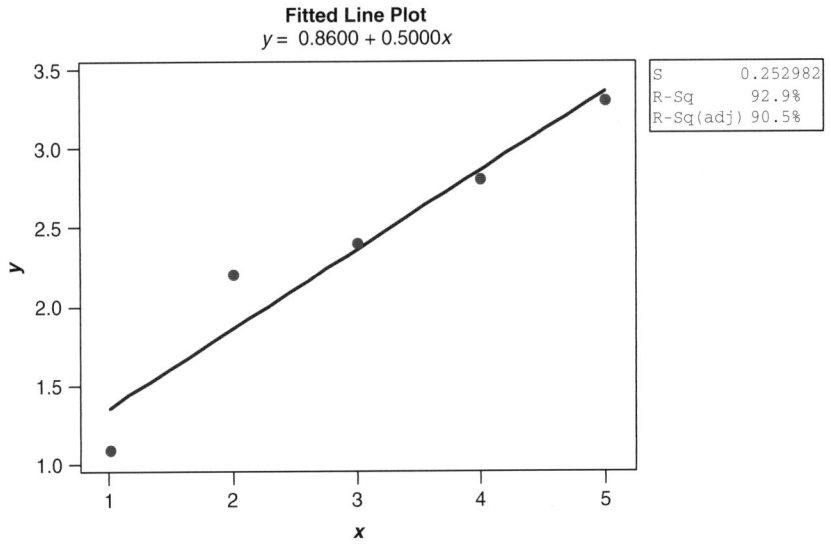

FIGURE 6.42(b)
Scatterplot Showing Effect on Regression Line of Outlier.

Data

x	1	2	3	4	5
y	1.1	2.2	5.0	2.8	3.3

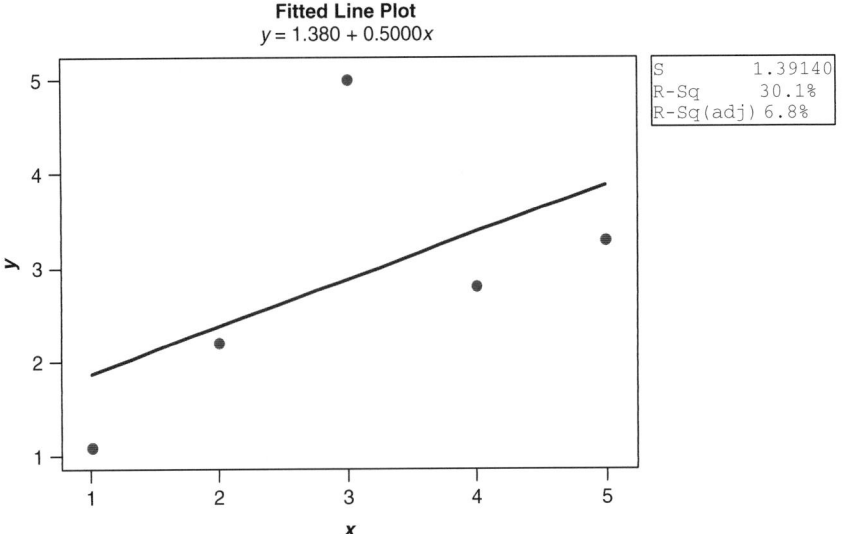

FIGURE 6.43(a) Scatterplot Showing Effect of Leverage Point in Line with Other Data.

Data						
x	1	2	3	4	5	10
y	1.1	2.2	2.4	2.8	3.3	6.0

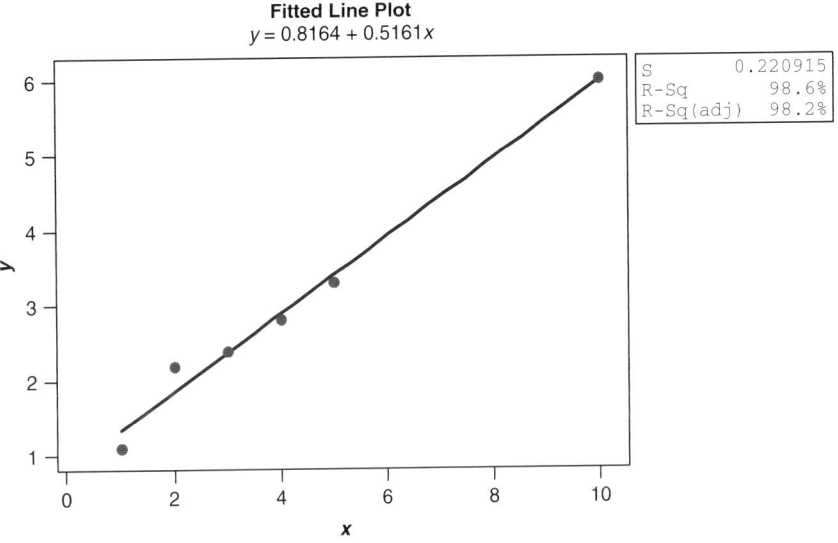

FIGURE 6.43(b) Scatterplot Showing Effect of Moving Leverage Point to Alternate Position.

Data						
x	1	2	3	4	5	10
y	1.1	2.2	2.4	2.8	3.3	2.0

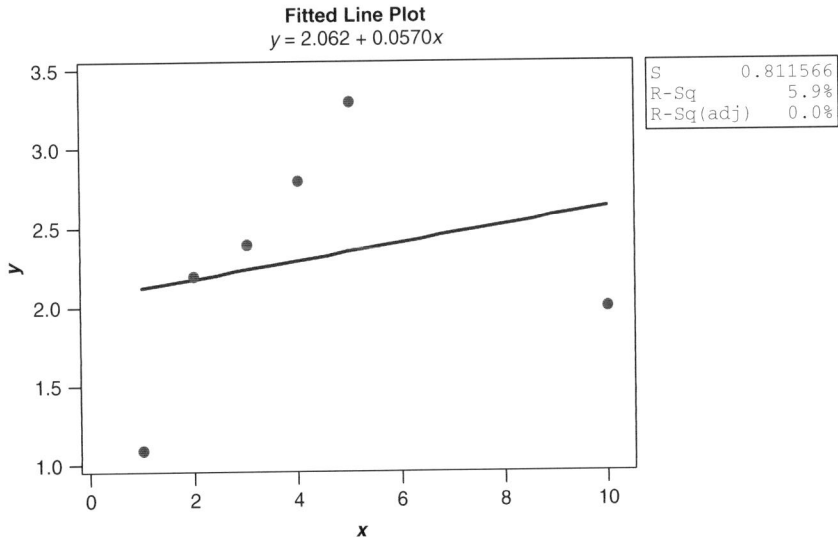

points [see Figure 6.42(a)]. In Figure 6.43(b), the placement of the additional point has drastically changed the slope of the line. In fact, the slope of the line appears to be determined almost entirely by this one point. This sixth observation is said to have high leverage and is referred to as a *leverage point*. The term "leverage point" means that the point is placed in such a way that it has the *potential* to affect the regression line.

In both Figure 6.43(a) and 6.43(b), the added point is a leverage point. This is due to its extreme placement on the x axis. The x value for this point is much different from the x values for the other sample points. The point, however, does not exert much influence in Figure 6.43(a) because it is in line with all the other points in terms of the position of its y value relative to its x value. It is important to recognize any leverage points due to their possible effect on the regression line. As with outliers, leverage points may be good or bad depending on whether they add information about the process under study or mask information that otherwise would be obtained.

Finally, it is important to note that an observation can be both a leverage point and an outlier.

6.7.2 IDENTIFYING OUTLIERS

The use of the standardized residuals has already been discussed. The standardized residuals are computed by dividing the raw residual, $\hat{e}_i = y_i - \hat{y}_i$, by the standard deviation of $\hat{e}_i$:

$$\hat{e}_{is} = \frac{\hat{e}_i}{\text{stdev}(\hat{e}_i)}$$

where $\hat{e}_{is}$ indicates the standardized residual. The variance of the standardized residuals is 1 (note that in this chapter it is necessary to distinguish between the raw residual $\hat{e}_i$ and the standardized residual $\hat{e}_{is}$).

If the residuals come from a normal distribution, then a standardized residual with an absolute value larger than 2 is expected only about 5% of the time. Thus, any observation with a standardized residual larger than 2 in absolute value might be classified as an outlier. MINITAB, for example, indicates any such observations in a table of Unusual Observations following the regression results.

Another measure sometimes used in place of the standardized residual is the standardized residual computed after deleting the ith observation. This measure is called the *studentized residual* or *studentized deleted residual*. To compute the studentized deleted residual, the residual, $\hat{e}_i$, is again standardized, but the divisor is different from that used to compute the standardized residual. The standard deviation of the ith residual is computed from the regression with the ith observation deleted. By doing this, the ith observation exerts no influence over the value of the standard deviation. If the ith observation's y value is unusual, this is reflected in the residual but not in its divisor. Thus, unusual y values should stand out. Also, because of the way they are computed, the studentized deleted residuals are known to follow a t distribution with $n - K - 1$ degrees of freedom. The studentized deleted residuals can be compared to a value chosen from the t table to determine whether they should be classified as outliers. (The standardized residuals do not follow a t distribution.) But this approach should not be used as a test of significance to determine whether the observation should be discarded. What to do about outliers once they are identified will be discussed later in this section.

6.7.3 IDENTIFYING LEVERAGE POINTS

Leverage was previously defined as the potential of an observation to affect the regression line. As shown in Figure 6.43, a point can possess leverage without significantly altering the position of the regression line. On the other hand, given sufficient leverage, a single point can significantly affect the slope of the regression line.

The leverage of the ith point in a sample is denoted h_i and is computed by some regression software packages. Leverage is a measure of how extreme the point is in terms of the values of the explanatory variables. Observations with extreme x values possess greater leverage than observations with x values that are similar to the other sample points. In Figure 6.42, the observation that was changed has smaller leverage than the observation that was changed in Figure 6.43. Note that the slope of the regression line is affected more by changes in the point with greater leverage.

MINITAB indicates certain observations with very high leverage. Any data value with leverage greater than $2(K + 1)/n$ (where K is the number of explanatory variables and n is the sample size) is indicated in a table of Unusual Observations following the regression results along with observations that have large standardized residuals.

6.7.4 COMBINING MEASURES TO DETECT OUTLIERS AND LEVERAGE POINTS

The effect of an observation on the regression line is determined both by the y value of the point and the x value(s). As shown in Figures 6.42 and 6.43, an observation with an unusual y value has a much greater effect on the regression line if it also has high leverage. Several statistics have been developed that consider extremity in both the y and x dimensions in an attempt to determine which points are highly influential on the regression line. Two of these measures, the DFITS statistic and Cook's D statistic, are discussed in this section.

Both of these measures combine information from the residuals and the leverage of each observation to try to pick out observations that may have a large influence on the regression line. As with the individual measures, unusual values of the DFITS statistic or Cook's D statistic are not indications that anything is wrong with the particular data value or that it should be discarded. It does indicate that the value with the unusual statistic is somehow different from the remaining values in the data set and should be given additional consideration before accepting the regression model.

The DFITS statistic can be written as

$$\text{DFITS}_i = \hat{e}_i \sqrt{\frac{n - K - 2}{SSE(1 - h_i) - \hat{e}_i^2}} \sqrt{\frac{h_i}{1 - h_i}}$$

where $\hat{e}_i$ is the residual for the ith observation, h_i is the leverage value for the ith observation, and SSE is the error sum of squares for the regression.

Cook's D statistic is computed as

$$D_i = \frac{\hat{e}_i^2}{MSE(K + 1)} \left(\frac{h_i}{(1 - h_i)^2} \right)$$

where $\hat{e}_i$ and h_i are as defined for the DFITS statistic and MSE is the mean square error for the regression.

As can be seen, both statistics use the residuals and the leverage of each individual point. Since they use the values in different ways, however, different information may be obtained from each of the statistics. Cook's D statistic is usually thought to represent the combined impact on all the regression coefficients of the ith observation. DFITS represents the combined impact on the fitted values of the ith observation.

There are two different schools of thought about how the DFITS statistic and Cook's D statistic should be used:

1. The values should be compared to some absolute cutoff. For example, the Cook's D value often is compared to an $F(\alpha; K + 1, n - K - 1)$ value. The DFITS value is compared to $2\sqrt{(K + 1)/n}$. Values bigger than either of the numbers should be further examined. But these are not to be viewed as statistical tests to reject or throw out observations.

2. Do not use absolute cutoffs. Simply pick out those observations whose Cook's D or DFITS values (if any) differ appreciably from most of the values.

The second approach is used in this text. Much recent research has shown that comparison to absolute cutoffs is not as effective in identifying influential observations as examining observations with unusually large DFITS or Cook's D values.

When specific cutoffs are not used, a method is needed to compare the different values of each statistic to determine which observations may be unusual. One way of doing this is simply to graph the DFITS (or Cook's D) values for each observation on the vertical axis and the number of the observation on the horizontal axis. This can also be accomplished by doing a time-series plot of the DFITS (or Cook's D) values, remembering that the index on the horizontal axis refers to the number of the observation and not necessarily to a time period. This method of presenting the DFITS and Cook's D statistics is demonstrated in Example 6.8.

6.7.5 WHAT TO DO WITH UNUSUAL OBSERVATIONS

As noted earlier, the fact that an observation has been classified as unusual does not mean that it is useless or that it should be deleted from the analysis. It is merely a flag to indicate that the observation deserves further examination. This is true regardless of which measure of "unusualness" has been used (standardized residual, DFITS, Cook's D statistic, and so on).

There are many reasons an observation may appear unusual. If there is a violation of the linearity or constant variance assumptions, this can cause certain observations to appear unusual until the violation has been corrected by choosing an appropriate transformation of the data.

If a data value has been typed in incorrectly, the value may be flagged as unusual. This is useful since incorrectly coded values should not be included in the data set. The true data value should be located and used to replace the incorrect value. If the true value cannot be found, then this is one case when it is almost always better to omit the incorrect data value before running the analysis.

If the unusual value is not due to the violation of an assumption or incorrect coding but is a correct value that is simply unusual with respect to most of the values in the data set, then the choice of what to do with it is more difficult. There are certain cases when deleting the observation from the data set is appropriate. These cases occur when the unusual observation is somehow very different from the observations included in the analysis. For example, if a production process is being examined, unusual observations may occur at the beginning of the process due to start-up problems. When the process reaches a steady state of performance, the data values generated may be quite different from those generated initially. The initial observations may be deleted if the process in its steady state is to be studied.

Consider another example. Suppose data on price and size of houses are obtained to develop an equation to help set prices for the houses. If interest lies in pricing houses with 1500 to 2500 square feet of space, then it is proper to exclude a house if it contains 4500 square feet. If the x values for the unusual observation fall outside the range of interest, it may be best to discard the observation and run the regression without that value.

In any case, before deleting an observation from an analysis, the observation should be studied carefully to see whether deletion is an appropriate option. If the observation is somehow so different from the others in the data set that it is inappropriate to include it in the analysis or if the value has been coded incorrectly and the true value is unknown, then deletion is a viable option. However, this option should not be used indiscriminately.

EXAMPLE 6.8 **S&L Rate of Return (continued)**   Figure 6.44 shows a plot of DFITS for the S&L rate of return data. Figure 6.45 shows a plot of the Cook's D statistic. Both of these plots were created by doing a time-series plot of the statistics. The index on the horizontal axis refers to the number of the observation from the sample. From each of these plots, the 29th observation stands out as unusual relative to the rest of the observations in the sample. In Table 6.3, the 29th observation is Mercury Saving. Note that the return for Mercury Saving is much higher than that for the other S&Ls. Mercury Saving has a very unusual y value resulting in a standardized residual of 4.69, which is far greater than would be expected if the disturbances were normally distributed.

This observation should definitely be examined further. The question at this point is what to do with it. In the case of Mercury Saving, it is unclear whether the return of 13.05% is correct. In the article from which these data were taken, however, it appears that Mercury Saving has been deleted from the analysis. Because the true return is not known, Mercury Saving is excluded from the data set and the analysis redone. It is important to emphasize that casual deletion of points from the data set is not recommended. If there is a good reason for not including a particular value, then it can be deleted. If the return for Mercury Saving is incorrect and the correct value is unknown, it is better to delete this observation than to use the incorrect information. Or if Mercury Saving is believed to be so different from the other S&Ls that it should not be included in the analysis, this is a good reason for deletion. For example, perhaps Mercury Saving had undertaken a particularly risky line of investments, different from those of the other S&Ls, which led to its very high return. Because the

FIGURE 6.44 Plot of DFITS for S&L Rate of Return Example.

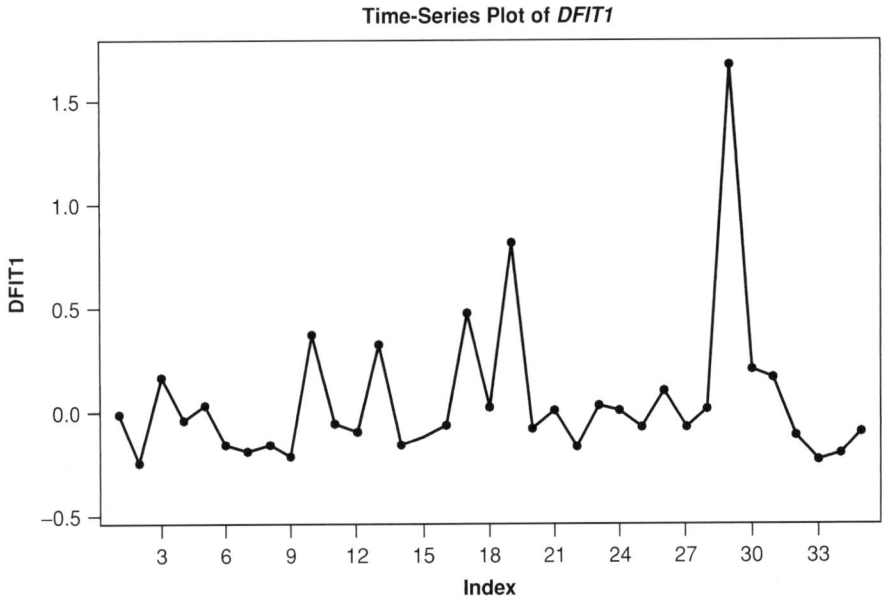

FIGURE 6.45 Plot of Cook's D Statistic for S&L Rate of Return Example.

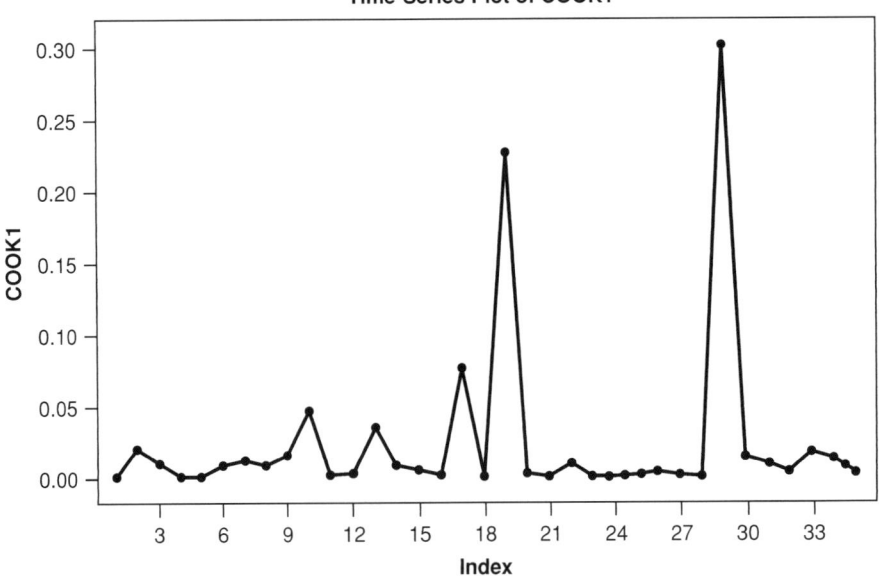

investment behavior of Mercury is extremely different from the other S&Ls in this case, there might be grounds for excluding it from the analysis. It is assumed here that there is sufficient reason for omitting Mercury Saving since that appears to be what was done in the original article. In a real application, it would be beneficial to study Mercury further to see what makes it so different from the other S&Ls.

6.7 Influential Observations 247

In the following example, note the difference between the analyses with and without Mercury Saving.

EXAMPLE 6.9 **S&L Rate of Return Without Mercury Saving** Figures 6.46 and 6.47 show the scatterplots with Mercury Saving omitted. The regression results are shown in

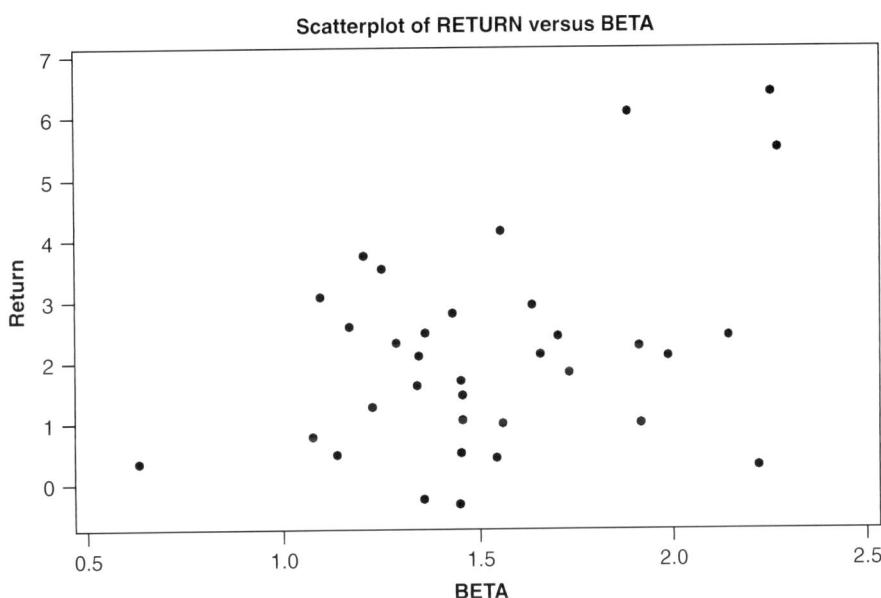

FIGURE 6.46
Scatterplot of RETURN versus BETA with Mercury Saving Omitted.

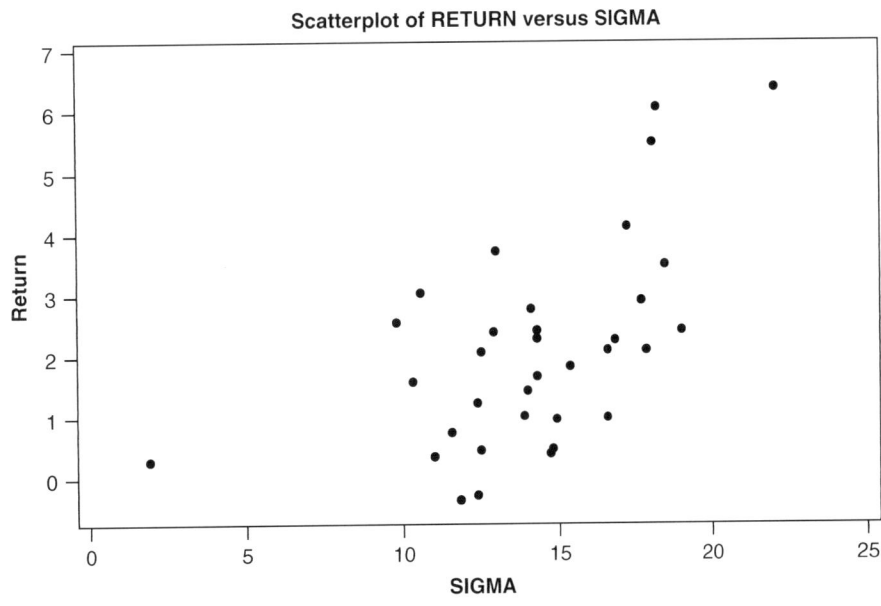

FIGURE 6.47
Scatterplot of RETURN versus SIGMA with Mercury Saving Omitted.

Figure 6.48. The residual plots are in Figures 6.49 to 6.51, and the plots of DFITS and Cook's D are in Figures 6.52 and 6.53.

The regression results differ considerably from the regression with Mercury Saving included. Compare the Figure 6.48 regression (without Mercury) to the Figure 6.37 regression (with Mercury). In the initial regression, neither variable is significant at the 5% level. In the regression without Mercury, SIGMA is now significant. Thus, the conclusions drawn from these two regressions are completely different because of the presence (or absence) of one influential observation. This

FIGURE 6.48 Regression for S&L Rate of Return Example with Mercury Saving Omitted.

```
Variable      Coefficient    Std Dev      T Stat      P Value
Intercept       -2.5103      1.1529       -2.18        0.037
BETA             0.8463      0.6843        1.24        0.225
SIGMA            0.2322      0.0714        3.25        0.003

Standard Error = 1.35165     R-Sq=37.2%    R-Sq(adj) = 33.1%

Analysis of Variance

Source         DF    Sum of Squares    Mean Square    F Stat    P Value

Regression      2        33.537           16.768       9.18      0.001
Error          31        56.635            1.827
Total          33        90.172
```

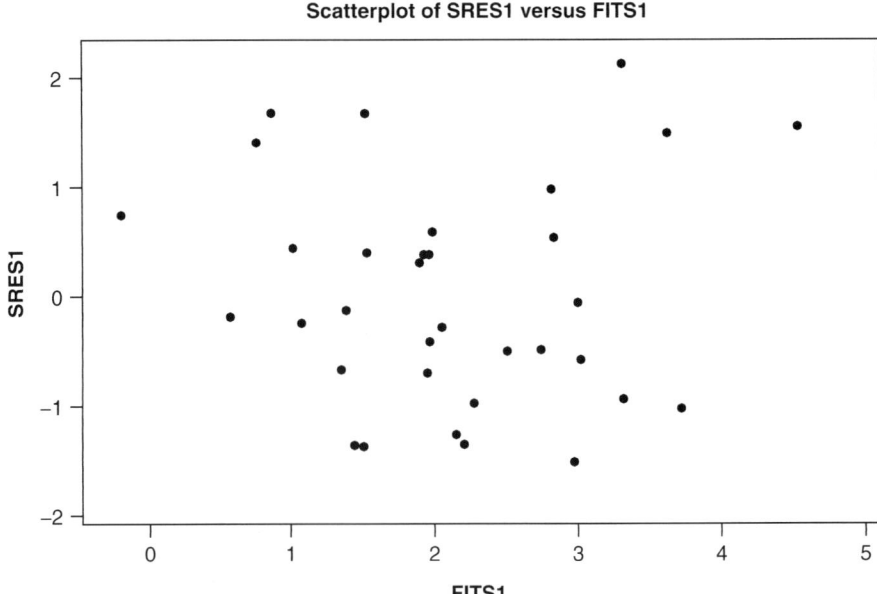

FIGURE 6.49 Plot of Standardized Residuals versus Fitted Values for S&L Rate of Return Example with Mercury Saving Omitted.

FIGURE 6.50 Plot of Standardized Residuals versus BETA for S&L Rate of Return Example with Mercury Saving Omitted.

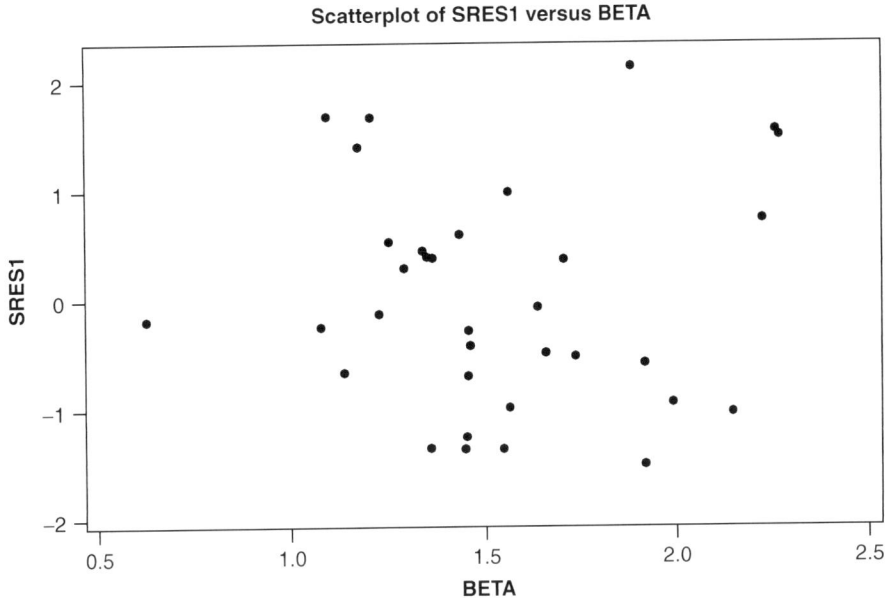

FIGURE 6.51 Plot of Standardized Residuals versus SIGMA for S&L Rate of Return Example with Mercury Saving Omitted.

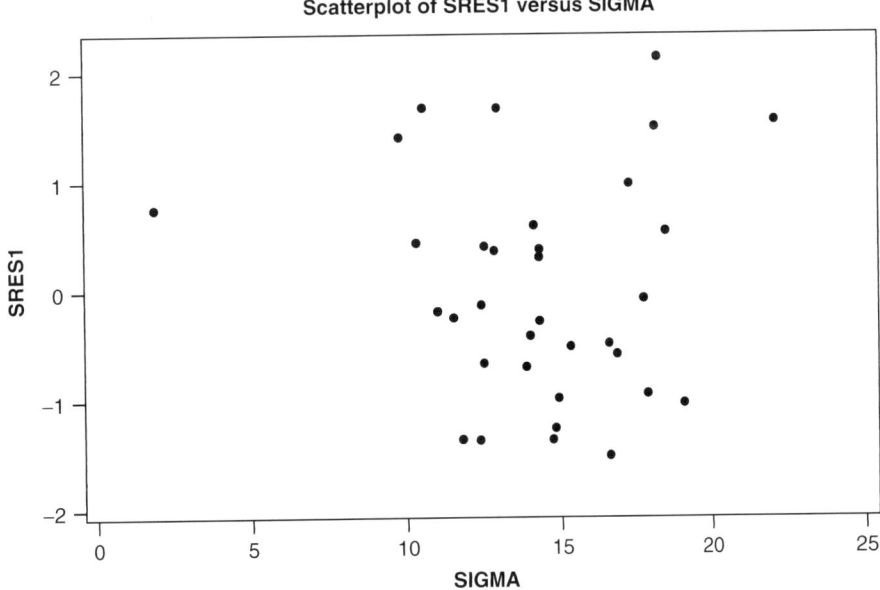

highlights the need to be careful when deleting influential observations. Knowing whether Mercury Saving belongs in our analysis is especially important because omitting it causes a reversal in our conclusion concerning the importance of the variable SIGMA.

250 Assessing the Assumptions of the Regression Model

FIGURE 6.52 Plot of DFITS for S&L Rate of Return Example with Mercury Saving Omitted.

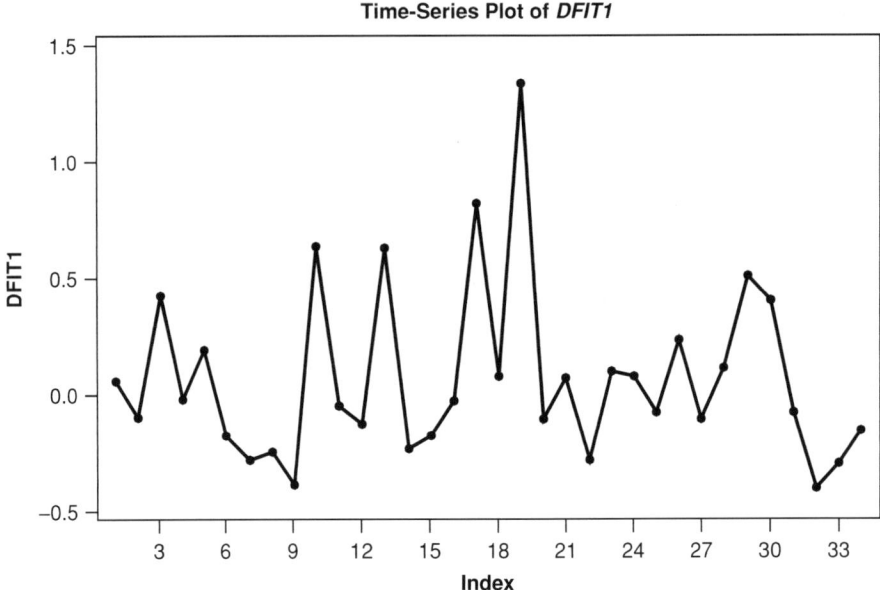

FIGURE 6.53 Plot of Cook's D Statistic for S&L Rate of Return Example with Mercury Saving Omitted.

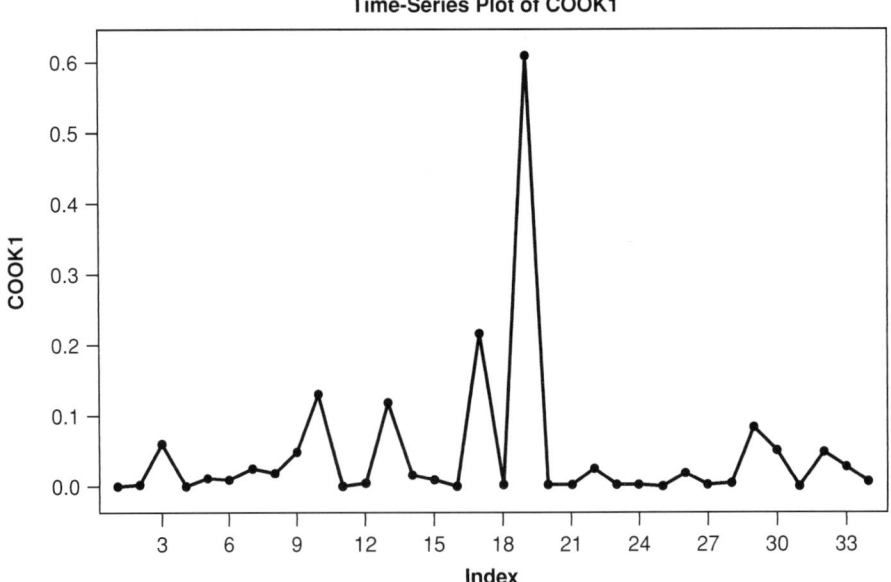

EXERCISES

4. **Petroleum Imports.** The data for U.S. monthly petroleum imports from 1991 through 1997 appear in Table 6.4. Figure 6.54 shows the regression results for the regression of monthly petroleum imports (in millions of barrels) on the one-period lagged imports. Figure 6.55 is a time-series plot of

6.7 Influential Observations

TABLE 6.4 Monthly Petroleum Imports (in millions of barrels)

Date	PETROIMP	Date	PETROIMP	Date	PETROIMP
Jan-91	180.1	May-93	211.3	Sep-95	244.9
Feb-91	163.6	Jun-93	222.5	Oct-95	226.4
Mar-91	169.2	Jul-93	229.5	Nov-95	231.6
Apr-91	177.8	Aug-93	204.1	Dec-95	235.7
May-91	215.2	Sep-93	206.7	Jan-96	239.0
Jun-91	199.2	Oct-93	223.4	Feb-96	198.1
Jul-91	201.4	Nov-93	228.3	Mar-96	201.7
Aug-91	220.2	Dec-93	203.1	Apr-96	238.2
Sep-91	190.3	Jan-94	206.0	May-96	261.6
Oct-91	189.3	Feb-94	176.9	Jun-96	253.0
Nov-91	182.9	Mar-94	219.8	Jul-96	275.2
Dec-91	182.8	Apr-94	217.2	Aug-96	251.0
Jan-92	186.7	May-94	216.6	Sep-96	260.4
Feb-92	155.0	Jun-94	248.1	Oct-96	250.7
Mar-92	172.4	Jul-94	245.6	Nov-96	217.0
Apr-92	186.0	Aug-94	242.5	Dec-96	247.8
May-92	195.5	Sep-94	260.4	Jan-97	224.1
Jun-92	193.1	Oct-94	221.0	Feb-97	211.1
Jul-92	112.7	Nov-94	229.2	Mar-97	245.6
Aug-92	201.4	Dec-94	220.9	Apr-97	250.9
Sep-92	190.0	Jan-95	212.3	May-97	278.2
Oct-92	216.5	Feb-95	195.9	Jun-97	254.2
Nov-92	193.2	Mar-95	239.7	Jul-97	266.3
Dec-92	192.0	Apr-95	212.0	Aug-97	280.9
Jan-93	212.0	May-95	240.5	Sep-97	274.6
Feb-93	175.9	Jun-95	242.0	Oct-97	280.6
Mar-93	206.1	Jul-95	245.4	Nov-97	252.8
Apr-93	220.3	Aug-95	240.8	Dec-97	250.1

Source: Business Statistics of the United States. Lanham, MD: Berman Press.

FIGURE 6.54 Regression of Petroleum Imports on the Lagged Value of Petroleum Imports.

```
Variable            Coefficient     Std Dev      T Stat     P Value

Intercept           63.087          17.044       3.70       0.00
PETROIMPL1          0.716           0.077        9.30       0.00


Standard Error = 22.2218        R-Sq = 51.7%    R-Sq(adj) = 51.1%


Analysis of Variance

Source          DF      Sum of Squares     Mean Square     F Stat     P Value

Regression      1       42754              42754           86.58      0.000
Error           81      39999              494
Total           82      82753
```

FIGURE 6.55
Time-Series Plot of Standardized Residuals for Petroleum Imports Regression.

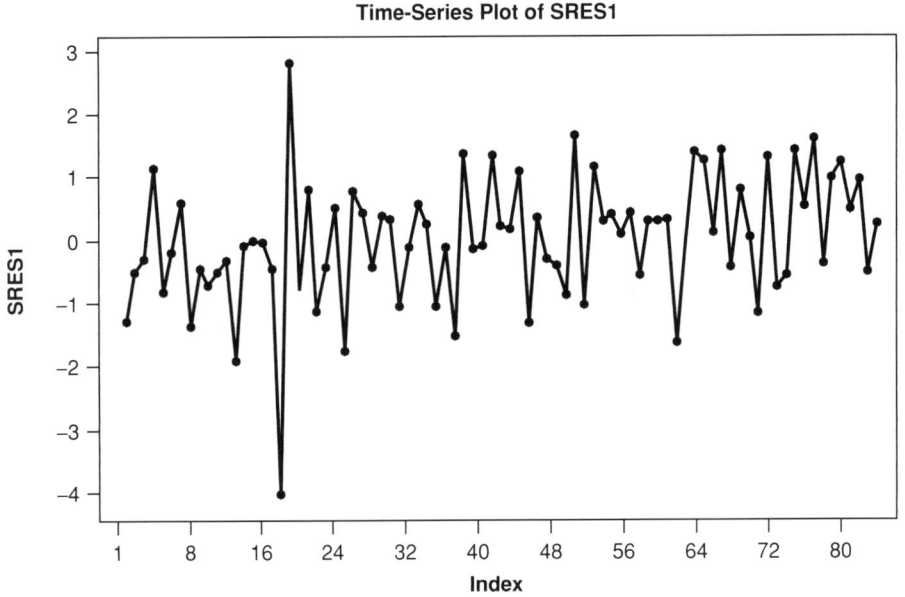

FIGURE 6.56
Plot of Standardized Residuals versus Fitted Values for Petroleum Imports Regression.

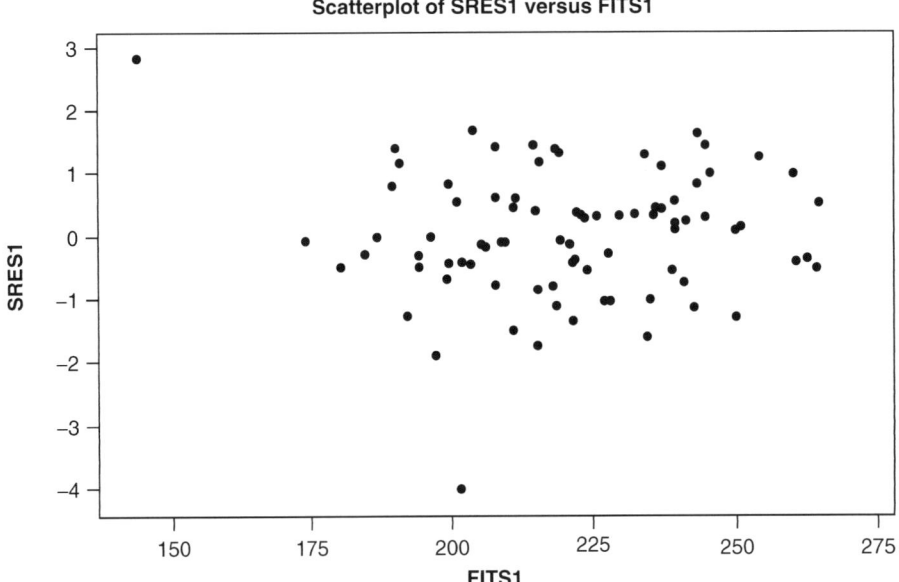

the standardized residuals from this regression. Figure 6.56 shows the plot of the standardized residuals versus the fitted values. Figures 6.57 and 6.58 show, respectively, plots of the DFITS and Cook's D statistics. These data are available in a file named PETRO6 on the CD.

FIGURE 6.57 Plot of DFITS Statistic for Petroleum Imports Regression.

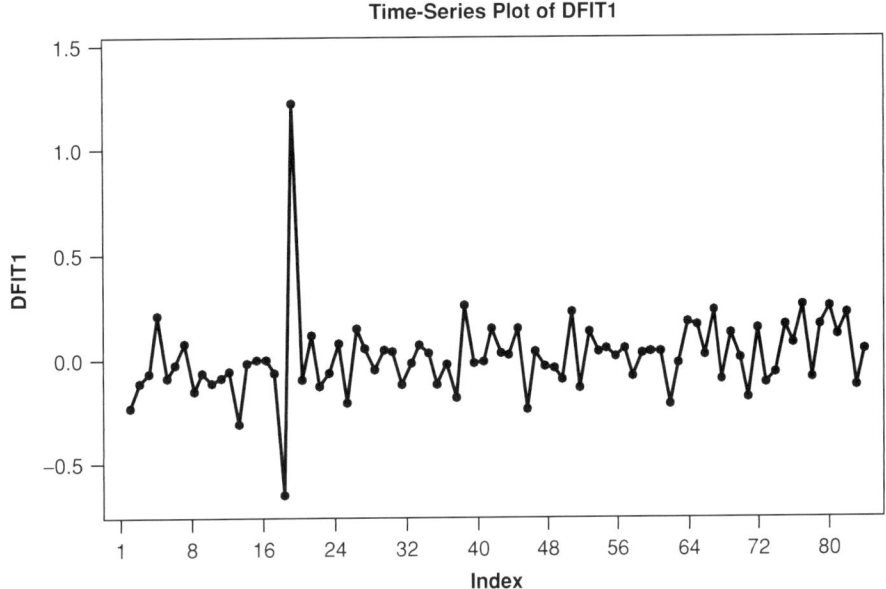

FIGURE 6.58 Plot of Cook's D Statistic for Petroleum Imports Regression.

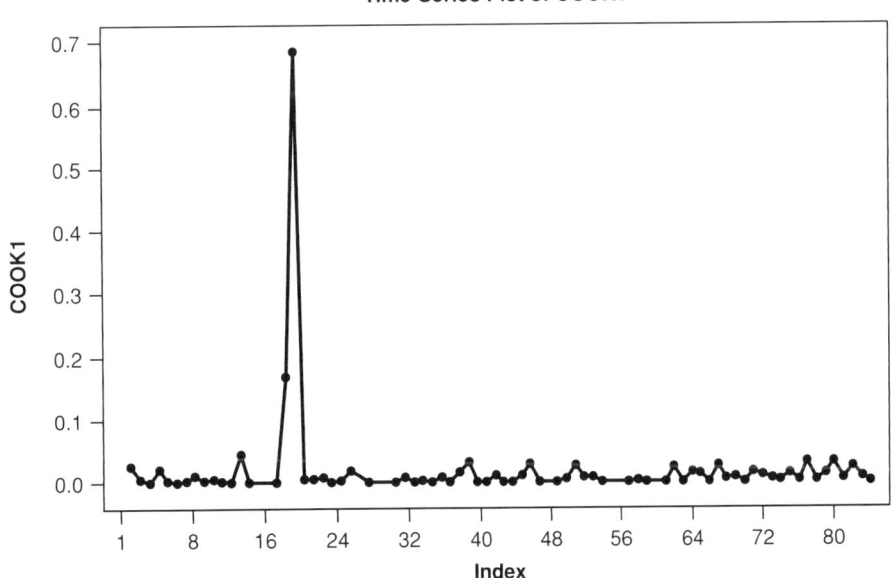

Are there any unusual observations that should be checked before accepting these regression results? If so, which observations? Can you determine what might be causing certain observations to appear unusual? Justify your answers.

6.8 ASSESSING THE ASSUMPTION THAT THE DISTURBANCES ARE INDEPENDENT

6.8.1 AUTOCORRELATION

One assumption that is frequently violated when using time-series data is that of independence of the disturbances, e_i. Disturbances in adjacent time periods are often correlated because an event in one time period may influence an event in the next time period. The relationship between disturbances in adjacent time periods is often represented by

$$e_i = \rho e_{i-1} + u_i$$

In this equation, e_i is the disturbance in the ith time period, e_{i-1} is the disturbance in the previous time period, ρ is called the *serial correlation coefficient* or *autocorrelation coefficient*, and u_i represents a disturbance that meets the assumption of independence. The regression relationship can be written as

$$y_i = \beta_0 + \beta_1 x_i + e_i$$

where $e_i = \rho e_{i-1} + u_i$.

When disturbances exhibit autocorrelation, least-squares estimates of the regression coefficients are unbiased, but the estimated standard errors of the coefficients are biased. As a result, confidence intervals and hypothesis tests do not perform as expected. The estimated regression coefficients also have larger sampling variance than certain other estimators that correct for autocorrelation.

The autocorrelation coefficient, ρ, determines the strength of the relationship between disturbances in successive time periods. Like any other correlation coefficient, it varies between -1 and $+1$, with values close to ± 1 indicating very strong relationships and values close to 0 indicating weak relationships. Ideally, a value of $\rho = 0$ is desired since this means that the disturbances are independent and the independence assumption has not been violated.

To determine whether autocorrelation is present, the Durbin–Watson test is used. This test is discussed in the next section.

6.8.2 A TEST FOR FIRST-ORDER AUTOCORRELATION

A well-known and widely used test for first-order autocorrelation is the *Durbin–Watson test*. When autocorrelation is present in business and economic data, it is typically positive autocorrelation ($\rho > 0$). For this reason, we test for positive autocorrelation.

The hypotheses to be tested may be written as follows:

$$H_0: \rho = 0$$
$$H_a: \rho > 0$$

where ρ is the first-order autocorrelation coefficient. If the null hypothesis is not rejected, the correlation, ρ, of adjacent disturbances is zero, and no problem of first-order autocorrelation exists. If the null hypothesis is rejected, the disturbances are correlated, and some correction for autocorrelation needs to be made.

The Durbin–Watson statistic is computed by first using least squares to estimate the regression equation and then by computing the residuals

$$\hat{e}_i = y_i - \hat{y}_i$$

where y_i represents one of the sample y values and $\hat{y}_i$ is the corresponding predicted y value. The residuals are used to compute the Durbin–Watson statistic, d:

$$d = \frac{\sum_{i=2}^{n}(\hat{e}_i - \hat{e}_{i-1})^2}{\sum_{i=1}^{n}\hat{e}_i^2}$$

When the disturbances are independent, d should be approximately equal to 2. When the disturbances are positively correlated, d tends to be smaller than 2.

The decision rule for the test is

Reject H_0 if $d < d_L(\alpha; n, K)$

Do not reject H_0 if $d > d_U(\alpha; n, K)$

Here, $d_L(\alpha; n, K)$ and $d_U(\alpha; n, K)$ are the critical values, which can be found in Table B.7 in Appendix B. The critical values depend on the level of significance of the test, α, the sample size, n, and the number of explanatory variables in the equation, K. For the Durbin–Watson test, there is a range of values for the test statistic where the test is said to be inconclusive. The inconclusive range of values for d is

$$d_L \leq d \leq d_U$$

If the test statistic d falls in this region, there is some question as to how to proceed.

Rejection of the null hypothesis suggests that a correction for autocorrelation is necessary. Failing to reject the null hypothesis means that no correction is necessary. But what should be done when d falls in the inconclusive region? There have been some additional procedures developed to further examine these cases, but they have not been incorporated into many computer regression routines. Without easy access to the additional procedures, one possibility is to treat values of d in the inconclusive region as if they suggested autocorrelation. If the regression results after correction for autocorrelation differ from those prior to the correction, then conclude that the correction was necessary. If the results are similar, then the correction was unnecessary, and the original uncorrected results can be used.[3]

EXAMPLE 6.10 **Sales and Advertising** Table 6.5 shows data on sales (in millions) and advertising (ADV) (in thousands) for the ABC Company (see the file SALESADV6 on the CD). These are annual data covering the period from 1967 to 2002. The regression of SALES on ADV is shown in Figure 6.59. The Durbin–Watson statistic is shown in

---

[3] To test $H_0: \rho = 0$ versus $H_a: \rho < 0$, the decision rule would be: Reject H_0 if $d > 4 - d_L$; Do not reject H_0 if $d < 4 - d_U$; Inconclusive if $4 - d_U \leq d \leq 4 - d_L$. A two-sided test can be performed by using both the upper- and lower-tailed tests separately (adjusting for the desired level of significance).

TABLE 6.5 Sales and Advertising Data

Year	SALES	ADV
1967	381.0	5316.8
1968	383.9	5413.2
1969	384.4	5486.9
1970	370.5	5537.8
1971	396.4	5660.6
1972	421.8	5750.8
1973	379.2	5782.2
1974	390.9	5781.7
1975	420.9	5821.9
1976	408.8	5892.5
1977	407.2	5950.2
1978	408.4	6002.1
1979	444.2	6121.8
1980	437.2	6201.2
1981	376.1	6271.7
1982	454.6	6383.1
1983	459.2	6444.5
1984	478.2	6509.1
1985	492.8	6574.6
1986	541.2	6704.2
1987	512.0	6794.3
1988	562.0	6911.4
1989	590.1	6986.5
1990	617.7	7095.7
1991	629.3	7170.8
1992	653.9	7210.9
1993	698.6	7304.8
1994	707.8	7391.9
1995	735.9	7495.3
1996	748.3	7629.2
1997	755.4	7703.4
1998	762.0	7818.4
1999	794.3	7955.0
2000	815.5	8063.4
2001	840.9	8170.8
2002	820.8	8254.5

the output. A time-series plot of the standardized residuals is shown in Figure 6.60. To test for positive first-order autocorrelation, the following decision rule is used:

Reject H_0 if $d < d_L\,(0.05;\,36,\,1) = 1.41$

Do not reject H_0 if $d > d_U\,(0.05;\,36,\,1) = 1.52$

Inconclusive if $1.41 \leq d \leq 1.52$

Because $d = 0.47$, the null hypothesis of no autocorrelation is rejected. First-order autocorrelation is a problem and should be corrected before inferences or forecasts are made.

FIGURE 6.59
Regression of SALES on ADV Including Durbin–Watson Statistic.

```
Variable         Coefficient       Std Dev        T Stat      P Value

Intercept        -632.694476       47.276973      -13.38      0.000
ADV                 0.177233        0.007045       25.16      0.000


Standard Error = 36.4920    R-Sq = 94.9%    R-Sq(adj) = 94.8%

Analysis of Variance

Source           DF       Sum of Squares    Mean Square    F Stat    P Value

Regression        1           842685           842685      632.81    0.000
Error            34            45277             1332
Total            35           887961


Durbin-Watson statistic = 0.467294
```

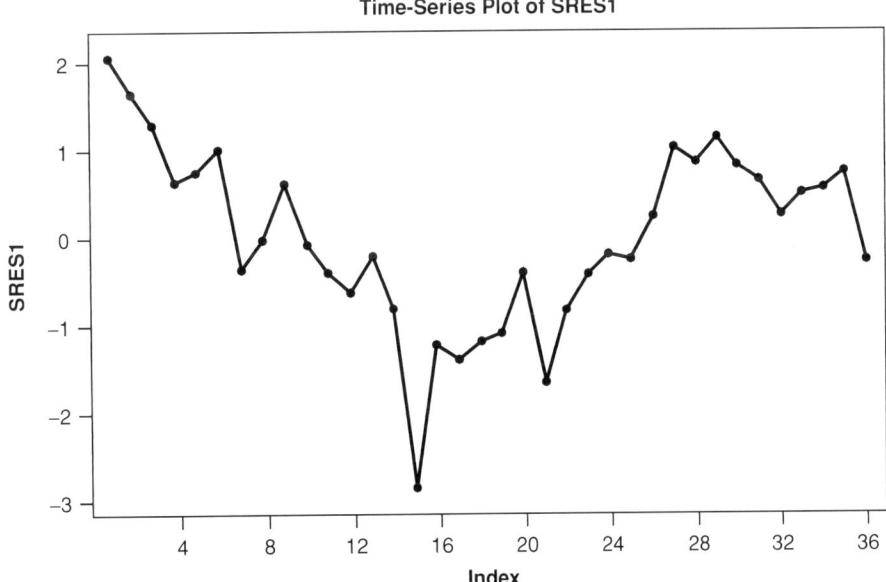

FIGURE 6.60 Time-Series Plot of Standardized Residuals for Sales and Advertising Example.

6.8.3 CORRECTION FOR FIRST-ORDER AUTOCORRELATION

When disturbances are autocorrelated, it may be due to the omission of an important variable from the regression. If the time-ordered effects of such a missing variable are positively correlated, then the disturbances in the regression tend to be positively correlated. The remedy to this problem is to locate the missing variable and include it in the regression equation, although this is easier said than done in many cases.

Another possible correction for first-order autocorrelation transforms the original time-series variables in the regression so that a regression using the transformed variables has independent disturbances. The original regression model for the time period i can be written

$$y_i = \beta_0 + \beta_1 x_i + e_i$$

where the disturbances, e_i, have first-order autocorrelation

$$e_i = \rho e_{i-1} + u_i$$

To remove the autocorrelation, the following transformations are used. Create new dependent and explanatory variable values y_i^* and x_i^* using

$$y_i^* = y_i - \rho y_{i-1} \quad \text{and} \quad x_i^* = x_i - \rho x_{i-1}$$

for time periods $i = 2, 3, \ldots, n$.

In addition, transform the first time-period observation as

$$y_1^* = \sqrt{1 - \rho^2} \, y_1 \quad \text{and} \quad x_1^* = \sqrt{1 - \rho^2} \, x_1$$

The new regression can be written as

$$y_i^* = \beta_0 + \beta_1 x_i^* + u_i$$

The disturbances u_i are independent. Now regress y_i^* on x_i^* to obtain estimates of β_0 and β_1.

In practice, there are various refinements to this process. In choosing a statistical package to perform these transformations and run the regressions, one of the most important things to keep in mind is that the transformation should include the first observation (x_1^*, y_1^*). Many statistical packages incorporated routines that simply dropped the first observation rather than transforming it as shown and including it in the new regression. Recent research shows that dropping this observation results in the loss of important information and adversely affects the results of the new regression. One common estimation procedure that drops the first observation is called the Cochrane–Orcutt method. This method should be avoided. Procedures that incorporate the first observation include the Prais–Winsten method and full maximum likelihood. When using an automatic method from a statistical package to correct for autocorrelation, only those methods incorporating the initial observation should be used. Statistical packages such as SAS and SHAZAM have single commands that transform the data appropriately, rerun the regression on the transformed data, and print out the results so that correcting for autocorrelation in this case is a simple matter. MINITAB and Excel have not incorporated any of these automatic procedures, so this type of correction is difficult to perform.

When the Prais–Winsten transformation is used, forecasts are computed in a slightly different manner than shown for a regression without autocorrelation. The T-period-ahead forecast can be written in general as

$$\hat{y}_{n+T} = \hat{\beta}_0 + \hat{\beta}_1 x_{n+T} + \rho^T \hat{e}_n$$

where $\hat{\beta}_0$ and $\hat{\beta}_1$ are used to represent the estimates of β_0 and β_1 from the transformed regression model (rather than the least-squares estimates b_0 and b_1), x_{n+T} is the value of the explanatory variable in the period to be forecast, ρ is the autocorrelation coefficient, and $\hat{e}_n$ is the residual from the last sample time period. Note that only the last residual in the sample contains any information about the future. Also note that the value of this information declines as forecasts are generated further into the future. The term ρ^T can be seen to decrease (because $-1 < \rho < 1$) as T increases.

A third option to correct for autocorrelation is to add a lagged value of the dependent variable as an explanatory variable. This is a viable option especially if building an extrapolative model and if the number of observations is reasonably large (since one observation is lost because of the lagged variable).

EXAMPLE 6.11 **Sales and Advertising (again)** A lagged value of the dependent variable will be introduced to try to correct for first-order autocorrelation in the model for sales. The new equation can be written

$$y_i = \beta_0 + \beta_1 y_{i-1} + \beta_2 x_i + e_i$$

where y_i is sales in time period i, y_{i-1} is the lagged value of sales, and x_i is ADV. The regression is shown in Figure 6.61. Evaluating whether this model is an improvement over the previous model is discussed in the next section.

6.8.4 H TEST FOR AUTOCORRELATION

When lagged values of the dependent variable are used as explanatory variables, the Durbin–Watson test is no longer appropriate. An alternative test typically recommended in this case is Durbin's h. The hypotheses to be tested are

$$H_0: \rho = 0$$
$$H_a: \rho > 0$$

The test statistic is

$$h = r\left(\frac{n}{1 - ns_{b_1}^2}\right)^{1/2}$$

FIGURE 6.61 Regression of SALES on ADV and a Lagged Dependent Variable.

```
Variable            Coefficient      Std Dev        T Stat        P Value

Intercept           -234.4752        78.0688        -3.00         0.005
SALESL1             0.6751           0.1123         6.01          0.000
ADV                 0.0631           0.0202         3.12          0.004

Standard Error = 24.1223             R-Sq=97.8%     R-Sq(adj)=97.7%

Analysis of Variance

Source              DF               Sum of Squares    Mean Square    F Stat    P Value

Regression          2                841098            420549         722.74    0.000
Error               32               18620             582
Total               34               859718

Durbin-Watson statistic = 2.33302
```

where r is an estimate of the first-order autocorrelation coefficient, n is the sample size, and $s_{b_1}^2$ is the estimated variance of the regression coefficient of the lagged dependent variable (y_{i-1}). If the null hypothesis is true, h has a standard normal distribution. The decision rule to conduct the test is

Reject H_0 if $h > z_\alpha$

Do not reject H_0 if $h \leq z_\alpha$

Note that the test statistic h cannot be computed if $1 - ns_{b_1}^2 < 0$ because this results in the need to take the square root of a negative number. (Alternative test procedures are available in this case. The reader is referred to Judge et al., *The Theory and Practice of Econometrics*, pp. 326–327, for an example.) In practice, a quick way to estimate the autocorrelation correlation coefficient is to use $r = 1 - d/2$, where d is the Durbin–Watson statistic.

Finally, it should be noted that other procedures are available for analyzing time-series data. Certain of these procedures are designed especially for extrapolative models using lagged dependent variables, autocorrelated errors, or both. For more detail on such time-series forecasting methods, the reader is referred to Bowerman and O'Connell, *Forecasting and Time Series: An Applied Approach*.

EXAMPLE 6.12 **Sales and Advertising (again)** In Example 6.10, a regression of SALES on ADV was run. The Durbin–Watson test indicated that the disturbances from this regression were not independent. In Example 6.11, a lagged dependent variable was added to the regression to try to correct for the first-order autocorrelation. The appropriate test to determine whether the introduction of the lagged variable has eliminated the autocorrelation is the h test. The quantities necessary to compute the h statistic are available in the regression output in Figure 6.61.

Using a 5% level of significance, the decision rule to conduct the test is

Reject H_0 if $h > 1.645$

Do not reject H_0 if $h \leq 1.645$

The estimate of the autocorrelation coefficient can be computed as

$$r = 1 - d/2 = 1 - 2.33/2 = -0.165$$

The sample size used in the regression is 35 (one of the original observations was lost because of the use of the lagged variable), and the standard deviation of the coefficient of the lagged variable is 0.1123, so the h statistic is

$$h = -0.165\sqrt{\frac{35}{1 - 35(0.1123)^2}} = -1.31$$

Our decision is that H_0 cannot be rejected. Conclude that first-order autocorrelation is not a problem. Adding the lagged variable proved effective in correcting for autocorrelation.

EXAMPLE 6.13 **Sales and Advertising (for the last time)** Figure 6.62 shows the resulting output from SAS when PROC AUTOREG is used (see Using the Computer section for

FIGURE 6.62
SAS Regression of SALES on ADV Using Prais–Winsten Correction for Autocorrelation.

The AUTOREG Procedure

Dependent Variable   SALES

Ordinary Least Squares Estimates

SSE	45276.5625	DFE	34
MSE	1332	Root MSE	36.49197
SBC	366.263543	AIC	363.096505
Regress R-Square	0.9490	Total R-Square	0.9490
Durbin-Watson	0.4673		

Variable	DF	Estimate	Standard Error	t Value	Approx Pr > |t|
Intercept	1	-632.6945	47.2770	-13.38	<.0001
ADV	1	0.1772	0.007045	25.16	<.0001

Estimates of Autocorrelations

Lag	Covariance	Correlation	-1 9 8 7 6 5 4 3 2 1 0 1 2 3 4 5 6 7 8 9 1
0	1257.7	1.000000	|********************|
1	891.8	0.709089	|************** |

Preliminary MSE   625.3

Estimates of Autoregressive Parameters

Lag	Coefficient	Standard Error	t Value
1	-0.709089	0.122745	-5.78

Algorithm converged.

The AUTOREG Procedure

Yule-Walker Estimates

SSE	19487.808	DFE	33
MSE	590.53964	Root MSE	24.30102
SBC	340.302604	AIC	335.552047
Regress R-Square	0.8057	Total R-Square	0.9781
Durbin-Watson	2.0797		

Variable	DF	Estimate	Standard Error	t Value	Approx Pr > |t|
Intercept	1	-577.8041	97.8730	-5.90	<.0001
ADV	1	0.1696	0.0145	11.70	<.0001

details). As used here, this is the same as the Prais–Winsten correction discussed in the text. SAS first estimates the relationship using least squares and reports the results (Ordinary Least Squares Estimates). The usual coefficient estimate, standard error, t value, and p value are reported along with a number of other statistics, some of which were discussed in this text (*SSE*, *MSE*, R-square, etc.) and some of which were not (AIC and SBC). The estimate of the first-order autocorrelation coefficient is $r = 0.709089$. The results under "The AUTOREG procedure: Yule-Walker estimates" show the model estimates after using the Prais–Winsten correction for first-order autocorrelation. The coefficient estimate, standard error, t value and p value are reported and are used as with least squares. In this example the revised equation would be written

$$SALES = -577.8041 + 0.1696 ADV$$

In Section 6.8.3, the coefficient estimates after the Prais–Winsten transformation were written as

$$\hat{\beta}_0 = -577.8041 \text{ and } \hat{\beta}_1 = 0.1696$$

to distinguish them from the original least-squares estimates.

EXERCISES

5. Cost Control (reconsidered). The file COST6 on the CD contains data on three variables:

COST is total manufacturing cost per month in thousands of dollars (the dependent variable)

PAPER is total production of paper per month in tons

MACHINE is total machine hours used per month

The regression of COST on PAPER and MACHINE is shown in Figure 6.63. The Durbin–Watson statistic is included in the output. A time-series plot of the standardized residuals is shown in Figure 6.64.

Test whether the disturbances are autocorrelated. Use a 5% level of significance. Be sure to state the hypotheses to be tested, the decision rule, the test statistic, and your decision.

FIGURE 6.63 Regression of COST on PAPER and MACHINE Including Durbin–Watson Statistic.

Variable	Coefficient	Std Dev	T Stat	P Value
Intercept	59.4318	19.6388	3.03	0.006
PAPER	0.9489	0.1101	8.62	0.000
MACHINE	2.3864	0.2101	11.36	0.000

Standard Error = 10.9835   R-Sq = 99.9%   R-Sq(adj) = 99.9%

Analysis of Variance

Source	DF	Sum of Squares	Mean Squares	F Stat	P Value
Regression	2	2271227	1135613	9413.48	0.000
Error	24	2895	121		
Total	26	2274122			

Durbin-Watson statistic = 2.14197

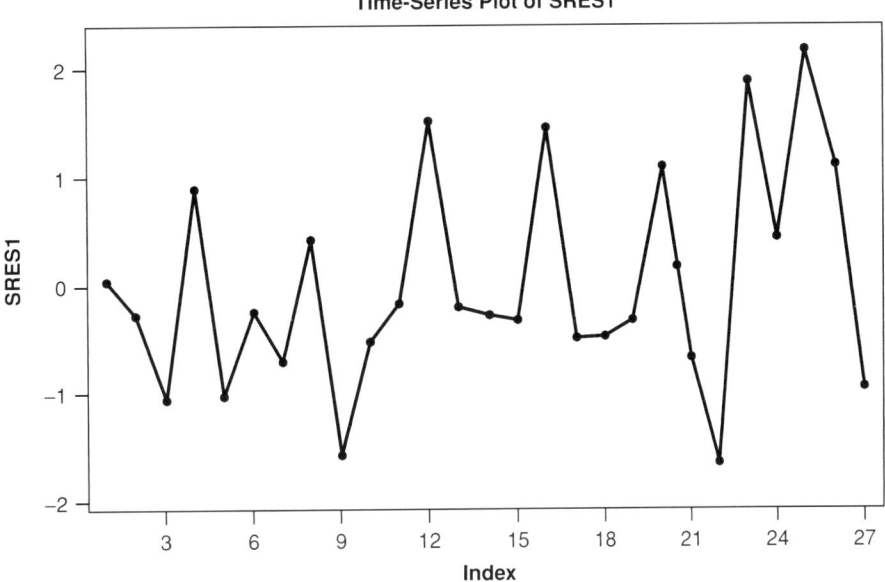

FIGURE 6.64
Time-Series Plot of Standardized Residuals from Regression of COST on PAPER and MACHINE.

ADDITIONAL EXERCISES

6. Major League Baseball Salaries. The owners of Major League Baseball (MLB) teams are concerned with rising salaries (as are owners of all professional sports teams). Table 3.9 provides the average salary (AVESAL) of the 30 MLB teams for the 2002 season. Also provided is the number of wins (WINS) for each team during the 2002 season. In Chapter 3 you were asked to run the regression of WINS on AVESAL to determine whether there is evidence that teams with higher total payrolls tend to be more successful. Rerun this regression now. The data are in the file BBALL6 on the CD. Examine the standardized residuals. Are there any teams that appear to be winning more games than expected given the size of their payroll? Justify your answer. Which team or teams, if any, did you identify?
(*Source*: Data courtesy of the *Fort Worth Star-Telegram.*)

7. Imports. The gross domestic product (GDP) and imports (IMPORTS) for 25 countries are available in a file named IMPORTS6 on the CD. Construct the scatterplot of IMPORTS versus GDP. Run the regression using IMPORTS as the dependent variable and GDP as the explanatory variable. Plot the standardized residuals versus the fitted values and the explanatory variable. Use the results to help answer the following questions:

a. Do any of the assumptions of the regression model appear to be violated? If so, which one (or ones)? Justify your answer.

b. Construct the scatterplot and rerun the regression with the United States omitted. Construct the residual plots for this new regression. Do the results appear any different from the original regression results? If so, how do they differ? Do you prefer the original results or the results with the United States omitted? On what do you base your choice? Do there still appear to be problems with this regression?

c. Try to develop a curvilinear model using the original data (with the United States included) that provides improved results over the linear model. Be sure to examine the residual plots from the curvilinear model to see if any regression assumptions are violated for this model.

8. Outlier. Consider the following time-series regression model:

$$y_i = \beta_0 + \beta_1 x_i + e_i$$

Suppose that the y value in one time period can be regarded as an outlier.

a. Indicate how you might be able to detect the presence of the outlier from any computer output you request or receive while analyzing the data.

b. Suppose the outlier is detected and the person in charge of the analysis decides to delete the point from the analysis and to rerun the regression without this observation. Are there alternate suggestions you would make prior to taking this course of action, or do you believe deletion of the observation is the best course of action?

9. **Consumer Credit.** Consider the time series of consumer installment credit (U.S.) in billions of dollars. Two different models are being considered:

MODEL 1: CREDIT $= \beta_0 + \beta_1 \text{TREND} + e$

and

MODEL 2: $\log(\text{CREDIT}) = \beta_0 + \beta_1 \text{TREND} + e$

where CREDIT is billions of dollars of consumer installment credit at the end of each month, TREND is a linear trend component, and log(CREDIT) is the natural logarithm of consumer installment credit. The regressions for these two models are shown in Figures 6.65 and 6.66, respectively. The researcher wishes to compare these two models and choose the best one. Can the choice be made on the basis of the two outputs shown? If yes, state how and state which model you would choose. If no, state why not and state what you would need to make the comparison.

10. **ABX Company.** The file named ABSENT6 on the CD contains data on two variables. The dependent variable is absenteeism among employees at the ABX Company (ABSENT). The explanatory variable to be considered is seniority (SENIOR), the employee's time with the company. Our goal is to express the relationship between absenteeism and seniority in the form of an equation. Start with the linear regression using absenteeism as the dependent variable and seniority as the independent variable.

Based on an examination of scatterplots and residual plots, do any assumptions of the linear regression model appear to be violated? If so, which one (or ones)? If any violations are detected, suggest possible corrections. Rerun the regression with the suggested corrections and compare your results to the original results. Be sure to do residual plots for the model using the suggested correction. Which model do you prefer and why?

11. **Coal Mining Fatalities.** The file CUTTING6 on the CD contains data on the following two variables:

FATALS: the annual number of fatalities from gas and dust explosions in coal mines for the years 1915 to 1978

CUTTING: the number of cutting machines in use

(*Source*: These data are from K. D. Lawrence and L. C. Marsh, "Robust Ridge Estimation Methods for Predicting U.S. Coal Mining Fatalities." *Communications in Statistics*, 13 (1984): pp. 139–149. Used by permission.)

Run the regression using FATALS as the dependent variable and CUTTING as the independent variable.

FIGURE 6.65 Regression Results for Consumer Credit Model 1.

Variable	Coefficient	Std Dev	T Stat	P Value
Intercept	737.933	4.650	158.71	0.000
TREND	9.763	0.165	59.10	0.000

Standard Error = 15.86   R-Sq = 98.7%   R-Sq(adj) = 98.7%

Analysis of Variance

Source	DF	Sum of Squares	Mean Square	F Stat	P Value
Regression	1	878056	878056	3492.70	0.000
Error	46	11564	251		
Total	47	889621			

FIGURE 6.66
Regression Results for Consumer Credit Model 2.

```
Variable      Coefficient    Std Dev      T Stat      P Value

Intercept     6.62890        0.00441      1504.82     0.000
TREND         0.01004        0.00016      64.14       0.000


Standard Error = 0.01502    R-Sq = 98.9%     R-Sq(adj) = 98.9%

Analysis of Variance
Source        DF     Sum of Squares    Mean Square    F Stat     P Value

Regression    1      0.92843           0.92843        4114.28    0.000
Error         46     0.01038           0.00023
Total         47     0.93881
```

Based on an examination of scatterplots and residual plots, do any assumptions of the linear regression model appear to be violated? If so, which one (or ones)? If any violations are detected, suggest possible corrections. Rerun the regression with the suggested corrections and compare your results to the original results. Be sure to do residual plots for the model using the suggested correction. Which model do you prefer and why?

12. **Piston Corporation (reconsidered).** Reexamine the data from the PISTON exercise in Chapter 5. The original variables in this problem were

   COST, the dependent variable; a measure of cost for the company
   PROD, production level
   INDEX, a cost index for the industry

   The data for these variables are in the file PISTON6 on the CD.

   Use the variables 1/PROD and INDEX as explanatory variables. Test to see whether the disturbances are positively autocorrelated. State the hypotheses to be tested, the decision rule, the test statistic, and your decision. On the basis of the test result, what action should be taken? Use a 0.05 level of significance.

13. **Computer Repair.** A computer repair service is examining the time taken on service calls. The data obtained for 30 service calls are in the file named COMPREP6 on the CD. Information obtained includes:

   number of machines to be repaired (NUMBER)
   years of experience of service person (EXPER)

   time taken (in minutes) to provide service (TIME)

   Develop a model to predict average time on the service calls using EXPER and NUMBER as explanatory variables. Use scatterplots and residual plots to determine whether any of the assumptions of the linear regression model have been violated.

   If any of the assumptions have been violated, state which one or ones and suggest possible corrections. Try the new model to see if it is an improvement over the original one. Be sure to examine residual plots from the corrected model (or models) that you try. Indicate your choice for the best model.

14. **U. S. Population.** The data file USPOP6 on the CD contains the population of the United States for the years 1930 through 1999. Fit a linear trend to these data.

   a. What is the resulting regression equation?
   b. What percentage of the variation in y has been explained by the regression?
   c. Test to see whether the disturbances are positively autocorrelated. State the hypotheses to be tested, the decision rule, the test statistic, and your decision. On the basis of the test result, what action should be taken? Use a 0.05 level of significance.
   d. Add a lagged value of the dependent variable to the equation. Your equation will now have a term for linear trend as well as the lagged dependent variable. What is the resulting regression equation?
   e. Test to see whether the disturbances are positively autocorrelated. (What test should you

use for this model?) State the hypotheses to be tested, the decision rule, the test statistic, and your decision. Use a 0.05 level of significance. On the basis of the test result, what action should be taken?

f. Add a two-period lagged value of the dependent variable to the equation. Your equation will now have a term for linear trend as well as the one- and two-period lagged dependent variables. What is the resulting regression equation?
(*Source*: U.S. Bureau of the Census.)

g. Test to see whether the disturbances are positively autocorrelated. (What test should you use for this model?) State the hypotheses to be tested, the decision rule, the test statistic, and your decision. Use a 0.05 level of significance. On the basis of the test result, what action should be taken?

h. Use your choice of the "best" model from among the three examined in this exercise to predict the U.S. population for 2000 and 2001. Do you encounter any problems with the use of this model to generate these forecasts? If so, what can you do to overcome these problems?

15. Major League Baseball Wins. What factor is most important in building a winning baseball team? Some might argue for a high batting average. Or it might be a team that hits for power as measured by the number of home runs. On the other hand, many believe that it is quality pitching as measured by the earned run average of the team's pitchers. The file MLB6 on the CD contains data on the following variables for the 30 major league baseball teams during the 2002 season:

WINS = number of games won
HR = number of home runs hit
BA = average batting average
ERA = earned run average

Using WINS as the dependent variable, run the regression relating the three explanatory variables to WINS. Examine the standardized residuals from this regression and any other regression diagnostics you believe might be useful. Do any of the regression assumptions appear to be violated? Justify your answer.
(*Source*: Data courtesy of the *Fort Worth Star-Telegram*.)

16. Byron Nelson Donations. Since 1982, the year that PGA Tour officials began tracking charitable donations from its tournaments on an annual basis, no event has contributed more to charities in its community than the Byron Nelson tournament. The amount donated each year from 1982 through 2002 (in millions of dollars) is provided in the file named DONATIONS6 on the CD. Fit a linear trend to these data.

a. What is the resulting regression equation?

b. What percentage of the variation in *y* has been explained by the regression?

c. Test to see whether the disturbances are positively autocorrelated. State the hypotheses to be tested, the decision rule, the test statistic, and your decision. Use a 0.05 level of significance. On the basis of the test result, what action should be taken?

d. Add a lagged value of the dependent variable to the equation. Your equation will now have a term for linear trend as well as the lagged dependent variable. What is the resulting regression equation?

e. Test to see whether the disturbances are positively autocorrelated. (What test should you use for this model?) State the hypotheses to be tested, the decision rule, the test statistic, and your decision. Use a 0.05 level of significance. On the basis of the test result, what action should be taken?

f. Use your choice of the "best" model to predict donations for 2003. Find a point prediction and a 95% prediction interval.
(*Source*: Data courtesy of the *Fort Worth Star-Telegram*.)

17. Estimating Residential Real Estate Values. The Tarrant County Appraisal District must appraise properties for the entire county. The appraisal district uses data such as square footage of the individual houses as well as location, depreciation, and physical condition of an entire neighborhood to derive individual appraisal values on each house. This avoids labor-intensive reinspection each year.

Regression can be used to establish the weight assigned to various factors used in assessing values. For example, the file REALEST6 on the CD contains the value (VALUE), size in square feet (SIZE), a physical condition index (CONDITION), and a depreciation factor (DEPRECIATION) for a sample of 100 Tarrant County houses (in 1990). Using these data, develop an equation that might be useful to the appraisal district in assessing values.

If any of the assumptions have been violated, state which one or ones and suggest possible corrections. Try the new model to see if it is an improvement over the original one. Be sure to examine residual plots from the corrected model (or models) that you try.

Discuss how the equation developed here could be used to value houses. What would be the value assigned to a 1400-square-foot house with physical condition index 0.70 and depreciation factor 0.02?

18. **Criminal Justice Expenditures.** The file CRIMSPN6 on the CD contains the following data for each of the 50 states:

> total expenditures on a state's criminal justice system (in thousands of dollars) (EXPEND)
> total number of police employed in the state (POLICE)
> population of the state (in thousands) (POP)
> total number of incarcerated prisoners in the state (PRISONER)

State governments must try to project spending in many areas. Expenditure on the criminal justice system is one area of continually rising cost. Your job is to build a model that can be used to forecast spending on a state's criminal justice system. Any of the three possible explanatory variables can be used. Be sure to consider violations of any assumptions in building your model and correct for any violations. Once your model is complete, predict expenditures for a state that plans to hire 10,000 police personnel, has a population of 3 million, and expects 600 prisoners. Find a point prediction and a 95% prediction interval.
(*Source*: These data were obtained from the U.S. Department of Criminal Justice web site and are for the year 1999.)

19. **Intersections.** One factor related to the number of accidents at an intersection is the peak-hour volume of traffic. Data on both the volume of traffic during the peak hour (VOLUME) and the total number of accidents (ACCIDENT) are available for 62 intersections in Fort Worth, Texas. These data are available in a file named TRAFFIC6 on the CD.
(*Source*: City of Fort Worth)

The city wants to identify intersections that have an unusual number of accidents. Its definition of unusual involves first establishing a base-level forecast of the number of accidents, taking account of the peak-hour volume of traffic, and then judging which intersections still have an unusual number of accidents. As a consultant to the city, your job is to determine which, if any, of these 62 intersections appear to have an unusually high number of accidents given the volume of traffic.

20. **Fuel Consumption.** Table 5.4 showed the fuel consumption (FUELCON) in gallons per capita for each of the 50 states and Washington, D.C. The following variables are also shown: the population of the state (POP), the area of the state in square miles (AREA). From these data, create the variable population density (DENSITY) defined as population divided by area. The file FUELCON6 on the CD contains this information as well as data for the following variables:

> DRIVERS, the ratio of licensed drivers to private and commercial motor vehicles registered
> HWYMILES, the number of federally funded miles of highway
> GASTAX, the tax on a gallon of gas
> INCOME, the average personal income

The object is to develop a regression equation to predict fuel consumption. Any of the variables in the file can be used as explanatory variables. Find what you believe is the best equation to explain fuel consumption. Be sure to check for violations of any of the regression assumptions.
(*Source*: The population data and average personal income are from the U.S. Bureau of the Census web site. All other data are from the Federal Highway Administration: Office of Highway Information Management.)

USING THE COMPUTER

The Using the Computer section in each chapter describes how to perform the computer analyses in the chapter using Excel, MINITAB, and SAS. For further detail on Excel, MINITAB, and SAS, see Appendix C.

268 Assessing the Assumptions of the Regression Model

EXCEL

Storing Standardized Residuals and Fitted Values

In the Regression dialog box, check Standardized Residuals, as shown in Figure 6.67. Checking the box for the standardized residuals saves the standardized residuals and the fitted values (as well as the raw residuals).

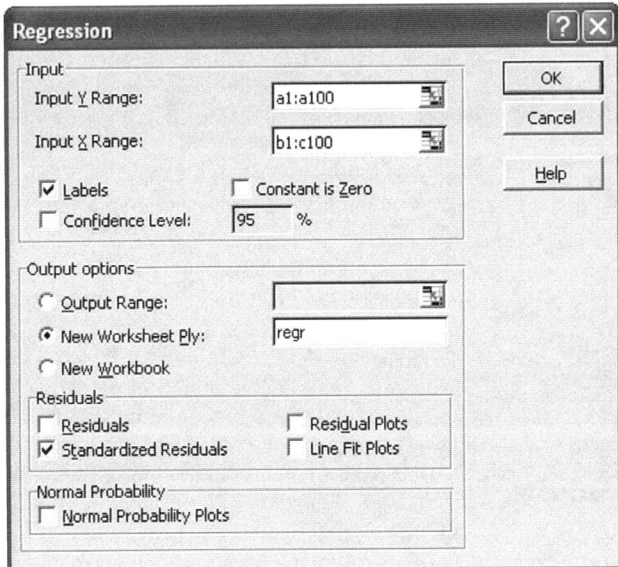

FIGURE 6.67 Saving Fitted Values and Standardized Residuals with Excel.

Requesting Other Influence Statistics and the Durbin–Watson Statistic

Excel does not provide options for automatically producing the other influence statistics discussed in the text or the Durbin–Watson statistic. Although these could be computed by creating formulas, this option is not discussed in this text.

A regression add-in is available that does compute a number of these additional statistics. See the Excel section in Appendix C for more information on the add-in.

Variable Transformations

See Variable Transformations in the Using the Computer section at the end of Chapter 5.

Testing for Normality

Although there is a normal probability plot option in the Excel Regression dialog box, it does not do you much good. It provides a normal probability plot of the dependent variable, when in fact, what you really want is a normal probability plot of the standardized residuals. Plus, I find the plot difficult to interpret given the way it is produced in Excel.

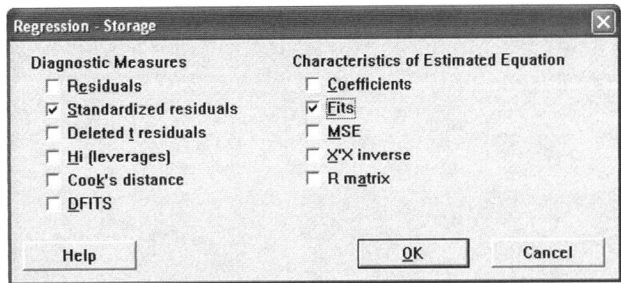

FIGURE 6.68 MINITAB Regression—Storage Dialog Box for Saving Fitted Values, Standardized Residuals, and Influence Statistics.

MINITAB

Storing Standardized Residuals and Fitted Values

STAT: REGRESSION: REGRESSION: STORAGE

Store standardized residuals and fitted values by using the Storage button in the Regression dialog box. Then click "Fits" and "Standardized residuals" as shown in Figure 6.68. The fitted values and standardized residuals are stored, respectively, in the next two free columns in the MINITAB worksheet.

Requesting Other Influence Statistics

Refer again to Figure 6.68. Requests for other influence statistics are made by checking the appropriate boxes in the Regression—Storage dialog box. Other influence statistics available in MINITAB include leverage values (Hi), Cook's Distance, and DFITS. These values are stored in a column in the worksheet by simply clicking the box next to the desired statistic.

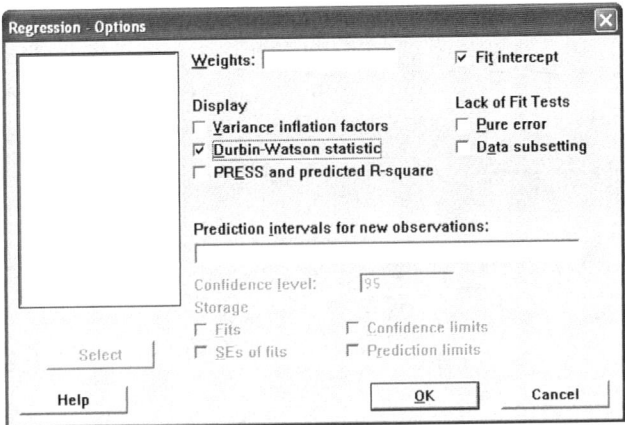

FIGURE 6.69 MINITAB Regression—Options Dialog Box for Requesting Durbin–Watson Statistic.

Requesting the Durbin–Watson Statistic
STAT: REGRESSION: REGRESSION: OPTIONS

To request the Durbin–Watson statistic, click the Durbin–Watson statistic box on the Regression—Options screen, as shown in Figure 6.69.

Variable Transformations

See Variable Transformations in the Using the Computer section at the end of Chapter 5.

Testing for Normality of Disturbances
STAT: REGRESSION: REGRESSION: GRAPHS

To request a test for the normality of the regression disturbances, click the Graphs button in the Regression Dialog Box. Then check the "normal plot of residuals" option, as shown in Figure 6.70. Note that "Residuals for Plots" allows a choice of which residuals to use in the test. The Anderson–Darling test statistic is reported along with a p value.

STAT: BASIC STATISTICS: NORMALITY TEST

To request a test for normality (for the disturbances or any other variable), use the Normality Test option on the Basic Statistics menu. Fill in the variable to be tested

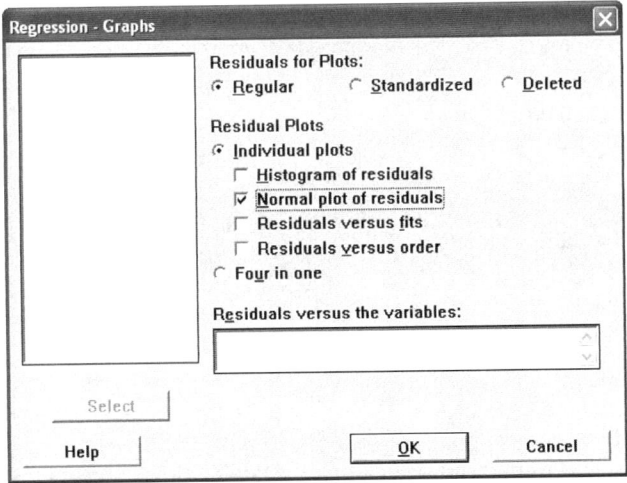

FIGURE 6.70 MINITAB Normality Test for Disturbances.

(the standardized residuals from a regression are used in the example) and the test desired, as shown in Figure 6.71. You have a choice of the Anderson–Darling test, the Ryan–Joiner test, or the Kolmogorov–Smirnov test. Regardless of which test is chosen, a p value will be reported, which makes conducting any of the tests very easy.

Szroeter's Test for Nonconstant Variance

Although this test cannot be requested with a single command, it can be computed through a sequence of commands. This will be demonstrated for MINITAB. (I used commands here rather than menu items because I find it easier when there is a sequence

FIGURE 6.71 MINITAB Normality Test Dialog Box.

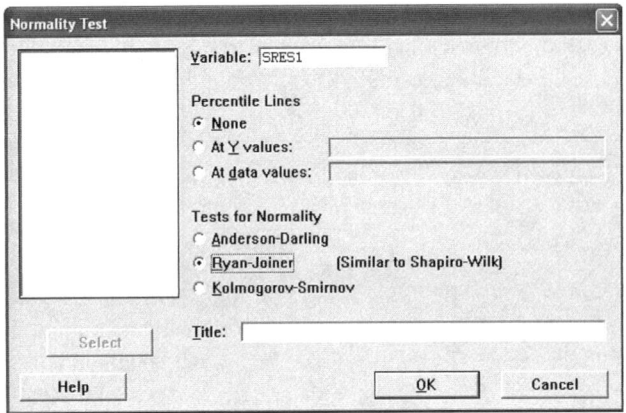

of operations to be performed. Obviously, these same procedures could be accomplished with menus.) Assume that the dependent (y) variable is in C1 and the (one) explanatory variable (x) is in C2. First, recall that the data must be ordered according to increasing values of x. The Sort command accomplishes this. The standardized residuals from the regression of y on x are then stored in C10.

```
SORT C2, CARRY C1, PUT IN C20, C21
REGR Y IN C21 ON 1 PRED IN C20, STORE IN
   C10
```

A patterned Set command is used to put integers 1 through n (the sample size) in C25.

```
SET IN C25
1:n
END
```

Now compute the quantities needed to compute h in Szroeter's test statistic:

```
LET C16 = C10*C10
LET C17 = C16*C25
SUM C17 K2
SUM C16 K1
LET K3 = K2/K1
```

The value of K3 printed will be h.

```
PRINT K3
```

Szroeter's Q statistic can then be computed using the formula

$$Q = \left(\frac{6n}{n^2 - 1}\right)^{1/2}\left(h - \frac{n+1}{2}\right)$$

SAS

Storing Standardized Residuals and Fitted Values

The following command sequence produces a regression of SALES on two independent variables, ADV and BONUS, saves the fitted values and standardized residuals, and then plots the standardized residuals versus the fitted values and each of the independent variables.

```
PROC REG;
MODEL SALES = ADV BONUS;
OUTPUT PREDICTED = FITS STUDENT = STRES;
PROC GPLOT;
PLOT STRES*FITS STRES*ADV STRES*BONUS;
```

In the model statement, the dependent variable (SALES) is listed first; the independent variables (ADV and BONUS) follow the equal sign. In the output statement, the fitted (predicted) values are requested and labeled FITS. The standardized residuals (STUDENT) are also requested and labeled STRES. PROC GPLOT is used to produce residual plots. As shown, residual plots of the standardized residuals versus the fitted values and each of the two explanatory variables are generated.

Requesting Other Influence Statistics

Influence statistics other than standardized residuals and fitted values can be requested in SAS as follows:

```
PROC REG;
MODEL Y = X/INFLUENCE;
OUTPUT COOK = COOKD;
```

The INFLUENCE option requests that SAS print out the leverage values, the studentized deleted residuals, and the DFITS statistics (among others not discussed in this text). The OUTPUT command requests that Cook's D statistic be computed, stored, and labeled COOKD.

Requesting the Durbin–Watson Statistic

```
PROC REG;
MODEL SALES = ADV/DW;
```

requests that the Durbin–Watson statistic be printed. SAS also prints out an estimate of the first-order autocorrelation statistic.

Variable Transformations

See Variable Transformations in the Using the Computer section in Chapter 5.

Testing for Normality

In SAS, a normal probability plot for the variable STRES can be produced using the commands:

```
PROC UNIVARIATE PLOT NORMAL;
VAR STRES;
```

A variety of descriptive statistics on the variable STRES are printed along with a stem-and-leaf plot, a box plot, and a normal probability plot. Four tests for normality are performed: Shapiro–Wilk (essentially the same test as the Ryan–Joiner test produced in MINITAB), Kolmogorov–Smirnov, Cramer–von Mises, and Anderson–Darling. In all cases, a p value corresponding to the observed statistic is computed and printed.

Prais–Winsten Transformation

The Prais–Winsten transformation to correct for first-order autocorrelation was discussed earlier in this chapter.

```
PROC AUTOREG;
MODEL SALES = ADV/NLAG = 1 ITER;
```

These commands request the Prais–Winsten transformation through the PROC AUTOREG command. The dependent variable here is SALES and the explanatory variable is ADV (more than one explanatory variable can be included if necessary). The NLAG = 1 option tells the procedure that first-order autocorrelation is to be corrected (AUTOREG is a general procedure that can correct for higher-order autocorrelation not discussed in this text). The ITER option requests an iterative procedure, which has in general been shown superior in various studies.

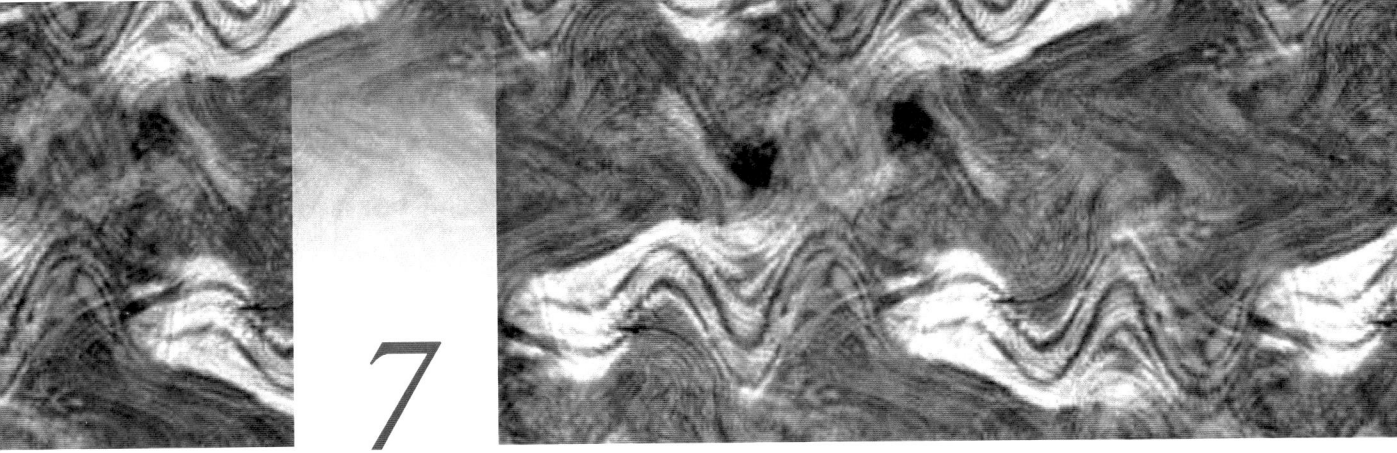

7 Using Indicator and Interaction Variables

7.1 USING AND INTERPRETING INDICATOR VARIABLES

Indicator variables or *dummy variables* are a special type of variable used in a variety of ways in regression analysis. Indicator variables take on only two values, either 0 or 1. They can be used to indicate whether a sample unit either does (1) or does not (0) belong in a certain category. For example, a dummy variable could be used to indicate when an individual in the sample was employed by constructing the variable as

D_{1i} = 1 if individual i is employed
  = 0 if individual i is not employed

To indicate when an individual is unemployed, the variable could be constructed as

D_{2i} = 1 if individual i is unemployed
  = 0 if individual i is not unemployed

Obviously, any type of split into two groups can be easily represented by indicator variables.

If there are more than two groups into which individuals may be classified, this simply requires the use of additional indicator variables. Suppose firms in a sample are to be categorized according to the exchange on which they are listed: NYSE,

AMEX, or NASDAQ. This could be accomplished by constructing the following variables:

D_{1i} = 1 if firm i is listed on the NYSE
    = 0 if firm i is not listed on the NYSE
D_{2i} = 1 if firm i is listed on the AMEX
    = 0 if firm i is not listed on the AMEX
D_{3i} = 1 if firm i is listed on the NASDAQ
    = 0 if firm i is not listed on the NASDAQ

Thus far, when sample individuals could belong to one of m different groups, m indicator variables were constructed, one for each group. When indicator variables are used in a regression analysis, however, only $m - 1$ of the indicator variables are included in the regression because only $m - 1$ indicator variables are needed to indicate m groups. The one group whose indicator is omitted serves as what might be called a *base-level* group. Consider the following example to clarify this point.

EXAMPLE 7.1 **Employment Discrimination** Regression analysis has been used extensively in employment discrimination cases. The desire in such cases is typically to compare mean salaries of two groups of employees (say, male and female employees) to determine whether one group has significantly lower salaries than the other group. Evidence of lower average salaries can provide some support for a discrimination suit against the employer. It is recognized that a simple two-sample comparison of mean salaries is not sufficient to conclude that one group has been discriminated against. Obviously, there are many factors other than discrimination that affect salary to which differences in average salary might be attributed. Regression is used to adjust for the effects of these other factors before the two groups are compared. An indicator variable is added to the regression to separate the employees into two groups: male and female.

Table 7.1 presents data from the case of *United States Department of the Treasury* v. *Harris Trust and Savings Bank* (1981). The data, contained in the file HARRIS7 on the CD, include the salary of 93 employees of the bank (SALARY), their educational level (EDUCAT), and an indicator variable (MALES) signifying whether the employee is male (1) or female (0). Figure 7.1 shows a scatterplot of salary versus education for all 93 employees. Figure 7.2 shows the same plot but with the two groups indicated by different symbols (male = ■ and female = ●). There is some indication from the plot in Figure 7.2 that male salaries are higher than female salaries, even when differing education levels have been taken into account. To obtain a better sense of the magnitude of these differences and to provide a test for whether the differences are significant or whether they are small enough that they could have occurred by chance, the regression of SALARY on the two explanatory variables EDUCAT and MALES is shown in Figure 7.3. The resulting equation is:

$$\text{SALARY} = 4173 + 80.7\text{EDUCAT} + 692\text{MALES}$$

The next question is: How do we interpret this equation?

TABLE 7.1 Data for Employment Discrimination Example

SALARY	EDUCAT	MALES	SALARY	EDUCAT	MALES	SALARY	EDUCAT	MALES
3900	12	0	5220	12	0	5040	15	1
4020	10	0	5280	8	0	5100	12	1
4290	12	0	5280	8	0	5100	12	1
4380	8	0	5280	12	0	5220	12	1
4380	8	0	5400	8	0	5400	12	1
4380	12	0	5400	8	0	5400	12	1
4380	12	0	5400	12	0	5400	12	1
4380	12	0	5400	12	0	5400	15	1
4440	15	0	5400	12	0	5400	15	1
4500	8	0	5400	12	0	5700	15	1
4500	12	0	5400	12	0	6000	8	1
4620	12	0	5400	12	0	6000	12	1
4800	8	0	5400	15	0	6000	12	1
4800	12	0	5400	15	0	6000	12	1
4800	12	0	5400	15	0	6000	12	1
4800	12	0	5400	15	0	6000	12	1
4800	12	0	5520	12	0	6000	12	1
4800	12	0	5520	12	0	6000	15	1
4800	12	0	5580	12	0	6000	15	1
4800	12	0	5640	12	0	6000	15	1
4800	12	0	5700	12	0	6000	15	1
4800	16	0	5700	12	0	6000	15	1
4980	8	0	5700	15	0	6000	16	1
5100	8	0	5700	15	0	6300	15	1
5100	12	0	5700	15	0	6600	15	1
5100	12	0	6000	12	0	6600	15	1
5100	15	0	6000	15	0	6600	15	1
5100	15	0	6120	12	0	6840	15	1
5100	16	0	6300	12	0	6900	12	1
5160	12	0	6300	15	0	6900	15	1
5220	8	0	4620	12	1	8100	16	1

Source: These data were obtained from D. Schafer, "Measurement-Error Diagnostics and the Sex Discrimination Problem," *Journal of Business and Economic Statistics*. Copyright 1987 by the American Statistical Association. Reprinted with permission. All rights reserved.

Consider a regression such as the one in Example 7.1 with one quantitative explanatory variable (x_1) and one indicator variable (D)

$$y = \beta_0 + \beta_1 x_1 + \beta_2 D$$

The indicator variable D is coded as 1 if an item in the sample belongs to a certain group and as 0 if the item does not belong to the group. The equation can be separated into two parts as follows. If the sample item is in the indicated group ($D = 1$):

$$y = \beta_0 + \beta_1 x_1 + \beta_2(1) = (\beta_0 + \beta_2) + \beta_1 x_1$$

If the sample item is not in the indicated group ($D = 0$):

$$y = \beta_0 + \beta_1 x_1 + \beta_2(0) = \beta_0 + \beta_1 x_1$$

FIGURE 7.1
Scatterplot of Salary versus Education for Employment Discrimination Example.

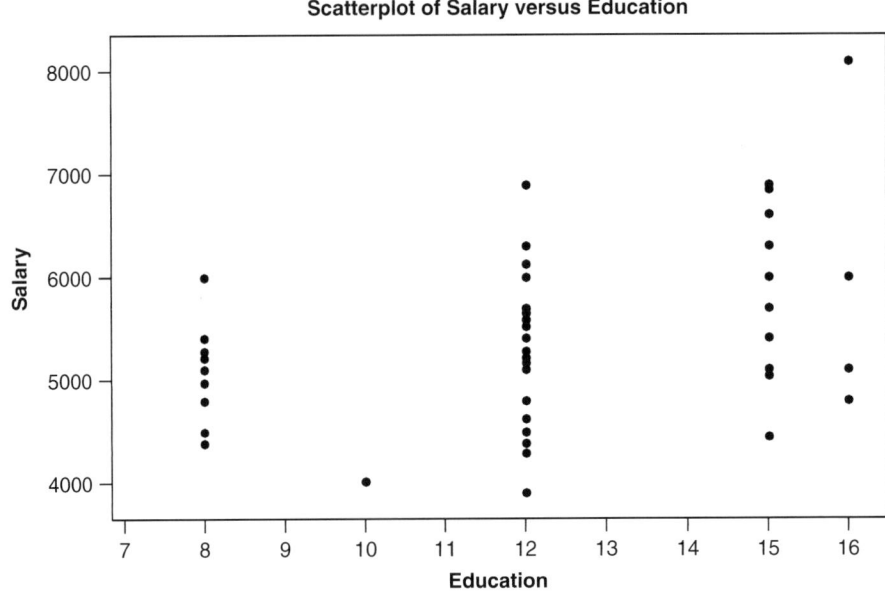

FIGURE 7.2
Scatterplot of Salary versus Education with Males (■) and Females (●) Indicated.

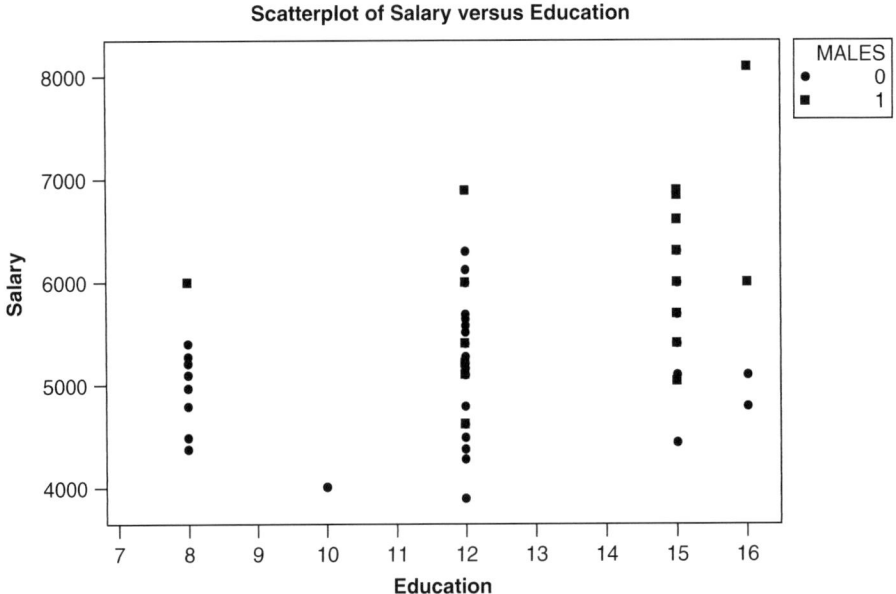

Using the indicator variable results in one equation for the indicated group and another for the other group. Note that the equation for the indicated group has been rewritten with two components making up the intercept term: the original intercept, β_0, and the coefficient of the indicator variable, β_2. Two lines have been fit with the same slope but different intercepts, even though only one

FIGURE 7.3
Regression of SALARY on EDUCAT and MALES for Employment Discrimination Example.

```
Variable      Coefficient    Std Dev    T Stat    P Value

Intercept       4173.1        339.2     12.30     0.000
EDUCAT            80.7         27.7      2.92     0.004
MALES            691.8        132.2      5.23     0.000

Standard Error = 572.437    R-Sq = 36.3%   R-Sq(adj) = 34.9%

Analysis of Variance
Source         DF    Sum of Squares    Mean Square    F Stat    P Value

Regression      2        16831744         8415872      25.68     0.000
Error          90        29491546          327684
Total          92        46323290
```

FIGURE 7.4 Graph Showing Relative Placement of Regression Lines If β_2 Is Assumed to Be Positive.

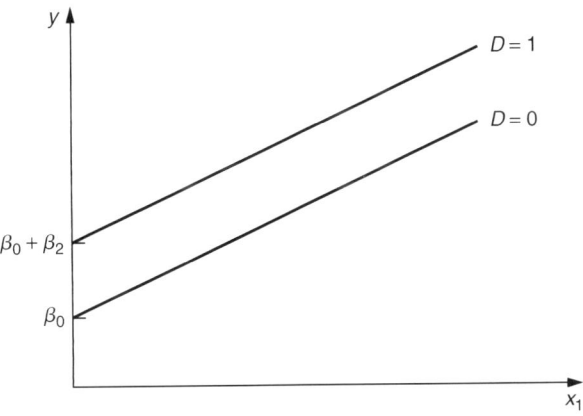

regression has been run. Figure 7.4 shows an example of how we might draw the two estimated lines. The difference in the intercepts is given by β_2 (which is assumed to be positive in the graph). This process allows us to answer the question of whether there is a difference in the average value of the y variable for the two groups after adjusting for the effect of the quantitative variable (or variables if there is more than one) and how much the average difference is. The adjusted difference in the averages is represented by the difference in the intercepts (that is, by the coefficient of the indicator variable). A t test on this coefficient will help decide whether the difference is large enough to be considered statistically significant.

EXAMPLE 7.2 **Employment Discrimination (continued)** The equation in the employment discrimination example can be interpreted as follows: Salary increases by $80.70 for each year of education. Males earn, on average, $692 more per year than females (this was in 1977, so $692 a year was more than it sounds like now).

The next question is whether $692 is large enough to be considered statistically significant. Or could this difference have occurred purely by chance? To determine this, the following hypotheses will be tested.

$$H_0: \beta_2 \leq 0$$
$$H_a: \beta_2 > 0$$

If the null hypothesis is not rejected, then there is not enough evidence to conclude that the average salary for males is higher (after taking into account education). If the null hypothesis is rejected, we conclude that the difference observed is too large to have occurred by chance. There is evidence that the average salary for males is higher, even after taking into account the effect of education. The test performed here is a one-tailed test because of the question we are asking: Is there evidence that the average salary for males is higher?

Decision rule: Reject H_0 if $t > 1.645$

Do not reject H_0 if $t \leq 1.645$

Because there are a large number of degrees of freedom (90) we use the z value, 1.645, as the critical value.

Test Statistic: t value = 5.23

Decision: Reject H_0

Our conclusion is that males, on average, appear to be earning significantly more than females, even after we take into account education. There is evidence of employment discrimination. Legal counsel for Harris Trust and Savings Bank in this case would point out that education is not the only factor that could cause a difference in salary. Factors like previous experience and time on the job might also have an effect. These factors, if measurable, could be included in the regression as well (and typically are in such applications).[1]

One concern that arises is whether the way the two groups were coded matters. Would the results be the same in our example if males were coded 0 and females were coded 1? Yes, they would. Figure 7.5 shows the regression of SALARY on EDUCAT and a new variable called FEMALES. This new variable is coded 0 for males and 1 for females. Note that the coefficient of EDUCAT is exactly the same as it was for the regression in Figure 7.3. The coefficient of the indicator variable is the same except for the sign. In the original regression, it was positive and now it is negative. The intercept of the new equation is larger by $692. However, the interpretation of the results is exactly the same. When the variable MALES was used, our interpretation was that males earned, on average, $692 per year *more* than females. Using the FEMALES variable, our interpretation is that females earned, on average, $692 per year less than males. Also, the t test again tells us that the difference is statistically significant.

---

[1] For a more detailed examination of the use of regression in employment discrimination cases, see D. A. Conway and H. V. Roberts, "Regression Analyses in Employment Discrimination Cases," in *Statistics and the Law*, New York: Wiley, 1986.

FIGURE 7.5
Regression of SALARY on EDUCAT and FEMALES for Employment Discrimination Example.

Variable	Coefficient	Std Dev	T Stat	P Value
Intercept	4864.9	387.9	12.54	0.000
EDUCAT	80.7	27.7	2.92	0.004
FEMALES	-691.8	132.2	-5.23	0.000

Standard Error = 572.437   R-Sq = 36.3%   R-Sq(adj) = 34.9%

Analysis of Variance

Source	DF	Sum of Squares	Mean Square	F Stat	P Value
Regression	2	16831744	8415872	25.68	0.000
Error	90	29491546	327684		
Total	92	46323290			

In the employment discrimination example, an indicator variable was used to indicate when sample observations fell into one of two groups. The groups represented qualitative variables (that is, variables that could not be quantified in a meaningful way). For example, there is no meaningful way to quantify the difference between males and females. An indicator variable was used to separate the two groups for separate analyses, but no numeric values were assigned to express the difference between being male or female. Indicator variables can be used in this manner to incorporate qualitative information into the regression equation. The next example illustrates the use of indicator variables to represent qualitative information when there are more than two categories.

EXAMPLE 7.3   **Meddicorp** In the Meddicorp example in Chapter 4, the relationship between the dependent variable sales (SALES) and two explanatory variables, advertising (ADV) and bonus (BONUS), was examined. It also was noted in the example that Meddicorp sold in three different regions of the country: South, West, and Midwest. A variable has been added to the data set to indicate each of these three regions. The variable is called REGION and is coded as follows:

REGION = 1 if South
= 2 if West
= 3 if Midwest

Data for the new variable and the original variables are shown in Table 7.2 (see the MEDDICORP7 file on the CD). The new variable is included in a regression along with the ADV and BONUS variables. The results are shown in Figure 7.6.

The estimated regression equation is

$$\hat{y} = -84 + 1.55\text{ADV} + 1.11\text{BONUS} + 119\text{REGION}$$

The coefficient of REGION indicates that as the value assigned to the region increases, so does the amount of sales. On average, there is a difference of 119 units between sales in the territories of the three regions. Each unit is $1000. Sales in

TABLE 7.2 Meddicorp Data Including the Variable REGION and Three Resulting Indicator Variables SOUTH, WEST, and MIDWEST

SALES	ADV	BONUS	REGION	SOUTH	WEST	MIDWEST
963.50	374.27	230.98	1	1	0	0
893.00	408.50	236.28	1	1	0	0
1057.25	414.31	271.57	1	1	0	0
1183.25	448.42	291.20	2	0	1	0
1419.50	517.88	282.17	3	0	0	1
1547.75	637.60	321.16	3	0	0	1
1580.00	635.72	294.32	3	0	0	1
1071.50	446.86	305.69	1	1	0	0
1078.25	489.59	238.41	1	1	0	0
1122.50	500.56	271.38	2	0	1	0
1304.75	484.18	332.64	3	0	0	1
1552.25	618.07	261.80	3	0	0	1
1040.00	453.39	235.63	1	1	0	0
1045.25	440.86	249.68	2	0	1	0
1102.25	487.79	232.99	2	0	1	0
1225.25	537.67	272.20	2	0	1	0
1508.00	612.21	266.64	3	0	0	1
1564.25	601.46	277.44	3	0	0	1
1634.75	585.10	312.35	3	0	0	1
1159.25	524.56	292.87	1	1	0	0
1202.75	535.17	268.27	2	0	1	0
1294.25	486.03	309.85	2	0	1	0
1467.50	540.17	291.03	3	0	0	1
1583.75	583.85	289.29	3	0	0	1
1124.75	499.15	272.55	2	0	1	0

FIGURE 7.6 Regression of SALES on ADV, BONUS, and REGION.

Variable	Coefficient	Std Dev	T Stat	P Value
Intercept	-84.219	177.907	-0.47	0.641
ADV	1.546	0.306	5.05	0.000
BONUS	1.106	0.573	1.93	0.067
REGION	118.899	28.687	4.14	0.000

Standard Error = 68.8881   R-Sq = 92.0%   R-Sq(adj) = 90.9%

Analysis of Variance

Source	DF	Sum of Squares	Mean Square	F Stat	P Value
Regression	3	1149317	383106	80.73	0.000
Error	21	99657	4746		
Total	24	1248974			

territories of Region 2 (West) would be $119,000 more (on average) than sales in territories of Region 1 (South), and sales in territories of Region 3 (Midwest) would be $119,000 more than sales in Region 2. Note that the difference in sales between regions is forced to be $119,000. Although this situation may not be realistic, it is required by the direct use of the variable REGION in the regression equation. A more flexible representation of the changes in average sales between regions is allowed by using indicator variables.

Note also that the coding of the variable REGION is arbitrary. The coding could be done as follows: Midwest = 1, West = 2, and South = 3. The numbers assigned to the regions are merely naming devices. They represent qualitative categories rather than quantitative data. Changing the order of the number of the categories does change the results, however, when using the REGION variable. This does not happen if indicator variables are used.

The REGION variable can be transformed into three indicator variables. An indicator variable SOUTH can be developed that indicates whether or not a territory is in the South. SOUTH is made to take on the value 1 whenever the REGION variable is 1, and SOUTH is given the value 0 if REGION is 2 or 3. An indicator variable WEST can be developed that indicates whether or not a territory is in the West. WEST takes on the value 1 whenever the REGION variable is 2 (that is, when the territory is in the West) and is given the value 0 if REGION is 1 or 3. Similarly, an indicator variable for the Midwest can be developed. The three indicator variables are defined as follows:

SOUTH = 1 if the territory is in the South
= 0 otherwise

WEST = 1 if the territory is in the West
= 0 otherwise

MIDWEST = 1 if the territory is in the Midwest
= 0 otherwise

These three indicator variables—SOUTH, WEST, and MIDWEST—indicate into which of the three mutually exclusive regions each territory falls. Table 7.2 shows the SALES, ADV, BONUS, and REGION variables and the resulting indicators. As stated previously, only two of the three indicators need to be used in the regression. The territories indicated by the third indicator serve as a base-level group.

If the MIDWEST is used as the base-level group, the regression results shown in Figure 7.7 are obtained. The regression equation is

$$\hat{y} = 435 + 1.37\text{ADV} + 0.975\text{BONUS} - 258\text{SOUTH} - 210\text{WEST}$$

The interpretation of this equation is similar to that of the equation developed in the employment discrimination example. Here, however, the territories have been separated into three groups through the use of the indicator variables. The coefficient of each indicator variable represents the difference in the intercept between the indicated group and the base-level (MIDWEST) group. This can be expressed through the use of three separate equations:

SOUTH $\quad \hat{y} = 435 + 1.37\text{ADV} + 0.975\text{BONUS} - 258$
$\quad\quad\quad = 177 + 1.37\text{ADV} + 0.975\text{BONUS}$

WEST $\quad \hat{y} = 435 + 1.37\text{ADV} + 0.975\text{BONUS} - 210$
$\quad\quad\quad = 225 + 1.37\text{ADV} + 0.975\text{BONUS}$

MIDWEST $\quad \hat{y} = 435 + 1.37\text{ADV} + 0.975\text{BONUS}$

FIGURE 7.7
Regression of SALES on ADV, BONUS, and the Indicator Variables SOUTH and WEST.

Variable	Coefficient	Std Dev	T Stat	P Value
Intercept	435.099	206.234	2.11	0.048
ADV	1.368	0.262	5.22	0.000
BONUS	0.975	0.481	2.03	0.056
SOUTH	-257.892	48.413	-5.33	0.000
WEST	-209.746	37.420	-5.61	0.000

Standard Error = 57.6254   R-Sq = 94.7%   R-Sq(adj) = 93.6%

Analysis of Variance

Source	DF	Sum of Squares	Mean Square	F Stat	P Value
Regression	4	1182560	295640	89.03	0.000
Error	20	66414	3321		
Total	24	1248974			

The slopes of the equations are constrained to be the same, but the intercepts are allowed to differ. As a further interpretation, consider the values of $\hat{y}$ for a given level of ADV and BONUS—say, ADV = 500 and BONUS = 250. These values are

SOUTH	$\hat{y} = 177 + 1.37(500) + 0.975(250) = 1105.8$
WEST	$\hat{y} = 225 + 1.37(500) + 0.975(250) = 1153.8$
MIDWEST	$\hat{y} = 435 + 1.37(500) + 0.975(250) = 1363.8$

The conditional mean sales for advertising equal to 500 and bonus payment equal to 250 are shown by these computations. The sales figures differ according to the coefficients of the indicator variables: $1,105,800 for SOUTH; $1,153,800 for WEST; $1,363,800 for MIDWEST.

To determine whether there is a significant difference in sales for territories in different regions, the following hypotheses should be tested:

$H_0: \beta_3 = \beta_4 = 0$

H_a: At least one of β_3 and β_4 is not equal to zero

The model hypothesized is

$$y = \beta_0 + \beta_1 \text{ADV} + \beta_2 \text{BONUS} + \beta_3 \text{SOUTH} + \beta_4 \text{WEST}$$

so the null hypothesis states that the coefficients of the indicator variables are both zero. If this hypothesis is not rejected, then no difference between the various regions exists, and the indicator variables can be dropped from the model. The simpler model

$$y = \beta_0 + \beta_1 \text{ADV} + \beta_2 \text{BONUS}$$

explains just as much variation in sales. This hypothesis can be tested using the partial F test discussed in Chapter 4. The full model contains the indicator variables; the reduced model does not. The test statistic to be used is computed exactly as discussed in Chapter 4:

$$F = \frac{(SSE_R - SSE_F)/(K - L)}{MSE_F} = \frac{(181{,}176 - 66{,}414)/2}{3321} = 17.3$$

using the reduced model regression results in Figure 7.8 to get SSE_R and the full model results in Figure 7.7 to get SSE_F.

The decision rule is

Reject H_0 if $F > 3.49$

Do not reject H_0 if $F \leq 3.49$

where 3.49 is the 5% F critical value with 2 numerator and 20 denominator degrees of freedom.

The null hypothesis is rejected because $17.3 > 3.49$. Thus, at least one of the coefficients of the indicator variables is not zero. This means that there are differences in average sales levels between the three regions in which Meddicorp does business.

When using indicator variables, the partial F statistic is used to test whether the variables are important as a group. The t test on individual coefficients should not be used to decide whether individual indicator variables should be retained or dropped from the equation (except when there are two groups represented and therefore only one indicator variable). The indicator variables are designed to have a particular meaning as a group. They are either all retained in the equation or all dropped from the equation as a group. Dropping individual indicators changes the meaning of the coefficients of the remaining indicators. In the Meddicorp example, each indicator

FIGURE 7.8
Regression of SALES on ADV and BONUS.

Variable	Coefficient	Std Dev	T Stat	P Value
Intercept	-516.444	189.876	-2.72	0.013
ADV	2.473	0.275	8.98	0.000
BONUS	1.856	0.716	2.59	0.017

Standard Error = 90.7485   R-Sq = 85.5%   R-Sq(adj) = 84.2%

Analysis of Variance

Source	DF	Sum of Squares	Mean Square	F Stat	P Value
Regression	2	1067797	533899	64.83	0.000
Error	22	181176	8235		
Total	24	1248974			

coefficient represents the difference in sales between the indicated group and the base level group (MIDWEST). If one of the other indicators is dropped, say, WEST, the remaining coefficients then represent the difference in sales between the indicated group and the new base-level group, which now becomes the MIDWEST and WEST regions combined. The interpretation of the coefficients is totally different due to the change in the base-level group. To answer the question of whether there is a difference in the intercepts for the groups involved, the indicators must be retained and tested as a group (although there is not universal acceptance of this point of view).

Recall that in the discrimination example, a t test was used to determine whether the intercepts for the males and females differed. Because there were only two groups and therefore only one indicator, the partial F test and the t test were equivalent. If more than two indicators are used, the partial F test is required.

Figure 7.9 shows the regression of SALES on ADV, BONUS, and the indicators WEST and MIDWEST. The base-level group is now the SOUTH region. The coefficients of the indicator variables measure the difference in sales between the base-level group and the indicated group. Although the regression coefficient estimates have different values from the previous regression (see Figure 7.7), the results are the same. For example, consider the value of $\hat{y}$ for ADV = 500 and BONUS = 250 for the SOUTH region:

$$\hat{y} = 177 + 1.37(500) + 0.975(250) = 1105.8$$

This is the same value determined from the regression using the WEST and SOUTH indicator variables. Comparisons for the other regions also show the same values regardless of which set of indicators is used. Any one of the three indicators can be omitted, and the omitted group simply serves as the base-level group. The values of the remaining indicator coefficients equal the difference between the indicated group and the chosen base-level group.

FIGURE 7.9
Regression of SALES on ADV, BONUS, and the Indicator Variables WEST and MIDWEST.

Variable	Coefficient	Std Dev	T Stat	P Value
Intercept	177.207	170.116	1.04	0.310
ADV	1.368	0.262	5.22	0.000
BONUS	0.975	0.481	2.03	0.056
WEST	48.146	32.801	1.47	0.158
MIDWEST	257.892	48.413	5.33	0.000

Standard Error = 57.6254 R-Sq = 94.7% R-Sq(adj) = 93.6%

Analysis of Variance

Source	DF	Sum of Squares	Mean Square	F Stat	P Value
Regression	4	1182560	295640	89.03	0.000
Error	20	66414	3321		
Total	24	1248974			

EXERCISES

1. **Discrimination.** Data for the following variables for 93 employees of Harris Bank Chicago in 1977 are available in a file named HARRIS7 on the CD:

    y = beginning salaries in dollars (SALARY)
    x_1 = years of schooling at the time of hire (EDUCAT)
    x_2 = number of months of previous work experience (EXPER)
    x_3 = number of months after January 1, 1969, that the individual was hired (MONTHS)
    x_4 = indicator variable coded 1 for males and 0 for females (MALES)

    (*Source*: These data were obtained from D. Schafer, "Measurement-Error Diagnostics and the Sex Discrimination Problem," *Journal of Business and Economic Statistics*. Copyright 1987 by the American Statistical Association. Used with permission. All rights reserved).

    The results for the regression of y on all four explanatory variables are shown in Figure 7.10. In this example, we are still concerned with whether there is evidence of discrimination, but we are now taking into account two other potentially important variables besides education. Use the results to answer the following questions:

    a. Conduct the F test for the overall fit of the regression. State the hypotheses to be tested, the decision rule, the test statistic, and your decision. Use a 5% level of significance.

    b. What conclusion can be drawn from the test result in part a?

    c. Is there a difference in salaries, on average, for male and female workers after accounting for the effects of the three other explanatory variables? Use a 5% level of significance to answer this question. State the hypotheses to be tested, the decision rule, the test statistic, and your decision.

    d. Is there evidence that Harris Bank discriminated against female employees?

    e. What salary would you forecast, on average, for males with 12 years education, 10 years of experience, and with time hired equal to 15? A point forecast is sufficient. What salary would you forecast, on average, for females if all other factors are equal?

2. **Graduation Rate.** The following data for 195 colleges were obtained from *Kiplinger's Personal Finance* and are available in a file named COLLEGE7 on the CD:

    y = the percentage of students who earned a bachelor's degree in four years (GRADRATE4)

FIGURE 7.10 Regression Results for Discrimination Exercise.

Variable	Coefficient	Std Dev	T Stat	P Value
Intercept	3526.422	327.725	10.76	0.000
EDUCAT	90.020	24.694	3.65	0.000
EXPER	1.269	0.588	2.16	0.034
MONTHS	23.406	5.201	4.50	0.000
MALES	722.461	117.822	6.13	0.000

Standard Error = 507.422   R-Sq = 51.1%   R-Sq(adj) = 48.9%

Analysis of Variance

Source	DF	Sum of Squares	Mean Square	F Stat	P Value
Regression	4	23665351	5916338	22.98	0.000
Error	88	22657939	257477		
Total	92	46323290			

FIGURE 7.11
Regression Results for Graduation Rate Exercise.

Variable	Coefficient	Std Dev	T Stat	P Value
Intercept	0.58944	0.04034	14.61	0.000
ADMISRATE	-0.35044	0.05759	-6.09	0.000
PRIVATE	0.28196	0.02399	11.75	0.000

Standard Error = 0.139754   R-Sq = 65.5%   R-Sq(adj) = 65.1%

Analysis of Variance

Source	DF	Sum of Squares	Mean Square	F Stat	P Value
Regression	2	7.1215	3.5608	182.31	0.000
Error	192	3.7500	0.0195		
Total	194	10.8715			

x_1 = admission rate expressed as a percentage (ADMISRATE)
x_2 = an indicator variable coded as 1 for private schools and 0 for public schools (PRIVATE)

(*Source*: Used by permission from the November and December 2003 issues of *Kiplinger's Personal Finance*. © 2003 The Kiplinger Washington Editors, Inc. Visit our website at www.kiplingers.com for further information.)

Results for the regression of y on x_1 and x_2 are shown in Figure 7.11. Use the results to answer the following questions:

a. Conduct the F test for the overall fit of the regression. State the hypotheses to be tested, the decision rule, the test statistic, and your decision. Use a 5% level of significance.

b. What conclusion can be drawn from the test result in part a?

c. Is there a difference in graduation rate, on average, for public and private schools after the effect of admission rate is taken into account? State the hypotheses to be tested, the decision rule, the test statistic, and your decision. Use a 5% level of significance to answer this question.

d. What graduation rate would you forecast, on average, for a private school with a 15% admissions rate? A point forecast is sufficient. What graduation rate would you forecast, on average, for a public school with a 15% admissions rate?

7.2 INTERACTION VARIABLES

Another type of variable that is used in regression is called an *interaction* variable. An interaction variable is formed as the product of two (or more) variables. To illustrate the effect of using an interaction variable, consider a regression equation with dependent variable y and independent variables x_1 and x_2. Construct the interaction variable x_1x_2, which is the product of the two explanatory variables. Now consider two possible regression models, one with the interaction term and one without it:

$$y = \beta_0 + \beta_1 x_1 + \beta_2 x_2 + e$$

and

$$y = \beta_0 + \beta_1 x_1 + \beta_2 x_2 + \beta_3 x_1 x_2 + e$$

Now determine the change in y given a one-unit change in x_1 with each of these models. For the model without the interaction term, a one-unit change in x_1 produces a change in y of β_1 units. For the model with the interaction term, rewrite the equation as

$$y = \beta_0 + (\beta_1 + \beta_3 x_2)x_1 + \beta_2 x_2 + e$$

Then a one-unit change in x_1 produces a change of $\beta_1 + \beta_3 x_2$ units in y. As shown, the change in y resulting from a one-unit change in x_1 also depends on the value of the variable x_2. If x_2 is small, smaller changes result; if x_2 is large, larger changes result. Thus, the effect of movements in x_1 cannot be judged independently of the value of x_2 (and the effect of movements in x_2 cannot be judged independently of movements in x_1).

An important application of interaction variables is in testing for differences in the slopes of two regression lines. This is done in a manner similar to the procedure to test for differences in the intercepts. Consider a regression with one quantitative explanatory variable (x_1) and one indicator variable (D):

$$y = \beta_0 + \beta_1 x_1 + \beta_2 D$$

The indicator variable D is coded as 1 if an item in the sample belongs to a certain group and as 0 if the item does not belong to the group. Now add the variable representing the interaction between x_1 and D, $x_1 D$:

$$y = \beta_0 + \beta_1 x_1 + \beta_2 D + \beta_3 x_1 D$$

This equation can be separated into two parts as follows. If the sample item is in the indicated group ($D = 1$):

$$y = \beta_0 + \beta_1 x_1 + \beta_2(1) + \beta_3 x_1(1) = (\beta_0 + \beta_2) + (\beta_1 + \beta_3)x_1$$

If the sample item is not in the indicated group ($D = 0$):

$$y = \beta_0 + \beta_1 x_1 + \beta_2(0) + \beta_3 x_1(0) = \beta_0 + \beta_1 x_1$$

Using the indicator variable allows us to examine differences in the intercepts for the two groups. Using the interaction variable allows us to examine differences in the slopes. Note that the equation for the indicated group has been rewritten with two components making up the intercept term (the original intercept, β_0, and the coefficient of the indicator variable, β_2), and two components making up the slope term (the original slope, β_1, and the coefficient of the interaction term, β_3). Two lines have been fit with different slopes and different intercepts, even though only one regression has been run. The difference in the intercepts is given by β_2; the difference in the slopes by β_3.

Not only can we determine whether there is a difference in the average value of the y variable for the two groups after adjusting for the effect of the quantitative variable, but we can also tell whether there is a difference in the slopes for the two groups. A t test for whether β_3 is equal to zero helps to determine whether the slopes differ. Also, a partial F test of the hypotheses

$H_0: \beta_2 = \beta_3 = 0$

H_a: At least one of β_2 and β_3 is different from zero

can be used to tell us whether there is any difference in the regression lines for the two groups (intercept or slope). The following example illustrates.

EXAMPLE 7.4 **Employment Discrimination (again)** In Examples 7.1 and 7.2, we concluded that there is evidence of employment discrimination at Harris Bank even after taking into account education. Now suppose the following question is considered: Does the difference in average salaries increase between the two groups as education increases? This is one question an interaction term allows us to investigate. The equation can be written as

$$\text{SALARY} = \beta_0 + \beta_1 \text{EDUCAT} + \beta_2 \text{MALES} + \beta_3 \text{MSLOPE}$$

where MSLOPE represents the interaction between EDUCAT and MALES. Thus,

$$\text{MSLOPE} = \text{EDUCAT} * \text{MALES}$$

The regression results for this equation are shown in Figure 7.12.

We ask the question: Is there *any* difference between the two groups (males and females)? To answer this, the following hypotheses can be tested:

$H_0: \beta_2 = \beta_3 = 0$

$H_a:$ At least one of β_2 and β_3 is different from zero

To test these hypotheses, a partial F test should be used. The reduced model has only the variable EDUCAT. This regression is shown in Figure 7.13. The F statistic is

$$F = \frac{(38{,}460{,}756 - 29{,}054{,}426)/2}{326{,}454} = 14.41$$

Using a 5% level of significance, the decision rule is

Reject H_0 if $F > 3.15$

Do not reject H_0 if $F \leq 3.15$

FIGURE 7.12
Regression Results for Discrimination Example with Interaction Variable to Represent Different Slopes.

Variable	Coefficient	Std Dev	T Stat	P Value
Intercept	4395.32	389.21	11.29	0.000
EDUCAT	62.13	31.94	1.95	0.055
MALES	-274.86	845.75	-0.32	0.746
MSLOPE	73.59	63.59	1.16	0.250

Standard Error = 571.362   R-Sq = 37.3%   R-Sq(adj) = 35.2%

Analysis of Variance

Source	DF	Sum of Squares	Mean Square	F Stat	P Value
Regression	3	17268865	5756288	17.63	0.000
Error	89	29054426	326454		
Total	92	46323290			

FIGURE 7.13
Regression Results for Reduced Model in Discrimination Example with Interaction Variable.

```
Variable      Coefficient     Std Dev      T Stat      P Value

Intercept     3818.56         377.44       10.12       0.000
EDUCAT         128.09          29.70        4.31       0.000

Standard Error = 650.112      R-Sq = 17.0%    R-Sq(adj) = 16.1%

Analysis of Variance

Source        DF     Sum of Squares     Mean Square    F Stat    P Value

Regression     1        7862534           7862534      18.60     0.000
Error         91       38460756            422646
Total         92       46323290
```

The critical value used is the $F(0.05; 2,60)$ value since the value for $F(0.05; 2,89)$ is not in the tables. The decision is to reject H_0. (We should already have guessed that this would be the decision since the test in Example 7.2 showed that the coefficient of the indicator variable was not zero).

The coefficients can be tested individually to see whether one or both are different from zero. Note the conflicting results when we do this. The p values (or t ratios) on each of the individual coefficients for MALES and MSLOPE suggest that the coefficients are equal to zero. But the F test just told us that at least one of the coefficients is different from zero. The reason for these conflicting results is the high correlation between MALES and MSLOPE (0.986). This is an example of the multicollinearity problem discussed in Chapter 4. Because we already have concluded that the indicator variable is important in the regression and the addition of the interaction variable does not add much (R^2 only increases by 1%), it might be best in this example to stick to the simpler model in Example 7.2.

Example 7.4 illustrates one caution in using interaction variables. If the correlation is high between interaction variables and the original variables in the regression, multicollinearity problems can result. In a regression with several variables, the number of interaction variables that could be created is very large and the likelihood of multicollinearity problems is high. Therefore, it is wise not to use interaction variables indiscriminately. There should be some good reason to suspect that two variables might be related or some specific question that can be answered by an interaction variable before this type of explanatory variable is used.

EXERCISES

3. **More on Possible Discrimination.** Suppose that legal counsel representing Harris Bank suggests that an interaction exists between education and experience and that the introduction of this term into the regression may account for the difference in average salaries. The interaction term

$$EDUCEXPR = EDUCAT * EXPER$$

is created and introduced into the regression. The regression results are in Figure 7.14. Use the results to help answer the following questions:

a. What is the adjusted R^2 for this regression? Compare this value to the adjusted R^2 for the regression without the interaction variable (see Figure 7.10). Which model appears to be the best choice based on the adjusted R^2?

b. Test to see whether the interaction term is important in this regression model. Use a 5% level of significance. State the hypotheses to be tested, the decision rule, the test statistic, and your decision.

c. Does the interaction variable seem to be important in explaining SALARY?

d. How would you respond to the suggestion that introduction of the interaction term into the regression may account for the difference in average salaries?

4. **Graduation Rate (continued).** Consider again the data from Exercise 7.2:

y = the percentage of students who earned a bachelor's degree in four years (GRADRATE4)

x_1 = admission rate expressed as a percentage (ADMISRATE)

x_2 = an indicator variable coded as 1 for private schools and 0 for public schools (PRIVATE)

We define the interaction variable:

ADSLOPE = ADMISRATE * PRIVATE

When ADSLOPE is included in the regression, it allows the slopes of the regression lines for public and private schools to differ. The regression model can be written

$$\text{GRADRATE4} = \beta_0 + \beta_1 \text{ ADMISRATE} + \beta_2 \text{ PRIVATE} + \beta_3 \text{ ADSLOPE} + e$$

The regression results for this model are shown in Figure 7.15. Figure 7.16 contains the results for the following model:

$$\text{GRADRATE4} = \beta_0 + \beta_1 \text{ ADMISRATE} + e$$

Use the outputs to help answer the following questions:

a. Is there *any* difference between the regression lines for public and private schools? State the hypotheses to be tested, the decision rule, the test statistic, and your decision. Use a 5% level of significance.

b. Is there a difference, on average, in graduation rates for public and private schools? State the hypotheses to be tested, the decision rule, the test statistic, and your decision. Use a 5% level of significance.

c. Is there a difference between the slopes of the regression lines representing the relationship

FIGURE 7.14 Regression Results for Discrimination Exercise Using Interaction Term Between Education and Experience.

Variable	Coefficient	Std Dev	T Stat	P Value
Intercept	3006.167	490.660	6.13	0.000
EDUCAT	134.470	39.814	3.38	0.001
EXPER	5.679	3.164	1.79	0.076
MONTHS	22.421	5.218	4.30	0.000
MALES	687.630	119.697	5.74	0.000
EDUCEXPR	-0.364	0.257	-1.42	0.160

Standard Error = 504.530   R-Sq = 52.2%   R-Sq(adj) = 49.4%

Analysis of Variance

Source	DF	Sum of Squares	Mean Square	F Stat	P Value
Regression	5	24177362	4835472	19.00	0.000
Error	87	22145928	254551		
Total	92	46323290			

FIGURE 7.15
Regression Results for Graduation Rate Interaction Variable Exercise.

Variable	Coefficient	Std Dev	T Stat	P Value
Intercept	0.63554	0.06200	10.25	0.000
ADMISRATE	-0.42084	0.09213	-4.57	0.000
PRIVATE	0.21666	0.07088	3.06	0.003
ADSLOPE	0.11560	0.11800	0.98	0.329

Standard Error = 0.139769   R-Sq = 65.7%   R-Sq(adj) = 65.1%

Analysis of Variance

Source	DF	Sum of Squares	Mean Square	F Stat	P Value
Regression	3	7.1403	2.3801	121.83	0.000
Error	191	3.7313	0.0195		
Total	194	10.8715			

FIGURE 7.16
Regression Results for Reduced Model for Graduation Rate Interaction Variable Exercise.

Variable	Coefficient	Std Dev	T Stat	P Value
Intercept	0.93429	0.03620	25.81	0.000
ADMISRATE	-0.72332	0.06285	-11.51	0.000

Standard Error = 0.182775   R-Sq = 40.7%   R-Sq(adj) = 40.4%

Analysis of Variance

Source	DF	Sum of Squares	Mean Square	F Stat	P Value
Regression	1	4.4241	4.4241	132.43	0.000
Error	193	6.4475	0.0334		
Total	194	10.8715			

between GRADRATE and ADMISRATE for public and private schools? State the hypotheses to be tested, the decision rule, the test statistic, and your decision. Use a 5% level of significance.

7.3 SEASONAL EFFECTS IN TIME-SERIES REGRESSION

Seasonal effects are fairly regular patterns of movement in a time series, repeating within a 1-year period. For example, sales of swimsuits are expected to be higher in spring and summer months and lower in fall and winter. Although the influence of seasonal effects is not expected to be exactly the same every year, the same general pattern is expected to persist.

Seasonal patterns can be modeled in a regression equation by using indicator variables, which can be created to indicate the time period to which each observation belongs. For example, if quarterly data are being analyzed, the following indicator variables could be created:

$Q1 = 1$ if the observation is from the first quarter of any year
$ = 0$ otherwise
$Q2 = 1$ if the observation is from the second quarter of any year
$ = 0$ otherwise
$Q3 = 1$ if the observation is from the third quarter of any year
$ = 0$ otherwise
$Q4 = 1$ if the observation is from the fourth quarter of any year
$ = 0$ otherwise

Four indicator variables have been created to denote the quarter to which each observation belongs, but only three of the variables should be included in the regression equation. The excluded quarter simply serves as a base-level quarter from which changes in the seasonal levels are measured. The interpretation of the coefficients of the indicator variables is the same as for any set of indicator variables. To illustrate, consider the following estimated regression equation with only quarterly indicator variables:

$$\hat{y} = b_0 + b_1 Q1 + b_2 Q2 + b_3 Q3$$

Note that the fourth quarter serves as the base-level quarter. Predictions of observations in the fourth quarter would be computed using the equation

$$\hat{y} = b_0$$

because $Q1 = Q2 = Q3 = 0$. This point has been located on the graph in Figure 7.17.

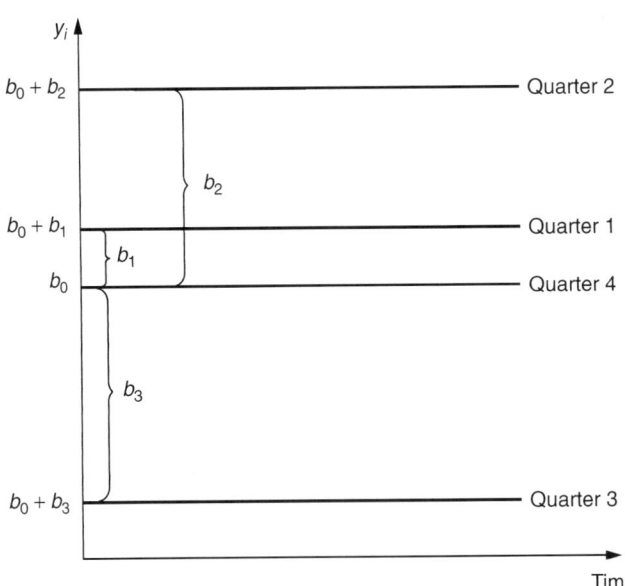

FIGURE 7.17 Graph Showing Placement of Seasonal Levels and Interpretation of Seasonal Indicator Coefficients.

Predictions of observations in the first quarter would be computed using the equation

$$\hat{y} = b_0 + b_1$$

because $Q1 = 1$ and $Q2 = Q3 = 0$.

Similarly, predictions of observations in the second quarter would be computed using the equation

$$\hat{y} = b_0 + b_2$$

and in the third quarter by

$$\hat{y} = b_0 + b_3$$

The coefficients b_1, b_2, and b_3 represent the differences between the fitted values for the indicated quarter (first, second, and third, respectively) and the base-level quarter. The graph in Figure 7.17 has been constructed with the assumption that b_1 and b_2 are positive and b_3 is negative. The lines shown for each quarter have a zero slope because no term has been included in the regression except the indicator variables. Differences in the overall level of the series in different quarters are shown by the quarterly indicators. The regression coefficients estimate the differences in the mean levels of y in the indicated quarters. Obviously, variables could be used in the regression in addition to the seasonal indicator variables. This is illustrated in the following example.

EXAMPLE 7.5 **ABX Company Sales** Consider again the ABX Company from Example 3.11 (ABXSALES7 on the CD). The ABX Company sells winter sports merchandise including skis, ice skates, sleds, and so on. Quarterly sales in thousands of dollars for the ABX Company are shown in Table 7.3. The time period represented starts in the first quarter of 1994 and ends in the fourth quarter of 2003.

A time-series plot of sales is shown in Figure 7.18. In Chapter 3, it was decided that a strong linear trend in sales appeared in this plot. The regression with the linear trend variable was estimated, and the results are shown again in Figure 7.19. The residual plot of the standardized residuals versus the fitted values is shown in Figure 7.20. From this plot, no obvious violations of assumptions can be observed.

A time-series plot of the standardized residuals is shown in Figure 7.21. Here a clear pattern emerges. The residuals tend to be higher in the first and fourth quarters and lower in the second and third quarters. This is not unexpected, because the ABX Company sells winter sports merchandise. This pattern is the result of seasonal variation in the data that has not been accounted for in the regression. Note that the pattern can also be seen in the time-series plot of the original data in Figure 7.18. Again, the residual plot emphasizes the pattern and points out some systematic variation in the data that we should try to model with our regression.

Figure 7.22 shows the regression of sales on the linear trend variable and indicator variables designed to indicate the first, second, and third quarters. The

TABLE 7.3 Data for ABX Company Sales Example Showing Seasonal Indicator

SALES	TREND	QUARTER	Q1	Q2	Q3	Q4
221.0	1	1	1	0	0	0
203.5	2	2	0	1	0	0
190.0	3	3	0	0	1	0
225.5	4	4	0	0	0	1
223.0	5	1	1	0	0	0
190.0	6	2	0	1	0	0
206.0	7	3	0	0	1	0
226.5	8	4	0	0	0	1
236.0	9	1	1	0	0	0
214.0	10	2	0	1	0	0
210.5	11	3	0	0	1	0
237.0	12	4	0	0	0	1
245.5	13	1	1	0	0	0
201.0	14	2	0	1	0	0
230.0	15	3	0	0	1	0
254.5	16	4	0	0	0	1
257.0	17	1	1	0	0	0
238.0	18	2	0	1	0	0
228.0	19	3	0	0	1	0
255.0	20	4	0	0	0	1
260.5	21	1	1	0	0	0
244.0	22	2	0	1	0	0
256.0	23	3	0	0	1	0
276.5	24	4	0	0	0	1
291.0	25	1	1	0	0	0
255.5	26	2	0	1	0	0
244.0	27	3	0	0	1	0
291.0	28	4	0	0	0	1
296.0	29	1	1	0	0	0
260.0	30	2	0	1	0	0
271.5	31	3	0	0	1	0
299.5	32	4	0	0	0	1
297.0	33	1	1	0	0	0
271.0	34	2	0	1	0	0
270.0	35	3	0	0	1	0
300.0	36	4	0	0	0	1
306.5	37	1	1	0	0	0
283.5	38	2	0	1	0	0
283.5	39	3	0	0	1	0
307.5	40	4	0	0	0	1

original sales data, the trend variable, a variable representing the number of the quarter (1 through 4) labeled QUARTER, and four indicator variables, labeled $Q1$, $Q2$, $Q3$, and $Q4$ are shown in Table 7.3. Remember that only three of the four indicator variables should be used in the regression. (Also, note that the variable QUARTER is not used in the regression; it is merely in the table to indicate the number of each quarter.)

7.3 Seasonal Effects in Time-Series Regression 295

FIGURE 7.18 Time-Series Plot of ABX Sales.

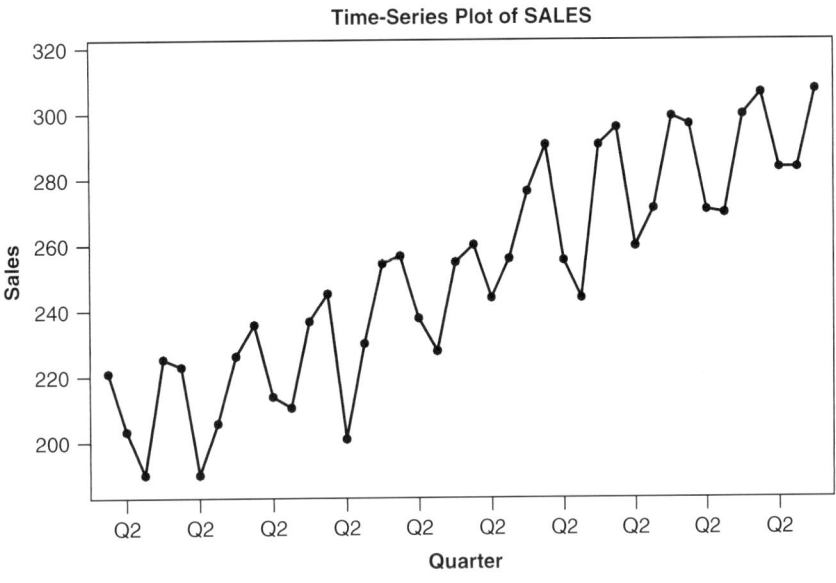

FIGURE 7.19 Regression of Sales on Linear Trend Variable for ABX Sales Example.

```
Variable        Coefficient      Std Dev        T Stat       P Value
Constant         199.017         5.128          38.81        0.000
TREND              2.556         0.218          11.73        0.000

Standard Error = 15.9126       R-Sq = 78.3%    R-Sq(adj) = 77.8%

Analysis of Variance

Source         DF     Sum of Squares    Mean Square    F Stat    P Value
Regression      1         34818            34818       137.50    0.000
Error          38          9622              253
Total          39         44440
```

FIGURE 7.20 Plot of Standardized Residuals versus Fitted Values for ABX Sales Example.

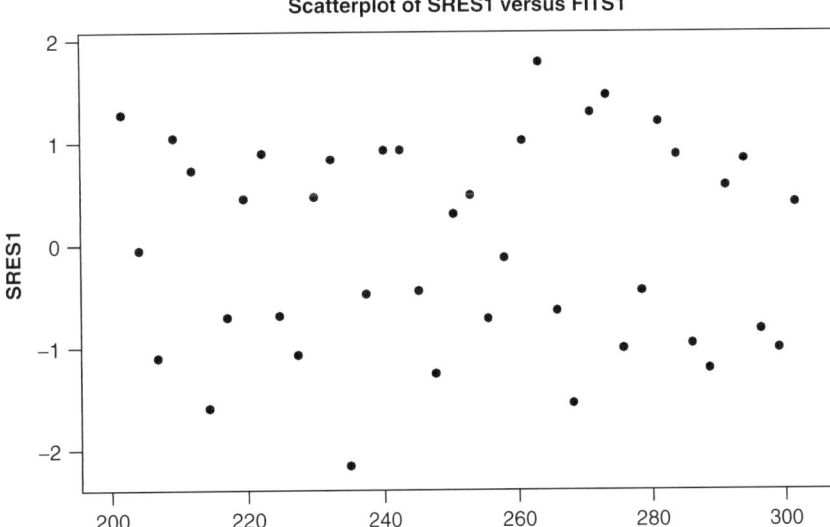

FIGURE 7.21
Time-Series Plot of Standardized Residuals for ABX Sales Example.

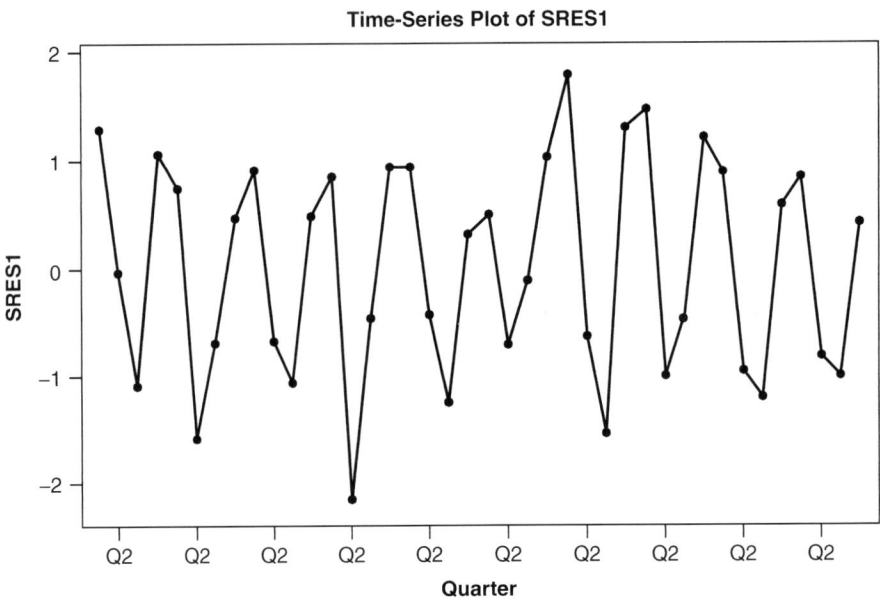

FIGURE 7.22
Regression of Sales on Linear Trend Variable and Quarterly Indicator Variables for ABX Sales Example.

```
Variable        Coefficient     Std Dev     T Stat      P Value

Intercept       210.846         3.148       66.98       0.000
TREND             2.566         0.099       25.93       0.000
Q1                3.748         3.229        1.16       0.254
Q2              -26.118         3.222       -8.11       0.000
Q3              -25.784         3.217       -8.01       0.000

Standard Error = 7.19028      R-Sq = 95.9%   R-Sq(adj) = 95.5%

Analysis of Variance

Source          DF      Sum of Squares      Mean Square     F Stat      P Value

Regression       4           42630              10658        206.14      0.000
Error           35            1810                 52
Total           39           44440
```

The residual plot of the standardized residuals versus the fitted values for the regression with the quarterly indicators is shown in Figure 7.23. A time-series plot of the standardized residuals is shown in Figure 7.24. No further violations of any assumptions appear in either of these plots.

An important question to be asked in seasonal time-series models is, "Are there seasonal differences in the level of the dependent variable?" Another way of asking

FIGURE 7.23 Plot of Standardized Residuals versus Fitted Values for ABX Sales Example with Quarterly Indicator Variables.

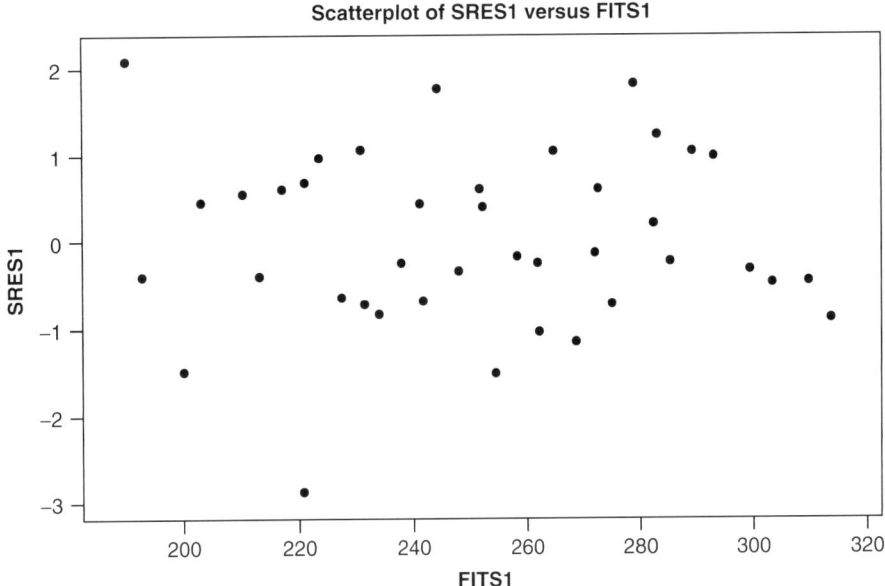

FIGURE 7.24 Time-Series Plot of Standardized Residuals for ABX Sales Example with Quarterly Indicator Variables.

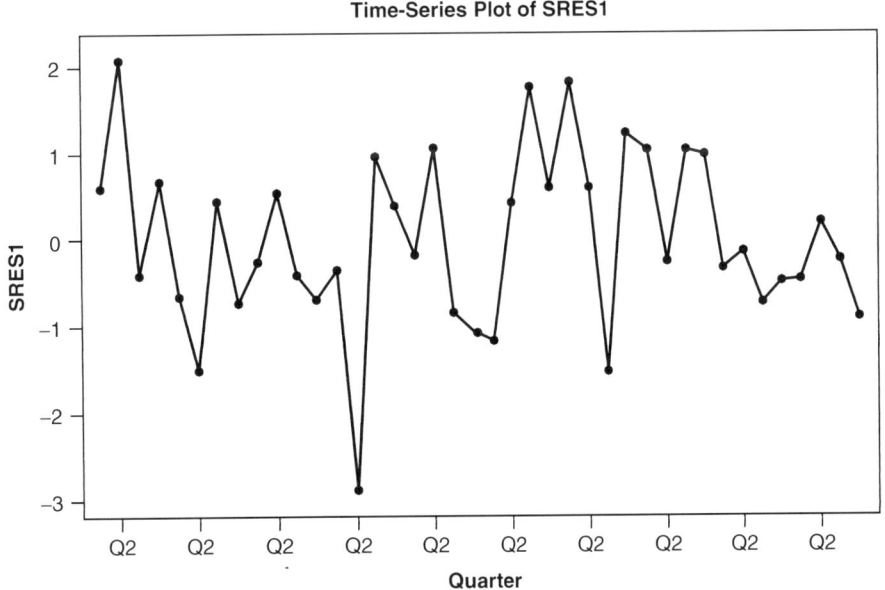

this question is, "Are the seasonal indicator variables necessary in the model?" The partial F test discussed in Chapter 4 can be used to answer the questions.

Consider the following model with explanatory variable x and quarterly seasonal indicators $Q1$, $Q2$, and $Q3$:

$$y = \beta_0 + \beta_1 x + \beta_2 Q1 + \beta_3 Q2 + \beta_4 Q3 + e$$

This is referred to as the full model. The hypotheses to be tested are

H_0: $\beta_2 = \beta_3 = \beta_4 = 0$

H_a: At least one of the coefficients β_2, β_3, and β_4 is not zero

If the null hypothesis is not rejected, the seasonal indicator variables add nothing to the model and can be removed. In other words, there are no seasonal differences. In this case, the following reduced model is adopted:

$$y = \beta_0 + \beta_1 x + e$$

If the null hypothesis is rejected, then there are seasonal differences, and the quarterly indicators should remain in the model.

To conduct the test, the partial F test statistic is computed as

$$F = \frac{(SSE_R - SSE_F)/(K - L)}{MSE_F}$$

where SSE_R is the error sum of squares from the reduced model, SSE_F is the error sum of squares from the full model, and MSE_F is the mean square error from the full model. Because the hypothesis test determines whether three coefficients are equal to zero, a divisor of 3 is used in the numerator in place of $K - L$ ($K = 4$, $L = 1$). The decision rule for the test is

Reject H_0 if $F > F(\alpha; K - L, n - K - 1)$
Do not reject H_0 if $F \le F(\alpha; K - L, n - K - 1)$

EXAMPLE 7.6 **ABX Company Sales (continued)** In Example 7.5, the following model was examined for ABX Company sales:

$$\text{SALES} = \beta_0 + \beta_1 \text{ TREND} + \beta_2 Q1 + \beta_3 Q2 + \beta_4 Q3 + e$$

To determine whether there are seasonal components affecting sales, the following hypotheses should be tested:

H_0: $\beta_2 = \beta_3 = \beta_4 = 0$

H_a: At least one of the coefficients β_2, β_3, and β_4 is not zero

To test this hypothesis, the partial F test is used. The test statistic is

$$F = \frac{(9622 - 1810)/3}{52} = 50.1$$

Note that the reduced model regression is in Figure 7.19, and the full model regression is in Figure 7.22. Using a 5% level of significance, the decision rule for the test is

Reject H_0 if $F > F(0.05; 3.35) = 2.92$ (approximately)
Do not reject H_0 if $F \le 2.92$

The null hypothesis is rejected. Thus, the conclusion is that seasonal components do affect sales and should be taken into account, as was done in Example 7.5.

When computing forecasts with seasonal models, the coefficients of the seasonal indicators are used to adjust the level of the forecast in the appropriate time periods. The quarterly forecasts for the year 2004 are as follows, using the seasonal model with trend:

Time Period	Point Forecast
2004 Q1	211 + 2.57(41) + 3.75 = 320.12
2004 Q2	211 + 2.57(42) − 26.1 = 292.84
2004 Q3	211 + 2.57(43) − 25.8 = 295.71
2004 Q4	211 + 2.57(44) = 324.08

Throughout this section, the use of quarterly indicator variables has been discussed. If monthly instead of quarterly data are used, the applications are similar. Instead of four quarterly indicators, twelve monthly indicators are created. Eleven of the twelve monthly indicators are used in the regression with the estimated coefficients interpreted in a manner similar to those of the quarterly coefficients. Tests for seasonal variation involve the set of eleven indicator variable coefficients. Since the use of monthly indicators is so similar to quarterly indicators, the demonstration of their use is reserved for the exercises.

EXERCISES

5. Furniture Sales. The file named FURNSALE7 on the CD contains data on monthly furniture sales (in millions of dollars) for the United States from January 1992 through December 2002. These data were obtained from the web site www.econo-magic.com. Figure 7.25 shows the time-series plot of the data. An extrapolative model to forecast furniture sales for each month in 2003 was developed. Figure 7.26 shows the regression results for the model. The model can be written as follows:

SALES = $\beta_0 + \beta_1$ TREND + β_2 LAGSALES
+ β_3 JAN + β_4 FEB + β_5 MARCH
+ β_6 APRIL + β_7 MAY + β_8 JUNE
+ β_9 JULY + β_{10} AUG + β_{11} SEPT
+ β_{12} OCT + β_{13} NOV + e

FIGURE 7.25
Time-Series Plot of Furniture Sales.

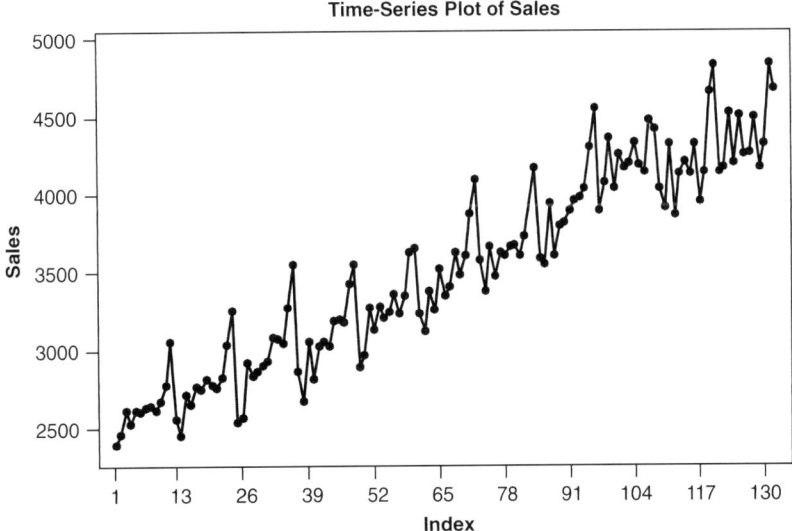

FIGURE 7.26
Regression Results with Seasonal Indicators for Furniture Sales Exercise.

Variable	Coefficient	Std Dev	T Stat	P Value
Intercept	1552.151	228.680	6.79	0.000
TREND	7.903	1.298	6.09	0.000
LAGSALES	0.484	0.083	5.84	0.000
JAN	−643.717	42.576	−15.12	0.000
FEB	−404.481	53.915	−7.50	0.000
MARCH	−80.369	56.921	−1.41	0.161
APRIL	−457.714	42.852	−10.68	0.000
MAY	−178.202	53.127	−3.35	0.001
JUNE	−316.796	45.155	−7.02	0.000
JULY	−278.410	47.612	−5.85	0.000
AUG	−183.319	47.143	−3.89	0.000
SEPT	−357.504	43.251	−8.27	0.000
OCT	−239.408	48.319	−4.95	0.000
NOV	6.826	45.791	0.15	0.882

Standard Error = 93.2329   R-Sq = 98.0%   R-Sq(adj) = 97.8%

Analysis of Variance

Source	DF	Sum of Squares	Mean Square	F Stat	P Value
Regression	13	49839825	3833833	441.06	0.000
Error	117	1017007	8692		
Total	130	50856832			

FIGURE 7.27
Regression Results with Seasonal Indicators for Furniture Sales Exercise.

Variable	Coefficient	Std Dev	T Stat	P Value
Intercept	1977.945	215.778	9.17	0.000
TREND	12.483	1.427	8.75	0.000
LAGSALES	0.201	0.087	2.32	0.022

Standard Error = 195.734   R-Sq = 90.4%   R-Sq(adj) = 90.2%

Analysis of Variance

Source	DF	Sum of Squares	Mean Square	F Stat	P Value
Regression	2	45952938	22976469	599.72	0.000
Error	128	4903894	38312		
Total	130	50856832			

TREND is a linear trend variable, LAGSALES is SALES lagged one period, and JAN through NOV are 11 monthly seasonal indicators. Note that December has been used as the base-level month.

Figure 7.27 shows the regression results for the model without the seasonal indicators.

Use the regression results to help answer the following questions.

a. Is there seasonal variation in furniture sales? State the hypotheses to be tested, your decision rule, the test statistic, decision, and conclusion. Use a 5% level of significance.

b. Are the TREND and LAGSALES variables important to the equation explaining furniture sales? Test the coefficient of each of these variables individually to answer this question. State the hypotheses to be tested, your decision rule, the test statistic, decision, and conclusion. Use a 5% level of significance.

c. Based on the test result in part a, use either the full or reduced model (with or without seasonal indicators) to develop a forecast of furniture sales for each month in 2003.

ADDITIONAL EXERCISES

6. **Absenteeism.** Data on 77 employees of the ABX Company have been collected. The dependent variable is absenteeism (ABSENT). The possible explanatory variables are

   COMPLX = measure of job complexity
   SENIOR = seniority
   SATIS = response to "How satisfied are you with your foreman?"

   These data are available in a file named ABSENT7 on the CD. In this exercise, use SENINV = 1/SENIOR, which is the reciprocal of the seniority variable, and COMPLX as two of the explanatory variables. The variable SATIS should be transformed into indicator variables as follows:

   FS1 = 1 if SATIS = 1 (very dissatisfied)
       = 0 otherwise
   FS2 = 1 if SATIS = 2 (somewhat dissatisfied)
       = 0 otherwise
   FS3 = 1 if SATIS = 3 (neither satisfied nor dissatisfied)
       = 0 otherwise
   FS4 = 1 if SATIS = 4 (somewhat satisfied)
       = 0 otherwise
   FS5 = 1 if SATIS = 5 (very satisfied)
       = 0 otherwise

   Five indicator variables are created to represent all five supervisor satisfaction categories. Recall that only four need to be used in the regression. Run the regression with the explanatory variables described here. Answer the following questions:

   a. Is there a difference in average absenteeism for employees in different supervisor satisfaction groups? Perform a hypothesis test to answer this question. State the hypotheses to be tested, the decision rule, the test statistic, and your decision. Use a 5% level of significance.

   b. Using the model chosen in part a (and keeping the variables COMPLX and SENINV in the model), what would be your estimate of the average absenteeism rate for all employees with COMPLX = 60 and SENIOR = 30 who were very dissatisfied with their supervisor? What if they were very satisfied with their supervisor, but COMPLX and SENIOR were the same values?

   c. How do you account for the differences in the estimates in part b?

   d. How could this equation be used to help identify employees who might be prone to absenteeism?

7. **Work-Order Closing.** Management at the Texas Christian University (TCU) Physical Plant is interested in reducing the average time to completion of routine work orders. The time to completion is defined as the difference between the date of receipt of a work order and the date closing information is entered. The number of labor hours charged to each work order and the cost of materials are two variables believed to be related to the time to completion of the work order. Management wants to know if there is any difference in the time to completion of work orders, on average, for different types of buildings. Buildings are classified into four types on the TCU campus: residence halls, athletic, academic, and administrative. In answering the question, take into account the possible effect of labor hours charged and materials cost. The data for a random sample of 72 work orders (chosen from a population of 11,720) are available in a file named WORKORD7 on the CD. The variables are as follows:

   y = DAYS = number of days to complete each work order
   x_1 = HOURS = number of hours of labor charged to each work order
   x_2 = MATERIAL = cost of materials charged to each work order
   x_3 = BUILDING = 1 for residence halls
                     2 for athletic buildings
                     3 for academic buildings
                     4 for administrative buildings

8. **Beer Production.** The file named BEER7 on the CD contains monthly U.S. beer production in millions of barrels for January 1982 through December 1991. Develop an extrapolative model for these data and use it to examine whether there is seasonal variation in beer production and whether beer production seems to be increasing, decreasing, or staying fairly constant over this time period. Use the model you select as best for

beer production to forecast monthly production for each month in 1992.

(*Source*: Data were obtained from *Business Statistics 1963–91* and *Survey of Current Business*.)

9. **Monthly Temperatures.** Lone Star Gas recognizes that one of the simplest and most effective ways to forecast natural gas use is with average monthly temperatures. A model can be developed that relates gas usage to average temperature, and then forecasts can be made based on forecasts of average temperatures in the area of interest. Lone Star first needs a model to forecast average temperatures in the Dallas–Fort Worth area. The file named TEMPDFW7 on the CD contains the average monthly temperatures for January 1978 to December 2002 for the Dallas–Fort Worth area. Develop an extrapolative model for these data and use the model to forecast average monthly temperatures for each month in the years 2003 and 2004. Data were obtained from the web site www.srh.noaa.gov/fwd/clmdfw.html.

10. **BigTex Services.** BigTex Services is undergoing scrutiny for a possible wage discrimination suit. As consulting statistician hired by the corporate lawyer, you are to examine data on BigTex employees to further investigate the charges. The following data are available in a file named BIGTEX7 on the CD:

>monthly salary for each employee (SALARY)
>years with the company (YEARS)
>position with the company (POSITION)
>>coded as
>>>1 = manual labor
>>>2 = secretary
>>>3 = lab technician
>>>4 = chemist
>>>5 = management
>
>amount of education completed (EDUCAT)
>>coded as
>>>1 = high school degree
>>>2 = some college
>>>3 = college degree
>>>4 = graduate degree
>
>gender (GENDER) coded as
>>>0 = female
>>>1 = male

What would you conclude from the data? Should BigTex Services be worried about possible wage discrimination charges? Why or why not?

11. **FOC Sales.** Techcore is a high-tech company located in Fort Worth, Texas. The company produces a part called a fibre-optic connector (FOC) and wants to generate reasonably accurate but simple forecasts of the sales of FOC's over time. The company has weekly sales data for 265 weeks starting in January of 1995. The weekly sales data are in a file named FOC7 on the CD. You are to build an extrapolative regression model to forecast FOC sales. Do you detect any "seasonal" patterns where the use of indicator variables might be useful?

12. **Fort Worth Crime.** In the early 1990s, a concerted effort was put into place to revitalize downtown Fort Worth, Texas. This was a combined effort involving both the city and private investors. The city police and private security coordinated their communications networks. New entertainment and dining establishments were located in the area. Buildings were refurbished and remodeled. Prior to this time, the downtown area had experienced high crime rates. Because of police and private security efforts beginning in the summer of 1992, there is interest in determining whether these interventions had a significant effect in decreasing crime in downtown Fort Worth. You are to use an extrapolative regression model to examine the data and to produce a report for the city of Fort Worth outlining your findings regarding any changes in the downtown crime rate. The report should be written in two parts: an executive summary outlining your findings and a technical report to support those findings.

The data are in a file named FWCRIME7 on the CD. The variables included are YR/MONTH, which gives the year and month of the observation and the number of monthly occurrences of crimes in the categories shown: MURDER, RAPE, ROBBERY, ASSAULT, BURGLARY, LARCENY and AUTOTHEFT.

13. **Fund Returns.** An investment company has two different strategies that it follows resulting in hedge funds that will be called LONG/SHORT and EQUITY. The equity fund mirrors the S&P 500 whereas the LONG/SHORT strategy is intended to provide returns that are relatively uncorrelated with the market. Daily returns for these two strategies are available from January 3, 2002, through July 16, 2003. On November 1, 2002, a

change in the investment strategies is made for both of these funds. Management wants to know if the beta coefficients of these two funds changed as a result. The beta coefficient is the slope of the regression line relating the fund return to the return on the market, in this case, the S&P 500. The beta coefficient represents the systematic risk of the fund, which means that the beta indicates the amount of change in the fund return, on average, that will occur when the market return changes by 1%. If the beta coefficient is 2 then a 1% increase (decrease) in the market return will be mirrored, on average, by a 2% increase (decrease) in the fund return. The file FUND7 on the CD contains the following variables:

RET1, daily returns for the EQUITY strategy
RET2, daily returns for the LONG/SHORT strategy
SP500, daily returns for the S&P 500

Use the data to estimate the beta coefficient for the two funds and to determine whether the change in investment strategy resulted in a shift in the beta coefficient. (These data have been disguised at the request of the company to provide confidentiality.)

14. **Productivity.** A company that owns and operates 12 hospitals is interested in developing a measure of staff productivity. It would like to use a measure called EEOB, "equivalent employees per occupied bed." EEOB represents how many full-time employee equivalents (40 hours per week) are used for each patient. Corporate headquarters is set on using EEOB to measure productivity. It knows that average length of stay (ALOS) has an inverse impact on EEOB. In the last two years, the length of stay in hospitals has decreased drastically. As a result, headquarters believes that it can expect the EEOB to increase. Another factor that may influence productivity is outpatient factor (OUTFAC), which takes into account time spent on various outpatient processes.

You have been hired to investigate the relationships between EEOB and the two explanatory variables ALOS and OUTFAC. You have monthly data from January 2002 through May 2003 for 12 different hospitals. The variable labeled HOSP numbers each hospital 1 through 12. The relationships may differ between hospitals. You might want to use indicator variables in your regression to allow for these differences.

Examine the relationships using the data in the file HOSP7 on the CD. What conclusions can you make based on the results? Be sure to examine residual plots. Do the assumptions of the regression appear satisfied? Based on the residual plots, what questions might you have about the data? (These data have been disguised at the request of the company to provide confidentiality.)

15. **Predicting Movie Grosses.** The file named MOVIES7 on the CD contains data on movies released in the United States during the calendar year 1998. The variables in this file are

TDOMGROSS	the total domestic gross revenue (dependent variable)
BACTOR	the number of "best actors" in the movie; to qualify as a best actor, the person must be listed in *Entertainment Weekly*'s lists of the 25 Best Actors and 25 Best Actresses of the 1990s
TDACTOR	the number of actors or actresses appearing in the movie who were among the top 20 actors and top 20 actresses in average box office gross per movie in their careers ("top dollar actors"), according to The Movie Times web site at the beginning of the 1998 movie season
GENRE	classification of movie types coded as 1 = action, 2 = drama, 3 = children's, 4 = comedy, 5 = documentary, 6 = thriller, 7 = horror, 8 = science fiction
MPAA	MPAA rating coded as 1 = G, 2 = PG, 3 = PG13, 4 = R, 5 = NC17, 6 = unrated
COUNTRY	where the move was made coded as 1 = USA, 2 = English-speaking country other than USA, 3 = non-English speaking country
CHRISTMAS	coded as 1 if the movie was to be released during the

HOLIDAY	Christmas season, 0 otherwise
	coded as 1 if the movie was released before any holiday weekend (President's Day, Memorial Day, Independence Day, Labor Day, Thanksgiving, or the Christmas season), 0 otherwise
SUMMER	coded as 1 if the movie was released during the summer (Memorial Day through Labor Day), 0 otherwise
SEQUEL	coded as 1 if the movie was a sequel, 0 otherwise

We would like to find an equation to predict the total domestic gross revenue of movies based on variables available prior to the release of the films. All of the variables in the data file are known or could be determined prior to release. Use regression to examine the relationships and determine what factors might be useful in helping to predict gross revenues. Use the log of TDOMGROSS as the dependent variable. Some of the explanatory variables are qualitative in nature and will need to be transformed into indicators. What relationships do you find that would be helpful in predicting revenues? What variables are not important? What recommendations would you make on the basis of your results?

These data are discussed in the article "Predicting Movie Grosses: Winners and Losers, Blockbusters and Sleepers," by Jeffrey S. Simonoff and Ilana R. Sparrow, *Chance*, Vol. 13, 2000, 15–24 and were obtained from Dr. Simonoff's web site.

16. **Rangers' Attendance.** The Texas Rangers major league baseball team needs your help. The team is interested in determining key identifiable variables that affect attendance at games played at the Ballpark in Arlington. It has provided you with an extensive data set, including the following variables:

DATE:	The date of the game played
WEEKDAY:	The day of the week, coded as 1 = Monday, 2 = Tuesday,..., 7 = Sunday
PROMOTION:	Equal to 1 if there was a special promotion (Dollar Decker Dog Night, Half Price Group Night, and so on); equal to 0 if there was no special promotion
WINS:	The cumulative number of wins for the Rangers prior to playing the game
AHEAD:	Equal to 1 if the Rangers were ahead in their division when the game was played; equal to zero otherwise
ATTENDANCE:	The attendance at all home games for the years 1994 through 1998; this provides a total of 375 observations
SCHOOL:	Equal to 1 if public schools were in session; equal to zero otherwise
OPPONENT:	Coded as 1 = New York 2 = Baltimore 3 = Cleveland 4 = Chicago 5 = Anaheim 6 = Minnesota 7 = Any National League team (coded this way since the teams played vary from year to year) 8 = Oakland 9 = Detroit 10 = Seattle 11 = Toronto 12 = Milwaukee 13 = Kansas City 14 = Boston
NIGHT:	Equal to 1 for a night game; 0 for a day game

These data are in a file named RANGERS7 on the CD.

Prepare a report summarizing the data and providing any information concerning what you see as factors that influence attendance. Recognize that the Rangers organization has little or no control over certain factors. It cannot, for example, schedule all the games with the Yankees, even if playing the Yankees produces higher attendance levels than any other team. Still, the team would be interested in finding out whether the opponent has an influence on attendance for planning purposes.

On the other hand, the Rangers do have control over some factors. If you find any controllable factors that might be of interest, be sure to single those out. Your report should consist of two parts: (1) an executive summary outlining your findings—any recommendations should be based on the results shown by the data; (2) a technical section including any analyses you need to support your position.

Data used courtesy of the Texas Rangers organization and Major League Baseball. Major League Baseball trademarks and copyrights are used with permission of Major League Baseball Properties, Inc.

17. **Book Cost.** A major publishing company would like to develop an equation that will help it in determining the cost of books that it publishes. It has a sample of 207 books that have been published recently. Of the 207 books in the sample, 83 are hardcover and 124 are softcover. Hardcover books obviously are priced at a premium, so some adjustment for this will need to be made. The variables in the data file named BOOKCOST7 on the CD are as follows:

COST	= the cost of producing the book
PAGES	= the number of pages in the book
SOFTCOVER	= 1 if the book is softcover and 0 if it is hardcover

Develop an equation that will help the publisher to predict cost for books to be published in the future.

USING THE COMPUTER

The Using the Computer section in each chapter describes how to perform the computer analyses in the chapter using Excel, MINITAB, and SAS. For further detail on Excel, MINITAB, and SAS, see Appendix C.

EXCEL

Creating and Using Indicator Variables

Indicator variables have values of either 0 or 1. Obviously, one way to create an indicator variable (or variables) is simply to type it into a data set. Logical if statements can be used to simplify the creation of indicators in some cases.

Consider the data in Table 7.2 for the Meddicorp example. The variable REGION was typed into the original data set. This variable numbers the three regions as 1, 2, or 3. It can be used to create the indicator variables for the three regions after they are defined as follows:

SOUTH	= 1 whenever REGION is 1 and 0 otherwise
WEST	= 1 whenever REGION is 2 and 0 otherwise
MIDWEST	= 1 whenever REGION is 3 and 0 otherwise

Suppose the REGION variable is in column D, as shown in Figure 7.28. The indicator SOUTH can be created as follows. In the first cell of the column where the SOUTH indicator is desired, type in the formula = if (d2 = 1,1,0). This formula puts 1 in the SOUTH column if the entry in cell d2 is 1; it puts 0 in the SOUTH column otherwise. Copy this formula down the SOUTH column. Now type the formula = if (d2 = 2,1,0) in the first cell of the WEST column. This formula puts 1 in the WEST column if the entry in cell d2 is 2; it puts 0 in the WEST column otherwise. Copy this formula down the WEST column. Now type the formula = if (d2 = 3,1,0) in the first cell of the MIDWEST column. This formula puts 1

306 Using Indicator and Interaction Variables

FIGURE 7.28 Creating Indicator Variables in Excel.

	A	B	C	D	E	F	G
1	SALES	ADV	BONUS	REGION	SOUTH	WEST	MIDWEST
2	963.50	374.27	230.98	1	=IF(D2=1,1,0)		
3	893.00	408.50	236.28	1			
4	1057.25	414.31	271.57	1			
5	1183.25	448.42	291.20	2			
6	1419.50	517.88	282.17	3			
7	1547.75	637.60	321.16	3			
8	1580.00	635.72	294.32	3			
9	1071.50	446.86	305.69	1			
10	1078.25	489.59	238.41	1			
11	1122.50	500.56	271.38	2			
12	1304.75	484.18	332.64	3			
13	1552.25	618.07	261.80	3			
14	1040.00	453.39	235.63	1			
15	1045.25	440.86	249.68	2			

FIGURE 7.29 Creating Interaction Variables in Excel.

	A	B	C	D
1	SALES	ADV	BONUS	INTERACT
2	963.50	374.27	230.98	=B2*C2
3	893.00	408.50	236.28	
4	1057.25	414.31	271.57	
5	1183.25	448.42	291.20	
6	1419.50	517.88	282.17	
7	1547.75	637.60	321.16	
8	1580.00	635.72	294.32	
9	1071.50	446.86	305.69	
10	1078.25	489.59	238.41	
11	1122.50	500.56	271.38	
12	1304.75	484.18	332.64	
13	1552.25	618.07	261.80	
14	1040.00	453.39	235.63	
15	1045.25	440.86	249.68	

in the MIDWEST column if the entry in cell d2 is a 3; it puts 0 in the MIDWEST column otherwise. Copy this formula down the MIDWEST column. This creates the three indicator variables.

Creating and Using Interaction Variables

Interaction variables are created by multiplying one variable by another. This can be done in Excel by using formulas. Consider the example shown in Figure 7.29. To create an interaction variable between ADV and BONUS, type the formula= $b2*c2$ into the first cell of the interaction variable column. Then copy this formula down the column.

Creating and Using Seasonal Indicators for Time-Series Regression

To create quarterly seasonal indicators, follow this sequence: First, type in a variable that numbers the quarters 1 through 4 for each year. Then use the logical if statement to create the indicators. Consider the example shown in Figure 7.30. The column labeled "quarter" numbers the quarters of each year. In the column labeled Q1, type in the formula = if($c2 = 1,1,0$). If cell c2 is 1, this results in 1 in the column for Q1. Otherwise, 0 is entered. Copy this formula down the column for Q1. In the column labeled Q2, type in the formula = if($c2 = 2,1,0$). If cell c2 is 2, this results in 1 in the column for Q2. Otherwise, 0 is entered. Copy this formula down the column for Q2. Continue this process for each indicator variable. You can alter this process for monthly data by numbering the months from 1 to12 and then creating 12 monthly indicators.

FIGURE 7.30 Creating Quarterly Seasonal Indicator Variables in Excel.

	A	B	C	D	E	F	G
1	SALES	TREND	QUARTER	Q1	Q2	Q3	Q4
2	221.00	1	1	=IF(C2=1,1,0)			
3	203.50	2	2				
4	190.00	3	3				
5	225.50	4	4				
6	223.00	5	1				
7	190.00	6	2				
8	206.00	7	3				
9	226.50	8	4				
10	236.00	9	1				
11	214.00	10	2				
12	210.50	11	3				
13	237.00	12	4				
14	245.50	13	1				
15	201.00	14	2				

In cell D2: IF(logical_test, [value_if_true], [value_if_false])

FIGURE 7.31 Make Indicator Variables Dialog Box in MINITAB.

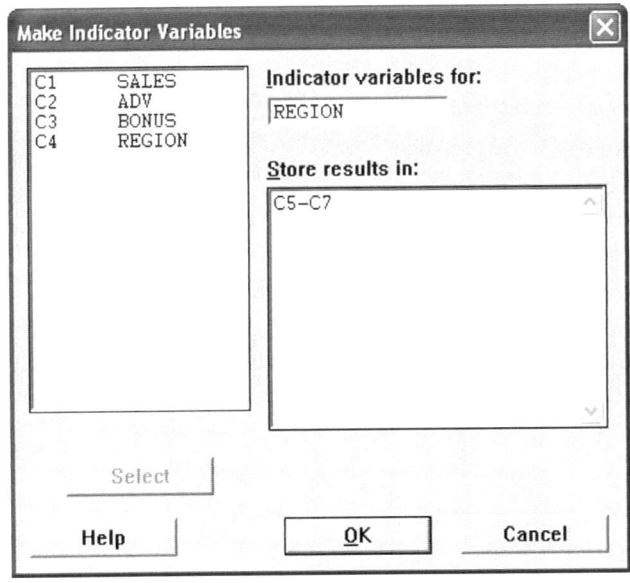

MINITAB

Creating and Using Indicator Variables

CALC: MAKE INDICATOR VARIABLES

Indicator variables have values of either 0 or 1. Obviously, one way to create an indicator variable (or variables) is simply to type it into a data set. Make Indicator Variables on the CALC menu can be useful in creating indicator variables more quickly in some cases.

Consider the data in Table 7.2 for the Meddicorp example. The variable REGION was typed into the original data set. This variable numbers the three regions as 1, 2, or 3. It can be used to create the indicator variables for the three regions after they are defined as follows:

SOUTH   = 1 whenever REGION is 1 and 0 otherwise
WEST   = 1 whenever REGION is 2 and 0 otherwise
MIDWEST = 1 whenever REGION is 3 and 0 otherwise

To create the indicator variables SOUTH, WEST, and MIDWEST, click CALC and then Make Indicator Variables. The Make Indicator Variables dialog box is shown in Figure 7.31. Fill in the "Indicator variables for" box with the name of the variable to be transformed into indicators (REGION in this case). Then list the locations of the indicator variables in the "Store results in" box. Be sure to list locations for all m indicators (m being the number of categories) even though only $m - 1$ indicators are used in the regression.

Creating and Using Interaction Variables

CALC: CALCULATOR

Interaction variables are created by multiplying one variable by another. This can be done in MINITAB by using the CALCULATOR on the CALC menu. Put the

FIGURE 7.32 Using the Calculator to Create Interaction Variables in MINITAB.

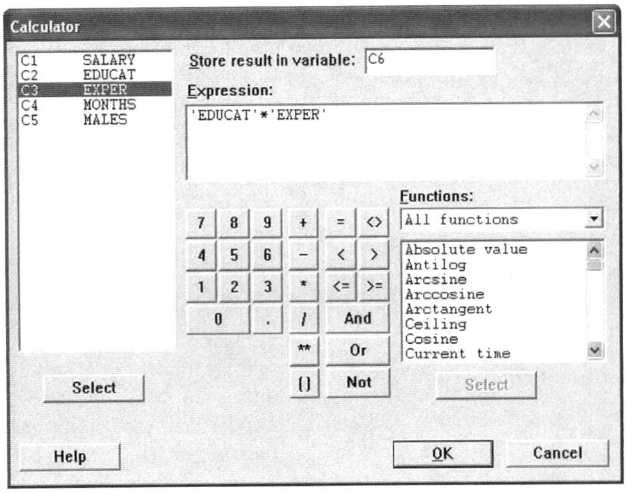

FIGURE 7.33 Step One in Creating Seasonal Indicators in MINITAB.

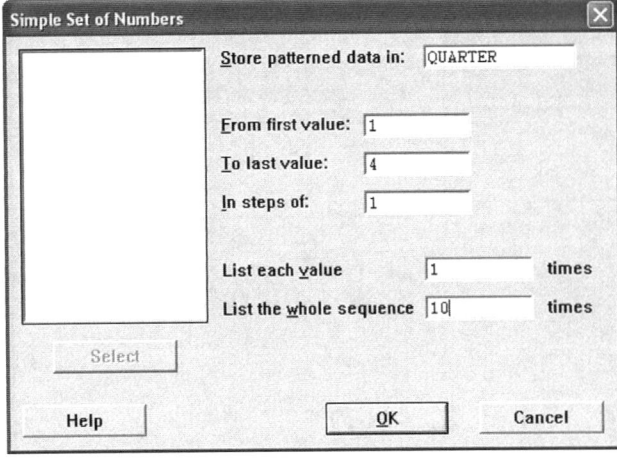

interaction variable column number in the "Store result in variable" box and put the action to be taken in the "Expression" box (see Figure 7.32). For example, put C6 in the "Store result in variable" box and 'EDUCAT'*'EXPER' in the "Expression" box and MINITAB multiplies C2 by C3 and stores the resulting interaction variable in C6. The interaction variable can then be used just like other variables.

Creating and Using Seasonal Indicators for Time-Series Regression

To create quarterly seasonal indicators, follow this sequence: First, create a variable that numbers each quarter 1 through 4: CALC: MAKE PATTERNED DATA: SIMPLE SET OF NUMBERS. In the Simple Set of Numbers dialog box, fill in a column number in the "Store patterned data in" box, put 1 in the "From first value" box, 4 in the "To last value" box, and 1s in the "In steps of" and "List each value" boxes. In the "List the whole sequence" box, put the number of years of data you have available. This creates a variable with the sequence of numbers 1,2,3,4 for each year of data, thus numbering the quarters in each year 1 through 4 (see Figure 7.33). Now click CALC: MAKE INDICATOR VARIABLES. Put the column number of the variable you just created in the "Indicator variables for" box and list four columns in the "Store results in" box. Click OK and the seasonal indicators are created. You can alter this process for monthly data by numbering the months from 1 to 12 and then creating 12 monthly indicators.

SAS

Creating and Using Indicator Variables

Indicator variables have values of either 0 or 1. Obviously, one way to create an indicator variable (or variables) is simply to type it into a data set. IF/THEN statements can be used to simplify the creation of indicator variables in some cases.

Consider the data in Table 7.2 for the Meddicorp example. The variable REGION was typed into the original data set. This variable numbers the three regions as 1, 2, or 3. It can be used to create the indicator variables for the three regions after they are defined as follows:

SOUTH   = 1 whenever REGION is 1 and 0 otherwise
WEST   = 1 whenever REGION is 2 and 0 otherwise
MIDWEST = 1 whenever REGION is 3 and 0 otherwise

The three indicators can be created during the Input phase in SAS as follows:

```
INPUT SALES ADV BONUS REGION;
IF REGION = 1 THEN SOUTH = 1;
ELSE SOUTH = 0;
IF REGION = 2 THEN WEST = 1;
ELSE WEST = 0;
IF REGION = 3 THEN MIDWEST = 1;
ELSE MIDWEST = 0;
```

After creating the three indicator variables, any two of them can be used in PROC REG. For example,

```
PROC REG;
MODEL SALES = ADV BONUS SOUTH WEST;
```

runs a regression with SALES as the dependent variable and ADV, BONUS, SOUTH, and WEST as independent variables.

Creating and Using Interaction Variables

Interaction variables are created by multiplying one variable by another. In SAS, such variable transformations are performed during the data input phase. For example, the following commands create the interaction variable called EDUCEXPR, which is the product of EDUCAT and EXPER:

```
INPUT SALARY EDUCAT EXPER MONTHS MALES;
EDUCEXPR = EDUCAT*EXPER;
```

The interaction variable, EDUCEXPR, can then be used in PROC REG, PROC PLOT, and so on, just like other variables.

Creating and Using Seasonal Indicators for Time-Series Regression

Seasonal indicators are simply indicator variables, and their creation is like that of indicator variables discussed previously. The only difference is that, instead of the variable representing REGION in the example given, a variable numbering the quarters or the months in the seasonal cycle is needed.

For quarterly data, a variable numbering the quarters 1, 2, 3, or 4 is used with the following command sequence in the input phase. Call the variable that numbers the quarters QUARTER and proceed as follows:

```
IF QUARTER = 1 THEN Q1 = 1;
ELSE Q1 = 0;
IF QUARTER = 2 THEN Q2 = 1;
ELSE Q2 = 0;
IF QUARTER = 3 THEN Q3 = 1;
ELSE Q3 = 0;
IF QUARTER = 4 THEN Q4 = 1;
ELSE Q4 = 0;
```

Q1, Q2, Q3, and Q4 are the four quarterly indicator variables. Three of these four can then be used in PROC REG.

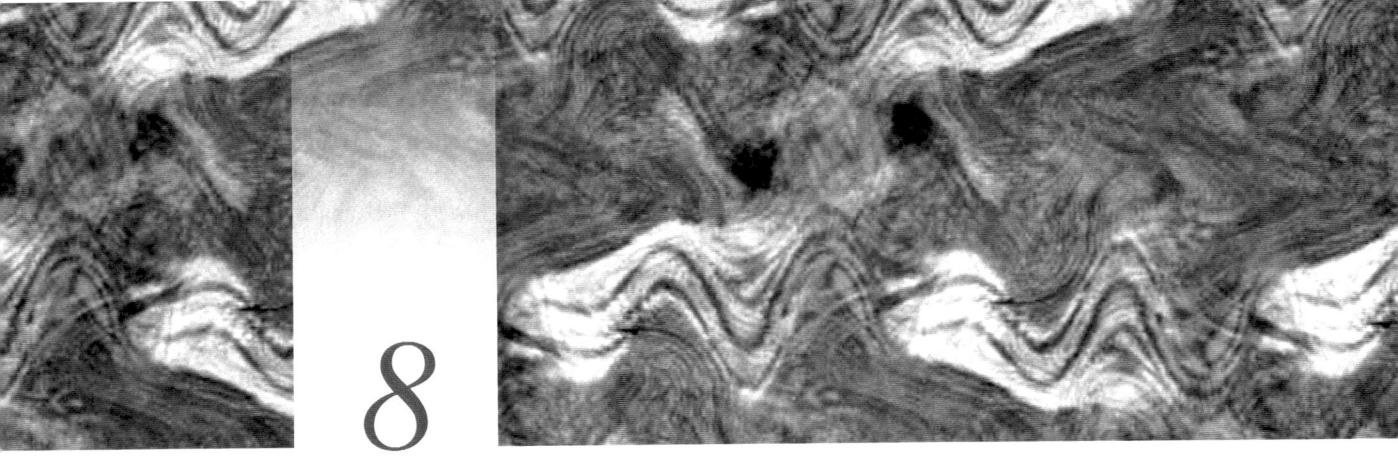

8 Variable Selection

8.1 INTRODUCTION

One of the primary tasks discussed in this text has been choosing which variables to include in the regression equation. Several hypothesis tests have been suggested to aid in this task. The F test for overall fit of the regression, the partial F test, and the t test are all designed to help decide whether certain variables should be included in the regression.

Several additional procedures, called *variable selection techniques*, also can be used to help choose which variables are important. These procedures are discussed in this chapter. The importance of choosing the correct variables is highlighted by examining what happens when either (a) important variables are omitted from the regression equation or (b) unimportant variables are included in the equation.

If an important variable is omitted from the regression, the effect of this variable is not taken into account. The estimates of the other regression coefficients become biased (systematically either too high or too low). Forecasts generated by the regression are also biased. If an unimportant variable is included in the regression equation, the standard errors of the coefficients and forecasts become inflated. Thus, forecasts and coefficient estimates are more variable than they would be with a proper choice of variables. The larger standard errors may also make results from inferences less dependable.

Whether identifying important variables to be included or unimportant variables to be deleted, variable choice is an important aspect of regression analysis. As a result, considerable work has been done in developing methods to help choose the "best" group of variables to include in the regression equation. A word of caution is in order before these techniques are introduced, however. None of these methods is guaranteed to automatically pick the best of all regression models. Otherwise, this text could have started with these procedures and stopped after one chapter.

Any regression analysis requires considerable input from the researcher performing the analysis. This person must define the problem to be solved, determine what variables might be useful in the regression equation, obtain data on these variables, and set up the data to be analyzed in a data file. During the analysis itself, the researcher must check the regression assumptions to ensure that none has been violated. The automatic methods do not do that. For example, if $\ln(y)$ rather than y is the appropriate dependent variable for analysis, the variable selection procedures cannot determine this. The functional form of the relationship must be determined separately from the use of these procedures. In fact, any correction for a violation of one of the assumptions of regression must be determined by the researcher. These methods also cannot suggest explanatory variables to add to the regression unless they have been initially determined by the researcher and included in the data set.

Finally, none of these automatic procedures has the ability to use the researcher's knowledge of the business or economic situation being analyzed. This knowledge should be used to help establish what form the regression model takes, what variables should be considered, and so on. This knowledge of the theory associated with the subject being analyzed is important. In many cases in business and economics, theoretical results are available that suggest what variables should be included in a relationship or what functional form the relationship should have. This theory will be provided throughout other courses in your business or economics major. Be sure to examine the theory in that field for suggestions that may help in building a model for your data.

The variable selection techniques discussed here are tools to aid the researcher in sifting through a number of explanatory variables to determine which ones should be included in the regression equation. With the knowledge that the techniques cannot be reliably applied without the judgment of the person researching the problem and the use of subject matter theory, the following variable selection procedures are discussed in this chapter:

1. all possible regressions (along with several criteria to choose which is the best regression)
2. backward elimination
3. forward selection
4. stepwise regression

8.2 ALL POSSIBLE REGRESSIONS

One of the variable selection techniques suggested to aid in choosing the best regression model is called *all possible regressions*. As the name suggests, the procedure is designed to run all possible regressions between the dependent variable and all possible subsets of explanatory variables. For example, if the three possible explanatory variables identified for consideration in the problem are denoted x_1, x_2, and x_3, then a total of eight possible regressions may be best. The possible regressions include the following subsets of the three explanatory variables:

1. no variables
2. x_1
3. x_2
4. x_3
5. x_1 and x_2
6. x_1 and x_3
7. x_2 and x_3
8. all three variables

The all possible regressions procedure evaluates each of these regressions and prints out summary statistics to aid in choosing which of the eight possibilities is best. The choice of the criterion to use and the final choice of a model are then up to the researcher. Commonly used criteria to help in choosing between the alternative regressions include:

1. R^2, adjusted or unadjusted, as the researcher prefers
2. C_p
3. s_e, the standard error of the regression

The R^2, R^2_{adj} and s_e have been discussed extensively in previous chapters, but the statistic C_p has not. C_p measures the total mean square error of the fitted values of the regression. The total mean square error involves two components: one resulting from random error and one resulting from bias. When there is no bias in the estimated regression model, the expected value of C_p is equal to p, which, in the notation of this text, is equal to $K + 1$, the number of coefficients to be estimated. When evaluating which regression is best, it is recommended that regressions with small C_p values and those with values near $K + 1$ be considered. If the value of C_p is large, then the mean square error of the fitted values is large, indicating either a poor fit, substantial bias in the fit, or both. If the value of C_p is much greater than $K + 1$, then there is a large bias component in the regression, usually indicating omission of an important variable.

The formula for computing C_p is

$$C_p = \frac{SSE_p}{MSE_F} - (n - 2p)$$

where SSE_p is the error sum of squares for the regression with p ($p = K + 1$) coefficients to be estimated and MSE_F is the mean square error for the model with all possible explanatory variables included.

Although the C_p measure is highly recommended as a useful criterion in choosing between alternate regressions, keep in mind that the bias is measured with respect to the total group of variables provided by the researcher. This criterion cannot determine when the researcher has forgotten about some variable not included in the total group. In other words, the input of the researcher is still important. The all possible regressions procedure is illustrated in the following example.

EXAMPLE 8.1 **Meddicorp Revisited** Consider again Example 4.1, the study of sales in the Meddicorp Company (see the MEDDICORP8 file on the CD). The complete data set for this example is shown in Table 8.1 with descriptions of each variable as follows:

y, Meddicorp's sales (in thousands of dollars) in each territory for 2003 (SALES)

x_1, the amount (in hundreds of dollars) that Meddicorp spent on advertising in each territory in 2003 (ADV)

x_2, the total amount of bonuses paid (in hundreds of dollars) in each territory in 2003 (BONUS)

x_3, the market share currently held by Meddicorp in each territory (MKTSHR)

TABLE 8.1 Data for Meddicorp Example Territory

	SALES	ADV	BONUS	MKTSHR	COMPET	REGION
1	963.50	374.27	230.98	33.	202.22	1
2	893.00	408.50	236.28	29.	252.77	1
3	1057.25	414.31	271.57	34.	293.22	1
4	1183.25	448.42	291.20	24.	202.22	2
5	1419.50	517.88	282.17	32.	303.33	3
6	1547.75	637.60	321.16	29.	353.88	3
7	1580.00	635.72	294.32	28.	374.11	3
8	1071.50	446.86	305.69	31.	404.44	1
9	1078.25	489.59	238.41	20.	394.33	1
10	1122.50	500.56	271.38	30.	303.33	2
11	1304.75	484.18	332.64	25.	333.66	3
12	1552.25	618.07	261.80	34.	353.88	3
13	1040.00	453.39	235.63	42.	262.88	1
14	1045.25	440.86	249.68	28.	333.66	2
15	1102.25	487.79	232.99	28.	232.55	2
16	1225.25	537.67	272.20	30.	273.00	2
17	1508.00	612.21	266.64	29.	323.55	3
18	1564.25	601.46	277.44	32.	404.44	3
19	1634.75	585.10	312.35	36.	283.11	3
20	1159.25	524.56	292.87	34.	222.44	1
21	1202.75	535.17	268.27	31.	283.11	2
22	1294.25	486.03	309.85	32.	242.66	2
23	1467.50	540.17	291.03	28.	333.66	3
24	1583.75	583.85	289.29	27.	313.44	3
25	1124.75	499.15	272.55	26.	374.11	2

x_4, the largest competitor's sales (in thousands of dollars) in each territory (COMPET)

x_5, a variable coded to indicate the region in which each territory is located:

1 = SOUTH, 2 = WEST, and 3 = MIDWEST (REGION)

The REGION variable was transformed to a set of three possible indicator variables—SOUTH, WEST, and MIDWEST—in Chapter 7. The interpretation of the coefficients of these variables was preferred to the single REGION variable. Recall from Chapter 7, however, that indicator variables should be treated as a group rather than individually. The all possible regressions technique combines each indicator variable with each other possible combination of variables. To keep the indicator variables grouped together, they are not included as possible explanatory variables in the all possible regressions procedure. Instead, they are examined later as a group (this also greatly reduces the amount of computation necessary and simplifies this example).

Figure 8.1 presents a summary of results for all possible regressions. Figure 8.2 shows the result as it would appear in SAS. In Figure 8.1 the summary measures included are: R^2, R^2_{adj}, C_p, and the standard error of the regression, s_e. These are descriptive measures that are often used to evaluate individual regression equations and to compare different equations. The R^2 does not compensate for the number of variables in the model, but the other three measures do. For this reason, the latter three measures are often considered more reliable for comparing equations with different numbers of variables.

Note that only 15 of the 16 ($2^4 = 16$) possible regressions are shown in the summary. The missing one is the "regression" with no variables included, which would be chosen as best only if none of the possible explanatory variables were linearly related to the dependent variable.

The researcher now must use the summary measures along with subject matter knowledge to choose from among the possible regressions. Recall that small values of C_p and values close to p are of interest in choosing good sets of explanatory variables. The smallest C_p value is for a two-variable regression. This is the regression with ADV and BONUS as explanatory variables; it has a C_p value of 1.61 and

FIGURE 8.1 Summary Results for All Possible Regressions.

Variables in the Regression	R^2	R^2_{adj}	C_p	s_e
ADV	81.1%	80.2	5.90	101.42
BONUS	32.3%	29.3	75.19	191.76
COMPET	14.2%	10.5	100.85	215.83
MKTSHR	0.0%	0.0	120.97	232.97
ADV, BONUS	85.5%	84.2	1.61	90.75
ADV, MKTSHR	81.2%	79.5	7.66	103.23
ADV, COMPET	81.2%	79.5	7.74	103.38
BONUS, COMPET	38.7%	33.2	68.03	186.51
BONUS, MKTSHR	32.8%	26.7	76.46	195.33
MKTSHR, COMPET	16.1%	8.5	100.18	218.20
ADV, BONUS, MKTSHR	85.8%	83.8	3.11	91.75
ADV, BONUS, COMPET	85.7%	83.6	3.33	92.26
ADV, MKTSHR, COMPET	81.3%	78.6	9.59	105.52
BONUS, MKTSHR, COMPET	40.9%	32.5	66.95	187.48
ADV, BONUS, MKTSHR, COMPET	85.9%	83.1	5.00	93.77

FIGURE 8.2 SAS All Possible Regressions Output.

```
                        The REG Procedure
                         Model: MODEL1
                     Dependent Variable: SALES

                     R-Square Selection Method

       Number in
        Model       R-Square         C(p)        Variables in Model

          1          0.8106         5.9046       ADV
          1          0.3228        75.1869       BONUS
          1          0.1422       100.8498       COMPET
          1          0.0005       120.9698       MKTSHR
       ----------------------------------------------------------------
          2          0.8549         1.6052       ADV BONUS
          2          0.8123         7.6612       ADV MKTSHR
          2          0.8117         7.7404       ADV COMPET
          2          0.3873        68.0320       BONUS COMPET
          2          0.3280        76.4591       BONUS MKTSHR
          2          0.1610       100.1763       MKTSHR COMPET
       ----------------------------------------------------------------
          3          0.8585         3.1054       ADV BONUS MKTSHR
          3          0.8569         3.3270       ADV BONUS COMPET
          3          0.8128         9.5912       ADV MKTSHR COMPET
          3          0.4090        66.9457       BONUS MKTSHR COMPET
       ----------------------------------------------------------------
          4          0.8592         5.0000       ADV BONUS MKTSHR COMPET
```

explains 85.5% of the variation in sales ($R^2 = 85.5\%$). As competing models, for example, there are two three-variable models with relatively small C_p values:

Variables	R^2	R^2_{adj}	C_p	s_e
ADV,BONUS,COMPET	85.7%	83.6%	3.33	92.26
ADV,BONUS,MKTSHR	85.8%	83.8%	3.11	91.75

Note that only modest increases in R^2 are achieved in these models. The adjusted R^2 is highest for the two-variable model, again supporting this model as best. Other models with small C_p values could be examined, but there appears to be little bias in the two-variable model with ADV and BONUS (the small deviation of C_p from p probably results from random variation), and it has the smallest C_p value, largest adjusted R^2, and smallest standard error. Therefore, the all possible regressions procedure suggests using this model.

The number of computations involved in the all possible regressions technique is very large. (For example, with $K = 10$ explanatory variables, there are $2^{10} = 1024$ possible models). This is a limiting factor in using this procedure. With very large numbers of candidate explanatory variables, the number of possible models to investigate becomes unwieldy. As a result, several alternatives have been suggested. One is called best subsets regression.

Best subsets regression can be described as follows for subsets of size 2: Among all one-predictor regression models, the two models giving the largest R^2 are found and summary information is printed. Then the two models with the largest R^2 using two predictors are found and summary information is printed. This process continues until the model containing all predictors is examined. The number of subsets

8.2 All Possible Regressions 317

examined is chosen by the researcher. If all subsets are examined then the procedure is identical to all possible regressions.

EXAMPLE 8.2 **Meddicorp Revisited Again** The MINITAB best subsets procedure applied to the Meddicorp data produces the results shown in Figure 8.3. By default, MINITAB only prints out the regressions for the two subsets with the highest R^2 values for each number of explanatory variables. Requests for as many as five subsets are allowed, however. The result of the best subsets procedure asking for the five best subsets to be shown is in Figure 8.4. Note that five best subsets are not available for all combinations of explanatory variables. Note also that this procedure would be the same as all possible regressions if there were three or fewer explanatory variables.

FIGURE 8.3 MINITAB Best Subsets Results Using Default Number (2) of Regressions.

```
Best Subsets Regression: SALES versus ADV, BONUS, MKTSHR, COMPET
Response is SALES
                                                      M C
                                                    B K O
                                                    O T M
                                                    A N S P
                                         Mallows    D U H E
  Vars    R-Sq   R-Sq(adj)     C-p         S       V S R T
    1     81.1      80.2       5.9       101.42    X
    1     32.3      29.3      75.2       191.76      X
    2     85.5      84.2       1.6        90.749   X X
    2     81.2      79.5       7.7       103.23    X     X
    3     85.8      83.8       3.1        91.751  X X X
    3     85.7      83.6       3.3        92.255  X X   X
    4     85.9      83.1       5.0        93.770  X X X X
```

FIGURE 8.4 MINITAB Best Subsets Results Using Maximum Number (5) of Regressions.

```
Best Subsets Regression: SALES versus ADV, BONUS, MKTSHR, COMPET
Response is SALES
                                                      M C
                                                    B K O
                                                    O T M
                                                    A N S P
                                         Mallows    D U H E
  Vars    R-Sq   R-Sq(adj)     C-p         S       V S R T
    1     81.1      80.2       5.9       101.42    X
    1     32.3      29.3      75.2       191.76      X
    1     14.2      10.5      100.8      215.83          X
    1      0.1       0.0      121.0      232.97            X
    2     85.5      84.2       1.6        90.749   X X
    2     81.2      79.5       7.7       103.23    X     X
    2     81.2      79.5       7.7       103.38    X       X
    2     38.7      33.2      68.0       186.51      X X
    2     32.8      26.7      76.5       195.33      X   X
    3     85.8      83.8       3.1        91.751   X X X
    3     85.7      83.6       3.3        92.255   X X   X
    3     81.3      78.6       9.6       105.52    X   X X
    3     40.9      32.5      66.9       187.48      X X X
    4     85.9      83.1       5.0        93.770   X X X X
```

SAS will also perform best subsets regressions. The output uses the same format as that shown for all possible regressions. See the Using the Computer section for more information.

8.3 OTHER VARIABLE SELECTION TECHNIQUES

As noted in the previous section, the all possible regressions technique becomes unwieldy when the number of possible explanatory variables is large. The techniques examined in this section attempt to cut down on the computational expense and the difficulty of examining a large number of potential models while still choosing variables that are important in explaining variation in the dependent variable.

Three procedures are discussed in this section:

1. backward elimination
2. forward selection
3. stepwise regression

Again, it should be stressed that none of these procedures is guaranteed to produce the best possible regression equation. The judgment of the researcher as well as careful examination of scatterplots, residual plots, and regression diagnostics is vital in choosing an appropriate model. The stepwise techniques are merely tools to help the researcher sort through a large number of possible explanatory variables. They help identify some important variables but by themselves are not sufficient to produce a good regression model.

8.3.1 BACKWARD ELIMINATION

The *backward elimination* procedure begins with a regression on all possible explanatory variables. After this regression is run, the explanatory variables are examined to determine which one has the smallest partial F statistic value. Calling this variable x_k, the following hypothesis test is performed:

$$H_0: \beta_k = 0$$
$$H_a: \beta_k \neq 0$$

The decision rule is

Reject H_0 if $F > F_c$
Do not reject H_0 if $F \leq F_c$

where F_c represents some critical value chosen as a cutoff for the test. If H_0 is rejected, the coefficient is judged to be nonzero, and the variable is considered important in the relationship. Because the partial F statistics for all other coefficients are known to exceed the partial F statistic for β_k, the null hypothesis is rejected for these coefficients also. The backward elimination procedure terminates at this point and produces summary statistics of the chosen regression.

On the other hand, if the null hypothesis is not rejected, then the variable is deleted from the equation, and a new regression is run with one less explanatory variable. The procedure is repeated until the null hypothesis is rejected, at which point the procedure terminates.

In this way, the backward elimination procedure sorts through the list of possible explanatory variables, eliminating those that are of little importance in explaining the variation in y and keeping those that are important. Importance is judged by the size of the partial F statistic for testing H_0: $\beta_k = 0$ relative to some critical value.

Two additional aspects of this procedure to note are

1. Although the test was described as a partial F test, it can just as easily be thought of as a t test. Rather than the partial F statistic, the t statistic is used and the decision rule is

   Reject H_0 if $t > t_c$ or $t < -t_c$
   Do not reject H_0 if $-t_c \leq t \leq t_c$

   where t_c is the chosen critical value. The t test and the partial F test for a *single* coefficient are equivalent, as was discussed in Chapter 4, as long as the same levels of significance are used for both tests.

2. Another way the test could be performed is by using p values. Whether the t or F statistic is computed, the p value could be calculated and used in the decision rule:

   Reject H_0 if p value $< \alpha$
   Do not reject H_0 if p value $\geq \alpha$

   where α is the chosen level of significance. When using p values, the critical value is α, while with the t or F tests it is a value chosen from the t or F tables. The p value form of the test is a better choice, if available, because the level of significance used to determine whether a variable stays or goes is the same regardless of the sample size or the number of variables in the equation. When a single t or F critical value is chosen, level of significance varies depending on the sample size and number of variables.

Example 8.3 demonstrates the backward elimination procedure as well as the other procedures. First, however, the remaining techniques are discussed.

8.3.2 FORWARD SELECTION

Forward selection starts by examining the list of possible explanatory variables and computing a simple regression for each one. The partial F statistic (or t statistic or the p value) is computed for the slope coefficient in each of these regressions, and the variable with the largest partial F statistic is noted. The hypothesis test

$$H_0: \beta_k = 0$$
$$H_a: \beta_k \neq 0$$

is conducted just as was done in the backward elimination procedure. If the null hypothesis is not rejected, then the conclusion is that x_k is of no importance to the regression, and none of the other variables are important because they have smaller partial F statistics. The forward selection procedure terminates at this point.

If the null hypothesis is rejected, then x_k is judged important and is retained in the regression. Next, each remaining variable is examined to determine which one will have the largest partial F statistic if added to the regression that already contains x_k. The hypothesis test is performed for the added variable, and the decision is made either to keep the variable in the regression or to discard it. When no more variables are judged to have nonzero coefficients, the procedure terminates and summary statistics are printed.

8.3.3 STEPWISE REGRESSION

The *stepwise regression* procedure combines elements of both backward elimination and forward selection. It begins like forward selection by examining the list of all possible explanatory variables in simple regressions and choosing the one with the largest partial F statistic. The hypothesis test for significance is performed, and if the variable is judged important, this variable is added to the model. Each of the remaining variables is then examined. The variable with the largest partial F statistic is chosen, and the hypothesis test for significance is performed on the coefficient of this variable to determine whether it should be added to the model. If the variable is judged important, it is added as in the forward selection procedure.

At this point, however, the stepwise procedure begins to act like the backward elimination procedure. After adding a new variable to the model, the stepwise procedure retests the coefficients of the previously added variables, deleting these variables if the test judges them to be unnecessary and retaining them otherwise. Because the addition of one variable can result in a change in the partial F statistic associated with another variable, it is possible for the stepwise procedure to allow a variable to enter the equation at one step, delete the variable at a later step, and allow the variable to reenter at an even later step.

Once none of the remaining out-of-equation variables test as significant and all of the variables in the equation are judged to be necessary, the stepwise regression procedure terminates.

8.4 WHICH VARIABLE SELECTION PROCEDURE IS BEST?

In an attempt to identify the best set of variables for a regression model, several techniques have been examined. This leaves the user to decide which technique is best for his or her purposes. As noted, none of the techniques examined is guaranteed to find the best possible regression model. The judgment of the researcher, including incorporation of relevant theory and subject matter knowledge, and careful examination of scatterplots, residual plots, and regression diagnostics are vital in choosing an appropriate model. With this caveat in mind, several trade-offs must be considered when choosing the variable selection technique to be used.

The all possible regressions technique is considered the best because it examines every possible model, given a certain list of variables. From the summary statistics such as R^2 and C_p, the researcher can decide which model is best. Note that even with the all possible regressions technique, a single best model might not be identifiable. There may be several competing models that have nearly identical summary statistics, leaving the researcher with the task of using judgment in choosing between these similar best models. This should not be looked upon as a drawback, however, but as a benefit. With a variety of models from which to choose, the researcher has more freedom to pick the one, say, with the most easily obtainable data or the simplest interpretation. The best subsets procedure is very similar to all possible regressions and would be a close second in its ability to identify the best possible models.

The remaining procedures are not as highly favored, but they do have one advantage over all possible regressions: computational cost. The all possible regressions procedure can be expensive in terms of the computer time needed to produce a solution and may result in too many models to consider if a large number of explanatory variables are used. If the variable list is a very large one, the researcher may be forced to avoid all possible regressions in favor of one of the computationally less expensive methods. As will be shown in Example 8.3, the forward selection, backward elimination, and stepwise regression procedures can be useful in identifying important variables. Research has shown, however, that these procedures can also choose unimportant variables for inclusion in the regressions by chance and may miss important variables. Thus, some caution must be exercised in their use. Among the three procedures, the stepwise regression and backward elimination procedures are very similar. The forward selection procedure is generally considered the least reliable of the techniques.

EXAMPLE 8.3 **Meddicorp Once Again** Figure 8.5 shows the MINITAB backward elimination output when applied to the Meddicorp data. The forward selection output is shown in Figure 8.6, and the stepwise output is in Figure 8.7. The equivalent outputs for SAS are shown in Figures 8.8, 8.9, and 8.10. The indicator variables have not been included in these analyses because these variables are treated as a group rather than individually.

The MINITAB outputs summarize the results of each step of these variable selection procedures in a column. For example, in the backward elimination output shown in Figure 8.5, the variables included in the model are in the left-hand column. In the column numbered Step 1, a summary of the regression equation at the first step is given. The estimated equation is

$$\text{SALES} = -593.5 + 2.51\text{ADV} + 1.91\text{BONUS} + 2.70\text{MKTSHR} - 0.12\text{COMPET}$$

Below the estimated coefficients in the columns are the t ratios and P values for testing $H_0: \beta_k = 0$. At the bottom of the column are the standard error of the regression, the R^2 (both unadjusted and adjusted) and the C_p. The variable with the smallest partial F statistic (or t ratio) is chosen: COMPET. The hypothesis test

$$H_0: \beta_k = 0$$
$$H_a: \beta_k \neq 0$$

FIGURE 8.5 MINITAB Output for Backward Elimination.

```
Stepwise Regression: SALES versus ADV, BONUS, MKTSHR, COMPET

Backward elimination. Alpha-to-Remove: 0.1

Response is SALES on 4 predictors, with N = 25
```

Step	1	2	3
Constant	-593.5	-620.6	-516.4
ADV	2.51	2.47	2.47
T-Value	8.00	8.87	8.98
P-Value	0.000	0.000	0.000
BONUS	1.91	1.90	1.86
T-Value	2.57	2.62	2.59
P-Value	0.018	0.016	0.017
MKTSHR	2.7	3.1	
T-Value	0.57	0.72	
P-Value	0.574	0.478	
COMPET	-0.12		
T-Value	-0.32		
P-Value	0.749		
S	93.8	91.8	90.7
R-Sq	85.92	85.85	85.49
R-Sq(adj)	83.10	83.82	84.18
Mallows C-p	5.0	3.1	1.6

FIGURE 8.6 MINITAB Output for Forward Selection.

```
Stepwise Regression: SALES versus ADV, BONUS, MKTSHR, COMPET

Forward selection. Alpha-to-Enter: 0.25

Response is SALES on 4 predictors, with N = 25
```

Step	1	2
Constant	-157.3	-516.4
ADV	2.77	2.47
T-Value	9.92	8.98
P-Value	0.000	0.000
BONUS		1.86
T-Value		2.59
P-Value		0.017
S	101	90.7
R-Sq	81.06	85.49
R-Sq(adj)	80.24	84.18
Mallows C-p	5.9	1.6

is performed, and the null hypothesis is not rejected. COMPET is removed from the regression, and a new equation is estimated. This regression is summarized in the Step 2 column. This process continues until the null hypothesis H_0: $\beta_k = 0$ is rejected for all remaining variables. The last column (Step 3 in this example) shows the result of the final regression. The summaries are similar for forward selection and stepwise regression.

8.4 Which Variable Selection Procedure Is Best?

FIGURE 8.7 MINITAB Output for Stepwise Regression.

```
Stepwise Regression: SALES versus ADV, BONUS, MKTSHR, COMPET

Alpha-to-Enter: 0.15    Alpha-to-Remove: 0.15

Response is SALES on 4 predictors, with N = 25

Step                 1         2
Constant         -157.3    -516.4

ADV                2.77      2.47
T-Value            9.92      8.98
P-Value           0.000     0.000

BONUS                        1.86
T-Value                      2.59
P-Value                     0.017

S                   101      90.7
R-Sq              81.06     85.49
R-Sq(adj)         80.24     84.18
Mallows C-p         5.9       1.6
```

FIGURE 8.8 SAS Output for Backward Elimination.

The STEPWISE Procedure
Model: MODEL1
Dependent Variable: SALES

Backward Elimination: Step 0

All Variables Entered: R-Square = 0.8592 and C(p) = 5.0000

Analysis of Variance

Source	DF	Sum of Squares	Mean Square	F Value	Pr > F
Model	4	1073119	268280	30.51	<.0001
Error	20	175855	8792.75990		
Corrected Total	24	1248974			

Variable	Parameter Estimate	Standard Error	Type II SS	F Value	Pr > F
Intercept	-593.53745	259.19585	46107	5.24	0.0330
ADV	2.51314	0.31428	562260	63.95	<.0001
BONUS	1.90595	0.74239	57955	6.59	0.0184
MKTSHR	2.65101	4.63566	2875.57485	0.33	0.5738
COMPET	-0.12073	0.37181	927.06773	0.11	0.7488

Bounds on condition number: 1.4799, 20.838

(Continued).

FIGURE 8.8
(*Continued*).

Backward Elimination: Step 1

Variable COMPET Removed: R-Square = 0.8585 and C(p) = 3.1054

Analysis of Variance

Source	DF	Sum of Squares	Mean Square	F Value	Pr > F
Model	3	1072191	357397	42.46	<.0001
Error	21	176782	8418.20313		
Corrected Total	24	1248974			

Variable	Parameter Estimate	Standard Error	Type II SS	F Value	Pr > F
Intercept	-620.63774	240.10769	56245	6.68	0.0173
ADV	2.46979	0.27839	662567	78.71	<.0001
BONUS	1.90030	0.72620	57643	6.85	0.0161
MKTSHR	3.11646	4.31354	4394.15366	0.52	0.4780

Bounds on condition number: 1.2212, 10.325

Backward Elimination: Step 2

Variable MKTSHR Removed: R-Square = 0.8549 and C(p) = 1.6052

Analysis of Variance

Source	DF	Sum of Squares	Mean Square	F Value	Pr > F
Model	2	1067797	533899	64.83	<.0001
Error	22	181176	8235.29179		
Corrected Total	24	1248974			

Variable	Parameter Estimate	Standard Error	Type II SS	F Value	Pr > F
Intercept	-516.44428	189.87570	60924	7.40	0.0125
ADV	2.47318	0.27531	664572	80.70	<.0001
BONUS	1.85618	0.71573	55389	6.73	0.0166

Bounds on condition number: 1.2126, 4.8502

All variables left in the model are significant at the 0.1000 level.

Summary of Backward Elimination

Step	Variable Removed	Number Vars In	Partial R-Square	Model R-Square	C(p)	F Value	Pr > F
1	COMPET	3	0.0007	0.8585	3.1054	0.11	0.7488
2	MKTSHR	2	0.0035	0.8549	1.6052	0.52	0.4780

FIGURE 8.9
SAS Output for Forward Selection.

```
                          The STEPWISE Procedure
                  Model: MODEL1, Dependent Variable: SALES
                         Forward Selection: Step 1

           Variable ADV Entered: R-Square = 0.8106 and C(p) = 5.9046

                            Analysis of Variance

                              Sum of       Mean
        Source         DF    Squares      Square      F Value    P > F

        Model           1    1012408     1012408       98.43     <.0001
        Error          23     236566       10285
        Corrected Total 24   1248974


                         Parameter    Standard
         Variable         Estimate      Error     Type II SS   F Value   Pr > F

         Intercept       -157.33011   145.19120      12077       1.17    0.2898
         ADV                2.77212     0.27941    1012408      98.43    <.0001

                      Bounds on condition number: 1, 1
--------------------------------------------------------------------------------
                         Forward Selection: Step 2

          Variable BONUS Entered: R-Square = 0.8549 and C(p) = 1.6052

                            Analysis of Variance

                              Sum of       Mean
        Source         DF    Squares      Square      F Value    Pr > F

        Model           2    1067797      533899       64.83     <.0001
        Error          22     181176     8235.29179
        Corrected Total 24   1248974


                         Parameter    Standard
         Variable         Estimate      Error     Type II SS   F Value   Pr > F

         Intercept       -516.44428   189.87570      60924       7.40    0.0125
         ADV                2.47318     0.27531     664572      80.70    <.0001
         BONUS              1.85618     0.71573      55389       6.73    0.0166

                   Bounds on condition number: 1.2126, 4.8502
--------------------------------------------------------------------------------
                         Forward Selection: Step 3

          Variable MKTSHR Entered: R-Square = 0.8585 and C(p) = 3.1054

                            Analysis of Variance

                              Sum of       Mean
        Source         DF    Squares      Square      F Value    Pr > F

        Model           3    1072191      357397       42.46     <.0001
        Error          21     176782     8418.20313
        Corrected Total 24   1248974
```

(*Continued*).

FIGURE 8.9
(*Continued*).

Variable	Parameter Estimate	Standard Error	Type II SS	F Value	Pr > F
Intercept	-620.63774	240.10769	56245	6.68	0.0173
ADV	2.46979	0.27839	662567	78.71	<.0001
BONUS	1.90030	0.72620	57643	6.85	0.0161
MKTSHR	3.11646	4.31354	4394.15366	0.52	0.4780

Bounds on condition number: 1.2212, 10.325

No other variable met the 0.5000 significance level for entry into the model.

Summary of Forward Selection

Step	Variable Entered	Number Vars In	Partial R-Square	Model R-Square	C(p)	F Value	Pr > F
1	ADV	1	0.8106	0.8106	5.9046	98.43	<.0001
1	BONUS	2	0.0443	0.8549	1.6052	6.73	0.0166
2	MKTSHR	3	0.0035	0.8585	3.1054	0.52	0.4780

FIGURE 8.10
SAS Output for Stepwise Regression.

The STEPWISE Procedure
Model: MODEL1
Dependent Variable: SALES

Forward Selection: Step 1

Variable ADV Entered: R-Square = 0.8106 and C(p) = 5.9046

Analysis of Variance

Source	DF	Sum of Squares	Mean Square	F Value	P > F
Model	1	1012408	1012408	98.43	<.0001
Error	23	236566	10285		
Corrected Total	24	1248974			

Variable	Parameter Estimate	Standard Error	Type II SS	F Value	Pr > F
Intercept	-157.33011	145.19120	12077	1.17	0.2898
ADV	2.77212	0.27941	1012408	98.43	<.0001

Bounds on condition number: 1, 1

FIGURE 8.10
(*Continued*).

```
                        Stepwise Selection: Step 2

         Variable BONUS Entered: R-Square = 0.8549 and C(p) = 1.6052

                            Analysis of Variance

                         Sum of         Mean
Source           DF      Squares        Square      F Value    Pr > F

Model             2      1067797        533899       64.83     <.0001
Error            22       181176       8235.29179
Corrected Total  24      1248974


                Parameter      Standard
Variable        Estimate       Error      Type II SS    F Value    Pr > F

Intercept      -516.44428     189.87570       60924       7.40     0.0125
ADV               2.47318       0.27531      664572      80.70     <.0001
BONUS             1.85618       0.71573       55389       6.73     0.0166

              Bounds on condition number: 1.2126, 4.8502
```

All variables left in the model are significant at the 0.1500 level.

No other variable met the 0.5000 significance level for entry into the model.

```
                        Summary of Stepwise Selection

       Variable   Variable   Number   Partial    Model
Step   Entered    Removed    Vars In  R-Square   R-Square   C(p)     F Value   Pr > F

1      ADV                      1     0.8106     0.8106     5.9046    98.43    <.0001
1      BONUS                    2     0.0443     0.8549     1.6052     6.73    0.0166
```

SAS also summarizes each step of the variable selection procedures, but prints out a more complete version of the regression results. Consider the backward elimination output in Figure 8.8. In the output labeled Step 0, the regression including all the potential explanatory variables is run. In Figures 8.9 (forward selection) and 8.10 (stepwise) the output labeled Step 1 is the regression using the variable with the largest partial F statistic. An ANOVA table is printed for the regression at each step. The table is the same as the standard ANOVA table for SAS. Below that, information about the regression coefficients is printed. This table contains the parameter estimate and the standard error for each variable in the regression as usual. However, an additional column is added: Type II SS. This column contains the sum of squares used to compute the partial F statistic to test if the coefficient of the variable in that row of the table is zero. The partial F statistic is given in the next column (rather than the usual t statistic) and the associated p value follows. The F statistic or the associated p value can be used as described earlier in the chapter to perform the tests needed for the variable selection procedures. SAS proceeds through the steps necessary to complete the various procedures and ends with a summary of the variables that were either omitted from or added to the regression equation at each step.

Regardless of the procedure or statistical package used, the result is the same in all but one instance. The equation chosen is SALES = −516.4 + 2.47ADV + 1.86BONUS. The one exception is the SAS output for forward selection. In this case the equation chosen is SALES = −620.64 + 2.47ADV + 1.90BONUS + 3.12MKTSHR. The reason that MKTSHR is included in the equation developed by the SAS forward selection procedure is that different cutoffs for the hypothesis tests to choose the variables are used. SAS uses p values to determine what variables should and should not be included in the equations. For the forward selection procedure, the p value for inclusion of a variable is 0.5. MKTSHR is included in the final equation because the p value for its coefficient (0.4780) is less than 0.5. In contrast, MINITAB's forward selection procedure uses a p value of 0.25 as a cutoff and MKTSHR is not included in the resulting equation. For more on choosing cutoffs for these procedures, see the Using the Computer section at the end of this chapter.

Combining all information gathered from the all possible regressions procedure and the various stepwise procedures, the best model appears to be

$$SALES = -516.4 + 2.47ADV + 1.86BONUS$$

Before concluding, recall that the indicator variables for the region have not been included in any of this analysis. They could be added to the regression at this point to see whether they improve the model. The regression with the indicators included is shown in Figure 8.11. The partial F test to test the hypotheses

$$H_0: \beta_3 = \beta_4 = 0$$
$$H_a: \text{At least one of } \beta_3 \text{ and } \beta_4 \text{ is not equal to zero}$$

can be used to determine whether the indicator variables improve the model. The null hypothesis is rejected, and the conclusion is to retain the indicator variables as well.

FIGURE 8.11 Regression Results for Model Including Indicator Variables.

Variable	Coefficient	Std Dev	T Stat	P Value
Intercept	435.099	206.234	2.11	0.048
ADV	1.368	0.262	5.22	0.000
BONUS	0.975	0.481	2.03	0.056
SOUTH	−257.893	48.413	−5.33	0.000
WEST	−209.746	37.420	−5.61	0.000

Standard Error = 57.6254   R-Sq = 94.7%   R-Sq(adj) = 93.6%

Analysis of Variance

Source	DF	Sum of Squares	Mean Square	F Stat	P Value
Regression	4	1182560	295640	89.03	0.000
Error	20	66414	3321		
Total	24	1248974			

EXERCISES

1. **Cost Control.** Exercise 1 in Chapter 4 discussed data available for a firm that produces corrugated paper for use in making boxes and other packing materials. The variables discussed were

    y, total manufacturing cost per month in thousands of dollars (COST)

    x_1, total production of paper per month in tons (PAPER)

    x_2, total machine hours used per month (MACHINE)

    x_3, total variable overhead costs per month in thousands of dollars (OVERHEAD)

    x_4, total direct labor hours used each month (LABOR)

    The data, available in a file named COST8 on the CD, are monthly and refer to the time period from January 2001 to March 2003. Use the backward elimination procedure to analyze these data, then answer the following questions:

    a. What is the regression equation chosen by the backward elimination procedure?

    b. What is the R^2 for the chosen equation?

    c. What is the adjusted R^2 for the chosen equation?

    d. What is the standard error of the chosen equation?

    e. What variables were omitted? Do you feel these variables are unrelated to COST? Why or why not? Do you feel the omitted variables are necessary in the regression equation? Why or why not?

2. **Sales Force Performance.** Data on the following variables were obtained for a random sample of 25 sales territories for a company (the data have been transformed to preserve confidentiality).

    y, sales in units for the territory (SALES)

    x_1, length of time territory salesperson has been with the company (TIME)

    x_2, industry sales in units for the territory (POTENT)

    x_3, dollar expenditures on advertising (ADV)

    x_4, weighted average of past market share for 4 previous years (SHARE)

    x_5, change in market share over the 4 years before the time period analyzed (SHARECHG)

    x_6, total number of accounts assigned to salesperson (ACCTS)

    x_7, average workload per account using a weighted index based on annual purchases of accounts and concentration of accounts (WORKLOAD)

    x_8, an aggregate rating on a 1–7 scale on eight dimensions of performance by applicable field sales manager (RATING)

    The data are available in a file named TERRITORY8 and were analyzed in D. W. Cravens, R. B. Woodruff, and J. C. Stamper, "An Analytical Approach for Evaluating Sales Territory Performance," *Journal of Marketing* 36 (1972): 31–37.

    The goal of the study is to identify factors that influence territory sales (y). The equation to be developed will be used to assess whether salespersons in respective territories are performing up to standard.

    Develop an appropriate model to explain sales territory performance. Use any of the techniques discussed to select appropriate variables. Be sure to examine scatterplots and residual plots for violations of assumptions and to correct for any such violations.

    a. For the model you select, report the estimated regression equation. Be sure to define the variables used.

    b. For the model you select, report any corrections for violations of assumptions.

    c. For the model you select, report the R^2, R^2_{adj}, standard error, and C_p value.

    d. Discuss how this equation could be used to set a performance standard for sales territories. How would average performance be determined? Below-average performance? What limitations would this approach have for setting performance standards?

3. **2003 Cars.** Data on 147 cars were obtained from the October 2002 issue of *Road & Track: The New Cars*. The following data are available in a file named CARS8 on the CD:

    name of car
    weight, in pounds (WEIGHT)
    mileage in city driving (CITYMPG)
    mileage in highway driving (HWYMPG)

horsepower, @ 6300 rpm (HP)
number of cylinders (CYLIN)
displacement, in liters (LITER)

Using the available data, try to determine what factors involved in the construction of a car affect either mileage in city driving or mileage in highway driving. (Choose either CITYMPG or HWYMPG as your dependent variable. If you choose CITYMPG, do not use HWYMPG as a possible explanatory variable, and vice versa.) Use any of the techniques discussed to select appropriate variables. Be sure to examine scatterplots and residual plots for violations of assumptions and to correct for any such violations.

   a. For the model you select, report the estimated regression. Be sure to define the variables used.
   b. For the model you select, report any corrections for violations of assumptions. Explain why the correction was needed and justify the correction you used.
   c. Justify your choice of variables from both a statistical and a practical standpoint.
   d. State any limitations of the model.

(*Source*: From *Road & Track: The New Cars.* Copyright 2002 by Hachette Filipacchi Magazines, Inc. Used with permission.)

4. **1998 American League Pitchers.** Data on 44 American League pitchers were obtained for the 1998 season. The pitchers included in the data set must have pitched in at least 40 innings to be listed here. Also, these pitchers were used only as starting pitchers (pitchers with even one appearance in relief were not included). The data are available in a file named ALPITCH8 on the CD as follows:

   name of pitcher
   team: coded as

   1 = Baltimore
   2 = Boston
   3 = Cleveland
   4 = Detroit
   5 = Anaheim
   6 = Chicago
   7 = Kansas City
   8 = Milwaukee
   9 = Minnesota
   10 = Seattle
   11 = New York
   12 = Oakland
   13 = Texas
   14 = Toronto
   15 = Tampa Bay

   number of wins (W)
   number of losses (L)
   earned run average (ERA)
   innings pitched (IP)
   hits allowed (H)
   home runs allowed (HR)
   bases on balls (BB)
   strikeouts (SO)

As a consultant to the Texas Rangers' coaching staff, you have been hired to determine what makes a starting pitcher successful. The Rangers are painfully aware of the need for good starting pitchers. They would like to improve their starting pitching before the next season. You have data available for American League starting pitchers in the 1998 season. Your goal is to determine what factors might be important in the success of a starting pitcher during a particular season. First, you must define success (the dependent variable). Then decide which of the variables available might make sense in evaluating a pitcher's success.

Write up a report to the Rangers with your recommendations. Your report should consist of two parts: (a) an executive summary with a brief nontechnical discussion of your recommendations and (b) a technical report to support your suggestions. The technical report should include a discussion of the analysis that led to your conclusions. This should include regression results, model validation, corrections for violations of assumptions, justification of your choice of variables, and an explanation of how this choice will help the Rangers in their decision making.

(*Source*: Used courtesy of the *Fort Worth Star-Telegram.*)

5. **FOC Sales.** Techcore is a high-tech company located in Fort Worth, Texas. The company produces a part called a fibre-optic connector (FOC) and wants to generate forecasts of the sales of FOCs over time. The company has weekly sales data for the past 265 weeks. The data are in the file FOC8 on the CD in the following columns: SALES, MONTH (numbers the month in which the observations were taken), FOV, COMPOSITE, INDUSTRIAL, TRANS, UTILITY, FINANCE, PROD, and HOUSE. (The SALES and FOV data have been disguised to provide confidentiality.) The variables are defined as follows:

FOV: Sales of a complementary product; sales of FOV are much easier to forecast than FOC sales
COMPOSITE: Friday close of the NYSE Composite Index
INDUSTRIAL: Friday close of the NYSE Industrial Stocks
TRANS: Friday close of the NYSE Transportation Stocks
UTILITY: Friday close of the NYSE Utility Stocks
FINANCE: Friday close of the NYSE Financials
PROD: Industrial Production—computers, communications equipment, and semiconductors, not seasonally adjusted
HOUSE: Monthly housing permits in thousands, seasonally adjusted rates

You have been hired as a consultant to Techcore to help build the forecasting model. Create a two-part report for Techcore that includes an executive summary with the essential nontechnical results of your study and a technical report that contains the details of your model building process.

USING THE COMPUTER

The Using the Computer section in each chapter describes how to perform the computer analyses in the chapter using Excel, MINITAB, and SAS. For further detail on Excel, MINITAB, and SAS, see Appendix C.

FIGURE 8.12 MINITAB Best Subsets Dialog Box.

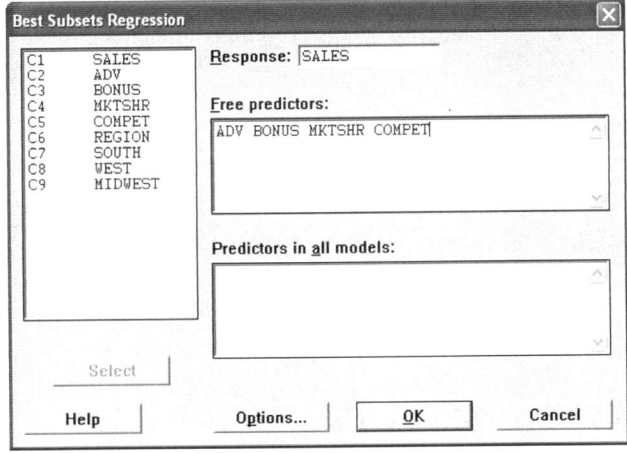

EXCEL
Excel does not come with any variable selection techniques built in. However, Appendix C describes an add-in that makes it possible to perform certain of the procedures. See the Appendix C section on Excel for more information.

MINITAB

Best Subsets Regression
STAT: REGRESSION: BEST SUBSETS

To use the best subsets procedure, click STAT, then REGRESSION from the STAT menu, and then BEST SUBSETS. The Best Subsets dialog box is shown in Figure 8.12. Fill in the "Response" variable (y) and the "Free predictors" (x variables) and click OK. Use "Predictors in all models" if you want certain x variables to be included in all regressions. The Options screen allows, among other things, the choice of from one to five subsets to be shown in the results.

STEPWISE REGRESSION, BACKWARD ELIMINATION, AND FORWARD SELECTION

Backward elimination, forward selection, and stepwise regression are all run from the STEPWISE Methods on the REGRESSION menu. Figure 8.13 shows the stepwise regression dialog box.

The MINITAB Stepwise—Options Dialog Box in Figure 8.14 allows the user to choose either F values or p values to be used in conducting the tests described in

FIGURE 8.13 MINITAB Stepwise Regression Dialog Box.

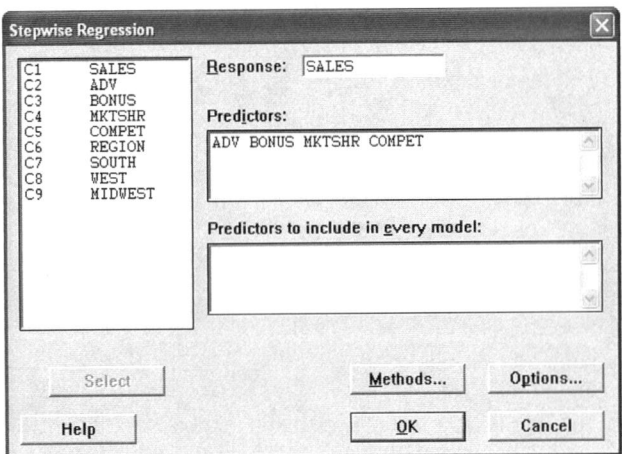

FIGURE 8.14 MINITAB Stepwise—Options Dialog Box.

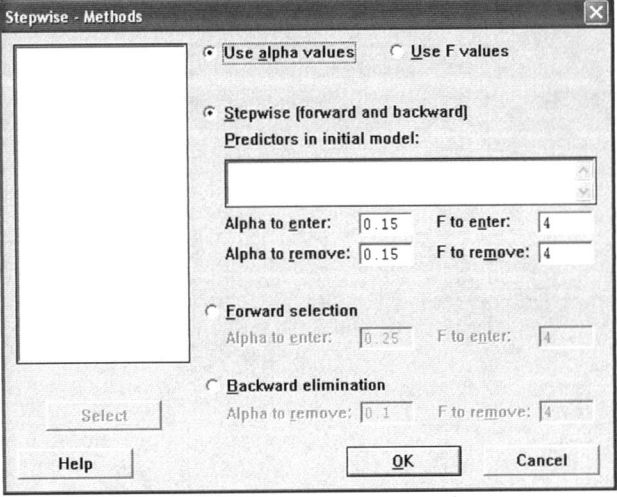

this chapter. If F values are chosen, the default critical value will be 4. The justification for the partial F test cutoff value of 4 is that this results in a test with a 5% level of significance (approximately) if a large number of observations are available. When using any of these three procedures, it is often recommended that a smaller cutoff value be used, say, $F_c = 2$ or $F_c = 1$. After examining the variables chosen by the procedure(s) at this critical value, the researcher is left to make the final choice about which variables should remain in the model. These critical values can be changed in the "F to enter" and/or "F to remove" boxes.

If p values are chosen, the default values are 0.15 both to enter and remove variables for stepwise regression, 0.25 to enter in forward selection, and 0.1 to remove in backward elimination. These default values can be changed at the discretion of the user.

Stepwise Regression

STAT: REGRESSION: STEPWISE

For stepwise regression, fill in the "Response" variable (y) and the names of the x variables in the "Predictors" box (Figure 8.13). Then click OK. Click Methods and the dialog box shown in Figure 8.14 opens. Be sure the stepwise button is clicked. Stepwise is the default method in MINITAB so it should be set unless it has been changed in a previous analysis.

Backward Elimination

STAT: REGRESSION: STEPWISE

For backward elimination, fill in the "Response" variable (y) and the names of the x variables in the "Predictors" box (Figure 8.13). Now click Methods, and the dialog box shown in Figure 8.14 opens. The backward elimination option is requested by clicking the button for that option.

Forward Selection

STAT: REGRESSION: STEPWISE

For forward selection, fill in the "Response" variable (y) and the names of the x variables in the "Predictors" box (Figure 8.13). Click Methods and the dialog box shown in Figure 8.14 opens. The forward selection option is requested by clicking the button for that option.

SAS

In SAS, the PROC REG command can be used to request all possible regressions, best subsets, backward elimination, forward selection, and stepwise. The choice of variable selection procedure is made using the SELECTION option in the MODEL statement. Each of the choices is demonstrated here.

All Possible Regressions and Best Subsets

```
PROC REG CP;
MODEL SALES=ADV BONUS MKTSHR COMPET/SELECTION=RSQUARE;
```

This command sequence produces all possible regressions output for the dependent variable SALES and independent variables ADV, BONUS, MKTSHR, and COMPET. The CP option requests that the C_p value associated with each regression be printed. In addition, the R^2 value (unadjusted) is printed. The MODEL statement works just like it does for PROC REG. The dependent variable is listed to the left of the equal sign, and the possible explanatory variables are listed to the right. The SELECTION = RSQUARE option requests all possible regressions. If BEST = k is added after SELECTION = RSQUARE, then the k best subsets are summarized rather than all possible regressions. Here, k is replaced by the number of desired subsets.

Stepwise Regression

```
PROC REG;
MODEL SALES=ADV BONUS MKTSHR COMPET/SELECTION=STEPWISE;
```

Control of the variables to enter or leave the regression is through the p values for the partial F test (rather than the F critical values). SLENTRY represents the maximum p value the coefficient of a variable can have and still have the variable enter the model. SLSTAY represents the maximum p value the coefficient of a variable can have for the variable to remain in the model. For stepwise regression, these values are set at SLENTRY = 0.15 and SLSTAY = 0.15. These default values can be changed. For example, to change SLENTRY and SLSTAY each to 0.2, the following sequence could be used:

```
PROC REG;
MODEL SALES=ADV BONUS MKTSHR COMPET/SELECTION=STEPWISE/SLENTRY=0.2 SLSTAY=0.2;
```

Backward Elimination

```
PROC REG;
MODEL SALES=ADV BONUS MKTSHR COMPET/SELECTION=BACKWARD;
```

Backward elimination uses SLSTAY = 0.1 in SAS.

Forward Selection

```
PROC REG;
MODEL SALES=ADV BONUS MKTSHR COMPET/SELECTION=FORWARD;
```

Forward selection uses SLENTRY = 0.5 in SAS.

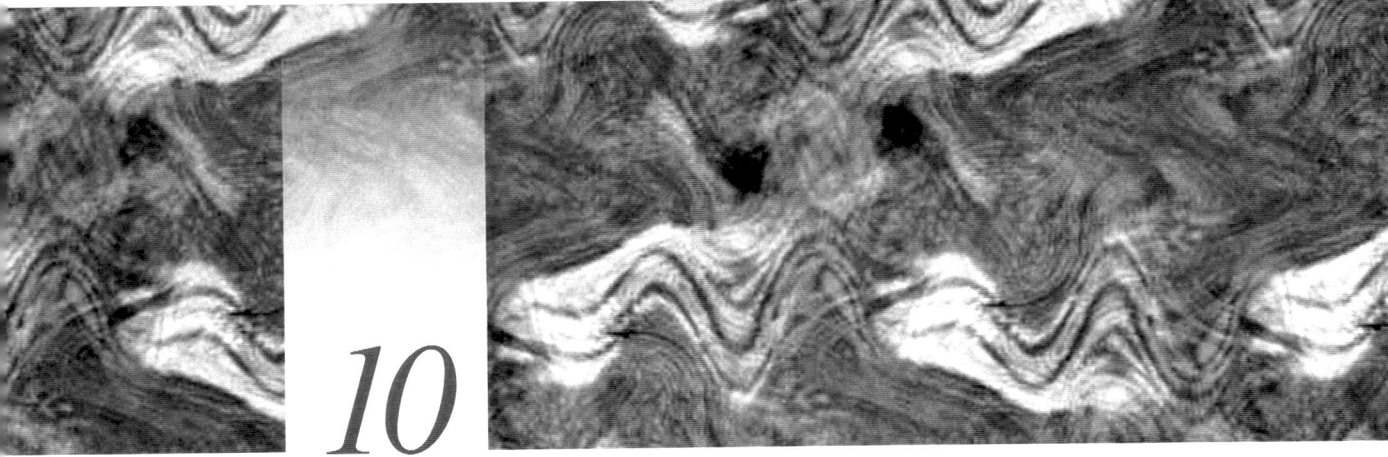

10
Qualitative Dependent Variables: An Introduction to Discriminant Analysis and Logistic Regression

10.1 INTRODUCTION

A bank is interested in determining whether certain borrowers are creditworthy. Should loans be made to these potential customers? The bank has a variety of quantitative information about each potential borrower and would like some way of classifying them into groups of qualifying and nonqualifying candidates.

An investment company is trying to determine the likelihood that certain firms will end up in bankruptcy. The investment company has quantitative information on the firms it is considering and wants to classify them into two groups: those that will go bankrupt and those that will not.

The director of personnel for a large corporation would like to know which of a group of new trainees will be successful in a certain position at the company. Using demographic information and the results of certain aptitude tests, the director wants to classify trainees into two groups: those who will succeed and those who will not.

The marketing department of a retail firm is interested in predicting whether people will buy a new product within the next year. The marketing department will have the results from surveys given to a sample of potential buyers and a quantitative score representing propensity to buy as a summary measure from these surveys. The department wants to use this score to divide the people into groups of potential buyers and nonbuyers so it can better target future advertising.

Each of these examples represents a case in which certain observations (people, firms, and so on) are to be divided into two groups: firms that do and do not qualify for credit; firms that will and will not go bankrupt; people that will and will not be successful in new positions; people who will and will not purchase a new product. The goal is to try to pick which of the two groups each observation will fall into based on available information. This type of problem can be placed in a regression setting. Write the regression model as

$$y = \beta_0 + \beta_1 x_1 + \cdots + \beta_K x_K + e$$

where

$y = 1$ if the observation falls into the group of interest

$y = 0$ if the observation does not fall into the group of interest

In Chapter 7, methods were discussed for dealing with explanatory variables representing qualitative information. Examples used were employed/unemployed, male/female, and so on. The way this information was incorporated into the regression was with indicator variables. These variables took on values of either 0 or 1 to indicate whether an item was or was not in a certain group (male = 1, female = 0, for example). In this chapter, we adopt a similar approach for situations where the dependent variable of interest is qualitative in nature. Situations are examined where the task is to try and determine whether items do or do not fall into a certain group. The dependent variable used to represent this situation will be a variable that has either the value 1 if the item is in the group or the value 0 if the item is not in the group.

Standard linear regression analysis is not designed for directly analyzing this situation. In the regression applications examined so far in this text, the dependent variable was a variable that could be effectively modeled as a continuous variable. The assumptions of the regression model are designed for such a situation. When a 0/1 dependent variable is used, several problems occur. To discuss the first problem, it is important to note that the conditional mean of y given x has a different interpretation when the dependent variable is a 0/1 variable than it did in the previous case when the dependent variable was modeled as continuous. It can be shown that $\mu_{y|x}$ is equal to the probability that the observation belongs to the indicated group: $\mu_{y|x} = p$, where $p = P(Y = 1)$. Probabilities must be between 0 and 1, of course, but when the regression is estimated, there is nothing to guarantee that the predictions from the estimated regression equation fall between 0 and 1. The actual predictions can vary considerably, with values above 1 or even negative values occurring.

Second, certain assumptions of the regression model will be violated. For example, the disturbances are not normally distributed and the variance around the regression line is not constant.

As a result of the problems that occur with applying standard regression techniques to situations with 0/1 dependent variables, alternative methods of analysis have been proposed. Two methods are discussed in this chapter: discriminant analysis and logistic regression. Both techniques have characteristics similar to the regression models that have been presented thus far in this text. Even though linear regression, discriminant analysis, and logistic regression differ, their similarities will help the reader understand the use of these procedures.

10.2 DISCRIMINANT ANALYSIS

EXAMPLE 10.1 **Employee Classification** The personnel director for a firm that manufactures computers has classified the performance of each of the employees in a certain position as either satisfactory or unsatisfactory. The director has two tests that she would like to use to help determine which future employees will perform in a satisfactory manner and which will not before they are assigned to the position. With this knowledge, she will be better able to suggest jobs within the firm at which each employee will have a greater chance of success. To help determine whether the tests will be useful, she administers them to the current employees and records their scores. The resulting data are shown in Table 10.1 and in the file TRAIN10 on the CD. The first column notes whether the employee is currently classified as satisfactory (1) or unsatisfactory (0). The next two columns show the test score results on each of the two tests. The director's task is now to determine how to use the test scores to predict the correct classification of each employee.

Figure 10.1 shows a scatterplot of each employee's scores on the two tests. The employees classified as successful are denoted with ■ and those classified as unsuccessful as ●. From the plot, it does appear that knowledge of the test score may provide information useful in classifying the employees. There is no exact cutoff to separate the two groups precisely, but there is enough separation of the members to indicate that the information provided on the tests may be useful.

Discriminant analysis can be used to help solve the type of problem posed in Example 10.1. Discriminant analysis can be described as follows: Let y represent the dependent variable, defined as

$y = 1$ if the observation falls into the group of interest

$y = 0$ if the observation does not fall into the group of interest

Let $x_1, \ldots, x_K$ be explanatory variables that are used to help predict into which of the two groups each of the observations in the sample should be classified. The explanatory variables are assumed to be approximately normally distributed. Although discriminant analysis can be used when the explanatory variables are not normally distributed, it is not guaranteed to be optimal in such cases and may not provide good results. This is one of the more serious limitations of this technique because it limits the kinds of variables that can be used as explanatory variables in the equation. It should be noted, however, that a study by Amemiya and Powell[1] showed that discriminant analysis does well in prediction and estimation even when the explanatory variables are nonnormal if sample sizes are large. Discriminant analysis also assumes that the variation of the explanatory variables is the same for each group. Write the equation representing the relationship between y and the explanatory variables as

$$y = \beta_0 + \beta_1 x_1 + \cdots + \beta_K x_K + e$$

---

[1] See T. Amemiya and J. Powell, "A Comparison of the Logit Model and Normal Discriminant Analysis When Independent Variables Are Binary." See References for complete publication information.

TABLE 10.1 Data and Discriminant Analysis Results for Employee Classification Example

GROUP	TEST1	TEST2	Predicted Group	Discriminant Score
1	96	85	1	0.99694
1	96	88	1	1.04291
1	91	81	1	0.64276
1	95	78	1	0.83111
1	92	85	1	0.76262
1	93	87	1	0.85185
1	98	84	1	1.09878
1	92	82	1	0.71666
1	97	89	1	1.11681
1	95	96	1	1.10691
1	99	93	1	1.29526
1	89	90	1	0.66350
1	94	90	1	0.95639
1	92	94	1	0.90052
1	94	84	1	0.86446
1	90	92	1	0.75272
1	91	70	0*	0.47421
1	90	81	1	0.58418
1	86	81	0*	0.34986
1	90	76	0*	0.50757
1	91	79	1	0.61211
1	88	83	0*	0.49766
1	87	82	0*	0.42376
0	93	74	1*	0.65266
0	90	84	1*	0.63014
0	91	81	1*	0.64276
0	91	78	1*	0.59679
0	88	78	0	0.42105
0	86	86	0	0.42647
0	79	81	0	−0.06020
0	83	84	0	0.22009
0	79	77	0	−0.12149
0	88	75	0	0.37508
0	81	85	0	0.11825
0	85	83	0	0.32192
0	82	72	0	−0.02236
0	82	81	0	0.11554
0	81	77	0	−0.00433
0	86	76	0	0.27325
0	81	84	0	0.10293
0	85	78	0	0.24531
0	83	77	0	0.11283
0	81	71	0	−0.09626

* Indicates a misclassification

FIGURE 10.1
Scatterplot of TEST1 versus TEST2 for Satisfactory (■) and Unsatisfactory (●) Employees.

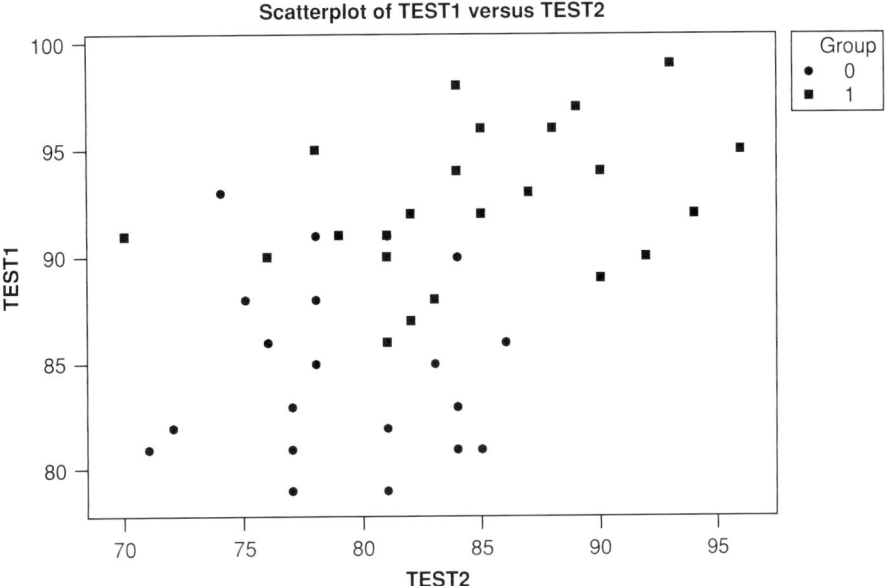

Apply linear regression to the data and estimate the previous equation. The estimated equation is written

$$d = b_0 + b_1 x_1 + \cdots + b_K x_K$$

where d is called the discriminant score. The *discriminant score* is just the predicted value from the estimated regression equation. These discriminant scores are used to classify each of the observations in the sample. A cutoff value is chosen, call it c, and the following classification rule is used:

If $d \leq c$, assign the observation to group 0

If $d > c$, assign the observation to group 1

How is the cutoff value in the classification rule chosen? Our real goal in such a situation is not to classify the items in the sample correctly; we know what group these items are in. The goal is to classify future observations correctly. In the employee classification example, this means being able to correctly classify future employees as satisfactory or unsatisfactory in a particular position on the basis of their test scores. So ideally, the estimates $b_0, b_1, \ldots, b_K$ should be chosen so that the number of misclassified future observations is minimized.

In the problem considered here, this can be done as follows: Estimate the equation for the discriminant score with least-squares regression using the 0/1 dependent variable. Record the predicted or fitted values from this estimated equation. These are the discriminant scores, d. Choose a value, c, as a cutoff in such a way that the probability of misclassification for future observations is minimized. This is done by choosing c halfway between the average discriminant scores for the two groups if the sample sizes are equal. If we write

$\overline{d}_1$ = average discriminant score for the 1 group
$\overline{d}_0$ = average discriminant score for the 0 group

then

$$c = \frac{\overline{d}_0 + \overline{d}_1}{2}$$

If the sample sizes are not equal, then a weighted average of the two average discriminant scores can be used:

$$c = \frac{n_0 \overline{d}_0 + n_1 \overline{d}_1}{n_0 + n_1}$$

where n_0 is the sample size for the group labeled 0 and n_1 is the sample size for the group labeled 1.

EXAMPLE 10.2 **Employee Classification (continued)** Continuing with Example 10.1, the discriminant scores for the employees in the sample in Table 10.1 have been computed and are shown in the fifth column of the table. Column 4 shows the results of a classification rule applied to the discriminant scores. The employees who were incorrectly classified have been denoted with an *. Note that five employees who were in group 1 have been classified as being in group 0, and four employees who were in group 0 have been classified as being in group 1. These errors are termed misclassifications. The percentage of correct classifications in this example is 79.1%.

The average discriminant score computed from the linear regression for the satisfactory (1) group can be found from the data in column 5. This average is 0.7848. The average for the unsatisfactory (0) group is 0.2475. The weighted average of these two values is given by

$$\frac{20(0.2475) + 23(0.7848)}{20 + 23} = 0.5349$$

This number can be used to classify each of the observations in Table 10.1 and to check the classifications in column 4. Whenever the d score in column 5 is less than or equal to 0.5349, the observation should be classified as belonging to group 0. When the d score in column 5 is greater than 0.5349, the observation should be classified as belonging to group 1. This is how the classifications in the table were determined.

The regression used to determine the d scores is shown in Figure 10.2. For future employees, this regression can be used to determine the d scores for classification.

Even though linear regression was used to perform the discriminant analysis, this is not typically the method of choice. There are specific computer routines available for discriminant analysis. Figure 10.3 shows the MINITAB output for the

FIGURE 10.2
Regression Used to Perform Discriminant Analysis for Employee Classification Example.

```
Variable      Coefficient   Std Dev    T Stat    P Value

Intercept     -5.9291       0.9633     -6.15     0.000
TEST1          0.0586       0.0112      5.23     0.000
TEST2          0.0153       0.0010      1.54     0.132


Standard Error = 0.351797    R-Sq = 53.7%  R-Sq(adj) = 51.4%

Analysis of Variance
Source        DF    Sum of Squares    Mean Square    F Stat    P Value

Regression     2    5.7472            2.8736         23.22     0.000
Error         40    4.9505            0.1238
Total         42   10.6977
```

employee classification problem when a procedure designed specifically for discriminant analysis is used. Figure 10.4 shows the SAS discriminant analysis output. The results of the classification are identical to those from the regression approach, but the method used is somewhat different and the output differs considerably.

The primary reason that specific discriminant analysis routines are used rather than the regression approach is that discriminant analysis can be extended to more than two groups. When more than two groups are included in the analysis, the regression approach is not as straightforward. The case for more than two groups can be handled easily by discriminant analysis routines, however. Each group is given a different number. The numbers assigned do not have to include 0 and 1. (In fact, in the two-group case, the numbers that identify the two groups can be any integers, not necessarily 0 and 1. The values 0 and 1 were used here for convenience in extending the discussion to the topic of logistic regression.)

Discriminant analysis routines proceed by computing the means of each of the different groups for each explanatory variable. Then a measure of the distance from each observation to each set of means is computed. The observation is classified into the group whose set of means is closest. Example 10.3 illustrates a discriminant analysis routine applied to the employee classification data.

EXAMPLE 10.3 **Employee Classification (continued)** Figure 10.3 shows the MINITAB discriminant analysis output applied to the employee classification data. In this example, the group counts are noted. There are 20 in the group denoted 0 and 23 in the group denoted 1. A Summary of Classification table is provided next. For example, for those employees who were actually unsatisfactory (True Group 0), 16 were classified as unsatisfactory (Put into Group 0) and 4 were classified as satisfactory (Put into Group 1) out of the total of 20. The 16 correct classifications for this group yield a proportion correct of 0.800 (80.0%). The same information is given for each group. Overall, there were 43 employees, 34 were correctly categorized, and the overall proportion correct was 0.791 (79.1%).

FIGURE 10.3
MINITAB Discriminant Analysis Output for Employee Classification Example.

```
Linear Method for Response: Group

Predictors: TEST1, TEST2

Group          0         1
Count         20        23

Summary of classification

                  True  Group
Put into Group     0      1
0                 16      5
1                  4     18
Total N           20     23
N correct         16     18
Proportion      0.800  0.783

N = 43            N Correct = 34        Proportion Correct = 0.791

Squared Distance Between Groups

         0         1
0  0.00000   4.44946
1  4.44946   0.00000


Linear Discriminant Function for Groups

                 0         1
Constant    -298.27   -351.65
TEST1          5.20      5.68
TEST2          1.97      2.10

Summary of Misclassified Observations

                True   Pred          Squared
Observation    Group  Group  Group  Distance  Probability
      17**       1      0      0     6.657      0.586
                               1     7.352      0.414
      19**       1      0      0     0.1911     0.799
                               1     2.9454     0.201
      20**       1      0      0     2.561      0.518
                               1     2.703      0.482
      22**       1      0      0     1.034      0.538
                               1     1.340      0.462
      23**       1      0      0     0.5264     0.682
                               1     2.0567     0.318
      24**       0      1      0     6.438      0.244
                               1     4.177      0.756
      25**       0      1      0     2.2891     0.280
                               1     0.4008     0.720
      26**       0      1      0     2.6389     0.259
                               1     0.5416     0.741
      27**       0      1      0     2.900      0.339
                               1     1.564      0.661
```

The Squared Distance Between Groups table is not discussed in this text.

The next item in the output is the Linear Discriminant Function for Groups. When a discriminant analysis procedure is used, a separate equation is computed for each group. Note that these equations are not the same as the discriminant function computed using the regression approach. The two equations can be combined in a certain way to produce the overall discriminant function, however. The way this combination is achieved is not discussed here, but the use of the equations for the separate groups is demonstrated:

$$\text{Group 0: } -298.27 + 5.20\text{TEST1} + 1.97\text{TEST2}$$
$$\text{Group 1: } -351.65 + 5.68\text{TEST1} + 2.10\text{TEST2}$$

These equations are applied to each employee. The employee is then classified into the group for which his or her score is the highest. These equations can also be used to classify any future applicants. Administer the tests, record the test scores, and compute the values for each equation. The applicant is then assigned to the group for which the value from the equations is the highest. Although this method differs from the way the linear regression approach classified the employees, the results are the same, as can be seen from the Summary of Misclassified Observations. This is always true in the two-group case. For three or more groups, however, discriminant analysis procedures should be used.

In the Summary of Misclassified Observations, the number of each employee misclassified is shown along with the True Group and the Pred (predicted) Group. The Squared Distance column shows a measure of the distance computed from each observation to the means of each group. Each employee has been classified into the group with the smaller distance. In the last column, a Probability has been computed that can be thought of as the predicted probability that the employee belongs to a particular group. The employee has been classified into the group that has the highest probability.

The SAS output in Figure 10.4 provides information similar to MINITAB. Class Level Information provides the number in each group (and the proportion as well). Prior probability provides the researcher opportunity to weight the likelihood that individuals will fall into each category. The default probability is that the likelihood in each group is equal (0.5 when there are two groups). Under Linear Discriminant Function, the equations used to classify observations into groups are provided. The use of these equations was described earlier in this example. Number of Observations and Percent Classified into GROUPS provides a breakdown of the performance of the discriminant analysis in terms of correct and incorrect classifications. Error Count Estimates for Groups summarizes the incorrect classifications.

In linear regression, an equation was developed using a certain set of observations. It is typically not these observations for which predictions are desired, however. The quality of predictions is important for observations not included in the original sample. This situation is the same in discriminant analysis. What really matters in discriminant analysis is how well the discriminant equations classify future observations. The percentage of correctly classified observations given in the

FIGURE 10.4
SAS Discriminant Analysis Output for Employee Classification Example.

The DISCRIM Procedure

Observations	43	DF Total		42
Variables	2	DF Within Classes		41
Classes	2	DF Between Classes		1

Class Level Information

GROUPS	Variable Name	Frequency	Weight	Proportion	Prior Probability
0	_0	20	20.0000	0.465116	0.500000
1	_1	23	23.0000	0.534884	0.500000

Pooled Covariance Matrix Information

Covariance Matrix Rank	Natural Log of the Determinant of the Covariance Matrix
2	6.05465

Generalized Squared Distance to GROUPS

From GROUPS	0	1
0	0	4.44946
1	4.44946	0

Linear Discriminant Function

Linear Discriminant Function for GROUPS

Variable	0	1
Constant	−298.26618	−351.64588
TEST1	5.19936	5.68452
TEST2	1.97076	2.09766

Number of Observations and Percent Classified into GROUPS

From GROUPS	0	1	Total
0	16	4	20
	80.00	20.00	100.00
1	5	18	23
	21.74	78.26	100.00
Total	21	22	43
	48.84	51.16	100.00
Priors	0.5	0.5	

Error Count Estimates for GROUPS

	0	1	Total
Rate	0.2000	0.2174	0.2087
Priors	0.5000	0.5000	

discriminant analysis output can be used as a guide for this, but this percentage will likely overstate the quality of future classifications. The same is true of the R^2 for a linear regression. The R^2 represents a measure of fit of the sample data, but may not reflect how well the equation will do in classifying future data.

There are other methods to assess the quality of future classifications when using discriminant analysis. One method is to split the original sample into two parts (if sample size is large enough) called an *estimation sample* and a *validation sample*. Use the estimation sample to determine the discriminant equations. Then use the discriminant equations to classify the items in the validation sample. The percentage of correct classifications in the validation sample should provide a better indication of how discriminant analysis will perform on future observations. After this has been done, the two samples can then be combined to compute the discriminant function for future use.[2]

10.3 LOGISTIC REGRESSION

Discriminant analysis is known to be statistically valid when we can assume that the independent variable in the regression equation is normally distributed. When the equation has more than one independent variable, the assumption is that the x variables have a multivariate normal distribution, a strong assumption that is not discussed in detail here. Suffice it to say that this normality assumption excludes many possible variables as explanatory variables in the discriminant function equation. *Logistic regression* is another procedure for modeling a 0/1 dependent variable that does not depend on the assumption that the independent variables are normally distributed. As a result, many other types of variables, including indicator variables, are in the possible set of explanatory variables.

The logistic regression approach does have its own set of assumptions, however. To briefly describe logistic regression, the notion presented earlier that the conditional mean of y given x has a different interpretation when the dependent variable is a 0/1 variable must be reconsidered. When the dependent variable is either 0 or 1, it can be shown that the conditional mean of y given x, $\mu_{y|x}$, is equal to the probability that the observation belongs to the indicated group:

$$\mu_{y|x} = p = P(Y = 1)$$

Probabilities must be between 0 and 1; thus, to model the conditional mean of y, a function that is restricted to lie between 0 and 1 must be used. The function considered in logistic regression is called the *logistic function* (what a coincidence!) and can be written as follows:

$$\mu_{y|x} = \frac{1}{1 + e^{-(\beta_0 + \sum_{j=1}^{K} \beta_j x_j)}}$$

---

[2] For more detail on discriminant analysis, including a discussion of discriminant analysis with more than two groups, see C. T. Ragsdale and A. Stam, "Introducing Discriminant Analysis to the Business Statistics Curriculum." See References for complete publication information.

FIGURE 10.5
The S-Shaped Curve of the Logistic Function.

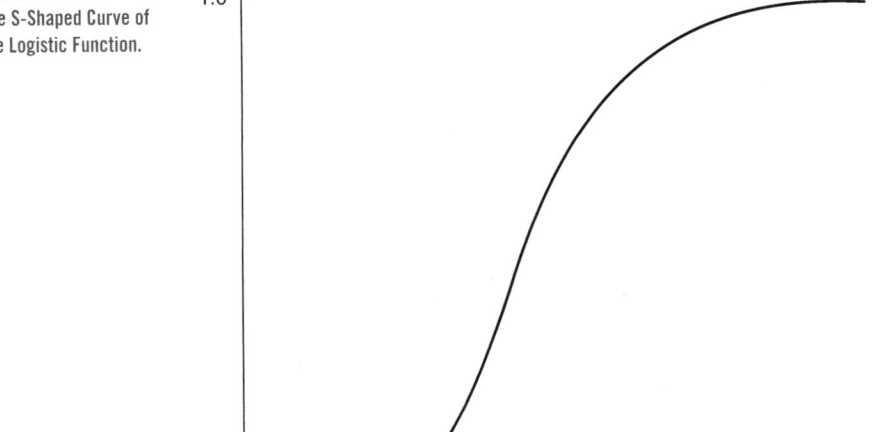

where the x_j are explanatory variables, K is the number of explanatory variables, and β_0 and the β_j are coefficients to be estimated.

This function works well for modeling probabilities because it is restricted to be between 0 and 1. The function forms an S-shaped curve such as the one in Figure 10.5. The logistic function is a nonlinear function of the regression coefficients and must be solved by a nonlinear regression routine. This makes the description of the solution process more complicated than that for linear least squares. However, logistic regression routines are available in certain statistical software packages and usually use a procedure called *maximum likelihood estimation* to estimate the regression coefficients in the logistic regression function. The following example illustrates.

EXAMPLE 10.4 **Logistic Regression and Employee Classification** Consider again the employee classification problem discussed in Example 10.1. Consider trying to estimate the probability that each employee belongs to the satisfactory group (the group coded 1). A possible nonlinear model to estimate this probability using only the result from TEST1 can be written:

$$\mu_{y|x} = \frac{1}{1 + e^{-(\beta_0 + \beta_1 x_1)}}$$

where x_1 is the TEST1 result.

The MINITAB logistic regression output used to estimate this equation is shown in Figure 10.6. The Response Information table shows the number of observations

FIGURE 10.6
MINITAB Logistic Regression Output for Employee Classification Example Using Only TEST1 as an Independent Variable.

```
Link Function: Logit

Response Information

Variable  Value  Count
Group     1      23     (Event)
          0      20
          Total  43

Logistic Regression Table

                                              Odds      95% CI
Predictor      Coef    SE Coef      Z      P  Ratio  Lower  Upper
Constant   -43.3684   12.9243   -3.36  0.001
TEST1       0.489722  0.144998   3.38  0.001   1.63   1.23   2.17


Log-Likelihood = -15.585
Test that all slopes are zero: G = 28.232, DF = 1, P-Value = 0.000


Goodness-of-Fit Tests
Method            Chi-Square   DF     P
Pearson              9.19454   17  0.934
Deviance             9.52955   17  0.922
Hosmer-Lemeshow      2.28161    8  0.971

Table of Observed and Expected Frequencies:
(See Hosmer-Lemeshow Test for the Pearson Chi-Square Statistic)

                              Group
Value   1    2    3    4    5    6    7    8    9   10   Total
1
  Obs   0    0    1    2    4    3    4    4    4    1    23
  Exp  0.1  0.2  1.0  1.6  3.2  3.8  4.3  3.8  3.9  1.0
0
  Obs   6    4    4    2    1    2    1    0    0    0    20
  Exp  5.9  3.8  4.0  2.4  1.8  1.2  0.7  0.2  0.1  0.0
Total   6    4    5    4    5    5    5    4    4    1    43

Measures of Association:
(Between the Response Variable and Predicted Probabilities)

Pairs        Number  Percent  Summary Measures
Concordant      411    89.3   Somers' D               0.82
Discordant       35     7.6   Goodman-Kruskal Gamma   0.84
Ties             14     3.0   Kendall's Tau-a         0.42
Total           460   100.0
```

in each group (satisfactory = 1, unsatisfactory = 0). The Logistic Regression Table shows information pertinent to the estimation of β_0 (Constant row) and β_1 (TEST1 row). Estimates of the coefficients are in the Coef column, and standard errors of these estimates are in the SE Coef column. A Z statistic and the associated p value are shown. These are used just as in simple linear regression to test whether

β_0 or β_1 are equal to zero as follows, using β_1 to illustrate:

Hypotheses: $H_0: \beta_1 = 0$
$H_a: \beta_1 \neq 0$

Decision rule: Reject H_0 if $Z > 1.96$ or $Z < -1.96$ (using a 5% level of significance)

Do not reject H_0 if $-1.96 \leq Z \leq 1.96$

Test statistic: $Z = 3.38$

Decision: Reject H_0

Conclusion: TEST1 is useful in this model

The p value can also be used to conduct the test in the usual manner:

Decision rule: Reject H_0 if p value < 0.05

Do not reject H_0 if p value ≥ 0.05

Test statistic: p value $= 0.001$

The same decision and conclusion are reached whether the Z statistic or its p value is used.

Information in the first row of the table can be used to test hypotheses about β_0, although interest is generally centered on the coefficients of the explanatory variables, as in linear regression.

Below this table is a statistic to test whether all slopes are zero. This is similar to the overall-fit F test in linear regression. The p value provided can be used in the usual way to conduct this test. This test is obviously of more interest when there is more than one explanatory variable. The remainder of the output is not discussed in this text.

The SAS logistic regression is in Figure 10.7. The Analysis of Maximum Likelihood Estimates table shows information pertinent to the estimation of β_0 (Intercept row) and β_1 (TEST1 row). Estimates of the coefficients are in the Estimate column, and standard errors of these estimates are in the Standard Error column. A chi-square statistic and the associated p value are shown. These are used to test whether β_0 or β_1 are equal to zero as follows, using β_1 to illustrate:

Hypotheses: $H_0: \beta_1 = 0$
$H_a: \beta_1 \neq 0$

Decision rule: Reject H_0 if $\chi^2 > \chi^2(0.05,1) = 3.841$

Do not reject H_0 if $\chi^2 \leq \chi^2(0.05,1) = 3.841$

where χ^2 is the test statistic and $\chi^2(0.05,1)$ is a chi-square critical value chosen from the chi-square table in Appendix B. One degree of freedom is used in selecting the proper critical value for the α level of significance.

Test statistic: $\chi^2 = 11.407$

Decision: Reject H_0

Conclusion: TEST1 is useful in this model

FIGURE 10.7
SAS Logistic Regression Output for Employee Classification Example Using Only TEST1 as an Independent Variable.

The LOGISTIC Procedure

Model Information

Data Set	WORK.TRAIN
Response Variable	GROUP
Number of Response Levels	2
Number of Observations	43
Model	binary logit
Optimization Technique	Fisher's scoring

Response Profile

Ordered Value	GROUP	Total Frequency
1	1	23
2	0	20

Probability modeled is GROUP = 1.

Model Convergence Status

Convergence criterion (GCONV = 1E − 8) satisfied.

Model Fit Statistics

Criterion	Intercept Only	Intercept and Covariates
AIC	61.401	35.169
SC	63.162	38.692
−2 Log L	59.401	31.169

Testing Global Null Hypothesis: BETA = 0

Test	Chi-Square	DF	Pr > ChiSq
Likelihood Ratio	28.2321	1	<.0001
Score	21.9226	1	<.0001
Wald	11.4070	1	0.0007

The LOGISTIC Procedure

Analysis of Maximum Likelihood Estimates

Parameter	DF	Estimate	Standard Error	Wald Chi-Square	Pr > ChiSq
Intercept	1	−43.3684	12.9243	11.2599	0.0008
TEST1	1	0.4897	0.1450	11.4070	0.0007

Odds Ratio Estimates

Effect	Point Estimate	95% Wald Confidence Limits	
TEST1	1.632	1.228	2.168

Association of Predicted Probabilities and Observed Responses

Percent Concordant	89.3	Somers' D	0.817
Percent Discordant	7.6	Gamma	0.843
Percent Tied	3.0	Tau-a	0.416
Pairs	460	c	0.909

The p value can also be used to conduct the test in the usual manner:

Decision rule: Reject H_0 if p value < 0.05
Do not reject H_0 if p value ≥ 0.05

Test statistic: p value $= 0.0007$

The Z test in MINITAB and the χ^2 test in SAS are essentially equivalent, and serve the same purpose.

Above this table in the Testing Global Null Hypothesis: BETA = 0 table, information is provided for a test of whether all slopes are zero. This is similar to the overall-fit F test in linear regression. There are three different test statistics provided (Likelihood Ratio, Score and Wald) which can be used to conduct the test. For any of the three options, the p value provided can be used in the usual way to perform this test. The Likelihood Ratio appears to be one of the more dependable approaches for this test.

The remainder of the output is not discussed in this text.

When using logistic regression, the goal of the analysis may be to classify the observations into a particular group (as in discriminant analysis). In this case, some rule must be designed to help decide into which group each observation should be classified. Predicted values of the probability of group membership in the indicated group can be computed from the logistic regression. A rule using these predicted probabilities can then be designed. The form of such a rule is

Classify observations into the

$y = 0$ group if the predicted value is below the cutoff
$y = 1$ group otherwise

A cutoff value of 0.5 is reasonable when the 0 and 1 outcomes are equally likely and the costs of misclassification into each group are about equal. In other cases, a different cutoff may be considered superior.[3]

EXERCISES

1. **Employee Classification.** Figure 10.8 shows summary logistic regression output for the employee classification data from Example 10.4 using both test results as explanatory variables. Use the output shown to answer the following questions. The data are in a file named TRAIN10 on the CD.

    a. Which variable or variables appear to be useful in the logistic regression function? Justify your answer. Use a 5% level of significance for any hypotheses tests.

    b. Using a cutoff probability of 0.5, classify potential employees whose test scores were as follows:
    1. TEST1 = 94   TEST2 = 88
    2. TEST1 = 80   TEST2 = 87
    3. TEST1 = 82   TEST2 = 74
    4. TEST1 = 90   TEST2 = 80

---

[3] For a more complete presentation of logistic regression, see C. E. Lunneborg, *Modeling Experimental and Observational Data*, Chapters 16, 17, and 18, or J. Neter, W. Wasserman, and M. Kutner, *Applied Linear Regression Models*, Chapter 16.

FIGURE 10.8
Logistic Regression
Output for Employee
Classification Example
Using TEST1 and TEST2
as Independent
Variables.

```
Response Information

Variable  Value  Count
Group       1     23   (Event)
            0     20
          Total   43

Logistic Regression Table

                                              Odds    95% CI
Predictor   Coefficient   StdDev     Z     P   Ratio  Lower  Upper

Constant     -56.1704    17.4516   -3.22  0.001
TEST1          0.483314   0.157779  3.06  0.002  1.62  1.19   2.21
TEST2          0.165218   0.102070  1.62  0.106  1.18  0.97   1.44

Log-Likelihood = -13.959
Test that all slopes are zero: G = 31.483, DF = 2, P-Value = 0.000
```

2. **Harris Salaries.** In Exercise 1 in Chapter 7, data from Harris Bank were examined to test for possible discrimination. In that exercise, the dependent variable was salary and one of the explanatory variables was an indicator variable to separate the employees into male and female groups. The coefficient of the indicator variable served as a measure of whether males earned more (or less), on average, than females.

Another way of examining this problem might be to use the male/female indicator variable as the dependent variable and see if group membership can be predicted from knowledge of salary. This can be done with either discriminant analysis or logistic regression. Try discriminant analysis and/or logistic regression and see how well these methods do in predicting whether employees are male or female based only on knowledge of their salary. Does your result support the claim that Harris Bank discriminated by underpaying female employees?[4]

The data are defined as follows and are in the file named HARRIS10 on the CD:

x_1 = beginning salaries in dollars (SALARY)
x_2 = years of schooling at the time of hire (EDUCAT)
x_3 = number of months of previous work experience (EXPER)
x_4 = number of months after January 1, 1969, that the individual was hired (MONTHS)
y = variable coded 1 for males and 0 for females (MALES)

(*Source*: These data were obtained from D. Schafer, "Measurement-Error Diagnostics and the Sex Discrimination Problem," *Journal of Business and Economic Statistics*. Copyright 1987 by the American Statistical Association. Used with permission. All rights reserved.)

3. **Computer Purchase.** The Daleway Corporation owns stores throughout the United States that carry its brand of computer. Management would like some information on the purchasing practices of the American public. Specifically, it would like to know if certain factors affect the decision to upgrade a system with the purchase of a new computer. The company surveyed 40 recent customers who were considering upgrading and collected information on the following variables:

PURCHASE, coded as 1 if the customer purchased a computer, 0 if the customer did not
INCOME, the household income of the customer (in thousands of dollars)
AGE, the age of the customer's current computer

---
[4] Note: Using discriminant analysis or logistic regression in this manner would probably not be the preferred method of examining this question in a legal proceeding. The regression approach discussed in Chapter 7 would be preferred, but this makes an interesting exercise.

Are either of the variables INCOME or AGE of use in predicting whether the customer will purchase a new computer? Justify your answer. Use a 5% level of significance for any hypothesis tests. The file containing these data is named COMPPURCH10 on the CD.

4. **March Madness (Men's Tournament).** Each March the games in the men's NCAA (National Collegiate Athletic Association) basketball tournament begin. Sixty-four teams are selected for the tournament. Each team is given a number from 1 to 16, called the *seed*. There are 4 teams assigned to each seed number (4 number 1s, 4 number 2s, etc). The lower the seed number, the higher the perceived quality of the team. In the first round of the tournament these 64 teams are paired according to their seed. Number 1 seeds play number 16 seeds; number 2 seeds play number 15 seeds; number 3 seeds play number 14 seeds and so on. The games are distributed in brackets with 16 teams (consisting of 1 through 16 seeds) in each bracket. The seed numbers are assigned by a committee formed by the NCAA. It is a commonly accepted fact that the seed numbers are of use in predicting the winner of the games. For example, a number 16 seed has never beaten a number 1 seed. This seems reasonable. Number 1 seeds, the teams of highest quality, should have a high probability of beating number 16 seeds. But do the seed numbers contain all the information useful in predicting the winners, or are there other team characteristics that might be of use? The file NCAAMEN10 on the CD contains data on the following variables:

$$\text{WIN} = 1 \text{ if the higher seed won the game}$$
$$= 0 \text{ otherwise}$$
$$\text{DIFF} = \text{lower seed number–higher seed number; note that DIFF will always be negative (or zero if two teams of equal seed play)}$$
$$\text{PCTHIGH} = \text{winning percentage for the higher-seeded team}$$
$$\text{PCTLOW} = \text{winning percentage for the lower-seeded team}$$

There is one additional column in the file labeled ROUND. This column indicates in which round of the tournament the game took place. Although it might be of interest as an explanatory variable, it was included primarily for information.

$$\text{ROUND} = 1, \text{ first-round game}$$
$$\text{(32 total games)}$$
$$= 2, \text{ second-round game (16 games pairing the 32 teams who survived the first round)}$$
$$= 3, \text{ third-round games (8 games—these 16 teams are often referred to as the Sweet 16)}$$
$$= 4, \text{ fourth-round games (4 games—these 8 teams are often referred to as the Elite 8 or the Great 8)}$$
$$= 5, \text{ fifth-round games (2 games—Final 4 teams)}$$
$$= 6, \text{ championship game}$$

Use logistic regression to determine whether DIFF, PCTHIGH, or PCTLOW are useful in predicting the winners of games in the NCAA basketball tournament. Use a 5% level of significance in any hypothesis tests. Are there other variables that might be useful in addition to ones given in this problem?

5. **March Madness (Women's Tournament).** Each March the games in the women's NCAA (National Collegiate Athletic Association) basketball tournament begin. Sixty-four teams are selected for the tournament. Each team is given a number from 1 to 16, called the *seed*. There are 4 teams assigned to each seed number (4 number 1s, 4 number 2s etc). The lower the seed number, the higher the perceived quality of the team. In the first round of the tournament these 64 teams are paired according to their seed. Number 1 seeds play number 16 seeds; number 2 seeds play number 15 seeds; number 3 seeds play number 14 seeds, and so on. The games are distributed in brackets with 16 teams (consisting of 1 through 16 seeds) in each bracket. The seed numbers are assigned by a committee formed by the NCAA. It is a commonly accepted fact that the seed numbers are of use in predicting the winner of the games. But do the seed numbers contain all the information useful in predicting the winners, or are there other team characteristics that might be of use? The file NCAAWOMEN10 on the CD contains data on the following variables:

$$\text{WIN} = 1 \text{ if the higher seed won the game}$$
$$= 0 \text{ otherwise}$$

DIFF = lower seed number–higher seed number; note that DIFF will always be negative (or zero if two teams of equal seed play)
PCTHIGH = winning percentage for the higher-seeded team
PCTLOW = winning percentage for the lower-seeded team

There is one additional column in the file labeled ROUND. This column indicates in which round of the tournament the game took place. Although it might be of interest as an explanatory variable, it was included primarily for information.

ROUND = 1, first-round game (32 total games)
= 2, second-round game (16 games pairing the 32 teams who survived the first round)
= 3, third-round games (8 games—these 16 teams are often referred to as the Sweet 16)
= 4, fourth-round games (4 games—these 8 teams are often referred to as the Elite 8 or the Great 8)
= 5, fifth-round games (2 games—Final 4 teams)
= 6, championship game

Use logistic regression to determine whether DIFF, PCTHIGH, or PCTLOW are useful in predicting the winners of games in the NCAA basketball tournament. Use a 5% level of significance in any hypothesis tests. Are there other variables that might be useful in addition to ones given in this problem?

6. **Loan Performance.** The National Bank of Fort Worth, Texas, wants to examine methods for predicting sub-par payment performance on loans. It has data on unsecured consumer loans made over a 3-day period in October 1994 with a final maturity of 2 years. There are a total of 348 observations in the sample. The data, which have been transformed to provide confidentiality, include the following:

PASTDUE: Coded as 1 if the loan payment is past due and zero otherwise.
CBSCORE: Score generated by the CSC Credit Reporting Agency. Values range from 400 to 8390, with higher values indicating a better credit rating.
DEBT: This is a debt ratio calculated by taking required monthly payments on all debt and dividing it by the gross monthly income of the applicant and coapplicant. This ratio represents the amount of the applicant's income that will go toward repayment of debt.
GROSSINC: Gross monthly income of the applicant and coapplicant.
LOANAMT: Loan amount.

You have been asked to examine the feasibility of predicting past-due loan payment. Describe your results to the bank in a two-part report. The report should include an executive summary with a brief nontechnical description of your results and an accompanying technical report with the details of your analysis. The data are in a file named LOAN10 on the CD.

USING THE COMPUTER

The Using the Computer section in each chapter describes how to perform the computer analyses in the chapter using Excel, MINITAB, and SAS. For further detail on Excel, MINITAB, and SAS, see Appendix C.

EXCEL

Discriminant Analysis and Logistic Regression

There are currently no options for either discriminant analysis or logistic regression in Excel.

MINITAB

Discriminant Analysis

STAT: MULTIVARIATE: DISCRIMINANT ANALYSIS

Performs a discriminant analysis. The "Groups:" (dependent) variable, which designates the two (or more) groups into which the observations are classified, and the explanatory or predictor variables are requested in the dialog box. See Figure 10.9.

Logistic Regression

STAT: REGRESSION: BINARY LOGISTIC REGRESSION

FIGURE 10.9 MINITAB Dialog Box for Discriminant Analysis.

Performs a logistic regression. Binary logistic regression assumes the sample observations come from two groups. More than two groups can be considered by using ORDINAL LOGISTIC REGRESSION (categories are ordinal in nature) or NOMINAL LOGISTIC REGRESSION (categories have no natural ordering). Figure 10.10 shows the dialog box for a binary logistic regression.

SAS

Discriminant Analysis

PROC DISCRIM;
CLASS VAR1;
VAR VAR2;

Performs a discriminant analysis. The dependent variable, which designates the two (or more) groups into which the observations are classified, is denoted here as VAR1 and is listed on the CLASS statement. The explanatory or predictor variables are listed on the VAR statement. In this example, only one predictor variable, VAR2, is used.

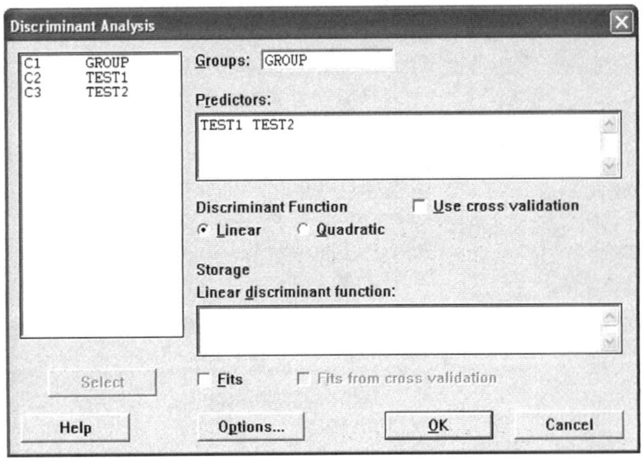

FIGURE 10.10 MINITAB Dialog Box for Binary Logistic Regression.

Logistic Regression

PROC LOGISTIC DESCENDING;
MODEL VAR1 = VAR2;

Performs a logistic regression. VAR1 is assumed to be the dependent variable. This variable is a 0/1 variable (as discussed in this chapter). The DESCENDING option is used so that probabilities for the group labeled 1 (the highest value of the dependant variable) are predicted. If you do not use this option, the only difference will be that the signs of the coefficients will be reversed. The explanatory variables are listed on the right-hand side of the equality in the MODEL statement. In this example, the only variable listed is VAR2. The MODEL statement setup is similar to PROC REG.

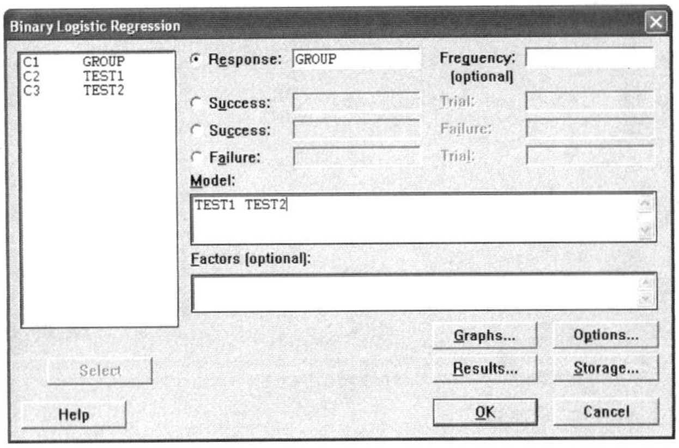